BAYESIAN STATISTICS 7

BAYESIAN STATISTICS 7

Proceedings of the Seventh Valencia International Meeting

Dedicated to Dennis V. Lindley

June 2–6, 2002

Edited by

J. M. Bernardo
M. J. Bayarri
J. O. Berger
A. P. Dawid
D. Heckerman
A. F. M. Smith
and
M. West

CLARENDON PRESS · OXFORD
2003

OXFORD
UNIVERSITY PRESS

Great Clarendon Street, Oxford OX2 6DP

Oxford University Press is a department of the University of Oxford.
It furthers the University's objective of excellence in research, scholarship,
and education by publishing worldwide in

Oxford New York

Auckland Bangkok Buenos Aires Cape Town Chennai
Dar es Salaam Delhi Hong Kong Istanbul Karachi Kolkata
Kuala Lumpur Madrid Melbourne Mexico City Mumbai Nairobi
São Paulo Shanghai Taipei Tokyo Toronto

Oxford is a registered trade mark of Oxford University Press
in the UK and in certain other countries

Published in the United States
by Oxford University Press Inc., New York

First published 2003

A catalogue record for this title is available from the British Library

Library of Congress Cataloging in Publication Data
(Data available)

ISBN 0-19-852615-6

10 9 8 7 6 5 4 3 2 1

Typeset by the editors using TEX
Printed in Great Britain
on acid-free paper by T. J. International Ltd., Padstow, Cornwall

OPENING ADDRESS BY THE FORMER CONFERENCE PRESIDENT

Sunday, June 2nd, Mare Nostrum Resort, Tenerife*

Thank you so much for remembering me at this meeting. It is a great sadness to me that I cannot be with you, both for the pleasure of meeting old friends and making new ones, and because I shall miss hearing about exciting developments in Bayesian ideas as the message continues to gather strength. It is only 18 years until, according to de Finetti, we shall all be Bayesians and our distinctive adjective superfluous.

When the first of these Valencia conferences was being planned, I have to admit that I had strong reservations about the wisdom of our meeting. This was because Bayesian statistics is a way of looking at the whole of statistics and not just a branch of statistics, so that our ideas should not be confined within a group. After the first meeting, and the feeling has grown stronger since, it was clear that the conference was a great success and the advantages that came from talking about ideas without the influence of frequentists, outweighed any sectarian considerations. Since then the conference has grown both in quantity and quality, influencing the dispersal of our ideas into mainstream thinking so that Bayesian papers occur more frequently. Nevertheless, there remain pockets of statistical thought that are unaware of the coherent approach and the errors of the frequentist viewpoint. For example, workers in multiple comparisons ignore the role of a probability description of parameters in making multiple statement. So there is still a way to go in spreading our ideas into other parts of our subject, let alone the world at large. At an earlier conference I gave a paper entitled "Is our view of Bayesian statistics too narrow?" I thought it was then and feel even more strongly today.

The description of uncertainty through probability, which I take to be the defining idea behind the Bayesian viewpoint, is an important, indeed essential, tool in all walks of life, including the activities of governments, and this because uncertainty is present everywhere. To quote de Finetti again: "Our world, our life, our destiny, are dominated by Uncertainty; this is perhaps the only statement we may assert without uncertainty." It is my very sincere hope that the splendid ideas that will be expressed here will embrace many aspects of society today, and I wish you well in your endeavours.

Dennis V. Lindley

---

* Read by Professor Adrian F. M. Smith.

PREFACE

The four-yearly Valencia International Meetings on Bayesian Statistics meetings are firmly established as the premier conferences for the exchange and advancement of Bayesian ideas.* By a quirk of Bayesian geography, the Seventh Valencia Meeting was held in the Spanish Canary Islands, at the Mare Nostrum Resort, Tenerife, from 2 to 6 June 2002. Over 440 delegates, from 35 different countries, were present. The meeting was dedicated to Professor Dennis V. Lindley, pioneer and champion of modern Bayesian Statistics, a founding member of the Valencia organizing committee and former Conference President.

For this meeting the organizing committee invited 25 leading experts in the field to present papers, followed by discussion. The International Society for Bayesian Analysis selected another 50 contributed papers for oral presentation. A further 290 papers were presented in three poster sessions.

These Proceedings contain 23 invited papers with their discussions, as well as 31 contributed papers (of which 10 were presented orally and 21 as posters), selected after a rigorous refereeing process. Together they provide a definitive overview of current concerns and activity in Bayesian Statistics, encompassing a wide range of theoretical and applied research.

Three papers treat topics related to the fundamental concept of exchangeability. **Lauritzen** characterizes random Rasch-type models for binary data in terms of sufficiency and row-column exchangeability; **Arellano-Valle, Iglesias and Vidal** study elliptical symmetry and its implications for inference; while **Esteves, Wechsler, Iglesias and Pereira** introduce a mixture extension of Polyá's urn scheme and discuss its predictive properties.

Problems in nonparametric Bayesian inference are considered by **Salinetti**, who describes the use of the hypo-strong law of large numbers to show consistency; and **Gutiérrez-Peña and Nieto-Barajas**, who analyze a nonparametric mixed Poisson process using an independent increment process prior for the mean function.

Other theoretical topics are treated by **Fraser, Reid, Wong and Yi**, who construct a third-order accurate likelihood function free of nuisance parameters; and by **Pettit and Sugden**, who consider estimation of a finite population total while allowing for outliers.

The objective Bayesian school is represented by **Bernardo and Juárez**, who treat the problem of point estimation in invariant fashion; **Rodríguez, Álvarez and Sansó**, who develop Bayesian reference analyses and intrinsic Bayes factors for a model motivated by a problem in oil exploration; and **Girón, Martínez, Moreno and Torres**, who provide a default model choice perspective on the analysis of matched pairs in the presence of covariates.

Two papers deal with Bayesian approaches to inverse problems. **Wolpert, Ickstadt and Hansen** develop a nonparametric analysis for ill-posed problems, given the values of certain integrals with respect to an unknown measure; while **Higdon, Lee and Holloman** construct methodology for estimating spatially dependent inputs from outputs generated by a complex physical or computational model.

The growing interest in causal inference is represented by three papers. **Chib** presents a Bayesian analysis, within the "potential outcomes" framework, for estimating treatment effects in the presence of unobserved confounders; **Daneshkhah and Smith** relate independence and consistency properties of prior distributions for a causal Bayesian network; while **Leucari and Consonni** apply reference conditioning to define and study compatibility of priors across a collection of such networks.

---

* The Proceedings of previous meetings have been published: the first by the University Press, Valencia (1980); the second by North-Holland, Amsterdam (1985); and the third, fourth, fifth and sixth by Clarendon Press, Oxford (1988, 1992, 1996, 1999). The editors in each case were the members of the organizing committee.

Simulation-based Bayesian computation is both a practically important and a theoretically challenging topic. **Mengersen and Robert** introduce the "pinball sampler" for generating Monte Carlo samples with an exact target distribution; **Papaspiliopoulos, Roberts and Sköld** compare the effects of centered and non-centered hierarchical model parametrizations on the convergence of MCMC routines; **Lavine** supplies a theorem allowing the identification of an ergodic chain for a subvector within a non-convergent Markov chain; and **Rasmussen** reduces the computational demands of hybrid Monte Carlo methods by incorporating a Gaussian process model.

Variational approximations are a useful fast alternative to simulation. **Beal and Ghahramani** develop variational bounds on marginal model likelihoods in the presence of incomplete data; while **Blei, Jordan and Ng** apply variational methods to a hierarchical model of a large-scale text collection.

Bayesian approaches to multivariate and regression analysis form a popular topic. **Liu, Zhang, Palumbo and Lawrence** apply Markov chain Monte Carlo methods to select and transform variables for cluster analysis; **Chipman, George and McCulloch** develop Bayesian analyses of treed generalized linear models, allowing local specification of the response model; **Neal** introduces "Dirichlet diffusion trees," combining exchangeability with hierarchical clustering, to structure multivariate prior distributions; **Peña, Rodríguez and Tiao** apply the SAR procedure to identify data heterogeneity and fit piecewise regression models; and **Smith and Walshaw** take some bivariate steps towards a Bayesian analysis of multivariate extreme values.

Three paper deal with gambling and finance. **Schervish, Seidenfeld and Kadane** use the gambling set-up of de Finetti's "dutch book" coherence argument to introduce measures of incoherence; **Quintana, Lourdes, Aguilar and Liu** consider the relevance of dynamic Bayesian modelling to investment in financial markets; and **Polson and Stroud** develop Bayesian methodology for combining data on financial derivatives and on their underlying assets.

A number of papers treat various aspects of temporal and spatial modelling and analysis. **Virto, Martín, Ríos-Insua and Moreno-Díaz** explore the applicability of MCMC methods to dynamic programming; **Zohar and Geiger** discuss problems of tracking groups of moving objects; **Scott and Smyth** use a hidden Markov model representation to analyze a Markov modulated Poisson process; **Lefebvre, Gadeyne, Bruyninckx and de Schutter** extend Kalman filter methods to online estimation of nonlinear systems; **Dethlefsen** applies state-space models and Kalman smoothing to analyze spatial Markov random fields; **Garside and Wilkinson** develop computational estimation tools for a lattice model with a latent spatio-temporal process; **Tamminen and Lampinen** use hierarchical modelling to aid feature location in computer vision; and **Carlin and Banerjee** develop a hierarchical spatio-temporal model for multivariate survival data.

In other fruitful applications of hierarchical modelling, **Davy and Godsill** analyze harmonic musical sounds; **Johnson, Graves, Hamada and Reese** consider the reliability of multi-component systems; and **Ferreira, West, Lee, Higdon and Bi** use multi-scale time-series models to estimate hydrological permeability fields.

Several papers are motivated by genetic microarray technology, where many variables may be measured on few individuals. In this context, **Genovese and Wasserman** compare Bayesian and frequentist approaches to multiple testing; while **West** discusses Bayesian factor regression models. **Mertens** applies Bayesian logistic regression models to microarray discrimination problems; while **Wakefield, Zhou and Self** use a hierarchical model to cluster temporally ordered gene expression data.

Other applied and methodological contributions include: **Newton, Yang, Gorman, Tomlinson and Roylance**, who describe a modelling strategy for analyzing genomic aberrations;

Gebouský, Kárný and Quinn, who exploit prior information to make inferences from sparse medical diagnostic data; **Choy, Chan and Yam**, who analyze salamanders with heavy tails; **Jamieson and Brooks**, who investigate density dependence in duck dynamics; and **Ausín, Lillo, Ruggeri and Wiper**, who attempt to optimize hospital bed provision. **Vehtari and Lampinen** use cross-validation to estimate the predictive utility of a model; **Erosheva** develops a Bayesian analysis of the "grade of membership" model; **van der Linde and Osius** consider uses of the odds ratio parametrization; **Zheng and Marriott** analyze a linear trend model with smooth transitions in intercept and slope; and **Dobra, Fienberg and Trottini** present a Bayesian framework for thinking about the confidentiality of individual data in a contingency table.

The task of deciding which of the many excellent conference papers should be included in the limited space of this volume was not an easy one. The organizers would like to express their deep appreciation for the assistance and advice of the associate editors: Bradley P. Carlin, Siddhartha Chib, Stephen Fienberg, Edward George, Simon Godsill, Eduardo Gutiérrez-Peña, David Higdon, Valen Johnson, Steffen Lauritzen, Jun Liu, Angelika van der Linde, Kerrie Mengersen, Michael Newton, Daniel Peña, Gareth O. Roberts, Gabriella Salinetti and Mark J. Schervish.

We are indebted to the Conference sponsors, the *Universitat de València* and the *International Society for Bayesian Analysis*, as well as to *Iberia Airlines*, *Microsoft Corporation*, the *Universidad de La Laguna, Canarias* and the *United States National Science Foundation*, for their generous financial support.

We are also most grateful to Mailo Albiach, Miguel Juárez, Carlos Perez and Dolores Tortajada for their invaluable administrative and technical assistance, and in particular to Dolores Tortajada for preparing the final TEXversion of these Proceedings. The Eighth Valencia International Meeting on Bayesian Statistics is planned to take place in the early summer of 2008.

M. J. Bayarri
J. O. Berger
J. M. Bernardo
A. P. Dawid
D. Heckerman
A. F. M. Smith
M. West

CONTENTS

I. INVITED PAPERS (with discussion)

II. CONTRIBUTED PAPERS

INVITED PAPERS

BAYESIAN STATISTICS 7, pp. 3–24
J. M. Bernardo, M. J. Bayarri, J. O. Berger, A. P. Dawid,
D. Heckerman, A. F. M. Smith and M. West (Eds.)

Bayesian Inference for Elliptical Linear Models: Conjugate Analysis and Model Comparison

REINALDO B. ARELLANO-VALLE, PILAR L. IGLESIAS and IGNACIO VIDAL
Pontificia Universidad Católica de Chile, Chile
reivalle@mat.puc.cl pliz@mat.puc.cl ividal@mat.puc.cl

SUMMARY

In this work, we review and extend some results presented in the literature related to Bayesian analysis in elliptical (symmetric) linear models. A new class of prior distributions is considered, which generalizes the normal-chi-squared family. It is shown that for this class, the posterior analysis is simple to perform under some conditions, and conjugacy is achieved for ϕ. One aspect of the models studied in detail is the invariance of some distributions with respect to changes in the generator. In particular, it is shown that some Bayes estimators and Bayes factors do not depend on the generator function of an elliptical model. We also discuss practical implications of these results.

Keywords: LINEAR REGRESSION MODELS; ELLIPTICAL DISTRIBUTIONS; SQUARED RADIAL DISTRIBUTIONS; INTRINSIC BAYES FACTOR; FRACTIONAL BAYES FACTOR; UTILITY FUNCTION.

1. INTRODUCTION

Bayesian inference for normal regression models, including sensitivity analysis, model comparison and error in variables under non-informative and conjugate prior for the model parameters has received considerable attention in the last decades. From a distributional point of view the results can be extended in several directions. One is by considering a wider class of prior distributions for the parameters of the model. Another, is by considering alternative distributions for the error terms.

Usually, the results with non-conjugate priors rely heavily on MCMC methods. On the other hand, many extensions has been obtained by considering the so called dependent elliptical model, which is often used in linear regression analysis to accommodate the kurtosis of the error terms and to accommodate outliers. Bayesian inference with multivariate elliptical models was initially presented in Chu (1973). Posteriorly, Meinhold and Singpurwalla (1989) consider a robustification of Kalman filters by using the multivariate Student-t distribution. This distribution was also used by Zellner (1976), who considered a Bayesian treatment of linear regression models under non-informative prior distributions.

The results of Zellner (1976) were extended to the case where the errors are modelled as scale mixtures of normal distributions by Jammalamadaka *et al.* (1987) and Chib *et al.* (1988) and to the entire class of multivariate elliptical distributions by Osiewalski and Steel (1993). See also, Branco *et al.* (2001) and Arellano-Valle *et al.* (2000) for connections to calibration problems and diagnostic in elliptical regression models.

In this paper, we extend the previous results by considering proper prior distributions for the location and dispersion parameters involved in the multivariate elliptical regression models.

The consequences of these extensions are analyzed in the context of the calibration problem, diagnostic and model comparison.

The paper is organized as follows. In Section 2, we review some results related to elliptical distributions. In Section 3, we present a Bayesian analysis of elliptical linear models by considering different specifications for the prior distribution and we examine the consequences for the calibration problem and diagnostic. The main result of this section is that many solutions to inference problems on elliptical regression models coincide with those obtained under normal regression models.

Finally, in Section 4 we compute default Bayes factors to compare elliptical linear models including a discussion of the performance of these measures as an objective criteria for elliptical model comparison. Conclusions and final remarks are presented in Section 5.

In this paper, we will employ the usual symbols $\|\|$, $\amalg$ and $\stackrel{d}{=}$ to denote the length of a vector, the (conditional) independence of two random vectors (given a third), and the equality in distribution, respectively.

2. PRELIMINARIES: ELLIPTICAL DISTRIBUTIONS

In this section, we summarize the basic properties of elliptical distributions. Roughly speaking, a random real variable Z has a spherical distribution if $Z \stackrel{d}{=} -Z$. In this work we restrict the study to the case when the c.d.f. of Z is absolutely continuous, so that the spherical symmetry implies that Z has density given by

$$f_Z(z) = h(z^2) I_{\Re}(z), \tag{1}$$

where $\int_0^\infty u^{-1/2}\, h(u)\, du = 1$.

The function h is called the density generator and we write $Z \sim S(h)$. For example, if $h(u) = (1/\sqrt{2\pi})\, e^{-u/2}$ then we obtain the standard normal distribution $Z \sim N(0,1)$, and if $h(u) = c\{\nu + u\}^{-(\nu+1)/2}$, for some constant c, then the standard Student-t distribution with ν degrees of freedom, say $Z \sim t(0,1,\nu)$, follows. For exponential power distribution, we have $h(u) = k \exp(-\frac{1}{2}\,|\,u\,|^{\,s})$, $s > 0$, for some constant k. The class of spherical distributions is a large family and contains symmetric distributions with heavier and lighter tails than those of the normal distribution. We note also that if $Z \sim S_1(h)$ then $T = Z^2$ has density given by

$$f_T(t) = t^{-1/2}\, h(t)\, I_{(0,\infty)}(t). \tag{2}$$

T is usually termed radial random variable and is denoted by $T \sim \mathcal{R}^2(h)$. Now, let $Y = \mu + \phi^{1/2} Z$ then $f_Y(y) = \phi^{1/2} h((y-\mu)^2\phi) I_{\Re}(y)$, where h satisfies (1) and $T = \phi(Y-\mu)^2$ has density given by (2). We say that the random variable Y has elliptical distribution with parameters μ (location) and ϕ (precision), with $\mu \in \Re$ and $\phi > 0$, and we write $Y \sim EL_1(\mu, \phi, h)$.

Multivariate distributions with spherical univariate marginal distributions can be constructed in several ways. The simplest procedure is to consider $\boldsymbol{Z} = (Z_1, \ldots, Z_n)^t$ a random vector with $Z_i \stackrel{iid}{\sim} S(h)$. In this case, we say that $\boldsymbol{Z}$ has poly-spherical distribution. On the other hand, we can construct a multivariate distribution with constant density function over spheres, that is, $f_{\boldsymbol{Z}}(\boldsymbol{z}) = h^n(\|\boldsymbol{z}\|^2)$, where

$$\int_0^\infty \frac{\pi^{n/2}}{\Gamma(n/2)}\, u^{(n/2)-1}\, h^n(u)\, du = 1. \tag{3}$$

In this case, we say that $\boldsymbol{Z}$ has a multivariate spherical distribution and we write $\boldsymbol{Z} \sim S_n(h^n)$. Note that $T = \|\boldsymbol{Z}\|^2$ has density given by $f_T(t) = \frac{\pi^{n/2}}{\Gamma(n/2)} t^{(n/2)-1} h^n(t) I_{(0,\infty)}(t)$.

For simplicity, the upper index n will be omitted when it is not necessary. We say here that T has radial squared distribution with n degrees of freedom and density generator function h and we write this as $T \sim \mathcal{R}_n^2(h)$. Thus, $\boldsymbol{Z} \sim S_n(h)$ if and only if $T = \|\boldsymbol{Z}\|^2 \sim \mathcal{R}_n^2(h)$. Note also that the random variable $S = T^{-1}$ has density function $\frac{\pi^{n/2}}{\Gamma(n/2)}(1/s)^{(n/2)-1} h^n(1/s) I_{(0,\infty)}(s)$,which is referred to as the inverted radial-squared distribution with n degrees of freedom and density generator h and we write this as $S \sim \mathcal{IR}_n^2(h)$.

All marginal and conditional distributions of a multivariate spherical distribution are also spherical (see, for example, Fang *et al.*, 1990). Any linear combination $W = \boldsymbol{a}^t \boldsymbol{Z}$ is spherically distributed too. Moreover, if $\boldsymbol{Z} = (Z_1, \ldots, Z_n)^t \sim S_n(h)$ then $Z_1, \ldots, Z_n$ are independent if and only if $S_n(h)$ is the normal spherical distribution (Kelker, 1970). Thus, except in the normal case, the poly-spherical and multivariate spherical distributions are different classes. Both contain distributions that are long-tailed and short tailed relative to the normal distribution, but the multivariate spherical approach seems to be a more realistic model because the independence assumption is relaxed. For example, in the context of the Student-t model, we have that:

$$\boldsymbol{Z} \sim \text{ Poly-}t(0,1,\nu) \leftrightarrow f_{\boldsymbol{Z}}(\boldsymbol{z}) = k \prod\nolimits_{i=1}^{n} \{1 + z_i^2/2\}^{-(\nu+1)/2},$$

and

$$\boldsymbol{Z} \sim t_n(0, \boldsymbol{I}_n, \nu) \leftrightarrow f_{\boldsymbol{Z}}(\boldsymbol{z}) = c\{1 + \frac{1}{\nu} \sum\nolimits_{i=1}^{n} z_i^2\}^{-(n+\nu)/2}.$$

Note that the poly-Student-t distribution is not spherically symmetric. This distribution remains invariant only under change of sign. On the other hand, the multivariate Student-t distribution remains invariant under all orthogonal transformations and has Student-t univariate marginal distributions, but the elements of $\boldsymbol{Z}$ are not independent. A justification from a predictivistic point of view of the dependence structure in the multivariate Student-t model is given by Loschi *et al.* (2002); see also Arellano-Valle *et al.* (1994). Table 4.1 contains different families of spherical generators function.

In this work, we deal only with multivariate spherical distributions, more generally with multivariate elliptical distributions. An $n \times 1$ random vector $\boldsymbol{Y}$ is said to have multivariate elliptical distribution with parameters $\boldsymbol{\mu}$ (the location vector) and Σ (the dispersion matrix) of dimensions $n \times 1$ and $n \times n$, respectively, with Σ being positive definite ($\Sigma > 0$), if $\boldsymbol{Y}$ has density function of the form

$$|\Sigma|^{-1/2} h^n((\boldsymbol{y} - \boldsymbol{\mu})^t \Sigma^{-1} (\boldsymbol{y} - \boldsymbol{\mu})),$$

where h^n satisfies (3).

In this case, we write $\boldsymbol{Y} \sim El_n(\boldsymbol{\mu}, \Sigma; h)$, which is equivalent to $\boldsymbol{Z} = \Sigma^{-1/2}(\boldsymbol{Y} - \boldsymbol{\mu}) \sim El_n(0, \boldsymbol{I}_n; h) = S_n(h)$ and therefore to $(\boldsymbol{Y} - \boldsymbol{\mu})^t \Sigma^{-1} (\boldsymbol{Y} - \boldsymbol{\mu}) = \|\boldsymbol{Z}\|^2 = T \sim \mathcal{R}_n^2(h)$. From the above results, we can show also that if $\mathrm{E}(T) < \infty$, then $\mathrm{E}(\boldsymbol{Y}) = \boldsymbol{\mu}$ and $\mathrm{Var}(\boldsymbol{Y}) = \alpha_h \Sigma$, where $\alpha_h = \mathrm{E}(n^{-1}T)$ is the variance parameter associated with the density generator $h = h^n$.

Now, let $\boldsymbol{Z}_k$ be a k-dimensional $(1 \le k < n)$ random sub-vector of $\boldsymbol{Z} \sim S_n(h)$. Then, $\boldsymbol{Z}_k \sim S_k(h)$ and has density function $h^k(\|\boldsymbol{Z}_k\|^2)$, $\boldsymbol{Z}_k \in \Re^k$, where

$$h^k(u) = \int_0^\infty \frac{\pi^{(n-k)/2}}{\Gamma((n-k)/2)} v^{(n-k)/2-1} h^n(u+v)\, dv,$$

so that $T_k = \|\boldsymbol{Z}_k\|^2 \sim \mathcal{R}_k^2(h)$, see Fang *et al.* (1990). Moreover, provided that the required moments exist, we have that $\mathrm{E}(k^{-1}T_k) = \alpha_h$ and $\mathrm{Var}(k^{-1}T_k) = \{k^{-1}(k+2)(\kappa_h + 1) - 1\}\alpha_h^2$, where $\kappa_h = \alpha_h^{-2} E[\{n(n+2)\}^{-1} T^2] - 1$ is the kurtosis parameter of the elliptical family with density generator $h = h^n$. Similar results hold for the inverse radial squared random variable $S_k = T_k^{-1}$.

Let us now consider the partition $\boldsymbol{Z} = (\boldsymbol{Z}_k^t, \boldsymbol{Z}_{(k)}^t)^t \sim S_n(h)$. Thus, the conditional distribution of $\boldsymbol{Z}_k$ given $\boldsymbol{Z}_{(k)} = \boldsymbol{z}_{(k)}$ is such that

$$\boldsymbol{Z}_k \,|\, \boldsymbol{Z}_{(k)} = \boldsymbol{z}_{(k)} \overset{d}{=} \boldsymbol{Z}_k \,|\, \|\boldsymbol{Z}_{(k)}\|^2 = t \sim S_k(h_t), \tag{4}$$

where $t = \|\boldsymbol{z}_{(k)}\|^2$ and, for each $t \geq 0$,

$$h_t^k(u) = \frac{h^n(u+t)}{h^{n-k}(t)}, \qquad u \geq 0, \tag{5}$$

is the conditional density generator function. Moreover, by noting that $T_k = \|\boldsymbol{Z}_k\|^2 \sim \mathcal{R}_k^2(h)$, $T_{(k)} = \|\boldsymbol{Z}_{(k)}\|^2 \sim \mathcal{R}_{n-k}^2(h)$ and $T = \|\boldsymbol{Z}\|^2 \sim \mathcal{R}_n^2(h)$, we obtain the following relationship: $\mathcal{R}_n^2(h) \overset{d}{=} \mathcal{R}_k^2(h) + \mathcal{R}_{n-k}^2(h)$, since $\|\boldsymbol{z}\|^2 = \|\boldsymbol{z}_k\|^2 + \|\boldsymbol{z}_{(k)}\|^2$. From (4), it also follows that $T_k \,|\, T_{(k)} = t \sim \mathcal{R}_k^2(h_t)$, so that the variance and kurtosis parameters, $\alpha_{h_t^k}$ and $\kappa_{h_t^k}$, respectively, associated with the conditional elliptical model in (4) are functions of $\boldsymbol{z}_{(k)}$ through $t = \|\boldsymbol{z}_{(k)}\|^2$.

More details about the relationship between elliptical and radial-squared distributions can be found in Arellano-Valle *et al.* (2002a).

3. BAYESIAN INFERENCE FOR ELLIPTICAL LINEAR MODELS

In this section, we consider the elliptical regression linear model

$$\boldsymbol{y} \,|\, \boldsymbol{X}, \boldsymbol{\beta}, \phi \sim EL_n(\boldsymbol{X}\boldsymbol{\beta}, \phi^{-1}\boldsymbol{I}_n, h_{a_0\phi}^n), \tag{6}$$

where, from (5),

$$h_{a_0\phi}^n(u) = \frac{h^{n+d_0}(u + a_0\phi)}{h^{d_0}(a_0\phi)},$$

a_0 is known $(a_0 > 0)$ and h^{n+d_0} is a generator function of a $(n+d_0)$-dimensional elliptical distribution.

If we adopt the convention that $h^0(u) = 1$, $h^k(0) = c$, for some constant c, and $h_0^k = h^k$, then, when $d_0 = a_0 = 0$, (6) yields the standard elliptical model $EL_n(\boldsymbol{X}\boldsymbol{\beta}, \phi^{-1}\boldsymbol{I}_n, h)$, with $h = h^n$. In the latter case and under the non-informative prior distribution

$$\pi(\boldsymbol{\beta}, \phi) \propto \phi^{-1}. \tag{7}$$

Osiewalski and Steel (1993) have shown that the posterior of $\boldsymbol{\beta}$ is the same for all density generators of elliptical distributions h, and therefore, for the normal linear model. Similar results hold for the predictive analysis. Only the posterior distribution of ϕ is affected by departures from normality within the class of elliptical distributions. Some results related to posterior moments of ϕ are given in Osiewalski and Steel (1996) by considering the conditional distribution of ϕ given the location parameters $\boldsymbol{\beta}$ and the data $(\boldsymbol{y}, \boldsymbol{X})$. Arellano-Valle *et al.* (2000) provide an alternative proof of this fact, and determine the posterior distribution of ϕ explicitly, obtaining a convenient formula for examining the effects of departures of normality, which are reflected on the posterior of ϕ. Proposition 3.1 extends the previous results by considering a more general class of priors distributions for $(\boldsymbol{\beta}, \phi)$.

Specifically, we consider

$$\phi \mid h \sim a_0^{-1} \mathcal{R}^2_{d_0}(h) \tag{8}$$

which yields

$$\pi(\boldsymbol{\beta}, \phi \mid h) = \frac{(a_0 \pi)^{d_0/2}}{\Gamma(d_0/2)} \phi^{d_0/2-1}\, h^{d_0}(a_0 \phi)\, \pi(\boldsymbol{\beta} \mid \phi, h). \tag{9}$$

The dependence on h in (8) is reasonable, since in the present context the interpretation of the scale parameter changes with the density generator. We will interpret $d_0 = 0$ in (8) as the non-informative prior $\pi(\phi \mid h) \propto \phi^{-1}$, so that (9) is reduced to (7) when $\pi(\boldsymbol{\beta} \mid \phi, h)$ is constant.

Proposition 3.1. *Under* (6) *and* (9) *with* $\boldsymbol{\beta} \amalg \phi \mid h$, *we have*

$$\pi(\boldsymbol{\beta} \mid \boldsymbol{X}, \boldsymbol{y}, h) \propto \{a_{q(\boldsymbol{y})} + \|\boldsymbol{X}\boldsymbol{\beta} - \boldsymbol{X}\hat{\boldsymbol{\beta}}\|^2\}^{-(p+d)/2} \pi(\boldsymbol{\beta} \mid h) \tag{10}$$

and,

$$\pi(\phi \mid \boldsymbol{X}, \boldsymbol{y}, h) \propto \phi^{(p+d)/2-1} \int_{\Re^p} h^{p+d}(\{a_{q(\boldsymbol{y})} + \|\boldsymbol{X}\boldsymbol{\beta} - \boldsymbol{X}\hat{\boldsymbol{\beta}}\|^2\}\phi)\, \pi(\boldsymbol{\beta} \mid h)\, d\boldsymbol{\beta}$$

where $d = n - p + d_0$ *are the remaining degrees of freedom,* $a_{q(\boldsymbol{y})} = a_0 + q(\boldsymbol{y})$ *with* $q(\boldsymbol{y}) = \|\boldsymbol{y} - \boldsymbol{X}\hat{\boldsymbol{\beta}}\|^2 = (n-p)S^2$.

Assuming h *as given and* $\pi(\boldsymbol{\beta} \mid h)$ *to be constant, then*

$$\boldsymbol{\beta} \mid \boldsymbol{X}, \boldsymbol{y} \sim t_p(\hat{\boldsymbol{\beta}}, (\boldsymbol{X}^t \boldsymbol{X})^{-1}; a_{q(\boldsymbol{y})}, d) \tag{11}$$

and,

$$\phi \mid \boldsymbol{X}, \boldsymbol{y}, h \sim \frac{1}{a_{q(\boldsymbol{y})}} \mathcal{R}^2_d(h) \tag{12}$$

Proof. See Arellano-Valle *et al.* (2002a). □

Remark 3.1. Note from (10) that if the prior distribution of $\boldsymbol{\beta}$ does not depend on h, then the posterior distribution of $\boldsymbol{\beta}$ is invariant on the class of elliptical models under consideration and can be obtained under the normality assumption. In particular, applying (11) and the well-known properties of the Student-t distribution we get

$$\mathrm{E}(\boldsymbol{\beta} \mid \boldsymbol{X}, \boldsymbol{y}) = \hat{\boldsymbol{\beta}} \quad (n > p+2), \quad \mathrm{Var}(\boldsymbol{\beta} \mid \boldsymbol{X}, \boldsymbol{y}) = \frac{a_{q(\boldsymbol{y})}}{d-2} (\boldsymbol{X}^t \boldsymbol{X})^{-1} \quad (n > p+4)$$

and, provided that they exist,

$$\mathrm{E}(\phi \mid \boldsymbol{X}, \boldsymbol{y}, h) = \frac{d\alpha_h}{a_{q(\boldsymbol{y})}}, \quad \mathrm{Var}(\phi \mid \boldsymbol{X}, \boldsymbol{y}, h) = \left\{ \frac{d+2}{d}(\kappa_h + 1) - 1 \right\} \left\{ \frac{d\alpha_h}{a_{q(\boldsymbol{y})}} \right\}^2.$$

In particular, if $d_0 = a_0 = 0$, $d = n-p$ and $a_{q(\boldsymbol{y})} = (n-p)S^2$, where $S^2 = \|\boldsymbol{y} - \boldsymbol{X}\hat{\boldsymbol{\beta}}\|^2/n-p$, we have

$$(\boldsymbol{\beta} \mid \boldsymbol{X}, \boldsymbol{y}) \sim t_p(\hat{\boldsymbol{\beta}}, S^2(\boldsymbol{X}^t \boldsymbol{X})^{-1}, \quad (\phi \mid \boldsymbol{X}, \boldsymbol{y}, h) \sim (n-p)^{-1} S^{-2} \mathcal{R}^2_{n-p}(h)$$

These results reduce to those in Osiewalski and Steel (1993), see also Zellner (1976) .

Assuming in Proposition 3.1, h as given and $\pi(\boldsymbol{\beta}\,|\,h)$ to be constant, then (12) implies that the posterior distribution of $\sigma^2 = \phi^{-1}$ is

$$\sigma^2\,|\,\boldsymbol{X},\boldsymbol{y},h \sim a_{q(\boldsymbol{y})}\mathcal{IR}^2_d(h),$$

that is,

$$\pi(\sigma^2\,|\,\boldsymbol{X},\boldsymbol{y},h) = \frac{\pi^{1/2}}{\Gamma(d/2)}\,a^{d/2}_{q(\boldsymbol{y})}\left(\frac{1}{\sigma^2}\right)^{d/2+1} h^d\left(\frac{a_{q(\boldsymbol{y})}}{\sigma^2}\right).$$

Example 3.1. For the case of Student-t distribution we have $\pi(\phi\,|\,\boldsymbol{X},\boldsymbol{y}) = a^{-1}_{q(\boldsymbol{y})}\,F_{d,\nu}$; thus,

$$\mathrm{E}(\phi\,|\,\boldsymbol{X},\boldsymbol{y}) = \frac{\nu}{(\nu-2)a_{q(\boldsymbol{y})}},\quad (\nu>2),$$

$$\mathrm{Var}(\phi\,|\,\boldsymbol{X},\boldsymbol{y}) = \left\{\frac{(d+2)(\nu-2)}{d(\nu-4)}-1\right\}\left\{\frac{\nu}{(\nu-2)a_{q(\boldsymbol{y})}}\right\}^2,\quad (\nu>4).$$

Also, from (12), we can obtain an $100(1-\alpha)\%$ central posterior interval for ϕ and its general form is $a^{-1}_{q(\boldsymbol{y})} = [\mathcal{R}^2_d(h)_{;\frac{\alpha}{2}}, \mathcal{R}^2_d(h)_{;1-\frac{\alpha}{2}}]$. For Student-$t$ case we have $a^{-1}_{q(\boldsymbol{y})}[F_{d,\nu;\frac{\alpha}{2}}, F_{d,\nu;1-\frac{\alpha}{2}}]$.

In the next examples we discuss some consequences of the previous results related to the calibration problem and diagnostic in elliptical linear models. The problem of model comparison is approached in section 4.

Example 3.2 (Calibration problem). We consider the linear calibration model specified by $y_i = \beta_1+\beta_2x_i+\epsilon_i$; $i=1,\ldots,n$ and $y_0 = \beta_1+\beta_2x_0+\epsilon_0$ where β_1,β_2,x_0 are known quantities. The objective is to make inference on x_0 (the calibration parameter) based on $y_0,y_1,\ldots,y_n$ and $x_1,\ldots,x_n$ which are observed.

Let $\boldsymbol{y} = (y_1,\ldots,y_n)^t$ denote the observed response vector corresponding to the first step of the calibration experiment and $\boldsymbol{y}_* = (y_0,y^t)^t$. Similarly, we define the vectors $\boldsymbol{X} = (x_1,\ldots,x_n)^t$, $\boldsymbol{X}_* = (x_0,x^t)^t$ and the matrices $\boldsymbol{X} = [1_n\,;\boldsymbol{X}]$ and $\boldsymbol{X}_* = [1_{n*}\,;\boldsymbol{X}_*]$ where 1_k denotes the k-dimensional vector of ones. Assuming that $\boldsymbol{y}_*\,|\,\boldsymbol{X}_*,\boldsymbol{\beta},\phi \sim EL_{n*}(\boldsymbol{X}_*\boldsymbol{\beta},\,\phi^{-1}\boldsymbol{I}_{n*},h^{n*}_{a_0\phi})$, where $\boldsymbol{\beta} = (\beta_1,\beta_2)^t$, $n^* = n+1$ and $h^{n*}_{a_0\phi}$ is as in (5), and the prior distribution is given by (9) with $(\boldsymbol{\beta},x_0)\amalg(\phi,h)\,|\,\boldsymbol{X}$. It can be shown that the posterior distribution of x_0 does not depend on h, in fact, $\pi(x_0\,|\,\boldsymbol{X},\boldsymbol{y}_*) \propto f(\boldsymbol{y}_*\,|\,\boldsymbol{X},x_0)\pi(x_0\,|\,\boldsymbol{X})$, with

$$f(\boldsymbol{y}_*\,|\,\boldsymbol{X},x_0) = \int_{\boldsymbol{\beta}}\int_0^\infty f(\boldsymbol{y}_*\,|\,\boldsymbol{X}_*,\boldsymbol{\beta},\phi)\pi(\boldsymbol{\beta},\phi\,|\,\boldsymbol{X}_*)\,d\phi\,d\boldsymbol{\beta}.$$

Following the same ideas of Proposition 3.1, it can be shown that $f(\boldsymbol{y}_*\,|\,\boldsymbol{X}_*)$ is independent of h. Consequently, the posterior distribution of x_0 does not depend on h.

Remark 3.2. If in the above example we suppose $a_0 = d_0 = 0$ and $x_0\,|\,\boldsymbol{X} \sim t(\bar{x}, \frac{n+1}{n-3}S^2_x; n-3)$, $n>3$, then we obtain the result derived by Hoadley (1970), who proved that the inverse estimator is a Bayes estimator of x_0, when $\epsilon_* \sim N(0,\phi^{-1}\boldsymbol{I}_{n+1})$. For details see Branco *et al.* (2001).

Example 3.3 (Influence measures). Here we consider the question about how to assess influence of a subset of observations under the models of the form (6), with respect to the deletion of sets of observations. As usual in the literature, I denotes a subset with k elements of the set $N = \{1,\ldots,n\}$. Also, if a subset I of size k has been deleted, then $(\boldsymbol{y}_I,\boldsymbol{X}_I)$ and $(\boldsymbol{y}_{(I)},\boldsymbol{X}_{(I)})$

are the corresponding reduced data associated with the subsets I and $N - I$, respectively. Under assumptions of proposition 3.1, we have that the posterior distribution $\pi(\boldsymbol{\beta} \mid \boldsymbol{X}_{(I)}, \boldsymbol{y}_{(I)})$ does not depend on h. Thus, the influence of observations on the posterior of $\boldsymbol{\beta}$ is similar to that for the normal model as long as the prior distribution on $(\boldsymbol{\beta}, \phi)$ has the form (9) with $\boldsymbol{\beta} \amalg (\phi, h) \mid \boldsymbol{X}$. However, the posterior distribution of $\pi(\phi \mid \boldsymbol{X}_{(I)}, \boldsymbol{y}_{(I)}, h)$ depends on h (similarly $\pi(\boldsymbol{\beta}, \phi \mid \boldsymbol{X}_{(I)}, \boldsymbol{y}_{(I)}, h)$)

The influence of observations can be also approached (Kempthorne, 1986) in a Bayesian decision-theory framework. That is, by examining the influence of the observations as its impact on the posterior Bayes Risks. For example, if the interest focuses on the estimation of ϕ, with $\pi(\boldsymbol{\beta}) \propto$ constant, then the impact of the observations $(\boldsymbol{y}_{(I)}, \boldsymbol{X}_{(I)})$ measured as the ratio of risks is given by

$$R_{\phi}(I) = \frac{\mathrm{Var}(\phi \mid \boldsymbol{X}_{(I)}, \boldsymbol{y}_{(I)}, h)}{\mathrm{Var}(\phi \mid \boldsymbol{X}, \boldsymbol{y})} = \left\{ \frac{((d+2)/d)(\kappa_h + 1) - 1}{(d_{(I)} + 2)/d_{(I)})(\kappa_h + 1) - 1} \right\} \left(\frac{d\, a_{q(\boldsymbol{y})}}{d_{(I)} a_{q(\boldsymbol{y}_{(I)})}} \right)^2,$$

where

$$d_{(I)} = n - k + d_0, \quad a_{q(\boldsymbol{y}_{(I)})} = a_0 + (n-k-p) S^2_{(I)}, \quad S_{(I)} = (n-k-p)^{-1} \|\boldsymbol{y}_{(I)} - \boldsymbol{X}_{(I)} \hat{\boldsymbol{\beta}}\|^2.$$

A review of this topic appears in Arellano-Valle *et al.* (2002a).

The results presented in this section can be also extended to the problem of Bayesian inference on elliptical measurement error model (see Arellano *et al.*, 2002b).

4. DEFAULT BAYES FACTORS FOR ELLIPTICAL MODELS

In this section, we consider the q alternative standard elliptical linear models

$$M_j : \boldsymbol{y} = \boldsymbol{X}\boldsymbol{\beta} + \boldsymbol{\epsilon}_j, \quad \boldsymbol{\epsilon}_j \sim El_n\left(0, \phi^{-1} \boldsymbol{I}_n; h_j\right), \tag{13}$$

$j = 1, \ldots, q$, where the h_j's are n-dimensional generators, $\boldsymbol{\beta} \in \Re^k$ $(n > k)$, and the non-informative prior distribution given by (7).

As is well known, the usual Bayes factors based on non-informative or default improper priors, do not work, because the resulting Bayes factors are undetermined. Several solutions to this difficulty have been proposed and discussed by Berger and Pericchi (2001), they are called objective Bayes model selection methods.

In this section we discuss the model comparison problem under improper prior by using the Intrinsic Bayes Factors (IBF) (Berger and Pericchi, 1996b) and Fractional Bayes Factors (FBF) (O'Hagan, 1995).

Let us consider the following partition $\boldsymbol{y} = \left(\boldsymbol{y}^t_{(1)}, \boldsymbol{y}^t_{(2)}\right)^t$ and $\boldsymbol{X} = \left(\boldsymbol{X}^t_{(1)}, \boldsymbol{X}^t_{(2)}\right)^t$ where $\boldsymbol{y}_{(i)} \in \Re^{n_i}$ $(i = 1, 2)$, $n = n_1 + n_2$ and the matrix $\boldsymbol{X}_{(i)}$ has dimension $n_i \times k$.

Proposition 4.1. *If rank*$(\boldsymbol{X}_{(1)}) = k < n_1$ *then, for each model* M_j *in* (13)*, the marginal density of the sub-vector* $\boldsymbol{y}_{(1)}$ *is given by*

$$m_j^N\left(\boldsymbol{y}_{(1)} \mid \boldsymbol{X}_{(1)}\right) = \frac{\Gamma((n_1 - k)/2)}{(\sqrt{\pi})^{n_1 - k} \left(|\boldsymbol{X}^t_{(1)} \boldsymbol{X}_{(1)}| \right)^{\frac{1}{2}}} \left\| \boldsymbol{y}_{(1)} - \boldsymbol{X}_{(1)} \hat{\boldsymbol{\beta}}_{(1)} \right\|^{-(n_1 - k)},$$

where $\hat{\boldsymbol{\beta}}_{(1)} = \left(\boldsymbol{X}^t_{(1)} \boldsymbol{X}_{(1)}\right)^{-1} \boldsymbol{X}^t_{(1)} \boldsymbol{y}_{(1)}$.

Proof. When $a_0 = d_0 = 0$ we can show that

$$f(\boldsymbol{y}, \boldsymbol{\beta} \,|\, \boldsymbol{X}, h) = \int_0^\infty \phi^{n/2-1}\, h\left(\phi \left\|\boldsymbol{y} - \boldsymbol{X}\boldsymbol{\beta}\right\|^2\right) d\phi = \frac{\Gamma(n/2)}{\pi^{n/2}} \left\|\boldsymbol{y} - \boldsymbol{X}\boldsymbol{\beta}\right\|^{-n},$$

which does not depend on h, and can be rewritten as

$$f\left(\boldsymbol{y}_{(1)}, \boldsymbol{y}_{(2)}, \boldsymbol{\beta} \,|\, \boldsymbol{X}\right) = \frac{\Gamma(n_1/2)\, t_{n_2}\left(\boldsymbol{y}_{(2)} \,|\, \boldsymbol{X}_{(2)}\boldsymbol{\beta}, n_1 \left\|\boldsymbol{y}_{(1)} - \boldsymbol{X}_{(1)}\boldsymbol{\beta}\right\|^{-2} \boldsymbol{I}_{n_2}, n_1\right)}{\pi^{n_1/2} \left\|\boldsymbol{y}_{(1)} - \boldsymbol{X}_{(1)}\boldsymbol{\beta}\right\|^{n_1}},$$

where $t_k(\boldsymbol{y} \,|\, \boldsymbol{\mu}, \Sigma, \nu)$ is density of $t_k(\boldsymbol{\mu}, \Sigma, \nu)$ distribution. If now integrating out $\boldsymbol{y}_{(2)}$ we obtain

$$f\left(\boldsymbol{y}_{(1)}, \boldsymbol{\beta} \,|\, \boldsymbol{X}_{(1)}\right) = \frac{\Gamma(n_1/2)}{\pi^{n_1/2}} \left\|\boldsymbol{y}_{(1)} - \boldsymbol{X}_{(1)}\hat{\boldsymbol{\beta}}_{(1)}\right\|^{-n_1} \left[1 + \frac{\left\|\boldsymbol{X}_1\boldsymbol{\beta} - \boldsymbol{X}_1\hat{\boldsymbol{\beta}}_1\right\|^2}{S_{(1)}^2(n_1 - k)}\right]^{-n_1/2},$$

where $S_{(1)}^2 = \frac{1}{n_1 - k} \left\|\boldsymbol{y}_{(1)} - \boldsymbol{X}_{(1)}\hat{\boldsymbol{\beta}}_{(1)}\right\|^2$. Since the last factor is the kernel of the Student-t density $t_k\left(\boldsymbol{\beta} \,|\, \hat{\boldsymbol{\beta}}_{(1)}, S_{(1)}^{-2}\boldsymbol{X}_{(1)}^t\boldsymbol{X}_{(1)}, n_1 - k\right)$ and $n_1 > k$, the result follows. □

Remark 4.1. From Proposition 4.1, the marginal density of any sub-vector $\boldsymbol{y}_{(1)}$, with $n_1 > k$, does not depend on the specific elliptical model under consideration. A similar result is obtained by Berger*et al.* (1998) to compare models of the form (13), but for a wider class of models. However, their result is not valid for $n_1 > k + 1$.

4.1. *Intrinsic Bayes Factor*

The general strategy for computing IBF's begins with the determination of a proper and minimal training sample. It is known that (Berger and Pericchi, 1996a) the minimal training sample for the elliptical models in (13) is a sub-vector $\boldsymbol{y}(l)$ of size $n_1 = k + 1$ such that the corresponding sub-matrix $\boldsymbol{X}(l)$ is of full rank. Computation to compare two elliptical models M_1 and M_2, yields the following PBF for data $\boldsymbol{y}$:

$$B_{12}(l) = \frac{m_1^N(\boldsymbol{y} \,|\, \boldsymbol{X})}{m_2^N(\boldsymbol{y} \,|\, \boldsymbol{X})} \cdot \frac{m_2^N(\boldsymbol{y}(l) \,|\, \boldsymbol{X}(l))}{m_1^N(\boldsymbol{y}(l) \,|\, \boldsymbol{X}(l))} = 1.$$

Therefore the IBF's would not be useful in order to compare the models in (13). However, the IBF's are useful and easy to calculate upon comparing elliptical linear models with different design matrices.

We now consider the comparison between the elliptical lineal models

$$M_j : \boldsymbol{y} = \boldsymbol{X}_j\boldsymbol{\beta}_j + \boldsymbol{\epsilon}_j, \quad \boldsymbol{\epsilon}_j \sim El_n\left(0, \phi^{-1}\boldsymbol{I}_n; h_j\right) \tag{14}$$

$j = 1, \ldots, q$ where the h_j's are the generators, $n > \max_j\{k_j\}$ and $\boldsymbol{\beta}_j \in \Re^{k_j}$.

Proposition 4.2. *The IBF's in order to compare any two models M_1 and M_2 of type* (14) *do not depend on h_1 and h_2.*

Proof. Note from Proposition 4.1 that the PBF does not depend on the generators. □

The previous result is very useful since it allows to calculate IBF's to compare elliptical linear models with different design matrices using the results of Berger and Pericchi (1996a) relating to the IBF's for normal linear models.

4.2. *Fractional Bayes Factor*

Another alternative approach to compare models is the FBF developed in O'Hagan (1995). As mentioned by this author, the FBF has a series of advantages over the IBF, for example it is easier to compute than IBF. It is necessary to note that those models specified by (14) differ in two aspects, the design matrix $\boldsymbol{X}_j$ and the generator h_j. The next results are related to the comparison of two models $M_1 = (\boldsymbol{X}_1, h_1)$ and $M_2 = (\boldsymbol{X}_2, h_2)$.

Proposition 4.3. *The FBF to compare two models M_1 and M_2 in (14) with design matrices of full rank is*

$$B_b(\boldsymbol{y}) = \frac{\Gamma\left(\frac{n-k_1}{2}\right)\Gamma\left(\frac{bn-k_2}{2}\right)\left(\left\|\boldsymbol{y}-\boldsymbol{X}_2\hat{\boldsymbol{\beta}}_2\right\|\right)^{n(1-b)}\int_0^\infty u^{\frac{bn}{2}-1}h_2^b(u)\,du}{\Gamma\left(\frac{n-k_2}{2}\right)\Gamma\left(\frac{bn-k_1}{2}\right)\left(\left\|\boldsymbol{y}-\boldsymbol{X}_1\hat{\boldsymbol{\beta}}_1\right\|\right)^{n(1-b)}\int_0^\infty u^{\frac{bn}{2}-1}h_1^b(u)\,du},$$

where $0 < b < 1$, $bn > \max_j\{k_j\}$, and $\hat{\boldsymbol{\beta}}_j = \left(\boldsymbol{X}_j^t\boldsymbol{X}_j\right)^{-1}\boldsymbol{X}_j^t\boldsymbol{y}$, $j = 1, 2$.

Proof. In this case the FBF is given by $B_b(\boldsymbol{y}) = \frac{q_1(b,\boldsymbol{y})}{q_2(b,\boldsymbol{y})}$, where

$$q_j(b,\boldsymbol{y}) = \frac{\int \pi(\boldsymbol{\beta}_j,\phi)\, f(\boldsymbol{y}\,|\,\boldsymbol{X}_j,\boldsymbol{\beta}_j,\phi,h_j)\, d\phi d\boldsymbol{\beta}_j}{\int \pi(\boldsymbol{\beta}_j,\phi)\, f^b(\boldsymbol{y}\,|\,\boldsymbol{X}_j,\boldsymbol{\beta}_j,\phi,h_j)\, d\phi d\boldsymbol{\beta}_j}$$

and $j = 1, 2$. Using the change of variables $u_j = \phi\left\|\boldsymbol{y}-\boldsymbol{X}_j\boldsymbol{\beta}_j\right\|^2$ and integrating out $\boldsymbol{\beta}_j$ we obtain that the denominator of $q_j(b,\boldsymbol{y})$ is given by

$$\frac{\Gamma\left(\frac{bn-k_j}{2}\right)\pi^{k_j/2}}{\Gamma\left(\frac{bn}{2}\right)\left(|\,\boldsymbol{X}_j^t\boldsymbol{X}_j\,|\right)^{1/2}}\left(\left\|\boldsymbol{y}-\boldsymbol{X}_j\hat{\boldsymbol{\beta}}_j\right\|\right)^{-(bn-k_j)}\cdot\int u_j^{\frac{bn}{2}-1}h_j^b(u_j)\,du_j.$$

Now, the numerator of $q_j(b,\boldsymbol{y})$ is just the predictive density of $\boldsymbol{y}$ under model j. Thus, from Proposition 4.1 we have that the numerator is given by

$$\frac{\Gamma\left(\frac{n-k_j}{2}\right)}{(\sqrt{\pi})^{n-k_j}\left(|\,\boldsymbol{X}_j^t\boldsymbol{X}_j\,|\right)^{1/2}}\left(\left\|\boldsymbol{y}-\boldsymbol{X}_j\hat{\boldsymbol{\beta}}_j\right\|\right)^{-(n-k_j)}.$$

Thus,

$$q_j(b,\boldsymbol{y}) = \frac{\Gamma\left(\frac{n-k_j}{2}\right)\Gamma\left(\frac{bn}{2}\right)\left(\left\|\boldsymbol{y}-\boldsymbol{X}_j\hat{\boldsymbol{\beta}}_j\right\|\right)^{n(b-1)}}{\Gamma\left(\frac{bn-k_j}{2}\right)(\sqrt{\pi})^n\int u_j^{\frac{bn}{2}-1}h_j^b(u_j)\,du_j},$$

as required. □

The FBF for especial cases are presented in what follows.

Corollary 4.1 *The FBF for comparing two models M_1 and M_2 of type (13) is given by*

$$B_b(\boldsymbol{y}) = \frac{\int_0^\infty u^{\frac{bn}{2}-1} h_2^b(u)\,du}{\int_0^\infty u^{\frac{bn}{2}-1} h_1^b(u)\,du}.$$

This corollary shows the lack of sensibility of the FBF when distinguishing between two different elliptical linear models, because the FBF depends on the data through the sample size n only. We note also that in order to compare models with different design matrices by using the result of Proposition 4.3 it is necessary to know $I(b, h_j) = \int_0^\infty u^{\frac{bn}{2}-1} h_j^b(u)\,du$. Table 4.1 shows the value of $I(b, h_j)$ for different generator density functions.

Table 4.1. *Some subclasses of n-dimensional spherical distributions, their density generator function, radial squared distributions and $I(b, h)$.*

Distribution	Density generator function	$I(b, h)$
Normal	$h_\phi(u) := (2\pi)^{-n/2} \exp\{-u/2\}$	$\frac{\Gamma\left(\frac{bn}{2}\right)}{(b\pi)^{bn/2}}$
Contaminated normal	$(1-\varepsilon)\,h_\phi(u) + \varepsilon\sigma^{-\frac{n}{2}} h_\phi\left(\frac{u}{\sigma}\right)$, $0 < \varepsilon < 1, \sigma > 0$	$\frac{\Gamma\left(\frac{bn}{2}\right)\left(1-\varepsilon+\varepsilon\sigma^{n(b-1)/2}\right)}{(b\pi)^{bn/2}}$
Student-t	$\frac{\Gamma\left(\frac{n+\nu}{2}\right)\nu^{\nu/2}}{\Gamma\left(\frac{\nu}{2}\right)\pi^{n/2}} \{\nu + u\}^{-\frac{(n+\nu)}{2}}, \nu > 0$	$\frac{\Gamma^b\left(\frac{n+\nu}{2}\right)\Gamma\left(\frac{bn}{2}\right)\Gamma\left(\frac{b\nu}{2}\right)}{\Gamma^b\left(\frac{\nu}{2}\right)\Gamma\left(b\frac{n+\nu}{2}\right)\pi^{bn/2}}$
Generalized Student-t	$\frac{\Gamma\left(\frac{n+\nu}{2}\right)\lambda^{\nu/2}}{\Gamma\left(\frac{\nu}{2}\right)\pi^{n/2}} \{\lambda + u\}^{-\frac{(n+\nu)}{2}}, \nu, \lambda > 0$	$\frac{\Gamma^b\left(\frac{n+\nu}{2}\right)\Gamma\left(\frac{bn}{2}\right)\Gamma\left(\frac{b\nu}{2}\right)}{\Gamma^b\left(\frac{\nu}{2}\right)\Gamma\left(b\frac{n+\nu}{2}\right)\pi^{bn/2}}$
Pearson Type II	$\frac{\Gamma\left(\frac{n+\nu}{2}\right)}{\Gamma\left(\frac{\nu}{2}\right)\pi^{n/2}} (1-u)^{\frac{\nu}{2}-1}, \nu > 0$	$\frac{\Gamma^b\left(\frac{n+\nu}{2}\right)\Gamma\left(\frac{bn}{2}\right)\Gamma\left(b\frac{\nu-2}{2}+1\right)}{\Gamma^b\left(\frac{\nu}{2}\right)\Gamma\left(b\frac{n+\nu-2}{2}+1\right)\pi^{bn/2}}$
Power exponential	$\frac{\Gamma\left(\frac{n}{2}\right)\nu}{\Gamma\left(\frac{n}{2\nu}\right)\pi^{\frac{n}{2}} 2^{\frac{n}{2\nu}}} \exp\{-u^\nu/2\}, \nu > 0$	$\frac{\Gamma^b\left(\frac{n}{2}\right)\Gamma\left(\frac{bn}{2\nu}\right)\nu^{b-1}}{\Gamma^b\left(\frac{n}{2\nu}\right)\pi^{\frac{bn}{2}} b^{\frac{bn}{2\nu}}}$
Kotz type	$\frac{\Gamma\left(\frac{n}{2}\right)\nu\rho^{\frac{2\lambda+n}{2\nu}}}{\Gamma\left(\frac{2\lambda+n}{2\nu}\right)\pi^{\frac{n}{2}} 2^{\frac{2\lambda+n}{2\nu}}} u^\lambda \exp\{-\rho u^\nu/2\}$, $\rho, \nu > 0, 2\lambda + n > 0$	$\frac{\Gamma^b\left(\frac{n}{2}\right)\Gamma\left(b\frac{2\lambda+n}{2\nu}\right)\nu^{b-1}}{\Gamma^b\left(\frac{2\lambda+n}{2\nu}\right)\pi^{\frac{bn}{2}} b^{b\frac{2\lambda+n}{2\nu}}}$

From Table 4.1, note that for the generalized Student-t distribution the value of $I(b, h)$ does not depend on the parameter λ, and therefore this value is the same for the Student-t distribution. Similarly, the value of $I(b, h)$ for the Kotz-type distribution does not depend on the parameter ρ.

The next corollary shows that for comparing two elliptical linear models with different design matrices and common generator function, it is enough to compare two normal linear models with different design matrices, see O'Hagan (1995).

Corollary 4.2. *The FBF for comparing two linear models $M_1 := (\boldsymbol{X}_1, h)$ and $M_2 := (\boldsymbol{X}_2, h)$ is given by*

$$B_b(\boldsymbol{y}) = \frac{\Gamma\left(\frac{n-k_1}{2}\right)\Gamma\left(\frac{bn-k_2}{2}\right)}{\Gamma\left(\frac{n-k_2}{2}\right)\Gamma\left(\frac{bn-k_1}{2}\right)}\left(\frac{\left\|\boldsymbol{y} - \boldsymbol{X}_2\hat{\boldsymbol{\beta}}_2\right\|}{\left\|\boldsymbol{y} - \boldsymbol{X}_1\hat{\boldsymbol{\beta}}_1\right\|}\right)^{n(1-b)},$$

where $0 < b < 1$, $bn > \max\{k_j\}$, and $\hat{\boldsymbol{\beta}}_j = \left(\boldsymbol{X}_j^t \boldsymbol{X}_j\right)^{-1} \boldsymbol{X}_j^t \boldsymbol{y}$, $j = 1, 2$.

Observe that, under conditions of the previous corollary, both the IBF and FBF remain invariant for the class of elliptical distributions.

4.3. *Model Comparison As a Decision Problem*

A more general approach is to consider the problem of model comparison within the decision theory framework, as described in Bernardo and Smith (1994). Following the notation used by those authors, we will call $\boldsymbol{\omega}$ the unknown state of the nature. In our case, the objective could be inference about $(\boldsymbol{\beta}, \phi)$, $(y_{n+1}, \ldots, y_m)$, etc. Thus, we would like to obtain the conditional distribution of $\boldsymbol{\omega}$ given $\boldsymbol{y}$ under the true model, by assuming that this model is contained in the class of models that we are comparing. Also, we will consider that, m_i means that, given data $\boldsymbol{y}$, we choose model M_i and a_j, $j \in J_i$ is some report of beliefs about $\boldsymbol{\omega}$ assuming model M_i.

Appropriate utility functions for these cases must be smooth, proper and local score functions; their definitions and more details can be found in Bernardo and Smith (1994). Under these conditions, these authors show that for proper score functions, $u_i(a_j, \boldsymbol{\omega})$, the optimal choice of a_j, $j \in J_i$, is $a_i^* = f(\boldsymbol{\omega} \mid \boldsymbol{y}, m_i)$ and, therefore the utility function would be $u(m_i, a_i^*, \boldsymbol{\omega}) = u_i(f(\boldsymbol{\omega} \mid \boldsymbol{y}, m_i), \boldsymbol{\omega}), i = 1, \ldots, q$. But also, under the assumption of local score function, we have that $u(m_i, a_i^*, \boldsymbol{\omega}) = A \log f(\boldsymbol{\omega} \mid \boldsymbol{y}, m_i) + B(\boldsymbol{\omega}), i = 1, \ldots, q$, where $A > 0$ is a constant and $B(\cdot)$ is a function of $\boldsymbol{\omega}$. Therefore, the optimal model is that which maximizes the expected utility function

$$\bar{u}(m_i \mid \boldsymbol{y}) = \int \{A \log f(\boldsymbol{\omega} \mid \boldsymbol{y}, m_i) + B(\boldsymbol{\omega})\} f(\boldsymbol{\omega} \mid \boldsymbol{y})\, d\boldsymbol{\omega}, \tag{15}$$

provided that this exists. In our case, if $P(Mi) = 1/q$ for all $i = 1, \ldots, q$, then

$$P(M_j \mid \boldsymbol{y}) = \left(\sum_{i=1}^{q} \frac{m_i^N(\boldsymbol{y} \mid \boldsymbol{X}_i)}{m_j^N(\boldsymbol{y} \mid \boldsymbol{X}_j)}\right)^{-1}.$$

If additionally all design matrices are equal then $P(M_j \mid \boldsymbol{y}) = 1/q$ and

$$f(\boldsymbol{\omega} \mid \boldsymbol{y}) = \frac{1}{q}\sum_{j=1}^{q} f(\boldsymbol{\omega} \mid \boldsymbol{y}, m_j). \tag{16}$$

That means that, except for different design matrices, that there is no posterior preference for any model. In such a case, the expected utility $\bar{u}(m_i \mid \boldsymbol{y})$ depends on the model through $f(\boldsymbol{\omega} \mid \boldsymbol{y}, m_j)$.

In what follows, we present results for computing $\bar{u}(m_i \mid \boldsymbol{y})$ to compare models $M_i = (\boldsymbol{X}_i, h_i)$ and $M_j = (\boldsymbol{X}_j, h_j)$ for different choices of $\boldsymbol{\omega}$. We will also assume that $n > \max_j\{k_j\}$, such that if $\boldsymbol{\omega} = (\boldsymbol{\beta}, \phi)$ then $f(\boldsymbol{\omega} \mid \boldsymbol{y}, m_j)$ and, therefore, $f(\boldsymbol{\omega} \mid \boldsymbol{y})$ are proper.

Proposition 4.4. *If* $\omega = (\boldsymbol{\beta}, \phi)$ *and* $\boldsymbol{X}_i = \boldsymbol{X}\ \forall\, i = 1, \ldots, q$,

$$\bar{u}\left(m_i \,|\, \boldsymbol{y}\right) = \frac{\pi^{\frac{n}{2}} A}{q\Gamma\left(\frac{n}{2}\right)} \sum_{j=1}^{q} \int \log\left[v^{\frac{n}{2}-1} h_i\left(v\right)\right] v^{\frac{n}{2}-1} h_j\left(v\right) dv$$
$$- A\mathrm{E}_{t_k}\left[\log\left(\|\boldsymbol{y} - \boldsymbol{X}\boldsymbol{\beta}\|^{n-2}\right)\right] - A\log\left[m^N\left(\boldsymbol{y} \,|\, \boldsymbol{X}\right)\right] + \mathrm{E}\left[B\left(\boldsymbol{\beta}, \phi\right) \,|\, \boldsymbol{y}\right],$$

where the expected value $\mathrm{E}_{t_k}(\cdot)$ *is calculated with respect to the Student-t distribution* $t_k\left(\hat{\boldsymbol{\beta}}, S^{-2}\boldsymbol{X}^t\boldsymbol{X}, n-k\right)$.

Proof. Since in this case we are assuming that $\omega = (\boldsymbol{\beta}, \phi)$ and $\boldsymbol{X}_i = \boldsymbol{X}$, $i = 1, \ldots, q$, it follows that $m_i^N\left(\boldsymbol{y} \,|\, \boldsymbol{X}\right) = m^N\left(\boldsymbol{y} \,|\, \boldsymbol{X}\right)$, $i = 1, \ldots, q$, so that

$$f\left(\boldsymbol{\beta}, \phi \,|\, \boldsymbol{y}, m_i\right) = \frac{\phi^{n/2-1}\, h_i\left(\phi\, \|\boldsymbol{y} - \boldsymbol{X}\boldsymbol{\beta}\|^2\right)}{m^N\left(\boldsymbol{y} \,|\, \boldsymbol{X}\right)}.$$

Then, from (16) and the previous equation,

$$\bar{u}\left(m_i \,|\, \boldsymbol{y}\right) = \frac{A\int \log\left[\phi^{\frac{n}{2}-1} h_i\left(\phi\, \|\boldsymbol{y} - \boldsymbol{X}\boldsymbol{\beta}\|^2\right)\right] \sum_{j=1}^{q} \phi^{\frac{n}{2}-1} h_j\left(\phi\, \|\boldsymbol{y} - \boldsymbol{X}\boldsymbol{\beta}\|^2\right) d\left(\boldsymbol{\beta}, \phi\right)}{q m^N\left(\boldsymbol{y} \,|\, \boldsymbol{X}\right)}$$
$$- A\log\left[m^N\left(\boldsymbol{y} \,|\, \boldsymbol{X}\right)\right] + \mathrm{E}\left[B\left(\boldsymbol{\beta}, \phi\right) \,|\, \boldsymbol{y}\right].$$

In the previous integral, the usual change of variable $v = \phi\, \|\boldsymbol{y} - \boldsymbol{X}\boldsymbol{\beta}\|^2$, yields

$$\bar{u}\left(m_i \,|\, \boldsymbol{y}\right) = \frac{A}{q m^N\left(\boldsymbol{y} \,|\, \boldsymbol{X}\right)} \int \|\boldsymbol{y} - \boldsymbol{X}\boldsymbol{\beta}\|^{-n} d\boldsymbol{\beta} \sum_{j=1}^{q} \int \log\left[v^{\frac{n}{2}-1} h_i\left(v\right)\right] v^{\frac{n}{2}-1} h_j\left(v\right) dv$$
$$- \frac{A}{q m^N\left(\boldsymbol{y} \,|\, \boldsymbol{X}\right)} \int \frac{\log\left(\|\boldsymbol{y} - \boldsymbol{X}\boldsymbol{\beta}\|^{n-2}\right)}{\|\boldsymbol{y} - \boldsymbol{X}\boldsymbol{\beta}\|^{n}} d\boldsymbol{\beta} \sum_{j=1}^{q} \int v^{\frac{n}{2}-1} h_j\left(v\right) dv$$
$$- A\log\left[m^N\left(\boldsymbol{y} \,|\, \boldsymbol{X}\right)\right] + \mathrm{E}\left[B\left(\boldsymbol{\beta}, \phi\right) \,|\, \boldsymbol{y}\right].$$

Since

$$\|\boldsymbol{y} - \boldsymbol{X}\boldsymbol{\beta}\|^{-n} = \left\|\boldsymbol{y} - \boldsymbol{X}\hat{\boldsymbol{\beta}}\right\|^{-n} \left[1 + \frac{\left\|\boldsymbol{X}\boldsymbol{\beta} - \boldsymbol{X}\hat{\boldsymbol{\beta}}\right\|^2}{S^2\left(n-k\right)}\right]^{-(n-k+k)/2},$$

and the last factor is the kernel of a Student-t distribution, the proof is concluded. □

Remark 4.2. From the above proposition and the previous results it follows that if $\omega = \boldsymbol{\beta}$ and $\boldsymbol{X}_i = \boldsymbol{X}$ for all $i = 1, \ldots, q$, then

$$\bar{u}(m_i \,|\, \boldsymbol{y}) = A\mathrm{E}_{t_k}[\log f(\boldsymbol{\beta} \,|\, \boldsymbol{y}) \,|\, \boldsymbol{y}] + \mathrm{E}_{t_k}[B(\boldsymbol{\beta}) \,|\, \boldsymbol{y}],$$

which do not depends on h.

Proposition 4.5. *If* $\omega = \phi$ *then*

$$\bar{u}\left(m_i \mid \boldsymbol{y}\right) = \frac{A}{\sum_{r=1}^{q} m_r^N\left(\boldsymbol{y} \mid \boldsymbol{X}_r\right)} \int \log\left[(n-k_i)\, s_i^2 v^{\frac{n-k_i}{2}-1} h_i(v)\right] v^{-1}$$

$$\sum_{j=1}^{q} \frac{(\pi v)^{\frac{n-k_j}{2}}}{\Gamma\left(\frac{n-k_j}{2}\right)} \left(\frac{(n-k_j)\, s_j^2}{(n-k_i)\, s_i^2}\right)^{\frac{n-k_j}{2}} h_j\left(\frac{(n-k_j)\, s_j^2}{(n-k_i)\, s_i^2} v\right) m_j^N\left(\boldsymbol{y} \mid \boldsymbol{X}_j\right) dv$$

$$+\,\mathrm{E}\left[B(\phi) \mid \boldsymbol{y}\right] + A \log\left[\frac{\pi^{(n-k_i)/2}}{\Gamma\left(\frac{n-k_i}{2}\right)}\right].$$

Proof. This follows from Proposition 3.1, by noting that when $a_0 = d_0 = 0$ then

$$f\left(\phi \mid \boldsymbol{y}, m_i\right) = \frac{\pi^{(n-k_i)/2}}{\Gamma\left(\frac{n-k_i}{2}\right)} \left((n-k_i)\, S_i^2 \phi\right)^{(n-k_i)/2} \phi^{-1} h_i\left((n-k_i)\, S_i^2 \phi\right)$$

where $S_i^2 = (n-k_i)^{-1} \left\| \boldsymbol{y} - \boldsymbol{X}_i \hat{\boldsymbol{\beta}}_i \right\|^2$. The change of variable $v = (n-k_i)\, S_i^2 \phi$ establishes the desired result. □

Remark 4.3. If in Proposition 4.5 we set $\boldsymbol{X}_i = \boldsymbol{X}$ for all $i = 1, \ldots, q$, then

$$\bar{u}\left(m_i \mid \boldsymbol{y}\right) = \frac{A\pi^{(n-k)/2}}{q\Gamma\left(\frac{n-k}{2}\right)} \sum_{j=1}^{q} \int \log\left[v^{\frac{n-k}{2}-1} h_i(v)\right] v^{\frac{n-k}{2}-1} h_j(v)\, dv$$

$$+A \log\left[\frac{\pi^{(n-k)/2} (n-k)\, s^2}{\Gamma\left(\frac{n-k}{2}\right)}\right] + \mathrm{E}\left[B(\phi) \mid \boldsymbol{y}\right],$$

which depends on h_i.

Let us suppose now that our interest is to select models to make inference about future observations $y_{n+1}, \ldots, y_m$. Thus, we will assume that the vector $\boldsymbol{y}$, as well as the matrix $\boldsymbol{X}$, are partitioned as $\boldsymbol{y} = \left(\boldsymbol{y}_{(n)}^t, \boldsymbol{y}_{(m-n)}^t\right)^t$ and $\boldsymbol{X}_i = \left(\boldsymbol{X}_{i(1)}^t, \boldsymbol{X}_{i(2)}^t\right)^t$ with $\boldsymbol{y}_{(n)} = (y_1, \ldots, y_n)^t$ and $\boldsymbol{y}_{(m-n)} = (y_{n+1}, \ldots, y_m)^t$ and $\boldsymbol{X}_{i(n)}, \boldsymbol{X}_{i(m-n)}$ are $n \times k$ and $(m-n) \times k$ dimensional known design matrices. Also, $\boldsymbol{y} \sim El_m\left(\boldsymbol{X}_i \boldsymbol{\beta}, \phi^{-1} \boldsymbol{I}_m; h_i\right)$ and we are comparing the models $M_i = (\boldsymbol{X}_i, h_i)$ and $M_j = (\boldsymbol{X}_j, h_j)$.

Proposition 4.6. *If* $\boldsymbol{\omega} = (y_{n+1}, \ldots, y_m)$

$$\bar{u}\left(m_i \mid \boldsymbol{y}_{(n)}\right) = \frac{A}{\sum_{r=1}^{q} m_r^N\left(\boldsymbol{y}_{(n)} \mid \boldsymbol{X}_r\right)} \int \log\left[f\left(\boldsymbol{y}_{(m-n)} \mid \boldsymbol{y}_{(n)}, \boldsymbol{X}_i\right)\right]$$

$$\sum_{j=1}^{q} f\left(\boldsymbol{y}_{(m-n)} \mid \boldsymbol{y}_{(n)}, \boldsymbol{X}_j\right) m_j^N\left(\boldsymbol{y}_{(n)} \mid \boldsymbol{X}_j\right) d\boldsymbol{y}_{(m-n)}$$

$$+\,\mathrm{E}\left[B\left(\boldsymbol{y}_{(m-n)}\right) \mid \boldsymbol{y}_{(n)}\right],$$

where

$$\left(\boldsymbol{y}_{(m-n)} \mid \boldsymbol{y}, \boldsymbol{X}_i\right) \sim t_{m-n}\left(\boldsymbol{X}_{i(m-n)} \hat{\boldsymbol{\beta}}_i, S_i^{-2} \boldsymbol{W}_i^{-1}, n-k_i\right)$$

with $\boldsymbol{W}_i = \boldsymbol{X}_{i(m-n)} \left(\boldsymbol{X}_i^t \boldsymbol{X}_i\right)^{-1} \boldsymbol{X}_{i(m-n)}^t + \boldsymbol{I}_{m-n}$.

Proof. The result follows from Osiewalski and Steel (1993), where it is shown that

$$\boldsymbol{y}_{(m-n)} \,|\, \boldsymbol{y}_{(n)}, M_i \sim t_{m-n}\left(\boldsymbol{X}_{i(m-n)}\hat{\boldsymbol{\beta}}_i, S_i^{-2}\mathbf{W}_i^{-1}, n-k_i\right),$$

which depends on M_i through $\boldsymbol{X}_i$ only. □

Remark 4.4 In the above proposition, if $\boldsymbol{X}_i = \boldsymbol{X}$ for all $i = 1, \ldots, q$, then

$$\bar{u}\left(m_i \,|\, \boldsymbol{y}_{(n)}\right) = A\mathrm{E}_{t_{m-n}}\left[\log\left[f\left(\boldsymbol{y}_{(m-n)} \,|\, \boldsymbol{y}_{(n)}, \boldsymbol{X}\right)\right] \,|\, \boldsymbol{y}_{(n)}\right] + \mathrm{E}_{t_{m-n}}\left[B\left(\boldsymbol{y}_{(m-n)}\right) \,|\, \boldsymbol{y}_{(n)}\right],$$

where the expected value $\mathrm{E}_{t_{m-n}}(\cdot)$ is taken with respect to the Student-t distribution $t_{m-n}\left(\boldsymbol{X}_{(m-n)}\hat{\boldsymbol{\beta}}, S_i^{-2}\boldsymbol{W}^{-1}, n-k\right)$.

We note from Propositions 4.4 and 4.5 that the expected utility function $\bar{u}\,(m_i \,|\, \boldsymbol{y})$ depends on the model m_i through $I(h_i) = \int \log\left[v^{(n-k)/2-1}h_i\,(v)\right] v^{(n-k)/2-1}h_j\,(v)\,dv$, which depends on the data only through the sample size n.

In general, the shape of the density (16) together with the fact that $f\,(\boldsymbol{\beta} \,|\, \boldsymbol{y}, M_i)$ and $f\,(y_{n+1}, \ldots, y_m \,|\, \boldsymbol{y}, M_i)$ do not depend on the elliptical model h_i, and

$$f\,(\boldsymbol{\beta}, \phi \,|\, \boldsymbol{y}, M_i) \propto v_1^{n/2-1}h_i\,(v_1), \quad f\,(\phi \,|\, \boldsymbol{y}, M_i) \propto v_2^{(n-k_i/2-1}h_i\,(v_2),$$

with $v_1 = \phi\,\|\boldsymbol{y} - \boldsymbol{X}\boldsymbol{\beta}\|^2$ and $v_2 = \phi\|\boldsymbol{y} - \boldsymbol{X}\hat{\boldsymbol{\beta}}\|^2$, respectively, is not useful when selecting the most appropriate model after having observed the data $\boldsymbol{y}$.

The comparison of elliptical models for the errors using the marginal densities, in linear models with prior distribution $\pi\,(\boldsymbol{\beta}, \phi) \propto \phi^{-1}$, should be reexamined. On the other hand, if we have chosen an elliptical model, that is to say h_i is fixed, then the comparison is centered in the design matrices, and the comparison could be carried out satisfactorily using the IBF, the FBF or maximizing (15): in the case of the IBF's, the well-known results of Berger and Pericchi (1996a) could be used, and in the case of the FBF's convenient formulas can be obtained for many models.

5. CONCLUSIONS

In this paper we present a Bayesian analysis of the elliptical linear model under different prior specifications for the parameters. We show that when using squared-radial distributions for ϕ with $\boldsymbol{\beta} \amalg \phi \,|\, h$ and $\pi(\boldsymbol{\beta} \,|\, h)$ the posterior of $\boldsymbol{\beta}$ does not depend on h. Hence, the inference on $\boldsymbol{\beta}$ and other related problems like calibration, sensitivity analysis is the same as the one obtained under normality. Only the posterior of ϕ depends on h, even under improper prior considered here.

Moreover, the IBF to compare two elliptical linear models (with common design matrix) does not work, because the predictive distributions are the same for the models under comparison. On the other hand, even though the FBF depends on h, it depends on the data only through the sample size. Similar results are obtained when we adopt the perspective of decision analysis for model comparison. Other alternative methods for nested hypothesis testing that must be explored are presented by Bernardo (1999) and Pereira and Stern (1999), because these procedures involve all parameters in the models being compared.

Thus, many results obtained under the normal model remain valid under dependent elliptical models. In general, the results derived here for the dependent elliptical models do not hold for poly-elliptical models. However, in the representable case, the Gibbs sampler provides a natural framework for obtaining approximations to the posterior distributions (see Arellano-Valle *et al.*, 2000)

ACKNOWLEDGEMENTS

The authors thank Dr. Heleno Bolfarine and Dr. Wilfredo Palma for helpful discussions and important contributions to develop this work. We also would like to acknowledge Juan Rossel for computational support, Fondecyt Líneas Complementarias 8000004 and Programa MECE de Educación Superior 0103, Chile.

REFERENCES

Arellano-Valle, R. B., Bolfarine, H. and Iglesias, P. (1994). A predictivistic interpretation to multivariate t-distribution. *Test* **3**, 221–236.

Arellano-Valle, R. B., Galea-Rojas, M. and Iglesias, P. (2000). Bayesian sensitivity analysis in elliptical linear regression models. *J. Statist. Plann . Inference* **86**, 175–199.

Arellano-Valle, R. B., del Pino, G. and Iglesias, P. (2002a). Bayesian analysis for the spherical linear model. *Tech. Rep.*, Pontificia Universidad Católica de Chile.

Arellano-Valle, R. B., Iglesias, P., San Martín, E. and Silva, A. J. (2002b). Ultrastructural models: ellipticity and invariance. *Tech. Rep.*, Pontificia Universidad Católica de Chile.

Berger, J. O. and Pericchi, L. R. (1996a). The intrinsic Bayes factor for linear models. *Bayesian Statistics 5* (J. M. Bernardo, J. O. Berger, A. P. Dawid and A. F. M. Smith, eds). Oxford: Oxford University Press, 25–44 (with discussion).

Berger, J. O. and Pericchi, L. R. (1996b). The intrinsic Bayes factor for model selection and prediction. *J. Am. Statist. Ass.* **91**, 109–122.

Berger, J. O. and Pericchi, L. R. (2001). Objective Bayesian methods for model selection. *Model Selection* (P. Lahiri, ed.). IMS: Lecture Notes **38**.

Berger, J. O., Pericchi, L. R. and Varshavsky, J. (1998). Bayes factor and marginal distributions in invariant situations. *Sankhyā A* **60**, 307–321.

Bernardo, J. M. (1999). Nested hypothesis testing: The Bayesian reference criterion. *Bayesian Statistics 6* (J. M. Bernardo, J. O. Berger, A. P. Dawid and A. F. M. Smith, eds). Oxford: Oxford University Press, 101–130 (with discussion).

Bernardo, J. M. and Smith, A. F. M. (1994). *Bayesian Theory*. Chichester: Wiley.

Branco, M., Bolfarine, H., Iglesias, P. and Arellano-Valle, R. B. (2001). Bayesian analysis of the calibration problem under elliptical distributions. *J. Statist. Plann . Inference* **90**, 69–95.

Chib, S., Tiwari, R. C. and Jammalaladaka, S. R. (1988). Bayes prediction in regressions with elliptical errors. *J. Econometrics* **38**, 349–360.

Chu, K. (1973). Estimation and decision for linear systems with elliptical random processes. *IEEE Trans. Automatic Control* **18**, 499–505.

Fang, K. T., Kotz, S. and Ng, K. W. (1990). *Symmetric Multivariate and Related Distributions*. London: Chapman and Hall.

Hoadley, B. (1970).A Bayesian look at inverse linear regression. *J. Am. Statist. Ass.* **65**, 356–369.

Jammalamadaka, S. R., Tiwari, R. C. and Chib, S. (1987). Bayes prediction in the linear model with spherically symmetric errors. *Econometric Letter* **24**, 39–44.

Kelker, D. (1970). Distribution theory of spherical distributions and a location-scale parameter generalization. *Sankhyā A* **32**, 419–430.

Kempthorne, P. J. (1986). Decision-theoretic measures of influence in regression. *J. R. Statist. Soc. B* **48**, 370–378.

Loschi, R., Iglesias, P. and Arellano-Valle, R. B. (2002). Predictivistic characterization of t distribution with centered spherical symmetry. *Tech. Rep.*, Pontificia Universidad Católica de Chile.

Meinhold, R. J. and Singpurwalla, N. D. (1989). Robustification of kalman filter models. *J. Am. Statist. Ass.* **84**, 479–486.

O'Hagan, A. (1995). Fractional Bayes factors for model comparison. *J. R. Statist. Soc. B* **57**, 99–138.

Osiewalski, J. and Steel, M. F. J. (1993). Robust Bayesian inference in elliptical regression models. *Econometrics J.* **57**, 345–363.

Osiewalski, J. and Steel, M. F. J. (1996). Posterior moments of scale parameters in elliptical regression models. *Bayesian Analysis in Statistics and Econometrics: Essays in Honor of Arnold Zellner* (D. A. Berry, K. M. Chaloner and J. K. Geweke, eds). New York: Wiley, 323–335.

Pereira, C. and Stern, J. M. (1999). Evidence and credibility: full bayesian test of precise hypothesis. *Entropy* **1**, 104–115.

Zellner, A. (1976). Bayesian and non-Bayesian analysis of the regression model with multivariate Student-t error terms. *J. Am. Statist. Ass.* **71**, 400–405.

DISCUSSION

JUDITH ROUSSEAU (*Université Paris 5 and CREST, France*)

Models: This paper considers elliptical models in a regression context defined by

$$Z \sim h^n_{a_0\phi}(Z^t Z) = \frac{h^{n+d_0}(Z^t Z + a_0\phi)}{h^{d_0}(a_0\phi)}, \quad Z = \sqrt{\phi}(Y - X\beta), \tag{1}$$

where h is called the density generator and $h^n_{a_0\phi}$ is the conditional generator.

This paper proposes a Bayesian inference on this type of model, extending the results of Osiewalski and Steel.

Osiewalski and Steel (1993) have proposed the prior $\pi(\phi, \beta) \propto \phi^{-1}$ as a non-informative prior for the model with $a_0 = d_0 = 0$, since it corresponds to Bernardo's reference prior associated with the order $\{\beta\}, \{\phi\}$. They have shown that, under such a prior, the posterior of β was independent of the choice of h, and the resulting analysis was thus robust to the model.

Arellano-Valle *et al.* extend these results. As a natural extension of Osiewalski and Steel prior they propose the following class of priors: $a_0\phi$ corresponds to the norm of an elliptical vector with generator h in $\mathbb{R}^{d_0}$, as was suggested by the model and β can have any prior. When β is independent of ϕ they prove that the posterior of β is independent of the model h and a bayes estimator would then be robust.

Obviously, model selection in this case becomes a problem, since the posterior is independent of the model. Moreover, they prove that Bayes factors are useless to discriminate between different $h's$ and so is Bernardo and Smith method.

• It is very nice to have finite sample results such as these, as they give an insight in the structure of such elliptical models. Also, conjugate or pseudo-conjugate priors can be very useful when using hierarchical models, as they simplify the implementation of the posterior.

• The general class of models considered by Arellano-Valle *et al.* and defined by (1) has a nice interpretation. Indeed, $h^n_{a_0\phi}$ is a conditional generator, *i.e.*, it is the generator of the vector Z conditional on some extra vector U such that $U^2 = a_0$ and $U \in \mathbb{R}^{d_0}$. This is strongly related to James-Stein estimators when the scale parameter is unknown, see for instance Brandwein and Strawderman (1991).

• In this paper, these models are used to describe a vector $Y^n = (Y_1, ..., Y_n)$ of observations. As such, it is common to assume that Y^n is a stochastic process as defined by Kolmogorov, in other words the number of observations is not an active factor in the generation of Y^n. In this case, the only elliptical models that should be considered are the scale mixtures of normals:

$$h(u) = \int_0^\infty \left(\frac{c}{2\pi}\right)^{n/2} e^{-uc/2} F(dc), \tag{2}$$

where F is a distribution function that satisfies $F(0) = 0$.

• The representation (2) is quite useful to understand some characteristics of the process Y^n. Indeed, we thus obtain a hierarchical model : $Y = \phi^{-1}(Z - X\beta)$, where Z satisfies,

$$Z|c \sim \mathcal{N}(0, c^{-1}I_n), \quad c \sim F.$$

The distribution of Y^n conditionally on X, β and ϕ thus appears as a marginal distribution of the observations. Hereafter, we only deal with such models.

Asymptotics and matching priors: Another way to study the structure of such models is to analyze its asymptotic behavior. It can also lead to the determination of matching priors. The asymptotic study of any estimator of the parameters of interest should be greatly simplified by the hierarchical approach. However the model is non-identifiable in (c, ϕ), only in the product of those two terms; therefore, classical asymptotic expansions based on the full conditional likelihood:

$$f_n(Y^n \mid X, \beta, \phi, c) = \frac{(c\phi)^{n/2}}{2\pi} e^{-c\phi \mid y^n - X\beta \mid^2},$$

cannot be obtained directly. However, assume that β is the parameter of interest and consider the following parametrization: $\theta = (\beta, \sigma^2, c)$ where $\sigma^2 = (c\phi)^{-1}$. The full conditional likelihood, then only depends on β and σ^2. Let $\pi(\beta, \phi)$ be any prior density on (β, ϕ), we thus obtain that the posterior density of θ has the form:

$$\pi(\beta, \sigma^2 \mid Y^n, X) \propto f(Y^n \mid X, \beta, \sigma^2)\tilde{\pi}(\beta, \sigma^2),$$

where

$$\tilde{\pi}(\beta, \sigma^2) = \int_0^\infty \pi(\beta, 1/(\sigma^2 c))(\sigma^2 c)^{-1} dF(c).$$

Hence, a change in the model corresponds to a change in the prior, with a gaussian model. This explains, in the special case of representable density generator h, why β is so robust to the choice of h and why things are so similar to the gaussian case.

One way to construct non-informative priors is then to construct $\tilde{\pi}$ in a gaussian context and then to determine π using the above equality. This can be done, for instance, when constructing reference priors. For matching priors, one has to be slightly more careful, since the underlying distributions of the observations is not the conditional distribution of Y^n given (β, ϕ, c) as would classically be the case but the marginal distribution of Y^n given (β, ϕ). However, if β is the parameter of interest, then first-order matching priors, as defined by Welch and Peers (1963) are obtained in the usual way: If π is a matching prior for β, then when $\beta_0(Y^n)$ is the α-posterior quantile of β, we have:

$$F_n^\pi(\beta_0(Y^n)) = P^\pi[\beta \leq \beta_0(Y^n) \mid Y^n, X] = \alpha,$$

$$P_{\phi,\beta}^n[\beta \leq \beta_0(Y^n)] = \alpha + O(n^{-1}). \quad (3)$$

As in Welch and Peers (1963),

$$F_n^\pi(\beta) = \frac{\int_{-\infty}^\beta \int_0^\infty f(Y^n \mid X, \beta, \sigma^2)\, d\tilde{\pi}(\beta, \sigma^2)}{\int_{\mathbb{R}} \int_0^\infty f(Y^n \mid X, \beta, \sigma^2)\, d\tilde{\pi}(\beta, \sigma^2)}.$$

Thus

$$F_n^\pi(\beta) = \Phi\left(h_n\left[1 + \frac{Q_1(h_n)}{\sqrt{n}} + \frac{R_n}{n}\right]\right),$$

where Φ is the cdf of a standard gaussian random variable, $h_n = \sqrt{n}(\beta - \hat{\beta})\hat{\sigma}^{-1}$, Q_1 is a polynomial function of h_n whose coefficients are normalized derivatives of the log-likelihood,

R_n is the remaining term, $\hat{\beta} = (X'X)^{-1}X'Y^n$ and $\hat{\sigma}^2$ is the MLE. Under a usual iid model (or independent but non identically distributed model, as in Yin, 1998), the remaining term and the coefficients in Q_1 would be a $O_{P_{\beta,\sigma^2}}(1)$ term, so that the above expansion would be a valid asymptotic expansion in probability under P_{β,σ^2}. Here, the main difference is due to the fact that the underlying distribution is $P^n_{\beta,\phi}$, in other words, this is a marginal density. To make it simpler, we now suppose that the Y_i's have the same mean β. Then (3) is equivalent to:

$$\begin{aligned}
P^n_{\phi,\beta}&\left[h_n\left(1+\frac{Q_1(h_n)}{\sqrt{n}}+\frac{R_n}{n}\right)\le\Phi^{-1}(\alpha)\right]\\
&=\int_0^\infty P_{\beta,\sigma}\left[h_n\left(1+\frac{Q_1(h_n)}{\sqrt{n}}+\frac{R_n}{n}\right)\le\Phi^{-1}(\alpha)\right]dF(c)\\
&=\alpha+\frac{1}{2\sqrt{\phi}\sqrt{n}}\int_0^\infty\sqrt{c}D_\beta\log\tilde{\pi}(\beta,1/\phi c)dF(c)+O(n^{-1}).
\end{aligned}$$

Actually to prove that the last equality is true, we have to prove that, in the Edgeworth expansion of

$$h_n + Q_1(h_n)/\sqrt{n} + R_n/n,$$

we can integrate out the terms, see for instance Ghosh *et al.* (1982) or Guihenneuc and Rousseau (2002).

Matching priors are then solutions of:

$$\int_0^\infty \sqrt{c}D_\beta \log\tilde{\pi}(\beta, 1/\phi c)\, dF(c) = 0$$

in particular, when restricting attention to priors in the form

$$\pi(\beta,\phi) = \pi_1(\beta)\pi_2(\phi),$$

the solutions are then $\pi(\beta,\phi) \propto \pi(\phi)$ thus the non informative prior proposed by Osiewalski and Steel (1993) and Arellano *et al.* is also a matching prior.

• The study of the posterior distribution of ϕ is much more involved as ϕ is strongly related to the mixture structure (1) and it is most likely that the posterior distribution of ϕ is not gaussian. This would be worth investigating.

Extensions to time series: This family of models could also be used with a covariance matrix that has a time series structure, so that the covariance matrix is the Toeplitz transform of some spectral density. This would be of special interest for time series models that have no explicit likelihood outside the gaussian case, as is typically the case for long-memory processes.

MANUEL MENDOZA (*Instituto Tecnológico Autónomo de México, Mexico*)

In Section 3 the authors show that for a wide class of priors the corresponding posterior distribution of β does not depend on the density generator h and thus, the inferences are the same to be obtained under a normal distribution. In addition, for a prior belonging to that class it is also shown, in connection to the calibration problem, that the posterior distribution of the calibration parameter x_0 again does not depend on h. I would like to point out that the calibration problem belongs to a more general family of problems where the parameter of interest can be expressed as a ratio of two linear combinations of the coefficients of a linear model. This class includes the slope ratio and the parallel lines bioassays as well as the problem of making inferences on the interception of two simple regression regimes, among many others. Moreover, in the normal case, a rather flexible family of prior distributions has been proposed to deal with this general structure (Mendoza, 1988). It seems to me that this family of priors satisfies conditions similar to those of Proposition 3.1 in the paper and then inferences concerning a much more general ratio-type parameter can be obtained that do not depend on the density generator h.

REPLY TO THE DISCUSSION

First of all, we would like to thank the discussants for their outstanding work. Besides enriching substantially our results, they pose interesting related problems that should be tackled in future studies.

Professor Rousseau's comments provide a deeper understanding of how the robustness results in our paper originated, and suggest different directions for extending them. Similarly, Professor Mendoza remarks that the robustness results in linear calibration discussed in Example 3.2 may be generalized to inferences on the ratios of two linear combinations of regression coefficients.

For convenience, we separate our reply in different sections.

Reference priors: Professor Rousseau remarks that the prior $\pi(\phi, \boldsymbol{\beta}) \propto \phi^{-1}$ corresponds to the reference prior of Bernardo, with the order $\{\boldsymbol{\beta}\}, \{\phi\}$. This is indeed the case, as proved by Fernández and Steel (1999) for location and scale families. Furthermore, this prior does not depend on the density generator function (h in the elliptical model). However, the results obtained in their paper are obtained under the assumption that asymptotic normality conditions needed to confirm those priors as reference priors hold. Under the extendibility condition, which guarantees the representation in (1) as a mixture of normals, Padilla *et al.* (2002) show that these conditions are assured. The key result is, in fact, related to the equivalence established by Professor Rousseau in the context of Asymptotic and Matching priors. We will reexamine this point in another section.

Robustness: Our paper presents a class of priors that make the posterior $\boldsymbol{\beta}$ independent of the choice of h. Although it is not established explicitly, the same invariance holds for the predictive distribution of $\boldsymbol{y}$, as pointed out in Osiewalski and Steel (1993). In our paper, we specify a conditional distribution for $\boldsymbol{y} \mid \boldsymbol{\beta}, \phi, h$ and a prior for $(\boldsymbol{\beta}, \phi) \mid h$, in such a way that $\boldsymbol{y} \perp\!\!\!\perp h$, considering h as random (*i.e.*, h is marginally ancillary). Thus, any procedure for model comparisons that is based on the predictive distributions would not be useful for discrimination among different density generators. Even if we introduce a prior for h, this would not be updated under the hypotheses imposed in this paper. Thus, we would like to put a stronger emphasis on the care needed in the modelling process and the choice of the model comparison procedure, rather than on robustness. On the one hand, hierarchical modelling does not allow the dependence relationships between model subcomponents to be seen clearly. On the other hand, it becomes clear that Bayesian model comparisons should include not only the predictive distribution, but also all the model components. Iglesias and Quintana (2002) point out that the comparisons should refer to alternatives for the joint distribution of $(\boldsymbol{y}, \boldsymbol{\theta})$. In particular, it is suggested there to study the ratio

$$\mathrm{BF}(\boldsymbol{\theta}) = \frac{f_1(\boldsymbol{y}, \boldsymbol{\theta})}{f_2(\boldsymbol{y}, \boldsymbol{\theta})}$$

as a function of $\boldsymbol{\theta}$, which is called Global Bayes Factor (GBF) and may be rewritten as

$$\mathrm{BF}(\boldsymbol{\theta}) = \frac{\pi_1(\boldsymbol{\theta} \mid \boldsymbol{y})}{\pi_2(\boldsymbol{\theta} \mid \boldsymbol{y})}\mathrm{BF},$$

where BF is the usual Bayes factor. Thus

$$\mathrm{E}_{\pi_2(\cdot\,|\,\boldsymbol{y})}\left[\mathrm{BF}\left(\boldsymbol{\theta}\right)\right] = \mathrm{BF}.$$

Another procedure, which eliminates some deficiencies of Bayes factors has been suggested by Bernardo (1999), in the framework of decision theory, whose aim is to choose one of the models under comparison, without any subsequent action.

Representable elliptical models: Professor Rousseau argues that it is usually assumed that the density generator of the elliptical distribution is representable as

$$h\left(u\right) = \int_0^{\infty} \left(\frac{c}{2\pi}\right)^{\frac{n}{2}} e^{-\frac{uc}{2}} dF\left(c\right), \tag{1}$$

where F is a c.d.f. with $F\left(0\right) = 0$.

Definition. *A random vector $\boldsymbol{y}$ has a distribution $EL_n\left(\boldsymbol{\mu}, \Sigma, h\right)$ representable by F if h may be written as (1), where $c\,\Pi\left(\boldsymbol{\mu}, \Sigma\right)$.*

As stated by Professor Rousseau, this is equivalent to a two-stage model specification:

(i) $\boldsymbol{y}\,|\,\boldsymbol{\mu}, \Sigma, \omega \sim \mathrm{N}\left(\boldsymbol{\mu}, \omega\Sigma\right)$

(ii) $\omega \sim F$ and $\omega\,\Pi\left(\boldsymbol{\mu}, \Sigma\right)$.

The advantage of this representation is that it facilitates the treatment of this model, particularly the implementation of computational strategies of the MCMC type, as shown in Arellano-Valle *et al.* (2000). As will be seen in the next point, it is also very useful in the study of asymptotic normality of the posteriors for the parameters involved. Nevertheless, there are some important distributions that do not satisfy the previous definition, such as Pearson type II distribution, which is a valid model for inference on finite populations (see Bolfarine *et al.*, 2002).

Asymptotic and matching priors: Professor Rousseau gives two beautiful results for representable elliptical models. One of them shows that the class of priors considered in our work, are also matching priors and its proof rests on a link established between inference in the normal model and on the representable elliptical model. This is a great contribution, since it helps both in the interpretation of the model and in studying asymptotic normality of the posterior distributions. To see this, we state her result as a Lemma:

Lemma. *Let $\boldsymbol{y}\,|\,\boldsymbol{X}, \boldsymbol{\beta}, \phi \sim EL_n\left(\boldsymbol{X}\boldsymbol{\beta}, \phi\boldsymbol{I}_n, h\right)$ representable by F and let $\pi\left(\boldsymbol{\beta}, \phi\right)$ be any prior density for $\left(\boldsymbol{\beta}, \phi\right)$. Then, there exists a prior density $\tilde{\pi}\left(\boldsymbol{\beta}, \sigma^2\right)$, with $\sigma^2 = \left(c\phi\right)^{-1}$ and $c \sim F$, satisfying*

$$\pi\left(\boldsymbol{\beta}, \sigma^2\,|\,\boldsymbol{y}, \boldsymbol{X}\right) \propto f\left(\boldsymbol{y}\,|\,\boldsymbol{X}, \boldsymbol{\beta}, \sigma^2\right) \tilde{\pi}\left(\boldsymbol{\beta}, \sigma^2\right),$$

where

$$f\left(\boldsymbol{y}\,|\,\boldsymbol{X}, \boldsymbol{\beta}, \sigma^2\right) = \left(\frac{\sigma^2}{2\pi}\right)^{n/2} \exp\left(-\frac{1}{2}\sigma^2 \left\|\boldsymbol{y} - \boldsymbol{X}\boldsymbol{\beta}\right\|^2\right).$$

This Lemma assures the asymptotic normality for the location parameters in representable elliptical models. For instance, if $\boldsymbol{y}\,|\,\mu, \phi \sim EL_n\left(\mu 1_n, \phi\boldsymbol{I}_n, h\right)$, where $\mu \in \Re$ and h is representable by F, then $\mu\,|\,\boldsymbol{y}$ is asymptotically normal with mean $\bar{y}$ and variance σ_0^2/n, where $\sigma_0^2 = lim_n\left(S_n^2\,|\,\sigma^2 = \sigma_0^2\right)$ and $S_n^2 = n^{-1}\sum_{i=1}^n \left(y_i - \bar{y}\right)^2$.

In fact, choosing $\boldsymbol{X} = 1_n$ in the Lemma, we get

$$\pi\left(\mu, \sigma^2 \,|\, \boldsymbol{y}, \boldsymbol{X}\right) = \tilde{\pi}\left(\mu, \sigma^2 \,|\, \boldsymbol{y}, \boldsymbol{X}\right),$$

where $\tilde{\pi}\left(\mu, \sigma^2 \,|\, \boldsymbol{y}, \boldsymbol{X}\right)$ is obtained under the assumption $\boldsymbol{y} \,|\, \boldsymbol{X}, \mu, \sigma^2 \sim \mathrm{N}_n\left(\mu 1_n, \sigma^2 \boldsymbol{I}_n\right)$ and the corresponding priors are related through

$$\tilde{\pi}\left(\mu, \sigma^2\right) = \int_0^{\infty} \pi\left(\mu, (c\sigma^2)^{-1}\right)(c\sigma^4)^{-1} dF(c),$$

taking $\pi\left(\mu, \sigma^2\right)$ to be continuous. Applying the Berstein Von-Mises Theorem we get

$$|\, \tilde{\pi}\left(\mu, \sigma^2 \,|\, \boldsymbol{y}, \boldsymbol{X}\right) - \mathrm{N}_2\left[\left(\mu, \sigma^2\right)^t \,|\, \left(\bar{y}, S_n^2\right)^t, n^{-1}\mathrm{diag}\left(\sigma_0^2, 2\sigma_0^2\right)\right] \,| \to 0.$$

For more details see Padilla *et al.* (2002).

Extensions to time series: Professor Rousseau suggests the use of elliptical distributions to model long memory processes. Some work along this line has been done in Iglesias and Palma (2002) and Bahamonde and Meza (2002), in the context of regression models with ARFIMA(d) errors, This work is motivated by the statistical analysis of air pollution data for Santiago de Chile and employs the $t_n\left(0, \Sigma, \nu\right)$ model, which is representable and allows the use of MCMC techniques.

Ratio of linear combination: Professor Mendoza points out an important fact about inferences for the ratio ρ of linear combinations of the regression coefficients (see Mendoza, 1988). A nice aspect of this formulation is that the calibration problem becomes a particular of our setting. His conjecture of the validity of the results to the class of elliptical models appears to be true. In fact, if in the regression model we choose a prior of the form $\pi\left(\boldsymbol{\beta}, \phi\right) \propto \pi\left(\boldsymbol{\beta}\right)\pi\left(\phi\right)$ with $\phi \sim \mathcal{R}_n^2\left(h\right)$, then the posterior of $\rho = g\left(\boldsymbol{\beta}\right)$ will not depend on h, since the posteriori of $\boldsymbol{\beta}$ does not.

In particular, consider the prior

$$\pi\left(\gamma, \phi\right) \propto \pi\left(\rho\right)\sigma^{-r}$$

with $\gamma_1 = \rho$, $\gamma_i = \beta_i$, $i = 1, \ldots, k$ used in Mendoza (1988). Taking $r = 2$ and applying a suitable reparametrization, his (explicit) results are indeed valid for the whole class of elliptical models. Furthermore, the reference prior derived when ρ is the parameter of interest might not depend upon h.

The formulation of the problem by Professor Mendoza opens up an interesting research area, to be explored in the framework of elliptical distributions.

ADDITIONAL REFERENCES IN THE DISCUSSION

Bahamonde, N. and Meza, C. (2002). *Contaminación Ambiental en Santiago. Análisis Bayesiano en Modelos de Regresión con Errores ARFIMA*. Master Thesis, Pontificia Universidad Católica de Chile.

Bolfarine, H., Iglesias, P. and Gasco, L. (2002). Bayesian operational approach for regression models in finite populations. *System and Bayesian Reliability 5: Essays in Honor of Professor Richard E. Barlow on his 70th Birthday* (Y. Hayakawa, T. Irony and M. Xie, eds). 375–390.

Brandwein, A.C. and Strawderman, W. E. (1991) Generalizations of James Stein estimators under spherical symmetry. *Ann. Statist.* **19**, 1639-1650.

Fernández, C. and Steel, M. F. J. (1999). Reference priors for the general location-scale model. *Statist. Prob. Lett.* **43**, 377–384.

Ghosh, J.K., Sinha, B. and Joshi, S. N. (1982). Expansions for posterior probability and integrated Bayes risk. *Statistical Decision Theory and Related Topics III* **1** (S. S. Gupta and J. O. Berger, eds). New York: Academic Press, 403-456.

Guihenneuc, C. and Rousseau, J. (2002). Laplace expansions in MCMC algorithms for latent variable models. *Tech. Rep.*, CREST, France.

Iglesias, P. and Palma, W. (2002). Analysis of air pollution in Santiago de Chile. *Tech. Rep.*, Pontificia Universidad Católica de Chile.

Iglesias, P. and Quintana, F. (2002). A global Bayes factor. *Tech. Rep.*, Pontificia Universidad Católica de Chile.

Mendoza, M. (1988). Inferences about the ratio of linear combinations of the coefficients in a multiple regression model. *Bayesian Statistics 3* (J. M. Bernardo, M. H. DeGroot, D. V. Lindley and A. F. M. Smith, eds). Oxford: Oxford University Press, 705–712.

Padilla, O., Iglesias, P. and Arellano-Valle, R. B. (2002). Reference priors for elliptical models. *Tech. Rep.*, Pontificia Universidad Católica de Chile.

Welch, B. and Peers, H. (1963). On formulae for confidence points based on intervals of weighted likelihoods. *J. R. Statist. Soc. B* **25**, 318-329.

Yin, M. (1998). Asymptotic expansions for posterior probability in regression model. *Statistics and Decisions* **16**, 349-368.

BAYESIAN STATISTICS 7, pp. 25–43
J. M. Bernardo, M. J. Bayarri, J. O. Berger, A. P. Dawid,
D. Heckerman, A. F. M. Smith and M. West (Eds.)

Hierarchical Bayesian Models for Applications in Information Retrieval

DAVID M. BLEI, MICHAEL I. JORDAN and ANDREW Y. NG
University of California, Berkeley, USA
blei@cs.berkeley.edu jordan@cs.berkeley.edu ang@cs.berkeley.edu

SUMMARY

We present a simple hierarchical Bayesian approach to the modeling collections of texts and other large-scale data collections. For text collections, we posit that a document is generated by choosing a random set of multinomial probabilities for a set of possible "topics," and then repeatedly generating words by sampling from the topic mixture. This model is intractable for exact probabilistic inference, but approximate posterior probabilities and marginal likelihoods can be obtained via fast variational methods. We also present extensions to coupled models for joint text/image data and multiresolution models for topic hierarchies.

Keywords: VARIATIONAL INFERENCE METHODS; HIERARCHICAL BAYESIAN MODELS; EMPIRICAL BAYES; LATENT VARIABLE MODELS; INFORMATION RETRIEVAL.

1. INTRODUCTION

The field of information retrieval is broadly concerned with the problem of organizing collections of documents and other media so as to support various information requests on the part of users. The familiar problem of returning a subset of documents in response to a query lies within the scope of information retrieval, as do the problems of analyzing cross-referencing within a document collection, analyzing linguistic structure so as to be able to say who did what to whom, automatic compilation of document summaries, and automatic translation. Another class of problems involve analyzing interactions between users and a collection, including the "collaborative filtering" problem in which suggestions are made to new users of a system based on choices made by previous users.

Clearly there is much grist for the mill of Bayesian statistics in the information retrieval problem. One can view the information retrieval system as uncertain about the needs of the user, an uncertainty that can be reduced in an ongoing "learning" process via a dialog with the user or with a population of users. There are also many modeling issues—particularly those surrounding the appropriate level of resolution at which to view a document collection—where the tools of hierarchical Bayesian modeling are clearly appropriate. The problem of formulating a response to a query can be viewed in Bayesian terms: one can model the conditional distribution of queries given documents, and in conjunction with a model of documents treat the problem of responding to a query as an application of Bayes' theorem (Zhai and Lafferty, 2001). Finally, many problems in information retrieval involve preferences and choices, and decision-theoretic analysis is needed to manage the complexity of the trade-offs that arise.

Despite these natural motivations, it is generally not the case that current information retrieval systems are built on the foundation of probabilistic modeling and inference. One important reason for this is the severe computational constraints of such systems. Collections involving tens or hundreds of thousands of documents are commonplace, and the methods that are used by information retrieval systems—reduction of a document to a "vector" of smoothed frequency counts, possibly followed by a singular value decomposition to reduce dimensionality, and computations of inner products between these "vectors"—have the important virtue of computational efficiency. If probabilistic modeling is to displace or augment such methods, it will have to be done without a major increase in computational load—and there are questions about whether this can be done with the current arsenal of Bayesian tools. A user who has sent a query to a search engine is generally not willing to wait for a Markov chain Monte Carlo simulation to converge.

In the current paper, we discuss a class of hierarchical Bayesian models for information retrieval. While simple, these models are rich enough as to yield intractable posterior distributions, and to maintain computational efficiency, we make use of *variational inference methods*. Variational methods yield deterministic approximations to likelihoods and posterior distributions that provide an alternative to Markov chain Monte Carlo. They are particularly apt in a domain such as information retrieval in which a fast approximate answer is generally more useful than a slow answer of greater fidelity.

The models that we discuss are all instances of so-called "bag-of-words" models (Baeza-Yates and Ribeiro-Neto, 1999). Viewing the words in a document as random variables, these are simply models in which the words are exchangeable. While clearly a drastic simplification, this assumption is generally deemed necessary in information retrieval because of the computational constraints; moreover, it has been found in practice that viable information retrieval systems can be built using such an assumption. In any case, by building up a sufficiently detailed model for the mixture underlying the word distribution, we hope to ameliorate the effect of the exchangeability assumption, and capture some of the latent structure of documents while maintaining computational tractability.

Finally, although we believe that probabilistic methods have an important role to play in information retrieval, full Bayesian computations are often precluded by the size of the problems in this domain. We therefore make significant use of empirical Bayesian techniques, in particular fixing hyperparameters via maximum likelihood. Some comfort is provided in this regard by the large scale of the problems that we study, but it is also important to acknowledge that all of the models that we study are very inaccurate reflections of the underlying linguistic reality.

2. LATENT VARIABLE MODELS OF TEXT

In any given document collection, we envision a number of underlying "topics"; for example, a collection may contain documents about novels, music or poetry. These topics may also viewed at varying levels of resolution; thus, we have have documents about romance novels, jazz music or Russian poetry. A traditional approach to treating such "topics" is via hierarchical clustering, in which one represents each document as a vector of word counts, defines a similarity measure on the vectors of word counts, and applies a hierarchical clustering procedure in the hopes of characterizing the "topics." A significant problem with such an approach, however, is the mutual exclusivity assumption—a given document can only belong to a single cluster. Textual material tends to resist mutual exclusivity—words can have several different (unrelated) meanings, and documents can be relevant to different topics (a document can be relevant to both jazz music and Russian poetry). If we are to use clustering procedures—and indeed computational concerns lead us to aim towards some form of divide-and-conquer strategy—care must be taken to define clustering procedures that are suitably flexible.

To provide the requisite flexibility, we propose a hierarchical Bayesian approach. The basic scheme is that of a mixture model, corresponding to an exchangeable distribution on words. The mixture has two basic levels. At the first level, we have a finite mixture whose mixture components can be viewed as representations of "topics." At the second level, a latent Dirichlet variable provides a random set of mixing proportions for the underlying finite mixture. This Dirichlet variable can be viewed as a representation of "documents"; *i.e.*, a document is modeled as a collection of topic probabilities. The Dirichlet is sampled once per document, and the finite mixture is then sampled repeatedly, once for each word within a document. The components that are selected during this sampling process are a multiset that can be viewed as a "bag-of-topics" characterization of the document.

We describe this basic model in detail in the following section, and treat various extensions in the remainder of the paper.

3. LATENT DIRICHLET ALLOCATION

The basic entity in our model is the *word*, which we take to be a multinomial random variable ranging over the integers $\{1, \ldots, V\}$, where V is the *vocabulary size*, a fixed constant. We represent this random variable as a V-vector w with components $w^i \in \{0, 1\}$, where one and only one component is equal to one.

A *document* is a sequence of N words, denoted by $\boldsymbol{w}$, where w_n is the n-th word. We assume that we are given as data a *corpus* of M documents: $\mathcal{D} = \{\boldsymbol{w}_d : d = 1, \ldots, M\}$.

We refer to our model as a *Latent Dirichlet allocation (LDA)* model. The model assumes that each document in the corpus is generated as follows:

(1) Choose $N \sim p(N \mid \xi)$.
(2) Choose $\theta \sim \mathrm{Di}(\alpha)$.
(3) For each of the N words w_n:
  (a) Choose a topic $z_n \sim \mathrm{Mult}(\theta)$;
  (b) Choose a word w_n from $p(w_n \mid z_n, \beta)$, a multinomial probability conditioned on the topic z_n.

Several simplifying assumptions are made in this basic model, some of which we remove in subsequent sections. First, the dimensionality k of the Dirichlet distribution (and thus the topic variable z) is assumed known and fixed. Second, the word probabilities are parameterized by a $k \times V$ matrix β where $\beta_{ij} = p(w^j = 1 \mid z^i = 1)$, which for now we treat as a fixed quantity that is to be estimated. Finally, note that we have left the document length distribution unspecified—in most applications we work conditionally on N, and indeed in the remainder of the paper we generally omit reference to N.

Given this generative process, the joint distribution of a topic mixture θ, N topics $\boldsymbol{z}$, and an N word document $\boldsymbol{w}$ is:

$$p(\theta, \boldsymbol{z}, \boldsymbol{w}, N \mid \alpha, \beta) = p(N \mid \xi)\, p(\theta \mid \alpha) \prod_{n=1}^{N} p(z_n \mid \theta)\, p(w_n \mid z_n, \beta), \tag{1}$$

where $p(z_n \mid \theta)$ is simply θ_i for the unique i such that $z_n^i = 1$. This model is illustrated in Figure 1 (left).

Note the distinction between LDA and a simple Dirichlet-multinomial clustering model. In the simple clustering model, the innermost plate would contain only w, the topic node would be sampled only once for each document, and the Dirichlet would be sampled only once for the whole collection. In LDA, the Dirichlet is sampled for each document, and the multinomial topic node is sampled *repeatedly* within the document.

4. INFERENCE

Let us consider the problem of computing the posterior distribution of the latent variables θ and z given a document (where we drop reference to the randomness in N for simplicity):

$$p(\theta, \boldsymbol{z} \mid \boldsymbol{w}, \alpha, \beta) = \frac{p(\theta, \boldsymbol{z}, \boldsymbol{w} \mid \alpha, \beta)}{p(\boldsymbol{w} \mid \alpha, \beta)}.$$

We can find the denominator—a marginal likelihood—by marginalizing over the latent variables in Eq. (1):

$$p(\boldsymbol{w} \mid \alpha, \beta) = \frac{\Gamma\left(\sum_i \alpha_i\right)}{\prod_i \Gamma(\alpha_i)} \int \Big(\prod_{i=1}^{k} \theta_i^{\alpha_i - 1}\Big)\Big(\prod_{n=1}^{N} \sum_{i=1}^{k} \prod_{j=1}^{V} (\theta_i \beta_{ij})^{w_n^j}\Big) d\theta. \tag{2}$$

This is an expectation under an extension to the Dirichlet distribution that can be represented with special hypergeometric functions (Dickey, 1983; Dickey *et al.,* 1987). Unfortunately, this function is infeasible to compute exactly, due to the coupling between θ and β inside the summation over latent factors.

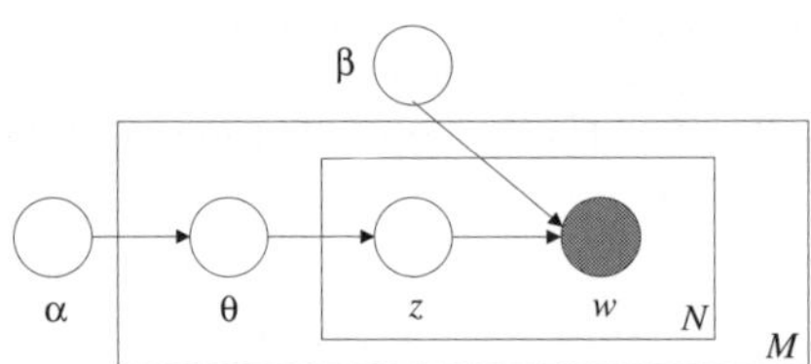

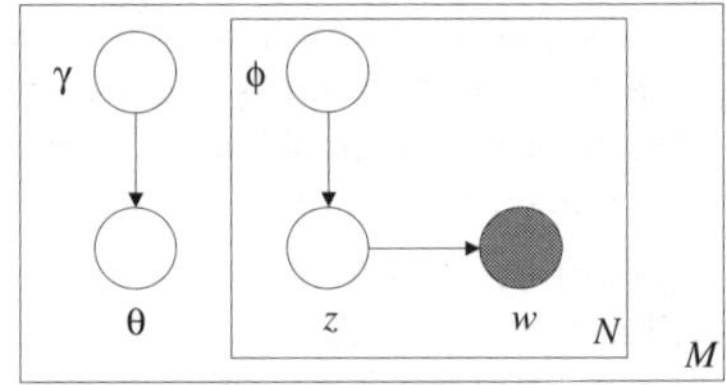

Figure 1. *(Left) Graphical model representation of LDA. The box is a "plate" representing replicates. (Right) Graphical model representation of the variational distribution to approximate the posterior in LDA.*

Note that we are treating the parameters α and β as fixed constants in Eq. (2); we show how to estimate their values using an empirical Bayes procedure in this section. In Section 6 we consider a fuller Bayesian model in which β is endowed with a Dirichlet distribution.

To approximate the posterior probability in a computationally efficient manner, we make use of *variational inference algorithms* (Jordan, *et al.,* 1999). Variational inference is related to importance sampling in its use of a simplified distribution to approximate the posterior. Rather than sampling from such a distribution, however, we consider a family of simplified distributions, parameterized by a set of *variational parameters*. Ranging over these parameters, we find the best approximation in the family, measuring approximation quality in terms of KL divergence. We essentially convert the inference problem (a problem of computing an integral) into an optimization problem (a problem of maximizing a function).

In the context of graphical models, the simplified distributions that provide the approximations used in a variational inference framework are generally obtained by omitting one or more edges in the graph. This decouples variables and provides a more tractable approximation to the posterior.

Figure 1 (Left) illustrates the LDA model for a corpus of documents. In this graph, the problematic coupling between θ and β is represented as the arc between θ and $\boldsymbol{z}$. We develop a variational approximation by defining an approximating family of distributions $q(\theta, \boldsymbol{z} \mid \boldsymbol{w}, \gamma, \phi)$, and choose the variational parameters γ and ϕ to yield a tight approximation to the true posterior.

In particular, we define the factorized variational distribution:

$$q(\theta, \boldsymbol{z} \mid \boldsymbol{w}, \gamma, \phi) = p(\theta \mid \boldsymbol{w}, \gamma) \prod_{n=1}^{N} p(z_n \mid \boldsymbol{w}, \phi_n),$$

as illustrated in Figure 1 (Right). This distribution has variational Dirichlet parameters γ and variational multinomial parameters ϕ_n. Note that all parameters are conditioned on $\boldsymbol{w}$; for each document, there is a different set of Dirichlet and multinomial variational parameters.

With this new model in hand, we can obtain an approximation to $p(\theta, \boldsymbol{z} \mid \boldsymbol{w}, \alpha, \beta)$ via a minimization of the KL divergence between the variational distribution and the true posterior:

$$(\gamma^*, \phi^*) = \arg\min_{(\gamma,\phi)} D(q(\theta, \boldsymbol{z} \mid \boldsymbol{w}, \gamma, \phi) \parallel p(\theta, \boldsymbol{z} \mid \boldsymbol{w}, \alpha, \beta)). \tag{3}$$

As we show in the following section, we can take decreasing steps in the KL divergence and converge to (locally) optimizing parameters (γ^*, ϕ^*) by alternating between the following pair of update equations:

$$\phi_{ni} \propto \beta_{iw_n} \exp\{\mathrm{E}_q[\log(\theta_i \mid \gamma)]\} \tag{4}$$

$$\gamma_i = \alpha_i + \textstyle\sum_{n=1}^{N} \phi_{ni}, \tag{5}$$

where a closed-form expression for the expectation in Eq. (4) is given in Section 4.1. Note again that we obtain variational parameters (γ^*, ϕ^*) for each document $\boldsymbol{w}$.

Equations (4) and (5) have an appealing intuitive interpretation. The Dirichlet update in Eq. (5) yields a posterior Dirichlet given expected observations taken under the variational distribution. The multinomial update in Eq. (4) is akin to using Bayes' theorem, $p(z_n \mid w_n) \propto p(w_n \mid z_n)p(z_n)$, where $p(z_n)$ is approximated by the exponential of the expected value of its log under the variational distribution.

Each iteration of the algorithm requires $O(Nk)$ operations. Empirically, we find that the number of iterations required for a single document scales linearly in the number of words in the document. This yields a total number of operations that scales empirically as $O(N^2k)$.

4.1. *Variational inference*

In this section, we derive the variational inference equations in Equations (4) and (5).

We begin by noting that the transformed parameters, $\log \theta_i$, are the natural parameters in the exponential family representation of the Dirichlet distribution, and thus the expected value of $\log \theta_i$ is given by $\mathrm{E}[\log \theta_i \mid \alpha] = \Psi(\alpha_i) - \Psi\left(\sum_{i=1}^{k} \alpha_i\right)$, where $\Psi(x)$ is the digamma function.

To derive the variational inference algorithm, we begin by bounding the marginal likelihood of a document using Jensen's inequality (Jordan *et al.,* 1999):

$$\begin{aligned}
\log p(\boldsymbol{w} \mid \alpha, \beta) &= \log \int_\theta \sum_{\boldsymbol{z}} p(\theta, \boldsymbol{z}, \boldsymbol{w} \mid \alpha, \beta)\, d\theta \\
&= \log \int_\theta \sum_{\boldsymbol{z}} \frac{p(\theta, \boldsymbol{z}, \boldsymbol{w} \mid \alpha, \beta) q(\theta, \boldsymbol{z})}{q(\theta, \boldsymbol{z})} d\theta \\
&\geq \int_\theta \sum_{\boldsymbol{z}} q(\theta, \boldsymbol{z}) \log p(\theta, \boldsymbol{z}, \boldsymbol{w} \mid \alpha, \beta)\, d\theta - \int_\theta \sum_{\boldsymbol{z}} q(\theta, \boldsymbol{z}) \log q(\theta, \boldsymbol{z})\, d\theta \\
&= \mathrm{E}_q[\log p(\theta, \boldsymbol{z}, \boldsymbol{w} \mid \alpha, \beta)] - \mathrm{E}_q[\log q(\theta, \boldsymbol{z})].
\end{aligned} \tag{6}$$

It is straightforward to show that the difference between the left-hand side and the right-hand side of this equation is precisely the KL divergence in Eq. (3), and thus minimizing that KL divergence is equivalent to maximizing the lower bound on the marginal likelihood in Eq. (6).

Letting $\mathcal{L}(\gamma, \phi; \alpha, \beta)$ denote the right-hand side of Eq. (6), we have:

$$\begin{aligned}\mathcal{L}(\gamma, \phi; \alpha, \beta) =& \mathrm{E}_q[\log p(\theta \mid \alpha)] + \mathrm{E}_q[\log p(\boldsymbol{z} \mid \theta)] + \mathrm{E}_q[\log p(\boldsymbol{w} \mid \boldsymbol{z}, \beta)] \\ & - \mathrm{E}_q[\log q(\theta)] - \mathrm{E}_q[\log q(\boldsymbol{z})].\end{aligned} \tag{7}$$

Next, we write Eq. (7) in terms of the model parameters (α, β) and variational parameters (γ, ϕ). Each of the five lines expands one of the five terms:

$$\begin{aligned}\mathcal{L}(\gamma, \phi; \alpha, \beta) = & \log \Gamma\left(\textstyle\sum_{j=1}^k \alpha_j\right) - \sum_{i=1}^k \log \Gamma(\alpha_i) + \sum_{i=1}^k (\alpha_i - 1)\left(\Psi(\gamma_i) - \Psi\left(\textstyle\sum_{j=1}^k \gamma_j\right)\right) \\ & + \sum_{n=1}^N \sum_{i=1}^k \phi_{ni}\left(\Psi(\gamma_i) - \Psi\left(\textstyle\sum_{j=1}^k \gamma_j\right)\right) \\ & + \sum_{n=1}^N \sum_{i=1}^k \sum_{j=1}^V w_n^j \phi_{ni} \log \beta_{ij} \\ & - \log \Gamma\left(\textstyle\sum_{j=1}^k \gamma_j\right) + \sum_{i=1}^k \log \Gamma(\gamma_i) - \sum_{i=1}^k (\gamma_i - 1)\left(\Psi(\gamma_i) - \Psi\left(\textstyle\sum_{j=1}^k \gamma_j\right)\right) \\ & - \sum_{n=1}^N \sum_{i=1}^k \phi_{ni} \log \phi_{ni}.\end{aligned} \tag{8}$$

In the following sections, we take derivatives with respect to this expression to obtain our variational inference algorithm.

Variational multinomial. We first maximize Eq. (8) with respect to ϕ_{ni}, the probability that the n-th word was generated by latent topic i. Observe that this is a constrained maximization since $\sum_{i=1}^k \phi_{ni} = 1$.

We form the Lagrangian by isolating the terms that contain ϕ_{ni} and adding the appropriate Lagrange multipliers. Let β_{iv} refer to $p(w_n^v = 1 \mid z^i = 1)$ for the appropriate v (and recall that each w_n is a V-vector with exactly one component equal to 1; we can select the unique v such that $w_n^v = 1$). We have:

$$\mathcal{L}_{[\phi_{ni}]} = \phi_{ni}\left(\Psi(\gamma_i) - \Psi\left(\textstyle\sum_{j=1}^k \gamma_j\right)\right) + \phi_{ni} \log \beta_{iv} - \phi_{ni} \log \phi_{ni} + \lambda_n \left(\textstyle\sum_{j=1}^k \phi_{nj} - 1\right).$$

Taking derivatives with respect to ϕ_{ni}, we obtain:

$$\frac{\partial \mathcal{L}}{\partial \phi_{ni}} = \Psi(\gamma_i) - \Psi\left(\textstyle\sum_{j=1}^k \gamma_j\right) + \log \beta_{iv} - \log \phi_{ni} - 1 + \lambda_n.$$

Setting this derivative to zero yields the maximized ϕ_{ni} (cf. Eq. (4)):

$$\phi_{ni} \propto \beta_{iv} \exp\left(\Psi(\gamma_i) - \Psi\left(\textstyle\sum_{j=1}^k \gamma_j\right)\right). \tag{9}$$

Variational Dirichlet. Next, we maximize Eq. (8) with respect to γ_i, the ith component of the posterior Dirichlet parameter. The terms containing γ_i are:

$$\mathcal{L}_{[\gamma]} = \sum_{i=1}^{k}(\alpha_i - 1)\left(\Psi(\gamma_i) - \Psi\left(\sum_{j=1}^{k}\gamma_j\right)\right) + \sum_{n=1}^{N}\phi_{ni}\left(\Psi(\gamma_i) - \Psi\left(\sum_{j=1}^{k}\gamma_j\right)\right)$$
$$- \log\Gamma\left(\sum_{j=1}^{k}\gamma_j\right) + \log\Gamma(\gamma_i) - \sum_{i=1}^{k}(\gamma_i - 1)\left(\Psi(\gamma_i) - \Psi\left(\sum_{j=1}^{k}\gamma_j\right)\right).$$

This simplifies to:

$$\mathcal{L}_{[\gamma]} = \sum_{i=1}^{k}\left(\Psi(\gamma_i) - \Psi\left(\sum_{j=1}^{k}\gamma_j\right)\right)\left(\alpha_i + \sum_{n=1}^{N}\phi_{ni} - \gamma_i\right) - \log\Gamma\left(\sum_{j=1}^{k}\gamma_j\right) + \log\Gamma(\gamma_i).$$

We take the derivative with respect to γ_i:

$$\frac{\partial\mathcal{L}}{\partial\gamma_i} = \Psi'(\gamma_i)\left(\alpha_i + \sum_{n=1}^{N}\phi_{ni} - \gamma_i\right) - \Psi'\left(\sum_{j=1}^{k}\gamma_j\right)\sum_{j=1}^{k}\left(\alpha_j + \sum_{n=1}^{N}\phi_{nj} - \gamma_j\right). \quad (10)$$

Setting this equation to zero yields a maximum at:

$$\gamma_i = \alpha_i + \sum_{n=1}^{N}\phi_{ni}. \quad (11)$$

Since Eq. (11) depends on the variational multinomial ϕ, full variational inference requires alternating between Eqs. (9) and (11) until the bound on $p(\boldsymbol{w} \mid \alpha, \beta)$ converges.

4.2. *Estimation*

In this section, we discuss approximate maximum likelihood estimation of the parameters α and β. Given a corpus of documents $\mathcal{D} = \{\boldsymbol{w}_1, \boldsymbol{w}_2, \ldots, \boldsymbol{w}_M\}$, we use a variational EM algorithm (EM with a variational E step) to find the parameters α and β that maximize a lower bound on the log marginal likelihood:

$$\ell(\alpha, \beta) = \sum_{d=1}^{M}\log p(\boldsymbol{w}_d \mid \alpha, \beta).$$

As we have described above, the quantity $p(\boldsymbol{w} \mid \alpha, \beta)$ cannot be computed efficiently. However, we can bound the log likelihood using:

$$p(\boldsymbol{w}_d \mid \alpha, \beta) = \mathcal{L}(\gamma, \phi; \alpha, \beta) + D(q(\theta, \boldsymbol{z} \mid \boldsymbol{w}_d, \gamma, \phi) \parallel p(\theta, \boldsymbol{z} \mid \boldsymbol{w}_d, \alpha, \beta)), \quad (12)$$

which exhibits $\mathcal{L}(\gamma, \phi; \alpha, \beta)$ as a lower bound since the KL term is positive.

We now obtain a variational EM algorithm that repeats the following two steps until Eq. (12) converges:

(E) Find the setting of the variational parameters $\{\gamma_d, \phi_d : d \in \mathcal{D}\}$ that tighten the bound in Eq. (12) as much as possible. This is simply variational inference for each training document as described in the previous section.

(M) Maximize Eq. (12) with respect to the model parameters α and β. This corresponds to finding the maximum likelihood estimates with the approximate expected sufficient statistics computed in the E step.

To maximize with respect to β, we isolate terms and add Lagrange multipliers:

$$\mathcal{L}_{[\beta_{ij}]} = \sum_{d=1}^{M}\sum_{n=1}^{N_d}\sum_{i=1}^{k}\sum_{j=1}^{V}\phi_{dni}w_n^j \log\beta_{ij} + \sum_{i=1}^{k}\lambda_i\left(\sum_{j=1}^{V}\beta_{ij} - 1\right).$$

We take the derivative with respect to β_{ij}, set it to zero, and find:

$$\beta_{ij} \propto \sum_{d=1}^{M}\sum_{n=1}^{N_d}\phi_{dni}w_n^j.$$

The terms that contain α are:

$$\mathcal{L}_{[\alpha]} = \sum_{d=1}^{M}\log\Gamma\left(\sum_{j=1}^{k}\alpha_j\right) - \sum_{i=1}^{k}\log\Gamma(\alpha_i) + \sum_{i=1}^{k}(\alpha_i - 1)\left(\Psi(\gamma_{di}) - \Psi\left(\sum_{j=1}^{k}\gamma_{dj}\right)\right)$$

Taking the derivative with respect to α_i gives:

$$\frac{\partial\mathcal{L}}{\partial\alpha_i} = M\left(\Psi\left(\sum_{j=1}^{k}\alpha_j\right) - \Psi(\alpha_i)\right) + \sum_{d=1}^{M}\Psi(\gamma_{di}) - \Psi\left(\sum_{j=1}^{k}\gamma_{dj}\right).$$

Given the coupling between the derivatives for the different α_j, we use Newton-Raphson to find the maximal α. The Hessian has the following form:

$$\frac{\partial\mathcal{L}}{\partial\alpha_i\alpha_j} = M\left(\Psi'\left(\sum_{j=1}^{k}\alpha_j\right) - \delta(i,j)\Psi'(\alpha_i)\right)$$

and this form allows us to exploit the matrix inversion lemma to obtain a Newton-Raphson algorithm that requires only a linear number of operations (Ronning, 1989).

5. EXAMPLE

We illustrate how LDA works by examining the variational posterior parameters γ and ϕ_n for a document in the TREC AP corpus (Harman, 1992). Recall that ϕ_{nj} is an approximation to the posterior probability associated with the i-th topic and the n-th word. By examining $\max_i \phi_{ni}$, we obtain a proposed allocation of words to unobserved topics.

Furthermore, we can interpret the i-th Dirichlet parameter (maximized by Eq. (5)) as the i-th Dirichlet parameter for the model plus the expected number of instances of topic i that were seen in the given document. Therefore, subtracting the posterior Dirichlet parameters from the model Dirichlet parameters we obtain an indication of the degree to which each factor is present in a document.

We trained a 100-factor LDA model on a subset of the TREC AP corpus. The following is an article from the same collection on which we did not train:

> The William Randolph Hearst Foundation will give $1.25 million to Lincoln Center, Metropolitan Opera Co., New York Philharmonic and Juilliard School. "Our board felt that we had a real opportunity to make a mark on the future of the performing arts with these grants an act every bit as important as our traditional areas of support in health, medical research, education and the social services," Hearst Foundation President Randolph A. Hearst said Monday in announcing the grants. Lincoln Center's share will be $200,000 for its new building, which will house young artists and provide new public facilities. The Metropolitan Opera Co. and New York Philharmonic will receive $400,000 each. The Juilliard School, where music and the performing arts are taught, will get $250,000. The Hearst Foundation, a leading supporter of the Lincoln Center Consolidated Corporate Fund, will make its usual annual $100,000 donation, too.

"Arts"	"Budgets"	"Children"	"Education"
NEW	MILLION	CHILDREN	SCHOOL
FILM	TAX	WOMEN	STUDENTS
SHOW	PROGRAM	PEOPLE	SCHOOLS
MUSIC	BUDGET	CHILD	EDUCATION
MOVIE	BILLION	YEARS	TEACHERS
PLAY	FEDERAL	FAMILIES	HIGH
MUSICAL	YEAR	WORK	PUBLIC
BEST	SPENDING	PARENTS	TEACHER
ACTOR	NEW	SAYS	BENNETT
FIRST	STATE	FAMILY	MANIGAT
YORK	PLAN	WELFARE	NAMPHY
OPERA	MONEY	MEN	STATE
THEATER	PROGRAMS	PERCENT	PRESIDENT
ACTRESS	GOVERNMENT	CARE	ELEMENTARY
LOVE	CONGRESS	LIFE	HAITI

Figure 2. *The four factors with largest expected counts for an article from the AP corpus. We show the 15 words with the largest probability, $p(w \mid z)$, for each of these factors.*

If we examine γ for this article, we find that most of the factors are very close to α while four of the factors achieve significant expected counts. Looking at the distribution over words, $p(w \mid z)$, for those four factors, we can identify the topics that mixed via the θ random variable to form this document (Figure 2).

6. SMOOTHING AND LDA

The large vocabulary size that is characteristic of many information retrieval problems creates serious problems of sparsity. A new document is very likely to contain words that did not appear in any of the documents in a training corpus. Maximum likelihood estimates of the multinomial parameters assign zero probability to such words, and thus zero probability to new documents. The standard approach to coping with this problem is to "smooth" the multinomial parameters, assigning positive probability to all vocabulary items whether or not they are observed in the training set (Jelinek, 1997). Laplace smoothing is commonly used; this corresponds to placing a uniform Dirichlet prior on the multinomial parameters.

Though it is often implemented in practice (*e.g.*, Nigam *et al.*, 1999), simple Laplace smoothing does not correspond to formal integration in the mixture model setting. In fact, exact integration is intractable for mixtures for the same reasons that exact inference is intractable under the LDA model. However, we can again utilize the variational framework and approximate the posterior Dirichlet given the data. We present this variational approximation in the remainder of this section.

We elaborate the basic LDA model to place Dirichlet priors on the parameters β_i, for $i \in \{1, \ldots, k\}$, where β_i are the multinomial probabilities $p(w \mid z^i = 1)$ of words given topics. Again making an exchangeability assumption, we have $\beta_i \sim \mathrm{Di}(\eta, \eta, \ldots, \eta)$. The probability of the data becomes:

$$p(\mathcal{D} \mid \alpha, \eta) = \int \prod_{i=1}^{k} p(\beta_i \mid \eta) \prod_{d=1}^{M} p(\boldsymbol{w}_d \mid \alpha, \beta)\, d\beta,$$

where $p(\boldsymbol{w} \mid \alpha, \beta)$ is simply the LDA model as described above.

This integral is intractable but we can again lower bound the log probability using a variational distribution:

$$\log p(\mathcal{D} \mid \alpha, \eta) \geq \mathrm{E}_q[\log p(\beta \mid \eta)] + \sum_{d=1}^{M} \mathrm{E}_q[\log p(\boldsymbol{w}_d, \boldsymbol{z}, \theta \mid \alpha, \beta)] + \mathrm{H(q)},$$

where the variational distribution takes the form:

$$q(\beta_{[1:K]}, z_{[1:D]}, \theta_{[1:D]} \mid w_{[1:D]}, \lambda, \phi, \gamma) = \prod_{i=1}^{K} \mathrm{Di}(\beta_i \mid \lambda_i) \prod_{d=1}^{M} q_d(\theta_d, z_d \mid w_d, \phi_d, \gamma_d).$$

Note that the variational parameter λ_i is a V vector (even though η is a scalar) and that q_d is the variational distribution for LDA as defined above.

Variational inference chooses values for the variational parameters so as to minimize the KL divergence between the variational posterior and the true posterior. The derivation is similar to the earlier derivation, and yields the following updates:

$$\lambda_{ij} = \eta + \sum_{d=1}^{M} \sum_{n=1}^{N_d} w_n^j \phi_{dni}$$

$$\gamma_{di} = \alpha_i + \sum_{n=1}^{N_d} \phi_{dni} \tag{13}$$

$$\phi_{dni} \propto \exp\{\mathrm{E}[\log \beta_{iv} \mid \lambda_i]\} \exp\{\mathrm{E}[\log \theta_i \mid \gamma_d]\}, \tag{14}$$

where v in Eq. (14) refers to the unique index for which $w_{dn}^v = 1$.

To estimate values for the hyperparameters, we again maximize the lower bound on the log likelihood with respect to the expected sufficient statistics taken under the variational distribution. The maximization for α is the same as above. We calculate the following derivatives:

$$\frac{d \log p(\mathcal{D} \mid \alpha, \eta)}{d\eta} = \sum_{i=1}^{K} \sum_{j=1}^{V} \mathrm{E}_q[\log \beta_{ij} \mid \lambda_i] + KV\Psi(V\eta) - KV\Psi(\eta)$$

$$\frac{d^2 \log p(\mathcal{D} \mid \alpha\eta)}{d\eta^2} = KV^2\Psi'(V\eta) - KV\Psi'(\eta),$$

and maximize η by Newton's method.

To compute the probability of a previously unseen document, we again form the variational lower bound:

$$\begin{aligned}\log p(w_{\mathrm{new}} \mid \alpha, \eta, \mathcal{D}) &\geq \mathrm{E}_q[p(\beta, \theta_{\mathrm{new}}, z_{\mathrm{new}}, w_{\mathrm{new}} \mid \alpha, \eta, \mathcal{D})] \\ &= \mathrm{E}_q[p(\beta \mid \alpha, \eta, \mathcal{D})] + \mathrm{E}_q[p(\theta_{\mathrm{new}}, z_{\mathrm{new}}, w_{\mathrm{new}})].\end{aligned}$$

The optimizing parameters in the first term are exactly the λ_i computed in the empirical Bayes parameter estimation procedure. The optimizing parameters in the second term are found by simply iterating Eqs. (13) and (14) for the data in the new document.

7. EMPIRICAL RESULTS

In this section, we present an empirical evaluation of LDA on the benchmark CRAN corpus (van Rijsbergen and Croft, 1975), containing 2630 medical abstracts with 7747 unique terms, and a subset of the benchmark TREC AP corpus, containing 2500 newswire articles with 37871 unique terms. In both cases, we held out 10% of the data for test purposes and trained the models on the remaining 90%. Finally, note that in preprocessing all the data, we removed a standard list of stop words. (Further experimental details are provided in Blei *et al.*, 2002).

We compared LDA to the following standard models: "unigram," "mixture of unigrams," and "pLSI." The "unigram" model is simply a single multinomial for all words, irrespective of the document. The "mixture of unigrams" is a finite mixture of multinomials (Nigam *et al.*, 1999). The "probabilistic latent semantic indexing" (pLSI) model is a precursor of LDA in which the Dirichlet distribution of LDA is replaced with a list of multinomial probability vectors, one for each document in the corpus (Hofman, 1999). This is an over-parameterized model, and a "tempering" heuristic is used in practice to smooth the (maximum likelihood) solution. We fit all of the latent variable models using EM (variational EM for LDA) with exactly the same stopping criteria (the average change in expected log marginal likelihood is less than 0.001%).

To evaluate the predictive performance of these methods, we computed the *perplexity* of the held-out test set. The perplexity, used by convention in language modeling, is monotonically decreasing in the likelihood of the test data, and can be thought of as the inverse of the per-word likelihood.

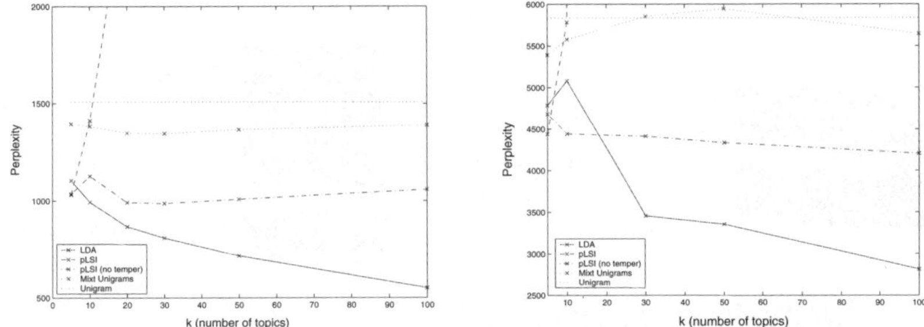

Figure 3. *Perplexity results on the CRAN (Left) and AP (Right) corpora for LDA, pLSI, mixture of unigrams, and the unigram model. Unigram is the higher dotted line; mixture of unigrams is the lower dotted line; untempered pLSI is dashed; tempered pLSI is dash-dot; and LDA is solid.*

Figure 3 illustrates the perplexity for each model and both corpora for different values of k. The latent variable models generally do better than the simple unigram model. The pLSI model severely overfits when not tempered (the values beyond $k = 10$ are off the graph) but manages to outperform the mixture of unigrams when tempered. LDA performs consistently better than the other models.

In Blei *et al.* (2002), we present additional experiments comparing LDA to related mixture models in the domains of text classification and collaborative filtering.

8. EXTENSIONS

The LDA model is best viewed as a simple hierarchical module that can be elaborated to obtain more complex families of models for increasingly demanding tasks in information retrieval.

In applications to corpora involving sets of images that have been annotated with text, we have studied a model that we refer to as "Corr-LDA," consisting of two coupled LDA models (Blei and Jordan, 2002). As shown in Figure 4, the model has two kinds of observables—"words" and "image blobs"—and also has latent "topics" for both kinds of variables. Briefly, a Dirichlet variable is used to parameterize the mixing proportions for a latent topic variable for images. A correspondence between topic variables for images and words is enforced via an explicit "translational" conditional probability. This yields a model that can associate particular words with particular regions of the image. We show annotations in Figure 5, comparing Corr-LDA to "GM-Mixture," a joint mixture model for words and images, and "GM-LDA," a pair of LDA models without the translational conditional. As suggested anecdotally by the annotations in the figure, and substantiated quantitatively in Blei and Jordan (2002), the Corr-LDA model

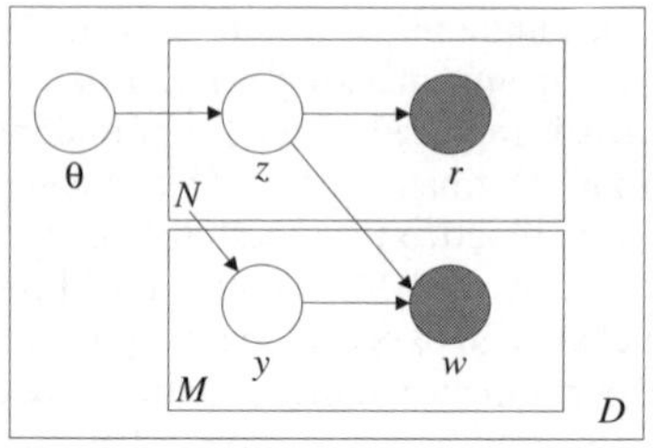

Figure 4. *The Corr-LDA model. The Gaussian random variable r encodes image blobs, while w encodes words. The variables z and y are latent "topics" for images and words, respectively. Note the "translational" conditional probability (the link from z to w)(that enforces a correspondence between image topics and word topics.*

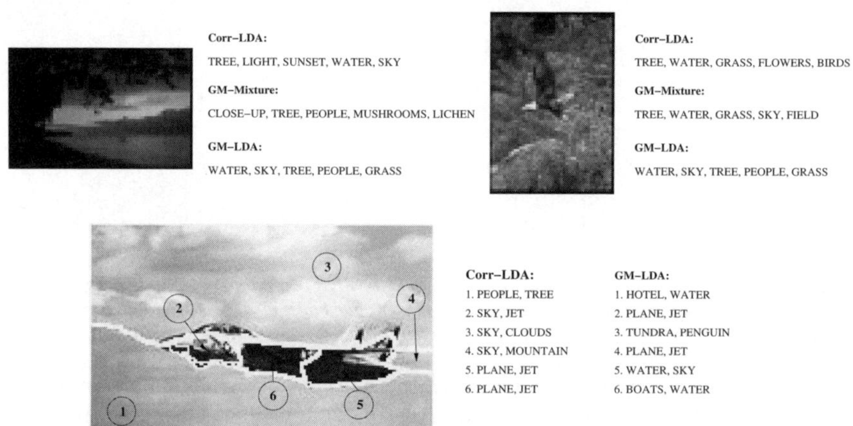

Figure 5. (Top) *Example images and their automatic annotations under different models.* (Bottom) *A segmented image and its labeling.*

is the most successful annotator of the models studied.

Another important extension of LDA involves an attempt to capture a notion of "resolution." Thus, documents might discuss "music" in a broad sense, might specialize to talk about "classical music," and might further specialize to talk about "chamber music." As in our earlier discussion, we do not want to impose any mutual exclusivity assumption between such levels of resolution—we want to allow a single document to be able to utilize different levels at different times. Thus, some of the words in the document may be viewed as associated with very specific topics, while other words are viewed as more generic. Elaborating the LDA model, we obtain such a multiresolution model by considering a tree in which the nodes correspond to latent topics. The generative process for generating a document chooses a path in this tree, chooses a level along the path, and selects a topic from the corresponding nonterminal node. Under this process, documents will tend to share topics as they do in the basic LDA model, but will be particularly likely to share topics that are high in the topic hierarchy. Thus, the hierarchy provides a flexible way to "share strength" between documents.

Both the pair LDA and the multiresolution LDA model are intractable for exact inference, but variational approximations are readily developed in both cases.

9. RELATED WORK

The LDA model and its hierarchical extension were inspired by the "probabilistic latent semantic indexing (pLSI)" model of Hoffman (1999). The pLSI model also assumes a mixture-of-multinomial model for the words in a document, but replaces the Dirichlet of LDA with a list of multinomial probability vectors, one for each document in the corpus. This can be viewed as a highly nonparametric version of LDA. As such, pLSI suffers from overfitting problems and is unable to assign probability mass to documents outside of the training corpus.

LDA is closely related to the Bayesian approach to mixture modeling developed by Diebolt and Robert (1994) and the Dirichlet multinomial allocation (DMA) model of Green and Richardson (2001). The DMA model posits a Dirichlet prior on a set of mixing proportions and a conjugate prior on the parameters associated with each mixture component. The mixing proportions and component parameters are drawn once. Each data point is assumed to have been generated by first drawing a mixture component, and then drawing a value from the corresponding parameters.

The main differences between LDA and DMA arise from the role of the Dirichlet random variable as a representation of "documents" in the LDA setting. In LDA, we draw a set of mixing proportions multiple times from the Dirichlet; in DMA, the mixing proportions are drawn only once. For DMA, new data points are assumed to have been drawn from a single mixture component; in LDA, new data points are collections of words, and each word is allocated to an independently drawn mixture component.

10. DISCUSSION

We have presented a simple hierarchical approach to modeling text corpora and other large-scale collections. The model posits that a document is generated by choosing a random set of multinomial probabilities for a set of possible "topics," and then repeatedly generating words by sampling from the topic mixture. An important virtue of this model is that a given document can be characterized by several topics—we avoid the mutual exclusivity assumption that is made by most clustering models.

One limitation of our current work on LDA is that we assume that the number of topics is a user-defined parameter. While our empirical work has shown a lack of sensitivity to this parameter, it clearly is of interest to study inference methods (variational or MCMC; see, *e.g.*, Green and Richardson, 1998; Attias, 2000) that allow this parameter to be inferred from data.

While we have focused on applications to information retrieval, problems with modeling collections of sequences of discrete data arise in many other areas, notably bioinformatics. While clustering methods analogous to those used in information retrieval have been usefully employed in bioinformatics, the mutual exclusivity assumption underlying these methods is particularly unappealing in the biological setting, and it seems likely that LDA-style models based on "topics" can play a useful role in these problems.

ACKNOWLEDGEMENTS

We would like to acknowledge support from NSF grant IIS-9988642 and ONR MURI N00014-00-1-0637.

REFERENCES

Attias, H. (2000). A variational Bayesian framework for graphical models. *Advances in Neural Information Processing Systems* **12** (S. Solla, T. Leen and K-R. Mueller, eds.). Cambridge: MIT Press, 209–215.

Baeza-Yates, R. and Ribeiro-Neto, B. (1999). *Modern Information Retrieval*. New York: ACM Press.

Blei, D. and Jordan, M. (2002). Modeling annotated data. *Tech. Rep.*, University of California, Berkeley, USA.

Blei, D., Jordan, M., and Ng. A. (2002). Latent Dirichlet allocation. *J. Machine Learning Res.* (submitted).

Dickey, J. M. (1983). Multiple hypergeometric functions: Probabilistic interpretations and statistical uses. *J. Am. Statist. Ass.* **78**, 628–637.

Dickey, J. M., Jiang, J. M., and Kadane, J. B. (1987). Bayesian methods for censored categorical data. *J. Am. Statist. Ass.* **82**, 773–781.

Diebolt, J. and Robert, C. P. (1994). Estimation of finite mixture distributions through Bayesian sampling. *J. R. Statist. Soc. B* **56**, 363–375.

Green, P. and Richardson, S. (2001). Modelling heterogeneity with and without the Dirichlet process. *Scand. J. Statist.* **28**, 355–377.

Harman, D. (1992). Overview of the first text retrieval conference (TREC-1). *Proceedings of the First Text Retrieval Conference* (D. Harman, ed). NIST Special Publication, 1–20.

Hofmann, T. (1999). Probabilistic latent semantic indexing. *Proceedings of the 22nd Annual International SIGIR Conference* (M. Hearst, F. Gey and R. Tong, eds). New York: ACM Press, 50–57.

Jelinek, F. (1997). *Statistical Methods for Speech Recognition*. Cambridge, MA: MIT Press.

Jordan, M. I., Ghahramani, Z., Jaakkola, T. S., and Saul, L. K. (1999). Introduction to variational methods for graphical models. *Machine Learning* **37**, 183–233.

Nigam, K., Lafferty, J. and McCallum, A. (1999). Using maximum entropy for text classification. *IJCAI-99 Workshop on Machine Learning for Information Filtering* (T. Joachims, A. McCallum, M. Sahami and L. Ungar, eds). San Mateo: CA Morgan Kaufmann, 61–67.

Ronning, G. (1989). Maximum likelihood estimation of Dirichlet distributions. *J. Statist. Comput. Simulation* **34**, 215–221.

van Rijsbergen, C. and Croft, W. (1975). Document clustering: An evaluation of some experiments with the Cranfield 1400 collection. *Inform. Processànd Mgm.* **11**, 71–182.

Zhai, C. and Lafferty, J. (2001). Document language models, query models, and risk minimization for information retrieval. *SIGIR Conference on Research and Development in Information Retrieval* (W. Croft, D. Harper, D. Kraft and J. Zobel, eds). New York: ACM Press, 111–119.

DISCUSSION

STEVEN N. MACEACHERN (*The Ohio State University, USA*)

In this paper, Blei, Jordan and Ng have done a favor to the Bayesian statistical community by introducing the variational inference paradigm and by illustrating its benefits in modelling documents. The further illustration in the talk, of how well the technique works for the description of pictures in terms of a small set of words, was impressive. I suspect that this paradigm will become one of the standard methods for Bayesian analysis in problems where speed of computation is essential. My comments will focus on two issues—namely an overview of the authors' modelling strategy and a recommendation for what I call directional assessment and adjustment for Bayesian analysis. I look forward to the authors' views on whether a directional adjustment is appropriate in this context, whether such an adjustment would be generally appropriate for variational inference, and also whether such adjustment is compatible with variational inference.

The authors implement a three step-strategy for their analysis: First, they choose simple, quickly computed statistics to monitor (each document is summarized by its word counts); second, they develop a simple model that generates these statistics (the multiple topics per document model, with conditionally independent choice of words drawn from the bag-of-words for each topic); third, they approximate the fit of the simple model (through the variational approximation). This strategy of writing a simplified model and then fitting it either exactly or by approximation is a mainstay of Bayesian inference. In many applications, one can provide a qualitative critique of a model. A simple model is retained for analysis, either because it simplifies computation or because specification of the structure and prior distribution for a more complex model would be difficult or not accepted by others. A great strength of this paper lies in its focus on computational algorithms that scale well and on the variational approximation used to fit the model. The next few paragraphs examine use of the strategy in a simplified context.

As an illustration of the contrast between simple and more complex models, consider two hierarchical models which have, at the middle stage, either a conditionally i.i.d. normal component or a stationary AR(1) component. The two models induce the same marginal distribution for each of the $\boldsymbol{Y}_i$. In the context of a larger model, with fairly large sample sizes, one would essentially learn the marginal distribution of the $\boldsymbol{Y}_i$, although their joint distribution could never be discovered from the simple analysis.

The simple model leads to inference for $\boldsymbol{\mu}$ based upon $\bar{\boldsymbol{Y}}$. This model has $\boldsymbol{\mu} \sim N(0, \boldsymbol{\sigma}_\mu{}^2)$; $\boldsymbol{\theta}_i \mid \boldsymbol{\mu} \sim N(\boldsymbol{\mu}, \boldsymbol{\sigma}_\theta{}^2)$; $\boldsymbol{Y}_i \mid \boldsymbol{\theta}_i \sim N(\boldsymbol{\theta}_i, \boldsymbol{\sigma}_Y{}^2)$, where $i = 1, \ldots, n$. The more complex model replaces the distribution on the $\boldsymbol{\theta}_i$ with a dependence structure, yielding $\boldsymbol{\theta}_0 \mid \boldsymbol{\mu} \sim N(\boldsymbol{\mu}, \boldsymbol{\sigma}_\theta{}^2)$; $\boldsymbol{\theta}_i \mid \boldsymbol{\mu}, \boldsymbol{\theta}_{<i} \sim N(\boldsymbol{\rho}\boldsymbol{\theta}_{i-1} + (1-\boldsymbol{\rho})\boldsymbol{\mu}, (1-\boldsymbol{\rho}^2)\boldsymbol{\sigma}_\theta{}^2)$, where $i = 1, \ldots, n$.

Since the sampling distribution of $\bar{\boldsymbol{Y}} \mid \boldsymbol{\mu}$ is normal under both models, the posterior distribution for $\boldsymbol{\mu} \mid \bar{\boldsymbol{Y}}$ can be found in closed form in both cases. In the simple model, the distribution of $\bar{\boldsymbol{Y}} \mid \boldsymbol{\mu}$ has mean $\boldsymbol{\mu}$ and variance $n^{-1}(\boldsymbol{\sigma}_\theta{}^2 + \boldsymbol{\sigma}_Y{}^2)$ while for the complex model it has mean $\boldsymbol{\mu}$ and variance $n^{-1}(\boldsymbol{\sigma}_\theta{}^2 + \boldsymbol{\sigma}_Y{}^2 + 2\boldsymbol{\sigma}_\theta{}^2\boldsymbol{\rho}/(1-\boldsymbol{\rho})) + O(n^{-2})$. The posterior distributions for $\boldsymbol{\mu} \mid \bar{\boldsymbol{Y}}$ under the models reflect this difference forever, with the ratio of posterior variances never tending to 1.

This overdispersion is not tied to the particulars of our simple and complex models. Instead, it is a general feature that appears whenever, loosely speaking, the $\boldsymbol{\theta}_i$ exhibit positive dependence. Thus, a qualitative assessment of the differences between the simple model and the complex model may lead us to conclude that the actual distribution of the statistic used in the Bayesian update has larger spread than the distribution we have used for formal calculation. As Bayesians, we are compelled to consider this assessment of our model and, if we judge the more complex posterior distribution to lie in a particular direction, to adjust our formally calculated posterior distribution.

There are a number of different perspectives that generate directional adjustments. A popular approach to adjustment is to flatten the likelihood either by raising it to a fractional power or by inflating the variance of a statistic. In the context of normal theory models, the approaches coincide. In the broader context of the exponential family, the adjustment can often be viewed as replacing the actual sample size with a smaller effective sample size while retaining the same values for the sufficient statistics. In early work on information processing, Zellner describes such adjustment, while Ibrahim and Chen (2000) describe it as a means of downweighting prior experiments.

There is a long tradition in survey sampling of directional adjustment. When a complex survey is administered, calculation of a standard error for an estimate can be difficult. The design based standard error accounts for features of the sampling design such as stratification and cluster sampling. Often, the very information needed for calculation of a standard error is unavailable to the analyst due to confidentiality constraints. In many surveys, standard errors are larger than those calculated by treating the data as a simple random sample. These difficulties have led to the notion of the "design effect", an inflator for the standard error. A design effect may be laboriously calculated for a number of estimators, taking into account the entire sample design. This design effect is then applied to inflate the standard errors of other estimators. The same notion of selecting an adjustment or of creating a distribution of adjustments can be ported over to Bayesian statistics through the device of a fractional likelihood. The analyst elicits, through whatever means, the fraction or the adjustment to posterior sample size.

Consider a directional assessment for the latent Dirichlet model, as applied to the information retrieval problem. The simple model is designed to work with the word counts for each of the documents. As a first step, we look at features of the problem which either suggest overdispersion or underdispersion of the word counts, relative to the behavior expected under the simple model.

To examine overdispersion/underdispersion in the context of the multinomial distribution, we first match the distributions on the cell means. Here, this is the number of words in the document times the cell probabilities. Overdispersion is naturally produced through a hierarchical structure where a vector of cell probabilities is drawn from some distribution and words are conditionally independent draws from these cell probabilities. Underdispersion is naturally obtained through stratification, where individual words are drawn as multinomials with differing vectors of cell probabilities. For stratification to reduce the dispersion, the different vectors of cell probabilities would be used in fixed proportion.

Turning to the documents, there are two main features that suggest overdispersion. The first is that language is an individual construct. Different individuals, even when writing on the same collection of topics, will consistently make different word choices. This effect is strong enough that it has been used to attribute authorship of documents to individuals. Mosteller and Wallace's (1984) examination of the Federalist papers relies on such a strategy.

The second feature which suggests overdispersion is the dynamic, ever-changing nature of language. New words enter the language (a check in the Merriam-Webster lists the word 'Bayesian' as only dating from 1961). Old words disappear, and usage frequency changes. Thus, word use provides information for dating documents. For collections of documents written over an extended period of time, changes in language lead to overdispersion. In the context of scientific documents, authors writing with different backgrounds will use a different vocabulary to describe many of the same concepts.

Certain features of the documents also suggest movement toward underdispersion. The main effects in this direction are tied to the structure of written language. The effects appear on the scale of entire documents, of paragraphs, and of sentences. At the level of the document, the classic training for writing an essay suggests that one introduce the topic, put in the guts of the essay, and then wrap things up with conclusions. Word choice is presumably different in these three parts of a document. This indicates the presence of stratification which leads in the direction of underdispersion. At a mid-level, a paragraph generally begins with a topic sentence, with details filling out the remaining sentences. Again, with differential word choice in the initial and remaining sentences of a paragraph, this suggests stratification. At the lowest level, English (and I believe most languages) tends to place the main information content of a sentence toward its beginning. Again, we have evidence of stratification. The grammatical structure of language, with a need for nouns, verbs and adjectives, also suggests stratification.

In my experience, the presence of both sorts of effects, those that move toward overdispersion and those that move toward underdispersion, is common. To complete a directional assessment, one must judge the relative sizes of the competing effects. My impression is that, in the document context, an individual's word choice will provide by far the strongest effect, leading to a net overdispersion. With this in mind, my belief would be that an exact analysis based on the simple model would produce likelihoods that are too sharp, leading to a posterior distribution that is too concentrated.

The next step in a directional assessment of the latent Dirichlet analysis is to judge the impact of the variational approximation to the posterior on the analysis. Here, my intuition is much weaker, as I have only worked through the approximation in some very simple cases. However, in the cases I have examined, the approximation systematically results in a distribution which is more concentrated than the distribution being approximated. The difference is tied to the asymmetry of the Kullback-Liebler divergence. The variational approximation reverses the roles of the usual "true" and "near" distribution that appear in asymptotic theory.

Taken together, the full model and the variational approximation to the simple model suggest that some adjustment to the posterior distribution is appropriate. Directional adjustment suggests flattening all, or part, of the approximate posterior. A systematic examination across problems might suggest how much flattening is appropriate.

Acknowledgement. This material is based upon work supported by the National Science Foundation under Award No. DMS-0072526.

REPLY TO THE DISCUSSION

We agree with Steven MacEachern that overdispersion is an important issue for the line of research that we have presented, and an important issue for the field of probabilistic information retrieval. There are at least three distinct reasons why the posterior that we obtain is likely to be overly concentrated with respect to the distributions of words that we attempt to model. First, our model is exceedingly simple, ignoring many linguistic phenomena that will tend to yield larger variability than our model accounts for, in particular the sequential phenomena referred to by MacEachern. Second, the empirical Bayes methodology that we use is known to yield underestimates of posterior variance. Third, the variational approach that we utilize for inference tends to yield approximating distributions that are overly concentrated. Let us briefly discuss each of these phenomena and outline possible solutions.

The exchangeability assumptions underlying LDA are aimed at computational simplicity, but clearly they are overly strong. A general goal of the field of information retrieval is to relax "bag-of-words" assumptions, and develop models that are closer to linguistic reality. We view hierarchical Bayesian methodology as providing a natural upgrade path. By introducing latent variables for linguistic concepts such as "sense" and "style," we can begin to capture sources of variability that are currently outside of our model. In particular, the latter concept might be naturally introduced as a discrete multinomial that conditions the Dirichlet in LDA—yielding a mixture of Dirichlet distributions in the topic simplex, and allowing us to move beyond the restrictive assumption that the corpus is captured by a single Dirichlet in the topic space. We can also introduce Markovian structure; for example, the topic variable for each word can be conditioned on the topic variable for the preceding word. Alternatively, we can consider Dirichlet/multinomial mixtures on subsequences of words ("n-grams") rather than single words. Finally, we can also make use of information available from parsing algorithms to provide conditioning variables for the LDA model, thereby capturing some of the stratification that, as discussed by MacEachern, is likely to lead to underdispersion. In all of these cases, however, while it is easy to specify natural extensions of LDA within the hierarchical Bayesian formalism, computational issues are of major concern. Any model that is aimed at applications in information retrieval must scale to tens or hundreds of thousands of documents, and inferential procedures must run in seconds.

Despite the large scale of problems in information retrieval, the data are also often sparse. Indeed, as we discussed in Section 6, document-by-word matrices tend to be sparse, and a new document is very likely to contain words that were not seen in the training corpus. This problem has been the subject of intense study, often inspired by theoretical work on species-sampling models and generally studied within a frequentist framework (Chen and Goodman, 1996). The variational smoothing method outlined in Section 6 is an approximate Bayesian solution to this problem.

While the discussion in Section 6 provided an example of the advantage of moving beyond an empirical Bayes inference method to a fuller hierarchical Bayesian approach (for the parameter β), we also find that computational considerations often weigh in favor of the simplicity offered by the empirical Bayes approach (cf. the parameter α in the LDA model). We expect that these considerations will be of increasing importance as we consider richer and more complex hierarchical Bayesian models, and thus we expect that empirical Bayes will continue to play an important role in this line of research. In this regard, it is important to note that the well-known fact that empirical Bayes leads to underestimates of the posterior variance (Carlin and Louis, 2000). In the context of overdispersion, this further reduction in variability may be particularly problematic. The corrections discussed in the empirical Bayes literature (*e.g.*, Kass and Steffey, 1989) may provide some relief.

Finally, the convexity-based variational techniques presented in Section 4 are known to be overly concentrated relative to the posterior distribution that they approximate. This difficulty has been addressed in a number of ways. One approach involves using higher-order approximations–Leisink and Kappen (2002) have presented a general methodology for converting low-order variational lower bounds into higher-order variational bounds. It is also possible to achieve higher accuracy by dispensing with the requirement of maintaining a bound, and indeed Minka and Lafferty (2002) have shown that improved inferential accuracy can be obtained for the LDA model via a higher-order variational technique known as "expectation propagation." Another general approach involves combining variational methods with sampling techniques to improve the accuracy of the variational distribution while maintaining its simple form (Ghahramani and Beal, 2000; de Freitas *et al.*, 2001). For example, the variational distribution can serve as a proposal distribution for an importance sampler.

Table 1. *Perplexity of a held-out test set as a function of the scaling factor.*

Scaling	Perplexity
0.1	2046.0
0.2	1818.6
0.3	1741.7
0.4	1661.1
0.5	1649.2
0.6	1652.1
0.7	1654.3
0.8	1654.0
0.9	1658.5
1.0	1660.5

While we believe that research in all of these areas—extended hierarchical Bayesian modeling of documents, corrections to empirical Bayes, and more accurate variational approximations—will eventually lead to well-motivated, computationally-efficient approximation procedures, the complexity of calibrating these various contributions to inaccurate assessment of variability also suggests that the kinds of "directional adjustments" suggested by MacEachern will play an important role. In an initial experiment, we adopted MacEachern's suggestion and rescaled the sample size while keeping the sufficient statistics fixed. Using a corpus of 740 documents, we estimated a five-factor LDA model, starting each run of variational EM from the same parameter setting. The results in Table 1 show the perplexity of a held-out set of 240 documents as a function of the scaling factor. The unscaled model was poorer than some of the scaled models, with the scaling of 0.5 yielding the best performance. Although these are preliminary results on a small corpus, they do indicate that simple directional adjustments may be useful in the Bayesian modeling of text corpora.

ADDITIONAL REFERENCES IN THE DISCUSSION

Carlin, B. P. and Louis, T. A. (2000). *Bayes and Empirical Bayes Methods for Data Analysis*. London: Chapman and Hall.

Chen, S. and Goodman, J. (1996). An empirical study of smoothing techniques for language modeling. *Proceedings of the Thirty-Fourth Annual Meeting of the Association for Computational Linguistics* (A. Joshi and M. Palmer, eds). San Mateo, CA: Morgan Kaufmann, 310–318.

de Freitas, N., Højen-Sørensen, P., Jordan, M. I., and Russell, S. (2001). Variational MCMC. *Uncertainty in Artificial Intelligence* (J. Breese and D. Koller, eds). San Mateo, CA: Morgan Kaufmann, 120–127.

Ghahramani, Z. and Beal, M. (2000). Variational inference for Bayesian mixtures of factor analyzers. *Advances in Neural Information Processing* **12**. Cambridge, MA: MIT Press, 449–455.

Ibrahim, J. G. and Chen, M.-H. (2000). Power prior distributions for regression models. *Statist. Sci.* **15**, 46–60.

Kass, R. E. and Steffey, D. (1989). Approximate Bayesian inference in conditionally independent hierarchical models (parametric empirical Bayes models). *J. Am. Statist. Ass.* **90**, 773–795.

Leisink, M. and Kappen, H. (2002). General lower bounds based on computer generated higher order expansions. *Uncertainty in Artificial Intelligence* (A. Darwiche and N. Friedman, eds). San Mateo, CA: Morgan Kaufmann, 293–300.

Minka, T. and Lafferty, J. (2002). Expectation-propagation for the generative aspect model, *Uncertainty in Artificial Intelligence* (A. Darwiche and N. Friedman, eds). San Mateo, CA: Morgan Kaufmann, 352–359.

Mosteller, F. and Wallace, D. L. (1984). *Applied Bayesian and Classical Inference: The Case of 'The Federalist' Papers*. New York: Springer.

BAYESIAN STATISTICS 7, pp. 45–63
J. M. Bernardo, M. J. Bayarri, J. O. Berger, A. P. Dawid,
D. Heckerman, A. F. M. Smith and M. West (Eds.)

Hierarchical Multivariate CAR Models for Spatio-Temporally Correlated Survival Data

BRADLEY P. CARLIN and SUDIPTO BANERJEE
University of Minnesota, USA
brad@biostat.umn.edu sudiptob@biostat.umn.edu

SUMMARY

Survival models have a long history in the biomedical and biostatistical literature, and are enormously popular in the analysis of time-to-event data. Very often these data will be grouped into strata, such as clinical sites, geographic regions, and so on. Such data will often be available over multiple time periods, and for multiple diseases. In this paper, we consider hierarchical spatial process models for multivariate survival data sets which are spatio-temporally arranged. Such models must account for correlations between survival rates in neighboring spatial regions, adjacent time periods, and similar diseases (say, different forms of cancer). We investigate Cox semiparametric survival modeling approaches, adding spatial and temporal effects in a hierarchical structure. Due to data limitations and computational complexity issues, we avoid geostatistical (kriging) models, and instead handle spatial correlation by placing a particular multivariate generalization of the conditionally autoregressive (CAR) distribution on the region-specific frailties. Exemplification is provided using time-to-event data for various cancers from the National Cancer Institute's Surveillance, Epidemiology, and End Results (SEER) database.

Keywords: CANCER SURVIVAL DATA; GEOGRAPHIC INFORMATION SYSTEM (GIS); LATTICE DATA; MARKOV CHAIN MONTE CARLO (MCMC).

1. INTRODUCTION

The analysis of spatially referenced data has been an increasingly active area of both methodological and applied statistical research. As with many fields, advances in computing and software have been largely responsible. Modern database engines, often widely and publicly available to anyone with a web browser, provide an enormous supply of georeferenced data. Sophisticated computer programs known as geographic information systems (GISs) have revolutionized the analysis and display of such data sets, through their ability to "layer" multiple data sources over a common study area. Finally, Markov chain Monte Carlo (MCMC) algorithms enable the fitting of complex hierarchical models in a Bayesian framework, permitting full posterior inference for underlying parameters in even complex model settings (*e.g.*, those involving data that are highly multivariate, temporally correlated, and perhaps even spatially or temporally misaligned).

In the fields of medicine and public health, a very common application of such models and methods is to the study of geographical patterns of disease. In the US, publicly available data on precise locations of disease cases is relatively rare due to strict confidentiality regulations. However, summaries of disease counts at a regional level (county, census tract, zip code, etc.) are often easy to obtain. As such, the most common model assumes that Y_i is the observed number of cases of a certain disease in region i, $i = 1, \ldots, p$, while E_i is the expected number of cases in this same region. The Y_i are thought of as random variables, while the E_i are thought of as fixed and known (and are often simply taken as proportional to the number of persons at risk in the region). If the E_i are not too large (i.e, if the disease is rare or the regions are sufficiently small), the model is then a spatial Poisson regression of the form

$$Y_i \mid \mu_i \overset{iid}{\sim} \text{Po}\,(E_i\, e^{\mu_i}), \quad \mu_i = x_i'\boldsymbol{\beta} + \theta_i + \phi_i\,. \tag{1}$$

The x_i are explanatory, region-level spatial covariates, having parameter coefficients $\boldsymbol{\beta}$. The θ_i capture region-wide *heterogeneity* via an exchangeable normal prior:

$$\theta_i \overset{iid}{\sim} N(0\,,\,1/\tau)\,,$$

where the precision τ controls the magnitude of the θ_i. These random effects capture extra-Poisson variability in the log-relative risks that varies "globally" (*i.e.*, over the entire study region).

Finally, the ϕ_i are the parameters that make this a truly spatial model by capturing regional *clustering*. While common geostatistical (e.g. exponential, spherical, Gaussian, etc.; see Cressie, 1993) models could be used as priors here, the most common approach is to adopt a *conditionally autoregressive* (CAR) prior (Besag, 1974; Besag *et al.,* 1991) of the form

$$\phi_i \mid \boldsymbol{\phi}_{-i} \sim N(\bar{\phi}_i\,,\,1/(\lambda m_i))\,, \tag{2}$$

where $\boldsymbol{\phi}_{-i} \equiv (\phi_1, \ldots, \phi_{i-1}, \phi_{i+1}, \ldots, \phi_p)^T$, m_i is the number of "neighbors" (adjacent regions) of region i, and $\bar{\phi}_i = m_i^{-1}\Sigma_{j\,adj\,i}\,\phi_j$, the average of the neighboring values. This model corresponds to a multivariate normal distribution for $\boldsymbol{\phi} \equiv (\phi_1, \ldots, \phi_p)^T$ with a less-than-full-rank covariance matrix. This arises due to the translation invariance clearly visible in (2); the prior is a *pairwise difference prior* (Besag *et al.,* 1995), identified only up to an additive constant. This impropriety is typically ignored, since the posterior for $\boldsymbol{\phi}$ will typically emerge as proper even though this prior is not. In practice, the constraint $\sum_{i=1}^{p} \phi_i = 0$ is added so that an overall intercept term in model (1) can be identified. This constraint is easily imposed *numerically* by recentering each sampled $\boldsymbol{\phi}$ vector around its own mean following each MCMC iteration; indeed, this procedure is now automated within WinBUGS and GeoBUGS (http://www.mrc-bsu.cam.ac.uk/bugs/welcome.shtml), software packages now routinely used by practicing biostatisticians and spatial epidemiologists.

In this paper we explore the extension of this methodology in three distinct directions. First, we wish to consider the case of *multivariate* CAR models. Such models are necessary to analyze more than one disease simultaneously, since a number of different diseases (say, cancer types) may share the same set of (spatially distributed) risk factors. While the vast majority of Bayesian CAR spatial modeling has concerned models for a single disease, some recent work in multivariate settings has appeared. Kim *et al.* (2001) presented a "twofold CAR" model to model counts for two different types of disease over each areal unit. However, their method seems specifically adapted for two diseases, and generalization of their method for a larger number of diseases seems prohibitively complicated. Similarly, Knorr-Held and Best (2000) have developed a "shared component" model for the above purpose, but their methodology too seems specific to the bivariate situation. Knorr-Held and Rue (2002) illustrate sophisticated MCMC blocking approaches in a model placing three conditionally independent CAR priors on three sets of spatial random effects in a shared component model setting. Most recently, Gelfand and Vounatsou (2002) investigated propriety conditions for the multivariate CAR model of Mardia (1988), and illustrated their use with a five-dimensional multinomial likelihood.

Gamerman *et al.* (2002) employ a Gaussian Markov random field (GMRF) model, a multivariate generalization of the pairwise difference CAR model above, and compare various MCMC blocking schemes for sampling from the posterior that results under a Gaussian multiple linear regression likelihood. They also investigate a "pinned down" version of this model that resolves the impropriety problem by centering the ϕ_i vectors around some mean location. These authors also place the spatial structure on the spatial regression coefficients themselves, instead of on extra intercept terms (*i.e.*, in (1) we would drop the ϕ_i, and replace $\boldsymbol{\beta}$ by $\boldsymbol{\beta}_i$, which would now be assumed to have a CAR structure). Assunção *et al.* (2002) refer to these models as *space-varying coefficient* models, and illustrate in the case of estimating fertility schedules. Assunção (2002) offers a nice review of the work to date in this area.

Our second extension concerns moving beyond models for mere counts in each region, to models for spatially referenced *survival times*. Again, previous Bayesian work in this context is rather limited. Crook *et al.* (2001) analyze spatially arranged time-to-event data by restructuring it as binary longitudinal, by combining risk sets containing failed and censored events within small time intervals. The BayesX software package (www.stat.uni-muenchen.de/ĩang/bayesx/bayesx.html) is used to obtain semiparametric posterior estimates from a probit conditional survival time model wherein a constant hazard is assumed over the short time intervals. Banerjee *et al.* (2002) use a parametric (Weibull) formulation for the baseline hazard and compare geostatistical and CAR priors for spatial frailties. Banerjee and Carlin (2002a) extend this approach to a semiparametric setup under the usual Cox proportional hazards model. Both of these papers reject the geostatistical frailty model, since it is quite time consuming yet produces answers that differ little from those using the CAR formulation. Finally, Henderson *et al.* (2001) use a Cox survival model with conditionally independent gamma frailties whose means are assumed to have a spatial correlation structure. They compare a variety of geostatistical models (powered exponential, Matérn, etc.) with the usual CAR model on the frailties, specified at either the regional or the individual location level. All of these papers use the DIC criterion (Spiegelhalter *et al.,* 2002) as their Bayesian model comparison tool, perhaps due to the very complex nature of the models.

Our third and final extension is to the *spatio-temporal* case. Bayesian disease mapping work in this area dates at least to that of Waller *et al.* (1997), where a subscript indexing time is added to the model, extending the hierarchical structure by one level and enabling the borrowing of estimative power across both time and space. More recently, Assunção *et al.* (2001) use the spatio-temporal model of Bernardinelli *et al.* (1995) and place conditionally independent CAR priors on the components of a log-relative risk (1) that is a polynomial in time. Banerjee and Carlin (2002b) extend the spatial frailty model of Banerjee and Carlin (2002a) to the spatio-temporal case, adding year of diagnosis to a space-only model for breast cancer incidence.

The remainder of our paper is organized as follows. Section 2 reviews the matrix formulation of the standard CAR model. Then in Section 3 we formulate the multivariate CAR (MCAR) model, which we derive from previous work by Mardia (1988). Section 4 then tailors the MCAR to fit our spatial frailty setting. In particular, we use a semiparametric model, and consider MCAR structure on both residual (spatial frailty) and regression (space-varying coefficient) terms. We also extend to the spatio-temporal case by including temporally correlated cohort effects (say, one for each year of initial disease diagnosis) that can be summarized and plotted over time. Section 5 illustrates the utility of our approach in an analysis of survival times of patients suffering from one or more types of cancer. We obtain posterior estimates of key fixed effects, smoothed maps of both frailties and spatially varying coefficients, and compare models using the DIC criterion. Finally, Section 6 briefly discusses our findings and suggests avenues for further research in this area.

2. MATRIX FORMULATION FOR CAR MODELS

Before we develop the multivariate CAR model, we first review the matrix formulation of the standard CAR, as described for example by Besag and Kooperberg (1995) and Best *et al.* (1999). Consider a vector $\boldsymbol{\phi} = (\phi_1, ..., \phi_p)^t$ of p components that follows a multivariate Gaussian distribution with mean 0 and B as the inverse of the dispersion matrix, so that B is $p \times p$ symmetric and positive definite. The density for $\boldsymbol{\phi}$ is given by

$$p(\boldsymbol{\phi}) = (2\pi)^{-p/2} |B|^{1/2} \exp\Big(-\frac{1}{2}\boldsymbol{\phi}^t B \boldsymbol{\phi}\Big) . \tag{3}$$

For such a distribution it is of interest to look at the conditional distribution of a particular component given the remaining components. In terms of the elements of the matrix $B = ((b_{ij}))$, it is well known from normal theory that ϕ_i has full conditional distribution

$$p\left(\phi_i|\boldsymbol{\phi}_{-i}\right) \propto \exp\Big(-\frac{1}{2}b_{ii}\Big(\phi_i - \sum_{j\neq i} \frac{-b_{ij}}{b_{ii}}\phi_j\Big)^2\Big) , \tag{4}$$

which is normal $N(\sum_{j\neq i} -b_{ij}b_{ii}^{-1}\phi_j, b_{ii}^{-1})$ distribution. To see this, note that the quadratic form $\boldsymbol{\phi}^t B\boldsymbol{\phi}$ can be expressed as $\sum_{i=1}^{p} b_{ii}\phi_i^2 + \sum\sum_{i\neq j} b_{ij}\phi_i\phi_j$. Collecting only the terms involving ϕ_i and remembering that $b_{ij} = b_{ji}$, we see that $p\left(\phi_i|\boldsymbol{\phi}_{-i}\right) \propto \exp\left(-\frac{1}{2}\{b_{ii}\phi_i^2 + 2\phi_i \sum_{j\neq i} b_{ij}\phi_j\right)$. The result in (4) now follows by simply completing the square.

A fundamental result in the understanding of CAR priors is due to Besag (1974), in which, using a remarkably simple proof of Brooks' Lemma, the conditions under which the specification of the full conditional distributions in (4) uniquely determine the joint distribution in (3) are explored. Thus suppose we now specify all the ϕ_i full conditional distributions as $N\left(\sum_{j\neq i} c_{ij}\phi_j, \sigma_i^2\right)$, so that

$$p(\phi_i|\boldsymbol{\phi}_{-i}) \propto \exp\Big(-\frac{1}{2\sigma_i^2}\Big(\phi_i - \sum_{j\neq i} c_{ij}\phi_j\Big)^2\Big). \tag{5}$$

When compared with (4), this reveals that $c_{ij} = -b_{ij}/b_{ii}$ and $b_{ii} = 1/\sigma_i^2$. Now form a matrix C with $c_{ii} = 0$ and $c_{ij} = -b_{ij}/b_{ii}$, and another matrix $M = \text{Diag}\left(\sigma_i^2\right)$ (so that $M^{-1} = \text{Diag}(b_{ii})$). Then B is related to M and C as

$$B = M^{-1}\left(I - C\right) , \tag{6}$$

where I is the identity matrix. As a result, the joint distribution of $\boldsymbol{\phi}$ is $N\left(0, (I - C)^{-1} M\right)$. Note that for modeling purposes the C matrix is often modeled directly, and so M must be specified so that $M^{-1}\left(I - C\right)$ is symmetric. The condition $c_{ij}\sigma_j^2 = c_{ji}\sigma_i^2$ guarantees this symmetry.

The C matrix may involve actual distances, between either regional centroids or actual event locations, if available (see, *e.g.*, Kaiser *et al.*, 2002). However as already mentioned, the most popular CAR model (Besag *et al.*, 1991) involves only the *adjacency* relationships amongst the various regions. From (5), we can identify the C matrix for this model as $c_{ii} = 0$, and $c_{ij} = 1/m_i$ if j is adjacent to i and 0 otherwise, while M^{-1} becomes the diagonal matrix $\text{Diag}\left(\lambda m_i\right)$. Thus, if W denotes the adjacency matrix of the map (*i.e.*, $w_{ii} = 0$, and $w_{ij} = 1$ if j is adjacent to i and 0 otherwise), then $C = W_s$ where $W_s = \text{Diag}\left(1/m_i\right)W$. That is, W_s is a scaled adjacency matrix, the i-th row being scaled by the number of neighbors of region i. By the discussion above (6), the above expressions for the elements of C and M translate to the following specifications for the inverse covariance matrix B: $b_{ii} = \lambda m_i$, and $b_{ij} = -\lambda$ if j is adjacent to i and 0 otherwise. Thus, B is symmetric and may be expressed as $B = \lambda(\text{Diag}(m_i) - W)$.

The conditional distribution specifications through C and M will provide a valid joint distribution only if $M^{-1}(I - C)$ is positive definite. However in many practical modeling situations this is not ensured. For example, with the above modeling of $C = W_s$ we actually have the aforementioned singularity in $M^{-1}(I - C)$ arising from the fact that $W_s 1 = 1$ (*i.e.*, the row sums of W_s all add up to 1). A possible repair for this situation is to include a "propriety parameter" α in the precision matrix B, modifying (6) to

$$B = M^{-1}(I - \alpha C)\,. \tag{7}$$

If $|\alpha| < 1$, then with M diagonal and C the scaled adjacency matrix, the matrix $M^{-1}(I - \alpha C)$ is diagonally dominant and symmetric. But symmetric diagonally dominant matrices are positive definite (see *e.g.*, Harville, 1997), thus resolving the impropriety problem.

Unfortunately, a difficulty with this solution is that this new prior may not deliver enough spatial similarity unless α is quite close to 1, in which case of course the impropriety problem returns (at least numerically). Some authors thus recommend an informative prior for α (say, a $\text{Beta}(18, 2)$) that insists on larger α's, but this is somewhat controversial.

3. MCAR MODEL DEVELOPMENT

We now develop the MCAR model. While most of the theoretical details can be extended from the results in Besag (1974) and have been explicitly derived by Mardia (1988), this distribution has rarely been employed in practice, perhaps due to its opaqueness in its original form, which we now try to elucidate. Let $\Phi^t = (\phi_1^t, \ldots, \phi_p^t)$ where Φ is $np \times 1$ with each ϕ_i being an n-dimensional vector. Once again consider a multivariate Gaussian distribution for Φ of the form

$$p(\Phi) = (2\pi)^{-np/2}\,|B|^{1/2} \exp\left(-\frac{1}{2}\Phi^t B \Phi\right)\,.$$

Here B is an $np \times np$ symmetric positive definite matrix. In fact it is easier to visualize B as a $p \times p$ block matrix with $n \times n$ blocks B_{ij}. Analogous to the univariate situation, it then follows that the full conditional distributions $p(\phi_i|\phi_{-i})$ are proportional to

$$\exp\Big(-\frac{1}{2}\Big(\phi_i - B_{ii}^{-1}\sum_{j\neq i}\big(-B_{ij}\big)\phi_j\Big)^t B_{ii}\Big(\phi_i - B_{ii}^{-1}\sum_{j\neq i}\big(-B_{ij}\big)\phi_j\Big)\Big)\,, \tag{8}$$

which means that $p(\phi_i|\phi_{-i})$ is $N_n\left(B_{ii}^{-1}\sum_{j\neq i}(-B_{ij})\,\phi_j, B_{ii}^{-1}\right)$. This is easily derived by noting that $\Phi^t B \Phi = \sum_{i=1}^p \phi_i^t B_{ii} \phi_j + \sum_{i=1}^p \sum_{j\neq i} \phi_i^t B_{ij} \phi_j$. Given a particular i, collecting only the terms involving ϕ_i, and using the fact that $\phi_i^t B_{ij} \phi_j = \phi_j^t B_{ij}^t \phi_i$, we find that

$$p(\phi_i|\phi_{-i}) \propto \exp\Big(-\frac{1}{2}\Big\{\phi_i^t B_{ii} \phi_i + 2\sum_{j\neq i} \phi_i^t B_{ij} \phi_j\Big\}\Big)\,.$$

The result in (8) now follows from the standard multivariate completion of the square technique.

We now turn to the identification of the joint distribution through conditional specifications. Following Mardia (1988), for a zero-centered MCAR we set

$$p(\phi_i|\phi_j, j \neq i, \Sigma_i) = N_n\Big(\sum_{j\neq i} C_{ij}\phi_j, \Sigma_i\Big), \quad i = 1, \ldots, p\,, \tag{9}$$

where each C_{ij} is a $n \times n$ matrix, as is each Σ_i. In addition, the Σ_i are each positive definite matrices and represent the conditional variance matrix. We immediately identify (with relation to the B matrix) $C_{ij} = -B_{ii}^{-1} B_{ij}$ and $B_{ii} = \Sigma_i^{-1}$. Setting Σ to be a block diagonal matrix with Σ_i as blocks (denoted as BlockDiag (Σ_i)) and C as a partitioned matrix with blocks C_{ij} (denoted as *Block* (C_{ij})) and with $C_{ii} = 0_{n \times n}$, we see that, analogous to (7),

$$B = \Sigma^{-1} (I - \alpha C) , \tag{10}$$

once we add the propriety parameter α. Note that both C and Σ are $np \times np$ matrices. Analogous to the univariate case, symmetry conditions for $\Sigma^{-1} (I - \alpha C)$ require $C_{ij} \Sigma_j = \Sigma_i C_{ji}^t$. We will refer to this joint distribution as an MCAR (C, Σ). We address the important issue of positive definiteness of $\Sigma^{-1} (I - \alpha C)$ when we discuss the modeling of the C matrix in the next section.

4. MCAR SPATIAL FRAILTY MODELING

4.1. *Static Spatial Survival Data with Multiple Causes of Death*

We now turn to the multivariate survival setting. Let t_{ijk} denote the time to death or censoring for the k-th patient having the j-th type of primary cancer living in the i-th county, $i = 1, \ldots, p$, $j = 1, \ldots, n$, $k = 1, \ldots, s_{ij}$, and let δ_{ijk} be the corresponding death indicator. Let us write x_{ijk} as the vector of covariates for the above individual, and let z_{ijk} denote the vector of cancer indicators for this individual. That is, $z_{ijk} = \left(z_{ijk1}, z_{ijk2}, ..., z_{ijkn}\right)^t$ where $z_{ijkl} = 1$ if patient ijk suffers from cancer type l, and 0 otherwise (note that $z_{ijkj} = 1$ by definition). Then we can write the likelihood of our proportional hazards model $L(\boldsymbol{\beta}, \boldsymbol{\theta}, \Phi; t, x, \gamma)$ as

$$\prod_{i=1}^{p} \prod_{j=1}^{n} \prod_{k=1}^{s_{ij}} \left\{h\left(t_{ijk}; x_{ijk}, z_{ijk}\right)\right\}^{\delta_{ijk}} \times \exp\left\{-H_{0i}\left(t_{ijk}\right) \exp\left(x_{ijk}^t \boldsymbol{\beta} + z_{ijk}^t \boldsymbol{\theta} + \phi_{ij}\right)\right\} , \tag{11}$$

where

$$h\left(t_{ijk}; x_{ijk}, z_{ijk}\right) = h_{0i}\left(t_{ijk}\right) \exp\left(x_{ijk}^t \boldsymbol{\beta} + z_{ijk}^t \boldsymbol{\theta} + \phi_{ij}\right) , \tag{12}$$

$$H_{0i}\left(t_{ijk}\right) = \int_0^{t_{ijk}} h_{0i}(u)\, d\boldsymbol{\phi}_i = \left(\phi_{i1}, \phi_{i2}, ..., \phi_{in}\right)^t ,$$

$$\boldsymbol{\beta}, \boldsymbol{\theta} \sim \text{ flat}, \quad \text{and} \quad \Phi \equiv \left(\boldsymbol{\phi}_1^t, \ldots, \boldsymbol{\phi}_p^t\right)^t \sim \text{MCAR}\,(C, \Sigma) .$$

The region-specific baseline hazard functions $h_{0i}\left(t_{ijk}\right)$ are modeled using the beta mixture approach (Gelfand and Mallick, 1995; Carlin and Hodges, 1999) in such a way that the intercept in $\boldsymbol{\beta}$ remains estimable. We note that we could extend to a county *and* cancer-specific baseline hazard h_{0ij}; however, preliminary exploratory analyses of our data suggest such generality is not needed here.

Several alternatives to model formulation (12) immediately present themselves. For example, we could convert to a space-varying coefficients model (Assunção, 2002), replacing the log-relative hazard $x_{ijk}^t \boldsymbol{\beta} + z_{ijk}^t \boldsymbol{\theta} + \phi_{ij}$ in (12) with

$$x_{ijk}^t \boldsymbol{\beta} + z_{ijk}^t \boldsymbol{\theta}_i , \tag{13}$$

where again $\boldsymbol{\beta}$ has a flat prior, but now $\Theta \equiv \left(\boldsymbol{\theta}_1^t, \ldots, \boldsymbol{\theta}_p^t\right)^t \sim$ MCAR (C, Σ). In Section 5 we apply this method to our cancer data set; we defer mention of still other log-relative hazard modeling possibilities until Sections 5 and 6.

4.2. MCAR *Specification, Simplification, and Computing*

To efficiently implement the MCAR (C, Σ) as a prior distribution for our spatial process, we make several simplifying assumptions. First, because the areal units may have different numbers of neighbors, we assume that the conditional variance at site i depends upon the number of neighbors, i.e. $\Sigma_i = m_i^{-1}\Lambda^{-1}$, where m_i is again the number of neighbors for areal unit i. From (9), the matrix Λ^{-1} describes the relative variability and covariance relationships between the different diseases given the neighboring sites. This assumption provides a mechanism for allowing extra variability at "edge" sites. Combining the above assumption with the symmetry restriction $C_{ij}\Sigma_j = \Sigma_i C_{ji}^t$, we obtain $m_i^{-1}C_{ij}\Lambda^{-1} = m_i^{-1}\Lambda^{-1}C_{ji}^t$.

We may also simplify the covariance structure by assuming the spatial influence of the neighbors j of county i is the same for all neighbors. Hence, we assume $C_{ij} = C_i$ if j is a neighbor of i, and $0_{n\times n}$ otherwise.

As a final simplification, we assume $C_i = m_i^{-1}I$. Looking again at (9), this final assumption implies that, for each disease, the conditional mean is simply the unweighted average of the neighboring values, as in the standard CAR model (2). We note for later use that, under the above simplifications, we may express the matrix B in terms of the $p \times p$ adjacency matrix W as

$$B = (\text{Diag}(m_i) - \alpha W) \otimes \Lambda ,$$

where we have again added a propriety parameter α. Note that this is a Kronecker product of a $p \times p$ and an $n \times n$ matrix, thereby rendering B as $np \times np$ as required. In fact, B may be looked upon as the Kronecker product of two partial precision matrices: one for the spatial components, $(\text{Diag}(m_i) - \alpha W)$ (depending upon their adjacency structure and number of neighbors), and another for the variation across diseases, given by Λ.

Also as a consequence of this simplified form, a sufficient condition for positive definiteness of the dispersion matrix for the MCAR$(\boldsymbol{C}, \Sigma)$ becomes $|\alpha| < 1$ (as in the univariate case). Negative smoothness parameters are not desirable, so we typically take $0 < \alpha < 1$. Thus we may now simplify our MCAR $(\boldsymbol{C}, \Sigma)$ notation to MCAR (α, Λ). We can now complete the Bayesian hierarchical formulation by placing appropriate priors on α (say, a $U(0, 1)$ or Beta$(18, 2)$) and Λ (say, a Wishart (ρ, Λ_0)).

The Gibbs sampler is the MCMC method of choice here, particularly because, as in the univariate case, it takes advantage of the MCAR's conditional specification. Adaptive rejection sampling may be used to sample the regression coefficients $\boldsymbol{\beta}$ and $\boldsymbol{\theta}$, while Metropolis steps with (possibly multivariate) Gaussian proposals may be employed for the spatial effects Φ. The full conditional for α is nicely suited for slice sampling (see *e.g.*, Neal, 2002), given its bounded support. Finally, the full conditional for Λ^{-1} emerges in closed form as an inverted Wishart distribution.

We conclude this subsection by generalizing our model to admit different propriety parameters α_j for different diseases (*cf.* Gelfand and Vounatsou, 2002).

Consider a permutation of the Φ vector, say $\Psi^t = (\boldsymbol{\psi}_1^t, \boldsymbol{\psi}_2^t, \ldots, \boldsymbol{\psi}_n^t)$, where $\boldsymbol{\psi}_j = (\phi_{1j},\ldots,\phi_{pj})^t$. That is, the vector of spatial effects, which were originally arranged as side-by-side vectors of the regional effects, are now permuted as side-by-side vectors of the disease effects. Moreover, since a permutation is an orthogonal transformation, $\Psi = P\Phi$ for some orthogonal matrix P. It then follows that the inverse covariance matrix of this transformed MCAR variable, PBP^t, which we denote by $B_{(P)}$, becomes

$$B_{(P)} = \Lambda \otimes (\text{Diag}(m_i) - \alpha W) . \tag{14}$$

This structure suffers from two obvious limitations. First, the spatial covariances among counties are proportional across diseases, *i.e.*, $\text{Cov}(\phi_{ij}, \phi_{i'j}) = K\text{Cov}(\phi_{ij'}, \phi_{i'j'})$ for some $K > 0$.

Second, the cross-covariances among counties for different diseases must obey a unnatural symmetry property, *i.e.*,

$$\text{Cov}(\phi_{ij}, \phi_{i'j'}) = \text{Cov}(\phi_{ij'}, \phi_{i'j})$$

Allowing different α_j for each disease j is one way of avoiding these difficulties. To illustrate this extension, consider the case of two diseases. We now want to model the above inverse dispersion matrix with possibly two different propriety parameters, say α_1 and α_2, so that Λ is 2×2. Since $(\text{Diag}(m_i) - \alpha_j W)$ is positive definite for each α_j, it has a Cholesky factorization $R_j^t R_j$, where R_j is $p \times p$. Gelfand and Vounatsou (2002) suggest the inverse dispersion matrix

$$B_{(P)} = \begin{pmatrix} \Lambda_{11} R_1^t R_1 & \Lambda_{12} R_1^t R_2 \\ \Lambda_{21} R_2^t R_1 & \Lambda_{22} R_2^t R_2 \end{pmatrix} = R^t \left(\Lambda \otimes I_{p \times p}\right) R$$

where $R = \text{BlockDiag}\,(R_j)$. Clearly the above formulation renders a positive definite $B_{(P)}$ as long as Λ is positive definite.

4.3. *Spatio-temporal survival data*

Here we extend the model of Section 4.1 to allow for cohort effects. Let r index the year in which patient ijk entered the study (*i.e.*, the year in which the patient's primary cancer was diagnosed). Extending model (6) we obtain the log-relative hazard

$$x_{ijkr}^t \boldsymbol{\beta} + z_{ijkr}^t \boldsymbol{\theta} + \phi_{ijr} \;, \tag{15}$$

with the obvious corresponding modifications to the likelihood (11). Here, $\boldsymbol{\phi}_{ir} = (\phi_{i1r}, \phi_{i2r}, ...,$ $\phi_{inr})^t$ and $\Phi_r = \left(\boldsymbol{\phi}_{1r}^t, \ldots, \boldsymbol{\phi}_{pr}^t\right)^t \stackrel{iid}{\sim} \text{MCAR}\,(\alpha_r, \Lambda_r)$. This permits addition of an exchangeable prior structure

$$\alpha_r \stackrel{iid}{\sim} \text{Beta}(a, b) \quad \text{and} \quad \Lambda_r \stackrel{iid}{\sim} \text{Wishart}(\rho, \Lambda_0) \;,$$

where we may choose fixed values for a, b, ρ, and Λ_0, or place hyperpriors on them and estimate them from the data. Note also the obvious extension to disease-specific α_{jr}, as mentioned at the end of the previous Section 4.2.

5. APPLICATION TO MULTIPLE CANCER SURVIVAL DATA

The National Cancer Institute's SEER program (http://seer.cancer.gov) is the most authoritative source of cancer data in the US, offering county-level summaries on a yearly basis for several states in various parts of the country. SEER currently includes cancer incidence and survival data from 11 population-based cancer registries and three supplemental registries, covering approximately 14% of the US population. Survival information for different types of cancer, individual level covariate information, and geographical information (county of residence) is available.

We illustrate our methods with an analysis of SEER data on 17,146 patients from the 99 counties of the state of Iowa who have been diagnosed with cancer between 1992 and 1998, and who have a uniquely identified primary cancer. Our covariate vector x_{ijk} consists of a constant (intercept), a gender indicator, the age of the patient, indicators for race with "white" as the baseline, indicators for the stage of the primary cancer with "local" as the baseline, and indicators for year of primary cancer diagnosis (cohort) with the first year (1992) as the baseline. The vector z_{ijk} comprises the indicators of which cancers the patient has; the corresponding parameters will thus capture the effect of these cancers on the hazards regardless of whether they emerge as primary or secondary.

We began by running five parallel MCMC chains from our basic MCAR spatial frailty model (6) for $15,000$ iterations each. After a burn-in period of 10,000 iterations, the remaining output was thinned to every 5 iterations, yielding a final posterior sample of size 5000 for computing posterior summaries. Our model used five separate (cancer-specific) propriety parameters α_j having an exchangeable $\text{Beta}(18, 2)$ prior, and a vague Wishart with $\rho = 5$, and $\Lambda_0 = \text{Diag}(0.01, 0.01, 0.01, 0.01, 0.01)$ for Λ. (Results for $\boldsymbol{\beta}, \boldsymbol{\theta}$, and Φ under a $U(0, 1)$ prior for the α_j were broadly similar.) Table 1 gives posterior summaries for the main effects $\boldsymbol{\beta}$ and $\boldsymbol{\theta}$; note that $\boldsymbol{\theta}$ is estimable despite the presence of the intercept since many individuals have more than one cancer. No race or cohort effects emerged as significantly different from zero, so they have been deleted; all remaining effects are shown here. All of these effects are significant and in the directions one would expect. In particular, the five cancer effects are consistent with results of previous modeling of this and similar data sets, with pancreatic cancer emerging as the most deadly (posterior median log relative hazard 1.701) and colorectal and small intestinal cancer relatively less so (0.252 and 0.287, respectively).

Table 1. *Posterior quantiles for the fixed effects $\boldsymbol{\beta}$ and $\boldsymbol{\theta}$ in the* MCAR *frailty model.*

Variable	2.5%	50%	97.5%
Intercept	0.102	0.265	0.421
Sex (female = 0)	0.097	0.136	0.182
Age	0.028	0.029	0.030
Stage of primary cancer (local = 0)			
Regional	0.322	0.373	0.421
Distant	1.527	1.580	1.654
Type of cancer			
Colorectal	0.112	0.252	0.453
Gallbladder	1.074	1.201	1.330
Pancreas	1.603	1.701	1.807
Small intestine	0.128	0.287	0.445
Stomach	1.005	1.072	1.141

Table 2 gives posterior variance and correlation summaries for the frailties ϕ_{ij} amongst the five cancers for two representative counties, Dallas (urban; Des Moines area) and Clay (rural northwest). Note that the correlations are as high as 0.528 (pancreas and stomach in Dallas County), suggesting the need for the multivariate structure inherent in our MCAR frailty model. Note also that summarizing the posterior distribution of Λ^{-1} would be inappropriate here, since despite the Kronecker structure in (14), Λ^{-1} cannot be directly interpreted as a primary cancer covariance matrix across counties.

Turning to geographic summaries, Figure 1 shows ArcView maps of the posterior means of the MCAR spatial frailties ϕ_{ij}. Recall that in this model, the ϕ_{ij} play the role of spatial residuals, capturing any spatial variation not already accounted for by the spatial main effects $\boldsymbol{\beta}$ and $\boldsymbol{\theta}$. The apparent lack of spatial pattern in these maps suggest there is little additional spatial "story" in the data beyond what is already being told by the fixed effects. However, the map scales reveal that one cancer (gallbladder) is markedly different from the others, both in terms of total range of the mean frailties (rather broad) and their center (negative; the other four are centered near 0).

Table 2. *Posterior variances and correlation summaries, Dallas and Clay counties,* MCAR *spatial frailty model. Diagonal elements are estimated variances, while off-diagonal elements are estimated correlations.*

Dallas County	colorectal	gallbladder	pancreas	intestine	stomach
Colorectal	0.852	0.262	0.294	0.413	0.464
Gallbladder		1.151	0.314	0.187	0.175
Pancreas			0.846	0.454	0.528
Small intestine				1.47	0.413
Stomach					0.908

Clay County	colorectal	gallbladder	pancreas	intestine	stomach
Colorectal	0.903	0.215	0.273	0.342	0.352
Gallbladder		1.196	0.274	0.128	0.150
Pancreas			0.852	0.322	0.402
Small intestine				1.515	0.371
Stomach					1.068

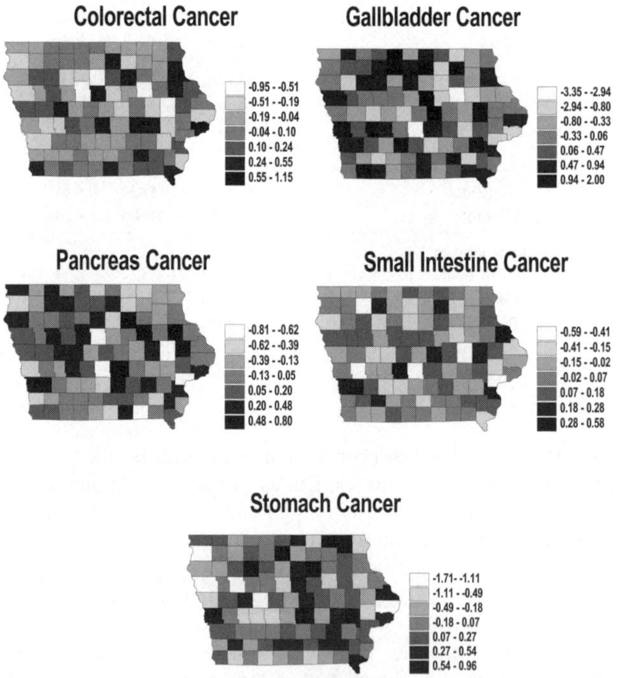

Figure 1. *Posterior mean spatial frailties, Iowa cancer data, static spatial* MCAR *model.*

Next, we change from the MCAR spatial frailty model to the MCAR spatially varying coefficients model (13). This model required a longer burn-in period (20,000 instead of 10,000), but otherwise our prior and MCMC control parameters remain unchanged. Figure 2 shows ArcView maps of the resulting posterior means of the spatially varying coefficients θ_{ij}. Unlike

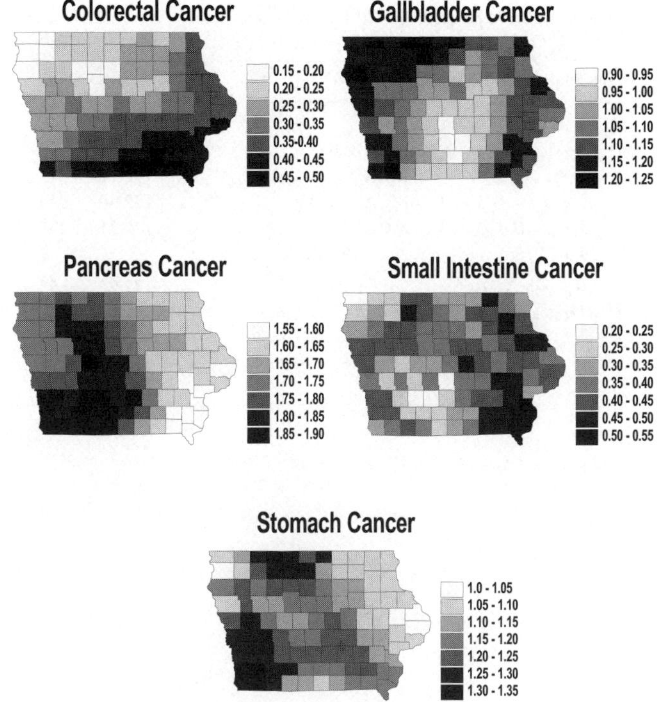

Figure 2. *Posterior mean spatially varying coefficients, Iowa cancer data, static spatial* MCAR *model.*

the ϕ_{ij} in the previous model, these parameters are not "residuals," but the effects of the presence of the primary cancer indicated on the death rate in each county. Clearly these maps show a strong spatial pattern, with (for example) southwestern Iowa counties having relatively high fitted values for pancreatic and stomach cancer, while southeastern counties are faring relatively poorly with respect to colorectal and small intestinal cancer. The overall levels for each cancer are consistent with those given for the corresponding fixed effects $\boldsymbol{\theta}$ in Table 1 for the spatial frailty model.

Table 3 gives the effective model sizes p_D and DIC scores for a variety of spatial survival models. The first two listed (fixed effects only and standard CAR frailty) have few effective parameters, but also rather poor (large) DIC scores. The MCAR spatial frailty models (which place the MCAR on Φ) fare better, especially when we add the disease-specific α_j (the model summarized in Tables 1 and 2 and Figure 1). However, adding heterogeneity effects ϵ_{ij} to this model adds essentially no extra effective parameters, and is actually harmful to the overall DIC score (since we are adding complexity for little or no benefit in terms of fit). Finally, the two spatially varying coefficients models enjoy the best (smallest) DIC scores, but only by a small margin over the best spatial frailty model.

In an attempt to validate the fit of some of these models, Figure 3 shows the Kaplan-Meier curve from the raw data (irregular dashed line) along with the Kaplan-Meier curves based on samples from the posterior predictive distributions for our "top three" (in terms of DIC) models: the spatially varying coefficients model (13) with cancer-specific α_j (solid line), the spatially

Table 3. *DIC comparison, spatial survival models for the Iowa cancer data.*

Log-relative hazard model	p_D	DIC
$x_{ijk}^t\boldsymbol{\beta} + z_{ijk}^t\boldsymbol{\theta}$	10.97	642
$x_{ijk}^t\boldsymbol{\beta} + z_{ijk}^t\boldsymbol{\theta} + \phi_i,\ \boldsymbol{\phi} \sim CAR(\alpha, \lambda)$	103.95	358
$x_{ijk}^t\boldsymbol{\beta} + z_{ijk}^t\boldsymbol{\theta} + \phi_{ij},\ \Phi \sim \text{MCAR}(\alpha = 1, \Lambda)$	172.75	247
$x_{ijk}^t\boldsymbol{\beta} + z_{ijk}^t\boldsymbol{\theta} + \phi_{ij},\ \Phi \sim \text{MCAR}(\alpha, \Lambda)$	172.40	246
$x_{ijk}^t\boldsymbol{\beta} + z_{ijk}^t\boldsymbol{\theta} + \phi_{ij},\ \Phi \sim \text{MCAR}(\alpha_1, \ldots, \alpha_5, \Lambda)$	175.71	237
$x_{ijk}^t\boldsymbol{\beta} + z_{ijk}^t\boldsymbol{\theta} + \phi_{ij} + \epsilon_{ij},\ \Phi \sim \text{MCAR}(\alpha_1, \ldots, \alpha_5, \Lambda),$ $\epsilon_{ij} \overset{iid}{\sim} N(0, \tau^2)$	177.25	255
$x_{ijk}^t\boldsymbol{\beta} + z_{ijk}^t\boldsymbol{\theta}_i,\ \Theta \sim \text{MCAR}(\alpha, \Lambda)$	169.42	235
$x_{ijk}^t\boldsymbol{\beta} + z_{ijk}^t\boldsymbol{\theta}_i,\ \Theta \sim \text{MCAR}(\alpha_1, \ldots, \alpha_5, \Lambda)$	171.46	229

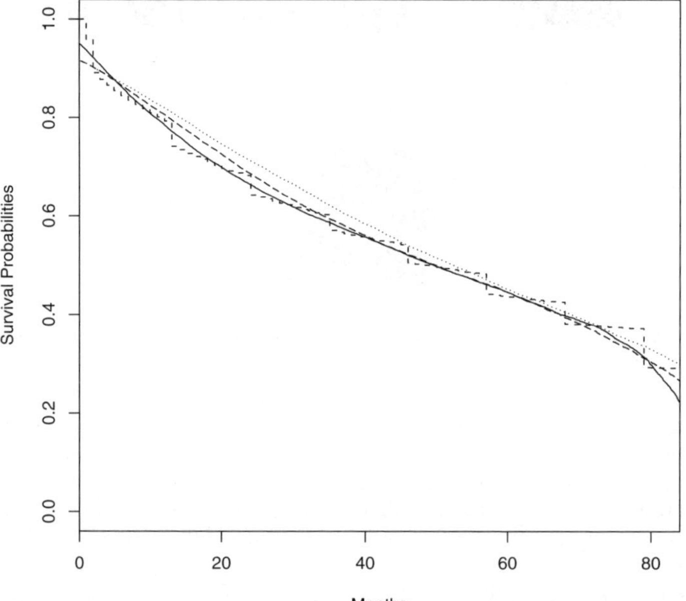

Figure 3. *Kaplan-Meier curves based on the raw data (irregular dashed line) and posterior samples from three* MCAR *spatial survival models (see text for legend).*

varying coefficients model (13) with a single α (dashed line), and the spatial frailty model (6) with cancer-specific α_j (dotted line). The first of these three models does seem to offer the best fit to the raw data over the range of survival times observed. This further augments this model's appeal (alongside its attractive Figure 2 maps and good DIC score), though we caution a final decision as to "best" model must of course take into account the use to which the chosen model will be put.

Finally, we fit the spatio-temporal extension (15) of our MCAR frailty model to the data where the cohort effect (year of study entry r) is taken into account. Year-by-year boxplots of the posterior median frailties (not shown) indicate a slightly decreasing trend in the final 2-3 years for all of the cancers, though this might be mostly an artifact of the paucity of observed survival times for these most recent cohorts. We are currently investigating the spatio-temporal extension of the spatially-varying coefficients model (13) (*i.e.*, $x^t_{ijkr}\boldsymbol{\beta} + z^t_{ijkr}\boldsymbol{\theta}_{ir}$) to see if the results are temporally more interesting.

6. DISCUSSION

Our use of propriety parameters α_j likely remains controversial, since $\alpha_j < 1$ implies that the mean of the ϕ_{ij} is a *shrunk* version of the average of the neighboring $\phi_{i'j}$. However, such parameters do allow the computation of the posterior variances and correlations in Table 2, while still permitting a reasonable degree of smoothing (visible, for example, in Figure 2).

We have investigated only fairly basic MCAR spatial survival models, and data complexities may well cloud matters considerably. For instance, the primary cancer j need *not* be the first cancer discovered in a patient; some patients may also have more than one primary. We may also have access to *cause of death* data $v_{ijk} = \left(v_{ijk1}, v_{ijk2}, ..., v_{ijkn}\right)^t$, where $v_{ijkl} = 1$ if patient ijk is observed to die from cancer type l, and 0 if not. In this case, we might augment model (6) to

$$x^t_{ijk}\boldsymbol{\beta} + z^t_{ijk}\boldsymbol{\theta} + v^t_{ijk}\boldsymbol{\gamma} + \phi_{ij} \ ,$$

where the frailties $\Phi \sim \text{MCAR}\,(C, \Sigma)$ as before. Alternatively, model (13) could augment to

$$x^t_{ijk}\boldsymbol{\beta} + z^t_{ijk}\boldsymbol{\theta}_i + v^t_{ijk}\gamma_i \ ,$$

where $\Theta \sim \text{MCAR}\,(C, \Sigma)$ and $\Gamma \equiv \left(\gamma^t_1, \ldots, \gamma^t_p\right)^t \sim \text{MCAR}(C^*, \Sigma^*)$. This "doubly MCAR" does not emerge as practically identifiable for our data (the causes of death are nearly always equal to the primary cancers), but it does suggest a possibly fruitful avenue for further model generalization.

Finally, in the future we also look toward extending our work from the single endpoint, multiple cause case to the *multiple* endpoint, multiple cause case—say, for analyzing times until diagnosis of each cancer (if any), rather than merely a single time until death.

REFERENCES

Assunção, R. M. (2002). Space-varying coefficient models for small area data. *Environmetrics* (to appear).

Assunção, R. M., Reis, I. A. and Oliveira, C. D. L. (2001). Diffusion and prediction of Leishmaniasis in a large metropolitan area in Brazil with a Bayesian space-time model. *Statistics in Medicine* **20**, 2319–2335.

Assunção, R. M., Potter, J. E. and Cavenaghi, S. M. (2002). A Bayesian space varying parameter model applied to estimating fertility schedules. *Statist. Med.* (to appear).

Banerjee, S. and Carlin, B. P. (2002a). Spatial semiparametric proportional hazards models for analyzing infant mortality rates in Minnesota counties. *Case Studies in Bayesian Statistics VI* (C. Gatsonis, R. Kass, A. Carrriquiri, A. Gelman, D. Higdon, D. Pauler and I. Verdinelli, eds). New York: Springer (to appear).

Banerjee, S. and Carlin, B. P. (2002b). Semiparametric spatio-temporal frailty modeling. *Environmetrics* (to appear).

Banerjee, S., Wall, M. M. and Carlin, B. P. (2002). Frailty modeling for spatially correlated survival data, with application to infant mortality in Minnesota. *Biostatistics* (to appear).

Bernardinelli, L., Clayton, D., Pascutto, C., Montomoli, C., Ghislandi, M. and Songini, M. (1995). Bayesian analysis of space-time variation in disease risk. *Statist. Med.* **14**, 2433–2443.

Besag, J. (1974). Spatial interaction and the statistical analysis of lattice systems. *J. R. Statist. Soc. B* **36**, 192–236. (with discussion).

Besag, J. and Kooperberg, C. (1995). On conditional and intrinsic autoregressions. *Biometrika* **82**, 733–746.

Besag, J., York, J. C. and Mollié, A. (1991). Bayesian image restoration, with two applications in spatial statistics (with discussion). *Ann. Inst. Statist. Math.* **43**, 1–59.

Besag, J., Green, P., Higdon, D. and Mengersen, K. (1995). Bayesian computation and stochastic systems. *Statist. Sci.* **10**, 3–66 (with discussion).

Best, N. G., Arnold, R. A., Thomas, A., Waller, L. A. and Conlon, E. M. (1999). Bayesian models for spatially correlated disease and exposure data *Bayesian Statistics 6* (J. M. Bernardo, J. O. Berger, A. P. Dawid and A. F. M. Smith, eds). Oxford: Oxford University Press, 131–156 (with discussion).

Carlin, B. P. and Hodges, J. S. (1999). Hierarchical proportional hazards regression models for highly stratified data. *Biometrika* **55**, 1162–1170.

Cressie, N. A. C. (1993). *Statistics for Spatial Data* (revised ed). New York: Wiley.

Crook, A. M., Knorr-Held, L. and Hemingway, H. (2001). Measuring spatial effects in time to event data: A case study using months from angiography to coronary artery bypass graft (CABG). *Tech. Rep.*, Imperial College, London, UK.

Gamerman, D., Moreira, A. R. B., and Rue, H. (2002). Space-varying regression models: Specifications and simulation. *Comput. Statist. Data Anal.* (to appear).

Gelfand A. E. and Mallick, B. K. (1995). Bayesian analysis of proportional hazards models built from monotone functions. *Biometrika* **51**, 843–852.

Gelfand, A. E. and Vounatsou, P. (2002). Proper multivariate conditional autoregressive models for spatial data analysis. *Biometrika* (to appear).

Harville, D. D. (1997), *Matrix Algebra from a Statistician's Perspective*. New York: Springer.

Henderson, R., Shimakura, S. and Gorst, D. (2001). Modelling spatial variation in leukemia survival data. *Tech. Rep.*, Lancaster University, UK.

Kaiser, M., Daniels, M., Furakawa, K., and Dixon, P. (2002). Analysis of particulate matter air pollution using Markov random field models of spatial dependence. *Environmetrics* (to appear).

Kim, H., Sun, D. and Tsutakawa, R. K. (2001). A bivariate Bayes method for improving the estimates of mortality rates with a twofold conditional autoregressive model. *J. Am. Statist. Ass.* **96**, 1506–1521.

Knorr-Held, L. and Best, N. G. (2000). A shared component model for detecting joint and selective clustering of two diseases. *J. R. Statist. Soc. A* **164**, 73–85.

Knorr-Held, L. and Rue, H. (2002). On block updating in Markov random field models for disease mapping. *Scand. J. Statist.* (to appear).

Mardia, K. V. (1988). Multi-dimensional multivariate Gaussian Markov random fields with application to image processing. *J. Multiv. Anal.* **24**, 265–284.

Neal, R. M. (2002). Slice sampling. *Ann. Statist.* (to appear).

Spiegelhalter, D. J., Best, N., Carlin, B. P. and van der Linde, A. (2002). Bayesian measures of model complexity and fit. *J. R. Statist. Soc. B* **64**, 1–34 (with discussion).

Waller, L. A., Carlin, B. P., Xia, H., and Gelfand, A. E. (1997). Hierarchical spatio-temporal mapping of disease rates. *J. Am. Statist. Ass.* **92**, 607–617.

DISCUSSION

DEBORAH ASHBY (*Queen Mary, University of London, UK*)

Cancer is a major public health problem: in the Western world, one in three of us can expect to get it at some point in out lives, one in four of us expect to die from it. Understanding patterns of both incidence and survival is vital to prevention and cure. To this end cancer registration systems have been set up across the globe. Such routine data is essentially multivariate in nature, as all cancers are captured by the data collection systems, but most analyses are univariate, with comparisons across cancers happening at an informal comparative level. This paper builds on previous work looking at spatial and temporal patterns in survival from cancer, and takes up the challenge of studying survival from different cancers simultaneously.

The paper is mainly concerned with the survival of different individuals diagnosed with different cancers. However a feature of the data is that some individuals are recorded as having more than one cancer. The authors claim that this contributes to the identifiability of $\boldsymbol{\theta}$ in the presence of the intercept. Before considering the technical aspects of how this is handled, what is going on from a biological perspective?

Multiple cancers. There are two senses in which a patient may get 'secondary' cancers. A new cancer is known as a 'primary' cancer, and this is what we are referring to when we say someone has, say, lung cancer, or breast cancer. A primary cancer starts off well-defined within its organ of origin, it may then grow and invade neighboring tissues, and it may throw off 'seeds', which lead to secondary cancers ('metastases') in other parts of the body. Each primary cancer has its own patterns of spread, but typically secondary cancers are found in lungs, bones, the brain, and the liver. In this data set, this information is captured by stage of cancer, coded as local, regional or distant.

Patient may also get subsequent primaries, which are new tumors. These may arise through bad luck: if one in three get cancer, then assuming independence one in nine might expect a second, provided they survive long enough to see it. But cancers are not independent, and risk factors such as smoking increase the risk of several cancers. In particular genetic predisposition may lead to multiple primaries: somebody with retinoblastoma (cancer of the eye) is at high risk of other cancers in the eye and other organs. A further reason for multiple primaries is that radiotherapy and chemotherapy may induce further cancers, for example, following treatment for Hodgkin's disease patients are at an increased risk of leukemia (Ashby *et al.,* 1993). All of this implies complex dependencies within and between patients, which are not currently accounted for in this modelling. In addition, multiple primaries can be difficult to classify. Different registries may operate different procedures, complicating the analysis, and producing artifactual spatial variations in incidence rates (Filali *et al.,* 1996).

Modelling. The basic model used is proportional hazards, with survival time t_{ijk} (k-th patient with j-th type of primary cancer in i-th county). This is fine if there is only one cancer, but if a patient has two primary cancers then there are actually two survival times, starting at different time points, ending at the same time point. At the moment the modelling only uses one of theses survival times, which seems rather odd. It effectively assumes that the second cancer is somehow known about at baseline, when in essence it is a time-dependent covariate. If there is any structure to which cancers diagnosed in which order (for example pancreas likely to be the last cancer a patient has diagnosed) then it could also introduce bias.

Depending on how the data are coded, a patient with cancers 1 and 2 could appear with t_{i1k} and t_{i2k}, so that both survival times are used, with the dependence captured to some extent by use of indicator z_{ijk}. This would not capture the individual frailties unless this was explicitly modelled in. I am also not clear why there is a $z_{ijk}\boldsymbol{\theta}$ term, but not $z_{ijk}\boldsymbol{\phi}$?

Further questions on the modelling: Can you put any interpretation on baseline hazard (*i.e.*, Figure 3)? Is it averaged across cancers? In Table 1, are $\boldsymbol{\beta}$'s really constant across all cancers *i.e.*, do sex, age and stage of cancer really have same impact on survival for five different sites? Sex and age are person-level covariates, but stage is a cancer level covariate. How do you code the person with local colorectal cancer who is later discovered to have pancreatic cancer with regional spread? And even for the person with just one cancer, progression from local to regional to distant is a time-dependent covariate, rather than one uniquely identified at baseline. How is this handled under the current model?

What does it all mean?. Spatial correlations in 'non-contagious' diseases can be thought of as reflections of unmeasured covariates. For cancer incidence these include smoking, drinking, social factors such as child-bearing patterns.

What might be the unmeasured covariates for survival? These could include the tendency to self-refer in timely fashion, which is socially determined, for example, by health beliefs and access to healthcare. Diagnostic practices vary between hospitals. There is some arbitrariness about the date of diagnosis, and registration practices can differ. Clinical care may vary, as may patient factors affecting survival such as age, sex, smoking, diet and compliance with treatment. Some of these may vary smoothly geographically, but others may not.

To what use might they be put? Their role ought to be descriptive, and challenge those working in health to understand the system better, but I can foresee ranking of frailties as in Goldstein and Spiegelhalter (1996). Without due account of the kind of factors outlined, they may be subject to serious misinterpretation. But those that are not due to artifacts are indicators of inequalities in health between populations, and should be taken seriously. An understanding of which factors are common to several cancers, and which are distinctive to a particular cancer will be invaluable.

Future Possibilities. This work is already complex, but can be developed further. There is a need to include covariates to explore the possibilities I have just raised. Despite the title, there is little presented in this paper on temporal patterns and it probably needs a longer time frame to make worthwhile. Some of these trends are likely to be subtle, but in respect of changed registration practices, or changes in screening or treatment policy it may be worth considering change point models. In addition the ambiguity in the start date of a cancer may be construed as a measurement error problem, and this could be incorporated (Hirst, 1998).

I spent many years working with Mersey Regional Cancer Registry in the UK. I am acutely aware of the enormous work that goes into collecting and collating such data, and of its importance in the fight against cancer. What I find most exciting about this paper is that it demonstrates the potential of Bayesian statistics to exploit the full complexities of these data, enabling us to compare and contrast survival experiences from different cancers for geographically diverse groups, which should ultimately improve survival for those with cancer.

AKI VEHTARI (*Helsinki University of Technology, Finland*)

My comments concern the use of DIC for spatial models and model comparison.

DIC is Bayesian extension of Akaike's (1973) AIC using posterior mean instead of maximum likelihood estimate and DIC also has a more elaborate way of estimating the model complexity than the AIC. Although DIC is more sophisticated I think that same dependency assumptions hold for AIC and DIC.

AIC was derived assuming independent data samples. Related FPE criteria by Akaike (1970), which is based on squared error instead of predictive likelihood, may be used for for time series with dependent data if errors are such that $E(\epsilon_i \epsilon_j | X_1, \ldots, X_j) = 0$ for $i < j$. This holds, for example, for finite parameter Markov processes.

Burman and Nolan (1992) postulate that functionals other than squared error may also be successfully used, provided that they may be well approximated by a quadratic form. I do not know if these results have been generalized for spatial models. DIC is not based on squared error and above condition for errors does not necessarily hold in complex real life spatial modelling problems. Vehtari (2001) demonstrates (although not in spatial modelling problem) that if there are dependencies in the data DIC does not work correctly.

Do the authors have some comments on how the dependencies in their data affect the model comparison and selection based on DIC? How do the authors estimate what is significant difference between DIC values in Table 3?

REPLY TO THE DISCUSSION

We thank Prof. Ashby and Dr. Vehtari for their insightful remarks. Both discussions reflect extensive thought on key issues arising in our paper. We will begin by replying to Dr. Vehtari's remarks, and then segue to the longer discussion by Prof. Ashby.

Dr. Vehtari is concerned with our use of the DIC criterion for model choice. While we are not certain, we think the data dependencies that hurt DIC performance in the example to which he refers are dependencies that are *not captured by the model.* While our data obviously feature a variety of spatially induced dependencies, we believe that all are accounted for by the MCAR and other aspects of our model, so under this "good model" assumption we believe DIC should perform reasonably well. Indeed, the "main example" in the original DIC paper by Spiegelhalter *et al.* (2002) is a disease mapping application (the oft-analyzed Scottish lip cancer data set) where the most promising models use the standard CAR prior. Nothing in this example's results suggests use of DIC would be inappropriate in our multivariate CAR setting. Regarding estimation of a "significant difference" in DIC score, here unfortunately we lack a good answer; attempts to date to estimate the variance of a DIC estimate (e.g. Zhu and Carlin, 2000) have met with limited success at best. In our work we typically resort to recomputing DIC a few times with different random number seeds, in order to get some idea of its posterior sampling variability for each model.

Turning to the comments of Prof. Ashby, we are at the same time delighted and terrified to have as our primary discussant someone with her understanding of cancer registries and their proper analysis! Our paper is only a first step in applying multivariate CAR models to cancer registry data; indeed, we were initially thrilled that our complex algorithms would even run at all! But clearly the detailed nature of the modelling does not excuse us from thinking more carefully about the data, and we intend to continue refining our approach in order to heighten its realism for cancer registry data sets like ours.

The most serious issue raised by Prof. Ashby involves what constitutes a 'primary' cancer. We confess to being somewhat confused here, a condition exacerbated by ambiguity in both the SEER database and the advice of various experts we consulted. While our paper referred to 'primary' and 'secondary' cancers, we now believe that *all* the cancers in the SEER database are primary (*i.e.*, physiologically independent, and not the result of metastasis from some other existing tumor). The cancer identified as primary, then, is presumably merely the *first* one diagnosed, for those having more than one. This then implies (as Prof. Ashby has pointed out) that one should be able to compute multiple times to death for these patients (one for each primary cancer diagnosis), but sadly we have not yet been able to cull this information from SEER. Thus, all of our event times refer to time from first primary until death. By the same token, stage refers to stage of the first primary at baseline (though Prof. Ashby is right to wonder about modelling stage as a time-varying covariate). Given the complex dependencies among the cancers, we think our univariate response model has merit, though as we mention in our discussion we do hope ultimately to extend our models to the multiple response setting. Along similar lines, we agree that there are many other potentially important covariates (smoking, drinking, self-referral, local diagnostic practices, etc.) which could be easily modelled if such data were available. Interactions (say, between age group and primary cancer) of the sort Prof. Ashby mentions could also be incorporated and estimated.

Prof. Ashby asks why we use a $z_{ijk}\boldsymbol{\theta}$ but not a $z_{ijk}\boldsymbol{\phi}$ term in model (12). We did this so that each individual (instead of each tumor) would contribute precisely one frailty term to the model. But of course we could certainly go the other way, and indeed this is precisely what model (13) in effect does. After hearing an earlier talk on this paper, Prof. Scott Zeger of Johns Hopkins University pointed out that there must be a lot of patients with more than one cancer, or else Figures 1 and 2 would merely be "shifted" versions of each other, showing the same basic spatial patterns (which they currently do not). We investigated this by fitting the model

$$x_{ijk}^t\boldsymbol{\beta} + z_{ijk}^t\boldsymbol{\theta} + z_{ijk}^t\boldsymbol{\phi}_i \ ,$$

where $\boldsymbol{\beta}$ and $\boldsymbol{\theta}$ have flat priors and the ϕ_i are distributed MCAR.

Table 4. *Proportion of patients in each group, Iowa cancer data.*

	number of primary cancers				
	1	2	3	4	5
County 5	0.51	0.23	0.13	0.10	0.03
all counties	0.94	0.04	0.01	0.01	0.00

Table 5. *Proportion of cancers in each group, Iowa cancer data.*

	number of primary cancers				
	1	2	3	4	5
County 5	0.27	0.24	0.20	0.22	0.07
all counties	0.87	0.07	0.04	0.02	0.01

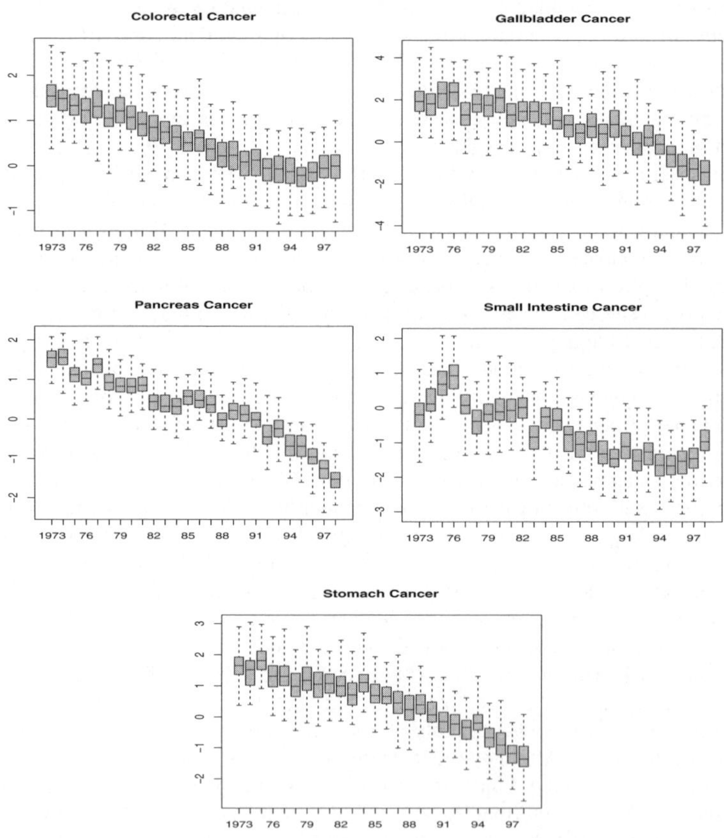

Figure 4. *Boxplots of posterior median spatial frailties over the 99 Iowa counties for each year, from 1973 to 1998.*

This model would be nothing but a "centered" version of model (13) if every patient had only one cancer. Our results indicated no significant change in the spatial story from that shown in Figure 2, suggesting that the effect of multiple cancers must indeed be significant. To check this, we created Table 4, which gives the proportion of SEER patients in County 5 (a small, rural county) and the entire state having 1, 2, 3, 4, or 5 primary cancers. Table 5 repeats this process using cancer (instead of patient) as the unit of analysis. Individuals with more than one primary cancer account for 6% of the patients and more than 13% of the cancers overall. Moreover, the totals for County 5 (49% and 73%, respectively) show that these totals can be much more extreme for thinly populated counties. These tables clearly show why Figures 1 and 2 need not be shifted versions of each other, and why better understanding of how 'primary' cancer is defined and modelled is critical to our ongoing analytic efforts.

Prof. Ashby also wonders about an interpretation of our baseline hazard $h_{0i}(t)$. As the notation suggests, this baseline hazard is permitted to vary across counties i, but not across primary cancers j nor individuals k within each county; a richer model might well be justified. The baseline hazard captures a county's "inherent" time to cancer death having accounted for the differential sex, age, stage, and primary cancer mix across counties. If plotted, the fitted baseline hazards would allow identification of outlying counties (say, counties with an overabundance of early deaths or long-term survivors). By contrast, Figure 3 shows Kaplan-Meier curves based either on the raw data or on "fake" data generated from our fitted models. These do not estimate a baseline hazard but rather a "marginal" hazard, averaged over cancers, counties, individuals, and the various covariate levels in our data set.

Finally, we agree that our paper does relatively little on the temporal side, and that our original data do not cover a long enough time period to reveal meaningful trends anyway. However, we have recently fit our spatio-temporal model (15) to SEER data from diagnosis years 1973 through 1998. Boxplots of the posterior medians of the frailties ϕ_{ijr} (Figure 4) reveal the expected steadily decreasing trend for all five cancers, though it is not clear how much of this decrease is simply an artifact of the censoring of survival times for patients in more recent cohorts. Incorporating change points, cancer start date measurement errors, and other model enhancements (say, interval censoring) will also be important aspects of future investigations.

ADDITIONAL REFERENCES IN THE DISCUSSION

Akaike, H. (1970). Statistical predictor identification. *Ann. Inst. Statist. Math.* **22**, 203–217.

Akaike, H. (1973). Information theory and an extension of the maximum likelihood principle. *Second International Symposium on Information Theory* (B. N. Petrov and F. Csaki, eds). Budapest: Academiai Kiado. Reprinted in *Breakthroughs in Statistics* **1** (S. Kotz and N. L. Johnson, eds). Berlin: Springer, 610–624.

Ashby, D., Hutton, J. L. and McGee, M. A. (1993). Simple Bayesian analyses for case-control studies in cancer epidemiology. *The Statistician* **42**, 385–397.

Burman, P. and Nolan, D. (1992). Data dependent estimation of prediction functions. *J. Time Ser. Anal.* **13**, 189–207.

Filali, K., Hedelin, G., Schaffer, P., Esteve, J., Arveux, P., Bouchardy, C., Exbrayat, C., Faivre, J., Levi, F., Mace-Lesech, J., Pottier, D. and Torhorst, J. (1996). Multiple primary cancers and estimation of the incidence rates and trends *Eur. J. Cancer* **32A**, 683–690

Goldstein, H. and Spiegelhalter, D. J. (1996). League tables and their limitations - statistical issues in comparisons of institutional performance. *J. R. Statist. Soc. A* **159**, 385–409 (with discussion).

Hirst, W. M. (1998). *Outcome Measurement Error in Survival Analysis*. Ph.D. Thesis, Univ. Liverpool, UK.

Vehtari, A. (2001). *Bayesian Model Assessment and Selection Using Expected Utilities.* Ph.D. Thesis, Helsinki University of Technology. `http://lib.hut.fi/Diss/2001/isbn9512257653`.

Zhu, L. and Carlin, B. P. (2000). Comparing hierarchical models for spatio-temporally misaligned data using the Deviance Information Criterion. *Statist. Med.* **19**, 2265–2278.

BAYESIAN STATISTICS 7, pp. 65–84
J. M. Bernardo, M. J. Bayarri, J. O. Berger, A. P. Dawid,
D. Heckerman, A. F. M. Smith and M. West (Eds.)

On Inferring Effects of Binary Treatments with Unobserved Confounders

SIDDHARTHA CHIB
Washington University, St. Louis, USA
chib@olin.wustl.edu

SUMMARY

This paper is concerned with the following problem. Suppose x is a binary $\{0, 1\}$ indicator of a treatment and y is a continuous response. The goal is to estimate the effect, say the average treatment effect (ATE), of x on y when there are unobserved confounders, even after conditioning on covariates, say because of unmeasured or unmeasurable covariates that affect both the treatment and the outcome. To deal with this problem, we analyze a modelling and inferential framework that has been little explored from the Bayesian perspective. In this framework, one requires a covariate (an instrument) that is correlated with the treatment and uncorrelated with the unobservables affecting the response. Such a covariate is available by design in so-called "quasi-experiments," in "natural experiments," and in other settings. We provide a Bayesian analysis of models with instruments and potential outcomes focusing on how prior-posterior inference may be conducted under a set of reasonable assumptions. Because identification of the ATE is fragile if the instrument is weak, and/or the degree of confounding is high, we discuss how relevant posterior distributions can be used to aid study conclusions. The framework is extended to univariate discrete outcomes where one important difference is the manner in which the ATE must be calculated. Estimation of the models is conducted by Markov chain Monte Carlo methods and models with and without confounding on unobservables are compared in terms of marginal likelihoods.

Keywords: ASSIGNMENT MECHANISM; AVERAGE TREATMENT EFFECT; BINARY OUTCOMES; CONFOUNDING; IDENTIFICATION; INSTRUMENTAL VARIABLE; MARGINAL LIKELIHOOD; MARKOV CHAIN MONTE CARLO; POTENTIAL OUTCOMES; STRUCTURAL MODEL.

1. INTRODUCTION

One of the most pervasive problems in statistics is the following. Suppose x is a binary $\{0, 1\}$ indicator of a treatment (where the term treatment is intended to embrace a covariate of interest, not necessarily one arising in a medical or epidemiological setting) and y is a response and the objective is to isolate the effect ("treatment effect") of x on y. To fix ideas, x may be an indicator of smoking status taking the value one if the subject is a current smoker and the value zero otherwise and y may be some measure of subject health. For simplicity we assume that both the treatment and response are univariate although this can obviously be relaxed. It is well understood that outside of the experimental setting inferring effects of this kind raise a multitude of challenges that are not easy to address (even in a randomized treatment setting, inference is difficult when human subjects are involved due to dropouts, non-compliance and other such complications). The problem is that when the treatment intake is non-random, as in an observational setting where subjects self-select into a treatment state, the choice of treatment may be influenced by unmeasured or unmeasurable or unobservable covariates that also affect the response.

In the presence of such confounding the effect of the treatment is, in general, only identifiable in conjunction with additional assumptions about the precise nature of the confounding. For example, in the work of Rosenbaum and Rubin, (Rosenbaum and Rubin, 1983; Rosenbaum, 2002), it is assumed that there exist observable covariates (say z_1) such that any unobserved determinant of the response is independent of the treatment, conditioned on the covariates z_1.

A more general setting is one in which the latter conditional independence assumption does not hold, and unobservable confounders remain, even after conditioning on a set of covariates. This setting has its origins in the structural equation models of econometrics developed by Haavelmo (1943, 1947) (see also Zellner,1971; Goldberger, 1972), later extended by Heckman (1978) to the case of binary treatments (Pearl, 2000) provides a directed acyclic graphical view of certain structural models). A necessary condition for identification of the treatment effect in this setting is the existence of a covariate that is correlated with the treatment assignment but uncorrelated with the response. Such a covariate, called an instrumental variable, or instrument for short, is available in so-called "encouragement designs" where subjects are randomly urged to take the treatment but actual compliance with treatment (the treatment intake), which is observed, is left to the discretion of the subject; in so-called "natural experiments" where one or more of the covariates that determine treatment assignment changes randomly due to a change in a government policy or some other intervention, and in other settings. When the validity of the instrument can be reasonably certified, the structural approach provides the means to infer the treatment effect without assuming that confounding on unobservables is absent, conditional on covariates.

Even though there is by now a large body of work dealing with these problems, much of the literature on structural models is non-Bayesian and prior to work of Heckman was entirely concerned with models with continuous responses and continuous treatments. The discrete-treatment structural models were initially specified without sufficient justification for the choice of instrument and without explicit discussion of treatment effects. Matters have changed considerably over the last decade although almost all the discussion is cast in frequentist terms, for example, Angrist *et al.* (1996), although Imbens and Rubin (1997) and Chib and Hamilton (2000, 2002) are exceptions. The purpose of this paper is to remedy this deficiency by taking up the question of discrete treatments in the context of first continuous responses and then discrete responses. We are particularly interested in developing a framework for analyzing these models that is concerned not just with the determination of the average treatment effect (ATE) but also with other inferences that are helpful in such problems. For example, it has been widely demonstrated (albeit in the context of continuous treatments, for example, Staiger and Stock, 1997, and Woglom, 2001) that an instrument that is weakly correlated with the treatment causes problems for a standard frequentist estimator of the parameters. Thus, it of interest to have a measure of such correlation in the discrete treatment context and to then study its posterior distribution. The extent of confounding is another parameter of interest whose posterior distribution can be derived and analyzed. We are also interested in the value of the Bayesian approach for comparing models with and without confounding on unobservables. For obvious reasons, all the calculations in the paper rely on Markov chain Monte Carlo methods. Fortunately, the fitting of the models raises no particular problems.

The rest of the paper is organized as follows. In Section 2,, the modelling and inferential framework for dealing with unobserved confounders is developed in the context of continuous outcomes. This section contains useful background information about the confounding problem and motivation for the structural model. A Markov chain Monte Carlo (MCMC) algorithm for estimating the parameters is provided and two examples are considered. The MCMC algorithm is also used to estimate the marginal likelihood of the model. In Section 3, the modelling and inferential framework is extended to binary responses. In the last section, we discuss briefly how the ideas in the paper can be extended even further.

2. MODELLING AND INFERENTIAL FRAMEWORK: CONTINUOUS UNIVARIATE RESPONSES

To begin with consider the setting of a univariate continuous response y and suppose that each subject i, $i \leq n$, is characterized by the potential or counterfactual outcomes (y_{i0}, y_{i1}) corresponding to the two possible levels $(0, 1)$ of the treatment x_i. Depending on the level of the treatment received, the observed outcome y_i is one of the two potential outcomes and is given by

$$y_i = y_{i0} + (y_{i1} - y_{i0})x_i \tag{2.1}$$

which exemplifies the fundamental missingness that is present in such settings (and, for Dawid, 2000, the principal reason for his misgivings about the potential outcomes paradigm). Letting $E(y_{ij}) = \mu_j$ $(j = 0, 1)$, one central objective of the analysis is to estimate the average treatment effect $\mu_1 - \mu_0$ given that only one of the two potential outcomes is ever observed for any subject.

In order to make progress with this problem, write the potential outcomes as

$$y_{i0} = \mu_0 + \eta_{i0} \tag{2.2}$$

$$y_{i1} = \mu_1 + \eta_{i1}$$

where η_{i0} and η_{i1} are unobserved random variables with mean zero. Let us assume throughout that $\eta_{i0} = \eta_{i1} = \eta_i$. Now substituting the potential outcomes into (2.1) gives rise to

$$y_i = \mu_0 + x_i\beta + \eta_i$$

where $\beta = (\mu_1 - \mu_0)$ is the average treatment effect. It would now be possible to estimate β by a regression of y_i on x_i provided the term $(\mu_0 + x_i\beta)$ is equal to the expected value of y_i given x_i or in other words if $E(\eta_i|x_i) = 0$. The latter condition is not likely to be satisfied in the context of observational data because the factors included in η_i are likely to be correlated with x_i. One way to deal with this problem is to assume that x_i is independent of η_i conditional on a set of covariates $\boldsymbol{z}_1$ (whose first element is unity) which we express in the form $\eta_i \perp\!\!\!\perp x_i|\boldsymbol{z}_{1i}$. This assumption is commonly associated with the approach of Rosenbaum and Rubin (1983). If we let $E(\eta_i|\boldsymbol{z}_{1i})$ be given by $\boldsymbol{z}_{1i}'\boldsymbol{\gamma}_1^*$ it follows that $y_i = \mu_0 + x_i\beta + \eta_i$ can be re-expressed as

$$y_i = \boldsymbol{z}_{1i}'\boldsymbol{\gamma}_1 + x_i\beta + \varepsilon_i \tag{2.3}$$

where $\boldsymbol{\gamma}_1$ is identical to $\boldsymbol{\gamma}_1^*$ except that the first element of $\boldsymbol{\gamma}_1$ is equal to the first element of $\boldsymbol{\gamma}_1^*$ plus μ_0. It is easy to check that under the assumption $\eta_i \perp\!\!\!\perp x_i|\boldsymbol{z}_{1i}$, we have that $\varepsilon_i \perp\!\!\!\perp x_i|\boldsymbol{z}_{1i}$, and hence a regression analysis applied to the preceding equation will deliver consistent inferences about β.

The conditional independence assumption provided in the preceding paragraph may also be interpreted in another way. Because x_i is a binary random-variable, suppose that treatment assignment is determined according to the probability model $x_i = I(\boldsymbol{z}_{1i}'\boldsymbol{\gamma}_{21} + u_i > 0)$, where $I(A)$ is the indicator function that takes the value one if the condition A is satisfied and the value zero otherwise, and u_i is a mean zero, variance one random-variable. Then, the conditional independence assumption amounts to saying that u_i is independent of ε_i, conditioned on $\boldsymbol{z}_{1i}$. Thus, in this sense, the conditional independence assumption rules out confounding on unobservables.

But suppose, more generally, that the unobservables (ε_i, u_i) are correlated even after conditioning on covariates, say because of unmeasured or unmeasurable covariates that affect both the treatment and the outcome. In this case, the effect of x on y is confounded with that of u on y (through ε). Suppose now that we have available, in addition to the covariates $\boldsymbol{z}_1$, at least one additional covariate $\boldsymbol{z}_2$ (instrument) that satisfies the twin conditions

A1 $\boldsymbol{z}_{2i}$ is a determinant of the treatment x_i given $\boldsymbol{z}_{1i}$
A2 $\boldsymbol{z}_{2i} \perp\!\!\!\perp \varepsilon_i | \boldsymbol{z}_{1i}$

In words, the instrument is a covariate that is correlated with the treatment but uncorrelated with the unobservables affecting the response, conditioned on $\boldsymbol{z}_{1i}$. Assumption A1 is sometimes referred to as an "exclusion restriction" because the covariate is a determinant of the treatment but not of the response given $\boldsymbol{z}_{1i}$. Assumption A2 is a statement that the model generating $\boldsymbol{z}_{2i}$ is ignorable. The directed acyclic graph of the model with the instrument is given in Figure 1.

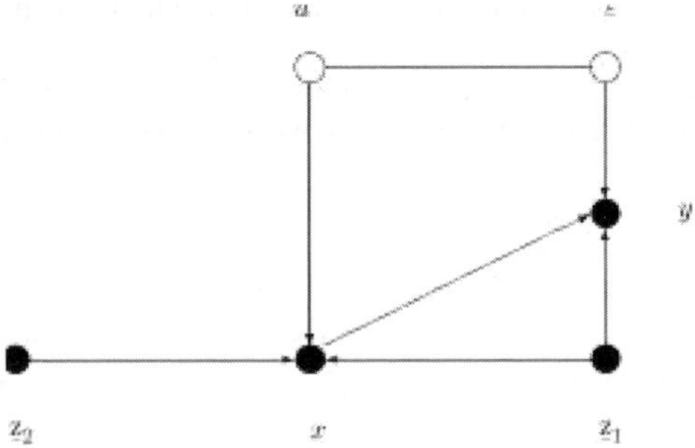

Figure 1. *Directed graph of the interactions with instrument variable* $\boldsymbol{z}_2$

It turns out that under weak additional assumptions related to the manner in which $\boldsymbol{z}_1$ and $\boldsymbol{z}_2$ determine the treatment assignment, the ATE is identified given an instrument. Although weaker assumptions can be made, we assume that the distribution of the unobservables is bivariate normal and the effect of the covariates on the treatment probability is given by the cdf of a standard normal distribution evaluated at a point (index) that is additive in $\boldsymbol{z}_1$ and $\boldsymbol{z}_2$. Specifically, we assume that $x_i = I(\boldsymbol{z}_i'\boldsymbol{\gamma} + u_i > 0)$, where $\boldsymbol{z}_i'\boldsymbol{\gamma}$ is the index function that is equal to $\boldsymbol{z}_{1i}'\boldsymbol{\gamma}_{21} + \boldsymbol{z}_{2i}'\boldsymbol{\gamma}_{22}$. Neither assumption is required for identification. Heckman and Vytlacil (1999) show that one only needs to assume that the joint distribution of the unobservables is continuous and $\boldsymbol{z}_i'\boldsymbol{\gamma} | \boldsymbol{z}_{1i}$ is a non-degenerate random variable. Under our assumptions, the joint model for the response and the treatment assignment mechanism is given by

$$\begin{aligned} y_i &= \boldsymbol{z}_{1i}'\boldsymbol{\gamma}_1 + x_i\beta + \varepsilon_i \\ x_i &= I(\boldsymbol{z}_i'\boldsymbol{\gamma} + u_i > 0), \quad i \leq n \end{aligned} \tag{2.4}$$

where (ε_i, u_i) is $N(0, \Omega)$ and $\Omega = (\omega_{ij})$ is a 2×2 matrix with $\omega_{22} = 1$.

We next present some implications of the model structure to enhance understanding of subsequent developments. First, the outcome model is a convenient way of expressing the marginal distribution of the counterfactual variables y_{i0} and y_{i1}. Thus, the marginal density of y_{i0} is Gaussian

$$f_0(y_i | \boldsymbol{z}_{1i}, \boldsymbol{\psi}) = N(y_i | \boldsymbol{z}_{1i}'\boldsymbol{\gamma}_1, \omega_{11}) \,, \tag{2.5}$$

where $N(\cdot|a, b)$ is the normal density with mean a and variance b and the notation f_0 is used to emphasize that this is the density of the counterfactual y_{i0}. Equivalently, if we were to apply the do-calculus framework of Pearl (2000), and intervene on x_i, with $do(x_i = 0)$, then $f(y_i|\boldsymbol{z}_{1i}, do(x_i = 0), \boldsymbol{\psi}) = N(y_i|\boldsymbol{z}_{1i}'\boldsymbol{\gamma}_1, \omega_{11})$ because an intervention severs any link to the probability model that determines the treatment assignment. Similarly, the marginal density of y_{i1} is

$$f_1(y_i|\boldsymbol{z}_{1i}, \boldsymbol{\psi}) = N(y_i|\boldsymbol{z}_{1i}'\boldsymbol{\gamma}_1 + \beta, \omega_{11}) \tag{2.6}$$

and thus the marginal distributions of the potential outcomes differ solely in the mean by an additive term. This assumption is common in the treatment literature (see, for example, Rubin, 1991). As we have mentioned earlier, we never get a draw from both distributions for the same subject. Also note that the difference in the expected values of y_{1i} and y_{i0} is β, and since this does not depend on $\boldsymbol{z}_{1i}$, β is the ATE.

Another way to appreciate the point in the preceding paragraph is to note that the distribution of y_i conditional on x_i is not $N(y_i|\boldsymbol{z}_{1i}'\boldsymbol{\gamma}_1 + x_i\beta, \omega_{11})$ because this ignores the dependence between ε_i and u_i. To derive the form of conditional density it is necessary to utilize the joint outcome-treatment model. For instance, to calculate the distribution of y_i conditional on x_i, write ε_i in equivalent form as $\varepsilon_i = \omega_{12}u_i + v_i$, where v_i is $N(0, \sigma_{11} = \omega_{11} - \omega_{12}^2)$ and independent of u_i. Then, conditional on u_i and x_i, we have that

$$f(y_i|x_i, \boldsymbol{z}_i, u_i, \boldsymbol{\psi}) = N(y_i|\boldsymbol{z}_{1i}'\boldsymbol{\gamma}_1 + x_i\beta + \omega_{12}u_i, \sigma_{11}).$$

The required density is now obtained by marginalizing over u_i using the density of $u_i|x_i, \boldsymbol{z}_i, \boldsymbol{\psi}$ which is truncated standard normal; for example, $f(u_i|x_i = 1, \boldsymbol{z}_i, \boldsymbol{\psi}) = N(u_i|0, 1)/\Phi(\boldsymbol{z}_i'\boldsymbol{\gamma})$, for $u_i > -\boldsymbol{z}_i'\boldsymbol{\gamma}$, where Φ is the cdf of the standard-normal distribution. If we let $s_i = (2x_i - 1)$, then after some tedious calculations it emerges that

$$f(y_i|x_i, \boldsymbol{z}_i, \boldsymbol{\psi}) = N(y_i|\boldsymbol{z}_{1i}'\boldsymbol{\gamma}_1 + x_i\beta, A) \times \Phi\left(s_i\frac{\boldsymbol{z}_i'\boldsymbol{\gamma} + b_i}{\sqrt{B}}\right)\frac{1}{\Phi(s_i\boldsymbol{z}_i'\boldsymbol{\gamma})} \tag{2.7}$$

where $A = \omega_{12}^2 + \sigma_{11}$, $B = (1 + \omega_{12}^2/\sigma_{11})^{-1}$ and $b_i = B(\sigma_{11}^{-1}\omega_{12}(y_i - \boldsymbol{z}_{1i}'\boldsymbol{\gamma}_1 - x_i\beta))$.

Third, from the second equation of the model the treatment assignment mechanism is given by

$$\Pr(x_i = 1|\boldsymbol{z}_i, \boldsymbol{\psi}) = \Phi(\boldsymbol{z}_i'\boldsymbol{\gamma}) \tag{2.8}$$

which is in the probit form as a consequence of the Gaussian assumption for u_i.

Fourth, one can derive the probability that $x_i = 0$ given y_{i0}. Write u_i in equivalent form as $u_i = \delta^*\varepsilon_i + v_i^*$, where v_i^* is $N(0, \sigma_{22})$ and uncorrelated with ε_i, $\delta^* = \omega_{12}/\omega_{11}$ and $\sigma_{22} = \sigma_{11}/\omega_{11}$. Now after some simple calculations it follows that

$$\Pr_0(x_i = 0|y_i, \boldsymbol{z}_i, \boldsymbol{\psi}) = \Phi\left(\frac{1}{\sqrt{\sigma_{22}}}\left(-\boldsymbol{z}_i'\boldsymbol{\gamma} - \delta^*\left\{y_i - \boldsymbol{z}_{1i}'\boldsymbol{\gamma}_1\right\}\right)\right) \tag{2.9}$$

where we use the notation $\Pr_0$ to highlight the specific counterfactual sample-space in which the probability is being computed. By similar calculations, one finds that $\Pr_1(x_i = 1|y_i, \boldsymbol{z}_i, \boldsymbol{\psi})$, the probability that $x_i = 1$ given y_{i1}, is $\Phi(\frac{1}{\sqrt{\sigma_{22}}}(\boldsymbol{z}_i'\boldsymbol{\gamma} + \delta^*\{y_i - \boldsymbol{z}_{1i}'\boldsymbol{\gamma}_1 - \beta\}))$.

Finally, all of the marginal and conditional distributions given above are consistent in the sense that

$$f(y_i|x_i = 1, \boldsymbol{z}_i, \boldsymbol{\psi})\Pr(x_i = 1|\boldsymbol{z}_i, \boldsymbol{\psi}) = f_1(y_i|\boldsymbol{z}_{1i}, \boldsymbol{\psi})\Pr_1(x_i = 1|y_i, \boldsymbol{z}_i, \boldsymbol{\psi}) \tag{2.10}$$

and

$$f(y_i|x_i = 0, \boldsymbol{z}_i, \boldsymbol{\psi})\Pr(x_i = 0|\boldsymbol{z}_i, \boldsymbol{\psi}) = f_0(y_i|\boldsymbol{z}_{1i}, \boldsymbol{\psi})\Pr_0(x_i = 0|y_i, \boldsymbol{z}_i, \boldsymbol{\psi}) \tag{2.11}$$

The calculations needed to show these equalities are lengthy and are, therefore, suppressed.

2.1. *Identification of Parameters and Related Issues*

Under the assumptions that have been made, the parameter ψ is identified given a random-sample of data on (y_i, x_i). To see this, note that the parameter $\gamma = (\gamma_{21}, \gamma_{22})$ is identified from the marginal distribution of x_i in (2.8). The remaining parameters are identified from the probabilities $\Pr_0(x_i = 0|y_i, \boldsymbol{z}_i, \psi)$ and $\Pr_1(x_i = 1|y_i, \boldsymbol{z}_i, \psi)$ given above. Because γ_{22} is identified, the coefficient on z_2 from the latter probabilities identifies σ_{22} which, therefore, identifies δ^* from the coefficient on y_i. Immediately then, we also identify γ_1 and β. Finally, ω_{11} and ω_{21} are identified from the already identified parameters σ_{22} and δ^*. Thus, under the stated assumptions, β, the ATE, is identified.

Due to the importance of the instruments in the identification argument, it is helpful to infer to what extent the instruments $\boldsymbol{z}_2$ are correlated with the treatment, after controlling for the influence of $\boldsymbol{z}_1$. Because the treatment is binary, the partial correlation of $\boldsymbol{z}_2$ with x cannot be easily defined. We can, nonetheless, consider the quantity

$$R^2 = \frac{n^{-1}\sum_{i=1}^n \gamma_{22}'\tilde{z}_{2i}\tilde{z}_{2i}'\gamma_{22}}{n^{-1}\sum_{i=1}^n \gamma_{22}'\tilde{z}_{2i}\tilde{z}_{2i}'\gamma_{22} + 1} \tag{2.12}$$

where $\tilde{z}_2$ is the residual from a regression of $\boldsymbol{z}_2$ on $\boldsymbol{z}_1$, which is the partial correlation of $\boldsymbol{z}_2$ with x^* after removing the effect of $\boldsymbol{z}_1$, and x^* is a latent variable that is included as part of the MCMC sampling strategy outlined in the Section 2.2. The posterior distribution of this quantity is a useful indicator of the strength of the instruments. Another important parameter of interest is the correlation coefficient between ε_i and u_i given by

$$\rho = \omega_{12}/\sqrt{\omega_{11}}.$$

This parameter can be interpreted as a measure of the extent of confounding on unobservables.

2.2. *Fitting of the Model*

Bayesian estimation of the structural model presented above can be effectively done by MCMC methods. Let $\boldsymbol{y}_i = (y_i, x_i)$ denote the data on the ith subject and let $\boldsymbol{y} = (\boldsymbol{y}_1, ..., \boldsymbol{y}_n)$ denote the entire sample data collected by random sampling. Because the parameters ω_{11} and ω_{12} must satisfy the positive-definiteness constraint $\omega_{11} > \omega_{12}^2$, reparameterize these parameters as $\sigma_{11} = \omega_{11} - \omega_{12}^2$ and $\delta = \omega_{12}$. The positive-definiteness constraint is enforced by ensuring that σ_{11} is positive. As a prior on the parameters, let $\boldsymbol{\beta} = (\gamma_1, \beta, \gamma)$ and $\boldsymbol{\theta} = (\omega_{12}, \sigma_{11})$, assume $\boldsymbol{\beta} \perp\!\!\!\perp \boldsymbol{\theta}$ and let

$$\pi(\boldsymbol{\beta}, \boldsymbol{\theta}) = N(\boldsymbol{\beta}|\boldsymbol{\beta}_0, \boldsymbol{B}_0)\, \mathrm{IG}(\sigma_{11}\,|\, n_0/2, d_0/2) N(\omega_{12}|m_0, \sigma_{11}\boldsymbol{M}_0)$$

where $IG(\sigma_{11}|a, b) \propto \sigma_{11}^{-(a+1)} \exp(-b/\sigma_{11})$ is the inverse-gamma density and the values indexed by the subscript zero are hyperparameters. Note that the joint inverse-gamma-normal assumption for σ_{11} and ω_{12} is quite natural in this context. Under these prior assumptions, one can find the implied prior of $\rho = \omega_{12}/\sqrt{\omega_{11}}$ by simulation.

Let $f(\boldsymbol{y}|\boldsymbol{z}, \psi)$ denote the likelihood function and let $f(\boldsymbol{y}_i|\boldsymbol{z}_i, \psi)$ denote the contribution of the ith observation. There are two ways of finding the likelihood contribution. One is as the product of $[y_i|x_i, \boldsymbol{z}_i, \psi]$ and $[x_i|\boldsymbol{z}_i, \psi]$ and the other as the product of $[y_i|\boldsymbol{z}_{i1}, \psi]$ and $[x_i|y_i, \boldsymbol{z}_i, \psi]$, where square brackets are used to denote densities. Both are equivalent due to (2.10) and (2.11). Regardless of the decomposition adopted, the likelihood has two distinct pieces: $\prod f(y_i, x_i = 0|\boldsymbol{z}_i, \boldsymbol{\psi})$ for the $x_i = 0$ observations and $\prod f(y_i, x_i = 1|\boldsymbol{z}_i, \boldsymbol{\psi})$ for the $x_i = 1$ observations.

It is readily seen that the form of the likelihood precludes direct extraction of tractable full conditional posterior densities. Following the approach of Albert and Chib (1993), write the joint model in (2.4) in terms of latent variables x_i^* as

$$\begin{aligned} y_i &= \boldsymbol{z}_{1i}'\gamma_1 + x_i\boldsymbol{\beta} + \varepsilon_i \\ x_i^* &= \boldsymbol{z}_i'\boldsymbol{\gamma} + u_i \end{aligned} \tag{2.13}$$

with $x_i = I(x_i^* > 0)$. From properties of the conditional normal distribution (applied to the latter) it follows that conditioned on the parameters $\psi = (\gamma_1, \beta, \boldsymbol{\gamma}, \omega_{12}, \sigma_{11})$ and the observed data we sample $\boldsymbol{x}^* = (x_1^*, ..., x_n^*)$ from the truncated normal distributions

$$x_i^*|y_i, x_i, \boldsymbol{z}_i, \boldsymbol{\psi} \propto \begin{cases} N\left(\boldsymbol{z}_i'\boldsymbol{\gamma} + \delta^*(y_i - \boldsymbol{z}_{1i}'\gamma_1), \sigma_{22}\right) I_{(-\infty,0)} & \text{if} \quad x_i = 0 \\ N\left(\boldsymbol{z}_i'\boldsymbol{\gamma} + \delta^*(y_i - \boldsymbol{z}_{1i}'\gamma_1 - \beta), \sigma_{22}\right) I_{(0,\infty)} & \text{if} \quad x_i = 1 \end{cases}$$

Next one samples $\boldsymbol{\beta}$ jointly by noting that despite the presence of the term $I(x_i^* > 0)$ on the right-hand side of the equation generating y_i, the joint density of $\boldsymbol{y}_i^* = (y_i, x_i^*)$ has the form of a bivariate normal density

$$f(\boldsymbol{y}_i^*|\psi) \propto \exp\Big(-\frac{1}{2}(\boldsymbol{y}_i^* - \boldsymbol{X}_i\boldsymbol{\beta})'\Omega^{-1}(\boldsymbol{y}_i^* - \boldsymbol{X}_i\boldsymbol{\beta})\Big), \quad i \leq n$$

where

$$\boldsymbol{X}_i = \begin{pmatrix} \boldsymbol{z}_{1i}' & x_i & 0' \\ 0' & 0 & \boldsymbol{z}_i' \end{pmatrix}$$

and $\Omega = (\omega_{ij})$ with $\omega_{22} = 1$. Thus, conditioned on $\boldsymbol{y}_i^* = (y_i, x_i^*)$, $i = 1, 2, ..., n$, and $\boldsymbol{\theta}$, we sample $\boldsymbol{\beta}$ from the density

$$\pi(\boldsymbol{\beta}|\{\boldsymbol{y}_i^*\}, \boldsymbol{z}, \boldsymbol{\theta}) = N(\boldsymbol{\beta}|\hat{\boldsymbol{\beta}}, \boldsymbol{B}) \tag{2.14}$$

where $\boldsymbol{B} = (\boldsymbol{B}_0^{-1} + \sum \boldsymbol{X}_i'\Omega^{-1}\boldsymbol{X}_i)^{-1}$ and $\hat{\boldsymbol{\beta}} = \boldsymbol{B}(\boldsymbol{B}_0^{-1}\boldsymbol{\beta}_0 + \sum \boldsymbol{X}_i'\Omega^{-1}\boldsymbol{y}_i^*)$ and all sums are over i.

Last, conditioned on the latent data and $\boldsymbol{\beta}$, we sample $\boldsymbol{\theta}$ by a Metropolis-Hastings step (Chib and Greenberg, 1995). Joint sampling of $\boldsymbol{\theta}$ is necessary; separate sampling although superficially attractive, because the full conditional distribution of each is tractable, worsens the performance of the sampling when measured in terms of the inefficiency factors (autocorrelation times). In particular, given $\boldsymbol{y}_i^* = (y_i, x_i^*)$, $i = 1, 2, ..., n$, and $\boldsymbol{\beta}$, $\boldsymbol{\theta}$ is sampled from

$$\begin{aligned} \pi(\boldsymbol{\theta}|\{\boldsymbol{y}_i^*\}, \boldsymbol{z}, \boldsymbol{\beta}) &\propto \mathrm{IG}(\sigma_{11} \mid n_0/2, d_0/2) N(\omega_{12}|m_0, \sigma_{11}\boldsymbol{M}_0) \\ &\quad \times \prod_{i=1}^{n} N(y_i|\boldsymbol{z}_{1i}'\gamma_1 + x_i\beta + \omega_{12}(x_i^* - \boldsymbol{z}_i'\boldsymbol{\gamma}), \sigma_{11}) \end{aligned}$$

using a tailored (Chib and Greenberg, 1998) bivariate-*t* proposal density $q(\boldsymbol{\theta}|\{\boldsymbol{y}_i^*\}, \boldsymbol{z}, \boldsymbol{\beta}) = T(\boldsymbol{\theta}|\boldsymbol{\mu}, \boldsymbol{V})$, where $\boldsymbol{\mu}$ and $\boldsymbol{V}$ are the mode and inverse of the observed information, respectively, of the function $\prod N(y_i|\boldsymbol{z}_{1i}'\gamma_1 + x_i\beta + \omega_{12}(x_i^* - \boldsymbol{z}_i'\boldsymbol{\gamma}), \sigma_{11})$. This completes the description of the MCMC simulation scheme.

2.3. *Model Comparisons: Confounding versus no Confounding*

In accordance with formal Bayesian precepts, it is possible to gauge the support for a model with confounding by computing its marginal likelihood and comparing it with that of the relevant benchmark model (one in which the parameter ω_{12} is zero). The marginal likelihood is easily computed by the method of Chib (1995). The basic idea is that on the log-scale the marginal likelihood $m(\boldsymbol{y}|\boldsymbol{z})$ can be written as

$$\ln m(\boldsymbol{y}|\boldsymbol{z}) = \ln f(\boldsymbol{y}|\boldsymbol{z}, \boldsymbol{\psi}^*) + \ln \pi(\boldsymbol{\psi}^*) - \ln \pi(\boldsymbol{\psi}^*|\boldsymbol{y}, \boldsymbol{z})$$

where $\boldsymbol{\psi}^*$ is (say) the posterior mean of the parameters and $f(\boldsymbol{y}|\boldsymbol{z}, \boldsymbol{\psi}^*)$ is the likelihood. The first two terms are clearly available directly. The third can be estimated efficiently by decomposing the posterior ordinate as

$$\pi(\boldsymbol{\psi}^*|\boldsymbol{y}, \boldsymbol{z}) = \pi(\boldsymbol{\beta}^*|\boldsymbol{y}, \boldsymbol{z})\pi(\sigma_{11}^*, \omega_{12}^*|\boldsymbol{y}, \boldsymbol{z}, \boldsymbol{\beta}^*)$$

where the first term is obtained by averaging the normal density in (2.4) at $\boldsymbol{\beta}^*$ over the draws from the MCMC run; the second ordinate is obtained by a refinement of the Chib method developed by Chib and Jeliazkov (2001). The marginal likelihood of the benchmark model is acquired in the same way after setting ω_{12} to zero in all expressions.

2.4. *Synthetic Data Example*

Consider a synthetic example in which the data are generated from the model $y_i = x_i\beta + \varepsilon$ and $x_i = I[\boldsymbol{z}_i\gamma + u_i > 0]$, where z_i is univariate. This model arises from a setting in which the expected value of y_{i0} (i.e., μ_0) is zero and the projection of z_1 on η_i in (2.2) is also zero. In this case, the parameter β is simply μ_1. We are primarily interested in the effect of ω_{12} on inferences about β and so for the purposes of generating the data, we set $\beta = 0.5$, $\omega_{11} = 0.5$, $z_i \sim$ Uniform$(-0.5, 0.5)$ and $R^2 = 0.4$, which determines the value of γ, and set the value of $\rho = \omega_{12}/\sqrt{\omega_{11}}$ to 0.7 to reflect a high level of confounding. This design is loosely based on that of Woglom (2001) who uses a similar strategy but in the context of a continuous treatment. Our simulated sample contains $n = 200$ observations.

The results from 5000 simulated draws under weak priors on the parameters are reported in Figure 2 for the three parameters of interest, β, R^2 and ρ (the posterior distribution of the last two parameters is derived from that of the other parameters of the model).

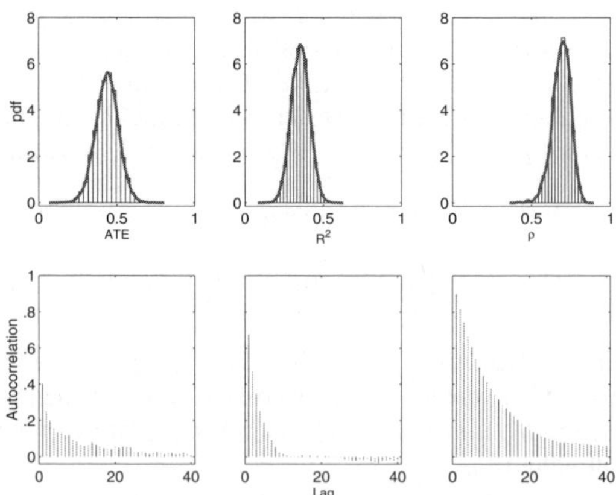

Figure 2. *Marginal posterior densities and output autocorrelations of* (i) *the ATE β (true value = 0.5);* (ii) *R^2 (true value = 0.4); and* (iii) *ρ (true value = 0.7) for simulated data example.*

What is interesting from these figures is that the posterior distribution concentrates quite accurately around the values that were used to generate the data. For instance, the posterior mean of β is 0.435 with a standard deviation of 0.072; for R^2 the corresponding values are 0.358 and 0.057 and for ρ the posterior mean is 0.691 and standard deviation is 0.059.

The fact that the posterior distributions provide accurate inferences about both the strength of the instrument and the degree of confounding is indicative of what is possible from the Bayesian perspective. Of course, the above results are based on a value of $R^2 = 0.4$. To demonstrate what happens when the instrument is weak, we also simulate data with $R^2 = 0.1$ keeping the level of confounding at $\rho = 0.7$. It is of interest now to study the joint posterior distribution of the ATE and ρ which we plot in Figure 3 for our data generated under the two settings of R^2. The top panel of Figure 3 shows that the joint posterior distribution is well behaved when $R^2 = 0.4$ even though the level of confounding is high. However, with the weaker instrument, as shown in the bottom panel of Figure 3, the two parameters are negatively correlated and the distribution has shifted away from the true values. Thus, our ability to learn about the parameters is compromised if the instrument is weak, a consequence of the fact that in such a case the parameters are poorly identified.

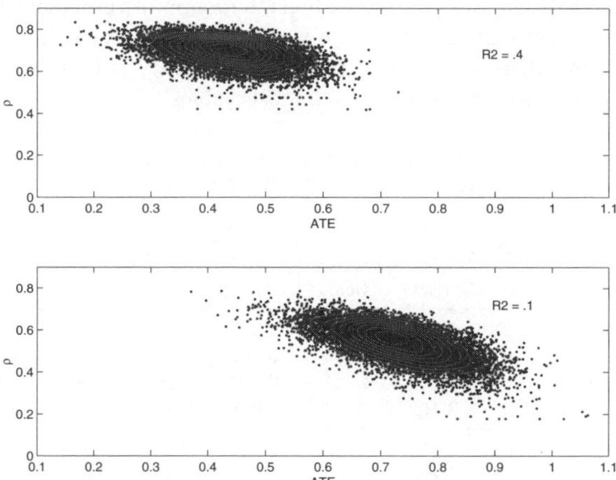

Figure 3. *Joint posterior densities of the ATE and ρ for data generated with $R^2 = 0.4$ (top panel) and $R^2 = 0.1$ (bottom panel) and $\rho = 0.7$.*

2.5. *Birthweight Data Example*

We now consider a real data example from a "quasi-randomized experiment" concerned with the effect of smoking in the third-trimester of a woman's pregnancy on the birthweight of the child. In this experiment, we have complete data on $n = 775$ pregnancies that resulted in child-birth. The data comes from a trial where, because smoking, the treatment, cannot be randomized, the women are randomly assigned to two groups in month four of the pregnancy: the first group is encouraged not to smoke while the other receives no such encouragement. Clearly, this randomized encouragement has no direct bearing on birthweights but can be expected to influence the treatment to the extent that women comply with the encouragement received, conditional on the other factors that influence the treatment. Our instrument, therefore, is z_{2i} which takes the value one if the woman is encouraged not to smoke, and the value zero otherwise. The actual treatment intake is x_i which is an indicator of the woman's smoking behavior in month eight of the pregnancy. The outcome variable y_i is birthweight measured in kilograms. In addition, the data contains information on the number of cigarettes smoked prior to the pregnancy (CIGSP) and the number of cigarettes smoked at the time of randomization (CIGSR). These form the vector $\boldsymbol{z}_{1i}$.

The setting described above is similar to that considered in a number of recent papers where z_{2i} is viewed as the treatment and x_i is an indicator of compliance with the assigned treatment; for example, (see Imbens and Rubin, 1997) where data from such a trial is analyzed by modelling the unobserved confounders as representing four discrete groups of subjects: "compliers," "defiers," "always takers," and "never takers." Instead of assuming that the unobservables are described in this way, the data can be analyzed by our structural model where the observed compliance to the encouragement received is determined partly by unobservable factors, such as the Imbens and Rubin (1997) subject type, but not restricted to such a factor. One generic confounder with compliance may be the health of the subject; those women that are healthy presumably care about the health of the child and hence do not smoke, or comply if encouraged not to. Those women who are in poor health may see no incentive to switch.

We fit our structural model to these data under a reasonable prior on the parameters (the first two moments of the prior appear in Table 1 along with a summary of the posterior moments, computed from 10000 MCMC draws). We see that intervention on treatment, that is the cessation of smoking in month eight, can be expected to increase the birthweight between 0.168 to 0.557 kg (the lower and upper ends of the 95% credibility interval). This is the interval estimate of the ATE after controlling for confounding on unobservables.

Table 1. *Prior-posterior summary for birthweight data based on a MCMC sample of 10,000 draws; L and U are the 2.5th and 97.5th percentiles and Ineff is the inefficiency factor*

	Prior		Posterior				
	Mean	Std dev	Mean	Std dev	L	U	Ineff
Const	0	3.162	3.434	0.062	3.311	3.556	8.166
CIGSP	0	3.162	0.004	0.002	0.000	0.008	1.000
CIGSR	0	3.162	0.001	0.003	-0.005	0.007	7.512
x	0	3.162	-0.354	0.098	-0.557	-0.168	13.536
const	0	3.162	0.100	0.129	-0.153	0.357	3.795
CIGSP	0	3.162	-0.002	0.006	-0.014	0.009	3.578
CIGSR	0	3.162	0.096	0.008	0.080	0.113	8.088
z_2	0	3.162	-0.791	0.110	-1.003	-0.577	4.531
ω_{11}	0.105	0.137	0.273	0.015	0.246	0.303	2.692
ω_{12}	0.050	0.163	0.072	0.058	-0.036	0.192	15.518
Log Marginal likelihood					-966.58		

The marginal posterior distributions of the ATE parameter and those of R^2 and ρ are reported in Figure 4. The posterior distribution of ρ indicates that the partial correlation of the instrument with the treatment is moderate and so identification of the parameters and the conclusions from the study are likely to be robust.

We finish the analysis by determining the support for the confounding model versus one in which confounding is absent, namely a model in which ω_{12} is zero. Under the same prior for the common parameters, we fit the outcome and assignment models independently and calculate the marginal likelihood as the sum of the log-marginal likelihoods from each of the sub-models. We find that the log marginal likelihood for the full model is -996.59 (numerical

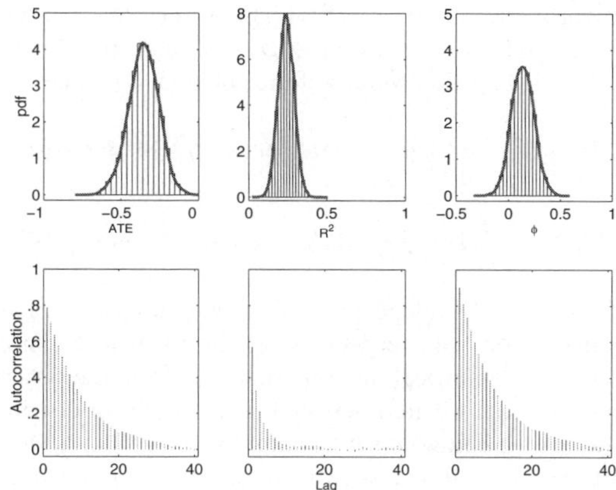

Figure 4. *Marginal posterior densities and output autocorrelations of* (i) *the ATE* β*;* (ii) R^2*; and* (iii) ρ *for birthweight data example.*

standard error of 0.05) while that of the no confounding model is -996.58 (with a numerical standard error of 0.01), as close to a virtual tie as one can possibly imagine. Thus, the data do not support the presence of confounding on unobservables, a conclusion that may have been partially anticipated from Table 1 but without the weight of evidence engendered by this subsequent marginal likelihood comparison.

3. BINARY OUTCOMES

In this section we expand the discussion to cover the case of binary outcomes. For example, suppose that y_i is a zero-one variable that represents post-surgical outcomes for hip-fracture patients (one if the surgery is successful, and zero if not) and x_i is an indicator representing the extent of delay before surgery (one if the delay exceeds say two days, and zero otherwise). In this case, it is quite likely that delay and post-surgical outcome are both affected by patient factors, *e.g.*, patient frailty, that are usually either imperfectly measured and recorded or absent from the data. A structural model that jointly models the outcomes and the treatment assignment is required to control for the unobservable confounders.

Following the notation introduced in Section 2, suppose that the outcome model is given by

$$y_i = I(\boldsymbol{z}_{1i}'\boldsymbol{\gamma}_1 + x_i\beta + \varepsilon_i), \quad i \leq n\,, \tag{3.15}$$

and suppose that the effect of the binary treatment x_i is confounded with that of unobservables that are correlated with ε_i, even after conditioning on $\boldsymbol{z}_{1i}$. To model this confounding, let the treatment assignment mechanism be given by $x_i = I(\boldsymbol{z}_i'\boldsymbol{\gamma} + u_i)$ where $\boldsymbol{z}_i = (\boldsymbol{z}_{1i}, \boldsymbol{z}_{2i})$ contains the instrumental variables $\boldsymbol{z}_{2i}$, and the confounder u_i is correlated with ε_i. For simplicity, we assume that the joint distribution of the unobservables (ε_i, u_i) is Gaussian with mean zero and covariance matrix Ω that is in correlation form due to the fact that the scale is unidentified. A little bit of reflection (starting with the joint distribution of (ε_i, u_i) followed by a change of variable) shows that even though the outcome from the second equation is present in the model for y_i, this set-up is a special case of the multivariate probit model analyzed in Chib and Greenberg (1998).

One change from the continuous outcomes case is the way in which the ATE is computed. By definition, the ATE is the difference, $E(y_{i1}|\boldsymbol{z}_{1i}, \boldsymbol{\gamma}_1, \beta) - E(y_{i0}|\boldsymbol{z}_{1i}, \boldsymbol{\gamma}_1, \beta)$, without involving the treatment assignment model. From (3.15) it can be seen that the two potential outcomes are $y_{i0} = I(z'_{1i}\gamma_1 + \varepsilon_i)$ and $y_{i1} = I(z'_{1i}\gamma_1 + \beta + \varepsilon_i)$ and, therefore, under the normality assumption, the difference in expected values of the potential outcomes conditioned on z_{1i} and ψ is

$$E(y_{i1}|z_{1i}, \psi) - E(y_{i0}|z_{1i}, \psi) = \Phi(z'_{1i}\gamma_1 + \beta) - \Phi(z'_{1i}\gamma_1)\,.$$

Thus, the ATE is

$$\text{ATE} = \int \{\Phi(z'_1\gamma_1 + \beta) - \Phi(z'_1\gamma_1)\}\,\pi(z_1)\,dz_1 \tag{3.16}$$

where $\pi(z_1)$ is the density of z_1 (independent of ψ by assumption). Calculation of the ATE, therefore, entails an integration with respect to z_1. In practice, a simple idea is to perform the marginalization using the empirical distribution of z_1 from the observed sample data. Of course, in our Bayesian context, we are interested in the posterior distribution of the ATE and this can be obtained (as shown below) by evaluating the integral as a Monte Carlo average, for each sampled value of (γ_1, β) from the MCMC algorithm. Identification of the parameters of the model and the ATE follows from Heckman and Vytlacil (1999) under our assumptions that z_{2i} is an instrument and the joint distribution of the unobservables is continuous.

3.1. *Model Fitting*

Because this model is a special case of the multivariate probit (MVP) model considered by Chib and Greenberg (1998), we can apply the estimation and marginal likelihood procedure developed for the MVP model. The estimation procedure, in the manner of Albert and Chib (1993), utilizes latent variables to represent the joint model as

$$\begin{pmatrix} y_i^* \\ x_i^* \end{pmatrix} = \begin{pmatrix} z'_{1i} & x_i & 0' \\ 0' & 0 & z'_i \end{pmatrix} \begin{pmatrix} \gamma_1 \\ \beta \\ \gamma \end{pmatrix} + \begin{pmatrix} \varepsilon_i \\ u_i \end{pmatrix} \tag{3.17}$$

where the observed outcomes arise as $y_i = I(z_i^* > 0)$ and $x_i = I(x_i^* > 0)$. The prior distribution on the parameters $\boldsymbol{\beta} = (\gamma_1, \beta, \gamma)$ is specified as $N(\boldsymbol{\beta}|\boldsymbol{\beta}_0, \boldsymbol{B}_0)$ and that of ω_{12} is taken to be $N(\omega_{12}|m_0, \boldsymbol{M}_0)$ truncated to $(-1, 1)$. The MCMC fitting algorithm consists of three blocks, one in which the latent data are sampled from truncated univariate normal distributions, a second in which $\boldsymbol{\beta}$ is sampled from a multivariate normal distribution conditioned on the sampled latent data and finally one in which ω_{12} is sampled by the Metropolis-Hastings algorithm. Briefly, the algorithm proceeds as follows.

Conditioned on the data and the parameters, we sample $(x_1^*, ..., x_n^*)$ from the truncated normal distributions

$$x_i^*|y_i^*, x_i, z_i, \psi \propto \begin{cases} N\left(z'_i\gamma + \omega_{12}(y_i^* - z'_{1i}\gamma_1), 1 - \omega_{12}^2\right) I_{(-\infty,0)} & \text{if} \quad x_i = 0 \\ N\left(z'_i\gamma + \omega_{12}(y_i^* - z'_{1i}\gamma_1 - \beta), 1 - \omega_{12}^2\right) I_{(0,\infty)} & \text{if} \quad x_i = 1 \end{cases}$$

and $(y_1^*, ..., y_n^*)$ from the truncated normal distributions

$$y_i^*|y_i, x_i, x_i^*, z_i, \psi \propto \begin{cases} N\left(z'_{1i}\gamma_1 + x_i\beta + \omega_{12}(x_i^* - z'_i\gamma), 1 - \omega_{12}^2\right) I_{(-\infty,0)} & \text{if} \quad y_i = 0 \\ N\left(z'_{1i}\gamma_1 + x_i\beta + \omega_{12}(x_i^* - z'_i\gamma), 1 - \omega_{12}^2\right) I_{(0,\infty)} & \text{if} \quad y_i = 1. \end{cases}$$

Next, conditioned on $\boldsymbol{y}_i^* = (y_i^*, x_i^*)$, $i = 1, 2, ..., n$, and ω_{12}, we sample $\boldsymbol{\beta}$ from the density

$$\pi(\boldsymbol{\beta}|\{\boldsymbol{y}_i^*\}, \boldsymbol{z}, \boldsymbol{\theta}) = N(\boldsymbol{\beta}|\hat{\boldsymbol{\beta}}, \boldsymbol{B})$$

where $\boldsymbol{B}$ and $\hat{\boldsymbol{\beta}}$ are as defined in Section 2. Finally, conditioned on $\boldsymbol{y}_i^* = (y_i^*, x_i^*), i = 1, 2, ..., n,$ and $\boldsymbol{\beta}$, we sample ω_{12} from the conditional posterior density

$$\pi(\omega_{12}|\{\boldsymbol{y}_i^*\}, \boldsymbol{z}, \boldsymbol{\beta}) \propto N(\omega_{12}|m_0, \boldsymbol{M}_0) \prod_{i=1}^{n} N\left(y_i^*|\boldsymbol{z}_{1i}'\gamma_1 + x_i\beta + \omega_{12}(x_i^* - \boldsymbol{z}_i'\gamma), 1 - \omega_{12}^2\right)$$

by the Metropolis-Hastings algorithm. Alternatively, ω_{12} can be sampled marginalized over $\{\boldsymbol{y}_i^*\}$, but conditioned on $\boldsymbol{\beta}$, from the posterior density that is proportional to the product of the prior and the likelihood. This is computationally more demanding since the likelihood evaluation requires the calculation of the bivariate normal cdf but this strategy does tend to improve the mixing of the chain, especially when the degree of confounding is high.

3.2. *Birthweight Data Example, Continued*

We return to the birthweight data example considered above by artificially modifying (entirely for illustrative purposes) the continuous birthweights into binary indicators representing normal versus below-normal weights. In practice, a weight below 2.5 kg is considered below-normal but with this cut-off, in our dataset, only 5% of the responses are below-normal; with a threshold of 3 kg, which we use, that figure rises to 29%.

The results from the MCMC fitting of a structural model to the binary responses are summarized in Table 2. The table shows that smoking in the third-trimester reduces birthweights; that encouragement not to smoke reduces the smoking propensity and that the support for confounding as reflected by the marginal posterior distribution of ω_{12} is weak. All these results are based on 10000 samples from our MCMC algorithm following a burn-in of 500 cycles. The inefficiency factors (autocorrelation times) reported in the last column of Table 2 indicate that the MCMC sampler is mixing well in general, but especially so for the parameters that are better supported by the data.

Table 2. *Prior-posterior summary for birthweight data with binary response based on a MCMC sample of 10000 draws*

	Prior		Posterior				
	Mean	Std dev	Mean	Std dev	L	U	Ineff
Const	0.000	10.000	0.713	0.166	0.373	1.030	11.115
CIGSP	0.000	10.000	0.011	0.005	0.000	0.022	3.183
CIGSR	0.000	10.000	0.004	0.009	-0.014	0.021	10.912
x	0.000	10.000	-0.612	0.280	-1.148	-0.046	17.096
const	0.000	10.000	0.101	0.127	-0.149	0.350	3.384
CIGSP	0.000	10.000	-0.003	0.006	-0.014	0.009	3.345
CIGSR	0.000	10.000	0.096	0.008	0.081	0.113	8.009
z_2	0.000	10.000	-0.784	0.110	-0.997	-0.570	4.284
ω_{12}	0.300	0.707	0.171	0.178	-0.181	0.512	16.012
Log-marginal likelihood				-865.37			

An interesting question in this case is about the posterior distribution of the ATE. We calculate this distribution by evaluating the integral in (3.16) using 200 draws on z_1 drawn from its empirical distribution. This is done for each value of the parameters (γ_1, β) from the

MCMC simulation. The resulting posterior sample of the ATE is plotted in Figure 5. The lower and upper values of the ATE are found to be -0.316 and -0.016, respectively, with a posterior median of -0.186. Thus, it is reasonable to believe that the ATE of smoking is negative.

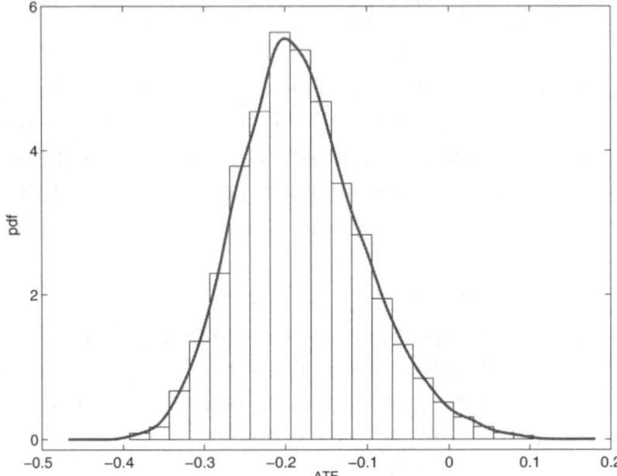

Figure 5. *Marginal posterior density of the ATE for birthweight data example with binary response;* z_1 *in* (3.16) *is marginalized using its empirical cdf, for each of 10,000 draws of the parameters from the posterior distribution.*

Finally, we compare the weight of evidence for the joint model versus a model that assumes that confounding is absent. We find that the log-marginal likelihood of the full model is -865.37 while that of the reduced model is -864.61, which indicates that the support for the full model is weak at best, which parallels the findings from the same data with a continuous response.

4. REMARKS AND EXTENSIONS

In this paper, we have considered the Bayesian analysis of a class of structural models that allow for confounding for unobservables even after conditioning on covariates. Our focus here has been on parametric structural models although it is possible to generalize the framework in the direction of semi-parametric structural models. In addition, it is possible to consider extensions to clustered data problems, for example, those arising in longitudinal settings with time-varying treatments, where most of the analysis to date has not permitted the kind of confounding on observables that we have outlined above. Robins in his extensive work in this area, in line with a large literature in econometrics, has primarily adopted the assumption of sequential ignorability, which amounts to assuming independence between the treatment and the potential outcomes, conditional on the treatment history.

In companion work by Chib and Greenberg (2002), both of the above extensions are discussed and the features emphasized in this paper, namely calculation of the average-treatment effect, measures of instrument strength and confounding, comparison of models via marginal likelihoods, are all discussed and illustrated. In that paper we also consider the computation of other treatment effects that are pertinent in the context of structural models, similar to the complier-averaged causal effect discussed by Imbens and Rubin (1997), which isolates the effect of the instruments on the outcomes, regardless of the treatments actually assigned.

REFERENCES

Albert, J. and Chib, S. (1993). Bayesian analysis of binary and polychotomous response data. *J. Am. Statist. Ass.* **88**, 669–679.

Angrist, J., Imbens, G. W. and Rubin, D.B. (1996). Identification of causal effects using instrumental variables. *J. Am. Statist. Ass.* **91**, 444–472 (with discussion).

Chib, S. (1995). Marginal likelihood from the Gibbs output. *J. Am. Statist. Ass.* **90**, 1313–1321.

Chib, S. and Greenberg, E. (1995). Understanding the Metropolis-Hastings algorithm. *Ann. Statist.* **49**, 327–335.

Chib, S. and Greenberg, E. (1998). Analysis of multivariate probit models. *Biometrika* **85**, 347–361.

Chib, S. and Greenberg, E. (2002). Inferring effects with unobserved confounders. *Tech. Rep.*, Washington University, St. Louis, USA.

Chib, S. and Hamilton, B (2000). Bayesian analysis of cross-Section and clustered data treatment models. *J. Econometrics* **97**, 25-50.

Chib, S. and Hamilton, B (2002). Semiparametric Bayes analysis of longitudinal data treatment models. *J. Econometrics* **97**, 67-89

Chib, S. and Jeliazkov, I. (2001). Marginal likelihood from the Metropolis-Hastings output. *J. Am. Statist. Ass.* **96**, 270–281.

Dawid, A. P. (2000). Causal inference without counterfactuals. *J. Am. Statist. Ass.* **95**, 407–424 (with discussion).

Goldberger, A. S. (1972). Structural equation methods in the social sciences. *Econometrica* **40**, 979–1001.

Haavelmo, T. (1943). The statistical implications of a system of simultaneous equations. *Econometrica* **11**, 1–12.

Haavelmo, T. (1947). Methods of measuring the marginal propensity to consume. *J. Am. Statist. Ass.* **42**, 105–122.

Heckman, J. J. (1978). Dummy endogenous variables in a simultaneous equation system. *Econometrica* **46**, 931–959.

Heckman, J. J and Vytlacil, E. J. (1999). Local instrumental variables and latent variable models for identifying and bounding treatment effects. *Proc. Nat. Acad. Sci.* **96**, 4730–4734.

Imbens, G. W. and Rubin, D. B. (1997). Bayesian inference for causal effects in randomized experiments with noncompliance. *Ann. Statist.* **25**, 305–327.

Pearl, J. (2000). *Causality*, Cambridge: Cambridge University Press.

Rosenbaum, P. R. and Rubin, D. B. (1983). The Central role of the propensity score in observational studies for causal effects. *Biometrika* **70**, 41–55.

Rosenbaum, P. R. (2002), *Observational Studies* (2nd ed). New York: Springer.

Rubin, D. B. (1991). Practical implications of modes of statistical inference for causal effects and the critical role of the assignment mechanism. *Biometrics* **47**, 1213–1234.

Staiger, D. and Stock, J. H. (1997). Instrumental variable regression with weak instruments. *Econometrica* **65**, 557–586.

Woglom, G. (2001). More results on the exact small sample properties of the instrumental variable estimator. *Econometrica* **69**, 1381–1390.

Zellner, A. (1971). *An Introduction to Bayesian Inference in Econometrics*. New York: Wiley

DISCUSSION

MING-HUI CHEN (*University of Connecticut, USA*)

I first congratulate the author for this excellent piece of work on the complete treatment of Bayesian analysis for inferring treatment effects when there are unobserved confounders. The paper considers an important problem of current interest. I particularly like the novel modelling strategy, which includes the introduction of model assumptions and the discussion of the relationship between the unmeasured covariates and the observed treatment and outcome, and a comprehensive elaboration of the properties of the proposed models. The paper presents a step-by-step development of the model and an excellent discussion on identification of the model parameters. The paper also addresses several other important issues such as model fitting and model comparisons for determining confounding versus no confounding. A simulation study and an illustration of proposed models using the birthweight data are nice additions to the paper as well.

Second, I would like to discuss the choice of prior distributions. Using the same notation as in the paper, for the continuous univariate response, the prior proposed in the paper takes the form $\pi(\boldsymbol{\beta}, \boldsymbol{\theta}) = \mathrm{N}(\boldsymbol{\beta}|\boldsymbol{\beta}_0, B_0)\mathrm{IG}(\sigma_{11}\,|\,n_0/2, d_0/2)\mathrm{N}(\omega_{12}|m_0, \sigma_{11}M_0)$, where $\boldsymbol{\beta} = (\gamma_1, \beta, \gamma)$, $\boldsymbol{\theta} = (\omega_{12}, \sigma_{11})$, and $\sigma_{11} = \omega_{11} - \omega_{12}^2$. This choice of the prior for $\boldsymbol{\beta}$, σ_{11}, and ω_{12} is reasonable, since it does ensure the positive-definiteness constraint. However, I do feel that there are other more direct approaches to specify the priors. One approach is to take $\pi(\omega_{11}, \omega_{12}) = \mathrm{IG}(\omega_{11}\,|\,n_0/2, d_0/2)\mathrm{TN}_{A_{\omega_{11}}}(\omega_{12}\,|\,m_0, \omega_{11}M_0)$, where $A_{\omega_{11}} = (-\sqrt{\omega_{11}}, \sqrt{\omega_{11}})$ and $\mathrm{TN}(\cdot|m_0, \omega_{11}M_0)$ denotes the truncated normal distribution with mean m_0 and variance $\omega_{11}M_0$ over the interval $A_{\omega_{11}}$. Let $\boldsymbol{\omega} = (\omega_{11}, \omega_{12})$. Then, the joint prior is given by $\pi(\boldsymbol{\beta}, \boldsymbol{\omega}) = \mathrm{N}(\boldsymbol{\beta}|\boldsymbol{\beta}_0, B_0)\pi(\omega_{11}, \omega_{12})$. We notice that the truncated normal prior specification ensures the positive-definiteness constraint.

Let $\rho = \omega_{12}/\sqrt{\omega_{11}}$, which is the correlation between ϵ_i and u_i. Then, $\sigma_{11} = \omega_{11}(1-\rho^2)$ and $\omega_{12} = \rho\sqrt{\omega_{11}}$. Instead of a truncated normal prior for ω_{12}, we may directly take a beta prior for ρ and assume ρ and ω_{11} independent *a priori*. Let $\mathrm{Be}(\rho|a_0, b_0)$ denote a beta prior for ρ and $\boldsymbol{\omega}^* = (\omega_{11}, \rho)$. Then, the joint prior is of the form $\pi(\boldsymbol{\beta}, \boldsymbol{\omega}) = \mathrm{N}(\boldsymbol{\beta}|\boldsymbol{\beta}_0, B_0)\mathrm{IG}\left(\omega_{11}\middle|\frac{n_0}{2}, \frac{d_0}{2}\right)\mathrm{Be}(\rho|a_0, b_0)$. With this choice of the prior, the conditional posterior distribution for ω_{11} is an inverse gamma after introducing the latent variables x_i^*.

In addition to the above choices of the prior for ψ, another "convenient" noninformative prior is also available. To fix the idea, we consider the case of the continuous univariate response and also let D denote the observed data. In this case, the likelihood function can be written as $L(\psi|D) = \prod_{i=1}^n f_{x_i}(y_i|z_{1i}, \psi)\,\mathrm{Pr}_{x_i}(x_i|y_i, z_i, \psi)$, where $\psi = (\gamma_1, \beta, \gamma, \boldsymbol{\theta})$. Observe that

$$\mathrm{Pr}_0(x_i = 0) = \Phi\left(\sigma_{22}^{-1/2}(-z_i'\gamma - \delta^*\{y_i - z_{1i}'\gamma_1\})\right)$$
$$\mathrm{Pr}_1(x_i = 1) = \Phi\left(\sigma_{22}^{-1/2}(z_i'\gamma + \delta^*\ \{y_i - z_{1i}'\gamma_1 - \beta\})\right),$$

where $\sigma_{22} = \sigma_{11}/\omega_{11}$ and $\delta^* = w_{12}/\omega_{11}$. Let $\gamma^* = \gamma/\sqrt{\sigma_{22}}$ and $\delta^{**} = \delta^*/\sqrt{\sigma_{22}}$. After some algebra, we observe that

$$\delta^{**} = \frac{\rho}{\sqrt{1-\rho^2}} \cdot \frac{1}{\sqrt{\omega_{11}}}.$$

Clearly, $-\infty < \delta^{**} < \infty$ and δ^{**} and ρ share the same sign. Also let $\psi^* = (\gamma_1, \beta, \omega_{11}, \gamma^*, \delta^{**})$. Then, ψ^* is a one-to-one transformation of ψ. Now, we can take an improper uniform distribution for ψ^*, i.e., $\pi(\psi^*) \propto 1/\omega_{11}$. With this "convenient" noninformative prior and the introduction of latent variables x_i^*, all conditional posterior distributions are either standard (truncated) normals or inverse gamma, which leads to straightforward implementation of the Gibbs sampling algorithm without using any Metropolis algorithms. I also notice that there are other "convenient" informative priors available. Here, the term "convenient" implies an easy implementation of the posterior computation.

Third, the results shown in Section 2.4 are very interesting. However, only one synthetic dataset was used. Therefore, it is difficult to know whether this happened by chance or by the nature of the problem. The use of multiple simulations may be one way to determine whether the partial correlation R^2 really serves as a device in identifying the model parameters. I would also like to know whether the results obtained in Section 2.4 can be theoretically justified. In any case, this is an interesting issue that is worthy of further investigation.

Fourth, the identification of parameters has been nicely discussed in Section 2.1. Also, the posterior distribution is always proper when a proper prior is used. However, it may be interesting to see whether the posterior distribution is still proper when an improper prior such as $\pi(\psi^*) \propto 1/\omega_{11}$ is used. In general, if the posterior is proper with an improper uniform prior, the posterior estimates are usually not too sensitive to the choice of hyperparameters when a proper prior is used. Chen and Shao (1999) and Chen *et al.* (2002) have examined this issue in detail for the generalized linear mixed models and the multivariate categorical response models. For example, in the case of binary outcomes, suppose we take an improper uniform prior for $\beta = (\gamma_1, \beta, \gamma)$. Let $w_{y_i} = 1 - 2y_i$ and $w_{x_i} = 1 - 2x_i$. Also, let $X^* = ((X_1^*)', \cdots, (X_n^*)')'$, where

$$X_i^* = \begin{pmatrix} w_{y_i} z_{1i}' & w_{y_i} x_i & 0' \\ 0' & 0' & w_{x_i} z_i' \end{pmatrix}, \quad i = 1, 2, \ldots, n.$$

Then, the resulting posterior is proper if (i) X_0^* is of full rank, and (ii) there exists a $2n \times 1$ positive vector a such that $a' X^* = 0$. It would be interesting to examine propriety conditions for the continuous response case when an improper uniform prior is used.

Fifth, In Section 2.5, Table 1 of Chib's paper indicates that a zero prior mean for the intercept was used. Since the intercept reflects the mean birthweight, a zero mean may not make much sense from the practical point of view.

Sixth, in the case of binary outcomes, Liu (2001) proposed a very efficient and clever CA-MCMC algorithm (see Chen *et al.,* 2000) for sampling from $(\gamma_1, \beta, \gamma, \omega_{12})$ together from the joint posterior distribution. His algorithm does not require any Metropolis steps and it has been proven that the CA-MCMC provides a good mixing and fast convergence. Certainly, it would be interesting to see how well the CA-MCMC performs for the birthweight data.

Seventh, in the case of binary outcomes,

$$\text{ATE} = \int \{\Phi(z_1'\gamma_1 + \beta) - \Phi(z_1'\gamma_1)\}\pi(z_1)\, dz_1.$$

I would like to point out that when the empirical distribution of z_1 from the observed sample data is used, the ATE automatically reduces to

$$\text{ATE} = \frac{1}{n}\sum_{i=1}^{n}\{\Phi(z_{1i}'\gamma_1 + \beta) - \Phi(z_{1i}'\gamma_1)\}.$$

Thus, there is no need to make additional draws on z_1 from its empirical distribution for each sampled value of (γ_1, β) from the MCMC algorithm as was done in the paper.

Finally, I would like to briefly discuss the issues involved in model comparisons using Bayes factor or marginal likelihoods. The main concerns with the decision on model choice based solely on the observed value of the Bayes factor or the marginal likelihoods are (i) uncertainty (a statistic after all); (ii) prior dependence; and (iii) extremely sensitive to imprecise priors. Therefore, Bayes factor or marginal likelihoods should be calibrated under prior predictive distributions and then the decision will be made based on their calibration distributions. Some attempts have been made recently in this regard. Vlachos and Gelfand (2002) considered the calibration of Bayesian model choice criteria; Ibrahim *et al.* (2001) examined the theoretical properties of calibration distributions of L-measures; and García-Donato and Chen (2002) proposed a model selection criterion using the calibrating value of Bayes factors based on the prior predictive distributions. To illustrate the calibrating value, we consider the following example. Suppose $y_1, y_2, \ldots, y_n$ are a sample from model M_i $(i = 1, 2)$, where M_1 is $N(\theta_0, \sigma^2)$ with θ_0 and σ^2 known and M_2 is $N(\theta, \sigma^2)$ with $\theta \sim N(0, \tau^2)$ with σ^2 and τ^2 known. The sampling distribution of Bayes Factor is related to chi-squared distribution (under both models), and any characteristic (means, probabilities, etc.) can be easily calculated. As an illustration, let $n = 20$, $\theta_0 = 0$, $\sigma^2 = 1$ and $\tau^2 = 5$. The expression for Bayes factor is

$$B_{12} = \exp\left\{\frac{1}{2}\left(\ln(1+a) - \frac{na}{\sigma^2(1+a)}\bar{y}^2\right)\right\},$$

where $a = n\tau^2/\sigma^2$. It is easy to observe that $\lim_{\tau^2\to\infty} B_{12} = \infty$ (Lindley's paradox). The range of B_{12} is $(0, (1+a)^{1/2})$. A value c is called the calibrating value if

$$\Pr(B_{12}(\mathbf{Y}) \geq c \mid M_2) = \Pr(B_{12}(\mathbf{Y}) \leq c \mid M_1),$$

where $\Pr(\cdot|M_i)$ denotes the prior predictive probability for $i = 1, 2$. García-Donato and Chen (2002) proposed that we accept model M_1 if $B_{12}(\boldsymbol{y}) > c$ and accept model M_2 otherwise. Figure 6 shows the calibration distributions of B_{12} under both models. The value c for this example is 3.08, which makes the black (M_2) and gray (M_1) areas equal. The other detailed discussions on the calibrating value can be found in García-Donato and Chen (2002).

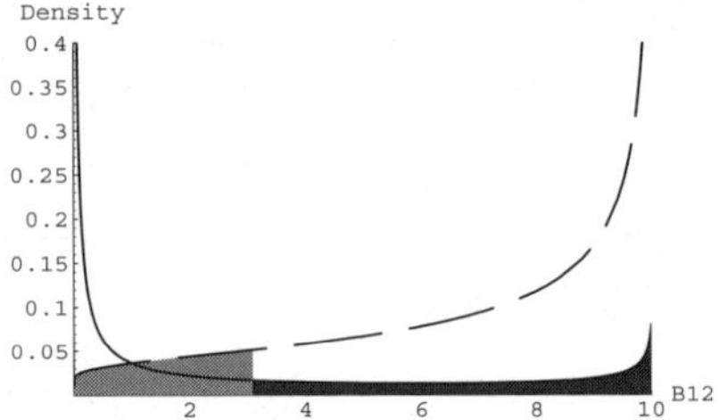

Figure 6. *Densities of* $B_{12}(\mathbf{Y})$ *under* M_1 *(dashed line) and under* M_2 *(solid line).*

DALE J. POIRIER and JUSTIN L. TOBIAS (*University of California-Irvine, USA*)

An essential ingredient of Rubin's potential outcome model in the presence of unobserved confounders is that only one of the two potential outcomes is observed for any one observation. This feature implies an inherent unidentifiability of the correlation between the two potential outcomes. For a Bayesian with a proper prior (and for some other partially informative priors) this causes no immediate difficulty. In fact Koop and Poirier (1997) show how learning about the unidentified parameter can occur, and suggest a MCMC algorithm for analyzing the model. Chib and Hamilton (2000) employ a dogmatic prior that sets the unidentified correlation equal to *zero*. In his invited paper, Chib's assumption that $\eta_{i0} = \eta_{i1} = \eta_i$ implicitly assumes a dogmatic prior that sets the unidentified correlation equal to *unity*, and in addition, equates the potential outcome variances. Such dogmatism strikes us as unnecessary.

This particular assumption also imposes strong and potentially inappropriate restrictions on key parameters of interest. To see this more clearly, let x(z) denote the binary treatment decision given that the instrument is set to z.

With this notation, note that the Average Treatment Effect: ATE $\equiv E(y_1 - y_0) = \beta$, the effect of Treatment on the Treated: $\text{TT}(z, x(z) = 1) \equiv E(y_1 - y_0|z, x(z) = 1) = \beta + E(\eta_1|z, x(z) = 1) - E(\eta_0|z, x(z) = 1) = \beta$ and the Local Average Treatment Effect: $\text{LATE}(z, \tilde{z}, x(z) = 1, x(\tilde{z}) = 0) \equiv E(y_1 - y_0|z, \tilde{z}, x(z) = 1, x(\tilde{z}) = 0) = \beta$ *must be the same* under the assumption $\eta_0 = \eta_1$. In other words, a necessary implication of this restriction is that the expected outcome gain from treatment (denoted $y_1 - y_0$) for an average person (ATE), for those actually taking the treatment at z (TT), and for those "complying" with the treatment at z but not at $\tilde{z}$ (LATE) must be equal. There seems to be no reason to require equality of these parameters *a priori*. Li, Poirier and Tobias (2002) do not impose this restrictive assumption and provide an observational study where these parameters clearly differ.

Furthermore, the focus of attention on the Average Treatment Effect (or average impacts generally) which are not affected by the unidentifiability, is modest in its goal. More interesting is the entire distribution of the outcome gain, not just its mean. Poirier and Tobias (2001) and Li *et al.* (2002) show how to analyze this entire distribution.

Finally, we note that when $\eta_0 = \eta_1$ is assumed the model reduces to the one described in Li (1998). Li also describes an MCMC estimation algorithm and discusses marginal likelihood calculations regarding unobservable confounding versus no confounding using comparatively simple Savage-Dickey density ratios.

REPLY TO THE DISCUSSION

Chen. We are grateful to Professor Chen for his detailed (and appreciative) reading of the paper and have no disagreement with any of the remarks. Professor Chen is particularly interested in the formulation of the prior and offers some alternatives to the one used in the paper. Each of these alternatives has merit and could be used easily without altering the estimation approach outlined in the paper. We have some reservations, however, about the use of noninformative priors in this context since an important goal is the formal comparison of models, for example, with and without confounding.

It is not quite correct to say that only one synthetic data set is used in the simulation study since, in fact, the data are generated for different values of the two key parameters R^2 and ρ. Anyway, we agree that a more extensive simulation study should be conducted (when one is not subject to the Valencia proceedings page constraint!).

The suggestion about trying alternative MCMC strategies is well taken and merits further study. It should be noted, however, that the MCMC procedure proposed in this paper is well tuned and produces low inefficiency factors.

We agree entirely with Professor Chen that the ATE in the binary response case can be computed without the simulation of z_{1i} from its empirical distribution.

Finally, it is always worthwhile to examine the sensitivity of the Bayes factors to the model inputs, including (but not restricted to) the prior. Albert and Chib (1997) provide a graphical approach for examining the effect of such changes in the inputs. In this context, the relevant question is whether reasonable changes in those inputs alters (not just the Bayes factor) but the ranking of the models. From our experience, this ranking is usually quite resistant to reasonable changes in the inputs.

Poirier and Tobias. We greatly appreciate the comments by Professors Poirier and Tobias. Professors Poirier and Tobias are uncomfortable with aspects of the modelling, in particular the assumption that the unobservables affecting the two potential outcomes are the same. This assumption should be viewed as asserting that the person-specific gain is zero, which is a reasonable assumption. This assumption can be relaxed. If the unobservables are different, however, one must be clear that it is not possible to estimate the joint distribution of the unobservables, despite the comment that one should, because any inferences about the correlation between the potential outcomes must necessarily depend on either specific parametric forms and/or other untestable modelling assumptions.

The second comment that the assumption that $\eta_{i0} = \eta_{i1} = \eta_i$ is responsible for the equality of the three treatment effects is not correct. If the outcome model has an interaction term between z_{1i} with x_i, and the distribution of z_{1i} differs between the treated and the untreated, then the treatment effects would differ, even under the assumption that $\eta_{i0} = \eta_{i1} = \eta_i$.

There is of course no reason to focus on the ATE but there is also no specific reason not to since the ATE is an important parameter. Chib and Hamilton (2000, 2002) provide posterior distributions of various other treatment effect parameters that measure the gain in different ways.

Finally, the paper by Li (1998) deals with a censored outcomes model that is different from the two models studied here. That paper is, however, not cast in the potential outcomes tradition and does not examine the identification issue or include an instrument in the modelling. We would not advocate using the Savage-Dickey method for finding the marginal likelihood because it is less general than the method discussed in this paper.

ADDITIONAL REFERENCES IN THE DISCUSSION

Albert, J. and Chib, S. (1997). Bayesian tests and model diagnostics in conditionally independent hierarchical models. *J. Am. Statist. Ass.* **92**, 916–925.

Chen, M.-H. and Shao, Q.-M. (1999). Properties of prior and posterior distributions for multivariate categorical response data models. *J. Multiv. Anal.* **71**, 277–296.

Chen, M.-H., Shao, Q.-M. and Ibrahim, J. G. (2000). *Monte Carlo Methods in Bayesian Computation*. New York: Springer.

Chen, M.-H., Shao, Q.-M. and Xu, D. (2002). Sufficient and necessary conditions on the propriety of posterior distributions for generalized linear mixed models. *Sankhyā A* **64**, 57–85.

García-Donato, G. and Chen, M.-H. (2002). Calibrating Bayes factor under prior predictive distributions. *Tech. Rep.*, University of Connecticut, USA.

Ibrahim, J. G., Chen, M.-H. and Sinha, D. (2001). Criterion based methods for Bayesian model assessment. *Statist. Sinica* **11**, 419–443.

Koop, G. and Poirier, D. J. (1997). Learning about the across-regime correlation in switching regression models. *J. Econometrics* **78**, 217–227.

Li, K. (1998). Bayesian inference in a simultaneous equation model with limited dependent variables. *J. Econometrics* **85**, 387–400.

Li, M., Poirier, D. J. and Tobias, J. L. (2002). Do dropouts suffer from dropping out? Estimation and prediction of outcome gains in generalized selection models. *Tech. Rep.*, University of California, Irvine, USA.

Liu, C. (2001). Bayesian analysis of multivariate probit models: Comments on "The art of data augmentation" by Van Dyk and Meng. *J. Comput. Graph. Statist.* **10**, 75–81.

Poirier, D. J. and Tobias, J. L. (2001). On the predictive distributions of outcome gains in the presence of an unidentified parameter. *J. Business Econom. Statist* (to appear).

Vlachos, P. K. and Gelfand, A. E. (2002). On the calibration of Bayesian model choice criteria. *J. Statist. Plann . Inference* (to appear).

BAYESIAN STATISTICS 7, pp. 85–103
J. M. Bernardo, M. J. Bayarri, J. O. Berger, A. P. Dawid,
D. Heckerman, A. F. M. Smith and M. West (Eds.)

Bayesian Treed Generalized Linear Models

HUGH A. CHIPMAN
University of Waterloo, Canada
hachipman@uwaterloo.ca

EDWARD I. GEORGE
University of Pennsylvania, USA
edgeorge@wharton.upenn.edu

ROBERT E. MCCULLOCH
University of Chicago, USA
robert.mcculloch@gsb.uchicago.edu

SUMMARY

For the standard regression setup, conventional tree models partition the predictor space into regions where the variable of interest Y, can be approximated by a constant. A treed model extends this idea by allowing a functional relationship between Y and the predictors within each region. As opposed to using a single model to describe the global variation of the response, treed models allow for local modelling across the predictor space. In this paper, we consider treed versions of generalized linear models (GLMs) and propose a Bayesian approach to finding and fitting such models. The potential of this approach is illustrated with a treed Poisson regression.

Keywords: BINARY TREES; LAPLACE APPROXIMATION; LOGISTIC REGRESSION; MARKOV CHAIN MONTE CARLO; MODEL SELECTION; POISSON REGRESSION; STOCHASTIC SEARCH.

1. INTRODUCTION

Consider the standard regression setup where Y a variable of interest, and $X_1, \ldots, X_p$ a set of potential explanatory variables or predictors are vectors of n observations. Suppose it is of interest to estimate the conditional distribution of $Y \mid X$ where $X = (X_1, \ldots, X_p)$. A standard modelling approach is to express this conditional distribution as a member of a single parametric family of models. If such an approach is inadequate, a better alternative may be to partition the predictor space so that within each subset of the partition, the parametric model accurately describes the conditional distribution of the response. Such an alternative can be accomplished by using a treed model.

A treed model is composed of two parts—a recursive partitioning of the predictor space using a tree structure and a distinct model for $Y \mid X$ associated with each subset of the partition. The binary tree T used as a recursive partition is inspired by earlier work (e.g. Morgan and Sonquist, 1963; Hawkins and Kass, 1982; Breiman *et al.*, 1984; Quinlan, 1986; Clark and Pregibon, 1992). An important distinction is that this earlier work assumes that within each subset of the partition, $E(Y \mid X)$ is constant. We shall refer to models making this assumption as "conventional trees." For the second component of the treed model, the subset models considered are all generalized linear models (Nelder and Wedderburn, 1972; McCullagh and

Nelder, 1989; Dey *et al.*, 2000). To fit the two components of the treed GLM, we consider a Bayesian approach. This entails the formulation of prior on the space of trees and on the parameters of the subset models. Efficient Metropolis-Hastings algorithms, obtained using Laplace approximations, are then used to stochastically search for high probability models.

The results in this paper extend the results of Chipman, George and McCulloch (1998, 2002) (hereafter CGM 1998 and CGM 2002) and Denison, Mallick and Smith (1998), (hereafter DMS 1998). Both CGM 1998 and DMS 1998 developed similar Bayesian approaches to conventional trees, the special case of treed models where $E(Y \mid X)$ is constant within each subset of the partition. CGM 2002 extended this Bayesian approach to treed regression models where $Y \mid X$ follows a normal linear model within the partition subsets.

The idea of treed models is not new. The simplest way to fit such models is to construct a conventional tree, and afterwards replace the piecewise constant model for $E(Y \mid X)$ with a richer parametric model. This approach is taken by Quinlan (1992) and Torgo (1997). A shortcoming of such approaches is that the partition is not optimized for the final parametric model. Partitioning the predictor space so that the response is as homogeneous as possible within each node makes linear models less likely to be useful. A better approach would be to grow the tree using a splitting criterion that reflects the model used in each partition. Several papers have suggested such a strategy: Karalič (1992) for multiple regression models, Alexander and Grimshaw (1996) for simple linear regression models, and Chaudhuri *et al.* (1994) for multiple polynomial regression models. This was extended to incorporate GLMs by Chaudhuri *et al.* (1995), and overdispersed treed logistic regressions by Ahn and Chen (1997). Treed models also bear some resemblance to hierarchical mixtures of experts models (Jordan and Jacobs 1994), in which a soft decision rule based on a linear combination of predictors is used in each interior node, and a logistic regression is performed in each terminal node. Further discussion of precedents appears in CGM 2002.

What distinguishes our work from these earlier papers is the Bayesian approach. Rather than use ad hoc penalty criteria for ranking models, the posterior distribution coherently ranks the models by using a likelihood to extract the information provided by the data. Rather than use an ad hoc greedy algorithm to find a model, which can be especially challenging with a treed model, our MCMC algorithm uses the posterior information to guide the stochastic search. Simulation studies in CGM 1998 illustrated that for Bayesian CART, our stochastic search found better trees than a greedy search. CGM 1999 further showed that this MCMC algorithm can find a wider variety of trees than a bootstrapped greedy grow/prune algorithm. Lutsko and Kuijpers (1994) considered a MCMC-like approach using simulated annealing, and also found improvements.

Although the formulation of our methods is entirely model based, in many ways it resembles a machine learning algorithm in the spirit of what Breiman (2001) calls "algorithmic modelling". Our formulation can be construed as an algorithm for discovering structure that is controlled by hyper-parameters that can be treated as tuning constants. An advantage of the Bayesian formulation is that it provides natural intrepretability of the hyperparameters thereby facilitating their calibration. It is interesting that Breiman distinguishes between two separate cultures of statistics, a culture that treats data as realizations from a model and a culture that gives primacy to out-of-sample prediction for algorithmic construction. We have used a throughly model-based approach to construct algorithms that provide excellent out-of-sample predictions.

2. TREED GENERALIZED LINEAR MODELS

2.1. *The General Model*

For the purpose of modelling the relationship between a variable of interest Y and a $p \times 1$ vector of predictor variables X, a treed model is a specification of the conditional distribution of $Y \mid X$. Such a model consists of two components—a binary tree T that partitions the domain of X, denoted $\mathcal{X}$, and a parametric model for Y associated with each subset of the partition.

The tree T partitions $\mathcal{X}$ as follows. Each interior node of the tree is associated with a single predictor splitting rule that assigns each (X, Y) observation to one of its two child nodes. For ordered predictors, the assignment is determined by whether or not the predictor is less than a fixed value. For categorical predictors, the assignment is determined by whether or not the predictor belongs to a particular subset of the possible categories. By successive assignments, beginning with the root node, T assigns each (X, Y) observation to one of the b terminal nodes, thereby partitioning the predictor space $\mathcal{X}$ into b disjoint sets.

The treed model then associates a parametric model for the distribution of $Y \mid X$ with each of the terminal nodes of T. More precisely, for X values that are assigned to the ith terminal node of T, the conditional distribution of Y is given by a parametric model $Y \mid X \sim p(y \mid X, \theta_i)$ indexed by θ_i. Letting $\Theta = (\theta_1, \ldots, \theta_b)$, a treed model is then fully specified by the pair (Θ, T). By treating the observed data as realizations from a treed model, we can then compute the posterior distribution over (Θ, T). This parametric modelling sets the stage for a Bayesian analysis. In contrast, early tree formulations were essentially proposed as data analysis tools rather than models.

CGM 1998 and DMS 1998 proposed Bayesian approaches to finding and fitting conventional trees. In contrast to treed models, conventional trees use terminal node distributions for $Y \mid X$ that are not functions of X. For example, with a continuous response, a conventional tree would assume $Y \mid X \sim \mathrm{N}(\mu_i, \sigma_i^2)$ for X values assigned to the ith terminal node of T. Such models correspond to step functions for the expected value of $Y \mid X$, and may require large trees to approximate an underlying distribution $Y \mid X$ whose mean is continuously changing in X. By using a richer structure at the terminal nodes, treed models can transfer structure from the tree to the terminal nodes. When such structure exists, smaller and hence more interpretable trees may be used to describe the distributions for $Y \mid X$.

Finally, we should point out that although we use the one symbol "X" for notational simplicity, one can decide to restrict attention to one subset of the components of X for the splitting rules in T and to a different subset for the terminal node models $p(y \mid X, \theta)$. These subsets need not be disjoint.

2.2. *Terminal Node GLMs*

Each tree T induces a partition $T_1, \ldots, T_b$ of the predictor space $\mathcal{X}$, where T_i is the subset of $\mathcal{X}$ corresponding to the ith terminal node of T. Note that $\bigcup T_i = \mathcal{X}$ and $T_i \bigcap T_{i'} = \emptyset$ for $i \neq i'$. For a given T, specification of the terminal node models for Y is facilitated by using a double indexing scheme where (x_{ij}, y_{ij}) denotes each of the $j = 1, \ldots, n_i$ observations of (X, Y) assigned to T_i. All the data assigned to T_i are denoted $x_i = (x_{i1}, \ldots, x_{in_i})$ and $y_i = (y_{i1}, \ldots, y_{in_i})$, and the entire data set is denoted $x = (x_1, \ldots, x_b)$ and $y = (y_1, \ldots, y_b)$. The overall sample size is denoted $n = \sum n_i$. This indexing scheme is conditional on T, and will be different for trees that induce a different partition of $\mathcal{X}$. Finally, in order to accommodate an intercept term in all of the terminal node models below, we shall assume throughout that the first component variable in every x_{ij} is identically equal to 1.

Perhaps the most natural and tractable case of a treed model is the treed regression model, CGM 2002, which is obtained by associating an independent normal linear model with each terminal node subset T_i. Using the notation above, this can be expressed as

$$Y_{ij} \mid x, \Theta, T \sim \mathrm{N}(x_{ij}^t \beta_i, \sigma_i^2) \tag{1}$$

with all Y_{ij} conditionally independent given (x, Θ, T). Here, β_i is an unknown $p \times 1$ vector of regression coefficients and $\theta_i = (\beta_i, \sigma_i^2)$. Under this model both the mean $E(Y_{ij} \,|\, x, \Theta, T)$ and the variance $Var(Y_{ij} \,|\, x, \Theta, T)$ functions can change across the terminal node subsets T_i.

In this paper, we consider generalized linear models (GLMs) as the terminal node models. For a given T, such models are of the form

$$p(y_{ij} \,|\, x, \Theta, T) = \exp\left\{\phi_i^{-1}[y_{ij}\eta_{ij} - \psi(\eta_{ij})] + c(y_{ij}, \phi_i)\right\}. \tag{2}$$

where for some strictly increasing function $h(\cdot)$,

$$\eta_{ij} = h(x_{ij}^t \beta_i), \tag{3}$$

$j = 1, \ldots, n_i$ and $i = 1, \ldots, b$. We will also assume throughout that all the Y_{ij} are conditionally independent given (x, Θ, T). Here, $\theta_i = \beta_i$ or $\theta_i = (\beta_i, \phi_i)$ according to whether the dispersion parameter ϕ_i is treated as known.

Letting $\mu_{ij} \equiv E(Y_{ij} \,|\, x, \Theta, T)$ denote the conditional mean of Y_{ij}, μ_{ij} is related to $x_{ij}^t \beta_i$ by $g(\mu_{ij}) = x_{ij}^t \beta_i$, where $g(\cdot)$ is called the link function. Because $\mu_{ij} = \psi'(\eta_{ij})$, the link function is implicitly determined by the relationship between η_{ij} and $x_{ij}^t \beta_i$. Indeed, $h(\cdot) = \psi'^{-1}(g^{-1}(\cdot))$ in (3). When $g^{-1} = \psi'$ so that $\eta_{ij} = x_{ij}^t \beta_i$, g is called the canonical link function. Note that the variance of Y_{ij}, $\sigma_{ij}^2 \equiv \text{Var}(Y_{ij} \,|\, x, \Theta, T) = \phi_i \psi''(\eta_{ij})$ depends on η_{ij} and ϕ_i.

The normal linear model (1) is the special case of (2) where g is the identity transform so that $\mu_{ij} = x_{ij}^t \beta_i$, $\sigma_{ij}^2 = \phi_i$ and $\psi(\eta_{ij}) = \eta_{ij}^2/2$. Other exponential family distributions for Y are easily subsumed by (2). When Y_{ij} are 0-1 random variables with means $\mu_{ij} = \pi_{ij}$, the logistic regression model is obtained with the logit transformation $g(\pi_{ij}) = \log[\pi_{ij}/(1 - \pi_{ij})] = x_{ij}^t \beta_i$, $\phi_i \equiv 1$ and $\psi(\eta_{ij}) = \log(1 + \exp(\eta_{ij}))$. In this case, the Y_{ij} are conditionally independent Bernoulli random variables with means $\pi_{ij} = g^{-1}(x_{ij}^t \beta_i) = \exp(x_{ij}^t \beta_i)/(1 + \exp(x_{ij}^t \beta_i))$. When Y_{ij} are counts with mean $\mu_{ij} = \lambda_{ij}$, the Poisson regression model is obtained with the log transformation $g(\lambda_{ij}) = \log \lambda_{ij} = x_{ij}^t \beta_i$, $\phi_i \equiv 1$ and $\psi(\eta_{ij}) = \exp(\eta_{ij})$. In this case, the Y_{ij} are conditionally independent Poisson random variables with means $\lambda_{ij} = g^{-1}(x_{ij}^t \beta_i) = \exp(x_{ij}^t \beta_i)$. In each of these cases, g is a canonical link and $\eta_{ij} = x_{ij}^t \beta_i$.

The assumption that dispersion is fixed at $\phi_i \equiv 1$ in some GLMs can render such models inadequate for data that exhibits greater dispersion. One solution might be to simply consider an overdispersed version of such models where $\phi_i > 1$. Another popular possibility is to go outside the GLM family by using mixture model elaborations such as the beta-binomial or the gamma-Poisson. Other useful alternatives have been proposed by Efron (1986), Jorgensen (1987) and Gelfand and Dalal (1990).

3. PRIOR SPECIFICATIONS FOR TREED GLMS

Since a treed model is identified by (Θ, T), a Bayesian analysis of the problem proceeds by specifying a prior probability distribution $p(\Theta, T)$. This is most easily accomplished by specifying a prior $p(T)$ on the tree space, a conditional prior $p(\Theta \,|\, T)$ on the parameter space, and then combining them via $p(\Theta, T) = p(\Theta \,|\, T)p(T)$.

3.1. *Specification of* $p(T)$

For $p(T)$, we recommend the specification proposed in CGM 1998 for conventional trees. This prior is implicitly defined by a tree-generating stochastic process that grows trees from a single-node tree by randomly splitting terminal nodes. A tree's propensity to grow under this process is controlled by a two-parameter node splitting probability

$$P(\text{node splits} \,|\, \text{depth} = d) = \alpha(1 + d)^{-\gamma},$$

where the root node has depth 0. The parameter α is a base probability of growing a tree by splitting a current terminal node and γ determines the rate at which the propensity to split diminishes as the tree gets larger. The specification of (α, γ) can be guided by the marginal prior distribution on the number of terminal nodes (the tree size), which can be easily simulated. For example, using such marginals, (α, γ) can be chosen to express the belief that reasonably small trees should yield adequate fits to the data. The tree prior $p(T)$ is completed by specifying a prior on the splitting rules assigned to intermediate nodes. We use a prior that is uniform on available variables at a particular node, and within a given variable, uniform on all possible splits for that variable.

3.2. *Specification of* $p(\Theta \mid T)$

Turning to the specification of $p(\Theta \mid T)$, we note that while $p(T)$ above is sufficiently general for all treed model problems, the specification of $p(\Theta \mid T)$ will necessarily be tailored to the particular form of the model $p(y \mid x, \theta)$ under consideration. However, some aspects of the $p(\Theta \mid T)$ specification should be generally considered. When reasonable, an assumption of iid components of Θ reduces the choice to that of a single prior $p(\theta)$ for $\theta_1, \ldots, \theta_b$. However, even with this simplification, the specification of $p(\theta)$ can be difficult and crucial. In particular, a key consideration is to avoid conflict between $p(\Theta \mid T)$ and the likelihood information from the data. On the one hand, if we make $p(\theta)$ too tight (i.e. with very small spread around the prior mean) the prior may be too informative and overwhelm the information in the data corresponding to a terminal node. On the other hand, if $p(\theta)$ is too diffuse (spread out), $p(\Theta \mid T)$ will be even more so, particularly for large values of b (large trees) given our iid model for the θ_i. Excessively diffuse prior priors can "wash out" the likelihood in the sense of the Bartlett-Lindley paradox, Bartlett (1957), pushing the posterior of T towards concentration on very small trees. Such a relationship between tree size and dispersion of the θ prior is illustrated in Table 1.

For simplicity, we shall assume $\theta_i = \beta_i$ throughout, treating the dispersion parameter ϕ_i as known and to be used as a tuning parameter in applications. More elaborate techniques, such as those mentioned at the end of Section 2, would be required to deal with unknown ϕ_i. Instead we take $\phi_i = \phi$ and investigate the effect of various fixed ϕ values. This strategy is illustrated in Section 5.1 where ϕ_i is seen to be very influential in the modelling process, playing a similar role to the residual variance in least squares regression.

For the prior on the β_is, we consider the simple choice that they are iid multivariate normal conditionally on T,

$$\beta_1, \ldots, \beta_b \mid T \quad \text{iid} \quad \sim N_p(\bar{\beta}, A^{-1}). \tag{4}$$

This further reduces the specification problem to the choice of values for the hyperparameters $\bar{\beta}$ and the $p \times p$ inverse covariance matrix A. We recommended such a normal prior for treed normal regression models in CGM 2002 because of the relative transparency for hyperparameter selection and because it was conjugate, yielding an exact closed form expression for $p(y \mid T)$. Although (4) will not be conjugate for other GLMs, it does allow for relatively straightforward use of an effective Laplace approximation of $p(y \mid T)$ as will be seen in Section 4.1.

We use information about the distribution of the transformed mean, $g(\mu_{ij})$, to guide the choice of hyperparameter values for $\bar{\beta}$ and A. To simplify notation, we restrict attention here to canonical link models where $g(\mu_{ij}) = x_{ij}^t \beta_i = \eta_{ij}$. Suppose, for the moment, that plausible values were available for $\eta_{\min}, \bar{\eta}, \eta_{\max}$, the minimum, central, and maximum values for the η_{ij}. Automatic choices for such values are discussed at the end of the section.

To further simplify hyperparameter values selection, we also standardize the last $(p-1)$ components of x_{ij} to each have mean 0 and range 1. (Recall that the first component of x_{ij} is always 1 so that an intercept term is included in every terminal node model.)

We are now ready to consider the choice of $\bar{\beta}$. To get started, it may be useful to consider this choice under the assumption that a tree is not needed, and a single GLM model is appropriate for the complete data. In this case, a natural default choice is $\bar{\beta} = (\overline{\eta}, 0, \ldots, 0)^t$. This choice guards against an unreasonable value for the intercept while incorporating the neutral value 0 for the remaining components of β, indicating indifference between positive and negative values. Note that standardization of the predictors to have mean 0 decouples the global relationship between the intercept and the other coefficients. However, given T, the mean values of predictors will generally not be 0 within subsets of the partition.

Turning to the choice of A, we make the simplifying reduction that $A = \text{diag}(\ 1/\sigma_0^2, 1/\sigma_\beta^2, \ldots, 1/\sigma_\beta^2)$ where σ_0 is the prior standard deviation of the intercept, and σ_β is the prior standard deviation of the other regression coefficients. This reduces the specification to the choice of two scalars, σ_0 and σ_β. As noted above, the choice of these hyperparameters is crucial. Essentially, the challenge is to choose σ_0 and σ_β large enough to accommodate all plausible values of β, but no larger than that. For choosing σ_0, suppose that all slope coefficients except the intercept were zero. Then we would want σ_0 such that $6\sigma_0 \approx \eta_{\max} - \eta_{\min} \equiv \Delta$, corresponding to the belief that a substantial mass of the normal prior lies within $\overline{\eta} \pm 3\sigma_0$. With this in mind, we treat σ_0 as a tuning parameter and consider various values around $\Delta/6$ as default choices to be explored.

For choosing σ_β, note that by standardizing the range of the predictors to have range 1, a full-range increase in a predictor with a regression coefficient equal to Δ would lead to full-range increase in $g(\mu_{ij}) = x_{ij}^t \beta_i$ when all the other predictors remained unchanged. However, such reasoning is not completely satisfying in at least two ways. First, for a given tree T, the predictor observations will generally no longer be standardized within each subset of the partition (although they would have a range less than 1, and a mean between -1 and 1). Second, the presence of multicollinearity can necessitate substantially larger coefficient values. However, if severe multicollinearity is present we can usually shrink coefficients towards zero without appreciable damage to the fit. In fact, such shrinkage often stabilizes calculations and even improves predictions. Given these considerations, we also treat σ_β as a tuning parameter and consider various values around $\Delta/6$ as default choices to be explored. Generally, smaller σ_β values will result in estimated coefficients that are shrunk, and trees with fewer terminal nodes.

We conclude this section with a brief discussion of automatic choices of $\eta_{\min}$, $\overline{\eta}$ and $\eta_{\max}$. In the absence of prior information, a natural automatic choice for $\overline{\eta}$ is $\overline{\eta} = g(\overline{y})$ where $\overline{y}$ is the overall mean of the y_{ij} values. Choice of $\eta_{\min}$ and $\eta_{\max}$ is more challenging. We have found it reasonable to fit a single GLM to the data and to use the minimum and maximum of the MLEs $\hat{\eta}_{ij}$ to estimate $\eta_{\min}$ and $\eta_{\max}$ respectively. A potential drawback is that if the single GLM severely underfits the data, then the range of $\hat{\eta}_{ij}$ may be too small, yielding too tight a prior on β. Despite this drawback, using predictions from a GLM model is certainly superior to using the range of the observed y_{ij}, which can lead to unrealistic bounds. For example with Poisson regression, an observed count $y_{ij} = 0$ would yield $\eta_{\min} = \log(0) = -\infty$. Finally, we note that genuine prior information may well exist in many applications. For example, in modelling of insurance claim counts, actuaries have a good idea of the lowest and highest possible accident rates among all rating groups. Such information might be used exclusively, or combined with automatic choices.

4. POSTERIOR COMPUTATION AND EXPLORATION

Given a set of training data, a Bayesian analysis would ideally proceed by computing the entire posterior distribution $p(\Theta, T \mid y, x)$. Unfortunately, in problems such as this, the size of the model space is so large that exhaustive calculation of the posterior is simply not feasible. However, posterior information can still be obtained by using a combination of analytical simplification or approximation together with MCMC sampling from the posterior. For example, the general strategy used in CGM 1998, 2002 was to first eliminate Θ by obtaining a closed form expression for the marginal likelihood

$$p(y \mid x, T) = \int p(y \mid x, \Theta, T) p(\Theta \mid T) d\Theta, \tag{5}$$

and then to use a Metropolis-Hasting (MH) algorithm to simulate a Markov chain sample from $p(T \mid y, x) \propto p(y \mid x, T)p(T)$. Because the Markov chain simulation tends to gravitate towards higher posterior probability trees, it can effectively be used as a stochastic search algorithm.

CGM 1998, 2002 were able to analytically perform the integration in (5) because conjugate priors were used. However, for GLMs (2) other than the normal linear model, analytical integration is unavailable with the normal prior (4). Instead, we use a Laplace approximation described below to obtain $\tilde{p}(y \mid x, T) \approx p(y \mid x, T)$, and then apply an MH algorithm to simulate a Markov chain sample from

$$\tilde{p}(T \mid y, x) \propto \tilde{p}(y \mid x, T)p(T).$$

A similar strategy in the context of model averaging of survival models was successfully used by Raftery *et al.* (1996).

4.1. *Laplace Approximation of* $p(y \mid x, T)$

For our treed GLMs, we express the integral in (5) as

$$p(y \mid x, T) = \prod_{i=1}^{b} \int L(\beta_i \mid x_i, y_i, T) p(\beta_i \mid T) d\beta_i \tag{6}$$

where $L(\beta_i \mid x_i, y_i, T) = \prod_{j=1}^{n_i} p(y_{ij} \mid x_{ij}, \beta_i, T)$ is the likelihood of β_i from (2) and (3). (Recall that we treat ϕ_i as known). To approximate $p(y \mid x, T)$, it thus suffices to approximate each of the integrals in (6), and for this purpose we use a Laplace approximation, see Tierney and Kadane (1986).

Let $\mathcal{L}(\beta_i) \equiv \log[L(\beta_i \mid x_i, y_i, T)p(\beta_i \mid T)]$ denote the log posterior of β_i (up to a norming constant). For notational convenience, we suppress the dependence on x_i, y_i, T in $\mathcal{L}(\beta_i)$. Using a quadratic approximation of $\mathcal{L}(\beta_i)$ around the posterior mode β_i^*, each of the integrals in (6) can be approximated as

$$\begin{aligned} \int L(\beta_i \mid x_i, y_i, T) p(\beta_i \mid T) d\beta_i &= \int \exp\{\mathcal{L}(\beta_i)\} d\beta_i \\ &\approx \int \exp\left\{\mathcal{L}(\beta_i^*) - \frac{1}{2}(\beta_i - \beta_i^*)^t(-\mathcal{L}''(\beta_i^*))(\beta_i - \beta_i^*)\right\} d\beta_i \\ &= \exp\{\mathcal{L}(\beta_i^*)\}\ (2\pi)^{p/2}\ | - \mathcal{L}''(\beta_i^*)|^{-1/2} \end{aligned} \tag{7}$$

where $\mathcal{L}''(\beta_i^*)$ is the $p \times p$ matrix of second derivatives of $\mathcal{L}$ evaluated at β_i^*.

Now, under our normal prior $\beta_i \sim \mathrm{N}(\bar{\beta}, A^{-1})$ in (4), the log posterior $\mathcal{L}$ can be conveniently expressed, (up to a norming constant), as

$$\mathcal{L}(\beta_i) \equiv l(\beta_i) - \frac{1}{2}(\beta_i - \bar{\beta})^t A(\beta_i - \bar{\beta}) \tag{8}$$

where $l(\beta_i) \equiv \log L(\beta_i \mid x_i, y_i, T)$ denotes the log likelihood of β_i. It also follows that

$$\mathcal{L}'(\beta_i) = l'(\beta_i) - A(\beta_i - \bar{\beta}) \tag{9}$$

and

$$\mathcal{L}''(\beta_i) = l''(\beta_i) - A. \tag{10}$$

Using (8) and (10), the Laplace approximation (7) with the normal prior on β_i can be expressed, (up to a norming constant) as

$$\int L(\beta_i \mid x_i, y_i, T) p(\beta_i \mid T) d\beta_i \approx \exp\{\mathcal{L}(\beta_i^*)\} \, (2\pi)^{p/2} \, |-l''(\beta_i^*) + A|^{-1/2}$$

$$= \frac{|A|^{1/2}}{|-l''(\beta_i^*) + A|^{1/2}} \exp\left\{ l(\beta_i^*) - \frac{1}{2}(\beta_i^* - \bar{\beta})^t A(\beta_i^* - \bar{\beta}) \right\}. \tag{11}$$

This approximation depends on the data through β_i^*, $l(\beta_i^*)$ and $l''(\beta_i^*)$.

From (2) and (3), the general form for the log likelihood for each terminal node GLM is

$$l(\beta_i) = \phi_i^{-1} \sum_{j=1}^{n_i} \left[y_{ij} h(x_{ij}^t \beta_i) - \psi(h(x_{ij}^t \beta_i)) \right].$$

The first and second derivatives of $l(\beta_i)$ are

$$l'(\beta_i) = \phi_i^{-1} \sum_{j=1}^{n_i} \left[[y_{ij} - \psi'(h(x_{ij}^t \beta_i))] h'(x_{ij}^t \beta_i) \right] x_{ij}$$

and

$$l''(\beta_i) = \phi_i^{-1} \sum_{j=1}^{n_i} \left[[y_{ij} - \psi'(h(x_{ij}^t \beta_i))] h''(x_{ij}^t \beta_i) - \psi''(h(x_{ij}^t \beta_i))(h'(x_{ij}^t \beta_i))^2 \right] x_{ij} x_{ij}^t$$

Special cases of l, l' and l'' are easily obtained for particular models based on $\psi(\cdot), h(\cdot)$ and ϕ_i. For example, in the logistic regression setting where $\psi(\eta) = \log(1 + \exp(\eta))$, $h(x) = x$ and $\phi_i \equiv 1$, we obtain $l(\beta_i) = \sum_{j=1}^{n_i} [y_{ij}\eta_{ij} - \log(1 + \exp(\eta_{ij}))]$, $l'(\beta_i) = \sum_{j=1}^{n_i} [y_{ij} - \pi_{ij}] \, \eta_{ij} x_{ij}$, and $l''(\beta_i) = \sum_{j=1}^{n_i} -x_{ij} x_{ij}^t \pi_{ij}(1 - \pi_{ij})$, since $\psi'(\eta) = e^\eta/(1 + e^\eta) = \pi$, $\psi''(\eta) = e^\eta/(1 + e^\eta)^2$, $h'(x) \equiv 1$ and $h''(x) \equiv 0$. In the Poisson regression setting where $\psi(\eta) = e^\eta$, $h(x) = x$ and $\phi_i \equiv 1$, we obtain $l(\beta_i) = \sum_{j=1}^{n_i} [y_{ij}\eta_{ij} - \lambda_{ij}]$, $l'(\beta_i) = \sum_{j=1}^{n_i} [y_{ij} - \lambda_{ij}] \eta_{ij} x_{ij}$, and $l''(\beta_i) = \sum_{j=1}^{n_i} -\lambda_{ij} \; x_{ij} x_{ij}^t$, since $\psi'(\eta) = \psi''(\eta) = e^\eta = \lambda$, $h'(x) \equiv 1$ and $h''(x) \equiv 0$.

Finally, to compute (11) for each i using these expressions, the posterior mode β_i^* is needed. To find β_i^*, we use a simple Newton-Raphson algorithm

$$\beta^{(k+1)} = \beta^{(k)} - (\mathcal{L}''(\beta^{(k)}))^{-1} (\mathcal{L}'(\beta^{(k)}))$$

where $\mathcal{L}'$ and $\mathcal{L}''$ are obtained from (9) and (10).

One implementation detail is worth noting. Throughout this section, norming constants have been omitted for the sake of clarity. These constants can actually affect the posterior probability, since each terminal node GLM will have its own norming constants, and the number of terminal nodes varies across trees. All norming constants are retained in our program code except where they are known to cancel.

4.2. *Markov Chain Monte Carlo Posterior Exploration*

We use MCMC to stochastically search for high posterior trees T by using the following MH algorithm which simulates a Markov chain $T^0, T^1, T^2, \ldots$ with limiting distribution $\tilde{p}(T \mid y, x) \propto \tilde{p}(y \mid x, T)p(T)$, where $\tilde{p}(y \mid x, T)$ is the Laplace approximation to $p(y \mid x, T)$ proposed above. Starting with an initial tree T^0, this algorithm iteratively simulates the transitions from T^i to T^{i+1} by the two steps:

1. Generate a candidate value T^* with probability distribution $q(T^i, T^*)$.
2. Set $T^{i+1} = T^*$ with probability

$$\alpha(T^i, T^*) = \min\left\{\frac{q(T^*, T^i)}{q(T^i, T^*)}\frac{\tilde{p}(y \mid x, T^*)p(T^*)}{\tilde{p}(y \mid x, T^i)p(T^i)}, 1\right\}. \tag{12}$$

Otherwise, set $T^{i+1} = T^i$.

In (12), $q(T, T^*)$ is the kernel which generates T^* from T by randomly choosing among four steps: GROW, PRUNE, CHANGE, and SWAP. Details of these steps are given in CGM 1998, 2002. Although these moves better explore the posterior than a greedy grow/prune algorithm, the chain may get trapped in local maxima. Multiple restarts may be employed to efficiently explore the posterior on trees. See CGM 1998, 2002 for additional details.

5. AN APPLICATION

5.1. *A Wave Soldering Experiment*

To illustrate and assess our Bayesian treed GLM approach, we applied it to the Poisson regression dataset `solder2`, which is available in S and is described in Chapter 1 of Chambers and Hastie (1992). These data were originally collected by Comizzoli *et al.* (1990) as part of an experiment to investigate a wave soldering procedure for mounting electrical components on circuit boards. The response, `skips`, is a visual count of the number of skips in solder applied to a circuit board. In 623 of the 750 observations, `skips` had a value 0. The mean and maximum value of `skips` were 1.19 and 32, respectively. The five categorical predictors are:

- `Opening (S/M/L)` : amount of clearance around the mounting pad;
- `Solder (Thick/Thin)` : amount of solder;
- `Mask (5 levels)` : type and thickness of the material used for the solder mask;
- `PadType (10 levels)` : the geometry and size of the mounting pad; and
- `Panel (1/2/3)` : each board was divided into three panels, with three runs on a board.

We began by fitting two Poisson regressions with log link functions. The first model contained indicator variables for all main effects, while the second (suggested in Chambers and Hastie, 1992) contained main effects plus the three two-way interactions `Opening:Solder`, `Opening:Mask` and `Mask:Solder`. The mean deviance for the first model was 1.52 with 731 df and the mean deviance for the second was 1.24 with 719 df. Mean deviances greater than 1 suggest either lack of fit or overdispersion.

As an alternative to these Poisson GLMs, we proceeded to consider a treed Poisson GLM, and elected to consider all predictors in both the splitting rules of interior nodes and in the terminal node GLMs. As described in Section 3.2, we set the prior mean of the intercept to be $\hat{\beta}_0 = g(\bar{y}) = \log(\bar{y})$ and set the prior means of the slope components of β equal to 0. To gauge the choices for the hyperparameters σ_0 and σ_β, we fitted the main effects Poisson regression mentioned above and found the range of predicted values of $\hat{\eta}_{ij} = \log \hat{\mu}_{ij}$ to be $\Delta = 17.8$. This suggests $\sigma_0 = \sigma_\beta = \Delta/6 \approx 3$. We considered two settings for the prior standard deviation in a neighborhood of 3, namely slopes $\sigma_\beta = 2, 4$. The prior standard deviation of the intercept σ_0 was fixed at 4. A value of 4 was used instead of 3 because it was felt that additional dispersion in the intercept was less likely to have an impact on the treed model. Because of the overdispersion we observed in the Poisson GLMs, it seemed reasonable to consider that the dispersion parameter ϕ would exceed 1 (no overdispersion) and be no more than 3. With this in mind, we treated ϕ as a tuning parameter and considered three settings $\phi = 1.5, 2, 3$. As will be seen below, varying this dispersion parameter ϕ plays an important role in the analysis. The modelling procedure was run separately for the six combinations of σ_β and ϕ.

The MH search was run with tree prior parameters $\alpha = 0.25$ and $\gamma = 2$, giving prior mass of approximately (0.75, 0.22, 0.03) on trees with 1, 2, and ≥ 3 terminal nodes, respectively. We set the Markov transition kernel $q(T^i, T^*)$ to randomly choose one of the four steps with probabilities P(GROW) = P(PRUNE) = 0.1 and P(CHANGE) = P(SWAP) = 0.4. For each estimation, one chain with 2500 steps was used, taking approximately 11 minutes to execute on a PentiumIII / 1 GHz computer. Although many trees are visited by the MH algorithm, the "best" tree is determined as follows: the most frequently visited tree size is identified, and then for this size we choose the tree with the highest log integrated likelihood. While the alternative of ranking trees by log posterior probabilities may seem appealing, it suffers from the dilution phenomenon discussed in CGM 1998.

We also considered a conventional tree with constant means at terminal nodes, fit with a Poisson likelihood via greedy search and cross-validation. Compared to our treed Poisson regressions, this conventional model relies much more on the tree structure to explain the variation of the response `skips`. This tree was fit to the data with the `rpart` implementation (Therneau and Atkinson, 1997) in Splus and R. Sensitivity to overdispersion is not a substantial issue when cross-validation is used to determine tree complexity. For this reason, we did not consider different parameter settings of the `rpart` procedure.

To compare the different modelling methods, as well as the various tuning parameter choices, we used a repeated 10-fold cross-validation. Within each 10-fold cross-validation, each data point is predicted out-of-sample once (i.e., in 9/10 of the cases it will be used for training and 1/10 for testing). Five different replications of 10-fold cross-validation were carried out. All methods were compared with the same five replications, thereby removing blocked effects. For the `rpart` tree, two cross-validations are in fact being performed: cross-validation internal to the `rpart` code is determining an appropriate tree size, and our cross-validation algorithm is assessing out-of-sample performance.

Figure 1 compares the performance of the different methods using mean and median deviance contributions, where the deviance contribution from observed response y_i with out-of-sample prediction $\hat{\mu}_i$ given by $d_i = 2\left[(\hat{\mu}_i - y_i) + y_i(\log y_i - \log \hat{\mu}_i)\right]$. The log posterior could have also been used as a performance measure, but due to comparisons with likelihood based methods (`glm, rpart`), we used deviance measures instead.

We see in Figure 1 that the best performers in terms of mean deviance are treed models with $\phi = 2$ or $\phi = 3$. However, the distribution of the d_i's turned out to be very long-tailed, with a few influential large values. For this reason we also considered the median deviance. For this measure, we see an even more substantial difference, as well as reduced variability in the values across the five simulations.

An understanding of the relationship between tree size and the parameters ϕ and σ_β may be useful in interpreting Figure 1. Table 1 shows the mean size of the "best" tree reported in the 50 runs (5 permutations $\times$ 10 folds) for each of the six (ϕ, σ_β) settings considered. As overdispersion increases, the tree size decreases, since the log posterior is divided by the factor ϕ. Increasing σ_β also makes smaller trees more likely. Evidently, a tree of four or more

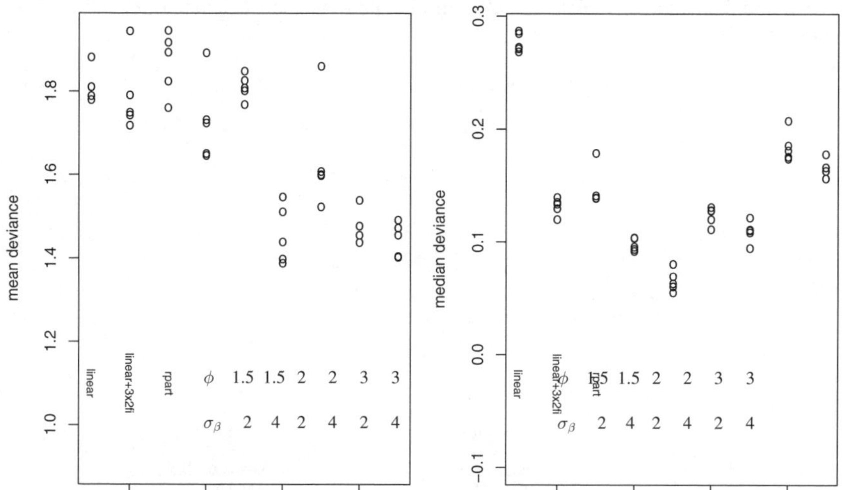

Figure 1. *Comparison of two GLMs (main effects models and main effects with three two-way interactions),* `rpart` *(conventional tree with Poisson data) and six treed models (with various parameter settings).*

nodes is overfitting, and a tree with two nodes may be slightly underfitting (as can be seen from the median deviance contribution).

Table 1. *Mean size of tree across 10 folds of cross-validation, and five permutations of data. Note that this "mean" is not across the posterior, since only one tree is reported for each of the 50 runs.*

ϕ	σ_β	Mean tree size
1.5	2	4.94
1.5	4	4.54
2.0	2	3.18
2.0	4	3.04
3.0	2	2.44
3.0	4	2.00

From the cross-validation results, it appeared that reasonably good and robust performance was obtained with the settings $\phi = 2, \sigma_\beta = 2$. We thus ran our treed model search algorithm using all the data with these settings. From a run of 2500 steps, we selected a "best" tree, which is given in Figure 2. This tree splits first on `Opening`, and subsequently on `PadType` in one node. Since the GLM with interactions involving `Opening` was an improvement over the main effects GLM, it is not surprising that this variable was split upon. Note however, that not all interactions with `Opening` are fit by this tree, since two categories (`Opening = middle, large`) are kept together. The subsequent split on PadType is suggestive of a three-way interaction between `Opening, PadType` and other variables.

In each terminal node a separate Poisson regression model is fit to the data. The degrees of freedom used by each model varies from one node to the next because some predictors are constant within nodes. This is obvious for variables used as splitting rules, such as `Opening` in Node 1, and the experimental design also eliminates some categorical variable levels in some of the terminal nodes.

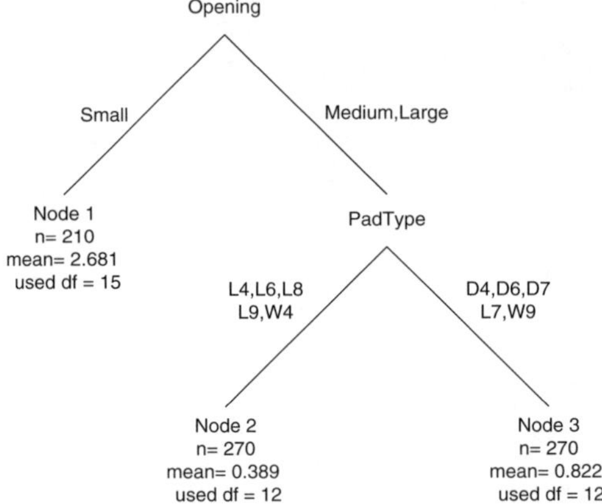

Figure 2. *The best tree found with $\phi = 2, \sigma_\beta = 2$.*

The coefficients of the GLMs in each node are plotted in Figure 3. Although the actual degrees of freedom used in each node varies, the informative prior makes it possible to calculate posterior means for all 19 regression coefficients, even when some predictors are constant in terminal nodes. This somewhat restricts interpretation of this plot, since within a node, some estimates are aliased with others. We can see however, that the effect of solder thickness is large when the opening is small and near zero otherwise, an interaction noted in the original analysis. We see also that in Node 3 (`Openings=Med/Large`, `PadType=D4,D6,D7,L7,W9`) coefficients for `L7` and `W9` are especially large. Mean numbers of `skips` and sample sizes for a partitioning of the data into four groups are given in Table 2.

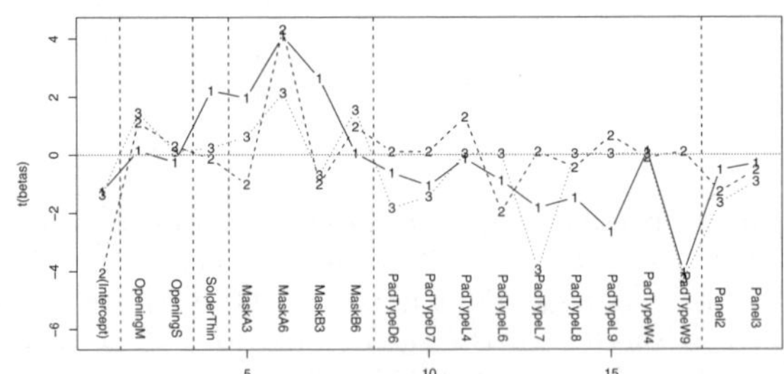

Figure 3. *Posterior mean regression coefficients for the best tree (tree given in Figure 2). Coefficients for each node are joined with a line, and the numeric plotting symbols correspond to the node numbers given in Figure 2.*

Table 2. *Summaries of the response for subsets of the data indicated by particular values of the predictors*

Group	Mean skips	n
Opening = S	2.6810	210
Opening = M,L, PadType = L7	0.0185	54
Opening = M,L, PadType = W9	0.0000	54
Opening = M,L, PadType ≠ L7,W9	0.7546	432
Total	1.1867	750

Table 2 indicates a very low rate of skips for PadType = L7,W9 and M or L levels of Opening. Evidently, the split on PadType was chosen because of quite different mean levels, but within nodes 2 and 3, there are also differences by PadType. The differing coefficients in nodes 2 and 3 for Mask suggest that this may be a secondary reason for the choice of this particular split.

5.2. *Simulation Study of the Null Case*

A potential criticism of any flexible model is that it finds complicated structure when there is none. In the solder2 example, out-of-sample validation indicates that complex structure actually is present. But how will the Bayesian search for treed models perform when the true model is a single GLM?

To study this null case, we simulated data from a single GLM, using the same predictor values as in the solder2 dataset. Response values were simulated using regression coefficients similar to the MLEs from a single GLM fit to the original data. An overdispersion component was incorporated into some of the simulated data sets by adding a random effect to η_{ij} before generating the observed response. We generated 60 data sets, 20 with no overdispersion, 20 with moderate overdispersion, and 20 with severe overdispersion. For each of these data sets, we ran our procedure with the same settings as in the previous section, except with four settings of $\phi = 1, D/2, D, 2D$ where D was the observed mean deviance of a Poisson GLM fit to the data. We considered these choices of ϕ to explore the effect of calibrating ϕ to the data.

In the interest of brevity, we give but a precis of our findings, which were very favorable. First of all, we were most interested to see how often our approach incorrectly partitioned the data by using a tree with more than one node. For data simulated from a single GLM with no overdispersion, the selected trees had a mean size of just over 1 for $\phi = 1, D, 2D$. With moderate overdispersion, average tree size was smallest at just over 1 when $\phi = 2D$ and around 2 when $\phi = D$. With severe overdispersion, average tree size was smallest, between 2 and 3, when $\phi = 2D$. In terms of fit to the data, our treed models were very competitive with a single GLM fit to the data, usually achieving a similar value for out-of-sample deviance. It is interesting to note that in most cases when the tree size was larger than 1, the treed model fits were not dramatically worse than those of a single GLM. Overfitting only became a problem when excessively small ϕ values were used.

REFERENCES

Ahn, H. and Chen, J. J. (1997). Tree-structured logistic model for over-dispersed binomial data with application to modelling developmental effects. *Biometrics* **53**, 435–455.

Alexander, W. P. and Grimshaw, S. D. (1996). Treed regression. *J. Comput. Graph. Statist.* **5**, 156–175.

Bartlett, M. (1957). A comment on D. V. Lindley's statistical paradox. *Biometrika* **44**, 533–534.

Breiman, L. (2001). Statistical modeling: the two cultures. *Statist. Sci.* **16** , 199–231 (with discussion).

Breiman, L., Friedman, J. Olshen, R. and Stone, C. (1984). *Classification and Regression Trees*. Pacific Drove, CA: Wadsworth.

Chipman, H. A., George, E. I. and McCulloch, R. E. (1998). Bayesian CART model search. *J. Am. Statist. Ass.* **93**, 935–960 (with discussion).

Chipman, H., George, E. I. and McCulloch, R. E. (2002). Bayesian treed models. *Machine Learning* **48**, 299–320.

Chambers, J. M. and Hastie, T. J. (eds) (1992) *Statistical Models in S*. Boca Raton: CRC Press.

Chaudhuri, P., Huang, M.-C., Loh, W.-Y., and Yao, R. (1994). Piecewise-polynomial regression trees. *Statist. Sinica* **4**, 143–167.

Chaudhuri, P., Lo, W.-D., Loh, W.-Y., and Yang, C.-C. (1995). Generalized regression trees. *Statist. Sinica* **5**, 641–666.

Clark, L. A. and Pregibon, D. (1992). Tree-Based Models. *Statistical Models in S*. (J. M. Chambers, and T. J. Hastie, eds.) Boca Raton: CRC Press, 377–420.

Comizzoli, R. B, Landwehr, J. M., and Sinclair, J. D. (1990). Robust materials and processes: key to reliability. *AT&T Tech. J.* **69**, 113–128.

Denison, D., Mallick, B. and Smith, A. F. M. (1998). A Bayesian CART algorithm. *Biometrika* **85**, 363–377.

Dey, D. K., Ghosh, S. K. and Mallick, B. K. (2000). *Generalized Linear Models: A Bayesian Perspective*. New York: Marcel Dekker.

Efron, B. (1986). Double exponential families and their use in generalized linear regression. *J. Am. Statist. Ass.* **81**, 709-721.

Gelfand, A. E. and Dalal, S. (1990). A note on overdispersed exponential families. *Biometrika* **77**, 55–64.

Hawkins, D. M. and Kass, G. V. (1982). Automatic interaction detection. *Topics in Applied Multivariate Analysis*. (D. M. Hawkins, ed). Cambridge: Cambridge University Press.

Jorgensen, B. (1987). Exponential dispersion models. *J. R. Statist. Soc. B* **49**, 127–162 (with discussion).

Jordan, M. I. and Jacobs, R. A. (1994). Hierarchical mixtures of experts and the EM algorithm. *Neural Comput.* **6**, 181–214.

Karalič, A. (1992). Employing linear regression in regression tree leaves. *Proceedings of ECAI-92*. Chichester: Wiley, 440–441.

Lutsko, J. F. and Kuijpers, B. (1994). Simulated annealing in the construction of near-optimal decision trees. *Selecting Models from Data: AI and Statistics IV* (P. Cheeseman and R. W. Oldford, eds). Berlin: Springer, 453–462.

McCullagh, P. and Nelder, J. A. (1989). *Generalized Linear Models*. London: Chapman and Hall.

Morgan, J. N. , and Sonquist, J. A. (1963). Problems in the analysis of survey data, and a proposal. *J. Am. Statist. Ass.* **58**, 415–434.

Nelder, J. A. and Wedderburn, R. W. M. (1972). Generalized linear models. *J. R. Statist. Soc. A* **135**, 370–384.

Quinlan, J. R. (1986). Induction of decision trees, *Machine Learning* **1**, 81–106.

Quinlan, J. R. (1992). Learning with continuous classes, in *Proceedings of the 5th Australian Joint Conference on Artificial Intelligence*. World Scientific, 343–348.

Raftery, A. E., Madigan, D. and Volinsky, C. T. (1996). Accounting for model uncertainty in survival analysis improves predictive performance. *Bayesian Statistics 5* (J. M. Bernardo, J. O. Berger, A. P. Dawid and A. F. M. Smith, eds). Oxford: Oxford University Press, 323–349.

Therneau, T. M. and Atkinson, E. J. (1997). An introduction to recursive partitioning using the RPART routines, *Tech. Rep.*, Mayo Clinic, Rochester, USA. `http://www.mayo.edu/hsr/techrpt/61.pdf`

Tierney, L. and Kadane, J. B. (1986). Accurate approximations for posterior moments and marginal densities. *J. Am. Statist. Ass.* **81**, 82–86.

Torgo, L. (1997). Functional models for regression tree leaves. *Proceedings of the International Machine Learning Conference (ICML-97)*. San Mateo, CA: Morgan Kaufmann, 385–393.

DISCUSSION

CHRISTOPHER M. BISHOP (*Microsoft Research, UK*)

In this stimulating paper, the authors have successfully exploited Markov chain Monte Carlo methods to explore the space of graphs for CART-like trees in which the terminal nodes represent generalized linear models (GLMs). Integration over the parameters of the terminal

GLMs, in order to compute the marginal likelihood (probability of data given the model) for the MCMC search, is accomplished using the Laplace approximation. Hyperparameters (such as those governing the GLM parameters) are either set by hand or fixed after a brief empirical search.

The underlying CART model on which this approach is based, however, suffers from some significant limitations, namely (i) the splits are axis-aligned, that is dependent on only one of the input variables at a time (this limitation is removed in some other variants of CART), (ii) the splits are binary, (iii) the splits are hard, so that each region of input space is associated with one, and only one, leaf node.

An alternative tree-based model, which avoids these limitations, is the *Hierarchical Mixtures of Experts* (HME) proposed by Jordan and Jacobs (1994). Each non-terminal node in the tree, called a *gating* node, corresponds to a multi-way indicator variable $Z = \{Z_1, \ldots, Z_M\}$, where $Z_i \in \{0, 1\}$ and $\sum_i Z_i = 1$. The conditional distribution of Z is given by a normalized exponential, or *softmax*, function

$$P(Z_i = 1 \mid V, X) = \frac{\exp(V_i^t X)}{\sum_{j=1}^{M} \exp(V_j^t X)}$$

where X is the vector of explanatory, or input, variables, and $\{V_i\}$ is a set of parameter vectors governing the orientation and steepness of the gating function.

In the case of binary splits this is equivalent to a single binary indicator variable Z with $P(Z = 1 \mid V, X) = \sigma(V^t X)$, where $\sigma(X) \equiv 1/(1+\exp(-X))$ is the logistic sigmoid function. For a given X, the probability of selecting a particular terminal node is obtained by multiplying together all of the conditional probabilities along the unique path from the root node to the terminal node. Thus each point of the input space is assigned probabilistically to each of the terminal nodes through a partition of unity.

The terminal nodes for the HME model follow those of Chipman *et al.*, namely softmax for multi-way classification, Gaussian for regression and so on. Jordan and Jacobs (1994) proposed an efficient EM algorithm for setting the model parameters of the HME, once the architecture of the tree has been prescribed.

It would be straightforward to use the MCMC approach of Chipman *et al.* to explore the tree structure of the HME model. While this could again be accomplished using a Laplace approximation to determine the likelihood score for each graph, a more appealing approach is to use variational methods (Jordan *et al.*, 1999) to marginalize over the model parameters.

Variational methods optimize an analytical approximation to the posterior distribution by maximization of a lower bound on the log marginal likelihood (in contrast to the Laplace approximation that simply fits the second-order moments at a mode of the distribution). The variational posterior distribution is chosen to have some factorization property with respect to the hidden variables, but is otherwise unconstrained.

The application of variational methods to the HME model is complicated by the fact that the softmax gating function does not lie within the conjugate exponential family. Previous attempts to apply variational methods to the HME have either resorted to mode fitting to circumvent this problem (Waterhouse *et al.*, 1996) thereby losing the appealing property of the lower bound, or else have modelled the joint distribution of input and output variables (Ueda and Ghahramani, 2002), which may be wasteful of resources and data particularly if the input space has high dimensionality.

Recently, Jaakkola and Jordan (2000) have developed a variational bound for the logistic sigmoid function which takes the form

$$\sigma(x) \geq F(x, \xi) \equiv \sigma(\xi) \exp\left\{(x - \xi)/2 - \lambda(\xi)(x^2 - \xi^2)\right\}$$

where $\lambda(\xi) = \tanh(\xi/2)/(4\xi)$, and ξ is a variational parameter. For any given value of x we can make this bound exact by an appropriate choice of the variational parameter ξ, namely $\xi = x$. (In fact the bound is exact at both $x = \xi$ and $x = -\xi$.) The bound is illustrated in Figure 1(a) for the case of $\xi = 2$, where the solid curve shows the logistic sigmoid function $\sigma(x)$, and the dashed curve shows the lower bound $F(x, \xi)$.

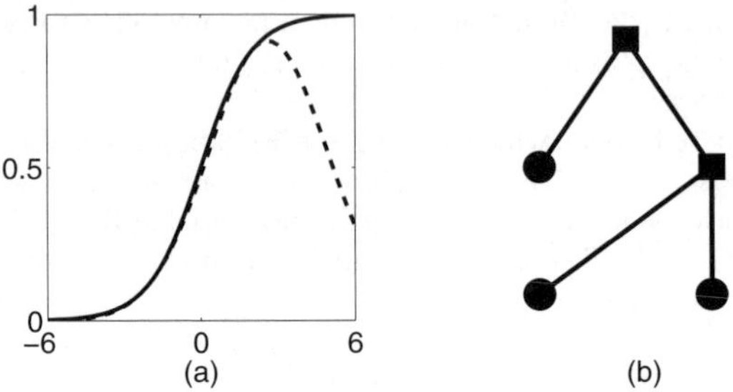

Figure 4. *(a) Logistic sigmoid function and variational bound. (b) Example HME model.*

We can use this bound to develop a fully Bayesian variational treatment of the HME model, for the case of binary gating nodes, which optimizes a rigorous lower bound on the marginal likelihood (Bishop, 2002). Since the bound transforms the logistic sigmoid into the exponential of a quadratic form over V, we can obtain a conjugate model by using Gaussian priors $V \sim \mathcal{N}(\mu, \Sigma)$, with conjugate hyperpriors for μ and Σ. Note that, for each gating node, there is a separate variational parameter ξ_n for each observation n.

Optimization of the ξ parameters is achieved by maximizing the lower bound on the marginal likelihood, leading to the re-estimation equations

$$\xi_n^2 = X_n^t \langle VV^t \rangle X_n$$

where $\langle \cdot \rangle$ denotes an average with respect to the variational posterior distribution. Re-estimation of ξ_n is interleaved with re-estimation of the factors in the variational posterior.

Unfortunately, the variational bound given by $F(x, \xi)$ does not extend to multi-way gating nodes governed by softmax functions. However, a complex, multi-way division of the input space can always be represented using binary splits provided the tree structure is sufficiently rich.

We illlustrate this approach using the simple HME model shown in Figure 4(b). Here, the two square nodes denote logistic sigmoid gating functions, and the three circular terminal nodes correspond to Gaussian conditional distributions over the output variable. In Figure 5(a) we show a simple data set with one input and one output variable, together with the means of the variational posterior distributions over the terminal node variables V. The corresponding outputs of the gating nodes are shown by the two curves in Figure 5(b). Finally, in Figure 5(c) we show the conditional distribution $P(Y \mid X)$ of the output variable given the input variable, as a function of the input variable, i.e. each vertical slice through this plot represents $P(Y \mid X)$ for the given value of X.

A key feature to note is that this conditional distribution is multi-modal. This is possible because the gating node outputs are smooth functions of the input variable. Such multi-modality could *not* be captured in a CART model, since it assigns each point of the input space to one, and only one, of the terminal nodes ("hard" splits).

We have illustrated the variational HME using a regression example with Gaussian terminal nodes. It is straightforward to apply this approach to two-way classification problems for a

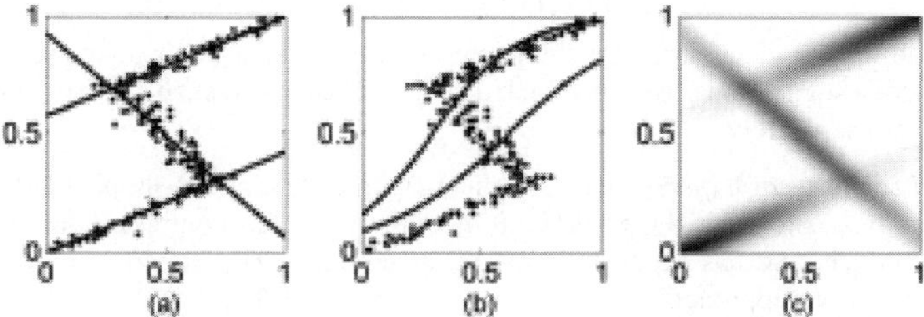

Figure 5. *(a) Means of terminal nodes, (b) means of gating nodes, (c) conditional distribution.*

model with logistic sigmoid terminal nodes simply by making further use of the variational bound $F(x, \xi)$.

It is also straightforward to evaluate the lower bound on the log marginal likelihood under the variational posterior distribution. This could readily be used in the Markov chain Monte Carlo scheme of Chipman *et al.* to sample from the posterior distribution over graphs, using opertors to add and remove nodes as before.

The use of MCMC methods to explore the space of graphs in tree structured classification and regression models is clearly a rich area for research. The paper of Chipman *et al.* provides some important tools for tackling such problems, as well as motivating a number of future research directions.

AKI VEHTARI (*Helsinki University of Technology, Finland*)

My comments concern the use of cross-validation for model comparison. Authors have used five-times-repeated 10-fold-CV (Algorithm 1) for model comparison, but they do not mention how the results in Figure 1 are used to tell whether the difference between two models is significant. Five repetitions in Algorithm 1 produce samples from the distribution in which the variability comes from the internal randomness in the algorithm (including such things as how sensitive the approach is to initial values and properties of the stochastic search algorithm) and the variability due to different random divisions in the 10-fold-CV. If the goal is to estimate which model would make best predictions for new data, it is important also to take into account the uncertainty from not knowing the distribution of the future data. Vehtari and Lampinen (2002) describe how to obtain samples from the distribution of the cross-validation estimated expected utility (e.g., deviance) estimate taking properly into account the uncertainty from the internal randomness in the algorithm, the variability due to different random divisions in the k-fold-CV, *and* the approximation of the future data distribution. Vehtari and Lampinen (2002) discuss and demonstrate that the uncertainty from the approximation of the future data distribution dominates and the other uncertainties are small (at least for stable models and algorithms). Using the obtained samples from the distributions of the expected utilities, models can be compared in Bayesian way by computing the probability of one model having a better expected utility than some other model (Vehtari and Lampinen, 2002). Also, when using k-fold-CV it is useful to use correction term. Since for each fold $1/k$ of the data is left out, the expected utility is estimated conditioning only on $1 - 1/k$ of the data and thus uncorrected k-fold-CV provides biased estimate of expected utility conditioned on the full data. This can be corrected using less well known first-order correction proposed by Burman (1989) and demonstrated for Bayesian models by Vehtari and Lampinen (2002). This correction is important in model assessment but also in model comparison if the models compared have different steepness of the learning curves.

REPLY TO THE DISCUSSION

We thank both of the discussants for their interest in our paper, their insightful remarks and their constructive suggestions. Because each discussant has raised such different issues, we will respond to them separately.

Hierarchical Mixtures of Experts. We very much appreciate Chris Bishop's kind remarks and his general enthusiasm about Bayesian treed modelling. He has contrasted our approach with the HME approach, and has suggested how the potential of HME models may be enhanced using elements of our approach.

As Bishop points out, the HME approach is a powerful tree-based modelling approach that differs from our approach in several ways: (i) we only split along coordinate axes (ii) we restrict attention to binary splits (iii) we use hard splits rather than soft splits. However, the first two differences are really not essential. To allow for splits along other directions, one can simply introduce new variables of the form $U^t X$, where X is the vector of explanatory variables and U is a vector of fixed constants. One could go further and treat U as a set of unknown parameters to be updated as part of our MCMC algorithm. Of course, the potential of such enhancements would have to be weighed against the costs in complexity of computation and interpretability. The restriction to binary splits is also not essential, since any multiway split can be recast as a sequence of binary splits.

What does fundamentally distinguish the HME approach from ours is the third difference above, the use of soft versus hard splits. Under an HME model, the distribution of each data point is modelled by a mixture of simple models, where the mixture weights are determined by the predictor values through the soft splitting functions. Under our model, the distribution of each data point is modelled by a single simple model that is determined by the predictor values through the hard splitting functions. Because of this difference, the HME approach can capture multimodal phenomena such as that which Bishop illustrates in Figure 5. But our approach cannot unless multimodal distributions are used at each terminal node.

Compared to our approach, the HME model provides a much richer structure for describing the data. But this richer structure comes at a cost in computational flexibility. Because each given HME model seems to require substantial computation for learning the parameters, the tree structure for each HME has been traditionally fixed in advance. This contrasts with our approach which is fundamentally about learning the tree structure. If MCMC methods coupled with new variational approximations, as Bishop suggests, could be used to enable automatic learning of tree structures for HMEs, that would be a powerful innovation which might greatly increase its flexibility.

The richer HME structure also comes at a cost in interpretability. Our approach is geared towards searching for simple local structure. When it exists, exposing such a simple structure can lead to a better understanding of the nature of the variation in the data. The HME model is instead geared towards describing complex variation with a rich mixture model. Although the HME approach might find simple structure, for example by finding close to hard splits, it probably would not unless the data provided overwheming evidence in that direction. In our opinion, to find a simple structure one needs to put more prior mass on simplicity.

When it does not find a simple structure, the HME approach will be more useful for prediction than for explanation. A straightforward way of enhancing our approach for improved prediction would be through model averaging. This could be done by averaging a subset of the stochastic search output with respect to the conditional posterior tree probabilities. Such a mixture of treed models would almost certainly yield improved predictions, but at a cost in interpretability. Interestingly, Bishop's insights suggest that there may be another route to enhancing our approach, namely by replacing our hard splitting rules with soft splitting rules such as the softmax gating functions. Although the implementation details of such an approach still need to be worked out, such soft Bayesian treed models could yield alternative mixture models that provide a smoother description of the variation across terminal nodes.

Statistical comparison of cross-validated results. Aki Vehtari provides a variety of interesting suggestions for ways in which cross-validation can be enhanced for model selection. However, he also expresses concern about the statistical significance of the cross-validation comparisons which appear in Figure 1. Although we had hoped that the visual comparisons would be sufficient to convince readers of the potential of our approach, we went ahead and performed a more careful classical ANOVA analysis.

For the mean deviance analysis, we compared each of the procedures in Figure 1 with the leftmost linear Poisson GLM. At the 0.05 level of significance and using a conservative Bonferroni adjustment, the performance of the rightmost four models were all significantly better than the linear Poisson GLM. For the median deviance analysis, we compared each of the procedures with the linear plus interaction Poisson GLM, since this model appeared to be the closest competitor to ours. At the 0.05 level of significance, again using a Bonferroni adjustment, the performance of both $\phi = 1$ models were significantly better, both $\phi = 2$ models were statistically equivalent and both $\phi = 3$ models were significantly worse. Every model was significantly better than the linear Poisson GLM. As we suspected, all of these ANOVA results conformed to our visual conclusions from Figure 1.

ADDITIONAL REFERENCES IN THE DISCUSSION

Bishop, C. M. (2002). Bayesian hierarchical mixtures of experts: A variational treatment. (in preparation).

Burman, P. (1989). A comparative study of ordinary cross-validation, v-fold cross-validation and the repeated learning-testing methods. *Biometrika* **76**, 503–514.

Ueda, N. and Ghahramani, Z. (2002). Bayesian model search for mixture models based on optimizing variational bounds. *Neural Networks* (to appear).

Jaakkola, T. and Jordan, M. I. (2000). Bayesian Parameter Estimation via Variational Methods. *Statist. Computing* **10**, 25–37.

Jordan, M. I., Ghahramani, Z., Jaakkola, T. S. and Saul, L. K. (1999). An introduction to variational methods for graphical models. *Learning in Graphical Models* (M. I. Jordan, ed). Cambridge, MA: MIT Press.

Vehtari, A. and Lampinen, J. (2002). Bayesian model assessment and comparison using cross-validation predictive densities. *Neural Comput.* **14** (to appear).

Waterhouse, S., MacKay, D. J. C. and Robinson, T. (1996). Bayesian methods for mixtures of experts. *Advances in Neural Information Processing Systems* **8**. Cambridge, MA: MIT Press, 351–357.

BAYESIAN STATISTICS 7, pp. 105–124
J. M. Bernardo, M. J. Bayarri, J. O. Berger, A. P. Dawid,
D. Heckerman, A. F. M. Smith and M. West (Eds.)

Bayesian Harmonic Models for Musical Signal Analysis

MANUEL DAVY
IRCCyN/CNRS, France
manuel.davy@irccyn.ec-nantes.fr

SIMON J. GODSILL
University of Cambridge, UK
sjg@eng.cam.ac.uk

SUMMARY

This paper is concerned with the Bayesian analysis of musical signals. The ultimate aim is to use Bayesian hierarchical structures in order to infer quantities at the highest level, including such things as musical pitch, dynamics, timbre, instrument identity, etc. Analysis of real musical signals is complicated by many things, including the presence of transient sounds, noises, and the complex structure of musical pitches in the frequency domain. The problem is truly Bayesian in that there is a wealth of (often subjective) prior knowledge about how musical signals are constructed, which can be exploited in order to achieve more accurate inference about the musical structure. Here we propose developments to an earlier Bayesian model that describes each component "note" at a given time in terms of a fundamental frequency, partials ("harmonics"), and amplitude. This basic model is modified for greater realism to include non-white residuals, time-varying amplitudes and partials "detuned" from the natural linear relationship. The unknown parameters of the new model are simulated using a variable dimension MCMC algorithm, leading to a highly sophisticated analysis tool. We discuss how the models and algorithms can be applied for feature extraction, polyphonic music transcription, source separation, and restoration of musical sources.

Keywords: MUSICAL ANALYSIS; AUTOMATIC PITCH TRANSCRIPTION; PITCH ESTIMATION; INSTRUMENT CLASSIFICATION; AUDITORY SCENE ANALYSIS.

1. INTRODUCTION

Inference about the high-level information contained in musical audio signals is complex, and requires sophisticated signal processing tools (Bregman, 1990). In this paper, we focus on the automatic interpretation of musical signals. Musical audio is highly structured, both in the time domain and in the frequency domain. In the time domain, tempo specifies the range of possible note transition rates. In the frequency domain, two levels of structure can be considered. First, each note is composed of a fundamental frequency (related to the "pitch" of the note), and partials whose relative amplitudes determine the timbre of the note. (This frequency domain description can be regarded as an empirical approximation to the true process, which is in reality a complex non-linear time-domain system; (McIntyre *et al.*, 1993; Fletcher and Rossing, 1998.) The frequencies of the partials are approximately integer multiples of the fundamental frequency. Second, several notes played at the same time form chords or polyphony. The fundamental frequencies of each note comprising a chord are typically related by simple multiplicative rules.

For example, a C major chord may be composed of the frequencies 523 Hz, 659 Hz $\approx 5/4\times523$ Hz and 785 Hz $\approx 3/2\times523$ Hz. An additional level of structure is the melody, which gives the frequency dependence of successive notes. A given melody is characterized by the succession of fundamental frequencies at specific time instants. Figure 1 shows a flute extract in waveform. Figure 2 shows a spectrogram analysis for a simple monophonic (single note) flute recording (this may be auditioned at `www-sigproc.eng.cam.ac.uk$jg/sounds/flute.wav`) In these both the temporal segmentation and the frequency domain structure are clearly visible on the plot. In polyphonic musical examples, several (possibly many!) such structures are superimposed, and the eye is then typically unable to separate the individual note structures from the spectrogram alone.

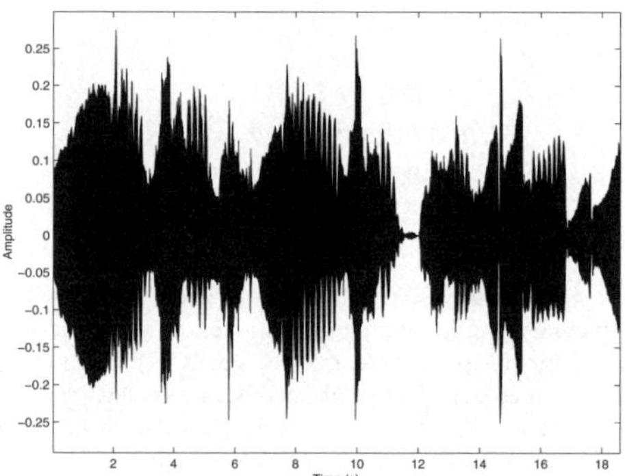

Figure 1. *Waveform: flute extract.*

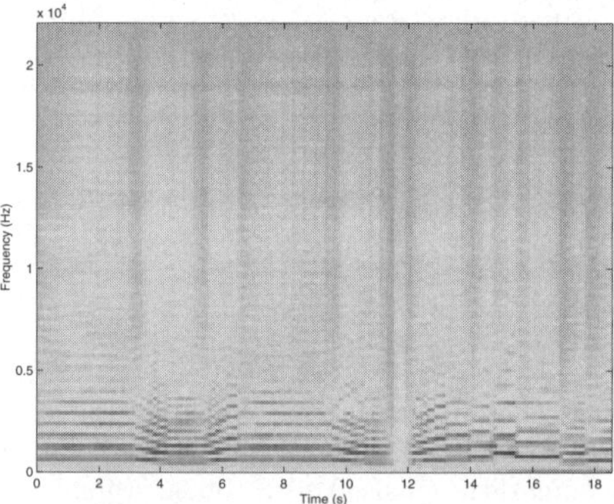

Figure 2. *Spectrogram: flute extract.*

The goals of a musical analysis can be manifold, and we seek to make our models general enough that they will fit with the inference requirements at hand. Some important goals and applications, which can all be expressed in probabilistic terms through the use of suitably chosen model structures, include automatic transcription (generation of a musical "score"), classification and search (*e.g.*, determining which instruments are playing or whether a particular tune is present), and source separation (separation of individual instruments from a polyphonic mix).

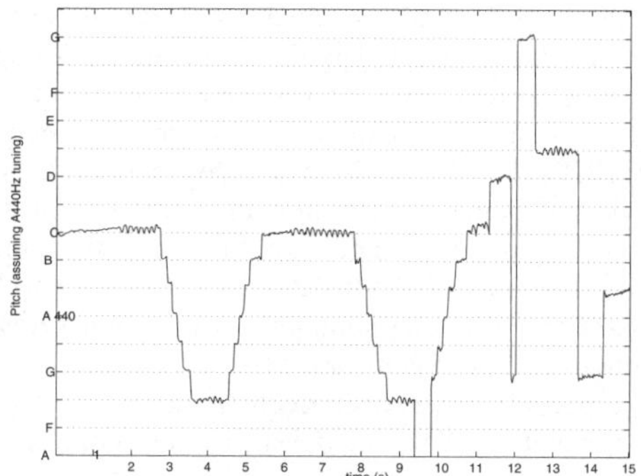

Figure 3. *Pitch estimation from flute extract.*

It should be clear from this discussion that musical audio provides an ideal structure for Bayesian modelling and inference: there is plenty of prior information available (both subjective and physically-based), there is a natural hierarchical structure (from individual partial frequencies through notes, chords and eventually whole melodies). All of these elements of the structure should ideally be estimated jointly in order to exploit the full power of the information available. This is a formidable task that none has successfully achieved to our knowledge. Most researchers have focused either on very high level, or very low level, modelling alone. Here we attempt partially to bridge the gap between these extremes, by exploring models which directly model musical signals in terms of their component "notes", while retaining a moderately realistic signal level model. Of course, in future work we would wish to see the whole task performed jointly, and we will expect to see dramatic performance improvements once this is properly achieved.

At the level of musical notes, two principal tasks may be identified for the analysis of musical audio: a segmentation step that identifies note transitions in time, and an estimation step in which the number of notes as well as their fundamental frequencies, their partial structure and other characteristics are estimated at any given time. We focus on the latter, since efficient music segmentation algorithms such as time-frequency (Laurent and Doncarli, 1998), support vector machines (Davy and Godsill, 2002b) or generalized likelihood ratio (Basseville and Nikiforov, 1993) techniques can be used for this step.

Numerous musical pitch estimation and analysis techniques can be found in the literature. Most apply only to monophonic (single note) recordings and rely on non-parametric signal analysis tools (local autocorrelation function, spectrogram, etc.). We do not have space to reference all approaches here. Certain authors have, however, adopted methods with a statistical modelling flavor, often using iterative procedures to estimate the individual components of a musical signal; see, for example, De Cheveigne (1993), De Cheveigne and Kahawara (1999), and Virtanen and Klapuri (2001). Bayesian approaches have been surprisingly rare, considering the large quantities of prior information available about musical signals. Notable exceptions include Kashino *et al.* (1995), Kashino and Murase (1999), who adopt a Bayesian hierarchical structure for high-level features in music such as chords, notes, timbre, etc., and Sterian *et al.* (1999), who adopt Bayesian tracking ideas for modelling of time-varying frequency partials. Bayesian models for polyphonic music have been proposed in Walmsley *et al.* (1998, 1999) and it is these that we extend and discuss here.

In this paper, we devise novel Bayesian models for periodic, or nearly periodic, components in a musical signal. The work develops upon models devised for automatic pitch transcription in

Walmsley *et al.* (1998, 1999) in which it is assumed that each musical note may be described by a fundamental frequency and linearly related partials with unknown amplitudes. The number of notes, and also the number of harmonics for each note are generally unknown and so a reversible jump MCMC procedure is adopted for inference in this variable dimension probability space; see Godsill and Rayner (1998a, 1998b), Andrieu and Doucet (1999), and Davy *et al.* (2002) for some relevant MCMC work in signal processing and audio. Use of these powerful inference methods allows estimation of pitch, harmonic amplitudes, and the number of notes/harmonics present at each time. The methods of Walmsley *et al.* (1998, 1999) have shown promise in highly complex problems with many notes simultaneously present. However, in the presence of non-stationary or ambiguous data, problems are expected in terms of large residual modelling errors and pitch errors (especially errors of $\pm$ one octave). Here we seek to address some of these shortcomings by elaboration of the model to include more flexibility in the modelling of non-stationary data, modelling of non-white residual noise, and also to allow the modelling of inharmonicity (or "detuning" of individual harmonics relative to the usual linear frequency spacing). As before, a variable dimension MCMC strategy is adopted for inference in the new model, and novel proposal mechanisms are developed for this purpose.

The paper is organized as follows. In Section 2, we present the basic harmonic model for the description of musical signals. Moreover, we specify the probabilistic framework, and give the parameter priors. In Section 3, we discuss estimation objectives and summarize the Bayesian computational method. Simulation results are presented in Section 4, and finally a discussion is given. Given space restrictions it has been impossible to describe in detail the prior models and exact MCMC implementation scheme used. In fact, we have implemented several different versions of the models and MCMC algorithms, and the code is still undergoing development as more sophisticated and realistic modelling assumptions are incorporated. A snapshot detailing the implementation that generated the simulation results for this paper can be found in Davy and Godsill (2002a).

2. BAYESIAN MODEL FOR MUSICAL ANALYSIS

Consider for the moment a short-time frame of musical audio data, denoted $y(\tau)$, in which note transitions do not occur. This would correspond, for example, to the analysis of a single musical chord. Throughout, we assume that the continuous time audio waveform $y(\tau)$ has been discretized with a sampling frequency ω_s rad s^{-1}, so that discrete time observations are obtained as $y_t = y(2\pi t/\omega_s),\ t = 0, 1, 2, \ldots, N-1$. We assume that $y(\tau)$ is bandlimited to $\omega_s/2$ rad s^{-1}, or equivalently that it has been prefiltered with an ideal low-pass filter having cut-off frequency $\omega_s/2$ rad s^{-1}.

In this section, we describe Bayesian models suited to musical audio analysis. We first introduce a robust monophonic (single note) model for music. We then explain how to expand it to a polyphonic (many note) model, and discuss the salient features of the approach. It is assumed throughout that the musical audio has been segmented such that no note transitions occur for $t \in \{0, 1, 2, ..., N-1\}$.

2.1. *Monophonic Case: Single Note Models*

In the *monophonic* case it is assumed that at any given time only one single musical pitch is sounding, for examples, solo trumpet, solo clarinet, etc. From this simple case we can build more sophisticated *polyphonic* (many note) structures by superposition of several monophonic units.

Physical considerations and empirical observation of spectrograms (see, *e.g.*, Figure 2—it is clear from this that there is a series of spectral "lines" corresponding to the fundamental and partials of each note) lead to the conclusion that the notes in musical signals are composed of a *fundamental frequency* ω_0 and a set of M *partials*. This classic model, see *e.g.*, Serra (1997), results from the approximate short-term periodicity of musical signals, which enables a Fourier series decomposition. In the simplest cases, this idealized assumption holds well and the following model can be applied for short time segments (as in Walmsley *et al.* , 1998, 1999):

$$y_t = \left\{ \sum_{m=1}^{M} \alpha_m \cos(m\omega_0 t) + \beta_m \sin(m\omega_0 t) \right\} + v_t \tag{1}$$

for $t \in \{0, \ldots, N-1\}$. Here, $M > 0$ is the number of partials present, α_m and β_m give the amplitudes of these partials, and v_t is a residual noise component. Note that $\omega_0 \in (0, \pi)$ is here scaled for convenience—its audible frequency is $\omega_0 \omega_s / (2\pi)$.

It turns out that this model is over-idealized for many realistic cases and must be modified in several ways to improve performance. In particular, partials can be expected to exhibit time-varying amplitudes, and the partials can be expected to deviate from the ideal frequency spacing. These two facts can be accommodated in a new model:

$$y_t = \left\{ \sum_{m=1}^{M} \alpha_{m,t} \cos\left[(m + \delta_m)\omega_0 t\right] + \beta_{m,t} \sin\left[(m + \delta_m)\omega_0 t\right] \right\} + v_t \tag{2}$$

where the partial amplitudes $\alpha_{m,t}$ and $\beta_{m,t}$ can now depend on time, and de-tuning parameters δ_m allow each partial to be offset from its nominal frequency of $m\omega_0$.

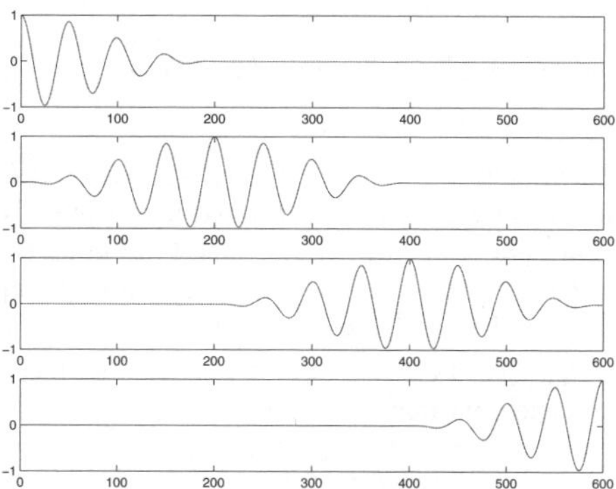

Figure 4. *Typical set of Gabor atoms $\phi_{i,t}\cos(.)$.*

Many evolution models are possible for the amplitude processes $\alpha_{m,t}$ and $\beta_{m,t}$, including random walks, autoregressions, etc., and most would be tractable within our Bayesian framework. It is important, however, to regularize the evolution of these components *a priori* in order that ambiguities between true frequency modelling and modelling of the time-varying amplitudes do not occur. We adopt a simple solution which consists of representing the amplitudes $\alpha_{m,t}$ and $\beta_{m,t}$ in terms of smooth basis functions ϕ_i, $i = 1, \ldots, I$ (with I fixed and known) such that

$$\alpha_{m,t} = \sum_{i=0}^{I} a_{m,i}\phi_{i,t}; \qquad \beta_{m,t} = \sum_{i=0}^{I} b_{m,i}\phi_{i,t} \tag{3}$$

There are many possible choices for the basis functions, and in practice any sufficiently smooth interpolation functions will do. The number of basis functions $(I+1)$ has to be upper bounded to avoid unidentifiability: low frequencies can actually be modelled by a time-varying amplitude as well as by a sinusoid. It is thus important to limit the number of basis functions such that $\omega_{\text{ampmax}} = (I+1)\omega_s/N$ is below the lowest note frequency in a given harmonic signal (*e.g.*, 20 Hz). Here we have implemented a simple scheme involving raised cosine functions (Hanning windows) with 50% overlap, see Figure 4. Since I is typically chosen much smaller than N, the reparametrization in terms of basis coefficients $a_{m,i}$ and $a_{m,i}$ is of much lower dimensionality than the original formulation in terms of $\alpha_{m,t}$ and $\beta_{m,t}$. The monophonic note model now becomes:

$$y_t = \left\{ \sum_{m=1}^{M} \sum_{i=0}^{I} a_{m,i}\phi_{i,t} \cos\left[(m+\delta_m)\omega_0 t\right] + b_{m,i}\phi_{i,t} \sin\left[(m+\delta_m)\omega_0 t\right] \right\} + v_t \quad (4)$$

We note that the model can now be seen as a representation of y_t in terms of a set of *Gabor atoms* (Flandrin, 1999). Here, each atom has a precise time-frequency location $(t_i, [m+\delta_m]\omega_0)$ and an amplitude $(a_{m,t}^2 + b_{m,t}^2)^{1/2}$ where t_i is the temporal center of ϕ_i. We will refer to each term $\phi_{i,t}\cos[\omega t]$ or $\phi_{i,t}\sin[\omega t]$ as an "atom" in the subsequent discussion.

2.2. *A Polyphonic Harmonic Model*

The above monophonic model can be easily expanded to the polyphonic case; that is, for signals composed of K concurrent notes. A suitable model is:

$$y_t = \left\{ \sum_{k=1}^{K} \sum_{m=1}^{M_k} \sum_{i=0}^{I} a_{k,m,i}\phi_{i,t} \cos\left[(m+\delta_{k,m})\omega_{0,k} t\right] + b_{k,m,i}\phi_{i,t} \sin\left[(m+\delta_{k,m})\omega_{0,k} t\right] \right\} + v_t \quad (5)$$

for $t = 0, \ldots, N-1$. Each note, $k = 1, \ldots, K$, now has its own set of parameters, and the notation is extended in an obvious way with an additional subscript k. The (unknown) parameters determining the polyphonic model of Eq. (5) are: the total number of notes K, the number of partials for each note $\boldsymbol{M} = [M_1, \ldots, M_K]^t$, the de-tuning parameters $\boldsymbol{\delta} = [\boldsymbol{\delta}_1, \ldots, \boldsymbol{\delta}_K]$ with $\boldsymbol{\delta}_k = [\delta_{k,1}, \ldots, \delta_{k,M_k}]^t$, the fundamental frequencies $\boldsymbol{\omega}_0 = [\omega_{0,1}, \ldots, \omega_{0,K}]^t$ and the amplitudes $\boldsymbol{\theta} = [a_{1,1,0}, b_{1,1,0}, \ldots, a_{K,M_K,I}, b_{K,M_K,I}]^t$. It is assumed that I is prespecified. In vector notation the model is now written as:

$$\boldsymbol{y} = \boldsymbol{D}\boldsymbol{\theta} + \boldsymbol{v} \quad (6)$$

where $\boldsymbol{y} = [y_0, \ldots, y_{N-1}]^t$, $\boldsymbol{v} = [v_0, \ldots, v_{N-1}]^t$, the matrix $\boldsymbol{D}$ contains the Gabor atoms stacked in columns, and $\boldsymbol{\theta}$ contains the amplitude parameters $a_{k,m,i}$ and $b_{k,m,i}$. The detailed expressions for $\boldsymbol{D}$ and $\boldsymbol{\theta}$ are given in Davy and Godsill (2002a).

In practice, musical signals also have non-harmonic components, such as emitted air sounds or aspiration noise. These components, in addition to any background noise, are subsumed in the noise term v_t in Equation (5). It is desirable that the noise v_t models accurately all the possible sources of model errors. A simple and general possibility is the *autoregressive* (AR) model of order p:

$$v_t = \gamma_1 v_{t-1} + \cdots + \gamma_p v_{t-p} + \epsilon_t \quad (7)$$

where ϵ_t is a zero mean Gaussian white noise of variance σ_ϵ^2. This introduces an additional set of parameters σ_ϵ^2, p and $\boldsymbol{\gamma} = [\gamma_1, \ldots, \gamma_p]^t$.

Given the linear model formulation of Eq. (5) and the assumption of i.i.d. Gaussian excitation for the AR process, we immediately obtain the likelihood function:

$$p(\boldsymbol{y} \mid \boldsymbol{\theta}, \boldsymbol{\omega}_0, \boldsymbol{\delta}, \boldsymbol{M}, K, \boldsymbol{\gamma}, \sigma_\epsilon^2) = \frac{1}{(2\pi\sigma_\epsilon^2)^{(N-p)/2}} \exp\left[-\frac{1}{2\sigma_\epsilon^2}(\boldsymbol{y} - \boldsymbol{D\theta})^t \boldsymbol{A}^t \boldsymbol{A}(\boldsymbol{y} - \boldsymbol{D\theta})\right] \quad (8)$$

The $(N-p) \times N$-dimensional matrix $\boldsymbol{A}$ is constructed by stacking the AR coefficients in rows, with appropriate zero padding, as detailed in Davy and Godsill (2002a).

2.3. *Features of the Model*

The models proposed in earlier subsections, both monophonic and polyphonic, all fall into the general category of the linear model. Under a Gaussian prior for $\boldsymbol{\theta}$ this will facilitate inference in the model by allowing exact simulation of $\boldsymbol{\theta}$ from its full conditional, and marginalization of $\boldsymbol{\theta}$ from the posterior distribution. This can provide an important dimensionality reduction in the model since $\boldsymbol{\theta}$ is often high-dimensional. This, however, is standard material, and we see the main interest of our model to be in the specific structure chosen for the $\boldsymbol{D}$-matrix and in the prior distributions of the unknown parameters, both of which are carefully tuned to subjective and objective information about musical signals.

The polyphonic model presented as Equation (5) has several features that distinguish it from other work in the area. First, it directly incorporates the frequency relationship between partials and fundamental frequencies. This is different from typical approaches in the literature to musical pitch transcription that estimate the frequencies of each line spectral component independently of the others, performing grouping into units such as chords and notes as a post-processing stage (but note that Gribonval and Bacry, 2001, goes some way towards integrating the harmonic structure directly into the model). This misses an opportunity for greater estimation accuracy through direct modelling of the waveform at the level of notes. We retain this feature in our model, based on the earlier model of Equation (1) (Walmsley *et al.,* 1998, 1999). An extension provided in the model of Eq. (5) is the incorporation of detuned harmonics with parameters $\delta_{k,m}$. Many researchers believe this to be an important component of any realistic musical instrument model (Fletcher and Rossing, 1998) and we plan to investigate this claim using the new model. Potential ambiguities can occur if the $\delta_{k,m}$ parameters allow one partial (harmonic) to stray into the frequency range of adjacent harmonics. However, this ambiguity is suppressed here by careful choice of priors that favor $\mid \delta_{k,m} \mid << 1$. Note that this model, which essentially includes random deviations from the natural harmonic positions, could easily be extended to model the more systematic trends in spacing of harmonics observed in, *e.g.*, string instruments (Fletcher and Rossing, 1998).

Secondly, instruments produce notes with time varying amplitude. This time variation is present throughout an individual note, but particularly evident around the start-up region, or *attack*. A constant amplitude model such as Eq. (1) is clearly unable to deal with such a case and will lead to misleading parameter inferences. The chosen decomposition of amplitudes in terms of a set of basis functions, Eq. (3), ensures smoothness and reduces the number of parameters in the model considerably.

Finally, the residual noise is an AR process that can model residual noise from instruments as well as general background noise.

Further improvements in modelling could be achieved by allowing the fundamental frequencies ω_0 to vary over the time-frame—however, this leads to a much more intractable model that we have avoided implementing thus far—provided frame sizes are kept short, the frequencies can usually be modelled as constant within a frame. See Walmsley *et al.* (1999) for some progress on models with time-varying frequency.

Even without time-varying frequencies the price of this more flexible model is a large number of unknown parameters. The principal unknowns are, then, the number of notes K, the fundamental frequency, number of harmonics and amplitudes for each note: $\{\omega_{0,k}, M_k, a_{k,m,i}, b_{k,m,i}\}$, the AR parameters γ and the variance σ_ϵ^2. This variable-dimension space of parameters is embedded in a Bayesian scheme as outlined in the next section. To our knowledge, this problem has never received a fully Bayesian treatment before. As will be seen, the Bayesian priors, in addition to the special structure of $\boldsymbol{D}$, play a key role in defining the model.

2.4. *Bayesian Model*

A natural hierarchical prior structure for the musical model is as follows:

$$\begin{aligned} &p(\boldsymbol{\theta}, \boldsymbol{\omega}_0, \boldsymbol{\delta}, \boldsymbol{M}, K, \gamma, \chi^2, \sigma_\epsilon^2) \\ &\quad = p(\boldsymbol{\theta} \mid \boldsymbol{\omega}_0, \boldsymbol{\delta}, \boldsymbol{M}, K, \sigma_\epsilon^2) p(\boldsymbol{\delta} \mid \boldsymbol{\omega}_0, \boldsymbol{M}, K)\, p(\boldsymbol{\omega}_0 \mid \boldsymbol{M}, K) p(\boldsymbol{M} \mid K)\, p(K)\, p(\gamma)\, p(\chi^2)\, p(\sigma_\epsilon^2) \end{aligned}$$

where χ is introduced later. The form of the prior distributions can be chosen to reflect prior beliefs about particular types of music, or particular instruments, and this is certainly an interesting line of future study. Here we adopt a generic approach in which the priors are designed to match the average character of musical notes. We consider the prior distributions one by one.

Prior for $\boldsymbol{\theta}$. The amplitudes of the partials determine the characteristic "timbre" of a musical note. Hence it is important to model these accurately in applications such as source separation and musical instrument classification. We adopt a zero-mean Gaussian prior for these parameters. This matches well the variability observed when the same note is played under slightly different conditions or on different instruments. It is also reasonable to assume the scale of the amplitudes is related to the scale of the AR residual process, since the AR residual models principally non-harmonic noises produced by the instruments. Thus we adopt a zero-mean Gaussian prior with covariance matrix $\sigma_\epsilon^2 \Sigma_\theta$. Choice of the matrix Σ_θ will then determine the properties of the prior. We have implemented a number of possibilities here, and it is clear that the prior would ideally be instrument-specific. However, it is possible to build in a certain amount of physical prior knowledge without limiting to very narrow classes of instrument. See, for example, Figure 5. This displays a single short-time Fourier magnitude spectrum for the flute extract in Figure 1. The spectrum is computed from approximately the first 0.25 s of the music where no note transitions occur. The fundamental frequency and partials are clearly visible, exhibiting a slow decay in amplitude with increasing frequency. This general observation applies to most acoustical sounds and so can be usefully incorporated into a prior. One successful implementation sets Σ_θ as diagonal with diagonal elements equal to χ^2/m^α, where m is the number of the partial, α is experimentally determined ($\alpha = 2$ is a good match to many signals we have analyzed), and χ^2 is an unknown scale parameter that is sampled in the MCMC scheme with an inverted gamma prior. This form of covariance matrix assumes joint prior independence of all notes, partials and atoms. While this functions well in practice, it may be argued that dependence should be modelled between partials and atoms within a particular note. There are many ways this could be achieved and we leave this as a future topic of research. As an alternative to these physically based priors, we have also implemented with some success the well known G-prior that has been found to be effective in similar contexts (Andrieu and Doucet, 1999; Walmsley *et al.,* 1999)). A full investigation of the relative merits of these choices is again left as a topic of future work.

Prior for $\boldsymbol{M}$ *and* K. The number of partials is an instrument- and realization-specific quantity. The precise distribution can be learned from training examples with different instruments. The general feature is that a particular instrument has a mean number of partials with a certain spread about this value. In order to model this, we have adopted an independent Poisson prior for each M_k, truncated to user-specified maximum and minimum limits. Similarly, the number of notes, K, has a truncated Poisson prior. These are specified vaguely for general musical extracts, but can be tuned more precisely when a particular instrument is known to be present.

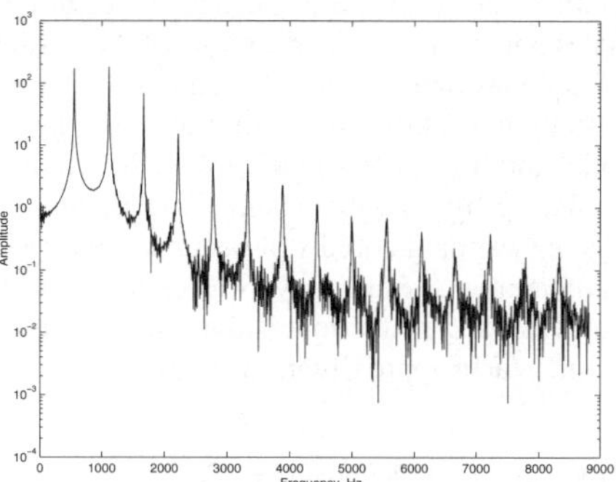

Figure 5. *Flute extract.*

Prior for $\boldsymbol{\delta}$. A third key parameter is the vector of detuning factors, $\boldsymbol{\delta}$, which is aimed at modelling slight harmonic de-tuning among the partials. Its value is expected to be close to zero, and here we assume no dependence between the δ parameters of different partials, although for certain instruments theory would dictate the general form of the δs (Fletcher and Rossing (1998)). Here a zero mean independent Gaussian is assumed, with variance σ_δ^2, fixed to a small value in order to favor small de-tuning parameters. The distribution can additionally be truncated in order that adjacent partials within a single note do not cross over one another in the frequency domain.

Prior for $\boldsymbol{\omega}_0$. For some instruments such as piano or organ, notes are tuned to a fixed grid of frequencies (one for each key of the instrument). In other instruments, the player will usually tune notes to be close to the fixed grid of note frequencies, see *e.g.*, the pitch transcription of the flute extract in Figure 3. A convenient and informative prior would thus favor these fixed frequencies above others. Moreover, when several notes are played at the same time (a chord), there exist simple relations between the fundamental frequencies. The prior density $p(\boldsymbol{\omega}_0 \mid \boldsymbol{M}, K)$ should reflect this prior information. However, these considerations only apply to tonal Western music, and for the sake of generality we here adopt a uniform prior over some region of interest, $\boldsymbol{\omega}_0 \in [\omega_{min}, \omega_{max}]^K$. Again, there are plenty of interesting possibilities for informative priors in future investigations.

Remaining parameters. Briefly, regarding the noise v_t, $p(\sigma_\epsilon^2)$ is inverted gamma with parameters α_ϵ and β_ϵ, $p(\boldsymbol{\gamma})$ is zero-mean Gaussian, with diagonal covariance matrix $\boldsymbol{I}_p$. p is fixed in our implementations thus far, but we note that standard samplers are readily available should this parameter become important (Troughton and Godsill, 1998, 2001; Godsill, 2001; Vermaak *et al.*, 2002).

Posterior distribution. Given the above prior structure, the posterior distribution may be computed. In particular, $\boldsymbol{\theta}$ and σ_ϵ are marginalized using standard linear model computations (Bernardo and Smith, 1994; West and Harrison, 1997) to give a reduced posterior $p(\boldsymbol{\omega}_0, \boldsymbol{\delta}, \boldsymbol{M}, K, \boldsymbol{\gamma}, \sigma_\epsilon^2, \chi \mid \boldsymbol{y})$. Full conditionals are also readily available for $\boldsymbol{\theta}$, σ_ϵ, $\boldsymbol{\gamma}$ and χ, owing to the conjugate prior structures chosen. These distributions are all employed in the variable dimension MCMC algorithms for computation, as summarized in the next section.

3. BAYESIAN COMPUTATIONS

The precise objectives of inference in these models is very much application driven. In the case of pitch estimation, for example, point estimates will typically be required for the fundamental frequencies ω_0. The posterior distribution of these parameters will also give a useful estimate of the uncertainty and multimodality in the posterior. In source separation and instrument classification tasks, estimates and posterior distributions of the partial amplitudes, the detuning parameters, fundamental frequency and number of harmonics are required for each note. There are, however, potential pitfalls in making these estimates, as the individual note ordering is not constrained in our model and it is quite conceivable that two notes are swapped during the MCMC simulation. An ordering by, say, increasing fundamental frequency or energy would help somewhat, although note swapping could still occur when large jumps in parameter values are made (*e.g.*, a change in fundamental frequency by an octave or more). We have avoided these ambiguities in this work by estimating functionals that do not depend upon the note labelling. These are computed as Monte Carlo approximations to posterior means:

$$I(f) = \int_\Omega f(\Phi)\, p(d\Phi \,|\, \boldsymbol{y}) \approx \frac{1}{L}\sum_{l=1}^{L} f(\widetilde{\Phi}^{(l)}) \tag{9}$$

where $\Phi = \{\boldsymbol{\theta}, \boldsymbol{\omega}_0, \boldsymbol{\delta}, \boldsymbol{M}, K, \gamma, \chi^2, \sigma_\epsilon^2\}$ is the collection of all unknowns in the model, $f(.)$ is a given integrable function with respect to the posterior, and Ω is the sample space for the posterior distribution. (The integral $I(f)$ over the discrete parameters should be seen as a discrete sum over all the possible values of these discrete parameters.) $\widetilde{\Phi}^{(l)}$ are (possibly dependent) Monte Carlo samples drawn from $p(\Phi \,|\, \boldsymbol{y})$.

A suitable family of functions to estimate are spectrogram-like representations. The short-time energy spectrum of the ith component in the model Eq. (5) is defined as

$$g_{i,\boldsymbol{\theta},\boldsymbol{\omega}_0,\boldsymbol{\delta},K}(\omega) = \Big|\sum_{k=1}^{K}\sum_{m=1}^{\boldsymbol{M}_k} a_{k,m,i}\Phi_{i,(m+\delta_{k,m})\omega_{0,k}}(\omega) + b_{k,m,i}\Phi_{i,(m+\delta_{k,m})\omega_{0,k}-\frac{\pi}{2}}(\omega)\Big|^2 \tag{10}$$

where $\Phi_{i,\omega'}(\omega)$ is the Fourier transform of $\phi_{i,t}\cos(\omega' t)$

The construction of the spectrogram-like representation consists in computing for $i = 0, \ldots, I$ and various values of ω in $[0, \pi]$

$$\widehat{g}_{i,\boldsymbol{\theta},\boldsymbol{\omega}_0,\boldsymbol{\delta},K}(\omega) = \frac{1}{L}\sum_{l=1}^{L} g_{i,\widetilde{\boldsymbol{\theta}}^{(l)},\widetilde{\boldsymbol{\omega}}_0^{(l)},\widetilde{\boldsymbol{\delta}}^{(l)},\widetilde{K}^{(l)}}(\omega) \tag{11}$$

As stated above, these computations require that a set of samples

$$\{\widetilde{\boldsymbol{\theta}}^{(l)}, \widetilde{\boldsymbol{\omega}}_0^{(l)}, \widetilde{\boldsymbol{\delta}}^{(l)}, \widetilde{\boldsymbol{M}}^{(l)}, \widetilde{K}^{(l)}, \widetilde{\gamma}^{(l)}, \widetilde{\sigma}_\epsilon^{2\,(l)}\}, \qquad l = 1, \ldots, L,$$

is available from the posterior $p(\boldsymbol{\theta}, \boldsymbol{\omega}_0, \boldsymbol{\delta}, \boldsymbol{M}, K, \gamma, \sigma_\epsilon^2 \,|\, \boldsymbol{y})$. The reversible jump MCMC algorithm for achieving this is summarized in the following paragraphs.

The simulation algorithm is a variable dimension MCMC procedure, using reversible jumps to achieve model space moves for both the number of harmonics in each note and the number of notes. Other parameters are updated using Metropolis-within Gibbs sampling moves for non-standard conditionals and Gibbs sampling where the conditionals are standard. The posterior distribution in this problem is highly multimodal and strongly peaked. This is partly as a result of ambiguities inherent in the model, and implies that the MCMC algorithms have to be carefully constructed in order to avoid getting stuck in local "traps" of the distribution. In fact, much of the innovative work in this project has been concerned with generation of effective proposal distributions for fast exploration of the parameter space. As well as standard random walk moves, these include independence proposals based upon the sample autocorrelation function and spectrum of the data, and octave/fifth-jumping moves aimed at moving rapidly between local maxima (these moves update the number of partials as well, preserving the spectral structure, and this leads to improved acceptance rates). The reversible jump proposals allow for several partials to be added/removed at once from a note, and notes may be split and merged in meaningful ways. Full details can be found in Davy and Godsill (2002a).

4. SIMULATION RESULTS

Results are presented based on two implementations. The first is a full implementation of the models as described in the paper with detuned partials and unknown numbers of partials M_k and number of notes K. The full details of this sampler can be found in Davy and Godsill (2002a). This first sampler is used to demonstrate the effectiveness of the model in analyzing short data sets containing isolated notes and chords. The second sampler is a reduced version of the first, in which the partials are not detuned (i.e. $\boldsymbol{\delta}$ is fixed to zero), and the number of notes K is specified *a priori* (but M_k is sampled using a reversible jump procedure). As a result, though less robust and general, the second sampler takes far less computational time, both per iteration and in terms of number of iterations to convergence. It is thus used for a rapid frame-based analysis of long data sets containing many different pitches and note transitions. In this way we hope to demonstrate both the potential of the full model and also the possibilities of realistic analysis for long monophonic and polyphonic musical extracts.

4.1. *Full Sampler*

The first example is a two-note mixture of a saxophone and trumpet, playing with fundamental frequencies 349 Hz (F) and 523 Hz (C), assuming A440 Hz tuning. This short extract is taken from example "Commit" at 2.5 s, see Section 4.2. Note that the number of notes, as well as the number of partials for each notes were unspecified for the MCMC simulations.

After fitting the model with MCMC to the data, the fitting error is almost perfect when viewed in the time domain. Figure 6 displays the spectra of the two estimated notes as well as the spectrum of the error signal. As can be seen, both the number of notes and the number of partials were accurately estimated (note, however, that the 10th partial of the note F was missed, but this had no major consequence for the inference). In addition, the fundamental frequencies are correctly estimated.

We have also plotted the spectrogram-like representation of the estimated notes, see Figure 7. As can be seen, the notes are very concentrated around the "true frequencies" 349 and 523 Hz, which shows that the posterior distribution is well concentrated around the true frequency values. Moreover, the spectrogram of the original data has also been computed (for equal comparison, we used $\phi_{i,t}$ as windowing function with 50% overlap). The frequencies and amplitudes of the line components in the two representations are very similar, which demonstrates again the accuracy of the approach. An audio animation of the MCMC procedure during convergence can be listened to at `http://www-sigproc.eng.cam.ac.uk/~md283/harmonic.html`.

4.2. *Reduced Sampler*

The reduced sampler is applied frame-wise to the data, as in the restoration processing of Godsill and Rayner (1998a). The waveform was arbitrarily segmented into blocks of duration

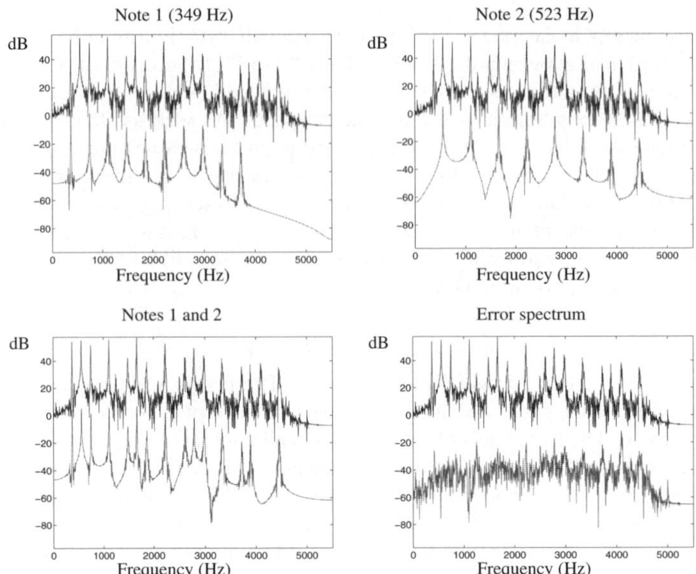

Figure 6. *Comparison of the original spectrum with MCMC-estimated spectra. In order to improve clarity, the original spectrum (top graph in each panel) has been translated by adding 50 dB to its amplitude in the four panels. (Top row) Estimated spectra of the individual notes. (Bottom Left) Spectrum of notes 1 and 2 superimposed. (Bottom Right) Spectrum of the residual error. Commit extract.*

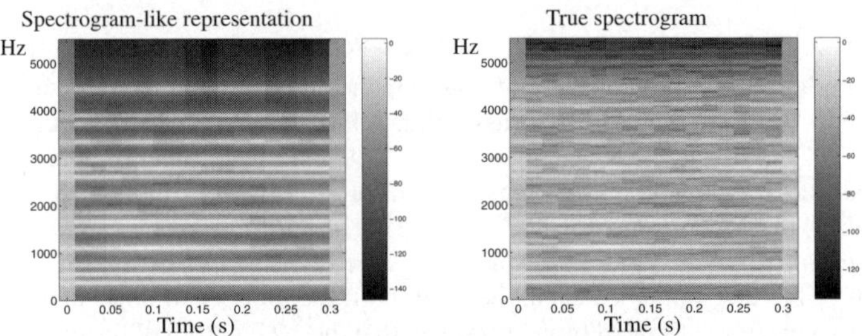

Figure 7. *(Left) Spectrogram-like representation inferred from the MCMC samples using Eq. (11). (Right) Spectrogram of the original time series, computed using the windows $\phi_{i,t}$.*

0.1 s with 50% overlap and the reduced MCMC sampler applied in turn to each block. First a short solo flute extract considered here is the opening of Debussy's Syrinx, downsampled to a 22,050 Hz sampling rate. This is monophonic throughout and hence processed with $K = 1$ throughout. The pitch estimates obtained are shown in Figure 3, corresponding to the waveform and spectrogram in Figures 1 and 2. Estimated pitches are plotted logarithmically with grid lines showing semitone steps relative to A440 Hz. The estimated pitch corresponds exactly to a manual transcription of the recording with the exception of the brief low G around 12 s. Close listening around 12 s shows that the flute plays a low distortion undertone in addition to the scored pitch at this point, and the algorithm is clearly modelling this undertone. The "drop-out" between 9 s and 10 s corresponds to a short period of silence. Informal examination of spectrograms indicates that the reversible jump algorithm for determining the number of harmonics is very successful. This demonstrates the high reliability and accuracy of the models for monophonic pitch estimation.

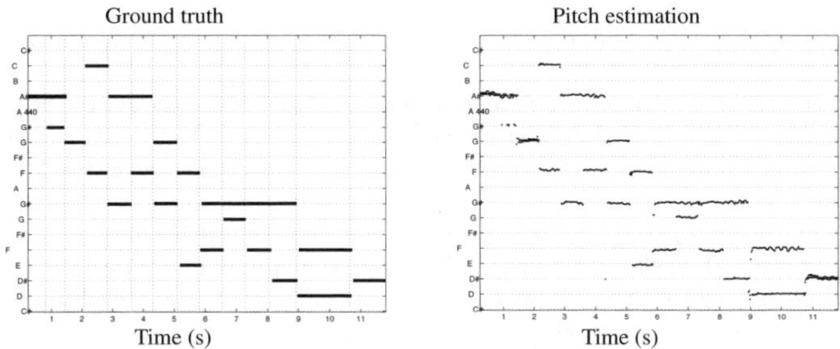

Figure 8. *Ground truth (manual transcription) and pitch extraction for Commit extract, assuming A440 Hz tuning.*

Next the example "Commit" (see full sampler above) is processed completely. There are two notes playing throughout, so $K = 2$ is used. Figure 8 displays the pitch estimation results for the extract. Note that the number of notes is known, and was provided for inference. Comparison with ground truth shows the pitch estimation accuracy in the presence of polyphonic music.

5. DISCUSSION

We have presented a small selection of results here. The methods have in fact been tested on a range of real audio material and found to be robust provided that there are no more than three notes playing simultaneously, in which case ambiguities can cause errors in the fundamental frequency estimation. This result is similar to those reported by other authors using other techniques, such as Kashino *et al.* (1995) and Virtanen and Klapuri (2001). However, these other methods integrate contextual information or more specific instrument-based knowledge into the processing, while we have specified prior distributions at a very generic level and have not integrated temporal information from surrounding data or the relationships that exist between notes at a particular time. These aspects can all be encoded within a Bayesian framework and we anticipate that future incorporation of ideas such as these into our problem will lead to significant enhancements in performance. Computation remains a concern, however, as the distributions involved are highly multimodal and intractable to more efficient analysis.

ACKNOWLEDGEMENTS

The work of both authors was partially supported by the EU project Models for Multimedia Information Retrieval. We would also like to thank Steve Hainsworth, Prof. Peter Rayner, Patrick Wolfe and Dr. Kunio Kashino for helpful discussions relating to this paper.

REFERENCES

Andrieu, C. and Doucet, A. (1999). Joint Bayesian detection and estimation of noisy sinusoids via reversible jump MCMC. *IEEE Trans. Signal Process.* **47**, 2667–2676.

Basseville, M. and Nikiforov, I. (1993). *Detection of Abrupt Changes: Theory and Application*. Englewood Cliffs, NJ: Prentice-Hall.

Bernardo, J. M. and Smith, A. F. M. (1994). *Bayesian Theory*. Chichester: Wiley

Bregman, A. (1990). *Auditory Scene Analysis*. Cambridge, MA: MIT Press.

Davy, M., Doncarli, C. and Tourneret, J. Y. (2002). Classification of chirp signals using hierarchical Bayesian learning and MCMC methods. *IEEE Trans. Signal Process.* **50**, 377–388.

Davy, M. and Godsill, S. (2002a). Bayesian harmonic models for musical pitch estimation and analysis. *Tech. Rep.*, University of Cambridge, UK.

Davy, M. and Godsill, S. (2002b). Detection of abrupt spectral changes using support vector machines. *Proc. IEEE ICASSP.*

De Cheveigne, A. (1993). Separation of concurrent harmonic sounds: Fundamental frequency estimation and a time-domain cancellation model for auditory processing. *J. Acoust. Soc. Am.* **93**, 3271–3290.

De Cheveigne, A. and Kawahara, H. (1999). Multiple period estimation and pitch perception model. *Speech Commun.* **27**, 175–185.

Flandrin, P. (1999). *Time-Frequency/Time-Scale Analysis*. New York: Academic Press

Fletcher, N. and Rossing, T. (1998). *The Physics of Musical Instruments*. Berlin: Springer.

Godsill, S. J. (2001). On the relationship between Markov chain Monte Carlo methods for model uncertainty. *J. Comput. Graph. Statist.* **10**, 230–248.

Godsill, S. J. and Rayner, P. J. W. (1998a). *Digital Audio Restoration: A Statistical Model-Based Approach*. Berlin: Springer

Godsill, S. J. and Rayner, P. J. W. (1998b). Robust reconstruction and analysis of autoregressive signals in impulsive noise using the Gibbs sampler. *IEEE Trans. Speech Audio Process.* **6**, 352–372.

Gribonval, R. and Bacry, E. (2001). Harmonic decomposition of audio signals with matching pursuit. *Tech. Rep.*, IRISA-INRIA.

Kashino, K. and Murase, H. (1999). A sound source identification system for ensemble music based on template adaptation and music stream extraction. *Speech Commun.* **27**, 337–349.

Kashino, K., Nakadai, K., Kinoshita, T. and Tanaka, H. (1995). Organisation of hierarchical perceptual sounds: Music scene analysis with autonomous processing modules and a quantitative information integration mechanism. *Proc. 14th Int. Joint Conf. on Artificial Intelligence*.

Laurent, H. and Doncarli, C. (1998). Stationarity index for abrupt changes detection in the time-frequency plane. *IEEE Signal Process. Lett.* **5**, 43–45.

McIntyre, M., Schumacher, R. and Woodhouse, J. (1993). On the oscillations of musical instruments. *J. Acoust. Soc. Am.* **74**, 1325–1345.

Serra, X. (1997). Musical sound modeling with sinusoids plus noise. *Musical Signal Processing* (C. Roads, S. T. Pope, A. Piccialli and G. De Poli, eds). Swets and Zeitlinger.

Sterian, A., Simoni, M. H. and Wakefield, G. H. (1999). Model-based musical transcription. *Proc. Int. Computer Music Conference*.

Troughton, P. and Godsill, S. J. (1998). A reversible jump sampler for autoregressive time series. *Proc. IEEE ICASSP.*

Troughton, P. and Godsill, S. J. (2001). MCMC methods for restoration of nonlinearly distorted autoregressive signals. *Signal Process.* **81**, 83–97.

Vermaak, J., Andrieu, C., Doucet, A. and Godsill, S. J. (2002). Bayesian model selection of autoregressive processes. *J. Time Ser. Anal.* (to appear).

Virtanen, T. and Klapuri, A. (2001). Separation of harmonic sounds using multipitch analysis and iterative parameter estimation. *Proc. IEEE Workshop on Audio and Acoustics*. Mohonk, NY.

Walmsley, P., Godsill, S. J. and Rayner, P. J. W. (1998). Multidimensional optimisation of harmonic signals. *Proc. EUSIPCO.*

Walmsley, P., Godsill, S. J. and Rayner, P. J. W. (1999). Polyphonic pitch tracking using joint Bayesian estimation of multiple frame parameters. *Proc. IEEE Workshop on Audio and Acoustics*. Mohonk, NY.

West, M. and Harrison, J. (1997). *Bayesian Forecasting and Dynamic Models*. New York: Springer

DISCUSSION

RAQUEL PRADO (*University of California, Santa Cruz, USA*)

I would like to begin by congratulating the authors for presenting a wonderful application of Bayesian hierarchical modelling to the challenging problem of analyzing musical signals. The superposition of different components (notes) that characterize musical signals and the enormous amount of prior information available, makes this problem an ideal setting for applying the Bayesian machinery. Although the models developed in the paper are somehow limited in the sense that they do not really handle signals in which there are note transitions, or signals in which there are more than three notes played simultaneously, the results presented here are rather promising.

My comments, questions and concerns are mainly centered on two issues: the choice of the basis functions used to model the non-stationarities in the amplitudes of the signal and the structure of the noise component.

The model described in Section 2.2 has the form

$$y_t = S_t + v_t = \sum_{k=1}^{K} \sum_{m=1}^{M_k} S_{k,m,t} + v_t,$$

for $t = 0, \ldots, N-1$, with K the number of notes and M_k the number of harmonics per note. It is assumed that K and M_k are unknown and that the noise component is modelled as an autoregression, $v_t \sim \text{AR}(p)$, with AR parameters $\boldsymbol{\gamma} = (\gamma_1, \ldots, \gamma_p)'$ and p fixed. Each component $S_{k,m,t}$ in the signal is the contribution of the m-th partial of the fundamental frequency $w_{0,k}$, detuned via $\delta_{k,m}$, that is

$$S_{k,m,t} = \alpha_{k,m,t} \cos[(m + \delta_{m,k})w_{0,k}t] + \beta_{k,m,t} \sin[(m + \delta_{m,k})w_{0,k}t],$$

with $\alpha_{k,m,t}$ and $\beta_{k,m,t}$, the parameters that describe the behavior of the amplitude of the component over time. These amplitude parameters are written as a linear combination of smooth basis functions $\phi_{i,t}$

$$\alpha_{k,m,t} = \sum_{i=1}^{I} a_{k,m,i}\phi_{i,t}, \qquad \beta_{k,m,t} = \sum_{i=1}^{I} b_{k,m,i}\phi_{i,t}.$$

The authors use Hanning windows (raised cosine functions) with 50% overlap and I, the number of basis functions, fixed, known and such that $I << N$. I will now discuss the impact that the choice of the basis functions has on modelling the signals. The model can be written in DLM form (West and Harrison, 1997) as

$$\begin{aligned} y_t &= \mathbf{F}_t'\boldsymbol{\theta}_t + v_t \\ \boldsymbol{\theta}_t &= \mathbf{G}_t\boldsymbol{\theta}_{t-1} + \boldsymbol{\omega}_t \end{aligned}$$

with $\mathbf{G}_t = \mathbf{I}$, $\boldsymbol{\omega}_t = 0$ for all t and $\mathbf{F}_t'$ and $\boldsymbol{\theta}_t$ given by $\mathbf{F}_t' = (\phi_{1,t}\cos[(1+\delta_{1,1})w_{0,1}t],$ $\phi_{1,t}\sin[(1+\delta_{1,1})w_{0,1}t], \ldots, \phi_{I,t}\cos[(M_K+\delta_{K,M_K})w_{0,K}t], \phi_{I,t}\sin[(M_K+\delta_{K,M_K})w_{0,K}t])$, and $\boldsymbol{\theta}_t = \boldsymbol{\theta} = (a_{1,1,1}, b_{1,1,1}, \ldots, a_{K,M_K,I}, b_{K,M_K,I})'$. It is worth looking at the form of the forecast function in order to understand what kind of structure is imposed through the prior and what type of behavior in the signal can be captured by the model. I will focus on the second issue here. For a given value of $\boldsymbol{\theta}$, the expected development of the series into the future $l > 0$ steps ahead is $E(y_{t+l}|\boldsymbol{\theta}) = \mathbf{F}_{t+l}'\mathbf{G}_{t+l}\ldots\mathbf{G}_{t+1}\boldsymbol{\theta} = \mathbf{F}_{t+l}'\boldsymbol{\theta}$, while the forecast function l steps ahead is $f_t(l) = \mathbf{F}_{t+l}'E(\boldsymbol{\theta}|D_t)$, where D_t denotes all the information available up to time t. For example, if we consider a model in which there is only one note, one harmonic and no detuning, i.e. $K = 1$, $M = 1$, and $\delta = 0$, we have $E(y_{t+l}|\boldsymbol{\theta}) = \sum_{i=1}^{I} E_{i,t,l}$ with

$$E_{i,t,l} = a_i\phi_{(t+l),i}\cos[w_0(t+l)] + b_i\phi_{(t+l),i}\sin[w_0(t+l)],$$

or equivalently, $E_{i,t,l} = A_{i,t,l}\cos[w_0(t+l) + \gamma_{i,t,l}]$, with $A_{i,t,l}$ the *amplitude* and $\gamma_{i,t,l}$ the *phase* given by

$$A_{i,t,l} = \phi_{(t+l),i}\sqrt{a_i^2 + b_i^2}, \qquad \gamma_{i,t,l} = \arctan\left(-\frac{b_i}{a_i}\right).$$

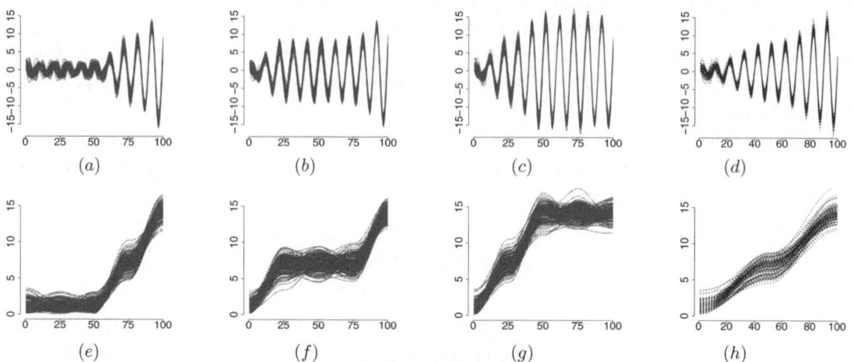

Figure 9. *The panels display the form of $E(y_l|\boldsymbol{\theta})$ (top panels) and $\sum_{i=1}^{I} A_{l,i}$ (bottom panels) using $I = 5$ (graphs (a), (b), (c), (e), (f) and (g)) and $I = 3$ (graphs (d) and (h)). In all the cases the amplitudes are modelled as increasing functions of time. Each picture displays 100 samples of $E(y_l|\boldsymbol{\theta})$ and $\sum_{i=1}^{I} A_{l,i}$.*

Figure 9 shows the form of $E(y_l|\boldsymbol{\theta})$ (graphs (a) – (d)) and $\sum_{i=1}^{I} A_{i,l}$ (graphs (e) – (h)) for $t = 0$, $l = 1, \ldots, 100$, $w_0 = \pi/10$, and two choices of I. Pictures (a), (b), (c), (e), (f) and (g) illustrate three different ways of modelling an increasing amplitude using $I = 5$ windows with 50% overlap. The pictures correspond to $\boldsymbol{\theta} \sim N(m, C)$ with $C = 1$ and $m = (0, 0, 0, 0, 0, 0, 5, 5, 10, 10)'$ for (a) and (e), $m = (0, 0, 5, 5, 5, 5, 5, 5, 10, 10)'$ for (b) and (f), while for (c) and (g) $m = (0, 0, 5, 5, 10, 10, 10, 10, 10, 10)'$ was used. The graphs (d) and (g) correspond to $I = 3$ windows, $C = 1$ and $m = (0, 0, 5, 5, 10, 10)'$. As it is illustrated with these pictures, the choice of the basis functions and the choice of the number of such functions are important modelling issues. Increasing I results in more flexible models—in the example, we clearly have more options to model an increase of the amplitude in time with $I = 5$ windows than with $I = 3$ windows—but also, this may result in models that are highly overparameterized. If the amplitudes change abruptly in time, then a large number of basis functions will be necessary to capture the non-stationarities, so it might be more appropriate to use basis functions that are not as smooth as the Hanning windows. If, on the other hand, the amplitudes change smoothly in time, a small I would be adequate. How do you choose I? Also, I suppose that there is no reason to use the same I for all the harmonics and all the fundamental frequencies. Have the authors considered other basis functions? Why is a 50% overlap between the basis functions a good choice?

The second point that I would like to discuss refers to the AR noise term v_t. How do you choose p, the order of the AR component? A related concern has to do with the structure of v_t. In previous work by Godsill and other authors, AR(p) models have been successfully used to describe audio signals (see for example Godsill and Rayner 1998a). I imagine that there are important identifiability issues related to which components are modelled by the signal S_t and which components are modelled by the noise v_t via the AR structure. An AR model whose characteristic polynomial has complex roots or "poles," will describe the periodicities in the series that are not explained by S_t. The authors use a zero-mean Gaussian prior on the AR parameters γ, with diagonal covariance matrix I_p. Therefore, no restrictions are imposed on the AR structure, and consequently, no restrictions are imposed on the frequency ranges that may be described by v_t. I believe this is a key issue that is not addressed in the paper.

Finally, I would like the authors to comment about the use of alternative parameterizations.

For example, parameterizing the model components in terms of $A_t \cos((m + \delta_m)w_0 t + \gamma_t)$ puts the focus on the amplitude and the phase and this may be more appropriate for incorporating subjective knowledge. Also, more realistic models should include a time-varying number of notes and harmonics per note. What are your thoughts on adapting your models to include such features without increasing too much the computational complexity, since this is already a major concern?

BRADLEY P. CARLIN (*University of Minnesota, USA*)

Congratulations to Dr. Godsill and Dr. Davy on a paper that was stimulating both in its content and in its non-traditional presentation! I would merely like to state my admiration for the work done so far, and encourage the authors to pursue the line of advanced modelling mentioned in Section 5. Specifically, the use of prior information regarding temporal patterns in pitch, tempo, and other aspects of the music seems likely to pay significant dividends. For example, in many jazz solos (especially those of beboppers like Charlie Parker), temporally adjacent notes are much more likely than not to be similar in pitch and length; indeed, the way lesser players (such as myself) "fake" such solos is to play only the most important notes, and "swallow" (half-play) the rest, according to the ascending or descending patterns they typically follow. In computer reconstruction of these solos, this suggests the use of adjacency priors, such as the conditionally autoregressive (CAR) priors (Besag *et al.*, 1991) often seen in spatial statistics, or perhaps actual mean "templates" reminiscent of those used in Bayesian image analysis (Phillips and Smith, 1994). Since such solos are notoriously difficult and laborious to transcribe, yet often feature musical structure we should be able to teach the computer, the authors' work and its extensions offer an important and valuable contribution to both art and science!

REPLY TO THE DISCUSSION

We would like to thank both discussants for their informed and useful comments on our work. Although the topic of automated musical transcription has received quite substantial research attention over several decades, results are as yet well below those that a trained human listener can achieve. We believe that Bayesian methods, with the possibility of including large amounts of expert musical prior knowledge, provide an excellent machinery for emulating and (in some cases) exceeding the performance of the human listener. At this level, however, the field is still in its infancy and has a great deal of ground to cover. In particular, as noted by Prado, the models proposed in our paper are still highly idealized and do not directly account for many features found in real musical signals. The principal reason for this over-idealized modelling is the computational cost associated with processing such large datasets, but a secondary reason is that we simply do not yet know what an appropriate general model for music would look like. Both of these factors can hopefully be overcome gradually over coming years as computational methods develop and our understanding of musical acoustics improves.

The first point raised by Prado is the choice of basis functions. In the most general form of our model, Eq. (2), a full time evolution is permitted for the amplitude parameters $\alpha_{m,t}$ and $\beta_{m,t}$. A suitably smooth dynamical model can then be set up which expresses the way these random amplitudes evolve over time. Such a model could then capture all possible amplitude fluctuations of a musical note, and Kalman filter/dynamic linear model type simulations could then be used to simulate from their conditional distributions in a (relatively) efficient manner. This would be our preferred approach given unlimited computational power. Here, however, we have substantially reduced the dimensionality of the problem, and hence the computational complexity, by expressing the amplitude variations in terms of a few smooth basis functions.

The choice of a Hanning window for these basis functions is largely arbitrary and could be replaced by any suitably smooth interpolation function.

Such an approach can be expected to capture the smoothness of a musical note in a parsimonious way. See, for example, the saxophone attack (the "attack" of a note is the initial start-up transient at the beginning of a note—we expect the most rapid amplitude changes during the attack) modelled in Figure 10. We can clearly see the effects of appropriate and inappropriate choice of basis functions. Using $I = 9$, i.e. a Hanning window with width 256 (bottom left), gives a near-perfect modelling (top right). However, using $I = 3$ and a Hanning window of 1024 width is completely inappropriate for the data. In this instance the fundamental frequency was correctly estimated in both cases, but in other instances, especially multi-pitch scenarios, this modelling problem can lead to pitch estimation errors. So, choice of basis functions is an important issue, as highlighted by Prado. Ideally, we would wish to choose the width and number of the basis functions automatically, so that rapid changes could be modelled around the start of a note (many narrow basis functions) and slow change in the steady state regime of a note (a few wide basis functions).

Such extensions, of course, fit elegantly into the Bayesian scheme and would be well worth pursuing in the future with multiscale representations based on continuous wavelets or multiresolution Gabor dictionaries, for example. One aspect of this that we have already tried successfully is to allow the number and width of the basis functions to depend on the fundamental frequency ω_0 (low notes have few long basis functions and *vice versa*). This is in keeping with the idea that low notes will naturally fluctuate less rapidly than high notes, and it also prevents some low-frequency over-modelling errors that sometimes occurred in our simulations. An alternative extension of the current framework includes the basis function width, or equivalently I as an unknown parameter of the system, to be integrated out from the posterior distribution. A reversible jump scheme is then implemented for the sampling of I, in much the same way as the number of harmonics M in the original scheme. Results from initial simulations are encouraging. A discrete set of possible window widths is specified *a priori*, $\{128, 180, 256, 361, 722, 1024, 1444, 2048\}$, and moves are randomly proposed to values within this set. A uniform prior is assumed for all candidate window widths. In the rapidly changing saxophone example illustrated below, for example, we obtain posterior probability of approximately 0.95 favoring a window width of 256, and 0.05 for a window width of 361. This agrees roughly with our previous simulations from the data in Figure 10. Conversely, a second data set in Figure 11 is from the "steady-state" region of the same note. Here the MCMC sampler favors a longer window width of 512, with probability 0.8, which again agrees with intuition for such a smooth signal.

In response to the remaining comments of Prado, we agree that there are potential ambiguities between the AR model and the sinusoidal basis components. We comment simply here that the prior structures we have implemented (scaling the sinusoidal magnitudes relative to the AR excitation variance) have not so far led to a case where harmonic components were falsely modelled by the AR residual. One way to guard against this happening, however, is to impose a stronger prior structure on the AR model to prevent such ambiguities from arising. A clear way to achieve this would be to assign zero probability mass to polynomial roots in some neighborhood of the unit circle. This could be implemented using rejection sampling and is certainly something that we plan to try. Finally, a more natural parameterization of the model is indeed in terms of amplitude and phase of the harmonics, A_t and ϕ_t. This would allow more informative modelling of the amplitudes than is currently possible with our zero-mean Gaussian priors on the α_t and β_t. Again, this would be a highly desirable development, although it takes us out of the conditionally linear Gaussian model framework that we currently benefit from.

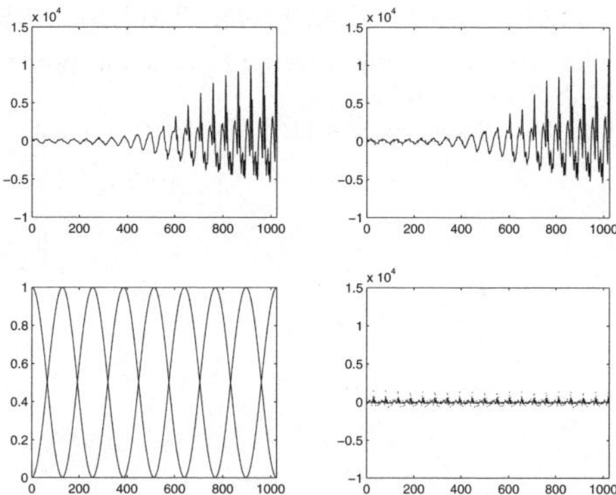

Figure 10. *Saxophone note modelling. Top left: original signal y_t. Bottom left: basis functions $\phi_{i,t}$ for $I = 9$. Top right: posterior mean (solid) and 5/95 percentiles (dotted) for signal estimate using $I = 9$. Bottom right: as top right with $I = 3$.*

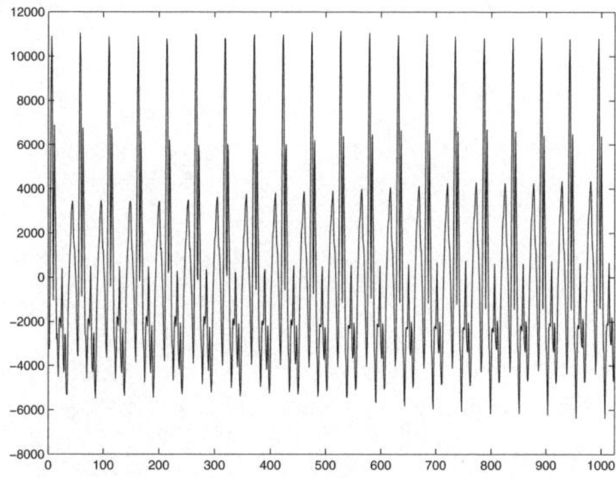

Figure 11. *Saxophone note–steady state region.*

Future developments of this work will include many of these natural extensions to the basic model. Most crucial as we see it is the modelling of large temporal sections of data jointly: after all the human brain retains a long memory (seconds at least of material) of the music in order to decipher what is happening at the present. We currently treat long datasets by splitting them into many shorter datasets, or frames, which are processed independently on one another (this technique generated the results of Figures 3 and 8, for example). This is highly sub-optimal and only adopted for computational reasons. Simple ways to incorporate interdependence between these sub-blocks include using inferences from previous frames to give "good" MCMC initializations, or even the generation of informative priors using the inference from previous frames. These techniques are not particularly robust, however, since an error in frame n can then propagate through subsequent frames $n+1, \ldots$. A more principled approach, and very much in line with the musical modelling suggestions of Carlin here, will most likely be required, with an over-arching hierarchical dynamical model describing the evolution of things such as musical pitch, number of harmonics, number of notes, etc., over time. In order to implement this we can use combinations of MCMC at the frame level with importance sampling (particle filtering/smoothing) for sequential propagation of inferences from frame to frame.

ADDITIONAL REFERENCES IN THE DISCUSSION

Besag, J., York, J. C. and Mollié, A. (1991). Bayesian image restoration, with two applications in spatial statistics. *Ann. Inst. Statist. Math.* **43**, 1–59 (with discussion).

Phillips, D. B. and Smith, A. F. M. (1994). Bayesian faces via hierarchical template modeling. *J. Am. Statist. Ass.* **89**, 1151–1163.

BAYESIAN STATISTICS 7, pp. 125–144
J. M. Bernardo, M. J. Bayarri, J. O. Berger, A. P. Dawid,
D. Heckerman, A. F. M. Smith and M. West (Eds.)

Assessing the Risk of Disclosure of Confidential Categorical Data

ADRIAN DOBRA
National Institute of Statistical Sciences, USA
adobra@niss.org

STEPHEN E. FIENBERG
Carnegie Mellon University, USA
fienberg@stat.cmu.edu

MARIO TROTTINI
Universitat de València, Spain
Mario.Trottini@uv.es

SUMMARY

Disclosure limitation involves the application of statistical tools to limit the identification of information on individuals (and enterprises) included as part of statistical databases such as censuses and sample surveys. We outline the major issues involved in assessing disclosure risk and assuring the protection of confidentiality for databases, especially those in the form of multi-way contingency tables, and we present a Bayesian framework for thinking about such problems both from the perspective of an intruder and the agency trying to protect its data.

Keywords: CONTINGENCY TABLES; DATA UTILITY; DIRICHLET PRIOR; DISCLOSURE LIMITATION; INTRUDER BEHAVIOR; LOG-LINEAR MODELS.

1. INTRODUCTION

Maintaining the confidentiality of statistical data is essential if government agencies are to collect and publish high-quality census and survey data. Typically, agencies promise respondents that their data will be kept confidential and used for statistical purposes only. For example, Title 13, Section 9 of the United States Code prohibits the US Census Bureau from publishing results in which an individual's or business's data can be identified. How can an agency comply with such legal strictures while at the same time provide public access to as much data as possible? This paper addresses this issue in the context of categorical data in the form of a cross-classification of counts.

Disclosure limitation is the process of protecting the confidentiality of statistical data. This paper focuses on *identity disclosure* where an intruder uses published statistical information to identify individual data provider. (For simplicity we set aside the issue of *attribute disclosure*, where an intruder learns that everyone in an identifiable group has a particular attribute.) Since virtually any form of data release contains some information about the individuals whose data are included in it, disclosure is not an all-or-none concept, but rather a probabilistic one seen differently from the eyes of the agency protecting the data, the individuals providing the data, and an "intruder" attempting to gain access to identifiable individual information (*cf.*, Lambert, 1993). In this sense, disclosure risk and the development of methods to limit disclosure are inherently Bayesian. For general introductions to some of the statistical aspects of confidentiality and disclosure limitation see Doyle *et al.* (2001), Fienberg (1994), and Willenborg and De Waal (1996, 2001). Early Bayesian contributions to the literature on disclosure limitation include Duncan and Lambert (1986, 1989) and Rubin (1993).

Disclosure limitation procedures alter or limit the data to be released, *e.g.*, by modifying or removing those characteristics that put confidential information at risk for disclosure. In the case of sample categorical data, a count of "1" can generate confidentiality concerns if that individual is also unique in the population. Much confidentiality research has focused on measures of risk that attempt to infer the probability that an individual is unique in the population given uniqueness in the sample (*e.g.*, see Chen and Keller-McNulty, 1998; Fienberg and Makov, 1998, 2001; Skinner and Holmes, 1998; Samuels, 1998). For simplicity, we focus on population tables of counts here and thus set aside this issue of making inferences from sample tables. But in either population or sample settings, small counts raise issues of disclosure risk.

In the next section we describe the identity disclosure problem in the context of a sequence of releases of marginal tables from a multi-way cross-classification. Then, in Sections 3 and 4, we outline a Bayesian approach to the balancing of disclosure risk and data utility, apply it to the case of tabular categorical data, and derive some commonly used risk measures for the release of a sequence of marginal tables. In Section 5, we adopt the perspective of the intruder and consider updating distributions over the space of possible tables subject to margin constraints. We illustrate the methodology using a $2 \times 3 \times 3$ contingency table drawn from the 1990 US decennial census. We conclude with a discussion of a number of unaddressed elements that need to be part of a full Bayesian approach to the problem.

2. DISCLOSURE LIMITATION FOR CATEGORICAL DATA

We think of the confidentiality problem as one involving three parties: a "statistical agency" that controls the data, "users" who wish to analyze all or perhaps subsets of the data, and an "intruder" who is attempting to identify one or more individuals in the data for some purpose.

Clearly, any release of data from a database increases the information available about individuals in the database and thus increases in some sense the probability that an individual in the database will become identifiable. Harm to such an individual occurs when an intruder matches the identifiable record to an existing database and learns information about the individual that was not previously available. Following Fienberg *et al.* (1997), we assume that the intruder acts as Bayesian updating his probabilities of identification of individuals in the database as more and more information becomes available. Further, we assume that the agency acts in a Bayesian fashion and makes a trade-off between the utility of the data were it to be released to the users and the disclosure risk associated with that release.

Our goal here is to outline a statistical framework for the release of cross-classified categorical data in the form of a contingency table. We are thinking in terms of requests from users for (marginal) sub-tables involving a subset of the variables. Potential responses include the release of the requested sub-table, the release of an "altered" or "masked" sub-table, or perhaps a refusal to release the sub-table. Note that there is statistical information for an intruder that comes from a refusal, although we have yet to see a Bayesian analysis that takes such information into account. We think in terms of a public system so that once a sub-table is released it is publicly available, and thus usable by an intruder. Clearly, the more sub-tables that are released, the more information we have about the full joint distribution of the cross-classifying variables.

The notion of data masking, introduced in Duncan and Pearson (1991), involves a transformation to the data so that individual records are altered to make them less identifiable. For categorical data, when releases consist of marginal tables, the types of masks suggested in the literature include stochastic perturbations subject to the constraint that the transformed data are consistent with the released marginals (*e.g.*, see Fienberg *et al.*, 1998; Duncan *et al.*, 2001).

How should the agency assess disclosure risk in this setting? What strategy should the intruder use to update his information about the individuals whose data are included in the full table? And, finally, given such choices, how should the agency respond to requests for specific marginal tables, given the set of tables already released?

We are unaware of any systematic and coherent statistical approach to the confidentiality problem as we have just outlined it, although Raghunathan and Rubin's (2001) multiple imputation strategy may provide a sensible Bayesian solution to it. Statistical agencies do in fact release sub-tables of very large contingency tables all of the time (*e.g.*, the website for the US Census Bureau's American Factfinder system releases selected three-way tables for various levels of geography: `http://factfinder.census.gov/`) and otherwise make judgments about the safety of releasing microdata files from sample surveys, but the judgments about the "safety" of such data releases is ad hoc at best. Recent efforts to study the release of margins of contingency tables have focused on the role of bounds on cell entries that result (*e.g.*, see Dobra and Fienberg, 2000, 2001; Dobra *et al.*, 2002), and on perturbations of data based on "exact" distributions for contingency tables under log-linear models given marginals corresponding to minimal sufficient statistics (*e.g.*, see Diaconis and Sturmfels, 1998; Fienberg *et al.*, 1998). This work offers a starting point for the present paper in which we attempt to outline some of the elements of a Bayesian approach.

3. A GENERAL FRAMEWORK FOR ASSESSING DISCLOSURE RISK

Let $\boldsymbol{f}$ represent the original data and $\mathcal{D}$ a set of candidate data *masks* or transformations of the data, typically stochastic in nature. Here we outline a general Bayesian framework, based on Trottini (2001) and Trottini and Fienberg (2002), to answer the question: "Which mask should the agency select?" In Section 4, we apply the framework to tabular categorical data.

The evaluation of a generic data mask $\tilde{\boldsymbol{f}}$ depends on the extent to which its release is beneficial for the users, (*data utility* of $\tilde{\boldsymbol{f}}$) and the extent to which its release can harm the agency or the data providers (*disclosure risk* of $\tilde{\boldsymbol{f}}$). For simplicity, we assume that there are only two users of the data: an *intruder* (I) who wants to "undo" the candidate mask $\tilde{\boldsymbol{f}}$ to disclose confidential information about the data provider, and a *scientist* (S) who wants to use the released data to infer some general feature of the population underlying $\tilde{\boldsymbol{f}}$. We denote the intruder's target by Θ_I and the scientist's target by Θ_S. We assume that user h ($h = I, S$) incurs a loss $L_h(e, \Phi_h)$, by using the estimate e for his target Θ_h, which depends on an unknown "state of the world" Φ_h. In most cases of interest, $\Phi_h = \Theta_h$. We denote by $\pi_h(\cdot)$ and $\pi_h(\cdot \mid \tilde{\boldsymbol{f}})$ the users prior and posterior distributions for Φ_h, $h = I, S$.

We assume that both S and I act in accord with the expected loss principle, *i.e.*, they estimate their target values by

$$\hat{\theta}_h = \arg\min_a \int L_h(a; \phi_h)\pi_h(\phi_h \mid \tilde{\boldsymbol{f}})d\phi_h, \qquad h = I, S.$$

Following DeGroot (1962), we define user h's uncertainty about the true value of the target as the expected loss associated with the optimal estimate of the target,

$$U_h(\tilde{\boldsymbol{f}}) = \int L_h(\hat{\theta}_h; \phi_h)\pi_h(\phi_h \mid \tilde{\boldsymbol{f}})d\phi_h, \qquad h = I, S.$$

We assume that the user stops trying to estimate Θ_h if his uncertainty is very large, in accord with the following decision rule:

User h's decision rule: *For a fixed threshold t_h, if $U_h(\tilde{\boldsymbol{f}}) \leq t_h$ then h takes action a_{1h} and estimates Θ_h by $\hat{\theta}_h$. If $U_h(\tilde{\boldsymbol{f}}) > t_h$ then h takes action a_{0h} and stops trying to estimate Θ_h.*

We assume that the loss that the agency incurs when user h takes action A_h, ($A_h \in \{a_{1h}, a_{0h}\}$) depends on an unknown state of the world $\Phi_A^{(h)}$ and is denoted by $L_A^{(h)}(\cdot,\cdot)$, $h = S, I$. Thus, $L_A^{(I)}(A_I, \phi_A^{(I)})$ quantifies, from the agency's perspective, the harm that the intruder's action A_I produces to the agency and the data providers when $\Phi_A^{(I)} = \phi_A^{(I)}$. In most of the cases it will be $\Phi_A^{(I)} = \Phi_I$. Similarly, $L_S^{(A)}(A_S, \phi_A^{(S)})$ quantifies, from the agency's perspective, the loss that the agency and the scientist incur if scientist takes action A_S and $\Phi_A^{(S)} = \phi_A^{(S)}$. In most of the cases this is just the loss that the scientist incurs by taking action A_S when $\Phi_A^{(S)} = \phi_h^{(S)}$, *i.e.*, $\phi_A^{(S)} = \phi_S$ and $L_A^{(S)}(a_{1S}, \phi_S) = L_S(a_{1S}, \phi_S)$. We denote by $\pi_A^{(h)}(\cdot)$ and $\pi_A^{(h)}(\cdot \mid \boldsymbol{f})$ the agency's prior and posterior distribution for $\Phi_A^{(h)}$.

We assume that the agency treats A_h and the "states of the world" $\Phi_A^{(h)}$ as random variables, and we propose to measure *disclosure risk*, DR, and *data utility*, DU, averaging losses with respect to the agency's joint posterior distribution for A_h and $\Phi_A^{(h)}$ given the original data $\boldsymbol{f}$,

$$\mathrm{DR}(\tilde{\boldsymbol{f}}) = E_{A_I, \Phi_A^{(I)} \mid \boldsymbol{f}} \{L_I^{(A)}(A_I, \phi_A^{(I)})\}, \quad \mathrm{DU}(\tilde{\boldsymbol{f}}) = -[E_{A_S, \Phi_A^{(S)} \mid \boldsymbol{f}} \{L_S^{(A)}(A_S, \phi_A^{(S)})\}].$$

We assume that the users' targets as well as the users' priors, $\pi_h(\cdot)$, and the users' loss functions, $L_h(\cdot,\cdot)$, are known to the agency. This implies that the agency knows the users' posterior distributions, users' optimal estimate of Θ_h, $\hat{\theta}_h$, and users' uncertainty, U_h, $h = S, I$. We make this assumption largely for convenience and extensions to classes of targets, priors, and loss functions are possible.

We further assume that the users' thresholds, t_h, are fixed but unknown to the agency, which thus treats them as random variables independent of the state of the world $\Phi_A^{(h)}$. The independence assumption is reasonable if user h fixes his threshold on the basis of what he knows about Θ_h but never on the basis of the agency's knowledge of $\Phi_A^{(h)}$. It follows that the agency's posterior distribution for A_h, $\{\Pr(a_{0h} \mid \boldsymbol{f}), \Pr(a_{1h} \mid \boldsymbol{f})\}$, depends only on the agency's distributions, $\pi_{T_h}(\cdot)$, for the users' thresholds and we can rewrite the disclosure risk and data utility as:

$$\mathrm{DR}(\tilde{\boldsymbol{f}}) = \sum_{j \in \{1,0\}} \Pr(a_{jI} \mid \boldsymbol{f}) \cdot E_{\Phi_A^{(I)} \mid \boldsymbol{f}} \{L_A^{(I)}(a_{jI}, \phi_A^{(I)})\}, \tag{1}$$

$$\mathrm{DU}(\tilde{\boldsymbol{f}}) = -\sum_{j \in \{1,0\}} \Pr(a_{1S} \mid \boldsymbol{f}) \cdot E_{\Phi_A^{(S)} \mid \boldsymbol{f}} \{L_A^{(S)}(a_{1S}, \phi_A^{(S)})\}. \tag{2}$$

In most of the cases the agency does not know the users' target but can only identify classes $\mathcal{Z}_h$ of possible targets, *i.e.*, $\Theta_h \in \mathcal{Z}_h = \{\Theta_h(1), \ldots, \Theta_h(r_h)\}$, and we can average (1) and (2) with respect to the probability that $\Theta_h = \Theta_h(j)$.

3.1. *The Utility–Risk Trade-off*

The most common criterion for the choice of the best mask in $\mathcal{D}$ consists of selecting the mask $\tilde{\boldsymbol{f}}$ that maximizes data utility subject to an upper bound for disclosure risk (Willenborg and de

Waal, 2001; Duncan *et al.*, 2001; Trottini and Fienberg, 2002). The optimal mask is the solution of the optimization problem:

$$\max\{\mathrm{DU}(\tilde{\boldsymbol{f}}) : \tilde{\boldsymbol{f}} \in \mathcal{D}, \text{ and } \mathrm{DR}(\tilde{\boldsymbol{f}}) \leq \alpha\}$$

where α is a threshold value for the maximum tolerable risk fixed by the statistical agency. Defining an optimality criterion corresponds to specifying suitable measures of disclosure risk and data utility. We believe that the framework outlined this section is the natural tool to define such measures. Once we have specified the users' targets, the information available about these targets prior to the release of the data, the estimation procedure used by the users, and the consequences for the agency of users' actions, then (1) and (2) automatically provide measures of disclosure risk and data utility coherent with these inputs.

One might argue that all these elements are mostly unknown to the agency and, as a result, that our framework is difficult to implement, and that heuristic measures could do a better job. In fact, the uncertainty about inputs is a major strength of our approach, since our framework allows us to incorporate this uncertainty in a natural way. Heuristic measures are not assumption-free. Rather the assumptions simply are not stated (and therefore not understood). We have been able to use our framework to produce most of the measures of disclosure risk and data utility proposed in the literature of statistical confidentiality for suitable choices of the input values. This allows us to understand whether these measures are statistically sensible.

In the next section, we apply the framework to tabular categorical data and, because of space limitations, we focus only on measures of disclosure risk. Similar results hold for data utility.

4. DISCLOSURE RISK FOR TABULAR CATEGORICAL DATA

Suppose that a statistical agency records the value of k categorical variables for each individual in a given population and summarizes the result in a frequency table $\boldsymbol{f}$ with m cells (corresponding to the possible cross-classifications of the k variables). Let $\mathcal{I} = \{1, 2, \ldots, m\}$. We assume that the table total (population size), n, is known *a priori* to the users, who view the original table as a random variable, $\boldsymbol{F}$. Before the generic masked data $\tilde{\boldsymbol{f}}$ is released, users know that $\boldsymbol{F}$ takes values in the set $\mathcal{X}$ of all non-negative integer m-vectors adding to n

$$\boldsymbol{F} \in \mathcal{X} = \{(x_1, \ldots, x_m) : x_i \text{ is a non-negative integer and } \sum\nolimits_{i=1}^{m} x_i = n\}.$$

We let $\mathcal{T}$ be the set of tables in $\mathcal{X}$ that are compatible with the candidate release $\tilde{\boldsymbol{f}}$ and $M(\mathcal{X})$ and $M(\mathcal{T})$ the cardinality of $\mathcal{X}$ and $\mathcal{T}$ respectively.

We now use the framework of Section 3 to define three measures of disclosure risk associated with the release of a generic mask, $\tilde{\boldsymbol{f}}$, which correspond to well-known ones proposed on an *adhoc* basis in the literature on statistical confidentiality. In all three examples, $\Phi_A^I = \boldsymbol{F}$ and, since the agency knows the original table, $\pi_A^{(I)}(\phi_A^{(I)} \mid \boldsymbol{f})$ is degenerate at $\boldsymbol{f}$. These examples illustrate how our approach can be used to assess effectiveness of existing criteria. We think of a measure of disclosure (data utility) as sensible if we can obtain it as a result of disclosure scenarios characterized by "natural" choices of users targets, priors, loss functions, etc. At least as important is the application of our framework to define new measures derived from equations (1) and (2) for suitable choices of the input values, but this goes beyond the goal of the present paper.

4.1. *Example 1: Disclosure Risk as Tightness of Bounds for Small Cell Counts*

Suppose that the intruder's target is the original table, $\Theta_I = \boldsymbol{F} = (F(1), \ldots, F(m))$, and let the intruder's action space for the problem "estimate $\boldsymbol{F}$" be the m-fold product space (for the purposes of the example we do not require intruder's estimates to lie on the simplex, although in general a rational intruder would include this constraint):

$$\mathcal{N}_I = \overbrace{\mathcal{N} \times \cdots \times \mathcal{N}}^{m\text{-times}}, \ \mathcal{N} = \{ [a,b] : \quad a \leq b, \quad a, b \text{ non-negative reals}\}.$$

Suppose that when the intruder can define tight bounds for all cells in the table his loss when estimating $\boldsymbol{F}$ is small, whereas if he cannot accurately estimate at least one cell, his loss is large. In particular for a generic $e = (e(1), \ldots, e(m)) \in \mathcal{N}_I$ and $f_j = (f_j(1), \ldots, f_j(m)) \in \mathcal{T}$ assume:

$$L_I(e, f_j) = \begin{cases} \sum_{i=1}^{m} \text{length}\, e(i), & \text{if } f_j(i) \in e(i), \quad i = 1, \ldots, m, \\ \infty, & \text{otherwise.} \end{cases}$$

Let $L(i)$ and $U(i)$ be the lower and upper bounds for the ith cell in the original table based on the candidate release $\tilde{\boldsymbol{f}}$, *i.e.*, $L(i) = min\{f_j(i) : f_j \in \mathcal{T}\}$ and $U(i) = max\{f_j(i) : f_j \in \mathcal{T}\}$. For this case, when $\pi_h(\cdot)$ has support $\mathcal{X}$, the intruder's optimal action and uncertainty are, respectively,

$$\hat{\theta}_I = ([L(1), U(1)], \ldots, [L(m), U(m)]), \qquad U_I = \sum_{i=1}^{m} U(i) - L(i).$$

Suppose now that the loss that the agency incurs when the intruder takes action a_{rI} takes its minimum when the intruder stops trying to identify the original table ($r = 0$) or when none of the intruder's set estimates of small cell counts in the true table contains the correct value of the cell. Further suppose that the loss increases as the bounds for small cell counts become tighter. This situation corresponds to:

$$L_I^{(A)}(a_{rI}, f_j) = \begin{cases} -\min\limits_{i \in \mathcal{Q}} \text{length}\, \hat{\theta}_I(i), & \text{if } r = 1 \text{ and } \mathcal{Q} \neq \emptyset \\ -n, & \text{otherwise} \end{cases}$$

where $\hat{\theta}_I(i)$ is the intruder's optimal (set) estimate of $F(i)$ and

$$\mathcal{Q} = \{i \in \{1, \ldots, m\} : \quad f_j(i) \in \hat{\theta}_I(i) \text{ and } 0 < f_j(i) < 3\}.$$

If the agency believes that the intruder never stops trying to estimate his target (*i.e.*, $\pi_{T_I}(\cdot)$ is degenerate at nm), then the disclosure risk in (1) becomes:

$$\text{DR}(\tilde{\boldsymbol{f}}) = -\min_i\{U(i) - L(i) : \quad 0 < f(i) < 3\}. \tag{3}$$

Choosing this degenerate distribution for the intruder's threshold does not necessarily imply that the agency believes that the intruder always tries to estimate the original table, regardless of his uncertainty, but rather may reflect a conservative attitude based on the worst-case scenario where an intruder always tries to make inference about $\boldsymbol{F}$. The measure in (3) has been discussed on an *adhoc* basis by several authors (*e.g.*, see Dobra *et al.*, 2002) and is a risk criterion used by many statistical agencies.

4.2. *Example 2: Disclosure Risk as Conditional Probability of the True Table*

Suppose that the intruder's target is the distribution of the original table $\boldsymbol{F}$ and that he uses a logarithmic utility function (Bernardo, 1979):

$$L_I(\hat{P}, f_j) = -log[\hat{P}(f_j)], \quad \hat{P} \in \mathcal{P}, \quad f_j \in \mathcal{X}, \tag{4}$$

where $\mathcal{P}$ denotes the class of all possible distributions with support $\mathcal{X}$. Thus, the loss that intruder pays for estimating the distribution of $\boldsymbol{F}$ by $\hat{P}$ when $\boldsymbol{F} = f_j$ (*i.e.*, when the original table is f_j) is a decreasing function of the probability of f_j under $\hat{P}$. Under the loss in (4), the intruder's optimal estimate of the distribution of $\boldsymbol{F}$ is his posterior distribution, and his uncertainty is the entropy of the posterior distribution.

Suppose that the agency pays no loss if the intruder stops trying to estimate the distribution of $\boldsymbol{F}$, and it pays a loss equal to the probability of the true table under the intruder's estimate otherwise,

$$L_I^{(A)}(a_{rI}, f_j) = \begin{cases} 0, & \text{if } r = 0, \\ \hat{\theta}_I(f_j) = \pi_I(f_j \mid \tilde{\boldsymbol{f}}), & \text{if } r = 1. \end{cases}$$

If the agency believes that intruder always tries to estimate the distribution of $\boldsymbol{F}$ no matter what his uncertainty (*i.e.*, if the agency's distribution for the intruder's threshold is degenerate at $log[M(\mathcal{T})]$), then the disclosure risk in (1) is just the (intruder's) posterior probability of the true table $\boldsymbol{f}$ given $\tilde{\boldsymbol{f}}$. If the intruder's prior for $\boldsymbol{F}$ is uniform on $\mathcal{X}$, then from Bayes' Theorem, his posterior given the released table $\tilde{\boldsymbol{f}}$ is uniform on $\mathcal{T}$ and (1) becomes $\text{DR}(\tilde{\boldsymbol{f}}) = \pi_I(\boldsymbol{f} \mid \tilde{\boldsymbol{f}}) = 1/M(\mathcal{T})$. Both measures of disclosure have been proposed on an *adhoc* basis in the literature of statistical confidentiality (*e.g.*, see Dobra, 2002). Since they correspond to different assumptions about the intruder's prior for $\boldsymbol{F}$ we can choose between them according to which prior is appropriate for a given problem.

4.3. *Example 3: Disclosure Risk as Fraction of Small Cells Values Correctly Identified*

Suppose that the agency knows that the intruder's target is to identify one cell of the original table, but it does not know which one. If we assume that each cell is equally likely to be the target, we have: $\Theta_I \in \{F(1), \ldots, F(m)\}$, and $\Pr(\Theta_I = F(i)) = 1/m$ for $i = 1, \ldots, m$.

Suppose further that the intruder uses a 0-1 loss function, *i.e.*, $L_{Ii}(e, \boldsymbol{f}_j) = 1 - I_{\boldsymbol{f}_j(i)}(e)$. Under these assumptions the intruder's optimal estimate of $F(i)$ is the permissible value $\hat{\theta}_{Ii}$ with highest posterior probability and the intruder's uncertainty is one minus this maximum (posterior) probability.

If the agency's distribution is degenerate at some value t_I^* then, from the agency perspective, the intruder's action is a degenerate random variable that takes values a_{0Ii} or a_{1Ii} depending on whether or not $\Pr(F(i) = \hat{\theta}_{Ii} \mid \tilde{\boldsymbol{f}}) < 1 - t_I^*$. Let δ be a threshold value and let $n_\delta(\boldsymbol{f}_j)$ be the number of cells in $\boldsymbol{f}_j$ such that $\boldsymbol{f}_j(i) < \delta$. Suppose that, when the intruder correctly estimates a small cell value $F(i)$, the agency incurs a loss that is a decreasing function of the number of "small" cell values in the true table and there is no loss if either $F(i)$ is "big" or the intruder's estimate is incorrect. This corresponds to:

$$L_{Ii}^{(A)}(a_{rIi}, \boldsymbol{f}_j) = \begin{cases} m/n_\delta(\boldsymbol{f}_j), & \text{if } r = 1, \hat{\theta}_{Ii} = f_j(i) \text{ and } f_j(i) < \delta, \\ 0, & \text{otherwise.} \end{cases}$$

Then, conditionally on $\Theta_I = F(i)$, the disclosure risk is:

$$\text{DR}_i(\tilde{\boldsymbol{f}}) = \begin{cases} m/n_\delta(\boldsymbol{f}), & \text{if } \hat{\theta}_{Ii} = f(i), f(i) < \delta \text{ and } \Pr(F(i) = \hat{\theta}_{Ii} \mid \tilde{\boldsymbol{f}}) > 1 - t_I^*, \\ 0, & \text{otherwise.} \end{cases}$$

and from the mixture version of (1) the (unconditional) disclosure risk is:

$$\text{DR}(\tilde{\boldsymbol{f}}) = \sum_{i=1}^{m} \Pr(\Theta_I = F(i)) \cdot \text{DR}_i(\tilde{\boldsymbol{f}}) = \frac{\#\,(f(i) < \delta \text{ correctly identified})}{n_\delta(\boldsymbol{f})}. \tag{5}$$

where in (5) a cell is correctly identified if $\hat{\theta}_{Ii} = f(i)$ and $\Pr(F(i) = \hat{\theta}_{Ii} \,|\, \tilde{\boldsymbol{f}}) > 1 - t^*_I$.

This measure of disclosure has been discussed on an *adhoc* basis by Dobra (2002) and Dobra *et al.* (2002). A similar version for microdata is also discussed in Lambert (1993) with $t^*_I = 1$. We next illustrate the implementation of (5) in an example where the released data consists of a set of marginal tables of the original table $\boldsymbol{f}$.

5. UPDATING POSTERIOR DISTRIBUTIONS OVER POSSIBLE TABLES

Now that we have criteria for assessing disclosure risk we can consider the problem posed originally in Section 2, deciding how the agency should respond to requests for specific marginal tables, given the set of tables already released. We do so by looking at the inferences made by the intruder about the possible tables that are consistent with the marginals released to date and we highlight the computational problems that characterize the evaluation of measures of disclosure risk from Section 4.

If the agency has released the l marginals $\mathcal{R} = \{\boldsymbol{f}_1, \boldsymbol{f}_2, \ldots, \boldsymbol{f}_l\}$, and no other information is available about $\boldsymbol{f}$, an intruder knows only that the table $\boldsymbol{f}$ belongs to the set of tables $\mathcal{T}$. [Here $\mathcal{R}$ is equivalent to the masked table $\tilde{\boldsymbol{f}}$ in Section 4.] We treat the population table observation $\boldsymbol{f} = \{f(i)\}_{i\in\mathcal{I}}$ as having been generated from a super-population specified by the random variable $\boldsymbol{F} = \{F(i)\}_{i\in\mathcal{I}}$. Evaluating the disclosure risk associated with releasing $\boldsymbol{f}_1, \ldots, \boldsymbol{f}_l$ by counting the number of tables in $\mathcal{T}$ could create a false sense of security if the probability $\Pr(\boldsymbol{F} = \boldsymbol{f} \,|\, \mathcal{R})$ is high, while $M(\mathcal{T})$ is very large. In this situation, there may be a reasonably substantial probability that the intruder could actually correctly identify the original table $\boldsymbol{f}$. Moreover, we can assess the level of protection for an individual cell count $f(i)$, $i \in \mathcal{I}$, by examining the feasibility interval $[L(i), U(i)]$, where $L(i) = \min\{F(i) : \boldsymbol{F} \in \mathcal{T}\}$, and $U(i) = \max\{F(i) : \boldsymbol{F} \in \mathcal{T}\}$. In many situations we can calculate these bounds directly or using relatively simple algorithms (*e.g.*, see Dobra, 2002, for a general algorithm and Dobra and Fienberg, 2000, 2001 for special cases).

The marginal distribution induced by $\Pr(\boldsymbol{F} = \boldsymbol{f} \,|\, \mathcal{R})$ on the possible values of a cell $i \in \mathcal{I}$, $q \in [L(i), L(i) + 1, \ldots, U(i) - 1, U(i)]$, is given by:

$$\Pr(F(i) = q \,|\, \mathcal{R}) = \sum\nolimits_{\{\boldsymbol{f}:\boldsymbol{f}\in\mathcal{T}, f(i)=q\}} \Pr(\boldsymbol{F} = \boldsymbol{f} \,|\, \mathcal{R}). \tag{6}$$

The intruder could infer that the "true" value of cell $i \in \mathcal{I}$ is the value q with the highest conditional probability $\Pr(F(i) = q \,|\, \mathcal{R})$. One could be misled by the fact that the feasibility interval $[L(i), U(i)]$ seems to be wide enough to guarantee the protection of cell count $f(i)$ because the probability of the "true" value $f(i)$ for cell i in (6) might be, in fact, very large and hence $f(i)$ might not be adequately protected.

5.1. *Conditional Distribution of a Table of Counts Under a Log-linear Model*

Suppose the distribution of the cell counts $\boldsymbol{f}$ is multinomial with a fixed total n:

$$\Pr(\boldsymbol{F} = \boldsymbol{f} \,|\, \boldsymbol{\theta}) = \frac{n!}{\prod\limits_{i\in\mathcal{I}} f(i)!} \exp\left[\sum_{i\in\mathcal{I}} f(i) \log \theta(i)\right],$$

where $\theta(i)$ is the probability that an individual cross-classified in table belongs to cell $i \in \mathcal{I}$. The cell probabilities $\boldsymbol{\theta} = \{\theta(i)\}_{i\in\mathcal{I}}$ are constrained to lie within the simplex

$$\Theta = \Big\{\boldsymbol{\theta} : \theta(i) > 0 \text{ for all } i \in \mathcal{I} \text{ and } \sum_{i\in\mathcal{I}} \theta(i) = 1\Big\}. \tag{7}$$

We are more accustomed to working with parameters associated with specific models. We therefore assume that the cell probabilities $\boldsymbol{\theta}$ lie in a space $\Theta_{\mathcal{A}}$ associated with a hierarchical log-linear model $\mathcal{A}$, given by

$$\Theta_{\mathcal{A}} = \Theta \cap \{\boldsymbol{\theta} : \quad \log \boldsymbol{\theta} = A \cdot \boldsymbol{\psi} \text{ for some } \boldsymbol{\psi} = \{\psi(i)\}_{i \in \mathcal{I}} \text{ with } \psi(i) > 0\},$$

where A is the design matrix of $\mathcal{A}$. The introduction of log-linear models for the cell probabilities here is a device and, at the end of this section, we suggest how the results for separate models should be combined.

If $\mathcal{A}$ is the saturated log-linear model, $\Theta_{\mathcal{A}}$ becomes Θ, see Eq. (7). The conditional distribution of $\boldsymbol{F} = \boldsymbol{f}$ given the released marginals $\mathcal{R}$ under model $\mathcal{A}$ is

$$\Pr(\boldsymbol{F} = \boldsymbol{f} \mid \mathcal{R}, \mathcal{A}) = \int_{\Theta_{\mathcal{A}}} \Pr(\boldsymbol{F} = \boldsymbol{f} \mid \boldsymbol{\theta}) \cdot \pi(\boldsymbol{\theta} \mid \mathcal{R}, \mathcal{A})\, d\boldsymbol{\theta}, \tag{8}$$

where $\pi(\boldsymbol{\theta} \mid \mathcal{R}, \mathcal{A})$ is the posterior distribution of cell probabilities given the released margins $\mathcal{R}$ under model $\mathcal{A}$.

Estimating $\Pr(\boldsymbol{F} = \boldsymbol{f} \mid \mathcal{R}, \mathcal{A})$ is difficult because the minimal sufficient statistics of the log-linear model $\mathcal{A}$ might be unknown if we are only provided with the set of marginals $\mathcal{R}$. We need to "augment" the observed data $\mathcal{R}$ to form a complete table $\boldsymbol{F} \in \mathcal{T}$ in order to obtain the minimal sufficient statistics of $\mathcal{A}$. This suggests a data augmentation approach for sampling from the joint density

$$\Pr(\boldsymbol{F}, \boldsymbol{\theta} \mid \mathcal{R}, \mathcal{A}) \propto \Pr(\boldsymbol{F} \mid \boldsymbol{\theta}) \pi(\boldsymbol{\theta} \mid \mathcal{R}, \mathcal{A}).$$

Start with $\boldsymbol{\theta}_0 \in \Theta_{\mathcal{A}}$. At the s-th step of the algorithm, do

1. Simulate $\boldsymbol{F}^{(s+1)} \propto \Pr(\boldsymbol{F} \mid \mathcal{R}, \mathcal{A}, \boldsymbol{\theta}^{(s)})$.
2. Simulate $\boldsymbol{\theta}^{(s+1)} \propto \Pr(\boldsymbol{\theta} \mid \mathcal{R}, \mathcal{A}, \boldsymbol{F}^{(s+1)})$.

If we are given the complete table with cell probabilities $\boldsymbol{\theta}^{(s)}$, it no longer makes sense to condition on the log-linear model $\mathcal{A}$. Similarly, given the complete table $\boldsymbol{F}^{(s+1)}$, conditioning on the observed data $\mathcal{R}$ becomes obsolete. Thus

$$\Pr(\boldsymbol{F} \mid \mathcal{R}, \mathcal{A}, \boldsymbol{\theta}^{(s)}) = \Pr(\boldsymbol{F} \mid \mathcal{R}, \boldsymbol{\theta}^{(s)}),$$
$$\Pr(\boldsymbol{\theta} \mid \mathcal{R}, \mathcal{A}, \boldsymbol{F}^{(s+1)}) = \Pr(\boldsymbol{\theta} \mid \mathcal{A}, \boldsymbol{F}^{(s+1)}).$$

We make use of the Markov chain Monte Carlo approach suggested by Diaconis and Sturmfels (1998) for generating draws from the posterior distribution $\Pr(\boldsymbol{F} = \boldsymbol{f} \mid \mathcal{R}, \boldsymbol{\theta}^{(s)})$. This sampling technique relies on the existence of a Markov basis–a finite set of moves or data swaps connecting any two tables with the same marginals.

We need to specify a prior distribution for the cell probabilities that is consistent with the constraints induced by the log-linear model. We take the prior density for $\boldsymbol{\theta}$ to be a constrained Dirichlet prior with hyperparameters $\boldsymbol{\alpha} = \{\alpha(i)\}_{i \in \mathcal{I}}$ (Schafer, 1997):

$$\pi_{\Theta_{\mathcal{A}}}(\boldsymbol{\theta}) \propto \prod_{i \in \mathcal{I}} \theta(i)^{\alpha(i)-1},$$

for $\boldsymbol{\theta} \in \Theta_{\mathcal{A}}$. It follows that the complete-data posterior density for $\boldsymbol{\theta}$ is

$$\Pr(\boldsymbol{\theta} \mid \mathcal{A}, \boldsymbol{F}^{(s+1)}) \propto \prod_{i \in \mathcal{I}} \exp\{[F^{(s+1)}(i) + \alpha(i) - 1] \cdot \log \theta(i)\},$$

for $\boldsymbol{\theta} \in \Theta_{\mathcal{A}}$ and zero otherwise. This is equivalent to the likelihood function for $\boldsymbol{\theta}$ given the table with cell entries $F^{(s+1)}(i) + \alpha(i) - 1$, for $i \in \mathcal{I}$. The constrained Dirichlet prior forms a conjugate class for the multinomial likelihood and hence the posterior of $\boldsymbol{\theta}$ is another constrained Dirichlet prior with hyper-parameters $\boldsymbol{F}^{(s+1)} + \boldsymbol{\alpha}$. We use Bayesian iterative proportional fitting (Gelman *et al.*, 1995; Schafer, 1997) for simulating random draws from the constrained Dirichlet posterior $\Pr(\boldsymbol{\theta} \mid \mathcal{A}, \boldsymbol{F}^{(s+1)})$.

By employing this data augmentation procedure, we can generate a sample $\boldsymbol{\theta}_1, \boldsymbol{\theta}_2, \ldots, \boldsymbol{\theta}_t$ from the posterior distribution $\pi(\boldsymbol{\theta} \mid \mathcal{R}, \mathcal{A})$ and estimate the conditional density of $\boldsymbol{F} = \boldsymbol{f}$ given data $\mathcal{R}$ under model $\mathcal{A}$ from (8) as

$$\Pr(\boldsymbol{F} = \boldsymbol{f} \mid \mathcal{R}, \mathcal{A}) \approx \frac{1}{t} \sum_{j=1}^{t} \Pr(\boldsymbol{F} = \boldsymbol{f} \mid \mathcal{R}, \boldsymbol{\theta}_j).$$

We are only looking at tables that are consistent with the marginals $\mathcal{R}$; hence we give zero probability to tables that are outside $\mathcal{T}$ by "normalizing" the posterior probabilities in (8) so that they add up to "1":

$$\Pr(\boldsymbol{F} = \boldsymbol{f} \mid \mathcal{R}, \mathcal{A}) \longleftarrow \frac{\Pr(\boldsymbol{f} \mid \mathcal{R}, \mathcal{A})}{\sum_{\boldsymbol{f}' \in \mathcal{T}} \Pr(\boldsymbol{f}' \mid \mathcal{R}, \mathcal{A})}. \tag{9}$$

5.2. *Example*

Table 1 gives a $2 \times 3 \times 3$ table drawn from the 1990 US decennial census public use sample for a local tract, and analyzed previously in Fienberg *et al.* (1998). Consistent with the discussion in Sections 1 and 4, we act *as if* Table 1 contains population counts. We focus on the four cells containing counts of "1" and "2."

Table 1. *Three-way cross-classification of Gender, Race, and Income for a selected US census tract. (Source: Fienberg et al., 1998). The bounds given in square brackets result from the release of a pair of two-way marginals: Race × Income and Income × Gender.*

		Income level		
Gender	Race	$\leq \$10,000$	$> \$10,000$ and $\leq \$25,000$	$> \$25,000$
Male	White	96 $[85, 107]$	72 $[64, 80]$	161 $[158, 169]$
	Black	10 $[0, 21]$	7 $[0, 14]$	6 $[0, 9]$
	Chinese	1 $[0, 1]$	1 $[0, 2]$	2 $[0, 2]$
Female	White	186 $[175, 197]$	127 $[119, 135]$	51 $[43, 54]$
	Black	11 $[0, 21]$	7 $[0, 14]$	3 $[0, 9]$
	Chinese	0 $[0, 1]$	1 $[0, 2]$	0 $[0, 2]$

Suppose that the agency releases a pair of two-way marginals: Race × Income and Income × Gender. Table 1 also includes in square brackets the bounds on the cell values resulting from the release of these marginals (Dobra and Fienberg, 2000). Because these marginals are the minimal sufficient statistics of a decomposable log-linear model, there exists a Markov basis that links all $2 \times 3 \times 3$ tables with these marginals (see Dobra, 2002). Table 2 reports the conditional marginal probabilities for the four cells containing small counts induced by conditioning on the saturated log-linear model $\mathcal{A}_1$. We assume a non-informative prior distribution with $\theta(i) = 0.5$ for every cell $i \in \mathcal{I}$. We marked by "–" the values outside the feasibility intervals. By employing the data augmentation algorithm outlined above, we generated a sample of size 500 from $\mathcal{T} \times \Theta_{\mathcal{A}_1}$.

The burn-in time for the Markov chain was $1,000,000$. To reduce the correlation between two consecutive draws, we discarded 1000 pairs $(\boldsymbol{F}, \boldsymbol{\theta})$ before selecting a new pair in the resulting sample. If the intruder were to "guess" that the true values of the entries for these cells are the values with the highest posterior probability, then his guess would be either incorrect or indecisive for each of the four cells.

Tables 2–4. *Marginal conditional probabilities under the log-linear models $\mathcal{A}_1$ (Table 2), $\mathcal{A}_2$ (Table 3)(and $\mathcal{A}_3$ (Table 4) for the cells containing small counts in Table 1 induced by releasing the Race × Income and Income × Gender marginals.*

	Table 2			Table 3			Table 4		
Cell	0	1	2	0	1	2	0	1	2
$(1,3,1)$	0.50	0.50	—	0.64	0.36	—	0.65	0.35	—
$(1,3,2)$	0.43	0.26	0.31	0.44	0.36	0.20	0.39	0.47	0.14
$(2,3,2)$	0.31	0.26	0.43	0.20	0.36	0.44	0.14	0.47	0.39
$(1,3,3)$	0.39	0.25	0.36	0.14	0.34	0.52	0.06	0.37	0.57

5.3. *Updating the Intruder's Posterior Distribution Over Permissible Tables*

In Section 5.1, we calculated the intruder's posterior distribution given a specific log-linear model and noted that we were introducing models as a device. To get rid of this conditioning, we now need to average over the model space, $\mathcal{H} = \{\mathcal{A}_1, \mathcal{A}_2, \ldots, \mathcal{A}_L\}$. The conditional distribution of $\boldsymbol{F} = \boldsymbol{f}$ given the released marginals $\mathcal{R}$ under the family of models $\mathcal{H}$ is (Kass and Raftery, 1995; Madigan and Raftery, 1994):

$$\Pr(\boldsymbol{F} = \boldsymbol{f} \mid \mathcal{R}, \mathcal{H}) = \sum_{l=1}^{L} \Pr(\boldsymbol{F} = \boldsymbol{f} \mid \mathcal{R}, \mathcal{A}_l) \cdot \Pr(\mathcal{A}_l \mid \mathcal{R}), \tag{10}$$

This is an average of the conditional probabilities of $\boldsymbol{F} = \boldsymbol{f}$ under each of the models, weighted by their posterior model probabilities. As we update $\mathcal{R}$ as a result of the release of additional margins, some of the terms in the sum on the r.h.s. of (10) are zero since we need to include only those log-linear models whose minimal sufficient statistics are the same as or include the released margins, $\mathcal{R}$. We note that the posterior probabilities $\Pr(\boldsymbol{F} = \boldsymbol{f} \mid \mathcal{R}, \mathcal{A}_l)$ in (10) are not "normalized" as in (9). After calculating $\Pr(\boldsymbol{F} = \boldsymbol{f}' \mid \mathcal{R}, \mathcal{H})$, $\boldsymbol{f}' \in \mathcal{T}$, however, we need to "normalize" them to give probability "0" to tables that are inconsistent with $\mathcal{R}$.

The probability of the data $\mathcal{R}$ given the model $\mathcal{A}_l$ is

$$\Pr(\mathcal{R} \mid \mathcal{A}_l) = \sum_{\boldsymbol{f}' \in \mathcal{T}} \Pr(\boldsymbol{F} = \boldsymbol{f}' \mid \mathcal{R}, \mathcal{A}_l),$$

and thus the posterior probability of model $\mathcal{A}_l$ given data $\mathcal{R}$ is

$$\Pr(\mathcal{A}_l \mid \mathcal{R}) = \frac{\Pr(\mathcal{R} \mid \mathcal{A}_l) \cdot \Pr(\mathcal{A}_l)}{\sum_{l'=1}^{L} \Pr(\mathcal{R} \mid \mathcal{A}_{l'}) \cdot \Pr(\mathcal{A}_{l'})}.$$

5.4. *Example Revisted*

We return to the data in Table 1. There are three log-linear models compatible with the agency release of the two two-way marginals, Race × Income and Income × Gender: (i) the saturated log-linear model $\mathcal{A}_1$, (ii) the log-linear model $\mathcal{A}_2$ of no second order interaction, and (iii) the decomposable log-linear model $\mathcal{A}_3$ for the conditional independence of Race and Gender given Income.

One of the minimal sufficient statistics of $\mathcal{A}_2$, namely Race $\times$ Gender, is not determined by fixing the other two two-way marginals of Table 1. Table 3 displays the posterior distribution for the possible values of the four cells containing small counts of "1" or "2" given the marginals Race $\times$ Income and Income $\times$ Gender under model $\mathcal{A}_2$. The count of "2" from cell $(1, 3, 3)$ has the largest posterior probability. Hence an intruder using the maximum posterior probability rule would correctly infer one out four values associated with the small counts cells in Table 1.

The minimal sufficient statistics for model $\mathcal{A}_3$ are just the released two-way marginals. We present the posterior probabilities of the four small count cells under $\mathcal{A}_3$ in Table 4. Here, the intruder would infer the correct value of cells $(1, 3, 2)$, $(2, 3, 2)$ and $(1, 3, 3)$.

The structure of the parameter space clearly makes a difference here since an intruder would not be able to correctly infer with any degree of accuracy the value of any of the four small count cells based on working with the saturated log-linear model $\mathcal{A}_1$. But under the no-second-order interaction model, $\mathcal{A}_2$, he could correctly guess one of the four counts and under the conditional independence model, $\mathcal{A}_3$, three of four counts.

Suppose we had assigned these three log-linear models a priori probabilities of 0.22, 0.67, and 0.11, respectively. Combining the results using the model averaging approach of (10), yields posterior probabilities of 0.18, 0.72, and 0.10, respectively. The released marginals (*i.e.*, our data) tend to give more probability to the model $\mathcal{A}_2$ of no-second-order interaction–not surprising, since this model fits the data reasonably well whereas the simpler model does not. Table 5 displays the posterior probabilities for the four cells and they are close to those for model $\mathcal{A}_2$. Thus, the intruder would not correctly identify the counts in the small cells except for the "2" in the $(1, 3, 3)$ cell.

Table 5. *Posterior conditional probabilities under the family of models $\mathcal{H} = \{\mathcal{A}_1, \mathcal{A}_2, \mathcal{A}_3\}$ for the possible values of the cells containing small counts in Table 1 induced by releasing the Race $\times$ Income and Income $\times$ Gender marginals.*

Cell	0	1	2
$(1, 3, 1)$	0.62	0.38	—
$(1, 3, 2)$	0.44	0.36	0.20
$(2, 3, 2)$	0.20	0.36	0.44
$(1, 3, 3)$	0.15	0.33	0.52

6. SUMMARY AND OPEN PROBLEMS

In this paper, we have attempted what we believe to be the first systematic Bayesian treatment of the problem of disclosure limitation for tables of counts, beginning with the trade-off between disclosure risk and data utility, and focusing on intruder efforts to identify small cell counts. The treatment is far from complete, however.

In Section 4, our discussion of data utility was restricted to a single user other than the intruder. But multiple users with differing analytical goals raise new issues. For example, releasing a high-dimensional margin requested by one user might well be "safe," but this action might preclude the release of many other lower-dimensional margins that would be of value to several other users.

Missing from Section 5 is an effort to address the information to the intruder when an agency chooses not to release a requested margin. If the agency is otherwise attempting to maximize the utility of the data for other users, the intruder should understand that the only reason not to release a margin is that when the information in it is combined with the other released margins, the intruder would be able to make "strong" inferences about small cell counts in the full table.

We have also not addressed the alternative strategy to not releasing a requested margin, *i.e.*, perturbing the table (subject of course to the constraints imposed by the already released margins) and releasing the margin from the perturbed table. This is a form of data transformation or mask in the spirit of Section 3. Clearly to do such perturbation in an efficient manner, the agency would do well to compute its posterior distribution over the parameters of the super-population space, and then draw tables from that distribution. This would be akin to the approach suggested in Fienberg *et al.* (1998) or the multiple imputation approach of Raghunathan and Rubin (2001). But then the intruder needs to update his distribution over the space of possible tables in a somewhat different fashion than that in Section 5.

As we noted in Section 1, small counts in a sample table may not necessarily correspond to small counts in a population table. Thus we need to adapt the strategies for assessing disclosure risk from Section 5 to deal with sample tables. Intuitively, as the sampling fraction gets smaller we expect disclosure risk to go down. But this may not be sufficient protection.

The US decennial census files from which that three-way table was extracted contain 53 categorical variables, cross-classified at multiple levels of geography. The kinds of computations we were able to implement on the three-way table in Section 5 do not necessarily scale well for such large class-classifications. Many of the calculations in Section 5 have a remarkable similarity to those involved in Bayesian model search with hierarchical log-linear models and especially the subclasses of decomposable and graphical models, just as the work on bounds for contingency tables in Dobra and Fienberg (2000, 2001) had intimate links to decomposable and graphical models. Thus tools such as those associated with the hyper-Markov laws in Dawid and Lauritzen (1993), and the suite of expert system tools in Cowell *et al.* (1999) may prove useful as we develop scalable approaches to disclosure limitation in large tables of counts.

ACKNOWLEDGEMENT

This research has been supported in part by National Science Foundation Grant No. EIA-9876619 to the National Institute of Statistical Sciences, and by a Marie Curie Fellowship of the European Community program "Improving The Human Research Potential" under the contract number HPMFCT-2000-00463 to the Universitat de València. We thank our colleagues on these projects, Susie Bayarri, George Duncan, Alan Karr, Steve Roehrig, and Ashish Sanil, for helpful ideas and comments.

REFERENCES

Bernardo, J. M. (1979). Expected information as expected utility, *Ann. Statist.* **7**, 686–690.

Chen, G. and Keller-McNulty, S. (1998). Estimation of identification disclosure risk in microdata, *J. Official Statist.* **14**, 79–95.

Cowell, R. G., Dawid, A. P., Lauritzen, S. L., and Spiegelhalter, D. J. (1999). *Probabilistic Networks and Expert Systems.* Berlin: Springer.

Dawid, A. P. and Lauritzen, S. L. (1993). Hyper Markov laws in the statistical analysis of decomposable graphical models. *Ann. Statist.* **21**, 1272–1317.

DeGroot, M. H. (1962). Uncertainty, information and sequential experiments, *Ann. Math. Statist.* **33**, 404–419.

Diaconis, P. and Sturmfels, B. (1998). Algebraic algorithms for sampling from conditional distributions. *Ann. Statist.* **26**, 363–397.

Dobra, A. (2002). Statistical Tools for Disclosure Limitation in Multi-way Contingency Tables. Ph.D. Thesis, Carnegie Mellon University, USA.

Dobra, A. and Fienberg, S. E. (2000). Bounds for cell entries in contingency tables given marginal totals and decomposable graphs. *Proc. Nat. Acad. Sci.* **97**, 11885–11892.

Dobra, A. and Fienberg, S. E. (2001). Bounds for cell entries in contingency tables induced by fixed marginal totals. *Statist. J. United Nations ECE* **18**, 363–371.

Dobra, A., Fienberg, S. E., Karr, A. and Sanil, A. (2002). Software systems for tabular data releases. *Int. J. Uncertainty, Fuzziness Knowledge Based Systems* (to appear).

Dobra, A., Tebaldi, C. and West, M. (2002). Reconstruction of contingency tables with missing data. *Tech. Rep.*, National Institute of Statistical Sciences, USA.

Doyle, P., Lane, J. Theeuwes, J., and Zayatz, L. (eds). (2001). *Confidentiality, Disclosure and Data Access: Theory and Practical Applications for Statistical Agencies*, Amsterdam: Elsevier.

Duncan, G. T., Fienberg, S. E., Krishnan, R., Padman, R. and Roehrig, S. F. (2001). Disclosure limitation methods and information loss for tabular data. *Confidentiality, Disclosure and Data Access: Theory and Practical Applications for Statistical Agencies* (P. Doyle, J. Lane, J. Theeuwes, and L. Zayatz, eds). Amsterdam: Elsevier, 135–166.

Duncan, G. T., Keller-McNulty, S., and Stokes, S. L. (2001). Disclosure risk vs. data utility: The R-U confidentiality map, *Tech. Rep.*, Los Alamos National Laboratory, USA.

Duncan, G. T. and Lambert, D. (1986). Disclosure-limited data dissemination (with discussion), *J. Am. Statist. Ass.* **81**, 10–28.

Duncan, G. T. and Lambert, D. (1989). The risk of disclosure for microdata, *J. Business EconStatist.* **7**, 207–217.

Duncan, G. T., and Pearson, R. B. (1991). Enhancing access to microdata while protecting confidentiality: Prospects for the future. *Statist. Sci.* **6**, 219–239 (with discussion).

Fienberg, S. E. (1994). Conflicts between the needs for access to statistical information and demands for confidentiality. *J. Official Statist.* **10**, 115–132.

Fienberg, S. E. and Makov, U. E. (1998). Confidentiality, uniqueness, and disclosure limitation for categorical data. *J. Official Statist.* **14**, 385–397.

Fienberg, S. E. and Makov, U. E. (2001). Uniqueness and disclosure risk: Urn models and simulation. *Research in Official Statist.* **4**, 23–40.

Fienberg, S. E., Makov, U. E., and Sanil, A. P. (1997). A Bayesian approach to data disclosure: Optimal intruder behavior for continuous data. *J. Official Statist.***13**, 75–89.

Fienberg, S. E. and Makov, U. E., and Steele, R. J. (1998). Disclosure limitation using perturbation and related methods for categorical data. *J. Official Statist.* **14**, 485–511 (with discussion).

Gelman, A., Carlin, J. B., Stern, H. S., and Rubin, D. B. (1995). *Bayesian Data Analysis*. London: Chapman and Hall.

Kass, R. E. and Raftery, A. E. (1995). Bayes factors. *J. Am. Statist. Ass.* **90**, 773–795.

Lambert, D. (1993). Measures of disclosure risk and harm, *J. Official Statist.* **9**, 313–334.

Madigan, D. and Raftery, A. E. (1994). Model selection and accounting for model uncertainty in graphical models using Occam's window, *J. Am. Statist. Ass.* **89**, 1535–1546.

Raghunathan, T. E. and Rubin, D. B. (2001). Multiple Imputation for statistical disclosure limitation. *Tech. Rep.*, Harvard University, USA.

Rubin, D. B. (1993). Satisfying confidentiality constraints through the use of synthetic multiply imputed microdata. *J. Official Statist.* **9**, 461–468.

Samuels, S. M. (1998). A Bayesian, species-sampling-inspired approach to the uniqueness problem in microdata disclosure risk assessment. *J. Official Statist.* **14**, 373–383.

Skinner, C. J. and Holmes, D. J. (1998). Estimating the re-identification risk per record in microdata. *J. Official Statist.* **14**, 361–372.

Schafer, J. L. (1997). *Analysis of Incomplete Multivariate Data.* London: Chapman and Hall.

Trottini, M. (2001). A decision-theoretic approach to data disclosure problems. *Res. Official Statist.* **4**, 7—22.

Trottini, M. and Fienberg, S. E. (2002). Modelling user uncertainty for disclosure risk and data utility. *Int. J. Uncertainty, Fuzziness Knowledge Based Systems* (to appear).

Willenborg, L. and De Waal, T. (1996). *Statistical Disclosure Control in Practice.* Berlin: Springer.

Willenborg, L. and De Waal, T. (2001). *Elements of Statistical Disclosure Control.* Berlin: Springer.

DISCUSSION

CHRISTOPHER MEEK (*Microsoft Research, USA*)

Congratulations to the authors for an interesting and thought-provoking paper. Methods for managing and assessing disclosure risks for information is an area of growing importance due to increasing amount of confidential information available electronically. It is useful to note that while the authors have used examples from census surveys for illustration, they could have chosen from a variety of alternative scenarios including disclosure of information from records and surveys on medical procedures, product purchases, internet usage, and subscriptions.

The paper provides a general decision-theoretic framework for assessing and managing disclosure risk. In addition, the authors describe techniques that begin to address the problem of inference about a joint table given released sub-tables; a problem that is central to using their framework. The goal of the general decision-theoretic framework is to allow an *agency* to balance the risks and benefits of disclosing information. The authors capture the benefit and risk with data-utility and data-risk functions, respectively, and evaluate the alternative potential releases by combining the data-utility and data-risk functions. The authors suggest combining these function with a thresholded decision rule. This suggestion seems odd; whereas this rule does utilize both the data utility and the data risk it does not balance them. If two sub-tables have the same data utility why not choose the one with the lower data risk?

It is most natural to consider the problem faced by such an agency as a multi-step decision problem or game-theoretic problem but, to simplify, the authors consider a myopic one-shot decision problem: should the agency release a sub-table $\mathcal{T}$ given previously released tables $\mathcal{R}$. The basic tactic employed by the authors in defining the benefits and risks is to equate the benefit (data utility) with the negative expected loss of the scientist and the data risk with the expected loss of the intruder. In their analysis, they ignore the game-theoretic aspects of the loss function; in actuality we would expect that the actions of the *intruders* and *scientists* are informed by their understanding of the system and their beliefs about the other participants, future requests and so on. The authors rather consider simplified loss functions that allow one to decompose the larger problem into more tractable sub-problems. Unfortunately, it is unclear whether this approach will allow one to create reasonable data utility and data risk functions. A fundamental question that the authors do not discuss is how one should go about improving the data utility and data risk functions. Should one interview intruders and scientists? Perhaps more importantly, how does one combine the loss functions of multiple intruders and scientists? One cannot simply add loss functions.

The authors demonstrate that they can use their framework to obtain data risk functions that have been used in the literature and thus ascribe a loss function to the putative intruder. This work demonstrates how extreme and unreasonable loss functions must be to reproduce the data risk functions, and points to the need to identify more realistic loss functions.

The development of more realistic loss functions in combination with the framework has other potentially useful applications. For instance, there are a number of alternative actions that the agency might take to protect data confidentiality when given a request for data. The authors focus on the case in which the agency simply must decide which table, if any, to release. Other methods for releasing data such as alteration, masking, or releasing ranges of values could be compared and evaluated using the framework when coupled with realistic loss functions.

Throughout the paper the authors equate small counts with high data risk. At an intuitive level this seems reasonable because counts of one or two might be identifiable and lead to the release of identifiable confidential information either directly through information released by the agency or indirectly through the linking of information release by the agency and other information sources. The following examples illustrate potential problems of equating small counts with high data risk.

First, consider a sub-table with attributes hair color, height and town. Let us suppose that there is one small town in which there is one person who has red hair and who is 7 feet tall. Clearly this person is identifiable (just go to the town and you can probably find the person!), however, this information alone is not confidential.

Second, consider a sub-table with attributes corresponding to a person's favorite movie, favorite book, and favorite color. In this situation, there are certainly going to be some entries that will have small counts. Unlike the previous case, these attributes do not lead to identifiability because we do not wear our preferences on our shirt sleeves.

Finally, consider the sub-table given in Table 6 of Income (I) and Hair Color (H) for town X. In this table, none of the entries is small in the sense discussed by the authors. Nonetheless, a release of such a table does seem to provide a release of information—that is, all blondes have high income in town X—and the blondes in town X are identifiable (again, just go to the town). If income is deemed to be confidential, then the release of this sub-table would be a release of identifiable confidential information. This example illustrates that the release of confidential information is not limited to the case of small counts.

Table 6. *Example of disclosure without small counts.*

	Town = X	
	I = High	I = Low
Black	4	4
Red	4	4
Blond	4	0

The last section of the paper is devoted to a specific technical problem central to the application of the general framework: the problem of computing a posterior over the joint table given a set of released sub-tables $\mathcal{R}$. This problem is not novel to the analysis of disclosure risk; it also arises in sociology, political science, spatial epidemiology and ecology where it is know as the ecological inference problem. The authors choose to attack the problem by considering the joint table to be a sample from an infinite super-population. In the simple case, the authors then consider a fixed log-linear model and utilize a two-step Gibbs sampler with one step requiring MCMC and the other a "Bayesian iterative proportional fitting" procedure. In the more complicated case, the author combines the Gibbs sampling approach with model averaging over a set of alternative log linear models.

For the MCMC step, the authors are to be congratulated on what is probably the first Bayesian use of the Markov basis results of Diaconis and Sturmfels (1998). They construct a Markov chain to sample tables from the posterior of tables that satisfy the released sub-tables. Unfortunately, the computation of the Markov basis is very expensive, and precious little is known about mixing times of Markov chains constructed from them. An alternative approach to sampling a table from the posterior of tables that satisfy the released sub-tables is based on sequential cell sampling (Dobra *et al.,* 2002). In this approach, one sequentially samples every cell and updates the upper and lower bounds. Clearly, this method can also be extremely expensive computationally (potentially exponential in cost). Given these computational difficulties, which of these methods do you think will be able to scale to large problems? Do we need to develop some new class of approaches?

The specific approach advocated by the authors is not completely specified. When performing model averaging, neither how one should go about selecting the set of alternative models, nor how one ought to compute the model posteriors for these alternatives is clear. What is the precise recipe? For some classes of models, for instance, non-decomposable and non-graphical models, the computation of the marginal likelihood is non-trivial even in the case of complete data. The situation in which one is computing the marginal likelihood of released sub-tables $\Pr(\mathcal{R} \mid \mathcal{A}_l)$ is certainly more challenging. If one wishes to include such alternative models, it may be useful to combine the model averaging with the Gibbs sampling by using a reversible jump MCMC approach.

In addition, given the focus on the small counts, it would seem sensible to carefully consider the sensitivity of the results.

Finally, the authors consider the problem of computing a posterior on the joint table given a set of released sub-tables when, in fact, there is more information to consider. Perhaps the most important information that is not accounted for is the refusal to release. Consider an agency that uses the rule that it will release no sub-table that makes a cell unique. In this case, if a request for a two-by-two table with a grand total of four is refused after both of the margins have been released, then there is only one table that is possible. Because the rule used by the agency is a function of the actual joint table, the tactic used by the authors of using a super-population to make inferences about the joint table complicates the inferential process of conditioning on the information provided by the refusal to release a sub-table. Whereas assumptions such as the super-population assumption allows one to use many standard statistical tools, perhaps it is worthwhile considering alternative approaches to this problem.

REPLY TO THE DISCUSSION

We thank Chris Meek for insightful comments and questions and for his clear appreciation of the complexity of the problem we are attempting to address. As he notes, issues of confidentiality and disclosure limitation arise in many different contexts. Most recently we have begun to consider them as part of a research effort in computer security. We organize our response according to the three components of the paper: understanding disclosure risk for categorical data, aspects of our decision-theoretic approach, and the computing the posterior distribution over feasible tables given data releases.

Understanding disclosure risk. Meek asks why we equate small counts with high data risk. As we noted in our introduction, we distinguish, as does the confidentiality literature, between *identity* disclosure (through uniqueness in the population) and *attribute* disclosure (see also Duncan *et al.,* 1993). His example of identity disclosure involves a small town with only one person with red hair and who is 7 feet tall and he notes that this information alone may well not be confidential. This is true as far as he goes. But if the two variables are only a subset of those involved in our multi-way table, then all of the values on the other variables in the database on this individual are disclosable and this may indeed be a serious problem. It is largely for this reason that statistical agencies and others who gather data promise those that supply it that the data will not be released in a form that allows individuals to be identified. Protecting identities then becomes by definition a primary confidentiality issue.

Meek's second example, of a small town where all blonds have high income, is an example of attribute disclosure—we do not learn the identity of any specific blond but we learn that they all have high income. We explicitly set such an issue aside in our paper, although it turns out that we could easily adapt the discussion of both decision-theoretic criteria and intruder identification to deal with this situation. For example, instead of looking at bounds on all cells and then focusing on small cells for identity disclosure, we can focus on bounds for and cells whose values almost equal one or more of the released marginal values to which it sums for attribute disclosure.

Aspects of our decision-theoretic approach. Meek asks about our use of the threshold rule which is widely used in the current literature on statistical confidentiality. Our simplified version implicitly assumes that for a fixed threshold α there is a unique data mask that maximizes data utility among all the available masks with disclosure risk below α (this is the most common case in applications). Clearly this simplification does not perform well when there are two or more masks that maximize data utility (the example raised by Meek). We implicitly assume in this case, however, that the mask with minimum risk should be selected. Thus, the rule actually balances disclosure risk and data utility. A general threshold rule, that solves the incoherence highlighted by Meek, could be stated as follows: *Let $\mathcal{D}$ be the class of all available masks and let $\mathcal{D}_\alpha$ be the subclass of $\mathcal{D}$ containing all masks with disclosure risk below α. Finally, let $\mathcal{D}_\alpha^{\mathrm{OPT}}$ be the class of masks in $\mathcal{D}_\alpha$ that maximize data utility. If $\mathcal{D}_\alpha^{\mathrm{OPT}}$ is empty, release no data. Otherwise release the mask in $\mathcal{D}_\alpha^{\mathrm{OPT}}$ with minimum disclosure risk.*

The suggestion that we formalize the disclosure limitation problem as a multi-step decision problem or game-theoretic problem is appealing at a first glance and it seemed to us the most natural choice when we first approached the problem. In the end, we preferred the one-shot decision problem formalization because: (i) the real-world disclosure limitation problem is actually a one-shot problem: the agency release the data, the users make their move (estimate of their targets); (ii) there is a strong asymmetry of information between the agency and the users. The agency knows the original data, the alternative masks available, the mask that has been used, the users' targets and loss functions, and it fixes the threshold value for the maximum tolerable disclosure risk. On the other hand, users only know that the released data have been obtained using a particular masking technique and sometimes they ignore the same threshold rule. Users' understanding of the system is quite limited and their beliefs about other participants (the agency in particular) are so vague that they are of little inferential value. Thus, the one-shot decision approach is not a myopic simplification of the problem but rather a very good approximation of how things works in reality. The key point is that in order to fill its mission of data dissemination, the agency does not need to reveal its strategy to the users but simply provide the data in a form that is useful for the *scientist* and safe from the attack of an *intruder*.

Meek questions how we should approach defining suitable loss functions for the intruder and the researcher. Many statistical agencies either ignore the uses of the data they release or claim that taking into account all users' targets is impossible. Instead, they use heuristic *adhoc* measures that try to preserve basic features of the original data while preserving confidentiality. Our framework clearly shows that defining measures in such a way does not avoid the problem of defining targets and loss functions. Rather, *adhoc* measures turn out to correspond to very specific (and often unrealistic) choices of them. While defining users' targets and loss functions is not easy, it is necessary and unavoidable if we want suitable measures of disclosure risk and data utility. The idea of interviewing the users is appealing but difficult to implement for the intruders that matter the most. For scientists the job is easier, however, since most statistical agencies already keep track of requests from users and design forms of data release that can meet these requests. Thus, if an agency knows that the principal use of the data is the estimation of a set of statistical models, then it can design suitable releases to preserve the statistical features of the parameters of these models. To define intruders' targets and loss functions the agency needs to reflect on what information in the original data needs to be protect and then consider a range of possible attacks an intruder might attempt to disclose this information.

Combining the loss functions of multiple users is also relevant. For disclosure risk, a reasonable solution is to use a worst case strategy and adopt the maximum disclosure risk measured for different targets on a comparable scale. The data utility case is more complex. We could consider linear combinations of users' loss, with different weights for different users and we are currently working on this problem. The current version of the framework nonetheless allows us to identify situations where there exists a suitable data mask that can meet the needs of several users of the data while preserving confidentiality and situations where the targets of the multiple scientists are not compatible (*i.e.*, masks suitable for one user are almost useless for another).

Computing the posterior distribution over feasible tables. As Meek notes this problem also arises in other contexts such as ecological inference (*e.g.*, see King, 1997) as well as tomography image reconstruction (*e.g.*, see Gelman, 1989), although in these other settings we are typically interested in some form of model whereas in the disclosure case we are not necessarily.

Meek asks which of the methods we explore holds the promise to scale to large problems and whether we should consider developing some new classes of approaches. The approach proposed by Diaconis and Sturmfels is not likely to scale because of the huge complexity of the computations required to compute a Markov basis. Although there exists a formula for dynamically generating a Markov basis in the decomposable case (Dobra, 2001) and ways to significantly speed up the computations in some other special cases (see, *e.g.*, the divide-and-conquer algorithm of Dobra and Sullivant, 2002), there is little chance that a Markov basis associated with a higher dimensional table and an arbitrary set of marginals will be easily computable in the near future. The sequential cell sampling approach of Dobra *et al.* (2002) offers a possible answer to this general problem, but this method relies on the computation of bounds, which is itself a hard problem for large tables. Therefore neither approach seems to scale to large tables and we need to explore new avenues as well as approximations that use reasonable amounts of computation.

Meek argues that we did not completely specify our approach in computing the posterior distribution over the space of feasible tables. Our example was small enough that, at least for it, we believe that there was no ambiguity, and we tried to make clear how one can compute or at least approximate the posterior distribution for each model. Nonetheless, there is clearly a major issue here for high-dimensional tables. Ideally we need to average over the space of all possible hierarchical log-linear models "consistent" with the released marginals, but, as Meek notes, the computations appear daunting without the added complexity of simultaneously using a reversible-jump MCMC approach. We are currently exploring approximations based on decomposable models but we also need to be able to compute the posterior distribution for the model whose minimal sufficient statistics correspond to the released margins, so we do not as yet have any recipe.

Meek suggested that it would be sensible to consider the sensitivity of our results to the specification of the model priors. We agree, and expect to explore alternatives such as the compatible priors associated with hyper-Markov laws.

Finally, we recognize the need to take into account the information supplied to an intruder through the refusal to release tables. Meek presents an extreme example, and the more likely scenario is that an agency is willing to release most lower-order margins and its refusals will occur as the dimensionality grows. Our bounds work (*e.g.*, see Dobra *et al.,* 2002) is clearly consistent with this scenario, as is the US Census Bureau's American FactFinder system, which has been constructed to release at most three-way tables. But this is one of the reasons why the option of perturbing a table before releasing a new "masked" margin is an option worthy of exploration.

ADDITIONAL REFERENCES IN THE DISCUSSION

Dobra, A. (2001). Markov bases for decomposable graphical models. *Tech. Rep.*, National Institute of Statistical Sciences, USA.

Dobra, A. and Sullivant, S. (2002). A divide-and-conquer approach for generating Markov bases of multi-way tables. *Tech. Rep.*, National Institute of Statistical Sciences, USA.

Duncan, G. T., Jabine, T. B., and de Wolf, V. A. (eds). (1993). *Private Lives and Public Policies: Confidentiality and Accessibility of Government Statistics*. Washington, DC: National Academy Press.

Gelman, A. (1989). Constrained maximum entropy methods in an image reconstruction problem, in *Maximum Entropy and Bayesian Methods* (J. Skilling, ed). Dordrecht: Kluwer, 429–435.

King, G. (1997). *A Solution to the Ecological Inference Problem: Reconstructing Individual Behavior from Aggregate Data*. Princeton: Princeton University Press.

BAYESIAN STATISTICS 7, pp. 145–161
J. M. Bernardo, M. J. Bayarri, J. O. Berger, A. P. Dawid,
D. Heckerman, A. F. M. Smith and M. West (Eds.)

Bayesian and Frequentist Multiple Testing

CHRISTOPHER GENOVESE and LARRY WASSERMAN
Carnegie Mellon University, USA
genovese@stat.cmu.edu larry@stat.cmu.edu

SUMMARY

We introduce a Bayesian approach to multiple testing. The method is an extension of the false discovery rate (FDR) method due to Benjamini and Hochberg (1995). We also examine the empirical Bayes approach to simultaneous inference proposed by Efron *et al.* (2001). We show that, in contrast to the single hypothesis case—where Bayes and frequentist tests do not agree even asymptotically—in the multiple testing case we do have asymptotic agreement.

Keywords: MULTIPLE TESTING; P-VALUES; FALSE DISCOVERY RATE; BOOTSTRAP.

1. INTRODUCTION

There has been renewed interest in simultaneous inference due to the abundance of large, complex data sets. As a motivating example, consider genetic microarrays. Suppose there are two groups (treatment and control) and m genes. The data, after non-trivial pre-processing, take the following form:

	Control				Treatment			
gene 1	X_{11}	X_{12}	$\cdots$	X_{1k}	Y_{11}	Y_{12}	$\cdots$	Y_{1k}
gene 2	X_{21}	X_{22}	$\cdots$	X_{2k}	Y_{21}	Y_{22}	$\cdots$	Y_{2k}
$\vdots$	$\vdots$	$\vdots$		$\vdots$	$\vdots$	$\vdots$		$\vdots$
$\vdots$	$\vdots$	$\vdots$		$\vdots$	$\vdots$	$\vdots$		$\vdots$
gene m	X_{m1}	X_{m2}	$\cdots$	X_{mk}	Y_{m1}	Y_{m2}	$\cdots$	Y_{mk}

Each row is a gene and each column is a microarray. Typically, m is around 5000, and newer arrays will have as many as 50,000 genes. Each data point represents the expression level of the gene. This is a measure of how active the gene is, i.e. how much protein is being produced by that gene. Let $H_i = 1$ if the treatment changes the distribution of the expression level for gene i and $H_i = 0$ otherwise. We want to test

$$H_i = 0 \text{ versus } H_i = 1 \qquad \text{for } i = 1, \ldots, m.$$

The microarray example is a prototype; the ideas that follow apply more generally. The key feature is that m is large but k is small.

2. MODERN FREQUENTIST MULTIPLE TESTING

Let $P^m = (P_1, \ldots, P_m)$ where P_i is a p-value for testing hypothesis H_i. Multiple testing methods involve choosing a threshold $T = T(P^m)$ and rejecting all hypotheses whose p-values are less than T. The most common methods are uncorrected testing: reject $H_i = 0$ if $P_i < \alpha$; and family-wise corrected testing such as the Bonferroni method: reject $H_i = 0$ if $P_i < \alpha/m$.

Uncorrected testing does not adequately control false positives. Bonferroni and its relatives control the probability of a single error which is too strict for large m. Two recent methods that avoid these extremes are the *false discovery rate* (FDR) method and the *empirical Bayes testing* (EBT) method. The FDR method is due to Benjamini and Hochberg (1995)—hereafter referred to as BH—and has since been extended by many others including Benjamini and Yekutieli (2002), Storey (2002) and Genovese and Wasserman (2001, 2002). The EBT method is due to Efron *et al.* (2001). Of course, empirical Bayes is not new but the particular way that Efron *et al.* use empirical Bayes for multiple testing has new twists. We would also like to mention that interesting connections between FDR and Bayes are discussed in Storey (2001).

2.1. *A Model for Multiple Testing*

Let $(P_1, H_1), \ldots, (P_m, H_m)$ be independent pairs such that $P_i|H_i = 0 \sim \text{Uniform}(0, 1)$ and $P_i|H_i = 1 \sim \Xi_i$ for some cdf Ξ_i which is stochastically smaller than the uniform cdf $U(t) = t$. Let $P^m = (P_1, \ldots, P_m)$ and $H^m = (H_1, \ldots, H_m)$. We further assume that $H_i \sim \text{Bernoulli}(a)$ and that the $\Xi_i' s$ are random distribution functions drawn from some distribution $\mathcal{L}$. The model may be written as

$$\begin{aligned} H_1, \ldots, H_m &\sim \text{Bernoulli}(a) \\ \Xi_1, \ldots, \Xi_m &\sim \mathcal{L}(d\xi) \\ P_i|H_i = 0, \Xi_i = \xi_i &\sim \text{Uniform}(0, 1) \\ P_i|H_i = 1, \Xi_i = \xi_i &\sim \xi_i. \end{aligned}$$

The marginal distribution of P_i is

$$P_1, \ldots, P_m \sim G = (1 - a)U + aF \tag{1}$$

where

$$F(t) = \int \xi(t) d\mathcal{L}(\xi). \tag{2}$$

In the non-parametric case we leave $\mathcal{L}$ unspecified and hence F is arbitrary except for the stochastic dominance condition.

The restriction to p-values is not necessary. In general, we let

$$D_i = (X_{i1}, \ldots, X_{ik}, Y_{i1}, \quad \ldots, Y_{ik})$$

denote all the data associated with H_i and we let V_i be some one-dimensional summary statistic or test statistic derived from D_i. The marginal model for V_i is $R = (1 - a)R_0 + aR_1$.

Example. (Normal Example). An example that will be useful for illustrative purposes is the following. Let $V_i \sim N(\theta_i, 1)$ where $\theta_i = 0$ when $H_i = 0$ and where $\theta_i = \theta$ when $H_i = 1$. Here, θ is an unknown parameter. Using a common (but unknown) alternative for each hypothesis makes this a toy example.

2.2. *FDR*

The Benjamini–Hochberg (BH) procedure rejects all null hypotheses for which $P_i \leq T \equiv P_{(j)}$ where

$$j = \max\left\{0 \leq i \leq m : \; P_{(i)} \leq \alpha\frac{i}{m}\right\}, \tag{3}$$

and $0 \equiv P_{(0)} < P_{(1)} < \cdots < P_{(m)}$ denote the ordered p-values. BH (1995) proved that

$$E(\text{FDR}) \leq (1-a)\alpha \leq \alpha, \tag{4}$$

where FDR is the *realized false discovery rate*, the number of false rejections divided by the number of rejections.[1] The BH result is remarkable: it holds regardless of how many nulls are true and regardless of the distribution of the p-values under the alternatives. In fact, they proved the stronger result

$$\sup_{h^m,\xi^m} \mathrm{E}_{J(h^m,\xi^m)}(\text{FDR}) \leq \alpha$$

where the supremum is over all binary vectors h^m, all vectors of distribution functions $\xi^m = (\xi_1, \ldots, \xi_m)$,

$$J(h^m, \xi^m) = \bigotimes_{i=1}^{m} U^{1-h_i}\xi_i^{h_i}.$$

There is an alternative way of viewing the BH procedure due to Storey (2002) and Genovese and Wasserman (2001, 2002). Consider rejecting all p-values less than some fixed threshold t. Genovese and Wasserman define the *realized false discovery rate process* by

$$\Gamma(t) = \frac{\sum_i \mathrm{I}(P_i \leq t)(1-H_i)}{\sum_i \mathrm{I}(P_i \leq t) + \prod_i \mathrm{I}(P_i > t)}, \qquad 0 \leq t \leq 1 \tag{5}$$

where the second term in the denominator forces $\Gamma(t)$ to be 0 when there are no rejections. Regarded as a function of the threshold t, this is a stochastic process. Typically, $\prod_i \mathrm{I}(P_i > t)$ is exponentially small and hence,

$$\Gamma(t) \approx \frac{\sum_i \mathrm{I}(P_i \leq t)(1-H_i)}{\sum_i \mathrm{I}(P_i \leq t)} \overset{d}{=} \frac{\text{Binomial}(m,(1-a)t)}{\text{Binomial}(m,G(t))}.$$

Thus, $\mathrm{E}[\Gamma(t)] = Q(t) + O(m^{-1/2})$ where

$$Q(t) = \frac{(1-a)t}{G(t)}. \tag{6}$$

If we want the expected FDR to be less than α, this suggests choosing a threshold t_* defined by

$$t_* = \max\{t : \quad Q(t) \leq \alpha\}. \tag{7}$$

It will then follow that $E(\Gamma(t_*)) = Q(t_*) + O(m^{-1/2}) \leq \alpha + O(m^{-1/2})$. Unfortunately, $t_* \equiv t_*(G, a)$ is a function of G and a which are unknown. An obvious thing to do is to find estimates $\hat{G}$ and $\hat{a}$ for G and a and then use the threshold $\hat{t} = t_*(\hat{G}, \hat{a})$. To be more explicit,

$$\hat{t} = \max\{t : \quad \hat{Q}(t) \leq \alpha\}. \tag{8}$$

where

$$\hat{Q}(t) = \frac{(1-\hat{a})t}{\hat{G}(t)}.$$

---

[1] BH call E(FDR) the false discovery rate. We call FDR the realized false discovery rate and E(FDR) the expected false discovery rate.

If we use the conservative estimator $\hat{a} = 0$, then we get back the BH procedure, as pointed out by Storey (2001). More power can be obtained by taking a less conservative estimator for a such as

$$\hat{a} = \frac{\hat{G}(1/2) - \frac{1}{2}}{1 - \frac{1}{2}}$$

which was suggested by Storey (2002). Other estimators of a are considered in Storey (2001) and Genovese and Wasserman (2001). Since $\hat{t}$ is obtained from (8) instead of (7), it is not obvious that $E[\Gamma(\hat{t})] \leq \alpha + O(m^{-1/2})$. After all, $\Gamma(\hat{t})$ is a random process $\Gamma(\cdot)$ evaluated at a random point $\hat{t}$ and, moreover, $\Gamma(\cdot)$ and $\hat{t}$ are correlated. However, Genovese and Wasserman (2001) showed that $E[\Gamma(\hat{t})] \leq \alpha + O(m^{-1/2})$ does hold assuming reasonable regularity conditions. A close inspection of their proof reveals that this result will continue to hold without the assumption of independence as long as the empirical distribution $\hat{G}$ is a consistent estimator of G.

The above procedures control the expected FDR. Genovese and Wasserman (2001) introduced *confidence thresholds* which control the realized FDR. Given c and α, a random variable $T = T(P^m)$ is level (c, α) *confidence threshold* if $P(\Gamma(T) < c) \geq 1 - \alpha$. In this paper, we introduce Bayesian procedures that control expected and realized FDR.

On the V scale, assuming that rejections correspond to large values of V, then all the above discussion applies with

$$Q(v) = \frac{(1-a)(1-R_0(v))}{(1-R(v))}. \tag{9}$$

2.3. *EBT*

The empirical Bayes approach for multiple testing, due to Efron *et al.* (2001), works as follows. First, suppose that a and $f = F'$ are known. Then, Bayes' theorem yields

$$P(H_i = 0|P^m) = P(H_i = 0|P_i) = \frac{1-a}{g(P_i)} \equiv q(P_i) \tag{10}$$

where $g(t) = (1-a) + af(t)$. (The density under the null is $f_0(t) = 1$.) In terms of V, we have $q(V_i) = (1-a)r_0(V_i)/r(V_i)$ where $r_0 = R_0'$ and $r = R'$.

When a and f are not known, we proceed as follows. Let $\hat{g}$ be an estimate of g. Since $f(t) \geq 0$ for all t, it follows that $g(t) \leq 1 - a$ for all t. Hence, $a \geq 1 - \min g(t)$ which suggests the estimator $\hat{a} = 1 - \min \hat{g}(t)$. Finally, define

$$\hat{q}(P_i) \equiv \frac{1-\hat{a}}{\hat{g}(P_i)} = \frac{\min_t \hat{g}(t)}{\hat{g}(P_i)}.$$

On the V scale we get $\hat{a} = 1 - \inf_v \hat{r}(v)/r_0(v)$ and

$$\hat{q}(V_i) = \frac{\min_v \frac{\hat{r}(v)}{r_0(v)}}{\hat{r}(V_i)}.$$

A similar approach is used in Efron *et al.* (2001). Actually, when r_0 and a are unknown, it is no longer the case that H_i is conditionally independent of V_j for $j \neq i$ so one should compute $P(H_i = 0|V^m)$ rather than $P(H_i = 0|V_i)$. However, $P(H_i = 0|V_i)$ can be regarded as an approximation to $P(H_i = 0|V^m)$.

An issue not addressed by Efron *et al.* – that we will address in this paper – is the accuracy of $\hat{q}$. This is important since the cases of interest are when V_i are large and the accuracy of $q(V_i)$ will then be driven by the tails of $\hat{r}$.

2.4. *Thresholds or Posterior Probabilities?*

Should we report $P(H_i = 0|V^m)$ or should we choose a threshold T to have a given FDR? Reporting $P(H_i = 0|V^m)$ is informative and intuitive but does not control the number of false positives. On the other hand, choosing an FDR-threshold automatically controls the errors but we lose the quantification of the strength of evidence provided by $P(H_i = 0|V^m)$. Our recommendation is to compute both. This provides the experimenter with the best of both worlds.

The FDR and the posterior probability are related. Consider a fixed threshold $T(P^m) \equiv t$ on the p-value scale. Recall that $E(\Gamma(t)) \approx Q(t) = (1-a)t/G(t)$ and $P(H_i = 0|P^m) = q(t) = (1-a)/g(t)$. Suppose that G is concave, as it is in most problems. Then we have the following FDR–EBT relation:

$$tg(t) \leq \frac{Q(t)}{q(t)} = \frac{tg(t)}{G(t)} \leq 1.$$

Now, consider a Bayesian who chooses to reject whenever $q(t)$ is less than some probability β. This defines a threshold rule $T_B(P^m) = \hat{q}^{-1}(\beta)$. Under appropriate conditions, $T \xrightarrow{P} q^{-1}(\beta) = t_*$ and thus, asymptotically, $Q(T) \approx Q(t_*) = \beta t g(t)/G(t) \leq \beta$ since $G(t) \geq tg(t)$. From an FDR perspective, the Bayesian is being conservative. Storey (2001) and Efron *et al.* (2001) discuss other relationships between Bayes and FDR. In particular, $Q(t) = P(H = 0|P \leq t)$.

Example. Return to the toy example. Figure 1 shows Q and q and illustrates the conservativeness of the using the posterior probability to define the rejection threshold.

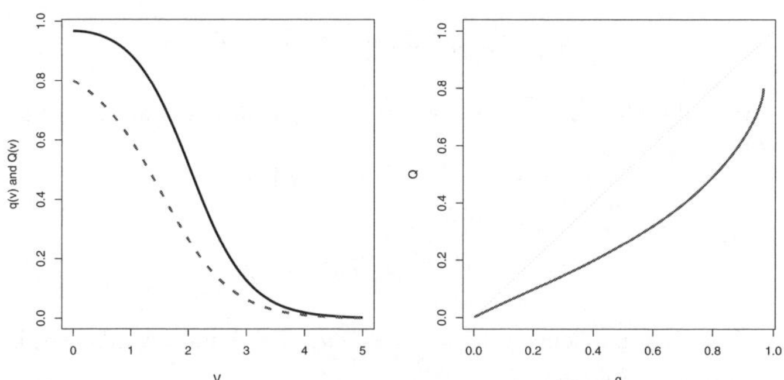

Figure 1. *Q and q.*

As we shall see, even if we want to focus on FDR instead of empirical Bayes, we will still need to use the quantity q. Hence, the accuracy of $\hat{q}$ is of importance from either point of view.

3. ASYMPTOTIC BEHAVIOR OF THE $\hat{Q}$ PROCESS

The Bayesian who wants to report posterior probabilities must use $\hat{q}$ in place of q since a and g are unknown. Moreover, the confidence threshold methods we describe later also depend on knowing q. The implications of having to estimate q are best understood by examining the asymptotics of $\hat{q}$ viewed as a stochastic process. For simplicity, first assume that a is known. In what follows we work on the V scale. As a direct consequence of the functional delta method we have the following result.

Theorem 1. *Let $\hat{r}(v)$ be an estimator of $r(v)$. Suppose that*

$$m^{\alpha}(\hat{r}(v) - r(v)) \overset{d}{\to} W$$

for some $\alpha > 0$, where W is a mean 0 Gaussian process with covariance kernel $\tau(v, w)$. Then

$$m^{\alpha}\ (\hat{q}(v) - q(v)) \overset{d}{\to} Z \tag{11}$$

where

$$Z(v) \overset{d}{=} -\frac{(1-a)r_0(v)W(v)}{r^2(v)}.$$

Hence, Z is a Gaussian process with mean 0 and covariance kernel

$$K_q(v, w) = \frac{(1-a)^2\tau(v, w)r_0^2(v)r_0^2(w)}{r(v)^4 r(w)^4}. \tag{12}$$

To explore this further, we need to say something specific about $\hat{r}$. We consider two cases: parametric models and kernel density estimators.

In the parametric case, $q(v) = (1-a)r_0(v)/((1-a)r_0(v) + ar_\theta(v))$ and let us assume that θ is a scalar parameter. Let $\hat{\theta}$ be a regular, $\sqrt{n}$-consistent estimator of θ and let $\hat{q}(v) = (1-a)r_0(v)/((1-a)r_0(v) + ar_{\hat{\theta}}(v))$. The asymptotic standard error of $\hat{q}(v)$ is

$$\mathrm{se}_v = \mathrm{se}(\hat{\theta})q(v)(1-q(v))|\dot{\ell}_\theta(v)|$$

where $\mathrm{se}(\hat{\theta})$ is the standard error of $\hat{\theta}$ and $\ell_\theta(v) = \log r_\theta(v)$. In the case where $V \sim N(\theta, \sigma^2)$,

$$\mathrm{se}_v = \frac{\sigma}{\sqrt{m}} q(v)(1-q(v))|v-\theta|.$$

Since we are especially interested in cases where $q(v)$ is small, it is more relevant to consider the relative error

$$\mathrm{rel}_v = \frac{\mathrm{se}_v}{q(v)} = \mathrm{se}(\hat{\theta})(1-q(v))|\dot{\ell}_\theta(v)|.$$

In the Normal case,

$$\mathrm{rel}_v = \frac{\sigma}{\sqrt{m}}(1-q(v))|v-\theta|.$$

In the tails, $1 - q(v) \approx 1$ and hence $\mathrm{rel}_v \approx \frac{\sigma}{\sqrt{m}}|v-\theta|$. This suggests that $\hat{q}(v)$ is reliable in a neighborhood of order $\sqrt{m}$ around θ.

Now consider kernel density estimation:

$$\hat{r}(v) = \frac{1}{m}\sum_{i=1}^{m}\frac{1}{h_m}K\left(\frac{v-V_i}{h_m}\right)$$

where K is a kernel and h_m is the bandwidth. The usual choice of bandwidth $h_m = O(m^{-1/5})$ yields an asymptotically biased estimator and hence (11) fails. Assume, therefore, that we undersmooth the density estimate, for example $h_m = O(m^{-1/4})$. The asymptotic variance of $\hat{r}(v)$ is $c^2 r(v)/(mh_m)$ where $c^2 = \int K^2(v)dv$. Hence, the standard error $\hat{q}(v)$ is

$$\mathrm{se}_v = \frac{c\,q(v)}{\sqrt{mh_m r(v)}}$$

with relative error

$$\text{rel}_v = \frac{c}{\sqrt{mh_m r(v)}}.$$

In particular, when $h_m = 1/m^\beta$, the relative error is

$$\text{rel}_v = \frac{c}{m^{(1-\beta)/2}\sqrt{r(v)}}.$$

This will be small when

$$r(v) > \frac{c'}{m^{1-\beta}}$$

for some c', suggesting unreliability in the tails as expected.

Example. Figure 2 shows the relative error when $r(v) = 0.5N(0,1) + 0.5N(1,1)$. The solid line is for a kernel density estimator (with $\beta = 1/4$) and the dashed line is under the parametric model. The shapes of the curves show the striking difference between the two in the tails.

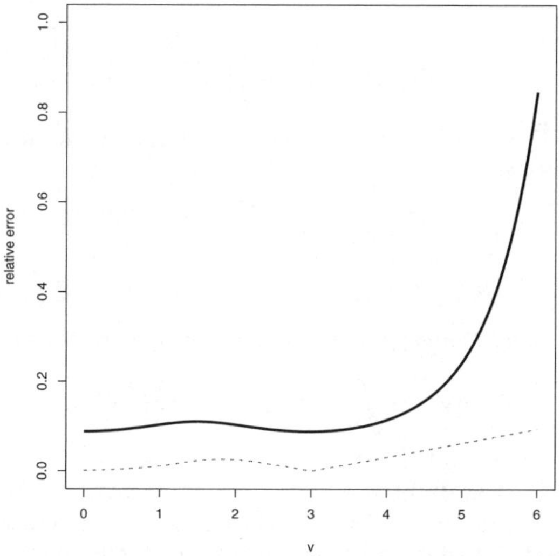

Figure 2. *Relative error in estimating $\hat{q}$. Solid line: non-parametric. Dashed line: parametric.*

When estimation of a is taken into account, things get more complicated. Recall that

$$\hat{q}(v) = \hat{r}(v)^{-1} \min_t \frac{\hat{r}(t)}{r_0(t)}.$$

In general, $\hat{q}$ is not a Hadamard differentiable function of $\hat{r}$ making it difficult to get a limit law without further assumptions. Results will be reported elsewhere.

4. BAYESIAN FDR

Let $\Gamma = \{\Gamma(t) : 0 \le t \le 1\}$ denote the entire realized FDR process. From the Bayesian point of view, the posterior of Γ is completely determined by the posterior for H^m. Recall that

$V_i \sim (1-a)R_0 + aR_1$ where R_1 depends on unknown parameters θ. In the non-parametric case, θ could be infinite dimensional. Let $\psi = (a, \theta)$. Note that the H^m are conditionally independent given V^m and ψ. If V_i is sufficient, H^m are conditionally independent given V^m and ψ. Let $\hat{\psi}$ be a consistent estimator of ψ. Then,

$$\begin{aligned} P(H^m = h^m | V^m) &= \int P(H^m = h^m | V^m, \psi) f(\psi | V^m) d\psi \\ &\approx P(H^m = h^m | V^m, \hat{\psi}) \\ &= \prod_i P(H_i = h_i | V_i, \hat{\psi}) \\ &\approx \prod_i (1 - \hat{q}(V_i))^{h_i} \hat{q}(V_i))^{1-h_i}. \end{aligned}$$

It follows that a simple method for generating random draws of Γ from the (approximate) posterior is:

$$H_i \sim \text{Bernoulli}(1 - \hat{q}(V_i)), \quad i = 1, \ldots, m$$

$$\text{set } \Gamma(v) = \frac{\sum_i I(V_i > v)(1 - H_i)}{\sum_i I(V_i > v) + \prod_i I(V_i < v)}.$$

Theorem 2. *Let $\tilde{\Gamma} \sim \Gamma | V^m$. Suppose that q is known. Under appropriate regularity conditions we have that*

$$\begin{aligned} \tilde{\Gamma}(v) | V^m &\approx N\left(\frac{\sum_i I(V_i > v) q(V_i)}{\sum_i I(V_i > v)}, \frac{\sum_i I(V_i > v) q(V_i)(1 - q(V_i))}{(\sum_i I(V_i > v))^2} \right) \\ &\equiv N\left(\mu_q(v), \tau_q^2(v)\right). \end{aligned}$$

When q is unknown, the limiting posterior is a mixture of normals, namely,

$$\tilde{\Gamma}(v) | V^m \approx \int N\left(\mu_q(v), \tau_q^2(v)\right) d(q | V^m).$$

Example. Figure 3 is based on 100 observations with $a = 0.5$ and $\theta = 3$. The first panel shows 1000 draws from the posterior for Γ. The second panel shows $\hat{q}$ as a function of v. Figure 4 shows the posterior for $\Gamma(2)$ and the Normal approximation for this posterior.

4.1. *Controlling Posterior* FDR

Rather than controlling E(FDR), a Bayesian would prefer to control E(FDR$|V^m$). Define

$$T_{Bayes} = \sup\{t : E(\Gamma(t) | V^m) \leq \alpha\}.$$

By definition, $E(\Gamma(T_{Bayes}) | V^m) \leq \alpha\}$ and clearly this rejects as many hypotheses as possible while controlling E(FDR$|V^m$).

Bayesian confidence thresholds are obtained as follows. Let

$$T = \sup\{t : P(\Gamma(t) > c | V^m) < \alpha\}. \tag{13}$$

It follows that $P(\Gamma(T) < c | V^m) \geq 1 - \alpha$ and hence T is a (c, α) posterior confidence threshold. But is T is a frequentist confidence threshold? If so, this represents an important instance of

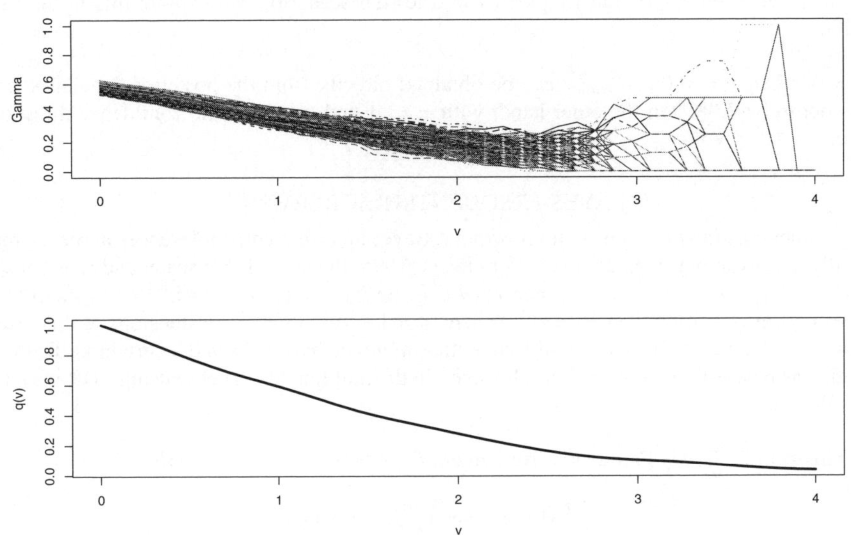

Figure 3. *Simulations of* Γ *from the posterior and* $\hat{q}$.

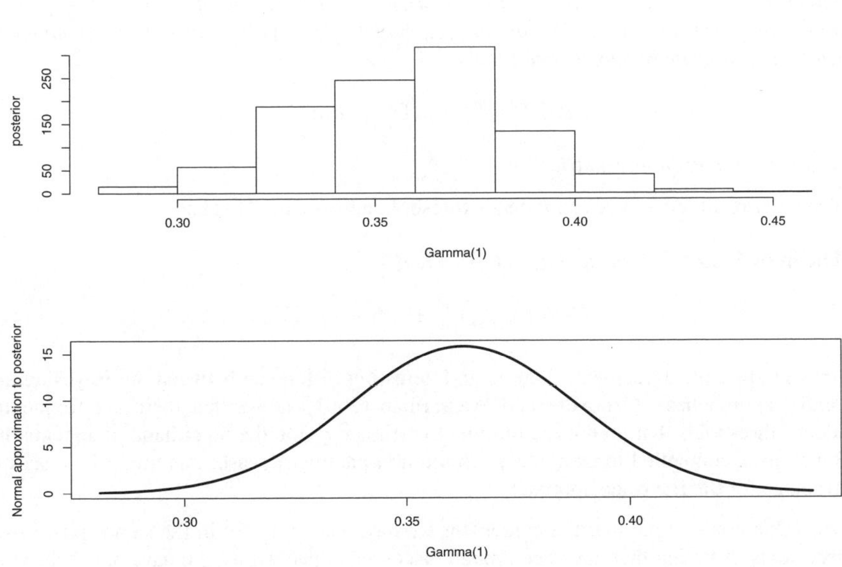

Figure 4. *Posterior of* $\Gamma(1)$ *from simulation and the Normal approximation.*

agreement between Bayes and frequentist in a testing scenario. We explore this in the next section.

Example. Let $\alpha = 0.05$. T_{Bayes} can be obtained directly from the posterior simulation and turns out to be 3.08. On the other hand, with $c = .1$ and $\alpha = .05$, the confidence threshold is 4.0.

5. BAYES-FREQUENTIST AGREEMENT

For parameter estimation, it is well known that Bayes and frequentist inferences agree asymptotically. For example, consider the Welch–Peers (1995) theorem. If θ is scalar and c_n is chosen to satisfy $P(\theta < c_n|D^n) = 1 - \alpha$, then $P_\theta(\theta \in (-\infty, c_n]) = 1 - \alpha + O(n^{-1/2})$. Moreover, if the Jeffreys prior is used then $P_\theta(\theta \in (-\infty, c_n]) = 1 - \alpha + O(n^{-1})$. Extensions of this result abound. In this sense, Bayesian and frequentist inference have achieved a certain unification. Testing has resisted such unification. However, in the multiple testing case using FDR we have the following.

Theorem 3. Bayes-Frequentist Agreement *Fix $t > 0$. Let c_m be such that*

$$P(\Gamma(t) \leq c_m|V^m) = 1 - \alpha.$$

Then, $P(\Gamma(t) \leq c_m) = 1 - \alpha + O(m^{-1/2})$.

A stronger result is that the law of the whole process agrees.

Theorem 4. *Let c be any finite constant. $\mathcal{L}(\Gamma|V^m)$ be the law of $\{\Gamma(v) : v \in [-c, c]\}$ under the posterior. Let $\mathcal{L}_P(\Gamma)$ be the frequentist law of $\{\Gamma(v) : v \in [-c, c]\}$ under P. Under appropriate regularity conditions,*

$$d(\mathcal{L}(\Gamma|V^m), \mathcal{L}_P(\Gamma)) = o_P(1)$$

where d is the Prohorov metric.

The latter result guarantees confidence threshold agreement. That is:

Theorem 5. *Let T be defined as in (13). Then,*

$$P(\Gamma(T) \leq c) \geq 1 - \alpha + o_P(1).$$

Remark: Despite the agreement, there is an interesting difference between the Bayesian and frequentist approaches. Genovese and Wasserman (2001) shows that there are frequentist confidence thresholds that do not require one to estimate q. On the other hand, it appears that the Bayesian is compelled to estimate q. In the non-parametric case, this might be a serious disadvantage for the Bayesian approach.

Remark: One could argue that the agreement we have shown is not in the same spirit as the disagreements in testing that have been much discussed. Specifically, we have not focused on measures of evidence in favor of or against a hypothesis. Whether the discussion should be framed this way in multiple testing is an interesting question.

6. FALSE CONFIDENCE RATES

In some cases, focusing attention on sharp null hypotheses may be inappropriate. Instead, interval nulls or confidence intervals may be more relevant. In the microarray example, we may be interested in genes whose expression levels have changed by a factor of 2. If θ_i denotes the difference on a log-scale, this means we are interested in genes for which $|\theta_i| > \log 2$.

More generally, suppose that we call an effect θ_i *interesting* if $|\theta_i| > \delta$ for some fixed $\delta > 0$. Let $C_i = (a_i, b_i)$ be a level $1-\beta$ confidence interval for θ_i. Let us declare that the i-th case is *significant* if $C_i \cap (-t, t) = \emptyset$ where t is some threshold to be chosen. Let R_i be the indicator for the event $\{C_i \cap (-t, t) = \emptyset\}$ and let $\Delta = (-\delta, \delta)$. We define the *false confidence rate* to be

$$\Lambda(t) = \frac{\sum_i R_i I(\theta_i \in \Delta)}{\sum_i R_i}.$$

We would like to choose T and β so that $E(\Lambda(T)) \leq \alpha$. It seems natural to choose $\beta = \alpha$ which leaves the problem of choosing T. Assume that $\theta_i \sim F$ for some arbitrary F and that $V_i \approx N(\theta_i, \sigma_n)$. Then,

$$\begin{aligned} \Lambda(t) &= \frac{\sum_i R_i I(\theta_i \in \Delta)}{\sum_i R_i} \approx \frac{\int_{-\Delta}^{\Delta} P(R_i = 1|\theta) dF(\theta)}{P(R_i = 1)} \\ &\leq \frac{\int_{-\Delta}^{\Delta} dF(\theta) \left[1 - \Phi\left(\frac{t-\Delta}{\sigma_n} - z_{\alpha/2}\right)\right]}{P(R_i = 1)} \\ &\leq \frac{\left[1 - \Phi\left(\frac{t-\Delta}{\sigma_n} - z_{\alpha/2}\right)\right]}{P(R_i = 1)} \approx \frac{\left[1 - \Phi\left(\frac{t-\Delta}{\sigma_n} - z_{\alpha/2}\right)\right]}{\hat{P}(R_i = 1)} \end{aligned} \tag{14}$$

where

$$\hat{P}(R_i = 1) = \frac{1}{m} \sum_i R_i.$$

We then select T to be the largest value of t for which the right-hand side of (14) is less than α. The procedure can also be based on V_i alone without the use of a confidence intervals. Bayesian and frequentist versions of this procedure will be discussed elsewhere.

7. MICROARRAY EXAMPLE

Data were obtained on 5355 mouse genes from 3T3L1 (fat) cells over 24 h after application of Troglitazone, which is used to treat diabetes and obesity.

The experiment was carried out by Dave Peters and Rob O'Doherty at the University of Pittsburgh. A full analysis of the data will be reported by our group elsewhere. For each gene we have 18 measurements over time. For each gene we computed its median expression level over time and recorded the sign of each measurement as being above or below its median. Let V_i be the longest run of 1's or -1's. In this case V has a discrete distribution on $\{1, \ldots, 18\}$. The null r_0 is know exactly. Figure 5 shows the p-values and 100 draws from the posterior of the Γ process. It is interesting that there is a very steep change in the posterior of Γ between $v = 7$ and $v = 9$. This is much easier to visualize from the posterior draws of Γ than from the p-value plot.

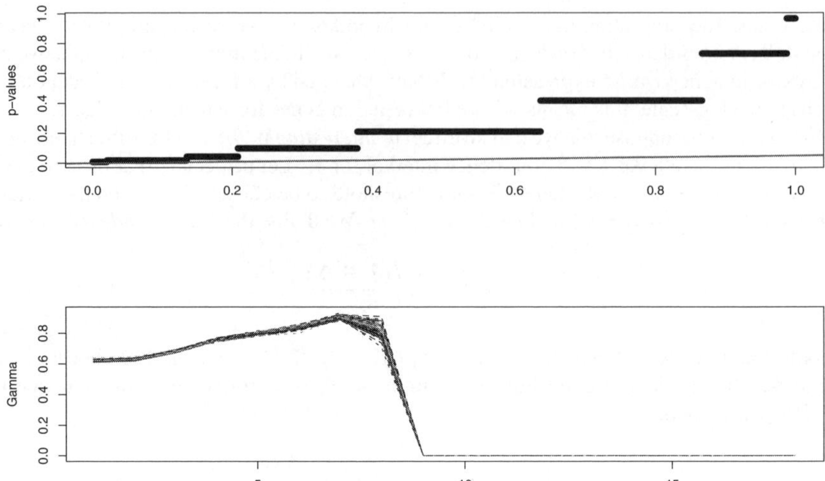

Figure 5. *p-value plot and posterior samples of the* Γ *process.*

8. CONCLUSION

As experiments and data sets get increasingly complex, simultaneous inference becomes more important and more common. We have discussed several frequentist and Bayesian methods for dealing with multiple testing problems that arise in these settings. We also briefly discussed interval estimation versions. Extensions are underway to deal with various complications. For example, dependence has been addressed in Benjamini and Yekutieli (2002), Storey and Tibshirani (2002) and Farcomeni *et al.* (2002). However, dependence can be complex and Bayesian hierarchical models may be useful here.

REFERENCES

Benjamini, Y. and Hochberg, Y. (1995). Controlling the false discovery rate: A practical and powerful approach to multiple testing. *J. R. Statist. Soc. B* **57**, 289–300.

Benjamini, Y. and Yekutieli, D. (2001). The control of the false discovery rate in multiple testing under dependency. *Ann. Statist.* **29**, 1165–1188.

Efron, B., Tibshirani R., Storey, J. and Tusher, V. (2001). Empirical Bayes analysis of a microarray experiment. *J. Am. Statist. Ass.* **96**, 1151–1160.

Farcomeni, A., Genovese, C. and Wasserman, L. (2002). FDR for correlated statistics (in preparation).

Genovese, C. R. and Wasserman, L. (2001). False discovery rates. *Tech. Rep.*, Carnegie Mellon University, USA.

Genovese, C. R. and Wasserman, L. (2002). Operating characteristics and extensions of the FDR procedure. *J. R. Statist. Soc. B* **64**, 499–515.

Storey, J. (2001). The positive False Discovery Rate: A Bayesian interpretation and the q-value. *Tech. Rep.*, Stanford University, USA. `www-stat.stanford.edu/~jstorey`.

Storey, J. (2002). A direct approach to false discovery rates. *J. R. Statist. Soc. B* **64**, 479–498.

Storey, J. and Tibshirani, R. (2002). Estimating false discovery rates under dependence with applications to DNA microarrays. *Tech. Rep.*, Stanford University, USA.

DISCUSSION

MERLISE A. CLYDE (*Duke University, USA*)

I would like to congratulate the authors for this interesting paper on False Discovery Rates. How multiplicity adjustments should be made in multiple testing and multiple comparison problems has been debated by frequentists and Bayesians (*e.g.*, Berry, 1988; Berry and Hochberg, 1997; Gopolan and Berry, 1998; Westfall *et al.*, 1997). The current work of Genovese and Wasserman (GW) presents a new Bayesian approach to multiple testing by controlling the posterior expected FDR.

The multiple testing problem can be addressed as a simultaneous solution of a multiple component decision problem. From a Bayesian perspective, one finds the set of component decisions that maximizes one's overall expected utility (Lindley 1990), ideally using subjective prior distributions and meaningful loss functions. If the loss function is additive in component loss functions for each hypothesis, then the joint Bayes rule for the simultaneous problem is comprised of the Bayes rule for each of the component problems considered on its own, and does not depend on the number of tests. While computations may not always be straightforward, there are no conceptual problems with dealing with general loss functions or dependent hypotheses.

At the heart of the Bayesian decision-theoretic approach are choice of joint prior distributions and utility functions. Difficulties may arise when one attempts to be "objective" about prior assessments. If the prior probability that each $H_i = 0$ is initially set to 1/2 in order to be "fair" and hypotheses are assumed to be independent, one may find that the prior probability assigned to the event that all nulls are true is too small for large m, and that the prior is not well calibrated. Westfall *et al.* suggest adjusting prior probabilities so that posterior probabilities correspond to Bonferroni adjusted p-values (which inherits problems of the conservative Bonferroni approach). Rather using a fixed prior probability on hypotheses, the hierarchical models implicit in Empirical Bayes model selection/multiple testing (George and Foster, 2000; Clyde and George, 2000; Efron *et al.* 2001) and hierarchical Bayes (Gopolan and Berry, 1998) provide automatic adjustments of probabilities.

While the posterior distribution over all hypotheses serves for inferential purposes, making decisions about which hypotheses are true requires specification of a loss or utility function. GW suggest that a Bayesian would prefer to control posterior expected FDR. Viewed as a decision problem, the actions are $\mathcal{A}_i = 1$ corresponding to rejecting $H_i = 0$ and $\mathcal{A}_i = 0$ for accepting $H_i = 0$. Ignoring the product term in the denominator of $\Gamma(t)$, the maximum expected utility corresponding to controlling $E(\text{FDR} \mid V^m)$ can be reformulated as

$$\mathcal{U}(\mathcal{A}^*) = \sup_{\mathcal{A}_i \in \{0,1\}} \sum_{i=1}^{m} \mathcal{A}_i \quad \text{subject to} \quad \frac{\sum \mathcal{A}_\mathrm{i} \mathrm{P}(\mathrm{H}_\mathrm{i} = 0 | \mathrm{V}^\mathrm{m})}{\sum \mathcal{A}_\mathrm{i}} \leq \alpha$$

and defines a threshold rule with optimal decisions $\mathcal{A}_i^* = I(P(H_i = 0|V^m) < \beta^{\text{FDR}})$ where β^{FDR} is the FDR threshold in terms of posterior probabilities. Controlling the posterior expected FDR constrains the average posterior probability that $H = 0$ over the set of rejected hypotheses to be less than α.

Under a simple additive loss function for testing, a loss of L_{i0} is incurred under decision $A_i = 1$ when $H_i = 0$ (cost of a false positive) and L_{i1} under decision $\mathcal{A}_i = 0$ if $H_i = 1$ (cost of a false negative), with no costs if the correct decision is made. If the L_{ij} are the same for all i, the component solution is take action $\mathcal{A}_i = 1$ if $P(H_i = 0|V^m) < L_1/(L_1 + L_0) = \beta$ which leads to a *fixed* threshold β that takes into account possible differential costs of false positives and false negatives.

To explore behavior of the Bayes FDR procedure and the Bayes rule under additive losses, data were simulated from an iid normal means model similar to the example in GW: $V_i|\theta_i \sim$

$N(\theta_i, 1)$, and $\theta_i|H_i \sim N(0, cH_i)$ (degenerate normal if $H_i = 0$), and $H_i \sim$ Bernoulli(a). Using proper prior distributions on the hyperparameters, $p(c) = (1+c)^{-2}$ and a uniform distribution for a, as in Bayarri and Berger (2002), the posterior probabilities

$$P(H_i = 0|V^m) = \frac{\int_0^1 \int_0^1 (1-a) \prod_{i \neq j}^m (1 - a + a\sqrt{1-w}\exp(0.5wV_j^2))da\, dw}{\int_0^1 \int_0^1 \prod_{j=1}^m (1 - a + a\sqrt{1-w}\exp(0.5wV_j^2))da\, dw}$$

can be obtained by numerical integration (after a change of variables $w = c/(1+c)$) using a common importance proposal distribution for (w, a) for $i = 1, \ldots m$. For large m, these can be approximated by Empirical Bayes estimates $1 - \hat{H}_i$ obtained via an EM algorithm (Clyde and George 2000), where one iterates among the solutions $\hat{H}_i = E(H_i|V^m, \hat{c}, \hat{a})$, $\hat{c} = \max(0, \sum \hat{H}_i V_i^2/(4+\sum \hat{H}_i)-1)$, and $\hat{a} = \sum \hat{H}_i/m$. The EB estimate of a corresponds to the average posterior probability that $H_i = 1$, and can be used with non-parametric estimates, $\hat{H}_i$. For moderate m, how sensitive are estimated posterior probabilities and FDR to the estimation of a; does the estimator given by GW provide robustness to prior misspecification?

Data were simulated for $m = 50, 500$, and 5000 with $c = 15$ and $a = 10/m$, with a total of 5000 simulations for each sample size. If a false positive is twice as bad as a false negative, the threshold for the Bayes rule, β, is 1/3. While the average of β^{FDR} over the simulations was similar to the fixed threshold (Table 1), the range of β^{FDR} varied greatly, ranging from 0.05 to 0.95. The posterior expected FDR under the fixed threshold procedure can be reported as a summary, and, of course, is higher, with an average posterior probability of rejected hypotheses around 0.2 to 0.3. These values will be affected by the choice of prior distributions and losses, but should these values be interpreted as a problem? Finally, Table 1 contrasts the percent of false positives (number of rejected nulls/number of actual nulls). While the Bayes FDR procedure is better with regard to this measure, the fixed threshold Bayes decision rule has good frequentist performance with a Type 1 rate less than 2%, and even smaller as m increases.

Table 1 *Comparisons of* FDR *and Bayes rules for testing in a normal means model with 10 non-zero means. The Bayes threshold $\beta = 1/3$. False positives correspond to the actual percent of nulls that were rejected under* FDR *and Bayes. Summaries are based on the mean over 5000 simulations.*

m	β^{FDR}	$E(\Gamma(\beta^{\mathrm{FDR}}\|V^m))$	$E(\Gamma(1/3\|V^m))$	% False Positives FDR	% False Positives Bayes
50	0.362	0.024	0.233	0.687	1.978
500	0.411	0.017	0.203	0.027	0.082
5000	0.367	0.016	0.301	0.002	0.013

In the simulation study, the Bayes FDR yielded thresholds over 0.90. Figure 6 illustrates such a situation, where in order to achieve an E(FDR $|V^m)$ of 0.05, $\beta^{\mathrm{FDR}} = 0.918$. This leads to 13 nulls out of the 5000 hypotheses having $P(H_i = 0|V^m)$ between 0.77 and 0.918. Strictly adhering to the FDR decision rule, however, would lead to the rejection of these 13 nulls. Bounding the posterior probability of the false discovery rate (as in the posterior confidence threshold) will not necessarily alleviate this problem. Is this a desirable posterior property for testing?

How should FDR be extended to more complex decision problems? If the FDR is phrased as a variable/model selection problem in terms of the vector H^m, does controlling the FDR require that a model is selected only if its posterior probability is greater than $1 - \alpha$? As the

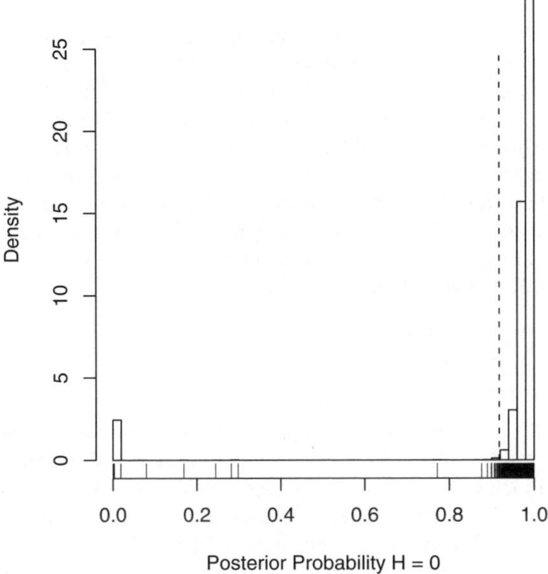

Figure 6. *Histogram of $P(H_i = 0|V^m)$, m = 5000 with threshold at $\beta^{\mathrm{FDR}} = 0.918$ (dashed line). The "rug plot" indicates locations of probabilities.*

number of models grows, posterior probabilities are often "diluted" (Chipman *et al.* 2001) so that this may not be feasible. In micro-array experiments, the hypotheses are not just simple tests for expression, but one may be interested in whether genes are up-regulated, down regulated, or have no change in expression, which genes are in the same pathway, or which genes are predictive of outcomes, such as survival. There are often known costs that limit the number of genes that can be used in a predictive model; should these costs and some measure of predictive accuracy (as in Draper and Fouskakis, 2001) limit the number of "significant" genes, rather than the expected FDR?

FDR procedures have provided real advancements over other frequentist approaches to multiple testing. Bayesian solutions (can) automatically adjust as the number of null components are added; the difficulty is in constructing objective Bayesian procedures that have good frequentists properties (Berger and Pericchi, 2001) when subjective prior information is unavailable. Is controlling the posterior expected FDR on top of a fully hierarchical analysis necessary? Or does the FDR provide some degree of protection for the Bayesian whose prior is really not well calibrated?

CHRISTIAN P. ROBERT and JUDITH ROUSSEAU (*CREST, France*)

This is a quite exciting paper about important problems and procedures, and we find it should be inspirational for (our) future work. The Bayesian flavor of the paper is, however, rather "mild" in that the authors start from frequentist (and partially *ad hoc*) procedures such as FDR and EBT, and try to draw Bayesian analyses of these, rather than positioning the problem within a more Bayesian and decision-theoretic framework, where a loss function would lead to an optimal Bayesian procedure, even in a non-parametric setting. For instance, the present formalization problem suffers from the (usual) frequentist imbalance between Types I and II errors.

Maybe paradoxically, we also consider that multiple testing should not be ranked as a core testing problem (*which imposes hard boundaries*) but rather as a milder generic *classification* problem. We have, on the other hand, no objection to the representation of the dataset as a collection of p-values P_i, since it often occurs that these are the only summary provided by the experimenters. (Even though care should be taken in making a provision for possible biases in the derivation of the P_i's, as, for instance, in the use of t statistics in microarray data.) The general assumption (1) is thus perfectly acceptable and we could consider starting from (1) to estimate the distribution G and then allocate the P_i's to the first or the second component of (1). More specifically, we can use a mixture representation of (1) as in Verdinelli and Wasserman (1998) and Robert and Rousseau (2002):

$$g(t) = (1-a) + a\sum_{i=1}^{\infty} \omega_i \frac{t^{\alpha_i - 1}(1-t)^{\beta_i - 1}}{B(\alpha_i, \beta_i)}, \qquad a, \alpha_i, \beta_i, \omega_i > 0, \qquad a < 1, \sum_i \omega_i = 1.$$

If we impose $\alpha_i < 1$, $\beta_i > 1$, the non-uniform part of the mixture stochastically dominates the uniform part *[although this is not a necessary condition]* and the uniform coefficient should be identifiable since all the other components of the density tend to 0 when t goes to 1. (*The number of components in the mixture can be evaluated from Richardson and Green (1997).*) The evaluation of $P(H_i|P^m)$ can then be changed into the probability of allocating P_i to the second component, using, for instance, the loss functions of Celeux *et al.* (2000).

REPLY TO THE DISCUSSION

We thank Merlise Clyde, Christian Robert, and Judith Rousseau for their comments. Merlise' discussion is a clear exposition of FDR from the Bayesian viewpoint. One of her main points is that when controlling FDR, the marginal posterior probability of the rejected hypotheses can be quite large. This is consistent with the results in our paper where we discuss the relationship between $Q(t)$ and $q(t)$. Should this be a concern? We think not.

Whether the FDR is a useful quantity to control is open to debate. In the end, the answer must depend on the specifics of a problem. But having chosen FDR as a target, we should not be concerned by potentially large values of the marginal quantity $P(H_i = 0|X)$ for rejected hypotheses. As Merlise pointed out, thresholding the $P(H_i = 0|X)$ does not control the posterior FDR either. Admittedly, in the example selected by Merlise, $P(H_i = 0|X)$ is quite large, but this example focuses on the sparse case ($a = 0.002$) where we would expect $d\beta_{\text{FDR}}/d\alpha$ to be large as well. We suspect that a slight change of α would drastically reduce the maximum $P(H_i = 0|X)$. A similar problem occurs with discrete test statistics, as in a permutation test. The p-values and the posterior probabilities in that case are such that achieving a specified criterion (FDR or marginal) can lead to a large change in the number of rejected hypothesis as a function of the target value. This sensitivity to the choice of target is not desirable in general but is most severe in the sparse case. It is an open question about how to "smooth" the method to control this sensitivity.

Christian Robert and Judith Rousseau suggest modelling g as a mixture of a uniform and a second component. Their model forces identifiability by making $g(1) = 0$. Indeed, this is a necessary and sufficient condition for identifiability (Genovese and Wasserman, 2001). But forcing this condition may be an error. Generally, the mixture will not satisfy this condition: even a two-sided Normal testing problem does not satisfy $g(1) = 0$. We do favor the non-parametric approach, but we suggest the approaches taken in Genovese and Wasserman (2001), Efron *et al.* (2001) and Storey (2002). These methods are fully non-parametric and they do not require heavy-duty non-parametric Bayesian machinery nor do they force identifiability on the model.

ADDITIONAL REFERENCES IN THE DISCUSSION

Bayarri, M. J. and Berger, J. O. (2002). Hypothesis testing and model selection. *Tech. Rep.*, Duke University, USA. `http://www.stat.duke.edu/~berger/talks/valencia7.pdf`

Berry, D. A. (1988). Multiple comparison, multiple tests, and data dredging: A Bayesian perspective. *Bayesian Statistics 3* (J. M. Bernardo, M. H. DeGroot, D. V. Lindley and A. F. M. Smith, eds). Oxford: Oxford University Press, 79–94 (with discussion).

Berry, D. A. and Hochberg, Y. (1999). Bayesian perspectives on multiple comparisons. *J. Statist. Plann . Inference* **82**, 215–227.

Berger, J .O. and Pericchi, L. R. (2001). Objective Bayesian methods for model selection: Introduction and comparison. *Model Selection* (P. Lahiri, ed). 135–193.

Celeux, G., Hurn, M. and Robert, C. P. (2000). Computational and inferential difficulties with mixtures posterior distributions. *J. Am. Statist. Ass.* **95**, 957–979.

Chipman, H., George, E. I. and McCulloch, R. E. (2001). The practical implementation of Bayesian model selection. *Model Selection* (P. Lahiri, ed). 65–134.

Clyde, M. A. and George, E. I. (2000). Flexible empirical Bayes estimation for wavelets. *J. R. Statist. Soc. B* **62**, 681–698.

Draper, D. and Fouskakis, D. (2001) Stochastic optimization: A review. *Tech. Rep.*, University of California, Santa Cruz, USA.

George, E. I. and Foster, D. P. (2000). Calibration and empirical Bayes variable selection. *Biometrika* **87**, 731–747.

Gopalan, R and Berry, D. A. (1998). Bayesian multiple comparisons using Dirichlet process priors. *J. Am. Statist. Ass.* **93**, 1130–1139

Lindley, D. V. (1990). The present position in Bayesian statistics. *Statist. Sci.* **5**, 44–65.

Richardson, S. and Green, P. J. (1997) On Bayesian analysis of mixtures with an unknown number of components. *J. R. Statist. Soc. B* **59**, 731–792 (with discussion).

Robert, C. P. and Rousseau, J. (2002) A mixture approach to Bayesian goodness of fit.*Tech. Rep.*, CREST, France.

Verdinelli, I. and Wasserman, L. (1998) Bayesian goodness-of-fit testing using infinite-dimensional exponential families. *Ann. Statist.* **26**, 1215–1241.

Westfall, P. H., Johnson, W. O. and Utts, J. M. (1997). A Bayesian perspective on the Bonferroni adjustment. *Biometrika* **84**, 419–427.

BAYESIAN STATISTICS 7, pp. 163–179
J. M. Bernardo, M. J. Bayarri, J. O. Berger, A. P. Dawid,
D. Heckerman, A. F. M. Smith and M. West (Eds.)

Bayesian Nonparametric Inference for Mixed Poisson Processes

EDUARDO GUTIÉRREZ-PEÑA
Universidad Nacional Autónoma de México, Mexico
eduardo@sigma.iimas.unam.mx

LUIS E. NIETO-BARAJAS
Instituto Tecnológico Autónomo de México, Mexico
lnieto@itam.mx

SUMMARY

This paper presents a Bayesian nonparametric approach to the analysis of two different types of non-homogeneous mixed Poisson processes. The unknown mean function is modelled *a priori* as a process with independent increments and the corresponding posteriors are derived. Posterior inferences are carried out via a Gibbs sampling scheme.

Keywords: BAYESIAN NONPARAMETRICS, EXTENDED GAMMA PROCESS, LÉVY PROCESS, LOG-BETA PROCESS, NEGATIVE-BINOMIAL PROCESS.

1. INTRODUCTION

In the theory of stochastic processes there are two that are fundamental. One is the Weiner model of Brownian motion; the other is the Poisson process. Both are important instances of exponential families of stochastic processes (Küchler and Sorensen, 1997). The Poisson process has several interesting properties and characterizations. For one thing, it is the only point process with stationary and independent increments. In addition, Poisson processes are ordinary renewal processes with exponentially distributed inter-occurrence times; they are the only ordinary renewal processes that are also stationary. For a thorough account of the structure and properties of general Poisson processes, the reader is referred to Kingman (1993).

Despite its nice properties, the Poisson process is often too simple to be useful in applications, and so several generalizations of it have arisen. Nonhomogeneous Poisson processes (NHPP), for instance, are widely used in reliability (see Kuo and Yang, 1996; Kuo and Ghosh, 1997; and the references therein). On the other hand, the mixed Poisson process (MPP) has played a prominent role in areas such as actuarial science (Grandell, 1991, 1997; and Albrecht, 1982) and flood frequency analysis in hydrology (*e.g.*, Lang, 1999). Non-homogeneous mixed Poisson processes are a natural generalization thereof (Grandell, 1997).

A further generalization, introduced by Cox (1955) and usually referred to as doubly stochastic Poisson processes, has also been extensively used in risk theory (Grandell, 1976, 1991). Such processes (in two dimensional spaces) are widely used as models for point patterns that are thought to reflect underlying environmental heterogeneity (Wolpert and Ickstadt, 1998a; Brix and Diggle, 2001).

Lo (1982) discusses Bayesian nonparametric inference for Poisson point processes based on i.i.d. multiple samples. Kuo and Yang (1996) and Kuo and Ghosh (1997) address Bayesian parametric and nonparametric inference, respectively, for non-homogeneous Poisson processes. In the latter paper, either the intensity or the cumulative intensity of the NHPP is assigned one of a number of alternative independent increment processes.

This paper is concerned with Bayesian nonparametric inference for nonhomogeneous mixed Poisson processes (NHMPP) of two different types, each defined by a specific mixing distribution or process (see Sections 2.2 and 2.3). For both types of NHMPP, we assign a general Lévy process prior to simple transformations of the mean function and derive the corresponding posteriors. Such priors exhibit a sort of partial conjugacy, which makes the analysis relatively simple. Posterior analysis is carried out via a Gibbs sampling scheme. Each type of NHMPP gives rise to one of two different types of negative binomial processes when the mixing distribution (process) is gamma. These are perhaps the most important instances of NHMPP and we will focus on them throughout the paper.

In the next section, we briefly review some basic facts concerning Poisson, mixed Poisson, Cox and Lévy Processes. Section 3 deals with Bayesian nonparametric inference for the class of NHPP (cf. Kuo and Ghosh, 1997). The results of Section 3 are then used in Section 4 in order to analyze the NHMPP of type 1 and type 2 via the Gibbs sampler. Section 5 describes in some detail how the required posterior simulations can be carried out, and the methods are illustrated using a data set of 31 failure times previously analyzed in the literature. Finally, Section 6 contains some concluding remarks.

2. PRELIMINARIES

In this section we review basic aspects of Poisson, mixed Poisson and Cox processes, and introduce some notation. A more detailed account can be found in Grandell (1997). We also briefly review the theory of Lévy processes. See Gikhman and Skorokhod (1965) and Sato (1999) for details.

2.1. *Poisson Processes*

A point process $\tilde{\mathcal{N}}(\cdot)$ is called a Poisson process with intensity λ, denoted $\tilde{\mathcal{N}}(t) \sim \text{Po}(\lambda t)$, if

(i) $\tilde{\mathcal{N}}(\cdot)$ has independent increments; and

(ii) $[\tilde{\mathcal{N}}(t) - \tilde{\mathcal{N}}(s)] \sim \text{Po}\{\lambda(t-s)\}$ for all $0 < s < t$.

A Poisson process is homogeneous, *i.e.*, $[\tilde{\mathcal{N}}(t+s) - \tilde{\mathcal{N}}(s)]$ has the same distribution for all $s \geq 0$. Moreover, Poisson processes are the only point processes with stationary and independent increments.

In the sequel, $\mathcal{A} = \{A(t) : t \geq 0\}$ will denote the set of all functions $A(\cdot)$ such that (i) $A(0) = 0$; (ii) $A(t) < \infty$ for all $0 \leq t < \infty$; and (iii) $A(\cdot)$ is non-decreasing and right continuous. Let $\Lambda(\cdot) \in \mathcal{A}$. A point process $\tilde{\mathcal{N}}(\cdot)$ is called a non-homogeneous Poisson process with mean function (intensity measure, cumulative intensity) $\Lambda(\cdot)$ if

(i) $\tilde{\mathcal{N}}(\cdot)$ has independent increments; and

(ii) $[\tilde{\mathcal{N}}(t) - \tilde{\mathcal{N}}(s)] \sim \text{Po}\{\Lambda(t) - \Lambda(s)\}$ for all $0 < s < t$.

In this case the increments are no longer stationary unless $\Lambda(t)$ is linear in t.

2.2. *Mixed Poisson Processes of Type 1*

Let θ be a positive random variable and let $\mathcal{N}(\cdot)$ be a Poisson process with intensity θ. Then the point process $\mathcal{N}(\cdot)$ is called a (homogeneous) mixed Poisson process. This process has stationary increments, but they will not be independent unless the distribution of θ is concentrated at one point.

Now, for a fixed function $\Lambda(\cdot) \in \mathcal{A}$, let θ be a positive random variable and let $\mathcal{N}(\cdot)$ be a NHPP with mean function $\mathcal{M}_\theta(\cdot) = \theta\Lambda(\cdot)$. Then the point process $\mathcal{N}(\cdot)$ is termed a non-homogeneous mixed Poisson process of type 1 (NHMPP1). Such processes have the following intuitive interpretation: first, a realization of a positive random variable θ is generated; conditional on θ, $\mathcal{N}(\cdot)$ is a Poisson process with mean function $\mathcal{M}_\theta(\cdot)$. As in the case of the NHPP, the increments of this process are not stationary unless $\Lambda(t)$ is linear in t. In addition, the increments of this process are not independent unless the distribution of θ is concentrated at one point.

Suppose, in particular, that $\theta \sim \text{Ga}(a, b)$. Then $\mathcal{N}(\cdot) \sim \text{Nb}_1\{a, p(\cdot)\}$ with $p(t) = \Lambda(t)/\{b + \Lambda(t)\}$ for all $t \geq 0$, *i.e.*, $\mathcal{N}(\cdot)$ is a negative binomial process of type 1. Let $\mu = \text{E}(\theta) = a/b$. Then $\text{E}\{\mathcal{N}(\cdot)\} = \mu\Lambda(\cdot)$ and $\text{Var}\{\mathcal{N}(\cdot)\} = \mu\Lambda(\cdot)\{1 + \Lambda(\cdot)/b\}$. Thus, $\mathcal{N}(\cdot)$ is overdispersed relative to the NHPP in the sense that $\text{E}\{\mathcal{N}(t)\} \leq \text{Var}\{\mathcal{N}(t)\}$ for all $t \geq 0$. For fixed μ, the parameter b controls both the degree of overdispersion and the strength of the dependence between the increments of $\mathcal{N}(\cdot)$.

2.3. *Mixed Poisson Processes of Type 2*

Let $\mathcal{N}(\cdot)$ be a NHPP with mean function $\Lambda(\cdot)$. When $\Lambda(\cdot)$ is assumed to be a random measure, we call $\mathcal{N}(\cdot)$ a non-homogeneous mixed Poisson process of type 2 (NHMPP2). Elsewhere in the literature, such processes are known as doubly stochastic Poisson Processes or Cox processes. We assume that the realizations of $\Lambda(\cdot)$ belong to the set $\mathcal{A}$ defined in Section 2.1 in order for $\mathcal{N}(\cdot)$ to be well defined. Similarly to the previous case, a NHMPP2 has the following interpretation: first, a realization of the stochastic process $\Lambda(\cdot)$ is generated; conditional on $\Lambda(\cdot)$, $\mathcal{N}(\cdot)$ is a Poisson process with mean function $\mathcal{M}_\Lambda(\cdot) = \Lambda(\cdot)$. If $\Lambda(\cdot)$ is assumed to have independent increments, then $\mathcal{N}(\cdot)$ will also have independent (but not stationary) increments.

Suppose, for example, that $\Lambda(\cdot) \sim \text{GaP}\{a(\cdot), b\}$, *i.e.*, $\Lambda(\cdot)$ is a gamma process (*e.g.*, Ferguson, 1973; van der Weide, 1997). Then $\mathcal{N}(\cdot) \sim \text{Nb}_2\{a(\cdot), 1/(b+1)\}$. In other words, $\mathcal{N}(\cdot)$ is a negative binomial process of type 2. In this case, $\text{E}\{\mathcal{N}(\cdot)\} = a(\cdot)/b$ and $\text{Var}\{\mathcal{N}(\cdot)\} = \{a(\cdot)/b\}\{1 + (1/b)\}$, and so $\mathcal{N}(\cdot)$ is also overdispersed relative to the NHPP. As in the case of the NHMPP1, b controls the degree of overdispersion.

As pointed out by Grandell (1997, Ch. 5), Cox processes, and not mixed Poisson processes, are actually the process version of the mixed Poisson distribution. Homogeneous mixed Poisson processes can be seen as Cox processes with underlying random measure $\Lambda(dt) = \theta dt$ where θ is a positive random variable. This is why we refer to Cox processes as mixed Poisson processes (of type two, so as to distinguish them from the usual MPP described in the previous section).

2.4. *Lévy Processes*

A continuous time process $L(\cdot)$ with independent stationary increments is called a *Lévy process* if its sample paths are right continuous with limits from the left and $L(0) = 0$. If the stationarity requirement is dropped, then $L(\cdot)$ is usually called an additive process (*e.g.*, Sato, 1999). Recently, however, especially in the Bayesian literature, the term "Lévy process" has been used to refer to the more general additive process. Throughout this paper we will use the latter terminology.

According to Gikhman and Skorokhod (1965), every Lévy process $L(\cdot)$ can be written as the sum of a fixed jump component and a 'continuous' component, *i.e.*, if $t_1, t_2, \ldots$ correspond to fixed points of discontinuity having independent non-negative jumps $L\{t_1\}, L\{t_2\}, \ldots$ (also independent of the rest of the process), then

$$L(t) = L_c(t) + \sum_j L\{t_j\} I(t_j \le t),$$

where $L_c(\cdot)$ is a non-decreasing process with no fixed points of discontinuity and therefore has a Lévy representation for the Laplace transform

$$\mathrm{E}\left\{e^{-\psi L_c(t)}\right\} = \exp\left\{-\int_0^\infty \left(1 - e^{-\psi\nu}\right) dN_t(\nu)\right\}, \tag{1}$$

with $N_t(\nu)$ a Lévy measure satisfying:

(1). For every Borel set B, $N_t(B)$ is continuous and nondecreasing as a function of t;

(2). For every real $t > 0$, $N_t(\cdot)$ is a measure on the Borel sets of $(0, \infty)$;

(3). $\int_0^1 \nu dN_t(\nu) < \infty$; and

(4). $\int_1^\infty dN_t(\nu) < \infty$.

3. NON-HOMOGENEOUS POISSON PROCESS

In this section, we present a detailed Bayesian analysis of NHPP, which will be used in Section 4 in order to study NHMPP.

3.1. *Likelihood and Prior*

Following Kuo and Ghosh (1997), let us consider a time-truncated model where the process is observed up to a fixed time τ. We denote the ordered epochs of the n observed jumps by $0 = x_0 < x_1 < x_2 < \cdots < x_n \le \tau$. If we allow ties in the observed jump times, then the probability of observing no jumps in the interval $(0, x_1)$, d_1 jumps at x_1, no jumps in (x_1, x_2), and so on up to no jumps in (x_n, τ), is given by

$$\mathrm{Lik}(\Lambda \mid D, \tau) = \left\{\prod_{i=1}^n [\Lambda(x_i) - \Lambda(x_i^-)]^{d_i} e^{-[\Lambda(x_i) - \Lambda(x_i^-)]} e^{-[\Lambda(x_i^-) - \Lambda(x_{i-1})]}\right\} e^{-[\Lambda(\tau) - \Lambda(x_n)]}.$$

Here, D denotes the data and d_i is the number of multiple jumps observed at time x_i. The total number of observed jumps in the interval $(0, \tau]$ is then $d = \sum_{i=1}^n d_i$.

We will assign a Lévy process prior to $\Lambda(\cdot)$. As pointed out in Section 2.4, the prior can then be characterized by

$$M = \{t_1, t_2, \ldots\}, \quad \{f_{t_1}, f_{t_2}, \ldots\},$$

the set of fixed points of discontinuity together with the corresponding density functions for the jumps, and $N_t(\cdot)$, the Lévy measure for the part of the process without fixed points of discontinuity. We assume the Lévy measure to be of the form

$$dN_t(\nu) = d\nu \int_0^t K(\nu, u) du. \tag{2}$$

Typically, $K(\cdot, \cdot)$ is parametrized in terms of a non-negative measure $\alpha(\cdot)$ and a non-negative, piece-wise continuous function $\beta(\cdot)$, *i.e.*,

$$\Lambda(\cdot) \sim \text{LévyProcess}\{\alpha(\cdot), \beta(\cdot)\}.$$

For instance, if

$$K(\nu,u)du = \nu^{-1}\exp\{-\nu\beta(u)\}d\alpha(u) \tag{3}$$

then $\Lambda(\cdot)$ is an extended gamma process (see Dykstra and Laud, 1981).

On the other hand, if

$$K(\nu,u)du = \{1-\exp(-\nu)\}^{-1}\exp\{-\nu\beta(u)\}d\alpha(u) \tag{4}$$

then $\Lambda(\cdot)$ is a log-beta process (see Walker and Muliere, 1997).

3.2. *Posterior Distributions*

Experience with updating Lévy processes for modelling cumulative hazard functions (see, for example, Walker and Muliere, 1997) and updating Lévy-driven processes for modeling hazard rate functions (see Nieto-Barajas, 2001) gives rise to the following posterior distributions.

Theorem 1. *Let $\Lambda(\cdot)$ be a Lévy process and let $x_1,\ldots,x_n$ be the ordered epochs of the observed jump times from a NHPP $\mathcal{N}(\cdot)$. Let $x_1^*,\ldots,x_m^*$ be the time epochs not equal to any of the prior fixed jumps. Then the posterior distribution of $\Lambda(\cdot)$ given the data is again a Lévy process with the following characteristics (we will use (*) to denote an updated parameter/function):*

$M^* = M\cup\{x_1^*,\ldots x_m^*\}$, *with*

$$f_j^*(\nu\,|\,D)\propto\begin{cases}\nu^{d_i}e^{-\nu}f_j(\nu) & \text{if}\quad t_j = x_i,\\ e^{-\nu}f_j(\nu) & \text{if}\quad t_j\neq x_i,\end{cases}$$

$$f_{x_i^*}(\nu\,|\,D)\propto\nu^{d_i}e^{-\nu}K(\nu,x_i),$$

for $i=1,\ldots,m$ and

$$K^*(\nu,u) = e^{-\nu}K(\nu,u).$$

Proof. Posterior distribution of the jumps at the prior fixed points of discontinuity are obtained by looking at the likelihood and using standard Bayesian updating. We will concentrate in obtaining the posterior distribution of the continuous part of the process for a single time epoch x_1. The idea is to characterize the posterior distribution by its Laplace transform. If we define

$$\phi(\psi,\Lambda,x_1,\epsilon) = \mathrm{E}\left[\exp\left\{-\int_0^\tau\psi(s)d\Lambda(s)\right\}\,\middle|\,\mathcal{N}(x_1-\epsilon,x_1]=1\right],$$

then our aim is to find

$$\lim_{\epsilon\to0}\phi(\psi,\Lambda,x_1,\epsilon).$$

Thus, $\phi(\psi,\Lambda,x_1,\epsilon)$ can be expressed as

$$\frac{\mathrm{E}\left[\int_{x_1-\epsilon}^{x_1}d\Lambda(s)\exp\left\{-\int_0^\tau\{\psi(s)+1\}d\Lambda(s)\right\}\right]}{\mathrm{E}\left[\int_{x_1-\epsilon}^{x_1}d\Lambda(s)\exp\left\{-\int_0^\tau d\Lambda(s)\right\}\right]}.$$

Splitting up the integral and using independence properties this becomes

$$H(x_1-\epsilon,x_1)\frac{\mathrm{E}\left[\exp\left\{-\int_{(0,x_1-\epsilon]\cup(x_1,\infty)}\{\psi(s)+1\}d\Lambda(s)\right\}\right]}{\mathrm{E}\left[\exp\left\{-\int_{(0,x_1-\epsilon]\cup(x_1,\infty)}d\Lambda(s)\right\}\right]},$$

where

$$H(x_1-\epsilon, x_1) = \frac{\mathrm{E}\left[\int_{x_1-\epsilon}^{x_1} d\Lambda(s) \exp\left\{-\int_0^T \{\psi(s)+1\} d\Lambda(s)\right\}\right]}{\mathrm{E}\left[\int_{x_1-\epsilon}^{x_1} d\Lambda(s) \exp\left\{-\int_0^T d\Lambda(s)\right\}\right]}$$

Taking limits, we can express $\lim_{\epsilon\to 0} \phi(\psi, \Lambda, x_1, \epsilon)$ as

$$H(x_1) \exp\left[-\int_0^T \int_0^\infty \left\{1 - e^{-\psi(s)\nu}\right\} e^{-\nu} K(\nu, s) d\nu ds\right],$$

with

$$H(x_1) = \frac{\int_0^\infty e^{-\psi(x_1)\nu} \nu e^{-\nu} K(\nu, x_1) d\nu}{\int_0^\infty \nu e^{-\nu} K(\nu, x_1) d\nu},$$

which can be written as

$$\mathrm{E}\left\{e^{-\psi(x_1)\Lambda^*\{x_1\}}\right\} \mathrm{E}\left[\exp\left\{-\int_0^T \psi(s) d\Lambda^*(s)\right\}\right],$$

with $\Lambda^*\{x_1\}$ and $\Lambda_c^*(t)$ as stated in the theorem. The general case for $x_1, \ldots, x_n$ time epochs can be obtained by a straightforward extension of this procedure. This completes the proof. $\square$

In particular, if $\Lambda(\cdot) \sim \mathrm{EGaP}_c\{\alpha(\cdot), \beta(\cdot)\}$, *i.e.*, $\Lambda(\cdot)$ is an extended gamma process without prior fixed points of discontinuity, then the posterior distribution of $\Lambda(\cdot)$ is again an extended gamma process with parameters $\alpha^*(\cdot) = \alpha(\cdot)$ and $\beta^*(\cdot) = \beta(\cdot) + 1$, and with fixed points of discontinuity at $M^* = \{x_1, \ldots, x_n\}$ with posterior distributions for the jumps $f_{x_i}^*(\nu) = \mathrm{Ga}(\nu \mid d_i, \beta(x_i) + 1)$, $i = 1, \ldots, n$.

Moreover, if $\Lambda(\cdot) \sim \mathrm{logBeP}_c\{\alpha(\cdot), \beta(\cdot)\}$, that is, if $\Lambda(\cdot)$ is a log-beta process without prior fixed points of discontinuity, then the posterior distribution of $\Lambda(\cdot)$ is again a Lévy process with a log-beta measure for the continuous part, with parameters $\alpha^*(\cdot) = \alpha(\cdot)$ and $\beta^*(\cdot) = \beta(\cdot) + 1$, and with fixed points of discontinuity at $M^* = \{x_1, \ldots, x_n\}$ with posterior distributions for the jumps $f_{x_i}^*(\nu) \propto \nu^{d_i} e^{-\nu} K(\nu, x_i)$, $i = 1, \ldots, n$, where $K(\nu, u)$ is given by (4).

In both cases, $\alpha(\cdot)$ is a non-negative measure and $\beta(\cdot)$ is a non-negative, piece-wise continuous function.

It must be pointed out that Kuo and Ghosh (1997) derive the posterior distributions for several particular priors, both on the cumulative intensity $\Lambda(t)$ and on the intensity function $\lambda(t) = d\Lambda(t)/dt$. Specifically, these authors assume gamma and beta (Hjort, 1990) process priors for $\Lambda(\cdot)$ and an extended gamma process prior on $\lambda(\cdot)$.

4. NONHOMOGENEOUS MIXED POISSON PROCESSES

4.1. *Mixed Poisson Processes of Type 1*

Let $\mathcal{N}(\cdot)$ be a NHMPP1 such that $[\mathcal{N}(\cdot) \mid \theta] \sim \mathrm{Po}\{\theta\Lambda(\cdot)\}$ and $\theta \sim f(\theta)$ for some distribution f. This implies that $\mathcal{N}(\cdot)$ has mean function $\mathcal{M}(\cdot) = \mu_\theta \Lambda(\cdot)$, where $\mu_\theta = \mathrm{E}_f[\theta]$. We can make Bayesian inferences about $\mathcal{M}(\cdot)$ by assigning a nonparametric prior to $\Lambda(\cdot)$. Because $\Lambda(\cdot)$ is a non-decreasing function, a reasonable choice is to use a Lévy process prior.

Posterior distributions. The corresponding posterior distributions are given by the next result.

Theorem 2. *Let $\Lambda(\cdot)$ be a Lévy process and let $x_1, \ldots, x_n$ be the ordered epochs of the observed jump times from a* NHMPP1 $\mathcal{N}(\cdot)$. *Let $x_1^*, \ldots, x_m^*$ be time epochs not equal to any of the prior fixed jumps. Then the posterior conditional distribution of $\Lambda(\cdot)$ given θ is again a Lévy process with $M^* = M \cup \{x_1^*, \ldots x_m^*\}$, where*

$$f_j^*(\nu \mid D, \theta) \propto \begin{cases} \nu^{d_i} e^{-\theta\nu} f_j(\nu) & \text{if } t_j = x_i, \\ e^{-\theta\nu} f_j(\nu) & \text{if } t_j \neq x_i, \end{cases}$$

$$f_{x_i^*}(\nu \mid D, \theta) \propto \nu^{d_i} e^{-\theta\nu} K(\nu, x_i),$$

for $i = 1, \ldots, m$, $K^*(\nu, u) = e^{-\theta\nu} K(\nu, u)$, and $(*)$ denotes an updated parameter or function. Furthermore, the posterior conditional distribution of θ given $\Lambda(\cdot)$ is given by

$$f^*(\theta \mid D, \Lambda) \propto \theta^{\sum_{i=1}^n d_i} e^{-\theta\Lambda(\tau)} f(\theta).$$

Proof. The posterior conditional distribution of $\Lambda(\cdot)$ comes from conditioning on θ and extending the proof given to Theorem 1. The posterior conditional distribution of θ is obtained by using standard Bayesian updating. □

In particular, if $\Lambda(\cdot) \sim \text{EGaP}_c\{\alpha(\cdot), \beta(\cdot)\}$, *i.e.*, if $\Lambda(\cdot)$ is an extended gamma process without prior fixed points of discontinuity, and if $\theta \sim \text{Ga}(a, b)$, then the conditional posterior distribution of $\Lambda(\cdot)$ given θ is again an extended gamma process with parameters $\alpha^*(\cdot) = \alpha(\cdot)$ and $\beta^*(\cdot) = \beta(\cdot) + \theta$. It has fixed points of discontinuity at $M^* = \{x_1, \ldots, x_n\}$ with posterior distributions for the jumps

$$f_{x_i}^*(\nu) = \text{Ga}(\nu \mid d_i, \beta(x_i) + \theta), \quad i = 1, \ldots, n.$$

Also, the posterior conditional of θ given $\Lambda(\cdot)$ is $\text{Ga}(a + d, b + \Lambda(\tau))$.

If, on the other hand, $\Lambda(\cdot) \sim \text{logBeP}_c\{\alpha(\cdot), \beta(\cdot)\}$, then the posterior conditional of $\Lambda(\cdot)$ given θ is again a Lévy process with a log-beta measure for the continuous part and a D-distribution (Walker, 1995) for the jumps. Also, provided that $\theta \sim \text{Ga}(a, b)$, the posterior conditional of θ given $\Lambda(\cdot)$ is $\text{Ga}(a + d, b + \Lambda(\tau))$.

Prior elicitation. An important issue when carrying out a Bayesian analysis is how to determine the hyper-parameters of the prior process. Recall that our interest lies on the mean function $\mathcal{M}(\cdot) = \mu_\theta \Lambda(\cdot)$. A simple idea is to "center" the process $\Lambda(\cdot)$ in such a way so that the first two moments of $\mathcal{M}(\cdot)$ equal two arbitrary functions $\mathcal{M}_0(\cdot)$ and $\Psi_0(\cdot)$, respectively (see, *e.g.*, Walker and Damien, 1998). By way of illustration, let us consider the case when $\Lambda(\cdot)$ is an extended gamma process.

Lemma 1. *Let $\Lambda(\cdot)$ be an extended gamma process without prior fixed points of discontinuity, i.e., $\Lambda(\cdot)$ has a representation of the form* (1) *and Lévy measure given by* (2) *and* (3). *Let $\mathcal{M}_0(\cdot)$ and $\Psi_0(\cdot)$ be non-negative and differentiable functions on $(0, \infty)$. Then, there exist functions $\alpha(\cdot)$ and $\beta(\cdot)$ such that, for all $t \geq 0$, $\mathcal{M}_0(t) = \text{E}\{\mathcal{M}(t)\}$, and $\Psi_0(t) = \text{Var}\{\mathcal{M}(t)\}$.*

Proof. We require

$$\mathcal{M}_0(t) = \text{E}\{\mathcal{M}(t)\} = \mu_\theta \int_0^t \frac{d\alpha(u)}{\beta(u)}$$

and

$$\Psi_0(t) = \text{Var}\{\mathcal{M}(t)\} = \mu_\theta^2 \int_0^t \frac{d\alpha(u)}{\beta^2(u)}.$$

Differentiating both equations with respect to t and solving the simultaneous equations we get

$$\beta(t) = \mu_\theta \left\{ \frac{d\mathcal{M}_0(t)}{d\Psi_0(t)} \right\}$$

and

$$d\alpha(t) = \frac{\{d\mathcal{M}_0(t)\}^2}{d\Psi_0(t)}.$$

This completes the proof. □

Following Walker and Damien (1998), we can center the extended gamma process at a Bayesian parametric model. For example, if we use the Weibull model $\mathcal{M}(t) = \gamma t^\delta$ with "prior" distribution $\gamma \sim \text{Ga}(p, q)$, and if $\theta \sim \text{Ga}(a, b)$, from Lemma 1 we obtain $\beta(t) = (aq/2b)t^{-\delta}$ and $d\alpha(t) = \delta(p/2)t^{-1}dt$.

4.2. *Mixed Poisson Processes of Type 2*

Suppose that $\mathcal{N}(\cdot)$ is a NHMPP2 such that

$$[\mathcal{N}(\cdot) \,|\, \Lambda(\cdot)\,] \sim \text{Po}\{\Lambda(\cdot)\}$$

and

$$\Lambda(\cdot) \sim \text{Lévy Process}\{a(\cdot), b(\cdot)\},$$

where $b(\cdot)$ is a fixed known function. For the sake of simplicity, we consider only Lévy processes without prior fixed points of discontinuity as distributions for $\Lambda(\cdot)$.

As in the previous case, we want to make inference about the mean function $\mathcal{M}(\cdot)$, which depends on $a(\cdot)$. In a Bayesian approach we can assign a Lévy process prior to $a(\cdot)$.

Posterior distributions. The following proposition provides the posterior distribution for the NHMPP2 model.

Proposition 1. *Let $a(\cdot)$ be a Lévy process such that $a(ds) \sim f\{a(ds)\}$, and let $x_1, \ldots, x_n$ be the ordered epochs of the observed jump times from a NHMPP2 $\mathcal{N}(\cdot)$. Then the posterior conditional distribution of $a(ds)$ given the data and $\Lambda(ds)$ can be obtained as (we will use an (*) to denote an updated parameter/function):*

$$f^*\{a(ds) \,|\, D, \Lambda(ds)\} \propto f\{\Lambda(ds) \,|\, a(ds), b(s)\} f\{a(ds)\}.$$

Furthermore, the posterior conditional distribution of $\Lambda(\cdot)$ given the data and $a(\cdot)$ is again a Lévy process with $M^* = \{x_1, \ldots x_n\}$, where

$$f^*_{x_i}(\nu \,|\, D, a) \propto \nu^{d_i} e^{-\nu} K(\nu, x_i),$$

for $i = 1, \ldots, n$, and $K^*(\nu, u) = e^{-\nu} K(\nu, u)$.

Proof. Straightforward, given the independence of the increments. □

If, in particular, we take

$$\Lambda(\cdot) \sim \text{EGaP}\{a(\cdot), b(\cdot)\} \tag{5}$$

and an extended gamma process prior for $a(\cdot)$, *i.e.*,

$$a(\cdot) \sim \text{EGaP}\{\alpha(\cdot), \beta(\cdot)\}, \tag{6}$$

then the mean function of $\mathcal{N}(\cdot)$ becomes

$$\mathcal{M}(t) = \int_0^t \frac{da(u)}{b(u)}.$$

It follows from Proposition 1 that the conditional posterior distributions are given by

$$[\,\Lambda(ds)\,|\,D, a\,] \sim \text{Ga}\left\{a(ds) + \sum_i d_i I(x_i \in ds), b(s) + 1\right\}$$

and

$$[\,a(ds)\,|\,D, \Lambda\,] \propto \frac{\{b(s)\Lambda(ds)\}^{a(ds)}}{\Gamma\{a(ds)\}}\, a(ds)^{\alpha(ds)-1} e^{-\beta(s)a(ds)} I\{a(ds) > 0\}.$$

Prior elicitation. Consider the special case where $\Lambda(\cdot)$ and $a(\cdot)$ are both extended gamma processes. We can center the prior process $a(\cdot)$ in such a way that the prior expected value and the prior variance of the mean function $\mathcal{M}(\cdot)$ be equal to two given functions $\mathcal{M}_0(\cdot)$ and $\Psi_0(\cdot)$, respectively.

Lemma 2. *Let $a(\cdot)$ be an extended gamma process without prior fixed points of discontinuity, as specified in Lemma 1. Let $\mathcal{M}_0(\cdot)$ and $\Psi_0(\cdot)$ be non-negative and differentiable functions defined on $(0, \infty)$. Then, there exist functions $\alpha(\cdot)$ and $\beta(\cdot)$ such that*

$$\mathcal{M}_0(t) = \text{E}\{\mathcal{M}(t)\} \quad \textit{and} \quad \Psi_0(t) = \text{Var}\{\mathcal{M}(t)\},$$

for all $t \geq 0$.

Proof. We require

$$\mathcal{M}_0(t) = \text{E}\{\mathcal{M}(t)\} = \int_0^t \frac{d\alpha(u)}{b(u)\beta(u)}$$

and

$$\Psi_0(t) = \text{Var}\{\mathcal{M}(t)\} = \int_0^t \frac{d\alpha(u)}{b^2(u)\,\beta^2(u)}.$$

Differentiating both equations with respect to t and solving the simultaneous equations we get

$$\beta(t) = \frac{d\mathcal{M}_0(t)}{b(t)\,d\Psi_0(t)}$$

and

$$d\alpha(t) = \frac{\{d\mathcal{M}_0(t)\}^2}{d\Psi_0(t)}.$$

This completes the proof. □

Again, if we use a Weibull model $\mathcal{M}(t) = \gamma t^\delta$ with "prior" distribution $\gamma \sim \text{Ga}(p, q)$ to center our prior process, then, using Lemma 2, we obtain $\beta(t) = [q/\{2b(t)\}]\,t^{-\delta}$ and $d\alpha(t) = (\delta p/2)\,t^{-1}dt$.

5. POSTERIOR SIMULATIONS

In this section, we describe in some detail how the required posterior simulations can be carried out. The methods described in the last section are illustrated using data previously analyzed by Kuo and Ghosh (1997) and other authors. The data set consists of the following 31 failure epochs in days: 9, 21, 32, 36, 43, 45, 50, 58, 63, 70, 71, 77, 78, 87, 91, 92, 95, 98, 104, 105, 116, 149, 156, 247, 249, 250, 337, 384, 396, 405 and 540, and relates to trouble reports for one module of the US Naval Tactical Data System.

5.1. *Example 1: NHMPP1*

First we briefly describe a way of simulating a Lévy process as defined in Section 2.4, since this will be required in order to sample from the posterior distribution of $\mathcal{M}(\cdot)$ under the NHMPP1 model. The method is based on a representation of a Lévy process derived by Ferguson and Klass (1972) and further discussed in Walker and Damien (2000). A related algorithm appears in Wolpert and Ickstadt (1998b).

Consider a Lévy process $L(\cdot)$, as defined in Section 2.4, and recall that the process is assumed to be observed up to a fixed time τ. Now let

$$M(x) = -N_\tau[x, \infty) = -\int_x^\infty dN_\tau(\nu),$$

where $N_t(\nu)$ denotes the Lévy measure of the process $L_c(t)$.

Define positive random variables $J_1 \geq J_2 \geq \cdots$ by

$$Pattn.\ Recogn.\ (J_1 \leq x_1) = \exp\{M(x_1)\}$$

and

$$Pattn.\ Recogn.\ (J_i \leq x_i \mid J_{i-1} \leq x_{i-1}) = \exp\{M(x_i) - M(x_{i-1})\}$$

for $x_i < x_{i-1}$.

We can obtain the J_i via $\omega_i = -M(J_i)$, where $\omega_1, \omega_2, \omega_3, \ldots$ are the jump times of a standard Poisson process (*i.e.*, with intensity 1). In other words, $\omega_1, \omega_2 - \omega_1, \omega_3 - \omega_2, \ldots$ are all iid $\text{Ga}(1, 1)$. The Ferguson-Klass representation of the process $L_c(t)$ is then given by

$$L_c(t) = \sum_i J_i\, I\{U_i \leq n_t(J_i)\},$$

where $U_1, U_2, \ldots$ are all iid $\text{Un}(0, 1)$, and $n_t(\nu) = \frac{dN_t}{dN_\tau}(\nu)$, for $t \in [0, \tau]$.

Posterior simulation of $\mathcal{M}(\cdot)$ can be carried out by means of a Gibbs sampling scheme (*e.g.*, Smith and Roberts, 1993) using the full conditionals given in Theorem 2. In order to sample from the full conditional of $\Lambda(\cdot)$, we use the algorithm described above based on the Ferguson–Klass representation of a Lévy process.

We now illustrate this using the data set described above. We take $\theta \sim \text{Ga}(5, 1)$, so that $\mathcal{N}(\cdot) \sim \text{Nb}_1(5, p(\cdot))$ with $p(t) = \Lambda(t)/\{1 + \Lambda(t)\}$. Also, we give $\Lambda(\cdot)$ an extended gamma process prior $\text{EGaP}\{\alpha(\cdot), \beta(\cdot)\}$. Hence the corresponding full conditionals are those described in Section 4.1. We centered the prior on a Weibull model, as discussed in Section 4.1, and took $p = q = \delta = 1$. We ran the Gibbs sampler for 10,000 iterations with a burn-in period of 1000 iterations. Convergence was assessed informally based on plots of ergodic averages for θ and $\Lambda(t)$ for a few selected values of t. Figure 1 shows the data together with the posterior mean (continuous line), the median (dotted line), and a 95% credible band (dashed lines) for the mean function $\mathcal{M}(\cdot)$. The straight line appearing on the left of the plot represents the function at which we centered the prior of $\mathcal{M}(\cdot)$, namely $\mathcal{M}_0(t) = t$.

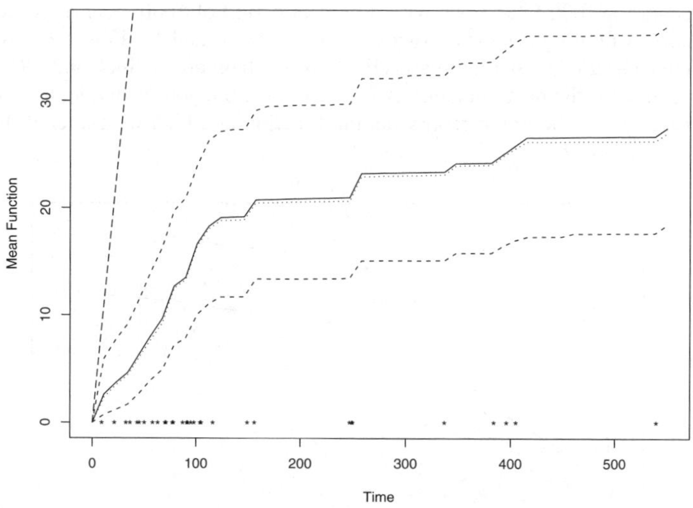

Figure 1. *Posterior mean, median, and 95% band for* $\mathcal{M}(\cdot)$ (NHMPP1; *extended gamma prior).*

5.2 *Example 2: NHMPP2*

In this example, we will illustrate the analysis of a NHMPP2 using the same data as in Example 1. For the mixing distribution we used a gamma process with parameters $a(\cdot)$ and b, as in (5) with $b(t) \equiv b$ for all t, and an extended gamma process prior for $a(\cdot)$ as in (6). Thus, $\mathcal{N}(\cdot)$ is a negative binomial process of type 2. We centered $a(\cdot)$ at a Weibull model as described in Section 4.2, and took $\delta = 1$, $p = 0.5$, $q = 0.5$ and $b = 100$. One way of approximating the posterior process $a^*(\cdot)$ is to take a fine partition, say $t_0, t_1, \ldots, t_R$ for R sufficiently large, and implement a Gibbs sampler.

Let $\Lambda_r = \Lambda(t_{r-1}, t_r]$, $a_r = a(t_{r-1}, t_r]$, $\alpha_r = \alpha(t_{r-1}, t_r]$ and $\beta_r = \beta(t_r)$, for $r = 1, \ldots, R$. Then the posterior conditional distributions of Λ_r and a_r are,

$$[\,\Lambda_r \,|\, D, a_r\,] \sim \text{Ga}\left\{a_r + \sum_i d_i\, I(t_{r-1} < x_i \le t_r), b+1\right\}$$

and

$$[\,a_r \,|\, D, \Lambda_r\,] \propto \frac{(b\Lambda_r)^{a_r}}{\Gamma(a_r)}\, a_r^{\alpha_r - 1} e^{-\beta_r a_r} I(a_r > 0),$$

for $r = 1, \ldots, R$.

We can simulate from the posterior conditional distribution of a_r introducing a Metropolis–Hastings step (see, *e.g.*, Tierney, 1994) in the following way: at iteration $(k+1)$, for each $r = 1, \ldots, R$, generate $a_r^\star \sim \text{Ga}(\alpha_r, \beta_r)$; then take $a_r^{(k+1)} = a_r^\star$ with probability $\pi(a_r^\star, a_r^{(k)})$ and $a_r^{(k+1)} = a_r^{(k)}$ with probability $\{1 - \pi(a_r^\star, a_r^{(k)})\}$, where

$$\pi(a_r^\star, a_r^{(k)}) = \frac{\Gamma(a_r^{(k)})}{\Gamma(a_r^\star)}\, (b\Lambda_r)^{(a_r^\star - a_r^{(k)})}.$$

We implemented this algorithm with $t_0 = 0$, $t_r = t_{r-1} + 0.54$ and $R = 1000$. We ran the Markov chain for 10000 iterations with a burn-in period of 1000. Convergence was also assessed visually by plotting ergodic averages of some of the a_r and Λ_r. Figure 2 shows the data together with the posterior mean (continuous line), the median (dotted line), and a 95% credible band (dashed lines) for the mean function $\mathcal{M}(\cdot)$. As in the previous example, the straight line appearing on the left of the graph represents the function at which we centered the prior of $\mathcal{M}(\cdot)$, namely $\mathcal{M}_0(t) = t$.

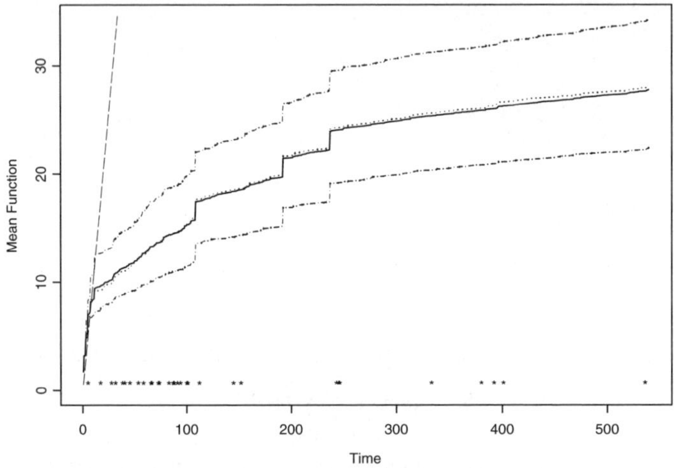

Figure 2. *Posterior mean, median and 95% band for* $\mathcal{M}(\cdot)$ (NHMPP2; *extended gamma prior*).

6. CONCLUDING REMARKS

In this paper, we have presented a relatively simple procedure to make Bayesian inferences for two types of non-homogeneous mixed Poisson processes. No parametric form for the mean functions was assumed; instead, the mean functions were given Lévy process priors 'centered' at suitable parametric models. Due to the nature of the parameter space, this approach can be regarded as nonparametric. For each of the two types of mixed Poisson processes, we showed how one can center the prior in order to reflect prior knowledge about the mean function.

Posterior inference for the NHMPP1 was achieved by implementing a Gibbs sampling scheme, using an efficient algorithm (Ferguson and Klass, 1972) which allowed us to simulate whole paths from the required Lévy processes (Section 5.1). However, for the NHMPP2, a partitioning simulation scheme proved to be a simpler way to carry out the required posterior simulations (Section 5.2). In both cases, the algorithms were easy to code, although it must be pointed out that in general the algorithm described in Section 5.1 is computationally more expensive. The methods were illustrated with the analysis of two different types of negative binomial processes.

One possible extension to the models discussed in this paper would be to use a Lévy-driven process prior (Nieto-Barajas, 2001; Nieto-Barajas and Walker, 2001) for the intensity function of a NHMPP. Such priors would be especially useful in cases where the mean function (cumulative intensity) is constrained to be continuous, and the required posterior simulations could be carried out without much additional effort with respect to the methods discussed here.

ACKNOWLEDGEMENTS

The first and second authors were supported by grants 32256-E and I39357-E, respectively, from CONACyT, Mexico. The first author wishes also to acknowledge partial support from the Sistema Nacional de Investigadores, Mexico.

REFERENCES

Albrecht, P. (1982). On some statistical methods connected with the mixed Poisson process. *Scand. Actuarial J.* 1–14.

Brix, A. and Diggle, P. J. (2001). Spatiotemporal prediction for log-Gaussian Cox Processes. *J. R. Statist. Soc. B* **63**, 823–841.

Cox, D. R. (1955). Some statistical methods connected with series of events (with discussion). *J. R. Statist. Soc. B* **17**, 129–164.

Dykstra, R. L. and Laud, P. W. (1981). A Bayesian nonparametric approach to reliability. *Ann. Statist.* **9**, 356–367.

Ferguson, T. S. (1973). A Bayesian analysis of some nonparametric problems. *Ann. Statist.* **1**, 209–230.

Ferguson, T. S. and Klass, M. J. (1972). A representation of independent increment processes without Gaussian components. *Ann. Math. Statist.* **43**, 1634–1643.

Gikhman, I. I. and Skorokhod, A. V. (1965). *An Introduction to the Theory of Random Processes*. New York: Dover.

Grandell, J. (1976). *Doubly Stochastic Processes*. Lecture Notes in Mathematics 529. New York: Springer.

Grandell, J. (1991). *Aspects of Risk Theory*. New York: Springer.

Grandell, J. (1997). *Mixed Poisson Processes*. London: Chapman and Hall.

Hjort, N. L. (1990). Nonparametric Bayes estimators based on beta processes in models for life history data. *Ann. Statist.* **18**, 1259–1294.

Kingman, J. F. C. (1993). *Poisson Processes*. Oxford: Oxford University Press

Küchler, U. and Sorensen, M. (1997). *Exponential Families of Stochastic Processes*. New York: Springer.

Kuo, L. and Ghosh, S. K. (1997). Bayesian nonparametric inference for non-homogeneous Poisson processes. *Tech. Rep.*, University of Connecticut, Storrs, USA.

Kuo, L. and Yang, T. Y. (1996). Bayesian computation for non-homogeneous Poisson processes in software reliability. *J. Am. Statist. Ass.* **91**, 763–773.

Lang, M. (1999). Theoretical discussion and Monte-Carlo simulations for the negative binomial process paradox. *Stoch. Environ. Res. Risk Assess.* **13**, 183–200.

Lo, A. Y. (1982). Bayesian nonparametric statistical inference for Poisson point processes. *Wahrscheinlichkeits-theorie und verwandte Gebiete* **59**, 55–66.

Nieto-Barajas, L. E. (2001). *Bayesian Nonparametric Survival Analysis via Markov Processes*. Ph.D. Thesis, University of Bath, UK.

Nieto-Barajas, L. E. and Walker, S. G. (2001). Bayesian nonparametric survival analysis via Lévy driven Markov processes. *Tech. Rep.*, University of Bath, UK.

Sato,K.-I. (1999). *Lévy Processes and Infinitely Divisible Distributions*. Cambridge: Cambridge University Press.

Smith, A. F. M. and Roberts, G. O. (1993). Bayesian computations via the Gibbs sampler and related Markov chain Monte Carlo methods. *J. R. Statist. Soc. B* **55**, 3–23.

Tierney, L. (1994). Markov chains for exploring posterior distributions. *Ann. Statist.* **22**, 1701–1762 (with discussion).

van der Weide, H. (1997). Gamma processes. In *Engineering Probabilistic design and Maintenance for Flood Protection* (R. Cooke *et al.*, eds). Dordrecht: Kluwer, 77–83.

Walker, S. G. (1995). Generating random variates from D-distributions via substitution sampling. *Statist. Comput.* **5**, 311–315.

Walker, S. G. and Damien, P. (1998). A full Bayesian nonparametric analysis involving a neutral to the right process. *Scand. J. Statist.* **25**, 669–680.

Walker, S. G. and Damien, P. (2000). Representation of Lévy processes without Gaussian components. *Biometrika* **87**, 477–483.

Walker, S. G. and Muliere, P. (1997). Beta-Stacy processes and a generalization of the Pólya-urn scheme. *Ann. Statist.* **25**, 1762–1780.

Wolpert, R. L. and Ickstadt, K. (1998a). Poisson/gamma random field models for spatial statistics. *Biometrika* **85**, 251–267.

Wolpert, R. L. and Ickstadt, K. (1998b). Simulation of Lévy random fields. *Practical Nonparametric and Semiparametric Bayesian Statistics* (D. Dey, P. Müller and D. Sinha, eds). New York: Springer, 227–242.

DISCUSSION

LUKE TIERNEY (*University of Minnesota, USA*)

I congratulate the authors on a very interesting paper. I have just a few technical comments and a few general remarks on the models used.

The NHMPP1 prior distribution is a semi-parametric model with a one-dimensional parametric component. One advantage of using a Lévy process for the nonparametric part is that one may be able to integrate out the nonparametric part to obtain a computable expression for the marginal density of the parametric component. This is the approach used by Kalbfleisch (1978) for the proportional hazards model. For an extended Gamma process prior distribution on Λ and assuming no ties in the data this yields

$$f(\theta|D) \propto \theta^n \left[\prod_{i=1}^{n} \frac{\alpha'(x_i)}{\theta + \beta(x_i)}\right] \exp\left\{-\int_0^T \log\left(1 + \frac{\theta}{\beta(u)}\right) d\alpha(u)\right\} f(\theta)$$

for the marginal posterior density of θ. The one-dimensional integral in this representation can easily be computed numerically. For the software reliability example the marginal posterior density of $\log\theta$ is shown as the solid curve in Figure 3. With a little experimentation one can easily find a scaled t distribution with 10 degrees of freedom that can be used as an envelope for rejection sampling from this marginal distribution; such an envelope is shown as the dashed curve in Figure 3. Thus a strategy for exact sampling of the joint posterior distribution is available: sample θ from its marginal distribution by rejection sampling, and then sample Λ from the extended Gamma process conditional posterior distribution using the Ferguson-Klass method.

In this particular example the envelope was determined empirically; it may be useful to explore methods for automatic determination of the envelope. Similar strategies may be useful in other nonparametric settings with a low dimensional parametric component.

A second technical point concerns the use of approximations in the sampling methods. Some form of approximation seems to be unavoidable when dealing with infinite dimensional parameters. The paper uses different approaches for the two settings considered: in the NHMPP1 model the Ferguson–Klass steps for the nonparametric component are truncated in some form (the paper does not appear to say explicitly how this is done); in the NHMPP2 model the infinite dimensional posterior distribution is replaced by a finite dimensional approximation. A concern with the NHMPP1 approach is that the use of an approximate conditional distribution within a Gibbs sampler may lead to a Markov chain that does not converge or converges to a distribution that is not close to the target distribution. Perhaps some theoretical results can be obtained to show that there is no need for this concern, but at this point I am not aware of such results. The NHMPP2 approach, while in some sense less sophisticated, is a bit safer in that the sampler is assured to have the approximate posterior distribution as its stationary distribution.

At the modeling level, one point on which I was not clear is how the additional hierarchy level in the two models can best be used to answer questions relevant to a particular application. For the software reliability example it seems that there would be two primary types of questions. The first concerns burn-in: does the data we have so far on a particular software system suggest

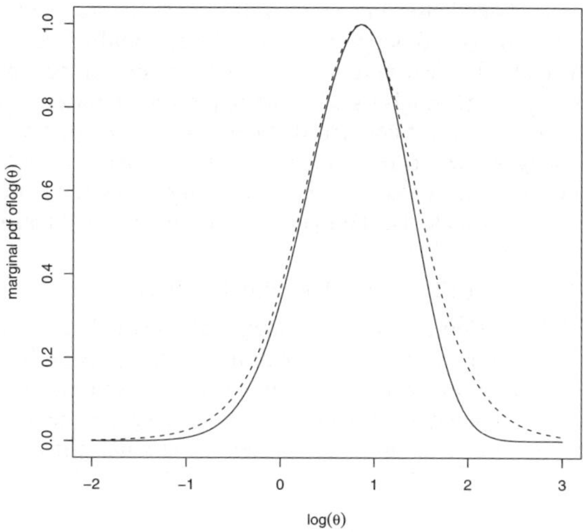

Figure 3. *Marginal posterior density of* $\log\theta$ *with scaled* t *envelope.*

that we have found enough bugs to warrant release of the software? The second type of question concerns generalization: what does our experience with this project tell us about other similar projects? For the first kind of question it would seem that we would be most interested in inferences about Λ, the rate process for the particular software system under study, rather than the mean rate process $\mathcal{M}$. In this context the added hierarchy provides a richer family of prior distributions for Λ than a simple Lévy process but at the cost of considerable added complexity. It also may not by itself provide enough structure for extrapolation beyond the period of observation, which would be quite important in this context.

For answering questions of generalization the NHMPP models can be viewed as hierarchical models representing variability of individual software projects from an overall mean rate $\mathcal{M}$ at the first stage, and uncertainty about the overall mean rate $\mathcal{M}$ at the second stage. To allow for learning about the magnitude of the first-stage variability it would be very useful to extend the analysis of the paper to allow for the use of data on multiple software projects, i.e. multiple realizations $\mathcal{N}$, and to allow for a prior distribution on the magnitude of the dispersion of the first stage. For the NHMPP1 model this might mean taking $f(\theta \mid \alpha)$ to have a $\mathrm{Ga}(\alpha, \beta)$ distribution and adding a prior distribution for α; for the NHMPP2 model it might involve adding a prior distribution for at least a parametric component of the first stage Lévy process dispersion $b(\cdot)$.

As a final comment, while Lévy processes are quite popular in the literature as formal prior distributions on functional parameters I am somewhat uneasy about their use in cases where the functions are directly interpretable and would typically be expected to have some smoothness properties. In practical Bayesian analyses it is rarely possible to completely elicit the actual prior distribution that should be used. Instead, a formal prior distribution is used that is matched to certain key features that have been elicited and that is hoped to be reasonably close in other features. In well-behaved finite dimensional problems the resulting formal posterior distributions can be expected to be close to those that would have been obtained with a more complete elicitation. As the elicitation sections of the paper show, Lévy processes can be chosen to match many features that one might like a formal prior distribution to have. However, they also have some peculiar properties, such as assigning probability one to pure jump functions with jumps that are everywhere dense. In applications such as the software reliability example my prior probability that the rate function is of this form would be zero. Given this wide separation between the real prior distribution and the formal Lévy process prior distribution, it is not clear whether the resulting formal posterior distribution will be close to my actual posterior distribution in the features I am most concerned about. The inconsistency results of Diaconis and Freedman (1986) suggest there is reason for concern. Formal prior distributions based on splines or Gaussian processes that produce smooth realizations seem less likely to create such obvious discrepancies, though such distributions may have their own undesirable features.

REPLY TO THE DISCUSSION

We would like to thank Prof. Tierney for his comments, all of which raise interesting issues.

Prof. Tierney's strategy for exact sampling will undoubtedly be useful in many nonparametric (or semiparametric) settings with a low-dimensional parametric component where the required integrals can be computed in closed form. For instance, for the log-beta process prior without prior fixed points of discontinuity, the posterior marginal distribution of θ would be given by

$$f(\theta \mid D) \propto \theta^n \left[\prod_{i=1}^{n} \alpha'(x_i)\psi'\{\theta + \beta(x_i)\} \exp\left\{ -\int_0^{\tau} (\psi\{\theta + \beta(u)\} - \psi\{\beta(u)\})\, d\alpha(u) \right\} \right],$$

where $\psi(x) = d \log \Gamma(x)/dx$ denotes the digamma function. This turns out to be very similar to the one obtained by Prof. Tierney for the extended gamma case, and could therefore be also sampled from using a rejection method.

Concerning the second technical point, we would like to emphasize that in the Ferguson and Klass's representation of a Lévy process as an infinite sum of random jumps at random locations (see Section 5.1), the sequence of jumps $J_1, J_2, \ldots$ is decreasing. Therefore, when truncating the sum in the representation of $L_c(t)$, the dropped terms always correspond to the smallest jumps. This is in contrast to alternative algorithms such as that of Wolpert and Ickstadt (1998b), for which truncation is not as straightforward.

As for the actual truncation method, one possibility is to monitor, at each iteration of the Gibbs sampler, the relative error

$$\text{RE}(r) = \left| \frac{S_r - S_{r-1}}{S_r} \right|$$

until $\text{RE}(r) \leq \epsilon$, where $S_r = \sum_{i=1}^{r} J_i$ is the cumulative sum of the jump sizes and ϵ is a small positive constant. Thus, at each iteration the truncation point would be given by the minimum value of r satisfying the last condition.

Alternatively, one could truncate the sum uniformly, fixing the number of terms at R, say, across all iterations of the Gibbs sampler. The value of R can be determined on the basis of a few preliminary runs. Actually, this was the procedure we followed in the example of Section 5.1, where we used $R = 500$.

As far as modelling is concerned, we regard both NHMPP1 and NHMPP2 as providing alternatives to the usual non-homogeneous Poisson process at the "likelihood level", and not necessarily as providing a richer family of prior distributions for $\Lambda(\cdot)$. Acknowledgedly our notation is slightly confusing, but what we are assuming is that the observed data are samples from the marginal distribution of $\mathcal{N}(\cdot)$ with respect to the corresponding mixture representation. We then put a prior on $\mathcal{M}(\cdot) = \mathrm{E}\{\mathcal{N}(\cdot)\}$ via

(i) a Lévy prior for $\Lambda(\cdot)$ in the NHMPP1 case, where $\mathcal{M}(t) = \mu_\theta \Lambda(t)$;

(ii) a Lévy prior for $a(\cdot)$ in the NHMPP2 case, where, for example, $\mathcal{M}(t) = a(t)/b$ if we let $\Lambda(\cdot)$ have a gamma process $\mathrm{GaP}\{a(\cdot), b\}$.

In this latter case we do not see the distribution on $\Lambda(\cdot)$ as a "prior", but rather as a mixing distribution which gives rise to a model that is overdispersed relative to the usual NHPP. Thus, in the example of Section 5.2, where we used a gamma process for $\Lambda(\cdot)$, we ended up with a negative binomial process (of type 2) for $\mathcal{N}(\cdot)$, a process parameterized by $a(\cdot)$.

In both cases the hierarchical representation is a convenient device allowing us to carry out the analysis via a Gibbs sampler. In view of the discussion above, we feel that the function of interest should be the mean $\mathcal{M}(\cdot)$ of the marginal process $\mathcal{N}(\cdot)$, and not $\Lambda(\cdot)$.

The hierarchical model discussed by Prof. Tierney, describing the variability of individual software projects, looks very interesting and certainly deserves further study.

We agree with Prof. Tierney that care must be taken when choosing nonparametric priors, and that consistency is a genuine concern in such cases. However, recent and ongoing work by several authors (Barron *et al.*, 1999; Kim and Lee, 2001; Walker and Hjort, 2001; Walker, 2002ab) shows that, under certain conditions on the prior processes, many of the priors commonly used in applications yield consistent posteriors.

Concerning the last point of the discussion about the continuity of the prior process, apart from the priors suggested by Prof. Tierney, as mentioned in the article one can also consider kernel mixture (Lévy-driven) priors of the form

$$\Lambda(t) = \int h(t, \nu)\, dL(\nu)$$

where $L(\nu)$ is the driving Lévy process and $h(\cdot, \cdot)$ is a suitable kernel function. Such priors have been successfully used by Nieto-Barajas (2001) and Nieto-Barajas and Walker (2001) in the context of survival analysis.

ADDITIONAL REFERENCES IN THE DISCUSSION

Barron, A., Schervish, M. J. and Wasserman, L. (1999). The consistency of posterior distributions in nonparametric problems. *Ann. Statist.* **27**, 536–561.

Diaconis, P. and Freedman, D. (1986). On inconsistent Bayes estimates of location. *Ann. Statist.* **14**, 68–87.

Kalbfleisch, J. D. (1978). Nonparametric Bayesian analysis of survival time data. *J. R. Statist. Soc. B* **40**, 31–38.

Kim, Y. and Lee, J. (2001). On posterior consistency of survival models. *Ann. Statist.* **29**, 666–686.

Walker, S. G. and Hjort, N. L. (2001). On Bayesian consistency. *J. R. Statist. Soc. B* **63**, 811–821.

Walker, S. G. (2002a). On sufficient conditions for Bayesian consistency. *Tech. Rep.*, University of Bath, UK.

Walker, S. G. (2002b). A new approach to Bayesian consistency. *Tech. Rep.*, University of Bath, UK.

BAYESIAN STATISTICS 7, pp. 181–197
J. M. Bernardo, M. J. Bayarri, J. O. Berger, A. P. Dawid,
D. Heckerman, A. F. M. Smith and M. West (Eds.)

Markov chain Monte Carlo-based approaches for inference in computationally intensive inverse problems

DAVE HIGDON
Los Alamos National Laboratory and Duke University, USA
dhigdon@lanl.gov

HERBIE LEE and CHRIS HOLLOMAN
Duke University, USA
herbie@stat.duke.edu chris@stat.duke.edu

SUMMARY

A typical setup for many inverse problems is that one wishes to update beliefs about a spatially dependent set of inputs x given rather indirect observations y. Here, the inputs and observed outputs are related by the complex physical relationship $y = \zeta(x) + \epsilon$. Applications include medical and geological tomography, hydrology, and the modelling of physical and biological systems. We consider applications where the physical relationship $\zeta(x)$ can be well approximated by detailed simulation code $\eta(x)$.

When the forward simulation code $\eta(x)$ is sufficiently fast, Bayesian inference can, in principle, be carried out via Markov chain Monte Carlo (MCMC). Difficulties arise for two main reasons:

- Even though the code may accurately represent the physical process, there are a large number of unknown, but required, inputs that must be calibrated to match the observed data y.
- The computational burden of the fastest available forward simulators is often large enough that approaches for speeding up the MCMC calculations are required.

This paper develops approaches for specifying effective low-dimensional representations of the inputs x along with MCMC approaches for sampling the posterior distribution. In particular we consider augmenting the basic formulation with fast, possibly coarsened, formulations to improve MCMC performance. This approach can be very easily implemented in a parallel computing environment. We give examples in single photon emission computed tomography and in hydrology.

Keywords: MULTIGRID MARKOV CHAIN MONTE CARLO; METROPOLIS COUPLED MARKOV CHAIN MONTE CARLO; SPATIAL STATISTICS; DISTRIBUTED COMPUTING.

1. INTRODUCTION

A typical setup for many inverse problems is that one wishes to update beliefs about a spatially dependent set of inputs x given indirect observations $y = (y_1, \ldots, y_n)^t$. Here the inputs and observed outputs are related by the complex physical relationship $y = \zeta(x) + \epsilon$ where $\zeta(x)$ denotes the actual physical system at the true, but unknown state $x = (x_1, \ldots, x_m)$, and ϵ denotes sampling error. Many such systems can be approximated by detailed computer simulation code $\eta(x)$. A very incomplete list of applications includes medical tomography (Weir, 1997), geological tomography (Andersen *et al.* 2001), hydrology (Lee *et al.*, 2002), petroleum engineering (Craig *et al.*, 2001; Hegstad and Omre, 2001), as well as a host of other physical, biological, or social systems. The observed data

$$y = \zeta(x) + \epsilon$$

are modelled statistically by

$$y = \eta(x) + e$$

where the discrepancy term e accounts for both sampling error and mismatch between the simulator $\eta(x)$ and reality $\zeta(x)$:

$$e = \zeta(x) - \eta(x) + \epsilon.$$

The goal is to use the observed data y to make inference about the spatial input parameters x; in particular, to characterize the uncertainty about x.

The likelihood $L(y|x, \theta_y)$, which may depend on additional parameters held in θ_y, is then specified to account for both mismatch and sampling error. It is worth noting here that the data come only from a single experiment. So there is no opportunity to obtain data from additional experiments for which some controllable inputs have been varied. Because of this, there is little hope of modelling the mismatch term $\zeta(x) - \eta(x)$ separately from the sampling error as is often done in the statistical analysis of complex computer code outputs (Kennedy and O'Hagan, 2001). Therefore, the likelihood specification will often need to be done with some care, incorporating the modeler's judgement about the appropriate size and nature of the mismatch term.

We consider systems for which the model input parameters x denote a spatial field or image. For example, in single photon emission computed tomography (SPECT) the image intensity x denotes blood flow within a region of the body; in a hydrologic application, x might give the spatial distribution of hydraulic conductivities or permeability. The simulator requires gridded inputs and the resolution of the grid is a pre-specified input to the simulator. The spatial prior for x, $\pi(x|\theta_x)$, will typically include an additional parameter vector θ_x to control x. The parameter θ_x may then be treated as fixed or have a prior of its own $\pi(\theta_x)$. Both modelling and computing considerations go into specification of $\pi(x|\theta_x)$, which is discussed in the following section.

The resulting posterior is then given by

$$\pi(x, \theta) \propto L(y|\eta(x), \theta_y) \times \pi(x|\theta_x) \times \pi(\theta)$$

where θ holds both nuisance parameters (θ_y, θ_x), This posterior can, in principle, be explored via Markov chain Monte Carlo (MCMC). However the combined effects of the high dimensionality of x and the computational demands of the simulator make implementation difficult in practice. By itself, the high dimensionality of x is not necessarily a problem. MCMC with single-site updating has been carried out with relative ease in large image applications. However, a high-dimensional input vector x does make it quite difficult to build any sort of statistical model $\hat{\eta}(x)$ to approximate the simulator as in Sacks *et al.* (1989) or Kennedy and O'Hagan (2001). Any MCMC implementation using a single-site updating scheme is impractical since it will require m forward runs for a single update scan through all the parameters. In addition, a simulator may require a fine grid to ensure satisfactory numerical performance, but the numerical error may unduly affect the small changes in output y when only a single component of x has been updated. The use of higher dimensional proposals has proven somewhat successful (Oliver *et al.*, 1997; Lee *et al.*, 2002), especially when some direct measurements on x are available as in Hegstad and Omre (2001). A similar strategy that we have found effective is to reparameterize x; this is discussed in Section 2.

To deal with the computational burden of the forward simulator $\eta(x)$, Section 3 lays out a Metropolis coupled MCMC (Geyer, 1991) implementation that simultaneously runs chains to sample multiple posterior formulations $\pi(x^1, \theta^1), \ldots, \pi(x^K, \theta^K)$ for which the spatial input parameters $x^1, \ldots, x^K$ are coarsened to varying degrees. Each formulation runs its simulator $\eta^k(x^k)$ at its own particular grid resolution. This MCMC scheme, which borrows from the work of Goodman and Sokal (1989) and Liu and Sabatti (1999), allows information from the faster running, but less accurate, coarse formulations to speed up the mixing for the fine scale chains. In addition, this scheme is relatively easy to implement on a parallel environment, without having to "parallelize" the actual simulator code. This distributed, coupled MCMC approach is discussed in Section 3. Section 4 follows giving a final discussion.

2. SPATIAL REPRESENTATIONS

The simulator typically requires that x be input over m regular grid points at spatial locations denoted by the set $s^x = \{s_1^x, \ldots, s_m^x\}$, which is contained in the spatial domain $\mathcal{S}$. Hence the actual input to $\eta(\cdot)$ requires x be restricted to the grid points $x_{s^x} = (x(s_1^x), \ldots, x(s_m^x))^t$. As regards to notation, we use x when the process is only considered at the set of spatial locations s^x; we take $x(s)$ to mean that the process is defined for all $s \in \mathcal{S}$. The grid size m can often be specified in the simulator $\eta(x)$, with fine grids typically giving more accurate results at the cost of increased computation. We note that recent literature has stressed the importance of specifying spatial models that are consistent under coarsening or aggregation schemes. Clearly, the single component $x_{s_i^x}$ of the input grid x is some form of aggregate of a continuously defined process in the neighborhood of the location s_i^x. However, in many applications involving simulation of physical systems, aggregation—equivalently, upscaling or closure—is a difficult, or even an ill-posed task by itself. So, even though issues regarding aggregation consistency can play an important role, especially when the aggregation process is well defined and data are sufficiently informative, the data are usually insufficient to resolve x in much detail in the applications we consider. Hence, we only require that the prior distribution for x infuse prior knowledge about its spatial distribution—at least at the resolution/level of detail that we expect from the data—as well as regularize the posterior for x.

In this paper we use both intrinsic Gaussian Markov random fields (MRFs) that model the m-dimensional process x at the spatial locations s^x, and standard Gaussian processes (GP) that define a process $x(s)$ over continuous space. The intrinsic Gaussian MRF has the form

$$\pi(x|\theta_x) \propto \theta^{m/2} \exp\left\{\tfrac{1}{2}\theta x^t W x\right\} \tag{1}$$

where θ_x controls the scale of x and the MRF precision matrix has the simple form

$$W_{ij} = \begin{cases} n_i & \text{if} \quad i = j \\ -1 & \text{if} \quad i \sim j \\ 0 & \text{otherwise,} \end{cases} \tag{2}$$

where n_i is the number of neighbors of site s_i^x and $i \sim j$ means locations s_i^x and s_j^x are neighbors of one another. With the regular grids considered in this paper, we specify two sites s_i^x and s_j^x to be neighbors if they are directly adjacent on the grid so that interior points of a 2-d rectangular grid have four neighbors; edge sites have three; and corner sites have two.

Gaussian process priors are typically specified through their mean and covariance function. We take the mean to be constant and define the covariance by

$$\text{Cov}(x(s_i^x), x(s_j^x)) = \theta_1 \rho \left(\frac{\|s_i^x - s_j^x\|}{\theta_2} \right)$$

where the correlation function $\rho(\cdot)$ must be positive definite and satisfy $\rho(0) = 1$. We typically take $\rho(d) = e^{-d^2}$, which leads to very smooth realizations for $x(s)$. By contrast, realizations under the locally planar MRF model (2) exhibit local roughness.

This distinction is important if one wishes to infer about the local nature of x and if the data are informative about the small scale nature of x. It is often the case in inverse problems that the indirectly observed data give no information regarding the small-scale behavior of x. Also, the input grid x can best be regarded as the aggregate of an underlying continuous process. For the two reasons above it is often impossible to distinguish between locally smooth and locally rough character of $x(s)$ from the data alone. When this is the case, as it is in the hydrology examples, computational considerations can lead us to favor models with smooth local behavior.

When we can get away with a smooth GP specification for $x(s)$, we can then efficiently represent $x(s)$ by convolving a white noise process $u(s)$, $s \in \mathcal{S}$ with a smoothing kernel $k(s)$ so that

$$x(s) = \int_{\mathcal{S}} k(\nu - s)u(\nu),\, d\nu \quad \text{for } s \in \mathcal{S}. \tag{3}$$

The resulting covariance function for $x(s)$ depends on the displacement vector $d = s - s'$ and is given by

$$\text{Cov}(x(s), x(s')) \propto \rho(d) \propto \int_{\mathcal{S}} k(\nu - s)k(\nu - s')\, d\nu = \int_{\mathcal{S}} k(\nu - d)k(\nu)\, d\nu. \tag{4}$$

The proportionality depends on the scale of the white noise process $u(s)$ and on $\int_{\mathcal{S}} k^2(\nu)\, d\nu$. We typically take $\mathcal{S}$ to be $R^{1,2,\text{ or }3}$, and $k(\cdot)$ to be a normal density with independence between the coordinate component directions. This is an equivalent representation of a mean 0 GP with $\rho(d) = e^{-d^2}$, possibly after rescaling the coordinate axes. By restricting the latent process $u(s)$ to coarse lattice locations $s_1^u, \ldots, s_\ell^u$, a small number of parameters effectively control the entire process $x(s)$. Now with a discrete white noise process

$$u = (u(s_1^u), \ldots, u(s_\ell^u))^t \sim N(0, I_\ell/\theta_u)$$

$x(s)$ can be represented by the discrete analog of (3)

$$x(s) = \sum_{k=1}^{\ell} u_k k(s - s_k^u) \tag{5}$$

where $k(\cdot - s_k^u)$ is the smoothing kernel centered at s_k^u. Figure 1 shows three successively coarsened white noise realizations u, their induced processes $x(s)$ from convolving u with the kernel shown in the upper left of the top row of figures; the bottom row of figures shows $\text{Cor}(x(s_0), x(s))$ as a function of s. The dotted black line gives the ideal covariance function obtained via (4). For this smooth process, u can undergo substantial coarsening before the induced process begins to substantially deviate from the ideal one obtained from continuous white noise.

Before moving on, we note there are alternative lower dimensional representations of $x(s)$ that one may consider such as Cholesky, SVD, or Fourier. Taking x to be discrete, in each case we can express $x = Ku$ so that x is the weighted sum of bases given by the columns of K. The difference between the approaches is in the specification of K. We favor the moving average representation because of its local nature as well as the simplicity of its basis representation. Its local nature meshes well with MCMC in which a simple Metropolis update of individual u_k will influence a local region of $x(s)$. The simplicity of this basis representation easily allows for extending the basic model for which $u \sim N(0, I_\ell/\theta_u)$. By allowing more general dependence within u the model can be extended to account for non-stationarity or time dependence; see Calder *et al.* (2002) for example.

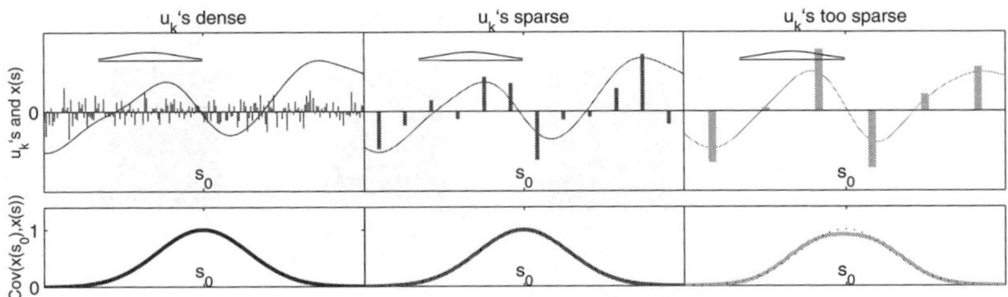

Figure 1. *A stationary spatial process $x(s)$ can be generated by smoothing white noise. The top frames show the induced process $x(s)$ obtained by smoothing the white noise shown by vertical lines using the kernel shown in the top left of the figure. Moving from left to right, the underlying white noise process becomes successively coarser. Below each of the top frames is a function showing Cov$(x(s_0), x(s))$ as a function of s; the location of s_0 is marked in the figures. In the rightmost frame the u_k's are so sparse that the covariance of the induced process begins to deviate from the ideal covariance function it is trying to match, which is shown by the black dotted line.*

Example 1. Studying the flow of water underground is of great interest to engineers, with important applications to cleanup of contaminated soil and petroleum exploration and production. A statistically interesting component of this problem is the inverse problem of inferring soil structure (*e.g.*, permeability) from flow data. Further details and references can be found in Lee *et al.* (2002).

The data presented here are from a larger study (Annable *et al.* 1998) at the Hill Air Force Base in Utah where the ground contains a number of contaminants. We look only at conservative tracer data, an experiment that yields information only on the permeabilities and not on the contamination. The site is 14 feet by 11 feet, with four injection wells along one edge and three production (extraction) wells along the opposite edge. Water is pumped continuously through the field, and then a tracer is added and the time of travel is measured for the tracer from the injection wells to five measurements sites (sampling wells) in the field. This time of travel is referred to as the breakthrough time for each sampling location. Since water flows faster through regions of higher permeability, one can learn about the underlying permeabilities through the breakthrough times. The upper left plot of Figure 2 shows the locations of the injectors, producers, and samplers, with the breakthrough times shown for the sampling locations.

Permeabilities vary spatially and are typically considered to be log-normally distributed. Thus, all our priors for permeabilities are stated on the log scale. We use a 42 by 33 grid of square cells, one-third of a foot on each side. For notational convenience, we represent the unknown (log) permeabilities as a $m = 42 \times 33 \times 1$ lattice x. Conditional on a specified permeability field x, the breakthrough times are found from the solution of differential equations given by physical laws, *i.e.*, conservation of mass, Darcy's Law, and Fick's Law. We do this using the S3D stream-tube computer code of King and Datta-Gupta (1998) and find the $n = 5$ fitted breakthrough times, $\hat{y} = \eta(x)$.

We consider two formulations—one for which x is modelled as a 2-d MRF prior using four nearest neighbors on a $m = 42 \times 33$ lattice; and one for which x is parameterized as a GP via (5), where the $\ell = 72$ kernel locations are shown by the dots in the bottom left frame of Figure 2 and the kernels are bivariate normal with a one sd ellipse shown in the bottom left frame of Figure 2. The resulting posteriors are

$$\pi(x, \theta_x|y) \propto \exp\{\tfrac{1}{2}\lambda(y - \eta(x))^t(y - \eta(x))\} \times \theta_x^{-m/2} \exp\{\tfrac{1}{2}\theta_x x^t W x\} \times \theta^{\alpha_x - 1} e^{-\beta_x \theta}$$
$$\pi(u, \theta_u|y) \propto \exp\{\tfrac{1}{2}\lambda(y - \eta(x))^t(y - \eta(x))\} \times \theta_u^{-\ell/2} \exp\{\tfrac{1}{2}\theta_u u^t u\} \times \theta_u^{\alpha_u - 1} e^{-\beta_u \theta_u}.$$

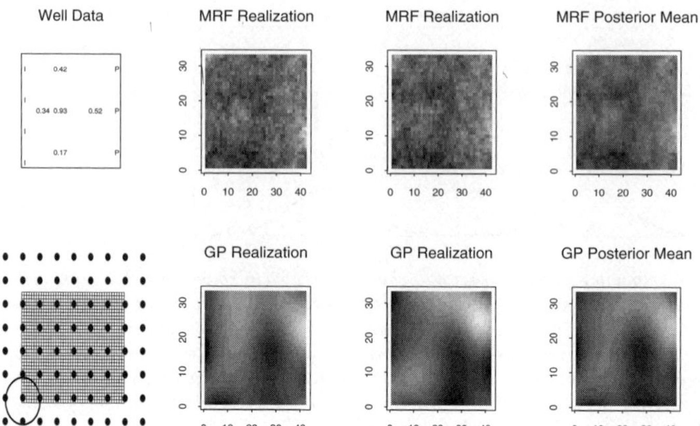

Figure 2. *Layout of wells, posterior realizations, and posterior means for both an MRF model and a moving average Gaussian Process model for the Hill Air Force Base data. In the upper left plot, the wells are labeled "I" for injectors, "P" for producers, and the samplers are shown with numbers where the value is the breakthrough time (in days) for each well. For the permeability plots, darker regions correspond to higher permeability values.*

In the MCMC implementations, x is updated via multivariate Hastings steps (Lee *et al.* 2002) and u is updated via single site Metropolis steps.

Figure 2 shows the results from the two formulations. The top row shows two realizations from the posterior and the posterior mean for the MRF prior, while the bottom row shows analogous plots for the GP prior. Both models fit the observed data well. In particular, both show a region of lower permeability in front of (relative to the injectors) the central well, which has the latest breakthrough time.

3. COMPUTATION

3.1. *Linking Coarse and Fine Formulations*

In many applications, the computational demands of the simulator greatly restrict the number of simulator runs that can be carried out, making posterior exploration via standard MCMC difficult or even impractical. An alternative is to formulate a coarsened version of the problem. Under this coarse specification, a coarsened counterpart for the input x is defined by $\tilde{x} = (\tilde{x}_1, \ldots, \tilde{x}_{\tilde{m}})^t = Cx$, where C is the coarsening operation which maps a m-vector to a lower dimensional $\tilde{m}$-vector. We use $s^{\tilde{x}} = \{s_1^{\tilde{x}}, \ldots, s_{\tilde{m}}^{\tilde{x}}\}$ to denote the spatial locations associated with this coarse grid. Typically, C is a $\tilde{m} \times m$ matrix so that Cx is a simple linear transformation, such as averaging or summing groups of fine-scale pixels to make coarse pixels. However, coarsening, or upscaling, could conceivably be a more complicated operation, depending on the application. Depending on the problem, y, θ_y, and θ_x might also require coarsened counterparts $\tilde{y}$, $\tilde{\theta}_{\tilde{y}}$, and $\tilde{\theta}_{\tilde{x}}$, which are modifications of their original form. In addition, the likelihood and priors under the coarsened formulation may also differ. The net result is two separate posterior distributions—one fine and one coarse:

$$\text{Fine:} \quad \pi(x, \theta|y) \propto L(y|\eta(x), \theta) \times \pi(x|\theta_x) \times \pi(\theta)$$
$$\text{Coarse:} \quad \tilde{\pi}(\tilde{x}, \tilde{\theta}|\tilde{y}) \propto \tilde{L}(\tilde{y}|\eta(\tilde{x}), \tilde{\theta}) \times \tilde{\pi}(\tilde{x}|\tilde{\theta}_{\tilde{x}}) \times \tilde{\pi}(\tilde{\theta}).$$

In order to link the coarse and fine-scale formulations, we make use of *Metropolis coupled* MCMC (Geyer, 1991). Now, instead of running two separate MCMC chains, one on the fine posterior and one on the coarse posterior, a single chain is run on the product distribution. This coupled chain has stationary distribution $\pi(x, \theta|y) \times \tilde{\pi}(\tilde{x}, \tilde{\theta}|\tilde{y})$. Because of the coarsened input $\tilde{x}$ to the simulator, the chain sampling the coarse-scale posterior will run more quickly. In addition, the coarse-scale posterior is typically smoother and easier to sample via MCMC as compared to its fine-scale counterpart. Hence, an efficient coupling scheme will allow information to move between the two formulations.

One possible implementation of such a coupled chain alternates standard within-scale updates with "swapping" updates that allow information to move between the two scales as shown below:

$$\begin{array}{ccccccccccc} (x,\theta)^1 & \overset{\text{MCMC}}{\longrightarrow} & (x,\theta)^2 & \overset{\text{SWAP}}{\longrightarrow} & (x,\theta)^3 & \overset{\text{MCMC}}{\longrightarrow} & (x,\theta)^4 & \overset{\text{SWAP}}{\longrightarrow} & (x,\theta)^5 & \cdots \\ (\tilde{x},\tilde{\theta})^1 & \overset{\text{MCMC}}{\longrightarrow} & (\tilde{x},\tilde{\theta})^2 & & (\tilde{x},\tilde{\theta})^3 & \overset{\text{MCMC}}{\longrightarrow} & (\tilde{x},\tilde{\theta})^4 & & (\tilde{x},\tilde{\theta})^5 & \cdots \end{array}$$

Here the updates denoted by $\overset{\text{MCMC}}{\longrightarrow}$ affect parameters within a given scale, while the updates denoted by $\overset{\text{SWAP}}{\longrightarrow}$ are a Hastings update that proposes new candidates $(x^*, \theta^*, \tilde{x}^*, \tilde{\theta}^*)$ according to the proposal kernel

$$q((x, \theta, \tilde{x}, \tilde{\theta}) \to (x^*, \theta^*, \tilde{x}^*, \tilde{\theta}^*)),$$

which is accepted according to the Hastings rule with probability

$$1 \wedge \frac{\pi(x^*, \theta^*|y)\tilde{\pi}(\tilde{x}^*, \tilde{\theta}^*|\tilde{y}) \times q((x^*, \theta^*, \tilde{x}^*, \tilde{\theta}^*) \to (x, \theta, \tilde{x}, \tilde{\theta}))}{\pi(x, \theta|y)\tilde{\pi}(\tilde{x}, \tilde{\theta}|\tilde{y}) \times q((x, \theta, \tilde{x}, \tilde{\theta}) \to (x^*, \theta^*, \tilde{x}^*, \tilde{\theta}^*))} \tag{6}$$

where $a \wedge b$ is the minimum of a and b.

We now describe some specific swapping proposals $q((x, \theta, \tilde{x}, \tilde{\theta}) \to (x^*, \theta^*, \tilde{x}^*, \tilde{\theta}^*))$ for the applications we consider. It is often convenient to break the swapping proposal kernel into the product

$$q((x, \theta, \tilde{x}, \tilde{\theta}) = q((x, \theta) \to (\tilde{x}^*, \tilde{\theta}^*)) \times q((\tilde{x}, \tilde{\theta}) \to (x^*, \theta^*))$$

where $q((x, \theta) \to (\tilde{x}^*, \tilde{\theta}^*))$ generates a coarse-scale proposal $(\tilde{x}^*, \tilde{\theta}^*)$ from the current fine-scale state (x, θ), and the kernel $q((\tilde{x}, \tilde{\theta}) \to (x^*, \theta^*))$ generates a fine-scale proposal (x^*, θ^*) from the current coarse-scale state $(\tilde{x}, \tilde{\theta})$.

Swapping proposals for MRF priors. When we use the MRF prior for x and $\tilde{x}$ (1), we generate the coarse-scale proposal by deterministically coarsening the fine-scale state and then generating a candidate value $\tilde{\theta}^*$ by simulating from its full conditional distribution (under the coarse-scale posterior) given the new proposed value $\tilde{x}^*$. This proposal kernel can be written

$$q((x, \theta) \to (\tilde{x}^*, \tilde{\theta}^*)) = I[\tilde{x}^* = Cx] \times \tilde{\pi}(\tilde{\theta}^*|\tilde{x}^*, \tilde{y})$$

where $I[\cdot]$ is the indicator function, Cx is the coarsening operation applied to the fine-scale x, and $\tilde{\pi}(\tilde{\theta}|\tilde{x}^*, \tilde{y})$ is the full conditional distribution of $\tilde{\theta}$ under the coarse formulation. If $\tilde{\theta}$ is given a conjugate $\Gamma(\alpha_x, \beta_x)$ prior, then its full conditional also has a gamma form. Also for the applications we consider, C is a simple summing or averaging operation.

The fine-scale candidate (x^*, θ^*) given the current coarse-scale state $(\tilde{x}, \tilde{\theta})$ is generated by drawing from the prior distribution $\pi(x^*|\theta^\dagger)$ subject to the constraint $Cx^* = \tilde{x}$. The value $\theta^\dagger$ is a deterministic function of $\tilde{\theta}$ chosen so that the candidate x^* most nearly matches the properties of typical fine-scale realizations. In the 1-d application of Example 2, we take $\theta^\dagger = \frac{1}{8}\tilde{\theta}$; in the

2-d SPECT application of Example 3, we take $\theta^\dagger = \frac{1}{32}\tilde{\theta}$. Once x^* has been generated, θ^* can then be drawn from its full conditional given the candidate value x^*. Hence

$$q((\tilde{x}, \tilde{\theta}) \to (x^*, \theta^*)) \propto \pi(x^*|\theta^\dagger, y) I[\tilde{x} = Cx^*] \times \pi(\theta^*|x^*, y).$$

Note that when the prior $\pi(x|\theta)$ has a multivariate normal form and C is a matrix, then the proposal x^* can be generated directly. This update is more problematic when the prior for x is not normal.

Example 2. Before considering swapping updates for formulations involving moving average specifications for x, we first consider a synthetic blur free, 1-d imaging example. A smooth, 1-d source is emitting according to a Poisson process with intensity given by the smooth, solid line(s) in Figures 3a and 3b. Under the fine-scale formulation of the problem is a 1-d array of $n = 40$ detectors recording emissions from the source; the count for each detector is shown in Figure 1(a). A 1-d Gaussian MRF prior over the $m = 40$ detector sites is assigned to the unknown fine-scale image x, with a $\Gamma(\alpha, \beta)$ prior for the precision parameter θ_x. A coarse-scale formulation is obtained by combining adjacent detector pairs so that the coarsened data consist of $\tilde{n} = 20$ counts (Figure 1b). Similarly, a MRF prior is assigned to the coarsened image $\tilde{x}$, which is divided into $\tilde{m} = 20$ sites, one for each coarse detector.

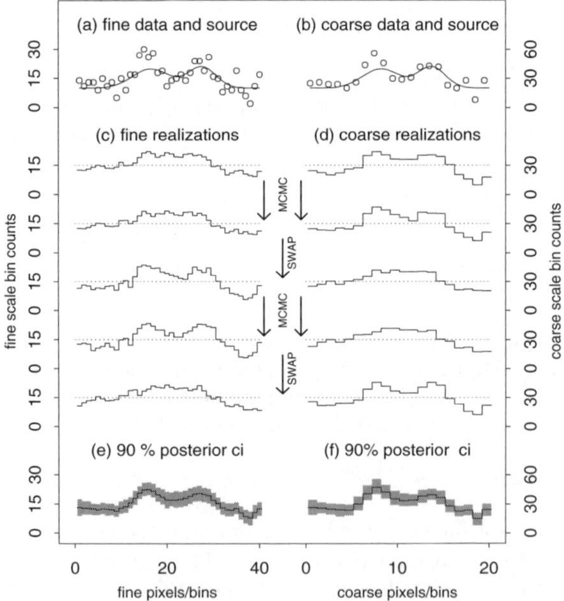

Figure 3. *Data, posterior realizations, and posterior summary for the coupled MCMC scheme: (a and b) Data and true image intensity under the fine and coarse formulations; (c and d) a sequence of four updates under the coupled MCMC scheme; (e and f) pointwise posterior 90% credible intervals for the image intensities x and $\tilde{x}$ under the fine and coarse formulations.*

The fine and coarse formulations are given by

$$
\begin{array}{ll}
\text{fine} & \text{coarse} \\
L(y|x) \propto \prod_{i=1}^{n} x_i^{y_i} \exp\{-x_i\} & \tilde{L}(\tilde{y}|\tilde{x}) \propto \prod_{i=1}^{\tilde{n}} \tilde{x}_i^{\tilde{y}_i} \exp\{-\tilde{x}_i\} \\
\pi(x|\theta) \propto \theta^{\frac{m}{2}} \exp\{-\frac{1}{2}\theta x^t W x\} & \tilde{\pi}(\tilde{x}|\tilde{\theta}) \propto \tilde{\theta}^{\frac{\tilde{m}}{2}} \exp\{-\frac{1}{2}\tilde{\theta}\tilde{x}^t \tilde{W} \tilde{x}\} \\
\pi(\theta) \propto \theta^{\alpha-1} e^{-\beta\theta} & \tilde{\pi}(\tilde{\theta}) \propto \tilde{\theta}^{\alpha-1} e^{-\beta\tilde{\theta}}
\end{array}
$$

where W and $\tilde{W}$ are given by (2) and adjacent detectors are defined to be neighbors. The swapping updates are carried out as described previously, with $\theta^\dagger = \frac{1}{8}\tilde{\theta}$. Figures 3c and 3d show four successive updates from this coupled MCMC scheme. The resulting posterior pointwise 90% credible intervals are shown in Figures 3e and 3f under both the fine and coarse formulations. In this application, the swap proposals were accepted about 14% of the time.

Example 3. In SPECT the goal is to estimate a photon emission intensity map using photon emissions from an object detected by a gamma camera. Figure 4 diagrams the information obtained during a SPECT scan. As the object emits photons, the gamma camera records the locations of photon hits along the camera array. The gamma camera array can rotate completely around the object. At a given camera position, photon emissions are recorded as counts at each of 128 bins indexed by b. This accumulation of counts is repeated at each of 120 rotation angles indexed by angle a.

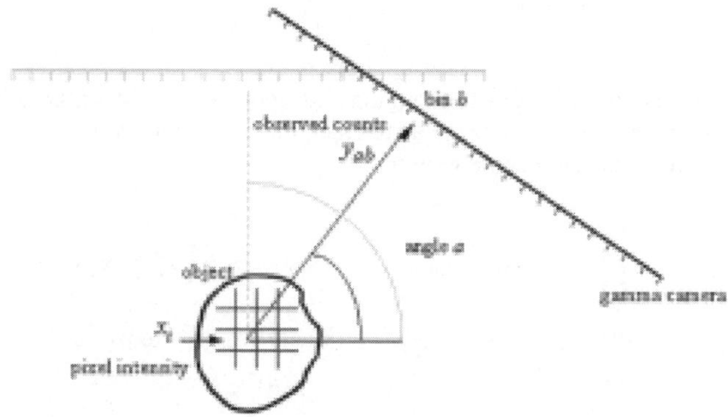

Figure 4. *SPECT: An object emits photons with location dependent intensity $x(s)$. The gamma camera obtains binned counts of photon emissions from various different positions controlled by the angle a. The counts from each angle a and each bin of the gamma camera b are recorded as y_{ab}.*

The data consist of counts y_{ab} obtained from bin b of the gamma camera while it was positioned at angle a. Lead columnators on the camera ensure that photons hit the camera at nearly right angles. Since a photon may be scattered, absorbed, miss the gamma camera, or otherwise fail to be detected, the probability map p_{abi} gives the probability of an emission from pixel i being detected at angle a and bin b.

We specify a fine-scale formulation that divides the emission source into a $m = 128 \times 128$ lattice of pixels; the coarse-scale formulation divides the emission source into a $\tilde{m} = 64 \times 64$

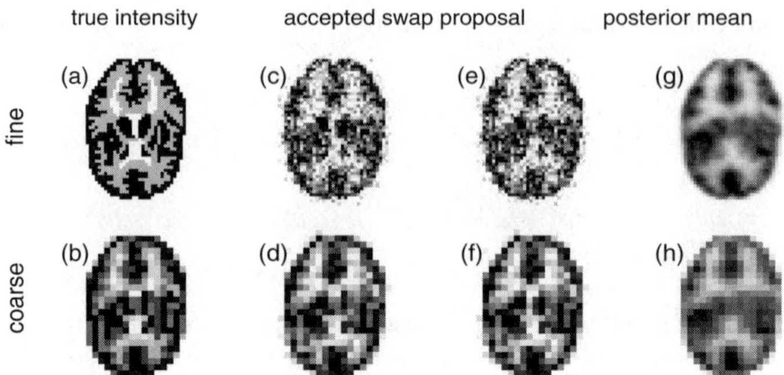

Figure 5. *Coupled fine and coarse-scale MCMC for a SPECT example.* (a) *true emission intensities;* (b) *coarsened version of the true intensities;* (c) *and* (d) *current values for x and $\tilde{x}$ during the coupled MCMC run;* (e) *and* (f) *proposed fine and coarse images x^* and $\tilde{x}^*$ after swapping an interior patch of the images in* (c) *and* (d)*;* (g) *posterior mean for x;* (h) *posterior mean for $\tilde{x}$.*

lattice of pixels. Hence, the fine formulation requires a $120 \times 128 \times 128^2$ probability map p_{abi}, and the coarse formulation requires a $120 \times 128 \times 64^2$ probability map $\tilde{p}_{abi}$, The counts y_{ab} then have a Poisson distribution with mean λ_{ab} under the fine-scale formulation, and mean $\tilde{\lambda}_{ab}$ under the coarse-scale formulation where

$$\lambda_{ab} = \sum_{i=1}^{m} x_i p_{abi} \quad \text{and} \quad \tilde{\lambda}_{ab} = \sum_{i=1}^{\tilde{m}} \tilde{x}_i \tilde{p}_{abi}.$$

Hence computing changes in λ_{ab} due to changing a component of x requires four times as much effort as does computing changes in $\tilde{\lambda}_{ab}$ due to changing a component of $\tilde{x}$.

The two formulations can then be written

$$\begin{array}{cc}
\text{Fine} & \text{Coarse} \\
L(y|x) \propto \prod_{a,b} \lambda_{ab}^{y_{ab}} \exp\{-\lambda_{ab}\} & \tilde{L}(y|\tilde{x}) \propto \prod_{a,b} \tilde{\lambda}_{ab}^{y_{ab}} \exp\{-\tilde{\lambda}_{ab}\} \\
\pi(x|\theta) \propto \theta^{m/2} \exp\{\tfrac{1}{2}\theta x^t W x\} & \tilde{\pi}(\tilde{x}|\tilde{\theta}) \propto \tilde{\theta}^{\tilde{m}/2} \exp\{\tfrac{1}{2}\tilde{\theta}\tilde{x}^t \tilde{W} \tilde{x}\} \\
\pi(\theta) \propto \theta^{\alpha-1} e^{-\beta\theta} & \tilde{\pi}(\tilde{\theta}) \propto \tilde{\theta}^{\alpha-1} e^{-\beta\tilde{\theta}}
\end{array}$$

where W and $\tilde{W}$ are given by (2) with vertically and horizontally adjacent pixels defined as neighbors. Note that the data are not coarsened in this example. Within-scale MCMC is carried out as described in Weir (1997).

We originally used swaps as described in Section 3.1. However we found that the fine-scale proposals were not sufficiently accurate near the edges of the emission phantom (Figure 5a). Instead, we proposed to swap only interior pieces of the fine and coarse images. Figure 6 shows how this is carried out. To construct the proposal, the same interior regions of the two images are exchanged. The region within the coarse-scale exterior is then deterministically coarsened; the region within the fine-scale exterior is then refined, conditioned on matching its coarse values and conditioned on the fine-scale pixels neighboring the region. This gives the proposal a better chance of being accepted—about one in eight swap proposals are accepted. Figure 5 shows an accepted swap along with coarse and fine-scale posterior mean images.

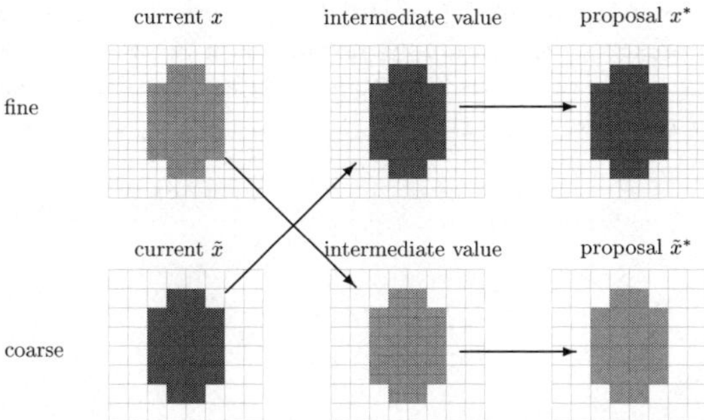

Figure 6. *A proposal that swaps only a piece of the image between the coarse and fine scales. Given the current values for x and $\tilde{x}$, the shaded regions of the two images are exchanged giving the intermediate values. The coarse shaded piece is refined to give a fine proposal x^* and the fine shaded piece is coarsened to give a coarse proposal $\tilde{x}^*$. The stochastic refining of the coarse shaded piece conditions on its previous coarse value as well as its neighboring fine-scale pixels.*

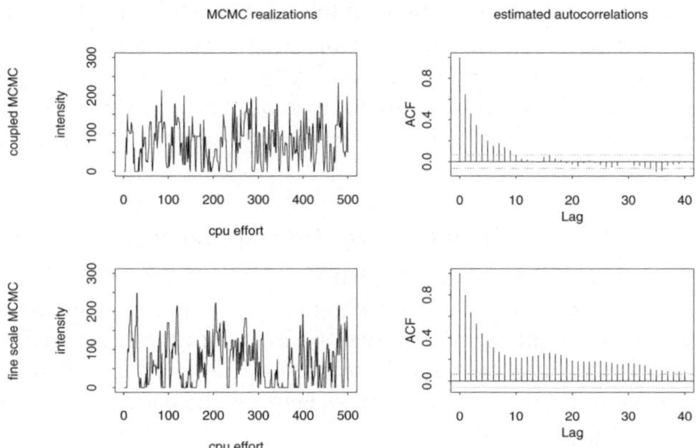

Figure 7. *MCMC trace plots and autocorrelation plots for the intensity of an interior pixel in the SPECT application under the coupled MCMC approach (top row) and standard MCMC within the fine-scale only (bottom row). The trace plots are standardized to comparable CPU time. The coupled MCMC is about three times as efficient when standardized to CPU effort.*

MCMC trace plots are shown in Figure 7 for an interior pixel under the two posterior sampling schemes. The coupled MCMC yields estimated autocorrelation times that are about a third of those obtained under the standard fine-scale MCMC algorithm.

Swapping proposals for continuous spatial priors. In the case when x and $\tilde{x}$ are both modelled *a priori* as restrictions of an identically distributed continuous processes $x(s)$ and $\tilde{x}(s)$ the swapping is trivial. These processes are constructed via (5) using independent copies u and $\tilde{u}$ with common spatial locations $s^u = \{s_1^u, \ldots, s_\ell^u\}$ so that

$$x(s) = \sum_{k=1}^{\ell} u_k k(s - s_k^u) \quad \text{and} \quad \tilde{x}(s) = \sum_{k=1}^{\ell} \tilde{u}_k k(s - s_k^u),$$

with u and $\tilde{u}$ modelled as independent $N(0, I_\ell/\theta)$ and $N(0, I_\ell/\tilde{\theta})$ draws, respectively. The gridded x essentially represents the continuous process as a piecewise constant over pixels centered at the locations s^x. Likewise, $\tilde{x}$ represents $x(s)$ as piecewise constants over larger pixels centered at the coarse locations $s^{\tilde{x}}$ (Figure 8).

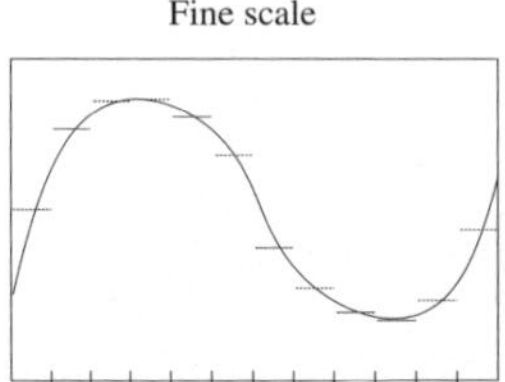

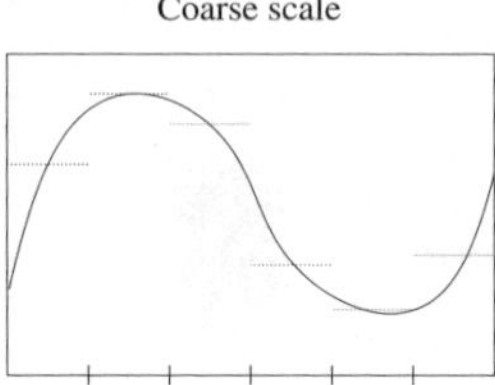

Figure 8. *Deterministic coarsening and refining in the case when x and $\tilde{x}$ are modelled as restrictions of identically distributed continuous processes $x(s)$ and $\tilde{x}(s)$. Given a realization of the underlying continuous process, the restriction of the process to the fine locations s^x or the coarse locations $s^{\tilde{x}}$ is completely determined.*

A swap between x and $\tilde{x}$ can be carried out by simply exchanging the values of (u, θ) and $(\tilde{u}, \tilde{\theta})$. Hence, coarsening x amounts to evaluating $x(s)$ at the coarse locations $s^{\tilde{x}}$; refining $\tilde{x}$ amounts to evaluating $\tilde{x}(s)$ at the fine locations s^x. Since this swap transition is symmetric and deterministic, the acceptance probability of (6) simplifies to a Metropolis acceptance rule. We defer to Section 3.2 to show an example of swapping using the continuous formulation for multiple levels of coarsening.

3.2. *Multi-processor Implementation*

Perhaps the most appealing aspect of this coupled MCMC approach is that it is readily amenable to multiprocessor implementation, without having to "parallelize" the simulator code. Multi-processor implementation is most easily carried out by running separate chains on the various processors, each exploring its own, possibly coarsened, posterior formulation. These chains are then coupled by periodically proposing swaps between the parameter values of the various chains as described in the previous section.

Table 1. *Acceptance rates of swap proposals.*

	28×28	24×24	20×20	16×16	12×12	8×8
32×32	0.86	0.70	0.39	0.13	0.01	0.00
28×28	-	0.80	0.47	0.22	0.01	0.00
24×24	-	-	0.69	0.30	0.03	0.00
20×20	-	-	-	0.56	0.11	0.00
16×16	-	-	-	-	0.21	0.00
12×12	-	-	-	-	-	0.01

As an example, we consider a synthetic application similar to the 2-d application of Section 2 where wells are laid out in an inverted nine spot pattern with a single injection well in the center surrounded by eight production wells. After a shock of tracer is introduced at the central production well, tracer breakthrough times are recorded at the eight production wells (Figure 9). Figure 10 shows an example of a three processor implementation with each processor

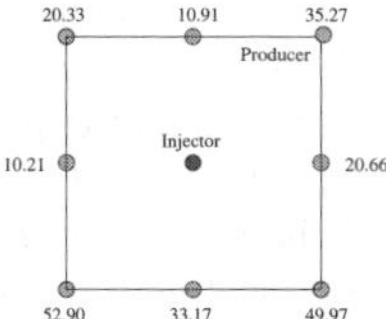

Figure 9. *Data for the multiprocessor sampler shown in Figure 10. An inverted nine spot pattern of a single injection well surrounded by 8 production wells. A shock of tracer is introduced at the injection well and the tracer concentration is recorded as a function of time at the production wells. The tracer breakthrough times are shown for each of the production wells. The likelihood is based on the breakthrough times.*

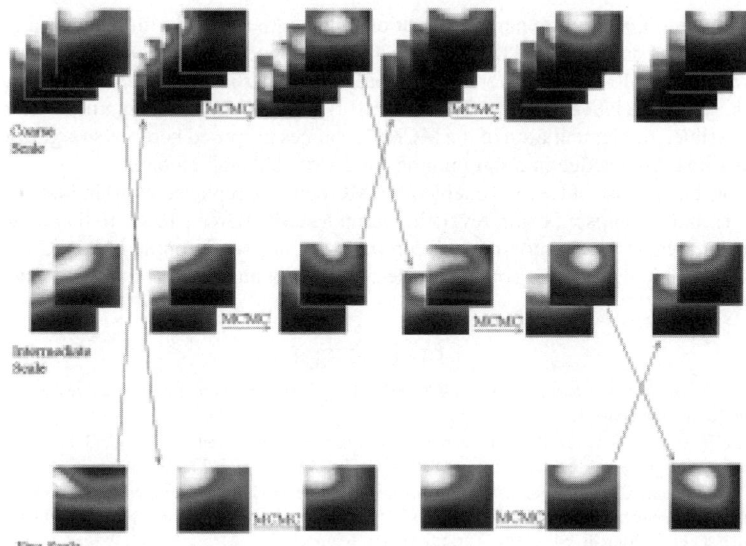

Figure 10. *Running formulations at different scales on different processors. Three distinct posterior distributions are obtained for the hydrology application of Section 2.2 by using different grid sizes in the flow simulator* (16×16, 20×20, *and* 24×24). *The three resulting posteriors are then sampled on three distinct processors. After a within-scale update scan consisting of four MCMC scans on the coarse-scale formulation, two on the intermediate-scale, and one on the fine-scale, metropolis swaps are proposed between the current realizations at each processor. This figure shows realizations at each level of coarseness for successive within-scale updates along with the result of the metropolis swaps between scales. Three such sequences are shown. In the first, the arrows denote an accepted swap between the coarse and fine-scales, in the second, a coarse-intermediate swap is accepted, in the third, an intermediate-fine swap is accepted. The* $\stackrel{\text{MCMC}}{\longrightarrow}$ *symbol denotes three additional within-scale update scans.*

running its own chain—one sampling a coarse-scale posterior, one sampling an intermediate-scale posterior, and one sampling a fine-scale posterior. The multiprocessor sampler alternates between within-scale updates and swapping updates. The within-scale updates consist of four MCMC scans of the permeability image at the coarse-scale, two scans at the intermediate-scale, and one scan at the fine level. The swapping scans consist of proposing swaps between current permeability images for each of the three possible scale pairings (coarse-intermediate, coarse-fine, and intermediate-fine). An implementation involving seven different levels of resolution was also carried out where swaps were attempted between all levels of coarseness. The proportion of accepted swaps are summarized in Table 1.

Though this example demonstrates a practical parallel implementation of a multi-grid MCMC scheme, clearly a number of questions loom regarding: allocation of processors to formulations; the choice of levels of coarseness in the auxiliary formulations; and appropriate swapping strategies, just to name a few. We have found the guidelines in Geyer and Thompson (1995) and Liu and Sabatti (1999) regarding constructing augmented chains relevant here. Future work and additional experience will give us a better handle on such questions.

4. DISCUSSION

Distributed computing, stingy parameterization, and augmentation with additional fast, coarsened formulations has expanded the universe of inverse/model calibration problems that can be handled using MCMC for posterior exploration. This is particularly relevant since distributed machines, such as relatively cheap clusters of workstations, are becoming more common and more accessible. Implementation of the MCMC schemes proposed here are straightforward and require minimal knowledge in programming for distributed architectures.

We note that the use of Geyer's coupled MCMC could be replaced with simulated tempering as in Geyer and Thompson, using reversible jump MCMC (Green 1995) to handle the change in dimension that comes in moving between scales. Our use of coupled MCMC allows us to control the amount of processing on each scale and makes it unnecessary to compute normalizing constants.

REFERENCES

Andersen, K. E., Brooks, S. P. and Hansen, M. B. (2001). Bayesian inversion of geoelectrical resistivity data. *Tech. Rep.*, Aalborg University, Denmark.

Annable, M. D., Rao, P. S. C., Hatfield, K., Graham, W. D., Wood, A. L. and Enfield, C. G. (1998). Partitioning tracers for measuring residual napl: field-scale test results. *J. Env. Eng.* **124**, 498–503.

Calder, C., Holloman, C. and Higdon, D. (2002). A space-time model for ozone concentration using process convolutions. *Case Studies in Bayesian Statistics VI* (C. Gatsonis, R. Kass, A. Carrriquiri, A. Gelman, D. Higdon, D. Pauler and I. Verdinelli, eds). New York: Springer (to appear).

Craig, P. S., Goldstein, M., Rougier, J. C. and Seheult, A. H. (2001). Bayesian forecasting using large computer models. *J. Am. Statist. Ass.* **96**, 717–729.

Geyer, C. J. (1991). Monte carlo maximum likelihood for dependent data. *Comp. Sci. and Statist.: Proc. 23rd Symp. Interface* (E. Keramidas, ed), 156–163.

Geyer, C. J. and Thompson, E. A. (1995). Annealing Markov chain Monte Carlo with applications to ancestral inference. *J. Am. Statist. Ass.* **90**, 909–920.

Goodman, J. and Sokal, A. D. (1989). Multigrid Monte Carlo method. *Phys. Rev. Let. D* **40**, 2035–2072.

Green, P. J. (1995). Reversible jump Markov chain Monte Carlo computation and Bayesian model determination. *Biometrika* **82**, 711–732.

Hegstad, B. K. and Omre, H. (2001). Uncertainty in production forecasts based on well observations, seismic data and production history. *Soc. Petrol. Eng. J.* 409–424.

Kennedy, M. and O'Hagan, A. (2001). Bayesian calibration of computer models (with discussion). *J. R. Statist. Soc. B* **68**, 425–464.

King, M. and Datta-Gupta, A. (1998). Streamline simulation: A current perspective. *In Situ* **22**, 91–140.

Lee, H., Higdon, D. M., Bi, Z., Ferriera, M. and West, M. (2002). Markov random field models for high-dimensional parameters in simulations of fluid flow in porous media. *Technometrics*, **44**, 230–241.

Liu, J. and Sabatti, C. (1999). Simulated sintering: Markov chain Monte Carlo with spaces of varying dimensions. *Bayesian Statistics 6* (J. M. Bernardo, J. O. Berger, A. P. Dawid and A. F. M. Smith, eds). Oxford: Oxford University Press, 386–413 (with discussion).

Oliver, D., Cunha, L. and Reynolds, A. (1997). Markov chain Monte Carlo methods for conditioning a permeability field to pressure data. *Math. Geol.* **29**, 61–91.

Sacks, J., Welch, W. J., Mitchell, T. J. and Wynn, H. P. (1989). Design and analysis of computer experiments. *Statist. Sci.* **4**, 409–423 (with discussion).

Weir, I. (1997). Fully Bayesian reconstructions from single photon emission computed tomography. *J. Am. Statist. Ass.* **92**, 49–60.

DISCUSSION

TONY O'HAGAN (*University of Sheffield, UK*)

I congratulate the authors on their fine paper. It contains many innovative and ingenious ideas, and is an excellent advertisement for Bayesian methods applied to difficult problems. Two important themes, in particular, emerge from this paper. The first is the use of coupled chains to make use of multi-level computer codes, and the second is the parsimonious representation of spatial structure. I will discuss each of these ideas following some description of the context in which this work sits.

Background. There are many important problems that arise in the use of complex computer simulation models. Users of such models are increasingly concerned to quantify the uncertainties that arise in their use. Can they trust the outputs of the model?

Uncertainties arise from a number of sources; see Kennedy and O'Hagan (2001). Of these, one that is brought home to model users quite clearly when they try to run the model is that there is uncertainty in the model inputs. Models typically demand very many inputs, some of which describe the particular real-world context in which the model is to be run; others may be generic constants or "tuning parameters".

Problems that confront models with observational data are particularly important, and difficult. This paper is concerned with such a problem, which the authors describe as the inverse problem, whereby we wish to use the observed data to infer an underlying true image. The true image is actually part of the model input, and is very much uncertain, while the observed data relate directly to model outputs. Thus, in the case of the hydrology examples, the permeabilities of the various cells in the grid are model inputs, while the flows received at the sampling wells are model outputs.

I think it is helpful to distinguish a pure inverse problem from one of calibration. I would describe the hydrology example as one where the primary interest is calibration. That is, we wish to learn about the true values of the model inputs that describe the real-world context of the application *in order to* then run the model in the same context to make predictions. For instance, we wish to know the permeability field *in order to* then predict where contaminants will have dispersed to. The SPECT example, on the other hand, is a pure inverse problem. We wish to learn about the true emission intensity image *for its own sake*, and not because we then want to predict other SPECT data.

The situations are different in their objectives, and this affects the detail with which we need the inverse problem solution. For calibration, we can often get away with rather poor inversion, because the detail does not affect subsequent prediction.

This paper is further evidence of the power of Bayesian methods in this field. However, appropriate methods depend heavily on the complexity (and particularly on the run-times) of the model. Simulation models vary enormously in their resource demands. Often, a run takes only a fraction of a second on a reasonably fast computer, so that methods demanding many thousands of runs are feasible. More expensive models, though, can take hours or even days for a single run. As they stand, the methods in this paper will not be practical for such expensive models.

Multi-level codes. The authors very nicely exploit the possibility to run these models at different resolutions. This feature is often available, but has been made little use of in the literature. The method presented here relies on explicit modelling of the relationship between the models at different resolutions, and it is their careful formulation of these relationships that is impressive. In other contexts, it may not be so obvious how to relate fast and slow versions of the model. Kennedy and O'Hagan (2000) model the difference between outputs at different code levels as another Gaussian process. A fast, approximate model is exploited differently by Craig *et al.* (1996) with the same objective of calibration (i.e. inversion, also known in their context as history matching).

The computational exploitation of the linkage between resolutions via coupled chains is, of course, also impressive.

Parsimonious modelling. The authors emphasize the value of parsimonious modelling of the Gaussian process, and this is also a very nice feature of the paper. I would like to touch on parsimony in a different sense.

In the hydrological example analyzed as Example 1 in Section 2 of the paper, we have essentially just five observations, i.e. the times to breakthrough at the five sampling wells. In this context, it might seem excessively cavalier to try to learn about the permeabilities on a 42×33 grid. We can basically learn about five things; everything else comes from prior information.

Now this is not as disastrous as it might appear. In most complex computer models, despite there often being huge numbers of inputs required, the output basically depends on only a few functions of those inputs. Identifying the "active" dimensions in a high-dimensional input space can be both instructive and a way to increase computational efficiency. In this case, I suspect that the five breakthrough times depend most on the average permeabilities over quite large blocks of the image. So why not consider representing the image as the mean permeabilities on a 2×2 grid (geometric means are probably the most meaningful), then deviations from these means on a 4×4 grid, and so on. After these first two levels of the hierarchy, the rest probably has very little impact on the five breakthrough times. Prior smoothness (and smallness) of those higher order deviations is reasonable and will have little impact on predictions. Thinking about prior information in the active dimensions can be profitable.

Remember that this is more a calibration problem than a pure inversion problem. It may not really matter what the detailed permeability map looks like when it comes to predicting something like contaminant transport.

Model inadequacy. Before finishing, I would like to mention another important source of uncertainty in the use of complex simulation models, which is model inadequacy. All models are wrong, so the model outputs will be different from the true real-world values even using the "best" settings of the inputs. This is important because it means there is extra noise in the observations, and this noise is spatially correlated, whereas the authors assume independent observation errors. The degree of correlation will depend on the model resolution, too. Honest modelling of prior beliefs about the model inadequacy term is necessary to avoid over-fitting. Kennedy and O'Hagan (2001) employ yet another Gaussian process for this.

I would have liked to see model inadequacy acknowledged in the examples in the paper.

Conclusion. I greatly enjoyed reading this fine paper. There were many things I liked a lot, and only a very few that I found myself able to criticize at all!

REPLY TO THE DISCUSSION

Thanks to Professor O'Hagan, for his thoughtful discussion. Anyone who has bothered to read this far into this article must know of the discussant's key role in developing methodology for statistical analysis based on both simulator output and observed data. We certainly feel that Bayesian methodology will continue to play a leading role in the growing field of statistical analysis of complex computer code outputs. Professor O'Hagan brings up a number of good points; we comment only on points for which we feel we have something to add.

If applications involving very slow simulators were the norm, perhaps our title should have been "MCMC-based approaches for inference in *easy* inverse problems." Though the methodology presented here cannot be applied directly in cases where a slow simulator is involved, it could conceivably be used if the analysis were augmented with faster, coarse versions of the simulator. Here MCMC could be used at resolutions that run sufficiently fast, lending information about the more refined resolutions. We also note that an explicit model of the difference between fast and slow simulator versions (as in Kennedy and O'Hagan, 2000) could be employed to give more efficient swap proposals for moves between resolutions. This would be particularly effective if this difference were sufficiently large and could be effectively modelled.

We certainly agree that model inadequacy is a key ingredient in applications such as these. For the applications we considered, the data were insufficient to estimate any sort of model inadequacy, or discrepancy, term. Hence, we implicitly built this discrepancy into the likelihood. In the hydrology example, had the wells been more numerous and less evenly distributed we would have opted for an explicit discrepancy term that depended on spatial location. Other applications we have worked with do include Gaussian process models for this model inadequacy.

Though we gave little ink to explicitly modelling the discrepancy between the simulator and reality, it has certainly been on our minds. A model inadequacy term is often essential for using the (calibrated) simulator for forecasting. However this can be quite problematic if the data used for calibration are for one type of model output and we wish to forecast/predict a different type of model output. An example might be using tracer concentration data in a hydrology application to predict flow rates within the region of interest. In this case one has to construct a discrepancy model for which the discrepancy of one type of output informs about the discrepancy of another type of output. Here one may need to take care that the crude inversion that worked fine for one type of calibration data is sensible for forecasting a different type of data/output.

Finally, we note that rather than directly modelling the discrepancy between observed data and simulator output, one could embed discrepancy terms within the simulator itself. This could allow for slight state modifications of the simulator at time steps, or within separate components of the model. Such an approach requires that one have access to the simulator code but has definite advantages: this more directly models the discrepancy between the simulator and reality; discrepancies are typically simpler as compared to errors propagated through the entire simulation; the form of such a discrepancy model could be suggested by taking account of the math model and physical process being simulated; and such discrepancy terms have the potential to deal with forecasting problems mentioned in the above paragraph. An example of recent work along these lines is Wikle *et al.* (2001).

ADDITIONAL REFERENCES IN THE DISCUSSION

Craig, P. S., Goldstein, M., Seheult, A. H. and Smith, J. A. (1996). Bayes linear strategies for matching hydrocarbon reservoir history. *Bayesian Statistics 5* (J. M. Bernardo, J. O. Berger, A. P. Dawid and A. F. M. Smith, eds). Oxford: Oxford University Press, 69–95 (with discussion).

Kennedy, M. C. and O'Hagan, A. (2000). Predicting the output from a complex computer code when fast approximations are available. *Biometrika* **87**, 1–13.

Wikle, C., Millif, R. F., Nychka, D. and Berliner, L. M. (2001). Spatio-temporal hierarchical Bayesian modelling: tropical ocean surface winds. *J. Am. Statist. Ass.* **98**, 382–397.

BAYESIAN STATISTICS 7, pp. 199–213
J. M. Bernardo, M. J. Bayarri, J. O. Berger, A. P. Dawid,
D. Heckerman, A. F. M. Smith and M. West (Eds.)

A Hierarchical Model for Estimating the Reliability of Complex Systems

VALEN E. JOHNSON
University of Michigan, USA
valenj@umich.edu

TODD L. GRAVES MICHAEL S. HAMADA
Los Alamos National Laboratory, USA
tgraves@lanl.gov hamada@lanl.gov

C. SHANE REESE
Brigham Young University, USA
reese@statmail.byu.edu

SUMMARY

We describe a hierarchical model for assessing the reliability of multi-component systems. Novel features of this model are the natural manner in which failure data collected at either the component or subcomponent level is aggregated into the posterior distribution, and pooling of failure information between similar components. Binary regression models are used to augment the model to account for the degradation of system performance with respect to time or other environmental factors. An example involving the performance of an anti-aircraft missile defense system illustrates the methodology.

Keywords: AGGREGATION ERROR, FAILURE MODEL, MULTI-COMPONENT SYSTEM, SYSTEM RELIABILITY.

1. BACKGROUND

This paper addresses the integration of component, subsystem and system data, and prior expert opinion to assess system reliability as it changes over time. Two problems in reliability which separately have received much attention in the literature are thus combined: (a) the integration of available information at various levels to assess system reliability and (b) estimating reliability growth or degradation. Methodology for integrating available information in a consistent fashion has proven problematic, and this paper describes a Bayesian hierarchical model that resolves this difficulty. For simplicity, we restrict discussion to systems in which components or subcomponents may be regarded as either functional or not.

To provide context, it is useful to begin with a review of related research in Bayesian system reliability. Most relevant to the model considered here are the papers by Martz *et al.* (1988) and Martz and Waller (1990), where complex systems, comprised of series and parallel subcomponents, were modelled using beta priors and binomial likelihoods at component, subsystem and system levels. Within this framework, an "induced" higher-level prior was obtained by propagating lower-level posteriors up through the system fault diagram, and combining these posteriors with "native" higher-level priors to obtain an induced prior at the next system level.

The induced priors were then approximated by beta distributions using a methods-of-moments type procedure. The combination of native priors and posterior distributions obtained from lower-level system data, both of which were expressed as beta distributions, was accomplished by expressing the induced priors as a beta distributions with parameters representing a weighted average of the constituent beta densities. This process was propagated through subsequent system levels to obtain an approximation to the joint posterior distribution on the total system reliability.

Many other reliability models are not able to account for prior expert opinion and data when such information is simultaneously obtained at several levels within a system. However, Springer and Thompson (1966, 1969), and Tang *et al.* (1994, 1997) have provided exact, and in complicated settings, approximate system reliability distributions based on binomial data by propagating component posteriors through the system's fault diagram. Thompson and Chang (1975), Chang and Thompson (1976), Lampkin and Winterbottom (1983) and Winterbottom (1984) employed approximations for system reliability distributions based on exponential lifetimes rather than binomial data. Others have proposed methods for evaluating or bounding moments of the system reliability posterior distribution (Cole, 1975; Mastran, 1976; Dostal and Iannuzzelli, 1977; Mastran and Singpurwalla, 1978; Barlow, 1985; Natvig and Eide, 1987; Soman and Misra, 1993); the first moment provides an estimate of system reliability. Moment estimators have also been used in the beta approximations employed by Martz etal (1988) and Martz and Waller (1990). In a somewhat different approach, Soman and Misra (1993) proposed distributional approximations based on maximum entropy priors.

Numerous models have, of course, also been proposed for modelling non-binomial data. Thompson and Chang (1975), Chang and Thompson (1976), Mastran (1976), Mastran and Singpurwalla (1978), Lampkin and Winterbottom (1983), and Winterbottom (1984) considered models for exponential lifetime data, while Hulting and Robinson (1990, 1994) examined Weibull models. Poisson count data, where the number of units failing in a specified period, are discussed in Hulting and Robinson (1990, 1994), Sharma and Bhutani (1992, 1994), and Martz and Baggerly (1997). Currit and Singpurwalla (1988) and Bergman and Ringi (1997a) considered dependence between components introduced through common operating environments. Bergman and Ringi (1997b) incorporated data from non-identical environments.

In many previously defined reliability models, a logical difficulty arises when prior information and data are combined at distinct component levels. Bier (1994) discussed this difficulty, which arises because data integration may be accomplished in one of several ways. In one approach, component-level priors and data are propagated upward to higher system levels in order to obtain a system-level posterior. In another, component-level priors are propagated to the system level, where they are combined (only) with system-level data to obtain a posterior distribution on the system reliability. Unfortunately, these approaches generally yield different posterior distributions on the system reliability. This effect is known as aggregation error. The model proposed here eliminates this problem.

Previous degradation models for system reliability have typically restricted attention to settings in which only system-level data are available (*e.g.*, Fries and Sen, 1996; Nolander and Dietrich, 1994; Sohn, 1996). An exception to this trend is Robinson and Dietrich (1988), who modelled component-level data collected during system development using exponential lifetime assumptions and decreasing failure rates. Our approach differs from that taken by Robinson and Dietrich in that we utilize binary regression models obtained at multiple component levels to model aging processes.

An outline for the remainder of this article is as follows. In Section 2, we review the baseline reliability model. That model is illustrated with an application to anti-aircraft missile system data in Section 3. The extension of the model to account for time degradation (or other component-level covariates) is described in Section 4. We conclude with a summary of results and suggestions for future work in Section 5.

2. METHODOLOGY

To illustrate the baseline model, consider Figure 1, which depicts a fault tree for an anti-aircraft missile system. The general features illustrated in this figure include the composition of a system by multiple subsystems, and the composition of these subsystems by further subsystems and components. In general, we assume that binomial data and prior expert opinion are available at different system levels, and that our primary goal in modelling such systems is the evaluation of the probability that a system (missile) drawn at random from the stockpile functions. Secondary goals might involve advising stockpile/inventory managers of the utility of conducting full-system or component-level tests to evaluate this probability, and to identify subsystems for which additional data might best be collected to improve estimation of overall system reliability.

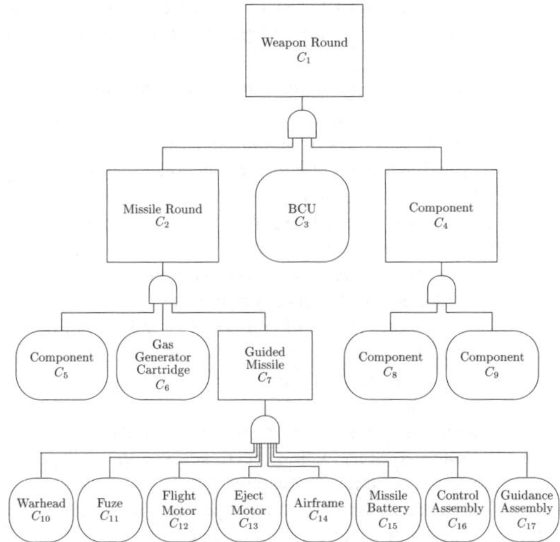

Figure 1. *Reliability Fault Tree Diagram for an Anti-Aircraft Missile System*

Several sources of information relevant to estimating system reliability are considered. The first is binomial data collected from actual component or subsystem tests. In the augmented degradation model, the age of the component at the time of the tests or other covariate information is also assumed to be available. The second source of information takes the form of expert opinion regarding the probability that a specific component or subsystem fails. This information is accompanied by relevant covariate values in the augmented degradation model. A third, less precise source of information is expert opinion regarding the similarity of the failure probabilities of groups of components within the system or across different systems. For example, in the missile system depicted above, an expert may assert that the reliability of the missile battery is similar to the reliability of a battery in a related missile system, or that reliabilities of the eject and flight motors are similar. However, the expert may not have knowledge regarding the specific probability that any component within a group of similar components functions. Finally, we incorporate the statistical notion that terminal nodes (*i.e.*, components in the fault tree having no subcomponents themselves) may also be grouped into sets of comparably reliable components without the guidance of actual expert opinion. In the baseline model, such information is modelled via an exchangeability assumption on the terminal probabilities themselves, while in the degradation model this assumption takes the form of an exchangeability assumption on binomial regression coefficients.

To model these sources of information, we first assume that the failure probabilities of components in distinct branches of the fault tree are conditionally independent, and that the success of the system requires successful functioning of all components. Extensions to systems that include redundant components, or in which component failures are not independent, are discussed in the summary. Nodes in the reliability diagram are labeled C_i, where i indicates the component or subcomponent index. The function $a(i)$ provides the parent component (or system) containing (sub)component i, while $g(i, m)$ indicates the group of components that expert m asserts have similar failure rates to component i. We let p_i denote the probability that component C_i functions when the missile is fired. The set of components for which test data is available is denoted by S_0, and within this set x_i denotes the number of times component i functioned successfully in n_i trials. In the baseline model, aging effects are not considered, making a simple binomial likelihood appropriate for modelling (x_i, n_i).

Table 1. *Notation used in model definition.*

Symbol	Description
C_i	Component i in system fault diagram
p_i	Probability that component i functions
$a(i)$	Parent of component i
$g(i, m)$	The group to which expert m assigns component i
S_0	Set of components for which test data is available
x_i	Number of successful trials of component i
n_i	Total number of trials of component i
p_{ij}	Success probability of component i under conditions z_{ij}
x_{ij}	Number of successful trials of component i under z_{ij}
n_{ij}	Total trials of component i under conditions z_{ij}
z_{ij}	Covariate vector
$\boldsymbol{\beta}_i$	Regression coefficient for ith terminal node
S_1	Components for which specific expert opinion is available
$\pi_{i,m}$	Expert m's point estimate of p_i
N_m	Beta parameter describing expert m's precision
$\alpha_m,\ \beta_m$	Gamma distribution parameters in prior on N_m
S_2	Components for which grouping, information is available
$\rho_{m,g}$	Central value of beta density on success probabilities for components in group g defined by expert m
K_m	Beta dispersion parameter for component probabilities in group g around $\rho_{m,g}$
$\delta_{g,m},\ \epsilon_{g,m}$	Beta hyperparameters in prior for $\rho_{m,g}$
$\zeta_m,\ \eta_m$	Gamma hyperparameters in prior for K_m
S_3	Set of terminal nodes
ϱ_0	Central value of beta density assumed for terminal nodes
J_0	Prior beta dispersion parameter for ϱ_0
$\tau_0,\ \phi_0$	Gamma hyperparameters in prior for K_m
$\psi_0,\ \omega_0$	Beta hyperparameters in prior on ϱ_0

In many actual applications, expert opinion plays a potentially important role in assessing system reliability, particularly in large complex systems for which data collected on individual subcomponents may be sparse. Furthermore, expert opinion may be available from several experts, and the quality of information obtained from each expert may vary. In the baseline model, we therefore assume that the prior density obtained from expert m concerning a specific value of p_i takes the form of a beta density, and we let the set of combinations of (i, m) for which expert opinion is available be denoted by S_1. More specifically, we assume that the net contribution in the joint posterior density arising from such prior information is

$$\begin{aligned} &\mathrm{B}(p_i \,|\, N_m\pi_{i,m} + 1, N_m(1 - \pi_{i,m}) + 1) \\ &\quad \equiv \frac{\Gamma(N_m + 2)}{\Gamma(N_m\pi_{i,m} + 1)\Gamma[N_m(1 - \pi_{i,m}) + 1]}\, p_i^{N_m\pi_{i,m}}(1 - p_i)^{N_m(1-\pi_{i,m})}. \end{aligned} \tag{1}$$

In (1), $\pi_{i,m}$ represents expert m's point estimate of p_i, and N_m represents the precision of expert m. For concreteness, we assume that each expert precision parameter N_m is drawn from a gamma density with known parameters α_m and β_m, parameterized here as

$$\mathrm{G}(N_m \,|\, \alpha_m, \beta_m) = \frac{\beta_m^{\alpha_m}}{\Gamma(\alpha_m)} N_m^{\alpha_m - 1} \exp(-\beta_m N_m).$$

Note that expert opinion is assumed to take the form of a binomial likelihood with a maximum at $\pi_{i,m}$—this convention eliminates the possibility that the joint density specified on all model parameters is improper, and also implicitly handles the aggregation problem identified by Bier (1994) by simply treating expert opinion as data.

When prior information regarding component success probabilities is unknown, but expert groupings of components are available, (1) is augmented in the baseline model by assuming that $\pi_{i,m}$ is replaced by $\rho_{m,g}$, where $\rho_{m,g}$ represents the common, but unknown, success probability assigned by expert m to components in group g (*i.e.*, components for which $g(i, m) = g$). The contribution to the joint posterior distribution on model parameters from such information is assumed to take the form

$$\prod\nolimits_{(i,m)\in S_2} \mathrm{B}(p_i \,|\, K_m\rho_{m,g} + 1, K_m(1 - \rho_{m,g}) + 1). \tag{2}$$

Here, S_2 denotes the combinations of (i, m) for which such grouping information is available.

As in (1), the parameter K_m is assumed to be drawn *a priori* from a gamma density having parameters ζ_m and η_m. The prior success parameter $\rho_{m,g}$ is assumed to be drawn from a beta density with known parameters $\delta_{g,m}$ and $\epsilon_{g,m}$, respectively.

Finally, for terminal nodes in the fault tree a hierarchical prior specification may be obtained by further assuming that each terminal node's success probability is drawn from a beta density with parameters $J_0\varrho_0$ and $J_0(1 - \varrho_0)$. The set of terminal nodes is denoted by S_3 (see Table 1).

For notational simplicity, we assume that all terminal nodes are, *a priori*, exchangeable, but this restriction may be relaxed by using expert judgment to group the terminal nodes in a manner similar to that used in the specification of (2). In that case, J_0 and ϱ_0 would be subscripted with the appropriate prior group. The parameter J_0 is assumed drawn from a gamma density with parameters τ_0 and ϕ_0; ϱ_0 is assumed *a priori* to be drawn from a beta density with parameters ψ_0 and ω_0.

As discussed in the previous section, combining data and prior information at different levels within a reliability diagram has often proven problematic, both from the perspectives of computational tractability and model consistency. Our solution to this conundrum is to simply re-express non-terminal node probabilities in terms of terminal node probabilities using deterministic relations derived from an examination of the system reliability diagram. For example, from Figure 1, it is evident that the probability that the guided missile component functions, p_7, is equal to the product of the probabilities that the warhead (p_{10}), fuze (p_{11}), flight motor (p_{12}), eject motor (p_{13}), airframe (p_{14}), missile battery (p_{15}), control assembly (p_{16}), and guidance assembly(p_{17}) all function. Thus,

$$p_7 = \prod_{i=10}^{17} p_i \tag{3}$$

and, for example, the prior specification on p_7 is interpreted as a prior specification on this product:

$$\begin{aligned} f_{7,m}(p_7 \mid \pi_{7,m}, K_m) &\equiv f_{7,m}(\prod_{i=10}^{17} p_i \mid \pi_{7,m}, N_m) \\ &\propto \left[\prod_{i=10}^{17} p_i\right]^{N_m \pi_{7,m}} \left[1 - \prod_{i=10}^{17} p_i\right]^{N_m(1-\pi_{7,m})}. \end{aligned}$$

Note that variable substitutions based on the reliability diagram do not uniquely identify a joint distribution on the terminal node probabilities, in this case p_{10} through p_{17}. However, together with the assumption that the distributions of these probabilities are defined with respect to Lebesgue measure on the unit interval and the given hierarchical specification, such substitutions do yield a uniquely defined joint distribution on these parameters.

Combining these assumptions leads to a joint posterior distribution on the baseline model parameters proportional to

$$\begin{aligned} &[\boldsymbol{p}, N, \boldsymbol{\rho}, K, \boldsymbol{\varrho}, J \mid x, n, \boldsymbol{\pi}, \boldsymbol{\alpha}, \boldsymbol{\beta}, \boldsymbol{\zeta}, \boldsymbol{\eta}, \boldsymbol{\delta}, \boldsymbol{\epsilon}, \boldsymbol{\tau}, \boldsymbol{\phi}, \boldsymbol{\psi}, \boldsymbol{\omega}] \propto \\ &\times \prod_{i \in S_0} p_i^{x_i}(1-p_i)^{n_i} \times \prod_{m:\exists(i,m)\in S_2} \mathrm{G}(K_m \mid \zeta_m, \eta_m) \\ &\times \prod_{(i,m)\in S_1} \mathrm{B}(p_i \mid N_m \pi_{i,m} + 1, N_m(1-\pi_{i,m}) + 1) \qquad (4) \\ &\times \prod_{(i,m)\in S_2} \mathrm{B}(p_i \mid K_m \rho_{m,g} + 1, K_m(1-\rho_{m,g}) + 1) \qquad (5) \\ &\times \prod_{i\in S_3} \mathrm{B}(p_i \mid J_0 \varrho_0, J_0(1-\varrho_0) + 1) \qquad (6) \\ &\times \prod_{m:\exists(i,m)\in S_2} \mathrm{B}(\rho_{m,g} \mid \delta_m, \epsilon_m) \times \mathrm{B}(\varrho_0 \mid \psi_0, \omega_0) \\ &\times \mathrm{G}(J_0 \mid \tau_0, \psi_0) \times \prod_{m:\exists(i,m)\in S_1} \mathrm{G}(N_m \mid \alpha_m, \beta_m). \qquad (7) \end{aligned}$$

In this expression, values of non-terminal node probabilities are assumed to be expressed in terms of the appropriate functions of terminal node probabilities, as defined from the system fault diagram.

An examination of the contributions to the joint posterior distribution arising from the three types of prior information, (4–6), reveals obvious similarities, but there are also important distinctions between these parameterizations. For example, in (4), the value of N_m represents the precision of the expert's opinion, while in (5) and (6), K_m and J_0 describe the similarity of item reliabilities within a grouping.

2.1. *Hierarchical Prior Model*

The hierarchical prior model on the terminal node probabilities plays a crucial role in rendering estimates of the overall system reliability insensitive to the level of detail included in the system fault diagram. As an illustration of this point, consider a simple system comprised of three components, and suppose that a single binomial observation with four successes and one failure is observed at the system level. Then without a hierarchical specification on the component probabilities and under the model assumptions stated above, the posterior distribution on the system reliability would be proportional to where the system reliability, p_1, is assumed to equal $p_2 p_3 p_4$.

With the implied uniform distribution on p_2–p_4, the posterior mean of p_1 in this model is 0.507; when the system is not decomposed into subsystems and a uniform prior is assumed on p_1, the posterior mean on p_1 (with a uniform prior) is 0.714. Furthermore, under such naive model specifications, the bias attributable to adding subcomponents to the fault tree becomes more severe as the number of subcomponents in the system increases.

In contrast, the hierarchical prior specification on p_2–p_4 with $\psi_0 = \omega_0 = 0.5$ results in a posterior mean of 0.718 for p_1, while the same specification with $\psi_0 = \omega_0 = 1.0$ results in a posterior mean of 0.687. Both estimates are largely insensitive to the number of subcomponents specified for the system.

2.2. *Estimation Strategies for the Baseline Model*

The joint distribution on model parameters specified in (7) does not lend itself to analytical evaluation of the system or component reliabilities. However, a component-wise Metropolis–Hastings algorithm can be implemented in a relatively straightforward way. In our version of such a scheme, we used a random-walk Metropolis–Hastings algorithm with Gaussian proposal densities specified on the logistic scale for the terminal node probabilities, as well as for ϱ_0 and $\rho_{m,g}$. Precision parameters were similarly updated using a random-walk Metropolis–Hastings scheme with Gaussian increments specified on the logarithmic scale. The resulting Metropolis–Hastings algorithm was implemented using a general-purpose Java MCMC system developed at Los Alamos National Laboratory (Graves, 2001).

3. ANALYSIS OF ANTI-AIRCRAFT MISSILE DATA

Anti-aircraft missiles are intended to provide defense from attacking enemy aircraft. Currently, the United States has over 15 different anti-aircraft missiles in its arsenal.

Each of these weapons is comprised of numerous components and subsystems, many of which are depicted schematically for a selected weapon system in Figure 1. Data available for assessing the reliability of this particular system include 45 observations on each of the components $C_4, C_5, C_6, C_{11}, C_{12}, C_{13}, C_{15}$ and C_{16}, and 126 tests of component C_3. The bulk of the test data, however, was performed at the system level, where over 1400 tests were performed. This situation is atypical of most applications in which component-level tests dominate, but this feature of the data offers an ideal opportunity for us to test our reliability model by comparing results obtained both with and without the full-system data. No component-level tests were performed on components $C_2, C_7, C_8, C_9, C_{10}, C_{14}$, and C_{17}.

Upon conferring with an expert, reliability classes were formed as follows. Components 2–4 were assigned to Group 1, Components 5–9 to Group 2, Components 10–17 to Group 3, and Components 5, 6, and 8–17 to Group 4. Informative priors with common, fixed group means (π_1–π_3) and a single, common precision parameter ($N_{1,2,3}$) were assumed for each of Groups 1–3. A common precision parameter was incorporated for each group owing to the fact that a single expert provided all prior information. A hierarchical prior with unknown mean and precision parameter (ϱ_0 and J_0, respectively) was assumed for components in Group 4. Also, gamma priors with parameters (5,1) were posited for both precision parameters ($N_{1,2,3}$ and J_0), and a non-informative prior ($\psi_0 = \omega_0 = 0.5$) was assumed for ϱ_0.

Applying the model discussed in Section 2, we obtained the posterior distributions on the component reliabilities for each of the components and the expert precision parameters. The

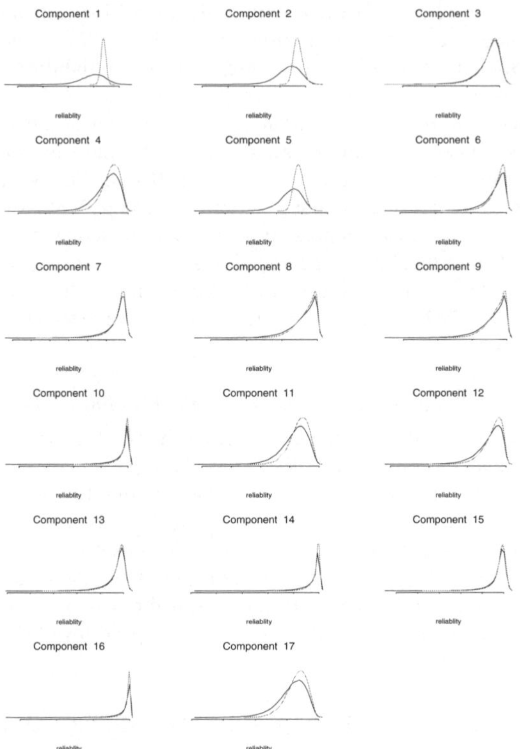

Figure 2. *Posterior distributions for the reliability of the system represented in Figure 1. In each pair of plots, the more peaked curve represents the marginal posterior density based on all test data, while the more dispersed curves represent the marginal posterior density using only component-level data (i.e., excluding system-level tests).*

system reliability posterior distributions with the system data included and system data excluded are plotted in Figure 2. We note the agreement between the two posterior distributions (full system tests included versus full system tests excluded). In every case, the 95% posterior probability region calculated by excluding full system data includes the posterior distribution obtained with the full system data. The scales from these plots have been removed due to classification concerns, although it can be noted that each plot contains a subinterval of $(0, 1)$ of length 0.1 or less. Also of interest is the posterior distribution for the expert precision parameter $N_{1,2,3}$. The posterior mean for this distribution is 12.2. This suggests that the expert's opinion is worth approximately 12 full system tests. Given the prior mean of 5, we conclude that the expert was reasonably well calibrated with the system structure and data.

3.1. *Diagnostics*

Two concerns commonly encountered in modelling system-level reliabilities using fault diagrams like that depicted in Figure 1 involves the extent to which different components function independently and whether system (or subsystem) reliability decreases when components are assembled. A simple cross-validation diagnostic useful for assessing the importance of these influences can be constructed by iteratively omitting data collected at each node from the estimation procedure, and then examining the predictive distribution for the omitted datum.

Such a procedure was applied to data obtained for this missile system, and resulted in an estimate of 0.83 for the predictive probability of observing fewer successes at the system-level than were actually observed. It therefore seems that there is little evidence to support the notion that the reliability of the system was degraded as components were assembled and required to operate as a unit.

There was, however, some indication of model lack-of-fit at the subcomponent level. For components 10 and 17, the predictive probability for observing fewer successes than were obtained at these nodes was approximately 0.035. The same number of failures was observed at each of these components, and these two components had the highest failure rate of any of the components in the system. Model lack-of-fit in this instance might thus be attributed to the fact that the hierarchical mean estimated for the terminal nodes, ϱ_0, increased substantially when the datum for either of these nodes was omitted, resulting in an overly optimistic estimate of this probability. Possible remedies for such model inadequacy would be to stochastically decrease the prior assigned to the value of J_0, or to introduce a separate hierarchical group for these nodes. In this case, neither remedy appeared to substantially affect estimates of system reliability in follow-on sensitivity analyses.

4. EXTENSION TO DEGRADATION MODELS

In many complex systems, reliabilities of system components degrade with age, and such degradation processes can be modelled naturally within the hierarchical framework described above. For purposes of illustration, we describe this extension within the context of logistic regression models on component reliabilities; generalizations to broader classes of regression models follow along similar lines.

In the baseline model, all component reliabilities are specified in terms of the reliabilities of terminal nodes, thus making it possible to consistently incorporate information collected at multiple component levels. A similar approach is adopted for modelling the degradation of component reliabilities over time. Specifically, for each terminal node in the system, say node i, we assume that

$$\log\left(\frac{p_{ij}}{1-p_{ij}}\right) = z'_{ij}\boldsymbol{\beta}_i,$$

where z_{ij} denotes a known vector of covariates relevant for predicting p_{ij} (the success probability of component i under conditions z_{ij}) and $\boldsymbol{\beta}_i$ denotes a regression coefficient specific to terminal node i. In the case of the missile data discussed later, the vectors z contain a constant term (intercept) and component age at time of testing or prior specification.

The hierarchical structure assumed for the component reliabilities in the baseline model is extended to the regression setting by assuming that, for terminal nodes within a common grouping, the vectors β_i are drawn from a multivariate normal distribution with mean, say, α and covariance matrix C. In this application, vague priors are assumed for α and C.

Specification of prior expert opinion is also modified to account for changes in component reliabilities over time. This is accomplished by substituting

$$\frac{\exp(z'_{ij}\boldsymbol{\beta}_i)}{1+\exp(z'_{ij}\boldsymbol{\beta}_i)}$$

for p_i in Eqs. (1) and (2) for an assumed value of the covariate vector z_{ij}. Often, z_{ij} is chosen to correspond to the state of a component at time 0.

Estimation of model parameters proceeds as in the baseline model, except that a random-walk Metropolis–Hastings update for the values of $\boldsymbol{\beta}_i$ replaces the corresponding updates of the values of p_{ij}.

4.1. *Application to Missile Data*

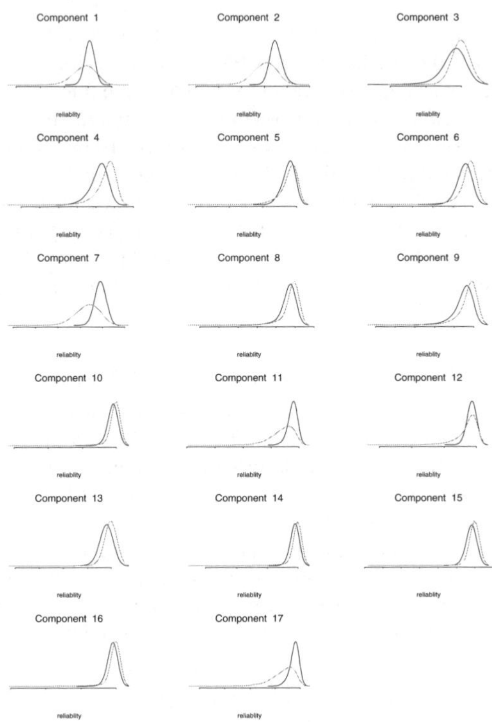

Figure 3. *Posterior distributions for the reliability of the missile-system components at time 0 and 120 months. In each case, the more disperse density corresponds to the posterior density estimated for 120 months.*

The success/failure data described in Section 3 were accompanied by the age, in months, of each subsystem at the time of the tests. Tests were conducted at 19 distinct times for components $C_4, C_5, C_6, C_{11}, C_{12}, C_{13}, C_{15}$ and C_{16}, while the 126 tests of Component 3 were conducted at 23 distinct times. The 1,400 system-level tests were performed at 106 distinct system ages. Ages at which system tests were conducted ranged from 0 to 143 months, though most system-level data was collected from systems less than 3 years old.

A plot of the marginal posterior densities on the reliabilities of components at different levels within the system is depicted in Figure 3. These plots were obtained by assuming that all prior opinion elicited in the baseline model applied at time 0 and using all system- and component-level data. As expected, greater uncertainty is associated with the reliability of most estimates at 10 years, owing to the comparatively high posterior uncertainty in many of the values of the regression coefficient corresponding to system age ($\boldsymbol{\beta}_i$ in (8)). Note also that

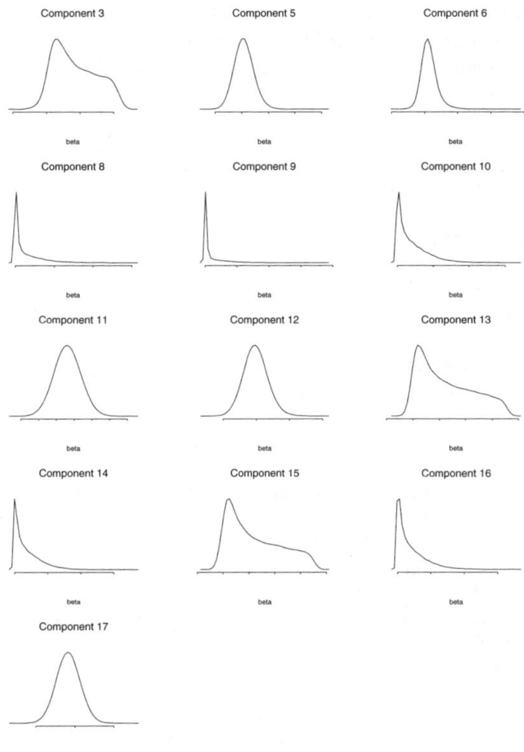

Figure 4. *Posterior distributions for the logistic slope parameters for terminal nodes. Note that Components 3, 13, and 15 experienced no failures in component-level testing.*

the posterior mean of the reliability of most components decreases gradually over time, again as is expected.

Plots of the marginal posterior distributions of the slope parameters of the terminal nodes are depicted in Figure 4. These plots highlight the extent to which many of the individual regression parameters are only weakly identifiable. In particular, for terminal nodes and subcomponents for which data were collected at a limited number of time points, and for which no failures were observed, many logistic regression curves provide nearly equivalent fits in the vicinity of the observed data. This phenomenon is further exacerbated at the subsystem level when data at lower-level components is sparse.

5. CONCLUSIONS

The proposed hierarchical model offers several advantages over existing models for system reliabilities. Among these are the ease of including diverse sources of information at different levels of the system in the model for overall system reliabilities, a coherent framework for incorporating multiple sources of prior expert opinion through the treatment of expert opinion as (imprecisely-observed) data, and the natural elimination of aggregation errors through the definition of non-terminal probabilities using the assumed structure of the system fault tree and terminal node probabilities.

A simplistic form of our hierarchical model for reliability was described in this paper. In future work we plan to extend this framework to include non-serial systems and extensions of the model to account for dependencies between component failures within a system or subsystem.

REFERENCES

Barlow, R. E. (1985). Combining Component and System Information in System Reliability Calculation. *Probabilistic Methods in the Mechanics of Solids and Structures* (S. Eggwertz and N.L. Lind, eds). Berlin: Springer, 375–383.

Bergman, B. and Ringi, M. (1997a). Bayesian system reliability prediction. *Scand. J. Statist.* **24**, 137–143.

Bergman, B. and Ringi, M. (1997b). System reliability prediction using data from non-identical environments. *Reliability Eng. System Safety*, **58**, 183–190.

Bier, V. M. (1994). On the concept of perfect aggregation in Bayesian estimation. *Reliability Eng. System Safety* **46**, 271–281.

Chang, E. Y. and Thompson, W. E. (1976). Bayes analysis of reliability of complex systems. *Oper. Res.* **24**, 156–168.

Cole, P. V. Z. (1975). A Bayesian reliability assessment of complex systems for binomial sampling. *IEEE Trans. Reliab.* **R-24**, 114–117.

Currit, A. and Singpurwalla, N. D. (1988). On the reliability function of a system of components sharing a common environment. *J. Appl. Probab.* **26**, 763–771.

Dostal, R. G. and Iannuzzelli, L. M. (1977). Confidence limits for system reliability when testing takes place at the component level. *The Theory and Applications of Reliability* **2**. New York: Academic Press, 531–552.

Fries, A. and Sen, A. (1996). A survey of discrete reliability-growth models. *IEEE Trans. Reliability* **45**, 582–604.

Graves, T. L. (2001). YADAS: An object-oriented framework for data analysis using Markov chain Monte Carlo. *Tech. Rep.*, Los Alamos National Laboratory, USA.

Hulting, F. L. and Robinson, J. A. (1990). A Bayesian approach to system reliability. *Tech. Rep.*, General Motors Research Laboratories, Warren, USA.

Hulting, F. L. and Robinson, J. A. (1994). The reliability of a series system of repairable subsystems: A Bayesian approach. *Naval Res. Logistics* **41**, 483–506.

Lampkin, H. and Winterbottom, A. (1983). Approximate Bayesian intervals for the reliability of series systems from mixed subsystem test sata. *Naval Res. Logistics* **30**, 313–317.

Martz, H. F. and Baggerly, K. A. (1997). Bayesian reliability analysis of high-reliability systems of Binomial and Poisson subsystems. *Internat. J. Reliability, Qual. Safety Eng.* **4**, 283–307.

Martz, H. F. and Waller, R. A. (1990). Bayesian reliability analysis of complex series/parallel systems of Binomial subsystems and components. *Technometrics*, **32**, 407–416.

Martz, H. F., Waller, R. A. and Fickas, E. T. (1988). Bayesian reliability analysis of series systems of Binomial subsystems and components. *Technometrics* **30**, 143–154.

Mastran, D. V. (1976). Incorporating component and system test data Into the same assessment: A Bayesian approach. *Oper. Res.* **24**, 491–499.

Mastran, D. V. and Singpurwalla, N. D. (1978). A Bayesian estimation of the reliability of coherent structures. *Oper. Res.* **26**, 663–672.

Natvig, B. and Eide, H. (1987). Bayesian estimation of system reliability. *Scand. J. Statist.* **14**, 319–327.

Nolander, J. L. and Dietrich, D. L. (1994). Attribute data reliability decay models. *Microelectronics Reliability* **34**, 1565–1596.

Robinson, D. and Dietrich, D. (1988). A system-level reliability-growth model. *Proc. Annual Reliability Maintainability Symposium*, 243–247.

Sharma, K.K. and Bhutani, R. K. (1994). Bayesian reliability analysis of a parallel system. *Microelectronics Reliability* **34**, 761–763.

Sharma, K. K. and Bhutani, R. K. (1992). Bayesian reliability analysis of a series System. *Reliability Eng. System Safety* **53**, 227–230.

Soman, K. P. and Misra, K. B. (1993). On Bayesian estimation of system reliability. *Microelectronics Reliability* **33**, 1455–1459.

Sohn, S. Y. (1996). Growth curve analysis applied to ammunition deterioration. *J. Qual. Technol.* **28**, 71–80.

Springer, M. D. and Thompson, W. E. (1966). Bayesian confidence limits for the product of n Binomial parameters. *Biometrika* **53**, 611–613.

Springer, M. D. and Thompson, W. E. (1969). Bayesian confidence limits for system reliability. *Proc. 1969 Annual Reliability and Maintainability Symposium*. New York: IEEE, 515–523.

Tang, J., Tang, K. and Moskowitz, H. (1994). Bayes credibility intervals for reliability of series systems with very reliable components. *IEEE Trans. Reliability* **43**, 132–137.

Tang, J., Tang, K. and Moskowitz, H. (1997). Exact Bayesian estimation of system reliability from component test data. *Naval Res. Logistics* **44**, 127–146.

Thompson, W. E. and Chang, E. Y. (1975). Bayes confidence limits for reliability of redundant systems. *Technometrics* **17**, 89–93.

Winterbottom, A. (1984). The interval estimation of system reliability from component test data. *Oper. Res.* **32**, 628-640.

DISCUSSION

DONGCHU SUN (*University of Missouri-Columbia, USA*)

It is my pleasure to congratulate Drs. Johnson, Groves, Hamada and Reese on a very promising hierarchical model to estimate the reliability of complex systems. Such a method could be extremely important in designing and predicting complex defense systems. I would also like to thank Valen for showing me his most recent work after the draft was written.

The paper contains many useful ideas. Given vast amounts of messy data, the authors want to estimate the reliability of complex systems consisting of many components. They propose an excellent method to combine data with three types of information: accurate information such as expert opinion, less accurate information (vague information) such as information from subsystems, and information on terminal nodes. I only have a few comments.

First, the way they combine prior distributions from the three types of prior information is based on the assumption of independence. I wonder if the prior information might be correlated. Some expert opinion might be effected by test results from components, subsystems and the whole system.

Next, it follows from Section 3.1 that the hierarchical prior model for terminal node probabilities performs better than a prior on a system or subsystem without a hierarchical specification on the component probabilities. For the example of a simple system with three terminal modes, the posterior mean of the system probability is 0.718 under the same Jeffreys priors for component reliabilities. This changes to 0.698, an 8.6% decrease, if constant priors are used for component reliabilities. In a really complex system, I wonder if the estimate is really insensitive to changes of hyperparameters in the beta priors for the reliabilities of hundreds or thousands of terminal nodes. My other question is how a node is classified as a terminal node.

The binomial assumption is based on independence and identical trails. For a system with a high reliability, I would imagine that the manufacturing cost is rather high. Will all these several hundred new systems be based on exactly the same design? Or would there be similar designs with minor changes, tested sequentially over time? Models might be adapted from spatial-temporal analysis, where independent observations are assumed for the data, and dependent priors are then used for the Bayesian analysis. Hierarchical priors are used, resulting in dependent priors for the success probabilities of the nodes. In such cases, perhaps some robustness studies would helpful.

If the system reliability is high or close to 1, the posterior distribution for terminal nodes and sub-system reliability could be quite skewed. Perhaps reparameterizations or transformations of such reliabilities could be helpful.

The hierarchical beta priors are useful. However, it is rather hard to include explanatory variables directly, such as location of tests, weather conditions including wind speed, temperature, etc. Such covariates may be useful in predicting the system reliability. The authors' recent development after the current paper on logistic models for system reliabilities is quite interesting. Perhaps one could use a generalized linear mixed prior for logistic or probit transformation of node reliabilities. Such a linear mixed prior could include fixed effects from various explanatory variables, random effects such as experts opinions, and possible spatial-temporal effects. The expert opinions could then be independent or correlated normal or other distributions. Linear or polynomial time trends, or autoregressive priors would be natural for the time effects, and various conditional autoregressive priors and intrinsic autoregressive priors could be used for spatial effects. Furthermore, based on my experience (which could be wrong in this context), the posterior computation based on such a linear mixed prior would be more efficient in the sense that the Markov chain for the Bayesian computation might converge faster compared to computation based on hierarchical beta priors.

Finally, might the problem be related to the competing risk problem? I imagine in certain tests, the cause of failure is not clear or is from two or more terminal nodes. I am not sure about the context in the complex systems discussed by the authors. If there are such cases, the results from the competing risk literature could be helpful.

TERJE AVEN (*Stavanger University College, Norway*)

First, I would like to congratulate the authors for an interesting contribution to the field of reliability quantification of complex systems. The work is in line with what is the prevailing Bayesian tradition on system reliability analysis; subjective probabilities are used to quantify uncertainties of the presumed, underlying reliability, p (say), which is defined as the probability of a component (or the system) to function as planned. But is such a starting point in accordance with the Bayesian paradigm; a paradigm which sees probability as a subjective measure of uncertainty? Is it appropriate to talk about *the* probability? No, I would answer. Such a terminology is confusing in a Bayesian setting. It is better to use terms such as *chance* or *propensity* (cf. Lindley, 1985, 2000). And an interpretation is in place: the reliability p expresses the proportion of units (components, systems) functioning when considering an infinite population of "similar" units. Thus, the reliability is an observable (or potentially observable) quantity. The reliability is then an objective property of the constructed sequence or population of units—but it is not a probability for the assessor, though were p known to the assessor it would be the assessor's probability. The quantity p is unknown, it is a random quantity, and the assessor's uncertainty related to its value is specified through a prior (posterior) distribution $H(p)$ (say). Formally, we may introduce a sequence of exchangeable binary random quantities and interpret p as the long run proportion of these taking value 1.

In many cases it is obvious what is meant by the infinite population of units. In other cases, however, such a population is difficult to think of. What is generating the variations within the infinite population? Often the population at hand is rather small, say 10 units, and extending this population to the infinite case is not always straightforward. And without a clear understanding of what the infinite population represents, it would be rather difficult to obtain meaningful uncertainty assessments (cf. Aven, 2001, 2002).

It is also possible to give an interpretation of p which is not based on the introduction of an infinite population. Let A be the event of interest, such that $P(A) = p$, given the background information and let A_i be a sequence of similar events. Then p is considered a candidate for our subjective probability for the event A, and $H(p)$ is a confidence measure, reflecting for a given value of p, the confidence we have in p for being able to predict the proportion of A_is occurring. If p is our choice, we believe that about $p \cdot 100\%$ of the A_is will occur, and $H(p)$ reflects for a given value of p, the confidence we have in p for being able to predict the number of A_is occurring. Using this interpretation, there is no true value of p. Thus it would be inappropriate to talk about estimation and uncertainty assessments of p.

REPLY TO THE DISCUSSION

The discussants raise several interesting and penetrating points. In response to those elucidated by Professor Aven, we would simply state that the probabilities discussed in this article, the p_i's, represent our subjective assessment of the probability that the system or component in question functions when called upon, given information available before the time of the test. Whether the unit under consideration represents a unit selected at random from an infinite population or is the last unit remaining from a stockpile does not affect our assessment of this probability. The deeper questions to which Professor Aven alludes would, of course, require careful consideration of the conditional independence assumptions made within the model framework and the extent to which exchangeability assumptions would hold for future units selected for testing.

Professor Sun's concern over the independence of prior assessments elicited from different experts is well-founded. Because of the similarity in training and experience of weapons scientists at Los Alamos National Laboratory, this particular issue complicates many analyses conducted at the lab. One common approach taken towards overcoming it involves the use of Delphi-type elicitation instruments to arrive at a "single" expert opinion. But further research into this problem is clearly warranted.

A terminal node in our system is defined as a component that itself has no sub-components. The value of the hierarchical treatment of terminal nodes in this context is that it renders the assessment of the system reliability insensitive to the resolution at which the system diagram is specified. This value is clearly diminished if large numbers of dissimilar components are grouped into the same hierarchical classification.

Changes in manufacturing processes over the lifetime of a system certainly do occur, and, as Professor Sun points out, should be accounted for when covariates involving manufacturing date or process are available. Such information can and should be incorporated into the model framework using the regression methodology described.

In regard to the skewness of the posterior estimates of system and component reliabilities, we do not view this as a drawback of our approach since the beta distributions assumed within the hierarchical specification are naturally equipped to handle this feature of the data. Along a similar line, the computational requirements for model estimation and convergence of the Metropolis-Hasting chain have not posed difficulties for systems of the complexity that we have considered. For the baseline model without covariates, calculation of posterior samples takes only a few seconds on a Sun workstation; calculations in the regression model are substantially more expensive but can still be accomplished within an hour for systems of complexity comparable to those examined in the paper.

Finally, the connection to competing risk models is an interesting one that we soon hope to investigate.

ADDITIONAL REFERENCES IN THE DISCUSSION

Aven, T. (2001). On the practical implementation of the Bayesian paradigm in reliability and risk analysis. *System and Bayesian Reliability. Essays in Honor of Professor Richard E. Barlow* (Y. Hayakawa, and M. Xie, eds). London: World Scientific, 269–286.

Aven, T. (2002). *How to Approach Risk and Uncertainty to Support Decision-Making*. New York: Wiley(to appear).

Lindley, D.V. (1985). *Making Decisions*. New York: Wiley.

Lindley, D.V. (2000). The Philosophy of Statistics. *The Statistician* **49**, 293–337.

BAYESIAN STATISTICS 7, pp. 215–232
J. M. Bernardo, M. J. Bayarri, J. O. Berger, A. P. Dawid,
D. Heckerman, A. F. M. Smith and M. West (Eds.)

Rasch Models with Exchangeable Rows and Columns

STEFFEN L. LAURITZEN
Aalborg Universitet, Denmark
steffen@math.auc.dk

SUMMARY

The article studies distributions of doubly infinite binary matrices with exchangeable rows and columns which satisfy the further property that the probability of any $m \times n$ submatrix is a function of the row and column sums of that matrix. We show that any such distribution is a (unique) mixture of random Rasch distributions. The non-degenerate elements of these distributions were introduced by Rasch (1960). We investigate the relationship between these random Rasch distributions and a problem in visual perception, the characters of a certain Abelian semigroup, and the problem of existence of measures with given marginals.

Keywords: DE FINETTI'S THEOREM; EXTREME POINT MODELS; INTELLIGENCE TESTING; MAJORIZATION; MARGINAL PROBLEM; PARTIAL EXCHANGEABILITY.

1. INTRODUCTION

This article is concerned with the dichotomous Rasch model for item analysis (Rasch, 1960). The model was developed to describe outcomes of psychological testing experiments. An item (question, problem) labeled i is presented to a person labeled j and a binary response X_{ij} is recorded. The Rasch model asserts that responses are independent, and that there are parameters α_i ("easinesses") characteristic for the items and parameters β_j ("abilities") characteristic for the persons so that

$$P(X_{ij} = 1 \mid \alpha_i, \beta_j) = \frac{\alpha_i \beta_j}{1 + \alpha_i \beta_j}. \tag{1}$$

The model and its variants has been the subject of intensive study and interest in the psychometric literature. But the Rasch model is also of fundamental interest in many other contexts and in itself. For example, X_{ij} could indicate whether or not a batter i is getting a hit when matched with a pitcher j (Gutmann *et al.*, 1991), X_{ij} could indicate presence or absence of a given species i of bird on island j (Wilson, 1987), or X_{ij} could denote the success or failure of mating when a female salamander i is paired with male salamander j (McCullagh and Nelder, 1989, pp. 439). An overview of literature related to the Rasch model can be found in Fischer and Molenaar (1995).

The Rasch model can in some sense be seen as the fundamental model of randomness for (0,1)-matrices and much effort has also been devoted to derivations of the Rasch model as the unique model satisfying certain fundamental principles. Rasch favored deriving it from his principle of *specific objectivity* (Rasch 1967, 1977), but, for example, he also derived it from sufficiency arguments, using the basic assumption that the probability of any binary $m \times n$

matrix should only depend on the row and column sums of this matrix (Rasch, 1971), a property clearly satisfied by the Rasch model, since

$$P_{\alpha\beta}\left\{(X_{ij}=x_{ij})_{i=1,\ldots,m;j=1,\ldots,n}\right\}=\prod_{i=1}^{m}\prod_{j=1}^{n}\frac{(\alpha_i\beta_j)^{x_{ij}}}{1+\alpha_i\beta_j}=\frac{\prod_{i=1}^{m}\alpha_i^{r_i}\prod_{j=1}^{n}\beta_j^{c_j}}{\prod_{i=1}^{m}\prod_{j=1}^{n}(1+\alpha_i\beta_j)}, \tag{2}$$

where $r_i=\sum_j x_{ij}$ and $c_j=\sum_i x_{ij}$. Other derivations (Andersen, 1973) assume sufficiency of the column sums c_j when item parameters α_i are arbitrary but known and show that this leads to the Rasch model. The derivations usually have an implicit or explicit assumption of *independence* and *regularity* ($0<P(X_{ij}=1)<1$); see Fischer (1995) for a survey. We note that the proof and theorem supplied by Rasch (1971) is inaccurate as it stands, but can be modified to become correct.

In the present paper we attempt to identify Rasch models as *extreme point models* (Lauritzen, 1988). More precisely, we replace the assumption of independence with the *exchangeability* of rows and columns and show (Theorem 2) that (randomized) Rasch models are extreme points in the simplex of row-column exchangeable binary matrices with distributions only depending on row and column sums. This yields an extension of de Finetti's theorem for binary sequences which supplements the results of Aldous (1981); see also Dawid (1982).

The article is composed as follows. In Section 2, we recapitulate some basic concepts and results on exchangeability and convex sets of measures. Section 3 reviews and extends the main results about random binary matrices, and Section 4 places the results in a slightly wider perspective.

2. PRELIMINARIES

2.1. *Exchangeable Sequences and Summarization*

We begin by rephrasing some classical results. A sequence $X_1,\ldots,X_n,\ldots$ is said to be *exchangeable* if for all n

$$P\left\{(X_i=x_i)_{i=1,\ldots,n}\right\}=P\left\{(X_i=x_{\pi(i)})_{i=1,\ldots,n}\right\},\qquad\text{for all }\pi\in S(n),$$

i.e., if its distribution is invariant under finite permutations. Clearly, if $X_1,\ldots,X_n,\ldots$ are independent and identically distributed, they are exchangeable, but not conversely.

A statistic $t(x)$ is *summarizing* for a discrete probability distribution P (Freedman, 1962) if $P(X=x)=p(x)=h(t(x))$ for some function h. Note that if $t(x)$ is summarizing for $P\in\mathcal{P}$, it is sufficient for $\mathcal{P}$ and for all $P\in\mathcal{P}$, $p(x\,|\,t)$ is uniform on $\{x:t(x)=t\}$.

For binary variables, $X_1,\ldots,X_n,\ldots$ is exchangeable if and only if for all n the sum $t_n(x)=\sum_i x_i$ is summarizing for $p(x_1,\ldots,x_n)$:

$$P(X_1=x_1,\ldots,X_n=x_n)=h_n(\textstyle\sum_i x_i)$$

because the group of permutations $S(n)$ *acts transitively on binary n-vectors with fixed sum, i.e.*, if x and y are two such vectors, there is a permutation which sends x into y, and thus t_n is a *maximal invariant*.

In general, a statistic t is summarizing for P if and only if P is invariant under the group of transformations that leaves t unchanged. Thus, the term *partial exchangeability* has often been used for this more general concept.

de Finetti (1931) shows that all exchangeable sequences are mixtures of Bernoulli sequences:

Theorem 1 (de Finetti) *A binary sequence $X_1, \ldots, X_n, \ldots$ is exchangeable if and only if there exists a distribution function F on $[0,1]$ such that for all n*

$$p(x_1, \ldots, x_n) = \int_0^1 \theta^{t_n(x)}(1-\theta)^{n-t_n(x)}\, dF(\theta).$$

It further holds that F is the distribution function of the limiting frequency:

$$Y = \lim_{n\to\infty} \frac{1}{n} \sum_i X_i, \qquad P(Y \leq y) = F(y) \tag{3}$$

and the Bernoulli distribution is obtained by conditioning with $Y = \theta$:

$$P(X_1 = x_1, \ldots, X_n = x_n \,|\, Y = \theta) = \theta^{t_n}(1-\theta)^{n-t_n}.$$

de Finetti's Theorem, with its generalizations (Hewitt and Savage, 1955) and variants (Diaconis, 1977; Diaconis and Freedman, 1980), has received much attention in the literature on probability and mathematical statistics (Kingman, 1978; Aldous, 1985).

2.2. *Convex Sets, Mixtures and Extreme Points*

Theorem 1 gives an *integral representation* of exchangeable measures. To pursue this perspective on de Finetti's Theorem, we need some basic facts about convex sets of measures.

In the following $\mathcal{P}$ denotes a set of probability measures on a space $\mathcal{X}$ which is the countable product of finite sets. Such a set $\mathcal{P}$ is *convex* if

$$P_1, P_2 \in \mathcal{P} \text{ and } 0 < \alpha < 1 \Longrightarrow \alpha P_1 + (1-\alpha)P_2 \in \mathcal{P}$$

and Q is an *extreme point* of $\mathcal{P}$ if

$$Q = \alpha P_1 + (1-\alpha)P_2 \text{ with } 0 < \alpha < 1, \quad P_1, P_2 \in \mathcal{P} \Longrightarrow Q = P_1 = P_2.$$

If $\mathcal{P}$ is equipped with the weak topology and $\mathcal{A}$ is a Borel subset of $\mathcal{P}$, P is a *mixture* of elements in $\mathcal{A}$ if there is a probability measure μ on $\mathcal{A}$ such that for all Borel subsets B of $\mathcal{X}$

$$P(B) = \int_{\mathcal{A}} A(B)\mu(dA).$$

A fundamental result is Choquet's Theorem: *If $\mathcal{P}$ is compact, the set of extreme points $\mathcal{E}$ of $\mathcal{P}$ is a non-empty Borel subset of $\mathcal{P}$, and any element of $\mathcal{P}$ is a mixture of the extreme points:*

$$P(B) = \int_{\mathcal{E}} E(B)\mu_P(dE).$$

A convex set $\mathcal{P}$ is a *simplex* if the mixing measure μ_P is uniquely determined by P. de Finetti's Theorem can alternatively be formulated as: *The exchangeable measures $\mathcal{P}_{\mathrm{E}}$ form a compact simplex, with Bernoulli measures as extreme points.*

A compact simplex is a *Bauer simplex* if the extreme points $\mathcal{E}$ form a closed (and therefore compact) subset of $\mathcal{P}$. The simplex $\mathcal{P}_{\mathrm{E}}$ is a Bauer simplex since the extreme points can be identified with the interval $[0, 1]$.

3. PARTIALLY EXCHANGEABLE BINARY MATRICES

This section first reviews results about binary matrices with exchangeable rows and columns, then binary matrices where the row and column sums are summarizing statistics, and finally gives new results about matrices with both properties.

3.1. *Row–column Exchangeable Matrices*

The distribution P is said to be *row-column exchangeable* (RCE) if for all permutations $\pi \in S(m)$ and $\rho \in S(n)$ we have

$$P\{(X_{ij} = x_{ij})_{i=1,\ldots,m;j=1,\ldots,n}\} = P\{(X_{ij} = x_{\pi(i)\rho(j)})_{i=1,\ldots,m;j=1,\ldots,n}\}.$$

We denote the corresponding group of transformations by $\mathcal{G}_{\mathrm{RC}}(m,n)$ and as before $\mathcal{G}_{\mathrm{RC}} = \cup_{m,n}\mathcal{G}_{\mathrm{RC}}(m,n)$ denotes the similar group acting on infinite matrices.

The set $\mathcal{P}_{\mathrm{RCE}}$ of RCE distributions was e.g. studied by Aldous (1981), Diaconis and Freedman (1981), Hoover (1982) and Lynch (1984), and the main results have been collected and extended in Aldous (1985).

There seems to be no simple expression for a statistic which is maximal invariant under the action of $\mathcal{G}_{\mathrm{RC}}(m,n)$ and thus there is no simple description of the sufficient (and summarizing) statistic.

Two σ-fields are particularly important for the study of $\mathcal{P}_{\mathrm{RCE}}$ and $\mathcal{P}_{\mathrm{RCS}}$. These are the *tail-field* $\mathcal{T}$ and *shell-field* $\mathcal{S}$ where

$$\mathcal{T} = \bigcap_{n=1}^{\infty} \sigma\{X_{ij}, \min(i,j) \geq n\}, \quad \mathcal{S} = \bigcap_{n=1}^{\infty} \sigma\{X_{ij}, \max(i,j) \geq n\}.$$

We exploit the following which is Proposition 14.8 of Aldous (1985):

Proposition 1. *If X has distribution $P \in \mathcal{P}_{\mathrm{RCE}}$ then the following are equivalent:*
1. *P is extreme in $\mathcal{P}_{\mathrm{RCE}}$;*
2. *$\mathcal{T}$ is P-trivial;*
3. *X is P-dissociated.*

Here a σ-field $\mathcal{A}$ is said to be *P-trivial* if $P(A) \in \{0,1\}$ for all $A \in \mathcal{A}$ and X is *P-dissociated* if for all A_1, A_2, B_1, B_2 with $A_1 \cap A_2 = B_1 \cap B_2 = \emptyset$,

$$\{X_{ij}\}_{i\in A_1, j\in B_1} \text{ and } \{X_{ij}\}_{i\in A_2, j\in B_2} \text{ are independent w.r.t. } P.$$

Diaconis and Freedman (1981) introduce the notion of a *ϕ-matrix*. This is constructed from a measurable function $\phi : [0,1]^2 \rightarrow [0,1]$ and independent sequences $U = (U_i)_{i=1,\ldots}$ and $V = (V_i)_{i=1,\ldots}$ of independent random variables, uniformly distributed on the unit interval $[0,1]$, by letting X_{ij} be conditionally independent given U and V and

$$P(X_{ij} = 1 \,|\, \mathcal{F}) = \phi(u_i, v_j),$$

where $\mathcal{F} = \sigma(U_i, V_i, i = 1, \ldots)$ is the *effect field* of the ϕ-matrix. Clearly ϕ-matrices are necessarily dissociated. Following Lynch (1984), a ϕ-matrix X is said to be *canonical* if $X \perp\!\!\!\perp \mathcal{S} \,|\, \mathcal{F}$, *i.e.*, if $\mathcal{F}$ captures the whole effect of $\mathcal{S}$ on X. Proposition 1 in combination with Corollary 2.4 of Lynch (1984) now yields:

Proposition 2. *P is extreme in $\mathcal{P}_{\mathrm{RCE}}$ if and only if P is the distribution of a canonical ϕ-matrix.*

Although this proposition gives a relatively simple description of $\mathcal{E}_{\mathrm{RCE}}$, it is still too implicit to be useful for statistical purposes. In particular, it is difficult to get a handle on the ambiguity of the function ϕ as many different ϕ-functions yield the same distribution of its ϕ-matrix.

3.2. *Summarized Matrices*

We consider distributions P of doubly infinite matrices $X = \{X_{ij}\}_{i,j=1,2,\ldots}$ of binary random variables. P is said to be *row–column summarized* (RCS) if the probability of any $m \times n$ (initial) submatrix, depends only on the row and column sums, *i.e.*, if for all m and n

$$P\left\{(X_{ij} = x_{ij})_{i=1,\ldots,m;j=1,\ldots,n}\right\} = P\left(\{(X_{ij} = y_{ij})_{i=1,\ldots,m;j=1,\ldots,n}\right\}.$$

whenever $\sum_{j=1}^{n} x_{ij} = \sum_{j=1}^{n} y_{ij}$ for all i and $\sum_{i=1}^{m} x_{ij} = \sum_{i=1}^{m} y_{ij}$ for all j.

For $r = (r_1, \ldots, r_m)$ and $c = (c_1, \ldots, c_n)$, we let $\mathcal{M}(r, c)$ denote the set of $m \times n$-matrices with row and column sums equal to (r, c). We then introduce the group $\mathcal{G}_S(m, n)$ of *switches.* This is the group of one-to-one transformations of $\mathcal{M}(r, c)$ generated by *simple switches,* where a simple switch changes a specified 2×2-submatrix of $x \in \mathcal{M}(r, c)$ as

$$\begin{pmatrix} 1 & 0 \\ 0 & 1 \end{pmatrix} \to \begin{pmatrix} 0 & 1 \\ 1 & 0 \end{pmatrix} \quad \text{and} \quad \begin{pmatrix} 0 & 1 \\ 1 & 0 \end{pmatrix} \to \begin{pmatrix} 1 & 0 \\ 0 & 1 \end{pmatrix},$$

and otherwise leave the entries of x invariant.

Theorem 3.1 of Ryser (1957) says that if x and y are matrices in $\mathcal{M}(r, c)$, there exists a switch $g \in \mathcal{G}_S(m, n)$ such that $x = gy$. In other words:

Lemma 1. *The group $\mathcal{G}_S(m, n)$ of switches acts transitively on $\mathcal{M}(r, c)$.*

It thus follows that (r, c) is a maximal invariant under the action of $\mathcal{G}_S(m, n)$. If we let $\mathcal{G}_S = \cup_{m,n}\mathcal{G}_S(m, n)$ denote the similar group acting on infinite matrices we thus have:

Corollary 1. *A distribution P on the set of doubly infinite binary matrices is row-column summarized if and only if it is $\mathcal{G}_S$-invariant.*

Lemma 1 and its corollary was also exploited by Besag and Clifford (1989) to construct a Markov chain Monte Carlo algorithm for simulating from the uniform distribution on $\mathcal{M}(r, c)$; see also Holst (1995), Rao *et al.* (1996) and Ponocny (2001).

We let $\mathcal{P}_{\mathrm{RCS}}$ denote the set of RCS-distributions. Lauritzen (1988) showed that $\mathcal{P}_{\mathrm{RCS}}$ is a convex simplex and found partially the extreme points $\mathcal{E}_{\mathrm{RCS}}$ of this simplex, in particular that $\mathcal{E}_{\mathrm{RCS}} \neq \mathcal{P}_{\mathrm{R}}$, where $\mathcal{P}_{\mathrm{R}}$ is the set of *Rasch distributions* given by (2). More precisely it was shown (Propositions 9.2 and 9.3, p. 250) that $P_{\alpha,\beta} \in \mathcal{E}_{\mathrm{RCS}}$ if

$$\sum_{i=1}^{\infty} \frac{\alpha_i \beta_i}{(1+\alpha_i)(1+\beta_i)(1+\alpha_i\beta_i)} = \infty \tag{4}$$

and $P_{\alpha,\beta} \notin \mathcal{E}_{\mathrm{RCS}}$ unless

$$\sum_{i=1}^{\infty} \frac{\alpha_i}{(1+\alpha_i)^2} = \sum_{i=1}^{\infty} \frac{\beta_i}{(1+\beta_i)^2} = \infty. \tag{5}$$

Roughly speaking, the conditions are preventing that the α and β sequences vary too rapidly with i. It is therefore natural to expect simpler results when extra symmetry, such as exchangeability, is assumed. The condition (4) implies (5), but it is not known whether any of these two conditions are both necessary and sufficient for $P_{\alpha,\beta}$ to be in $\mathcal{E}_{\mathrm{RCS}}$. Note that $\mathcal{P}_{\mathrm{RCS}}$ is not a Bauer simplex. To see this, define $P^n_{\alpha,\beta}$ by

$$\alpha_i = \beta_i = \begin{cases} 2^{-i} & \text{if } i < n, \\ 1 & \text{otherwise.} \end{cases}$$

Then $P^n_{\alpha,\beta} \in \mathcal{E}_{\mathrm{RCS}}$ for all n as it satisfies (4), but as n tends to infinity it converges to a measure which violates (5), hence $\mathcal{E}_{\mathrm{RCS}}$ is not closed.

3.3. *Summarized and Exchangeable Matrices*

This section deals with the set of distributions $\mathcal{P}_{\text{RCES}} = \mathcal{P}_{\text{RCE}} \cap \mathcal{P}_{\text{RCS}}$, which are both RCE and RCS. We let $\mathcal{G}_{\text{RCS}}(m,n)$ denote the group of transformations generated by row–column permutations $\mathcal{G}_{\text{RC}}(m,n)$ and switches $\mathcal{G}_S(m,n)$ and $\mathcal{G}_{\text{RCS}}$ the corresponding group of transformations on infinite matrices.

Lemma 2. *The pair of empirical measures induced by the row and column sums* $t_{mn}(x) = (\sum_{i=1}^m \delta_{r_i}, \sum_{j=1}^n \delta_{c_j})$ *is a maximal invariant for the action of* $\mathcal{G}_{\text{RCS}}(m,n)$.

Proof. Clearly, t_{mn} is invariant so we just have to show that $\mathcal{G}_{\text{RCS}}(m,n)$ acts transitively on the set of matrices with a given value of t_{mn}. So assume $t_{mn}(x) = t_{mn}(y) = t$. We first permute the rows and columns of x and y to form $g_1 x$ and $g_2 x$ with increasing row and column sums using $g_1, g_2 \in \mathcal{G}_{\text{RC}}(m,n)$. But then $g_1 x$ and $g_2 y$ have identical row and column sums and Lemma 1 yields the existence of $g \in \mathcal{G}_S(m,n)$ so that $gg_1 x = g_2 y$. Then, $g^* = g_2^{-1} g g_1 \in \mathcal{G}_{\text{RCS}}(m,n)$ has $g^* x = y$ as desired. □

The set $\mathcal{P}_{\text{RCES}}$ is a proper subset of $\mathcal{P}_{\text{RCE}}$. This is because the group $\mathcal{G}_{\text{RC}}(m,n)$ does not act transitively on sets of matrices x with a fixed value $t_{mn}(x) = t$. For example, if we let

$$x = \begin{pmatrix} 0 & 0 & 1 \\ 1 & 0 & 0 \\ 0 & 1 & 1 \end{pmatrix}, \qquad y = \begin{pmatrix} 0 & 0 & 1 \\ 0 & 0 & 1 \\ 1 & 1 & 0 \end{pmatrix},$$

then $\det x = 1$ and $\det y = 0$. Since the absolute value of the determinant is invariant under permutation of rows and columns, there is no element $g \in \mathcal{G}_{\text{RC}}(3,3)$ so that $x = gy$. Indeed, for a ϕ-matrix with $\phi(u,v) = uv$, we have $p(x) = 665/2985984$ but $p(y) = 1/4096$, so this distribution is in $\mathcal{P}_{\text{RCE}} \setminus \mathcal{P}_{\text{RCS}}$.

General results (Lauritzen, 1988) imply that $\mathcal{P}_{\text{RCES}}$ is a simplex. To identify the extreme points $\mathcal{E}_{\text{RCES}}$ of $\mathcal{P}_{\text{RCES}}$ we need the following lemma:

Lemma 3. *If* $P \in \mathcal{P}_{\text{RCES}}$ *and* $E \in \mathcal{T}$ *with* $P(E) > 0$, *then* $P(\cdot \,|\, E) \in \mathcal{P}_{\text{RCES}}$.

Proof. Let $E \in \mathcal{T}$ with $P(E) > 0$. By Lemma 2 we must just show that $P(\cdot \,|\, E)$ is $\mathcal{G}_{\text{RCS}}$-invariant. So let $g \in \mathcal{G}_{\text{RCS}}$. When $E \in \mathcal{T}$ it is clearly invariant both under switches, row and column permutations so $gE = E$. Let D be an arbitrary measurable subset of infinite binary matrices. We thus have

$$\begin{aligned} P(gX \in D \,|\, X \in E) &= \frac{P(gX \in D \wedge X \in E)}{P(X \in E)} = \frac{P(gX \in (D \cap gE))}{P(X \in E)} \\ &= \frac{P(X \in D \cap E)}{P(X \in E)} = P(X \in D \,|\, X \in E) \end{aligned}$$

and thus $P(\cdot \,|\, E) \in \mathcal{P}_{\text{RCES}}$. □

Lemma 4. $\mathcal{E}_{\text{RCES}} = \mathcal{P}_{\text{RCES}} \cap \mathcal{E}_{\text{RCE}}$.

Proof. The inclusion $\mathcal{E}_{\text{RCES}} \supseteq \mathcal{P}_{\text{RCES}} \cap \mathcal{E}_{\text{RCE}}$ is obvious. To show the reverse inclusion, we assume that $P \in \mathcal{P}_{\text{RCES}} \subseteq \mathcal{E}_{\text{RCE}}$ and show that $P \notin \mathcal{E}_{\text{RCES}}$. Since $P \notin \mathcal{E}_{\text{RCE}}$, $\mathcal{T}$ is not P-trivial. Thus there exists $E \in \mathcal{T}$ with $0 < P(E) < 1$. We may now write

$$P(\cdot) = P(\cdot \,|\, X \in E)P(E) + P(\cdot \,|\, X \notin E)(1 - P(E)).$$

By Lemma 3 this expresses P as a non-trivial convex combination of two elements of $\mathcal{P}_{\text{RCES}}$ implying that $P \notin \mathcal{E}_{\text{RCES}}$. □

Next we must realize that $\mathcal{P}_{\mathrm{RCS}}$ is stable under shell-conditioning:

Lemma 5. *If* $P \in \mathcal{P}_{\mathrm{RCS}}$ *then* $P(\cdot \mid \mathcal{S}) \in \mathcal{P}_{\mathrm{RCS}}$.

Proof. Using Corollary 1 this follows as in the proof of Lemma 3 because G_S leaves sets in $\mathcal{S}$ invariant. □

Note that $P(\cdot \mid \mathcal{S})$ is typically *not* RCE. Next we define ϕ to be of *Rasch type* if it satisfies the functional equation:

$$\phi(u,v)\bar{\phi}(u,v^*)\bar{\phi}(u^*,v)\phi(u^*,v^*) = \bar{\phi}(u,v)\phi(u,v^*)\phi(u^*,v)\bar{\phi}(u^*,v^*), \tag{6}$$

where we have let $\bar{\phi} = 1 - \phi$. If P is the distribution of a ϕ-matrix with ϕ of Rasch type, $P(\cdot \mid \mathcal{F})$ is $\mathcal{G}_S$ invariant. This therefore also holds for its unconditional distribution P, implying that $P \in \mathcal{P}_{\mathrm{RCES}}$. In fact, below we show that such ϕ-matrices exactly correspond to the extreme points $\mathcal{E}_{\mathrm{RCES}}$ of $\mathcal{P}_{\mathrm{RCES}}$.

Theorem 2. *If* $P \in \mathcal{P}_{\mathrm{RCES}}$, *it is in* $\mathcal{E}_{\mathrm{RCES}}$ *if and only if* P *is the distribution of a* ϕ*-matrix of Rasch type.*

Proof. If X is a ϕ-matrix, its distribution P is extreme in $\mathcal{P}_{\mathrm{RCE}}$ and *a fortiori* extreme in $\mathcal{P}_{\mathrm{RCES}} \subset \mathcal{P}_{\mathrm{RCE}}$. Thus, we only need to show the converse.

Assume P is an extreme point of $\mathcal{P}_{\mathrm{RCES}}$. From Lemma 4 and Proposition 2 we get that P is the distribution of some canonical ϕ-matrix X. Lemma 5 implies

$$P\left(\left\{\begin{matrix} X_{ij}=1 & X_{ij^*}=0 \\ X_{i^*j}=0 & X_{i^*j^*}=1 \end{matrix}\right\} \,\middle|\, \mathcal{S}\right) = P\left(\left\{\begin{matrix} X_{ij}=0 & X_{ij^*}=1 \\ X_{i^*j}=1 & X_{i^*j^*}=0 \end{matrix}\right\} \,\middle|\, \mathcal{S}\right).$$

Since X is canonical, $P(X \in A \mid \mathcal{S}) = P(X \in A \mid \mathcal{F})$ and thus

$$\begin{aligned} &\phi(U_i,V_j)\{1-\phi(U_i,V_{j^*})\}\{1-\phi(U_{i^*},V_j)\}\phi(U_{i^*},V_{j^*}) \\ &\quad = \{1-\phi(U_i,V_j)\}\phi(U_i,V_{j^*})\phi(U_{i^*},V_j)\{1-\phi(U_{i^*},V_{j^*})\}, \end{aligned}$$

i.e., it is a ϕ-matrix of Rasch type. □

If $0 < \phi < 1$ the solutions of (6) all have the special form

$$\phi(u,v) = \frac{a(u)b(v)}{1 + a(u)b(v)} \tag{7}$$

where a and b map the unit interval into the positive half-line. This is seen by letting $\rho = \phi/\bar{\phi}$ and fixing (u^*, v^*) whereby (6) can be manipulated to

$$\rho(u,v) = \frac{\rho(u,v^*)\rho(u^*,v)}{\rho(u^*,v^*)}$$

so that we may let $a(u) = \rho(u,v^*)/\rho(u^*,v^*)$ and $b(v) = \rho(u^*,v)$ to satisfy (7). Indeed we may without loss of generality assume that a and b are determined from distribution functions A and B on the positive halfline as $a = A^{-1}$ and $b = B^{-1}$ so that A and B are the distributions of $\alpha_i = a(U_i)$ and $\beta_j = b(V_j)$ respectively.

A *regular random Rasch distribution* is now defined to be the distribution of a ϕ-matrix with ϕ of the form (7) and the set of such distributions is denoted by $\mathcal{P}_{\mathrm{RR}}$.

Note that there is ambiguity between a and b in the sense that one can be transformed by multiplication with a positive constant and the other with division without changing the ϕ-matrix. But modulo this, the pairs (A, B) are in a one-to-one correspondence with the elements of $\mathcal{P}_{\mathrm{RR}}$.

It follows that the extreme points of $\mathcal{P}_{\mathrm{RCES}}$ which are non-degenerate, in the sense that they correspond to truly non-deterministic matrices, are regular random Rasch distributions. More precisely, if we say that P is *regular* when

$$0 < P(X_{ij} = 1 \mid \mathcal{S}) < 1, \qquad \text{for all } i, j,$$

we have:

Corollary 2. *P is a regular extreme point of $\mathcal{P}_{\mathrm{RCES}}$ if and only if $P \in \mathcal{P}_{\mathrm{RR}}$.*

There are many non-regular extreme points of $\mathcal{P}_{\mathrm{RCES}}$, essentially corresponding to all non-regular solutions of the functional equation (6). An example of such a solution is

$$\phi(u, v) = \chi_{\{u \leq v\}} = \begin{cases} 1 & \text{if } u \leq v, \\ 0 & \text{otherwise.} \end{cases}$$

This is an example of a ϕ-matrix of Rasch type which is "deterministic" in the sense that it is $\mathcal{S}$-measurable. In the context of intelligence tests, the interpretation of this model is that a person with ability v solves a problem of difficulty u with certainty if $v \geq u$ but never if $u > v$. Variants of the model appear for h and k being monotone functions from the unit interval to itself and then

$$\phi^*(u, v) = \chi_{\{h(u) \leq k(v)\}}. \tag{8}$$

It seems plausible that these are the only ϕ-matrices of Rasch type which are essentially $\mathcal{S}$-measurable. Proposition 3.6 of Aldous (1981) implies that this is true if and only if all solutions to the functional equation (6) with $\phi \in \{0, 1\}$ had the form (8).

There are many more ϕ-matrices of Rasch type. Consider for example

$$\phi(u, v) = \begin{cases} \dfrac{a(u)b(v)}{1 + a(u)b(v)} & \text{if } 1/3 < u, v < 2/3 \\ \chi_{\{u \leq v\}} & \text{otherwise} \end{cases}$$

which can be seen as dividing difficulties and abilities into three classes of equal size: *low, medium, high*, so that an ordinary Rasch model prevails when persons of medium ability solve questions of medium difficulty whereas other combinations yield a deterministic outcome.

There are similar models with more than three groups. For example, one can keep cutting out middle thirds of the unit interval as above to get

$$\phi(u, v) = \begin{cases} \dfrac{a(u)b(v)}{1 + a(u)b(v)} & \text{if } 1/9 < u, v < 2/9 \\ \dfrac{a(u)b(v)}{1 + a(u)b(v)} & \text{if } 1/3 < u, v < 2/3 \\ \dfrac{a(u)b(v)}{1 + a(u)b(v)} & \text{if } 7/9 < u, v < 8/9 \\ \chi_{\{u \leq v\}} & \text{otherwise,} \end{cases}$$

and so on. Since the simplex $\mathcal{P}_{\mathrm{RCES}}$ is a Bauer simplex (Ressel, 2002, personal communication), the set of its extreme points is closed. Thus the sequence of distributions of ϕ-matrices defined by this procedure will converge to what could be termed a *Cantor–Rasch distribution* with an infinite number of groups. Although these non-regular Rasch models are unusual, they are by no means counterintuitive.

4. OTHER PERSPECTIVES

4.1. *The Julesz Conjecture*

Diaconis and Freedman (1981) discuss a conjecture of Julesz (1975, 1980) in visual perception saying that two "random patterns" (*i.e.*, binary matrices) with the same first- and second-order statistics (joint distributions of singletons and pairs) cannot be visually distinguished. Indeed, they give several examples of ϕ-matrices with the same first- and second-order statistics as a purely random "coin tossing" matrix which are visually very different from such a matrix.

Here we show that such counterexamples cannot be of Rasch type. This implies that certain types of deviations from the Rasch model may indeed be visually detected from inspecting large matrices.

Say a binary matrix is *purely random* if X_{ij} are all independent and $P(X_{ij} = 1) = 1/2$ for all i, j.

Theorem 3. *Let X be a ϕ-matrix of Rasch type with the same first- and second-order statistics as a purely random matrix. Then it is a purely random matrix.*

Proof. Theorem (3.8) of Diaconis and Freedman says that a ϕ-matrix has the same first- and second-order statistics as a purely random matrix if

$$\int \phi(u,v)\,du = 1/2 \text{ a.e. } (u) \quad \text{and} \quad \int \phi(u,v)\,dv = 1/2 \text{ a.e. } (v), \tag{9}$$

so this is what we assume. We will show that if ϕ is of Rasch type, it is a.e. constant and thus equal to $1/2$.

Since ϕ is of Rasch type it satisfies (6). If we expand and reduce the terms in this equation we find that it is equivalent to

$$\begin{aligned}&\phi(u,v)\phi(u^*,v^*) + \phi(u,v)\phi(u,v^*)\phi(u^*,v) + \phi(u,v^*)\phi(u^*,v)\phi(u^*,v^*)\\ &= \phi(u,v^*)\phi(u^*,v) + \phi(u,v)\phi(u,v^*)\phi(u^*,v^*) + \phi(u,v)\phi(u^*,v)\phi(u^*,v^*)\end{aligned}$$

Integrating this equation with respect to u and using (9) yields that for almost all v and almost all v^*

$$\begin{aligned}&\phi(u^*,v^*)/2 + I(v,v^*)\phi(u^*,v) + \phi(u^*,v)\phi(u^*,v^*)/2\\ &= \phi(u^*,v)/2 + I(v,v^*)\phi(u^*,v^*) + \phi(u^*,v)\phi(u^*,v^*)/2,\end{aligned}$$

where $I(v,v^*) = \int \phi(u,v)\phi(u,v^*)\,du$. Reducing and rearranging the terms leads to

$$\{\phi(u^*,v^*) - \phi(u^*,v)\}\{I(v,v^*) - 1/2\} = 0. \tag{10}$$

Next let $A = \{v \mid I(v,v^*) = 1/2\}$, then for almost all $v \notin A$ we have $\phi(u,v) = \phi(u,v^*)$ and hence also $I(v,v^*) = I(v^*,v^*)$ for almost all $v \notin A$, whereby

$$\int I(v,v^*)\,dv = \int_A I(v,v^*)\,dv + \int_{A^C} I(v,v^*)\,dv = \lambda(A)/2 + I(v^*,v^*)\{1 - \lambda(A)\},$$

where λ is Lebesgue measure. Using now that

$$I(v^*,v^*) = \int \phi(u,v^*)^2\,du \geq \left\{\int \phi(u,v^*)\,du\right\}^2 = 1/4$$

and

$$\int I(v,v^*)\,dv = \int\int \phi(u,v)\phi(u,v^*)\,du\,dv = \int\int \phi(u,v)\phi(u,v^*)\,dv\,du = 1/4,$$

we find

$$1/4 \geq \lambda(A)/2 + \{1 - \lambda(A)\}/4 = 1/4 + \lambda(A)/4$$

whereby $\lambda(A) = 0$. Hence, $\phi(u,\cdot)$ is constant almost everywhere. By symmetry, $\phi(\cdot,v)$ is also constant so ϕ must be constant and equal to $1/2$. □

Note that we have not proved the somewhat stronger statement saying that two ϕ-matrices of Rasch type which have identical first- and second order statistics, have identical distributions, and indeed this does not hold in general.

4.2. *Analytic Properties of* $\mathcal{P}_{\text{RCES}}$

In several papers, Ressel (1985,1988,1994) has studied convex sets of measures with symmetry properties from an analytic point of view. For example, he has considered simplices of probability distributions which are summarized by additive statistics with values in Abelian semigroups. The case of $\mathcal{P}_{\text{RCES}}$ is such an example, where the semigroup S is the subsemigroup of pairs of measures on the non-negative integers generated by the summarizing statistics

$$t_{mn}(x) = (\sum_{i=1}^{m} \delta_{r_i}, \sum_{j=1}^{n} \delta_{c_j})$$

for $m, n = 1, 2, \ldots$. This family of statistics can be shown (Ressel, 2002, personal communication) to be "strongly almost additive" and thus "strongly positivity forcing" (Ressel, 1994), which implies that $\mathcal{P}_{\text{RCES}}$ is a Bauer simplex and the extreme points $\mathcal{E}_{\text{RCES}}$ are determined by normalized characters $\sigma \in \hat{S}$ so that for a binary $m \times n$ matrix x it holds

$$p_\sigma(x) = \sigma(t_{mn}(x)),$$

where $\sigma(t \oplus s) = \sigma(t)\sigma(s)$ for all $s, t \in S$. Theorem 2 thus identifies the characters of this semigroup in terms of solutions to the functional equation (6), albeit in a rather implicit fashion.

To describe the characters in more detail, we may represent the elements of the semigroup S by vectors (r, c) with elements ordered so that $r_1 \leq \cdots \leq r_m$ and $c_1 \leq \cdots \leq c_n$. Then $(r, c) \in S$ if and only if the set of matrices $\mathcal{M}(r, c)$ with row sums r and column sums c is non-empty. Gale (1957) and Ryser (1963) have shown that $\mathcal{M}(r, c) \neq \emptyset$ *if and only if* $r \preceq c^*$, where c^* is the *conjugate* sequence of c

$$c_j^* = |\{l \mid c_l \geq j\}|$$

and $\preceq$ denotes *majorization:* $a \preceq b \Leftrightarrow \sum_{i=1}^{k} a_i \leq \sum_{j=1}^{k} b_j$ for all $k = 1, \ldots, m$.

The value of the character ρ_ϕ where ϕ satisfies (6) is then given as

$$\rho_\phi(r, c) = \int \cdots \int \prod_i \prod_j \phi(u_i, v_j)^{x_{ij}} \{1 - \phi(u_i, v_j)\}^{1 - x_{ij}} \, du_i \, dv_j,$$

where x is an arbitrary element of $\mathcal{M}(r, c)$. The description is still somewhat implicit since many choices of ϕ lead to the same character ρ_ϕ.

4.3. *Marginal Problems*

A problem related to the Rasch model was investigated by Gutmann *et al.* (1991). Simulation models for baseball were considered in which a random batter of batting average Y was confronted with a random pitcher of pitching average Z. If we let $W = \psi(Y, Z)$ denote the probability of a hit, we must have

$$\mathrm{E}(W \mid Z) = Z; \quad \mathrm{E}(W \mid Y) = Y; \quad 0 \leq W \leq 1. \tag{11}$$

Dawid *et al.* (1995) discuss the related problem of coherent combination of experts' opinions. Here Y and Z are experts' opinions in the form of their subjective probabilities for some event A. Then, $W = \psi(Y, Z)$ is a *coherent combination* of the experts' opinions if and only if (11) holds.

If F and G are the distribution functions of Y and Z such a function ψ exists if and only if it holds for all $s, t \in [0, 1]$ that

$$\int_s^1 x\, F(dx) + \int_t^1 y\, G(dy) \leq \int_0^1 x\, F(dx) + \{1 - F(s)\}\{1 - G(t)\}. \tag{12}$$

This was shown as Theorem 4 of Gutmann *et al.* (1991), using classical results of Kellerer (1961) and Strassen (1965). Gutmann *et al.* (1991) also show that if this condition is met, ψ can be chosen to be increasing in each of its arguments, and ψ can also be chosen to be the indicator of a set, although not always both simultaneously, see Proposition 4 below. Note that for $0 \leq \psi \leq 1$, (11) may also be written as

$$\int_0^1 \psi(x, y)\, F(dx) = y \text{ a.e. (F)}, \qquad \int_0^1 \psi(x, y)\, G(dy) = x \text{ a.e. (G)}. \tag{13}$$

Clearly, if F and G are such a pair and we let

$$\phi(u, v) = \psi(F^{-1}(u), G^{-1}(v)),$$

we obtain a ϕ-matrix of batting outcomes.

The results of Gutmann *et al.* (1991) can be seen as a continuous analogue of the Gale–Ryser theorem. To make this more precise, we define the *conjugate* F^* of a distribution function F on the unit interval by

$$F^*(x) = 1 - F^{-1}(1 - x),$$

where F^{-1} is the left-continuous inverse of F:

$$F^{-1}(x) = \sup\{y \mid F(y) \leq x\}.$$

As in the discrete case we say that G *majorizes* F and write $F \preceq G$ if

$$\int_0^s F(x)\, dx \leq \int_0^s G(x)\, dx, \qquad \text{for all } s \in [0, 1].$$

Proposition 3. *Let F and G be two distribution functions on $[0, 1]$. Then there exists a function ψ satisfying (13) if and only if $F \preceq G^*$.*

Proof. We simply show that (12) holds if and only if $F \preceq G^*$. Partial integration in (12) yields

$$\int_0^s F(x)\, dx \leq sF(s) + tG(t) - F(s) - G(t) + F(s)G(t) + \int_t^1 G(x)\, dx. \tag{14}$$

A small picture makes it clear that

$$\int_t^1 G(x)\, dx = \int_0^{1-G(t)} G^*(x)\, dx + (1 - t)G(t).$$

Letting $u = 1 - G(t)$ and inserting the above into (14) yields that (11) holds for all s, t if and only if it holds for all s, u that

$$\int_0^s F(x)\, dx \leq (s - u)F(s) + \int_0^u G^*(x)\, dx. \tag{15}$$

If we assume (15), we may let $u = s$ and deduce that $F \preceq G^*$. Conversely, if we assume $F \preceq G^*$, we have

$$(s-u)F(s) + \int_0^u G^*(x)\,dx \geq (s-u)F(s) + \int_0^u F(x)\,dx \geq \int_0^s F(x)\,dx \tag{16}$$

because differentiation w.r.t. u shows that the right-hand side of (16) is at minimum for $u = s$. Thus we have shown (15), as needed. □

The proposition on p. 1793 of Gutmann *et al.* (1991) can now be rephrased as

Proposition 4. *If* F *and* G *are continuous, there exists* $\psi \in \{0,1\}$ *which is increasing in each of its arguments and satisfies* (13) *if and only if* $F = G^*$.

The analogy with the Gale–Ryser theorem becomes clearer if we let

$$F_{mn}(x) = \frac{1}{m}\sum_1^m \delta_{r_i/n}([0,x]), \quad G_{mn}(y) = \frac{1}{n}\sum_1^n \delta_{c_j/m}([0,y]) \tag{17}$$

whereby some manipulation shows that

$$r \preceq c^* \iff F_{mn} \preceq G^*_{mn}.$$

If we consider a random Rasch ϕ-matrix, given by distributions (A, B) of row and column sums, we get for the infinite row and column averages

$$\bar{X}_{i\infty} = \mathrm{E}(\bar{X}_{i\infty} \mid \mathcal{S}) = \lim_{n\to\infty} \frac{1}{n}\sum_{j=1}^n \frac{\alpha_i\beta_j}{1+\alpha_i\beta_j} = \int_0^\infty \frac{\alpha_i\beta}{1+\alpha_i\beta} B(d\beta) = \breve{B}(\alpha_i),$$

where $\breve{B}$ is what we choose to call the *Rasch transform* defined as

$$\breve{B}(x) = \int_0^\infty \frac{xy}{1+xy} B(dy).$$

Similarly, we get $\bar{X}_{\infty j} = \breve{A}(\beta_j)$. Thus if we let $F(x)$ denote the distribution function of the row average $\bar{X}_{i\infty}$, we have

$$F(x) = P(\bar{X}_{i\infty} \leq x) = P(\breve{B}(\alpha_i) \leq x) = A(\breve{B}^{-1}(x)) \tag{18}$$

and similarly

$$G(x) = B(\breve{A}^{-1}(x))$$

where G is the distribution function of $\bar{X}_{\infty j}$.

Clearly, we may consider the pair $t^*_{mn} = (F_{mn}, G_{mn})$ in (17) of empirical distributions of the row and column averages as the summarizing statistic for $\mathcal{P}_{\mathrm{RCES}}$. In analogy with (3) of de Finetti's theorem, we then have that for any $P \in \mathcal{P}_{\mathrm{RCES}}$ this pair converges to a pair (F, G) of distributions satisfying $F \preceq G^*$ and the mixing measure μ_P on $\mathcal{E}_{\mathrm{RCES}}$ is the distribution of this pair; we refrain from giving the details of the argument.

An obvious question to ask next is whether to any given *subconjugate pair* (F, G) of distributions, *i.e.*, pair of distributions satisfying $F \preceq G^*$, one can find a ϕ-matrix of Rasch type, so that $\mathcal{E}_{\text{RCES}}$ can be identified with the set of subconjugate pairs.

So consider a pair (F, G). From (18) it follows that these are the distributions of row and column averages of a regular random Rasch model if and only if there exist distributions A and B on $(0, \infty)$ so that

$$F(\check{B}(x)) = A(x) \quad \text{and} \quad G(\check{A}(y)) = B(y) \text{ for all } x \text{ and } y. \tag{19}$$

In the case where (F, G) are empirical distributions of the form (17), (19) is easily seen to be equivalent to the equation system

$$\frac{r_i}{n} = \frac{1}{n}\sum_j \frac{\alpha_i\beta_j}{1+\alpha_i\beta_j} \quad \text{and} \quad \frac{c_j}{m} = \frac{1}{m}\sum_i \frac{\alpha_i\beta_j}{1+\alpha_i\beta_j}, \tag{20}$$

where then A and B are the empirical distributions of $\{\alpha_i\}$ and $\{\beta_j\}$. This fact is most directly seen when row sums and column sums are all different and ordered to be increasing, since then

$$F_{mn}(\check{B}(\alpha_i)) = F_{mn}(r_i/n) = i/n = A(\alpha_i)$$

and similarly with β_j.

The equation system (20) is exactly the maximum likelihood equations for estimation of the parameters in the Rasch model and these are known to have a solution (Fischer, 1981) if and only if $r \prec s^*$ where $\prec$ denotes strict majorization

$$a \prec b \Leftrightarrow \sum_{i=1}^{k} a_i < \sum_{j=1}^{k} b_j \text{ for all } k = 1, \ldots, m,$$

and the solution is unique up to multiplication of α_i with a positive constant c and division of β_j with the same constant. Thus, if we say that (F, G) are *strictly subconjugate* if $F \prec G^*$, where $\prec$ means strict majorization:

$$F \prec G \Longleftrightarrow \int_0^s F(x)\,dx < \int_0^s G(x)\,dx \text{ for all } 0 < s < 1$$

it seems natural to conjecture:

Conjecture. *Let (F, G) be a pair of distribution functions on $[0, 1]$. Then there exists a ϕ-matrix of Rasch type with distributions of asymptotic marginal row and column averages given by F and G if and only if $F \preceq G^*$. Moreover, the distribution of the ϕ-matrix is injectively parametrized by (F, G) and the corresponding ϕ-matrix is regular if and only if $F \prec G^*$.*

However, at present it is not clear to the author how to prove this, although part of the conjecture should follow from a suitable limiting argument, using the result about existence and uniqueness of the maximum likelihood estimates. Note that the case $F = G^*$ of Proposition 4 indeed corresponds to the non-regular Rasch-matrix determined by (8).

ACKNOWLEDGEMENTS

The results in the present paper have their origin in my visit to London in 1979 and I am indebted to inspiration from discussions with Phil Dawid during that visit, where I first learned about subconjugate distributions and majorization and the results later published in Aldous (1981). Ever since, I have from time to time worked on the issues in the present paper. Significant progress was made in 1997 while I was a Fellow at the Center for Advanced Study in Behavioral Sciences at Stanford, USA, and I gratefully acknowledge financial support from NSF grant # SBR–9022192. Finally, a visit to Klitgaarden—a refuge in Skagen, Denmark, for scientists and artists—provided the necessary peace to complete the paper and consequently I am deeply indebted to this institution.

REFERENCES

Aldous, D. (1981). Representations for partially exchangeable random variables. *J. Multiv. Anal.* **11**, 581–598.

Aldous, D. (1985). Exchangeability and related topics. *École d'Été de Probabilités de Saint–Flour XIII (1983)*, (P. Hennequin, ed). Berlin: Springer, 1–198.

Andersen, E. B. (1973). Conditional inference for multiple-choice questionnaires. *BrJ. Math. Statist. Psychol.* **26**, 31–44.

Besag, J. and Clifford, P. (1989). Generalized Monte Carlo significance tests. *Biometrika* **76**, 633–642.

Dawid, A. P. (1982). Intersubjective statistical models. *Exchangeability in Probability and Statistics* (G. Koch and F. Spizzichino, eds). Amsterdam: North-Holland, 217–232.

Dawid, A. P., DeGroot, M. H., and Mortera, J. (1995). Coherent combination of experts' opinions. *Test* **4**, 263–313 (with discussion).

de Finetti, B. (1931). Funzione caratteristica di un fenomeno aleatorio. *Atti della R. Academia Nazionale dei Lincei, Serie 6. Memorie, Classe di Scienze Fisiche, Mathematice e Naturale* **4**, 251–299.

Diaconis, P. (1977). Finite forms of de Finetti's theorem on exchangeability. *Synthése* **36**, 271–281.

Diaconis, P. and Freedman, D. (1980). de Finetti's theorem for Markov chains. *Ann. Prob.* **8**, 745–764.

Diaconis, P. and Freedman, D. (1981). On the statistics of vision: the Julesz conjecture. *J. Math. Psych.* **24**, 112–138.

Fischer, G. H. (1981). On the existence and uniqueness of maximum-likelihood estimates in the Rasch model. *Psychometrika* **46**, 59–77.

Fischer, G. H. (1995). Derivations of the Rasch model. *Rasch Models. Foundations, Recent Developments, and Applications* (G. H. Fischer and I. W. Molenaar, eds). New York: Springer, 15–38.

Fischer, G. H. and Molenaar, I. W. (eds.) (1995). *Rasch Models. Foundations, Recent Developments, and Applications*. New York: Springer.

Freedman, D. (1962). Invariants under mixing which generalize de Finetti's theorem. *Ann. Math. Statist.* **33**, 916–923.

Gale, D. (1957). A theorem on flows in networks. *Pacific J. Math.* **7**, 1073–1082.

Gutmann, S., Kemperman, J. H. B., Reeds, J. A., and Shepp, L. A. (1991). Existence of probability measures with given marginals. *Ann. Prob.* **19**, 1781–1797.

Hewitt, E. and Savage, L. J. (1955). Symmetric measures on Cartesian products. *Trans. Am. Math. Soc.* **80**, 470–501.

Holst, C. (1995). *Item Response Theory*. Ph.D. Thesis, University of Copenhagen, Denmark.

Hoover, D. N. (1982). Row–column exchangeability and a general model for exchangeability. *Exchangeability in Probability and Statistics*, (G. Koch and F. Spizzichino, eds). Amsterdam: North-Holland, 281–91.

Julesz, B. (1975). Experiments in the visual perception of textures. *Sci. Am.* **232**, 34–43.

Julesz, B. (1980). Spatial nonlinearities in the instantaneous perception of textures with identical power spectra. *Phil. Trans. Roy. Soc. London B* **294**, 83–94.

Kellerer, H. G. (1961). Funktionen auf Produkträumen mit vorgegebenen Marginal-Funktionen. *Math. Annalen* **144**, 323–344.

Kingman, J. F. C. (1978). Uses of exchangeability. *Ann. Prob.* **6**, 183–197.

Lauritzen, S. L. (1988). *Extremal Families and Systems of Sufficient Statistics*. Berlin: Springer.

Lynch, J. (1984). Canonical row-column-exchangeable arrays. *J. Multiv. Anal.* **15**, 135–140.

McCullagh, P. and Nelder, J. A. (1989). *Generalized Linear Models*, 2nd edn. London: Chapman and Hall.

Ponocny, I. (2001). Nonparametric goodness-of-fit tests for the Rasch model. *Psychometrika* **66**, 437–460.

Rao, A. R., Jana, R., and Bandyopadhyay, S. (1996). A Markov chain Monte Carlo method for generating random (0,1)-matrices with given marginals. *Sankhyā A* **58**, 225–242.

Rasch, G. (1960). *Probabilistic Models for Some Intelligence and Attainment Tests*. Studies in Mathematical Psychology **1**. Copenhagen: Danmarks Pædagogiske Institut.

Rasch, G. (1967). An informal report on a theory of objectivity in comparisons. *Psychological Measurement Theory* (L. J. T. van der Kamp and C. A. J. Vlek, eds). Leyden: University Press, 1–19.

Rasch, G. (1971). Proof that the necessary condition for the validity of the multiplicative dichotomic model is also sufficient (unpublished manuscript).

Rasch, G. (1977). On specific objectivity. an attempt at formalizing the request for generality and validity of scientific statements. *The Danish Yearbook of Philosophy* (M. Blegvad, ed). Copenhagen: Munksgaard, 58–94.

Ressel, P. (1985). de Finetti-type theorems: An analytical approach. *Ann. Prob.* **13**, 898–922.

Ressel, P. (1988). Integral representations for distributions of symmetric stochastic processes. *Prob. Theo. Rel. Fields* **79**, 451–467.

Ressel, P. (1994). Non-homogeneous de Finetti-type theorems. *J. Theor. Prob.* **7**, 469–482.

Ryser, H. J. (1957). Combinatorial properties of matrices of zeros and ones. *Can. J. Math.* **9**, 371–377.

Strassen, V. (1965). The existence of probability measures with given marginals. *Ann. Math. Statist.* **36**, 423–439.

Wilson, J. B. (1987). Methods for detecting non-randomness in species co-occurrences: A contribution. *Oecologica* **73**, 579–582.

DISCUSSION

MICHAEL GOLDSTEIN (*University of Durham, UK*)

This paper contains results which are both deep and elegant. Are they important? I am reminded of the following quote from the introduction to the collection "Studies in subjective probability" (1964), in which Kyburg and Smokler write

> In some ways, the most important concept of the subjectivistic theory is that of exchangeable events. Until this notion was introduced by de Finetti in 1931, the subjectivistic theory of probability remained pretty much of a philosophical curiosity. None of those for whom probability theory was a means of livelihood or knowledge paid much attention to it.

Why is exchangeability so important? It will be helpful to have a story to hang this discussion on, so let us suppose that a new television program is created, called First Kiss. In this program, a group of men and women compete as follows. Each man and each woman kiss exactly once. Each kiss is determined to be either "good" or "bad." This determination is made by a strictly objective method which, due to space limitations, I will not be able to describe here. Each player is scored by the number of good kisses that they achieve, and the winners go on to compete in further stages of the competition.

Suppose that we want to carry out a statistical analysis of a round of the game. We have a table with rows corresponding to men, columns to women. The (i, j)-th entry, X_{ij}, is 1 for a good kiss, and zero for a bad kiss. While there are many ways that we might choose to analyze such a table, a simple and fairly standard approach would be to apply a linear log-odds model, representing the process generating the table as

$$L_{ij} = \log \frac{\mathrm{P}(\mathrm{X_{ij}} = 1)}{\mathrm{P}(\mathrm{X_{ij}} = 0)} = r_i + c_j \tag{21}$$

where r_i, c_j are row and column constants representing the ability of each contestant. We might then fit (21) to the table of data using our favorite Bayes or likelihood approach, and then carry out some form of diagnostic check for model fit. However, this sidesteps the fundamental question as to why we should entertain a model such as (21), in the first place.

One of de Finetti's fundamental contributions was to show how beliefs about underlying and unobservable parameters could be inferred strictly from beliefs expressed over observable quantities. In our problem, the results of this paper assure us of the following. Suppose that we can view our individual matrix as a sub-matrix of a (hypothetical) infinite matrix (i) with exchangeable rows and exchangeable columns, and (ii) for which row sums and column sums for any sub-matrix are sufficient statistics for that sub-matrix. Then the Rasch representation theorem tells us that our beliefs over our matrix must be exactly as if we believed (i) that each row has a true value r_i and each column has a true value c_j, satisfying (1) for all i, j; (ii) we don't know what values r_i and c_j are, but we believe that the sequence $r_1, r_2, \ldots$ is iid with probability distribution $\mathrm{P_R}$, and the sequence $c_1, c_2, \ldots$ is iid with probability distribution $\mathrm{P_C}$; (iii) we do not know what $\mathrm{P_R}, \mathrm{P_C}$ are, but we have a prior distribution P_{RC} over possible choices of $\mathrm{P_R}, \mathrm{P_C}$. Therefore, we see that the Bayesian analysis over (21) is indeed a necessary consequence of certain beliefs over the observables. Further, the diagnostic analysis of the model that we might carry out is precisely that which critically scrutinizes the generalized constraints on our beliefs which we require in order to apply the Rasch representation.

This is an important and useful result, partly in giving meaning to our analysis and partly in directing us to the diagnostic testing which is appropriate to use of the model. However, there is a further consideration which I believe that we must apply before we can claim that this paper offers genuine insights for the subjectivistic theory. The representation theorem argues that our beliefs must be as if there were true underlying probability distributions generating true underlying parameter values. But what is it about our beliefs over the kisses which compels us to believe in these underlying parameter distributions? The central result of this paper is a deep one, whose proof winds its way through various other deep results from a variety of sources. Therefore, it is difficult to see whether the representation is based on natural, finite considerations, or whether at some point in the development a step has been introduced which only makes sense within an infinite collection and which has no meaningful finite counterpart.

I shall now suggest that the result is indeed a consequence of natural and finite considerations. First, let us recall how de Finetti's representation theorem works for coin tosses. If we judge that coin tosses are exchangeable, then we may consider the outcomes of a large, but finite, collection of tosses. We may imagine filling a bucket with tokens, where the i-th token is marked heads or tails depending on the result of the i-th toss. Suppose that the proportion of heads in the bucket is p. As the tosses are exchangeable, our beliefs, given p, about the outcome of tossing the coin k times is exactly as though we were to make k independent selections of tokens from the bucket without replacement. If the number of tokens in the bucket is large compared to k, then we may view the selections as almost independent, each with probability p for heads. Of course, we do not know what the value of p will be, and therefore we have a prior distribution over this value. Thus, our beliefs will be exactly as described by de Finetti's representation theorem, up to the approximation arising from the finite nature of the bucket. Thus, there is a final book-keeping step of allowing the size of the bucket to tend to infinity, and showing that the limit is smoothly and consistently achieved, but this argument is sufficient to show that the representation is really concerned with our beliefs over large finite collections of tosses.

For the Rasch representation, the argument is more complicated, but similar finite arguments show why the representation holds. To simplify the discussion, suppose that we consider that there are three levels of ability for the men, namely Superb (S), Acceptable (A) and Terrible (T), and similarly for the women. We do not know *a priori* how many people fall into each category, nor do we know the quantitative differences between the groups and nor do we know which category each individual should fall into.

However, now suppose that we envisage a large, but finite, array of outcomes of the game. The row sums allow us to classify the men into their appropriate groups to an arbitrary level of accuracy, as the array size increases. Similarly, the column sums allow us to classify the women. Therefore, we may consider that we have nine buckets filled with kisses. In each bucket, some kisses are good and some are bad. Let p_{mw} be the proportion of good kisses among group mw, where each ability of the men, m, and of the women, w, is one of S, A, T. Our probability that an individual pair i, j of people have a good kiss, conditional on the row and column sums for the layout, comes from using the i-th row sum and the j-th column sum to allocate the pair to the appropriate groups m_i and w_j and then, from the row and column exchangeability, viewing the probability that the couple have a good kiss as $p_{m_i w_j}$ independently of all other kisses.

Given the nine values p_{mw}, we now fit the Rasch model

$$P(X_{ij} = 1) = \frac{\alpha_i \beta_j}{1 + \alpha_i \beta_j}, i, j = S, A, T \tag{22}$$

As it stands, the model is non-identifiable, so we nominate an individual to be the standard against which all others are judged. Suppose that we assign β_S, the score for superb women, to be one. This then fixes the scores for all men as $p_{mS} = (\alpha_m/[1 + \alpha_m]), m = S, A, T$. This now fixes each of the remaining scores for women, for example, looking at the groups with $m = S$ gives $p_{Sw} = (\alpha_S \beta_w/[1 + \alpha_S \beta_w]), w = A, T$. (This argument breaks down if any of the p_{mw} values are zero, which is why a separate argument is required in the general statement of the theorem for the non-regular case.) We have now fixed all of the values α_i, β_j and we must check that (22) is indeed satisfied over all subgroups. This follows as all the information that we have used is based on conditioning on row and column sums. Such conditioning preserves row–column summarizability (as, conditional on the row and column sums, all configurations with these row and column sums have the same probability, so that any two sub-matrices with the same row and column sums must have the same conditional probability as each can be embedded in exactly the same number of configurations for the full matrix with the given row and column sums). Therefore, consider, for example, our assessment for $P(X_{TT} = 1)$. This is uniquely determined, as by row–column summarizability, we must assign the same probability to each of the events $[X_{TT} = 1, X_{TS} = 0, X_{ST} = 0, X_{SS} = 1]$ and $[X_{TT} = 0, X_{TS} = 1, X_{ST} = 1, X_{SS} = 0]$. Equivalently, we must assign

$$\frac{\alpha_S}{1 + \alpha_S} \frac{1}{1 + \alpha_S \beta_T} \frac{1}{1 + \alpha_T} p_{TT} = \frac{1}{1 + \alpha_S} \frac{\alpha_S \beta_T}{1 + \alpha_S \beta_T} \frac{\alpha_T}{1 + \alpha_T} (1 - p_{TT})$$

from which $p_{TT} = (\alpha_T \beta_T/[1 + \alpha_T \beta_T])$ as required. We therefore see that the exchangeability construction for Rasch matrices corresponds in this case to a mixture of our uncertainties as to the relative proportions of each of the groups and our beliefs over the magnitudes of the effects within each group, as expressed by our beliefs over the nine probability values p_{mw} constrained by the Rasch relations (22). The book-keeping that is required to produce the general result is far more detailed than for the classic de Finetti representation, as we not only have to let the size of each bucket tend to infinity but we also need to let the number of buckets tend to infinity, reversing the argument that I gave where we started by knowing the number of groups and instead defining the groups through observed similarity of row or column sums.

The technical difficulties in such an explicit construction are considerable. However, the above argument is, I hope, sufficient to suggest that the reason that the Rasch representation, so expertly presented in this paper, does indeed offer powerful, practical insights into the treatment of binary layouts is that it is a genuinely subjectivistic result which is based on intuitive and finite considerations.

REPLY TO THE DISCUSSION

First I would like to thank Michael Goldstein for his positive reaction to this paper. Although mating of salamanders is a potential application of the Rasch model, I admit that the First Kiss program is much more fascinating! The description of the nature and genesis of the Rasch model given by Michael Goldstein is both very illuminating, accurate, and on the point.

Indeed, it was appropriate to mention that the non-degenerate Rasch model is nothing but an additive model for the log-odds, a model which is more familiar to statisticians today than it was in 1960, when Rasch introduced it.

It would be very valuable to have a derivation of the random Rasch model from finite considerations, as suggested by the discussant. Diaconis and Freedman (1980) give finite versions of de Finetti's classical theorem, with an explicit bound on the distance in total variation between the distribution of the first k of a sequence of exchangeable variables with a given finite length n, and the closest mixture of Bernoulli distributions. The bound, $4k/n$, originates from approximating the hypergeometric distribution with the binomial. Generalizations of this type of argument has, for example, been made by Diaconis *et al.* (1992), and the corresponding infinite versions of de Finetti type theorems then usually follow by a simple limit argument.

The problem here is that the bookkeeping associated with deriving such bounds and controlling their asymptotic behavior in the case of binary matrices is particularly difficult. Whereas there are efficient and well-known asymptotic results for the number $\binom{n}{x}$ of binary sequences of length n with sum x, it seems to be extremely hard to control $N(r,c) = |\mathcal{M}(r,c)|$, the number of binary matrices with rowsums equal to $r = (r_1, \ldots, r_m)$ and column-sums equal to $c = (c_1, \ldots, c_n)$. The combinatorial literature has only sporadic results; see for example O'Neil (1969), Békéssy *et al.* (1972), Bender (1974), Mineev and Pavlov (1976), and McKay (1984, 1985). The structure of the degenerate RCE matrices of Rasch type also indicates that the situation is quite complex.

ADDITIONAL REFERENCES IN THE DISCUSSION

Békéssy, A., Békéssy, P., and Komlós, J. (1972). Asymptotic enumeration of regular matrices. *Studia Scientiarum Mathematicarum Hungaria* **7**, 343–353.

Bender, E. A. (1974). The asymptotic number of non-negative integer matrices with given row and column sums. *Discrete Math.* **10**, 217-233.

Diaconis, P., Eaton, M. L. and Lauritzen, S. L. (1992). Finite forms of de Finetti theorems in linear models and multivariate analysis. *Scand. J. Statist.* **19**, 289–315.

Diaconis, P. and Freedman, D. (1980). Finite exchangeable sequences. *Ann. Prob.* **8**, 745–764.

McKay, B. (1984). Asymptotics for 0–1 matrices with prescribed line sums. *Enumeration and Design* (D. M. Jackson and S. A. Vanstone, eds). New York: Academic Press, 225–228.

McKay, B. (1985). Asymptotics for 0–1 matrices with prescribed row sums. *Ars Combinatoria* **19A**, 15–25.

Mineev, M. P. and Pavlov, A. I. (1976). On the number of (0-1) matrices with prescribed sums of rows and columns. *Dokl. Akad. Nauk SSSR* **230**, 1286-1290 (Russian original has page numbers 271–274).

O'Neil, P. E. (1969). Asymptotics and random matrices with row-sum and column-sum restrictions. *Bull. Am. Math. Soc.* **75**, 1276–1282.

BAYESIAN STATISTICS 7, pp. 233–248
J. M. Bernardo, M. J. Bayarri, J. O. Berger, A. P. Dawid,
D. Heckerman, A. F. M. Smith and M. West (Eds.)

Discrimination Based on an Odds Ratio Parameterization

ANGELIKA VAN DER LINDE and GERHARD OSIUS
Universität Bremen, Germany
avdl@math.uni-bremen.de osius@math.uni-bremen.de

SUMMARY

It is argued that the association of any two random elements with positive joint probability density function is characterized by its odds ratio function. The impact of this fundamental result is explored in two applications. In Bayesian analyses it is the association between an observable random variable and a random parameter that is of primary interest. In multivariate analysis it is the association between two random vectors that is investigated. If two random elements are strongly related the corresponding conditional distributions can well be separated. A concept of dependence therefore implies an approach to solving problems of discrimination. Discrimination based on the characterization of association by the odds ratio function is exemplified in Bayesian inference and in multivariate analysis.

Keywords: ASSOCIATION; ODDS RATIO; KULLBACK–LEIBLER DISTANCE; LOGISTIC REGRESSION; MODEL CHOICE; MODEL COMPLEXITY; LOGARITHMIC SCORE FUNCTION; CANONICAL CORRELATION; DISCRIMINANT ANALYSIS.

1. ASSOCIATION AND ODDS RATIOS

1.1. *Motivation*

How do you present concepts of dependence in your introductory course to statistics?

The best known definitions certainly are those of a correlation coefficient and a covariance matrix. In fact, for jointly Gaussian vectors the covariance matrix does characterize the joint distribution if the marginal distributions are specified. And in multivariate analysis the investigation of structures of dependence is widely based on covariance or correlation matrices.

With respect to discrete variables, for example referring to a 2×2-table of binary random variables X and Y, statisticians are also familiar with the odds ratio defined by

$$OR := \frac{P(Y=1\,|\,X=1)}{P(Y=0\,|\,X=1)} \Big/ \frac{P(Y=1\,|\,X=0)}{P(Y=0\,|\,X=0)}. \tag{1}$$

More generally, in a $(M+1) \times (K+1)$-contingency table where (X, Y) takes values in $\{0, 1, ..., M\} \times \{0, 1, ..., K\}$, the odds ratio can be "moved around" yielding a matrix with entries

$$OR(m,k) := \frac{P(Y=k\,|\,X=m)}{P(Y=0\,|\,X=m)} \Big/ \frac{P(Y=k\,|\,X=0)}{P(Y=0\,|\,X=0)}.$$

Given the marginal distributions of X and Y, any matrix with positive entries $OR(m,k)$ defines a unique joint probability distribution. (See the proof by Plackett, 1974, based on a result by Sinkhorn, 1967.) Hence, the odds ratio matrix like the covariance matrix for jointly Gaussian vectors describes that part of the joint distribution that is left if the information inherent in both marginal distributions P_X, P_Y is removed. We call that part the "*association* of X and Y". Is there a unique characterization of a joint probability distribution P_{XY} by a triple (P_X, P_Y, "*association*") in general ? How can association be formally defined ?

The analysis of a "mixed" example with binary Y and arbitrary random vector X again suggests that it might be an *odds ratio function* (with reference value x_0)

$$OR^0(x) := \frac{P(Y=1\,|\,X=x)}{P(Y=0\,|\,X=x)} \Big/ \frac{P(Y=1\,|\,X=x_0)}{P(Y=0\,|\,X=x_0)} \tag{2}$$

that captures the association between X and Y. In this case

$$\log\frac{P(Y=1\,|\,X=x)}{P(Y=0\,|\,X=x)} = \log\frac{P(Y=1\,|\,X=x_0)}{P(Y=0\,|\,X=x_0)} + \log OR^0(x) = \alpha + \log OR^0(x)$$

(say). Hence

$$P(Y=1\,|\,X=x) = (1+\exp\{-\alpha - \log OR^0(x)\})^{-1}, \tag{3}$$

and

$$P(Y=1) = E_x[(1+\exp\{-\alpha - \log OR^0(x)\})^{-1}]. \tag{4}$$

This expectation is a strictly increasing function of α and approaches the limits 1, respectively, 0 as $\alpha \to \infty$, respectively, $\alpha \to -\infty$. Hence for $0 < P(Y=1) < 1$ there exists a unique $\infty < \alpha < \infty$ such that (4) holds. This shows that for fixed marginal distribution P_Y the joint distribution of (X, Y) is uniquely determined by the log-odds ratio function. Furthermore, for fixed marginal distributions and a given log-odds ratio function a joint distribution is defined by (3) with α obtained from (4).

1.2. *Main Results*

Consider two random elements X and Y and assume their joint probability distribution P_{XY} to have a positive density $p(x, y)$ w.r.t. a *product* measure Q. Denote the log-density by $\phi(x, y)$ and define the log-odds ratio function ψ as a function of two pairs (x, y) and (x', y'),

$$\psi(x, y\,|\,x', y') := \phi(x, y) - \phi(x, y') - \phi(x', y) + \phi(x', y'). \tag{5}$$

Under mild regularity conditions the (log-)odds ratio function characterizes the association between X and Y, that is, P_{XY} is characterized by the triple (P_X, P_Y, ψ). We only sketch the ideas of proof here and refer to Osius (2000) for full details. Uniqueness: Two joint distributions P_{XY} and P'_{XY} with common marginal distributions P_X, P_Y and common log-odds ratio function ψ coincide, if their log-densities are integrable. Indeed, let $p(x, y)$ and $p'(x, y)$ denote the densities of two such joint distributions. Evaluation of the Kullback–Leibler distance

$$I(p, p') = \int \log\frac{p(x, y)}{p'(x, y)}\, dP_{XY}(x, y)$$

yields $I(p, p') = 0$ implying $P_{XY} = P'_{XY}$. Existence: Given P_X and P_Y and a function ψ, a joint probability density exists (under regularity conditions) such that P_X and P_Y are the corresponding marginal distributions and ψ is the log-odds ratio function. More precisely, the joint density p is obtained as a limit of a sequence of densities (p_n) where the corresponding joint distributions P^n_{XY} are constructed as follows. Start with any P^0_{XY} the log-odds ratio function of which is ψ. Then given P^n_{XY}, replace the marginal distribution P^n_X of X (conditioning) by the wanted margin P_X and obtain P'^n_{XY}. Next, replace the other margin P'^n_Y by P_Y to obtain P^{n+1}_{XY} which has the same log-odds ratio function ψ as P^n_{XY}. The sequence (p_n) of densities converges to a density p of the required joint distribution P_{XY} (and does not depend on the starting distribution P^0_{XY}). A sufficient condition (which can be weakened) for this reasoning to hold is the integrability of the odds ratio function $\exp\psi$ which can be used as the density of P^0_{XY} after normalization. This iterative procedure of "marginal fitting" generalizes the method used by Sinkhorn (1967) for marginals with finite support. His proof of convergence however exploits unique features of distributions with finite support which cannot be referred to in the general set-up considered here.

The required joint distribution P_{XY} can also be characterized in another way. For a density $p'(x, y)$ w.r.t. a product measure Q consider the functional

$$l(p') := \int p(x)p(y) \log \frac{p'(x,y)}{p(x)p(y)} dQ(x,y) = -I(p(x)p(y), p'(x,y)).$$

The required joint density then is given as the unique density $p(x, y)$ that maximizes $l(p')$ within the space of all log-integrable densities having ψ as their log-odds ratio function. The functional $l(p')$ is strictly concave and generalizes the log-likelihood used by Haberman (1974, Th. 2.6) for distributions with finite support.

1.3. *Properties of the Odds Ratio Function*

The log-odds ratio function has some desirable properties like the compatibility with 1:1-transformations of X and Y and the invariance under a change of the dominating (product) measure. The log-odds ratio function is already determined by any partial function ψ^0 with fixed reference values (x_0, y_0),

$$\psi^0(x, y) := \psi(x, y \mid x_0, y_0), \tag{6}$$

which can be regarded as representative of ψ. ψ^0 is also already defined using one of the conditional log-densities instead of $\phi(x, y)$ in Eq. (5).

1.4. *The Odds Ratio Parameterization*

Using ψ^0, the log-density ϕ can be decomposed analogously to a control parameterization in an ANOVA-model,

$$\phi(x, y) = \alpha + \beta(x) + \gamma(y) + \psi^0(x, y), \tag{7}$$

where

$$\alpha = \phi(x_0, y_0),$$
$$\beta(x) = \phi(x, y_0) - \phi(x_0, y_0), \quad \beta(x_0) = 0,$$
$$\gamma(y) = \phi(x_0, y) - \phi(x_0, y_0), \quad \gamma(y_0) = 0,$$

and

$$\psi^0(x_0, y) = \psi^0(x, y_0) = \psi^0(x_0, y_0) = 0.$$

Thus $\psi^0(x, y)$ corresponds to an interaction term. The log-odds ratio parameterization of conditional and marginal densities then is

$$\log p(y \mid x) = \log p(y_0 \mid x) + \gamma(y) + \psi^0(x, y), \tag{8}$$

$$\log p(y) = \alpha + \gamma(y) - \log p(x_0 \mid y). \tag{9}$$

1.5. *Bi-affine Log-Odds Ratio Functions*

In standard examples the log-odds ratio function exhibits a simple structure. Let T as superscript denote the transpose of a vector or a matrix.

Example 1. For a Normal conditional distribution $Y \mid x \sim N(Bx, \Sigma)$, ψ^0 is in general bi-affine,

$$\psi^0(x, y) = (y - y_0)^t \Sigma^{-1} B(x - x_0). \tag{10}$$

For reference values $x_0 = 0$, $y_0 = 0$ this reduces to the bi-linear form

$$\psi(x, y \mid 0, 0) = y^t \Sigma^{-1} Bx = y^t cov(y \mid x)^{-1} E(y \mid x). \tag{11}$$

Hence if the joint distribution of X and Y is Gaussian the conditional distributions are Gaussian as well, and the association between X and Y is described by a bi-linear function ψ^0.

Example 2. If the conditional distribution of Y given x belongs to an exponential family,

$$p(y \mid x) = a(y) \exp\{x^t t(y) - nM(x)\}, \qquad (say), \tag{12}$$

then

$$\psi^0(x, y) = (x - x_0)^t (t(y) - t(y_0)). \tag{13}$$

Example 3. For simple random variables with finite range bi-affinity holds, too, in a general sense. For X taking values in $\{0, 1, ..., M\}$ and Y taking values in $\{0, 1, ..., K\}$ define transformations

$$g(m) = e_{m+1} \in \Re^{M+1} \quad , \quad h(k) = e'_{k+1} \in \Re^{K+1}, \tag{14}$$

where e_i, e'_j denote the i-th respectively j-th unit vector. Choosing $x_0 = 0$, $y_0 = 0$,

$$\psi^0(m, k) = (e_{m+1} - e_1)^t ((\log p(m \mid k)))(e'_{k+1} - e'_1) \tag{15}$$

$$= (g(m) - g(0))^t ((\log p(m \mid k)))(h(k) - h(0)).$$

A similar result is obtained for transformations $g_0(m) = e_m \in \Re^M$, $g_0(0) = 0_M$ and $h_0(k) = e'_k \in \Re^K$, $h_0(0) = 0_K$ corresponding to the non-degenerate representation of the multinomial distribution. The entries of the matrix in the inner product then are directly the values $\psi^0(m, k)$ instead of $\log p(m \mid k)$.

Motivated by these examples we call ψ^0 bi-affine if there are transformations $g : \mathcal{X} \rightarrow \Re^{k_x}$, $h : \mathcal{Y} \rightarrow \Re^{k_y}$ and a $k_x \times k_y$-matrix A such that

$$\psi^0(x, y) = (g(x) - g(x_0))^t A(h(y) - h(y_0)). \tag{16}$$

1.6. *Logistic Regression*

If Y takes values in $\{0, 1, ..., K\}$ which are coded using h_0 as in Example 3 and if ψ^0 is bi-affine, the log-odds ratio parameterization of the conditional density (8) induces a logistic regression model

$$\log \frac{p(k \mid x)}{p(k_0 \mid x)} = \gamma'(k) + g(x)^t a_k, \qquad \text{(say)}. \tag{17}$$

In this case, the model is linear in the transformed values of x and A is a matrix of regression coefficients. Any modelling assumption specifying a structure of the log-odds ratio function corresponds to a (logistic) regression model, and reversely any regression model specifies a structure of ψ^0.

1.7. *Measures of Association and Dependence*

So far we have described the association of two random elements X and Y by a function, ψ^0. How can the strength of association be quantified ? In general, a measure of association (or dependence) is given by a functional that assigns to a (log-)odds ratio function a real value, for example an integral. If such a functional does not involve the marginal distributions, we call it a measure of association. If it does, for example, if expectations are taken w.r.t. a marginal distribution, we call it a measure of dependence.

A measure of the strength of the relation between X and Y defined in the same spirit as the association is the "mutual information"

$$I(X,Y) := \int p(x,y) \log \frac{p(x,y)}{p(x)p(y)} dQ(x,y). \tag{18}$$

In the log-odds ratio parameterization

$$I(X,Y) = -\alpha + E_x \log p(y_0 \,|\, x) + E_y \log p(x_0 \,|\, y) + E_{xy}\psi^0(x,y). \tag{19}$$

The measure is symmetric in X and Y, but it is not a symmetric distance between $p(x,y)$ and $p(x)p(y)$. The symmetrized distance

$$J(X,Y) := \int (p(x,y) - p(x)p(y)) \log \frac{p(x,y)}{p(x)p(y)} dQ(x,y) \tag{20}$$

can be written as

$$J(X,Y) = E_{xy}\psi^0(x,y) - E_x E_y \psi^0(x,y) \tag{21}$$

(for all reference values x_0, y_0) and thus can be expressed in terms of integrals of the log-odds ratio function ψ^0.

2. ASSOCIATION IN BAYESIAN ANALYSIS

In this section, we are interested in the association of an observed random vector Y and a random vector Θ induced by a prior distribution on a parameter θ that determines the conditional density $p(y \,|\, \theta)$. In our notation we switch from x to θ throughout this section. Details of the approach sketched here are given in van der Linde (2002b).

2.1. *Measures of Dependence of Interest*

We investigate measures of dependence defined by average Kullback–Leibler distances between probability densities q_1 and q_2. The Kullback–Leibler distance is also called the *directed divergence* of q_1 and q_2, and the *divergence* is the symmetrized measure

$$J(q_1,q_2) = I(q_1,q_2) + I(q_2,q_1). \tag{22}$$

The "mutual information" $I(\Theta, Y)$ is seen to be a special application to joint probability density functions of (θ, y) (cf. (18)). As

$$I(\Theta, Y) = E_y I(p(\theta \,|\, y), p(\theta)), \tag{23}$$

it is important in Bayesian inference describing the learning process about Θ using y, the transition from the prior to the posterior distribution. We do not focus on the mutual information

$I(\Theta, Y)$ though, but want to draw attention to measures of discriminative information, the (directed) divergences between conditional densities.

$$I(p(y \mid \theta), p(y \mid \theta')) = E_{y \mid \theta} \log \frac{p(y \mid \theta)}{p(y \mid \theta')} \tag{24}$$

as an average likelihood ratio describes the difference between probability models (sampling distributions) and is of interest in hypothesis testing and model selection. The symmetrized distance $J(p(y \mid \theta), p(y \mid \theta'))$ may be interpreted as a measure of variation of sampling distributions (centered at $p(y \mid \theta')$), thus quantifying the range of modelling assumptions for the distribution of the observed random vector. Dually $J(p(\theta \mid y), p(\theta \mid y'))$ quantifies the sensitivity of the posterior distribution to data.

The discriminative information is related to the "global" measure of (symmetrized) mutual information as indicated in (23). For example, in conjugate exponential families the prior distribution and posterior distribution are of the same type, but they differ in (hyper-)parameters. For $p(\theta) = p(\theta \mid \lambda)$ and $p(\theta \mid y) = p(\theta \mid \lambda(y))$, say,

$$J(\Theta, Y) = E_y J(p(\theta \mid \lambda(y)), p(\theta \mid \lambda)). \tag{25}$$

Dually also

$$J(\Theta, Y) = E_\theta J(p(y \mid \theta), p(y)), \tag{26}$$

and in conjugate exponential families the reference density $p(y)$ can be replaced by $p(y \mid E\Theta)$, that is,

$$J(\Theta, Y) = E_\theta J(p(y \mid \theta), p(y \mid E\Theta)). \tag{27}$$

Example 1. (continued). With Gaussian distributions $Y \mid \theta \sim N(\theta, \Sigma)$, $\Theta \sim N(0, K)$ for $\theta_0 = E\Theta$,

$$\psi^0(\theta, y) = tr(\Sigma^{-1}(\theta - E\Theta)(y - y_0)^t). \tag{28}$$

Hence

$$J(\Theta, Y) = tr(\Sigma^{-1}K), \tag{29}$$

and $E_\theta J(p(y \mid \theta), p(y \mid E\Theta))$ is given by

$$tr(E_\theta[\Sigma^{-1}(\theta - E\Theta)(\theta - E\Theta)^t]) = tr(\Sigma^{-1}K) = E_{\theta y}\psi(\theta, y \mid E\Theta, y_0).$$

The relation

$$J(\Theta, Y) = E_{\theta y}\psi(\theta, y \mid E\Theta, y_0), \tag{30}$$

observed in the example, can be proven to hold in general for exponential families.

2.2. *Model Complexity*

Turning from prior to posterior expectations with given data y', it can be shown that an equation like (30) still holds approximately. Thus

$$c(\psi, y') := E_{\theta \mid y'} E_{y \mid \theta} \psi(\theta, y \mid E(\theta \mid y'), y') \tag{31}$$

can be addressed as a measure of variation of sampling distributions or as a measure of model complexity (given y').

Let $\widetilde{E}$, $\widetilde{cov}$ denote expectation and covariance of Θ referring to either the prior or the posterior distribution. Further, let $\mathcal{I}(\theta)$, $\mathcal{I}_y(\theta)$ denote an expected, respectively, observed

Fisher information matrix and abbreviate the log-likelihood function $\log p(y \mid \theta) =: L_y(\theta)$. The argument to establish in general

$$\widetilde{E}_\theta E_{y \mid \theta}\psi(\theta, y \mid \widetilde{E}\Theta, y_0) \approx \widetilde{E}_\theta J(p(y \mid \theta), p(y \mid \widetilde{E}\Theta)), \tag{32}$$

which is an exact equality in the Gaussian example as well as for conjugate exponential families, is based on approximations derived from second-order Taylor expansions. It can be briefly sketched as follows:

$$\begin{aligned}
&\widetilde{E}_\theta E_{y \mid \theta}\psi(\theta, y \mid \widetilde{E}\Theta, y_0)\\
&\qquad = -\widetilde{E}_\theta E_{y \mid \theta}[L_y(\widetilde{E}\Theta) - L_y(\theta)] + (-\widetilde{E}_\theta[L_{y_0}(\theta) - L_{y_0}(\widetilde{E}\Theta)])\\
&\qquad \approx \frac{1}{2} tr(\mathcal{I}(\widetilde{\mathrm{E}}\Theta)\widetilde{\mathrm{cov}}\Theta) + \frac{1}{2}\mathrm{tr}(\mathcal{I}_{\mathrm{y}_0}(\widetilde{\mathrm{E}}\Theta)\widetilde{\mathrm{cov}}\Theta)\\
&\qquad \approx tr(\mathcal{I}_{y_0}(\widetilde{E}\Theta)\widetilde{cov}\Theta)\\
&\qquad \approx 2\widetilde{E}_\theta I(p(y \mid \theta), p(y \mid \widetilde{E}\Theta))\\
&\qquad \approx \widetilde{E}_\theta J(p(y \mid \theta), p(y \mid \widetilde{E}\Theta)).
\end{aligned} \tag{33}$$

A derivation of (33) is given in Kullback, (1959/1968), p. 26). Spiegelhalter *et al.* (2002) suggested

$$p_D(y') := 2E_{\theta \mid y'}[L_{y'}(E(\theta \mid y')) - L_{y'}(\theta)] \tag{34}$$

as measure of model complexity and studied its properties and interpretation as "effective degrees of freedom" in many examples. The derivation above shows that $c(\psi, y') \approx p_D(y')$ and thus theoretically substantiates the interpretation of $c(\psi, y')$, respectively, $p_D(y')$ as measures of model complexity. The dual structure of the association between Θ and Y and the corresponding duality of measures of dependence furthermore links the Bayesian measure $c(\psi, y')$ to the frequentist approach to model complexity in terms of sensitivity (e.g. Efron, 1986; Ye, 1998). $c(\psi, y')$ and $p_D(y')$ may be regarded as estimates of the measure of model complexity

$$c(\psi) := E_{y'}c(\psi, y') \tag{35}$$

which does not depend on the data and therefore is more appropriate as general definition. For computational purposes though, $p_D(y')$ is most advantageous.

Example 1. (continued). In this case for all y'

$$\begin{aligned}
c(\psi) &= c(\psi, y') = p_D(y')\\
&= tr(\Sigma^{-1}\mathrm{cov}(\theta \mid y')) = \mathrm{tr}(I + K^{-1}\Sigma)^{-1}.
\end{aligned} \tag{36}$$

If $K = \tau^2 K'$ and $\tau^2 \to \infty$, then $c(\psi) \to tr\ I_q$ where q is the dimensionality of θ ($\theta \in \Re^q$), that is, the number of unknown parameters in the sampling model.

Example 2. (continued). For exponential families we have

$$\begin{aligned}
c(\psi, y') &= n\, E_{\theta \mid y'}[(\theta - E(\theta \mid y'))^t \nabla M(\theta)]\\
&= n\,(\,\mathrm{tr}\,[\mathrm{cov}_{\theta \mid y'}(\theta, E_{\theta \mid y'}[t(y)])]),
\end{aligned}$$

and

$$p_D(y') = 2nE_{\theta \mid y'}[M(\theta) - M(E(\theta \mid y')].$$

The approximation suggests

$$c(\psi, y') \approx p_D(y') \approx \mathrm{tr}(\mathcal{I}_{\mathrm{y}_0}(\mathrm{E}(\theta \,|\, \mathrm{y}'))\mathrm{cov}(\theta \,|\, \mathrm{y}')). \tag{37}$$

2.3. *Expected Utilities*

The relation (cf. (31) and (33))

$$2E_{\theta \,|\, y'} I(p(y \,|\, \theta), p(y \,|\, E(\theta \,|\, y'))) \approx c(\psi, y')$$

and the very definition of $\psi(\theta, y \,|\, E(\theta \,|\, y'), y_0)$ indicate that $c(\psi, y')$ may serve as a correction term for imputing a representative value like $E(\theta \,|\, y')$ for θ in $\log p(y' \,|\, \theta)$ and similarly in expected utilities based on the logarithmic score function for probability densities as belief functions for a quantity of interest. In particular, in criteria for model choice the decomposition into "model fit" and "model complexity" results from such a corrected imputation.

For example, the Deviance Information Criterion (DIC) introduced by Spiegelhalter *et al.* (2002) is based on

$$-2E_{\theta \,|\, y'} E_{y \,|\, \theta} \log p(y \,|\, E(\theta \,|\, y')) \approx -2 \log p(y' \,|\, E(\theta \,|\, y')) + 2p_D(y'). \tag{38}$$

DIC is defined by

$$\mathrm{DIC} = D(\overline{\theta}) + 2p_D(y'),$$

where D denotes the deviance and the bar indicates the posterior expected value. Consider for illustration the expected loss $-2E_{\theta \,|\, y'} E_{y \,|\, \theta} \log p(y \,|\, \theta)$. Using this shorthand notation a first imputation step

$$-2E_{\theta \,|\, y'} E_{y \,|\, \theta} \log p(y \,|\, \theta) \approx -2E_{\theta \,|\, y'} E_{y \,|\, \theta} \log p(y \,|\, \overline{\theta}) - c(\psi, y') \tag{39}$$

can be derived directly from (33), and a second imputation step also used in the derivation of DIC is

$$\begin{aligned} -2E_{y \,|\, \theta} \log \frac{p(y \,|\, \overline{\theta})}{p(y' \,|\, \overline{\theta})} \\ &= -2E_{y \,|\, \theta} \log \frac{p(y \,|\, \overline{\theta})}{p(y \,|\, \theta)} - 2E_{y \,|\, \theta} \log \frac{p(y \,|\, \theta)}{p(y' \,|\, \theta)} - 2E_{y \,|\, \theta} \log \frac{p(y' \,|\, \theta)}{p(y' \,|\, \overline{\theta})} \\ &= 2E_{y \,|\, \theta} \psi(\theta, y \,|\, \overline{\theta}, y') - 2E_{y \,|\, \theta} \log \frac{p(y \,|\, \theta)}{p(y' \,|\, \theta)}. \end{aligned} \tag{40}$$

Neglecting the second term of expectation zero (under $E_{y' \,|\, \theta}$), we obtain

$$-2E_{\theta \,|\, y'} E_{y \,|\, \theta} \log p(y \,|\, \overline{\theta}) \approx -2 \log p(y' \,|\, \overline{\theta}) + 2c(\psi, y'). \tag{41}$$

Thus, $c(\psi, y')$ serves as correction term for imputing y' in $\log p(y \,|\, \overline{\theta})$ which amounts to imputing θ in $\log p(\widetilde{y} \,|\, \overline{\theta})$ twice for $\widetilde{y} \in \{y, y'\}$. The imputation is corrected by $c(\psi, y') \approx E_{\theta \,|\, y'} J(p(y \,|\, \theta), p(y \,|\, \overline{\theta}))$, the posterior average symmetrized distance between the corresponding densities. Combining (39) and (41),

$$-2E_{\theta \,|\, y'} E_{y \,|\, \theta} \log p(y \,|\, \theta) \approx -2 \log p(y' \,|\, \overline{\theta}) + c(\psi, y'). \tag{42}$$

The argument shows that it is the odds ratio parameterization that provides the formal access to useful approximations of expected utilities. It may lead to special definitions adapted to a particular application like that of $c(\psi, y')$ or $p_D(y')$ in model choice or it may help to link and to compare different approximations.

3. ASSOCIATION IN MULTIVARIATE ANALYSIS

We consider random vectors X and Y and measure the strength of their relation by $J(X, Y)$. We now assume ψ^0 to be bi-affine, referring to the discussion in Section 1.5. From (16) and (21) we obtain

$$J(X,Y) = \text{tr}(A\Sigma_{HG}), \tag{43}$$

where $\Sigma_{HG} = cov_{xy}(h(y), g(x))$. In order to simplify notations we will denote mean vectors by μ and covariance matrices by Σ with appended subscripts in capital letters indicating the random variables. In particular, the random vectors resulting from (coding) transformations (cf. 1.5, Example 3) will be denoted by (G_0) G and (H_0) H, respectively. The matrix $A\Sigma_{HG}$ is a quadratic $(k_x \times k_x)$-matrix but not necessarily symmetric.

3.1. *Linear Discriminant Functions*

We aim at a decomposition of $J(X, Y)$ analogously to a principal component analysis (PCA), that is, we want to decompose the trace into components of decreasing importance and would like to relate these components to functions that capture the relation between X and Y in decreasing amounts. To this end we use a singular value decomposition (SVD) of $A\Sigma_{HG}$, which provides rank optimal approximations to $A\Sigma_{HG}$. The eigenvectors of $\Sigma_{GH}A^t A\Sigma_{HG}$ occurring in this SVD then are used as coefficients to form linear combinations of the random variables in G, which we call "linear discriminant functions." The procedure is slightly modified due to the requirement that (the variance of) such discriminant functions should be standardized. Formally therefore we use the following definition.

Definition. Let $\widetilde{\Sigma}$ be a positive definite $k_x \times k_x$-matrix and $\widetilde{\Sigma}^{-1/2}$ its (e.g. Gramian) square root. Let r_x denote the rank of $A\Sigma_{HG}$, $r_x \leq \min\{k_x, k_y\}$. Let the SVD of

$$\widetilde{C}_x := \widetilde{\Sigma}^{1/2} A\Sigma_{HG}\widetilde{\Sigma}^{-1/2} \tag{44}$$

be given by $\widetilde{C}_x = \widetilde{U}_x\Lambda_x\widetilde{V}_x^t$, where Λ_x is a $r_x \times r_x$ diagonal matrix with entries λ_j^x, $j = 1, ..., r_x$ which are the square roots of the eigenvalues of $\widetilde{C}_x^t\widetilde{C}_x$ in decreasing order. Finally denote the columns of $V_x := \widetilde{\Sigma}^{-1/2}\widetilde{V}_x$ by $v_{x,j}$, $j = 1, ..., r_x$. Then the j-th linear discriminant function of X w.r.t. $\widetilde{\Sigma}$ is defined by

$$L_j^x(x) := v_{x,j}^t g(x). \tag{45}$$

The term "linear discriminant function" is chosen based on the intuition that functions, say of X, which capture the relation between X and Y can be expected to have a potential for discriminating y's given their values. For example, if Y is a group indicator, then L_1^x is supposed to be useful in grouping observed units for which X has been recorded. Kullback (1959/1968) used the term in a similar spirit but under special distributional assumptions, and his work motivated our approach (cf. Kullback, 1959/1968, Ch. 9.4). Kullback's interest was in testing equality of (Gaussian) sampling distributions in different groups, and he suggested divergences as measures of variation to be used as test statistics. His linear discriminant functions can be shown to result from a special application of our more general definition.

3.2. *Standard Situations*

In standard situations where X and Y are (conditionally) Gaussian or multinomial, bi-affinity of the log-odds ratio function can be checked directly as exemplified in Section 1.5. The matrices A and Σ_{HG} in these cases are obtained explicitly and so are the linear discriminant functions. We summarize some results for these classical set-ups. Proofs and further details are given in (van der Linde, 2002a).

(a) *Joint Gaussian distribution of X and Y.* Under the assumption of a joint Normal distribution of X and Y,

$$J(X,Y) = E_y[(\mu_{X\,|\,y} - \mu_X)^t \Sigma_{X\,|\,y}^{-1} (\mu_{X\,|\,y} - \mu_X)]. \tag{46}$$

The j-th linear discriminant function w.r.t. Σ_X equals the j-th canonical variate in X. Thus, in this case the decomposition of $J(X,Y)$ is equivalent to that in canonical correlation analysis (CCA).

(b) *Y a grouping indicator, X Gaussian in each group.* Assume Y to take values k in $\{0, 1, ..., K\}$ and $X\,|\,k \sim N(\mu_{X\,|\,k}, \Sigma_k)$. If the group-specific distributions of X have equal covariance matrices, $\Sigma_k = \Sigma$, say, then

$$J(X,Y) = tr(\Sigma^{-1}B), \tag{47}$$

where $B = \sum_{k=0}^{K} p_Y(k)(\mu_{X\,|\,k} - \mu_X)(\mu_{X\,|\,k} - \mu_X)^t$. The linear discriminant functions in this case coincide with Fisher's discriminant functions. Thus, the classical linear discriminant analysis (LDA) proves to be a special case of our approach.

If the conditional covariance matrices are heterogeneous we have the set-up leading to classical quadratic discriminant functions. In fact, ψ^0 is bi-affine but w.r.t. a transformation $g(x)$ that involves not only components X_i but also products $X_i X_j$ of components of X. The bi-affine form is given explicitly, and the use and performance of the linear discriminant functions is illustrated in a numerical example in van der Linde, (2002a). The heterogeneous set-up is of particular interest in applications and used, for instance, in model-based clustering Yeung *et al.* (2001). A common recommendation to check a cluster analysis (*e.g.*, Seber, 1988, p. 390) is to use a biplot based on PCA. Linear discriminant functions instead are optimally targeted to grouping, and we suggest to use a biplot based on them for model checking.

(c) *X and Y taking finitely many values.* In Example 3 of Section 1.5, in particular in (15) the log-odds ratio function ψ^0 for simple random functions was shown to be bi-affine, and the matrix A for the non-degenerate transformation h_0 to have entries $\psi^0(m,k)$. The use of linear discriminant functions in this case is illustrated in a problem of seriation in van der Linde (2002a).

(d) *Use of linear discriminant functions.* As indicated in the discussion of standard situations we suggest to use linear discriminant functions for dimension reduction rather than allocation. Their derivation requires the matrix A which is a matrix of regression coefficients under the (linear logistic) regression model corresponding to the assumption of bi-affinity of ψ^0. This assumption is frequently made for general distributions of X when Y is a grouping variable, and allocation can then be based on the logistic regression model. Dimension reduction is useful e.g. to obtain graphical displays that help to solve problems of seriation, identification and interpretation of groups or in model checking.

3.3. *Duality*

Interchanging X and Y directly yields dual linear discriminant functions, and the special case of CCA illustrates this feature of duality. Qualitatively, however, X and Y can be rather different.

In discriminant analysis, for example, Y as a grouping variable typically is discrete and of low dimension whereas X as a vector of many feature variables is continuous or mixed and of high dimension. Hastie *et al.* (1994, 1995) discussed this problem in the context of LDA and suggested solutions in classical terms.

A switch of variables in our approach yields

$$\psi^0(x, y) = (h(y) - h(y_0))A^t(g(x) - g(x_0)), \tag{48}$$

and hence $J(X, Y) = \text{tr } A\Sigma_{HG} = \text{tr } A^t\Sigma_{GH}$, where now $A^t\Sigma_{GH}$ is a $k_y \times k_y$-matrix and typically $k_y << k_x$. A (standardized) SVD of $A^t\Sigma_{GH}$ yields dual linear discriminant functions,

$$L_j^y := w_j^t H \quad \text{(say).} \tag{49}$$

If H is a coding transformation, then L_j^y represents an "optimal scoring variable" assigning real values w_{jk} to realizations k of Y. The key idea for applications is to define and use approximate dual functions

$$L_j^{y\,|\,x} := w_j^t E(H \,|\, X = x) \tag{50}$$

which are functions of x again. In general, $E(H \,|\, X = x)$ is not linear in x but according to the related regression model $\log E(H \,|\, X = x)$ is linearly related to $g(x)$.

4. DISCUSSION

In this section, we want to (re-)emphasize some prominent features of the odds ratio parameterization that determine its applications and to point to some experiences using it that suggest further work.

We introduced the odds ratio parameterization as a *universal* formal language which can be useful in any investigation of association or dependence of two random elements. We illustrated its potential in two major fields, both analyses basically referring to the symmetrized mutual information as measure of dependence. Beyond these applications though, its impact has hardly been elaborated. For example, different measures of association and dependence may be of interest, the ideas of sufficiency and ancillarity or, for instance, of "copulas" should be related. In multivariate analysis there appears to be a dominance of "Normal thinking" as most analyses are based on covariance matrices. Alternative approaches in discrete multivariate analysis might be interpreted in terms of association. Therefore, our presentation is to be understood as an invitation and stimulation to work out the concept of association, in particular, the benefits of an odds ratio parameterization in other fields of research.

The association between two random elements X and Y is defined *symmetrically* and thus induces a *dual* theory. Hence, parameterizing with the log-odds ratio function ψ^0 is particularly compatible with a symmetric concept like $J(X, Y)$ but less so with asymmetric or directed analyses like those based on $I(X, Y)$, although these can be expressed in terms of the odds ratio parameterization. Similarly, the dual theory can be very helpful in a symmetric setting as demonstrated in discriminant analysis, or it may reveal different features of probability distributions as, for example, in sensitivity analysis as compared to hypothesis testing. Yet, it is certainly worth elaborating deliberately the dual theories to explore their potential whenever an odds ratio parameterization is used.

We emphasized issues of parameterization and the identification of quantities of interest like measures of dependence. We did not even mention how to eventually *estimate* these quantities, and we would like to make some points in this respect. Association is characterized probabilistically, it is inherent in both conditional distributions. Thus, the association between Θ and Y is best accessible in a Bayesian approach. The association between X and Y is estimable already using sampling schemes based on one conditional distribution only, a fact that has been much discussed in epidemiology (see the famous paper by Prentice and Pyke, 1978). Measures of dependence like $J(X, Y)$ may require knowledge of the joint distribution, however. The odds ratio parameterization can be helpful in sorting out problems of estimability in restricted sampling schemes. The relation is one-to-one for a third component in the triple (P_X, P_Y, ψ^0) if the two remaining components are fixed, but this relation is unfortunately not explicit in terms of the parameterization. Yet—as is well known in epidemiology and classical discriminant analysis—the key to estimation is the induced regression model.

REFERENCES

Efron, B. (1986). How biased is the apparent error rate of a prediction rule? *J. Am. Statist. Ass.* **81**, 461–470.

Haberman, S. J. (1974). *The Analysis of Frequency Data.* Chicago: The University of Chicago Press.

Hastie, T., Tibshirani, R. and Buja, A. (1994). Flexible discriminant analysis by optimal scoring. *J. Am. Statist. Ass.* **89**, 1255–1270.

Hastie, T., Buja, A. and Tibshirani, R. (1995). Penalized discriminant analysis. *Ann. Statist.* **23**, 73–102.

Kullback, S. (1959). *Information Theory and Statistics.* Reprinted New York: Dover, 1968.

van der Linde, A. (2002a). Dimension reduction and linear discriminant functions based on an odds ratio parameterization. *Tech. Rep.*, Univ. of Bremen, Germany, `www.math.uni-bremen.de/~avdl`.

van der Linde, A. (2002b). On the association between a random parameter and an observable. *Tech. Rep.*, Univ. of Bremen, Germany, `www.math.uni-bremen.de/~avdl`.

Osius, G. (2000). The association between two random elements: A complete characterization in terms of odds ratios. *Tech. Rep.*, Univ. of Bremen, Germany, `www.math.uni-bremen.de/~osius`.

Plackett, R. L. (1974). *The Analysis of Categorical Data.* London: Griffin.

Prentice, R. L. and Pyke, R. (1979). Logistic disease incidence models and case–control studies. *Biometrika* **66**, 403–412.

Seber, G. A. F. (1984). *Multivariate Observations.* New York: Wiley

Sinkhorn, R. (1967). Diagonal equivalence to matrices with prescribed row and column sums. *Am. Math. Mon.*, **74**, 402–405.

Spiegelhalter, D. J., Best, N. G., Carlin, B. P. and van der Linde, A. (2002). Bayesian measures of model complexity and fit. *J. R. Statist. Soc. B* **64**, 1–34 (with discussion).

Ye, J. (1998). On measuring and correcting the effects of data mining and model selection. *J. Am. Statist. Ass.* **93**, 120–131.

Yeung, K. Y., Fraley, C., Murua, A., Raftery, A. E., Ruzzo, W. L. (2001). Model-based clustering and data transformations for gene expression data. *Tech. Rep.*, University of Washington, USA.

DISCUSSION

ROBERT E. KASS (*Carnegie Mellon University, USA*)

I would like to preface my discussion by saying how pleased I am to be contributing to this volume in honor of Dennis Lindley. Professor Lindley played a crucial role in the development of Bayesian methods, serving as their chief champion for many years, and gave great encouragement to many aspiring young Bayesians, myself included.

Now, concerning the paper, I must admit I find it difficult. It combines a seemingly technical result, that (P_X, P_Y, ψ) characterizes P_{XY} under weak conditions, with a sweeping conceptual vision, of "[the] odds ratio parameterization as a universal formal language." On the one hand, it is easy to agree that the log-odds ratio is important, indeed, fundamental; but that by itself is hardly new. There are, here, a series of potentially interesting observations, including remarks about a formal duality between some aspects of frequentist and Bayesian inference (in certain cases), and about some aspects of multivariate analysis. But it is not easy to appreciate the importance of these observations in the absence of some interesting new consequences. Reinterpretation, by itself, is at best tantalizing, and does not necessarily constitute progress.

The authors move from the log-odds ratio to some discussion of its expectation, and the expectation of various log densities. This raises the general question, When is it Bayesianly interesting to consider expectations over the sample space?

I have come across this question, and puzzled over it a bit, in the context of the use of mutual information in the analysis of neuronal data. Figure 1 displays the firing times of a single neuron under two different experimental conditions (see Olson *et al.,* 2000; Ventura *et al.,* 2002). The differential firing rates under the two conditions was the subject of the experiment, and it may be seen that the firing rate is somewhat elevated in the "pattern" condition compared to that in the "spatial" condition toward the end of the given epoch. To be specific, the neuron appears to discriminate between the two conditions in the 200 ms time interval $(400, 600)$ but not in the interval $(0, 200)$. In widely cited work, Optican and Richmond (1987) suggested that mutual information provides a useful measure for quantifying such temporal contrasts. In the context of this application, let θ be a dichotomous indicator of experimental condition and Y be the data collected over a particular interval, which, for reasons that will become clear in a moment, I would like to denote by e. Optican and Richmond's suggestion is that we evaluate the amount of information (about the condition) provided by the neuron during the interval e using the mutual information

$$I(\theta, Y \mid e) = \text{Entropy}(\theta \mid \text{e}) - \text{Entropy}(\theta \mid \text{Y}, \text{e}). \tag{51}$$

Many neurophysiologists like this idea, and having thought about it, I do too. But this begs the question, In what sense is mutual information interesting from a Bayesian point of view?

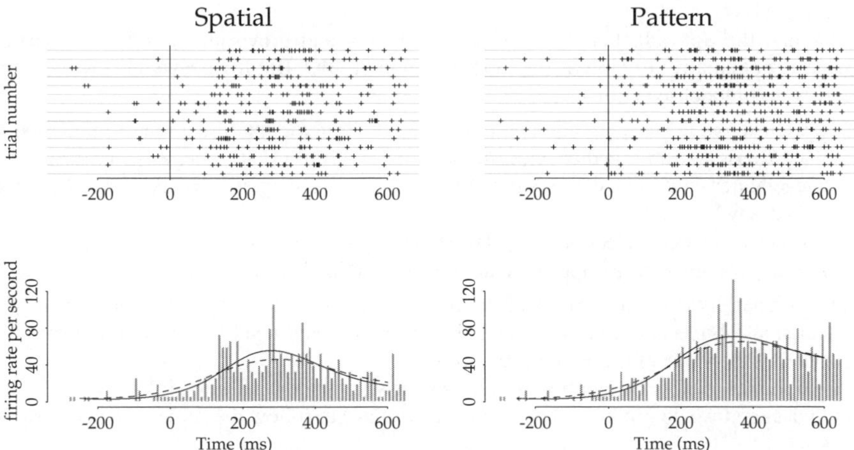

Figure 1. *Firing times of a neuron in repeated experimental trials under two experimental conditions. The data are shown in the top portions of the figure as hash marks along lines, each line representing a distinct trial (experimental replication). The bottom portions of the figure display results of pooling the data into 10 ms time bins, pooling across the trials. Smooth curves (using two alternative smoothing methods) are overlaid on the binned-data histograms.*

An answer was suggested by Lindley (1956) and Bernardo (1979), who showed that mutual information could be regarded as a Bayesian experimental design criterion. Specifically, for an experimental design e they proposed and studied the criterion of choosing the design to maximize $I(\theta, Y \mid e)$ given in (51). Thus, evaluating informativeness of data according to $I(\theta, Y \mid e)$ has a well-established Bayesian interpretation when we consider the alternatives e to amount, essentially, to alternative experimental designs. In the neurophysiological context the analogy with experimental design works well: when we choose alternative intervals of time, we are effectively choosing alternative data to examine. That is, time plays the role of the design variable.

Returning to the more general question, experimental design is also a leading example of a situation in which it is Bayesianly interesting to examine an expectation over the sample space. A second leading example involves the evaluation of Bayes risk. This brings me to my final substantive comment.

My impression was that a major motivation for this paper was the decision-theoretic interpretation of the DIC criterion, as discussed by Spiegelhalter *et al.* (2002). The essential result, as I understand it, is

$$\text{DIC} \approx \text{E}_{\Theta\,|\,y'}\, \text{E}_{Y\,|\,\theta}[-2\log p(y\,|\,\bar{\theta})] \tag{52}$$

where $\bar{\theta} = E(\theta\,|\,y)$ is the posterior mean, here used as a "plug-in" estimator. The approximation is not asymptotic, but rather uses some asymptotics while also invoking the *possibility* that a particular term (having zero expectation) is small. I am grateful to Professor van der Linde for emphasizing to me, in personal communication, that the right-hand side of (52) should be considered the risk (under the logarithmic loss) in using the sampling distribution $p(y\,|\,\bar{\theta})$ for the future data Y. In connecting this with the frequentist view, it is worth noting that the right-hand side of (52) has the form $E_{\Theta\,|\,y'}R(\Theta)$ where $R(\theta) = E_{Y\,|\,\theta}\left(-2\log p(y\,|\,\bar{\theta})\right)$ is the usual frequentist risk, so that the predictive criterion on the right-hand side of (52) is the posterior mean of the risk.

I found the Spiegelhalter *et al.* paper stimulating and provocative, and think that (52) provides a potentially very interesting interpretation of DIC. But if we take this predictive risk seriously as a model selection criterion, we should ask

- How close is DIC to the predictive risk in (52)? From the interpretation of predictive risk as a posterior mean together with the analysis of Efron (1986) one sees immediately (as Spiegelhalter *et al.* pointed out) that DIC will behave asymptotically like AIC. What more can one say?
- How close is model selection using DIC to model selection based on risk?
- How can we numerically approximate model selection based on risk?

In summary, while I find many of the results, including those on the formal "duality" of frequentist and Bayesian inference, intriguing, I feel I would need to see more in order to be convinced of the value of these interpretations.

I would like to close by mentioning my own view that model selection is hard when we have limited data and we do not know how (or are unwilling) to follow the subjectivist prescription, that is,

- We use default priors *that matter* (they matter much more than in estimation problems);
- And/or we use default utilities *that matter*.

The second item makes it particularly hard to do numerical comparisons: we are continually facing the tautology that a model selection method will perform well according to the criterion that defines it. Thus, we continue to be presented with competing model selection criteria, with each appearing sensible to its proponents, and we are unable to find any basis for reaching a consensus.

My guess is that this is an inherently insoluble problem. Is there any way forward? Perhaps case studies might help. There, one would have to present a specific scientific problem with well-justified statistical goals that require some kind of evaluation of alternatives models. In such a specific context it may be possible to argue convincingly that a particular method of model selection is more helpful than others in achieving those goals. An attempt of this sort was made in Viele *et al.* (2002), but I hope others will present different, and more informative case studies in the future.

REPLY TO THE DISCUSSION

We agree that the paper is difficult because it briefly indicates rather than spells out the impact of our results. We did elaborate on our findings in subsequent papers but take the opportunity to point again to some applications.

First of all we do not think that the characterization of P_{XY} by (P_X, P_Y, ψ^0) is technical. The log-odds ratio is indeed fundamental, and the result shows to what extent. In contrast to (previously) widespread beliefs it turns out that the odds ratio is *the* parameter of interest whenever the association between two random elements is to be studied in terms of (conditional) densities. Furthermore, we demonstrate under which modelling assumptions (*logistic* regression models) and sampling schemes (*conditional* and *joint* sampling) it is estimable. These results are valid for rather general distributions but have been known and used in restricted set-ups only, namely for (simple) random variables X and Y with finite range (i.e. contingency tables). For a random vector X and simple Y (finite range) log-odds ratios have been used in logistic regression to model the conditional distribution $P_{Y|X}$. But even in this special case our characterization of the joint distribution P_{XY} in terms of odds ratios has not been given so far. And in the general situation with arbitrary random vectors (or even arbitrary random elements, cf. Osius, 2000) X and Y the characterization of P_{XY} and the resulting odds ratio models like, for example, bi-affine models have not been given before.

In consequence, we see the benefit of the odds ratio parameterization in its potential for *distributionally* adequate generalizations and the reinterpretation of known results as guidance to such generalizations. For example, the definition of linear discriminant functions is validated by the identification of CCA and LDA as special cases, but it provides a general approach for arbitrary distributions with bi-affine log-odds ratio functions, which was exemplified for multinomial distributions. Hence, for example, new diagnostic (bi-)plots are suggested which take into account the distributional assumptions of the model. Similarly, identifying $J(X, Y)$ as a (transformed) coefficient of determination in Gaussian multiple regression guides its generalization to non-Gaussian regression and hence induces procedures of variable selection that are based on the contribution of each variable X_i to $J(X, Y)$.

Also, the discussion of model complexity offers clarity and mathematical foundation that was missing in the introduction of $p_D(y')$. In this way some issues of the related discussion can be settled. $p_D(y')$ turns out to be an *estimate* of a well interpretable quantity (posterior version of symmetrized mutual information) which can be derived independently and is invariant under one-to-one transformations of the parameter of interest θ. Related quantities like the one suggested by M. Plummer (in the discussion of Spiegelhalter et al., 2002) based on intuition can easily be qualified as being equal to $c(\psi, y')$ in exponential families.

Turning to the comparison of DIC with the predictive risk in (52) of the discussion we can only give a partial answer. Two "errors" can cause a difference: (i) the neglected term

$$-2E_{\theta|y'}E_{y|\theta}\log\frac{p(y|\theta)}{p(y'|\theta)}$$

(cf.(40)) and (ii) the estimation error $c(\psi, y') - p_D(y')$. It is certainly worth studying for which type of distribution these terms may effect the decision.

Although we agree that to a pragmatic statistician "progress" may not be obvious, we insist on and claim mathematical progress. Establishing an "economy of thought" derived from general mathematical structures has always been a genuine task for mathematicians. Foundational mathematical views allow for structurally well-justified results and procedures in future work. For example, establishing links between reference priors on model (hyper-)parameters and model priors as decreasing functions of model complexity requires a formalization in which parameters of interest and their approximations, respectively, estimates can be well separated. We do believe that the log-odds ratio parameterization provides such a tool for proofs.

We agree that this point of view aims at objective Bayesian procedures (targeting parsimony of a model in model choice, for example) rather than subjectivist specifications "that matter". While we prefer substantial prior specifications whenever possible, we also see a need for well-founded default priors (e.g. on model (hyper-)parameters).

Finally, we would like to (re-)emphasize that the duality of the frequentist and Bayesian approach results in coherent inference just about the association between Θ and Y respectively about ψ^0 obtainable from both conditional distributions, that is, from the likelihood as well as the posterior distribution. Discarding P_Y and P_Θ means not to refer to sampling expectations and not to invoke a(n informative) prior.

ADDITIONAL REFERENCES IN THE DISCUSSION

Bernardo, J. M. (1979). Expected information as expected utility. *Ann. Statist.* **7**, 686–690.

Lindley, D. V. (1956). On the measure of information provided by an experiment *Ann. Statist.* **27**, 986–1005.

Olson, C. R., Gettner, S. N., Ventura, V., Carta, R. and Kass, R. E. (2000). Neuronal activity in macaque supplementary eye field during planning of saccades in response to pattern and spatial cues. *J. Neurophysiol.* **84**, 1369–1384.

Optican, L. M. and Richmond, B. J. (1987). Temporal encoding of two-dimensional patterns by single units in primate inferior temporal cortex. III: Information-theoretic analysis. *J. Neurophysiol.* **57**, 162–178.

Ventura, V., Carta, R., Kass, R. E., Gettner, S. N. and Olson, C. R. (2002). Statistical analysis of temporal evolution in single-neuron firing rates. *Biostatistics* **3**, 1–20.

Viele, K., Kass, R. E., Tarr, M. J., Behrmann, M., and Gauthier, I. (2002). Recognition of faces versus Greebles: A case study in model selection. *Case Studies in Bayesian Statistics VI* (C. Gatsonis, R. Kass, A. Carrriquiri, A. Gelman, D. Higdon, D. Pauler and I. Verdinelli, eds). New York: Springer, 91–136 (with discussion).

BAYESIAN STATISTICS 7, pp. 249–275
J. M. Bernardo, M. J. Bayarri, J. O. Berger, A. P. Dawid,
D. Heckerman, A. F. M. Smith and M. West (Eds.)

Bayesian Clustering with Variable and Transformation Selections

JUN S. LIU
Harvard University, USA
jliu@stat.harvard.edu

JUNNI L. ZHANG
*Peking University, China**
zjn@gsm.pku.edu.cn

MICHAEL J. PALUMBO   CHARLES E. LAWRENCE
The Wadsworth Center, USA
palumbo@wadsworth.org   lawrence@wadsworth.org

SUMMARY

The clustering problem has attracted much attention from both statisticians and computer scientists in the past 50 years. Methods such as hierarchical clustering and the K-means method are convenient and competitive first choices off the shelf for the scientist. Gaussian mixture modelling is another popular but computationally expensive clustering strategy, especially when the data are high-dimensional. We propose to first conduct a principal component analysis (PCA) or correspondence analysis (CA) for dimension reduction, and then fit Gaussian mixtures to the data projected to the several major PCA or CA directions. Two technical difficulties of this approach are: (a) the selection of a subset of the PCA factors that are informative for clustering, and (b) the selection of a proper transformation for each factor. We propose a Bayesian formulation and Markov chain Monte Carlo strategies that overcome the two difficulties and examine the performances of the new method by both simulation studies and real applications in molecular imaging analysis and DNA microarray analysis.

Keywords: GAUSSIAN MIXTURES; GIBBS SAMPLER, MICROARRAY; SIMULATED TEMPERING.

1. INTRODUCTION

Clustering objects into homogeneous groups is an important step in many scientific investigations. Recently, good clustering techniques are of special interest to biologists because of the availability of large amounts of high-dimensional data resulting from the biotechnology revolution. These data include, for example, measurements of mRNA levels in the cell by microarray experiments, single-particle electron micrographs of macromolecules, high-throughput biological sequences of many species, protein–protein interaction data, etc. Although techniques for clustering high-dimensional observations have been subjected to active research for many years, traditional methods such as hierarchical clustering and K-means clustering are still top choices for scientists despite their various limitations in the analysis of complex data.

Due to the recent advances in Markov chain Monte Carlo (MCMC; see Liu, 2001, for a recent overview), the Bayesian clustering approach via mixture models has been shown to be

* The work of J. L. Zhang was done while she was a Ph.D. student at Harvard University, USA.

attractive in many applications (Celeux and Govaert, 1995; Richardson and Green, 1997; Fraley and Raftery, 1999; Moss *et al.*, 1999; Ishwaran *et al.*, 2001; Yeung *et al.*, 2001; Ghosh and Chinnaiyan, 2002; Kim *et al.*, 2002; McLachlan *et al.*, 2002; Sansó *et al.*, 2002). However, the use of Bayesian clustering methods in high-dimensional data has been hindered by the very high computational cost and instability of the generic Gaussian mixture models. To overcome this difficulty, Banfield and Raftery (1993) proposed a general framework for directly modelling/constraining the covariance matrices of the mixture components. Here we recommend to first conduct a principal component analysis (PCA) or correspondence analysis (CA) for data reduction and then fit a mixture model to the factors resulting from these analyses. Indeed, if the original data come from a mixture Gaussian distribution and the estimated PCA or CA directions can be treated as known, then the new data vectors resulting from the projection of the original data onto these directions still follow a mixture Gaussian distribution. A similar approach has been applied to a character recognition problem (Kim *et al.*, 2002).

When PCA is used in clustering problems, it can select classification-related directions if these directions are associated with the differences in the locations of the means of different clusters. It can also pick up some artificial directions resulting from certain unusually noisy components or highly correlated components. Consequently, a potential problem with the PCA–Gaussian-mixture approach is the determination of an appropriate set of the PCA or CA factors useful for clustering. A common practice is to choose the factors corresponding to the few large principal components. But it is not clear where to stop and whether some of these eigen-directions are caused by some artifact or noises in the data unrelated to the clustering task.

In this article, we propose a novel procedure called Bayesian clustering with variable selection (BCVS), which can simultaneously cluster the objects and select "informative" variables, or factors, for the clustering analysis. Since many real data do not fit the multivariate Gaussian or mixture Gaussian models well, it is a common practice to first transform certain variables (using logarithm or some power functions) and then do the model fitting. These transformation steps are often carried out by the investigator based on a certain exploratory pre-processing of the data. We note that if the transformations are indexed as in Box and Cox (1964), each factor can be associated with a transformation variable and a full Bayesian model can be set up to include all the variables. Consequently, the BCVS procedure can be automated to select both informative factors and proper transformations for these factors. The advantage of this type of full Bayesian models is its ability to treat all involved variables in a coherent framework, to combine different sources of information, and to reveal subtle patterns by properly averaging out noise.

Section 2 presents two examples that motivated our development of the method: image clustering and microarray analysis. Section 3 first describes the full Bayesian Gaussian mixture model with variable selection. After the illustration in Section 3.1 of a standard Gibbs sampler for the model, Section 3.2 prescribes a more efficient predictive updating strategy for simultaneous clustering and variable selection. Section 3.3 details a tempering strategy for improving the convergence of the BCVS sampler. Section 4 formulates the transformation selection problem based on the framework of Box and Cox (1964). Section 5 tests the BCVS method on a series of simulated data sets, a micrograph image clustering problem and two microarray studies for cancer patient clustering. Section 6 concludes with a brief discussion.

2. MOTIVATING EXAMPLES

2.1. *Image Analysis for Electron Micrographs*

In single-particle electron micrograph imaging, each macromolecule lies randomly on the specimen support in a limited number of ways (*i.e.*, onto a few different "faces" of the molecule).

A large number of images of the identical molecule are observed, and these images should fall into a few classes corresponding to the different characteristic orientations (Sansó *et al.*, 2002). Because particles within a class are still randomly rotated in the plane, they must also be aligned. We focus here only on the classification of previously aligned images. The goals are to identify classes and to infer average images using differences in the appearance of particles. Figure 3a shows images of *E. Coli* ribosome with four different high-tilt angles, and Figure 3b shows images with four low-tilt angles. It is seen that after adding noises, the four classes of images are difficult to distinguish. Our goal here is to use a model-based method to automatically cluster these images into different classes.

In most single-particle classification techniques, the images are first subject to CA or PCA (Frank and van Heel, 1982). The factors produced by CA or PCA are prioritized according to the eigenvalue weights that account for the variance contribution of each factor. Not all factors carry meaningful or signal-related information, since noise or artifacts that are unrelated to the shape of the macromolecule can also contribute to the variance associated with a factor. Although there are some previous researches that address this issue (Frank, 1996), these time-consuming methods necessitate an extensive knowledge of the system, are somewhat subjective, and tend to break down when the signal to noise ratio (SNR) is low. It is thus desirable to develop a method that can automatically select the factors to be used in clustering and improve the clustering of the images with low SNR.

2.2. *Patients Clustering Based on Gene Expression Microarrays*

The recent developments in gene chip or microarray technologies allow the scientist to observe simultaneously the expression levels of many genes in a cell at a given time, condition, or developmental stage (Schena *et al.*, 1995). Because genes with similar or related functions often behave similarly under various conditions, biologists can discover novel gene–gene relationships and transcriptional regulatory signals by analyzing gene clustes based on the similarities of their expression patterns (Roth *et al.*, 1998). It has also been reported that the gene expression profiles can be used to discriminate different cell types and predict patients' responses to certain drug treatment. It is also possible and important to cluster different cell types (or patients) based on their global gene expressions since the resulting gene clusters often correspond to clinically important subgroups. In this latter task, each cell type or patient is associated with the measurements of mRNA levels for thousands to tens of thousands of genes, a very high-dimensional vector, whereas the total number of patients is only in the range of hundreds or fewer. Alizadeh *et al.* (2000) used hierarchical clustering to divide 96 lymphoma patients into two homogeneous groups based on 4026 expression values of each individual. These two groups correspond to two subgroups of patients who respond differently to current therapy. Golub *et al.* (1999) used the self-organizing map (SOM) to cluster 38 leukemia samples into two groups based on the microarray values of 6817 genes for each individual. These two clusters again coincide well with the two important subtypes of the leukemia, ALL and AML. We show in the application section that BCVS can be successfully applied to these data to produce as good or better clustering results.

3. VARIABLE SELECTION IN MIXTURE MODELING

It is noticed that if we project a random Gaussian vector to a particular direction $\boldsymbol{v}$, then the resulting random variable also has a Gaussian distribution. Consequently, if we project observations from a mixture Gaussian distribution, the projected vectors should also follow mixture Gaussian. Although the PCA or CA directions need to be estimated from the data in all applications, it does not seem to make any material difference by treating these directions as given, as long as the number of factors extracted from the data is substantially smaller than the original dimension of the data. Thus, in practice we obtain the first k_0 principal vectors ordered as $V = (\boldsymbol{v}_1^t, \ldots, \boldsymbol{v}_{k_0}^t)$, and project the data onto these directions so as to form n vectors of k_0 dimensions. The data vectors that will be subject to BCVS analysis are $\boldsymbol{x}_i = (x_{i1}, \ldots, x_{ik_0})$, $i = 1, \ldots, n$, where each $\boldsymbol{x}_i$ is generated by multiplying the ith original data vector by projection matrix V. Each of the $(x_{1j}, x_{2j}, \ldots, x_{nj})^t$ will be called a factor throughout the paper.

It is reasonable to assume that each observation follows a mixture Gaussian distribution, of which each Gaussian component has the mean vector $\boldsymbol{\mu}_j$ and the covariance matrix Σ_j for $j = 1, \ldots, J$. The fractions of each component are $(p_1, \ldots, p_J)$. Notation-wise, we can write

$$\boldsymbol{x}_i \sim p_1 \mathrm{N}(\boldsymbol{\mu}_1, \Sigma_1) + \cdots + p_J \mathrm{N}(\boldsymbol{\mu}_J, \Sigma_J), \quad i = 1, \ldots, n.$$

Without having any constraints, each $\boldsymbol{\mu}_j$ is a k_0-dimensional vector and Σ_j a $k_0 \times k_0$ positive-definite matrix. A membership labeling variable J_i for each observation can be introduced so that

$$\boldsymbol{x}_i \mid J_i = j \sim \mathrm{N}(\boldsymbol{\mu}_j, \Sigma_j).$$

For the most part of this article, we assume that J is known in advance. There is a whole body of literature discussing how to choose a proper J in practice, and we will defer this issue to the discussion section.

We further assume that only a subset of the k_0 factors are informative for clustering. In the *anchor mode* model, we assume that this subset consists of the first K factors, where K is a random variable with a prior distribution $K \sim f(k)$. Thus, the data $\boldsymbol{x}_i$ has its first k components to follow a mixture Gaussian and its remaining components to follow a simple Gaussian distribution. Thus,

$$\boldsymbol{x}_i \mid J_i = j,\ \boldsymbol{\mu},\ K = k,\ \Sigma \sim N(\boldsymbol{\mu}_j, \Sigma_j) \times N(\boldsymbol{\mu}_0, \Sigma_0),$$

where $\boldsymbol{\mu}_j$ is a vector of length k and Σ_j is a $k \times k$ matrix, both are specific to cluster j; $\boldsymbol{\mu}_0$ is a vector of length $k_0 - k$ and Σ_0 is a $(k_0 - k) \times (k_0 - k)$ covariance matrix, common to all the observations.

Let $\boldsymbol{X} = (\boldsymbol{x}_1^t, \ldots, \boldsymbol{x}_n^t)^t$ and let $\boldsymbol{J} = (J_1, \ldots, J_n)$. Then

$$P(\boldsymbol{X} \mid \boldsymbol{J} = \boldsymbol{j}, \boldsymbol{\mu}, K = k) = \prod_{i=1}^{n} \mathrm{N}(\boldsymbol{x}_{i[1:k]} \mid \boldsymbol{\mu}_{j_i}, \Sigma_{j_i}) \prod_{i=1}^{n} \mathrm{N}(\boldsymbol{x}_{i[k+1:k_0]} \mid \boldsymbol{\mu}_0, \Sigma_0),$$

where $\mathrm{N}(\boldsymbol{x} \mid \boldsymbol{\mu}, \Sigma)$ is the multivariate Gaussian density. The joint posterior distribution is then

$$P(\boldsymbol{J}, K = k, \boldsymbol{\mu}, \Sigma \mid \boldsymbol{X}) \propto f(k) \pi(\boldsymbol{\mu}, \Sigma) \prod_{i=1}^{n} \{p_{j_i} \mathrm{N}(\boldsymbol{x}_{i[1:k]} \mid \boldsymbol{\mu}_{j_i}, \Sigma_{j_i}) \mathrm{N}(\boldsymbol{x}_{i[k+1:k_0]} \mid \boldsymbol{\mu}_0, \Sigma_0)\}.$$

Note that the dimensionality of $\boldsymbol{\mu}_j$, Σ_j, $\boldsymbol{\mu}_0$, and Σ_0 will change when k changes.

We first give some detailed calculation for the derivation of a MCMC algorithm for the anchor mode BCVS. The procedure can be easily generalized to the non-anchor mode, in which BCVS can select any combination of any number of the k_0 factors for the clustering.

3.1. *A Gibbs Sampling Algorithm*

Markov chain Monte Carlo algorithms for the estimation in mixture models have been an active topic in statistical research. Some of the recent articles include Diebolt and Robert (1994),

Richardson and Green (1997), Neal (2000), Brooks (2001), Ishwaran *et al.* (2001), just to start a list. The new feature in our algorithm is its variable/factor selection step.

Assume *a priori* that $[\boldsymbol{\mu}_j \mid \Sigma_j] \sim N(\boldsymbol{x}_0, \Sigma_j/\rho_0)$ and $\Sigma_j \sim \text{Inv-W}_{\nu_0}(S_0^{-1})$ (see the appendix for its density form). Let $\mathrm{N}(\cdot)$ denote the Gaussian density function. Then

$$P(J_i = j \mid \boldsymbol{\mu}_1, \ldots, \boldsymbol{\mu}_J, \Sigma; \boldsymbol{X}) = \frac{p_j \mathrm{N}(\boldsymbol{x}_{i[1:k]} \mid \boldsymbol{\mu}_j, \Sigma_j)\mathrm{N}(\boldsymbol{x}_{i[k+1:k_0]} \mid \boldsymbol{\mu}_0, \Sigma_0)}{\sum_{l=1}^{J} p_l \mathrm{N}(\boldsymbol{x}_{i[1:k]} \mid \boldsymbol{\mu}_l, \Sigma_l)\mathrm{N}(\boldsymbol{x}_{i[k+1:k_0]} \mid \boldsymbol{\mu}_0, \Sigma_0)}$$
$$= \frac{p_j \mathrm{N}(\boldsymbol{x}_{i[1:k]} \mid \boldsymbol{\mu}_j, \Sigma_j)}{\sum_{l=1}^{J} p_l \mathrm{N}(\boldsymbol{x}_{i[1:k]} \mid \boldsymbol{\mu}_l, \Sigma_l)}$$

and

$$[\boldsymbol{\mu}_j \mid \Sigma, \boldsymbol{J}, \boldsymbol{X}] \sim N\left(\frac{\rho_0 \boldsymbol{x}_{0[1:k]} + \sum_{i=1}^{n} \boldsymbol{x}_{i[1:k]} \times I_{\{J_i=j\}}}{\rho_0 + \sum_{i=1}^{n} I_{\{J_i=j\}}}, \frac{\Sigma_j}{\rho_0 + \sum_{i=1}^{n} I_{\{J_i=j\}}}\right),$$

$$[\boldsymbol{\mu}_0 \mid \Sigma, \boldsymbol{X}] \sim N\left(\frac{\rho_0 \boldsymbol{x}_{0[1:k]} + \sum_{i=1}^{n} \boldsymbol{x}_{i[k+1:k_0]}}{\rho_0 + n}, \frac{\Sigma_0}{\rho_0 + n}\right).$$

The conditional distribution for Σ_j is then

$$[\Sigma_j \mid \boldsymbol{X}, \boldsymbol{\mu}_j, \boldsymbol{J}, K = k] \propto |\Sigma_j|^{-(\nu_0+k+n_j+1)/2} e^{-\frac{1}{2}\operatorname{tr}[\Sigma_j^{-1}(S_0 + SS_j(\boldsymbol{\mu}_j))]},$$

where n_j is the number of observations in the jth cluster and $SS_j(\boldsymbol{\mu}_j) = \sum_{i:j_i=j}(\boldsymbol{x}_i - \boldsymbol{\mu}_j)^t(\boldsymbol{x}_i - \boldsymbol{\mu}_j)$ is the mean-corrected sum of squares for the observations in the jth cluster.

A prior $\mathrm{Di}(a_1, \ldots, a_J)$ distribution can be employed for the cluster proportions $(p_1, \ldots, p_J)$. Then, given the clustering indicators $\boldsymbol{J}$, it is straightforward to update the proportion vector by drawing from $\mathrm{Di}(n_1 + a_1, \ldots, n_J + a_J)$, where n_j is the size of the jth cluster.

Updating K in our model setting is not as trivial as the previous steps because once the $\boldsymbol{\mu}_j$ and Σ_j are fixed, the factor number K, which underlies the dimensionality of the mean vectors and covariance matrices, cannot be moved any more. It is possible to propose a change for all the $\boldsymbol{\mu}_j$, the Σ_j and K jointly and use the reversible jumping rule to guide for the acceptance/rejection decision. However, this type of proposals often encounter a high rejection rate, rendering the algorithm inefficient. Here we adopt a more effective alternative: marginalizing the $\boldsymbol{\mu}_j$ and Σ_j analytically. More precisely, using the standard Bayesian Gaussian inference results summarized in the Appendix, we can derive that

$$P(\boldsymbol{X}_{[1:k]} \mid K = k, \boldsymbol{J}) = \prod_{j=1}^{J} \frac{Z(\nu_0, S_0, k)}{Z(n_j + \nu_0, S_0 + SS_j, k)} (2\pi)^{-n_j k/2} \left(\frac{n_j + \rho_0}{\rho_0}\right)^{-k/2}, \qquad (1)$$

where SS_j is defined as in (11) for observations in the jth cluster. This calculation indicates that a more efficient approach, iterative predictive updating (Liu, 1994, Chen and Liu, 1996), is possible for our MCMC computation.

3.2. *Predictive Updating*

Instead of doing the full Gibbs, a convenient alternative is to iteratively draw the label of each $\boldsymbol{x}_i$ from its predictive distribution conditional on the labels of the remaining observations and then update the component number K conditional on the labels. This strategy not only saves our effort in updating the mean vectors and covariance matrices, but also improves the convergence rate of the sampler (Liu, 1994). More precisely, conditional on K and $\boldsymbol{J}$ and with conjugate priors, we can compute the analytical form of

$$P(\boldsymbol{X} \mid K = k, \boldsymbol{J}) = P(\boldsymbol{X}_{[1:k]} \mid K = k, \boldsymbol{J}) \times P(\boldsymbol{X}_{[k+1:k_0]}), \tag{2}$$

which leads to an iterative sampling of $J_i = j$ with probability proportional to the prior fraction p_j times a multivariate-t density function, and an update of K from $[K = k \mid \boldsymbol{J}, \boldsymbol{X}]$. We can also marginalize p_j if a prior $\mathrm{Di}(a_1, \ldots, a_J)$ has been used for the proportions, in which case we use in the place of p_j,

$$\hat{p}_j = (n'_j + a_j)/(n + a_1 + \cdots + a_J - 1),$$

where n'_j is the total number of objects in the jth cluster excluding the ith observation.

In order to compute the Metropolis ratio for changing the variable component number from k to $k+1$, we compute the related Bayes factors (normalizing constants) as in the standard Bayesian Gaussian inference (see Appendix). Let the normalizing function $Z(\nu, S, k)$ be defined as in (10). With known clustering information, we can compute (2) in two steps:

- The easier one (the Bayes factor $P(\boldsymbol{X}_{[k+1:k_0]} \mid \boldsymbol{J})$) is

$$P(\boldsymbol{X}_{[k+1:k_0]}) = \frac{Z(\nu_0, S_0, k_0 - k)}{Z(n + \nu_0, S_0 + SS_0, k_0 - k)} (2\pi)^{-n(k_0-k)/2} \left(\frac{n + \rho_0}{\rho_0}\right)^{-(k_0-k)/2},$$

where SS_0 is the sum of square matrix (as defined in (11)) computed from $\boldsymbol{X}_{[k+1:k_0]}$, from the prior mean $\boldsymbol{x}_0[k+1:k_0]$ of $\boldsymbol{\mu}_{[k+1:k_0]}$. If we model the columns of $\boldsymbol{X}$ from $k+1$ to k_0 as independent, then the above marginal likelihood can be modified as

$$P(\boldsymbol{X}_{[k+1:k_0]}) = \prod_{j=k+1}^{k_0} \frac{Z(\nu_0, s_0^2, 1)}{Z(n + \nu_0, s_0^2 + s^2(\boldsymbol{x}_j), 1)} (2\pi)^{-n/2} \left(\frac{n + \rho_0}{\rho_0}\right)^{-1/2},$$

where $s^2(\boldsymbol{x}_j)$ is the modified total sum of squares of the jth column. The prior for the unknown mean and variance of each independent component takes the same form as in the multi-dimensional case (see Appendix), but with s_0^2 replacing S_0.

- Compute the modified sum of square matrix SS_j (by formula (11) for observations in the jth cluster, with data $\{\boldsymbol{x}_{i[1:k]}, i \in \text{Cluster } j\}$. Here n_j is the cluster size. Then

$$P(\boldsymbol{X}_{[1:k]} \mid K = k, \boldsymbol{J}) = \prod_{j=1}^{J} \frac{Z(\nu_0, S_0, k)}{Z(n_j + \nu_0, S_0 + SS_j, k)} (2\pi)^{-n_j k/2} \left(\frac{n_j + \rho_0}{\rho_0}\right)^{-k/2}. \tag{3}$$

The Metropolis ratio for $k \to k'$ is then

$$r = \min\left\{1, \frac{f(k')P(\boldsymbol{X} \mid K = k', \boldsymbol{J})T(k' \to k)}{f(k)P(\boldsymbol{X} \mid K = k, \boldsymbol{J})T(k \to k')}\right\}. \tag{4}$$

If k_0 is small, we can draw K from its conditional distribution $P(K = k \mid \boldsymbol{J}, \boldsymbol{X})$ directly.

To summarize, we have the following predictive updating iterations to replace the regular Gibbs sampler:

1. Sample the group indicator variable of the ith observation from the t-distribution:

$$[J_i = j \mid \boldsymbol{J}_{[-i]}, K = k, \boldsymbol{X}] \propto \hat{p}_j \, f_t \left(\boldsymbol{x}_{i[1:k]};\, \nu_j, \hat{\boldsymbol{\mu}}_j, \frac{(n'_j + \rho_0 + 1)(S_0 + SS_k)}{(n'_j + \rho_0)(n'_j + \nu_0 - k + 1)} \right)$$

   where $\nu_j = n'_j + \nu_0 - k + 1$, n'_j is the current size of cluster j excluding the ith observation, and $\hat{\boldsymbol{\mu}}_j$ is a weighted combination of prior mean $\boldsymbol{x}_0$ and the sample mean as in (12). The weight for the prior is $\rho_0/(n'_j + \rho_0)$ and for the sample mean is $n'_j/(n'_j + \rho_0)$.
2. Propose a move for k to k' according to a transition rule $T(k \to k')$; accept the move with probability (4).

This algorithm can be easily modified to accommodate the non-anchor mode. Without the restriction of having to include the first k factors, we can choose a factor at random and ask whether we should treat it as an informative factor or not, *i.e.*, turn it on or not. Conditional on each factor's on-off states, $\boldsymbol{O}$, we can compute $P(\boldsymbol{X} \mid \boldsymbol{O}, \boldsymbol{Y})$ the same way as in (3). The Metropolis-ratio similar to (4) can be used to guide the transition from $\boldsymbol{O}$ to a new vector of the on-off states, $\boldsymbol{O}'$.

3.3. *Parallel Tempering*

It has been observed that the MCMC samplers for fitting a mixture model tend to be very "sticky." With the inclusion of the variable selection and transformation indicators, the MCMC algorithm tends to perform even worse. Parallel tempering (Geyer, 1991) seems to be an effective means to improve the mixing of a MCMC sampler.

To implement a tempering sampler conditional on $K = k$, we define the target distribution as

$$\pi(\boldsymbol{J}) \propto P(\boldsymbol{Y} \mid K = k, \boldsymbol{J}) P(\boldsymbol{J}),$$

where $P(\boldsymbol{J})$ is the prior distribution for $\boldsymbol{J}$. Then $\pi(\boldsymbol{J})$ can be evaluated as described in Section 3.2, up to a normalizing constant. In order to carry out the tempering idea, we construct a temperature ladder, $1 = t_1 < t_2 < \cdots < t_L$, and define $\pi_l(\boldsymbol{J}) \propto \{\pi(\boldsymbol{J})\}^{1/t_l}$.

The sample space of the tempering sampler is the product space of the $\boldsymbol{J}$. In other words, the new target distribution is

$$\Pi(\boldsymbol{J}_1, \ldots, \boldsymbol{J}_L) = \pi_1(\boldsymbol{J}_1) \times \cdots \times \pi_L(\boldsymbol{J}_L),$$

for which our tempering sampler will converge to. Here we let $\boldsymbol{J}_l = (J_{l,1}, \ldots, J_{l,n})$. The tempering process can be implemented as follows:

Tempering Sampler

1. Iterative classification (independently) at each level. For levels $l = 1, 2, \ldots, L$:
   - For $i = 1, \ldots, n$, compute

$$[J_{l,i} = j \mid \boldsymbol{J}_{l,[-i]}] \propto \left\{ p_j \, f_t \left(\boldsymbol{y}_{i[1:k]};\, \nu_j, \hat{\boldsymbol{\mu}}^*_j, \frac{(n_j + \rho_0 + 1)(S_0 + SS_k)}{(n_j + \rho_0)(n_j + \nu_0 - k + 1)} \right) \right\}^{1/t_k}$$

     for all possible j (in our example, $j = 1, 2, 3, 4$).
   - Update $J_{l,i}$ by a random draw from the above distribution.
2. For every N_0 (say, 10) cycles of iterative updating, we conduct one cycle of level exchange, starting from the highest-temperature configuration. For $k = 1, \cdots, L - 1$:

- compute the ratio

$$r = \frac{\pi_{L-k}(\boldsymbol{J}_{L-k+1})\pi_{L-k+1}(\boldsymbol{J}_{L-k})}{\pi_{L-k}(\boldsymbol{J}_{L-k})\pi_{L-k+1}(\boldsymbol{J}_{L-k+1})} = \left\{\frac{\pi(\boldsymbol{J}_{L-k+1})}{\pi(\boldsymbol{J}_{L-k})}\right\}^{t_{L-k}^{-1}-t_{L-k+1}^{-1}}$$

- exchange $\boldsymbol{J}_{L-k}$ and $\boldsymbol{J}_{L-k+1}$ with probability $\min\{1, r\}$.

3. Go back to step 1.

4. SELECTING A PROPER TRANSFORMATION

For ease of presentation, we give only the details for deciding whether logarithm transformations should be applied to certain factors. The formulas for the more general selection from a continuum of transformations using the Cox–Box formulation is presented with details omitted. We first consider the univariate case, and then generalize to consider several variables.

Suppose we have a set of univariate observations $x_1, \ldots, x_n$. Suppose the prior distribution for μ and Σ is as given in (8) and (9), we have

$$P(x_1, \ldots, x_n \mid \nu_0, s_0, x_0, \rho_0) = \frac{Z(\nu_0, s_0, 1)}{Z(n+\nu_0, SS+s_0, 1)}(2\pi)^{-n/2}\left(\frac{n+\rho_0}{\rho_0}\right)^{-1/2},$$

where $\sigma^2 \sim \text{Inv-W}_{\nu_0}(s_0^{-1})$, $\mu \mid \sigma^2 \sim N(x_0, \sigma^2/\rho_0)$, and SS is as defined in (11). If the data are to be log-transformed, then the likelihood is

$$P_l(x_1, \ldots, x_n \mid \mu, \sigma^2) = \left(\frac{1}{\sqrt{2\pi}\sigma}\right)^n \exp\left\{-\sum_{i=1}^n \frac{(\log x_i - \mu)^2}{2\sigma^2}\right\}\prod_{i=1}^n x_i^{-1}.$$

Thus, under the log-transformation model and a set of different prior parameters (indicated by "$*$"), we have

$$P_l(x_1, \ldots, x_n \mid \nu_0^*, s_0^*, x_0^*, \rho_0^*) = \frac{Z(\nu_0^*, s_0^*, 1)}{Z(n+\nu_0^*, SS^*+s_0^*, 1)}(2\pi)^{-n/2}\left(\frac{n+\rho_0^*}{\rho_0^*}\right)^{-1/2}\prod_{i=1}^n x_i^{-1},$$

where SS^* is the corresponding residual sum of squares for the $\log(x_i)$.

More generally, suppose we have n iid observations, $\boldsymbol{x}_i = (x_{i1}, \ldots, x_{ik})$, $i = 1, \ldots, n$, from a k-dimensional distribution. Let $\boldsymbol{X} = (\boldsymbol{x}_1^t, \ldots, \boldsymbol{x}_n^t)^t$. We notice what when considering the logarithm transformation for one of the variables, x_l, say, the computation of the data likelihood is almost unchanged. More precisely, we have

$$P_l(\boldsymbol{X} \mid \nu_0^*, S_0^*, \boldsymbol{x}_0^*, \rho_0^*) = \frac{Z(\nu_0^*, S_0^*, k)}{Z(n+\nu_0^*, SS^*+S_0^*, k)}(2\pi)^{-nk/2}\left(\frac{n+\rho_0^*}{\rho_0^*}\right)^{-k/2}\prod_{i=1}^n x_{il}^{-1},$$

where SS is the sum of square matrix of the original data as defined in (11), and SS^* is the sum of square matrix of the transformed data.

It is not very sensible to use the same prior for the original and the transformed data, but it is also difficult to decide what corresponding priors should be applied to the transformed data. Thus, noninformative priors seem to be desirable, although they will typically result in improper (infinite) Bayes factors. We note, however, that if we let $\rho_0 = \rho_0^*$, $\nu_0 = \nu_0^*$, $S_0 = S_0^*$, and let all of them converge to zero, we still have a proper ratio:

$$\frac{P(\boldsymbol{X})}{P_l(\boldsymbol{X})} = \frac{Z(n, SS^*, k)}{Z(n, SS, k)}\prod_{i=1}^n x_{il} = \left(\frac{|SS^*|}{|SS|}\right)^{n/2}\prod_{i=1}^n x_{il}. \tag{5}$$

For a transformation of the Box–Cox form $f_\alpha(x) = (x^\alpha - 1)/\alpha$, $\alpha > 0$, the ratio is

$$\frac{P(\boldsymbol{X})}{P_\alpha(\boldsymbol{X})} = \left(\frac{|SS^*|}{|SS|}\right)^{n/2} \prod_{i=1}^{n} x_{il}^{1-\alpha}. \tag{6}$$

Note that $\alpha = 0$ corresponds to the logarithm transformation. To insert the variable selection step into the predictive updates, we sample conditional on $\boldsymbol{J}$ for each variable whether a logarithm transformation should be applied, according to the ratio (5).

When the observations involve negative values (or are very large), it is sometimes useful to consider the transformation of the form $f_{\alpha,m} = [(x+m)^\alpha - 1]/\alpha$, where m is some constant to be added to the observation and can be estimated as well. We see from the foregoing derivations that the Bayes ratio should take a similar form:

$$\frac{P(\boldsymbol{x})}{P_{\alpha,m}(\boldsymbol{x})} = \left(\frac{|SS^*|}{|SS|}\right)^{n/2} \prod_{i=1}^{n} (x_{il} + m)^{1-\alpha}. \tag{7}$$

In practice, we may start m with $1 - \min\{x_{1i}, i = 1, \ldots n\}$ and update m along with the MCMC iterations.

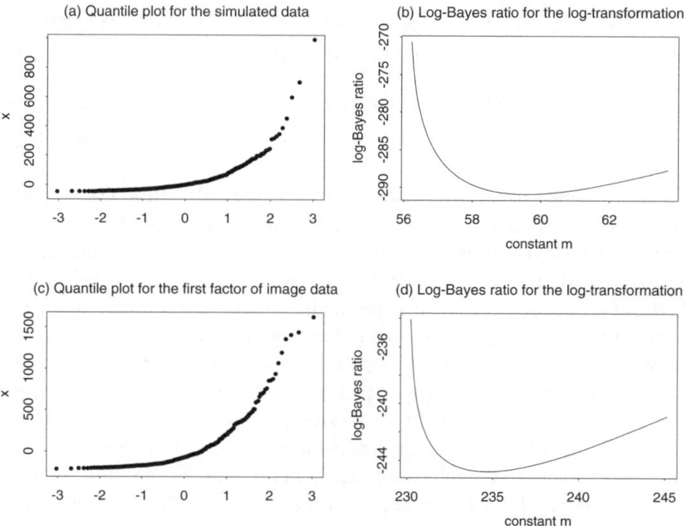

Figure 1. *Deciding whether a logarithm transformation should be used for* (a) *a simulated dataset and* (c) *the first factor of a micrograph image dataset.* (b) *and* (d): *the logarithm of the Bayes ratio as computed by* (7) *with* $\alpha = 0$.

We simulated four hundred observations from $\exp(Z + 4) - 60$, where Z is the standard Gaussian random variable. Its qq-plot is seen in Figure 1a. The log-Bayes ratio (the untransformed likelihood versus the logarithm transformation) as shown in Figure 1b clearly indicates that a constant around 50–60 should be added and the logarithm transformation is necessary. The qq-plot in Figure 1c shows the long-tailness of the first factor in the micrograph image example. By applying the transformation-selection procedure just described, we see from Figure 1d that a number around 235 should be added before the logarithm transformation. The Bayes ratio strongly indicates the use of log-transformation for a wide range of constant m. As shown later, this transformation significantly improved the clustering result for a molecular micrograph application.

5. PERFORMANCE EVALUATION OF THE BCVS

5.1. *A Simulation Study*

In order to investigate the usefulness of variable selection in clustering analysis, we simulated observations from a mixture of two bivariate Gaussian distributions,

$$\alpha \mathrm{N}(\boldsymbol{\mu}^{(1)}, \Sigma^{(1)}) + (1-\alpha)\mathrm{N}(\boldsymbol{\mu}^{(2)}, \Sigma^{(2)}),$$

then we add some factors which are just independent Gaussians. Thus, in these data, only the first two factors are "informative," and the other dimensions are noises. We compare the results of the following: clustering using only the first two factors; using BCVS with the anchored mode, using BCVS with non-anchored mode, and using all the factors indiscriminately.

We first simulated 100 data sets with four factors, for each data set, the parameters for the bivariate mixture Gaussian distribution were drawn according to

$$\begin{aligned}
&&&\alpha \sim \mathrm{Un}(0.3, 0.7),\\
\sigma_{11}^{(1)} &\sim Inv-\chi^2(2, 4.0), & \sigma_{22}^{(1)} &\sim Inv-\chi^2(8, 5.0), & \rho^{(1)} &\sim \mathrm{Un}(0, 0.6),\\
&& \mu_1^{(1)}|\sigma_{11}^{(1)} &\sim \mathrm{N}(3, \sigma_{11}^{(1)}), & \mu_2^{(1)}|\sigma_{22}^{(1)} &\sim \mathrm{N}(1, \sigma_{22}^{(1)}),\\
\sigma_{11}^{(2)} &\sim Inv-\chi^2(8, 5.0), & \sigma_{22}^{(2)} &\sim Inv-\chi^2(2, 4.0), & \rho^{(2)} &\sim \mathrm{Un}(0, 0.6),\\
&& \mu_1^{(2)}|\sigma_{11}^{(2)} &\sim \mathrm{N}(0, \sigma_{11}^{(2)}), & \mu_2^{(2)}|\sigma_{22}^{(2)} &\sim \mathrm{N}(0, \sigma_{22}^{(2)}),
\end{aligned}$$

where $\sigma_{ll}^{(j)}$ is the variance of the lth variable and $\rho^{(j)}$ the correlation coefficient for cluster j. Each noise factor has mean 0 and variance τ^2 drawn from Inv-$\chi^2(8, 5.0)$.

After the parameters were drawn, 200 observations were then simulated with these parameters. This setup resulted in a wide range of data sets, some of which had well-separated clusters, and some had close clusters. We also generated 100 data sets of size 200 with nine factors (seven noisy factors). For each data set, the parameters for the bivariate mixture Gaussian distribution and those for each of the seven noise factors were drawn similarly as described above. In each dimension setting (4 and 9, respectively), we stratified the 100 simulated datasets into three groups (easy, median, and difficult) of about equal sizes based on the performances of the "gold standard", *i.e.*, the clustering algorithm using only the two informative factors. Figure 2 compares the performances of the three clustering approaches, BCVS with anchor mode, BCVS with non-anchor mode, and clustering using all the factors, with the "gold standard", in each data set group as well as all data sets.

In the MCMC implementations, we assumed *a priori* that for each cluster j, $[\boldsymbol{\mu}_j \mid \Sigma_j] \sim \mathrm{N}(\boldsymbol{x}_0, \Sigma_j/\rho_0)$ and $\Sigma_j \sim$ Inv-W$_{\nu_0}(S_0^{-1})$, where $\boldsymbol{x}_0$ is the vector of sample means of the k factors, $\rho_0 = 0.01$, $v_0 = k$, and S_0 is a diagonal matrix with sample variances of the k factors as the diagonal elements. We also assumed uniform prior for the number of factors used for clustering in anchor mode and for the combination of factors used for clustering in non-anchor mode.

Figure 2 showed the differences in the number of correctly clustered objects for each method. Suppose the true clustering answer groups the observations into $\boldsymbol{T} = (T_1, T_2)$. Then the number of correctly clustered objects of a clustering result $\boldsymbol{C} = (C_1, C_2)$ is defined as the number of matched objects for the best match, *i.e.*,

$$N(\boldsymbol{C}) = \max\{|C_1 \cap T_1| + |C_2 \cap T_2|, |C_2 \cap T_1| + |C_1 \cap T_2|\}.$$

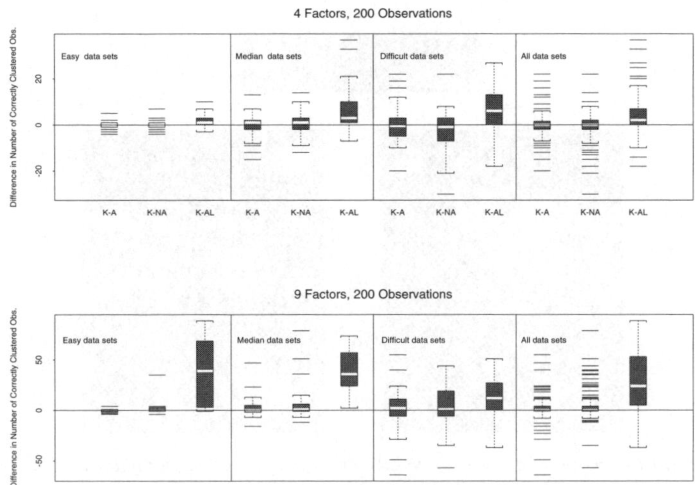

Figure 2. *The number of correctly clustered objects of the three clustering strategies (anchor, non-anchor, and all factors) were each compared with that of the clustering using the two known factors. A: BCVS with anchor mode; NA: non-anchor mode; K: clustering using only the first two factors; AL: using all the factors. The 100 simulated datasets for each setting are stratified according to the clustering performances when the two informative factors are known* (K).

A completely random assignment will result in a $N(\boldsymbol{C})$ slightly greater than $n/2$.

It is interesting to note from the boxplots that for the datasets with four factors the BCVS with both the anchor and non-anchor modes performed almost as good as the method with the two informative factors known, and all the three are better than the method using all the four factors indiscriminately. The datasets with nine factors showed more striking differences. Note that when all the nine factors are used in fitting a mixture of two Gaussians, we have to entertain 109 free parameters: two nine-by-nine dimensional covariance matrices, two mean vectors, and the mixture proportion. It is not surprising that such a method performed poorly. It is somewhat surprising, however, that BCVS performed so much better, suggesting that we do not lose much by not knowing which factors to use. A careful examination shows that the BCVS with anchor mode performed slightly better than that with non-anchor mode.

5.2. *Classification of Images from Electron Micrographs*

In a recent study (Sansó *et al.*, 2002), we applied the BCVS to classify the electron micrograph images and compared its performances with the popular hierarchical clustering (HAC) method. The signal components of these images were generated by projecting a volume of the 50S ribosomal subunit from *E. coli* reconstructed from electron micrographs of a negatively stained sample (Radermacher *et al.*, 1987). A set of four images, as shown in Figure 3a, was created by tilting four projections by different angles. A set of lower-tilt images were also produced for closer comparisons but is not shown here.

One hundred copies of each projection were generated and noise was added to each of them, yielding four tilt groups of 100 images each. Two different sources of noise were used. In the first series, Gaussian noise with standard deviation from 1 to 20 was added to each projection.

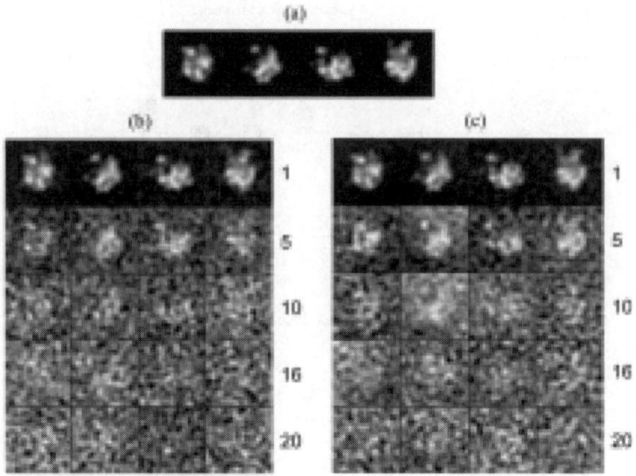

Figure 3. *(a) The images of the 50S ribosomal subunit from E. coli at four tilting angels. (b) Images with Gaussian noises at different SNR levels. (c) Images with real noises.*

In a second series, the "real noise" obtained by windowing 400 regions of an image of carbon film obtained in the electron microscope was added. The real-noise windows were also scaled to obtain a wide range of SNR. Examples of the resulting individual images from selected data sets are shown in Figures 3b and 3c. These data sets differ from those obtained in real experiments where particles are found in random orientations. However, these partially simulated data allow the comparison of different classification methods without confounding the comparison with alignment issues.

Each image data were submitted to either CA or PCA dimension reduction, and the first eight factors were saved. These eight-dimensional data sets were classified by both the BCVS and HAC with "complete-link" merging criterion. Since methods for factor selections in HAC are subjective, all eight factors were employed with this procedure. The HAC dendrogram tree was cut at a threshold giving four classes.

Results. The BCVS yielded a smooth sigmoidal decline of the success rate of classification with increased noise (Figure 4). HAC showed a more erratic pattern, *i.e.*, from noise 4 to 5 there is a sudden decrease of classification from 100% to 75.5%, Figure 4a, resulting from the merging of nearly all of the observations of two tilt groups (94% and 96%) into a single class, leaving a fourth class almost empty. At this noise level, BCVS classified 100% correctly. Up to and including noise level 14, the BCVS anchor and non-anchor modes produced similar results. At noise levels greater than 14, the anchor mode gave the best results.

For real-noise data sets, at high-tilt, both BCVS and HAC produce similar results through noise level 10 (Figure 4b). At noise level 12, HAC decreases its score suddenly (from 92% to 59%), whereas BCVS continues to return favorable scores through noise level 14, at which a score of over 79% is seen. The BCVS outperformed HAC more significantly in all low-tilt datasets (*i.e.*, 81% versus 48% at noise level 4, as shown by dashed lines in Figure 4c).

PCA versus CA. CA is historically preferred to PCA in image processing because PCA often results in factors whose scales differ in magnitudes, rendering the Euclidean distance-based HAC algorithm difficult to produce meaningfully clusters. However, since the BCVS adaptively adjusts scales of factors through inferences of variances, we examined its application under PCA

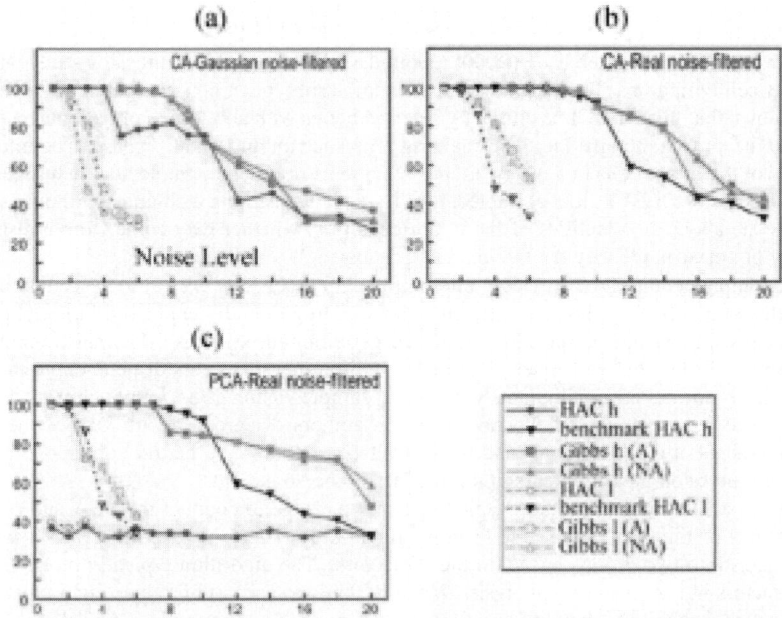

Figure 4. *The comparison results for the micrograph image data. Solid lines are for high-tilt datasets and the dashed lines for low-tilt datasets.*

data reduction. With well-behaved Gaussian noise we found that in all cases both BCVS and HAC perform better on PCA data than on CA data (Figure not shown).

HAC's results on real-noise data sets using PCA show very low scores (below 40%) for all noise levels (Figure 4c). Scores in this range indicate that the classification is not much better than a random assignment of particles into classes. Because HAC uses Euclidean distances, it is not well suited for cases in which the variance (*i.e.*, the spread) of a factor differs by an order of magnitude or more from the variance of other factors. These results support the accepted practice of using CA data reduction for HAC. In contrast, the BCVS algorithm is not hindered by data with such characteristics and can perform better than that using the CA.

Logarithm transformation of factors. We observed that, under PCA reduction, the first factor had a much larger variance than the other factors and the degree of this effect varies from class to class. The q–q plot for the first factor is displayed in Figure 1c, which shows a clear sign of long-tailness. We considered the transformation of the form $\log(x + m)$. As discussed in Section 4, the Bayes ratio of the original model versus the log-transformation data model for this factor can be computed. Figure 1d shows the logarithm of the ratio for a range of m values, which is overwhelmingly in favor of transformation (a value of -100 means that the logarithm transformation is e^{100} times more favorable than no-transformation).

With this factor log-transformed, the rate of correct classification is significantly improved for BCVS on real-noise, PCA-reduced data. For example, at the noise level 8, the correct-classification rate increases from 83.50 to 94.25 for the anchor mode, and from 82.75 to 94.25 for the non-anchor mode, and such a high success rate of more than 80% is maintained through noise level 18.

5.3. *Clustering of Microarray Data*

Lymphoma Data. Alizadeh *et al.* (2000) reported a microarray gene expression study on diffuse large B-cell lymphoma (DLBCL), the most common subtype of non-Hodgekin's lymphoma. It was known that this DLBCL is clinically heterogeneous with about 40% of patients responding well to the current therapy and the remaining 60% succumbing to the disease. The microarray studies of the tumor cells in these two types of patients revealed the molecular heterogeneity of the two types of DLBCL. It is of interest here to see if a clustering technique, without using any knowledge about the identities of the tumor cells (*i.e.*, whether they come from a responding patient or not), can identify the two distinct groups.

A complementary DNA (cDNA) microarray chip (Schena *et al.,* 1995) was constructed by Alizadeh *et al.* (2000), who selected genes that are preferentially expressed in lymphoid cells and genes with known or suspected roles in cancer or immune systems. The final "Lymphochip" consists of 18,000 cDNA clones. About 1.8 million measurements of gene expression were made in 96 normal and malignant lymphocyte samples using 128 Lymphochip microarrays. After some preprocessing of the raw data, the authors discarded about 75% of the mRNA measurements of all the patients and made available a 4026×96 matrix corresponding to the mRNA expression measurements of 4026 genes in the 96 patients.

We first submitted the data to Splus software package to conduct the PCA and return the projections of the data to the first 16 eigen directions. Then we asked the BCVS to cluster the objects into two groups based on the 16 factors. The algorithm typically picked seven to eight factors in the clustering analysis. When BCVS was asked to produce eight clusters with non-anchor mode, it resulted in very similar clusters as that reported in Alizadeh *et al.* (2000) who used HAC. On average 11 factors were selected, 1–9, 11 and 15. When four clusters were asked for, BCVS chose to use 10 factors, 1–9 and 12, and produced a group (47 members) with mostly DLCL types, a group that mixes CLL and FL types (24), a group of Blood B cell types (13), and a group of Blood T cell types (12). This result is in close agreement with Alizadeh *et al.* (2000). A small difference is that the two germinal center B cells were grouped with the CLL+FL cluster and DLCL-0009 and SUDHL5 were put into the DLCL cluster.

Leukemia data. Golub *et al.* (1999) applied gene expression microarray techniques to study human acute leukemias and discovered the distinction between acute myeloid leukemia (AML) and acute lymphoblastic leukemia (ALL). Distinguishing ALL from AML is crucial for successful treatment, since chemotherapy regimens for ALL can be harmful for AML patients. Their results demonstrated the feasibility of cancer classification based solely on gene expression monitoring and suggested a general strategy for discovering and predicting cancer classes for other types of cancer, independent of previous biological knowledge. They first constructed a classifier that can distinguish the two types of cancers using an initial collection of samples belonging to known classes. As a second task, they applied a two-cluster self-organizing map (SOM) to automatically group the 38 initial leukemia samples into two classes on the basis of the expression pattern of all 6817 genes. Their SOM results matched the known classes closely: Class A1 contained mostly ALL (24 of 25 samples) and class A2 contained mostly AML (10 of 13 samples). So SOM clustered three ALL samples with the AML class and one AML sample with ALL. A drawback of the SOM, however, is its lack of statistical interpretation and the uncertainty measurement of the results.

The dataset contains the expression levels of 6817 genes for 38 patients. There are actually 7129 probe sets–controls and gene redundancies bringing the 6817 up to 7129. The patient samples are known to come from two distinct classes of leukemia: 27 are ALL and 11 are AML. We submitted all the data (controls and redundancies) to PCA using the SPlus function. Recognizing that the total number of observations is very small, we chose to output the first eight eigenvectors from the PCA and use the BCVS with the anchor mode. The BCVS chose the first seven eigenvectors to include in the mixture modelling, and reported two classes: Class 1 contained 27 samples, 26 of which were ALL samples and Class 2 contained 11 samples, 10 of which were AML samples. This result is slightly better than that obtained by Golub *et al.* (1999) using SOM, illustrating that PCA plus BCVS produces competitive results for clustering. More importantly, in comparison with SOM, the PCA and Gaussian mixture models possess clear statistical interpretations and well-established mathematical properties, enabling the investigator to think further about the modelling issues.

6. DISCUSSION

Based on Gaussian mixture models, we propose a novel Bayesian method BCVS for clustering high-dimensional data. The new method has the following features: (a) factors informative to the clustering model are automatically selected; (b) transformations of the factors can be selected in an automatic and principled way; and (c) the new method combines the PCA with the formal Bayesian modelling. We have shown by simulation that the variable selection step in BCVS can significantly improve the clustering result especially when the number of factors in consideration is high. We have also shown by a few real applications that the BCVS produced as good or better results than the popular hierarchical clustering method.

What is lacking in our current method is a way to determine the total number of clusters for the mixture model. Conceptually, one can just give a prior for the clustering variables (total cluster number and memberships) and then proceed with the MCMC machinery. For example, a popular prior for clustering indicators is that derived from the Dirichlet process (Neal, 2000), which is most conveniently described conditionally: given the clustering of the first i observations, the $i+1$st observation can join an existing cluster of size n_j with probability $n_j/(q+i)$, and form a new cluster of its own with probability $q/(q+i)$. The prior expectation of the total number of clusters in this model is $O(\log(n))$, which may not be desirable in practice. Qin *et al.* (2002) proposed a modification: the $i+1$st observation joins an existing cluster with probability $1/(q+c_i)$ and forms a new cluster with probability $q/(q+c_i)$, where c_i is the total number of clusters formed by the first i observations. This prior gives an expected number of clusters of $O(\sqrt{n})$. One can also prescribe a prior for the clustering variable, as suggested by Richardson and Green (1997), by giving first a distribution on the number of clusters and then a distribution for the memberships of the n objects. A potential problem with this line of approach is the additional computation cost.

Another possible avenue to achieve the automatic selection of cluster number is to treat it as a model selection problem and use the Bayesian information criterion (BIC) to determine the number of Gaussian components (Yeung *et al.*, 2001). Although this approach is not directly based on a model, it is much cheaper computationally and gives satisfactory result in general. However, with the additions of variable and transformation selection variables, the BIC in Yeung *et al.* (2001) needs to be revised to suit for the new variable selection task.

ACKNOWLEDGEMENTS

This research was supported partly by NSF Grants DMS-0094613 and DMS-0204674 to JSL and the NIH Grant R21RR14036 and NSF Grant DBI-9515518 to CEL. We thank Epaminondas Sourlas in Harvard statistics department for some computational assistance.

REFERENCES

Alizadeh, A. A. *et al.* (2000). Distinct types of diffuse large B-cell lymphoma identified by gene expression profiling. *Nature* **403**, 503–511.

Banfield J. D. and Raftery, A. E. (1993). Model-based Gaussian and non-Gaussian clustering. *Biometrics* **49**, 803–821.

Box, G. E. P. and Cox, D. R. (1964). An analysis of transformations *J. R. Statist. Soc. B* **26**, 211–246 (with discussion).

Brooks, S. P. (2001). On Bayesian analyses and finite mixtures for proportions. *Statist. Comput.* **11**, 179–190.

Celeux, G. and Govaert, G. (1995). Gaussian parsimonious clustering models. *Pattern Recogn.* **28**, 781–793.

Chen, R. and Liu, J. S. (1996). Predictive updating methods with applications in Bayesian classification. *J. R. Statist. Soc. B* **58**, 397–415.

Diebolt, J. and Robert, C. P. (1994). Estimation of finite mixture distributions through Bayesian sampling. *J. R. Statist. Soc. B* **56**, 363–375.

Fraley, C. and Raftery, A. E. (1999). Mclust: software for model-based cluster analysis. *J. Classification* **16**, 297–306.

Frank, J. (1996). *Three-Dimensional Electron Microscopy of Macromolecular Assemblies.* New York: Academic Press.

Frank, J. and van Heel, M. (1982). Correspondence analysis of aligned images of biological particles. *J. Mol. Biol.* **161**, 134–137.

Gelman, A., Carlin, J. B., Stern, H. S., and Rubin, D. B. (1995). *Bayesian Data Analysis.* New York: Chapman and Hall.

Geyer, C. J. (1991). Markov chain Monte Carlo maximum likelihood. *Computing Science and Statistics: The 23rd Symposium on the Interface* (E. Keramigas, ed). Fairfax: Interface Foundation, 156–163.

Ghosh, D. and Chinnaiyan, A. M. (2002). Mixture modelling of gene expression data from microarray experiments. *Bioinformatics* **18**, 275–286.

Golub, T. R., Slonim, D. K., Tamayo, P., *et al.* (1999). Molecular classification of cancer: Class discovery and class prediction by gene expression monitoring. *Science* **286**, 531–537.

Ishwaran, H., James, L. F., and Sun J. Y. (2001). Bayesian model selection in finite mixtures by marginal density decompositions. *J. Am. Statist. Ass.* **96**, 1316–1332.

Kim, H. C., Kim, D. J., and Bang, S. Y. (2002). A numeral character recognition using the PCA mixture model. *Pattern Recogn. Lett.* **23**, 103–111.

Liu, J. S. (1994). The collapsed Gibbs sampler with applications to a gene regulation problem. *J. Am. Statist. Ass.* **89**, 958–966.

Liu, J. S. (2001). *Monte Carlo Strategies in Scientific Computing.* New York: Springer.

McLachlan, G. J., Bean, R. W., and Peel, D. (2002). A mixture model-based approach to the clustering of microarray expression data. *Bioinformatics* **18**, 413–422.

Moss, S., Wilson, R. C., and Hancock, E. R. (1999). A mixture model for pose clustering. *Pattern Recogn. Lett.* **20**, 1093–1101.

Neal, R. M. (2000). Markov chain sampling methods for Dirichlet process mixture models *J. Comput. Graph. Statist.* **9**, 249–265.

Qin, Z. S., McCue, L. A., Thompson, W., Mayerhofer, L., Lawrence, C. E. and Liu, J. S. (2002). Identification of co-regulated genes through Bayesian clustering of predicted regulatory binding sites. *Tech. Rep.*, Harvard University, USA.

Radermacher, M., Wagenknecht, T., Verschoor, A., and Frank J. (1987). Three-dimensional reconstruction from a single-exposure random conical tilt series applied to the 50S ribosomal subunit of Escherichia coli. *J. Microsc.* **146**, 113–136.

Richardson, S. and Green, P. J. (1997). On Bayesian analysis of mixtures with an unknown number of components. *J. R. Statist. Soc. B* **59**, 731–758.

Roth, F. P., Hughes, J. D., Estep, P. W., and Church, G. M. (1998). Finding DNA regulatory motifs within unaligned noncoding sequences clustered by whole-genome mRNA quantitation. *Nat. Biotechnol.* **16**, 939–945.

Samsó, M., Palumbo, M. J., Radermacher, M., Liu, J. S., and Lawrence, C. E. (2002). A Bayesian method for classification of images from electron micrographs. *J. Struct. Biol.* (to appear).

Schena, M., Shalon, D., Davis, R. W. and Brown, P. O. (1995). Quantitative monitoring of gene expression patterns with a complementary DNA microarray. *Science* **270**, 467–470.

Yeung, K. Y., Fraley, C., Murua, A. *et al.* (2001). Model-based clustering and data transformations for gene expression data. *Bioinformatics* **17**, 977–987.

APPENDIX: BAYESIAN MULTIVARIATE GAUSSIAN INFERENCE

For the consistency of notations and the self-containedness of the article, we here present the standard conjugate Bayesian analysis with Gaussian observations. More details can be found in, for example, Gelman *et al.* (1995). Suppose n iid realizations $\boldsymbol{X} = \{\boldsymbol{x}_1, \ldots, \boldsymbol{x}_n\}$ are observed from the k-dimensional Gaussian distribution $N(\boldsymbol{\mu}, \Sigma)$. It is of interest to make inference on $\boldsymbol{\mu}$, Σ, and a future observation $\boldsymbol{x}_{n+1}$.

The prior distribution for Σ is the Inverse-Wishart distribution, Inv-W$_{\nu_0}(S_0^{-1})$, which is of the form:

$$p_0(\Sigma) = c_0|\Sigma|^{-(\nu_0+k+1)/2} e^{-\frac{1}{2}\,\mathrm{tr}(\Sigma^{-1}S_0)}, \tag{8}$$

where c_0 is the normalizing constant, k is the dimensionality, and ν_0 and S_0 are two hyper-parameters to be given by the user. Conditional on Σ, the prior of $\boldsymbol{\mu}$ is

$$\boldsymbol{\mu} \,|\, \Sigma \sim N(\boldsymbol{x}_0, \Sigma/\rho_0). \tag{9}$$

To make our later analysis more convenient, we define the normalizing function

$$Z(\nu, S, k) = |S|^{\frac{\nu}{2}} \left\{ 2^{\frac{\nu k}{2}} \pi^{\frac{k(k-1)}{4}} \prod_{i=1}^{k} \Gamma\left(\frac{\nu+1-i}{2}\right) \right\}^{-1}. \tag{10}$$

Then $c_0 = Z(\nu_0, S_0, k)$.

In order to make p_0 a proper distribution, we need to have $\nu_0 > k-1$ and $|S_0| > 0$. With this distribution and $\nu_0 > k+1$, we have $\mathrm{E}(\Sigma) = S_0/(\nu_0 - k - 1)$.

If we have n iid observations $\boldsymbol{X} = \{\boldsymbol{x}_1, \ldots, \boldsymbol{x}_n\}$ from $N(\boldsymbol{\mu}, \Sigma)$, then data-parameter joint distribution is

$$\begin{aligned}
P(\boldsymbol{X}, \boldsymbol{\mu}, \Sigma) &= c_0(2\pi)^{-(n+1)k/2}|\Sigma|^{-(n+k+\nu_0+2)/2}\rho_0^{k/2} \\
&\quad \times \exp\Big[-\tfrac{1}{2}\{\mathrm{tr}(\Sigma^{-1}S_0) + \rho_0(\boldsymbol{x}_0 - \boldsymbol{\mu})\Sigma^{-1}(\boldsymbol{x}_0 - \boldsymbol{\mu})^t + \sum_{i=1}^{n}(\boldsymbol{x}_i - \boldsymbol{\mu})\Sigma^{-1}(\boldsymbol{x}_i - \boldsymbol{\mu})^t\}\Big] \\
&= c_0(2\pi)^{-(n+1)k/2}|\Sigma|^{-(n+k+\nu_0+2)/2}\rho_0^{k/2} \\
&\quad \times \exp\Big[-\tfrac{1}{2}\{\mathrm{tr}\{\Sigma^{-1}(S_0 + SS)\} + (\boldsymbol{\mu} - \bar{\boldsymbol{x}}^*)\Big(\frac{\Sigma}{n+\rho_0}\Big)^{-1}(\boldsymbol{\mu} - \bar{\boldsymbol{x}}^*)^t\}\Big]
\end{aligned}$$

where

$$SS = \sum_{i=1}^{n}(\boldsymbol{x}_i - \bar{\boldsymbol{x}})^t(\boldsymbol{x}_i - \bar{\boldsymbol{x}}) + \frac{n\rho_0}{n+\rho_0}(\bar{\boldsymbol{x}} - \boldsymbol{x}_0)^t(\bar{\boldsymbol{x}} - \boldsymbol{x}_0) \tag{11}$$

and

$$\bar{\boldsymbol{x}}^* = \frac{n}{n+\rho_0}\bar{\boldsymbol{x}} + \frac{\rho_0}{n+\rho_0}\boldsymbol{x}_0. \tag{12}$$

Thus, after integrating out $\boldsymbol{\mu}$, we have

$$\begin{aligned}
P(\boldsymbol{X}, \Sigma) &= c_0(2\pi)^{-nk/2}\rho_0^{k/2}(n+\rho_0)^{-k/2}|\Sigma|^{-(n+k+\nu_0+1)/2}\exp\left[-\tfrac{1}{2}\,\mathrm{tr}\{\Sigma^{-1}(S_0 + SS)\}\right] \\
&= c_0(2\pi)^{-nk/2}\rho_0^{k/2}(n+\rho_0)^{-k/2}c_1^{-1}\text{Inv-W}_{n+\nu_0}\{(S_0 + SS)^{-1}\}
\end{aligned}$$

where

$$c_1 = Z(n + \nu_0, S_0 + SS, k).$$

Hence, the model likelihood $P(\boldsymbol{X})$ is of the form

$$P(\boldsymbol{X} \mid n, k, \lambda_0) = \frac{Z(\nu_0, S_0, k)}{Z(n+\nu_0, S_0+SS, k)}(2\pi)^{-nk/2}\left(\frac{n+\rho_0}{\rho_0}\right)^{-k/2}. \tag{13}$$

Marginally, the distribution of $\boldsymbol{\mu}$ is

$$[\boldsymbol{\mu} \mid \boldsymbol{x}_1, \ldots \boldsymbol{x}_n] \sim t_{n+\nu_0-k+1}\left(\bar{\boldsymbol{x}}^*, \frac{S_0+SS}{(n+\rho_0)(n+\nu_0-k+1)}\right).$$

The predictive distribution of a new observation $\boldsymbol{x}$ is

$$[\boldsymbol{x} \mid \boldsymbol{x}_1, \ldots \boldsymbol{x}_n] \sim t_{n+\nu_0-k+1}\left(\bar{\boldsymbol{x}}^*, \frac{(n+\rho_0+1)(S_0+SS)}{(n+\rho_0)(n+\nu_0-k+1)}\right). \tag{14}$$

DISCUSSION

ADRIAN E. RAFTERY (*University of Washington, USA*)

Introduction. It is a pleasure to congratulate the authors on a paper that proposes promising solutions to several outstanding problems in model-based clustering. I prefer the term *model-based clustering* to *Bayesian clustering*, because many of the references that the authors cite as examples of Bayesian clustering in their introductory section are not specifically Bayesian, but are model-based, and because most clustering done in practice is heuristic rather than model-based. That is the big dichotomy in the clustering literature, rather than the Bayesian-frequentist one. For recent reviews of the literature on this topic, see McLachlan and Peel (2000) and Fraley and Raftery (2002).

I will start by summarizing the main features of the paper and highlighting what I see as some of its most important contributions. I will then address the issue of whether dimension reduction and clustering should be done separately (as the authors do), or together, and suggest that simultaneous solutions are possible. I will also discuss the nature of the hierarchical agglomerative clustering that the authors used as a comparison method. Finally, I will discuss the relative advantages and disadvantages of maximum likelihood via the EM algorithm, and fully Bayesian estimation via MCMC, for model-based clustering.

Highlights of the Paper. The paper addresses the important problem of clustering high-dimensional data, that is $n \times p$ data matrices containing J groups, where the dimension, p, is large relative to the number of data points, n. The work is motivated by three examples: clustering of electron microscope images, where $n = 400$, p is in the tens of thousands, and $J = 4$; lymphoma gene expression data, with $n = 96$, $p = 4026$ and there are two groups; and leukemia gene expression data, with $n = 38$, $p = 6817$ and there are two groups.

The method proposed is as follows, in brief. The number of groups, J, is taken to be known.

1. Extract the first k_0 principal components. In the examples, $k_0 = 8$ or 16.
2. Fit a mixture of multivariate normal distributions to a subset of the k_0 principal components. The covariance matrices are taken to be unconstrained.
3. The subset of the principal components to be used is treated as a parameter, and estimated using the Markov chain Monte Carlo model composition (MC^3) algorithm of Madigan and York (1995). This is a Metropolis–Hastings algorithm over the discrete space of possible subsets of the principal components in which the mixture model parameters are integrated out analytically.

4. Taking the first k principal components, rather than an arbitrary subset, is an option, called *anchor mode*.
5. Bayes factors can be used to choose a Box–Cox transformation of the variables.

A simulation study was carried out, in which $n = 200$, $p = 9$ and there were two groups. This gave good results, but in a situation very different from the motivating examples. The method was applied to the motivating data sets, with encouraging results.

The two biggest contributions of the paper are methods for variable selection and for the selection of transformations in model-based clustering. The results show clear gains when variable selection is used in clustering, over the approach when all the variables are used. This is similar to the situation in regression, where it has been shown that one gains by doing variable selection or model averaging in a Bayesian context (*e.g.*, Clyde, 1999; Hoeting *et al.,* 1999). The method proposed for selecting transformations makes a lot of sense.

Dimension reduction and clustering: Together or separate?. The authors first perform principal component analysis on the full data set. They then select the first k_0 principal components, where k_0 is very small relative to the total number of principal components, and they then do clustering on the resulting reduced data set. This is simple and appealing, and gives good results in their simulations and examples. It is worth pointing out that this is actually a very well known general approach in the clustering literature and goes back a long way (*e.g.*, Schnell, 1970; Tyron and Bailey, 1970; Everitt, 1974; Chien, 1978). The originality of the present approach is that after the first k_0 principal components have been extracted, further variable selection is carried out simultaneously with the clustering.

However, Chang (1983) has shown that the practice of reducing the data to the first principal components before clustering is not justified in general. He showed that the principal components with the larger eigenvalues do not necessarily contain the most information about the cluster structure, and that taking a subset of principal components can lead to a major loss of information about the groups in the data. Chang demonstrated this theoretically, by simulations, and also in applications to real data. Similar results have been found by other researchers, including Yeung and Ruzzo (2001) for clustering gene expression data, and Green and Krieger (1995) for market segmentation.

This point is illustrated in Figure 5. This is a simulated two-dimensional data set in which there are two clear groups. The first principal component is the diagonal line with equation roughly $y = -x$ that separates the two groups. This first principal component accounts for about 90% of the variance, so consideration of the eigenvalues of the covariance matrix would often lead to collapsing the data to the first principal component only. But if this were done, it is clear that all the cluster information would be lost. In fact, all the cluster information is contained in the second principal component, which accounts for only a small proportion of the variance.

Of course, Liu *et al.* do not use only one principal component; they use $k_0 = 8$ or 16. But they are reducing data with thousands of dimensions to 8 or 16; how can we be sure that something similar to Figure 1 is not happening, on a much larger scale? In their examples, this does not seem to be the case, but the other papers I have cited provide evidence that this can happen in practice.

This also suggests that the choice of k_0 is crucial in practice, and I wonder how the authors propose choosing it so as to avoid problems such as those I have mentioned.

Simultaneous parameter reduction and clustering: Is it possible? It is easy to criticize strategies such as those of the authors, but dimension reduction of some sort is clearly necessary in these very high-dimensional problems. The authors' strategy of reducing the data to the first few principal components is commonly used in high-dimensional clustering problems. So the

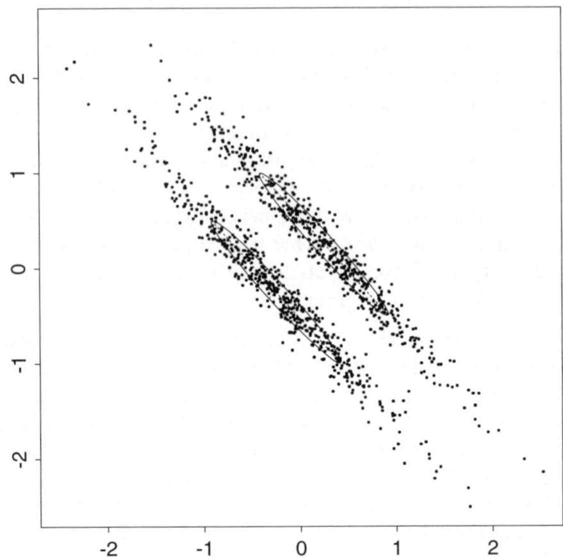

Figure 5. *Two groups in two dimensions. All cluster information would be lost by collapsing to the first principal component. The principal ellipses of the two groups are shown as solid curves.*

question is, what else can be done? Directly clustering data of such high dimension seems hopeless.

Some initial dimension reduction may be necessary and easy to do in problems like this. For example, in gene expression data, eliminating the unexpressed genes is standard practice, and this can reduce the number of dimensions from around 6,000 to a few hundred in typical examples. In medical image data, many of the pixels are of no interest because they belong to the background or to broad features, and eliminating them can reduce the effective dimension of the problem by 90% or more.

A general alternative strategy is to keep all the remaining dimensions in the analysis, but to use more parsimonious models for the covariance matrices in the multivariate normal mixture model. Various such parsimonious models have been proposed, and they allow surprising flexibility, as well as huge reductions in the number of parameters needed.

Perhaps the simplest such model is the diagonal covariance, or naive Bayes model, in which $\Sigma_j = \text{diag}(v_{1j}, \ldots, v_{pj})$. This is quite parsimonious relative to the Liu *et al.* model. For example, for a 400-dimensional data set (typical of gene expression microarrays when unexpressed genes have been removed), it has about the same number of parameters as the Liu *et al.* model with $k_0 = 18$ principal components. It is also quite flexible: it allows different volumes and shapes for each cluster, and also different orientations. The main restriction is that the orientations are axis-aligned. The big advantage here is that clustering is done simultaneously with parameter reduction. This model has performed well for clustering gene expression data (Yeung *et al.*, 2001).

A more general, but still parsimonious set of models arises from the volume-shape-orientation decomposition of Banfield and Raftery (1993), in which $\Sigma_j = \lambda_j D_j A_j D_j^t$. Here λ_j is a scalar that determines the volume of the jth cluster, $A_j$is a diagonal matrix of scaled eigenvalues determining its shape, and D_j is an orthogonal matrix determining its orientation. Each of the volume, shape and orientation may be constant across clusters, or allowed to vary between clusters.

If the orientation is the same across clusters, but the volume and shape are allowed to vary between clusters, a parsimonious but flexible model results. This generalizes the diagonal or naive Bayes model: the cluster orientations are still axis-aligned, but the axes are rotated, and the rotation is determined simultaneously with the clustering. The diagonal model would not work well in the situation of Figure 1, but this generalization would work very well.

Another parsimonious principal component-based approach is the mixture of probabilistic principal component analyzers of Tipping and Bishop (1999). This allows a different set of principal components for each group, and these are estimated simultaneously with the clustering. This seems likely to achieve much of the parsimony of the Liu *et al.* approach, but without the potential disadvantages pointed out by Chang (1983) and others. This is related to the mixture of factor analyzers approach of Ghahramani and Hinton (1997); see McLachlan and Peel (2000) for a review of these approaches.

Finally, I note that using a time series covariance structure for Σ_j may be very parsimonious and quite appropriate for some high dimensional situations that have a sequential structure, such as chemical spectra. The obvious covariance structures of this kind are those that arise from autoregressive-moving average (ARMA) models, and these are highly parsimonious. Model-based clustering has been applied with some success to the clustering of chemical spectra by Wehrens *et al.* (2002), but without using the specifically sequential nature of the data. It seems possible that exploiting this aspect of the data more fully might lead to more efficient and even more successful clustering in such applications.

The Hierarchical agglomerative clustering straw man. Liu *et al.* compare their methods with complete link hierarchical agglomerative clustering based on Euclidean distances. They refer to this as hierarchical agglomerative clustering, or HAC, but it is worth noting that this is only one of many possible kinds of HAC. It is related, although vaguely, to model-based clustering with the the model $\Sigma_j = \lambda I$, i.e. spherical, equal-volume clusters. This is likely often be to an inappropriate model for the data.

It is possible to carry out *model-based* hierarchical agglomerative clustering. One just uses the likelihood as merging criterion at each stage (Banfield and Raftery, 1993). This has been used recently with considerable success in the high-dimensional situation of text classification (Tantrum *et al.,* 2002).

Thus BCVS's advantage over HAC in the authors' simulation study may be due either to (i) being Bayes, or (ii) using a better model (the unconstrained covariance model) for the components. Which of (i) or (ii) is correct is an empirical question, and is not answered by the paper. It could be addressed by carrying out model-based HAC, and comparing thc result of this with BCVS. This can be done, for example, using the MCLUST software available at `www.stat.washington.edu/mclust`.

Label-Switching. The authors have laid out a Bayesian approach to the estimation of a mixture model using MCMC. As such, it would seem to be subject to the label-switching problem discussed, for example, by Richardson and Green (1997). This arises because one can change the labeling of the mixture components without changing the likelihood. Because there are J! labelings, it follows that there are J! components of the posterior distribution, which are identical except for the labeling, if the prior is symmetric with respect to labelings. This has various perverse consequences: for example, if the MCMC sampler does truly explore the posterior distribution, the posterior means of the means of the mixtures components will all be the same.

This problem is not easily diagnosed; for example, label switches do not necessarily correspond to sudden big jumps in the MCMC chain.

In the vanilla mixture model, various solutions to the label-switching problem have been proposed (Celeux *et al.,* 2000; Stephens, 2000), and these seem to work quite well.

I would be interested in the authors' comments on this problem. On the face of it, it seems as if it invalidates their method in its current form. I wonder if the authors feel that the approach could be rescued using some of the proposed solutions to the label-switching problem; could these be adapted to BCVS?

MLE via EM, or Bayes via MCMC? Several of the groups active in research on model-based clustering now focus more on maximum likelihood estimation via the EM algorithm than on Bayesian estimation via MCMC. These include the Grenoble group (*e.g.*, Celeux *et al.,* 1996), the Queensland group (*e.g.*, McLachlan and Peel, 2000), the Microsoft Research group (*e.g.*, Cadez *et al.,* 2000), and our own group at the University of Washington. This is surprising, because of the appeal of the Bayesian framework and because several of these researchers had earlier adopted Bayesian MCMC approaches to the model-based clustering problem (*e.g.*, Bensmail *et al.,* 1997). How can we explain this?

For estimation, the two approaches give similar results in many situations. Both have strengths and weaknesses. The EM approach is often simpler to implement, particularly when the components of the mixture are complex. The Bayesian approach requires the additional work involved in prior specification and the assessment of prior sensitivity, which may not seem very rewarding if similar results are obtained without using a prior. For example, Liu *et al.* use the prior

$$\mu_j \sim N(x_0, 100\Sigma_j).$$

This seems extremely spread out. How sensitive is the posterior for the principal components chosen to this choice, for example to the choice of the constant 100? The Bayesian approach also suffers from the label-switching problem mentioned earlier.

The Bayesian approach also has advantages. In model-based clustering, the MLE of Σ_j can be degenerate, with zero or near-zero eigenvalues or diagonal elements, yielding infinite or near-infinite likelihoods, corresponding to small and/or highly linear clusters. By specifying a prior for Σ_k, a Bayesian approach can alleviate this by effectively smoothing the likelihood so that its many uninteresting infinite spikes are removed. Similarly, the Bayes estimates of posterior cluster membership probabilities, $p(J_j|x_i)$, take account of parameter uncertainty, and so are more accurate and less extreme towards 0 or 1.

But the biggest advantage of the Bayesian approach lies in model selection, model averaging and hypothesis testing, rather than estimation. Liu *et al.* have shown this convincingly in their treatment of the selection of the principal components to be used. In this regard, the good performance of the anchor mode is striking. Do the authors have any thoughts on how general this is?

Throughout, the authors have taken the number of clusters to be known. They point out that a Bayesian approach could solve this also, and mention the use of BIC (Yeung *et al.,* 2001; Fraley and Raftery, 2002) as a possibility. As they say, this is cheap computationally and has given good results in many applications. It could be adapted to variable and transformation selection as well. Its use for transformation selection might require the calculation of a more exact Bayes factor.

Steele (2002) has derived a unit information prior for model selection in mixtures, and in simulation studies he found that Bayes factors based on it gave similar performance to BIC. This provides both some further justification for BIC, and a basis for possible further refinements of model selection methods in model-based clustering.

P. L. IGLESIAS and F. A. QUINTANA (*Pontificia Universidad Católica, Chile*)

This paper proposes a Bayesian clustering procedure after transforming the data via PCA. The number of factors to be included is considered random and therefore part of the inference problem. The goal is to search for homogeneous groups. Under BCVS the transformed vector via PCA is modelled as a mixture of gaussian distributions with Dirichlet-distributed weights and conjugate style priors for mean vectors and covariance matrices. We congratulate the authors for an excellent piece of work, which not only contains novel material, but also opens the door to a number of potentially fruitful research topics.

BCVS involves highly dimensional parameters and many variables. It is then natural to wonder about sensitivity on prior specification and robustness of the clustering method to departures from the gaussian mixture model. Concretely,

1. How would the clustering structure change under different choices of the prior $f(k)$ for the number of components? Would it be possible to elicit such prior in a fully Bayesian way, that is, without carrying out a pre-exploratory data analysis? Perhaps a natural alternative is to perform a reference analysis, which is not necessarily equivalent to a uniform prior on k.
2. Much of the structure in the gaussian mixture model is preserved for some elements in the class of multivariate elliptical distributions, without the need of using Box–Cox transformations. In this case, we anticipate some changes in the clustering structure. For instance, with heavy-tailed distributions we would expect some points to shift to different clusters, and maybe even less clusters than in the gaussian case.
3. An alternative way to produce clusters consists of using product partition models (PPM) after the k principal components are chosen. An advantage of this method is that the number of clusters J does not need to be known in advance.

DANIEL PEÑA (*Universidad Carlos III de Madrid, Spain*)

This is a very interesting paper with many insights and I would regret that the method proposed in the paper may fail because of the initial step of the procedure. Working with the first k_0 principal components of the data may lead to loss of very useful information, as the directions of maximum variability need not be the directions most useful for clustering.

An interesting direction for clustering is one in which the projected points cluster around well separated different means. Note that a univariate sample of zero-mean variables of size n will have maximum separation in two groups when it is composed of $n/2$ points equal to $-a$ and $n/2$ points equal to a, and then the kurtosis coefficient of the sample will take its minimum value, equal to one. Thus, directions of minimum kurtosis coefficient seems useful to show possible clusters.

Another interesting direction for clustering will be the one in which the majority of the points cluster around a common mean, but there are some small groups of points at both sides of the main group. In particular, we may have a central group and some outliers. In this case, the kurtosis coefficient of the distribution of the data will be large. Thus, interesting directions for clustering are those in which the projected data have either a small or large kurtosis coefficient and a powerful cluster method so that high dimensional data can be obtained by projecting the data in the directions with maximum or minimum kurtosis coefficient (see Peña and Prieto, 2001). It can be shown that this method has some optimal properties when the data have been generated by mixtures of elliptical distributions. For instance, when the data are generated by a mixture of two normal distributions with the same covariance matrix the direction which minimizes the kurtosis of the projection is the Fisher linear discriminant function (Peña and Prieto, 2000).

Given these results, I would suggest to the authors that instead of selecting the first k_0 principal components select those components in which the projected data have the largest or the smallest kurtosis coefficient. Alternatively, instead of working with principal components, they can directly take as variables the projections of the original data on the $k_0/2$ orthogonal directions of maximum and minimum kurtosis coefficient.

CHRISTIAN P. ROBERT (*Université Paris Dauphine, France*)

It is quite exciting to track the growing importance of mixture modelling in the Bayesian literature and, in particular, at the *Valencia 7* meeting. This paper is an exemplary illustration of the versatility of mixture modelling, since this modelling applies to many areas from clustering to nonparametric settings. My comments will mainly be limited to relevant works in the area, although I first want to point out that the computation of the Bayes factor (5) using improper priors is invalid, *stricto sensu* (Robert, 2000).

First, the authors rightly perceive the relevance of using *Principal Components* (PC) to reduce the dimension of the model and they introduce the compelling idea of *anchor*. In my opinion, using PC is an almost compulsory step as fitting a mixture model to a high-dimensional dataset will almost certainly lead to a large number of components, unless some strong and well-documented structure is available. During the talk, I was wondering, however, how the exploratory technique of PC (and the concept of anchor) could be embedded within a more Bayesian Decision Theory framework.

Second, the authors note that integrating the parameters out lead to improved performances of the sampler. There is more to this point: we showed in Casella *et al.* (2000) that this integration provides an easier exploration of the space of the missing variables (or of the corresponding sufficient statistics), and exhibited a quite peculiar phenomenon of *concentration*. In the cases we studied, it indeed appears that the marginal posterior distribution of the sufficient statistics, $(\sum_i I_j(J_i), \sum_i I_j(J_i)\, x_i, \sum_i I_j(J_i)\, x_i {x_i}^t)$ say, is highly concentrated on a few values, and thus that partitioned sampling is very helpful in uncovering the important parts of the missing variables space.

Third, Celeux *et al.* (2000) also implemented tempering in this setting, following Neal (1996). The algorithm in Section 3.2 of our paper is essentially the same as the *Tempering Sampler* of Liu *et al.* , with the difference that the level exchange is operated at every step of our algorithm. The motivation for using tempering there was to evacuate the *label switching* problem that plagues mixture posterior simulation, particularly in higher dimensions.

REPLY TO THE DISCUSSION

We thank all the discussants for their exciting ideas, sharp questions, and knowledge in the area of model-based clustering. As Raftery rightly pointed out, many of the methods we labeled as "Bayesian" are, in fact, EM-based likelihood methods based on a complete statistical model. To us, there is no major distinction between the two approaches except that the complete Bayesian approach often wins in giving one the full inference. Besides providing a principled approach to model selection, the Bayesian procedure is also more flexible in incorporating complex structures, such as the one for Box–Cox transformations.

Although the EM-based approach appears to have avoided the troubles of prior specifications and "label-switching" nuisance, it does so by sacrificing its inference power—now the inference has to be based on certain asymptotic results which may not hold particularly well for mixture model fitting. Being able to assess the sensitivity to prior specifications should be seen as a virtue of the Bayes approach, although it might give the researchers (us) some additional hassles. In other words, the EM approach in a way only hides the problems, not really solves them. If the Bayesians are willing to take an asymptotic approximation in the place of the (perhaps) more accurate MCMC computations, they will not have any label-switching problem at all—asymptotically the posterior modes corresponding to different labeling can never communicate. It is indeed true that in the finite-sample case the label-switching problem causes the Bayesians some headache, but we have not yet seen a problem where this difficulty causes any material damage. Furthermore, a cheap and effective way to get around the issue is to artificially impose an order among the clusters (*i.e.*, based on the ordering of the first component of the mean vector of each cluster).

Professor Robert objected to our use of improper priors in transformation selections. Although what Robert pointed out is a well-known and general phenomenon for Bayesian model selections, our situation is slightly different. As explained in Section 4, the ratio of the Bayes factors needed for the transformation selection is perfectly proper, as long as the priors for the parameters in the transformed and untransformed models are "equally" noninformative. One can probably also calibrate these priors to make them "equally" informative and proper. Using noninformative priors here, however, makes practical sense: if we are not sure whether we should impose a Gaussian model for the original data or the logarithm of the data, it is very likely that we are completely ignorant about the mean and variance parameters of the corresponding model.

Both Raftery and Robert pointed out the "incompatibility" and potential dangers of using the exploratory PCA method together with the general Bayesian framework. Although we agree with Robert that a coherent Bayesian model encompassing both PCA and mixture modelling would be a more aesthetic thing to do, we are skeptical whether such a full Bayesian paradigm exists that can completely dominate the approach we took. Regarding the counter-example shown by Raftery, we feel that if distinctions among different clusters cannot be discerned with the first 10 to 15 principle components, it is perhaps more fruitful for us to re-examine the underlying science and revise the model accordingly than to search for a more omnipotent model. Nevertheless, the suggestions made by Professors Peña and Raftery are very interesting and worth further exploration. In particular, Peña's idea of choosing judiciously additional "special" directions according to either kurtosis or other statistics can be an important addition to the current BCVS framework.

Microarray analysis provides for statisticians an excellent entry point to bioinformatics/ computational biology. We would like to take this opportunity to mention a few other important bioinformatics problems in which statistics is likely to play an important role. These include, by no means exclusively, the protein folding prediction, multiple sequence alignments, transcription factor binding-sites identification, gene regulatory network constructions, evolutionary analysis, and the analyses of single nucleotide polymorphisms (SNPs) in the human genome. Some related references can be found in Liu (2002).

The prediction of protein tertiary structures is of great importance to drug designs and the basic biochemistry. Although many proteins' structures have been worked out by X-ray crystallographers, these only account for a small part of the protein universe. Multiple sequence alignment is still the main tool for protein or DNA sequence analysis, which has been at the center of computational biology for about 30 years. With the completion of the human genome and genomes of many other species, the task of organizing and understanding the generated sequence data through multiple alignment becomes even more pressing and challenging. The control of genes' expressions is fundamental to cell survival, growth, and differentiation. An important form of control is exerted by interfering with genes' transcriptions by specialized proteins called the transcription factors, which recognizes short DNA segments in front of genes. Predicting novel transcription factor binding sites and deciphering genetic networks are all important steps towards the general goal of understanding gene regulation. As the great evolutionist T. Dobzhansky pointed out: nothing in biology made sense except in the context of evolution. Evolution study can not only help us understand where and how we come from, but also shed light on protein functions and cellular processes. The SNPs refer to frequently occurring (one in 1000 bases) single-base variations among the genomes of different individuals. Because of the availability of high throughput SNP detection and analysis tools and their great potential in mapping genes responsible for complex diseases, the SNPs have recently attracted much attention from scientists. Statistical modelling and computation are crucial to fully realize the power of SNPs.

Now is clearly an exciting time for statisticians, especially Bayesian statisticians. We are challenged by many important biological and other scientific problems through massive amount of data, which are often associated with in depth subject knowledge; yet, we are equipped with the powerful Bayesian modelling machine and unprecedented computational tools, such as MCMC and EM algorithms, and computer power. We hope that our paper and the discussions can get some of the readers interested in looking into a broader array of bioinformatics problems.

ADDITIONAL REFERENCES IN THE DISCUSSION

Bensmail, H., Celeux, G., Raftery, A. E. and Robert, C. P. (1997). Inference in model-based cluster analysis. *Statist. Comput.* **7**, 1–10.

Cadez, I., Heckerman, D., Meek, C., Smyth, P. and White, S. (2000). Visualization of navigation patterns on a web site using model-based clustering. *Tech. Rep.*, Microsoft Research, Redmond, USA.

Casella, G., Robert, C. P. and Wells, M. T. (2000). Mixture Models, Latent Variables and Partitioned Importance Sampling. *Tech. Rep.*, CREST, France.

Celeux, G., Chaveau, D. and Diebolt, J. (1996). Stochastic versions of the EM algorithm: An experimental study in the mixture case. *J. Statist. Comput. Simulation* **55**, 287–314.

Celeux, G., Hurn, M. and Robert, C. (2000). Computational and inferential difficulties with mixture posterior distributions. *J. Am. Statist. Ass.* **95**, 957–970.

Chang, W.-C. (1983). On using principal components before separating a mixture of tow multivariate normal distributions. *Appl. Statist.* **32**, 267–275.

Chien, Y. Y. (1978). *Interactive Pattern Recognition.* New York: Marcel Dekker.

Clyde, M. (1999). Bayesian model averaging and model search strategies. *Bayesian Statistics 6* (J. M. Bernardo, J. O. Berger, A. P. Dawid and A. F. M. Smith, eds). Oxford: Oxford University Press, 157–185 (with discussion).

Everitt, B. S. (1974). *Cluster Analysis*. New York: Wiley.

Fraley, C. and Raftery, A. E. (2002). Model-based clustering, discriminant analysis and density estimation. *J. Am. Statist. Ass.* **97**, 611–631.

Ghahramani, Z. and Hinton, G. E. (1997). The EM algorithm for factor analyzers. *Tech. Rep.*, University of Toronto, Canada.

Green, P. E. and Krieger, A. M. (1995). Alternative approaches to cluster-based market segmentation. *J. Market Res. Soc.* **37**, 221–239.

Hoeting, J. A., Madigan, D., Raftery, A. E. and Volinsky, C. T. (1999). Bayesian model averaging: A tutorial. *Statist. Sci.* **14**, 382–417 (with discussion). Corrected version at `www.stat.washington.edu/www/research/online/hoeting1999.pdf`.

Liu, J. S. (2002). Bioinformatics: Microarrays analyses and beyond. *Amstat News*, September, 59–67.

Madigan, D. and York, J. (1995). Bayesian graphical models for discrete data. *Internat. Statist. Rev.* **63**, 215–232.

McLachlan, G. J. and Peel, D. (2000). *Finite Mixture Models*. New York: Wiley.

Neal, R. (1996). Sampling from multimodal distributions using tempered transitions. *Statist. Comput.* **4**, 353–366.

Richardson, S. and Green, P. J. (1997). On Bayesian analysis of mixtures with an unknown number of components. *J. R. Statist. Soc. B* **59**, 731–758.

Robert, C. P. (2000). *The Bayesian Choice*. New York: Springer.

Peña, D. and Prieto, F. J. (2000). The kurtosis coefficient and the linear discriminant function. *Statist. Prob. Lett.* **49**, 257–261.

Peña, D. and Prieto, F. J. (2001). Clustering by projections. *J. Am. Statist. Ass.* **96**, 1433–1445.

Schnell, G. D. (1970). A phenetic study of the suborder *Lari Aves*: Methods and results of principal components analysis. *Systematic Zool.* **19**, 35–57.

Steele, R. (2002). *Practical Importance Sampling Methods for Mixture Models and Missing Data Problems*. Ph.D. Thesis, University of Washington, Seattle, USA.

Stephens, M. (2000). Dealing with label-switching in mixture models. *J. R. Statist. Soc. B* **62**, 795–809.

Tantrum, J. M., Murua, A. and Stuetzle, W. (2002). Hierarchical model-based clustering of large data sets through fractionation and refractionation. *Proceedings of the 8th International Conference on Knowledge Discovery and Data Mining* (KDD02), 183–190.

Tipping, M. E. and Bishop, C. M. (1999). Mixtures of probabilistic principal component analyzers. *Neural Comput.* **11**, 443–482.

Tyron, R. C. and Bailey, D. E. (1970). *Cluster Analysis*. New York: McGraw-Hill.

Wehrens, R., Simonetti, A. W. and Buydens, L. M. C. (2002). Mixture modelling of medical magnetic resonance data. *J. Chemometrics* **16**, 274–282.

Yeung, K. Y. and Ruzzo, W. L. (2001). Principal component analysis for clustering gene expression data. *Bioinformatics* **17**, 763–774.

BAYESIAN STATISTICS 7, pp. 277–292
J. M. Bernardo, M. J. Bayarri, J. O. Berger, A. P. Dawid,
D. Heckerman, A. F. M. Smith and M. West (Eds.)

IID Sampling using Self-Avoiding Population Monte Carlo: The Pinball Sampler

KERRIE L. MENGERSEN
University of Newcastle, Australia
kerrie.mengersen@newcastle.edu.au

CHRISTIAN P. ROBERT
Université Paris Dauphine and Insee, Paris, France
xian@ceremade.dauphine.fr

SUMMARY

Population Monte Carlo, in which a sample is generated at each iteration of a Monte Carlo algorithm, is now a popular and accepted practice. For example, particle filters are usually implemented in sequential settings or for processing and analysis of large datasets. We show that population Monte Carlo algorithms, or iterated importance sampling, can also produce iid samples for a given reference distribution π. The proposal chosen for the moves of the particle system is based on a self-avoiding random walk with a bouncing mechanism and a delayed rejection acceptance rule that avoids the immediate vicinity of the other particles, while preserving the stationary distribution π.

Keywords: POPULATION MONTE CARLO; ITERATED IMPORTANCE SAMPLING; ADAPTIVE SCHEME; DELAYED REJECTION; MCMC ALGORITHMS; PARTICLE SYSTEMS; PSEUDO-REFERENCE DISTRIBUTION; SELF-AVOIDING RANDOM WALK.

1. INTRODUCTION

Population Monte Carlo, or iterated importance sampling, involves simultaneous processing of a vector of random variables $\left(\theta_1^{(t)}, \ldots, \theta_M^{(t)}\right)$ towards simulation from a reference distribution π and computation of related integrals,

$$\int h(\theta)\, \pi(d\theta)\,. \tag{1}$$

This is now a popular method of choice for computation in many fields. For instance, particle filters are currently used in sequential settings such as tracking a moving target (see, *e.g.*, Gilks and Berzuini, 2000; Doucet *et al.* , 2001) or in static settings with large datasets as exploratory tools (see, *e.g.*, Berzuini and Gilks, 2001; Chopin, 2001). Various improvements have occurred in the past years with the introduction of additional moves based on MCMC techniques, as detailed in Doucet *et al.* (2001).

The current paper shows that such systems of particles also have an appeal from a purely Monte Carlo point of view in the sense that they can produce iid samples from the reference distribution π, a feature that is only achieved with difficulty via the complex perfect sampling technique in regular MCMC methodology. We present an updating system for the particle system that is based on a standard random walk with corrections to avoid the immediate vicinity of other particles. It, therefore, classifies as a self-avoiding random walk for the whole chain $\left(\theta_1^{(t)},\ldots,\theta_M^{(t)}\right)_t$.

In contrast to the current work on population Monte Carlo, we move away from the importance sampling justification and produce an iid and unbiased representation of the reference distribution. Paradoxically, the corresponding importance resampling particle system based on the same proposal enjoys poor properties like high degeneracy and low mixing. The method we propose thus pertains more closely to the theme of adaptive MCMC algorithms in that it takes advantage of the vector $\left(\theta_1^{(t)},\ldots,\theta_M^{(t)}\right)$ to speed up the exploration of the distribution of interest. Hence, it relates to earlier adaptive schemes such as Gilks *et al.* (1994), Haario *et al.* (1999) and Warnes (2000). In particular, it is similar to the multistate sampler of Warnes (2000) in that it simulates simultaneously a set of values in a dependent manner, two major differences being that, first, we do not aim at approximating the posterior distribution through a kernel estimator, as in Oudjane (2000) and Warnes (2000), and, second, the particles are connected via a "repulsive" proposal. The difference with adaptive proposals as in Haario *et al.* (2000) and Andrieu and Robert (2001) is that the chain we simulate is homogeneous and, hence, standard ergodic theorems apply.

The paper is organized as follows: we establish the general properties of our population Monte Carlo algorithm in Section 2, system algorithm, and a focus on the iid nature of the resultant samples. The bouncing mechanism is then detailed in Section 3 and justification is embedded in current concepts of delayed rejection. Some illustrations of the performance of the new algorithm are considered in Section 4.

2. POPULATION MONTE CARLO

A system of particles $\left(\theta_1^{(t)},\ldots,\theta_M^{(t)}\right)_t$ is a random vector evolving over time t according to a Markovian updating scheme. It is customarily associated with a vector of weights, $\left(\omega_1^{(t)},\ldots,\omega_M^{(t)}\right)$, $\sum_{i=1}^M \omega^{(t)} = 1$, that is used to approximate integrals such as (1) through importance sampling approximations

$$\sum_{k=1}^{M} \omega_k^{(t)} h(\theta_k^{(t)}) \,.$$

Therefore, the original particle filter algorithm, for instance, does not directly simulate the $\theta_k^{(t)}$'s, but rather updates the weights $\omega_k^{(t)}$ until the discrepancy between those weights is too high, in which case a resampling move based on the $\omega_k^{(t)}$'s is launched (and possibly supplemented with additional MCMC moves, to avoid degeneracy, as in Berzuini and Gilks, 2001, or Crisan and Doucet, 2000).

In this paper, we follow a completely different approach to particle filters in that (a) we do not use importance sampling schemes nor weights $\omega_k^{(t)}$, and (b) we resample the *whole* vector $\left(\theta_1^{(t)},\ldots,\theta_M^{(t)}\right)$ at each iteration of the algorithm. In this respect, the approach is one of iterated importance sampling and is closer to regular Monte Carlo simulation than to standard particle filtering, principally because we simulate a whole vector instead of a single variable θ. We can also draw a link with parallel MCMC algorithms where a thread of MCMC chains is generated, usually to assess convergence to the stationary distribution and mixing properties of the chain, as in Gelman and Rubin (1992) and subsequent literature. The difficulty with these early implementations is that the convergence and the dynamics of the various chains $\left(\theta_k^{(t)}\right)_t$ were considered individually, rather than as a whole: the distribution of $\theta_k^{(t)}$ is thus of the form $K_t\left(\theta_k^{(0)}, \theta_k^{(t)}\right)$ and convergence characteristics are dependent on the starting value $\theta_k^{(0)}$.

The algorithm developed in this paper introduces a kernel defined over the vector of particles; the corresponding stationary distribution is π^M, and we prove that convergence occurs for this vector. This approach also has its limitations in that the number of particles is necessarily fixed over the iterations to achieve convergence (unless some branching process extension applies or can be devised). Figure 1 shows nine successive iterations of 50 particles from a particle system as the particles explore the surface of a target distribution described in Section 4. This snapshot clearly indicates how the particles interact over the target surface, visiting the different modes in proportion to their amplitude.

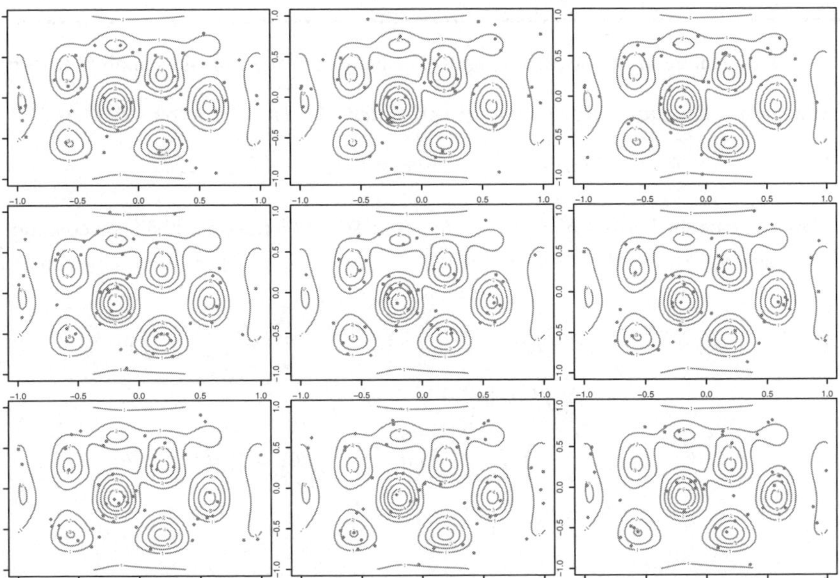

Figure 1. *Successive iterations of* 50 *particles over the contours of a two-dimensional target distribution.*

The move from the system at time t to the system at time $l + 1$ is Gibbs-like, in that the update of each particle is sequential and is conditional on the current value of the whole particle system (algorithm GPF, below) with proposals $K_k(\cdot|\theta_1, \ldots, \theta_M)$ that are associated with the stationary distribution π, e.g. that satisfy the detailed balance condition (see, *e.g.*, Robert and Casella, 1999)

$$\pi\left(\theta_k^{(t)}\right) K_k\left(\theta_k^{(t+1)} \middle| \theta_1^{(t+1)}, \ldots, \theta_{k-1}^{(t+1)}, \theta_k^{(t)}, \ldots, \theta_M^{(t)}\right)$$
$$= \pi\left(\theta_k^{(t+1)}\right) K_k\left(\theta_k^{(t)} \middle| \theta_1^{(t+1)}, \ldots, \theta_{k-1}^{(t+1)}, \theta_k^{(t+1)}, \ldots, \theta_M^{(t)}\right). \quad (2)$$

(A particular choice for K_k that pushes particles farther apart from one another will be detailed in the next section.) When the support of π is unbounded, the grid in the initial step of `[Al-`

`gorithm GPF]` can be constructed from a starting distribution μ or via a reparametrization of the distribution onto a compact support like $[-1,1]^p$ (see, *e.g.*, Robert, 2001) and a uniform grid.

[Algorithm GPF]

1. `Construct a grid` $\left(\theta_1^{(0)},\ldots,\theta_M^{(0)}\right)$ `of starting values over the support of` π
2. `For` $t=0,\ldots,T$,
   `for` $k=1,\ldots,M$,
   `simulate`

$$\theta_k^{(t+1)} \sim K_k\left(\theta \left|\theta_1^{(t+1)},\ldots,\theta_{k-1}^{(t+1)},\theta_k^{(t)},\ldots,\theta_M^{(t)}\right.\right)$$

A crucial feature of this sequential update is that it approximates iid sampling from π:

Theorem. *The stationary distribution associated with* `[Algorithm GPF]` *is*

$$(\theta_1,\ldots,\theta_M) \sim \pi(\theta_1)\times\ldots\times\pi(\theta_M) = \pi^M(\theta_1,\ldots,\theta_M)$$

Proof. Since the kernels $K_k(\cdot|\theta_1,\ldots,\theta_M)$ satisfy the (conditional) balance condition (2). the fact that $\left(\theta_1^{(t+1)},\ldots,\theta_M^{(t+1)}\right)\sim\pi^M$ if $\left(\theta_1^{(t)},\ldots,\theta_M^{(t)}\right)\sim\pi^M$ then follows by a chain rule:

$$\begin{aligned}
\left(\theta_1^{(t+1)},\ldots,\theta_M^{(t+1)}\right) \sim & \int \pi^M\left(d\theta_1^{(t)},\ldots,d\theta_M^{(t)}\right) K_1\left(\theta_1^{(t+1)}\left|\theta_1^{(t)},\ldots,\theta_M^{(t)}\right.\right) \\
& \cdots K_M\left(\theta_M^{(t+1)}\left|\theta_1^{(t+1)},\ldots,\theta_M^{(t)}\right.\right) \\
& = \pi\left(\theta_1^{(t+1)}\right)\int \pi^{M-1}\left(d\theta_2^{(t)},\ldots,d\theta_M^{(t)}\right) K_2\left(\theta_2^{(t+1)}\left|\theta_1^{(t+1)},\ldots,\theta_M^{(t)}\right.\right) \\
& \cdots K_M\left(\theta_M^{(t+1)}\left|\theta_1^{(t+1)},\ldots,\theta_M^{(t)}\right.\right) \\
& \vdots \\
& = \pi^M\left(\theta_1^{(t+1)},\ldots,\theta_M^{(t+1)}\right)
\end{aligned}$$

□

Note that, like the Gibbs sampler, `[Algorithm GPF]` does not satisfy the detailed balance condition on the whole vector $(\theta_1,\ldots,\theta_M)$ since it depends on the order in which the particles are updated. It could be transformed into a reversible algorithm by introducing a random permutation of the indices but we fail to see the gain in this additional step.

Provided the standard irreducibility conditions (see Robert and Casella, 1999) are satisfied by the componentwise transition kernels $K_k(\cdot|\theta_1,\ldots,\theta_M)$, the k-fold Markov chain $\left(\theta_1^{(t)},\ldots,\theta_M^{(t)}\right)_t$ is thus ergodic and positive recurrent with stationary distribution π^M. Therefore, once the particle system $\left(\theta_1^{(t)},\ldots,\theta_M^{(t)}\right)$ has eliminated its dependence on the initial value $\left(\theta_1^{(0)},\ldots,\theta_M^{(0)}\right)$, it can be considered as an iid sample from π at any given time t, rather than in the long run as in regular MCMC sampling. The result is therefore more interesting than usual ergodicity theorems in that a slice *(in time)* of the process provides an unbiased evaluation of the reference distribution π, once stationarity is achieved. An evaluation of the simulation output as in regular Monte Carlo experiments such as normal approximation confidence intervals can then be undertaken. The principal effort in implementing our method is expended on devising proposals that speed up mixing, as detailed in the following section.

3. THE PINBALL SAMPLER

The interest in using a population Monte Carlo sample $\left(\theta_1^{(t)}, \ldots, \theta_M^{(t)}\right)_t$, rather than a single chain $\left(\theta_1^{(t)}\right)_t$, is primarily to increase the possibilities in the exploration of the state space and to avoid trapping setups such as the usual concentration around a *single* mode of the reference distribution π that depends on the starting point for the chain. While the choice of the kernels K_k is quite open [for convergence to π^M to occur], we propose in this section a particular Metropolis-within-Metropolis set of kernels K_k that is designed to address these problems. The basic ideas behind our choice are

(1) to introduce a pseudo-reference distribution π^R that pushes particles farther apart from one another (hence the "pinball" label);
(2) to implement a Metropolis move calibrated for π^R;
(3) to increase the efficiency of this proposal via a (currently deterministic) delayed rejection mechanism;
(4) to validate the move with a final Metropolis acceptance probability calibrated for the true reference distribution π.

3.1. *Repulsive Proposal Distribution*

First, we introduce the *pseudo-reference distribution*

$$\pi_k^R(\theta) \propto \pi(\theta) \prod_{j \neq k} e^{-\xi/\pi(\theta_j)||\theta-\theta_j||^2}, \tag{3}$$

which is derived from the distribution of interest π by adding exponential terms that create holes around the other particles θ_j, since the j-th term $e^{-\xi/\pi(\theta_j)||\theta-\theta_j||^2}$ converges to 0 when θ goes to θ_j. These terms thus induce a repulsive effect (hence the R) around the other particles and simulating from π^R creates moves similar to those of a ball in a pinball machine. The factor $\pi(\theta_j)$ in the exponential moderates the repulsive effect in zones of high probability and enhances it in zones of low probability. (This effect can be further attenuated by replacing $\pi(\theta_j)$ with $\max(\zeta, \pi(\theta_j))$ so that points θ_j with almost zero density $\pi(\theta_j)$ do not induce a trapping effect of their own by excluding almost the entire state space.) The parameter ξ is a *tempering factor*, since the repulsive part in π_k^R can be seen as a term raised to the power ξ; this factor can be calibrated during the simulation against the number of particles M and the acceptance rate. Note also that the construction of π_k^R does not depend on the normalization constant of π, since such a dependence would jeopardize the implementation of the method in most MCMC setups: indeed, when the normalization constant in π is unknown, ξ can be rescaled as $\xi\pi(\theta_0)$, where θ_0 is an arbitrary value in the parameter space with $\pi(\theta_0) > 0$, or as the average of the $\pi(\theta_j)$'s over the other particles.

The move related to π_k^R maps $\theta_k^{(t)}$ to $\tilde{\theta}_k^{(t)}$ via a Metropolis–Hastings kernel

$$\tilde{K}_k\left(\theta_k \middle| \theta_1^{(t+1)}, \ldots, \theta_{k-1}^{(t+1)}, \theta_k^{(t)}, \ldots, \theta_M^{(t)}\right)$$

tuned for the pseudo reference distribution π_k^R. (Note that $\tilde{K}_k$ includes a Dirac mass at $\theta_k^{(t)}$). Then the value $\tilde{\theta}_k^{(t)}$ is evaluated again π, using the Metropolis–Hastings acceptance probability

$$1 \wedge \frac{\pi\left(\tilde{\theta}_k^{(t)}\right) \, \pi_k^R\left(\theta_k^{(t)}\right)}{\pi\left(\theta_k^{(t)}\right) \, \pi_k^R\left(\tilde{\theta}_k^{(t)}\right)}$$

as justified by the following lemma:

Lemma 1. *If θ' given θ is generated by a Metropolis–Hastings mechanism with stationary distribution π', the acceptance probability*

$$\alpha(\theta, \theta') = 1 \wedge \frac{\pi(\theta')}{\pi(\theta)} \frac{\pi'(\theta)}{\pi'(\theta')}$$

makes the move a Metropolis–Hastings mechanism with stationary distribution π.

Proof. Once again, the proof is based on the balance condition. If K' is the kernel associated with the original move,

$$\pi'(\theta)K'(\theta, \theta') = \pi'(\theta')K'(\theta', \theta)$$

and the overall kernel is

$$K^\star(\theta, \theta') = \alpha(\theta, \theta')K'(\theta, \theta') + \int (1 - \alpha(\theta, \xi))K'(\theta, d\xi)\delta_\theta(\theta'),$$

where $\delta_\theta(\theta')$ denotes the Dirac mass at θ. It therefore satisfies

$$\begin{aligned}
\pi(\theta)K^\star(\theta, \theta') &= \{\pi(\theta)/\pi'(\theta) \wedge \pi(\theta')/\pi'(\theta')\}\pi'(\theta)K'(\theta, \theta') \\
&\quad + \int (1 - \alpha(\theta, \xi))K'(\theta, d\xi)\pi(\theta)\delta_\theta(\theta') \\
&= \{\pi(\theta)/\pi'(\theta) \wedge \pi(\theta')/\pi'(\theta')\}\pi'(\theta')K'(\theta', \theta) \\
&\quad + \int (1 - \alpha(\theta', \xi))K'(\theta', d\xi)\pi(\theta')\delta_{\theta'}(\theta) \\
&= \pi(\theta')K^\star(\theta', \theta).
\end{aligned}$$

□

Notice that this correction for using the wrong target distribution can also be found in Neal (1996) and Celeux *at al.* (2000), in tempered moves.

3.2. *Avoiding the Other Particles: A Delayed Rejection Mechanism*

We now concentrate on the construction of the kernel $\tilde{K}_k(\cdot|\theta_1, \ldots, \theta_M)$, which is designed to approximate simulation from the repulsive distribution π_k^R.

To this end, we will make use of the delayed rejection method of Tierney and Mira (1999): when a move brings a particle θ_k too close to other particles, it is likely to be rejected because of the repulsive part in π_k^R. Rather than leaving θ_k in its current state, we propose a new move for θ_k which deflects it further away from those particles. This new move can be evaluated in its turn with a corrected acceptance probability, which takes into account the fact that the first move has been rejected. Continuing, if this second move is also rejected, further moves can be considered in the same way. As in Tierney and Mira (1999), these moves can depend on the previous [rejected] proposed values. The new feature is that, given that we are working with a system of particles, the moves can also depend on the other particles.

In order to achieve reversibility, a critical feature that ensures detailed balance, we use a random walk proposal, followed by iterated *deterministic* reflections if the current proposal is rejected. More precisely, given $\left(\theta_1^{(t+1)},\ldots,\theta_k^{(t)},\ldots,\theta_M^{(t)}\right)$, we generate the first proposal θ_{k0} from a symmetric distribution $q_\tau\left(||\theta-\theta_k^{(t)}||\right)$, with scale factor τ calibrated for the reference distribution π. This proposal is accepted with probability

$$\alpha_{k0}\left(\theta_k^{(t)},\theta_{k0}\right)=1\wedge\frac{\pi_k^R\left(\theta_{k0}\right)}{\pi_k^R\left(\theta_k^{(t)}\right)}\,.$$

If θ_{k0} is rejected, this may be for two reasons: either the new value θ_{k0} has fallen in a region of low density, and the rejection is governed by the first part of π_k^R, or θ_{k0} is too close to one or to several particles, and the rejection is governed by the second part of π_k^R. We then devise the delayed rejection for this second possibility, by introducing a *pinball effect*, i.e. by taking the particle away from the particle closest to θ_{k0}, θ_j say, to θ_{k1} through a symmetry of $\theta_k^{(t)}$ with respect to the (θ_j,θ_{k0}) line. Obviously, if the rejection is due to a low value for $\pi(\theta)$, the additional proposal may send the particle into a region of even lower dimension. This danger, however, can be partially alleviated in a compact space by devising reflection rules that bring back the particle to the center of the space after a series of moves toward the boundary of the space.

If $\sigma(\theta_0,\theta_1,\theta_2)$ denotes the density of the second move, the delayed rejection acceptance probability of Tierney and Mira (1999) is

$$\begin{aligned}&\alpha_{k1}\left(\theta_k^{(t)},\theta_{k0},\theta_{k1}\right)\\&=1\wedge\frac{\pi_k^R(\theta_{k1})q_\tau(||\theta_{k0}-\theta_{k1}||)\sigma\left(\theta_{k1},\theta_{k0},\theta_k^{(t)}\right)[1-\alpha_{k0}(\theta_{k1},\theta_{k0})]}{\pi_k^R\left(\theta_k^{(t)}\right)q_\tau\left(||\theta_{k0}-\theta_k^{(t)}||\right)\sigma\left(\theta_k^{(t)},\theta_{k0},\theta_{k1}\right)\left[1-\alpha_{k0}\left(\theta_k^{(t)},\theta_{k0}\right)\right]}\,.\end{aligned}$$

Since the first proposal is symmetric,

$$q_\tau\left(||\theta_{k0}-\theta_k^{(t)}||\right)=q_\tau(||\theta_{k0}-\theta_{k1}||)\,,$$

since $\sigma(\theta_0,\theta_1,\theta_2)$ reduces here to a Dirac mass, and since the first proposal was rejected, implying

$$\pi_k^R(\theta_{k0})<\pi_k^R\left(\theta_k^{(t)}\right)\,,$$

the delayed rejection acceptance probability simplifies into

$$\begin{aligned}\alpha_{k1}\left(\theta_k^{(t)},\theta_{k0},\theta_{k1}\right)&=1\wedge\left(\pi_k^R(\theta_{k1})\left[1-1\wedge\frac{\pi_k^R(\theta_{k0})}{\pi_k^R(\theta_{k1})}\right]\right)\Big/\left(\pi_k^R\left(\theta_k^{(t)}\right)\left[1-\frac{\pi_k^R(\theta_{k0})}{\pi_k^R\left(\theta_k^{(t)}\right)}\right]\right)\\&=1\wedge\frac{0\vee[\pi_k^R(\theta_{k1})-\pi_k^R(\theta_{k0})]}{\pi_k^R\left(\theta_k^{(t)}\right)-\pi_k^R(\theta_{k0})}\,.\end{aligned}$$

If θ_{k1} is rejected, the delayed rejection mechanism described above can be repeated, as a series of symmetries with respect to a line defined by the closest of the other particles. We can then generalize Mira's (1998) formula (5.13), obtained in the symmetric case, to our setting to establish the following result:

Lemma. *Delayed rejection associated with the acceptance probabilities* $(1 \le m \le H)$

$$\alpha_{km}(\theta_k^{(t)}, \ldots, \theta_{km}) = 1 \wedge \frac{0 \vee [\pi_k^R(\theta_m) - \max_{0 \le \ell < m} \pi_k^R(\theta_\ell)]}{\pi_k^R(\theta_k^{(t)}) - \max_{0 \le \ell < m} \pi_k^R(\theta_\ell)} \tag{4}$$

preserves π_k^R *as its stationary distribution.*

Proof. It is sufficient, as in Mira (1998, Section 5.3), to consider the third candidate, with acceptance probability

$$\begin{aligned}
\alpha_{k2}(\theta_k^{(t)}, \theta_{k0}, \theta_{k1}, \theta_{k2}) &= 1 \wedge \frac{\pi_k^R(\theta_{k2})}{\pi_k^R(\theta_k^{(t)})} \frac{1 - \alpha_{k0}(\theta_{k2}, \theta_{k1})}{1 - \alpha_{k0}(\theta_k^{(t)}, \theta_{k0})} \frac{1 - \alpha_{k1}(\theta_{k2}, \theta_{k1}, \theta_{k0})}{1 - \alpha_{k1}(\theta_k^{(t)}, \theta_{k0}, \theta_{k1})} \\
&= 1 \wedge \frac{\pi_k^R(\theta_{k2})}{\pi_k^R(\theta_k^{(t)})} \frac{1 - 1 \wedge \frac{\pi_k^R(\theta_{k2})}{\pi_k^R(\theta_{k1})}}{1 - \frac{\pi_k^R(\theta_{k0})}{\pi_k^R(\theta_k^{(t)})}} \frac{1 - 1 \wedge \frac{[\pi_k^R(\theta_{k0}) - \pi_k^R(\theta_{k1})] \vee 0}{[\pi_k^R(\theta_{k2}) - \pi_k^R(\theta_{k1})] \vee 0}}{1 - 1 \wedge \frac{[\pi_k^R(\theta_{k1}) - \pi_k^R(\theta_{k0})] \vee 0}{\pi_k^R(\theta_k^{(t)}) - \pi_k^R(\theta_{k0})}} .
\end{aligned}$$

This ratio is equal to 0 if $\pi_k^R(\theta_{k2}) < \pi_k^R(\theta_{k1})$. If $\pi_k^R(\theta_{k2}) > \pi_k^R(\theta_{k1})$,

$$\begin{aligned}
\alpha_{k2}(\theta_k^{(t)}, \theta_{k0}, \theta_{k1}, \theta_{k2}) &= 1 \wedge \frac{\pi_k^R(\theta_{k2}) - \pi_k^R(\theta_{k1})}{\pi_k^R(\theta_k^{(t)}) - \pi_k^R(\theta_{k0})} \frac{1 - 1 \wedge \frac{[\pi_k^R(\theta_{k0}) - \pi_k^R(\theta_{k1})] \vee 0}{\pi_k^R(\theta_{k2}) - \pi_k^R(\theta_{k1})}}{1 - 1 \wedge \frac{[\pi_k^R(\theta_{k1}) - \pi_k^R(\theta_{k0})] \vee 0}{\pi_k^R(\theta_k^{(t)}) - \pi_k^R(\theta_{k0})}} \\
&= 1 \wedge \frac{\pi_k^R(\theta_{k2}) - \pi_k^R(\theta_{k1}) - [\pi_k^R(\theta_{k0}) - \pi_k^R(\theta_{k1})] \vee 0}{\pi_k^R(\theta_k^{(t)}) - \pi_k^R(\theta_{k0}) - [\pi_k^R(\theta_{k1}) - \pi_k^R(\theta_{k0})] \vee 0}
\end{aligned}$$

Therefore,

$$\alpha_{k2}(\theta_k^{(t)}, \theta_{k0}, \theta_{k1}, \theta_{k2}) = \begin{cases} 0 & \text{if } \pi_k^R(\theta_{k2}) < \pi_k^R(\theta_{k1}), \\ 1 \wedge \frac{\pi_k^R(\theta_{k2}) - \pi_k^R(\theta_{k1})}{\pi_k^R(\theta_k^{(t)}) - \pi_k^R(\theta_{k0}) - \pi_k^R(\theta_{k1}) + \pi_k^R(\theta_{k0})} & \text{if } \pi_k^R(\theta_{k2}) > \pi_k^R(\theta_{k1}), \\ & \text{and } \pi_k^R(\theta_{k1}) > \pi_k^R(\theta_{k0}), \\ 1 \wedge \frac{\pi_k^R(\theta_{k2}) - \pi_k^R(\theta_{k1}) - \pi_k^R(\theta_{k1}) + \pi_k^R(\theta_{k0})}{\pi_k^R(\theta_k^{(t)}) - \pi_k^R(\theta_{k0})} & \text{if } \pi_k^R(\theta_{k2}) > \pi_k^R(\theta_{k1}), \\ & \text{and } \pi_k^R(\theta_{k1}) < \pi_k^R(\theta_{k0}), \end{cases}$$

$$= 1 \wedge \frac{[\pi_k^R(\theta_{k2}) - \max(\pi_k^R(\theta_{k1}), \pi_k^R(\theta_{k0}))] \vee 0}{\pi_k^R(\theta_k^{(t)}) - \max(\pi_k^R(\theta_{k1}), \pi_k^R(\theta_{k0}))}$$

□

The delayed rejection algorithm thus reads as follows:

[Algorithm DRPF]

---

1. Construct a grid $(\theta_1^{(0)}, \ldots, \theta_M^{(0)})$ of starting values over the support of π
2. For $t = 0, \ldots, T$,
   for $i = 1, \ldots, M$,
   a. Generate $\theta_{i0} \sim \mathcal{N}(\theta_i^{(t)}, \tau^2)$
   b. Determine the Metropolis--Hastings acceptance probability $\alpha_0(\theta_k^{(t)}, \theta_{i0})$
   c. While θ_{im} is rejected $(m = 0, \ldots, H-1)$,
      c1. Determine $\theta_m^\star$ the closest particle to θ_{im}
      c2. If $m = 0$, take θ_{i1} as the symmetric of θ_{i0} wrt the $(\theta_m^\star, \theta_k^{(t)})$ line
      Else, take $\theta_{i(m+1)}$ as the symmetric of $\theta_{i(m-1)}$ wrt the $(\theta_m^\star, \theta_{im})$ line
      c3. Compute the delayed rejection acceptance probability
      $$\alpha_{i(m+1)}\left(\theta_i^{(t)}, \ldots, \theta_{i(m+1)}\right)$$
   d. Take $\theta_i^{(t+1)} = \theta_{im}$

The deterministic move from θ_{im} to $\theta_{i(m+1)}$ can be replaced by a stochastic move centered at the symmetry, provided reversibility is preserved. Note that the overall scheme is dimension-free, and thus can be implemented in any dimension.

4. ILLUSTRATIONS

To illustrate how the particle system explores the whole space, we first chose the two-dimensional distribution

$$\pi(x,y) \propto ((x-1)(x+1)\sin(7y) + (y-1)(y+1)\sin(8x))^2 + ((x-1)(x+1)\cos(6y) - (y-1)(y+1)\sin(8x))^2,$$

defined on $[-1,1]^2$, and represented in Figure 2, which was chosen for its multimodal features.

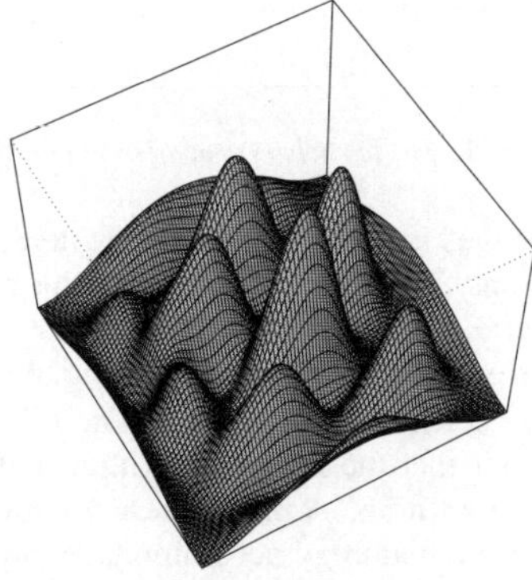

Figure 2. *Representation of the reference distribution $\pi(x,y)$ on $[-1,1]^2$.*

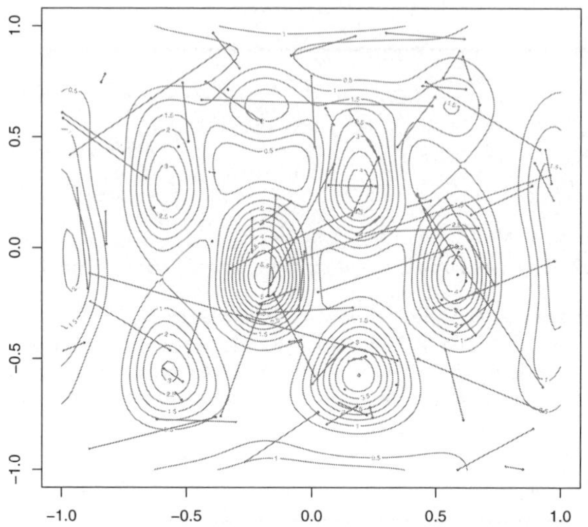

Figure 3. *One iteration of the particle system with moves indicated by lines.*

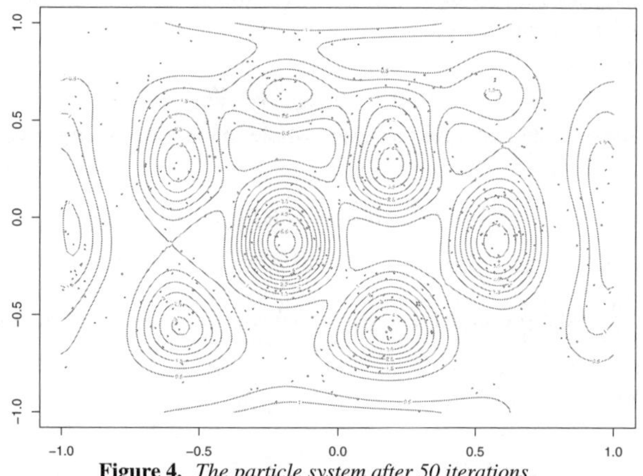

Figure 4. *The particle system after 50 iterations.*

Figure 3 represents the changes in the system after one iteration, while Figure 4 represents the particle filter after 50 iterations. The strong dispersion of the particles across the entire space is evident within and across iterations. A visually acceptable representation of the distribution is achieved as a result, with coverage of all modes according to the target density.

As a second illustration we consider a diffuse bivariate t_1 distribution with correlation of 0.4., based on the *a priori* belief that the DRPF algorithm will better describe tail behavior than the traditional Metropolis algorithm. A bivariate distribution was again adopted for ease of representation, although the methodology accommodates problems of general dimension. Figure 5 depicts perspective plots of the target distribution, typical samples arising from the

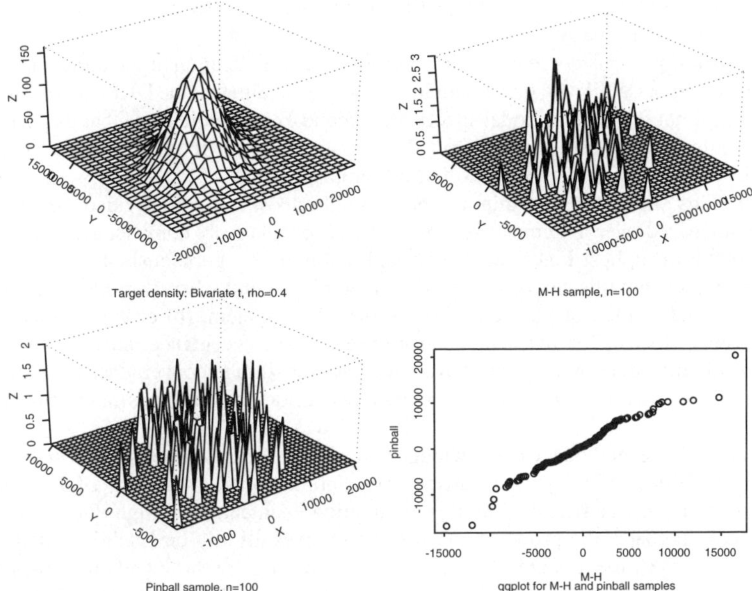

Figure 5. *Comparison of Metropolis and DRPF samples of a bivariate t distribution with one degree of freedom and correlation 0.4. The four plots represent the target density* (top left), *a typical sample of 100 iterations from a Metropolis algorithm* (top right), *a typical sample of 100 particles from a DRPF algorithm* (bottom left) *and a quantile–quantile plot of the Metropolis and pinball samples* (bottom right).

Metropolis and DRPF algorithms, and a corresponding quantile–quantile plot of the two sample outputs. Comparison of the ranges in the plots of the two sample outputs indicates that the DRPF algorithm may tend to push particles further, although the q–q plot confirms that this is a tenuous conclusion. Indeed, despite the disperse nature of the distribution, the attraction to the mode overwhelms the repulsive effect of neighboring particles. Interestingly, the dispersion is thus most evident in the upper tails rather than the outer tails of the distribution.

This latter tendency was further explored by comparing the proportion of the sample under each algorithm that lies outside defined "confidence bands," where the latter are defined in terms of distance from the center (0,0) of the target distribution. Thus, a 90% confidence band, for example, represents a boundary of the (two-dimensional) space within which 90% of the target probability lies in terms of Euclidean distance to the mean. Under this comparison, not dissimilar proportions of the Metropolis and DRPF samples were observed in the 95% to 70% bands and again in the 5% to 20% bands, but greater proportions of DRPF samples were seen in the 30%-60% bands.

5. CONCLUSION

This paper has two main aims. First, population Monte Carlo algorithms (including particle sampling algorithms) are demonstrated to produce iid samples from the target distribution. Second, a new algorithm is introduced that enhances exploration of a space and avoidance of trapping spaces by "bouncing" particles away from each other using a pseudo-reference distribution and a delayed rejection mechanism. The new approach is embedded in current population Monte Carlo and delayed rejection literature and its performance is examined both theoretically and empirically.

The illustrations considered in Section 4 provide some indication of the performance of the proposed algorithm. While the first example validates the claim that the DRPF algorithm indeed facilitates exploration of multimodal spaces, the second example indicates that its performance is at least equivalent to that of the traditional Metropolis algorithm in standard but difficult cases, with the major gain being the generation of iid samples.

These gains are obtained at minimal programming cost in that, provided the distribution π is reparametrized over a compact set like $[-1,1]^p$ (or relatively compact set like $(-1,1)^p$), the same program applies. Finer details like calibration of (ξ,τ) obviously need to be polished for new problems, but this can be achieved in a learning round where various values of (ξ,τ) are assessed against the acceptance rate (a) of the whole system, (b) of the various particles, with assessment focused on the maintenance of reasonable acceptance rates. In addition, the asymptotically iid nature of the particle system allows for simpler convergence assessment than in regular MCMC settings. For instance, nonparametric homogeneity tests can be run at various times.

Caveats with the method are acknowledged, including obvious issues such as the resurgence of the *curse of dimensionality*: the starting values are assumed to provide a fair coverage of the support of π. In a generic setting, this assumption is untenable in high dimensions. More fundamentally perhaps, the proposal and rejection moves all rely on Euclidean balls: as the dimension increases, the amount of space lost in a coverage by these balls increases as well. It may be that substituting balls based on the L_1 or maximum distances to the Euclidean balls brings better performance, but the implementation is likely to be more delicate.

The other difficulty is that the pseudo-distribution (3) has a complexity of order $O(M)$ and induces a complexity $O(M^2)$ for a step of `[Algorithm GPF]`. This constraint prevents the derivation of large samples when mixing is poor and convergence requires many iterations. It could be relaxed, however, by considering only certain neighbors in (3) at most steps and by updating the set of neighbors at fixed intervals. The attempts we made at implementing this method were, however, unsuccessful in that the definition of the neighborhoods cancels reversibility: using the same algorithm as above is then not valid, as shown by the behavior of the corresponding particle systems.

ACKNOWLEDGEMENTS

A conversation with Robert L. Wolpert along rue Estienne d'Orves in 1993 acted as a generator for this work. Thanks also to C. Andrieu, N. Chopin, P. Green, and M. Hurn for helpful discussions about this paper.

This work was supported by Center National de la Recherche Scientifique during a visit at CREST in April 2001 and by an ARC Large Grant for the first author, and by EU TMR network ERB–FMRX–CT96–0095 on *"Computational and Statistical Methods for the Analysis of Spatial Data"* for the second author.

REFERENCES

Andrieu, C. and Robert, C. P. (2001). Controlled MCMC for Optimal Sampling. *Tech. Rep.*, CEREMADE, Paris, France.

Berzuini, C. and Gilks, W. R. (2001). Resample-move filtering with cross-model jumps. *Sequential Monte Carlo in Practice* (A. Doucet, J. F. G. de Freitas and N. J. Gordon, eds).Berlin: Springer.

Cappé, O. and Robert, C. P. (2000). Marcok chain Monte Carlo: Ten years and still running!. *J. Am. Statist. Ass.* **95**, 1282–1286.

Celeux, G., Hurn, M. and Robert, C. P. (2000). Computational and inferential difficulties with mixtures posterior distribution. *J. Am. Statist. Ass.* **95**, 957–979.

Chopin, N. (2001). A sequential particle filter for static models. *Biometrika* (to appear).

Crisan, D. and Doucet, A. (2000). Convergence of sequential Monte Carlo methods. *Tech. Rep.*, University of Cambridge, UK.

Doucet, A., de Freitas, J. F. G. and Gordon, N. J. (2001). *Sequential Monte Carlo Methods in Practice*. New York: Springer.

Gelman, A. and Rubin, D. B. (1992). A single series for the Gibbs sampler provides a false sense of security. *Bayesian Statistics 4* (J. M. Bernardo, J. O. Berger, A. P. Dawid and A. F. M. Smith, eds). Oxford: Oxford University Press, 625–631.

Gilks, W. R., Best, N. G. and Tan, K. K. C. (1995). Adaptive rejection Metropolis sampling within Gibbs sampling. *J. R. Statist. Soc. C* **44**, 455–472.

Gilks, W. R. and Berzuini, C. (2001). Following a moving target: Monte Carlo inference for dynamic Bayesian models. *J. R. Statist. Soc. B* (to appear).

Gilks, W. R., Roberts, G. O. and George, E. I. (1994). Adaptive direction sampling. *The Statistician* **43**, 179–189.

Haario, H., Saksman, E. and Tamminen, J. (1999). Adaptive proposal distribution for random walk Metropolis algorithm. *Comput. Statist.* **14**, 375-395.

Neal, R. (1996). Sampling from multimodal distributions using tempered transitions. *Statist. Comput.* **4**, 353–366.

Oudjane, N. (2000). *Stabilité et Approximations Particulaires en Filtrage Non-Linéaire*. Ph.D. Thesis, INRIA, Rennes, France.

Robert, C. P. (2001). Discussion of a paper by G. Roberts. *Highly Structured Stochastic Systems* (P. G. Green, N. Hjört and S. Richardson, eds). London: Chapman and Hall.

Robert, C. P. and Casella, G. (1999). *Monte Carlo Statistical Methods*. New York: Springer.

Tierney, L. and Mira, A. (1999). Some adaptive Monte Carlo methods for Bayesian inference. *Statist. Med.* **18**, 2507–2515.

Warnes, G. R. (2000). *The Normal Kernel Coupler*. Ph.D. Thesis, University of Washington, Seattle, USA.

DISCUSSION

PETER MÜLLER (*University of Texas M.D. Anderson Cancer Center, USA*)

Mengersen and Robert (MR) propose a smart twist to parallel chain MCMC. The idea is simple, compelling and I believe very versatile. MR observe that in MCMC with parallel chains $\theta_1^{(t)}, \ldots, \theta_M^{(t)}$ transition probabilities need not be constrained to one chain at a time, as is traditionally done. Instead, they propose to focus on the joint distribution $h(\theta_1, \ldots, \theta_M) = \pi(\theta_1) \cdot \ldots \cdot \pi(\theta_M)$ of all M parallel chains, and devise a Markov chain for the joint state vector $\theta = (\theta_1, \ldots, \theta_M)$ with stationary distribution h. From this broader perspective it is now natural to consider transition probabilities that involve all M parallel chains jointly. This allows many variations of basic MCMC, in particular, the proposed "self-avoiding" sampler. The idea is so compelling that in hindsight I am surprised nobody had yet thought about it. A curious feature of the proposed scheme is that the stationary target distribution $h(\theta_1, \ldots, \theta_M)$ is independent, but the transition probability $p(\theta_1^{(t+1)}, \ldots, \theta_M^{(t+1)} | \theta_1^{(t)}, \ldots, \theta_M^{(t)})$ is highly dependent across θ_m. In fact, introducing this dependence is the central idea. The scheme proposed by MR combines this construction with another powerful mechanism to increase computational efficiency. They implement delayed rejection as proposed in Tierney and Mira (1999).

This raises my first question to the authors. The proposed approach combines two powerful mechanisms that are both designed to create faster-mixing Markov chains. I would like to see more discussion of the relative impact of both improvements on the reported gains in computational efficiency. A related concern is the fairness of the comparison in the example. If the comparison is meant to highlight the benefit of the self-avoiding proposal, then the competing MCMC implementation should also include delayed rejection. Confounding the benefits of both, delayed rejection and self-avoiding proposals, makes an easy interpretation of the comparison difficult.

Another related question is the need of the detour via the repulsive target distribution π^R. Is the introduction of π^R only a pedagogical device to facilitate the discussion? I feel it could be more misleading than helpful. In the end the proposed chain simply uses a Metropolis–Hastings transition probability with an acceptance probability to achieve the desired stationary distribution $h(\theta_1, \ldots, \theta_M)$.

Finally, I would like to point out some interesting connections with two other parallel MCMC schemes to highlight what I think is the important general theme in this proposal. Robert *et al.* (1999), Doucet *et al.* (2002) and others proposed a Markov Chain Monte Carlo scheme to evaluate marginal maximum *a posteriori* (MMAP) modes. Consider a parameter vector $\theta = (\eta, \lambda)$, and assume we are interested in the MMAP $\eta^* = \arg\max_\eta \int_\lambda p(\eta, \lambda | data)$. The central idea of the method is to construct a Markov chain for the target distribution $h(\eta, \lambda_1, \ldots, \lambda_M) \propto \prod_{i=1}^M p(\eta, \lambda_i | \text{data})$ (η is common across the factors, without subindex $_i$). In a simulated annealing like fashion M is increased over iterations. The chain can be shown to converge (in η) to the MMAP η^*. Another related parallel chain Monte Carlo algorithm is proposed in Müller (1999) and Müller *et al.* (2002). Consider the problem of expected utility maximization $d^* = \arg\max_d \int u(d, \theta) p_d(\theta) d\theta$. Here d is a decision to be determined, θ is a random variable (future data and parameters), $u(d, \theta)$ is the utility assuming θ, and $p_d(\theta)$ is the relevant probability model under d. We propose parallel chain Monte Carlo simulation from $h(\theta_1, \ldots, \theta_M, d) \propto \prod_{i=1}^M u(\theta_i, d)\pi(\theta_i)$, with M increasing between iterations. Under appropriate assumptions, the chain can be shown to converge (in d) to the optimal decision d^*.

NICOLAS CHOPIN (*CREST-ENSAE, France*)

I wish to congratulate both authors for their ingenious algorithm. I want to stress some key features. First, the basic structure provided by Theorem 1 for a "vectorized" MCMC sampler with stationary distribution π^M seems fairly general, and obviously calls for further research on this new flavor of i.i.d. samplers.

Second, the self avoiding mechanism introduced in Section 3.1 is quite appealing. Note the penalizing factor $\exp(-\xi/\pi(\theta_j)\|\theta - \theta_j\|^2)$ in (3) creates some repulsive effect around θ_j, in a small vicinity with radius r about $1/\pi(\theta_j)^{1/2}$ (up to a constant). This dependency in $\pi(\theta_j)$ is advocated by the authors to moderate the repulsive effect in high-probability regions. I propose somme additional justification for this dependency. The π-probability of appartenance to the ball of radius r and centered at point θ_j is approximated at first order as $\pi(\theta_j) v_d r^d$, where d is the dimension of sampling space, and v_d is the volume of a ball with unit radius, $v_d = \pi^{n/2}/\Gamma(1 + n/2)$. If r is set so that this probability equals $1/M$, where M is the number of particles, other particles should be restrained from entering this particular neighborhood, in order to avoid "over-representation" of this part of the space. We get as a rule of thumb to set r so that

$$r^d = \frac{1}{M\pi(\theta_j)v_d},$$

which is very similar to what is proposed by the authors (at least in dimension 2, and for a fixed M). Unfortunately, our rule of thumb is difficult to implement, since it depends on the (usually intractable) normalizing constant of π. Accordingly, a fine tuning of the repulsive factor proposed by the authors may amount in practice to (roughly) evaluate this normalizing constant, and therefore may be tricky in some settings.

Third, it is quite appreciable that this method barely depends on the structure of the target distribution, in contrast with Gibbs samplers for instance. This allows for more automated implementations. In complex settings, however, one may need to determine some distribution μ that covers the support of π, in order to sample initial particles. The choice of μ seems very akin to the specification of an instrumental distribution for an independent Metropolis–Hastings sampler, but my guess is that the choice of μ will have less impact on the behavior of the algorithm than in the independent MH case.

I wonder also if we really need to throw away everything but the M particles produced by the last stage of the algorithm (which are π-i.i.d. is stationarity is reached), rather than considering also values produced by (some) previous iterations, as in regular MCMC set-ups. While the i.i.d. argument obviously does not apply any more, ergodicity should allow for, say, computing Monte Carlo empirical averages over this larger set of particles.

Finally, do the authors have some evaluation of the marginal gain of their delayed rejection strategy, or to put it more bluntly, can they state the extent to which the improvement in mixing induced by these delayed rejections is worth their additional computational cost?

REPLY TO THE DISCUSSION

We thank the two contributors to the discussion of our paper and attempt to address their comments in turn.

The salient features of our algorithm are well identified by our main discussant, Peter Muller, who also provides welcome extensions to the potential of the method through connections with other parallel MCMC schemes. He raises two important points: First, what is the relative impact of the repulsion mechanism and the delayed rejection mechanism, and second, why is π^R employed?

Both the repulsion mechanism and delayed rejection are designed to enhance mixing. However, the former is focused on the "intrinsic" space, in that the aim is to push the particles themselves apart by creating "holes" around other particles in the space. The latter, on the other hand, is focused on the "extrinsic" space, in that the aim is to reduce the immobility of particles at any iteration and potentially propose their movement into hitherto unexplored areas of the space. The impact of delayed rejection has been explored by Mira (1998). The impact of the repulsion mechanism has yet to be fully explored. Furthermore, the questions of how many delayed rejections should be adopted and how the particles should "bounce" remain unanswered. The approach to bouncing taken in this paper is simplistic and deterministic. It is possible to design alternative moves that maintain detailed balance, but their relative advantages have not yet been considered.

In our algorithm, π^R acts as an intermediary between the distribution under which moves are proposed and the target distribution π. Hence, the approach might be considered as two-stage: first, moves are proposed according to a random walk and accepted according to π^R, and second, the emergent set of particles is corrected from π^R to π. It is not clear how this intermediary step would be abandoned in a straightforward manner.

Nicholas Chopin illuminates particular features of our algorithm and we thank him for the different perspectives that he gives to the approach. Indeed, the suggested "rule of thumb" in his discussion of dependency does amount to tuning the repulsive factor, and although the "trial and error" method adopted in our examples was suitable for the low-dimensional contexts, we acknowledge that in higher dimensions this may be more difficult. However, this is not dissimilar to regular MCMC settings, and it may be practicable to deconstruct the high-dimensional problem into lower-dimensional components in order to achieve this tuning.

The conjecture is made by Chopin that the choice of distribution from which to sample initial particles is less important in our algorithm than for a regular independent Metropolis–Hastings sampler. We have not investigated this, but we concur that the repulsive mechanism will indeed facilitate greater movement from the initial space to the entire target space. In an extension of the approach described in this paper, new particles could be born at some times $s, s > 0$; in this case, it may be possible to develop an adaptive independent distribution for initial particles, in which the appropriate space is "learnt" during time $t = 0, .., s$. Of course, adoption of an inappropriate independent sampler for all proposed moves would lead to the same convergence problems as in regular MCMC.

Chopin's observation about using values in previous iterations is valid. The usual trade-offs can be made between iidness and ergodic averages as in regular MCMC. It is interesting to consider, however, what these trade-offs would be in the current context. It is also interesting to consider how the algorithm could adapt to other purposes, for example constructing a negatively correlated sample from the target population. This is the subject of current research. The question of "when do we stop" naturally follows these lines of thought. In the current algorithm, this may be as soon as the influence of the initial value is erased, or when a test for iidness is passed.

Chopin's final question about the marginal gain of the delayed rejection is addressed by Mira (1998). However, optimizing the number of rejections in an iteration has not yet been explored. To some extent, comparison with other approaches is mitigated by Müller's query about the validity of such comparisons: if we use parallel chains or delayed rejection, should we also use these in the algorithms with which we are comparing? It is not clear how different MCMC approaches should be compared in general, although measures such as effective sample size, estimation and expectations of extremes (tail behavior) have been suggested. This issue of identifying the "best" method for a given situation is still open in our opinion.

ADDITIONAL REFERENCES IN THE DISCUSSION

Doucet, A., Godsill, S. J. and Robert, C. P. (2002). Marginal maximum a posteriori estimation using Markov chain Monte Carlo. *Statist. Comput.* (to appear).

Lee, S. J. (1998). *Semiparametric Bayesian Analysis: Selection Models and Meteorological Applications.* Ph.D. Thesis, Purdue University, USA.

Müller, P. (1999). Simulation based optimal design. *Bayesian Statistics 6* (J. M. Bernardo, J. O. Berger, A. P. Dawid and A. F. M. Smith, eds). Oxford: Oxford University Press, 459–474 (with discussion).

Müller, P., de Iorio, M. and Sansó B. (2002). Optimal Bayesian design by inhomogeneous Markov chain simulation. *Tech. Rep.*, University of Texas, USA.

Robert, C. P., Doucet, A. and Godsill, S. J. (1999). Marginal MAP estimation using Markov chain Monte Carlo. *Tech. Rep.*, CREST, INSEE, France.

BAYESIAN STATISTICS 7, pp. 293–305
J. M. Bernardo, M. J. Bayarri, J. O. Berger, A. P. Dawid,
D. Heckerman, A. F. M. Smith and M. West (Eds.)

A Statistical Approach to Modeling Genomic Aberrations in Cancer Cells

MICHAEL A. NEWTON HYUNA YANG
University of Wisconsin, Madison, USA
newton@stat.wisc.edu hyuna@stat.wisc.edu

PATRICIA A. GORMAN IAN TOMLINSON REBECCA R. ROYLANCE
Cancer Research, UK
ian.tomlinson@cancer.org.uk

SUMMARY

Whereas most cells in the body carry the normal complement of 23 chromosome pairs, the cells within a cancerous tumor very often present highly abnormal genomic structure. Deletions, amplifications, rearrangements and mutations are common at various scales and are highly variable amongst tumors, as indicated by molecular technologies which enable ever better measurement. It is an important statistical problem to separate sporadic abnormalities from those that may not be sporadic and that may have some biological significance. We discuss a modelling strategy for genomic aberration data, which allows us to to infer combinations of aberrations that together increase the chance that a precancerous cell will have a descendant tumor lineage. The likelihood component involves a network of pathway structures. Markov chain Monte Carlo is used to sample from the space of these oncogenic networks. We illustrate the methodology with comparative genomic hybridizations from a recent breast cancer study, and we derive a likelihood formula for the larger class of tree-like networks.

Keywords: ONCOGENIC PATHWAYS; GENETIC NETWORKS; MODEL-BASED CLUSTERING; INSTABILITY AND SELECTION.

1. BACKGROUND

In much the same way the detective uses clues to infer what has happened at a crime scene, the cancer biologist attempts to identify critical genomic changes that have carried a normal cell to its highly aberrant state in a cancerous tumor. The extent and variety of genomic aberration in tumor cells are striking (see *e.g.*, Knuutila *et al.,* 1998, 1999, for surveys of cytogenetic aberrations), and support a statistical approach to data analysis. Lengauer *et al.* (1998) comment that, "... all tumors are genetically unstable. Instability is the engine of both tumor progression and tumor heterogeneity, guaranteeing that no two tumors are exactly alike and that no single tumor is composed of genetically identical cells." Allowing that genetic instability somehow creates aberrations, it is equally important to know that a tumor cell lineage carrying some profile of aberrations is subjected to selective pressures which affect its fate (*e.g.*, Tomlinson *et al.,* 1996).

The concepts of instability and selection guide a general understanding of experimental results from cancer biology, but they also provide a framework for deriving probability models and statistical methodology to analyze measured aberrations (Newton *et al.,* 1998, 1999, 2000).

Briefly, the probability distribution for measured aberrations is built in two steps: first, one postulates random genomic damage in a progenitor cell—a cell that could become ancestral to observable tumor cells; second, one allows that certain damage (such as deletion of a suppressor gene) is beneficial to the tumor and increases the probability of selection, that is, the probability that descendents of the progenitor cell will populate an observable tumor. Interestingly, Bayes rule is used to derive the sampling distribution of the data since we measure aberrations conditionally on selection having occurred. The resulting statistical methods have been used to characterize significant changes in bladder cancer (Yeager *et al.,* 1998), prostate cancer (Jarrard *et al.,* 1999), hepatocellular carcinoma (Teeguarden *et al.,* 2000), colon cancer (Shoemaker *et al.,* 1998), and breast cancer (Haag *et al.,* 1996). A limitation of the methodology used in those studies is that it did not account for statistical dependence among unlinked genomic aberrations. More specifically, it did not deal with the fact that the co-occurrence of several aberrations could increase the probability of selection. The term *oncogenic pathway* has been used to describe such a combination of aberrations that are relevant to oncogenesis, and this concept is the focus of our present effort.

The importance of computational schemes to identify oncogenic pathways is well recognized, and the available tree-based methods provide the first quantitative, multivariate approach (Desper *et al.,* 1999, 2000; Jiang *et al.,* 2000). From the point of view of statistical inference, however, the available methods have several deficiencies. Notably, they rely on a high degree of initial data processing and they provide point estimates but no measures of confidence. In very recent work, it has been recognized that the instability-selection framework can incorporate oncogenic pathways (Newton, 2001). This finding provides a statistical basis to inference about oncogenic pathways. Here we review the *instability-selection-network* model from Newton (2001) and we illustrate the methodology on data from a recent breast cancer study. We also report a methodological extension of the work in Newton (2001) by solving the problem of how to calculate likelihoods for a class of overlapping-pathway models.

2. ILLUSTRATION

To be any more precise it is helpful to have an example. Figure 1 shows genomic aberrations taken on a set of 40 grade II invasive ductal breast cancers. The *profile* from each tumor is a vector $x = (x_1, \ldots, x_n)$ with binary elements $x_i \in \{0, 1\}$ that indicate whether or not aberration i occurs in that tumor. The number n of potential aberrations has to do with the resolution of the measurements, and turns out to be 82 in this study. This counts 41 possible amplifications and 41 possible deletions. Each of the 41 chromosome arms in the genome may exhibit a deletion of genomic material, an amplification of material (or both). (Recall that there are 23 chromosome pairs a normal cell. Except for several acrocentric chromosomes, there are two arms for each one – the short p arm and the long q arm, thus we measure 41 different arms.) The data in Figure 1 are obtained using comparative genomic hybridization (CGH), a general approach to assess DNA copy number variations in tumor cells (see the appendix for further details).

One of the first things to see from the data in Figure 1 is the extent of marginal heterogeneity among aberrations. This variability is much larger than one expects by chance alone, for example, if these sample frequencies are identically distributed Binomial counts (calculation not shown). Extra copies of DNA on the long arm of chromosome one, that is, $+1q$, is the most frequent aberration, occurring in 27 tumors. (On average, each aberration occurs in 5.5 tumors, and each tumor presents 11.2 aberrations.) Frequent aberrations are interesting for further study because they may be involved in the etiology of cancer. In considering oncogenic pathways, we are compelled to look beyond marginal frequencies towards sample correlations among aberrations. We performed a simple permutation test by shuffling within rows in Figure 1 and

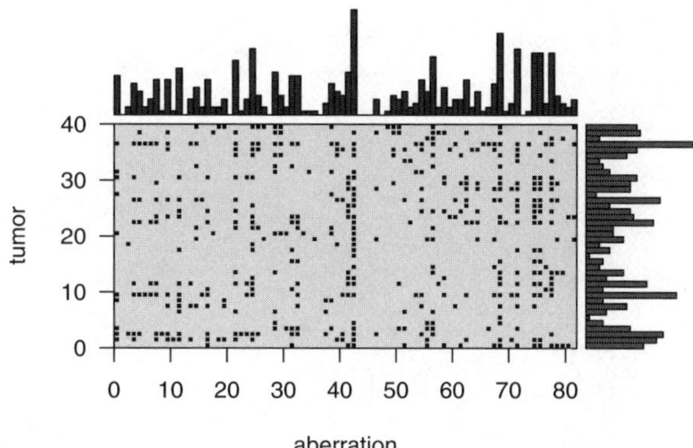

Figure 1. *Genomic aberrations (dark spots) in 40 grade II invasive ductal breast cancers (rows) for 82 potential aberrations (columns). The first 41 columns correspond to deletions on arms from* $1p$ *to* Xq*, and the next correspond to amplifications.*

recomputing the sample covariance matrix among aberrations and we found that the extent of correlations is quite significant (calculation not shown).

One result of model-based computations described in the next section is a posterior probability that a given aberration resides on some oncogenic pathway, as opposed to being neutral. In the model, a neutral aberration is one for which its occurrence in a progenitor cell does not influence selection. We presume that if the aberration is not neutral then it is somehow relevant to oncogenesis and thus we should like to be able to compute the probability of this event. Figure 2 compares the posterior probability of being relevant to the marginal empirical frequency of occurrence. Some 13 aberrations out of 82 are almost certainly relevant. Naturally, they tend to be the aberrations which occur most frequently, but the correspondence with empirical frequency is not perfect. For example, the deletion $-22q$ is probably relevant even though it occurs in only 8 of the 40 tumors; yet several other aberrations occur 10 times and they are probably neutral. Had we been processing only marginal frequencies then we would not expect this phenomenon; evidently, the way in which combinations of aberrations occur together is informative dependence.

Hierarchical clustering is a common tool in bioinformatics and provides a natural way to picture the dependencies between CGH aberrations. Figure 3 makes a comparison of two clusterings. Both use the default settings of the **hclust** function in **R** (Ihaka and Gentleman, 1996). In the first, we are clustering the raw profiles from Figure 1, using Euclidean distance between columns (aberrations) in Figure 1 and complete-linkage to build the tree. Perhaps not surprisingly, it is difficult to extract much information from the raw-data clustering. We find the second clustering rather more informative. Instead of clustering raw data, we are clustering certain posterior probabilities that come out of a Bayesian analysis of the data. More specifically, the distance between two aberrations is taken to be the posterior probability that they do not reside on a common oncogenic pathway, as calculated within the instability-selection-network model. A convenient feature of this clustering is that the many aberrations which are probably neutral (*i.e.*, , they are probably not on any oncogenic pathway), are compressed together on the right side of the plot. Dominant branches at height near 0 characterize collections of aberrations that probably constitute oncogenic pathways. This clustering provides one way to extract posterior summaries beyond point estimates of pathways. Furthermore, in our experience, the structure exposed in this clustering is rather insensitive to the prior distribution.

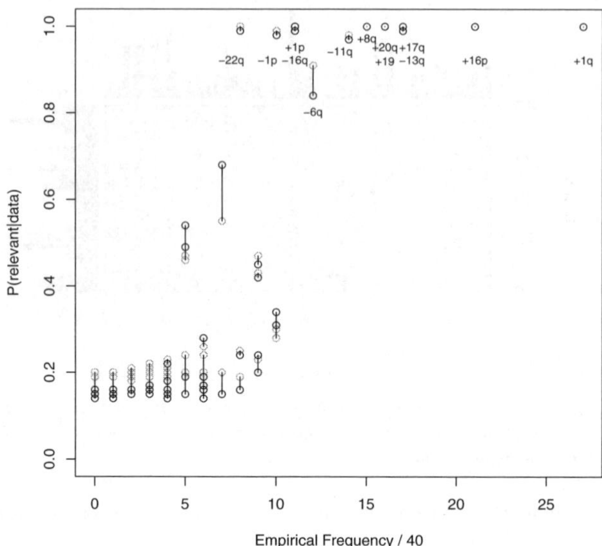

Figure 2. *Empirical aberration frequency and posterior probability that each aberration is relevant to oncogenesis. The posterior probabilities are computed twice, from independent MCMC runs, under the double Polya prior with* $\tau = 5$. *The level of Monte Carlo error is relatively low, as indicated by the fact that the two approximations (pairs of points connected by a line) are quite close.*

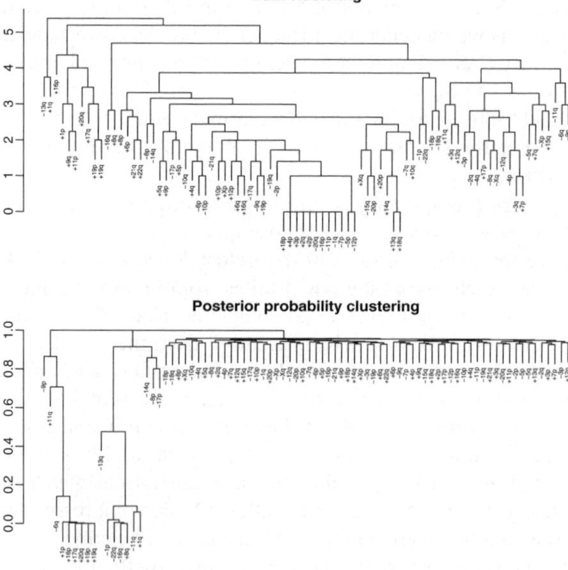

Figure 3. *Dependence among aberrations via hierarchical clustering.*

The summary calculations reviewed above are based on a specific probability model for measured aberrations, a prior distribution over the space of oncogenic pathways, and MCMC methods for posterior analysis. We consider these elements next.

3. A NETWORK MODEL

3.1. *Non-overlapping pathways*

Whether or not an aberration is neutral is unknown *a priori*. We represent these parameters with a vector $a = (a_1, \ldots, a_n)$ of binary indicators. The pathway structure amongst relevant (*i.e.*, , non-neutral) aberrations is represented in Newton (2001) as a set partition using a label vector $c = (c_1, \ldots, c_n)$. The ith label c_i has no meaning in isolation; but for relevant aberrations i and j, $c_i = c_j$ means that i and j reside on the same oncogenic pathway. An oncogenic pathway is a collection of relevant aberrations, and is also unknown. The assumption of non-overlapping pathways is not realistic but it leads to a feasible likelihood calculation.

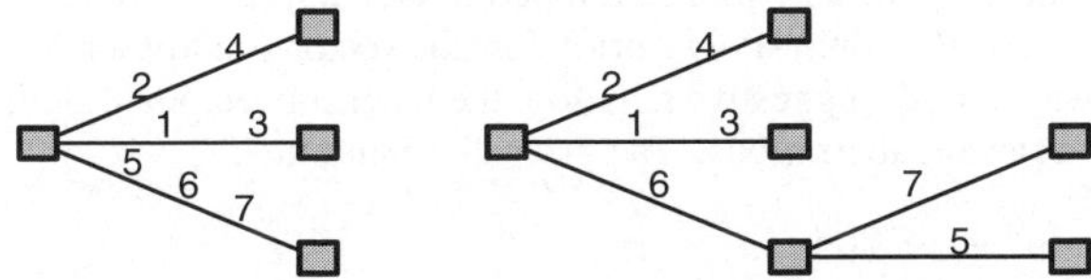

Figure 4. *Two hypothetical networks. The one on the left has three non-overlapping paths:* $\{2, 4\}$, $\{1, 3\}$ *and* $\{5, 6, 7\}$. *The one on the right has four paths:* $\{2, 4\}$, $\{1, 3\}$, $\{6, 7\}$ *and* $\{5, 6\}$. *The number of leaves on each tree corresponds to the number of paths. The node on the left in each case is the root.*

The pathway labels c_i for relevant aberrations ($a_i = 1$) describe some collection of pathways. For example, in Figure 4, the diagram on the left considers a hypothetical cancer comprised of three oncogenic pathways and seven relevant aberrations. In addition to measurable (*i.e.*, , overt) damage indicators x_i, the model envisions latent binary variables y_i that stand for covert aberrations which may be generated by genetic instability but which we are unable to measure. Genetic instability in the progenitor cell creates either overt damage $x_i = 1$ or covert damage $y_i = 1$ randomly: *i.e.*, $x_i \sim_{iid}$Bernoulli(α) and $y_i \sim_{iid}$Bernoulli(β), for unknown rates α and β. We say that instability has opened a pathway if all the aberrations on that pathway have occurred either overtly or covertly. Thus, the probability that pathway e, say, is open is, $p_e = \theta^{n_e}$, where $\theta = 1 - (1 - \alpha)(1 - \beta)$ is the chance that a given aberration occurs somehow, and n_e is the size of the oncogenic pathway. The progenitor cell, having incurred damage x and y, is selected by oncogenesis if there is at least one open pathway. In other words, the probability of selection is

$$\begin{aligned} P(\text{SEL}) &= 1 - P(\text{all pathways closed}) \\ &= 1 - \prod_e (1 - p_e). \end{aligned}$$

(The product form is valid only when two different pathways do not share a common aberration, but see the next section for an extension of this.) We can by a similar argument calculate

$$P(\text{SEL}|x) = 1 - \prod_e [1 - p_e(x)].$$

where $p_e(x) = P(\text{pathway } e \text{ is open}|x)$. If all the aberrations on e have occurred overtly, then $p_e(x) = 1$. Covert aberrations may be necessary to open the pathway, and thus, more generally,

$p_e(x) = \beta^{t_e}$ where $t_e = \sum_{i \in e}(1 - x_i)$. Combining the instability and selection terms, the likelihood contribution from a tumor presenting data x is:

$$P(x|\text{SEL}) = \alpha^{\sum_i x_i}(1-\alpha)^{\sum_i (1-x_i)} \left\{ \frac{1 - \prod_e [1 - p_e(x)]}{1 - \prod_e (1 - p_e)} \right\} \quad (1)$$

Newton (2001) first derived this model, showed some of its sampling properties, and presented Bayesian model fitting methods. It is interesting that the act of selection creates heterogeneity in marginal aberration rates, positive dependence between aberrations on the same pathway, negative dependence between aberrations on different pathways, and independence of all neutral aberrations.

Bayesian analysis was proposed in Newton (2001) to infer the unknown oncogenic pathways using the likelihood (1) and a prior distribution over rates α and β and, more importantly, over the network of oncogenic pathways. The number and composition of these pathways is unknown. The prior implemented by the software developed in Newton (2001) (and used in the example from Section 2) uses a Bernoulli–Polya prior for the vector a and then a second Polya-type prior for the pathways. With suggestive notation, the label subvector $c[a]$ indicates the pathway structure amongst relevant aberrations. The prior distribution is:

$$\begin{aligned} \pi(c[a], a) &= \pi(c[a]|a)\, \pi(a) \\ &= \left[\frac{\tau^K \Gamma(\tau) \left[\prod_e \Gamma(n_e)\right]}{\Gamma(\tau + m)} \right] \left[\frac{\Gamma(m+1)\Gamma(n-m+1)}{\Gamma(n+2)} \right] \end{aligned}$$

where K is the number of pathways, n_e is the length of pathway e, $m = \sum_e n_e$ is the total number of relevant aberrations, $n - m$ is the number of neutral aberrations, and $\Gamma()$ is the Gamma function. Among other things, the prior induces a uniform prior distribution over m. A single hyperparameter $\tau > 0$ affects the extent of clustering.

Markov chain Monte Carlo is the obvious approach to enable posterior analysis. The algorithm in Newton (2001) involves a collection of move types which alter the activation vector a, the cluster vector c and the unknown rates α and β. It is interesting to note that some simple move types are not very effective. For example, in changing a neutral aberration to a relevant aberration it seems reasonable to attempt to add a new pathway carrying that single aberration. Such a local change in network space may correspond to a very large change in likelihood (1), since short pathways are expected to present very particular statistical properties. On the other hand, two networks that differ by one long pathway may have very similar likelihood.

Figure 5 shows trace plots of log-posterior and network size m for two independent MCMC runs using the 40 breast cancer profiles. We show results for the particular hyperparameter $\tau = 5$, and we note that similar results obtained using $\tau = 1$ and $\tau = 10$. The mixing is reasonable but not ideal; it is encouraging that posterior summaries vary very little across replicate runs. The posterior mean for α is 0.094 and β is 0.085. The prior was uniform over the triangle $\alpha < \beta$.

A point estimate of the network is the one which achieves the highest posterior probability. For these data, we obtain two pathways of equal length:

e_1	$+1q$	$+8q$	$-16q$	$-13q$	$-22q$	$-1p$	$-11q$
e_2	$+19q$	$+20q$	$+16p$	$+17q$	$+19p$	$+1p$	$-6q$

(The MAP estimate was similar for the three hyperparameter values considered.)

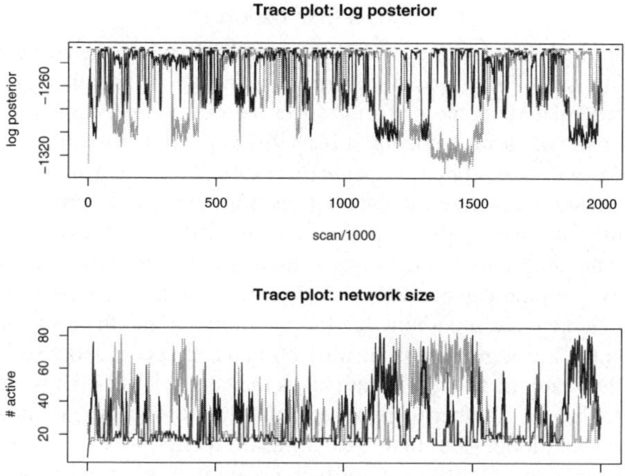

Figure 5. *Trace plots from MCMC: log posterior of each network* (top) *and number of relevant aberrations* (bottom). *Each chain started at a random network, proceeded for* 2×10^6 *scans, and was subsampled every 1000 scans.*

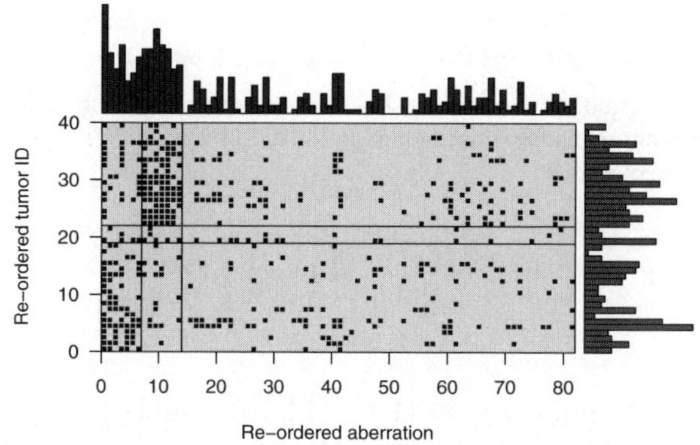

Figure 6. *Genomic aberrations like in Figure 1, but with rows and columns rearranged. The left block of columns corresponds to the first estimated pathway, the second block to the second estimated pathway, and the remainder are probably neutral. The rows (tumors) are organized by chances of following each pathway. The lower block of 19 tumors probably followed the first pathway* e_1*, the upper block of 18 tumors probably followed* e_2*, and it is a toss up for three of the tumors.*

Having an estimated network allows us to perform a model-based clustering of the tumors. The notion is that each tumor has traversed (at least) one of the pathways. Conditional on the data x, we can compute the probability that each of the estimated pathways was open for that tumor. Figure 6 shows a summary of this calculation. It is analogous to Figure 1, except that aberrations are rearranged according to the estimated pathways and tumors are reordered according to the pathway they probably followed. These pathway predictions might provide a useful biomarker to relate to other properties of the tumor or clinical outcomes.

3.2. *Tree-like Pathways*

A limitation of the methods developed in Newton (2001) and reviewed here is that pathways cannot overlap. This simplifying assumption makes it feasible to calculate the likelihood (1) of a network, but the general understanding from cancer biology compels us to do better. Rather than allow all possible pathway arrangements, it is useful to restrict attention to tree-like networks, as Desper and colleagues have done (Desper *et al.*, 1999, 2000; Jiang *et al.*, 2000). The question is can we develop instability-selection calculations for pathways arranged in a tree? Take the right panel in Figure 4 as an example. The tree is comprised of edges, as before, that are disjoint collections of relevant aberrations. No longer is there a 1–1 correspondence between edges and oncogenic pathways. A pathway is a series of edges moving from the root to a leaf node. To be more precise, each edge e has a parent edge $\mathrm{PA}(e)$, allowing that the root can be a parent edge also (*e.g.*, the parent of the edge containing aberration 6 is the root.) Some edges are not parents of any edge, and these are the leaves. What makes the network tree-like is the absence of loops in this graph. Note that the number of oncogenic pathways equals the number of leaves.

To calculate a likelihood, note that, as before, each edge e has a probability p_e of being open, and a conditional probability $p_e(x)$ of being open in light of overt damage x. Each edge e also represents the ancestral edge of a branch of the tree. For example (using the aberration label/s to label the edge), in Figure 4, e_6 begins a branch containing e_6, e_7 and e_5. Edges like $e_{1,3}$ constitute a single branch, as does any edge which is a leaf. Noting this special tree structure, we can define two other probabilities on each edge:

$$\tilde{p}_e = P(\text{branch starting at } e \text{ is open})$$

and

$$\tilde{p}_e(x) = P(\text{branch starting at } e \text{ is open}|x)$$

For leaf edges, $\tilde{p}_e = p_e$ and $\tilde{p}_e(x) = p_e(x)$. For non-leaf edges, the event that the branch starting at e is open is the event that some descendant pathway in that branch is open. We immediately obtain the recursive equations:

$$\tilde{p}_e = p_e \left[1 - \prod_{h:\mathrm{PA}(h)=e} (1 - \tilde{p}_h) \right]$$

and

$$\tilde{p}_e(x) = p_e(x) \left[1 - \prod_{h:\mathrm{PA}(h)=e} [1 - \tilde{p}_h(x)] \right]$$

For example, $\tilde{p}_6 = p_6[1-(1-p_5)(1-p_7)]$. Given any tree-like network we can compute $\tilde{p}_e$ and $\tilde{p}_e(x)$ for all edges by moving recursively from the leaves towards the root. The chance of selection and the likelihood involve $\tilde{p}_e$ and $\tilde{p}_e(x)$ for edges emanating from the root. Extending (1), we obtain

$$P(x|\mathrm{SEL}) = \alpha^{\sum_i x_i}(1-\alpha)^{\sum_i (1-x_i)} \left\{ \frac{1 - \prod_{e:[\mathrm{PA}(e)=\mathrm{root}]}[1 - \tilde{p}_e(x)]}{1 - \prod_{e:[\mathrm{PA}(e)=\mathrm{root}]}(1 - \tilde{p}_e)} \right\} \tag{2}$$

This formula allows us to extend the domain of instability-selection modelling to more realistic oncogenic pathway structures. It remains to develop a useful prior distribution over tree-like networks and to implement a posterior sampling algorithm.

APPENDIX

The data were obtained by using comparative genomic hybridization, a method of screening the entire genome for gains and losses of genetic material in a single experiment. Differentially labelled test or tumor DNA (green) and reference or normal DNA (red) are co-hybridized to normal metaphase spreads. Differences in the copy number between test and reference DNA are seen as differences in the ratio of green to red fluorescence intensity on the metaphase chromosomes. Images of the metaphases are captured using an epifluorescence microscope, equipped with a cooled CCD (charge-coupled device) camera. Quantification of the fluorescence ratios is performed using a digital image analysis system. For each tumor, 5–10 metaphases are analyzed and an average fluorescence ratio for each chromosome obtained. Regions of chromosomal gain are seen as an increased fluorescence ratio, while regions of loss are seen as a decrease in the fluorescence ratio. Conventionally, gains and losses are considered significant when the ratio is larger than $1.15 : 1$ and smaller than $0.85 : 1$, respectively.

REFERENCES

Desper, R., Jiang, F., Kallioniemi, O.-P., Moch, H., Papadimitriou, C. H. and Schäffer, A. A. (1999). Inferring tree models for oncogenesis from comparative genome hybridization data. *J. Comput. Biol.* **6**, 37–51.

Desper, R., Jiang, F., Kallioniemi, O. P., Moch, H., Papadimitriou, C. H. and Schäffer, A. A. (2000). Distance-based reconstruction of tree models for oncogenesis. *J. Comput. Biol.* **7**, 789–803.

Haag, J. D., L.-C. Hsu, Newton, M. A. and Gould, M.N. (1996). Allelic imbalance in mammary carcinomas induced by either 7,12-dimethylbenz[a]anthracene or ionizing radiation in rats carrying genes conferring differential susceptibilities to mammary carcinogenesis. *Mol. Carcinog.* **17**, 134–143.

Ihaka, R. and Gentleman, R. (1996). R: A language for data analysis and graphics. *J. Comput. Graph. Statist.* **5**, 299–314 (see `www.r-project.org`).

Jarrard, D. F., Sarkar, S., Shi, Y., Yeager, T. R., Magrane, G., Kinoshita, H., Nassif, N., Meisner, L., Newton, M. A., Waldman, F. M. and Reznikoff, C. A. (1999). p16/pRb pathway alterations are required for bypassing senescence in human prostate epithelial cells. *Cancer Res.* **59**, 2957–2964.

Jiang, F., Desper, R., Papadimitriou, C. H., Scäffer, A. A., Kallioniemi, O.-P., Richter, J., Schraml, P., Sauter, G., Mihatsch, M. J. and Moch, H. (2000). Construction of evolutionary tree models for renal cell carcinoma from comparative genomic hybridization data. *Cancer Res.* **60**, 6503–6509.

Knuutila S, Björkqvist A.-M., Autio K., Tarkkanen, M., Wolf, M., Monni, O., Szymanska, J., Larramendy, M. L., Tapper, J., Pere, H., El-Rifai, W., Hemmer, S., Wasenius, V.-M., Vidgren, V. and Zhu, Y. (1998). DNA copy number amplifications in human neoplasms. Review of comparative genomic hybridization studies. *Am. J. Pathol.* **152**, 1107–1123.

Knuutila, S., Aalto, Y., Autio, K., Björkqvist, A.-M., El-Rifai, W., Hemmer, S., Huhta, T., Kettunen, E., Kiuru-Kuhlefelt, S., Larramendy, M. L., Lushnikova, T., Monni, O., Pere, H., Tapper, J., Tarkkanen, M., Varis, A., Wasenius, V.-M., Wolf, M. and Zhu, Y. (1999). DNA copy number losses in human neoplasms: Review. *Am. J. Pathol. Online* **155**, 683–694.

Lengauer, C., Kinzler, K. W. and Vogelstein, B. (1998). Genetic instabilities in human cancers. *Nature* **396**, 643–649.

Newton, M. A. (2001). A statistical method to discover significant combinations of genetic alterations associated with cancer using comparative genomic hybridization profiles. *Tech. Rep.*, University of Wisconsin-Madison, USA.

Newton, M. A., Gould, M. N. Reznikoff, C. A. and Haag, J. D. (1998). On the statistical analysis of allelic-loss data. *Statist. Med.* **17**, 1425–1445.

Newton, M. A., Yeager, T. and Reznikoff, C. A. (1999). A statistical analysis of cancer genome variation. *Statistics in Genetics* (M. E. Halloran and S. Geisser, eds). Berlin: Springer, 223–236.

Newton, M. A. and Lee, Y. J. (2000). Inferring the location and effect of tumor suppressor genes by instability-selection modelling of allelic-loss data. *Biometrics* **56**, 1088–1097.

Shoemaker, A. R., Moser, C. A., Midgley, L., Clipson, M. A., Newton and W. F. Dove (1998). A resistant genetic background leading to incomplete penetrance of intestinal neoplasia and reduced loss of heterozygosity in $Apc^{Min/+}$ mice. *Proc. Natl. Acad. Sci. USA* **95**, 10826–10831.

Teeguarden, J. G., Newton, M. A., Dragan, Y. P. and Pitot, H. C. (2000). Genome-wide loss of heterozygosity analysis of chemically induced rat hepatocellular carcinomas reveals elevated frequency of allelic imbalances on chromosomes 1, 6, 8, 11, 15, 17, and 20. *Mol. Carcinog.* **28**, 51–61.

Yeager, T. R., DeVries, S., Jarrard, D. F., Kao, C., Nakada, S. Y., Moon, T. D., Bruskewitz, R., Stadler, W. M., Meisner, L. F., Gilchrist, K. W., Newton, M. A., Waldman, F. M. and Reznikoff, C. A. (1998). Overcoming cellular senescence in human cancer pathogenesis. *Genes Dev.* **12**, 163–174.

DISCUSSION

SCOTT C. SCHMIDLER (*Duke University, USA*)

I would like to begin by congratulating the authors on a very interesting paper, which describes an elegant stochastic model for genomic aberration data. This paper builds on a body of previous work by the authors, and as such represents a well-developed and relatively mature statistical methodology. Thus my comments will focus as much on the scientific questions under study as on the statistical methods, with comments on the latter restricted primarily to the methodological extensions introduced in this paper for incorporation of dependency structure between aberrations.

The data. Modern biologists often wish to ask a wide variety of questions of the data, and the model-based approach and simulation methods described in the paper are particularly attractive in such settings. However it is also worth asking what are likely to be the primary goals of the type of analysis described in the paper. Although this is not stated explicitly, one presumes that one such goal may be to identify aberrations implicated in tumorigenesis for more detailed study; another perhaps to find diagnostic markers for tumor identification. In both cases I wonder whether the data described are adequate to answer such questions in the absence of a control population. Figure 2 of the paper shows that high empirical frequency generally leads to high posterior probability of relevance, but perhaps some aberrations are more frequent in non-tumor cells as well. Do the authors have other data or domain knowledge which applies here?

A second question involves the interpretation of the "oncogenic pathways" under study. I find it somewhat difficult to ascribe a meaning to these pathways in terms of molecular events. This makes validation and interpretation of the inferred pathways unclear. Put another way, what does their inference tell us about the underlying biological system?

A minor point is that given the large amount of work invested here in modelling and computation, the reduction of the data to a set of directly observable $x_i \in \{0, 1\}$ seems undesirable. Might more quantitative information be obtained simply by treating the x_i's as missing data and writing a conditional density $\prod_k P(f_k \mid x_i)$ for the fluorescence ratio observations themselves?

The model. A few relatively minor questions regarding the model itself: Does x_i independent of x_j make sense when x_i denotes deletion of a site and x_j amplification of the same site? Are the inferences sensitive to such assumptions?

In reality, $\alpha_i \not\equiv \alpha$ and similarly for β_i. Moreover, these differences are likely to be of real interest. Is it possible with the currently available data to make inference about differences in these parameters?

The Bernoulli–Polya prior chosen has uniform marginal distribution on m. In previous work, the first author has shown that the resulting MAP inference can miss larger networks, suggesting that perhaps this prior over-penalizes large networks. In a related problem involving priors on set partitions, Schmidler (2002) found that such combinatorial priors which provide adequate marginal inferences often provide poor MAP estimates. This may be checked by performing the MCMC using a uniform prior $P(c \mid a) \propto const.$ This can be expected to give upwardly biased marginal probabilities of aberrations sharing a pathway, but the resulting MAP networks may be informative.

The results. A particularly attractive aspect of this paper is the elegant solution it provides to both pathway identification and tumor clustering using a common statistical model. The visualization of marginal posterior probabilities of dependence via hierarchical clustering is quite informative, and likely to have applications in other areas.

The posterior mean values obtained for α and β are 0.094 and 0.085, respectively. I find this slightly troubling in light of the prior used requiring $\beta < \alpha$. It seems to suggest that the "covert" aberrations in the model are playing nearly as big a role in inference as the aberrations actually observed. Perhaps the β's provide too much flexibility? Placing additional restrictions on the β's or directly on $\sum_i y_i$ may be desirable.

Finally, an important question not addressed in the paper is the validation of the inferred pathways. The MAP networks are interesting, but there will always be a MAP network—how can we verify whether it is correct? This relates back to the question of interpretation of oncogenic pathways—is there a measurable quantity that can provide ground truth?

Model extensions. The paper discusses but does not implement relaxing the assumption of non-overlapping pathways via a tree structure. Note that there is no inherent ordering among aberrations in a "pathway", so many network structures may also be rearranged to form trees as well. It should be noted that the recursive calculation given applies equally well to general network structures described in the paper, and the authors have indicated that they are aware of this as well.

The word "pathway" is a bit of a misnomer here, due to the lack of any real ordering among aberrations sharing a pathway. We may instead view pathways as simply a subset of aberrations all of which must occur, and therefore we may define a pathway as a new random variable formed by a logical AND, *e.g.*, $e = a_i \wedge a_j \wedge \ldots \wedge a_k$. This suggests more general logic statements, *i.e.*, use of logical OR and NOT. The trees described introduce a restricted form of OR, and networks extend this further. However use of negation may be valuable as well. Here I envision certain aberrations which may contribute to tumorigenesis, but which are lethal when occurring simultaneously. Such cases would best be modelled by statements of the form $a_i \oplus a_j = (a_i \vee a_j) \wedge \neg(a_i \wedge a_j)$.

General remarks. The approach developed here appears to have direct applications to related problems in genomics and bioinformatics, such as the discovery of genetic regulatory networks and metabolic pathways. Have the authors considered applying this approach to these other important and open problems? It is also worth noting the relation to the latent factor models of West *et al.* (2001) where the a_i's are not restricted to binary values. Another question is whether it is possible to leverage the large amounts of genomic sequence data recently deposited into public databases, perhaps in helping specify informative priors for aberrations occurring in the region of known genes.

Finally, I would like to congratulate the authors again on an elegant and practical model-based solution to an important applied problem in molecular biology. Too often problems in bioinformatics have been dominated by heuristic algorithmic formulation, ignoring the underlying uncertainty and implicit statistical assumptions. In recent years, a number of such problems have been treated from a formal (often Bayesian) statistical modelling perspective, including global and local sequence alignment (Lawrence *et al.*, 1993; Krogh *et al.*, 1994), protein structure prediction (Schmidler *et al.*, 2000), molecular structure analysis (Wu *et al.*, 1998; Schmidler, 2002), and an enormous literature on analysis of gene expression experiments (*e.g.*, Kerr and Churchill, 2000; Li and Wong, 2001). The authors have been among the pioneering statisticians in addressing important problems in molecular biology and genetics, and it is a great pleasure to be able to add oncogenic pathway discovery to this list of problems with proper statistical solutions.

REPLY TO THE DISCUSSION

The discussant has identified critical issues surrounding the proposed methodology and he has been generous in his overall assessment. We appreciate the informed comments.

To clarify our goals, the primary goal is to characterize potentially important combinations of genomic aberrations so as to guide future cancer studies. Such studies might derive useful diagnostic markers for clinical work or they could involve focused searches for the genes and molecules involved in the cancer process. Our approach is statistical, of course, and it is meant to provide a quantitative summary of the information collected in a given CGH study. One hopes that such an analysis can complement related investigations of tumor biology. Our calculations do not provide direct insight into molecular events, but that would be a lot to ask from the CGH data alone. Regarding controls, it is important to recognize that the genomic damage measured by CGH does not occur in normal tissue. That is, except for a very low rate of measurement error, Figure 1 would be all white (no aberrations) if non-tumor DNA was under investigation. In machine-learning terms, we have an unsupervised learning problem.

Fairly, the discussant criticizes our data reduction scheme. The binary data retain most of the signal, we suspect, but certainly one could attempt to model the whole intensity profile. Each binary arm-level indicator is an inference about whether or not some aberration has occurred in a large genomic region, so it is aggregating many intensity measurements together. This issue requires some serious thought for high-resolution array-based CGH measurements that are now becoming available. A brute-force model-based approach might be difficult to implement; it seems quite practical to segment the intensity profiles somehow for our downstream network calculations. Precisely how to do this is an open problem.

The discussant suggests that the marginal rates of occurrence of overt and covert aberrations ought to depend on the genomic location i. This may be so but it raises some interesting questions about the whole enterprise to look for cancer genes in aberration hot spots. The issue is investigated in Newton *et al.* 1998. Our model entails prior homogeneity (of aberration rates) that becomes heterogeneity in tumors because of selection. If genomic instability targets a location i, rather than being random (α_i is larger than other α_j, rather than complete equality), then aberration i may occur frequently in tumors, whether or not it is relevant to oncogenesis. Our premise, essentially, is that instability is less directed than that, implying that aberration hot spots in observed cells correspond somehow to relevant genes. One could allow independent but heterogeneous genomic instability. Then it would be the second-order statistical dependencies only, rather than the marginal frequencies that would deliver inferences about relevant aberrations. There seem to be clear benefits to the present model; for one thing, the estimated network provides a simple reorganization of the aberrations according to both marginal frequency and joint co-occurrence.

We are intrigued by discussant's related work suggesting that a flat prior over networks may be preferable when computing the MAP estimate, even though the informed prior produces reasonable marginal inferences. One might expect the opposite if shrinkage results from continuous parameter spaces provide any guidance.

Our prior constraint $\alpha > \beta$ was used to enforce some regularity since too many covert aberrations will quench the effect of selection. That the posterior pushes β near α suggests to us that the allowable pathways have not captured the dependence patterns effectively. This is why we have been so keen to develop tools for overlapping pathway models. As we have reported, we can compute the likelihood using a recursive algorithm, but we continue to build MCMC methods for this general case.

The potential for the instability-selection-network methodology to characterize statistical patterns in many forms of biological data has not gone unnoticed by us; indeed, we are excited about possible developments beyond chromosome-based CGH. The term "pathway" may be imprecise, as the discussant notes, since we use it to represent an unordered collection of markers; the term "ensemble" may be more appropriate (Newton, 2002). Whatever we call it, the framework is compelling. Data measuring aberrant features in the current state of a living system exhibit patterns; these patterns arose somehow prior to observation, and, insofar as they were favorable, they became observable. In allowing overlapping pathways, the class of models based on logical AND seems quite rich; we have not considered the more general class but the discussant's proposal ought to be followed up. With ever richer models the issue of identifiability seems to be important, in spite of the fact that a Bayesian analysis of some data with some model need not regard the issue. From the perspective of model development, we think it is useful to know what can be recovered in principle from a large data set.

Finally, we comment on "ground truth" and how to verify that an estimated network is correct in some sense. The best solution we can suggest is via prediction. Each CGH profile is a high-dimensional marker of tumor type; the modelling effort provides a clustering of tumors according to the probable pathways that each tumor has experienced (Figure 6). If the approach has any merit, it may be that the probable-pathway marker is an efficient low-dimensional predictor of other attributes, such as clinical outcome. Currently we are investigating the relationship between pathway predictions and clinical data.

ADDITIONAL REFERENCES IN THE DISCUSSION

Kerr, M. and Churchill, G. (2000). Analysis of variance for gene expression microarray data. *J. Comp. Biol.* **7**, 819–837.

Krogh, A., Brown, M., Mian, I. S., Sjolander, K. and Haussler, D. (1994). Hidden Markov models in computational biology: applications to protein modelling. *J. Mol. Biol.* **235**, 1501–1531.

Lawrence, C. E., Altschul, S. F., Boguski, M. S., Liu, J. S., Neuwald, A. F. and Wootton, J. C. (1993). Detecting subtle sequence signals: a Gibbs sampling strategy for multiple alignment. *Science* **262**, 208–214.

Li, C. and Wong, W. H. (2001). Model-based analysis of oligonucleotide arrays: expression index computation and outlier detection. *Proc. Natl. Acad. Sci. USA* **98**, 31–36.

Newton, M. A. (2002). Discovering combinations of genomic aberrations associated with cancer. *J. Am. Statist. Ass.* **97** (to appear).

Schmidler, S. C. (2002). *Statistical Models and Monte Carlo Methods for Protein Structure Prediction*. Ph.D. Thesis, Stanford University, USA.

Schmidler, S. C. (2002). Statistical shape theory and protein structure analysis. *Tech. Rep.*, Duke University, USA.

Schmidler, S. C., Liu, J. S. and Brutlag, D. L. (2000). Bayesian segmentation of protein secondary structure. *J. Comp. Biol.* **7**, 233–248.

West, M., Blanchette, C., Dressman, H., Huang, E., Ishida, S., Spang, R., Zuzan, H., Marks J. R. and Nevins J. R. (2001). Predicting the clinical status of human breast cancer using gene expression profiles. *Proc. Natl. Acad. Sci. USA* **98**, 11462–11467.

Wu, T. D., Schmidler, S. C., Hastie, T. and Brutlag, D. L. (1998). Regression analysis of multiple protein structures. *J. Comput. Biol.* **5**, 597–607.

BAYESIAN STATISTICS 7, pp. 307–326
J. M. Bernardo, M. J. Bayarri, J. O. Berger, A. P. Dawid,
D. Heckerman, A. F. M. Smith and M. West (Eds.)

Non-Centered Parameterizations for Hierarchical Models and Data Augmentation

OMIROS PAPASPILIOPOULOS GARETH O. ROBERTS
Lancaster University, UK
o.papaspiliopoulos@lancaster.ac.uk g.o.roberts@lancaster.ac.uk

MARTIN SKÖLD
Australian National University, Australia
martin.skold@anu.edu.au

SUMMARY

In this paper, we will compare centered and non-centered parameterizations for classes of hierarchical models. Our examples will include variance component models, random effect models, hidden Markov process models, and partially observed diffusion models. We will investigate the construction of non-centered methods by the use of state space expansion techniques, and will introduce methods for devising partially non-centered parameterizations, many of which are data-dependent.

Keywords: PARAMETERIZATION OF HIERARCHICAL MODELS; MISSING DATA PROBLEMS; CENTRED AND NON-CENTRED PARAMETRIZATIONS.

1. INTRODUCTION

For at least the last 15 year or so, hierarchical models in various guises have revolutionized Bayesian statistical methodology. Their key advantages have been their flexibility, interpretability and the ease of inference using MCMC techniques, see, for example, Smith and Roberts (1993) and Gilks *et al.* (1994). The simplest possible hierarchical model can be described by the directed graphical model in Figure 1a, where θ is a generic collection of hyperparameters, Y represents the observed data, and X can take the role of population parameters, missing data, hidden Markov process, or various other possibilities. Therefore, even this very simple setup encompasses a huge diversity of model types and statistical contexts. Most examples we consider here can be loosely described by Figure 1a, though there will often be other parameters present in the model not explicitly represented there. Convergence of MCMC methods, particularly when using the Gibbs sampler or related techniques, depends crucially on the parameterization used for the unknown quantities. From a modelling and interpretation perspective, the natural *centered parameterization* (denoted in this paper by CP) for Figure 1a is to use just $\theta,\ X$. This utilizes the conditional independence inherent within the model, which makes updating θ computationally less challenging in general. This is particularly true where conditional conjugacy is present, as is often the case by design. Thus, an algorithm for sampling from the joint distribution of θ, X, which we will call parameters and missing data, respectively, given the observed data Y might proceed by alternating between:

1. update θ from a Markov chain with stationary distribution $\theta \mid X$
2. update X from a Markov chain with stationary distribution $X \mid \theta, Y$ (1)

Algorithms like (1) will be called *Hastings-within-Gibbs* algorithms, the assumption being that each update is carried out using an appropriate Metropolis–Hastings update which preserves the relevant conditional distribution.

However, according to Figure 1a, θ and X exhibit *a priori* dependence and in many contexts this dependence is very strong. The presence of data tends to diminish the magnitude of that dependence, but the efficiency of the above successive substitution scheme will depend crucially on the extent to which this is the case, as we shall see in Section 2.

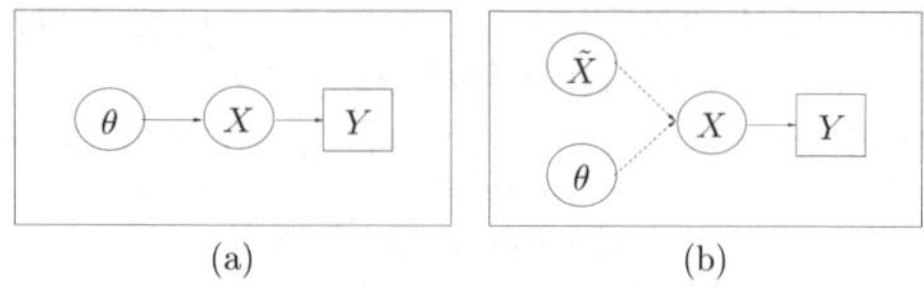

Figure 1. (a): *Graphical model of the centered hierarchical parameterization (CP),* (b): *Graphical model of the non-centered hierarchical parameterization (NCP).*

On the other hand, we might be able to find an alternative parameterization, $(X, \theta) \rightarrow (\tilde{X}, \theta)$, of the model in Figure 1a, *i.e.*, where the new missing data $\tilde{X}$ is some function of the previous missing data X and the parameters θ, such that $\tilde{X}$ is *a priori* independent of θ. This type of reparametrization of the hierarchical model in Figure 1a, is called *non-centered parameterization* (denoted by NCP) and the corresponding graphical model is given in Figure 1b. The MCMC algorithm corresponding to (1) for simulating from the posterior distribution of $(\tilde{X}, \theta)$ iterates

1. update θ from a Markov chain with stationary distribution $\theta \mid \tilde{X}, Y$
2. update $\tilde{X}$ from a Markov chain with stationary distribution $\tilde{X} \mid \theta, Y$ (2)

and the motivation behind the NCP is that in many contexts the convergence properties of (2) might be better than those of (1). Notice that where the conditional distributions in the above algorithms can be sampled directly, we shall refer to them as the *Gibbs sampler.*

1.1. *To Center or not to Center?*

NCPs have been used in many contexts dating back to the introduction of data-augmentation in Tanner and Wong (1987) and probably well before. On the other hand, Gelfand *et al.* (1996) give strong arguments for the adoption of CPs particularly for simple classes of hierarchical structures such as random effects models. Moreover, the conditional independence of θ and Y given X often means that Gibbs sampling can be implemented for the CP but not for the NCP, leading to a significant computational edge in favor of the CP. So is there any need to consider non-centered approaches at all?

In this paper, we shall argue that there is an important role for the NCP in many contexts. It provides a general reparametrization strategy to improve convergence of MCMC in cases where the latent process is relatively weakly identified by the data, since the prior independence between θ and $\tilde{X}$ will then ensure weak posterior dependence. We shall consider the construction of NCPs, and also classes of parameterizations which lie on continua between the CP and the NCP, the so-called partially non-centered parameterizations (PNCP). The first examples of the use of PNCP were introduced (originally in the context of the EM algorithm but subsequently for MCMC too) by Meng and van Dyk (1997, 1999).

Our work here should be taken as complementary to the marginal augmentation techniques introduced in Liu and Wu (1999) and Meng and van Dyk (1999). The marginal augmentation technique can be superior to both the CP and NCP in context where it can be applied, although it relies on implementation strictly by Gibbs sampling whereas our methodology is designed to be implemented in conjunction with Hastings-within-Gibbs strategies.

One problem with the NCP is the requirement of orthogonality between $\tilde{X}$ and θ. In many models this can be hard to achieve in practice, and this is therefore a major limitation on the use of non-centered methods. We will introduce state-space expansion techniques in this paper that allow easy implementation of the NCP in a wide variety of stochastic models.

Our three major examples will particularly emphasize the use of the CP, the NCP and the PNCP in the context where X is a hidden stochastic process. In our examples in Section 5, X either represents a Markov chain (or process) or in the geostatistical example of Section 5.1, X is a Gaussian random field.

1.2. *Example: Hierarchical Linear Models*

A toy example that serves to illustrate the main ideas in this paper and which is totally understood theoretically, is the following Normal hierarchical model (NHM), written as

$$\begin{aligned} Y_{ij} &= X_i + \sigma_y \epsilon_{ij}, \qquad j = 1, \dots, n \\ X_i &= \theta + \sigma_x z_i, \qquad i = 1, \dots, m. \end{aligned} \tag{3}$$

This model has also been used for pedagogical purposes in Liu and Wu (1999). Here ϵ_{ij} and z_i are independent standard normal random variables, θ is assigned a uniform improper prior and the variances are considered known for the time being at least. Due to the sufficiency of $\sum_j Y_{ij}/n$, which is again Gaussian, there is no loss of generality in assuming a single observation per random effect X_i and therefore from now on we will take $n = 1$ and drop the j subscript. The parameterization (θ, X), $X = (X_1, \dots, X_m)$ is known as the *centered parameterization* (CP), (see Gelfand *et al.*, 1995), and depicted graphically in Figure 1a illustrating the independence between θ and $Y = (Y_1, \dots, Y_m)$ conditional on X.

The name *non-centered parameterization* was originally used for the NHM, (see Gelfand *et al.*, 1995). In this context the NCP writes the model as

$$\begin{aligned} Y_i &= \tilde{X}_i + \theta + \sigma_y \epsilon_{ij} \\ \tilde{X}_i &= \sigma_x z_i, \; i = 1, \dots, m. \end{aligned} \tag{4}$$

Notice that $\tilde{X} = (\tilde{X}_1, \dots, \tilde{X}_m)$ and θ are *a priori* independent (see Figure 1b) but conditional on the data, they are dependent.

Gibbs sampling can be applied very easily in this context using either the CP or the NCP and it is of interest to know which gives the most rapidly convergent sampler. Fortunately, in this simple context, this question has a definitive answer as we shall see in Section 2.1. However, the result depends explicitly on the relative sizes of σ_x^2 and σ_y^2. We shall also investigate whether there are alternative parameterizations which are capable of improving on both the CP and NCP. Again in this simple context (but also for a very general form of (3)), we can give a definitive and positive answer to this - there always exists a *partially non-centered parameterization* (PNCP) which gives more rapid convergence than either of these two alternatives.

1.3. *Summary of Paper*

In this paper we will introduce, analyze and apply general forms of NCPs. In Section 2, we shall describe existing theory on rates of convergence for the Gibbs sampler, which allows us to compare different parameterization schemes, at least in the Gaussian context, and apply this theory to the linear Gaussian model example introduced in Section 1. Section 3 introduces NCPs in more generality, including methods which involve the expansion of the state space. Section 4 considers PNCPs, while Section 5 considers three examples of the effective use of NCPs: one involving a non-linear geostatistical model; another involving the problem of inference for partially observed diffusions; while the third considers a class of stochastic volatility models currently popular in finance.

2. RATES OF CONVERGENCE OF THE GIBBS SAMPLER

Let $Z = (Z_1, Z_2)$ denote a random variable with density π, partitioned into two components, Z_1, Z_2, of arbitrary dimension. We term a two-component Gibbs sampler on π, under the parameterization (Z_1, Z_2), the algorithm that iterates the following procedure:

1. sample Z_1 from the conditional distribution of $Z_1 \mid Z_2$
2. sample Z_2 from the conditional distribution of $Z_2 \mid Z_1$ (5)

It is well beyond the scope of this article to discuss rates of convergence of algorithms in any detail, though see Jones and Hobert (2001) and Roberts and Tweedie (2002) for recent summaries. However, where the two-component Gibbs sampler can be implemented, it admits a very complete (if not always practically useful) theory that we very briefly describe here. Let $P^n(x, \cdot)$ denote the distribution of the two-component Gibbs sampler after n iterations, where x denotes an arbitrary starting value for the (Z_1, Z_2) pair. Amit (1991) observed that the $\mathcal{L}_2$ distance from stationarity decays as $A(x)b(n)\rho^n$ for some function $b(n)$ which varies slower than an exponential function. The constant $\rho \leq 1$ is defined as

$$\rho^{1/2} = \sup \text{Corr}(f(Z_1), g(Z_2)) \tag{6}$$

and the supremum above is taken with respect to all real-valued non-constant functions f and g which admit finite variances under π. It turns out that other common norms (such as total variation distance) can also be shown to have this rate, at least for a large class of plausible target distributions (see Roberts and Tweedie, 2001). The covariance structure of the two-component Gibbs sampler has also been studied in detail by Liu *et al.* (1994).

It has long been recognized that the correlation structure of the target distribution determines the convergence behavior of the corresponding Gibbs sampler (see Hills and Smith, 1992; Gelfand *et al.*, 1995). In the two component case, this result makes the connection precise. The characterization (6) is of little practical use, since in general it will be impossible to evaluate the supremum in (6), although one important exception is for Gaussian π, when suprema of the kind appearing in (6) are achieved exclusively by linear functions. However, (6) has been used to compare different augmentation schemes (see *e.g.*, Meng and van Dyk, 1999; Liu and Wu, 1999).

For Gibbs samplers with larger numbers of components, it is not possible to find an explicit statement analogous to (6) which relates the rate of convergence of the algorithm to the target distribution correlation structure, though Amit (1991) does give some related inequalities. In the Gaussian target distribution case however, explicit formulae are available for rates of convergence of the sampler in terms of the target distribution correlation matrix (see Roberts and Sahu, 1997). We shall use these results to compare parameterization exactly for the NHM.

2.1. *Rates of Convergence for CP and NCP of the NHM*

The following results are taken directly from Roberts and Sahu (1997). We wish to sample from the joint posterior distribution of X and θ of the model (3) using the Gibbs sampler as in (1). Since this is a multivariate Gaussian distribution we can explicitly calculate the rate of convergence, denoted by ρ_c, using the results of Roberts and Sahu (1997),

$$\rho_c = 1 - \kappa, \tag{7}$$

$$\kappa = \frac{(\sigma_x^2 + \sigma_y^2)^{-1}}{(\sigma_x^2)^{-1}} = \frac{\sigma_x^2}{\sigma_x^2 + \sigma_y^2}\,. \tag{8}$$

The first expression for κ in (8) writes κ as the ratio of observed by augmented information for θ under the CP, defined as $1/\text{Var}(\theta \mid Y)$ and $1/\text{Var}(\theta \mid X, Y) = 1/\text{Var}(\theta \mid X)$, respectively. In this context, $1 - \kappa$ is the *Bayesian fraction of missing information* in the sense defined by Rubin (1987). The relationship between observed and augmented information and rates of convergence of algorithms was noted first in a very general framework for the EM algorithm (see Meng and van Dyk (1997)), but can be translated to the data augmentation methodology in this specialized linear model context (see Sahu and Roberts, 1999).

Therefore, the CP will perform well when $\kappa \to 1$, *i.e.*, when the data are relatively very informative in the sense that the observed data contain almost as much information about the parameter as the augmented. This result can also be expressed in terms of (6) since

$$\text{Corr}(\overline{X}, \theta \mid Y) = \sqrt{1 - \kappa}, \qquad \overline{X} = \sum_{i=1}^{m} X_i/m.$$

We return now to the NCP (4). In this case, we use the Gibbs sampler algorithm (2) and the rate of convergence, ρ_{nc}, turns out to be

$$\rho_{nc} = \kappa. \tag{9}$$

When one parameterization produces very slow mixing for the Gibbs sampler the other will be performing extremely well. For this model with the flat priors specified, the relative performance of the CP and NCP is explicit since $\rho_{nc} = 1 - \rho_c$. Note, however, that this relation does not hold when proper priors are used for θ, since in this case both algorithms have faster convergence than the rates given in (7) and (9).

2.2. *Example: State-space Models*

Another class of Gaussian models that we can analyze is that of linear state space models defined in the following way.

$$\begin{aligned} Y_i &= X_i + \sigma_y \epsilon_i \\ X_i &= \phi X_{i-1} + \theta(1 - \phi) + \sigma_x (1 - \phi^2)^{1/2} z_i, \qquad i = 1, \ldots, m \end{aligned} \tag{10}$$

where σ_x, σ_y and $\phi \in [0, 1]$ are assumed to be known. Here the stationary moments of $X = (X_1, \ldots, X_m)$ are $\text{E}(X_i) = \theta$, $\text{Var}(X_i) = \sigma_x^2$; therefore, (10) extends (3) to allow dependence among the X_is. The two-component Gibbs sampler alternates by updating θ given X, and X given θ and $Y = (Y_1, \ldots, Y_m)$, using, for instance, forward filtering–backward smoothing techniques as described in Carter and Kohn (1994). The NCP is derived by setting $\tilde{X}_i = X_i - \theta$.

For this example, explicit rates of convergence are easily calculable using the results of Roberts and Sahu (1997). It can be shown that as $m \to \infty$:

$$\frac{1 - \rho_{nc}}{1 - \rho_c} \approx \frac{\sigma_y^2}{\tau \sigma_x^2} \tag{11}$$

where τ is the integrated correlation time of X. A similar expression can be found in Pitt and Shephard (1999) together with considerably more detailed convergence rate analysis. Thus highly correlated hidden Markov models, with large marginal variance, favor the use of the CP over the NCP. Similar empirical results are obtained for a geostatistical model in Section 5.1.

In fact for both the CP and NCP, when X is updated as a block rather than by single-site updating, the corresponding rate of convergence does not converge to 1 for large m and thus suggests the use of forward–backward filtering as part of MCMC routines for general classes of hidden Markov processes (see Carter and Kohn, 1994) (without taking computational cost into account).

2.3. *Example: Linear Non-Gaussian Models*

The following toy example is very simple, but its results are quite striking. Suppose we modify the NHM (3) such that ϵ_{ij} has a standard Cauchy distribution, while z_i remains standard Gaussian. In this context, the heavy tailed nature of the Cauchy distribution makes the observation equation relatively un-informative for extremal values of X. Following the intuition gained from studying the NHM, we might expect the CP to perform poorly in some way in the tail regions in relation to the NCP. Figure 2 shows output from the Gibbs sampler for both CP and NCP (where $n = m = 1$) for different starting values for θ.

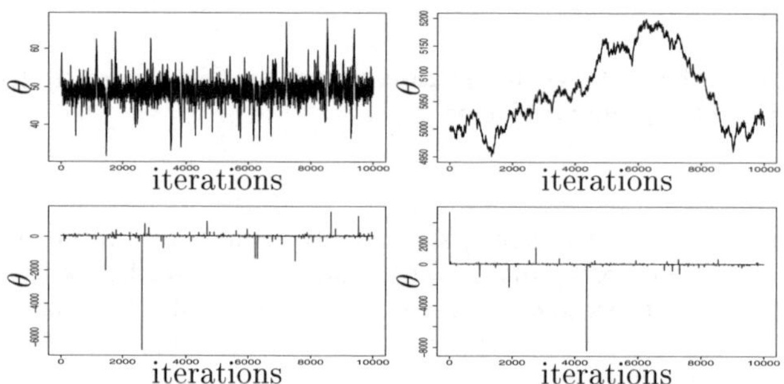

Figure 2. *Gibbs sampler output for θ in the Normal-Cauchy model described in Section 2.3 with $Y = 49.06$. Top: centered parameterization started from $\theta_0 = 50$ (left) and $\theta_0 = 5000$ (right). Bottom: non-centered parameterization for the same starting values. All chains were run for 10^4 iterations. Notice the different scales in the plots*

The CP exhibits unstable heavy-tailed excursions characteristic of algorithms which fail to be geometrically ergodic (see Roberts, 2003), while the NCP appears to return to the distribution mode very rapidly. In fact the following result (which we state without proof) holds. For the linear non-Gaussian model (3) with Cauchy observation equation and Gaussian hidden equation, the CP converges slower than geometrically, while the NCP converges uniformly quickly from all starting values (uniformly ergodic). This can be linked to the results obtained by Dawid (1973), where it is shown that, under the above setting, as $\theta \to \infty$, $\tilde{X}$ is independent of θ and Y and its distribution converges to the prior. Therefore, the NCP will have no difficulty making excursions far into the target distribution tails and swiftly come back to the modal area. For Cauchy hidden equation with Gaussian observation error (as used *e.g.*, in Wakefield *et al.*, 1994), the opposite result holds with the NCP failing to be geometrically ergodic while the CP is uniformly ergodic. Considerably more general results relating tail behavior to qualitative convergence properties will be found in Roberts (2002).

3. GENERAL NON-CENTERED PARAMETERIZATIONS

The important feature of the NCP for the NHM that can be extracted to a much more general context is the orthogonality of the prior structure. Specifically, we find $\tilde{X}$ which is *a priori* independent of θ and from which X can be constructed via a deterministic function

$$X = h(\tilde{X}, \theta). \tag{12}$$

In the Gaussian context under NCP (4), it is easy to identify h as $h(\tilde{x}, \theta) = \theta + \tilde{x}$. For the general hierarchical model in Figure 1a, such a function h always exists, but is not unique. However, it can be difficult to identify such a function h which is analytically sufficiently tractable to be of practical use.

The effect of this reparametrization is described by Figure 1b. The NCP does not exhibit conditional independence between θ and the data Y conditional on $\tilde{X}$. Once an NCP has been identified, it is therefore unlikely that Gibbs sampling in its pure form can be used to update the two components θ and $\tilde{X}$.

From experience in the NHM context, we would expect an NCP to be more effective than its CP rival when X is poorly identified by the data and remains highly correlated with θ. For this reason, it is particularly important in contexts where the dimensionality of X increases as the data set becomes larger, and this in turn explains its relevance to latent structure models such as hidden Markov models.

A technique that allows us to construct NCPs for a wide range of hierarchical models (for example, most of the models considered in Lee and Nelder, 1996) is the expansion of the state-space. In an NCP we allow the function $h(\cdot, \theta)$ to be non-invertible so that for example $\tilde{X}$ can exist on a higher dimensional space than X. As an example, suppose that X is distributed according to a Gamma with shape parameter α and scale parameter 1. We can take $\tilde{X}$ to be a standard Gamma process in $[0, \infty)$ and set $X = \tilde{X}(\alpha)$. The link between infinite divisible distributions and Lévy processes can be used for similar parameterizations. Due to lack of space, we will not give the details of this construction here, but the interested reader is referred to Papaspiliopoulos *et al.* (2002a). The implementation of a state-space expanded NCP is not more difficult than the corresponding CP, although it is more computationally intensive. We can also use these ideas to orthogonalize a stochastic process from its parameters and an example is given in Section 5.2.

4. PARTIALLY NON-CENTERED ALGORITHMS

We have already defined and analyzed the convergence properties of the CP and NCP for the normal hierarchical model in Section 1.2. Consider the following parameterization for the same model,

$$\begin{aligned} Y_i &= w\theta + \tilde{X}_i^w + \sigma_y \epsilon_i \\ \tilde{X}_i^w &= (1-w)\theta + \sigma_x z_i, \qquad i = 1, \ldots, m. \end{aligned} \tag{13}$$

where w is a fixed number in $[0, 1]$. It can easily be seen that

$$\tilde{X}_i^w = (1-w)X_i + w\tilde{X}_i$$

where X_i and $\tilde{X}_i$ are as defined in (3) and (4), respectively, and therefore (13) defines a continuum of parameterization strategies with the CP at one extreme, for $w = 0$ and the NCP at the other, for $w = 1$. The motivation behind this construction is the following: from the results of Section 2.1 we know that the CP is the optimal algorithm when the relative observation error σ_y/σ_x

tends to zero, *i.e.*, when the data are most informative, while the NCP is optimal when this error tends to infinity, *i.e.*, in absence of any data. We would therefore like to construct an algorithm that will adapt to the quantity of information present in the observed data, in order to provide samplers with superior convergence properties to both the CP and NCP. This can be achieved by (1), which we will call *partially non-centered* (PNCP), while it is the optimal Gibbs sampling algorithm for a specific choice of w.

The joint posterior distribution of $\tilde{X}^w = (\tilde{X}_1^w, \ldots, \tilde{X}_m^w)$ and θ is still Gaussian and therefore we can calculate the rate of convergence of the Gibbs sampler (1) under this parameterization which is given and plotted against w in Figure 3.

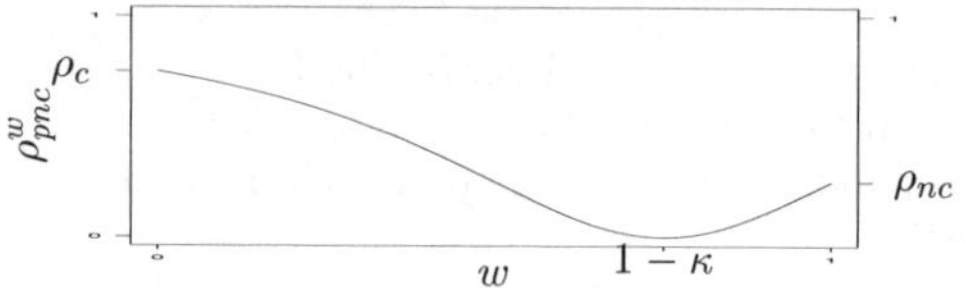

Figure 3. *Rate of convergence* $\rho^w_{pnc} = (w-(1-\kappa))^2/(w^2\kappa + (1-w)^2(1-\kappa))$ *for the PNCP on the NHM with* $\sigma_x = 1,\ \sigma_y = 3$.

We can easily derive that $\rho^w_{pnc} \leq \max(\rho_c, \rho_{nc}),\ \forall w \in (0,1)$, and $\rho^w_{pnc} = 0$ for $w = 1-\kappa$, which suggests not only that the PNCP algorithm (13) can outperform both CP and NCP, but also it can be tuned appropriately to produce IID samples, by setting $w = 1-\kappa$.

A similar result can be obtained for the general NHM (see Gelfand *et al.,* 1995)

$$\begin{aligned} Y_i &= C_{1i}X_i + (\sigma_y^2 I_{n_i})^{1/2}\epsilon_i \\ X_i &= C_2\theta + D^{1/2}z_i, \quad i = 1, \ldots, m \end{aligned} \tag{15}$$

where Y_i is an $n_i \times 1$ vector, X_i a $p \times 1$, θ a $q \times 1$, D is the prior $p \times p$ variance matrix, and $C_{1i},\ n_i \times p$ and $C_2,\ p \times q$ are design matrices, $C_2^t C_2$ assumed to be invertible. Then the NCP and PNCP are defined as

$$\begin{aligned} \tilde{X}_i &= X_i - C_2\theta \\ \tilde{X}_i^w &= X_i - W_i C_2\theta \end{aligned} \tag{16}$$

and IID samples can be produced by setting

$$W_i = B_i D^{-1}, \quad B_i^{-1} = \sigma_y^{-2} C_{1i}^t C_{1i} + D^{-1}. \tag{17}$$

The rate of convergence of the CP for (15) is given by the maximal modulus eigenvalue of $m^{-1}\sum_{i=1}^m W_i$. The importance of this matrix in assessing the convergence properties of the CP was observed by Gelfand *et al.* (1995).

Notice that in (16) the proportion of θ subtracted from each X_i varies with i unlike (13), reflecting the varying informativity of each Y_i about the underlying X_i present in (16).

4.1. *Partial Non-centering Outside the Gaussian Context*

Partial non-centering can be used for many models outside the Gaussian context. In general there is no unique way of defining a continuum of partial non-centering strategies. However often there will be a natural one suggested strongly by the model structure. Outside the Gaussian context, it is rare that pure Gibbs sampling can be used in conjunction with a PNCP, so that as with the NCP and often the CP, appropriate Hastings-within-Gibbs strategies will be necessary. Thus, optimal PNCP will not produce IID observations from the target distribution.

Sensitivity of algorithm performance to data is very common in many classes of hierarchical models, since it is often the case that the information about X_i contained in Y_i will depend on Y_i. Therefore, to extend the PNCP to other models in an efficient way, we will need to allow w to vary across $i = 1, \dots, m$, as in (16), and possibly be a function of the corresponding data Y_i.

To make this more concrete, consider the random effects model:

$$\begin{aligned} Y_i &\sim f(\cdot|X_i) \\ X_i &= \theta + \sigma_x z_i, \qquad i = 1, \dots, m \end{aligned} \tag{18}$$

for some class of densities $f(\cdot \mid \cdot)$.

A quadratic expansion of the log-likelihood, $\ell = \log f$ gives a rough indication into the information content present in Y_i about X_i. We set $I(y) = -\partial^2 \ell(y|x)/\partial x^2$ evaluated at the MLE $\hat{x}$ (ignoring the latent structure). Other approximations of information may be more appropriate in certain cases. In the NHM, $I(y) = (\sigma_y^2)^{-1}$, but more generally I will depend on y. Data-dependent non-centering then sets

$$\tilde{X}_i^w = X_i - w_i(y)\theta \tag{19}$$

where $w_i(y) = (1 + I(Y_i)/\sigma_x^2)^{-1}$.

A multivariate generalization of the technique outlined above is described in the geostatistical example of Section 5.1. A further spatial application of the PNC appears in Higdon (1998).

5. EXAMPLES

5.1. *Spatial GLMM*

In this section, we will consider a special case of a GLMM model introduced in Breslow and Clayton (1993) proposed in the spatial context by Diggle *et al.* (1998). Similar modelling approaches have received much attention recently but strong posterior correlation between parameters and latent variables makes a fully Bayesian approach difficult without the use of complex MCMC algorithms and careful reparameterizations. We will here consider a spatial Poisson log-Normal model also studied in Diggle *et al.* (1998) for modelling radioactive counts and in Christensen and Waagerpetersen (2002) for modelling counts of weed. For a more detailed description of the model refer to Diggle *et al.* (1998).

The data consists of recorded observations $Y = (Y_1, \dots, Y_m)$ with

$$\begin{aligned} Y_i &\sim \text{Po}(\exp(X_i)) \\ X &\sim \text{N}(\theta\mathbf{1}, \sigma^2 R) \end{aligned} \tag{20}$$

Here $X = (X_1, \dots, X_m) = (X(t_1), \dots, X(t_m))$ are (unobserved) values from a stationary isotropic Gaussian random field $\boldsymbol{X} = \{X(t), t \in \boldsymbol{R}^2\}$ with mean θ, standard deviation σ and correlation function $r(u) = \text{Corr}(X(s_1), X(s_2)) = \exp(-u/\alpha)$, $u = \|s_2 - s_1\|$ (Euclidean distance), and $\mathbf{1}$ is a $m \times 1$ vector of 1s.

Thus, the unknown components in this model are the parameters θ, α and σ, together with the underlying field X. It is beyond the scope of this article to fully describe partially non-centered methods which can be applied effectively to this problem. We shall instead concentrate on part of the algorithm in which the partial non-centering is applied to θ while the remaining parameters σ and α remain fixed. Similar non-centering strategies can be applied to σ and α

and also directly to X in order to break down the posterior correlation structure present in the field X. This will be reported in Christensen *et al.* (2002).

Both Diggle *et al.* (1998) and Christensen and Waagerpetersen (2002) use the NCP, *i.e.*, they alternate between updating $\tilde{X} = X - \mathbf{1}\theta$ and θ. Diggle *et al.* (1998) use single site Gibbs updating of $\tilde{X}$ and Christensen and Waagerpetersen (2002) use the Metropolis adjusted Langevin algorithm (MALA, see *e.g.*, Roberts and Tweedie, 1996), which can give considerable convergence advantages for large m. Here we shall extend the data-dependent partial non-centering ideas of Section 4.1 to the present case of spatially varying X.

In the absence of covariates, the partial centering parameters for the NHM with $m = 1$ are given by $W = BD^{-1} = (\text{diag}(\sigma_x^{-2}) + D^{-1})^{-1}D^{-1}$, where D is the $n \times n$ prior variance matrix and $\text{diag}(\sigma_x^{-2})$ is the $n \times n$ matrix with σ_x^{-2} on the diagonal and zeros elsewhere. The latter matrix being constant along the diagonal reflects the fact that the variance of the error distribution $Y|X$ is independent of X and thus the loss of information about θ is equal along the components of Y. This is a feature not shared by model (20) since here the error-distribution depends on X in a nonlinear way through the logarithmic link-function and large values in Y tend to be more informative about the mean than small ones.

Outside the Gaussian context, there is no direct analogue of the B_i matrices in (17), but a quadratic expansion of the likelihood as in Section 4.1 suggests we set

$$\hat{B}^{-1} := -\frac{\partial^2}{\partial X^2} \log \pi(X|Y,\theta)|_{X=\hat{X}} = -(\text{diag}(\exp(\hat{X})) + \sigma^{-2}R^{-1}) \approx -(\text{diag}(Y) + \sigma^{-2}R^{-1})$$

(where $\hat{X}$ is the MLE from the observation equation alone) leading to the partial centering for θ

$$\tilde{X}^w = X - W\theta.$$

with $W = \hat{B}\sigma^{-2}R^{-1}$. Where $\alpha = 0$ this reduces to the random effects case described by (19).

A simulation study involving 100 observations equally spaced on the unit square was carried out. It involved using two different combinations of θ and σ each under two levels of dependence. Although updates of $\tilde{X}^w$ need to be carried out by a suitable Hastings algorithm in this context for each of our parameterizations, our comparison is based on pure Gibbs updates (approximated by running multiple MALA steps for $\tilde{X}^w$ and θ). This eliminates the possibility that any difference between simulation performance could be due to varying efficiency of the MALA updates. ACFs summarizing our results are given in Figure 4.

The CP performs better in relation to the NCP as the dependence in $\boldsymbol{X}$ becomes stronger (as measured by α), which is in agreement with (11). Moreover, while the performance of the CP and NCP vary considerably as the parameters change, the data-dependent PNCP performs extremely well in all cases.

5.2. *Non-Gaussian Ornstein–Uhlenbeck Stochastic Volatility Models*

Non-centering was recently applied in Dellaportas *et al.* (2001) to construct efficient algorithms for Bayesian inference for the class of non-Gaussian stochastic volatility models introduced in Barndorff-Nielsen and Shephard (2001). The parameterization structures that we present for this model are more complicated than the CP and NCP shown in Figure 1; however, the general methodology of non-centering developed in the previous sections can also be extended in a natural way.

A stochastic volatility model describes the continuous time movement of the logarithm of an asset price, $Y(t)$, through the stochastic differential equation (SDE)

$$dY(t) = v(t)^{1/2} dB(t), \qquad t \in [0, T] \tag{21}$$

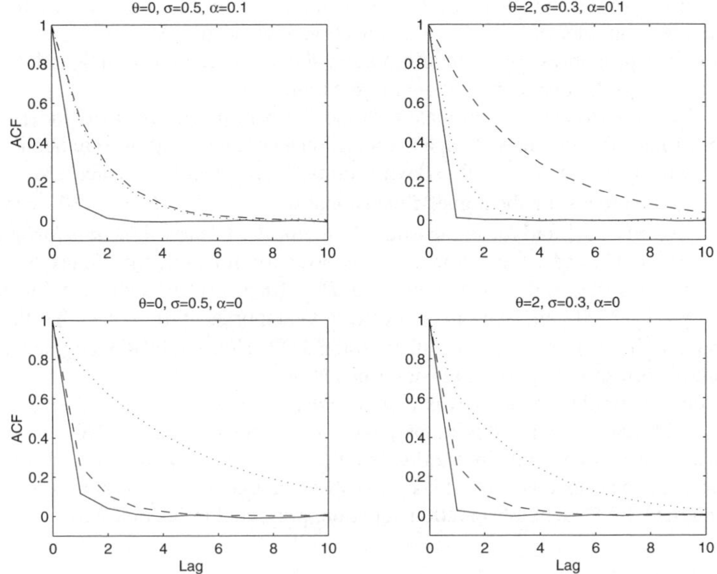

Figure 4. *The spatial GLMM of Section 5.1 and its special case, the random effects model of Section 4.1 (showed in the bottom two plots). ACF for θ using CP (dotted), NCP (dashed) and data-dependent PNCP (solid) for various parameter values.*

where $B(\cdot)$ is a standard Brownian motion. The volatility $v(t)$ is assumed to be unobserved, stochastic, and independent of $B(t)$. Defining $v^*(t_{i-1}, t_i)$ to equal volatility integrated over $[t_{i-1}, t_i]$, (20) implies that the conditional distribution of the increments of the data, $Y(t_i) - Y(t_{i-1})$, given v^* is $\mathrm{N}(0, v^*(t_{i-1}, t_i))$ and with conditional independence between disjoint time intervals. In Barndorff-Nielsen and Shephard (2001), the volatility process is modelled as a stationary non-Gaussian Ornstein–Uhlenbeck process described through an SDE with solution

$$v(t) = e^{-\mu t} v(0) + \sum_{j=1}^{N(t)} e^{-\mu(t-c_j)} \epsilon_j. \tag{22}$$

We will focus on the case where $N(t)$ is a Poisson process with rate λ and arrival times $c_1 < \cdots < c_{N(t)}$, and $\epsilon_j \sim \mathrm{Ex}(\phi)$ independently of the Poisson process so that (22) defines a process with stationary distribution $\mathrm{Ga}(\nu = \lambda/\mu, \phi)$. The data consist of discrete observations $Y = \{Y(t_1), \ldots Y(t_n)\}$. Although it is straightforward to write down the conditional density of the data given the corresponding integrated volatilities $v^*(t_{i-1}, t_i)$, $i = 1, \ldots, n$, the likelihood function $f(Y \mid \phi, \lambda, \mu)$ is unavailable since it involves integrating out the volatility process which is practically impossible (for details see Section 5.4 of Barndorff-Nielsen and Shephard, 2001). On the other hand a data augmentation method that augments the $v^*(t_{i-1}, t_i)$ is very inefficient, as shown in Barndorff-Nielsen and Shephard (2001). Instead, we explicitly augment $v(0)$, the jump times $\{c_j\}$ and the jump sizes $\{\epsilon_j\}$ and denote by Ψ the marked point process containing these pairs, since they uniquely determine v^*. Therefore, the missing data are $(\Psi, v(0))$, and conditionally on this λ, ϕ are independent of Y. Implementational details of the algorithm constructed for this parameterization appear in Dellaportas *et al.* (2001).

The algorithm described above is a CP for λ and ϕ. Therefore, from the intuition acquired from the simpler examples of Section 2.1 and Section 3, we would expect this method to exhibit poor convergence properties when the information about λ, ϕ contained in Ψ and $v(0)$ strongly dominates the marginal information about these parameters.

An NCP for this model, which makes Ψ and the parameters *a priori* independent, can naturally be produced using the state-space expansion technique mentioned in Section 3. Ψ is *a priori* a Poisson process on $[0,T]\times(0,\infty)$ with points $\{(c_j,\epsilon_j)\}$ and mean measure $\lambda\phi e^{-\phi\epsilon}dc\,d\epsilon$. Let $\tilde{\Psi}$ be a Poisson process on the higher dimensional space $[0,T]\times(0,\infty)\times(0,\infty)$ with points denoted by $\{(\tilde{c}_j,m_j,\tilde{\epsilon}_j)\}$ and mean measure $e^{-\tilde{\epsilon}}d\tilde{c}\,dm\,d\tilde{\epsilon}$. Clearly $\tilde{\Psi}$ is *a priori* independent of the parameters. Consider the following transformation illustrated in Figure 5. Choose all points of $\tilde{\Psi}$ with $m_j<\lambda$ and project them to $[0,T]\times(0,\infty)$. Denote those points by $(c_j,\tilde{\epsilon}_j)$. Then take $\epsilon_j=\tilde{\epsilon}_j/\phi$. It can easily be shown (see Dellaportas *et al.*, 2001) that the set of the resulting points $\{(c_j,\epsilon_j)\}$ has the same distribution as Ψ. Hence the NCP augments $\tilde{\Psi}$ and the transformation corresponding to (12) is described above.

An extensive simulation study was carried out in Dellaportas *et al.* (2001) to study the properties of CP and NCP for this model, revealing robustness of the NCP to a variety of different parameter values and time-series length. Unfortunately, due to lack of space, we cannot reproduce these results here. It was also observed that the CP has similar behavior as in the linear model with Cauchy observation equation, presented in Section 2.3.

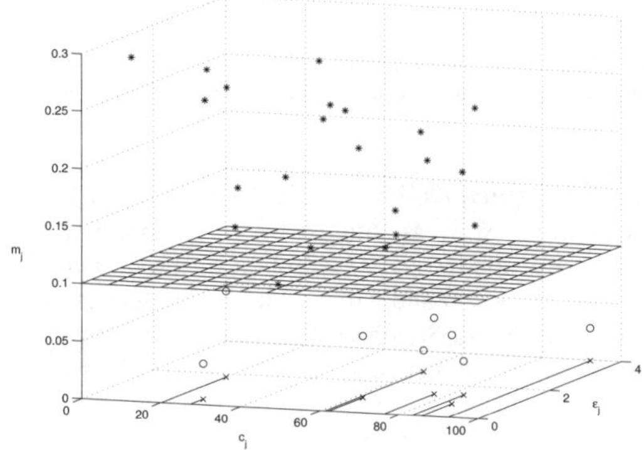

Figure 5. *Construction of the NCP of Section 5.2. Transformation of $\tilde{\Psi}$ to Ψ; using the current value of $\lambda=0.1$, choose all points of $\tilde{\Psi}$ with $m_j\leq\lambda$ (denoted by circles, where points with $m_j>\lambda$ by asterisks), project them to $[0,T]\times(0,\infty)$, where $T=100$; denote them as $\{c_j,\tilde{\epsilon}_j\}$. Set $\epsilon_j=\tilde{\epsilon}_j/\theta$ using the current value of $\theta=1$.*

5.3. *Partially Observed Diffusion Models*

Here we briefly review the techniques introduced in Roberts and Stramer (2001). We shall consider the simplest case of these models where the volatility is an unknown constant

$$dX(t)=\sigma dB(t)+\xi(\theta,X(t))\,dt\,. \tag{23}$$

We wish to make inference about σ and θ from observations $\{Y(t_1),\ldots,Y(t_n)\}$. The data are assumed to be insufficiently fine to allow useful approximations of the continuous time reality, so it is natural to treat this as a missing data problem.

Here we shall consider the theoretical algorithm that imputes the entire missing data, $X=\{X(s),t_i<s<t_{i+1},\ 1\leq i\leq n-1\}$. In practice, observations at a fine partition of time points need to be imputed, but here we shall assume that continuous imputation is feasible.

The CP that alternates updating the parameters and X is actually *reducible*, since σ never changes as a result of the quadratic variation identity

$$\lim_{m\to\infty}\sum_{i=2}^{m}(X(i(t_n-t_1)/m)-X((i-1)(t_n-t_1)/m))^2=\sigma^2(t_n-t_1)\,.$$

We call such an identity linking X and the parameters an *ergodicity constraint*.

A natural non-centering of the scale parameter σ sets

$$\tilde{X}(t)=X(t)/\sigma$$

(though this is not strictly a non-centering of σ without an associated transformation of the diffusion drift term). However, the corresponding NCP is again reducible. This time the problem is caused by the continuity of sample paths of (23) at the observed points $Y(t_j)$, which is violated if σ is updated.

Here there is essentially a unique piecewise linear transformation of $\tilde{X}$, giving a PNCP that leads to an irreducible algorithm. For $t_i<t<t_{i+1}$ it is described by

$$\tilde{\tilde{X}}(t)=\tilde{X}(t)-\frac{(t_{i+1}-t)Y(t_i)/\sigma+(t-t_i)Y(t_{i+1})/\sigma}{t_{i+1}-t_i}\,.$$

Much more detail and intuition into this construction can be obtained from Roberts and Stramer (2001). The problems of the CP are caused by the fact that the augmented data contain infinite information about σ while the observed finite data set can hardly do this. More complex SDE models can be treated by adaptations of the techniques outlined above.

6. DISCUSSION

In this paper, we have considered many strategies for constructing non-centered and partially non-centered parameterizations across wide-ranging classes of models. We have tried to provide a balance of theory, methodology, statistical insight and examples. Our motivation has been the search for effective parameterizations for the use in MCMC algorithms.

However, it has not been the purpose of this paper to promote unreservedly non-centered methods. In fact there are many situations where there is little reason to look further than the CP for the construction of effective algorithms. On the other hand, as we have seen in a number of examples, there are many situations where the posterior dependence between θ and X is prohibitively strong so that non-centered methods are needed. We have also shown how the tail behavior of the observation and hidden equations can affect the behavior of the CP and the NCP. Thus we have tried to present a balanced view of the available parameterizations. Moreover, we have discussed PNCP methods that offer a solution to finding algorithms with increased robustness to data.

Other work to which the ideas in this paper appear well-suited involves inference using stochastic epidemic models, and this is subject to ongoing work with Peter Neal (Neal *et al.*, 2002). Non-centering methods ought also to be well suited for problems in Bayesian non-parametrics and other situations where hidden structure is allowed to vary flexibly and we are currently working on this (see Papaspiliopoulos *et al.*, 2002b). We are also working on non-centering for multivariate θ, where there is the option of non-centering the parameters individually or simultaneously, and on hierarchical models with more than two stages.

ACKNOWLEDGEMENTS

All authors would like to acknowledge the support of TMR network FMRX-CT960095 on Spatial and Computational Statistics. The first author would like to thank the Onasis foundation and Lancaster University for support and similarly the third author thanks the Helmut Hertz and Wennergren foundations. We thank Petros Dellaportas for motivating discussions, and Ole Christensen and Paul Fearnhead for constructive comments on earlier drafts.

REFERENCES

Amit, Y. (1991). On rates of convergence of stochastic relaxation for Gaussian and non-Gaussian distributions. *J. Multiv. Anal.* **38**, 82–99.

Barndorff-Nielsen, O. and Shephard, N. (2001). Non-Gaussian Ornstein-Ulhenbeck-based models and some of their uses in financial economics. *J. R. Statist. Soc. B* **63**, 167–241.

Breslow, N. E. and Clayton, D. G. (1993). Approximate inference in generalized linear mixed models. *J. Am. Statist. Ass.* **88**, 9–25.

Carter, C. K. and Kohn, R. (1994). On Gibbs sampling for state space models. *Biometrika* **81**, 541–553.

Christensen, O. F., Roberts, G. O. and Sköld, M. (2002). Bayesian analysis of spatial GLMM using partially non-centered MCMC methods (in preparation).

Christensen, O. F. and Waagepetersen, R. P. (2002). Bayesian prediction of spatial count data using generalised linear mixed models. *Biometrics* **58**, 280–286.

Dawid, A. P. (1973). Posterior expectations for large observations. *Biometrika* **60**, 664–667.

Dellaportas, P., Papaspiliopoulos, O. and Roberts, G. O. (2001). Bayesian inference for non-Gaussian Ornstein–Uhlenbeck stochastic volatility processes (submitted for publication).

Diggle, P. J., Tawn, J. A. and Moyeed, R. A. (1998). Model-based geostatistics. *J. R. Statist. Soc. C* **47**, 299–350, (with discussion).

Gelfand, A. E., Sahu, S. K. and Carlin, B. P. (1995). Efficient parametrization for normal linear mixed models. *Biometrika* **82**, 479–488.

Gelfand, A. E., Sahu, S. K. and Carlin, B. P. (1996). Efficient parameterizations for generalised linear models. *Bayesian Statistics 5* (J. M. Bernardo, J. O. Berger, A. P. Dawid and A. F. M. Smith, eds). Oxford: Oxford University Press, 479–488.

Gilks, W. R., Thomas, A., and Spiegelhalter, D. J. (1994). A language and program for complex bayesian modelling. *The Statistician* **43**, 169–178.

Higdon, D. M. (1998). Auxiliary variable methods for Markov chain Monte Carlo with applications. *J. Am. Statist. Ass.* **93**, 585–596.

Hills, S. E. and Smith, A. F. M. (1992). Parameterization issues in Bayesian inference. *Bayesian Statistics 4* (J. M. Bernardo, J. O. Berger, A. P. Dawid and A. F. M. Smith, eds). Oxford: Oxford University Press, 227–246.

Jones, G. L. and Hobert, J. P. (2001). Honest exploration of intractable probability distributions via Markov chain Monte Carlo. *Statist. Sci.* **16**, 312–334.

Knuth, D. E. (1979). *TEX and METAFONT, New Directions in Typesetting*. New York: Digital Press.

Lee, Y. and Nelder, J. A. (1996). Hierarchical generalized linear models. *J. R. Statist. Soc. B* **58**, 619–656, (with discussion).

Liu, J. S. and Wu, Y. (1999). Parameter expansion for data augmentation. *J. Am. Statist. Ass.* **94**, 1264–1274.

Liu, J. S., Wong, W. H. and Kong, A. (1994) Covariance structure of the Gibbs sampler with applications to the comparisons of estimators and augmentation schemes. *Biometrika* **81**, 27–40.

Meng, X. and van Dyk, D. (1997). The EM algorithm — an old folk song sung to a fast new tune. *J. R. Statist. Soc. B* **59**, 511–567 (with discussion).

Meng, X. and van Dyk, D. (1999). Seeking efficient data augmentation schemes via conditional and marginal augmentation. *Biometrika* **86**, 301–320.

Neal, P., Roberts, G. O. and Viallefont, V. (2002). Robust MCMC algorithms for inference for stochastic epidemic models (in preparation).

Papaspiliopoulos, O., Roberts, G. O. and Sköld, M. (2002a). State-space expansion parameterizations for MCMC (in preparation).

Papaspiliopoulos, O., Roberts, G. O. and Sköld, M. (2002b). Non-centred MCMC methods for Bayesian non-parametrics (in preparation).

Pitt, M. K. and Shephard, N. (1999). Analytic convergence rates and parameterization issues for the Gibbs sampler applied to state space models. *J. Time Ser. Anal.* **20**, 63–85.

Roberts, G. O. (2002). Convergence of MCMC for hierarchical models with heavy tailed-links (in preparation).

Roberts, G. O. (2003). Linking theory and practice of MCMC. *Highly Structured Stochastic Systems* (P. J. Green, N. L. Hort and S. Richardson, eds). Oxford: Oxford University Press, 145-166

Roberts, G. O. and Sahu, S. K. (1997). Updating schemes, correlation structure, blocking and parameterization for the Gibbs sampler. *J. R. Statist. Soc. B* **59**, 291–397.

Roberts, G. O. and Stramer, O. (2001). Bayesian inference for incomplete observations of diffusion processes. *Biometrika* **88**, 203–221.

Roberts, G. O. and Tweedie, R. L. (1996). Exponential convergence of Langevin diffusions and their discrete approximations. *Bernoulli* **2**, 341–364.

Roberts, G. O. and Tweedie, R. L. (2001). Geometric $L2$ and $L1$ convergence are equivalent for reversible Markov chains. *J. Appl. Probab.* **38A**, 37–41.

Roberts, G. O. and Tweedie, R. L. (2002). *Understanding MCMC.* Berlin: Springer.

Rubin, D. B. (1987). *Multiple Imputation for Nonresponse in Surveys*. New York: Wiley.

Sahu, S. K. and Roberts, G. O. (1999). On convergence of the EM algorithm and the Gibbs sampler. *Statist. Comput.* **9**, 55–64.

Smith, A. F. M. and Roberts, G. O. (1993). Bayesian computation via the Gibbs sampler and related Markov chain Monte Carlo methods. *J. R. Statist. Soc. B* **55**, 3–24.

Tanner, M. A. and Wong, W. H. (1987). The calculation of posterior distributions by data augmentation. *J. Am. Statist. Ass.* **82**, 528–540.

Wakefield, J. C., Smith, A. F. M., Racine-Poon A. and Gelfand A. E. (1994). Bayesian analysis of linear and non-linear population models by using the Gibbs sampler. *Appl. Statist.* **43**, 201–221.

DISCUSSION

ALAN E. GELFAND (*Duke University, USA*)

Continuing the tradition of previous work by Roberts and his collaborators, this paper is clever, provocative, and potentially quite useful. Anyone who has invested time in implementing challenging MCMC model fitting will have appreciated the importance of choice of parametrization. This contribution provides some strategic, though generally terse, guidance for many hierarchical modelling settings.

My comments will be confined to spatial and nonparametric modelling contexts. For the former, consider first a Gaussian spatial process model of the form $Y(\boldsymbol{s}_i) = \boldsymbol{X}^t(\boldsymbol{s}_i)\boldsymbol{\beta}+\alpha(\boldsymbol{s}_i)+\epsilon(\boldsymbol{s}_i)$ where $\alpha(\boldsymbol{s}_i)$ is a 0 mean Gaussian process with variance σ^2 and arbitrary correlation function ρ while $\epsilon(\boldsymbol{s}_i)$ is a white noise process with variance τ^2. If σ^2, τ^2 and ρ are known and $\boldsymbol{X}(\boldsymbol{s}_i)\boldsymbol{\beta} = \mu$ then $(\mu, \boldsymbol{\alpha})$ is the non-centered parametrization, $(\mu, \boldsymbol{\eta})$ where $\eta_i = \mu + \alpha_i$ is the centered one and $(\mu, \boldsymbol{\eta}_{cent})$ where $\boldsymbol{\eta}_{cent} = \boldsymbol{\eta} - \mu W 1$ is partially non-centered (PNCP) . It is straightforward to compute that, provided var$(\mu|\boldsymbol{Y})$ exists,

$$\text{cov}(\boldsymbol{\eta}_{cent}, \mu|\boldsymbol{Y}) = \left[I - W - \left(I + \frac{\tau^2}{\sigma^2}R^{-1}\right)\right]^{-1} \text{var}(\mu|\boldsymbol{Y})$$

where R is such that $R_{ij} = \text{corr}(\alpha(\boldsymbol{s}_i), \alpha(\boldsymbol{s}_j))$. Hence, $\boldsymbol{\eta}_{cent}$ and μ are uncorrelated *a posteriori* if $W = (I + \sigma^2\tau^{-2}R)^{-1}$. This choice of W is exactly the one obtained by the authors in Section 5.1 to optimize the convergence rate, ρ_{NCP} derived under a normal hierarchical model (NHM).

Naturally arising questions are the following. Is the focus on finding a parametrization to achieve prior independence (as the introduction suggests) or on posterior uncorrelatedness? The foregoing calculation is done as a cross-correlation under the posterior while the rate of convergence is obtained as an L_2 distance between the distribution at the current iteration and the posterior. Is the former a legitimate way to obtain the PNCP under a NHM? The foregoing calculation produces the same PNCP for any prior on μ as long as var$(\mu|\boldsymbol{Y})$ exists. Is there a similar robustness to prior in rate of convergence? If we return to a general $\boldsymbol{X}^t(\boldsymbol{s}_i)\boldsymbol{\beta}$ with intercept, what sort of centering, partial centering, is sensible? If σ^2, τ^2 and ρ are not known would an estimated W using parameter estimates be appropriate to use?

Suppose in the above that we have a non Gaussian first stage with canonical link. That is, again taking $\boldsymbol{X}^t(\boldsymbol{s}_i)\boldsymbol{\beta} = \mu$, $f(Y(\boldsymbol{s}_i)|\mu, \alpha(\boldsymbol{s}_i)) = c\exp(Y(\boldsymbol{s}_i)\theta(\boldsymbol{s}_i) - b(\theta(\boldsymbol{s}_i)))$ where $\theta(\boldsymbol{s}_i) = \mu + \alpha(\boldsymbol{s}_i)$ with $\alpha(\boldsymbol{s}_i)$ as above. If we expand $b(\theta(\boldsymbol{s}_i))$ about μ to second order, we can again compute W to obtain $cov(\boldsymbol{\eta}_{cent}, \mu) \approx 0$. This W only involves $b''(\mu)$ rather than $\{b''(\mu+\alpha(\boldsymbol{s}_i))\}$ suggesting that in estimating W we would use only say $\overline{Y}$ rather than $\{Y(\boldsymbol{s}_i)\}$ as in the authors' expression in Section 5.1. Is it clear that the latter is a better data-based choice?

Next we consider two nonparametric illustrations. First, consider the quantal bioassay problem using Dirichlet processes as in, *e.g.*, Gelfand and Kuo (1991). Here $Y_i|p_i$ are independent $Bi(n_i, p_i)$, $i = 1, \cdots, k$ with $p_i = F(d_i)$ where d_i are increasing dosage levels and F is a random c.d.f. from a Dirichlet process, i.e., $F \sim DP(\alpha F_0)$ with F_0 a parametric family of c.d.f.'s indexed by say location μ and scale σ. In this situation there is misalignment between the likelihood, which is of the form $\prod_i p_i^{y_i}(1-p_i)^{n_i - y_i}$, and the prior, which is of the form $p_1^{\gamma_1 - 1}(p_2 - p_1)^{\gamma_2 - 1} \cdots (1-p_k)^{\gamma_{k+1}-1}$. The customary fitting approach introduces latent multinomial variables to break the misalignment but a slowly converging chain with high autocorrelation typically results.

Can a state space expansion approach work here? The centered parametrization is given by $((\alpha, \mu, \sigma), \{p_i\})$. If, analogous to the Gamma process example in section 3.1, we replace p_i with $F_i(.)$ we will be unable to ensure the hard order restrictions on the p_i. If we introduce a single $F(.)$ then given $\{d_i\}$ we can define $p_i = h_i(F(.), \mu, \sigma) = F((d_i - \mu)/\sigma)$ and convert to a non-centered parametrization $(F(.), \alpha, \mu, \sigma)$. (We can set $\sigma = 1$ for convenience of illustration). Would the size of the n_i determine which parametrization is preferred? Would the choice really matter since we still have the misalignment problem?

Lastly, consider a Dirichlet process mixing setting. Here $Y_i \sim f(Y_i|\theta_i, \phi)$ are conditionally independent and θ_i are conditionally independent from F. Finally, $F \sim DP(\alpha F_0)$ where, again F_0 is indexed by say μ and σ. See, *e.g.*, Gelfand and Kottas (2002) for details and computational discussion. Here, the centered parametrization is $((\alpha, \mu, \sigma), F, \{\theta_i\}, \phi)$. Customary marginalization over F still yields a centered parametrization, which, under MCMC, exhibits high auto- and cross-correlation.

Instead, we could consider marginalizing over $\{\theta_i\}$. Since F is a.s. discrete we could introduce a partial sum approximation $\tilde{F}$ to carry out the marginalization yielding the parametrization $((\alpha, \mu, \sigma), \tilde{F}, \phi)$. Is this parametrization partially non-centered? If Y_i does not inform well about θ_i, do we expect marginalizing over $\{\theta_i\}$ to be "better" than marginalizing over F? Are there other promising possibilities? Perhaps my real question is what suggestions the authors might have with regard to choice of parametrization in such nonparametric models.

OLE F. CHRISTENSEN (*Lancaster University, UK*)

I would like to congratulate the authors on providing a very useful characterization of different parameterizations whereby one can get an intuition on how they behave in practice. Basically, the hierarchical parameterization illustrated by Figure 1a works well when the information in data is relatively strong, and the *a priori* independence parameterization illustrated by Figure 1b works well when the information in the data is relatively weak. They also propose several strategies for constructing a robust parametrization that works well in both cases. The authors use the terminology "non-centered" and "partial non-centered" for the later two types of parameterization, which suggest that they find them un-natural. I think these two types of parameterization deserve better names.

My particular interest is the spatial example in Section 5.1 where I can confirm that the robust parameterization suggested by the authors performs well for all the data sets I have considered. Updating the covariance parameters σ^2 and α is more challenging since α enters the target density in a complicated way through the correlation matrix.

My last comment is about the possibility of marginalizing out parameters when using conjugate priors or limits of conjugate priors. Consider the simple example where the latent variable X is Gaussian with mean $D\beta$, covariance matrix Σ, and $\pi(\beta) \propto 1$. The marginal posterior distribution of X is

$$f(x \mid y) \propto f(y \mid x) \exp(-x^t(\Sigma^{-1} - \Sigma^{-1}D(D^t\Sigma^{-1}D)^{-1}D^t\Sigma^{-1})x/2),$$

and MCMC sampling can be done without updating β. In practice marginalization is an advantage, since one avoids having to tune the proposal variance for β. Could the authors comment on how well marginalization would work compared to their approach of updating both the latent variables and the parameters ?

DARREN J. WILKINSON (*University of Newcastle, UK*)

The authors are to be congratulated on a most interesting paper. They consider the common problem of inference for "latent process" models, with parameter θ, latent process X and data Y such that $\theta \perp\!\!\!\perp Y \mid X$.

The classic data augmentation scheme alternately samples $\theta \mid X, Y$ and $X \mid \theta, Y$ and often works well in the situation where the two steps can both be carried out exactly as pure Gibbs steps. Note, in particular, that the step $X \mid \theta, Y$ is often carried out as a sequence of smaller Gibbs moves but in many cases, such as the case of systems that are linear Gaussian conditional on θ, this is not necessary and block-updating schemes generally perform much better despite some increase in computational overhead (Rue, 2001; Wilkinson and Yeung, 2002a). Two-block updating schemes perform particularly well when there is high dependence within X and weak (posterior) dependence between θ and X. However, as noted in the paper, in the case of high posterior dependence between θ and X the classic data augmentation scheme can break down. The authors suggest that non-centered or partially non-centered parameterizations may provide a solution. It should be noted that there are often other possible strategies which can work even better and are often easier to implement. Like NCPs, such techniques involve sacrificing a Gibbs move for a Metropolis–Hastings scheme, but constructed in such a way as to get around the strong dependence between θ and X. The simplest way is to integrate X out of the problem completely, and simply construct a Metropolis–Hastings scheme for $\theta \mid Y$. If a new $\theta^\star$ is proposed from some kernel $q(\theta^\star \mid \theta)$, it is accepted with probability $\min\{1, A\}$ where

$$A = [\pi(\theta^\star)L(\theta^\star; Y)q(\theta \mid \theta^\star)]/[\pi(\theta)L(\theta; Y)q(\theta^\star \mid \theta)].$$

Note that this depends on the marginal likelihood of the data $L(\theta; Y)$, but this is usually tractable whenever sampling $X \mid \theta, Y$ in a single block is possible (Rue, 2001; Wilkinson and Yeung, 2002b). Even in cases where this is not tractable, effective updating schemes can be created by constructing "single-block" updating schemes where a new $(\theta^\star, X^\star)$ is sampled in two stages: first $\theta^\star$ is sampled from $q(\theta^\star \mid \theta)$, and then a new $X^\star$ is sampled from a tractable approximation to $X \mid \theta^\star, Y$ (Knorr-Held and Rue 2002). The acceptance probability is primarily related to the closeness of the approximation. By simulating $X^\star$ to be consistent with $\theta^\star$, the dependence between them is overcome without the need to resort to NCPs.

My second observation concerns the application of NCPs to discretely observed diffusions. This at first appears to be a "killer application" for NCPs, as CPs are well-known to exhibit pathologically poor mixing. However, despite a statement to the contrary in the discussion of Roberts and Stramer (2001), the techniques do not generalize easily from the univariate to the multivariate case. Even for the simple bivariate log-Gaussian stochastic volatility model

$$dX_t = \mu X_t\, dt + \exp\{Y_t/2\} X_t\, dW_t$$

$$dY_t = -\lambda(Y_t - \nu)\, dt + \tau\, dW_t'$$

it seems to be impossible (at least in practice) to find a transformation to constant volatility, a required step in the construction of an effective NCP.

REPLY TO THE DISCUSSION

We would like to thank the discussants for their insightful contributions. Many important points have been raised, and we will try to address most of these.

Alan Gelfand asks about what we should really be looking for in a suitable parameterization, and suggests that in many cases, it is sufficient to consider posterior correlation structure.

In this paper, our approach is different. We are primarily interested in constructing and assessing the performance of NCPs. Of course, there are many situations where neither of these methods is adequate, and when a CP and an NCP exist, we try to find ways to construct intermediate parameterizations that adapt according to the information in the data, so that the user does not have to choose *a priori* between the two extreme parameterizations. Actually , the PNCP in the NHM can also be obtained using a posterior uncorrelatedness argument, as Alan suggests, because in the Gaussian context the maximal correlation described in Amit (1991) is attained by linear functions. The same is true in the geostatistical example that mimics the construction for the NHM. However, outside the Gaussian context it is possible for posterior correlations to be very unreliable for the purposes of predicting convergence properties (see *e.g.*, Roberts, 1992). Moreover, for generalised hierarchical mixed models (such as those considered in Section 3 of Gelfand *et al.,* 1996), we have found examples where there exist parameterizations (essentially based on the weighted average form of the posterior expectation of X given θ; see formula 3.3 of Gelfand *et al.,* 1996), under which the missing data and the parameters are uncorrelated, but which possess convergence properties inferior to the CP (due to higher order dependence between the random effects and the parameters). Finally, it is unclear how to generalize the construction based on minimizing posterior correlation to models where the missing data live on non-Euclidean spaces (as in the example of Section 5.2).

Alan also asks about the effect of the prior on θ. It turns out that although the prior on θ affects the rates of convergence of all the competing algorithms, the optimal PNCP and the relative sizes of ρ_c and ρ_{nc} are both unaffected by the prior.

An important question raised in Alan's discussion asks how to proceed when multiple parameters are to be updated as part of the MCMC cycle. Using notation of the geostatistical example in Section 5.1, we can easily construct a multivariate NCP by setting $\tilde{X} = h(X, \theta, \sigma, \alpha) := (X - \theta 1)R(\alpha)^{-1/2}\sigma^{-1}$, making all parameters *a priori* independent. This construction also makes the components X_i, X_j of the latent field *a priori* independent, and potentially eases updating of X when this has to be performed with a Hastings step. The approach adopted in Christensen *et al.* (2002) proceeds by constructing a PNCP of the form $\tilde{X} = (X - \theta 1W)W^{-1/2}$, where $W = W(\sigma, \alpha)$ is chosen dynamically as a function of the current values of (σ, α). This works well in practise, although the extra computational expense involved means that for really large problems some shortcuts, involving less $W(\sigma, \alpha)$ evaluations, are necessary.

The idea behind the method sketched in Section 4.1 is to make a Gaussian approximation of the likelihood and then apply the PNCP given by (17). There are of course many competing strategies for partial non-centering, and even if we decide to determine the extent of non-centering on the basis of a Taylor series expansion, we still need to decide where to expand around. The most robust estimates of information content will be achieved by expanding about the posterior mode, but of course that information is unavailable *a priori*. While we carry out the Taylor expansion about $\hat{\boldsymbol{x}}$, Alan suggests performing the expansion about $\bar{Y}$, which in turn leads to a PNCP which non-centers each datum equally. Whilst this approach has the advantage of simplicity, it seems reasonable that the success of this approach depends on the quality of the Gaussian approximation and our experience has been that in situations like this where information contained in a particular datum depends strongly on the observed data value there is much to be gained by a heterogeneous degree of non-centering. However it might turn out that expanding about a weighted average of $\hat{\boldsymbol{x}}$ and $\bar{Y}$ (perhaps using some simple kriging estimate) would perform even better than $\hat{\boldsymbol{x}}$.

Ole Christensen asks to what extent marginalizing parameters will be beneficial to MCMC mixing. It turns out that this generally improves convergence when a pure Gibbs sampler can be applied (see *e.g.*, Liu, 1994). In practice, however, marginalization will sometimes complicate the likelihood surface of the latent variables X, and if X has to be updated with a Hastings step, this step can become very problematic after marginalization.

Bayesian non-parametric and semi-parametric analysis is ideally suited to the use of NCPs. We have been working on general non-centering constructions to Lévy processes which lead to natural applications in Bayesian non-parametrics. In particular, there are very natural ways of constructing non-centering of Dirichlet processes using their representation in terms of the Gamma process. Similarly, we can provide NCPs of Polya trees (and similar models) by working directly with the Gamma random variables used in their sequential construction. Thus, we intend to implement NCP in the Dirichlet mixture application suggested by Alan (see Papaspiliopoulos *et al.*, 2002b).

We also believe that the bioassay problem suggested by Alan is a natural one for our methods, and we look forward to implementing NCPs in this context. We agree strongly that the size of the n_i parameters here will be crucial in determining the effectiveness of the methods.

Darren Wilkinson is indeed correct that the generalization of the Roberts–Stramer methodology to multiple-dimensional problems in full generality is difficult. To see this consider a d-dimensional diffusion satisfying

$$d\boldsymbol{X}_t = \Sigma(\boldsymbol{X}_t)^{1/2} d\boldsymbol{B}_t + b(\boldsymbol{X}_t)dt$$

where Σ and b are functions of $\boldsymbol{X}_t$, t, and unknown parameters. In this context, the search for a non-centered parameterization decoupling the parameters governing Σ from $\boldsymbol{X}$ involves finding an invertible function $\boldsymbol{h} : \boldsymbol{R}^d \to \boldsymbol{R}^d$ which satisfies

$$\nabla \boldsymbol{h} (\nabla \boldsymbol{h})' = \Sigma^{-1} .$$

In full generality, this matrix differential equation is impossible to solve in practice, although there are important special cases which do admit solution very easily (for instance where Σ is diagonal and for all i, Σ_{ii} is not a function of X_j for any $j \neq i$). Unfortunately, this special case does not cover the stochastic volatility example cited by Darren.

ADDITIONAL REFERENCES IN THE DISCUSSION

Gelfand, A. E. and Kottas, A. (2002). A computational approach for full nonparametric Bayesian inference in single and multiple sampling problems. *J. Comput. Graph. Statist.* **11**, 289–305.

Gelfand, A. E. and Kuo, L. (1991). Nonparametric Bayesian bioassay including ordered polytomous response. *Biometrics* **78**, 657–666.

Knorr-Held, L. and H. Rue (2002). On block updating in Markov random fields for disease mapping. *Scand. J. Statist.* (to appear).

Liu J. S. (1994). The Collapsed Gibbs Sampler in Bayesian Computations With Applications to a Gene Regulation Problem *J. Am. Statist. Ass.* **89**, 958–966.

Roberts G. O. (1992). Discussion on Parameterization issues in Bayesian inference. *Bayesian Statistics 4* (J. M. Bernardo, J. O. Berger, A. P. Dawid and A. F. M. Smith, eds). Oxford: Oxford University Press, 227–246.

Rue, H. (2001). Fast sampling of Gaussian Markov random fields. *J. R. Statist. Soc. B* **63**, 325–338.

Wilkinson, D. J. and Yeung, S. K. H. (2002a). Conditional simulation from highly structured Gaussian systems, with application to blocking-MCMC for the Bayesian analysis of very large linear models. *Statist. Comput.* **12**, 287–300.

Wilkinson, D. J. and S. K. H. Yeung (2002b). A sparse matrix approach to Bayesian computation in large linear models. *Comput. Statist. Data Anal.* (to appear).

BAYESIAN STATISTICS 7, pp. 327–347
J. M. Bernardo, M. J. Bayarri, J. O. Berger, A. P. Dawid,
D. Heckerman, A. F. M. Smith and M. West (Eds.)

Identifying Mixtures of Regression Equations by the SAR procedure

DANIEL PEÑA JULIO RODRÍGUEZ
Universidad Carlos III de Madrid, Spain
dpena@est-econ.uc3m.es puerta@est-econ.uc3m.es

GEORGE C. TIAO
University of Chicago, USA
gct@gsb.uchicago.edu

SUMMARY

A procedure for identifying data heterogeneity when fitting regression models is presented. The method is based on the SAR procedure, developed by Peña and Tiao (2002), and has three steps. First, the sample is cleaned for large outliers by using a discrepancy measure based on predictive ordinates. Second, observations are split into small homogeneous groups by using a link function derived from cross-validation predictive distributions. Third, these small groups are then iteratively enlarged by incorporating observations homogeneous to those in the group. In this way the possible piece-wise regression equations are found. Examples are shown to illustrate the performance of the procedure for finding mixtures of regressions due to the presence of outliers or to switching regression models. A Monte Carlo study of the power of the proposed procedure is presented.

Keywords: OUTLIERS; PREDICTIVE DISTRIBUTION; STRUCTURAL CHANGE; SWITCHING REGRESSION.

1. INTRODUCTION

Data heterogeneity when a regression model is fitted implies that some kind of segmentation or clustering exists among the sample units. This problem has been studied under two main approaches: (1) outlier detection and/or robust regression estimation and (2) switching and/or structural change regression. From the Bayesian point of view these two situations can be formulated as mixture estimation problems with different types of mixture distributions. For the outlier problem, Box and Tiao (1968) proposed a scale contaminated normal model for the noise distribution and develop a Bayesian estimation procedure for this model. Several other outlier detection and robust Bayesian estimation methods have been proposed based on mixtures and heavy tails distributions. The estimation of these mixture models can be carried out by Markov Chain Monte Carlo (MC^2) methods, but the standard Gibbs sampling algorithm presents serious convergence problems when there are groups of masked outliers (see Justel and Peña, 1996).

A different source of heterogeneity was introduced by Quandt (1958) who proposed a model in which the data is assumed to follow the regression equation $y_i = \boldsymbol{x}_i'\boldsymbol{\beta}_1 + u_i$ for $i = 1, ..., n_1$ and a different regression equation $y_i = \boldsymbol{x}_i'\boldsymbol{\beta}_2 + u_i$ for $i = n_1 + 1, ..., n$ where $\boldsymbol{\beta}_1 \neq \boldsymbol{\beta}_2$. The key problem in this model is to identify the change point between the two regimes and estimate the parameters (see, for instance, Schweder, 1976) and this problem has been extensively studied under the name of regression under a structural change. A comparison of some of the procedures developed to estimate the change point can be found in Andrews et al (1996). A generalization of this model is to assume that each observation can be generated with some unknown probability by one of the two regressions models $y_i = \boldsymbol{x}_i'\boldsymbol{\beta}_j + \sigma_j u_i$, where $j = 1, 2$ and u_j is $N(0, 1)$. This situation may correspond to the case in which a categorical variable is omitted in the model, like gender, month, or day of the week. If the categorical variable only has effect on the intercept we have the standard missing attribute case, whereas if the regression coefficients depend on the omitted categorical variable we have the switching regression problem. In the general case of G possible regimes, or groups of data, and assuming normality, the model is

$$y_i \mid \boldsymbol{x}_i \sim \sum_{g=1}^{G} \alpha_g N(\boldsymbol{x}_i'\boldsymbol{\beta}_g, \sigma_g^2), \tag{1}$$

where $\alpha_g \geq 0$ and $\sum_{g=1}^{G} \alpha_g = 1$. This model has been studied extensively both from the Bayesian and the likelihood points of view. See Aitkin (2001) for some comparisons. When the number of groups is known and we have some reasonable initial estimate for the parameters in the different regimes the model can be estimated by MC^2 methods. However, when this information is not available the estimation of this model is a difficult problem. First, the components of the mixture have identification problems (Celeux *et al.*, , 2000). Second, when one of the groups has a much smaller variance than the others, and behaves like a set of high leverage outliers, a false convergence of the Gibbs sampling procedure may occur and some modifications are required to achieve convergence (see Justel and Peña, 2001). Third, more experience is required about the performance of the available procedures proposed to deal with the Bayesian estimation of the models when the number of parameters and the number of groups are large and we do not have good initial estimates to start the algorithm (see, for instance, Richarson and Green, 1997; Stephens, 2000). A recent analysis with Bayesian MC^2 methods can be found in Hurn *et al.* (2002).

In this paper, we present a different approach to solve the heterogeneity problem. The approach is exploratory and is based on the SAR procedure developed by Peña and Tiao (2002). It is designed to deal with any type of heterogeneity, including outliers and switching regressions, and it can be used either as a starting point in a MC^2 estimation algorithm or as a general model building procedure.

The paper is organized as follows. In Section 2, the SAR procedure applied to the regression model is briefly summarized. A key ingredient of the procedure is the link function, and the properties of this function in the regression case are studied in Section 3. Section 4 presents simple simulated examples to illustrate the algorithm and discusses its implementation. Section 5 shows that the proposed procedure can provide the same solution as sophisticated MC^2 algorithms; that it is able to succeed in cases in which MC^2 algorithms are found to fail; and that it can be applied in large data mining problems. Section 6 includes a simulation study about the size and power of the procedure. Section 7 contains some brief concluding remarks.

2. THE SAR PROCEDURE IN REGRESSION

Given a sample of n points $(y_i, \boldsymbol{x}_i')$, $i = 1, ..., n$, where $\boldsymbol{x}_i$ is a $p \times 1$ vector, let $\boldsymbol{Y} = (y_1, ..., y_n)'$ be the vector of responses and $\boldsymbol{X} = (\boldsymbol{x}_1, ..., \boldsymbol{x}_n)'$ the $n \times p$ full rank matrix of explanatory variables. The SAR (split and recombine) procedure is an iterative method to identify possible clusters in a sample. It can be applied for all sorts of data heterogeneity and includes three basic steps. First, large isolated outliers are identified and deleted from the sample. Second, the remaining data are split iteratively into homogeneous groups of some minimum fixed size, called basic sets. Third, each basic set is allowed to grow by recombining points one by one into it, as long as they are found homogeneous with the observations already in the group. The complement of the enlarged group will now be treated as a new sample for a repeated application of the three-step split and recombine procedure. This process will lead to a set of different grouping of the sample, called possible data configurations (PDCs). These solutions can be explored by a model selection criteria (see Peña *et al.*, 2002) or they can be used as starting points for a MC^2 algorithm for mixtures.

The SAR procedure is based on three statistics and a link function. The first statistic is based on the predictive ordinate, $p(y_i|\boldsymbol{Y}_{(i)})$, that has been used by several authors for outlier detection (see Box, 1980; Geisser, 1980, 1987; Pettit and Smith, 1985; Pettit, 1990; Peña and Tiao, 1992; Peña and Guttman, 1993, and others). The notation $\boldsymbol{Y}_{(i)}$ indicates that observation y_i has been deleted from the data set $\boldsymbol{Y}$. The justification of this measure is easy to see by introducing a dummy variable δ_i that is equal to 1 when observation y_i has been generated by the same model that has generated the data in $\boldsymbol{Y}_{(i)}$ and 0 otherwise. Then, we have that

$$P(\delta_i = 0|y_i, \boldsymbol{Y}_{(i)}) = 1 - kP(y_i|\delta_i = 1, \boldsymbol{Y}_{(i)})P(\delta_i = 1) \tag{2}$$

where P denotes the probability function and k is a constant. Calling $p(y_i|\boldsymbol{Y}_{(i)})$ the predictive density of y_i when it is generated by the same model that generates $\boldsymbol{Y}_{(i)}$ we see that the smaller the predictive ordinate the larger the probability that the observation has been generated by a model different from the one that generates the rest of the data. The first statistic we define is the standardized predictive ordinate $c_0(i)$, given by

$$c_0(i) = -2\log\left\{\frac{p(y_i|\boldsymbol{Y}_{(i)})}{p(\widehat{y}_{i(i)}|\boldsymbol{Y}_{(i)})}\right\}, \tag{3}$$

where $p(y_i|\boldsymbol{Y}_{(i)})$ is the predictive distribution of y_i given data $\boldsymbol{Y}_{(i)}$, and $\widehat{y}_{i(i)} = E(y_i|\boldsymbol{Y}_{(i)})$ is the expected value of this distribution. This standardized predictive ordinate is computed first to test for outliers, and second to test if a new point can be incorporated to a group formed by data $\boldsymbol{Y}_{(i)}$. The second statistic we define, $c_1(i)$, is given by

$$c_1(i) = \max_{y_j}(c_1(i|j)) = \max_{y_j}\left[-2\log\left\{\frac{p(y_i|\boldsymbol{Y}_{(ij)})}{p(\widehat{y}_{i(ij)}|\boldsymbol{Y}_{(ij)})}\right\}\right] \tag{4}$$

where $\boldsymbol{Y}_{(ij)}$ represents a set of data in which observations y_i and y_j are deleted from $\boldsymbol{Y}$, $p(y_i|\boldsymbol{Y}_{(ij)})$ is the predictive distribution of y_i given data $\boldsymbol{Y}_{(ij)}$, and $\widehat{y}_{i(ij)} = E(y_i|\boldsymbol{Y}_{(ij)})$. The maximum is taken with respect all the observations in $\boldsymbol{Y}_{(i)}$ and indicates the minimum standardized value of the predictive distribution of y_i that can be obtained by deleting a point from $\boldsymbol{Y}_{(i)}$. The third statistic we use is the maximum standardized change in the predictive distribution obtained by deleting an observation from the set $\boldsymbol{Y}_{(i)}$, and is given by $d_1(i) = c_1(i) - c_0(i)$ and this statistic is used jointly with $c_0(i)$ to identify masked outliers from the sample. Finally, the binary relationship implied by $c_1(i)$ provides a link function, $l(i)$, between observation y_i and the observation y_j which produces the maximum in (4): observation y_j is the one that, when deleted, will make y_i the most discrepant with the rest of the data. This link function is defined by

$$l(i) = j \quad \text{if} \quad y_j = \arg\max_{y_k}(c_1(i|k))$$

that is, the point $(y_i, \boldsymbol{x}_i')$ is linked to $(y_j, \boldsymbol{x}_j')$ if deleting the latter point from the sample produces the maximum discrepancy between the observation y_i and the remaining $n-2$ observations. We will call the point, $(y_j, \boldsymbol{x}_j')$, the discriminator of $(y_i, \boldsymbol{x}_i')$. Note that (1) each sample point must have a discriminator; (2) a discriminator is, in some metric implied by the model, an extreme point in the set $\boldsymbol{Y}_{(i)}$; and (3) the relationship defined by the link function is neither symmetric nor transitive.

In the linear regression case calling $\widehat{\boldsymbol{\beta}} = (\boldsymbol{X}'\boldsymbol{X})^{-1}\boldsymbol{X}'\boldsymbol{Y}$ and $s^2 = \boldsymbol{e}'\boldsymbol{e}/(n-p)$, where $\boldsymbol{e} = (\boldsymbol{I} - \boldsymbol{H})\boldsymbol{Y}$ and $\boldsymbol{H} = \boldsymbol{X}(\boldsymbol{X}'\boldsymbol{X})^{-1}\boldsymbol{X}'$ is the idempotent projection matrix, and assuming the standard non-informative prior for the vector of parameters $\boldsymbol{\theta} = (\boldsymbol{\beta}, \sigma)$, $p(\boldsymbol{\theta}) \propto \sigma^{-1}$, then, given $\boldsymbol{Y}$, the predictive distribution for a future observation y_f from the linear model (see *e.g.*, Box and Tiao, 1973), is the standard univariate t distribution with ν degrees of freedom

$$p(y_f|\boldsymbol{Y}) = \frac{1}{B\left(\frac{1}{2}, \frac{\nu}{2}\right)\sqrt{\nu}}\, s^{-1}(1+h_f)^{-1/2}\left(1 + \frac{t_f^2}{\nu}\right)^{-(\nu+1)/2}$$

where $t_f^2 = (y_f - \widehat{y}_f)^2/s^2(1+h_f)$, $\widehat{y}_f = \boldsymbol{x}_f'\widehat{\boldsymbol{\beta}}$, $h_f = \boldsymbol{x}_f'(\boldsymbol{X}'\boldsymbol{X})^{-1}\boldsymbol{x}_f$, and $\boldsymbol{x}_f$ is the future explanatory vector. For $c_0(i)$, we let $y_f = y_i$, $\boldsymbol{x}_f = \boldsymbol{x}_i$, and we have that

$$\log[p(y_i|\boldsymbol{Y}_{(i)})] = k - \frac{1}{2}\log[s_{(i)}^2(1 + h_{i(i)})] - \left(\frac{\nu_0 + 1}{2}\right)\log\left(1 + \frac{t_i^2}{\nu_0}\right)$$

where $\nu_0 = (n-p-1)$, $h_{i(i)} = h_f$, $t_i = t_f$ in which $\widehat{y}_{i(i)} = \boldsymbol{x}_i'\widehat{\boldsymbol{\beta}}_{(i)}$ is the predictive mean of y_i, and $\widehat{\boldsymbol{\beta}}_{(i)}$ and $s_{(i)}^2$ are the estimates based on $\boldsymbol{Y}_{(i)}$ and $\boldsymbol{X}_{(i)}$ where $\boldsymbol{X}_{(i)}$ is obtained from $\boldsymbol{X}$ by deleting $\boldsymbol{x}_i$. Since $\log[p(\hat{y}_{i(i)}|\boldsymbol{Y}_{(i)})] = k - (1/2)\log\left[s_{(i)}^2(1 + h_{i(i)})\right]$, we can write

$$c_0(i) = (\nu_0 + 1)\log\left(1 + \frac{t_i^2}{\nu_0}\right). \tag{5}$$

The statistic t_i is the studentized residual, or standardized predictive residual, that is normally used for testing an individual outlier in regression. Expression (5) shows that $c_0(i)$ is a monotonic transformation of t_i^2 and, for large n, these two measures will be equivalent.

In the same way, letting $y_f = y_i$, $\boldsymbol{x}_f = \boldsymbol{x}_i$, and deleting y_i and y_j from $\boldsymbol{Y}$ to become $\boldsymbol{Y}_{(ij)}$, we obtain $c_1(i)$ as

$$c_1(i) = (\nu_1 + 1)\max_{y_j}\left[\log\left\{1 + \frac{t_{i(ij)}^2}{\nu_1}\right\}\right] \tag{6}$$

Here $\nu_1 = (n-p-2)$ and $t_{i(ij)}^2$ is given by $t_{i(ij)}^2 = (y_i - \widehat{y}_{i(ij)})^2/s_{(ij)}^2(1 + h_{i(ij)})$, where $h_{i(ij)} = h_f$, $\widehat{y}_{i(ij)} = \boldsymbol{x}_i'\widehat{\boldsymbol{\beta}}_{(ij)}$ is the predictive mean of y_i, and $\widehat{\boldsymbol{\beta}}_{(ij)}$ and $s_{(ij)}^2$ are the estimates based on $\boldsymbol{Y}_{(ij)}$ and $\boldsymbol{X}_{(ij)}$. For large n, $c_1(i)$ will be equivalent to $\max_{y_j} t_{i(ij)}^2$.

The SAR procedure consists of the following three basic steps (see Peña and Tiao, 2002, for further justifications, tables of the statistics and further details):

1. Check the sample for outliers by computing $c_0 = \max c_0(i)$ and $d_1 = \max d_1(i)$ and test for outliers according to the null distributions of first c_0 and then d_1. The critical values of these statistics have been obtained by Monte Carlo. Delete the outliers and repeat the checking until either no more outliers are found or a set of size$p + h$, is obtained, where h is a parameter to control the degrees of freedom of the fitted model that is usually taken, for n small or moderate, equal to $\log(n - p)$.

2. Split the remaining sample by putting in the same group all the observations which share the same discriminator. Points that are either outliers or discriminators are considered as isolated observations and they are not incorporated into the groups. Thus: (1) If $l(i) = j$ and $l(k) = j$ then (i, k) are put in the same group. (2) discriminators and treated as isolated points. (3) observations in subgroups of sizes smaller than $p + h$ are treated as isolated points. Once the new groups are found, we go back to 1 and through repeated application of steps 1 and 2, we split the original sample S_0 into m basic sets $(B_1, ..., B_m)$ and isolated points consisting of outliers, discriminators and observations from undersized groups in the splitting process.
3. The recombining process starts by selecting a basic set, B_i, fitting the model to the data in this basic set, and checking the observations in the complementary set $\overline{B}_i$ for homogeneity with respect to the basic set one at a time. The checking is done with the statistic $c_0(i)$ conditional on data in B_i. If the minimum of $c_0(i)$ for all the points in $\overline{B}_i$ is below the 99th percentile of the distribution of this statistic, then the corresponding observation is incorporated into the basic set. Then the model is refitted to the newly enlarged basic set including a new observation and the process of checking the remaining observations is repeated. When no more observations can be incorporated into the initial set, the enlarged set is considered as a first level final homogeneous group F_i. This enlarging process is repeated for each of the m basic sets. The result will be a set of first level final groups $(F_1, ..., F_m)$. Now, select a first level final group F_i and conditional on this group the complementary set $\overline{F}_i$ is analyzed as a new sample and steps 1-3 are applied. Repeating this process, the algorithm is continued until we obtain all the non-redundant PDCs originated from the set of basic sets. Note that each PDC is a partition of the sample into disjoint subsets.

The key tool of the procedure is the link function, which defines the discriminator for each sample point. In the next section we analyze its properties in the regression case.

3. THE PROPERTIES OF THE LINK FUNCTION

Let $\boldsymbol{e} = \boldsymbol{Y} - \boldsymbol{X}\widehat{\boldsymbol{\beta}} = (e_1, ..., e_n)'$ be the residuals of the regression fitted to the sample $(\boldsymbol{Y}, \boldsymbol{X})$ and h_{ij} the ij-th element of the projection matrix $\boldsymbol{H} = \boldsymbol{X}(\boldsymbol{X}'\boldsymbol{X})^{-1}\boldsymbol{X}'$. It is shown in the appendix that we can write

$$t^2_{i(ij)} = \frac{e_i^2 + (1 - h_{jj})^{-1}[h_{ij}^2 e_j^2 (1 - h_{jj})^{-1} + 2e_i h_{ij} e_j]}{a_{ii} - c(1 - h_{jj})^{-1}[dh_{ij}^2 + b_{ii} e_j^2 + 2e_i h_{ij} e_j]}$$

where $c = (n - p - 2)^{-1}$ and $d = (n - p)s^2$ are constants and $b_{ii} = (1 - h_{ii})$, $a_{ii} = c[db_{ii} - e_i^2]$. This expression shows that for a given observation, $(y_i, \boldsymbol{x}_i')$, the statistic $t^2_{i(ij)}$ is maximized if the discriminator point $(y_j, \boldsymbol{x}_j')$ is chosen so that: (1) the product $e_i h_{ij} e_j$ is positivc, (2) the measure of leverage of the discriminator $(1 - h_{jj})^{-1}$ is as large as possible, (3) the predictive residual $e_j(1 - h_{jj})^{-1}$ is, in absolute value, as large as possible. Dividing both terms by $s^2(1 - h_{ii})$ and calling $z_{ij} = h_{ij}(1 - h_{jj})^{-1/2}(1 - h_{ii})^{-1/2}$ and $r_j = e_j[s^2(1 - h_{jj})]^{-1/2}$ to the standardized residuals, we have that

$$t^2_{i(ij)} = \frac{r_i^2 + [z_{ij}^2 r_j^2 + 2r_i r_j z_{ij}]}{c[(n - p) - r_i^2 - (n - p)z_{ij}^2 - r_j^2 - 2r_i r_j z_{ij}]}$$

and it is straightforward to check that, assuming $r_i r_j z_{ij}$ is positive, the partial derivatives of this function with respect to the new variables r_j and z_{ij}, which measure outlyingness and leverage, are positive. Thus, the discriminator should be a point with large standardized residual, r_j, and large leverage, as measured by z_{ij}. It is interesting to note that these measures appear in a natural way in the analysis of outlying observations in linear models using standard Bayesian procedures (see Peña and Guttman, 1993). Note that we can write, in the space of the explanatory variables, $h_{ij} = h_{ii}^{1/2} h_{jj}^{1/2} \cos\theta$, and, therefore, the cross leverage h_{ij} will be large for a fixed value of h_{ii} if the leverage of the discriminator, h_{jj}, is large and $\cos\theta$ close to one.

In order to understand better the behavior of the link function suppose that we want to find the discriminator of a point y_i with non-negative residual, $e_i \geq 0$. As the $t^2_{i(ij)}$ statistic is invariant to translations, suppose for simplification that all the variables are measured on deviations to the mean so that the regression equation has no constant term. We consider the following three possible cases:

(1) Suppose $x_i = 0$ so that $h_{ii} = 0 = z_{ij}$. Then $t^2_{i(ij)} = r_i^2/c[(n-p) - r_i^2 - r_j^2]$, and the discriminator will be the point with largest value of the standardized residual r_j^2.

(2) Suppose $x_i > 0$ but $e_i = 0 = r_i$ and the point analyzed lies on the regression line. Then $t^2_{i(ij)} = [(n-p) - (n-p)z_{ij}^2 - r_j^2]^{-1} z_{ij}^2 r_j^2/c$ and the point will be linked to the value that maximizes this expression. Note again that the discriminator will be a high leverage point with large standardized residual.

(3) Suppose $x_i \gg 0$ and $e_i \gg 0$. Then the discriminator must: (a) have $r_i r_j z_{ij} > 0$ which implies that either the residual is also positive and $\theta < \pi/2$ or the residual is negative and $\pi/2 < \theta < \pi$, and (b) have large values of the leverage and the standardized residual.

For instance, consider the simple regression through the origin, with $\sum x_i = \sum y_i = 0$ and $\sum x_i^2 = 1$ so that $h_{ij} = x_i x_j$, if an observation has an x value larger than the mean and the residual is positive the discriminator will be either: (1) a point with positive residual and also $x_j > 0$, or (2) a point with negative residual and $x_j < 0$.

Figure 1 shows a sample of simulated data from an homogeneous regression and the links obtained in this sample. The discriminators are indicated by: (1) an arrow and (2) the vertex of the cone of lines that go to all the points that are linked to the discriminator. Only eight points are discriminators. If we now put together points with the same discriminator and consider as a group a set of at least 3 points we will obtain four groups. Note that, as indicated in the previous rules, the discriminator is a point with either a large residual or a large leverage or a combination of both effects. For instance, points 8 and 24 have high leverage, whereas points 7 and 20 have a relatively large residual. Note that points are linked to discriminators located according to the previous rules.

We now illustrate the behavior of the link function in the four heterogeneous cases of simple regression considered in Figure 2: (a) concentrated contamination; (b) switching regression; (c) mixture of two regressions with omitted categorical variable, and (d) structural change. Figure 2 shows the data indicated by numbers and the results of applying the link function to the four data sets. Only a few points in each case are discriminators and they are indicated as before by: (1) an arrow and (2) the vertex of the cone of lines that go to all the points that are linked to the discriminator. In case (a) only five points are discriminators and putting together those observations with the same discriminator we obtain five basic sets. The link function puts all the outliers in the same basic set. Note that points are linked to discriminators that are extreme according to the previous rules. In case (b) the sample is split into five groups induced by the five discriminators. A discriminator is always an extreme point with large leverage and large residual, and points are linked to the discriminator with the same residual sign and at the same side of the mean of the explanatory variable. Again, the sample is split into homogeneous groups. A similar situation occurs in case (c) in which seven discriminators are found. Finally,

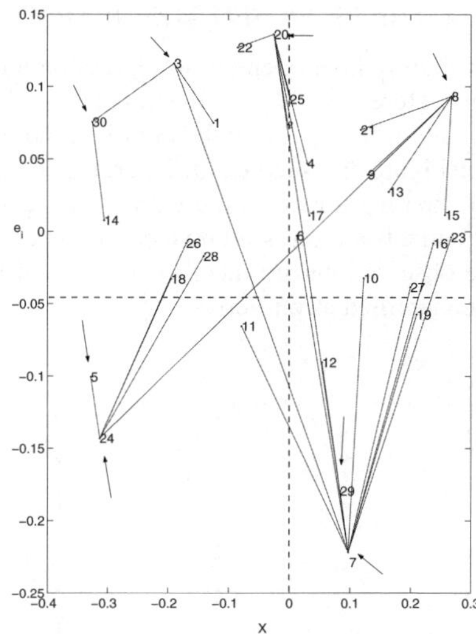

Figure 1. *Results of applying the link function in the homogeneous regression case.*

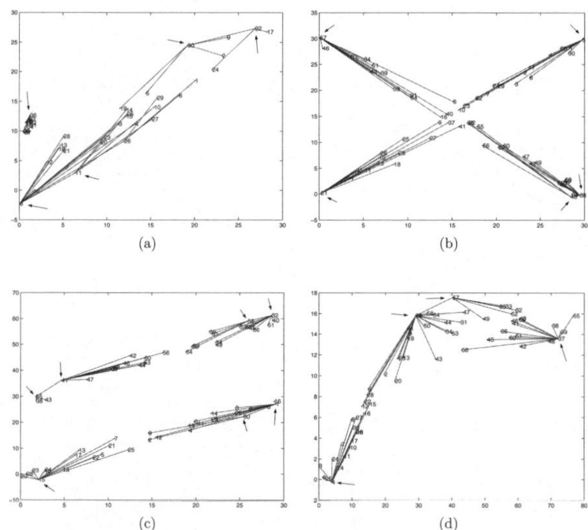

Figure 2. *Behavior of the link function in simple linear regression in the four cases: (a) group of masked outliers; (b) switching regression; (c) mixture of two regressions with omitted categorical variable; (d) regression under an structural change.*

in case (d) four discriminators are found and, as before, points are linked to extreme observations (high leverage and or high residual) located in the same region of the space as determined by the fitted regression line. In the four examples considered the sample is split into homogeneous groups and this is the pattern found in other SAR applications (see Peña and Tiao, 2002, and Peña, Rodríguez and Tiao, 2002).

4. THE PROPOSED ALGORITHM

The algorithm proposed for finding heterogeneity in regression models is the SAR algorithm briefly presented in Section 2. Here we will just illustrate it by using the data from Figure 2b. Some of the intermediate results when applying the algorithm are indicated in Figure 3. First, four basic stets are found (see Figure 3a); second, the enlarging of the first basic set incorporates all the observations from the first regime (see Figure 3b); third, conditioning on this result the complementary part is split and a basic set is found (see Figure 3c); fourth, the set is enlarged and the rest of the available observations are incorporated except for two points indicated by arrows in Figure 3d that are identified as outliers.

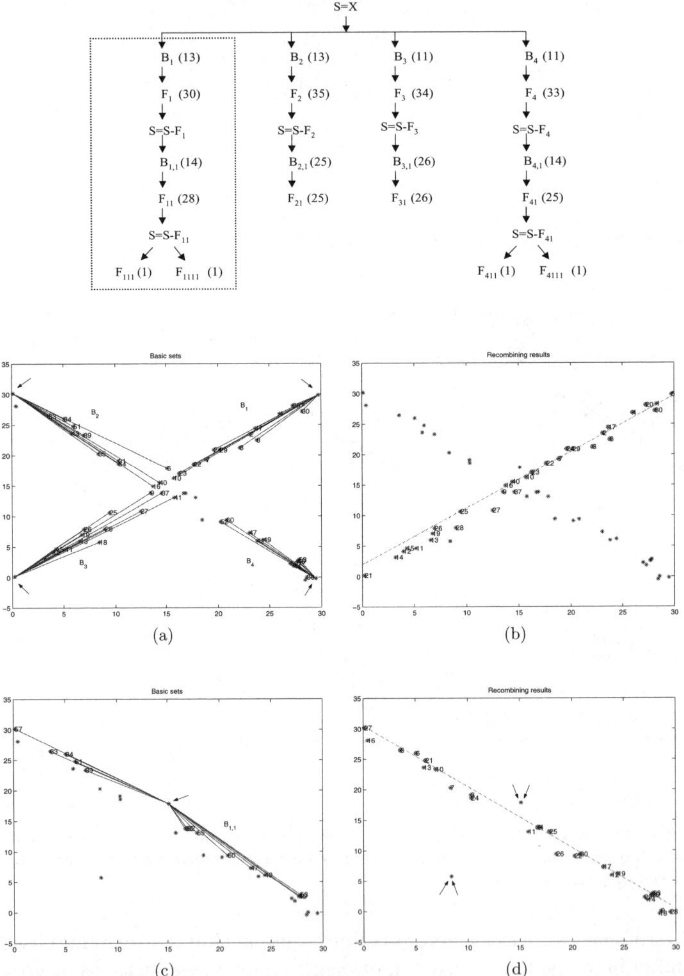

Figure 3. *Flow chart of the SAR procedure in the switching regression.* (a) *Basic sets;* (b) *enlarging of the first basic set;* (c) *Splitting of the complementary part, one basic set is found;* (d) *enlarging of the basic set and the two outliers detected, which are indicated by arrows.*

The two regression lines and the two outliers found in Figure 3 is one of the possible data configurations (PDCs), provided by the algorithm, because when starting from a different basic set some of the points may be allocated in a slightly different way. Figure 3, top, shows a flow chart of the four paths generated by each of the four basic sets. The path illustrated in Figure 3 is the first one, and it is presented in a dotted rectangle. Starting from the first basic set of 13 observations (B_1) the set is enlarged to include 30 observations (F_1). Then the complementary part of F_1 is analyzed and a basic set $B_{1,1}$ of 14 observations is found. This basic set grows to include 28 observations (F_{11}). The complementary part of this groups contains two observations that are considered as outliers (groups F_{111} and F_{1111}, each of one observation). The four PDCs finally found are shown in Figure 4. These four PDCs only differ in the way in which the doubtful observations in the intersection of the two lines are allocated. In the first PDC these observations are all included in the regression with positive slope and no outliers are found. In the second PDC, they are included in the regression with negative slope. In the third and fourth PDCs two of the points are considered as outliers and the remaining doubtful points are allocated: (1) in the third PDC to the regression with positive slope and (2) in the fourth to the one with negative slope. The selection of the best PDC can be made by: (1) Using the BIC criterion; (2) Fitting two regression lines to each of the two clear groups and then computing the posterior probabilities of each observation in the intersection belonging to each of the two regression lines.

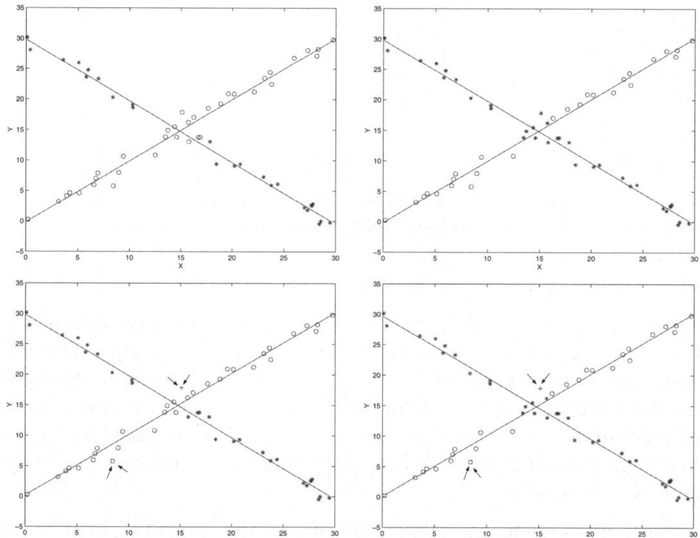

Figure 4. *The four PDCs found for data from Figure 2* (b) *in the switching regression problem.*

In the three other cases shown in Figure 1 also more than one PDC may be obtained. In case (a) two PDCs are found: the first considers the sample as homogeneous and the second splits the data into the homogeneous group and the outliers. In case (c) only one PDC is obtained and this is the correct solution. In case (d) three possible PDCs, shown in Figure 5, are found. The interesting ones are the second and the third, which differ in the way in which observations in the intersection of the two regimes are allocated. The method makes clear that some of these observations can be generated by both models. The true data generation pattern for this example is shown in Figure 5d.

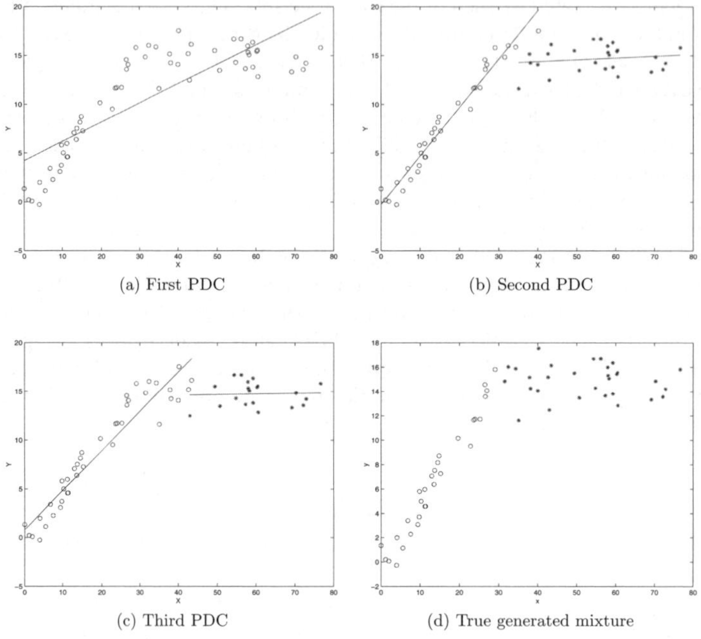

Figure 5. *PDCs for the structural change data from Figure 1d. The two interesting solutions (send and third) differ in the way the data around the change point are allocated between the two groups.*

Some remarks about the procedure are in order:

Remark 1. If p is large and n/p is small, let us say smaller than 20, many points can be extreme and many discriminators may be found, leading to many small initial groups. This will make the procedure slower and less powerful for the identification of small heterogeneous groups, as too many points will be deleted in the first split. In this case it is faster and safer to delete only discriminators of order m, that are defined as data points that are discriminators for at least m points in the sample. In Peña *et al.* (2002) these discriminators are defined and analyzed for cluster problems in high dimensions.

Remark 2. The procedure evaluates the relative size of statistics c_0 and d_1 by their sampling distribution which has been obtained by simulation. Although we believe that this type of crossbreeding between Bayesian and frequentist ideas enriches statistics, the procedure could be made easily completely Bayesian by working either with the posterior probabilities or the Bayes factors. In order to do so we have to introduce an alternative model to explain how the point y_i under consideration could be heterogeneous with the group. A simple solution is to assume as alternative distribution the scale normal contaminated model and use (2) to compute the posterior probabilities or the Bayes factors.

Remark 3. The solutions obtained by the SAR procedure are not very sensitive to the choice of the parameter h and therefore to the minimal size, $p + h$, of the basic set. When h is small the procedure obtains a large number of basic sets and this increases the power to find small heterogeneous groups of masking outliers, but the number of PDC increases making the procedure slower for large data sets. If we expect that heterogeneity will be due to the possible existence of two or more regressions, we can choose h moderately large.

5. EXAMPLES

In this section, three examples are shown to illustrate the performance of the SAR algorithm and to compare it with other approaches. The first one is a real data set that has already be analyzed in the Bayesian literature. We show that the SAR procedure leads to the same solution obtained by MC^2 methods. The second example is a masking outlier example, and there the SAR procedure succeeds where standard MC^2 methods may fail completely. The third example is presented to illustrate the usefulness of the SAR procedure as a data mining tool in large high dimensional regression data sets.

Example 1. Figure 6 shows the ethanol data set, which relates the equivalence ratio, a measure of the richness of the air–ethanol mix for burning ethanol in a single-cylinder automobile test to the concentration of nitric oxide in engine exhaust (normalized by engine work) (Brinkman, 1981). Hurvich *et al.* (1998) analyze this data set with nonparametric regression and Hurn *et al.* (2001) use Stephens (2000) birth-and-death proposal for carrying out MC^2 estimation of mixture switching regression models. Figure 6 shows the four PDCs found by the SAR procedure. The four PDCs differ in the way in which doubtful points are allocated: the two regimes are clearly identified in the four cases, but there is a large uncertainty on the allocation of the observations in the intersections of the two lines.

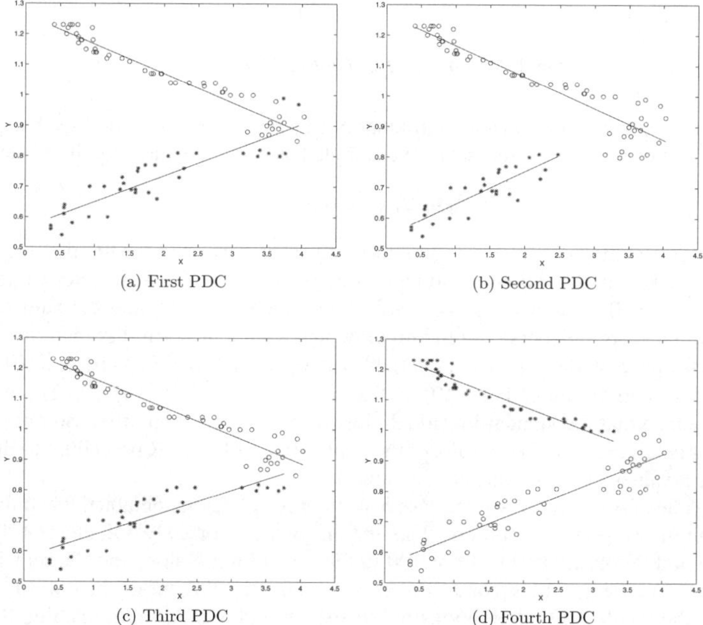

(a) First PDC (b) Second PDC

(c) Third PDC (d) Fourth PDC

Figure 6. *The four PDCs found for the ethanol data when y = equivalence ratio and x = nitric oxide concentration.*

Example 2. An interesting classical example of masking is the artificial data generated by Hawkins *et al.* (1984). The model includes 75 data points of one response and 3 explanatory variables. The data are generated in such a way that the first 10 data points are high leverage outliers, whereas the next four are good observations with high leverage. Traditional methods of outlier detection fail in this case due to the high leverage problem, and Justel and Peña (1996) showed that Gibbs Sampling also fails and is unable to identify the outliers, even after 30,000 iterations. These authors showed that the lack of convergence in the algorithm is not a problem of the outlier model considered, as the same lack of convergence was found in all the outliers models included in the study, including a nonparametric hierarchical model based on Dirichlet processes and later Justel and Peña (2001) introduced modifications in the MC chain to solve this problem. Figure 7 shows the two PDCs found with the SAR algorithm in this problem. The first one is the right one, and is the one chosen by the BIC criterion. The second one is the wrong one obtained by the standard application of the Gibbs Sampling algorithm.

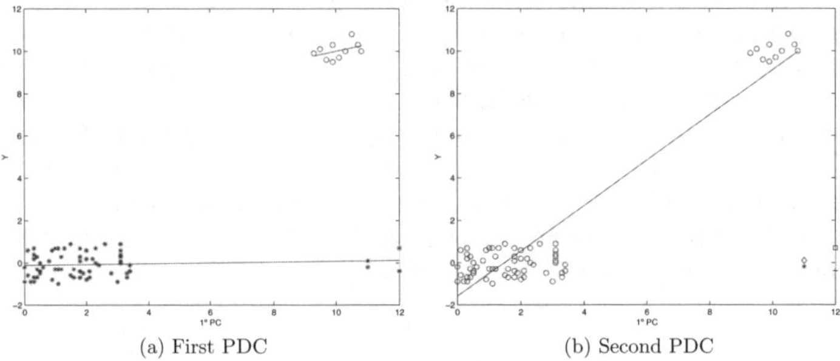

(a) First PDC (b) Second PDC

Figure 7. *The two PDCs for the HBK data.*

Example 3. In this example we consider a mixture of two regressions with omitted categorical variable in relatively large dimension data. The sample has been generated by the model

$$y = \beta_0 + \boldsymbol{\beta}_1' \boldsymbol{x} + \beta_2 z + u.$$

Here $\boldsymbol{x}$ has dimension 20, $u \sim N(0,1)$, and 400 values are generated for the first regression with $z = 0$, and 100 values for the second regression with $z = 1$. The parameter values have been chosen as $\beta_0 = 0$, $\boldsymbol{\beta}_1' = -1_{20}' = (-1, ..., -1)$, and $\beta_2 = 90$, and the values of the explanatory variables are independent random drawings from a uniform distribution. For the first regression the range of the explanatory variables is $(0, 10)$ so that $\boldsymbol{x}|(z = 0) \sim [U(0, 10)]^{20}$ whereas for the second the range is $(9, 10)$ so that $\boldsymbol{x}|(z = 1) \sim [U(9, 10)]^{20}$. These values have been chosen so that the standard residual plots from the fitted regression do not provide any evidence of heterogeneity. The results of the application of the SAR procedure to this data set with different values of h are indicated in Table 1.

If a small value of h is chosen, see the case $h = 10$, three PDCs are obtained that only differ in the allocation of two points as outliers. The first one assigns correctly 392 out of 400 to the first regression and 99 observations out of 100 to the second regression, and the remaining 9 points are isolated outliers. The second PDC assigns correctly 392 out of the 400 to the first regression and the whole 100 observations to the second and, again, the remaining 8 points are considered as outliers. Finally, the third PDC assigns correctly 391 out of 400 to the first regression and the 100 observations to the second and includes 9 isolated outliers. For moderate

Table 1. *PDCs for data in example 3 for several h values.*

h	PDC1			PDC2			PDC3			Time
	$z=0$	$z=1$	IA	$z=0$	$z=1$	IA	$z=0$	$z=1$	IA	
10	392	99	9	392	100	8	391	100	9	161 s
25	400	100	0	–	–	–	–	–	–	118 s
50	400	100	0	–	–	–	–	–	–	118 s
75	500	0	0	–	–	–	–	–	–	93 s

values of h, see the rows for $h = 25$ and/or 50, the SAR procedure obtains only one PDC with all observations correctly classified and without outliers. For large values of h, see the row of $h = 75$, the minimum size of the basic group is $75 + 21 = 96$, and the procedure obtains only one PDC with all observations in one group and the data set is found homogeneous. Note that this is to be expected when the size of the basic group is similar or larger than the group of heterogeneous observations. The table includes the running time in seconds of the program (written in Matlab) for several values of h. We have also tried $h = 6$, which corresponds to the rule $\log(n - p)$. Nine PDCs configuration are found. Eight of them identify clearly the two groups and include between 391 and 394 observation of the first group in the first regression and between 99 and 100 of the second in the second regression. These 8 PDCs only differ in the number of outliers that goes from 6 to 10. The 9th PDC also finds two groups of 393 and 105 observations plus 2 outliers, but the second group includes 5 points from the first that produce strong biases in the estimation of the regression coefficients. Thus, for large n we recommend that h should be chosen so that the size of the basic set is smaller than the expected size of the smallest heterogeneous groups we want to detect.

6. MONTE CARLO ANALYSIS FOR THE CHANGING REGRESSION PROBLEM

We have carried out a Monte Carlo experiment for different structures of mixture regressions. Samples (1000) for each combination of parameter values have been generated as follows. First, 50 uniform $U(0, 10)$ observations have been obtained and these values are used as the explanatory variable in the regression $y = 1.5x+1.5\epsilon$, where ϵ is $N(0, 1)$. Then, 30 observations are generated by $x = d + U(0, 15)$ and used as explanatory variables in the regression $y = a+bx+\sigma\epsilon$ where, as before, ϵ is $N(0, 1)$. Note that the important parameter d controls whether or not there is a horizontal gap ($H_g = d - 10$) and vertical gap ($V_g = a - 15 + db$) between the end point of the first and the beginning point of the two regression lines.

Table 2 shows the frequency of obtaining an approximate correct solution for 13 different parameter values. The first case corresponds to the size of the test, as all the data are generated for the same model. Cases 1–8 correspond to structural change, cases 9–10 to switching regression and cases 11–12 to mixture of two regressions with an omitted dummy variable. The size of the test has been computed with $h = 3$, similar values have been found for other values of $h \leq 10$. In the cases of two regression lines without a horizontal or vertical gap usually two solutions are found, but the points in the intersection are attributed either to the first or the second regression. There is no information to identify these points into one of the two groups unless we introduce some separation among them. Thus, in this case the proportion of points correctly identified is smaller than when a gap is introduced..

Table 2 shows a second Monte Carlo experiment to analyze the power of the procedure for finding a concentrated contamination. Samples (1000) for each combination of parameter values have been generated as follows. First n_1 uniform $U(0, 10)$ observations have been obtained and these values are used as the explanatory variable in the regression $y = 1.5x + \epsilon$, where ϵ

Table 2. *Power study for the two regimes and contaminated regression.*

Two regimes regression						
case	b	V_g	H_g	σ	95true	90true
0	1.5	0	0	1.5	0.979	1.00
1	-0.5	0	0	1	0.217	0.666
2	-0.5	0	0	0.5	0.512	0.929
3	-0.5	-2.5	5	1	0.836	0.894
4	-0.5	-2.5	5	0.5	0.906	0.971
5	-0.5	+2.5	5	1	0.904	0.958
6	-0.5	+2.5	5	0.5	0.980	1.000
7	0.5	-2.5	5	1	0.732	0.767
8	0.5	-2.5	5	0.5	0.827	0.869
9	-1.5	0	-10	1	0.220	0.808
10	-1.5	0	-10	0.5	0.435	0.921
11	1	0	-10	1	0.968	0.994
12	1	0	-10	0.5	0.975	0.996

Contaminated regression						
case	n_1	n_2	t_0	s_0	95true	90true
1	50	10	1	0.1	0.950	0.980
2	50	10	1	0.5	0.836	0.885
3	50	10	1	1	0.629	0.817
4	50	10	2	0.1	0.982	1.000
5	50	10	2	0.5	0.969	0.994
6	50	10	2	1	0.965	0.992
7	50	10	3	0.1	0.988	0.999
8	50	10	3	0.5	0.973	0.995
9	50	10	3	1	0.976	0.998
10	50	10	4	0.1	0.985	1.000
11	50	10	4	0.5	0.986	0.998
12	50	10	4	1	0.983	0.997

is $N(0,1)$. Then, n_2 observations are obtained from $N((x_0, y_0), s_0\boldsymbol{I})$, where x_0 is a random value from a uniform $U(-2.5, 12.5)$, and

$$y_0 = 1.5x + t_0 s_R \sqrt{(1 - x_0^2)},$$

where s_R is the residual standard deviation in the regression using the first n_1 observations. Note that t_0 represents the standardized size of the outlier and its location is made in agreement with this parameter. Finally, the parameter s_0 defines the dispersion of the outliers with respect to its center. Table 1 shows that the power of the SAR procedure in all cases is large, and the procedure seems to overcome the masking problem.

7. CONCLUDING REMARKS

The SAR procedure seems to offer a powerful method for the Bayesian analysis of regression mixture models. The method can be helpful in identifying patterns in heterogeneous regression data, including masked outliers, switching regression, change point problems and other multiple regime situations. Although the main contribution of the procedure is to find structure in the data as an exploratory tool, the selection between the possible data configurations (PDCs) can be done by applying a MC^2 algorithm for model estimation. An important difference of the SAR procedure with respect to alternative methods for finding heterogeneity in regression mixtures is that the mixture components do not have to compete to classify each observation. Thus, more than one possible solution may exist and the procedure will find all solutions coherent with the model structure, under different restrictions implied by the conditioning of the homogeneous enlarged basic sets. This property gives a high robustness to the SAR procedure because when there exist observations that could be assigned to various components of the mixture they are usually not split up between the different mixture components but are allocated in groups to the PDCs, avoiding the well-known masking problem in outlier detection. Finally, the ideas presented here can be easily extended to other regression situations as non-linear regression problems, heterocedastic regression models, generalized linear models, and multivariate regression models, including simultaneous equation econometric models.

ACKNOWLEDGEMENTS

Daniel Peña and Julio Rodríguez acknowledge support from GRANT BEC 2000-0167, MCYT, Spain, and by the Cátedra BBVA de Calidad.

REFERENCES

Aitkin, M. (2001). Likelihood and Bayesian Analysis of Mixtures. *Statist. Modelling* **1**, 287–304.

Andrews, D. W. K., Lee, I. and Ploberger, W. (1996). Optimal changepoint tests for normal linear regression. *J. Econometrics* **70**, 9–38.

Brinkman, N. D. (1981). Ethanol fuel—a single cylinder engine study of efficiency and exhaust emissions. *SAE Trans.* **90**, 1410–1427.

Box, G. E. P. (1980). Sampling and Bayesian inference in scientific modelling and robustness. *J. R. Statist. Soc. A* **143**, 383–430 (with discussion).

Box, G. E. P. and Tiao, G. C. (1968). A Bayesian approach to some outliers problems. *Biometrika* **55**, 119–129.

Box, G. E. P. and Tiao, G. C. (1973). *Bayesian Inference in Statistical Analysis*. Reading, MA: Addison-Wesley.

Celeux, G., Hurn, M. and Robert, C. (2000). Computational and inferential difficulties with mixture posterior distributions. *J. Am. Statist. Ass.* **95**, 957–970.

Cook, R. D. and Weisberg, S. (1982). *Residuals and Influence in Regression.* London: Chapman and Hall

Geisser, S. (1980). Discussion of a paper by G. E. P. Box. *J. R. Statist. Soc. A* **143**, 416–417.

Geisser, S. (1987). Influential observations, diagnosis and discordancy test. *J. Appl. Statist.* **14**, 133–142.

Hawkins, D. M., Bradu, D. and Kass, G. V. (1984). Location of several outliers in multiple regression data using elemental sets. *Technometrics* **26**, 197–208.

Hurn, M., Justel, A. and Robert, C. P. (2002). Estimating mixtures of regressions. *J. Comput. Graph. Statist.* (to appear).

Hurvich, C. M., Simonoff, J. S. and Tsai, C. (1998). Smoothing parameter selection in nonparametric regression using an improved Akaike information criterion. *J. R. Statist. Soc. B* **60**, 271–293.

Justel, A. and Peña, D. (1996). Gibbs sampling will fail in outlier problems with strong masking. *J. Comput. Graph. Statist.* **5**, 176–189.

Justel, A. and Peña, D. (2001). Bayesian Unmasking in Linear Models. *Comput. Statist. Data Anal.* **36**, 69-84.

Quandt, R. E. (1958). The Estimation of the Parameters of a Linear Regression System Obeying Two Separate Regimes. *J. Am. Statist. Ass.* **53**, 873–880.

Peña, D. and Guttman, I. (1993). Comparing Probabilistic Methods for Outlier Detection. *Biometrika* **80**, 603–610.

Peña, D. and Tiao, G. C. (1992). Bayesian Robustness Functions for Linear Models. *Bayesian Statistics 4* (J. M. Bernardo, J. O. Berger, A. P. Dawid and A. F. M. Smith, eds). Oxford: Oxford University Press, 365–388.

Peña, D. and Tiao, G. C. (2002). The SAR procedure: A diagnostic analysis of heterogeneous data (submitted).

Peña, D., Rodríguez, J. and Tiao, G. C. (2002). Cluster analysis by the SAR procedure. *Tech. Rep.*, Universidad Carlos III de Madrid, Spain.

Pettit, L. I. (1990). The Conditional Predictive Ordinate for the Normal Distribution. *J. R. Statist. Soc. B* **52**, 175–184.

Pettit, L. I. and Smith, A. F. M. (1985). Outliers and influential observations in linear models. *Bayesian Statistics 2* (J. M. Bernardo, M. H. DeGroot, D. V. Lindley and A. F. M. Smith, eds), Amsterdam: North-Holland, 473–494 (with discussion).

Richarson, S. and Green, P. J. (1997). On Bayesian analysis of mixtures with an unknown number of components. *J. R. Statist. Soc. B* **59**, 731–758.

Stephens, M. (2000). Bayesian analysis of mixture models with an unknown number of components–An alternative to reversible jump methods. *Ann. Statist.* **28**, 40–74.

Schweder, T. (1976). Some 'optimal' methods to detect structural shift or outliers in regression. *J. Am. Statist. Ass.* **71**, 491–450.

APPENDIX

Let $I = (i_1, ..., i_k)$ represent the indices for k observations with vector of responses $\boldsymbol{Y}_I$ and regressors $\boldsymbol{X}_I$. Calling $\widehat{\boldsymbol{\beta}}_{(I)} = (\boldsymbol{X}'_{(I)}\boldsymbol{X}_{(I)})^{-1}\boldsymbol{X}'_{(I)}\boldsymbol{Y}_{(I)}$ to the vector when these observations are deleted from the sample $\boldsymbol{X}, \boldsymbol{Y}$, and using the well known expression (see Cook and Weisberg, 1982) we have that,

$$\widehat{\boldsymbol{\beta}}_{(I)} = \widehat{\boldsymbol{\beta}} - (\boldsymbol{X}'\boldsymbol{X})^{-1}\boldsymbol{X}_I(\boldsymbol{I} - \boldsymbol{H}_I)^{-1}\boldsymbol{e}_I,$$

where $\boldsymbol{H}_I = \boldsymbol{X}_I(\boldsymbol{X}'\boldsymbol{X})^{-1}\boldsymbol{X}'_I$ and $\boldsymbol{e}_I = \boldsymbol{Y}_I - \boldsymbol{X}_I\widehat{\boldsymbol{\beta}}$. Now calling

$$\boldsymbol{e}_{I(I)} = \boldsymbol{Y}_I - \boldsymbol{X}_I\widehat{\boldsymbol{\beta}}_{(I)} = (\boldsymbol{I} - \boldsymbol{H}_I)^{-1}\boldsymbol{e}_I,$$

we have that

$$(n - p - I)s^2_{(I)} = (n - p)s^2 - \boldsymbol{e}'_I(\boldsymbol{I} - \boldsymbol{H}_I)^{-1}\boldsymbol{e}_I$$

and it is easy to show that $\boldsymbol{H}_{I(I)} = \boldsymbol{X}_I(\boldsymbol{X}'_{(I)}\boldsymbol{X}_{(I)})^{-1}\boldsymbol{X}'_{(I)}$ verifies $\boldsymbol{H}_{I(I)} = \boldsymbol{H}_I(\boldsymbol{I} - \boldsymbol{H}_I)^{-1}$. Now, making $\boldsymbol{Y}_I = (y_i, y_j)'$, $I = (i, j)$, using the previous equations and after some straightforward algebra, we found

$$t^2_{i(ij)} = \frac{(e_i + h_{ij}e_j/(1 - h_{jj}))^2}{a_{ii} - c/(1 - h_{jj})[dh^2_{ij} + b_{ii}e^2_j + 2e_ih_{ij}e_j]}$$

where $c = (n - p - 2)^{-1}$ and $d = (n - p)s^2$ are constant and $b_{ii} = (1 - h_{ii})$, $a_{ii} = c[db_{ii} - e^2_i]$ are only function of the i-th point.

To understand better this function, we consider the simple regression through the origin case with $\sum x_i = \sum y_i = 0$ and $\sum x^2_i = 1$ so that $h_{ij} = x_ix_j$. Then, we can write, for each (e_i, x_i), this function as

$$t^2_{i(ij)} = f(e_j, x_j) = \frac{(n - p - 2)(e_i + e_jx_ix_j/(1 - x^2_j))^2}{(d(1 - x^2_i) - e^2_i) - \left(1/(1 - x^2_j)\right)(dx^2_ix^2_j + (1 - x^2_i)e^2_j + 2e_ie_jx_ix_j)}.$$

It can be proved that if $e_i \neq 0$ and $x_i \neq 0$, then,

$$f(e_j, x_j) = f(-e_j, -x_j)$$

and $f(e_j, -x_j) = f(-e_j, x_j)$ and $(e_j = 0, x_j = 0)$ is a saddle point. If $e_i \neq 0$ or $x_i \neq 0$, then, $f(e_j, x_j) = f(-e_j, -x_j) = f(e_j, -x_j) = f(-e_j, x_j)$ and $(e_j = 0, x_j = 0)$ is a minimum.

DISCUSSION

HAL S. STERN (*University of California, Irvine, USA*)

Introduction. Peña, Rodríguez, and Tiao (PRT) present a novel exploratory approach for investigating data heterogeneity when fitting regression models. It is my pleasure as discussant to thank them on behalf of those attending the conference. This introduction provides a brief review of the PRT approach and some preliminary comments. The remaining sections provide more detailed discussion of some important issues.

The authors use the term data heterogeneity to refer to the existence of clusters or subgroups among the sample units. In the regression context the subgroups may be clusters of outlying observations in an otherwise homogeneous population or subpopulations best described by different linear models. The SAR (split and recombine) procedure, described more fully in Peña and Tiao (2002), is used to discover such heterogeneities. The key element in the approach is a three-step process that identifies a subgroup for which a single regression model seems appropriate. This process is repeated until all such subgroups are identified. The three steps are: (1) outliers are identified using the predictive ordinate $p(y_i|Y_{(i)})$ where $Y_{(i)}$ refers to the vector of responses without observation i; (2) the remaining data is split into small homogeneous groups known as basic sets (each containing at least a specified minimum number of observations) using the concept of "link"-ing or discriminator points; (3) these basic sets are enlarged by incorporating all of the observations in the data set that are consistent with the linear model that describes the homogeneous group. Once a basic set is selected and enlarged, the SAR procedure is restarted on the set of all observations that are excluded from the enlarged group. The outcome of the procedure is called a *possible data configuration* (PDC). The outcome depends on the group that is enlarged first, thus, there may be several different PDCs for a given data set.

The authors cite as their goal to "solve" (my quotes) the heterogeneity problem. Though I quibble below with the idea of solving the heterogeneity problem in this way, it is clear that SAR is a powerful exploratory technique that addresses a number of important heterogeneous regression models. It is interesting to note that the SAR procedure, though proposed by Bayesian researchers and motivated by Bayesian principles, will likely be welcomed by non-Bayesians. The cross-validatory ideas that motivate SAR are familiar in regression diagnostics and there is little for anyone to object to in terms of probability modelling. The remainder of this discussion considers four issues: the types of heterogeneous mixture models considered, appropriate alternatives for comparison with SAR, whether one can solve the heterogeneity problem in this way, and the nature of exploratory analysis.

Mixtures of regressions. Four heterogeneous regression models are used by the authors to demonstrate the utility of the SAR procedure. These are an outlier contamination model, a structural change model, a switching regression model (subgroups may differ in all coefficients), and a missing categorical variable model (subgroups differ in intercept only). Each of the four can be expressed as a probability model for scalar y_i conditional on covariate vector x_i of the form $y_i \mid x_i \sim \sum_j \alpha_j \mathrm{N}(x_i'\beta_j, \sigma_j^2)$. This common form of mixture model misses an important form of heterogeneity. The next paragraph describes a more general regression mixture model. It turns out that the SAR procedure is effective in identifying this form of heterogeneity as well.

One natural way to motivate mixture models is to introduce an indicator z_i for each case, with $z_i = j$ if the ith observation comes from subgroup j (or has level j for an unobserved categorical variable). Then the mixture model for y_i conditional on x_i is obtained by considering the joint model $y_i, z_i \mid x_i \sim p(z_i \mid x_i)p(y_i \mid z_i, x_i)$ and marginalizing over z_i to obtain

$$y_i \mid x_i \sim \sum_j \Pr(z_i = j \mid x_i) p(y_i \mid z_i, x_i).$$

The model considered by the authors is obtained if $\Pr(z_i = j \mid x_i) = \alpha_j$ independent of x_i and a normal linear regression model is used for y_i given z_i and x_i. When viewed in this way, however, it seems much more natural to allow for the possibility that the latent variable z_i depends on x_i, perhaps with a multinomial model $\Pr(z_i = j \mid x_i) = \exp(x_i'\delta_j)/\sum_k \exp(x_i'\delta_k)$. The multinomial-regression mixture model is used by Peng *et al.* (1996) as a "mixture of experts" and by Morduch and Stern (1997) as an approach to heterogeneity in data regarding family spending in Bangladesh.

The more complicated mixture poses some problems for inference in that the likelihood function may have multiple modes (including some degenerate cases with only enough observations in some subgroups to identify the regression parameters) and the multinomial parameters describing the latent indicators are not well identified. Example 3 in the paper is a mixture of this more complex type because the covariate distribution differs across the subgroups. That example proves that the SAR procedure can handle data heterogeneity of this type.

Alternatives to SAR. Several parts of the paper refer to Gibbs sampling or Markov chain Monte Carlo (MCMC) as an alternative to SAR. This is used as a shorthand way to reference a formal analysis of an apparently relevant model, say the outlier contamination model, using MCMC. Of course this is not really a fair comparison at all in the sense that the formal prior-to-posterior analysis provides a very different type of information than is provided by the SAR analysis. It may be difficult to get MCMC algorithms to converge for mixture models but once this is achieved the investigator is rewarded with the usual posterior inferences that Bayesians find so attractive.

As SAR does not offer such inferences it might be better compared to traditional and new clustering techniques. Especially noteworthy are the model-based clustering of Banfield and Raftery (1993) and the work by Liu in this volume. Model-based clustering allows for multivariate Gaussian clusters with varying shape and orientation and thus might pick up the mixtures of linear models contemplated here. Recent work on data visualization is also relevant (Sutherland *et al.*, 2000).

It might also be interesting to explore how well SAR does relative to the traditional approach of statisticians building up from simpler models with the aid of model diagnostics. For example, a traditional linear regression analysis of the authors' example 1 along with residual plots might suggest a quadratic model. Another model-based approach would be to search for modes of the likelihood under a hypothesized heterogeneity or mixture model. One wonders if the different PDCs might show up as different modes in the likelihood.

Solving the heterogeneity problem. The authors clearly identify their procedure as exploratory, suggesting that formal inference might be done after SAR helps to identify a suitable model. It is difficult to argue with such an approach. At times though there are claims about the SAR procedure's ability to find the "correct" solution (at one point noting that using BIC to choose among PDCs would have selected the right answer). There are at least three reasons it would be better to avoid such claims. First, the primary goal of such exploration should be to generate interesting views of the data not to identify the "true" data generating mechanism which is destined to remain unknown in most applications. Second, all of the simulated data sets are from linear regression models or mixtures of linear regressions. These simulations show that SAR can identify heterogeneity and that SAR will not always identify heterogeneity, but the simulations do not address how SAR might perform on data best described with a nonlinear relationship. The ethanol data of example 1 looks a bit like data from a quadratic model. Though the piecewise linear model may fit a bit better in that particular data set, this example raises the question as to whether all quadratics should be viewed as mixtures of regressions. A third argument for caution regarding SAR's performance concerns the effect of the minimum group size parameter. This is a key parameter and the advice provided here is that investigators should choose this parameter to match "the expected size of smallest heterogeneous groups we want to detect." Thus, the quality of SAR's results will depend mightily on the investigator's knowledge regarding the expected heterogeneity.

Concluding remarks on SAR and exploratory analysis. There are clear roles for exploratory tools in data analysis; such methods help find structure, suggest models, and critique models. The SAR approach seems to offer much in the first two of these roles. Exploratory techniques like SAR can help find structure in new and difficult settings. The mining of large databases and the analysis of large sets of biological data each demonstrate that there is a clear need for exploratory tools in the current scientific environment. The authors mention data mining as a possible application of SAR. The relationships sought in data mining applications are of all types, not just linear, so it would be interesting to see how SAR performs in the data mining context.

In some applications the exploratory analysis may reveal all that we need to know. It is much more common that the outcome of an exploratory analysis is the suggestion of a probabilistic model allowing formal inference regarding the data-generating process. In this regard SAR performs well. The end result of the SAR analysis will often be the suggestion of a suitable mixture model; one must be careful to avoid overstating the significance of such *post hoc* explanations of heterogeneity.

The authors' conclusions describe plans to extend the application of SAR to finding data heterogeneity in the fitting of generalized linear models, non-linear models, simultaneous equation models, etc. This is clearly an intriguing idea given the success they have had with linear models but there is some risk. One major appeal of the SAR approach to heterogeneity in linear models is that there is a simplicity and intuition which users are likely to find appealing. This may be more difficult to provide if the SAR building blocks are more complex models.

SAR is an exploratory tool worth having in the proverbial "statistician's toolbox". Though it does not "solve" the heterogeneity problem it seems capable of providing real insight into the structure of data for a wide range of applications.

P. IGLESIAS and R. B. ARELLANO-VALLE (*Universidad Católica, Chile*)

In this paper, an interesting procedure based on SAR is introduced for modelling heteroscedasticity in the context of regression analysis. Such a procedure is obtained from the predictive distribution of the normal regression model and the usual non-informative prior distribution. An interesting property is that if the normal regression model is replaced by an elliptical regression model, but the same non-informative prior distribution is considered, then such procedure will remain invariant (Osielwaski and Steel, 1993). On the other hand, if another prior distribution is adopted in order to obtain marginal equivalency with the normal model (Arellano-Valle etalc 2002), then the procedure obtained in this case is also invariant under elliptical models, but will yield different results than those obtained by considering the usual non-informative prior distribution. Although invariance (with respect to the likelihood in this case) can be a desirable property, we would expect that models with heavier or lighter tails than the normal model yield different clusters. We think that this "lack of robustness" is due to the fact that the procedure is based on the predictive distribution only, that is, the other components of the Bayesian model are not considered. It would be interesting to introduce in the procedure those parameters for which the posterior distributions is not invariant to departures from normality (the precision parameter in this case). Furthermore, if we consider that the usual non-informative prior distribution is not the only reference prior distribution, the following question arises naturally: what is the performance of the proposed procedure when using Jeffrey's prior or some other reference prior distribution that depends on the order of the model parameters (Bernardo and Smith, 1994)? What is the sensitivity of the procedure to changes on the prior distribution? Finally, there are alternative methods for modelling heteroscedasticity, like the product partition model introduced by Hartigan (1990), which considers all the components of the Bayesian model when obtaining the clustering. Quintana and Iglesias (2002) propose to adopt this approach within the context of decision theory. The idea is to clearly define the purpose of the study (such as estimation, hypothesis testing, outlier detection, etc.) and from this to develop a clustering algorithm that depends also on the loss function associated with the decision problem. An interesting problem is to consider justifying the proposed procedure from a decision theoretic viewpoint, doing at the same time a comparative study with other procedures considered in the literature.

CHRISTIAN P. ROBERT (*Université Paris Dauphine, France*)

While I find the diagnosis tools developed by the authors quite clever and apparently very efficient, I cannot but question the relevance of *exploratory* devices when "exact" procedures are available. Indeed, we studied in Hurn *et al.* (2002) the performances of a (fully) Bayesian approach to the estimation of the number of components in a mixture of standard [like (1)] and generalized linear models. Using Stephens's (2000) continuous time MCMC algorithm with a simple birth-and-death proposal, we found very satisfactory performances of the algorithm, with no obvious dependence on the starting values (as should be).

Given that such a (fully) Bayesian modelling is possible (and can be implemented in a fairly straightforward manner), it necessarily brings more information that the exploratory SAR procedure, since the later cannot, for instance, classify competing regression lines in terms of their posterior probability or eliminate dubious solutions. While Gibbs sampling indeed has difficulties to escape the "fatal attraction" of leverage points in outlier problems, more hybrid solutions using subsampling (in a spirit similar to the one developed in this paper) or tempering (Celeux *et al.,* 2000) should work better.

In addition, the SAR procedure does not strike me as being fundamentally Bayesian, since the predictive $p(y|Y)$ is built on a single normal regression model with conjugate priors, while the actual model is a mixture of normal regression models with or without conjugate priors. So the discriminating factors $c_0(i)$ are only formally related to the Bayesian approach. Also, for other generalized linear models, the construction of the predictive distribution $p(y|Y)$ is not possible in closed form. It thus seems to me that, unless the authors propose an alternative criterion, the necessary call to an approximative device of a numerical or simulational nature is not fundamentally different (in difficulty level) from the construction of the full MCMC apparatus. I nonetheless find the construction of the discriminators quite interesting in that they may serve as *anchors*, to borrow from Liu *et al.,* (2002) terminology, for constructing more adaptive or more efficient MCMC samplers.

REPLY TO THE DISCUSSION

We first want to thank our official discussant, Hal Stern, for his comments and to Christian Robert, Pilar Iglesias and Reinaldo Arellano-Valle for their contributions to the discussion of our paper.

Our reply to Hal Stern must be brief because we are in agreement with all the points he raised, and we fully share his point of view of the usefulness of the SAR procedure for exploratory data analysis. We appreciate his thoughtful and wise comments on our procedure, which will stimulate us to extend it for further applications. We agree that our method should be compared to clustering methods, and in fact we have carried out a comparison of the SAR procedure with some traditional and new clustering methods, including k-Means, Mclust, the Projection Pursuit method by Peña and Prieto (2001) and others (see Peña *et al.*, 2002). The result we have found by an extensive Monte Carlo study is that the SAR procedure seems to have the best performance according to standard criteria to compare cluster methods. In the present paper we have emphasized the exploratory role of the SAR procedure in regression and thus the comments of Hal Stern on our claims on solving the heterogeneity problem are right. However, we also believe that the SAR procedure can be extended to provide a formal structure for inference but this will be the subject of further research.

We agree with Christian Robert that if we had exact procedures we better use them. However the problem is that with a complex data set usually there is no exact procedure available. We may assume a model, run MCMC and get an answer but this type of "exact" procedure when applied with the wrong model can be misleading, as illustrated in the outlier problem we refer to in our paper. We believe that the SAR procedure can help in formulating a reasonable model for the data in hand. Regarding the second point, we assume that a regression model has been fitted to the data and we want to check for homogeneity. If the model was a mixture of regression we could, in principle, apply the SAR procedure to the mixture to check if it is homogeneous. We agree that a key advantage of the SAR procedure is its simplicity and small computational burden and we will try to keep this feature in its extension to generalized linear models.

The comments and suggestions by P. Iglesias and R.B. Arellano-Valle are very appropriate. We do not expect difference in behavior in the SAR procedure by moderate changes of the prior distribution, but this point deserves a careful investigation. Also, the suggestion to develop cluster algorithms as decision problems is attractive and we will be interested in developments in this area in the future.

ADDITIONAL REFERENCES IN THE DISCUSSION

Arellano-Valle, R. B., Iglesias, P. and Vidal, I. (2002). Bayesian inference for elliptical linear models: Conjugate analysis and model comparison (this volume).

Banfield, J. D. and Raftery, A. E. (1993). Model-based Gaussian and non-Gaussian clustering. *Biometrics* **49**, 803–821.

Bernardo, J. M. and Smith A. F. M. (1994). *Bayesian Theory*. Chichester: Wiley

Hartigan, J. A. (1990). Partition models. *Comm. Statist. Theory Meth.* **19**, 2745–2756.

Liu, J. S., Zhang, J. L., Palumbo, M. L. and Lawrence, C. E. (2002). Bayesian clustering with variable and transformation selections (this volume).

Morduch, J. J. and Stern, H. S. (1997). Using mixture models to detect sex bias in health outcomes in Bangladesh. *J. Econometrics* **77**, 259–276.

Osiewalski, J. and Steel, M. F. J. (1993). Robust Bayesian inference in elliptical regression models. *J. Econometrics* **57**, 345–363.

Peng, F., Jacobs, R. A. and Tanner, M. A. (1996). Bayesian inference in mixture-of-experts and hierarchical mixtures-of-experts models with an application to speech recognition. *J. Am. Statist. Ass.* **91**, 953–960.

Peña, D. and Prieto, F.J. (2001). Clustering by projections. *J. Am. Statist. Ass.* **96**, 1433–1445.

Peña, D., Rodríguez, J. and Tiao, G. C. (2002). The SAR Procedure for cluster analysis. *Tech. Rep.*, Universidad Carlos III de Madrid, Spain.

Quintana, F. and Iglesias, P. (2002). Nonparametric Bayesian clustering and product partition models. *Tech. Rep.*, Universidad Pontificia, Chile.

Sutherland, P., Rossini, A., Lumley, T., Lewin-Koh, N., Dickerson, J., Cox, Z. and Cook, D. (2000). Orca: A visualization toolkit for high-dimensional data. *J. Comput. Graph. Statist.* **9**, 509–529.

BAYESIAN STATISTICS 7, pp. 349–367
J. M. Bernardo, M. J. Bayarri, J. O. Berger, A. P. Dawid,
D. Heckerman, A. F. M. Smith and M. West (Eds.)

Global Gambling

JOSE MARIO QUINTANA
The Nikko Securities Co. Internat., Inc. and Bayesian Enhanced Strategic Trading, USA*
`JoseMarioQuintana@2BestSystems.Com`

VIRIDIANA LOURDES
ITAM, Mexico and Bayesian Enhanced Strategic Trading, USA
`vlourdes@itam.mx`

OMAR AGUILAR
Lehman Brothers, USA
`omar.aguilar@lehman.com`

JANE LIU
UBS Warburg, USA
`jane.liu@ubsw.com`

SUMMARY

The notion of gamble has played a central role from the advent of probability theory; to the first axiomatic definition, elicitation and revision of probabilities; to the concept of coherence, and the principle of maximizing expected utility. From a formal perspective, the largest stakes in the world are not raised at the Monte Carlo or Las Vegas casinos but at the Tokyo, Frankfurt, London, New York, and Chicago financial derivative markets. In this paper, we review the practical difficulties and advantages of applying the Bayesian paradigm, in general, and dynamic Bayesian modeling, in particular, to investment management.

Keywords: ASSET ALLOCATION; BAYESIAN DYNAMIC MODELING; STOCHASTIC DYNAMIC PROGRAMMING; GAMBLING; INVESTMENT MANAGEMENT.

1. INTRODUCTION

Phillip Johnson, a former chairman of the Commodity Futures Trading Commission, once remarked "If I buy stocks or bonds I am called an investor. If, on the other hand, I buy futures contracts, I am—at best—called a speculator or—worse—a gambler." He was suggesting that the colorful terminology of the futures markets might mislead the public. Nevertheless, the difference between investing, speculating and gambling, technically, blurs to variations in the objectives (*e.g.*, the fear and greed tradeoff attitude and the time horizon). Furthermore, financial derivatives contracts such as futures, forwards, options, etc. (as well as the traditional investment instruments) qualify as bets, stakes and gambles from a formal quantitative viewpoint. From this perspective, the cynics who assert that the financial derivatives markets are the greatest casinos of the world may have a point. The currency daily dealing alone is referred to as a trillion dollar market.

The concept of gamble, conversely, has played a remarkable role in the development of probability and the Bayesian methodology. Blaise Pascal and Pierre de Fermat initiated probability

* The author's views are personal and not necessarily shared by The Nikko Securities Co. International, Inc., its parent or any of its affiliates.

theory (circa 1650) when the Chevalier de Méré, a French nobleman gambler and philosopher, challenged Pascal to solve a puzzle involving a game of chance: How do you divide, fairly, the stakes in an uncompleted game? Bets and stakes are embedded in Daniel Bernoulli's (1738) pioneering work on risk and utility theory. The notion of gamble plays a central role in the legendary work of Bayes (1763) on the axiomatic definition, elicitation and revision of (subjective) probabilities. Indeed, he put gambling at the core of his "peculiar" definition of "The *probability of any event* is..." Detailed accounts of these early developments can be found in Stigler (1986) and Bernstein (1996). Gambling took a pivoting position during the last century in formalizing the Bayesian paradigms of maximizing the expected utility and coherence. The first aims to find the best actions of a gambler in terms of the uncertain consequences. The foundation of the second, de Finetti (1974), is the evasion, from a gambler's perspective, of a Dutch book (*i.e.*, the avoidance of becoming a potential sure loser). It is worth noting that there are alternative ways to justify, and even generalize, the Bayesian learning archetype (*e.g.*, Zellner, 1991, and references therein).

Coherence and maximizing the expected utility paradigm should, in principle, necessarily guide investment management; making it a Bayesian enterprise. However, avoiding being a potential sure loser and doing your best is not sufficient to become a winner. In fact, the proponents of the Efficient Market Hypothesis (EMH) state that beating 'the market' is an impossible mission. The EMH motivation is very appealing: if a market were inefficient, then relevant available information (*e.g.*, the price history) could be used to predict a future return (of an instrument or a portfolio of instruments) and one could profit from it; however, the resulting trading activity would then incorporate this information into the future prices making the market efficient. Information efficiency implies that the returns are independently distributed. Therefore, in an efficient market, making future gains is entirely random. Louis Bachelier (1900) in his, aptly named, doctoral dissertation initiated the EMH. As a result of considering futures, options, and "the mathematical expectation of a gambler" he developed the key Wiener process theory, predating Albert Einstein's work by five years. In one of its modern variants the EMH states that large series of small independently distributed random information shocks have a multiplicative effect in the price of securities, implying that the logarithm of the price follows a Wiener process (*i.e.*, the price follows a geometric Wiener process). These assumptions form the basis of standard models for pricing options and other derivative instruments; see, for example, Kolb (2000) and references therein. Notwithstanding, the EMH motivation is *strange* in the sense that if the argument is carried forward it also implies the opposite: if a market were efficient there would be no incentive for paying attention to market related information; therefore, it would not be assimilated by the market, making it inefficient! The apparent situation is that financial markets are in a persistent state of near efficiency, meaning that making excess money consistently is possible but very difficult.

It is in this extremely challenging, but arguably the most natural, environment where authors of this paper have put Bayesian methods into action. Our aim here, building on a previous short informal article (Quintana, 2000), is to outline the Bayesian technology; comment on its implementation and performance; and discuss the benefits, achievements, difficulties, and open problems. In Section 2, we outline a general Bayesian framework for investment management, and in Section 3, we discuss specific Bayesian dynamic models and potential new developments.

2. STOCHASTIC DYNAMIC PROGRAMMING

The only possible way to beat liquid financial markets (*i.e.*, generate excess returns over benchmarks) is by taking the risk of being beaten by the markets (i.e. the risk of underperforming the markets). A general financial strategy to outperform benchmarks is: (a) invest the capital in the benchmark, and (b) add an overlay portfolio of financial derivatives (*e.g.*, future and forward contracts) which virtually has no direct capital requirements (although the capital is indirectly needed to support these risky positions). From a pragmatic perspective, the overlay portfolio is essentially a set of (side) bets between gentlemen and gentlewomen. The investment problem is reduced to manage the overlay portfolio over time, aiming to produce consistent positive excess returns. This task, as most other tasks in life, involves making sequential decisions under uncertainty. Its best possible Bayesian resolution involves a stochastic version of Bellman's *Optimality Principle*. A derivation, which extends early proposals by Zellner and Chetty (1965) and Markowitz (1987), is found in Quintana and Putnam (1994) and it is available upon request. Here we merely outline the exposition as in Quintana *et al.* (1998).

In the above framework the investment manager is a Bayesian making multiple-period decisions on the size and directions (long and short positions are allowed) of the overlay financial derivatives positions (*i.e.*, the portfolio weights) where the excess return (*i.e.*, the overlay portfolio payoff normalized by the supporting capital) generating process is governed by a stochastic process involving endogenous and exogenous variables. We denote the excess returns of the financial derivatives (*i.e.*, the financial derivative payoffs of a long position normalized by the price to be paid in local currency) as $(\mathbf{y}_1, ..., \mathbf{y}_n)$, and their associated explanatory variables $(\mathbf{x}_1, ..., \mathbf{x}_n)$ by $(\mathbf{z}_1, ..., \mathbf{z}_n) = ((\mathbf{x}_1, \mathbf{y}_1), ..., (\mathbf{x}_n, \mathbf{y}_n))$. These joint variables follow a stochastic process defined by the joint probability density,

$$p(\mathbf{z}_1, ..., \mathbf{z}_n) = p(\mathbf{z}_1)p(\mathbf{z}_2 \mid \mathbf{z}_1)...p(\mathbf{z}_n \mid \mathbf{z}_1, ..., \mathbf{z}_{n-1}). \tag{2.1}$$

The densities appearing in (2.1) are generic and their form may change over time. The multiple-period goal is,

$$U^* = \max_{\mathbf{u}_1,...,\mathbf{u}_n} \operatorname*{E}_{\mathbf{z}_1,...,\mathbf{z}_n} U(\mathbf{z}_1, ..., \mathbf{z}_n; \mathbf{u}_1, ..., \mathbf{u}_n), \tag{2.2}$$

where $(\mathbf{u}_1, ..., \mathbf{u}_n)$ represent the decision variables (*e.g.*, the sets of portfolio weights and possible auxiliary decision variables). Each decision variable is set after observing its associated explanatory variable but before observing its corresponding excess return. The explanatory time index merely denotes its association with its excess return, thus the explanatory variables are relevant lagged variables, proxies of related contemporaneous variables, etc. Equations (2.1) and (2.2) summarize a general Bayesian investment management program in a predictivistic fashion (Geisser, 1988). The utility (2.2) should represent investors' attitudes implemented by portfolio managers. Ideally, according to Agency Theory, the portfolio managers (and the investment manager firm, etc.) interests should be aligned to those of the investors. This occurs, in practice, to some extent. Usually, it is reasonable to assume that the utility depends only on the excess returns, that is, $U(\mathbf{z}_1, ..., \mathbf{z}_n; \mathbf{u}_1, ..., \mathbf{u}_n) = U(\mathbf{y}_1, ..., \mathbf{y}_n; \mathbf{u}_1, ..., \mathbf{u}_n)$, and the extended form of the optimization (2.2) becomes,

$$U^* = \operatorname*{E}_{\mathbf{x}_1} \max_{\mathbf{u}_1} \operatorname*{E}_{\mathbf{y}_1|\mathbf{x}_1} ... \operatorname*{E}_{\mathbf{x}_n|\mathbf{x}_1,\mathbf{y}_1,...,\mathbf{x}_{n-1},\mathbf{y}_{n-1}} \max_{\mathbf{u}_n} \operatorname*{E}_{\mathbf{y}_n|\mathbf{x}_1,\mathbf{y}_1,...,\mathbf{y}_{n-1},\mathbf{x}_n} U(\mathbf{y}_1, ..., \mathbf{y}_n; \mathbf{u}_1, ..., \mathbf{u}_n). \tag{2.3}$$

This representation, a stochastic version of the conventional deterministic dynamic programming form, radically confines the search of the sequence of optimal weights. Equations (2.2) and (2.3) essentially state that *stochastic dynamic programming* is implied by the Bayesian paradigm of maximizing expected utility; related results are found in Raiffa and Schlaifer (1961), DeGroot (1970), Zellner (1971), Berger (1985), Bernardo and Smith (1994) and Quintana and Putnam (1994). Soyer and Tanyeri (2001) discuss an interesting simulation approach to solve a two-period Bayesian portfolio selection problem although for the case with no explanatory variables.

A major advantage of this formulation is that it is naturally suited to produce the customary simulations of portfolio (excess) returns with no hindsight. Regrettably, it is still necessary to consider the conditional expectations not only of all succeeding excess returns, but also of all subsequent explanatory variables. This is still a daunting modeling and computation enterprise.

2.1. *Myopic Mean–Variance Portfolio Efficiency*

Considerable research, in an effort to decompose the multiple-period optimization problem into a sequence of isolated single-period problems, focused on identifying terminal wealth utility functions that are myopic. It turns out that the Bernoulli (1738) logarithmic utility function is essentially the only myopic utility defined only in terms of the ending wealth; unfortunately, its single period optimization proves to be a dreadful problem (Hakansson, 1970; Dumas and Jacquillat, 1990). Fortunately, a myopic analysis follows simply by considering time additive utility functions in line with a suggestion by Markowitz (1959). By allowing the utility function to have the following form;

$$U(\mathbf{y}_1, ..., \mathbf{y}_n; \mathbf{u}_1, ..., \mathbf{u}_n) = U_1(\mathbf{y}_1; \mathbf{u}_1) + ... + U_n(\mathbf{y}_n; \mathbf{u}_n), \tag{2.4}$$

the optimization (2.2) is further reduced to,

$$U^* = \mathop{\mathrm{E}}_{\mathbf{x}_1} \max_{\mathbf{u}_1} \mathop{\mathrm{E}}_{\mathbf{y}_1|\mathbf{x}_1} U_1(\mathbf{y}_1; \mathbf{u}_1) + ... + \mathop{\mathrm{E}}_{\mathbf{x}_1,\mathbf{y}_1,...,\mathbf{y}_{n-1},\mathbf{x}_n} \max_{\mathbf{u}_n} \mathop{\mathrm{E}}_{\mathbf{y}_n|\mathbf{x}_1,\mathbf{y}_1,...,\mathbf{y}_{n-1},\mathbf{x}_n} U_n(\mathbf{y}_n; \mathbf{u}_n). \tag{2.5}$$

The multiple-period optimization problem is broken down into several isolated single-period optimizations, and each optimal set of weights can be determined on-line requiring only one-step ahead conditional predictions for the excess returns. The conditional expectations of the explanatory variables are irrelevant for deriving the on-line sequence of optimal portfolio weights, and only the following conditional excess return generating process needs to be specified by a model,

$$p(\mathbf{y}_1 \mid \mathbf{x}_1) p(\mathbf{y}_2 \mid \mathbf{x}_1, \mathbf{y}_1, \mathbf{x}_2) ... p(\mathbf{y}_n \mid \mathbf{x}_1, \mathbf{y}_1, ..., \mathbf{y}_{n-1}, \mathbf{x}_n). \tag{2.6}$$

The use of time-additive utility functions makes the problem more manageable and plausible. Indeed, most investors are concerned with the wealth of their investments all the way through the lives of their programs and do not only care about the final wealth. A reasonable choice for the single-period utility function is the following,

$$U_t(\mathbf{y}_t; \mathbf{u}_t) = p_t - \tfrac{1}{2}\lambda_t (y_t - p_t)^2, \tag{2.7}$$

such that $\mathbf{u}_t = (p_t, \mathbf{w}_t),\ p_t, \lambda_t > 0,\ y_t = \mathbf{y}_t' \mathbf{w}_t,\ t = 1, ..., n$. The overlay portfolio excess return target p_t (an auxiliary decision variable) represents the investor's greed at time t, the investor's fear is a function of the squared, actual overlay portfolio excess return y_t, deviation from the target, and λ_t stands for the greed-fear trade-off investor's tolerance. It can be shown (Quintana, 1992) that the single period optimal allocation weights $\mathbf{w}_t$ corresponding to (2.7) are the industry standard mean-variance efficient weights, in the sense of Markowitz (1987). That is, the optimal weights for (2.3) with the utilities prescribed by (2.4) and (2.7) correspond to the optimal weights in,

$$\max_{\mathbf{w}_t} \left(\left(\mathop{\mathrm{E}}_{\mathbf{y}_t|\mathbf{x}_1,\mathbf{y}_1,...,\mathbf{y}_{t-1},\mathbf{x}_t} y_t \right) - \tfrac{1}{2}\lambda_t \left(\mathop{\mathrm{Var}}_{\mathbf{y}_t|\mathbf{x}_1,\mathbf{y}_1,...,\mathbf{y}_{t-1},\mathbf{x}_t} y_t \right) \right), \tag{2.8}$$

for $t = 1, ..., n$. Although mean–variance efficiency also can be justified, from a Bayesian perspective, directly by means of quadratic utility functions of wealth, the resulting allocations can waste perfect arbitrage opportunities; this drawback does not apply to the solutions of the optimization program (2.8),

$$\mathbf{w}_t^* = \frac{1}{\lambda_t} \left(\mathop{\mathrm{Var}}_{\mathbf{y}_t|\mathbf{x}_1,\mathbf{y}_1,...,\mathbf{y}_{t-1},\mathbf{x}_t} \mathbf{y}_t \right)^{-1} \mathop{\mathrm{E}}_{\mathbf{y}_t|\mathbf{x}_1,\mathbf{y}_1,...,\mathbf{y}_{t-1},\mathbf{x}_t} \mathbf{y}_t. \tag{2.9}$$

An obvious difficulty for applying (2.8) and (2.9) occurs when the predictive distribution of the excess returns lacks the first two moments. Nevertheless, there is the temptation of using the moments of a very large simulated sample from the distribution since the limiting optimal weights (2.9) might still be determined. This poses *Challenge 1: Demonstrate the advantages (or pitfalls) of an indiscriminate use of mean-variance efficient portfolios based on very large distribution samples.*

2.2. *Mean–Variance–Turnover Portfolio Efficiency*

A significant defect of the assumption (2.4) is that it prevents accounting for the portfolio turnover (or transaction costs). A favorite practitioner's solution is to expand the usual two-way optimization (2.8) into a three-way (mean, variance, turnover) Pareto optimization by including a turnover (or transaction costs) penalty. Intuitively, one can wonder if there is a two-period myopic Bayesian argument to rationalize this approach. This brings *Challenge 2: Exhibit, or prove the inexistence of a Bayesian justification for mean–variance–turnover efficient portfolios.* Furthermore, the situation for the general multiple-period problem is similar to that of two-person zero-sum game theory: there is a theoretical solution but it is typically unworkable. However, there are practical approximate solutions, for example, for playing chess. This is *Challenge 3: Explore the merits of a limited tree expansion strategy coupled with a heuristic evaluation for the edging nodes (e.g., the mean–variance–turnover efficient rule).*

3. BAYESIAN DYNAMIC MODELING

Over the last decade and a half Bayesian dynamic (stochastic) models, as the conditional excess return generating process, have been put into action driving financial trading strategies (following the general framework described in Section 2). These statistical models have a common conditional probabilistic structure; Figure 1 depicts this (conditional) probabilistic influence. The system parameters $\boldsymbol{\theta}_t$ evolve randomly in a Markovian fashion, and the observations $\mathbf{y}_t$ (the excess returns) depend, in a probabilistic sense, only through their contemporaneous parameters. The basic premise is that the model structure, described by the values of the system parameters, is changing according to a stochastic process (as opposed to the traditional unlikely static formulations) but there is a degree of persistence making the sampling and filtering process worthwhile.

$$\begin{array}{ccccccc} \rightarrow & \boldsymbol{\theta}_{t-1} & \rightarrow & \boldsymbol{\theta}_t & \rightarrow & \boldsymbol{\theta}_{t+1} & \rightarrow \\ & \downarrow & & \downarrow & & \downarrow & \\ & \mathbf{y}_{t-1} & & \mathbf{y}_t & & \mathbf{y}_{t+1} & \end{array}$$

Figure 1. *Probabilistic influence diagram for Bayesian dynamic models.*

The model specification is defined by the set of stochastic Markovian parametric probability density functions and the set of observational probability density (*i.e.*, likelihood) functions:

$$\text{stochastic p.d.f.} \qquad p(\boldsymbol{\theta}_t \mid \boldsymbol{\theta}_{t-1}), \tag{3.1a}$$

$$\text{observational p.d.f.} \qquad p(\mathbf{y}_t \mid \boldsymbol{\theta}_t). \tag{3.1b}$$

For notational convenience we are omitting all the given available information (*e.g.*, $\mathbf{x}_1, \mathbf{y}_1, \ldots, \mathbf{y}_{t-1}, \mathbf{x}_t$). The initial parameter, without loss of generality, can be set $\boldsymbol{\theta}_0 = \mathbf{0}$. The updating recurrences for the evolution, prediction, and filtering steps are:

$$\text{Evolution:} \qquad p(\boldsymbol{\theta}_t) = \int_{\boldsymbol{\theta}_{t-1}} p(\boldsymbol{\theta}_t \mid \boldsymbol{\theta}_{t-1}) p(\boldsymbol{\theta}_{t-1}) d\boldsymbol{\theta}_{t-1}, \tag{3.2a}$$

Prediction: $$p(\mathbf{y}_t) = \int_{\boldsymbol{\theta}_t} p(\mathbf{y}_t \,|\, \boldsymbol{\theta}_t) p(\boldsymbol{\theta}_t) d\boldsymbol{\theta}_t, \tag{3.2b}$$

Filtering: $$p(\boldsymbol{\theta}_t \,|\, \mathbf{y}_t) = \frac{p(\mathbf{y}_t \,|\, \boldsymbol{\theta}_t) p(\boldsymbol{\theta}_t)}{p(\mathbf{y}_t)}. \tag{3.2c}$$

Although, in principle, this process can be generally performed, it is extremely difficult to implement in practice because, apart from specific cases, the absence of tractable sufficient statistics precludes the existence of close-form transformations to update a finite set of system hyper-parameters. There is an alternative, but equivalent, direct functional specification of the model; this form is particularly suitable to simulate the dynamic process,

stochastic equation: $$\boldsymbol{\theta}_t = g(\boldsymbol{\theta}_{t-1}, \mathbf{w}_t, t), \tag{3.3a}$$

observation equation: $$\mathbf{y} = f(\boldsymbol{\theta}_t, \mathbf{v}_t, t), \tag{3.3b}$$

where $\mathbf{v}_t$ and $\mathbf{w}_t$ denote independently generated random perturbations following fully–specified probability distributions (*i.e.*, independent of any other past, contemporaneous, or future system perturbation, parameter, or observation). These random perturbations can be thought of as the primary source of the system's uncertainty and can be set, if necessary, to be independent uniformly distributed random variables between zero and one (*i.e.*, $\mathbf{w}_t \sim \text{Un}(\mathbf{0}, \mathbf{1})$ and $\mathbf{v}_t \sim \text{Un}(\mathbf{0}, \mathbf{1})$ where $\mathbf{0} = (0, ..., 0)'$ and $\mathbf{1} = (1, ..., 1)'$). (The functions g and f unlike the system perturbation distributions, may generally depend on past observations.)

The representation (3.3) suggests an extended *functional* version of the *probabilistic* influence diagram in Figure 1; this functional influence diagram is shown in Figure 2. Henceforth, we use this representation to specify some Bayesian dynamic models that have been put into action. The reason is twofold: First, the functional specification is clearly unambiguous and provides the means for simulating the process. Second, there are important instances where the probabilistic representation (3.1) is intractable whereas there is a suitable functional form in terms of (3.3) as discussed in Section 3.5. In addition, the continuously compounded (c.c.) excess returns (as opposed to the plain excess returns) generating process, is the one being described ($\mathbf{y}_t$). All these models implicitly allow for variations of market efficiency (*e.g.*, full efficiency, near efficiency, etc.) to be revealed.

$$\begin{array}{ccccccc} & \mathbf{w}_{t-1} & & \mathbf{w}_t & & \mathbf{w}_{t+1} & \\ & \Downarrow & & \Downarrow & & \Downarrow & \\ \rightarrow & \boldsymbol{\theta}_{t-1} & \rightarrow & \boldsymbol{\theta}_t & \rightarrow & \boldsymbol{\theta}_{t+1} & \rightarrow \\ & \downarrow & & \downarrow & & \downarrow & \\ & \mathbf{y}_{t-1} & & \mathbf{y}_t & & \mathbf{y}_{t+1} & \\ & \Uparrow & & \Uparrow & & \Uparrow & \\ & \mathbf{v}_{t-1} & & \mathbf{v}_t & & \mathbf{v}_{t+1} & \end{array}$$

Figure 2. *Functional influence diagram for Bayesian dynamic models.*

3.1. *Stochastic Variance–Covariance Model*

The Stochastic Variance–Covariance Model (SVCM) was at the core of a risk management system to construct currency portfolios with attractive risk-adjusted returns. The aim was to beat a cash deposit by generating excess returns through an efficient overlay portfolio of currency forwards (the future cash flow of forward currency contract is the equivalent to borrowing money denominated in one currency and lending it in another). In the midst of the ERM (Exchange Rate Mechanism), it was reasonable to expect a cash flow equal to the corresponding interest rate differential (e.g, see Quintana and Putnam, 1996). The investment strategy was successfully

implemented to manage client investments, during 1990-1991, at Chase Investors, New York (a former investment management arm of The Chase Manhattan Bank).
The SVCM functional specification is,

$$\text{stochastic equation:} \qquad \Sigma_t = g(\Sigma_{t-1}, \mathbf{w}_t), \qquad \mathbf{w}_t \sim \text{Un}(\mathbf{0}, \mathbf{1}), \tag{3.4a}$$

$$\text{observation equation:} \qquad \mathbf{y}_t = \mathbf{m}_t + C(\Sigma_t)'\mathbf{v}_t, \qquad \mathbf{v}_t \sim \text{N}(\mathbf{0}, \mathbf{I}), \tag{3.4b}$$

where $C(\Sigma_t)$ is the Cholesky's transformation applied to Σ_t (*i.e.*, $C(\Sigma_t)$ is an upper triangular matrix and $\Sigma_t = C(\Sigma_t)'C(\Sigma_t)$). Note that $C(\Sigma_t)'\mathbf{v}_t \sim \text{N}(\mathbf{0}, \Sigma_t)$.

$$\begin{array}{ccccccc}
 & \mathbf{w}_{t-1} & & \mathbf{w}_t & & \mathbf{w}_{t+1} & \\
 & \Downarrow & & \Downarrow & & \Downarrow & \\
\rightarrow & \Sigma_{t-1} & \rightarrow & \Sigma_t & \rightarrow & \Sigma_{t+1} & \rightarrow \\
 & \downarrow & & \downarrow & & \downarrow & \\
 & \mathbf{y}_{t-1} & & \mathbf{y}_t & & \mathbf{y}_{t+1} & \\
 & \Uparrow & & \Uparrow & & \Uparrow & \\
 & \mathbf{v}_{t-1} & & \mathbf{v}_t & & \mathbf{v}_{t+1} &
\end{array}$$

Figure 3. *The SVCM functional influence diagram.*

The SVCM functional influence diagram is shown in Figure 3. The transformation in (3.4a) is complex and requires a special explanation. The goal of its definition is to provide for an inverted-Wishart conjugate evolution for the observation noise ($\mathbf{e}_t = \mathbf{y}_t - \mathbf{m}_t$) variance matrix Σ_t of dimension q. Thus, it is assumed that $\Sigma_{t-1} \sim \text{Wi}^{-1}(\mathbf{S}_{t-1}, d_{t-1})$, following Box and Tiao (1973), where $\text{E}\Sigma_{t-1} = (d_{t-1} - 2)^{-1}\mathbf{S}_{t-1}$. The stochastic equation (3.4a) consists of the following steps:

(a) Obtain $\Psi_{t-1} = (\mathbf{U}_{t-1}^{-1\prime}\Sigma_{t-1}\mathbf{U}_{t-1}^{-1})^{-1}, \mathbf{U}_{t-1} = C(\mathbf{S}_{t-1})$. Note that $\Psi_{t-1} \sim \text{Wi}(d_{t-1}+q-1, \mathbf{I})$ where $\text{E}\Psi_{t-1} = (d_{t-1} + q - 1)\mathbf{I}$ and $\text{E}\Sigma_{t-1}^{-1} = (d_{t-1} + q - 1)\mathbf{S}_{t-1}^{-1}$.

(b) Obtain $\Xi_{t-1} = C(\Psi_{t-1})$. Note that according to Bartlett's decomposition theorem (Muirhead, 1982) all the elements of Ξ_{t-1} are independently distributed, $\xi_{ii(t-1)}^2 \sim \chi^2_{d_{t-1}+q-i}$, and $\xi_{ij(t-1)} \sim \text{N}(0,1)$ for $1 \leq i < j \leq q$ (assuming the χ^2 distribution extension, to fractional degrees of freedom, induced by the Γ distribution).

(c) Obtain Ξ_t whose elements are formed by $\xi_{iit}^2 = \beta_{it}\xi_{ii(t-1)}^2$, $\xi_{ijt} = \xi_{ij(t-1)}$, $\beta_{it} = F^{\#}_{\text{Be}(\delta_{it}\eta_{it},(1-\delta_{it})\eta_{it})}(w_{it})$, where $w_{it} \sim \text{Un}(0,1)$ independently (*i.e.*, $\mathbf{w}_t \sim \text{Un}(\mathbf{0}, \mathbf{1})$), $\delta_{it} = (\delta_t d_{t-1} + q - i)/(d_{t-1} + q - i)$, $\eta_{it} = d_{t-1} + q - i$ and δ_t is a discount factor ($0 < \delta_t < 1$). Here $\text{Be}(\delta_{it}\eta_{it}, (1 - \delta_{it})\eta_{ij})$ denotes the beta distribution and $F^{\#}_{\text{Be}(\delta_{it}\eta_{it},(1-\delta_{it})\eta_{it})}$ denotes the inverse function of its cumulative distribution, so that $\beta_{it} \sim \text{Be}(\delta_{it}\eta_{it}, (1 - \delta_{it})\eta_{ij})$. Note that $\xi_{iit}^2 \sim \chi^2_{\delta_t d_{t-1}+q-i}$.

(d) Obtain $\Psi_t = \Xi_t'\Xi_t$. Note that, according to Bartlett's decomposition theorem, $\Psi_t \sim \text{Wi}(\delta_t d_{t-1} + q - 1, \mathbf{I})$.

(e) Obtain, at last, $\Sigma_t = (\delta_t^{1/2}\mathbf{U}_{t-1})'\Psi_t^{-1}(\delta_t^{1/2}\mathbf{U}_{t-1})$. Note that $\Sigma_t \sim \text{Wi}^{-1}(\mathbf{S}_t^*, d_t^*)$ where $\mathbf{S}_t^* = \delta_t\mathbf{S}_{t-1}$ and $d_t^* = \delta_t d_{t-1}$.

The hyperparameter evolution updating equations in step (e) validate the proposal by Quintana and West (1987) based on discounting information, described as well in West and Harrison (1997), as an authentic Bayesian dynamic model. See also the aforementioned references for the hyper-parameter observation close-form updating equations. The steps (a) and (b) are preparations for applying the key random shock in step (c). Beta random shocks are performed in the step (c), to the diagonal elements of Ξ_{t-1}, to discount the degrees of freedom of the

χ^2 distributions. It makes repeated use of an artifice employed by Shepard (1994) to justify the univariate variance discount model introduced by Harrison and Ameen (1985) as a genuine Bayesian dynamic model. Namely, if $\xi_{t-1}^2 \sim \chi_{d_t}^2$ and $\beta_t \sim \text{Be}(\delta_t d_t, (1-\delta_t)d_t)$ independently, then $\xi_t^2 = \beta_t \xi_{t-1}^2 \sim \chi_{\delta_t d_t}^2$ where $0 < \delta_t < 1$ (see DeGroot, 1970). The steps (d) and (e) reverse the preparations done in the steps (a) and (b); also in the step (e) the result includes a compensation factor δ_t, to discount the shape parameter $\mathbf{S}_{t-1}$, and to preserve a location measure of Σ_{t-1} provided by $\hat{\Sigma}_t = d_t^{-1}\mathbf{S}_t = d_{t-1}^{-1}\mathbf{S}_{t-1} = \hat{\Sigma}_{t-1}$. This estimate $\hat{\Sigma}_t$ becomes, over time, an exponentially weighted average of the past cross errors $\mathbf{e}_t\mathbf{e}_t'$ when the discount factor is set to be constant $\delta_t = \delta$.

Changing slightly the beta shocks and the compensation factor allows for preserving other measures of location of Σ_{t-1}, for example, its mean or its harmonic mean; see Liu (2000) for a definition of a stochastic equation to preserve the latter (although with some limitations) and a generalization to allow for singular Σ. Uhlig (1994) introduced a way to validate variance discounting in multivariate models using singular multivariate Beta distributions. However, a singular multivariate Beta distribution was only defined when both of the parameters were integers; as a result, discounting could only be justified in a very restricted way. In contrast, the stochastic equation (3.4a) allows for continuous flexibility over time.

3.2. *General Multivariate Dynamic Linear Model*

Although univariate instances of the General Multivariate Dynamic Linear Model (GMDLM) of West and Harrison (1997) have been put into action for driving trading strategies, the direct multivariate implementation has presented a serious obstacle. The main problem is the specification of the error variance matrix Σ. This matrix is not only unknown but also, arguably, changing randomly over time. Nevertheless, the GMDLM has been used as a building block for more suitable dynamic models. The GMDLM functional specification is,

$$\text{stochastic equation:} \quad \boldsymbol{\theta}_t = \mathbf{G}_t\boldsymbol{\theta}_{t-1} + C(\mathbf{W}_t)'\mathbf{w}_t, \qquad \mathbf{w}_t \sim \text{N}(\mathbf{0}, \mathbf{I}), \tag{3.5a}$$

$$\text{observation equation:} \quad \mathbf{y}_t = \mathbf{F}_t\boldsymbol{\theta}_t + C(\Sigma_t)'\mathbf{v}_t, \qquad \mathbf{v}_t \sim \text{N}(\mathbf{0}, \mathbf{I}). \tag{3.5b}$$

Figure 4 shows the GMDLM functional influence diagram.

$$\begin{array}{ccccccc}
 & \mathbf{w}_{t-1} & & \mathbf{w}_t & & \mathbf{w}_{t+1} & \\
 & \Downarrow & & \Downarrow & & \Downarrow & \\
\rightarrow & \boldsymbol{\theta}_{t-1} & \rightarrow & \boldsymbol{\theta}_t & \rightarrow & \boldsymbol{\theta}_{t+1} & \rightarrow \\
 & \downarrow & & \downarrow & & \downarrow & \\
 & \mathbf{y}_{t-1} & & \mathbf{y}_t & & \mathbf{y}_{t+1} & \\
 & \Uparrow & & \Uparrow & & \Uparrow & \\
 & \mathbf{v}_{t-1} & & \mathbf{v}_t & & \mathbf{v}_{t+1} &
\end{array}$$

Figure 4. *The GMDLM functional influence diagram.*

The tractable filtering and prediction implementation close-form algorithms for the GMDLM can be found in West and Harrison (1997).

3.3. *Stochastic Multiple-Factor Model*

The Stochastic Multiple-Factor Model (SMFM) was an enhancement of the SVCM introducing stock and government bond futures (a security future contract is essentially equivalent to a theoretical security forward contract) as dependent variables, and several explanatory "factors" as dependent variables (*e.g.*, short-term interest rates, money supply, bond yields, inflation proxies, dividend yields, etc). This model was implemented at The Chase Manhattan Bank, during

1992-1994, to manage proprietary and external client investments. A further improvement was achieved by structuring the appropriate functional equations to accommodate for pooling (shrinkage) techniques a la Zellner *et al.* (1991). This version of the model drove the asset allocation of client investments and funds at Bankers Trust Company during 1995–1997, and subsequently at CDC Investments, New York, during 1997–2002.

The SMFM functional specification is the same as that of the GMDLM with the additional SVCM stochastic equation stating an evolution for the variance matrix of the observation noise. That is, the SMFM stochastic equations are (3.5a) and (3.4a), and its observation equation is (3.5b). Figure 5 shows its functional influence diagram.

$$\begin{array}{ccccccc} & (\mathbf{w}_\theta, \mathbf{w}_\Sigma)_{t-1} & & (\mathbf{w}_\theta, \mathbf{w}_\Sigma)_t & & (\mathbf{w}_\theta, \mathbf{w}_\Sigma)_{t+1} & \\ & \Downarrow & & \Downarrow & & \Downarrow & \\ \rightarrow & (\boldsymbol{\theta}, \Sigma)_{t-1} & \rightarrow & (\boldsymbol{\theta}, \Sigma)_t & \rightarrow & (\boldsymbol{\theta}, \Sigma)_{t+1} & \rightarrow \\ & \downarrow & & \downarrow & & \downarrow & \\ & \mathbf{y}_{t-1} & & \mathbf{y}_t & & \mathbf{y}_{t+1} & \\ & \Uparrow & & \Uparrow & & \Uparrow & \\ & \mathbf{v}_{t-1} & & \mathbf{v}_t & & \mathbf{v}_{t+1} & \end{array}$$

Figure 5. *The SMFM functional influence diagram*

Approximate filtering and prediction implementation of the SMFM has been accomplished via proprietary methods involving running both, the SVCM and the GMDLM, updating algorithms. These are *adhoc* methods that share some similarities with MCMC procedures but take advantage of the structure due to the near efficiency of the markets (*i.e.*, the signal to noise ratio is relatively low for predicting excess returns in a short time period). Note that given Σ_t over time, the conditional model becomes a tractable GMDLM, conversely, given $\boldsymbol{\theta}_t$ over time, the conditional model becomes a tractable SVCM.

3.4. *Bayesian Dynamic Factor Model.*

In recent years, major developments in structured dynamic latent factor models and stochastic volatility have led to the introduction of various approaches to modeling dependencies in volatility processes. Pitt and Shephard (1999) and Aguilar and West (2000) develop and illustrate Multivariate Stochastic Volatility Models (MSVM) based on dynamic factor structure as direct generalizations of univariate stochastic volatility models in line with Kim etal (1998) and references therein.

The key motivating concept underlying interest in MSVM is to describe, explicitly, changes in covariance structure through patterns of time-variation in parameters driving underlying processes. Aguilar and West (2000) illustrate the potential of these models to provide practical improvements in short-term forecasting and portfolio construction in an application of multiple time series of international exchange rates. Aguilar and Lourdes (2001) extend such framework and associated computational methods to include dynamic regressions for the mean in line with Quintana and Putnam (1996) and references therein. This set of generalizations in the mean and the stochastic volatility processes enabled a class of dynamic factor models to be considered for practical implementation in forecasting and portfolio equity allocation and are currently used at Merrill Lynch Investment Managers.

The Bayesian Dynamic Factor Model (BDFM) functional specification, as appeared in Aguilar and Lourdes (2001), is: $\mathbf{y}_t = \mathbf{F}_t\boldsymbol{\theta}_t + \mathbf{v}_t$ where $\mathbf{y}_t$ is a q-vector time series; $\mathbf{F}_t$ is a matrix of econometric variables; $\boldsymbol{\theta}_t$ is a set of time-varying coefficients $\boldsymbol{\theta}_t = \boldsymbol{\theta}_{t-1} + C(\mathrm{W})'\mathbf{w}_{\theta,t}$ and $\mathbf{w}_{\theta,t} \sim \mathrm{N}(\mathbf{0}, \mathbf{I})$; $\mathbf{v}_t = \mathbf{X}\mathbf{f}_t + \epsilon_t$ is a vector with a dynamic factor structure; $\mathbf{f}_t$ are realizations of the k-vector latent factor process with a diagonal variance matrix $\mathbf{H}_t$ changing over time, that is, $\mathbf{f}_t = C(\mathbf{H}_t)'\mathbf{v}_{f,t}$ with $\mathbf{v}_{f,t} \sim N(\mathbf{0}, \mathbf{I})$; ϵ_t are a series-specific quantities with a diagonal matrix Ψ_t, and $\epsilon_t = C(\Psi_t)'\mathbf{v}_{\epsilon,t}$ and $\mathbf{v}_{\epsilon,t} \sim N(\mathbf{0}, \mathbf{I})$. Conditional on all quantities but the latent factor $\mathbf{f}_t$, the variance matrix for $\mathbf{v}_t$ can be decomposed as, $\Sigma_t = \mathbf{X}\mathbf{H}_t\mathbf{X}' + \Psi_t$ with possibly time-varying components.

Multivariate extensions of standard stochastic volatility models with dynamic factor structure were introduced and developed by Pitt and Shephard (1999) and Aguilar and West (2000) and extended and applied in Aguilar and Lourdes (2001). For instance, take $\mathbf{H}_t$ as diagonal with elements $\exp(\lambda_{jt})$ where λ_{jt} is the log of the instantaneous variance of the j-th latent factor at time t and write $\boldsymbol{\lambda}_t$ for the vector of the λ_{jt}. The model adopts a vector autoregression of order one for $\boldsymbol{\lambda}_t$, $\boldsymbol{\lambda}_t = \boldsymbol{\mu}_t + \Phi(\boldsymbol{\lambda}_{t-1} - \boldsymbol{\mu}_{t-1}) + \boldsymbol{\omega}_t$, with $\boldsymbol{\omega}_t = C(\mathbf{U})'\mathbf{w}_{\omega,t}$, $\boldsymbol{\mu}_t = \boldsymbol{\mu}_{t-1} + \boldsymbol{\zeta}_t$, $\boldsymbol{\zeta}_t = C(\mathbf{V})'\mathbf{w}_{\zeta,t}$ and $\mathbf{w}_{\omega,t} \sim N(\mathbf{0}, \mathbf{I})$, $\mathbf{w}_{\zeta,t} \sim N(\mathbf{0}, \mathbf{I})$. Note that if $\boldsymbol{\mu}_t = \mu_t\mathbf{1}$, μ_t would represent the equilibrium volatility level or global risk parameter interpreted as the risk associated to shocks outside the system that affect all factors and all series. Moreover, this structure explicitly allows contemporaneous dependencies between the innovations impacting the volatilities across factors through the general variance matrix $\mathbf{U}$. For the idiosyncratic variances, define $\eta_{jt} = \log(\psi_{jt})$ for each series j and write $\boldsymbol{\eta}_t$ for the vector of the η_{jt}. The model assumes standard univariate autoregressions of order one, namely, $\eta_{jt} = \alpha_j + \rho_j(\eta_{j,t-1} - \alpha_j) + \xi_{jt}$ with independent innovations ξ_{jt} with variance τ_j where $\xi_{jt} = \sqrt{\tau_j}\mathbf{w}_{\xi,t}$ and $\mathbf{w}_{\xi,t} \sim N(0, 1)$. Unlike the factor volatilities, the ξ_{jt} processes are mutually independent series; write $\boldsymbol{\xi}_t$ for the vector of the ξ_{jt}. Figure 6 shows the functional influence diagram for the BDFM.

$$
\begin{array}{ccccccc}
 & (\mathbf{w}_\theta, \mathbf{w}_\omega, \mathbf{w}_\zeta, \mathbf{w}_\xi)_{t-1} & & (\mathbf{w}_\theta, \mathbf{w}_\omega, \mathbf{w}_\zeta, \mathbf{w}_\xi)_{t} & & (\mathbf{w}_\theta, \mathbf{w}_\omega, \mathbf{w}_\zeta, \mathbf{w}_\xi)_{t+1} & \\
 & \Downarrow & & \Downarrow & & \Downarrow & \\
\rightarrow & (\boldsymbol{\theta}, \boldsymbol{\lambda}, \boldsymbol{\mu}, \boldsymbol{\eta})_{t-1} & \rightarrow & (\boldsymbol{\theta}, \boldsymbol{\lambda}, \boldsymbol{\mu}, \boldsymbol{\eta})_{t} & \rightarrow & (\boldsymbol{\theta}, \boldsymbol{\lambda}, \boldsymbol{\mu}, \boldsymbol{\eta})_{t+1} & \rightarrow \\
 & \downarrow & & \downarrow & & \downarrow & \\
 & \mathbf{y}_{t-1} & & \mathbf{y}_{t} & & \mathbf{y}_{t+1} & \\
 & \Uparrow & & \Uparrow & & \Uparrow & \\
 & (\mathbf{v}_f, \mathbf{v}_\epsilon)_{t-1} & & (\mathbf{v}_f, \mathbf{v}_\epsilon)_{t} & & (\mathbf{v}_f, \mathbf{v}_\epsilon)_{t+1} &
\end{array}
$$

Figure 6. *Functional influence diagram for BDFM.*

Model implementation requires prior distributions for the full set of parameters and the development of numerical methods. A set of standard reference priors is assumed for most model parameters, except for those subject to specific restrictions such as the autoregressive parameters on the stochastic volatility processes, and some of the loadings matrix elements. Moreover, informative proper prior distributions are assumed for the innovation variances $\mathbf{U}$ and τ_j to separate the sources of variability in the data that are confounded in the model as discussed in Aguilar and West (2000). Customized MCMC methods of model fitting and computation are described in Aguilar and Lourdes (2001) building on previous work on stochastic volatility models and Bayesian dynamic factor models as reviewed in Aguilar and West (2000). These models are likely to increase the accuracy in short-term forecasts of variance–covariance patterns in Σ_t in (3.4a) that are critical in portfolio equity allocation decisions, as discussed below. Furthermore, underlying factor processes can be identified as the main drivers for the structural changes in volatility and hence increase the opportunities for informed interventions based on econometric interpretations of such latent processes.

The latest application of the Bayesian Dynamic Factor Model (BDFM), illustrated in Aguilar and Lourdes (2001), is motivated by the recent focus on assessing a strategy aimed at optimally shifting equity exposure among various industrial sectors building on previous work by Beller, Kling and Levinson (1998). The models exploit the return predictability of the MSCI/S&P

industry composition of the S&P 500 index and their contribution to the equity market performance. This work illustrates findings in applying MSVM to different industry rotation strategies, including aspects of model performance in dynamic portfolio equity allocation that play a key role in the investment process of a dozen mutual funds managed at Merrill Lynch Investment Managers. In addition, the resulting industry forecast yields a natural style and size rotation strategies between growth, value, small and large investments. More importantly, the implications from those findings when extended to global industries are key to understanding the importance of country and industry effects in the practice of global equity investing. That is, by extending the analysis in a global industry rotation, one would simply refocus research efforts on industry effects and, in turn, portfolio construction to take this factor into account. The recommendation is that investors, instead of parsing the globe by political boundaries, adopt a different strategy to gain a different perspective and slice the global market place into distinct industrial sectors or a combination of both approaches. A global industry rotation product would seek excess returns by opportunistically allocating funds among global industries based on the aforementioned multivariate stochastic volatility models. In other words, instead of defaulting into global equity industrial allocation by focusing on country equity allocation, investors would optimize industrial equity allocation directly taking into account country effects in the modeling process.

3.5. *Bayesian Functionally Specified Stochastic Models.*

Although, both representations (3.1) and (3.3) are theoretically equivalent, there are important cases (at least in the field of investment management) where the associated probabilistic representation is virtually intractable (*i.e.*, there is no known close-form formulation) but there are sensible (computable) functional representations. Some examples follow (in all of them there is the assumption that the underlying (c.c.) excess returns generating process follows a multivariate Brownian motion model or the variants considered in the previous subsections):

1. Consider modeling the behavior of the payoffs of global stocks or bond future contracts. There is a well-known difficulty due to the asynchronous (or missing) nature of the observations (*i.e.*, a number of markets might open when the other markets have already been closed).
2. Consider the case when only the following specific partial information is available: open, high, low, and close prices of the markets sessions.
3. Consider the common case of modeling futures of indices taking into account their underlying structure.
4. Consider the option approach to model stocks, bonds, and mortgages (or any other kind of instruments with embedded options).

Challenge 4: Devise broad methods for Bayesian filtering and prediction for functionally specified stochastic models.

The following outlines an, admittedly, tentative effort of a scheme, inspired by particle filtering methods (Kitagawa, 1996; Pitt and Shephard, 1999) to face the above challenge:

(a) The process starts by sampling "particles" of $\boldsymbol{\theta}_{t-1}$ from an approximation of its posterior density, a mixture with a common elliptical kernel (*e.g.*, with a multivariate normal kernel or with a multivariate t kernel).

(b) The particles are projected one period ahead by applying the functional stochastic equation (3.3a) resulting in particles of $\boldsymbol{\theta}_t$.

(c) Corresponding particles for $\mathbf{y}_t$ are generated by applying the observation equation (3.3b).

(d) The joint distribution of these combined particles $(\boldsymbol{\theta}_t, \mathbf{y}_t)$ is approximated by a continuous mixture with a common elliptical kernel, preserving the first two moments, via a shrink-plus-stretch mixing procedure.
(e) An (approximate) marginal predictive density corresponding to $\mathbf{y}_t$ is readily available as the matching marginal from the joint mixture (which is another mixture of the same kind). Similar comments apply to the prior density of $\boldsymbol{\theta}_t$.
(f) Given the observation $\mathbf{y}_t$, a (approximate) posterior density for $(\boldsymbol{\theta}_t \mid \mathbf{y}_t)$ is readily available as the matching conditional from the joint mixture. This also turns out to be a mixture with a common elliptical kernel and the process can go forward to the next period.

The shrink-plus-stretch mixing step procedure (d) can be described, using the normal kernel as an example, via the following transformation: $\varphi_t \to \varphi_t^+ = \boldsymbol{\eta}_t + \xi_t$ where $\varphi_t = (\boldsymbol{\theta}_t, \mathbf{y}_t)$, $\varphi_t^+ = (\boldsymbol{\theta}_t^+, \mathbf{y}_t^+)$, $\boldsymbol{\eta}_t = \mathbf{m}_t + \delta_t(\varphi_t - \mathbf{m}_t)$, $\xi_t = \sqrt{1-\delta_t^2}C(\mathbf{C}_t)'\mathbf{u}_t$, $\mathbf{u}_t \sim \mathrm{N}(\mathbf{0}, \mathbf{I})$, $\mathrm{E}_{\varphi_t} \varphi_t = \mathbf{m}_t$, $\mathrm{Var}_{\varphi_t} \varphi_t = \mathbf{C}_t$, and $0 < \delta_t < 1$ is a mixing control parameter. Note that $\mathrm{E}_{\varphi_t^+} \varphi_t^+ = \mathbf{m}_t$ and $\mathrm{Var}_{\varphi_t^+} \varphi_t^+ = \mathbf{C}_t$ as well. Furthermore, when $\delta_t \to 1$ then, the shrink term $\boldsymbol{\eta}_t \to \varphi_t$, the stretch term (*i.e.*, the common kernel) $\xi_t \to$ a (normal) zero centered Dirac delta function, and $\mathrm{E}_{\varphi_t^+} h(\varphi_t^+) \to \mathrm{E}_{\varphi_t} h(\varphi_t)$ for any continuous function h.

Therefore, when φ_t is approximated by a cloud of particles (*i.e.*, when φ_t is assumed to be a finite discrete random variable taking the value of a particle $\varphi_t = \tilde{\varphi}_t$ with the corresponding probability $p(\tilde{\varphi}_t)$), φ_t^+ is distributed as the required approximating mixture, $p(\varphi_t^+) = \sum_{\varphi_t} p(\varphi_t^+ \mid \varphi_t) p(\varphi_t)$, and $\varphi_t^+ \mid \varphi_t \sim \mathrm{N}(\boldsymbol{\eta}_t, (1-\delta_t^2)\mathbf{C}_t)$ where the sum is over the set of particles of φ_t. The approximate marginal predictive density, in step (e), is, $p(\mathbf{y}_t^+) = \sum_{\varphi_t} p(\mathbf{y}_t^+ \mid \varphi_t) p(\varphi_t)$, and $\mathbf{y}_t^+ \mid \varphi_t$ is the corresponding marginal of $\varphi_t^+ \mid \varphi_t$. Similarly, the approximate posterior density, in step (f), is,

$$p(\boldsymbol{\theta}_t^+ \mid \mathbf{y}_t^+) = \sum\nolimits_{\varphi_t} p(\boldsymbol{\theta}_t^+ \mid \mathbf{y}_t^+, \varphi_t)\, p(\varphi_t \mid \mathbf{y}_t^+),$$

where $\boldsymbol{\theta}_t^+ \mid \mathbf{y}_t^+, \varphi_t$ is the corresponding conditional of $\varphi_t^+ \mid \varphi_t$ and

$$p(\varphi_t \mid \mathbf{y}_t^+) = \frac{p(\mathbf{y}_t^+ \mid \varphi_t)}{p(\mathbf{y}_t^+)}\, p(\varphi_t).$$

Preliminary testing results for the univariate SVCM are encouraging but it is very likely that the implementation of multivariate non-normal, non-linear dynamic models will pose serious difficulties. Even deterministic univariate evolutions can exhibit the fallacy that continuous random variables can be *effectively* simulated using digital computers. For example, the infamous tent function from non-linear dynamics, defined by the lines from (0,0) to (0.5,1) to (1,0), when it is applied repeatedly illustrates the problem. Indeed, if $\theta_{t-1} \sim \mathrm{Un}(0,1)$ and g is set as the tent function applied repeatedly any number of times, then theoretically $\theta_t = g(\theta_{t-1}) \sim \mathrm{Un}(0,1)$. However, a simulated particle cloud will converge to zero if the tent function is applied a sufficient number of times (*e.g.*, 64 times) as a consequence of the finite binary (*e.g.*, double precision) representation of "real" numbers, showing that the least significant bits can become all but insignificant.

It is interesting to note that the tent function is more than just a chaos theory artifice used to exhibit a pitfall. First, it is directly connected to the payoff of a combination of options, referred to as a "long *butterfly* spread" (Kolb; 2000). Second, it is a uniform randomizer in the sense that the uniform distribution $\mathrm{Un}(0,1)$ is not only a fixed point under the aforementioned map g but it is also an "attractor". Third, it is intrinsically related to a device often referred as

"Bayes' Billiard Table." Was the Reverend also a chaos theory frontrunner? Did the table have side cushions in his mind?

4. CONCLUSIONS

There is still a spectrum of opinions whether the overall trading activity in financial securities, in general, and derivative instruments, in particular, produce a real global economic benefit. Yet, there is a consensus that making excess returns in the global financial markets is not an easy task. Over the years, there have been suggestions for effectively using Bayesian methods in finance and econometric applications. In this paper, we reviewed the practical advantages and potential difficulties of applying Bayesian rationale to the investment management process. The sequential Bayesian approach for making decisions under uncertainty is *inherently* well suited for the investment management enterprise: Form a probabilistic predictive view of the markets and position the portfolio accordingly to take the best advantage, refresh repeatedly the view and the portfolio positioning, as developments occur and new information arrives. Furthermore, Bayesian dynamic modeling seems to represent adequately the ever-changing random structure of the markets; indeed, dynamic Bayesian formulations have produced consistent predictive distributions and improved portfolio performance. Although the Bayesian/non-Bayesian controversy, regarding models with unknown static parameters, might continue, when models with stochastic parameters are entertained the Bayesian way is unassailable.

For more than a decade, Bayesian ideas have been the building blocks of the investment process of multi-billion asset allocation management at major financial institutions such as Chase, Bankers Trust, CDC Investments, Deutsche Bank, Merrill Lynch Investment Managers, and The Nikko Securities Co. International, Inc. Bayesian models have been and are currently applied to global asset allocation, stock selection, industry, style and size rotation for a variety of asset classes and implemented via hedge funds, mutual funds and overlay strategies among others.

Additional formal research and empirical assessments of Bayesian methodology are relentlessly evaluated and new challenges are currently under investigation. Our experience to date leads us to believe that this kind of research will be fruitful and will support the outcomes in the last decade regarding the practical utility of Bayesian paradigm in the definitive (monetary) gambling game, and it is likely that further Bayesians will enter the field. This is our *Ultimate challenge to resolve: Can Bayesians consistently beat the financial markets?* ***You*** *bet!*

ACKNOWLEDGEMENTS

The authors acknowledge Mike West and Arnold Zellner for useful discussions that directly and indirectly have contributed to the development of this article and the Bayesian investment management accomplishments mentioned herein. We are also grateful to Diana Wyant for invaluable assistance in the completion of this paper.

REFERENCES

Aguilar, O. and West, M. (2000). Bayesian dynamic factor models and portfolio allocation. *J. Bus. Econom. Statist.* **18**, 338–357.

Aguilar, O. and Lourdes, V. (2001). Bayesian analysis of investment rotation strategies (submitted).

Bachelier (1900). *Theory of Speculation*. Annales de l'Ecole normale supérieure. Translated from French in *The Random Character of Stock Market Prices*. (P. H. Cootner, ed.). Cambridge, MA: MIT Press, 1964.

Bayes, T. (1763). An essay towards solving a problem in the doctrine of chances. Published posthumously in *Phil. Trans. Roy. Soc. London* **53**, 370–418 and **54**, 296–325. Reprinted in *Biometrika* **45** (1958), 293–315, with a biographical note by G. A. Barnard.

Beller, K. R., Kling, J. L. and Levinson, M. (1998). Are industry stock returns predictable? *Fin. Anal. J.*, **54**, 42–57.

Berger, J. O. (1985). *Statistical Decision Theory and Bayesian Analysis*, 2nd ed. New York: Springer.

Bernoulli, D. (1738). Specimen Theoriae Novae de Mensura Sortis (Exposition of a New Theory on the Measurement of Risk). Translated from the Latin by Louise Sommer in *Econometrica* **22**, 1954, 23–36.

Bernardo, J. M. and Smith, A. F. M. (1994). *Bayesian Theory*. Chichester: Wiley.

Bernstein, P. L. (1996). *Against the Gods, The Remarkable Story of Risk*. New York: Wiley.

Box, G. E. P. and Tiao, G. C. (1973). *Bayesian Inference in Statistical Analysis*. Reading, MA: Addison-Wesley.

DeGroot, M. H. (1970). *Optimal Statistical Decisions*. New York: McGraw-Hill.

de Finetti, B. (1974). *Theory of Probability*. Chichester: Wiley.

Dumas, B. and Jacquillat, B. (1990). Performances of currency portfolio chosen by a Bayesian technique: 1967–1985. *J. Banking Fin.* **14**, 539–558.

Geisser, S. (1988). The future of statistics in retrospect. *Bayesian Statistics 3* (J. M. Bernardo, M. H. DeGroot, D. V. Lindley and A. F. M. Smith, eds). Oxford: Oxford University Press, 147–158 (with discussion).

Hakansson, N. H. (1970). Optimal investment and consumption strategies under risk for a class of utility functions. *Econometrica* **38**, 587–607.

Harrison, P. J. and Ameen, J. R. M. (1985). Normal discount Bayesian models, *Bayesian Statistics 2* (J. M. Bernardo, M. H. DeGroot, D. V. Lindley and A. F. M. Smith, eds), Amsterdam: North-Holland, 271–298 (with discussion).

Kim, S., Shephard, N. and Chib, S. (1998). Stochastic volatility: likelihood Inference and comparison with ARCH Models. *Rev. Economic Studies* **65**, 361–393.

Kitagawa, G. (1996). Monte Carlo filter and smoother for non-Gaussian non-linear state space models. *J. Comput. Graph. Statist.* **5**, 1–25.

Kolb, R. W. (2000). *Futures, Options and Swaps*. Oxford: Blackwell.

Liu, J. (2000). *Bayesian Time Series: Analysis Methods Using Simulation Based Computation*. Ph.D. Thesis, ISDS, Duke University, USA.

Markowitz, H. M. (1959). *Portfolio Selection: Efficient Diversification of Investments*, 2nd ed. Oxford: Blackwell.

Markowitz, H. M. (1987). *Mean-Variance Analysis in Portfolio Choice and Capital Markets*. Oxford: Blackwell.

Muirhead, R. J. (1982). *Aspects of Multivariate Statistical Decision Theory*. New York: Wiley.

Pitt, M. K and Shephard, N. (1999). Filtering via simulation: Auxiliary particle filters. *J. Am. Statist. Ass.* **94**, 590–599.

Quintana, J. M. (1992). Optimal portfolios of forward currency contracts. *Bayesian Statistics 4* (J. M. Bernardo, J. O. Berger, A. P. Dawid and A. F. M. Smith, eds). Oxford: Oxford University Press, 753–762.

Quintana, J. M. (2000). Bayesian “betting” worldwide. *ISBA Bulletin* **7**

Quintana, J. M. and Putnam, B. H. (1994). Driving on a Hazardous Road with Limited Visibility. Unpublished presentation at the *5th Valencia International Meeting on Bayesian Statistics*, Alicante, Spain.

Quintana, J. M. and Putnam, B. H. (1996). Debating currency markets efficiency using dynamic multiple-factor models. *Proc. Bayesian Statistical Science*. Alexandria: American Statistical Association.

Quintana, J. M., Putnam B. H. and Wilford, D. S. (1998). Mutual and pension funds management: Beating the markets using a global Bayesian investment strategy. *Proc. Bayesian Statistical Science*. Alexandria: American Statistical Association.

Quintana, J. M. and West, M. (1987). Multivariate time series analysis: New techniques applied to international exchange rate data. *The Statistician* **36**, 275–281.

Raiffa, H. and Schlaifer, R. (1961). *Applied Statistical Decision Theory*. Harvard, MA: University Press.

Shephard, N. G. (1994). Local scale models: State space alternative to integrated GARCH processes. *J. Econometrics* **60**, 181–202.

Soyer, R. and Tanyeri, K. (2001). Bayesian portfolio selection in random variance models. *Bayesian Methods with Applications to Science, Policy and Official Statistics* (E. George, ed). Brussels: European Community, 527–535.

Stigler, S. M. (1986). *The History of Statistics: The Measurement of Uncertainty Before 1900*. Harvard, MA: University Press.

Uhlig, H. (1994). On singular Wishart and singular multivariate Beta distribution. *Ann. Statist.* **22**, 395–405.

West, M. and Harrison, P. J. (1997). *Bayesian Forecasting and Dynamic Models*, 2nd ed. New York: Springer.

Zellner, A. (1971). *An Introduction to Bayesian Inference in Econometrics*. New York: Wiley.

Zellner, A. (1991). Bayesian methods and entropy in economics and econometrics. *Maximum Entropy and Bayesian Methods* (W. T. Grandy and L. H. Schick, eds). Netherlands: Kluwer, 17–31.

Zellner, A. and Chetty, V. K. (1965). Prediction and decision problems in regression models from the Bayesian point of view. *J. Am. Statist. Ass.* **60**, 608-616.

Zellner, A., Hong, C. and Min, C. (1991). Forecasting turning points in international output growth rates using Bayesian exponentially weighted autoregression, time-varying parameter, and pooling techniques. *J. Econometrics* **49**, 275–304.

DISCUSSION

DALE J. POIRIER (*University of California-Irvine, USA*)

I wish to congratulate the authors for a thought-provoking narrative discussing their use of Bayesian procedures involving latent structure dynamic Bayesian modeling to address pragmatic issues in empirical finance. This paper has two main sections. Section 2 involves a dynamic stochastic programming model of an optimal asset portfolio. It presents a tractable formulation leading to allocation weights (2.9). Unfortunately, many strong assumptions are made (*e.g.*, quadratic utility, time separability, existence of moments, and zero transactions costs) to motivate the predictive moments it contains. Challenges 1–3 in part acknowledge these limitations. Section 3, where the main contribution of the paper lies, discusses four models described by the acronyms: SVCM, GMDLM, SMFM, and BDFM. I find the latter, Bayesian Dynamic Factor Models, the most interesting and the most challenging. The authors have "teased" us with an overview discussion in which many implementation details are omitted due to proprietary concerns. In some ways this invited paper fits awkwardly in an academic conference whose audience comprises cash-poor academics rather the cash-paying clients for whom the models are intended. Nonetheless, I wish to raise some questions that I hope can be answered in this public forum.

1. What diagnostic checks do you use to monitor model performance? Specifically, does the monetary diagnostic of whether you are making money supersede statistical diagnostics? Do model improvements arising out of statistical diagnostic checking lead to making more money?
2. How did your models react to the terrorist shock of 9/11? Did they quickly adapt to the increased volatility?
3. Your models are "parameter-rich" with many new parameters arriving with each new observation (see Figure 6). Clearly, priors matter quite a bit in such cases. You note the joint use of subjective informative, reference, and data-dependent priors. Such combinations are fairly rare. What guidelines (generic or application specific) did you follow in choosing your priors?

BRADLEY P. CARLIN (*University of Minnesota, USA*)

Congratulations to Dr. Quintana on a careful and thorough paper in a difficult field! I would merely like to point out some connections between the author's use of backward induction in Section 2 and some recent developments along these lines in the Bayesian clinical trial monitoring literature with which I am familiar. In the latter setting, it has long been recognized that Bayesian approaches offer the philosophically soundest solution, but the arduousness of both the computing and (especially) the bookkeeping inherent in backward induction rendered them essentially infeasible. Recently, however, Carlin *et al.* (1998) introduced a "forward" sampling algorithm that is able to control the book-keeping explosion, since its complexity is linear (not exponential) in the number of backward steps. However, this method (and its extension by Kadane and Vlachos, 2002) is limited to rather simple exponential family models.

Fortunately, Brockwell and Kadane (2002) have very recently provided a general method that uses a grid approximation to the expected loss at each decision time, working on a discretized version of the space of sufficient statistics. While the resulting algorithm is quite numerically intensive (generally requiring a network of parallel processors), it apparently offers complete generality regarding choice of model and utility function. I wonder if one of these approaches might free us from the "myopic" and time-additive assumptions employed by the author and others working in the financial modeling arena.

REPLY TO THE DISCUSSION

We wish to thank the discussants for their contributed questions, suggestions, comments, and criticisms.

Dale J. Poirier. Professor Poirier raised several interesting issues. His first concern is the potential model's strong assumptions in the stochastic programming yielding to optimal portfolios. We disagree with the criticism oriented to the quadratic utility (2.7) although we would agree if it were directed towards the conventional quadratic utility $U_t(\mathbf{y}_t; \mathbf{w}_t) = (y_t - p_t)^2$, where the target p_t is given. Although the utility (2.7) is not a panacea, it accommodates the fear-greed investor's dilemma, and yields a form of mean–variance optimization, when transaction costs concerns are ignored, that most investors find adequate (we have yet to meet an investor who would ask for a different one, in the absence of transaction costs). In the presence of transactions costs, the time additive utility functions assumption does not hold as discussed in Section 2. However, Professor Carlin has suggested a potential way out from an intractable cumbersome dynamic programming undertaking, which could improve on the conventional, yet unfulfilling, three-way Pareto optimization. Finally, the existence of predictive moments depends on the choice of the Bayesian dynamic models that generate the predictive distributions and not on the choice of utility functions. On the one hand, as the complexity of the models increases, the question of existence (or lack of existence) of moments becomes more difficult to answer, hence our preoccupation with *Challenge 1*. On the other hand, a cynical, but to some extent genuine attitude, is that continuous random magnitudes (*e.g.*, financial payoffs) are ideal approximations of actual random magnitudes that take only a finite number of values (whose moments existence is not in question). In the end, nobody is observing, storing, or processing an infinite number of bits.

Embedded in Professor Poirier's discussions is the comment of proprietary concerns and lack of transparency. Professor Poirier states that the authors have "teased" the spectators with a deliberately incomplete paper fitting awkwardly in an academic conference. We concede that, even if we had had enough space, we could not have disclosed, due to proprietary concerns, all the implementation information necessary to reproduce results (an academic ideal that is seldom accomplished). Yet, we largely disclosed the description of the Bayesian strategies and the Bayesian dynamic models in particular. Several years ago, during a presentation at a National Bureau of Economic Research (NBER) conference held at Duke University, Arnold Zellner suggested to the first author a potential forecasting improvement by means of utilizing information-pooling techniques. We feel obliged to report (in this and other academic public forums) the actual performance of this particular technique and the Bayesian paradigm in general, within a challenging environment. We believe that both, academia and the financial industry, benefit by this kind of interaction, regardless of their socio-economic status.

Further points in the discussion involve three sets of direct questions on several topics. A first set of key questions relates to diagnostic checks.

What diagnostic checks do you use to monitor model performance? We use minimal direct statistical diagnostic checks: *ex post* bias means absolute deviation and standard error. Specifically, does the monetary diagnostic of whether you are making money supersede statistical diagnostics? Yes. Do model improvements arising out of statistical diagnostic checking lead to making more money? The question is non-applicable. We do not make model improvements arising from direct statistical diagnostic checking.

The ultimate diagnostic of a trading strategy is its ability to make real excess returns in real time. The problem is how to choose strategies (and statistical models) in advance. On the one hand, one can calibrate, without major difficulties, a statistical model backwards using mean variance optimization systems. On the other hand, the success of a trading strategy depends on the forward performance of the statistical predictive model. We attempt to minimize the hindsight effect by considering only strategy improvements based on strong financial-economic, optimization, and statistical theoretical reasons. We use financial simulations of the performance of the strategies backwards to confirm improvements, when this process is carefully executed there are few major surprises, and the forward performance is in line with the backward (simulated) performance.

We are not overly concerned about using the *ex post* utility as the foremost diagnostic device. First, although the utility used is not a *proper scoring rule*, the optimal (expected) utility associated to a "wrong" model cannot exceed the optimal (expected) utility associated with "the true" model ("true" and "wrong" according to the analyst). Second, even if different forecasting models could produce the same optimal allocations (not to mention different dynamic models that could produce the same forecasts) hence the same ex-post utility, the statistical "adequacy" of the models would be, by definition, irrelevant from a utilitarian viewpoint.

A non-trivial implementation of the Bayesian paradigm virtually implies statistical modeling compromises. For example, Monte Carlo implementation methods typically utilize digital-computer generated *pseudo*-uniform random deviates. These methods are extremely useful while potentially treacherous, as noted in Section 3.5. Even overlooking the compromise due to fixed-precision computations, another issue still remains. It can be argued that the "probability" that binary digits of a uniform random deviate can be digitally computed, even in principle, is zero. Hence, the *pseudo* prefix is well deserved; there are too many (a continuum of) real numbers between zero and one and too few (countable) digital computer programs.

In our opinion, the key is to have a disciplined approach to minimize the hindsight effect, to be aware of potential modeling pitfalls, and to continuously monitor the *ex post* utility, rather than to try to avoid the use of a "wrong" model. We have, admittedly, a pragmatic viewpoint, as Box (1976) noted, "all models are wrong, but some are useful." We relentlessly but carefully search for the most useful Bayesian strategies. After all, investors would not give back profits generated by a useful strategy, merely because the underlying models were eventually found later to be "wrong" (or in particular, not "genuinely" Bayesian).

A second set of questions relates to the issue of model reaction to sudden external fierce shocks such as the tragic events of September 11th. In general, we have to acknowledge that a common feature of virtually all the models used in financial applications is their inability to anticipate the price of movements due to this kind of external shocks (*e.g.*, political events, war, and fraud amongst others). All these external shocks usually produce sharp market movements that are practically impossible to anticipate or hedge against using standard statistical tools. In practice, to avoid substantial losses in such cases, defensive mechanisms, such as reduction in position sizes, de-leverage are applied before rerunning or updating the models. In the particular case of September 2001, the number of models that were running at the time were affected by the lack of trading activity for a week on most capital markets and risk management mechanisms were immediately applied and positions were drastically reduced by the time markets reopened. From a model point of view, the level of observed and predicted volatility increased dramatically in the weeks following the tragic events yielding to more conservative portfolios.

A final set of questions relates to the number of parameters in the models and the choice of priors. First, regarding the issue of the "arrival" of new parameters with each new observation. We believe that this can be addressed pointing to standard Dynamic Linear Models (DLMs) theory (West and Harrison, 1997). As it is the case here, with each new observation, the posterior and predictive distribution of the model parameters are updated according to the evolution and observation equations without affecting the number of parameters in each model. In fact in all four models presented in the paper, the number of parameters is fixed and known. On a related note, as we acknowledge that the models are not necessarily parsimonious by nature, we tried to address the issues of model parameterization, parameter redundancy, and dimension reduction in the fourth model from Section 3.4 where we discuss a general class of Bayesian Dynamic Factor models. Finally, regarding priors, we chose initial vague priors for the strictly dynamic parameters whose forecasting influence wanes overtime anyway. In addition, we assume prior distributions that are either standard reference priors or proper priors that are relatively diffuse for some static parameters.

Bradley P. Carlin. Professor Carlin contributes some insightful comments on the potential connections between financial applications and recent developments in Bayesian clinical trials. In particular, he mentions the general methodology proposed by Brockwell and Kadane (2002) (henceforth BK2002) as a possible alternative to the backward induction algorithm (stochastic dynamic programming) presented in Section 2. The idea involves a smart algorithm to perform, approximately, optimal decision in Bayesian sequential decision problems where the computation time is linear in the number of stages. Their approximation goes beyond the exponential family models solutions introduced by Carlin *et al.* (1998) and Kadane and Vlachos (2002).

BK2002 provides a method that consists of a grid approximation to the expected utility at each decision time based on a discrete-space version of sufficient (or even near sufficient) statistics. As pointed out by the authors, their algorithm, together with the one proposed by Berry *et al.* (2000), are the only well-documented "general-purpose" algorithms for solving complicated sequential analysis problems, which do not suffer from exponentially growing computational times. Although these methodologies represent potentially superior substitutes for the myopic utility functions used in financial decision-making, we believe there are several key assumptions that could complicate the implementation in applied multiple financial time series.

First, in BK2002 the utility is a function of unknown but fixed parameters conditional on the observations and their associated cost. In our investment management context the utility is typically a function of the future excess returns (observations) and turnover (or transaction costs), driven by a model with underlying stochastic parameters. Nevertheless, it appears that their methods could be applied through the corresponding mappings since the concept of (near) sufficiency can be extended from static parameters, to dynamic parameters, to future observations.

Second, the backward induction model in BK2002 assumes a finite decision set. In portfolio management, the multiple-period goal is to maximize the utility of the wealth of the investor at the end of n periods by sequentially choosing the decision variables: $\mathbf{w}_t$, $t = 0, ..., n-1$ (this is already a simplification since most investments programs do not have an explicit termination date). In our context, the vector $\mathbf{w}_t = (w_t^1, ..., w_t^K)$ represents the allocations associated to each of the K risky securities for $t = 0, ..., n-1$. Therefore, the optimal allocation decisions are time-varying and are, in principle, continuous in nature. Furthermore, the possibility of having a terminal decision (*i.e.*, stop investing in risky assets) before time $t = n$ is not explicit, but it can be implicitly achieved by setting the weights (of the risky securities) to zero at a certain point in time and thereafter.

Third, the approximation in BK2002 relies on the construction of a grid defined on the space of the summary information statistic for the model parameters. In our financial applications, predictive distributions are derived from multivariate Bayesian dynamic models on a high-dimensional parameter space. Consequently, tractable (near) sufficient statistics are defined in high dimensions, and a complex grid might not be trivially constructed.

Fourth, embedded in BK2002's approximation there is a sampling mechanism of one-step-ahead observations $\mathbf{z}_{t+1}, t = 1, ..., n-1$. However, in the particular case of the models presented in Section 3, the predictive distributions are derived from multivariate dynamic regression schemes. This implies simulating from the one-step-ahead distribution of $\mathbf{z}_t = (\mathbf{x}_t, \mathbf{y}_t)$ where assumptions on the joint evolution of the explanatory variables $\mathbf{x}_t$ would be needed.

In summary, we believe that the innovative ideas recently developed in the Bayesian clinical trials arena could be difficult to implement, at this stage, in the context of the models provided in Section 3. There are a number of key assumptions and simplifications that would be needed to apply these methods to portfolio management, which might reduce the predictive power of dynamic time series models. Moreover, even if simulation is used, the decision space is made discrete and reduced, and a complex high-dimensional grid is constructed, the results could be computationally expensive, hence diminishing the appealing idea of controlling the bookkeeping explosion. Nevertheless, further research may overcome these obstacles.

ADDITIONAL REFERENCES IN THE DISCUSSION

Berry, D. A., Müller, P., Grieve, A. P., Smith, M., Parke, T., Blazek, R., Mitchard, N. Brearley, C. and Krams, M. (2000) Adaptive Bayesian designs for dose-ranging drug trials. *Case Studies in Bayesian Statistics V* (C. Gatsonis, R. Kass, B. Carlin, A. Carriquiry, A. Gelman, I. Verdinelli and M. West, eds). New York: Springer, 99-191 (with discussion).

Box, G. E. P. (1976). Science and statistics. *J. Am. Statist. Ass.* **71**, 791–799.

Brockwell, A. E. and Kadane, J. B. (2002). Sequential analysis by gridding sufficient statistics. *Tech. Rep.*, Carnegie Mellon University, USA.

Carlin, B. P., Kadane, J. B. and Gelfand, A. E. (1998). Approaches for optimal sequential decision analysis in clinical trials. *Biometrics* **54**, 964–975.

Kadane, J. B. and Vlachos, P. K. (2002). Hybrid methods for calculating optimal few-stage sequential strategies: Data monitoring for a clinical trial. *Statist. Comput.* **12**, 147–152.

BAYESIAN STATISTICS 7, pp. 369–384
J. M. Bernardo, M. J. Bayarri, J. O. Berger, A. P. Dawid,
D. Heckerman, A. F. M. Smith and M. West (Eds.)

New Tools for Consistency in Bayesian Nonparametrics

GABRIELLA SALINETTI
Università di Roma "La Sapienza", Italy
gabriella.salinetti@uniroma1.it

SUMMARY

Posterior consistency and the parallel behavior of consistency of maximum likelihood estimators is analyzed in nonparametric statistical problems. The framework is the hypo-Strong Law of Large Numbers, a form of "one-sided" Uniform Law of Large Numbers.

Keywords: POSTERIOR CONSISTENCY; HYPO-LAW OF LARGE NUMBERS; MAXIMUM LIKELIHOOD ESTIMATOR CONSISTENCY.

1. INTRODUCTION

Consistency in Bayesian nonparametric statistical problems continues to register increasing attention, motivated by various and easy to share reasons, widely illustrated in the rich literature on the subject starting from Diaconis and Freedman (1986) to the recent Ghosh and Ramamoorthi (2002).

The difficulty in nonparametrics is in the dimension of the problem: the "parameter" itself is typically a probability measure, the prior is a probability measure on a space of probability measures, and the resulting posterior, after seeing the data, is a random probability measure.

Posterior consistency at a given "parameter value" P_0 reduces then to almost sure (weak) convergence of the sequence of the posterior (random) probability measures to a probability measure concentrated in P_0. On the other hand, once the prior has been elicited, the posterior depends on the data through the likelihood function and then its asymptotic behavior depends on the asymptotic behavior of the likelihood, or the log of it, a functional of the empirical process determined by the observations.

With this view it is easy to realize that the standard Strong Law of Large Numbers, certainly a key tool in analyzing the asymptotic behavior of the likelihood, is too weak to determine consistency results; on the other hand Uniform Strong Laws of Large Numbers, often invoked in consistency of statistical functionals in nonparametric contexts, certainly produces consistency but they could reveal themselves too be demanding too mcuh in applications.

Hypo-Strong Law of Large Numbers, somehow between standard and uniform laws, seems to delineate a natural framework to deal with consistency of statistical functionals including posterior consistency.

Hypo-Strong Law of Large Numbers is based on the notion of hypo-convergence of functions; this is basically convergence of the hypographs of the functions (the set below the graph) and its relevance derives from the fact that it delineates the "minimal" setting for convergence of the suprema (values and solutions). In this role, almost sure hypo-convergence finds its natural and fruitful application in consistency of those statistical functionals which are, or can be expressed as, solutions of stochastic optimization problems; this is typically the case of Maximum Likelihood Estimators. The essential of hypo-convergence and hypo-Strong Law of Large Numbers is given in Section 3.

The parallel behavior of consistency of Maximum Likelihood Estimators and posterior consistency, often observed in the literature in consistency analyses, has been the initial motivation to analyze the role and the implications of the hypo-Strong Law of Large Numbers in posterior consistency. The main implication is posterior consistency on "compacts," a necessary prelude for posterior consistency; tightness of the posteriors is then a necessary and sufficient condition for posterior consistency; compactification or natural compact embedding can then become a way to verify posterior consistency. This is illustrated in Section 4.

Section 5 explores the parallelism between consistency of Maximum Likelihood Estimators and posterior consistency. Finally, in Section 6, together with conclusive remarks, further developments are delineated.

2. PROBLEM SETTING

Let E be the finite dimensional Euclidean space where the observations take value and $\mathcal{E}$ its Borel field. Let Θ be "parameter" space always assumed to be metric separable and complete, with its Borel field $\mathcal{B}(\Theta)$. For each $\theta \in \Theta$, P_θ is a probability measure on $\mathcal{E}$ and let $\mathcal{P} = \{P_\theta,\, \theta \in \Theta\}$. The prior beliefs are expressed by a prior probability measure Π on the Borel field $\mathcal{B}(\Theta)$. The observations $\{X_n, n \geq 1\}$ are random variables valued in E and modelled, given θ, as i.i.d. with probability law P_θ; it will be always convenient to look at the observations as co-ordinate processes on $(\Omega, \mathcal{A})$, $\Omega = E^\infty$, $\mathcal{A} = \mathcal{E}^\infty$.

The prior Π is updated by observing $X_1, ..., X_n$ in the posterior $\Pi(\cdot|X_1, ..., X_n)$, conditional distribution on $\mathcal{B}(\Theta)$ given $X_1, ..., X_n$. The sequence $\{\Pi(\cdot|X_1, ..., X_n),\, n \geq 1\}$ is a sequence of random probability measures on $\mathcal{B}(\Theta)$, simply denoted $\Pi_n(B, \cdot)$ with $B \in \mathcal{B}(\Theta)$.

The sequence $\{\Pi_n,\, n \geq 1\}$ is said to be *consistent at* θ_0, if there exists $\Omega_0 \subset \Omega$ with $P_{\theta_0}^\infty(\Omega_0) = 1$ such that if $\omega \in \Omega_0$ then for every open neighborhood U of θ_0 $\Pi_n(U, \omega) \to 1$ or, equivalently, $\Pi_n(U^c, \omega) \to 0$, where U^c denotes the complement of U in Θ.

Equivalently, posterior consistency at θ_0 holds if for all $\omega \in \Omega_0$, the sequence of probability measures $\{\Pi_n(\cdot, \omega),\, n \geq 1\}$ weakly converges to Π_0 where Π_0 is the probability measure giving mass 1 to θ_0.

Here the attention is restricted to the case where the probability measures P_θ are absolutely continuous with respect to the same σ-finite measure λ; correspondent densities are denoted $f(\cdot;\theta)$ and it is understood that $\theta \neq \theta'$ implies $f(\cdot;\theta) \neq f(\cdot;\theta')$; in fact, we will identify Θ with the set $\{f(\cdot;\theta);\, \theta \in \Theta\}$ when no confusion can arise. The likelihood, for the observations $X_1(\omega), \ldots, X_n(\omega)$ is $\prod_{i=1}^n f(X_i(\omega), \theta)$ and the posterior $\Pi_n(\cdot, \cdot)$ can now be expressed as

$$\Pi_n(B, \omega) = \frac{\int_B \prod_{i=1}^n f(X_i(\omega), \theta)\, \Pi(d\theta)}{\int_\Theta \prod_{i=1}^n f(X_i(\omega), \theta)\, \Pi(d\theta)}, \qquad B \in \mathcal{B}(\Theta).$$

The expression clarifies that the posterior depends only on the empirical measure determined by the observations. In fact, with standard manipulations, $P_{\theta_0}^\infty$-a.s., we have

$$\Pi_n(B, \omega) = \frac{\int_B \exp\left[nH_n(\omega, \theta)\right] \Pi(d\theta)}{\int_\Theta \exp\left[nH_n(\omega, \theta)\right] \Pi(d\theta)} \tag{1}$$

where

$$H_n(\omega, \theta) = \frac{1}{n} \sum_{i=1}^n \log \frac{f(X_i(\omega), \theta)}{f(X_i(\omega), \theta_0)} = \int_E \log \frac{f(x, \theta)}{f(x, \theta_0)} P_n(\omega, dx)$$

and $P_n(\omega,\cdot)$ is the empirical measure determined by the observations $X_1(\omega), ..., X_n(\omega)$. Statements involving the probability measure $P_{\theta_0}^{\infty}$ correspond "formally" to a situation where the observations $\{X_n,\, n \geq 1\}$ are i.i.d. P_{θ_0}; and from now on the underlying probability space is $(\Omega, \mathcal{A}, P_{\theta_0}^{\infty})$. Observe that

$$H_0(\theta) = \int \log \frac{f(x,\theta)}{f(x,\theta_0)} P_{\theta_0}(dx) = -\int f(x,\theta_0) \log \frac{f(x,\theta_0)}{f(x,\theta)} \lambda(dx) = -K(\theta_0,\theta)$$

where $K(\theta_0,\theta)$ is the Kullback–Leibler divergence of P_θ from P_{θ_0}. Once rephrased $\Pi_n(U^c,\omega)$ in the form of (1), consistency is expressed as

$$\Pi_n(U^c,\omega) = \frac{\int_{U^c} \exp\left[nH_n(\omega,\theta)\right]\Pi(d\theta)}{\int_{\Theta} \exp\left[nH_n(\omega,\theta)\right]\Pi(d\theta)} \to 0 \ ; \tag{2}$$

it is evident that the asymptotic behavior of Π_n depends on the asymptotic behavior of H_n and its relation with H_0; in fact, as it will be motivated next, on the hypo-convergence $P_{\theta_0}^{\infty}$-a.s. of $\{H_n\}$ to H_0.

The basic assumptions maintained in the rest of the paper are now introduced.

For every $\delta > 0$ let $K_\delta(\theta_0)$ be the Kullback–Leibler neighborhood of θ_0

$$K_\delta(\theta_0) = \{\theta \in \Theta : \, K(\theta_0,\theta) < \delta\}.$$

Even if main applications refer to the Hellinger metric, or the equivalent L_1-metric, the metric on Θ here will remain mostly not specified, the main goal being to identify situations of consistency or possible inconsistency issues depending on the metric adopted and the topological nature of the parameter space. It is also understood, until the metric is not specified, that we are inside "permissible" problems in the sense that when we write for example $\Pi(K_\delta(\theta_0))$ it is understood that $K_\delta(\theta_0) \in \mathcal{B}(\Theta)$.

As usually done in analyzing posterior consistency it will be assumed not only that $\Pi(U) > 0$ for every open neighborhood U of θ_0, a necessary condition for posterior consistency, but also that θ_0 is in the Kullback–Leibler support of the prior Π, *i.e.*,

$$(A) \qquad \Pi(K_\delta(\theta_0)) > 0 \qquad \forall \delta > 0.$$

We also tacitly assume that in the metric adopted on Θ, for every open neighborhood U of θ_0

$$(B) \qquad \inf\nolimits_{\theta \in U^c} K(\theta_0,\theta) > 0,$$

a condition satisfied in the cases of interest.

As consequence of (A), through the Strong Law of Large Numbers we have (see *e.g.*, Ghosh and Ramamoorthi, 2002, or Barron *et al.*, 1999, for the derivation)

$$\forall \delta > 0, \qquad \liminf_{n\to\infty} \; e^{n\delta} \int \prod_{i=1}^{n} \frac{f(X_i(\omega),\theta)}{f(X_i(\omega),\theta_0)} \Pi(d\theta) = \infty \quad P_{\theta_0}^{\infty}\text{-a.s.} \tag{3}$$

Condition (A) and its consequence (3) control the denominator in (2). We have for all $\omega \in \Omega_\delta$, $P_{\theta_0}^{\infty}(\Omega_\delta) = 1$, and n sufficiently large

$$\int \exp\left[nH_n(\omega,\theta)\right]\Pi(d\theta) > e^{-n\delta}\,. \tag{4}$$

In this case, as it will be seen later, it is easy to verify posterior consistency at θ_0 if $P_{\theta_0}^{\infty}$-a.s., for any open neighborhood U of θ_0

$$\sup_{\theta \in U^c} H_n(\omega, \theta) \longrightarrow \sup_{\theta \in U^c} H_0(\theta) \tag{5}$$

or, more generally

$$\limsup_{n \to \infty} \sup_{\theta \in U^c} H_n(\omega, \theta) < 0. \tag{6}$$

The convergence expressed by conditions (5) or (6), or variations of these, as a.s. convergence of suprema is directly connected with the a.s. hypo-convergence of the sequence $\{H_n(\omega, \cdot)\}$ to $H_0(\cdot)$ and it introduces a preliminary motivation to analyze posterior consistency in presence of hypo-convergence.

3. HYPO-CONVERGENCE AND HYPO-STRONG LAW OF LARGE NUMBERS

Hypo-convergence for a sequence of functions is basically set convergence of their hypographs. For the purposes of this paper it will suffice to restrict the attention to functions defined on a metric separable and complete space (Θ, d), even if some properties and results hold in more general settings.

For a function $H : \Theta \to \bar{\Re}$, the hypograph is the set hypo $H = \{(\theta, \alpha) \in \Theta \times \Re : H(\theta) \geq \alpha\}$, that is, all the points below the graph of H.

A sequence of functions $\{H_n, n \geq 1\}$ *hypo-converges* to the function H at θ if for every sequence $\theta_m \to \theta$ and any subsequence $\{n_m\}$

$$\limsup_{m \to \infty} H_{n_m}(\theta_m) \leq H(\theta) \tag{7}$$

and there exists $\theta_n \to \theta$ such that

$$H(\theta) \leq \liminf_{n \to \infty} H_n(\theta_n) \, . \tag{8}$$

If the two relations above hold for every $\theta \in \Theta$, the sequence $\{H_n, n \geq 1\}$ is said to *hypo-converge* to H, denoted $H_n \xrightarrow{h} H$, or $H = \text{hypo-lim}_{n \to \infty} H_n$.

The sequence $\{H_n, n \geq 1\}$ is said to *hypo-converge* to H *on a subset* C of Θ if relations (7) and (8) hold when restricted to C, that is, for every $\theta \in C$ and every sequence $\{\theta_n\}$ in C. When only condition (7) holds for every θ we simply write hypo-limsup$_{n \to \infty} H_n \leq H$.

Basic and extended references to hypo-convergence, its motivations, and its role in convergence of optimization problems are in Attouch (1984). Most of the literature on the subject concerns epi-convergence (convergence of epigraphs) and minimization problems, but any result on epi-convergence has its counterpart in the mirror setting of hypo-convergence. We refer to hypo-convergence, the choice being determined by the need to preserve the structure of the statistical problems where it will be used.

Hypo-convergence yields convergence of maximizers and optimal values in the sense stated below. The literature on the subject is quite rich, but here the basic aspects will suffice.

Proposition 3.1. *Suppose that $H_n \xrightarrow{h} H$. Then*

$$\sup_{\theta \in \Theta} H(\theta) \leq \liminf_{n \to \infty} \sup_{\theta \in \Theta} H_n(\theta) \, . \tag{9}$$

Moreover, if there is a subsequence $\{\theta_{n_m}\}$ such that for every m, $\theta_{n_m} \in$ argmax H_{n_m} and $\theta_{n_m} \to \hat{\theta}$ then $\hat{\theta} \in$ argmax H and $\sup_{\theta \in \Theta} H_n(\theta) \to \sup_{\theta \in \Theta} H(\theta)$.

This result is well known; an elementary proof (in the setting of epi-convergence) is given in Dong and Wets (2000). As immediate consequence of (7) we have

Proposition 3.2. *If* hypo-limsup$_{n\to\infty} H_n \leq H$ *then for every compact set* C

$$\limsup_{n\to\infty} \sup_{\theta\in C} H_n(\theta) \leq \sup_{\theta\in C} H(\theta).$$

Actually a form of characterization of convergence of suprema in terms of hypo-convergence can be given (see *e.g.*, Attouch, 1984, Theorem 2.11) as follows:

Proposition 3.3. *If* $H_n \xrightarrow{h} H$ *then the following are equivalent:*

(i) $\sup_{\theta\in\Theta} H_n(\theta) \to \sup_{\theta\in\Theta} H(\theta)$ *and argmax* $H \neq \emptyset$

(ii) $\exists \epsilon_n \downarrow 0$ *and a relatively compact sequence* $\{\theta_n\}$ *such that for all* n, $\theta_n \in \epsilon_n$*-argmax* $H_n = \{\theta : H_n(\theta) > \sup H_n - \epsilon_n\}$

(iii) $\exists \epsilon_n \downarrow 0$ *and a non empty relatively compact set* C *such that for all* n

$$\min\left\{\frac{1}{\epsilon_n}, \sup_{\theta\in\Theta} H_n(\theta) - \epsilon_n\right\} < \sup_{\theta\in C} H_n(\theta).$$

Remark. The sense of this result is that in the presence of hypo-convergence, convergence of suprema can fail only when ϵ-optimal solutions "move out" of the space, living on complements of compact sets.

We turn now to random functions $H_n(\cdot,\cdot)$ defined on a probability space $(\Omega, \mathcal{A}, \mu)$, which map every $\omega \in \Omega$ into a function $H_n(\omega, \cdot) : \Theta \to \bar{\Re}$.

For a sequence of such functions $\{H;\ H_n,\ n \geq 1\}$, μ*-a.s. hypo-convergence* of H_n to H means, as natural to expect, that for all $\omega \in \Omega \setminus N$, $\mu(N) = 0$,

$$H_n(\omega, \cdot) \xrightarrow{h} H(\omega, \cdot).$$

The theory of random semicontinuous functions, with upper (or lower) semicontinuous realizations, based on the topology generated by hypo- (respectively epi-) convergence, has received remarkable attention as the appropriate setting to deal with stochastic processes with semicontinuous realizations (Salinetti and Wets, 1986, 1990). In the specific setting of consistency of statistical estimators, as functionals of the empirical measures, it will suffice to refer to $P_{\theta_0}^{\infty}$-a.s. convergence of random functions generated by a sequence of i.i.d. random variables to which we refer as *hypo-Strong Law of Large Numbers*.

Let $\{X_n,\ n \geq 1\}$ be a sequence of random variables valued in $(E, \mathcal{E})$ i.i.d. with probability law P_{θ_0}, regarded as coordinate process on $(\Omega, \mathcal{A}, \mu)$ with $\Omega = E^{\infty}$, $\mathcal{A} = \mathcal{E}^{\infty}$ and $\mu = P_{\theta_0}^{\infty}$. For a function $h : E \times \Theta \to \bar{\Re}$ let

$$H_n(\omega, \theta) = \frac{1}{n}\sum_{i=1}^{n} h(X_i(\omega), \theta) = \int_E h(x, \theta) P_n(\omega, dx)$$

where $P_n(\omega, \cdot)$ is the empirical probability measure determined by $X_1, \ldots, X_n$ and let

$$H_0(\theta) = \int_E h(x, \theta) P_{\theta_0}(dx).$$

The sequence of random functions $\{H_n\}$ satisfies the *hypo-Strong Law of Large Numbers* (hypo-SSLN) if for all $\omega \in \Omega \setminus N$, $P_{\theta_0}^{\infty}(N) = 0$

$$H_n(\omega, \cdot) \xrightarrow{h} H_0(\cdot).$$

We basically refer here to the version reproduced in Dong and Wets (2000, Theorem A.4). A close derivation aiming at consistency of Maximum Likelihood Estimators (MLE's) is given in Hess (1996). The basic assumptions under which the hypo-SLLN holds require that the function $h : E \times \Theta \to \bar{\Re}$ is $\mathcal{E} \times \mathcal{B}(\Theta)$-measurable and for each $x \in E$, $h(x, \cdot)$ is upper semicontinuous on Θ, *i.e.*, h is a random upper semicontinuous function.

Theorem 3.4. *Let* (Θ, d) *be a Polish space,* $\{X_n, n \geq 1\}$ *a sequence of i.i.d. random variables valued in* $(E, \mathcal{E})$ *with common probability law* P_{θ_0} *defined on* $\mathcal{E}$ *with* $\mathcal{E}$ P_{θ_0}*-complete, and let* $h : E \times \Theta \to \bar{\Re}$ *be a random upper semicontinuous function. Suppose that for every* $\theta \in \Theta$ *there exists a neighborhood* U *of* θ *and a measurable function* $\alpha : E \to \bar{\Re}$ *with* $\int \alpha^+ dP_{\theta_0} < \infty$ *such that* P_{θ_0}*-a.s.* $h(\cdot, \theta') \leq \alpha(\cdot)$ *for all* $\theta' \in U$*. Then the hypo-SLLN holds, that is,* $P^{\infty}_{\theta_0}$*-a.s.*

$$H_n(\omega, \cdot) \xrightarrow{h} H_0(\cdot).$$

Remark. For the problems that will be approached in the next section often it will be actually sufficient a one-sided version of the hypo-SLLN in the form

$$P^{\infty}_{\theta_0}\text{-a.s. hypo-limsup}_{n\to\infty} H_n(\omega, \cdot) \leq H_0(\cdot).$$

In view of Propositions 3.2, 3.3, and Theorem 3.4, it is easy to realize the role of hypo-SLLN in consistency of statistical estimators when these are obtained, or can be expressed, as solution of optimization problems; this is quite frequent in the context of classical statistics and in fact a.s. hypo-convergence, and specifically hypo-SLLN, offers at the same time the theoretical setting (stochastic processes with semicontinuous realizations) for statistical functionals of empirical processes and a unified derivation of consistency properties. An approach of this type is pursued in Salinetti (1990, 2001) even if mainly restricted to the parametric case or regular "parameter" spaces.

Consistency of MLE's based on a.s. hypo-convergence is obtained in Hess (1996) under quite general assumptions. The same problem is approached in Dong and Wets (2000) in a more general setting where "prior" information, mainly in the form of model assumptions, is formalized as "constraints" of the maximization problem.

At the same time, also posterior consistency, when based on convergence of stochastic suprema as sketched in (5) and (6) takes advantage of the potential offered by hypo-convergence, as it will be seen in the next section.

The conditions under which hypo-SLLN holds do not appear particularly restrictive in applications, at least for MLE's cases; the uppersemicontinuity of the integrand is often a consequence of the metric adopted on Θ; examples are worked out in Dong and Wets (2000) and Hess (1996) and some easy applications are in the next section.

4. POSTERIOR CONSISTENCY AND HYPO-CONVERGENCE

In accordance to Section 2, posterior consistency requires that for every $\omega \in \Omega_0$, $P^{\infty}_{\theta_0}(\Omega_0) = 1$, and every open neighborhood U of θ_0,

$$\Pi_n(U^c, \omega) = \frac{\int_{U^c} \exp\left[nH_n(\omega, \theta)\right] \Pi(d\theta)}{\int_{\Theta} \exp\left[nH_n(\omega, \theta)\right] \Pi(d\theta)} \to 0.$$

A preliminary consideration easily leads to

Proposition 4.1. *If* θ_0 *belongs to the Kullback–Leibler support of* Π *and* $P^{\infty}_{\theta_0}$*-a.s. for every open neighborhood* U *of* θ_0

$$\limsup_{n\to\infty} \sup_{\theta \in U^c} H_n(\omega, \theta) \leq \sup_{\theta \in U^c} H_0(\theta) \tag{10}$$

then posterior consistency at θ_0 *holds.*

Proof. By (B) there exists $\epsilon > 0$ such that $\sup_{\theta \in U^c} H_0(\theta) < -\epsilon$. Let $\Omega_1 = \Omega \setminus N_1$, $P^{\infty}_{\theta_0}(N_1) = 0$, the set where (10) and the consequence (4) of (A) for $\delta < \epsilon$ hold. For $\omega \in \Omega_1$ and n sufficiently large we have

$$\Pi_n(U^c, \omega) \leq e^{-n(\epsilon-\delta)} \Pi(U^c)$$

and the result follows. □

The result, based on the convergence of suprema (10), induces to study posterior consistency in presence of a.s. hypo-convergence of the sequence $\{H_n(\omega, \cdot)\}$ to $H_0(\cdot)$ appealing to the recognized fact that hypo-convergence is the "minimal" setting for convergence of suprema (values and solutions).

However, the connection between hypo-convergence and posterior consistency is deeper and related to the tightness of the sequence of posteriors.

Since posterior consistency at θ_0 is also $P^{\infty}_{\theta_0}$-a.s. weak convergence of $\{\Pi_n(\cdot, \omega)\}$ to $\Pi_0(\cdot)$ and (Θ, d) is a Polish space, a necessary condition for posterior consistency at θ_0 is $P^{\infty}_{\theta_0}$-a.s. tightness of the sequence $\{\Pi_n(\cdot, \omega),\ n \geq 1\}$, *i.e.*, for every $\omega \in \Omega_0$, $P^{\infty}_{\theta_0}(\Omega_0) = 1$, and $\alpha > 0$ it has to exist a compact subset C_α of Θ such that $\Pi_n(C^c_\alpha, \omega) < \alpha$ for all n.

If $P^{\infty}_{\theta_0}$-a.s. tightness is guaranteed (for example, we know that Π has compact support), then for all $\omega \in \Omega_0$ and every neighborhood U of θ_0 we have

$$\Pi_n(U^c, \omega) = \Pi_n(U^c \cap C_\alpha, \omega) + \Pi_n(U^c \cap C^c_\alpha, \omega) < \Pi_n(U^c \cap C_\alpha, \omega) + \alpha\,.$$

Thus posterior consistency at θ_0 follows if we can prove that for every α and every neighborhood U of θ_0, $\Pi_n(U^c \cap C_\alpha, \omega) \to 0$.

Actually the sequence $\{\Pi_n,\ n \geq 1\}$ is posterior consistent at θ_0 *if and only if* there exists Ω_0, $P^{\infty}_{\theta_0}(\Omega_0) = 1$, such that for every $\omega \in \Omega_0$

(i) $\{\Pi_n(\cdot, \omega),\ n \geq 1\}$ is tight

(ii) for any open neighborhood U of θ_0 and any compact set C, $\Pi_n(U^c \cap C, \omega) \to 0$.

We now show that condition (ii) holds if $P^{\infty}_{\theta_0}$-a.s., the sequence $\{H_n(\omega, \cdot)\}$ hypo-converges to $H_0(\cdot)$. In fact just as consequence of Proposition 3.2 we have

Proposition 4.2. *Suppose that θ_0 belongs to the Kullback–Leibler support of Π and $P^{\infty}_{\theta_0}$-a.s. the following conditions hold*

(i) $\{\Pi_n(\cdot, \omega),\ n \geq 1\}$ *is tight;*

(ii) $H_n(\omega, \cdot) \xrightarrow{h} H_0(\cdot)$

then $\{\Pi_n\}$ is posterior consistent at θ_0.

Proof. Let Ω_0, $P^{\infty}_{\theta_0}(\Omega_0) = 1$, be such that on it (i) and (ii) hold. By tightness, for every $\alpha > 0$ there exists C_α compact such that $\Pi_n(C^c_\alpha, \omega) < \alpha$. Let U be an open neighborhood of θ_0. Let $\epsilon > 0$ such that $\sup_{U^c \cap C_\alpha} H_0(\theta) < -\epsilon$. Let $0 < \delta < \epsilon$ and Ω_δ, $P^{\infty}_{\theta_0}(\Omega_\delta) = 1$, such that, for all $\omega \in \Omega_\delta$ and n sufficiently large, (4) holds.

Hypo-convergence, by Proposition 3.2, yields

$$\limsup_{n\to\infty} \sup_{\theta \in U^c \cap C_\alpha} H_n(\omega, \theta) \leq \sup_{\theta \in U^c \cap C_\alpha} H_0(\theta) < -\epsilon. \tag{11}$$

For n sufficiently large the last relation with (4) yields

$$\Pi_n(U^c \cap C_\alpha, \omega) < e^{-n(\epsilon-\delta)}\, \Pi(U^c \cap C_\alpha);$$

it follows that

$$\Pi_n(U^c \cap C_\alpha, \omega) \to 0 \quad \text{and} \quad \limsup_{n\to\infty} \Pi_n(U^c, \omega) \leq \alpha.$$

The argument repeated for every α gives the conclusion. □

Remark. The role played by a.s. hypo-convergence in the proof is in obtaining (11). It is useful to observe that only the "hypo-limsup" part of it enters and actually it will suffice simply the "hypo-limsup" part on compact sets.

Relation (11) is evidently a key point, tightness being in any case a necessary condition. It is relevant to observe that under general conditions (see, *e.g.*, Barron *et al.*, 1999) $P_{\theta_0}^{\infty}$-a.s. we have

$$H_n(\omega,\cdot) \longrightarrow H_0(\cdot) \qquad \Pi\text{-a.s.},$$

however, "pointwise" convergence is not sufficient for convergence of suprema.

The argument used in the last Proposition actually allows to conclude

Proposition 4.3. *If θ_0 belongs to the Kullback–Leibler support of Π and $P_{\theta_0}^{\infty}$-a.s. $H_n(\omega,\cdot) \xrightarrow{h} H_0(\cdot)$, then for every compact subset C of Θ not containing θ_0*

$$\Pi_n(C,\omega) \to 0\,. \tag{12}$$

Moreover $\{\Pi_n\}$ is posterior consistent at θ_0 if and only if $P_{\theta_0}^{\infty}$-a.s., $\{\Pi_n(\cdot,\omega)\}$ is tight.

Proof. If C does not contain θ_0 there exists an open neighborhood U of θ_0 with $U \cap C = \emptyset$ so that $C \subset U^c$ and then, by (B), $\sup_{\theta \in C} H_0(\theta) < 0$. The argument used in the last proposition then shows (12). Tightness, as a sufficient condition, is Proposition 4.2; the necessary condition is obvious. □

Let us consider now the tightness condition. If θ_0 belongs to the Kullback–Leibler support of Π, a sufficient condition for $P_{\theta_0}^{\infty}$-a.s. tightness is that for every $\omega \in \Omega_0 = \Omega \setminus N_0$, $P_{\theta_0}^{\infty}(N_0) = 0$, there exists a compact subset C of Θ such that

$$\limsup_{n\to\infty} \sup_{\theta \in C^c} H_n(\omega,\theta) < 0. \tag{13}$$

The argument to show this is the usual one: $P_{\theta_0}^{\infty}$-a.s., appealing to (A) and (13), for $\epsilon > 0$ and $\delta_k \downarrow 0$ we get

$$\Pi_n(C^c,\omega) = \frac{\int_{C^c} \exp\left[nH_n(\omega,\theta)\right]\Pi(d\theta)}{\int \exp\left[nH_n(\omega,\theta)\right]\Pi(d\theta)} < e^{-n(\epsilon-\delta_k)}\Pi(C^c)$$

and this is enough to state the tightness for $\Pi_n(\cdot,\omega)$.

The argument above with Proposition 4.2 gives

Proposition 4.4. *If θ_0 belongs to the Kullback–Leibler support of Π and $P_{\theta_0}^{\infty}$-a.s. the following conditions hold*

(i) *there exists a compact C such that*

$$\limsup_{n\to\infty} \sup_{\theta \in C^c} H_n(\omega,\theta) < 0$$

(ii) $H_n(\omega,\cdot) \xrightarrow{h} H_0(\cdot)$

then $\{\Pi_n\}$ is posterior consistent at θ_0.

The development followed here aims at pointing out the role of hypo-convergence in posterior consistency: hypo-SLLN ensures the right posterior convergence on compact sets not containing θ_0; possible inconsistency can only derive from lack of tightness and this is recognized in the tail behavior of $H_n(\omega, \cdot)$ where tails are complements of compact sets.

Of course, if Θ is compact then the hypo-SLLN implies posterior consistency. This is also the case when Θ is locally compact and convex with convex metric, because in this case the concavity of $H_n(\omega, \cdot)$ and the strict concavity of $H_0(\cdot)$ recovers on the tightness.

The specific nature of these results, consistency on compacts and tail problems, where tails are complements of compacts, suggests to analyze posterior consistency in suitable compact embedding of the parameter space, where conditions for hypo-SLLN are easily satisfied. Two cases are briefly examined: the first one, relative to discrete observations is well known (Freedman, 1963). The revisitation aims at illustrating the procedure and the way it can be extended in the more general situation of the second example.

Example 1. Discrete observations. The observations take values in a countable set $I = \{x_1, x_2, ...\}$ and the parameter space consists of all the probability distributions on I. As in Freedman (1963), let $S = \{f : I \to [0, 1]\}$ with the product topology so S is compact and metrizable; $L = \{p \in S : \sum p(x_r) \leq 1\}$, with the relative topology, is compact too. The parameter space $\Theta = \{\theta \in L : \sum \theta(x_r) = 1\}$ is a G_δ set of L. Let Π be the prior on $\mathcal{B}(\Theta)$, the Borel field of Θ with the relative topology of L; suppose that θ_0 belongs to the Kullback–Leibler support of Π and $-\int \log \theta_0(x) P_{\theta_0}(dx) < \infty$. In accordance with Proposition 4.1 posterior consistency at every θ_0 follows if $P_{\theta_0}^{\infty}$-a.s., for every open neighborhood U of θ_0

$$\limsup_{n\to\infty} \sup_{\theta\in U^c} H_n(\omega, \theta) < 0. \tag{14}$$

We obtain it passing to the compact embedding L. Let us consider on the compact space L the functions

$$\hat{H}_n(\omega, p) = \int \log \frac{p(x)}{\theta_0(x)} P_n(\omega, dx), \quad \hat{H}_0(p) = \int \log \frac{p(x)}{\theta_0(x)} P_{\theta_0}(dx),$$

extensions of $H_n(\omega, \theta)$ and $H_0(\theta)$ defined on Θ. It is not difficult to prove that $\hat{H}_0$ is uppersemicontinuous on L and for every open neighborhood U_L of θ_0 in L, $\sup_{U_L^c} H_0 < 0$. It is also easy to verify that hypo-SLLN holds for the sequence $\{\hat{H}_n\}$; this follows after observing that the conditions of Theorem 3.4 are satisfied by the function $h(x, p) = \log[p(x)/\theta_0(x)]$; observe that $p(x) \leq 1$, so the dominant function α is $-\log \theta_0(x)$. Since U_L^c is compact, it follows by Proposition 3.2, $P_{\theta_0}^{\infty}$-a.s.

$$\limsup_{n\to\infty} \sup_{p\in U_L^c} \hat{H}_n(\omega, p) \leq \sup_{p\in U_L^c} \hat{H}_0(p) < 0.$$

On the other hand for any open neighborhood U of θ_0 in Θ, with U^c complement of U in Θ, we have $U^c \subset U_L^c$ where U_L is an open neighborhood of θ_0 in L; so $\sup_{\theta\in U^c} H_n(\omega, \theta) \leq \sup_{p\in U_L^c} \hat{H}_n(\omega, p)$ and (14) follows.

The argument above shows, as in Freedman (1963, Theorem 3.2), that *the posterior is consistent at every θ_0 in the Kullback–Leibler support of Π such that*

$$-\int \log \theta_0(x) P_{\theta_0}(dx) < \infty.$$

Example 2. Unimodal densities. Without entering into technical details, we consider the case where the parameter space consists of unimodal and uppersemicontinuous (u.sc) densities on the real line, dominated by an u.sc. function ψ. The mode is the same, conventionally put at 0, and the modal value is $+\infty$. Let $L_0 = \{f : \Re \to [0, \infty], f \text{ unimodal }, f(0) = \infty, \int f d\lambda \leq 1\}$, where λ is the Lebesgue measure, $\Theta_0 = \{\theta \in L_0 : \int f d\lambda = 1\}$ so the parameter space is $\Theta = \{\theta \in \Theta_0, \theta \leq \psi\}$. The space L_0 can be equipped with a metric based on the Levy distance for monotone functions; with this metric (L_0, d) is metric and compact as shown in Reiss (1973), Θ_0 is a G_δ set in L_0 and convergence $d(f_n, f) \to 0$ in L_0 means pointwise convergence on the continuity set of f.

Let Π be a prior on the Borel field of Θ_0 with $\Pi(\Theta) = 1$ and let θ_0 be in the Kullback–Leibler support of Π and such that $\int \log[\psi(x)/\theta_0(x)]\, P_{\theta_0}(dx) < \infty$. It is easy to realize that the topological structure of the problem is analogous to the previous case. Also in this case on the compact space $L = \{f \in L_0, f \leq \psi\}$ the extensions

$$\hat{H}_n(\omega, f) = \int \log \frac{f(x)}{\theta_0(x)} P_n(\omega, dx), \quad \hat{H}_0(f) = \int \log \frac{f(x)}{\theta_0(x)} P_{\theta_0}(dx),$$

satisfy the conditions of Theorem 3.4 for the hypo-SLLN; $\hat{H}_0$ is u.sc. on L and for every open neighborhood U_L of θ_0 in L, $\sup_{U_L^c} \hat{H}_0(f) < 0$. The argument above allows to conclude similarly to the previous case that *the posterior is consistent at any* θ_0 *in the Kullback–Leibler support of* Π *such that* $\int \log[\psi(x)/\theta_0(x)] P_{\theta_0}(dx) < \infty$.

It is relevant to observe that since in this class of unimodal functions, convergence in the metric d is equivalent to convergence in L_1, as shown in Reiss (1973), we actually get posterior consistency in the L_1-metric. Extensions to more general situations are possible, for example when the mode is not fixed but sits in a compact interval.

Remark. In both examples the parameter space has a global dominant function, constantly 1 in the first case and ψ in the second case. This is evidently more restrictive than that seen in Theorem 3.4; in this direction "sieves"-type constraints could lead to workable extensions.

5. POSTERIOR CONSISTENCY AND CONSISTENCY OF MLE'S

In this section the parallel behavior of posterior consistency and consistency of MLE's is analyzed and differences are highlighted. The parallelism between the two consistencies is delineated in Ghosh and Ramamoorthi (2002) and reference to consistency of MLE's often appears in the literature when dealing with posterior consistency. An explicit connection between posterior consistency and consistency of MLE's is, in fact, already contained in Strasser (1981). With the notations of Section 2 the MLE for θ_0, when existing, is $\hat{\theta}_n(\omega) = \arg\max_{\theta \in \Theta} H_n(\omega, \theta)$. Existence and measurability questions are not the main concern here; we will refer to Approximate Maximum Likelihood Estimators (AMLE) where the sets of α-approximate MLE's, $\alpha > 0$, are given by

$$A_{n,\alpha}(\omega) = \{\theta \in \Theta : H_n(\omega, \theta) > \sup H_n(\omega, \theta) - \alpha\};$$

AMLE's consistency for θ_0 means here that there exists Ω_0, $P_{\theta_0}^{\infty}(\Omega_0) = 1$, such that for all $\omega \in \Omega_0$ and any sequence $\alpha_n \downarrow 0$, for every open neighborhood U of θ_0

$$A_{n,\alpha_n}(\omega) \subset U$$

for n sufficiently large; in fact, a restrictive notion of consistency of AMLE's which however serves the purpose of comparison.

As easy completion of Proposition 4.2 we immediately have:

Proposition 5.1. *If θ_0 belongs to the Kullback–Leibler support of Π and $P_{\theta_0}^{\infty}$-a.s., for every open neighborhood U of θ_0 we have*

$$\limsup_{n\to\infty} \sup_{\theta\in U^c} H_n(\omega,\theta) < 0 \tag{15}$$

then AMLE's consistency and posterior consistency at θ_0 both hold.

Proof. Posterior consistency at θ_0 is Proposition 4.1. For $\omega \in \Omega_0$, with $P_{\theta_0}^{\infty}(\Omega_0) = 1$, $\{\alpha_n\}$ a sequence of positive numbers decreasing to 0, U any neighborhood of θ_0, since $H_n(\omega,\theta_0) = 0$, relation (15) implies that $A_{n,\alpha_n}(\omega) \subset U$ for n sufficiently large and AMLE's consistency follows. □

Beyond any definition of AMLE's it seems relevant to observe that most of the literature on consistency of AMLE's is based on a system of conditions where it is easy to recognize, in addition to the uppersemicontinuity of the maps $\theta \mapsto \log f(x,\theta)$, for all x, assumptions as "local dominance" of the type considered in Theorem 3.4 and "tail" assumptions which include global dominance, semi-dominance and global uniformity as in Perlman (1972) or, more recently, one-sided bracketing condition as in Dudley (1998). Uppersemicontinuity and local dominance imply hypo-SLLN; then, based on tail conditions, consistency of AMLE's for θ_0 is obtained showing that, $P_{\theta_0}^{\infty}$-a.s., (15) holds.

In all these cases then, which include Huber (1967) and Wald (1949), posterior consistency and AMLE's consistency both hold.

In general, however, consistency of AMLE's does not imply (15) unless additional regularity assumptions are introduced such as local compactness (Perlman, 1972); on the other hand consistency of AMLE's can be obtained without passing through (15), as for example in (Hess, 1996).

Of course the same parallel behavior is observed under the conditions of Proposition 4.4 as it can be easily proved. We have:

Proposition 5.2. *If θ_0 belongs to the Kullback–Leibler support of Π and $P_{\theta_0}^{\infty}$-a.s. the following conditions hold:*

(i) *there exists a compact C such that*

$$\limsup_{n\to\infty} \sup_{\theta\in C^c} H_n(\omega,\theta) < 0$$

(ii) $H_n(\omega,\cdot) \xrightarrow{h} H_0(\cdot)$

then posterior consistency and AMLE's consistency both hold.

The result suggests a final comment on the relation between the two consistencies. In view of all the above in dealing with AMLE's consistency, almost sure hypo-convergence is necessary as minimal requirement for convergence of stochastic suprema; conditions on tails based on the existence of convenient compact sets are sufficient. The situation is in a sense mirrored for posterior consistency: tightness is a necessary condition, almost sure hypo-convergence is sufficient, with the additional remark that, if tightness is guaranteed, almost sure hypo-convergence on compacts is sufficient.

6. SOME CONCLUSIONS AND POSSIBLE DEVELOPMENTS

Hypo-SLLN is a functional limit which preludes to convergence of stochastic suprema; then it is a preliminary requirement for consistency of AMLE's and a potential useful tool to verify posterior consistency through convergence of stochastic suprema. It does not imply, alone, posterior consistency; it implies consistency in presence of tightness and somehow it produces "approximate" consistency as consistency on compacts; it locates sources of inconsistency in the tails identified with complements of compact sets, thus alerting on the behavior of the prior on these sets. Moreover, it is a tool to obtain posterior consistency in those circumstances where there are "natural" compactifications and the crucial properties of the functionals involved are preserved in the extension to the compact embedding. The real point is the existence of suitable compactifications.

The case of unimodal u.sc. densities, regarded from a different perspective, indicates a way to proceed in this direction, relying on the topology generated by hypo-convergence. It is easy to show that convergence in the metric of unimodal u.sc. functions is equivalent to their hypo-convergence and actually many of the properties, compactness included, could be easily obtained as a consequence of this fact.

In fact, the topology generated by hypo-convergence is compact and metrizable (Dolecki *et al.,* 1983); thus when the prior information selects a.s. u.sc. densities, not necessarily unimodal, a compact embedding is available. Of course, in the Bayesian setting this way of proceeding has to be accompanied by a device to elicit a prior on the space of u.sc. functions. In this direction the theory developed for random u.sc. functions states that a probability measure on the Borel field generated by the hypo-topology determines and is uniquely determined by a Choquet capacity or hitting function (Salinetti and Wets, 1986); this function, with properties which are the natural extension of the properties of the distribution function of a random variable, can play the role of a general device to assign nonparametric prior on the space of u.sc. functions.

REFERENCES

Attouch, H. (1984). *Variational Convergence for Functions and Operators*. Boston, MA: Pitman.

Barron, A., Schervish, M. J. and Wasserman, L. (1999). The consistency of posterior distributions in nonparametric problems. *Ann. Statist.* **27**, 536–561.

Diaconis, P. and Freedman, D. (1986). On the consistency of Bayes estimates. *Ann. Statist.* **14**, 1–67 (with discussion). .

Dolecki, S., Salinetti, G. and Wets, R. J.-B. (1983). Convergence of functions: equi-semicontinuity. *Trans. Am. Math. Soc.* **276**, 409–430.

Dong, M. X. and Wets, R. J.-B. (2000). Estimating density functions: a constrained maximum likelihood approach. *J. Nonparametr. Statist.* **12**, 549–595.

Dudley, R. M. (1998). Consistency of M-estimators and one-sided bracketing. *High Dimensional Probability (Oberwolfach, 1996)*. Basel: Birkhäuser, 33–58.

Freedman, D. A. (1963). On the asymptotic behavior of Bayes estimates in the discrete case. *Ann. Math. Statist.* **34**, 1386–1403.

Ghosh, J. K. and Ramamoorthi, R. V. (2002). Bayesian nonparametrics (unpublished).

Hess, C. (1996). Epi-convergence of sequences of normal integrands and strong consistency of the maximum likelihood estimator. *Ann. Statist.* **24**, 1298–1315.

Huber, P. J. (1967). The behavior of maximum likelihood estimates under nonstandard conditions. *Proc. Fifth Berkeley Symp.* **1** (L. M. LeCam and J. Neyman, eds). Berkeley: Univ. California Press, 221–233.

Perlman, M. D. (1972). On the strong consistency of approximate maximum likelihood estimators. *Proc. Fifth Berkeley Symp.* **1** (L. M. LeCam and J. Neyman, eds). Berkeley: Univ. California Press,

Reiss, R.-D. (1973). On the measurability and consistency of maximum likelihood estimates for unimodal densities. *Ann. Statist.* **1**, 888–901.

Salinetti, G. (1990). On structural relationships between probability and statistics: Empirical processes and statistical functionals (in Italian). *Proc. Italian Statistical Society*.

Salinetti, G. (2001). Consistency of statistical estimators: The epigraphical view. *Stochastic Optimization: Algorithms and Applications (Gainesville, FL, 2000)*. Dordrecht: Kluwer, 365–383.

Salinetti, G. and Wets, R. J.-B. (1986). On the convergence in distribution of measurable multifunctions (random sets), normal integrands, stochastic processes and stochastic infima. *Math. Oper. Res.* **11**, 385–419.

Salinetti, G. and Wets, R. J.-B. (1990). Random semicontinuous functions. *Lectures in Applied Mathematics and Informatics*. Manchester: Manchester University Press, 330–353.

Strasser, H. (1981). Consistency of maximum likelihood and Bayes estimates. *Ann. Statist.* **9**, 1107–1113.

Wald, A. (1949). Note on the consistency of the maximum likelihood estimate. *Ann. Math. Statist.* **20**, 595–601.

DISCUSSION

JAYANTA K. GHOSH (*Purdue University, USA*)

This is a very clear, enjoyable paper. I begin by quickly summing up,

$$\Pi(\theta \in U^c | X_1, X_2, \ldots, X_n) = \frac{\int_{U^c} e^{nHn(\theta, X_1, \ldots, X_n)} \Pi(d\theta)}{\int_{\Theta} e^{nHn(\theta, X_1, \ldots, X_n)} \Pi(d\theta)}$$

where nHn is the log likelihood ratio as in the paper.

Under the relatively simple condition that θ_0 is in Kullback–Leibler support of Π, Schwartz (1965) showed the denominator is

$$0(e^{-n\delta}) \text{ a.s. } P_{\theta_0}, \tag{16}$$

for every positive δ, however small. So to show

$$\Pi(\theta \in U^c | X_1, X_2, \ldots, X_n) \to 1 \text{ a.s. } P_{\theta_0},$$

one has to show the numerator is

$$O(e^{-n\delta 0}) \text{ a.s. } P_{\theta_0} \text{ for some } \delta_0 > 0 \tag{17}$$

Since the numerator is bounded by

$$\exp\{n \sup_{\theta \in U^c} H_n(\theta, X_1, X_2, \ldots, X_n)\}$$

one way of showing posterior consistency is to study conditions under which the limit (or limsup) of

$$\sup_{\theta \in U^c} H_n(\theta, X_1, \ldots, X_n) \leq -\delta_0 \text{ a.s. } P_{\theta_0} \tag{18}$$

GS points out that a minimal and natural setting for studying (18) is "hypo-convergence" of H_n. This is well-established by her. She also exhibits several interesting implications of "hypo-convergence." From these follow her interesting observation that one needs hypo-convergence and tightness for (17) and hence for posterior consistency.

Comments, suggestions.

(i) The essence of this idea is well known but traditionally one uses "uniform convergence" instead of hypo-convergence, which is much weaker.

(ii) The snag is that GS has not told us how to replace the many Uniform Strong Laws of Large Number (USLLN) by Hypo-SLLN's with conditions that are easy to verify. Will bracketing entropy and uniform strong laws based on them help?

(iii) Also, to handle high or infinite dimensional Θ, one may have to develop hypo-convergence theorems in the context of sieves.

Finally, using (18) to prove (17) may not always be efficient for infinite dimensional (nonparametric) examples. Integrals are better behaved than maxima—a fact that is easy to accept for Bayesians. Bayesians prefer to integrate the likelihood instead of maximizing.

For example (as shown by Schwartz), if U is a weak neighborhood, then (17) holds *without any assumptions*, whereas we do not know any such general fact about (18).

Also if $\sup_{\theta\in U^c} H_n(\theta, X_1, X_2, \ldots, X_n)$ converges to infinity a.s. P_{θ_0}, the methods of hypo convergence will not work.

There is a famous counter-example of Bahadur with countable Θ where the MLE is not consistent. It is treated in detail in Lehmann's "Point Estimation." Ghosh and Ramamoorthi (forthcoming book on Bayesian Nonparametrics) establish $\sup_{\theta\in U^c} H_n(\theta, X_1, X_2, \ldots, X_n)$ converges to infinity a.s. P_{θ_0}. But by Doob's theorem the posterior is consistent at any θ_0 in the support of the prior.

SUBHASHIS GHOSAL (*North Carolina State University, U. S. A.*) and
AAD VAN DER VAART (*Free University Amsterdam, The Netherlands*)

This interesting paper attempts to relax the need for uniform convergence of log likelihood ratios in consistency theorems. Salinetti's proof (*e.g.*, Propositions 4.4 and 4.1) for posterior consistency involves two steps. First, it is noted that a standard consistency proof goes through if

$$\limsup_{n\to\infty} \sup_{\theta\in U^c} H_n(\omega, \theta) < 0. \tag{19}$$

She then shows that when the space is compact, hypo-convergence yields the condition above. In non-compact spaces, U^c must be intersected with a compact set C and the posterior probability of the remaining part must be bounded by other methods.

Below we try to argue why the notion of hypo-convergence is only moderately useful in the present context.

First, it does not yield global results, but must be complemented by showing tightness of the posterior sequence, which can be almost as hard as the original problem. Instead of a fixed compact, one could consider a sieve consisting of an increasing sequence of suitable compact sets. Then the posterior probability of the tail can be bounded by showing that its prior probability is exponentially small. However, hypo-convergence does not seem to yield the desired conclusion for the central part when it increases with n.

Secondly, because of averaging by the prior, convergence of the supremum of the log likelihood ratios is more than necessary for the consistency of the posterior. The inadequacy of the approach based on bounds for the supremum is particularly pronounced in a countable parameter space where posterior consistency holds under minimal conditions by Doob's (1948) theorem.

Thirdly, the concept of hypo-convergence appears conceptually more complicated than the condition(19) it is supposed to derive. The sufficient condition for hypo-convergence as in Theorem 3.4 is a device already available in Wald's consistency proof, and can be used directly to derive(19).

Finally, we note that(19) implies the classical testing condition of Schwartz (1965) for posterior consistency and hence it is not possible to derive any new consistency result using hypo-convergence. To see this, consider the test that accepts $H_0 : \theta = \theta_0$ when

$$\sup_{\theta\in U^c} H_n(\omega, \theta) < -\varepsilon$$

for some suitable $\varepsilon > 0$. Then (19) implies that the type I error probability is bounded away from 1. The type II error probability at any $\theta \in U^c$ is bounded by

$$P_\theta\Big(\sup_{\theta' \in U^c} H_n(\omega, \theta') < -\varepsilon\Big) \leq P_\theta\Big(H_n(\omega, \theta) < -\varepsilon\Big)$$
$$\leq P_\theta\Big(\prod_{i=1}^{n} \frac{f^{1/2}(X_i, \theta_0)}{f^{1/2}(X_i, \theta)} \geq e^{n\varepsilon/2}\Big) \leq e^{-n\varepsilon/2}$$

by Markov's inequality. The desired test can now be constructed using the i.i.d. structure.

The hypothesis of hypo-convergence does help to construct the desired tests. Because in a compact parameter space, the desired tests exist by the finiteness of the metric entropy (see *e.g.*, Ghosal *et al.*, 1999), this is again only of moderate usefulness.

REPLY TO THE DISCUSSION

First, I would like to thank the discussants for their useful suggestions and comments. The suggestions are all accepted and worth to be further developed; the comments, especially the critical ones, are all constructive and help to clarify potential and limits of hypo-convergence.

In answering the comments the role of a.s. hypo-convergence in convergence of stochastic suprema will be kept separated from its role in presence of tightness or in suitable compact embeddings. Preliminary to all that it has to be remarked that the approach to posterior consistency based on hypo-convergence does not replace or competes with uniformly consistent tests; these, when existing, provide a complete answer. Also, the reference to a compact parameter space as particular instance, can be of limited interest but this is not the main point as it will be tried to argue.

The fact that if a.s., for every open neighborhood U we have

$$\limsup_{n \to \infty} \sup_{\theta \in U^c} H_n(\omega, \theta) < 0 \tag{20}$$

then posterior consistency, as well as MLE's consistency hold, discloses the nature of the often observed parallel behavior of the two consistencies; since most of the literature on MLE's provides conditions under which (20) holds, posterior consistency, as already observed, can be obtained in all those circumstances where the prior information, including modelling assumptions, and the topology chosen, design a space Θ where these conditions are satisfied. In that, a.s. hypo-convergence is a preliminary "necessary" requirement, not sufficient in general, for the convergence of stochastic suprema. This will be the case if Θ is compact but that can be of limited interest in most applications. Weak compactness is easier to achieve either because it is provided by the metric adopted on Θ or requiring, for example, that the solutions of the optimization problem are bounded. In this case, a.s. Mosco-hypo-convergence, a somewhat stronger notion of hypo-convergence, provides convergence of suprema (Dong and Wets, 2000). This path parallels with the possibility offered by upper bracketing entropy with the consequent "upper" Glivenko–Cantelli classes which remains a key tool.

In the same direction to obtain posterior consistency through convergence of suprema, the suggestion to develop the hypo-convergence in the context of sieves is appealing; the a.s. convergence of $\sup_{\theta \in S_n} H_n(\omega, \theta)$ to $H_0(\theta_0)$ along the sieve $\{S_n\}$ relies on the hypo-convergence of $\hat{H}_n(\omega, \theta)$ to H_0, with $\hat{H}_n(\omega, \theta) = H_n(\omega, \theta)$ if $\theta \in S_n$ and $-\infty$ otherwise; it will provide also convergence of $\sup_{S_n \cap U^c} H_n(\omega, \theta)$ to $\sup_{U^c} H_0(\theta)$, hence posterior consistency if the prior $\Pi(S_n)$ is exponentially small. Some aspects concerning hypo-convergence in the context of sieves have been already touched in Dong and Wets (2000) in the form of penalized maximum likelihood, a sort of "dual" of the method of sieves.

Some additional comments are necessary on a.s. hypo-convergence, tightness and suitable compact embedding.

Hypo-convergence, when holding, highlights that possible sources of inconsistency are on the tails of $H_n(\omega, \cdot)$ identified with complements of compact subsets of Θ. Small prior probability on the tails can compensate the possible explosive behavior of these and accounts for those situations, as in Bahadur's example, where one has posterior consistency but not consistency of MLE's. On the other hand, a.s. tightness of the posteriors is a necessary condition; when holding, a.s. hypo-convergence on compacts is sufficient for posterior consistency. This delineates an approach based on conditions which jointly provide a.s. tightness and hypo-convergence. Almost sure hypo-convergence on compacts is not difficult to obtain; in general, tightness could be hard to verify but, since it is unavoidable, I am still convinced that it remains an interesting direction to work on; just as minor hint a possibility is offered by a revisitation, based on the prior Π, of the tail conditions of the type considered in Huber (1967).

The fact that inconsistency sits on complements of compact sets and the nature of inconsistency as highlighted in Freedman (1963) suggest an approach to the consistency problem based on a "suitable" compact embedding where, in the topology adopted, a.s. hypo-convergence is easily obtained. The way to proceed has been illustrated for two classes of problems, one well known in the context of posterior consistency and the other already considered in consistency of MLE's. Proceeding with this approach towards more general situations and considering the fact that consistency depends on the topology adopted, raise the questions of what is the "suitable" compact embedding, if there exists any, which is the relation with the metric in which the original consistency problem has been formulated and finally, in the Bayesian context, how to specify the prior. These questions are open, but in restricting the attention to uppersemicontinuous densities, as indicated in the conclusive remarks, the topology of hypo-convergence of the densities emerges as an appealing candidate.

ADDITIONAL REFERENCES IN THE DISCUSSION

Doob, J. L. (1948). Application of the theory of martingales. *Coll. Int. du CNRS, Paris*, 22–28.

Ghosal, S., Ghosh, J. K. and Ramamoorthi, R. V. (1999). Posterior consistency of Dirichlet mixtures in density estimation. *Ann. Statist.* **27**, 143–158.

Schwartz, L. (1965). On Bayes procedures. *Z. Wahrsch. Verw. Gebiete* **4**, 10–26.

BAYESIAN STATISTICS 7, pp. 385–401
J. M. Bernardo, M. J. Bayarri, J. O. Berger, A. P. Dawid, D. Heckerman, A. F. M. Smith and M. West (Eds.)

Measures of Incoherence: How not to Gamble if you Must

MARK J. SCHERVISH, TEDDY SEIDENFELD and JOSEPH B. KADANE
Carnegie Mellon University, USA
mark@stat.cmu.edu teddy@stat.cmu.edu kadane@stat.cmu.edu

SUMMARY

The degree of incoherence, when previsions are not made in accordance with a probability measure, is measured by the rate at which an incoherent bookie can be made a sure loser. We consider each bet from three points of view: that of the gambler, that of the bookie, and a neutral viewpoint. From each viewpoint, we define a normalization for each bet, and the sure loss for incoherent previsions is divided by the normalization to determine the rate of incoherence. Several different definitions of normalization are considered in order to determine plausible ranges for the degree of incoherence. We give examples of the measurement of incoherence of of some classical statistical procedures.

Keywords: BOOKIE; COHERENCE; ESCROW; GAMBLER.

1. INTRODUCTION

de Finetti (1974) describes the criterion of coherence for probabilities assigned to events and previsions assigned to general random variables. The idea is that, if one were to use these previsions as fair prices to pay for gambles on the random quantities, then the previsions are incoherent if and only if there exists a finite combination of the gambles that is guaranteed to lose at least some positive amount. To put this in mathematically precise language, let $X_1, \ldots, X_n$ be bounded random variables, that is bounded functions from a set of states of nature S to the reals. For $i = 1, \ldots, n$, let p_i be the prevision of X_i, so that $\alpha_i(X_i - p_i)$ is a fair gamble for all sufficiently small $|\alpha_i|$ values. The previsions are *incoherent* if there exist values $\alpha_1, \ldots, \alpha_n$ such that

$$\sup_s \sum_{i=1}^{n} \alpha_i(X_i(s) - p_i) < 0. \tag{1}$$

The previsions are *coherent* if they are not incoherent. If (1) holds, we say that *Dutch book* has been made against the person who offered the previsions.

de Finetti (1974) merely partitions all prevision assignments into two classes: coherent and incoherent. It seems reasonable, however, to expect that some collections of previsions are more incoherent than others. It is our goal in this paper to set up a framework in which one can measure how incoherent is a collection of incoherent previsions.

We begin by considering a slightly more general situation than that described above, both because it strengthens the results and because the most interesting examples are of the more general form. Think of previsions being assigned to random variables by a bookie who is then going to take bets from one or more gamblers. Suppose that the bookie chooses the prevision p for a bounded random variable X. The gain to the bookie when the state of nature is s and the bookie accepts the gamble $\alpha(X - p)$ from a gambler is $\alpha(X(s) - p)$. It is common in some gambling situations (horse racing is a common example) for bookies to offer only one-sided previsions. That is, the bookie offers a prevision p for X, but accepts only gambles of the form $\alpha(X - p)$ for $\alpha > 0$. Such a prevision p will be called a *lower prevision*. Similarly, if the bookie accepts only gambles of the form $\alpha(X - p)$ for $\alpha < 0$, we call p an *upper prevision* for X. It should be easy to see that, if a bookie is willing to accept the gamble $\alpha(X - p)$ with $\alpha > 0$, then the bookie should also be willing to accept the gamble $\alpha(X - x)$ for every $x < p$, since the payment to the bookie from this second gamble is greater than that of the first by $\alpha(p - x) > 0$ no matter what state of nature occurs. Similarly, if p is an upper prevision, the bookie should be willing to accept the gamble $\alpha(X - x)$ for all $x > p$ when $\alpha < 0$. In the future when we refer to a coefficient of the *appropriate sign*, we shall mean a positive α for a gamble based on a lower prevision and a negative α for a gamble based on an upper prevision.

Example 1. A simple example of incoherence occurs when a lower prevision is greater than an upper prevision. Suppose that $p > q$ where p is a lower prevision for X and q is an upper prevision for X. Then the bookie is willing to accept the following two gambles, $\alpha(X - p)$ and $-\alpha(X - q)$ for $\alpha > 0$. The sum of these two gambles is $\alpha(q - p) < 0$ no matter what state of nature occurs. There is a sense in which, the larger $p - q$ is, the more the bookie stands to lose from such gambles. Of course, the bookie loses more the larger α is, regardless of how large $p - q$ is. However, α is more a measure of how much the gambler wishes to bet than it is a measure of the bookie's incoherence.

We would like to measure the incoherence of the bookie's previsions by looking at how much the bookie can be forced to lose relative to how much the gambler needs to bet in order to force the loss. As noted in Example 1, if the gambler bets twice as much (double α) the bookie loses twice as much, even though the previsions have not changed. Normalization of the combinations of bets that a bookie is willing to accept would permit extraction of the degree of incoherence from the size of the loss that the bookie incurs.

In summary, incoherence arises when a bookie offers previsions such that there exists a combination of gambles for which the bookie is guaranteed to lose at least some positive amount. A gambler can increase the guaranteed loss by increasing the sizes of his/her bets with the bookie. We measure the incoherence of incoherent previsions by looking at how large the guaranteed loss can be made relative to some normalization for the collection of gambles that the gambler chooses.

2. NORMALIZATION

As noted earlier, normalizations measure the sizes of gambles. There are two important aspects of normalization that we have chosen to separate. The first is the normalization of a single gamble, and the second is the normalization of sums of gambles. Treating a sum of gambles as a single gamble obscures the fact that the coefficients can be chosen individually and that the previsions are given individually. For a single gamble, there are several possible ways to measure the size. Let X be a random quantity with prevision p, and let α be a constant with the appropriate sign. The normalization for $\alpha(X - p)$ is allowed to depend on all three of α, X, and p. (When we say that the normalization depends on X, we mean that it depends on the function X from states of nature to the real numbers, not on the unknown value $X(s)$. For example, the normalization can depend on the maximum and/or minimum value of X, etc.) Some examples of normalizations include the following three:

1. *The bookie's escrow:* $\max\{0, -\inf_s \alpha[X(s) - p]\}$.
2. *The gambler's escrow:* $\max\{0, \sup_s \alpha[X(s) - p]\}$.

3. *The neutral normalization:* $|\alpha|$.

The two escrows have interpretations in terms of betting. If the bookie requires each gambler to show that they have sufficient funds to pay off any bets they might lose, then the gambler's escrow, being the most the gambler can lose, will cover the bet. Similarly, the bookie's escrow will cover the the bookie's largest possible loss. Notice that all three of the normalizations listed above have the following two properties:

(i) The normalization is always nonnegative.
(ii) If α is changed to $c\alpha$ for $c > 0$, the normalization gets multiplied by c as well.

These are the important properties that we require of a normalization. Although there are many normalizations that satisfy these two conditions, we consider only the three described above in this paper.

When we combine several gambles, we need to be able to normalize the combination. We have chosen to require the normalization of a combination of gambles to depend solely on the normalizations of the individual gambles that go into the combination. The reason for this is that, once we start to combine gains and losses, we are starting to do some of the work of measuring incoherence. For example, consider $X_1 = I_A$, the indicator of an event with prevision p and $X_2 = I_{A^C}$, the indicator of the complement with prevision q. If we combine $(I_A - p) + (I_{A^C} - q)$, we get $1 - p - q$, which already tells us whether or not the previsions are coherent. Our assumption is designed to help us separate the normalization from the measurement of incoherence.

Schervish *et al.* (2002b) give a list of criteria that a normalization should satisfy, including the assumption just mentioned. We summarize these here. Let $f_n(x_1, \ldots, x_n)$ denote the normalization for a combination of n gambles whose individual normalizations are $x_1, \ldots, x_n$.

1. f_n is homogeneous of degree 1, that is, $f_n(cx_1, \ldots, cx_n) = cf_n(x_1, \ldots, x_n)$ for all $c > 0$.
2. f_n is invariant under permutations of its arguments.
3. f_n is nondecreasing in each argument.
4. $f_{n+1}(x_1, \ldots, x_n, 0) = f_n(x_1, \ldots, x_n)$.
5. The normalization is no larger than the sum of the individual normalizations.
6. f_n is continuous.
7. $f_1(x) = x$.

The first condition implies that scaling up a collection of gambles by the same amount does not change the rate of incoherence. The second condition expresses the fact that we do not care in what order the gambles are written. The third condition expresses the idea that bigger gambles should require larger normalization. The fourth condition says that if a gamble does not require any normalization, then the other gambles should determine the normalization for the combination. The fifth condition is like the triangle inequality for norms, saying that the whole is no greater than the sum of the parts. It makes particular sense if the normalization is thought of as an escrow. The sixth condition says that small changes in individual normalizations should produce small changes in the overall normalization. The seventh condition merely expresses the fact that the individual normalizations are just that.

Based on these criteria, Schervish, Seidenfeld and Kadane (2002b) characterize the collection of functions f_n that can be used for normalization. In particular, they find that the largest and smallest normalizations arise from the following two functions:

$$f_{0,n}(x_1, \ldots, x_n) = \max\{x_1, \ldots, x_n\}, \quad f_{1,n}(x_1, \ldots, x_n) = \sum_{i=1}^{n} x_i.$$

In this paper, we consider only normalizations based on these two functions, which we call the *max* and *sum* normalizations.

In summary, a normalization is defined by two choices. One is a choice of normalization for individual gambles, and the other is a choice of normalization function for combining individual normalizations. We have given specific examples of three individual normalizations and two combination functions. By pairing these, we produce six different examples of how to normalize combinations of gambles. We refer to these pairings with names such as neutral/sum, gambler's escrow/max, etc. We assume throughout this paper that the same individual normalization is chosen for all individual gambles when they are being combined.

3. RATE OF INCOHERENCE

We are now in position to say how we measure incoherence. Consider a finite collection of gambles $Y_i = \alpha_i(X_i - p_i)$ for $i = 1, \ldots, n$. The *guaranteed loss* from the combination $Y = \sum_{i=1}^{n} Y_i$ is

$$G(Y) = -\min\left\{0, \sup_s Y(s)\right\}.$$

Dutch book has been made with this combination if and only if $G(Y) > 0$. It is clear that if each Y_i is changed to cY_i for $c > 0$, then Y changes to cY and $G(cY) = cG(Y)$. We measure the rate of incoherence of Y by first dividing $G(Y)$ by a normalization. If $e(Y_i)$ is our chosen normalization for individual gamble Y_i, then our normalization for Y is $f_n(e(Y_1), \ldots, e(Y_n))$, for some function f_n satisfying the conditions mentioned earlier. The *rate of guaranteed loss* for Y is then

$$H(Y) = \frac{G(Y)}{f_n(e(Y_1), \ldots, e(Y_n))}.$$

For a fixed collection of random variables $X_1, \ldots, X_n$ and previsions $p_1, \ldots, p_n$, the *rate of incoherence* is the maximal value of $H(Y)$ over all choices of $\alpha_1, \ldots, \alpha_n$ that have the appropriate signs.

Example 2. Consider an event A and its complement A^C. Suppose that a bookie offers lower previsions of 0.6 for both of these events. These are clearly incoherent, and there are many combinations of gambles that make Dutch book. All such combinations must be of the form $Y = \alpha_1(I_A - 0.6) + \alpha_2(I_{A^C} - 0.6)$ for $\alpha_1, \alpha_2 > 0$. In this case, Y only takes two values, $0.4\alpha_1 - 0.6\alpha_2$ and $0.4\alpha_2 - 0.6\alpha_1$. Hence,

$$G(Y) = \begin{cases} 0 & \text{if} \alpha_1 \geq 1.5\alpha_2 \text{ or } \alpha_2 \geq 1.5\alpha_1, \\ 0.6\min\{\alpha_1, \alpha_2\} - 0.4\max\{\alpha_1, \alpha_2\} & \text{otherwise.} \end{cases}$$

That is, Dutch book is made if and only if both α_1/α_2 and α_2/α_1 are less than 1.5. For the neutral/sum normalization, the normalization equals $\alpha_1 + \alpha_2$. So, when Dutch book can be made,

$$H(Y) = \frac{0.6\min\{\alpha_1, \alpha_2\} - 0.4\max\{\alpha_1, \alpha_2\}}{\alpha_1 + \alpha_2} = r - 0.4,$$

where $r = \min\{\alpha_1, \alpha_2\}/(\alpha_1 + \alpha_2)$ can be any number between 0.4 and 0.5. Clearly, we maximize $H(Y)$ by choosing $r = 0.5$. This makes the rate of incoherence 0.1 in this example. We can achieve this rate by choosing $\alpha_1 = \alpha_2$.

If, instead of neutral/sum, we had used neutral/max, then the normalization would have been $\max\{\alpha_1, \alpha_2\}$ and then $H(Y) = 0.6r/(1 - r) - 0.4$, where r is the same as before. This is also maximized with $r = 0.5$, and the rate of incoherence is then 0.2. This illustrates a fact that is quite general. Since $f_{0,n}$ and $f_{1,n}$ are, respectively, the smallest and largest normalizations, they lead, respectively, to the largest and smallest rates of incoherence when combined with a common individual normalization. In the case of events, the neutral normalization is the sum of the gambler's escrow and the bookie's escrow whenever the prevision is between 0 and 1. Hence, rates of incoherence based on the two escrows are larger than those based on the neutral normalization.

4. CHOOSING AMONG NORMALIZATIONS

We have introduced at least six ways to normalize sums of gambles when computing rates of incoherence. Each has its advantages and disadvantages. There are, however, some properties that we have been able to determine for some of them. The two escrows have interpretations in terms of amounts needed to cover the bets. We have not yet found any operational interpretation for the neutral normalization. The escrows have the property that a prevision of p for X and a prevision of cp for cX (with $c > 0$) have equivalent effects on any rate of incoherence. This is not true of the neutral normalization. For example, suppose that we wish to include $\alpha(X - p)$ in a combination of gambles, where p is a coherent lower prevision for X and $\alpha > 0$. The three individual normalizations for $(\alpha/c)(cX - cp)$ are $\alpha(\sup X - p)$ (gambler's escrow), $-\alpha(\inf X - p)$ (bookie's escrow) and α/c (neutral). Notice that the first two are the same as they would be for the gamble $\alpha(X - p)$, whereas the third is different. One consequence of this is the following. If it turns out that $\alpha(X - p)$ is included in a combination of gambles Y that is incoherent, and if $c > 1$, then $H(Y)$ based on the neutral normalization will be larger if one uses $(\alpha/c)(cX - cp)$ instead, because this version requires smaller normalization and produces the same payoffs as $\alpha(X - p)$. We do not find this to be a troubling feature of the neutral normalization because incoherent agents are free to choose previsions other than cp for cX even after they have chosen p as the prevision of X. From this point of view, X and cX are different random variables, and there is no compelling reason that their effects on the rate of incoherence need be related.

Schervish, Seidenfeld and Kadane (2002b) have established some additional properties of the various normalizations. They have established conditions under which the rate of incoherence is continuous as a function of the random variables and their previsions. Continuity of the rate of incoherence means (loosely) if $\sup_s |X_1(s) - X_2(s)| < \epsilon$ and $|p_1 - p_2| < \epsilon$, then substituting X_2 with prevision p_2 for X_1 with prevision p_1 should make a small change to the rate of incoherence. Schervish, Seidenfeld and Kadane (2002b) show that the neutral/sum normalization gives a continuous rate of incoherence and all neutral normalizations give continuous rates of incoherence if only finitely many gambles have been assigned previsions. Normalizations based on the escrows give continuous rates of incoherence only under the assumption that each individual prevision is coherent by itself (and not just barely so). That is, if p is a lower prevision for X, then $p < \sup_s X(s)$ and if p is an upper prevision for X, then $\inf_s X(s) < p$. Equality in either of these would lead to 0 normalization for an individual gamble, and rates of incoherence are not continuous when 0 normalizations occur.

Example 3. Consider a constant random variable $X = c$. Suppose that an incoherent bookie specifies a lower prevision $p > c$. The gambler's escrow will be 0, since the gambler cannot lose a bet of the form $\alpha(X - p)$ with $\alpha > 0$. The rate of incoherence is ∞ in this case. The bookie's escrow is $\alpha(p - c)$ and so is $G(Y)$, hence the rate of incoherence is 1 no matter how far p is from c. If $p = c$, the rate of incoherence drops to 0 for both escrows. For the neutral normalization, the normalization is α and $G(Y) = \alpha(p - c)$, so the rate of incoherence is $p - c$, which increases as p increases, as intuition might suggest.

Another property considered by Schervish, Seidenfeld and Kadane (2002b) is dominance. Suppose that two different bookies offer previsions for the same random variables. Suppose that for a specific set of coefficients for the gambles that makes a Dutch book, the second bookie's losses are always larger than the first's. Then we say that the first bookie dominates the second with respect to those coefficients. Intuitively, one might expect the function H computed for the first bookie to be smaller than the function H computed for the second bookie. This is true whenever the normalization is neutral. For the bookie's escrow, we need to assume that none of the individual gambles be a sure winner for the bookie. For the gambler's escrow, we need to assume that none of the individual gambles be a sure winner for the gambler.

On balance, it might appear that the neutral normalizations have better mathematical properties than the others, but we continue to work with all of them for the remainder of this paper.

5. CLASSICAL INFERENCE

Inference techniques used by non-Bayesians have sometimes been criticized on the grounds that they are incoherent. If they are indeed incoherent, then we should be able to make Dutch book and measure the rate of incoherence. For example, suppose that one chooses to test all null hypotheses at level 0.05 regardless of how the data arose or how many observations are available. It is well-known (see Cox, 1958 and Lindley, 1972 for examples) that there are cases in which such behavior runs afoul of admissibility if not coherence. Two obvious stumbling blocks stand in the way of applying the concept of coherence to classical inferences. First, classical inferences do not provide previsions. In particular, they do not provide probabilities or expected values for unknown quantities. They are often based upon probabilities and expectations for random variables that used to be unknown, but have since been observed. The unknown quantities never become the subject of a probabilistic calculation. Secondly, classical statisticians are not prepared to gamble based on their inferences.

This last point suggests a fruitful avenue to pursue. If nothing is at stake, who cares what inference is made? So, suppose that there is a decision problem with a loss function. Classical statisticians are willing to talk about decision theory. Indeed, they have developed a theory of risk functions, admissibility, minimaxity, etc. When statisticians choose one decision rule δ_0 over another δ_1, they express a preference for suffering the risk function $R(\theta, \delta_0)$ to $R(\theta, \delta_1)$. We choose to interpret such a preference by saying that $R(\theta, \delta_1) - R(\theta, \delta_0)$ is a favorable gamble. If inferences are incoherent we should be able to combine such favorable gambles to make Dutch book.

5.1. *Simple Hypothesis Testing*

Consider first the case of testing a simple null hypothesis against a simple alternative hypothesis. (Schervish, Seidenfeld and Kadane, 2002a, consider this situation in more detail.) Suppose that the loss function has the simple 0–1 form. That is, the loss is 0 if the correct hypothesis is chosen, and the loss is 1 of the incorrect hypothesis is chosen.

Example 4. Suppose that X has a normal distribution with mean θ and variance σ^2, where σ^2 is known, but we want to test the null hypothesis $H_0 : \theta = 0$ versus the alternative $H_1 : \theta = 1$. Suppose that our classical statistician wants to use the most powerful level 0.05 test no matter what σ^2 is. For example, if $\sigma^2 = 1$, the most powerful level 0.05 test is to reject H_0 if $X > 1.645$. In general, we can write the risk function of the most powerful level α_0 test δ_{α_0} as

$$\begin{aligned} R(\theta, \delta_{\alpha_0}) &= \begin{cases} \alpha_0 & \text{if } \theta = 0, \\ \Phi(\Phi^{-1}(1-\alpha_0) - \sigma^{-1}) & \text{if } \theta = 1, \end{cases} \\ &= [\alpha_0 - \Phi(\Phi^{-1}(1-\alpha_0) - \sigma^{-1})] I_{\{0\}}(\theta) + \Phi(\Phi^{-1}(1-\alpha_0) - \sigma^{-1}). \end{aligned}$$

So, if the statistician prefers the level 0.05 test to the level 0.1 test when $\sigma = 1$, the difference in risk functions is

$$R(\theta, \delta_{0.1}) - R(\theta, \delta_{0.05}) = 0.1796[I_{\{0\}}(\theta) - 0.7217],$$

indicating that 0.7217 is a lower prevision for the event $\{\theta = 0\}$. (Technically, all we can really say is that (0.7217×0.1796) is a lower prevision for $0.1796 I_{\{0\}}$.) Suppose also that the statistician prefers the level 0.05 test to the level 0.01 test when $\sigma = 2$. Then the difference of the risk functions is

$$R(\theta, \delta_{0.01}) - R(\theta, \delta_{0.05}) = -0.1322[I_{\{0\}}(\theta) - 0.6975],$$

making it appear as if 0.6975 is an upper prevision for $\{\theta = 0\}$. An upper prevision that is lower than a lower prevision ought to be incoherent. If we combine the first gamble with $\alpha_1 = 1/0.1796$ and the second gamble with $\alpha_2 = 1/0.1322$, we get the combination with constant value -0.0242. Hence, we have made Dutch book. For the two gambles being discussed, the rate of incoherence using bookie's escrow/sum normalization is

$$\frac{0.0242}{0.7217 + (1 - .6975)} = 0.02363.$$

In Example 4, we could have chosen different risk functions to trade with the classical statistician. For example, when $\sigma = 1$, we could trade the risk function of the level 0.06 test instead of the risk function of the level 0.1 test. Similarly, when $\sigma = 2$, we could trade the risk function of the level 0.04 test instead of the level 0.01 test. Each of these alternative trades leads to two different gambles that provide different Dutch books, and different rates of incoherence. We might ask which trades, if any, lead to the largest rate of incoherence.

Schervish *et al.* (1997, 2002a) present a theorem that says essentially the following. Use bookie's escrow/sum normalization. For each σ, the most powerful level α_0 test is a Bayes rule with respect to a unique prior $p(\sigma) = \Pr(\theta = 0)$. Let σ_0 and σ_1 be two different possible variances. Let p_i $(i = 0, 1)$ be the two priors $p_i = p(\sigma_i)$ for $i = 0, 1$. Assume that $p_0 > p_1$. For each $i = 0, 1$, consider all possible trades of the risk function for some test δ for the risk function of the most powerful level α_0 test. The largest possible rate of incoherence from combining two such risk function trades is $(p_0 - p_1)/(1 - p_1 + p_0)$.

Example 5. We can return to Example 4 and ask what is the largest rate of incoherence. In that example, each of the two-level 0.05 tests is the Bayes rule with respect to a unique prior. When $\sigma = 1$, the level 0.05 test is the Bayes rule with respect to the prior $p_0 = 0.7586$, and when $\sigma = 2$, the level 0.05 test is the Bayes rule with respect to the prior $p_1 = 0.6676$. The theorem quoted above says that the maximum rate of incoherence for bookie's escrow/sum normalization is $(0.7586 - 0.6676)/(1 + 0.7586 - 0.6676) = 0.08341$.

5.2. *Minimax Estimation*

Hypothesis testing at a fixed level is not the only incoherent classical inference. Minimax estimation also suffers from the same malady. Consider a binomial random variable X with parameters n and θ. The minimax estimator of θ is (see Schervish, 1995, Example 3.62)

$$\delta_n(X) = \frac{\sqrt{n}/2 + X}{\sqrt{n} + n}.$$

For each sample size n, δ_n is the Bayes rule with respect to the beta distribution prior with parameters $\sqrt{n}/2$ and $\sqrt{n}/2$, as well as several others, but the prior must change as n changes.

If indeed it is incoherent to use the minimax rule for different sample sizes, we should be able to make Dutch book by trading risk functions.

Example 6. Suppose that we offer to trade the risk function of a Bayes rule with respect to a different beta distribution prior for the risk function of the minimax rule. The risk function of the the Bayes rule $\beta_{n,\gamma}$ with respect to a beta distribution with parameters γ and γ is

$$R(\theta,\beta_{n,\gamma}) = \frac{n\theta(1-\theta)+\gamma^2(1-2\theta)^2}{(n+2\gamma)^2}.$$

If $n=1$, the minimax rule is $\delta_1(X)=(0.5+X)/2$ and the minimax risk is $1/16$. Suppose that we offer to trade the risk function of the Bayes rule with respect to the beta distribution prior with parameters 1.3 and 1.3, $\beta_{1,1.3}(X)=(1.3+X)/3.6$. If $n=4$, the minimax rule is $\delta_4(X)=(1+X)/5$ with minimax risk $1/36$. In this decision problem, we might offer to trade the risk function of the rule $\beta_{4,0.8}$. If we make these two trades, we can then combine the two favorable gambles as follows:

$$\alpha_1[R(\theta,\beta_{4,0.8})-R(\theta,\delta_4)]+\alpha_2[R(\theta,\beta_{1,1.3})-R(\theta,\delta_1)]. \tag{2}$$

As a function of θ, (2) is a quadratic that is symmetric around $\theta=1/2$. The geometry of this quadratic shows that the maximum value is minimized when α_1 and α_2 are chosen to make the resulting function constant. Using neutral/sum normalization, the rate of incoherence in this case is 0.00032, and it is achieved with $\alpha_1=9.679$ and $\alpha_2=1$.

As in Example 5, we can replace the two trades in Example 6 by different trades that might achieve higher rates of incoherence. In general, the maximum value of

$$\alpha_1[R(\theta,\beta_{n_1,\gamma_1})-R(\theta,\delta_{n_1})]+\alpha_2[R(\theta,\beta_{n_2,\gamma_2})-R(\theta,\delta_{n_2})]$$

divided by $\alpha_1+\alpha_2$ is smallest when the function is constant. To achieve this, we need one of γ_i to be less than $\sqrt{n_i}/2$ and the other to be larger. Then, we can choose

$$\alpha_{3-i} = \left|\frac{4\gamma_i^2-n_i}{(n_i+2\gamma_i)^2}\right|,$$

for $i=1,2$. The constant value of the combination of gambles is then

$$\alpha_1\left(\frac{\gamma_1^2}{(n_1+2\gamma_1)^2}-\frac{1}{4+8\sqrt{n_1}+4n_1}\right)+\alpha_2\left(\frac{\gamma_2^2}{(n_2+2\gamma_2)^2}-\frac{1}{4+8\sqrt{n_2}+4n_2}\right).$$

Take the negative of this, normalize it, and then maximize the ratio by choice of γ_1 and γ_2. In Example 6, with neutral/sum normalization, the choices that provide the maximum are $\gamma_1=\gamma_2=0.653$, and the rate of incoherence for these choices is 0.0011.

We have not yet solved the general problem of finding the combination of risk function trades that leads to the largest rate of incoherence. To do that, we would have to consider the risk functions of all possible Bayes rules. Since every Bayes rule is the posterior mean of θ, we can write each Bayes rule as

$$\delta(x) = \frac{\int_{[0,1]}\theta^{x+1}(1-\theta)^{n-x}d\mu(\theta)}{\int_{[0,1]}\theta^{x}(1-\theta)^{n-x}d\mu(\theta)}, \tag{3}$$

where μ is an arbitrary prior on $[0,1]$. Notice that (3) depends only on the first $n+1$ moments of the prior distribution μ. Feller (1972, Section 7.3) shows that, for every possible set of first

$n+1$ moments, there is a prior concentrated on the points $k/(n+1)$, for $k = 0, \ldots, n+1$ that has those moments. Hence, the collection of all Bayes rules can be parameterized by the finite-dimensional collection of all prior distributions concentrated on $n+2$ equally spaced points. Although the resulting maximization problem is not likely to have a closed-form solution, a numerical algorithm should be possible to solve it.

5.3. *Testing a Sharp Null Hypothesis*

Suppose that X has a normal distribution with unknown mean θ and known variance σ^2, and we wish to test the null hypothesis $H_0 : \theta = 0$ versus the alternative $H_1 : \theta \neq 0$. Several authors have highlighted sharp differences between Bayesian methods and testing such hypotheses at fixed levels (see, *e.g.*, Berger and Sellke, 1987; Schervish, 1996). We can ask whether such testing strategies are incoherent. For example, suppose again that the loss function is of the following type:

$$L(\theta, a) = \begin{cases} c & \text{if } \theta = 0 \text{ and } a = 1, \\ 1 & \text{if } \theta \neq 0 \text{ and } a = 0, \\ 0 & \text{otherwise,} \end{cases} \tag{4}$$

where action $a = 1$ means reject H_0 and $a = 0$ means do not reject H_0. Then, the risk function of the uniformly most powerful unbiased (UMPU) level α_0 test δ_{α_0} has the following property for every $\alpha_0 < 1$:

$$\lim_{|\theta| \to \infty} R(\theta, \delta_{\alpha_0}) = 0.$$

It follows that, for every test δ,

$$\limsup_{|\theta| \to \infty} R(\theta, \delta) - R(\theta, \delta_{\alpha_0}) \geq 0.$$

Hence, if we combine finitely many such risk function trades, the combination will also have nonnegative lim sup as $|\theta| \to \infty$. It follows that we cannot make Dutch book by trading risk functions.

One could argue that, in practical problems, $|\theta|$ is bounded. Since UMPU tests remain UMPU when extreme portions of the parameter space are removed (as long as an interval around the null remains), we could try to make Dutch book with a bounded parameter space. Even this is not possible. The power function of every test is continuous, hence, for every test δ,

$$\lim_{\theta \to 0} R(\theta, \delta) - R(\theta, \delta_{\alpha_0}) = -\frac{1}{c}[R(0, \delta) - R(0, \delta_{\alpha_0})].$$

It follows that, for every finite combination of risk function trades, the value at 0 will have magnitude c times as big as but the opposite sign of the limit of the values as $\theta \to 0$. Hence, these two values cannot both be negative, and we still cannot make Dutch book. The arguments given here apply regardless of how one chooses the level α_0 as a function of σ^2. One could use the same level for all values of σ^2 or one could arbitrarily choose different values of α_0 as σ^2 changes.

If it is coherent to choose the level of the test as an arbitrary function of σ^2, one would expect that there must be at least one prior distribution such that these choices would be Bayes rules (or formal Bayes rules) with respect to each such prior. Indeed, we have been able to identify those prior distributions that have this property when the level is chosen to be the same value α_0 for all values of σ^2.

Theorem 1. *Let μ be a prior such that, for every test δ the trade of the risk function of δ for the risk function of the UMPU level α_0 test has nonnegative expected value for all σ. Then μ is finitely additive and has the following properties. Let $p_0 = \mu(\{0\})$, $q = \inf_{a>0} \mu(\{\theta : 0 < |\theta| < a\})$ and $r = \inf_{b>0} \mu(\{\theta : |\theta| > b\})$. Then $q = cp_0$ and $p_0 + q + r = 1$.*

Proof. Suppose that we trade the risk function of the UMPU level α_1 test for the risk function of the UMPU level α_0 test. The trade has the value $c(\alpha_1 - \alpha_0)$ for $\theta = 0$ and for $\theta \neq 0$, the value is

$$\Phi\left(\Phi^{-1}\left(1-\frac{\alpha_0}{2}\right)-\frac{\theta}{\sigma}\right)-\Phi\left(-\Phi^{-1}\left(1-\frac{\alpha_0}{2}\right)-\frac{\theta}{\sigma}\right)$$
$$-\Phi\left(\Phi^{-1}\left(1-\frac{\alpha_1}{2}\right)-\frac{\theta}{\sigma}\right)+\Phi\left(-\Phi^{-1}\left(1-\frac{\alpha_1}{2}\right)-\frac{\theta}{\sigma}\right). \tag{5}$$

For each $\theta \neq 0$, the value in (5) goes to 0 as $\sigma \to 0$ and it goes to $\alpha_0 - \alpha_1$ as $\sigma \to \infty$. From these facts, it follows that the limit as $\sigma \to 0$ of the expected value of (5) is $(\alpha_1 - \alpha_0)(cp_0 - q)$. Unless $q = cp_0$, this can be made negative by appropriate choice of α_1. Hence, $q = cp_0$ is necessary in order for the expected value to be nonnegative.

Next, notice that every prior as described in the theorem has the property that the expected value of the risk function trade is 0. Finally, suppose that $p_0 + q + r < 1$. Then there exists a bounded interval $[a, b]$ such that $\mu([a, b]) > 0$. Let $\alpha_1 > \alpha_0$. Then (5) is negative for all $\theta \neq 0$ and the expected value is strictly negative. Hence $1 = p_0 + q + r$. □

Although the prior in Theorem 1 gives non-negative expected value to every risk function trade of the form (5), it also gives expected value 0 to every risk function trade between any two UMPU tests regardless of what their levels are. That is, all UMPU tests are equally good.

5.4. *The Significance Level Depends on the Loss Function*

When asked "How should I choose the size of a test?" the classical statistician sometimes responds by saying that the size should depend on the costs of type I and/or type II errors. For example, the more serious is a type I error, the smaller the size should be. Consider loss functions of the form (4). The larger c gets, the smaller should be the significance level (size) of the preferred test. Theorem 1 suggests that even this interpretation will not stand up to Dutch book, if the parameter space is bounded. Notice that Theorem 1, when applied to decision problems with different loss functions (different values of c) says that the only prior distributions that support letting the size of the test depend only on c are the ones for which $\Pr(|\theta| > b) = 1$. If the parameter space is bounded, no prior will support such preferences.

Example 7. Suppose that we have two different decision problems both with loss functions of the form (4) but with two different values of c, c_0 and c_1. Call these problem 0 and problem 1 respectively. Suppose that the parameter space is bounded, that is, we know that $|\theta| < b$. Suppose that our classical statistician prefers the UMPU level α_i test in problem i for $i = 0, 1$, where α_0 is not necessarily the same as α_1. We can make Dutch book against such preferences. Without loss of generality, assume that $c_1 > c_0$. Suppose that we trade the risk function of a UMPU level β_i test for the risk function of the UMPU level α_i test in problem i. Then the values of these trades at $\theta = 0$ are $c_i(\beta_i - \alpha_i)$ for $i = 0, 1$. The values of such trades for $\theta \neq 0$ depend on σ, and they have a form similar to (5). The rate of incoherence depends on many factors, including the bound on the parameter space. Here, we merely illustrate how Dutch book can be made. Suppose that $\alpha_0 = 0.1$ and $\alpha_1 = 0.05$ while $c_0 = 1$ and $c_1 = 2$. In problem 0, our classical statistician prefers the level 0.1 test to all others regardless of σ, and in problem 1, he/she prefers the level 0.05 test to all others regardless of σ. Consider problem 0 with $\sigma = 1.234$ and $\beta_0 = 0.11$ and problem 1 with $\sigma = 1$ and $\beta_1 = 0.045$. Suppose that the parameter space is bounded at 4, that is $|\theta| \leq 4$. If we combine the two trades, described in this example with coefficients γ and $1-\gamma$, the value at $\theta = 0$ is $0.01\gamma - 0.01(1-\gamma) = 0.02\gamma - 0.01$. This will be negative for all $\gamma \in [0, .5)$. With $\gamma = 0.49223$, the value at $\theta = 0$ is -1.55×10^{-4}, and the max for $0 < |\theta| < 4$ is also -1.55×10^{-4}. The value $\gamma = 0.49223$ achieves the rate of incoherence for these two trades for all parameters spaces of the form $\{\theta : |\theta| \leq b\}$ for $b \in [2, 5.4]$. For b outside of this range, a different rate of incoherence occurs.

6. GENERAL RESULTS

Schervish *et al.* (1997, 2002a) prove several general results about incoherent previsions for elements of a partition. To summarize these, let $A_1, \ldots, A_n$ be a partition. That is, one of the events $A_1, \ldots, A_n$ must occur and no two of them can occur simultaneously. If upper previsions for these events add up to $q < 1$, then the rate of incoherence is $(1-q)/(n-q)$, $(1-q)/q$ and $(1-q)/n$ for bookie's escrow/sum, gambler's escrow/sum and neutral/sum normalizations, respectively. If lower previsions for these events add to $p > 1$, the rate of incoherence for bookie's escrow/sum normalization is $(p-1)/p$. For the other two normalizations based on sum, the rate of incoherence depends on whether any partial sums of the lower previsions are strictly more than 1.

Example 8. Let A_1, A_2, A_3 form a partition. Suppose that the lower previsions of 0.1, 0.8 and 0.7 are given for these events respectively. Suppose that we use neutral/sum normalization. The rate of guaranteed loss for the combination of gambles $(I_{A_2} - 0.8) + (I_{A_3} - 0.7)$ is 0.25. If we use all three gambles, the rate is 0.2. Hence, the maximal rate is achieved by combining fewer than three gambles.

Additional results concern a partition together with a simple random variable X measurable with respect to that partition. That is, $X = \sum_{i=1}^{n} x_i I_{A_i}$. Let p_i be both an upper prevision and a lower prevision for A_i for $i = 1, \ldots, n$. Let p_X be both an upper prevision and a lower prevision for X. Let $\mu = \sum_{i=1}^{n} x_i p_i$ and $\delta = p_X - \mu$. Schervish *et al.* (1997) address the question of how much, if any, the added prevision for X increases the rate of incoherence already computed for the previsions of the partition elements. Not surprisingly, the degree to which the added prevision increases the rate of incoherence depends on both δ and how incoherent the previsions are for the events in the partition. Indeed, there are cases in which the previsions of the events in the partition are sufficiently incoherent that the additional prevision might not increase the rate of incoherence at all if δ is small enough. To summarize these results, for bookie's escrow/sum normalization, let $q = \sum_{i=1}^{n} p_i$, and assume that $x_1 < x_2 < \cdots < x_n$. Then the values of p_X that do not increase the rate of incoherence above what it was for the previsions of the partition elements are as follows:

$$\mu + \frac{1-q}{n-1} \sum_{i=1}^{n-1} x_i \leq p_X \leq \mu + \frac{1-q}{n-1} \sum_{i=2}^{n} x_i \text{ if } q < 1,$$

$$\max\{x_1, \mu - (q-1)x_n\} \leq p_X \leq \min\{x_n, \mu - (q-1)x_1\} \text{ if } q > 1,$$

$$p_X = \mu \text{ if } q = 1.$$

For the gambler's escrow/sum normalization, when $q < 1$, the corresponding set of p_X is

$$\mu + (1-q)x_1 \leq p_X \leq \mu + (1-q)x_n.$$

This last range has an interesting interpretation. The p_X values that do not increase the rate of incoherence are those formed by computing the "expected value" of X using the incoherent previsions and then placing the remaining probability $1 - q$ anywhere else between the smallest and largest possible values of X. Schervish *et al.* (2000) describe how the incoherent previsions that add up to $q < 1$ are related to lower probabilities from ϵ-contamination models. In particular, the lower and upper expectations of the simple random variable X discussed above turn out to be $\mu + (1 - q)x_1$ and $\mu + (1 - q)x_n$.

7. DISCUSSION

In this article, we introduce several indices of incoherence of previsions, based on the gambling framework of de Finetti (1974). When a bookie is incoherent, a gambler can choose a collection of gambles acceptable to the bookie that result in a sure loss to the bookie (and a sure gain to the gambler). That is, the gambler can make a Dutch book against the bookie. Each of our indices of incoherence in the bookie's previsions is the maximum guaranteed rate of loss to the bookie that the gambler creates through his/her choice of coefficients, relative to a normalization. We introduced some properties that we want a normalization to have, and we identified largest and smallest normalizations amongst special classes of normalizations.

We then illustrated the methods for measuring incoherence with several examples of incoherent inferences. These included testing simple hypotheses at fixed levels, minimax estimation, and choosing the size of a test based solely on the loss function.

ACKNOWLEDGEMENT

This research was supported by NSF grant DMS-9801401.

REFERENCES

Berger, J. O. and Sellke, T. (1987). Testing a point null hypothesis: The irreconcilability of p-values and evidence. *J. Am. Statist. Ass.* **82**, 112–122 (with discussion).

Cox, D. R. (1958). Some problems connected with statistical inference. *Ann. Math. Statist.* **29**, 357–363.

de Finetti, B. (1974). *Theory of Probability* **1**. Chichester: Wiley.

Feller, W. (1972). *An Introduction to Probability Theory and its Applications* **2**. Chichester: Wiley.

Lindley, D. V. (1972). *Bayesian Statistics, A Review*. Philadelphia: SIAM.

Schervish, M. J. (1995). *Theory of Statistics*. New York: Springer.

Schervish, M. J. (1996). P-values: What they are and what they are not. *Am. Statist.* **50**, 203–206.

Schervish, M. J., Seidenfeld, T. and Kadane, J. B. (1997). Two measures of incoherence: How not to gamble if you must. *Tech. Rep.*, Carnegie Mellon University, USA.

Schervish, M. J., Seidenfeld, T. and Kadane, J. B. (2000). How sets of coherent probabilities may serve as models for degrees of incoherence. *Int. J. Uncertainty, Fuzziness Knowledge-Based Systems* **8**, 347–355.

Schervish, M. J., Seidenfeld, T. and Kadane, J. B. (2002a). How incoherent is fixed-level testing? *Proc. 2000 Meeting of the Philosophy of Science Association* (J. Barett, ed.) Chicago: Philosophy of Science Association. (to appear).

Schervish, M. J., Seidenfeld, T. and Kadane, J. B. (2002b). Measuring incoherence. *Sankhyā A* **64**, 561–587.

DISCUSSION

DEAN P. FOSTER (*University of Pennsylvania, USA*)

I would like to congratulate the authors on tackling the important problem of the philosophical differences between Bayesian and frequentist statistics. I feel that the actual practice of statistics is converging between the two camps, but the philosophical underpinings are as distant as they ever were. This paper continues the interesting conversation between Bayesian and frequentist statistics.

The first half of the paper, which discusses upper and lower probabilities, is a very nice contribution to the literature. However, the second half of the paper, which applies this reasoning to frequentist statistics, is based on an unstated assumption that relies on a fully specified prior distribution for its validity. The authors assume that

Assumption SSK. *If* $A \gg B$, *then* $A - B \gg 0$.

This assumption is true for probabilities but false for upper and lower previsions. Since algebra of this sort is only valid for traditional probabilities, the authors are tacitly assuming that the decision maker is a Bayesian. So when frequentists are later shown to be incoherent, this comes as no surprise.

Anyone who uses unequal upper and lower previsions can find a lottery A and a positive ϵ such that $-\epsilon \gg A$ and $-\epsilon \gg -A$. If they make assumption SSK they can be shown to be incoherent as follows.

From $-\epsilon \gg A$, apply assumption SSK to get $-\epsilon - A \gg 0$.

Adding ϵ to both sides yields $-A \gg \epsilon$, and hence $-\epsilon \gg -A \gg \epsilon$.

By transitivity $-\epsilon \gg \epsilon$, so that the decision maker is at least ϵ incoherent.

The only way to avoid this is to have ϵ be zero. In other words, one would have to have upper and lower previsions that were equal. The incoherence found in this paper is generic to all users of upper and lower previsions and not just to frequentists.

An economic look at upper and lower probabilities. This paper models frequentist statisticians as decision makers who use upper and lower previsions. In other words, the authors drop the axiom that decision makers should be willing to take either side of a bet at the reciprocal odds. Given how resistant frequentists are about assigning probabilities to unknown quantities, this is a good model for their behavior; it allows frequentists to simply refuse some bets. This should allow a system in which we can model both frequentist and Bayesian decision-making.

A good deal of my research has concerned theoretical economics. From that perspective, upper and lower previsions are equivalent to having a bid–ask spread in a market. Bid–ask spreads are now so important in economics that the question no longer is whether they should exist, but rather how large should they be. The intellectual basis for bid–ask spreads is the theory of asymmetric information for which George Akerlof, Michael Spence and Joseph Stiglitz were awarded the 2001 Nobel prize in economics. Recent work using these ideas to model markets can be found in Admati and Pfleiderer (1988) and Foster, Gervasi, Ramaswamy, 2002).

The motivation for bid–ask spreads in economics is private information. To see why, let me modify the usual story of a bookie posting odds. Suppose a fair coin is flipped and then before anyone can see it, the coin is covered with a hand. In a competitive market, with many bookies, the bid–ask spread would theorctically be driven towards zero. Thus, all bookies would give odds of nearly 1:1 on this coin. Mathematically this would be represented as:

$$\overline{P}(\mathcal{H}) - I_{\mathcal{H}} \gg 0, \quad \text{and} \quad (1 - \underline{P}(\mathcal{H})) - (1 - I_{\mathcal{H}}) \gg 0,$$

where $\mathcal{H}$ is the event that the coin lands heads, $I_{\mathcal{H}}$ is the indicator of $\mathcal{H}$, and the stated prices for both would be $\overline{P}(\mathcal{H}) = \underline{P}(\mathcal{H}) = 0.5$. Here, the bookie prefers the left-hand side of each of these equations to a payment of zero. For instance, the first equation says that if the bookie is paid $\overline{P}(\mathcal{H})$ in advance, then she is willing to pay out 1 when $\mathcal{H}$ occurs.

Now let us introduce private information. Suppose one of the customers, Mr. EagleEyes, can actually count the number of times that the coin flips in the air and thus has better than even odds of knowing if the coin has landed heads. If this coin is then tossed many times, over the course of several hours, he would gradually win money from any of the bookies that kept their odds at 1:1. But those bookies that increased their upper previsions would lose less and go home richer. Gradually competition would drive a bid–ask spread into the market by eliminating bookies who's spreads were too narrow. Here private information is the force that drives upper and lower previsions apart.

So economics motivates the size of the final bid–ask spread. If Mr. EagleEyes is the only bettor and has an accuracy of 75%, then any bookies that offers prices less than 0.75 will be run out of business. Further, any bookies that offers prices greater than 0.75 will be eliminated by competition with those who offer lower prices. So the equilibrium price is 0.75.

Relationship to frequentist statistics. Consider a bookie's relative preference of the following gambles:

$$\text{Gamble } A = 0, \qquad \text{Gamble } B = 0.6 - I_{\mathcal{H}}.$$

Given that Mr. EagleEyes has an accuracy of 75%, the bookie will consider gamble B as likely to lose. Thus, it is clear that $A \gg B$. Now if we use assumption SSK,

$$A - B = I_{\mathcal{H}} - 0.6 \gg 0,$$

which means that the bookie should be willing to take $A - B$. However, Mr. EagleEyes will only take the other side of $A - B$ when he believes $\mathcal{H}$ is unlikely to occur. Thus she will expect to lose 0.15 on this gamble with Mr. EagleEyes. So, "$A - B$" is not a gamble she should choose to accept in contradiction to what $A - B \gg 0$ which followed from assumption SSK.

Unfortunately, the paper relies on Assumption SSK for its results about frequentist statistics: "When statisticians choose one decision rule δ_0 over another δ_1 they express a preference for suffering the risk function $R(\theta, \delta_0)$ to $R(\theta, \delta_1)$. We choose to interpret such a preference by saying that $R(\theta, \delta_1) - R(\theta, \delta_0)$ is a favorable gamble." Let us look at this carefully. The authors are treating $R(\theta, \delta_0)$ and $R(\theta, \delta_1)$ as gambles where the payoff is determined by the value of θ. They assume that $R(\theta, \delta_0)$ is preferred to $R(\theta, \delta_1)$:

$$R(\theta, \delta_0) \gg R(\theta, \delta_1),$$

and then invoke assumption SSK to obtain

$$R(\theta, \delta_0) - R(\theta, \delta_1) \gg 0.$$

The rest of the paper relies on this definition of preferences between risk functions. Hence, all of Section 5 relies on assumption SSK. But as we saw before this reduction is only valid when there is a precise probability measure over θ, *i.e.*, when the statistician is a Bayesian.

Conclusions. Economics can describe situations where we should expect decision makers to have zero bid–ask spread and hence behave like Bayesians. The first class of these situations are "Robinson Crusoe" games, namely only one player is making decisions. In this simple setting the bid–ask should be zero. About the only realistic situation where this applies is whenone is making decisions that only have to do with oneself. In other words, where you are not having to communicate or interact with any other decision makers. The second class is when one decision maker believes that he knows every single fact that the other decision maker knows. Whether such a decision maker should be called pig-header or God depends on whether or not his belief of superior knowledge is delusional or correct. But in either case, he believes he does not need a bid–ask spread.

We can contrast the above two situation where a bid–ask spread is not necessary to the more realistic situation where it is necessary. Consider two decision makers who are interacting. Say the scientists publishing the study and the scientist reading the study. Suppose that each of these scientists has a domain of experties—that is, some private knowledge. (Say, one scientist is an expert in flowers that are pollinated by bees and the other is an expert in bees that pollinate flowers.) Then, a bid–ask spread is necessary to avoid placing losing bets. Elaborating these ideas to fit the real world is more suitable for a book rather than for a three page discussion, so I'll leave the details up to the imagination of the reader.

This division between when you should use bid–ask spreads and when you should not can be summarized as follows: To convince only yourself use Bayesian thinking, to convince someone else use frequentist behavior. Or as D. Berry put it more succinctly at Valencia 5:

> "Walk like an Egyptian,
> Talk like a Frequentist, but
> Think like a Bayesian."

REPLY TO THE DISCUSSION

We want to thank Professor Foster for his entertaining and illuminating comment on our paper. In our reply, we focus on three aspects of his commentary:

1. We explain in more detail the role of what he calls our assumption SSK in our application of rates of incoherence for assessing frequentist methods.
2. We discuss gamma-minimax as a rival to our theory for the coherent use of lower and upper previsions.
3. We distinguish between the possibility of a sure loss for a bookie and the advantage to a gambler who has "private" information when betting.

Assumption SSK. We agree that our calculations of measures of incoherence for frequentist procedures depend on our use of what Professor Foster calls our unstated assumption *Assumption SSK*. It seems odd to call this assumption "unstated" when Professor Foster himself quotes our very statement of the assumption later in his discussion. Dutch book calculations use only the set of what the bookie judges are *favorable gambles*. Those are gambles that the bookie strictly prefers to the status quo, which is treated as the 0 point for changes in the bookie's net wealth. (An *unfavorable gamble* is one that the bookie strictly disprefers to the status quo.) The Dutch book analysis does not consider, or even require, other of the bookie's preferences. The incoherent bookie may fail even to have strict preferences other than those reflected by the distinction between favorable and unfavorable gambles. This is the perspective that we use to interpret frequentist methods.

Loss functions (and so risks) as used in frequentist statistics are designed to be measured in (dis)utility units. For this reason we think we are justified assuming that, in the absence of any other frequentist principle guiding comparisons to the status quo, a preference for one risk function over another may be interpreted as a strict preference for the difference to the status quo.

Thus, we disagree with Professor Foster's characterization of how we model frequentists. At one point, he says that we model them as using upper and lower previsions. We do use Assumption SSK to convert those preferences (e.g. for level 0.05 tests over tests with other levels) that the frequentist has into favorable trades of risks, which are formally equivalent to upper and lower previsions. However, we do not attribute, e.g., a gamma-minimax decision rule to the frequentist, which would entail many other strict preferences not acknowledged by frequentist methods.

Later in his commentary, Professor Foster says that we do not allow our decision maker to be indifferent between options. We do better for the frequentist. We allow the decision maker to refuse to be indifferent without inferring that he/she has a strict preference. That is, he/she may refuse to make a choice based on a preference without that entailing indifference. For example, take the decision maker who always prefers a level 0.05 test to all others. The only choices that we assume this decision maker has made relate to this strict preference. For example, we make no assumption about how or even whether this decision maker would choose between a level 0.01 test and a level 0.10 test.

We also disagree with the final statement Professor Foster makes before his "Conclusions." It is consistent with the use of upper and lower previsions to define $A \gg B$ if and only if $A - B \gg 0$ (*i.e.*, if and only if, the trade is favorable). We developed such a theory in Seidenfeld *et al.* (1990). As is the case with the theory of upper and lower previsions that we presented in our paper, the theory of Seidenfeld *et al.* (1990) allows decision makers to not have a preference between some gambles without being indifferent.

Gamma-Minimax. Preference relations may violate reasonable axioms of rational decision making without leading to Dutch book. The case of gamma-minimax decision theory is one such theory, and it does not validate our Assumption SSK. A *gamma-minimax decision maker* (GM) is one who works with a set of probability distributions $\mathcal{P}$. This decision maker prefers A to B if and only if $\mathrm{E}(A) > \mathrm{E}(B)$, where $\mathrm{E}(\cdot)$ denotes lower prevision, namely

$$\mathrm{E}(A) = \inf_{P \in \mathcal{P}} \int A\, dP.$$

This is what we understand Professor Foster to mean by the decision rule that he describes early in his commentary. A GM will prefer a gamble A to status-quo if and only if $\mathrm{E}(A) > 0$. There are many examples of $\mathrm{E}(A) > \mathrm{E}(B)$ but $\mathrm{E}(A - B) \not> 0$. Indeed, it is straightforward to prove that it is impossible to make Dutch book against a GM, as we do next.

First, notice that $\mathrm{E}(A + B) \geq \mathrm{E}(A) + \mathrm{E}(B)$, hence a GM who prefers both A and B to 0, will prefer $A + B$ to 0. This means that a GM accepts combinations of favorable gambles. Furthermore, if there exist gambles $A_1, \ldots, A_n$ such that $\sup(A_1 + \cdots + A_n) < 0$, it follows that $\int (A_1 + \cdots + A_n)\, dP < 0$ for all $P \in \mathcal{P}$, hence $\mathrm{E}(A_1 + \cdots + A_n) < 0$ and at least one of $A_1, \ldots, A_n$ must have $\mathrm{E}(A_i) < 0$. Hence, at least one of $A_1, \ldots, A_n$ is not favorable.

For a GM, the relationship between strict preferences and favorable trades is subtle and, we think, counterintuitive. As an example, consider a GM who believes $A \gg B$ and $-\epsilon \gg A - B$. This sort of preference will be typical for a GM. Suppose also that this GM already has B in her portfolio. Then, even though she prefers A to B, she is willing to pay ϵ in order to avoid trading away B to get A.

At the risk of shooting ourselves in the feet, we admit that a minimax decision maker could (but not necessarily must) be interpreted as a GM. Such an interpretation would invalidate our use of Assumption SSK in measuring the incoherence of a minimax decision maker. The same argument cannot, however, be applied to the decision maker who always chooses level 0.05 tests. Consider the case of testing two simple hypotheses. A GM who is not a Bayesian would fix the ratio of type I and type II error probabilities from one problem to the next, instead of using tests with a fixed level.

Although gamma-minimax does not allow Dutch book, it does have some other unpleasant consequences. Consider the following sequential decision problem. There is an event E and its complement E^c. There is also a fair coin toss with $\mathcal{H}$ being the event that the coin lands heads. Assume that E and $\mathcal{H}$ are independent. The set $\mathcal{P}$ consists of all probabilities for E between 0 and 1. Let $X = I_{\mathcal{H}} I_E + I_{\mathcal{H}^c} I_{E^c}$, where I stands for the indicator of the event in the subscript. Consider how a GM would choose between the following two options. Under option A the GM pays 0.25 and receives her choice between X or 0. Because every element of $\mathcal{P}$ assigns prevision 0.5 to X, the GM would choose X and the lower prevision of option A is 0.25. In option B the GM pays nothing, but first learns whether $\mathcal{H}$ or $\mathcal{H}^c$ occurs, and then chooses between X and 0. After learning which of $\mathcal{H}$ or $\mathcal{H}^c$ occurs, the lower prevision of X is 0, just like status quo. So the lower prevision of option B is 0. Hence the GM will choose option A between these two. This exhibits a case of negative value of information. The GM actually pays 0.25 in order to avoid learning which of $\mathcal{H}$ or $\mathcal{H}^c$ occurs before choosing between X and 0.

Private information. Professor Foster's discussion of private information (and Mr. EagleEyes) is entertaining, but we think it misses the point of what Dutch book means. Being coherent is a very weak condition. It does not guarantee that nobody can win money from you with some additional information. On the other hand, being incoherent does guarantee, regardless of whether anyone has any additional private information, that someone could win money from you no matter what happens. There is nothing wrong with believing that a coin has equal odds of landing either side up while simultaneously believing that somebody else has good reason to think otherwise. We suspect that actually gambling with this other person would not maximize expected utility.

For example, suppose that we have watched Mr. EagleEyes long enough to believe that he has 75% accuracy. Consider the next flip with $\mathcal{H}$ denoting the event that the coin lands heads. If our bookie believes that the coin is equally likely to land either way, she would accept the bet $a(I_{\mathcal{H}} - 0.5)$ for all reasonably small values of a, positive or negative, chosen before any additional information becomes available. However, Mr. EagleEyes will choose the value of a after watching the coin being flipped. Let $A = 1$ if Mr. EagleEyes chooses the value $a = 1$ and let $A = -1$ if Mr. EagleEyes choose the value $a = -1$. Our bookie believes that $P(A = 1|\mathcal{H}) = 0.25$ and $P(A = 1\,|\,\mathcal{H}^c) = 0.75$. The gamble $A(I_{\mathcal{H}} - 0.5)$ has a prevision of -0.25 and would not be favorable. To make the gamble fair Mr. EagleEyes would have to pay the bookie 0.25 on each bet. The reader can easily check that this turns out to produce the same effect as having the bookie offer $\underline{P} = 0.25$ and $\overline{P} = 0.75$.

Conclusion. We thank Professor Foster again for the attention that he has given to our paper. We particularly appreciate the discovery that a minimax decision maker might be justified in rejecting Assumption SSK. Not all frequentist inferences are minimax, and minimax decision making suffers from its own set of ills. Although we agree with Professor Foster on the value of bid–ask spreads, we do not agree that frequentist behavior can be understood in that framework. We prefer to think *and* act as Bayesians. We need to work harder to make "others" understand how Bayesian methods can also be used to describe how they think.

ADDITIONAL REFERENCES IN THE DISCUSSION

Admati, A. and Pfleiderer, P. (1988). Selling and trading on information in financial markets. *Am. Econom. Rev.* **78**, 96–103.

Foster, D. and Gervasi, R. (2002). A liquity limited crossing market. *Tech. Rep.*, University of Pennsylvania, USA.

Seidenfeld, T. Schervish, M. J. and Kadane, J. B. (1990). Decisions without ordering. *Acting and Reflecting* (W. Seig, ed). Dordrecht: Kluwer, 143–170.

BAYESIAN STATISTICS 7, pp. 403–417
J. M. Bernardo, M. J. Bayarri, J. O. Berger, A. P. Dawid,
D. Heckerman, A. F. M. Smith and M. West (Eds.)

A Nonparametric Bayesian Approach to Inverse Problems

ROBERT L. WOLPERT
Duke University, USA
wolpert@stat.duke.edu

KATJA ICKSTADT
Universität Dortmund, Germany
ickstadt@statistik.uni-dortmund.de

MARTIN B. HANSEN
Aalborg Universitet, Denmark
mbh@math.auc.dk

SUMMARY

We propose a new method for making inference about an unknown measure $\Gamma(d\lambda)$ upon observing some values of the Fredholm integral $g(\omega) = \int k(\omega, \lambda)\Gamma(d\lambda)$ of a known kernel $k(\omega, \lambda)$, using Lévy random fields as Bayesian prior distributions for modelling uncertainty about $\Gamma(d\lambda)$. Inference is based on simulation-based MCMC methods. The method is illustrated with a problem in polymer chemistry.

Keywords: GAMMA PROCESS; LÉVY PROCESS; POLYMER; RANDOM FIELD; REVERSIBLE JUMP MCMC; RHEOLOGY.

1. INTRODUCTION

Fredholm (1900) initiated the formal study of integral equations of the first kind, in which we try to impute an unknown measure $\Gamma(d\lambda)$ from finitely many observed values of the integrals

$$G(\omega_i) = \int_\Lambda k(\omega_i, \lambda)\, \Gamma(d\lambda) \tag{1}$$

of a known kernel $k(\omega, \lambda)$. The problem is difficult in part because the integral operator $K : \Gamma \mapsto G$ is *smoothing*, making the "inverse problem" $K^{-1} : G \mapsto \Gamma$ ill-posed in the sense that small changes in G may be associated with large changes in Γ.

The most common approaches to solving Eq. (1) for the unknown Γ begin by approximating this infinite-dimensional continuous problem with the finite-dimensional discrete one

$$G_i = \sum_{j \in J} k_{ij}\, \Gamma_j. \tag{2}$$

The approximate solution of Eq. (2) is available from the normal equations as $\Gamma \approx [K'K]^{-1}K'G$ (see Kirsch, 1996) for a discussion of such discretization methods and of the numerical obstacles that arise in trying to solve the resulting linear systems). Typically, the matrix $[K'K]$ is ill-conditioned and so Eq. (2) remains ill-posed—if J is small there are no solutions, while if J is sufficiently large there are infinitely many which differ wildly. Commonly, this is addressed through the inclusion of some form of "roughness penalty" (see, *e.g.*, the *method of regularization* of Tikhonov, 1963).

2. A NEW BAYESIAN NONPARAMETRIC APPROACH

Our approach is to treat the solution of Eq. (1) as a Bayesian statistical inference problem, that of estimating the uncertain element Γ of the space $\mathcal{M}_+(\Lambda)$ of positive measures on a set Λ upon observing, perhaps with error, the quantities $G_i \approx G(\omega_i) \in \mathcal{G}$ at some finite set of points $\{\omega_i\}_{i\in I} \subset \Omega$. To complete the Bayesian model specification we must select a *prior distribution* $\pi(d\Gamma)$ on $\mathcal{M}_+(\Lambda)$, making Γ a random measure, and we must select a *measurement error model* for G_i given $g_i \equiv G(\omega_i)$, leading to a likelihood function $L(\Gamma)$.

In many applications (including ours in Section 3) localization arguments suggest that the uncertain positive measures $\Gamma(A)$ and $\Gamma(B)$ assigned to disjoint sets $A,\ B \subset \Lambda$ may be regarded as stochastically independent *a priori*. Under mild regularity conditions this leads to a Lévy–Khinchine-like representation for stochastic integrals $\Gamma[\phi] \equiv \int_\Lambda \phi(\lambda)\,\Gamma(d\lambda)$ of measurable functions $\phi : \Lambda \to \mathbb{R}$ of the form

$$\log \mathrm{E}\left[e^{i\Gamma[\phi]}\right] = \iint\limits_{\mathbb{R}_+\times\Lambda} \left(e^{iu\phi(\lambda)} - 1\right)\, \nu(du\, d\lambda)$$

for some positive measure $\nu(du\, d\lambda)$ on $\mathbb{R}_+\times\Lambda$ satisfying the integrability condition $\iint_{\mathbb{R}_+\times K}(1\wedge u)\nu(du\, d\lambda) < \infty$ for compact $K \subset \Lambda$. Jacod and Shiryaev (1987, Chapter 2, Section 4c) give details about this generalization of the usual Lévy–Khinchine formula to non-stationary processes and random fields. The Inverse Lévy Measure (ILM) algorithm of Wolpert and Ickstadt (1998a, 1998b) offers an explicit construction of such random fields, predicated on the representation

$$\Gamma[\phi] = \iint\limits_{\mathbb{R}_+\times\Lambda} u\phi(\lambda)\, H(du\, d\lambda) = \sum_{j\in J} u_j \phi(\lambda_j)$$

of $\Gamma(d\lambda)$ in terms of a Poisson measure $H(du\, d\lambda)$ on $\mathbb{R}_+\times\Lambda$ with Lévy mean measure

$$\mathrm{E}[H(du\, d\lambda)] = \nu(du\, d\lambda);$$

here $\{u_j, \lambda_j\}_{j\in J}$ represents an instance of the (at most countable) random support of $H(du\, d\lambda)$.

If $\nu(\mathbb{R}_+\times\Lambda) < \infty$ then $H(du\, d\lambda)$ (and hence $\Gamma(d\lambda)$) will have only finitely many points of support. If $\nu(du\, d\lambda)$ has a density function $\nu(u,\ \lambda)$ with respect to some finite reference measure $m(du\, d\lambda)$, then Γ will have a probability density function

$$\pi(\Gamma) = \left[\prod\nolimits_{j\in J} \nu(u_j,\ \lambda_j)\right] e^{m(\mathbb{R}_+\times\Lambda) - \nu(\mathbb{R}_+\times\Lambda)}$$

with respect to the random field with Lévy measure m.

If recorded measurements $G_i \in \mathcal{G}$ may be taken to differ only by independent measurement errors from the true values

$$g_i \equiv G(\omega_i) = \int_\Lambda k(\omega_i, \lambda)\, \Gamma(d\lambda) = \sum\nolimits_{j\in J} k(\omega_i, \lambda_j)\, u_j,$$

with probability density functions $f(G_i \mid g_i)$, then the likelihood function is simply

$$L(\Gamma) = \prod_{i\in I} f(G_i \mid g_i)$$

and by Bayes' theorem the posterior distribution for Γ has a probability density function

$$\pi(\Gamma \mid \{\vec{G}_i\}_{i\in I}) \propto \Big[\prod\nolimits_{j\in J} \nu(u_j, \lambda_j)\Big] \Big[\prod\nolimits_{i\in I} f(G_i \mid g_i)\Big] \, e^{m(\mathbb{R}_+\times\Lambda)-\nu(\mathbb{R}_+\times\Lambda)}. \tag{3}$$

This posterior distribution forms the basis for statistical inference about the solution Γ of the inverse problem in Eq. (1). Features of Γ that are well-determined by the data (or the prior) will show little posterior variation, while the system's ill-posedness will be expressed in wide posterior variability of features that are undetermined by the prior and data.

The role of the Lévy prior distribution is analogous to that of the roughness penalty in conventional regularization methods, resolving features left unspecified by the data, but with the important benefit of easy interpretability and coherence.

3. A RHEOLOGY EXAMPLE

A Newtonian fluid placed between two horizontal plates exhibits *viscous* behavior: a tangential force applied to one of the plates leads to a velocity gradient in the fluid proportional to the force per unit area, $\tau = \eta\frac{\partial v}{\partial y}$ (here τ represents the stress, or force per unit area, and v the horizontal velocity at any height y). The proportionality constant in this linear relationship, the viscosity η, is measured in pascal-seconds ("pascal" is the SI unit for pressure or stress, equal to one newton per square meter or one kilogram per meter per second squared, so one pound per square inch (p.s.i.) is about 6.89 kPa). The viscosity of familiar fluids ranges from about 10^{-5} Pa·s for air to 10^{-3} Pa·s for water to 1 Pa·s for glycerine.

A tangential force applied to one side of a springy or *elastic* substance induces a proportional deformation, $\tau = G\gamma$, where (unitless) γ represents the relative length change induced and the proportionality constant G (measured in Pa) is called the elastic modulus.

Polymers are gooey non-Newtonian materials whose behavior lies in between highly viscous Newtonian fluids, with viscosities in the range of 10^2–10^5 Pa·s, and elastic materials. Boltzmann (1876) had the idea of modelling these *visco-elastic* compounds by introducing time-dependence to the stress $\tau(t)$, elastic modulus $G(t)$, and deformation $\gamma(t)$ (with time-derivative $\dot\gamma(t)$), and relating all of them by a time-dependent extension of the elasticity equation of the form, $\tau(t) = \int_{-\infty}^{t} G(t-s)\dot\gamma(s)\,ds$ or, upon changing variables,

$$\tau(t) = \int_0^\infty G(\omega)\,\dot\gamma(t-\omega)\,d\omega. \tag{4}$$

Boltzmann took G to have a "fading memory," i.e., to be completely monotonically decreasing and so, by Bernstein's Theorem (see Feller, 1971, Section 13.4) representable in the form

$$G(\omega) = \int_0^\infty e^{-\omega/\lambda}\,\Gamma(d\lambda)/\lambda \tag{5}$$

as the Laplace transform of some positive measure $\Gamma(d\lambda)$ on $\mathbb{R}_+$ called the *relaxation spectrum*. For infinitesimal stresses the behavior is approximately that of a viscous fluid with *zero-shear viscosity* $\nu_0 = \int G(\omega)\,d\omega = \Gamma(\mathbb{R}_+)$. For periodic strains $\gamma(\omega) = \gamma_0\sin(\omega t)$ the solution to Eqs. (4) and (5) is available in closed form:

$$\tau(\omega) = \gamma_0\Big[G'(\omega)\sin(\omega t) + G''(\omega)\cos(\omega t)\Big]$$

where the elastic, in-phase, energy-conserving *storage modulus* and the viscous, out-of-phase, energy-dissipating *loss modulus* are given (respectively) by

$$G'(\omega) \equiv \int_0^\infty \frac{\omega^2\lambda}{1+\omega^2\lambda^2}\,\Gamma(d\lambda) \qquad G''(\omega) \equiv \int_0^\infty \frac{\omega}{1+\omega^2\lambda^2}\,\Gamma(d\lambda). \tag{6}$$

For small periodic strains both $G'(\omega)$ and $G''(\omega)$ can be measured experimentally using an oscillatory shear rheometer; our analysis below is based on measurements of Berger (1988), reproduced in Table 1 and plotted on both linear and logarithmic scale in Figure 1.

Table 1. *Experimental measurements of storage modulus $G'(\omega_i)$ and loss modulus $G''(\omega_i)$, both in Pa, at various frequencies ω_i (s^{-1}) for a polybutadiene melt at 23°C, from Berger (1988).*

ω	$G'(\omega)$	$G''(\omega)$	ω	$G'(\omega)$	$G''(\omega)$
2.493×10^0	2052	34526	7.680×10^1	432105	359952
3.670×10^0	4156	50445	1.144×10^2	534678	343388
5.373×10^0	8847	73294	1.654×10^2	619214	327629
7.864×10^0	18834	105329	2.433×10^2	701325	307419
1.144×10^1	37737	149699	3.539×10^2	772708	290069
1.695×10^1	74730	206936	5.238×10^2	841878	278292
2.451×10^1	136257	266220	7.529×10^2	897344	264055
3.608×10^1	223611	313420	1.114×10^3	956262	249131
5.218×10^1	324937	345321			

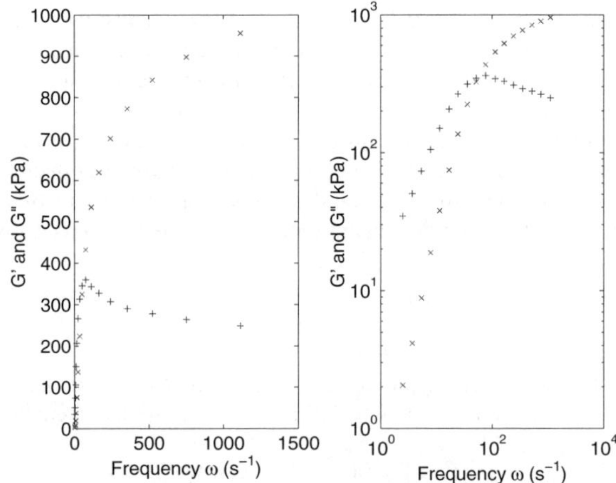

Figure 1. *Measured $G'(\omega_i)$ $(\times)$ and $G''(\omega_i)$ $(+)$ from Berger (1988).*

Eq. (6) is a two-dimensional Fredholm inverse problem of the form of Eq. (1) with $\omega \in \Omega \equiv \mathbb{R}_+$, $G(\omega) \equiv [G'(\omega), G''(\omega)] \in \mathcal{G} \equiv \mathbb{R}_+^2$, $\Lambda = \mathbb{R}_+$, $k : \Omega \times \Lambda \to \mathcal{G}$ with $k(\omega, \lambda) = [\omega^2\lambda, \omega](1+\omega^2\lambda^2)^{-1}$, and $\Gamma(d\lambda)$ unknown. To proceed with the Bayesian modelling approach of Section 2 we will need to specify a measurement-error model for $\{G_i = G(\omega_i)\}_{i\in I}$ and a prior distribution for Γ.

Berger's choice of sampling frequencies ω_i and the general form of both $G'(\omega)$ and $G''(\omega)$ in Figure 1 suggest that the logarithmic transformation simplifies the relation and may stabilize the variance, leading to our choice of a bivariate lognormal measurement error model

$$\log G_i' \sim \mathrm{N}(\log g_i', \sigma^2) \qquad \log G_i'' \sim \mathrm{N}(\log g_i'', \sigma^2)$$

or, more succinctly, $\vec{G}_i \sim \mathrm{LN}(\vec{g}_i, \sigma^2 I_2)$ for some fixed $\sigma^2 > 0$ with $\vec{G}_i \equiv [G_i', G_i'']$. In the absence of repeated measurements which might let us validate the log-normal model and help select a value for σ^2, we based our choices on an exploratory study of the residuals $\{(\vec{G}_i - \hat{G}_i)\}_{i\in I}$ from the best fit $\{\hat{G}_i\}_{i\in I}$ of Eq. (6) to Berger's data.

For a prior distribution we have chosen the Gamma random field with Lévy measure

$$\nu(du\, d\lambda) = \alpha\, u^{-1} \lambda^{-1}\, e^{-\beta u}\, du\, d\lambda$$

with uniform shape measure (on a logarithmic scale) on the interval $\Lambda \equiv (\lambda_-, \lambda^+)$ with $\lambda_- = e^{-7}$, $\lambda^+ = e^{-1}$, a convenient approximation to the interval $1/\omega_{\max} < \lambda < 1/\omega_{\min}$ that the localization principle argument of Davies and Anderssen (1997) suggests is the widest range on which we can hope to learn from the data about $\Gamma(d\lambda)$. The localization principle is also the basis for our choice of an independent-increment prior distribution.

With our choice of prior distribution the zero-shear viscosity $\int_0^\infty G(t)\, dt = \Gamma(\mathbb{R}_+)$ has a $\mathrm{Ga}(\alpha \log(\lambda^+/\lambda_-), \beta)$ distribution. Values for the parameters α, β were chosen to ensure that the mean and variance of $\Gamma(\mathbb{R}_+)$ would be approximately $6\alpha/\beta \approx 13,500$ and $6\alpha/\beta^2 \approx 50^2$, as suggested by the empirical evidence of Berger's observations (the zero-shear viscosity is also the slope $dG''/d\omega$ at $\omega = 0$). See Wolpert *et al.* (2003) for a wider range of prior distributions and measurement error models and see Anderssen and Hansen (2003) for a more specific rheological treatment and discussion of the sampling localization theorem in the context of relaxation spectral analysis.

In summary, the complete Bayesian model specification is:

$$\begin{array}{llcll}
\text{Data:} & \vec{G}_i & \sim & \mathrm{LN}(\vec{g}_i, \sigma^2 I_2) \text{ in } \mathcal{G} = \mathbb{R}_+^2 & \\
\text{Model:} & \vec{g}_i & = & \left[\int_0^\infty \frac{\omega_i^2 \lambda}{1+\omega_i^2\lambda^2}\,\Gamma(d\lambda),\ \int_0^\infty \frac{\omega_i}{1+\omega_i^2\lambda^2}\,\Gamma(d\lambda)\right] & \\
\text{Prior:} & \Gamma(d\lambda) & \sim & \mathrm{Levy}(\nu(du,\, d\lambda)), & u \in \mathbb{R}_+, \lambda \in \Lambda \equiv (\lambda_-, \lambda^+) \\
 & & = & \mathrm{Ga}(\alpha\lambda^{-1}\, d\lambda, \beta) &
\end{array}$$

and our goal is to estimate $\Gamma(d\lambda)$ upon observing $\vec{G}_i \equiv [G'(\omega_i),\ G''(\omega_i)] \approx \vec{g}_i$ for several frequencies ω_i.

4. COMPUTATIONS

Our choice of the Gamma Lévy prior features an infinite Lévy measure $\nu(\mathbb{R}_+ \times \Lambda) = \infty$ and, therefore, almost surely there are infinitely many terms in the representation

$$\vec{g}_i = \sum_{j\in J} [\omega_i^2 \lambda_j, \omega_i]\, (1 + \omega_i^2 \lambda_j^2)^{-1}\, u_j.$$

For any $\epsilon > 0$ the number $M_\epsilon = |J_\epsilon|$ of points with $u_j > \epsilon$ (indexed by $J_\epsilon = \{j \in J : u_j > \epsilon\}$) is a random variable whose prior distribution is Poisson with mean

$$\mathrm{E}[M_\epsilon] = \int_{\lambda_-}^{\lambda^+} \int_\epsilon^\infty \alpha\, u^{-1} \lambda^{-1}\, e^{-\beta u}\, du\, d\lambda = \alpha E_1(\beta\epsilon) \log \frac{\lambda^+}{\lambda_-} < \infty,$$

where $E_1(x) \equiv \int_x^\infty t^{-1} e^{-t}\, dt$ is the exponential integral function (Abramowitz and Stegun, 1964, Section 5.1). The expected total mass $\sum_{j\in J_\epsilon} \{u_j : u_j \le \epsilon\}$ of all points (u_j, λ_j) with

$u_j \le \epsilon$ is only

$$\mathrm{E}\Big[H\Big((0,\epsilon]\times\Lambda\Big)\Big] = \int_{\lambda_-}^{\lambda^+}\int_0^{\epsilon} \alpha\,\lambda^{-1}\,e^{-\beta u}\,du\,d\lambda = \alpha\beta^{-1}\log\frac{\lambda^+}{\lambda_-}(1-e^{-\beta\epsilon}),$$

a fraction $(1-e^{-\beta\epsilon})$ of the total prior expected mass $\mathrm{E}[\Gamma(\mathbb{R}_+)] = \alpha\beta^{-1}\log\frac{\lambda^+}{\lambda_-}$. In our implementation we select ϵ small enough that this represents 0.5% of the total mass, and include only the mass points $u_j \in U_\epsilon \equiv (\epsilon,\infty)$. The posterior distribution of M_ϵ is not Poisson, of course.

The space Θ of configurations we model may be represented as the countable union of Cartesian powers

$$\Theta = \cup_{M=0}^{\infty}(\mathbb{R}_+\times\Lambda)^M,$$

where the measure $\Gamma_\theta(d\lambda) \in \mathcal{M}_+(\Lambda)$ associated with index $\theta \in \Theta$ is

$$\Gamma_\theta(d\lambda) = \sum_{j=1}^{M} u_j\delta_{\lambda_j}(d\lambda),$$

the sum of M point masses of magnitudes $u_j \in \mathbb{R}_+$ at points $\lambda_j \in \Lambda$. From Eq. (3) we can compute the probability density function of the posterior distribution of $\theta \in \Theta$ (with respect to the Poisson random measure on Θ with rate $m(du\,d\lambda)$) upon observing the $N \equiv |I|$ vectors $\vec{G}_i = [G'_i,\,G''_i]$:

$$\begin{aligned}
\log\pi(\theta\,|\,\{\vec{G}_i\}_{i\in I}) &= c + \sum_{j\in J}\log\nu(u_j,\,\lambda_j) + \sum_{i\in I}\log f(\vec{G}_i\,|\,g_i)\\
&= c + M\log\alpha - \sum_{j\in J}\log u_j\lambda_j - \beta\sum_{j\in J}u_j\\
&\quad - N\log 2\pi\sigma^2 - \frac{1}{2\sigma^2}\sum_{i\in I}\left[\log^2\frac{G'_i}{g'_i} + \log^2\frac{G''_i}{g''_i}\right],
\end{aligned}$$

where $c = m(U_\epsilon\times\Lambda) - \nu(U_\epsilon\times\Lambda)$ does not depend on θ.

To implement the Metropolis-Hastings version of the Markov Chain Monte Carlo method (see, e.g., Tierney (1994)) we must select an irreducible transition probability distribution $Q(d\theta^*\,|\,\theta)$ on Θ. Our choice reflects our intention to *model* uncertainty about $\Gamma(d\lambda)$ using the Gamma random field with its infinite Lévy measure, even though our *implementation* permits us to simulate only the finite number M_ϵ Gamma masses of magnitude $u_j > \epsilon$.

The heuristic behind our proposal distribution is to imagine infinitely many particles at locations $(u_j,\lambda_j) \in \mathbb{R}_+\times\Lambda$ all undergoing simultaneous ergodic diffusion with the posterior as a stationary distribution. At any given time only finitely many points M_ϵ will lie above the line $u_j > \epsilon$; now and then one of these will diffuse below that line, causing M_ϵ to fall by one, while now and then one of the infinitely many points below the line will rise above it, increasing M_ϵ by one. Sampled at discrete times this would be a random walk similar to that we propose below, with three types of steps— those with M_ϵ unchanged ($\Delta M_\epsilon = 0$) and those where M_ϵ increases or decreases by one ($\Delta M_\epsilon = \pm 1$). For fixed logarithmic step size δ (we use $\delta = 0.25$), re-entry probability $0 < p < 1$ (we use $p = 0.01$), and re-entry distribution with density $f(u,\lambda)$ on $(0,\infty)\times\Lambda$ (see below), the move proposals are:

$\Delta M_\epsilon = 0$ With probability $1-p$, choose j uniformly from the integers $1{:}M$ and propose the lognormal step $u_* = u_j \exp(\delta Z_1)$, $\lambda_* = \lambda_j \exp(\delta Z_2)$ with $Z_1, Z_2 \sim \mathrm{N}(0,1)$. Reflect at the boundaries if necessary to ensure that $\lambda_* \in \Lambda = (\lambda_-, \lambda^+)$. If $u_* > \epsilon$, then the proposed new M_ϵ remains unchanged. Otherwise,

$\Delta M_\epsilon = -1$ If $u_* \le \epsilon$ above, remove j from J and decrease M_ϵ by one; the resulting proposal is to delete the single point (u_j, λ_j) from the ensemble θ.

$\Delta M_\epsilon = +1$ With probability p, increment M_ϵ by one and introduce a new index M to J and draw a new mass point $(u_M, \lambda_M) \sim f(u_M, \lambda_M)$ from the re-entry distribution.

For our re-entry distribution we draw λ from the uniform distribution on a logarithmic scale on Λ and, independently, draw u from the exponential distribution with mean $\mu = \mathrm{E}[H]/\mathrm{E}[M_\epsilon]$, conditioned to satisfy $u > \epsilon$, giving

$$f(u,\lambda) = (\lambda\mu \log\frac{\lambda^+}{\lambda_-})^{-1} \exp\,(\frac{\epsilon - u}{\mu}), \quad \text{on} \quad U_\epsilon \times \Lambda.$$

The conditional p.d.f. $Q(\theta_* \mid \theta)$ of the proposal transition probability distribution $Q(d\theta_* \mid \theta)$ (again, with respect to the $\mathrm{Po}(m(du\,d\lambda))$ Poisson random measure) is easily calculated from this prescription. Finally the MCMC algorithm proceeds as follows:

0. Initialize $t = 0$, $M^{(0)} \sim \mathrm{Po}(\mathrm{E}[M_\epsilon])$, $J = \{1, ..., M^{(0)}\}$, $\{(u_j, \lambda_j)\}_{j\in J} \sim f(u, \lambda)$, and set $\theta^{(0)} = \{(u_j, \lambda_j)\}_{j \in J}$.
1. Find a proposed new point $\theta_* \sim Q(d\theta_* \mid \theta^{(t)})$ and compute the Metropolis-Hastings log acceptance probability

$$\zeta^{(t+1)} = \log \pi(\theta_* \mid \{\vec{G}_i\}) + \log Q(\theta^{(t)} \mid \theta_*) - \log \pi(\theta^{(t)} \mid \{\vec{G}_i\}) - \log Q(\theta_* \mid \theta^{(t)}).$$

2. Generate a standard exponential random variable $Z \sim \mathrm{Ex}(1)$ and set

$$\theta^{(t+1)} = \begin{cases} \theta_* & \text{if } Z + \zeta^{(t+1)} \ge 0 \\ \theta^{(t)} & \text{if } Z + \zeta^{(t+1)} < 0 \end{cases}.$$

   Adjust $M^{(t)}$ and J if necessary. Increment $t \leftarrow t + 1$.
3. Periodically (e.g. at 100 evenly spaced times following "burn-in") store $\theta^{(t)}$.
4. If $t < TMAX$, repeat steps 1–4.

The parameters δ, μ, etc. are adjusted in trial runs to ensure that the rate of accepting proposed moves is approximately 20–50%. On contemporary small computers (2 GHz dual-processor Unix workstations), our MatLab implementation can complete approximately two million steps per hour.

5. RESULTS

Figure 2 shows a representation of the posterior distribution of the model's predictions $G(\omega) = [G'(\omega),\ G''(\omega)]$, with the measurements ($\times$ for $G'(\omega_i)$, $+$ for $G''(\omega_i)$); the figure shows the 25%, 50% and 75% percentile bands along with 100 MCMC iterations (equally spaced from among one million). These curves lie so close together that it is difficult to distinguish them in the plot, showing that there is little posterior uncertainty about $G(\omega)$.

Figure 3 shows the prior (dotted line) and posterior means of the spectral density $\Gamma(d\lambda)/d\lambda$, and Figure 4 the prior (dotted) and posterior cumulative spectral distribution $H[(0, \lambda)]$ with the

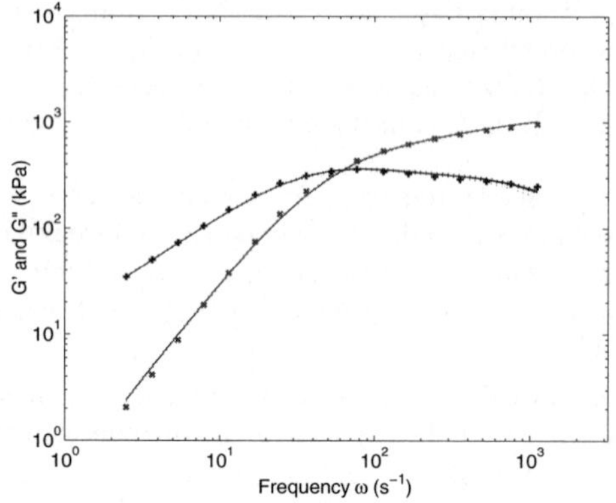

Figure 2. *Posterior distributions for* $G'(\omega_i)$ $(\times)$ *and* $G''(\omega_i)$ $(+)$.

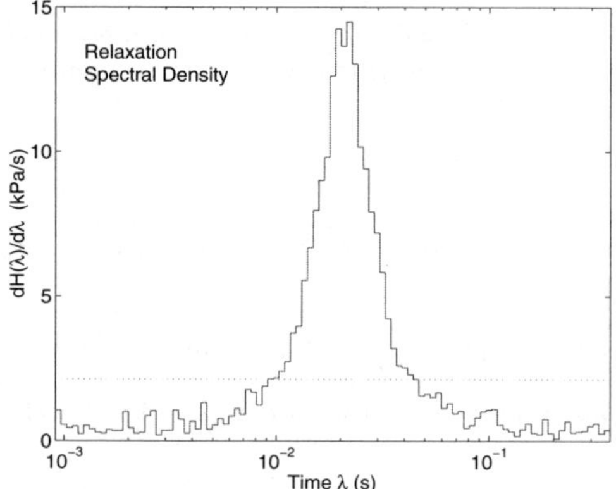

Figure 3. *Prior and posterior spectral density.*

25%, 50% and and 75% posterior percentiles and 100 MCMC iterations to illuminate the distribution. Evidently the spectral density is unimodal (or, if not, has no strong second mode), centered at about $\lambda \approx 0.020s$, with half of its mass in the interval $[0.015, 0.027]$ and 90% in the interval $[0.004, 0.086]$.

Figure 5 displays the posterior distribution of M_ϵ, the number of mass points for the spectral measure of magnitude $u > \epsilon$, with the posterior mean $\mathrm{E}[M_\epsilon \mid \{\vec{G}_i\}_{i \in I}] = 303.9$ indicated with a vertical line (the prior distribution was $M_\epsilon \sim \mathrm{Po}(118.0)$).

Figure 6 shows a single state $\theta^{(t)}$ from the simulation (the final one, with $t = 1,000,000$). This step features $M_\epsilon = 290$ mass points (u_j, λ_j) (dots represent recently-moved points, stars represent recently-added points from the re-entry distribution). Following the 100,000-step burn-in phase in this simulation run, M_ϵ ranged from a minimum of 247 to a maximum of 458, with a mean of 303.9. Magnitudes u_j are represented on a logarithmic scale, so a very large fraction of the mass is represented by the largest few points; the 30 points above $u > 100$ hold 50% of the total mass, and the 186 points above $u > 10$ hold about 95%. Here $\epsilon = 2.57$ was chosen to ensure that 99.5% of the prior mass exceeded ϵ.

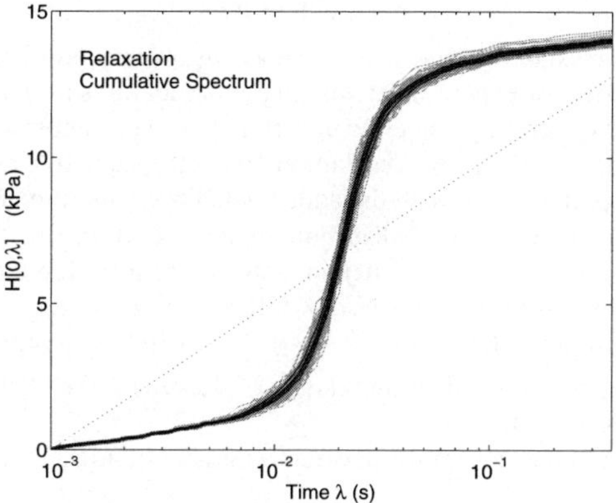

Figure 4. *Prior and posterior cumulative spectrum.*

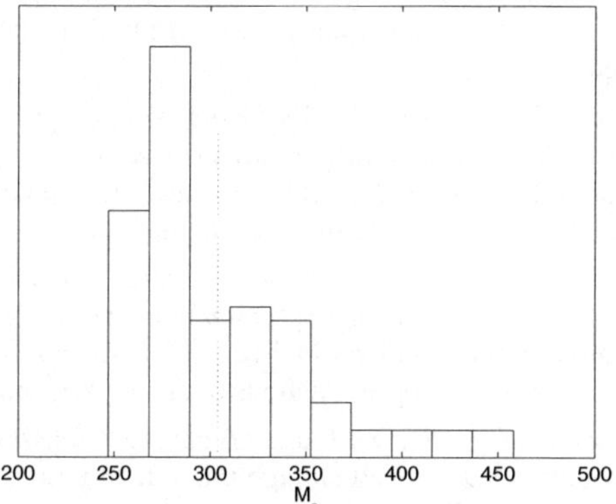

Figure 5. *Posterior distribution of M_ϵ.*

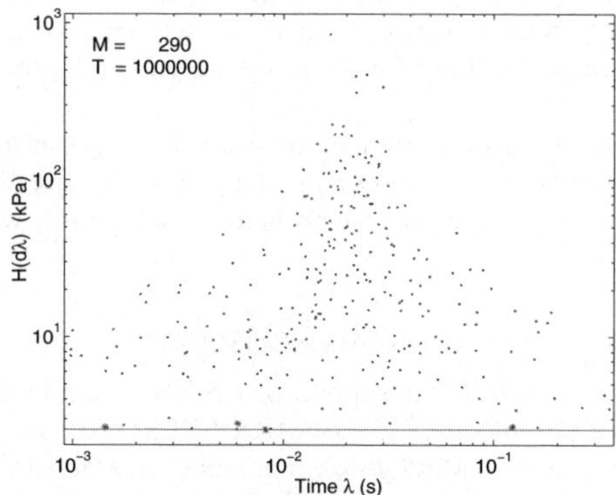

Figure 6. *A single configuration $\theta^{(t)}$ from the simulation.*

6. DISCUSSION

The nonparametric Bayesian approach to inference in inverse problems allows us to model explicitly our prior beliefs or expert understanding about features of the solution $\Gamma(d\lambda)$, and to represent honestly and coherently whatever uncertainty remains about these features following the observation of our data. The prior distribution serves the same role as the roughness penalty in the regularization approach, but with the added benefits of interpretability and coherence.

The specific choice of Lévy prior distributions leads both to (usually welcome) prior independence of the spectral measure of disjoint sets and to tractable computational problems. Note that the *posterior* distribution of $\Gamma(d\lambda)$, while still discrete, is *not* Lévy— the point process $H(du\, d\lambda)$ in the representation $\Gamma(d\lambda) = \int_{\mathbb{R}_+} u\, H(du, d\lambda)$ assigns independent random variables $H(A_k) \sim \text{Po}(\nu(A_k))$ to disjoint sets $A_k \subset \mathbb{R}_+ \times \Lambda$ but the likelihood function induces dependence among the $\{H(A_k)\}$.

Our nonparametric Bayesian approach offers a number of advantages over earlier methods, including

Flexibility : Different choices for the Lévy measure $\nu(du, d\lambda)$ will lead to "smooth" or "bumpy" measures, allowing the analysis to reflect any expert opinion about features of $\Gamma(d\lambda)$ such as smoothness, uni- or multi-modality, zero-shear viscosity $\Gamma(\mathbb{R}_+)$, etc.

Tractability : The MCMC approach described here works equally well for any of these choices of Lévy measure (and measurement-error model). Posterior distributions of any quantity $\Gamma[\phi] = \int \phi(\lambda)\, \Gamma(d\lambda)$, including interval measures $\Gamma(A)$ and zero-shear viscosity $\Gamma(\mathbb{R}_+)$, are easily computed.

Parsimony : Some choices of $\nu(du,\, d\lambda)$ will almost-surely have finite numbers M of mass points, and can even have $\text{E}[M]$ as small as two or three, leading to strikingly parsimonious representations of $\Gamma(d\lambda)$ as a sum of a small number of point masses, similar to the representations of Anderssen and Davies (2001).

In our work with the Gamma prior we found no apparent sensitivity to the choice of the cut-off $\epsilon > 0$. We settled on a value small enough that our truncated approximation includes 99.5% of the prior expected mass, but in a sensitivity analysis we varied ϵ over a wide range.

We also explored a similar algorithm modelling a fixed number M of mass points (u_j, λ_j) diffusing over $U_\epsilon \times \Lambda$ with *reflecting* boundary conditions at $u = \epsilon$, rather than the *free* boundary conditions of the present implementation. Model fit and posterior distributions of $\Gamma(A)$ for intervals A were very similar to those found in the present study, for a wide range of values of M.

We continue to explore the prior elicitation issues that arise in this modelling approach, studying a range of different Lévy measures and seeing how they affect posterior inference. We are also exploring inference for other inverse problems; some of this work will be described in Wolpert *et al.* (2003).

ACKNOWLEDGMENTS

The authors acknowledge gratefully the support of US NSF grant DMS-9626829, of US EPA grant R828686-01-0, of EC TMR grant ERB-FMRX-CT96-0096, and of the Centre for Mathematical Physics and Stochastics (MaPhySto), funded by the Danish National Research Foundation. We would like to thank Robert S. Anderssen for suggesting the problem and Jesper Møller and the Department of Mathematical Sciences of Aalborg University for facilitating our collaboration.

REFERENCES

Abramowitz, M. and Stegun, I. A. (1964). *Handbook of Mathematical Functions With Formulas, Graphs and Mathematical Tables*. Washington, DC: National Bureau of Standards.

Anderssen, R. S. and Davies, A. R. (2001). Simple moving-average formulae for the direct recovery of the relaxation spectrum. *J. Rheol.* **45**, 1–27.

Anderssen, R. S. and Hansen, M. B. (2003). Bayesian relaxation spectrum recovery. *Tech. Rep.*, Aalborg University, Denmark.

Berger, L. (1988). *Untersuchung zum Rheologischen Verhalten von Polybutadienen mit Bimodaler Molmassenverteilung*. Ph.D. Thesis, Eidgenössische Technische Hochschule (ETH), Zurich.

Boltzmann, L. (1876). Zur Theorie der elastischen Nachwirkung. *Ann. Phys. Chem.* **7**, 624–654.

Davies, A. R. and Anderssen, R. S. (1997). Sampling localization in determining the relaxation spectrum. *J. Non-Newtonian Fluid Mech.* **73**, 163–179.

Feller, W. (1971). *An Introduction to Probability Theory and its Application* **2**, 2nd ed. Chichester: Wiley.

Fredholm, E. I. (1900). Sur une nouvelle méthode pour la résolution du problèm de Dirichlet. *Œuvres complètes: publiées sous les auspices de la Kungliga svenska vetenskapsakademien par l'Institut Mittag-Leffler*, 61–68, reprinted 1955. Malmö: Litos reprotryck.

Jacod, J. and Shiryaev, A. N. (1987). *Limit Theorems for Stochastic Processes*. Berlin: Springer.

Kirsch, A. (1996). *An Introduction to the Mathematical Theory of Inverse Problems*. New York: Springer.

Tierney, L. (1994). Markov chains for exploring posterior distributions. *Ann. Statist.* **22**, 1701–1762 (with discussion).

Tikhonov, A. N. (1963). Solution of incorrectly formulated problems and the regularization method. *Sov. Doklady* **4**, 1035–1038.

Wolpert, R. L., Ickstadt, K. and Hansen, M. B. (2003) Inverse problems and a Lévy process solution. *Tech. Rep.*, Aalborg University, Denmark.

Wolpert, R. L. and Ickstadt, K. (1998a). Poisson/gamma random field models for spatial statistics. *Biometrika* **85**, 251–267.

Wolpert, R. L. and Ickstadt, K. (1998b). Simulation of Lévy random fields. *Practical Nonparametric and Semiparametric Bayesian Statistics* (D. Dey, P. Müller and D. Sinha, eds). New York: Springer, 227–242.

DISCUSSION

SUBHASHIS GHOSAL (*North Carolina State University, USA*)

First, I would like to congratulate the authors for an excellent paper. It once again demonstrates the usefulness of statistical, in particular Bayesian, methods in hard problems in numerical analysis. Statistical, and in particular Bayesian, methods have a lot to contribute in numerical analysis. Bayesian approaches to numerical analysis problems, in fact, go back to Poincaré. He considered a Bayes procedure for interpolation polynomials. His approach consisted of putting a prior distribution on the coefficients of the power series expansions of the interpolated functions. Despite the enormous potential of Bayesian approaches for numerical analysis problems, it is somewhat surprising that the methods did not receive a lot of attention. Notable exceptions are Bakhvalov (1959) and Diaconis (1988). The former author considered numerical quadrature problems and put a prior that is supported on finitely many functions. Diaconis (1988) formulated a Bayesian approach for a general numerical problem and considered several examples. In particular, he showed that for a quadrature problem, if the prior on the functions is taken to be a Brownian motion, then the trapezoid rule is the Bayes solution. The paper of Wolpert, Ickstadt and Hansen renews the hope of a revival of Bayesian methods for numerical analysis.

The authors consider certain inverse problems and offer Bayesian solutions. If variables are measured with errors, statistical methods considering the presence of random errors must be employed. Since an inverse problem is ill-conditioned, small changes in the observation lead to large changes in the solution. Only a small rounding error in the observations can also make a big difference in the solution. Therefore, even if the observations are deterministic functions of the parameters, Bayes solutions may offer the desired stability of the solution. Bayesian procedures are expected to be much more stable than many common frequentist methods such as the least square estimate in ill-conditioned problems. For instance, the ridge regression estimator, which is an approximate Bayes procedure, gives much more stable solutions than those given by the ordinary least squares.

In the continuous analogue of the regression problem considered by the authors, a nonparametric Bayesian approach is natural. The authors argue that a random field prior based on a Lévy process is natural for this problem.

It seems that the main question in this context is whether the Bayes procedure leads to a stable solution. We believe it does, since a Bayesian method has an inherent smoothing property. A formal verification would probably consist of the computation of the norm of the linear map given by the derivative, at a given vector of observations, of the map relating the observations to the Bayes estimate. This computation may turn out to be too difficult. In that case, at least the stability property should be checked by numerically computing the variations in the Bayes estimate for small perturbations of the observations.

The authors argue that a Lévy process is a natural candidate for the prior distribution on the space of the function to be found. We partly agree with the authors. The independence structure is certainly helpful for writing down the joint posterior density of the weights and the location of the support points up to a normalizing constant. Then it is relatively easy to apply the Metropolis-Hastings algorithm to calculate the posterior distribution. However, independence has a price to be paid in terms of the roughness of the sample paths of the underlying function.

Perhaps it is not unreasonable to think that a priori the sample paths of these functions are smooth. In particular, this means that the random measure $\Gamma(ds)$ considered by the authors is absolutely continuous and the density is in fact a smooth function. In that case, one should use a prior for $\Gamma(ds)$ that sits on the absolutely continuous smooth densities. If the same symbol $\Gamma(s)$ stands for the density of $\Gamma(ds)$, then $\Gamma(s)$ should be a random process that takes values in $(0, \infty)$ and has smooth paths. Then a possible alternative to the Lévy process prior could be a Gaussian process prior for $\log \Gamma(s)$, where the covariance kernel for the Gaussian process is chosen in such a way that the sample paths have the desired smoothness properties. Another attractive alternative is to consider a mixture prior of the form $\int \psi_k(s,z)\, dU(z)$, where $\psi_k(s,z)$ is a probability kernel, k is a smoothing parameter and $U(z)$ is an increasing process on $(0, \infty)$ without any smoothness restriction. Note that the total mass of $\Gamma(\cdot)$ on $(0, \infty)$ is equal to $U(\infty)$. Now a Lévy process prior can be used for $U(z)$. Using the notion of a Feller approximation, Petrone and Veronese (2002) argued that the gamma kernel is a natural kernel when the domain is $(0, \infty)$ as in this case. We agree that the kernel can be combined with the likelihood at the computation stage. However, in this case, one is not interested in the posterior distribution of $U(\cdot)$ but in that of $\int \psi_k(\cdot, z)\, dU(z)$, the smoothed out $U(\cdot)$. Smoothing is likely to have a useful consequence in consistency and rates of convergence. The situation is similar to the difference between the Dirichlet process and Dirichlet mixtures. Without the smoothing, one can only hope consistency in the weak-star topology or a Kolmogorov–Smirnov type norm, while with smoothing one would expect consistency in the stronger L_1 or L_2 distances. For instance, in the case of estimation of a probability measure that has a density, the Dirichlet prior only gives rise to consistency in the weak or Kolmogorov-Smirnov norm while Dirichlet mixtures are consistent under the variation norm (Ghosal *et al.*, 1999). The effect of smoothing is even more apparent in rates of convergence. For normal mixtures, one gets a very fast rate $\log n/\sqrt{n}$ of convergence (Ghosal and van der Vaart, 2001). On the other hand, the estimation of a mixing distribution is known to be a hard problem with only a logarithmic rate of convergence.

The Inverse Lévy Measure Algorithm comes handy in the computations. As it is common with most nonparametric Bayesian procedures, computations can be done only with the help of Markov chain Monte Carlo methods. The authors use the Metropolis algorithm by first reducing the infinitely many jump points of the Lévy process to finitely many jumps by ignoring all the remaining jump points. Perhaps this is a natural way to resolve the problem. However, one needs to be cautious about the number of terms to be taken. If one takes only a fixed number of terms independent of the sample size, then for large sample sizes, the approximate posterior will necessarily be very different from the actual posterior in the natural scale of accuracy for the sample size, since the former posterior is inconsistent (because the corresponding prior does not have the full support) while the latter is expected to be consistent — see the next paragraph. The small jumps ruled out in the prior stage may become important with increasing sample size. Therefore, it seems natural to ask the question how many support points are to be taken in the computation for a given sample size to guarantee a given level of accuracy. It seems that one needs to take enough terms so that the remaining part has a prior probability that is exponentially small with the increasing sample size.

It is important to know asymptotic properties such as consistency and rates of convergence of the Bayes solution. Consistency questions are important not only to objective Bayesians and frequentists, but also to subjective Bayesians since consistency is equivalent of merging of opinion; see Diaconis and Freedman (1986) for a discussion of this topic and a striking example of inconsistency. There seem to be two parameters — σ measuring the (lack of) precision and n, the number of observations taken, both of which may control the asymptotics. Letting $\sigma \to 0$ is the same as taking repeated measurements. Nevertheless, to estimate the whole function accurately, the number of points where observations are taken must increase and fill up the whole domain. Since a Lévy process can be chosen to have full weak-star support, consistency is expected to hold under the weak-star topology. In order to have this, the Lévy measure $\nu(\cdot)$ must necessarily have infinite total mass. To have a stronger form of consistency or a rate of convergence, one needs to use a prior that has smooth sample paths like the Gaussian or the mixture prior.

Finally, we wonder whether the log normal error distribution used in the computation is the most appropriate one. The log normal errors seem to have arisen because the measurements are positive and one switches to the logarithmic scale. Since a log normal variable can be incredibly large, and it is the actual variable rather than its logarithm that is measured, the log normal model for measurement errors may not well represent the uncertainty in the present context.

REPLY TO THE DISCUSSION

First, we would like to thank Professor Ghosal for his thorough discussion and encouraging remarks about this work, and for the additional perspective he offers us on the interplay between Bayesian statistics and numerical analysis.

Ghosal begins with a rousing endorsement of the application of Bayesian methods and principles to problems in numerical analysis, for which we thank him and with which we heartily concur. A large number of factors contribute to uncertainty in applied mathematical problems such as those we consider—measurement error is the principal one we consider here, but model simplifications or misspecification, minor recording, and transcription errors and, as Ghosal observes, even rounding can affect inference in ill-conditioned problems. The Bayesian approach is ideal for reflecting all the sources of uncertainty that affect inference in applied mathematics problems.

We agree with Ghosal that our Bayesian approach "leads to a stable solution" (we believe he is referring to the relative insensitivity of inference to small variations in the observations), but we would not characterize stability as "the main question." Earlier regularization methods also overcome the natural ill-conditioning of this and similar inverse problems by penalizing roughness, trading off a small degree of model fit for the comfort of a smooth (or *regular*) solution. What is new in the present approach is the *interpretability* of this trade-off (we express our anticipation of regularity or smoothness through a prior distribution with an easy probabilistic interpretation) and the opportunity to express coherently all the uncertainty in model predictions and inference that arises from measurement errors and other sources, an advantage inherent in our Bayesian formulation. The stability computation he suggests appears to be quite problem-specific, depending on details of the measurement error model used in specific applications; our goal here is to present a widely applicable methodology useful in many commonly studied inverse problems, with widely varying measurement error models.

The frequentist question of asymptotic consistency (as the number n of observations increases) is an interesting *mathematical* issue that would be fun to study further. We do not know precise conditions on the prior distribution (*i.e.*, on the Lévy measure), the likelihood (*i.e.*, on the measurement error model), and the specific problem (*i.e.*, on the kernel $k(t, s)$) that would guarantee consistency. That problem is less interesting *statistically*, however, since there is no opportunity here to acquire additional observations (the number $n = 17$ of data points is fixed). What *is* interesting in practice is how accurately and honestly we can represent the uncertainty about quantities of interest in applications, such as the elastic modulus $G(\omega) = \int_\Lambda \exp(-\omega/\lambda)\,\Gamma(d\lambda)/\lambda$ and integrals of the uncertain relaxation spectral measure $\Gamma(d\lambda)$ over intervals. We feel that the method succeeds in representing the evidence contained in the data and the prior about these issues of interest in this application.

Consistency, as Ghosal observes, can be expected to require an infinite Lévy measure $\nu(du\,d\lambda)$, hence a random field with infinitely many mass points, while any digital implementation must necessarily be finite. In some sense our approach actually models infinitely many mass points, though, explicitly keeping track of the locations and magnitudes of *all* of the $M_\epsilon < \infty$ points of macroscopic mass $u_j > \epsilon$ for some small $\epsilon > 0$ (these are the largest mass points), and also modelling the total mass $\sum\{u_j : u_j < \epsilon\}$ (but not the precise locations or individual masses) of the infinitely many points with mass below ϵ. When a Metropolis-Hastings move drops a mass point below the ϵ-boundary the number M_ϵ decreases by one and the total mass of small points experiences a corresponding increase; the re-entry distribution can be interpreted as the occasional elevation of a small-mass point above the ϵ-boundary, entering the explicitly modelled configuration and thus increasing M_ϵ by one.

Our choice of the gamma prior distribution (with infinite Lévy measure) and very small ϵ (ensuring that 99.5% of the prior Lévy mass will lie above the ϵ-boundary) lead to prior and posterior means of about $\mathrm{E}[M_\epsilon] \approx 118$ and $\mathrm{E}[M_\epsilon|\,\mathrm{data}] \approx 304$, respectively. For perspective, the optimal fit of the data with finitely many points M shows virtually no improvement beyond $M = 8$ or so (more precisely, there is negligible drop in deviance at the MLE for the $2M$-dimensional models with M masses $\{u_j\}$ and locations $\{\lambda_j\}$ for $M > 8$). The data alone offer no compelling reason to use more than a small handful of points— our implementation includes many more points than are needed.

Ghosal takes issue with our choice of a lognormal measurement error model. He suggests that such a model offers the possibility of huge positive deviations that do not appear in our dataset, and asks if the data themselves were not measured (with possible measurement error), rather than their logarithms, making a linear-scale measurement error model appear to be more appropriate than the log-scale one we used. His concern appears to be well founded and we will explore that issue further.

Finally, Ghosal suggests some alternatives to our approach, based on smoothed (or mixtures of) Lévy processes or on exponentiating Gaussian processes. Of course, there are many ways to model uncertainty about the relaxation measure $\Gamma(d\lambda)$ in our application or, more generally, about the measures arising in Fredholm integral equations of the first kind; we have offered and illustrated one, and Ghosal suggests a few more of varying degrees of tractability. Part of his motivation appears to be a concern with the discrete nature of the underlying measure $\Gamma(d\lambda)$ arising in our Lévy formulation, a consequence of our insistence on independent increments. The independent increment requirement was not made for our convenience, it followed from the localization result of Davies and Anderssen (1997). We know of no reason to expect $\Gamma(d\lambda)$ to have a density $\Gamma(\lambda)$, let alone a continuous one; the objects of apparent interest are not Γ itself but rather the elastic modulus $G(\omega) = \int_\Lambda \exp(-\omega/\lambda)\,\Gamma(d\lambda)/\lambda$ and the partial viscosity $\eta_{(a,b)} \equiv \int_{(a,b)} \Gamma(d\lambda)/\lambda$ associated with intervals $(a,b) \subset \Lambda$.

For investigators or problems where a continuous density function *is* required, it may be useful to note that the Fredholm equation $G_\epsilon(\omega) = \int_\Lambda k(\omega,\lambda)\Gamma_\epsilon(\lambda)d\lambda$ for a smoothed density $\Gamma_\epsilon(\lambda) \equiv \int \gamma_\epsilon(\lambda - t)\Gamma(dt)$ can be rewritten as $G_\epsilon(\omega) = \int_\Lambda k_\epsilon(\omega,\lambda)\Gamma(d\lambda)$ with a discrete Lévy measure $\Gamma(d\lambda)$ and a smoothed kernel $k_\epsilon(\omega,\lambda) \equiv \int k(\omega,t)\gamma_\epsilon(\lambda - t)dt$. Thus, the present formulation and implementation already includes the mixture-of-Lévy model, simply by convolving the kernel $k(\omega,\lambda)$ with a smoother $\gamma_\epsilon(\cdot)$.

We would like to thank the audience, organizers, and our discussant again for a lively session exploring new opportunities for Bayesian statisticians to make contributions in areas of applied mathematics where the role of uncertainty and the need for statistics have not yet fully been recognized. There are many more opportunities awaiting us.

ADDITIONAL REFERENCES IN THE DISCUSSION

Bakhvalov, N. S. (1959). Approximate computation of multiple integrals. *Vestnik Moskov. Univ. Ser. Mat. Mekh. Astr. Fiz. Khim.* **4**, 3–18.

Diaconis, P. (1988). Bayesian numerical analysis. *Statistical Decision Theory and Related Topics IV* **1** (S. S. Gupta and J. O. Berger, eds). Berlin: Springer, 163–175.

Diaconis, P. and Freedman, D. (1986). On the consistency of Bayes estimates. *Ann. Statist.* **14**, 1–67 (with discussion).

Ghosal, S., Ghosh, J. K. and Ramamoorthi, R. V. (1999). Posterior consistency of Dirichlet mixtures in density estimation. *Ann. Statist.* **27**, 143–158.

Ghosal, S. and van der Vaart, A. W. (2001). Entropies and rates of convergence for maximum likelihood and Bayes estimation for mixtures of normal densities. *Ann. Statist.* **29**, 1264–1280.

Petrone, S. and Veronese, P. (2002). Non parametric mixture priors based on an exponential random scheme. *J. It. Statist. Soc.* **11**, 1–20.

BAYESIAN STATISTICS 7, pp. 419–440
J. M. Bernardo, M. J. Bayarri, J. O. Berger, A. P. Dawid,
D. Heckerman, A. F. M. Smith and M. West (Eds.)

A Novel Framework for Tracking Groups of Objects

RON ZOHAR and DAN GEIGER
Technion, Israel
ronron@tx.technion.ac.il dang@cs.technion.ac.il

SUMMARY

We describe a novel framework for group tracking and identification termed *flow conservation group tracking*. Our framework integrates local kinematic measurements with flow conservation constraints so as to yield a robust estimation of the positions and types of targets across an arena. The framework is based on three components each of which is novel. The measurement-to-track component uses a reduction to minimum cost flow problems. The estimation of the number of objects on each track is based on estimation using flow networks, and the assessment of the types of vehicles is reduced to hierarchy matching algorithms.

Keywords: FLOW NETWORKS; GAUSSIAN NETWORKS; GROUP TRACKING; HIERARCHIES; TARGET IDENTIFICATION..

1. INTRODUCTION

Group tracking is an approach to tracking where groups of objects are being monitored as a single entity rather than monitoring each object individually. Group tracking is applied to tracking closely spaced objects with similar state vectors (Blackman and Popoli, 1999). For such closely spaced objects, it is often impossible to track each object separately due to sensors' limitations. Furthermore, in some applications, such as tracking a convoy of ground vehicles, maintaining individual tracks of each vehicle often presents an insurmountable computational burden.

On a battlefield, groups of vehicles may split or merge, and it is the responsibility of the group tracking system to continuously maintain the amount and types of targets within each group. An up-to-date text book of this field (Blackman and Popoli, 1999) comments: "Despite the intuitive appeal of group tracking, little development has been reported in the tracking literature since the classic work of Taenzer (1980) and the overview discussion given Blackman (1986, Ch. 11). It appears that the implementation problems resulting from splitting and merging of groups have discouraged the further development of group tracking methods. However, it seems clear that some form of group tracking is the best approach for the tracking environment of many closely spaced targets."

Standard tracking algorithms for single objects (Bar-Shalom and Li, 1995; Blackman and Popoli, 1999) associate measurements at sample time t_i with tracks estimated at time t_{i-1} (measurement-to-track algorithms). This association is mostly based on kinematic models, consisting of location, velocity, and sometimes acceleration. A well-known approach for extending this methodology to tracking groups is *centroid group tracking* (Blackman, 1986; Blackman and Popoli, 1999). Using this approach, only a centroid generated for the group is tracked where a *centroid* is a virtual object which summarizes the kinematics of the group. This virtual centroid object is treated in the tracking system as if it were a single real object.

The principle observation motivating this work is that centroid group tracking is a mis-generalization of the tracking framework from single objects to tracking groups of objects because it places the main emphasis on kinematics rather than on the principle of matter conservation, namely, that the true number of objects before and after a split is not changed. In single object tracking, matter is implicitly conserved by the mere existence or inexistence of a track; The association of a single observation to each existing track guarantees that objects disappear from the tracking system only when an explicit decision is made to drop a track. On the other hand, in centroid group tracking, each observation and each track may represent numerous objects, and the actual number of objects within each group may only be estimated. Consequently, a group tracking framework which does not insure that matter is conserved through time may yield erroneous decisions. This problem worsens when the frequency of sensor readings is low, yielding kinematics information that is not sufficiently accurate.

When groups of objects sampled at time t_i are associated with established tracks, one should use an estimation of the number of objects within the group. There is no point associating a large group of objects with a previously established track that contains a few objects just because the kinematic information matches. Similarly, when a group splits into several subgroups it is clear that the number of objects in the subgroups equals the number of objects in the original group. However, this important structural information is being neglected in current group tracking schemes, where the number of vehicles in each group and their type are ignored or estimated in a local fashion.

In this paper, we describe a novel framework for group tracking and identification termed *flow conservation group tracking*, which we believe to be a preferred extension of the methodology from tracking single objects to tracking groups of objects. Our framework integrates local kinematic measurements with flow conservation constraints so as to yield a more robust estimation of the position and types of objects across an arena. We base our framework to a large extent on the ability to estimate and match the number of targets in each group, in addition to kinematics.

An important component of our framework is the estimation of the types of objects in the arena given reports received at a group level. We present the notion of *hierarchical identification* where each report may be a list of labels of various specificity. For example (T55, T88, T88, Tank) is a report about four objects. We develop a framework and polynomial algorithms for the fusion of such consistent and inconsistent reports into a single summary report and for assessing the quality of the given reports.

The rest of this paper proceeds as follows: in Section 2, we describe the components of the flow conservation group tracking framework. In Sections 3 and 4, we describe in some detail two of the three components of our framework and conclude with a discussion of on-going research (Section 5).

2. THE FRAMEWORK

As in classic frameworks for multi-target tracking systems (*e.g.*, Blackman and Popoli, 1999) we assume iterative processing is being used so that tracks that have been formed on a previous scan are being updated at the next scan. Incoming observations are being associated with existing tracks using an association process which integrates kinematic estimation with flow conservation. Associated observations are used to update track parameters and may cause the splitting or the merging of tracks. Similarly, observations not assigned to existing tracks may initiate new tentative tracks. The tracks are updated and stored. Then, a flow estimation component uses the updated tracks information to estimate the updated number of objects on each track. Finally, an identification component uses the tracks information and the estimated number and types of objects on each track for the fusion of tracking reports into a single consistent summary report. This framework is depicted in Figure 1. In this paper we describe the guidelines of the first component and focus on the second and third component.

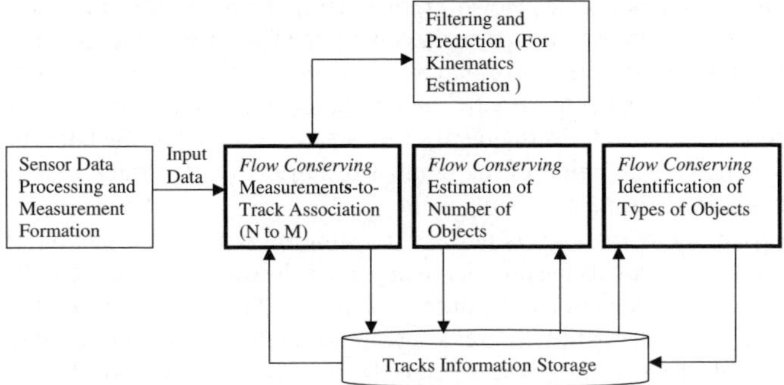

Figure 1. *Basic elements of a flow-conserving group-tracking system.*

The first component in our system performs *Flow Conserving Measurements-to-Track Association.* This component receives clusters from an external information source which samples groups in the arena at time t_i. The information provided for each reported cluster includes its approximate location and velocity, an estimated number of vehicles with error characterization and (possibly) aggregated information of the types of objects in the cluster. The measurements-to-track association procedure associates the clusters received at time t_i with tracks existing at time t_{i-1} where the information stored for each track includes the above-mentioned tracking information (location, velocity, estimated number of objects etc.).

The flow-conserving measurements-to-track association component has been formalized and efficiently solved as a minimum cost flow problem, where the association process is modelled as the flow of vehicles from tracks existing at time t_{i-1} to observations at time t_i. In particular, a bi-partite graph is generated with *track-nodes* for each track that existed at time t_{i-1} and with *cluster-nodes* for each cluster received at time t_i. Directed edges connect every track-node with every cluster-node to which it may be associated. Each edge (T_j, C_k) connecting track T_j with cluster C_k is assigned a value which represents the cost of flow through the edge. We select cost values that are proportional to the log probability for an object belonging to track T_j at time t_{i-1} to be detected at cluster C_k at time t_i. This probability may be determined using prediction of the track's location at time t_i, which can be done using existing approaches (*e.g.*, KF, EKF, IMM) and as proposed for centroid group tracking. Finally, a minimum cost flow algorithm is applied to this bi-partite graph yielding for each edge (T_j, C_k) the amount of vehicles from track T_j that are accounted for by cluster C_k. It should be noted that the suggested approach is capable of addressing the issues of low probability of detection and false detections, as well as arrival or departure of objects to and from the arena by a slightly more sophisticated modelling of the flow graph.

This approach for measurement-to-track association, which is currently being developed with Mark Matusevich, differs significantly from the component with a similar name used in classic frameworks for multi-target tracking systems. First, this component generates many-to-many association relations, where previous components typically generate one-to-one relations. Second, our component uses the estimation of number of objects to assure matter conservation.

The second component in our framework estimates the number of vehicles on each track. The accuracy of the estimation directly affects the performance of the measurement-to-track association component. In Blackman (1986), it is suggested to estimate the number of targets in a group by averaging the number of observations contained in several group detections associated with the track. Indeed, when a track is continuously maintained for a long period of time, this approach is appropriate. However, in general, the track splits and merges repeatedly such that the number of observations associated with the latest track does not provide sufficient information for number-of-objects estimation. It should be stressed that in reality, the merging and splitting of tracks may occur very frequently mainly due to clustering artifacts rather than actual operational maneuvers. Consequently, the samples received on the latest track alone may not suffice for reliable estimation of the number of objects on that track. This motivates our *Flow Conserving Flow Estimation* component.

When a group in the arena splits into several subgroups it is clear that the true number of objects in the subgroups equals the number of objects in the original group. Our flow conserving measurement-to-track association component ensures that this is also the case for the generated tracks. Consequently, it is possible to create a graph from the tracking information where edges represent tracks, internal nodes correspond to the merger or separation of groups of objects, such that the number of objects on each track is constant and the number of objects that enter an internal node equals the number of objects that leave it. Given a set of measurements of the number of objects on each edge the problem we address, which we call the *Most Probable Flow Estimation* problem (MPFE), is to estimate the most probable assignment of flow for every edge such that the conservation constraint is maintained. This problem is further described in Section 3.

Finally, the third component of our system performs *Flow Conserving Type Identification*. This component is applicable to systems which receive aggregated type information about groups in the arena. Once again, the frequent splitting and merging of groups requires incorporation of global information via flow conservation constraints on each type in order to yield a more accurate identification. In Section 4, we develop a framework and polynomial algorithms for the fusion of consistent and inconsistent reports on types of objects into a single summary report for each track separately. We present the notion of *hierarchical identification* where each report may be a list of labels of various specificity.

3. FLOW ESTIMATION

We now focus on the problem of accurately estimating the number of targets in an arena within our flow conservation group tracking framework. The estimated number of vehicles is an integral component in the measurement to track association procedure and therefore directly affects the quality of the group tracking system. When a group in the arena splits into several subgroups, it is clear that the number of targets in the subgroups equals the number of targets in the original group. Consequently, it is possible, given a set of measurements of the number of targets on each track, to globally estimate the most probable number of vehicles on each track by utilizing these flow conservation constraints. We model this problem using the notion of flow networks.

A *flow network* is a weighted directed graph $G = (V, E, w)$ where V is a set of nodes, E is a set of m directed edges and w is a function $w : E \to R$, which assigns a flow $w(l)$ for every edge l such that the following *flow conservation constraint* is maintained: the sum of flows into an internal node equals the sum of flows going out of an internal node. A *Hidden Flow Model* (HFM) is a family of conditional distributions of the measured flow for the edges in a flow network given the true flow in the network.

Given a set of measurements of the flow on each edge and assuming an independent Gaussian measurement error for each edge, the problem we address, which we called the MPFE problem, is to estimate the most probable assignment of flow for every edge such that the conservation constraint is maintained. In group tracking, edges represent tracks, and internal nodes correspond to the merger or separation of groups of objects. The number of objects on each track is constant and the number of objects that enter an internal node equals the number of objects that leave it. Various sensors are used to track the scene. Each sensor provides an estimate for the number of vehicles on each track. The goal is to improve the estimation of the number of vehicles on each track by utilizing the flow constraints on internal nodes.

Similar problems arise whenever an estimation is required in problems that can be modelled as a HFM. For example, estimation of currents in electrical networks, traffic estimation in roads, and throughput estimation in computer networks. Most probable flow estimation provides a mechanism for improving the estimation obtained by individual independent sensors by utilizing flow conservation constraints.

In this section, we present polynomial algorithms for the assignment of flow to each edge, such that the likelihood of this assignment is maximized given individual measurements with normally distributed measurement errors. An $O(m)$ algorithm is provided when the underlying undirected graph of G is a tree where m is the number of edges. Notably, a solution of this problem using Linear Minimum Variance Unbiased estimation (LMVU) (Kay, 1993) or Conditional Gaussian (CG) networks (Lauritzen, 1992; Cowell *et al.*, 1999) yields an $O(m^3)$ algorithm.

The rest of this section is organized as follows. We first provide a formal description of the flow estimation problem (Section 3.1) and discuss known solutions (Section 3.2). Then, we describe an $O(m)$ algorithm for the special case of binary polytrees, and analyze the algorithm (Section 3.3).

In a companion technical report (Zohar and Geiger, 2002a), we extend the algorithm to deal with general polytrees in time complexity of $O(m)$ and with unknown precision of the measuring devices and we describe several experiments that demonstrate the estimation improvement, and a sensitivity analysis of the assumption to normally distributed measurement errors.

3.1. *Problem Formulation*

A *flow network* is a weighted directed graph $G = (V, E, w)$ where V is a set of nodes, $E \subseteq V \times V$ is a set of m directed edges, and $w : E \rightarrow \mathrm{R}$, is a flow function that assigns a flow $w(l)$ to edge l. An *internal node* is a node that has some edges coming into it and some that are leaving it. A *source* is a node with a single outgoing edge and no incoming edges and a *sink* is a node with a single incoming edge and no outgoing edges. Let $A(v)$ and $B(v)$ denote the sets of edges into node v and out of node v, respectively.

For every internal node $v \in V$, the flow conservation constraint is given by

$$0 = \sum_{l \in A(v)} w(l) - \sum_{l \in B(v)} w(l) \tag{3.1}$$

These constraints are linear and can be represented using an $n \times m$ *constraint matrix* R via the equation $RX = 0$, where n is the number of rows (one per internal node), $m = |E|$ is the number of columns (one per edge), $X_j = w(l_j)$ is the flow on edge l_j, and $X = (X_1, ..., X_m)^T$. In particular, for every node v_i and edge l_j, we set $R_{ij} = -1$ if edge l_j points to v_i, $R_{ij} = 1$ if edge l_j points away from v_i, and $R_{ij} = 0$ otherwise.

We define the MPFE problem as follows:

Instance: Let $G = (V, E, w)$ be a flow network constrained by $RX = 0$, with X the flow on each edge and R a constraint matrix of the flows. Let e_i be the measured flow on the i-th edge.

Query: Compute

$$X^{\max} = \underset{\substack{X \\ s.t. RX=0}}{\arg\max} \prod_{i=1}^{m} P\left(e_i \left| X_i\right.\right), \tag{3.2}$$

where $P(e_i|X_i) = N(e_i|X_i, \tau_i)$ is a normal distribution function given by

$$N(e_i|X_i, \tau_i) \equiv \sqrt{\frac{\tau_i}{2\pi}} \exp\left\{-\frac{\tau_i \left(X_i - e_i\right)^2}{2}\right\}$$

with $\tau_i \triangleq \sigma_i^{-2}$ being the precision of the measurement device on the i-th edge.

Our formulation of the problem is based on several assumptions, which we now explicate. Let $e = \{e_1, ..., e_m\}$ where e_i is the measured value on the i-th edge. To find the most probable flow we compute using Bayes' rule

$$X^{\max} = \underset{X}{\arg\max}\, P\left(X \left| e\right.\right) = \underset{X}{\arg\max} \frac{P\left(e \left| X\right.\right) P\left(X\right)}{P\left(e\right)}$$

Assuming $P\left(X\right)$ is constant for all the consistent configurations of X and since $P(e)$ is also constant, we obtain

$$X^{\max} = \arg\max_{\{X; RX=0\}} P\left(e \left| X\right.\right) \tag{3.3}$$

Assuming the measurement on each edge is unbiased and independent of measurements on other edges given the true flow on that edge, we obtain Eq. (3.2).

3.2. *Various Solutions of MPFE*

The MPFE problem can be solved for arbitrary flow networks in $O(m^3)$ steps by various well known techniques which we now explicate.

One approach is to translate a given flow network into a Gaussian network such that each edge in the flow network becomes a variable in the Gaussian network and each flow constraint becomes a clique. Recall that in Gaussian networks each variable v is distributed normal with a mean that is a linear function of the vertices that point into it and with a conditional variance that is fixed given the variables that point into v (Shachter and Kenley, 1989; Lauritzen, 1992; Cowell *et al.,* 1999). This framework can be used in the limit to represent flow conservation constraints because when a variance of a variable tends to zero, the variable is simply a linear function of the variables that point into it. However, by transforming the flow network into a Gaussian network some independence assertions are lost and therefore the clique-tree algorithm (Jensen, 1996) yields an $O(m^3)$ algorithm even in cases where an $O(m)$ algorithm can be found.

A second relevant approach is a reduction of the MPFE problem to the *Minimal Cost Flow problem* (Ahuja *et al.* , 1993). The input of a Minimal Cost Flow problem is a directed graph with a cost function $G_e(f_e)$ of the flow f_e on each edge e. The output is an assignment of flows that minimizes the total cost $\sum_e G(f_e)$ and satisfies the flow conservation constraints. The MPFE problem is a special case of the Minimal Cost Flow problem because the cost function can be selected for edge l_i to be $-\log P(l_i|X_i)$. Also this approach does not yield an $O(m)$ algorithm for the problem we address herein. Furthermore, this approach does not support the computation of expected precision of estimation, or the convergence rate to the true flow values, or the estimation of the conditional distribution functions from data.

Another technique applicable for this problem is that of Linear Minimum Variance Unbiased estimation (LMVU). Although this technique is in general non-Bayesian, it identifies with the Bayesian approach given the prior we have chosen. The following theorem from (Kay, 1993) provides a unified approach for the formulation and solution of diverse problems including MPFE as a special case:

Theorem 3.1. *If the data can be modelled as $x = H\theta + w$ where x is an $N \times 1$ vector of observations, H is a known $N \times t$ observation matrix $(N > t)$ of rank t, θ is a $t \times 1$ vector of parameters to be estimated, and w is an $N \times 1$ noise vector with pdf $N(0, C)$ then the LMVU estimator is*

$$\hat{\theta} = (H^t C^{-1} H)^{-1} H^t C^{-1} x$$

and the covariance matrix is $C_{\hat{\theta}} = (H^t C^{-1} H)^{-1}$.

For the MPFE problem, the θ vector represents a set of $t = m - n$ independent flows. Since the flow conservation constraints are linear, the flow on each edge is represented using $H\theta$. Since the computation of $\hat{\theta}$ and $C_{\hat{\theta}}$ requires matrix inversions, it requires $O(m^3)$ time complexity and $O(m^2)$ space complexity. The main disadvantage of this solution for MPFE and other graph-based problems (such as inference in Bayesian networks) is the fact that the solution does not use the topology of the graph and is therefore less intuitive and, in some cases, less efficient, as we will demonstrate herein.

Our approach is that of bucket elimination (Dechter, 1996). It is also of $O(m^3)$ time complexity for general graphs. However, for polytrees we were able to develop an $O(m)$ algorithm. The principle ideas of this algorithm are developed in the next section. More details of our algorithm can be found in Zohar and Geiger (2002a).

3.3. *Solving MPFE on Binary Polytrees*

A *polytree* is a directed graph G such that the underlying undirected graph of G, namely, the graph where each directed edge is replaced with an undirected edge, is a tree. A *binary polytree* is a polytree such that each internal node has at most two outgoing or incoming edges. In this section we develop an $O(m)$ algorithm for the flow estimation problem, where the flow network G is a binary polytree. In the next section we extend the algorithm to arbitrary polytrees.

Foundations. We start with several lemmas describing relevant properties of a normal distribution $N(e_i| \; X_i, \; \tau_i)$. The first lemma restates a known property of normal distributions; its one-line proof is brought here for completeness. The lemma shows that to maximize Eq. (3.2) one can maximize an equivalent function in which several measurements with various precisions of the true flow X on a link are replaced with a single equivalent measurement of X. In the sequel we will always replace a set of measurement on an edge with its single equivalent measurement. It is convenient to denote the normal distribution $N(e_i|X_i, \tau_i)$ with $f_{e_i,\tau_i}(X_i)$.

Lemma 3.2.

$$\arg\max_X \prod_{j=1}^{k} f_{e_j,\tau_j}(X) = \arg\max_X f_{E,T}(X)$$

and where $\mathrm{E} = (\sum_{j=1}^{k} \tau_j e_j)/(\sum_{j=1}^{k} \tau_j)$ *and* $\mathrm{T} = \sum_{j=1}^{k} \tau_j$.

Proof.

$$\prod_{j=1}^{k} f_{e_j,\tau_j}\left(X\right) = C_1 \prod_{j=1}^{k} \exp\left\{-\frac{1}{2}\tau_j\left(X - e_j\right)^2\right\}$$

$$= C_2 \exp\left\{-\frac{1}{2}\left(\sum_{j=1}^{k}\tau_j\right)\left(X^2 - 2X\frac{\sum_{j=1}^{k}\tau_j e_j}{\sum_{j=1}^{k}\tau_j}\right)\right\}$$

$$= C_3 \exp\left\{-\frac{1}{2}\mathrm{T}\left(X - \mathrm{E}\right)^2\right\} = C f_{E,T}\left(X\right),$$

(where C_1 through C_3 are constants).

□

The second lemma provides a closed form maximum likelihood solution of the flow through an internal node which is adjacent to exactly three edges.

Lemma 3.3. *The maximum value for the product* $f_{\mu_l,\tau_l}(X_l)\, f_{\mu_r,\tau_r}(X_r)$ *subject to*

$$(-1)^{d_l} X_l + (-1)^{d_r} X_r = X_p \quad d_l, d_r \in \{0,1\} \tag{3.4}$$

equals

$$C f_{(-1)^{d_l}\mu_l + (-1)^{d_r}\mu_r, \frac{\tau_r\tau_l}{\tau_r+\tau_l}}(X_p)$$

for every X_p *where* C,d_l *and* d_r *are constants.*

Proof. It suffices to compute $L(X_p)$ where

$$L(X_p) = \max_{X_l} f_{\mu_l,\tau_l}(X_l)\, f_{\mu_r,\tau_r}\left((-1)^{d_r}\left(X_p - (-1)^{d_l} X_l\right)\right).$$

The value X_l^* which maximizes $L(X_p)$ is given by:

$$X_l^* = \min_{X_l} \tau_l (X_l - \mu_l)^2 + \tau_r\left((-1)^{d_r} X_p - (-1)^{d_l+d_r} X_l - \mu_r\right)^2. \tag{3.5}$$

Differentiating with respect to X_l and equating to zero yields

$$X_l^* = \frac{\tau_l\mu_l + (-1)^{d_l}\tau_r X_p - (-1)^{d_l+d_r}\tau_r\mu_r}{\tau_l + \tau_r} \tag{3.6}$$

which is a global minimum because 3.5 is quadratic. Consequently,

$$f_{\mu_l,\tau_l}(X_l^*) = C_1 \exp\left\{\frac{-\tau_l\tau_r^2}{2\left(\tau_l+\tau_r\right)^2}\left(X_p - \left((-1)^{d_r}\mu_r + (-1)^{d_l}\mu_l\right)\right)^2\right\}$$

$$= C_1' f_{(-1)^{d_r}\mu_r + (-1)^{d_l}\mu_l, \frac{\tau_l\tau_r^2}{(\tau_l+\tau_r)^2}}(X_p)$$

and

$$f_{\mu_r,\tau_r}\left((-1)^{d_r}\left(X_p - (-1)^{d_l} X_l^*\right)\right) = C_2' f_{(-1)^{d_l}\mu_l + (-1)^{d_r}\mu_r, \frac{\tau_r\tau_l^2}{(\tau_l+\tau_r)^2}}(X_p).$$

Lemma 3.2 yields

$$L(X_p) = C f_{(-1)^{d_l}\mu_l + (-1)^{d_r}\mu_r, \frac{\tau_r\tau_l}{\tau_r+\tau_l}}(X_p).$$

□

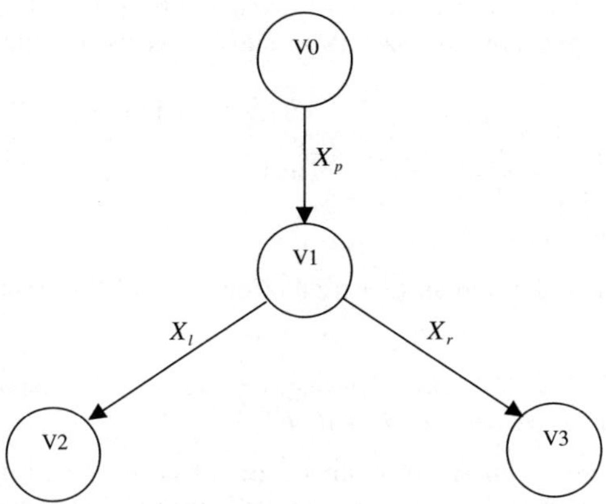

Figure 2. *A Simple Binary Tree.*

We now demonstrate the importance of these lemmas for solving the MPFE problem for the simple binary tree in Figure 2. Our algorithms generalize this example.

Let X_p, X_r and X_l be flows such that $X_p = X_r + X_l$. Let e_p, e_r, e_l be normally-distributed measurements of the flows on these edges with known precisions τ_p, τ_r and τ_l, respectively. Using Eq. (3.2), the most probable flow is given by:

$$\{X_p^*, X_l^*, X_r^*\} = \underset{\substack{\{\hat{X}_p, \hat{X}_l, \hat{X}_r\} \\ s.t. \hat{X}_p = \hat{X}_l + \hat{X}_r}}{\arg\max} \; f_{e_p, \tau_p}(\hat{X}_p) f_{e_r, \tau_r}(\hat{X}_r) f_{e_l, \tau_l}(\hat{X}_l)$$

in order to find X_p^* it suffices to compute

$$X_p^* = \underset{\hat{X}_p}{\arg\max} \, f_{e_p, \tau_p}(\hat{X}_p) \underset{\substack{\{\hat{X}_l, \hat{X}_r\} \\ s.t. X_p = \hat{X}_l + \hat{X}_r}}{\max} \; f_{e_r, \tau_r}(\hat{X}_r) f_{e_l, \tau_l}(\hat{X}_l). \tag{3.7}$$

using Lemma 3.3, Eq. (3.4) becomes:

$$X_p^* = \underset{\hat{X}_p}{\arg\max} \, f_{e_p, \tau_p}(\hat{X}_p) f_{e_l + e_r, \frac{\tau_r \tau_l}{\tau_r + \tau_l}}(\hat{X}_p)$$

using Lemma 3.2,

$$X_p^* = \underset{\hat{X}_p}{\arg\max} \, f_{\frac{e_p \tau_p + (e_l + e_r)\frac{\tau_r \tau_l}{\tau_r + \tau_l}}{\tau_p + \frac{\tau_r \tau_l}{\tau_r + \tau_l}}, \tau_p + \frac{\tau_r \tau_l}{\tau_r + \tau_l}}(\hat{X}_p).$$

Therefore,

$$X_p^* = \frac{e_p \tau_p + (e_l + e_r)\frac{\tau_r \tau_l}{\tau_r + \tau_l}}{\tau_p + \frac{\tau_r \tau_l}{\tau_r + \tau_l}},$$

and the precision τ_p^* of the estimator X_p^* is given by

$$\tau_p^* = \tau_p + \frac{\tau_r \tau_l}{\tau_r + \tau_l}.$$

After X_p^* is found, the optimal value for X_l is available from Eq. (3.6) and subsequently the optimal value for X_r is given by Eq. (3.4). Alternatively, we use Lemmas 3.2 and 3.3 to obtain

$$X_l^* = \arg\max_{\hat{X}_l} f_{\frac{e_l\tau_l + (e_p - e_r)\frac{\tau_r\tau_p}{\tau_r+\tau_p}}{\tau_l + \frac{\tau_r\tau_p}{\tau_r+\tau_p}}, \tau_l + \frac{\tau_r\tau_p}{\tau_r+\tau_p}}(\hat{X}_l) \quad \text{and therefore} \quad \tau_l^* = \tau_l + \frac{\tau_r\tau_p}{\tau_r + \tau_p},$$

and similarly, obtain X_r^* and τ_r^*.

The Algorithm. We now develop an $O(m)$ algorithm for MPFE on binary polytrees termed *bintree-mpfe*.

Definition *Let $G = (V, E, w)$ be a flow network and let $e \in E$ be the edge connected to a source or a sink in G. We call e an E-leaf of G.*

Recall that there are n linear constraints (one for each internal node). The algorithm consists of n steps. At each step, the algorithm uses Lemma 3.3 to eliminate two E-leaves with a common node v by adjusting the measurement and precision of measurement received on the third edge adjacent to v. The process iterates until a single edge remains for which the optimal flow estimation as well as its precision are given by the adjusted measurement and precision of measurement for that edge.

Suppose that l_j,l_k are E-leaves with a common node, and are also adjacent to l_h such that $X_j + X_k = X_h$. Let e_h, e_j, e_k and τ_h, τ_j, τ_k be the measurements and the precisions of measurements on the edges l_h, l_j, l_k, respectively. The elimination process consists of adjusting e_h and τ_h as follows:

$$e_h' \leftarrow \frac{e_h\tau_h + (e_j + e_k)\frac{\tau_j\tau_k}{\tau_j+\tau_k}}{\tau_h + \frac{\tau_j\tau_k}{\tau_j+\tau_k}} \qquad \tau_h' \leftarrow \tau_h + \frac{\tau_j\tau_k}{\tau_j + \tau_k} \tag{3.8}$$

The interpretation is that in order to find the optimal flow values for the rest of the graph, l_j and l_k may be eliminated by providing an additional measurement for l_h equal to $(e_j + e_k)$ with precision $(\tau_j\tau_k)/(\tau_j + \tau_k)$. This interpretation holds due to Lemma 3.3. The entire algorithm, which we call *bintree-mpfe*, is summarized in Figure 3.

The next theorem proves the correctness of *bintree-mpfe* and analyzes its complexity.

Theorem 3.4. *Let $G(V, E, w)$ be an instance of the MPFE problem where G is a binary polytree. bintree-mpfe correctly computes the most probable flow in time and space complexity of $O(|E|)$.*

Proof. Since G is a binary polytree, it is possible to find two E-leaves l_l and l_r adjacent to a third edge l_p such that $(-1)^{d_l}X_l + (-1)^{d_r}X_r = X_p$ with $d_l, d_r \in \{0, 1\}$. Equation 2 may then be rewritten as:

$$X^{\max} = \arg\max_{s.t. RX=0} f_{e_l,\tau_l}(X_l)\, f_{e_r,\tau_r}(X_r) \prod_{i\neq l,r} f_{e_i,\tau_i}(X_i)$$

Noticing that $(-1)^{d_l}X_l + (-1)^{d_r}X_r = X_p$ is the only constraint in $RX = 0$ that relates X_l with X_r, we write

$$X^{\max} \setminus \{X_l, X_r\} = \arg\max_{s.t. R'X=0} \prod_{i\neq l,r} f_{e_i,\tau_i}(X_i) \max_{s.t. (-1)^{d_l}X_l + (-1)^{d_r}X_r = X_p} f_{e_l,\tau_l}(X_l)\, f_{e_r,\tau_r}(X_r)$$

Algorithm *bintree-mpfe*
Input: An Instance of the MPFE problem. A queried edge l_q.
Output: The most probable flow on l_q and its precision (i.e. X_q^*, τ_q^*)
Initialization:
1. For each edge l_i, let e_i be the measurement of the flow on that edge and let τ_i be the precision of measurement.
2. Let $G' \leftarrow G$
Repeat
  Select two E-leaves, l_l, l_r, adjacent to a third edge, l_p such that $l, r \neq q$.
  Select $d_l, d_r \in \{0, 1\}$ such that $(-1)^{d_l} X_l + (-1)^{d_r} X_r = X_p$
  $e_{lr} \triangleq (-1)^{d_l} e_l + (-1)^{d_r} e_r$
  $\tau_{lr} \triangleq \frac{\tau_l \tau_r}{\tau_l + \tau_r}$
  $e'_p \leftarrow \frac{e_p \tau_p + e_{lr} \tau_{lr}}{\tau_p + \tau_{lr}}$
  $\tau'_p \leftarrow \tau_p + \tau_{lr}$
  $G' = G' \setminus l_l, l_r$
Until The single edge l_q remains in G'
$X_q^* \leftarrow e'_q$
$\tau_q^* \leftarrow \tau'_q$

Figure 3. *Algorithm bintree*-MPFE.

where R' is the constraints matrix excluding the constraint $(-1)^{d_l} X_l + (-1)^{d_r} X_r = X_p$. Using Lemma 3.3 we have:

$$X^{\max} \setminus \{X_l, X_r\} = \underset{s.t. R'X=0}{\arg\max}\, f_{e_p,\tau_p}(X_p)\, f_{e_{lr},\tau_{lr}}(X_p) \prod_{i \neq l,r,p} f_{e_i,\tau_i}(X_i)$$

with $\tau_{lr} = \frac{\tau_l \tau_r}{\tau_l + \tau_r}$ and $e_{lr} = (-1)^{d_l} e_l + (-1)^{d_r} e_r$. Using Lemma 3.2 the two terms depending on X_p can be merged yielding,

$$X^{\max} \setminus \{X_l, X_r\} = \underset{s.t. R'X=0}{\arg\max}\, f_{\frac{e_p \tau_p + e_{lr} \tau_{lr}}{\tau_p + \tau_{lr}}, \tau_p + \tau_{lr}}(X_p) \prod_{i \neq l,r,p} f_{e_i,\tau_i}(X_i). \tag{3.9}$$

Eq. (3.9) means that in order to find the optimal flow values for $X \setminus \{X_l, X_r\}$, it is sufficient to consider a smaller equivalent tree by removing the two E-leaves X_l, X_r (along with the sources/sinks connected to these edges), and by updating the parameters of $f_{e,\tau}(X_p)$ as specified by Eq. (3.9) and by *bintree-mpfe*.

Algorithm *bintree-mpfe* consists of n iterations. At each iteration the algorithm reduces the graph by two E-leaves l_l, l_r and updates the parameters of one remaining edge l_p which becomes an E-leaf (due to the elimination of its adjacent edges). After each iteration the graph remains a binary polytree. Therefore, this process may be iterated until a single edge remains. At each iteration the edges l_l, l_r are selected to be different than l_q, the queried edge. Consequently, after n iterations the remaining edge is l_q. The optimal flow value for this edge and its precision, X_q^*, τ_q^* are simply given by the parameters of the remaining distribution $f_{e'_q,\tau'_q}(X_q)$ as specified in *bintree-mpfe*.

It remains to analyze time and space complexity. Algorithm *bintree-mpfe* has n iterations. Each iteration requires $O(1)$ operations. Since $n \leq |E|/3$ the time complexity is $O(|E|)$. Space is used for storing $2n$ parameters, yielding space complexity $O(|E|)$. □

In order to find the most probable flow and its precision on every edge in the flow network, it is possible to run the above algorithm, each time for a different queried edge l_q. This results in time complexity of $O(|E|^2)$. However, an $O(|E|)$ algorithm is obtained with some additional bookkeeping and using a message-passing framework similar to the one suggested for inference in Bayesian networks in (Pearl, 1988). This modified algorithm defines for each internal node v two buckets for each of its adjacent edges (six buckets altogether). The "Up" bucket for an edge l saves the best estimation for the edge's flow and its precision using measurements from all the edges contained in the subtree rooted at l and not containing other edges adjacent to v. The "Down" bucket saves the best estimation for l using the measurements from all the other edges.

Initially, all the buckets are empty except for the "Up" buckets of internal nodes adjacent to E-leaves. These buckets are filled with the measurements received for the E-leaves and with their precisions. The algorithm proceeds in a distributed manner as follows: Each internal node monitors its "Up" buckets. For each pair of "Up" buckets that gets filled, the node fills the "Down" bucket of the third adjacent edge with the flow and precision of flow using Lemma 4.2. Independently, each non-E-leaf edge monitors the "Down" buckets of its adjacent nodes. Once a "Down" bucket gets filled, the edge combines the values from that bucket with the measurements received on the edge, using Lemma 4.1, and places the result on the "Up" bucket of the other node adjacent to the edge. Eventually, all the buckets get filled. The optimal values for each edge is then readily given by combining the values on the corresponding "Up" and "Down" buckets using Lemma 4.1.

Since each internal node performs at most three operations and each edge performs at most 2 operations, and since the final computation is $O(1)$ per edge, the time complexity of the algorithm is $O(|E|)$. Since there are less than $4|E|$ buckets, the space complexity of the algorithm is $O(|E|)$ as well.

3.4 *Solving MPFE on Polytrees*

In this section, we provide an $O(|E|)$ algorithm for finding the most probable flow on general polytrees. The algorithm consists of a construction of a binary polytree from a given general polytree and then application of algorithm *bintree-mpfe* on the constructed binary polytree.

Construction I: Let $G(V, E, w)$ be a polytree flow graph. For each node $v \in V$ of degree $d(v) > 3$, replace v by a directed chain of $d(v) - 2$ nodes as follows:

(i) connect two of the edges originally connected to v to the source of the chain;
(ii) connect two of the edges originally connected to v to the sink of the chain;
(iii) connect the remaining edges originally connected to v to the remaining nodes of the chain, one edge per node.

We denote the resulting graph by $\hat{G}$.

For an example of this construction, consider Figure 4. The original graph G as depicted in (a) consists of the nodes v_0 through v_6 and the edges e_1 through e_6. Note that $d(v_1) = 6 > 3$ and therefore, steps (i)–(iii) of construction I are activated to generate $\hat{G}$, as depicted in (b): The node v_1 is replaced by the directed chain $N0$ through $N3$. Edges e_1, e_2 are connected to the source of the chain (N0), edges e_5, e_6 are connected to the sink of the chain (N3) and the remaining edges, originally connected to v, are connected to the remaining nodes of the chain, one edge per node.

Algorithm *polytree-mpfe* is presented in Figure 5. The following theorem states the correctness of *polytree-mpfe* and its complexity. Due to lack of space, the proof is available in Zohar and Geiger (2002a).

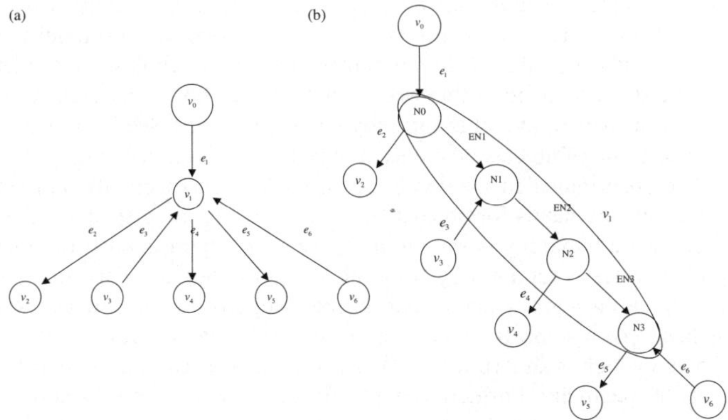

Figure 4. *Construction I example.*

Theorem 3.5. *Let $G(V, E, w)$ be an instance of the MPFE problem where G is a polytree. polytree-mpfe correctly computes the most probable flow in time and space complexity of $O(|E|)$.*

Suppose, for example, we wish to compute the most probable flow of edge l_1 in Figure 4a. Apply Construction I to obtain Figure 4b. Next apply algorithm *bintree-mpfe*. The first iteration yields: $e_{lr} = e_5 - e_6$ with $\tau_{lr} = \tau_5\tau_6/(\tau_5 + \tau_6)$. After the second iteration $e_{lr} = e_4 + e_5 - e_6$ with $\tau_{lr} = (\tau_4\tau_5\tau_6)/(\tau_4\tau_5 + \tau_4\tau_6 + \tau_5\tau_6)$. After the third iteration $e_{lr} = -e_3 + e_4 + e_5 - e_6$ with

$$\tau_{lr} = \frac{\tau_3\tau_4\tau_5\tau_6}{\tau_3\tau_4\tau_5 + \tau_3\tau_4\tau_6 + \tau_3\tau_5\tau_6 + \tau_4\tau_5\tau_6}$$

and finally, after the forth iteration: $e_{lr} = e_2 - e_3 + e_4 + e_5 - e_6$ with

$$\tau_{lr} = \frac{\tau_2\tau_3\tau_4\tau_5\tau_6}{\tau_2\tau_3\tau_4\tau_5 + \tau_2\tau_3\tau_4\tau_6 + \tau_2\tau_3\tau_5\tau_6 + \tau_2\tau_4\tau_5\tau_6 + \tau_3\tau_4\tau_5\tau_6}.$$

Hence, $X_1^* = (\tau_1 e_1 + \tau_{lr} e_{lr})/(\tau_1\tau_{lr})$, with $\tau_1^* = \tau_1 + \tau_{lr}$.

Note that our algorithm for efficiently solving polytree flow networks can be used as a preprocessing step by eliminating tree-like portions of the given flow network prior to activating a general approach such as *mpfe*. This preprocessing step improves the efficiency of such approaches.

Algorithm *polytree-mpfe*
$\hat{G} \leftarrow$ Apply Construction I to G
call *bintree-mpfe*($\hat{G}$), apply results to G

Figure 5. *Algorithm polytree-mpfe.*

4. HIERARCHICAL IDENTIFICATION

We now develop a framework and efficient algorithms for the conservative fusion of consistent and inconsistent reports received for a group of objects into a single summary report. We present the notion of *hierarchical identification* where each report may be a list of labels of various specificity. The number of vehicles on each track is assumed to have been estimated (say, as done in Section 3). These algorithms are developed as part of the flow-conserving identification component for the flow-conservation group tracking framework. However, unlike in Section 3, as a first step, the analysis in this section assumes that each track is solved independently of other tracks. A needed extension of this work which utilizes flow conservation constraints for each type along with an error model is currently being developed (see Section 5).

We consider a group of objects where each object can be described using labels of various specificity. Labels are arranged in a hierarchy according to their specificity. For example, Tank, IFV, Artillery Vehicle are labels for objects, and T88, T54 are more specific labels then Tank. An example for such a hierarchy is shown in Figure 6. A report is a list of labels associated with a group of objects, as reported by some identification device. For example, (T55, T88, T88, Artillery Vehicle) is a report about four objects. Reports do not associate which object in the group has been given a specific label. In our example, the report does not specify which object is a T55 and which is an Artillery Vehicle. The labels used in each report may be given in various levels of specificity. Furthermore, the information in the reports may be erroneous.

In this section, we first provide a formal description of the *Hierarchy Matching* problem (Section 4.1). Then we describe an intuitive $O(lm^2n)$ algorithm (Section 4.2) where m is the number of objects, n is the number of reports and l is the number of labels in the hierarchy.

In a companion technical report (Zohar and Geiger, 2002b), we provide a more efficient $O(n(l+m))$ algorithm for this problem, present the *Minimal Modification Hierarchy Matching* problem, and provide an algorithm for its solution for star hierarchies.

4.1. *Problem Formulation*

Suppose L is a finite set with a partial order $>$. Each element $l_i \in L$ is called a *label* and $l_1 > l_2$ is interpreted as l_2 is more *specific* than l_1, or equivalently, l_1 is more *general* than l_2. For example, l_1=Tank is more general than l_2=T88.

Definition 2. A *hierarchy* H is a directed tree (L, E) where each node in L is a label, and if (l_1, l_2) is an edge in E, then label l_2 is more specific than label l_1.

For example, consider the military-domain hierarchy in Figure 6. Tank, IFV, Artillery Vehicle are labels for objects, and T88, T54 are more specific labels then Tank.

We use the term *multiset* to denote a set with repetitions. A multiset A is *compatible* wrt a hierarchy $H = (L, E)$ if all elements of A belong to L and reside on a single directed path in H.

We define the *Hierarchy Matching* problem as follows.

Instance: Let $H = (L, E)$ be a hierarchy. Let $R_j = \{l_{j1}, ..., l_{jm}\}$, $j = 1, ..., n$, be multisets with elements from L and let R be the union $R_1 \cup R_2 \cup \ldots R_n$ (with repetitions) of the multisets R_i. In other words, the multiset R contains mn elements from L, not necessarily distinct.

Query: Partition R into disjoint multisets $A_1, ..., A_m$ each containing exactly one element from each R_j, $j = 1, ..., n$, such that for $1 \le i \le m$, A_i is compatible wrt H; or output a valid statement that no such partition exists.

When such a partition exists, the multiset R is said to be *consistent* wrt H and $A_1, \ldots, A_m$ is called a *consistent partition* of R. Otherwise it is *inconsistent* wrt H.

The *hierarchy matching* problem is motivated by the following interpretation of its components. Consider a group of objects, in which the number of objects m is known. Each object can be described using a label. The set of possible labels is denoted by L. Labels are arranged in a hierarchy tree as in Figure 6. A report R_j is a list of labels associated with a group of m objects, as reported by some identification device. For example, (T55, T88, T88, Artillery

Vehicle) is a report about four objects. Reports do not associate which object in the group has been given a specific label. In our example the report does not specify which object is a T55 and which is an Artillery Vehicle. The labels used in each report may be given in various levels of specificity.

The *hierarchy matching* problem seeks a consistent interpretation of the reports $R_1, \ldots, R_n$ if such exists by matching the labels of each object across the n reports. If the reports are consistent it is possible to summarize the information they convey using a *consensus report*

Definition 3. Given a consistent partition $A_1, ..., A_m$ we say that the *consensus report* is the multiset constructed of the most specific labels in each of the multisets $A_i, i = 1, ..., m$

Consider using Figure 6 the reports: $R_1 = (Equipment, Tank)$, $R_2 = (Military, \mathit{IFV})$. It is readily seen that these reports may be partitioned into the consistent partition: $A_1 = (Military, Tank), A_2 = (Equipment, \mathit{IFV})$. In this case, the consensus report is the most specific label from A_1 and the most specific label from A_2 that is (Tank, *IFV*).

Now consider a third report $R_3 = (Military, Truck)$. Although each pair of the reports R_1, R_2, R_3 is consistent, no consistent partition exists for the three reports. Consequently the reports are inconsistent and a consensus report does not exist.

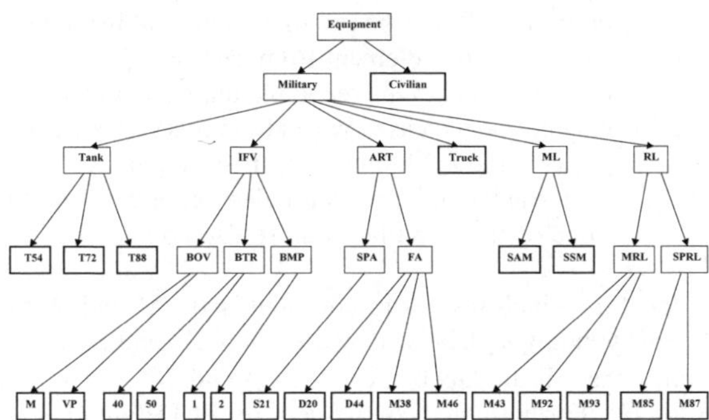

Figure 6. *A hierarchy of labels.*

The *hierarchy matching* problem is a restricted version of an n-*dimensional matching* problem, defined as follows. Let $G = (V, E)$ be a graph and let $V_1, \ldots, V_n$ be a partition of V into disjoint sets each containing exactly m nodes. The problem is to partition V into m disjoint sets $A_1, \ldots, A_m$ such that each A_i contains exactly one node from each set V_j (namely, n nodes) and for every two nodes u, v in A_i there exists an edge (u, v) in E. When $n = 2$, this becomes the bi-partite graph matching problem. When $n > 2$, the decision version of this problem is NP complete (Garey and Johnson, 1979). However, for the *Hierarchy Matching* problem, which can be viewed as a restricted version of the n-dimensional matching problem, we developed a polynomial algorithm with time complexity $O(n(l + m))$.

4.2. *Algorithm hierarchy-matching*

The algorithm described in this section is based on the fact that due to the constraint implied by a hierarchy the desired multisets A_i, $1 \leq i \leq m$ can be found sequentially in a greedy manner.

Definition 4. Let R be a multiset with elements from a set L with partial order $>$. An element l_1 is called a *least element* of R if $l_1 \in R$ and there is no other label $l_2 \in R$ such that $l_1 > l_2$.

Definition 5. Let l, l_1 be elements of a set L with partial order $>$, such that $l_1 \geq l$, and let R be a multiset with elements from L. The element l_1 is called a *least element* of R wrt l if $l_1 \in R$ and there is no other element $l_2 \in R$ such that $l_1 > l_2 \geq l$.

Definition 6. Let $(H, R_1, ..., R_n)$ be an instance of the *Hierarchy Matching* problem and let the multiset R be the union $R_1 \cup R_2 \cup \ldots R_n$ (with repetitions). Let $A = \{a_1, ..., a_n\}$ be a multiset with elements from R such that: (a) The multiset A is compatible wrt H; (b) there is an index $k \in [1, ..., n]$ such that a_k is a least element of R; (c) for $j = 1, ..., n$, label a_j is a least element of R_j wrt a_k. Then A is said to be a *reducible multiset* wrt to $(H, R_1, ..., R_n)$.

Lemma 4.1. *Let $(H, R_1, \ldots, R_n)$ be an instance of the hierarchy matching problem where R is the union $R_1 \cup R_2 \cup \ldots R_n$ (with repetitions). Let $A = \{a_1, \ldots, a_n\}$ be a reducible multiset wrt $(H, R_1, \ldots, R_n)$. Then R is consistent iff $R^* \equiv R \setminus A$ is consistent.*

Proof. First assume that the multiset R^* is consistent. Consequently it may be partitioned into disjoint compatible multisets wrt H. Since A is a reducible multiset it is also compatible wrt H by requirement (a) of Definition 6. Since $R = R^* \cup A$ with repetitions, the multiset R must be consistent as well.

Now assume that R is consistent. We will prove that R^* may be partitioned into $m-1$ disjoint compatible multisets. By definition, R may be partitioned into m disjoint compatible multisets $A_1, \ldots, A_m$ each containing exactly one element from each R_j, $j = 1, \ldots, n$. Clearly, if one of the multisets $A_1, \ldots, A_m$ is identical to the reducible multiset A then the proof is complete. However, in general, this may not be the case. We address this difficulty in the following.

Since $A = \{a_1, \ldots, a_n\}$ is a reducible multiset, then by requirement (b) of Definition 6 there is an index $k \in [1, \ldots, n]$ such that a_k is a least element of R. Since R is partitioned into the multisets $A_1, \ldots, A_m$, one of these multisets must also contain the label a_k. Denote this multiset $A^* = \{a_1^*, \ldots, a_n^*\}$.

Apart from the label a_k which resides in both multisets, A^* and A, the labels $a_j, j \neq k$ need not equal the corresponding a_j^* labels. However, both A^* and A are compatible multisets containing a_k, meaning that all the labels $\{a_1, \ldots, a_n\}$ and $\{a_1^*, \ldots, a_n^*\}$ must reside on the same directed path in H. Furthermore, requirement (c) of Definition 6 yields $a_j^* \geq a_j$ for $j = 1, \ldots, n$. Since a_j^* is more general than a_j it may replace a_j in any multiset without affecting this multiset compatibility. Consequently it is possible to locate for $j = 1, \ldots, n$ the label a_j in the reports $A_1, \ldots, A_m$ and swap it with a_j^*. The result is a consistent partition of R into m compatible multisets of which one is identical to A.

□

Lemma 4.1 implies that given an instance of the hierarchy matching problem it is possible to pursue equivalent problems with reduced size by removing reducible-multisets. The following Lemma proves that reducible-multisets can always be found in a consistent set of reports.

Lemma 4.2. *Let $(H, R_1, \ldots, R_n)$ be an instance of the hierarchy matching problem where R is the union $R_1 \cup R_2 \cup \ldots R_n$ (with repetitions). Let $\hat{a}$ be a least element of R. Then if R is consistent, there exists a reducible multiset $A = \{a_1, \ldots, a_n\}$ wrt $(H, R_1, \ldots, R_n)$ such that $\hat{a} \in A$.*

Proof. Since R is consistent it may be partitioned into m disjoint compatible multisets $A_1, \ldots, A_m$ each containing exactly one element from each R_j, $j = 1, \ldots, n$. Since R is partitioned into the multisets $A_1, \ldots, A_m$, one of these multisets must contain the label $\hat{a}$. Denote this multiset $A^* = \{a_1^*, \ldots, a_n^*\}$. The multiset A^* is compatible and contains a least element of R. Indeed, it is possible that for some report R_j, a_j^* is not a least element wrt $\hat{a}$. However, this may be easily fixed by replacing a_j^* with the least element of R_j wrt $\hat{a}$.

□

Lemmas 4.1 and 4.2 yield the following algorithm for hierarchy matching: At each iteration the algorithm finds a reducible-multiset. This can be done if the set of reports is consistent (Lemma 4.2). By Lemma 4.1 it is valid to remove the reducible multiset if it exists. This process is iterated. After m iterations either a consistent partition is found or such a partition does not exist and therefore inconsistency is declared. This algorithm is shown in Figure 7.

Algorithm *hierarchy-matching* $(H, R_1, \ldots, R_n)$
Main
  $R \leftarrow R_1 \cup R_2 \cup \ldots R_n$ (with repetitions)
  $R^{(0)} \leftarrow R$
  for i=1 to m
    $A_i \leftarrow$ FindReducibleMultiset($R^{(i-1)}$)
    $R^{(i)} \leftarrow R^{(i-1)} \setminus A_i$ {reduction}
FindReducibleMultiset(R)**:**
  $a_k \leftarrow$ find a least element in R
  $A \leftarrow a_k$
  for each report R_j **do**
      $j{\neq}k$
    $a_j \leftarrow$ find a least element in R_j wrt a_k
    if a_j is found **then**
      $A \leftarrow A \cup a_j$ (with repetitions)
    else declare *inconsistent*, halt
  return (A)

Figure 7. *Algorithm hierarchy-matching.*

Theorem 4.3. *Let $(H, R_1, \ldots, R_n)$ be an instance of the Hierarchy Matching problem where R is the union $R_1 \cup R_2 \cup \ldots R_n$ with repetitions. Then, algorithm hierarchy-matching solves the problem in time complexity $O(lm^2n)$ and space complexity $O(l + mn)$.*

Proof. Denote by $R^{(i)}$ the set of reports before the i-th iteration. At each iteration the algorithm attempts to find a reducible multiset in $R^{(i)}$. Recall that if $R^{(i)}$ is consistent it is always possible to find a reducible-multiset by Lemma 4.2. Therefore, if a reducible-multiset is not found, the multiset $R^{(i)}$ is inconsistent. By Lemma 4.1, if $R^{(i-1)}$ is consistent than so must be $R^{(i)}$. Consequently, the inconsistency of $R^{(i)}$ yields inconsistency of $R^{(i-1)}$ and so on until we obtain that $R^{(1)} = R$ is inconsistent as declared by the algorithm.

Consider now the case when the algorithm does not output a statement that R is consistent. This case occurs iff $R^{(m)}$ is found to be consistent which implies that $R^{(m-1)}$ is consistent and therefore that $R^{(m-2)}$ is consistent and thus $R^{(1)} \equiv R$ is consistent as well. Furthermore, a consistent partition to reducible multisets is readily given.

We now analyze time and space complexity. At each iteration of *FindReducibleMultiset* the algorithm finds a least element in R and then a least element in each of the remaining $n-1$ reports. Consequently, a naive implementation for *FindReducibleMultiset* yields time complexity $O(lmn)$ (where it is assumed that the query: given labels l_1, l_2 is $l_1 > l_2$? is $O(l)$). This function is called (at most) m times hence the entire algorithm has time complexity $O(lm^2n)$. Space is needed to store R and the multisets A_i as well as the hierarchy labels which yields a space complexity of $O(l + mn)$.

□

5. DISCUSSION AND FUTURE WORK

In this paper, we described a novel framework for group tracking and identification termed *flow conservation group tracking*, which we believe to be a preferred extension of the methodology for tracking single targets to tracking groups. Our framework integrates local kinematic measurements with flow conservation constraints and is based to a large extent on the ability to estimate and match the number and types of targets in each group, in addition to kinematics.

We envision several major improvements to our framework. First, we expect the framework to be augmented to support multiple hypothesis tracking. However, unlike the framework suggested in Reid (1979), each hypothesis must also include the estimation of number of vehicles on each of the tracks in addition to kinematics. Second, we believe that a tighter integration may be obtained between the measurement-to-track and the flow-estimation components. Finally, since flow conservation group tracking may be applied in several resolutions (depending on the clustering attributes) it will be interesting to characterize, given an arena, optimal attributes for clustering so as to provide a more accurate tracking procedure.

Each of the framework's components may also be improved independently. An improvement to the flow estimation component may be the development of a methodology that is not based on normal measurement errors. Our initial results reported in Zohar and Geiger (2002a) are very encouraging showing that our algorithm works even when data are generated from various distributions.

Our initial experiments show that flow conservation provides a constant factor asymptotic improvement versus sole local estimates regardless of sample size. Characterizing this improvement as a function of the topology may also be interesting. Also, it seems possible to improve the flow estimation algorithm for special graph topologies. We have already developed an algorithm that solves the MPFE problem in time complexity $O(|E|)$ provided that the flow network is a polytree and we believe it may be possible to extend this improvement to other network topologies as well for which our algorithm's current complexity is $O(|E|^3)$. This improvement may be helpful in the application of flow conservation group tracking in large scale real-time scenarios.

The main improvement we envision to the identification module is related to the exploration of both the *Hierarchy Matching* and the *Min Modification Hierarchy Matching* problems in the context of global estimation in flow networks. Instead of providing a summary report that is solely based on measurements received on a single edge, future algorithms will generate a summary report per edge in a global fashion that uses reports on other edges in the flow network to correct its estimates, analogous to the way we estimate the number of objects in *mfpe*.

Another improvement includes introducing an error model and a modified algorithm that provides the most likely summary report per edge in the flow network—both locally and globally.

ACKNOWLEDGEMENTS

The authors thank David Reinitz for useful conversations and Mark Matusevich for developing with us the use of minimum cost flow algorithms for the measurement-to-track component.

REFERENCES

Ahuja, R. K., Magnanti, T. L., and Orlin, J. B. (1993). *Network Flows: Theory, Algorithms and Applications*. Englewood Cliffs, NJ: Prentice-Hall

Bar-Shalom, Y. and Li, X. R. (1995). *Multitarget-Multisensor Tracking: Principles and Techniques*. Los Angeles: Bar-Shalom

Blackman, S. (1986). *Multiple Target Tracking with Radar Applications*. Boston: Artech House.

Blackman, S. and Popoli, R. (1999). *Design and Analysis of Modern Tracking Systems*. Boston: Artech House.

Cowell, R. G., Dawid, A. P., Lauritzen, S. L., and Spiegelhalter, D. J. (1999). *Probabilistic Networks and Expert Systems*. Berlin: Springer.

Dechter, R. (1996). Bucket elimination: A unifying framework for probabilistic inference algorithms. *Proceedings of Uncertainty in Artificial Intelligence*. San Fransisco: Morgan-Kaufmann Pub., 211–219.

Garey, M. R. and Johnson, D. S. (1979). *Computers and Intractability, A Guide to the Theory of NP-Completeness*. New York: Freeman.

Jensen, F. V. (1996). *An Introduction to Bayesian Networks*. London: UCL Press.

Kay, S. M. (1993). *Fundamentals of Statistical Signal Processing: Estimation Theory*. Englewood Cliffs, NJ: Prentice-Hall.

Lauritzen, S. L. (1992). Propagation of probabilities, means and variances in mixed graphical association models. *J. Am. Statist. Ass.* **87**, 1098–1108.

Pearl, J. (1988). *Probabilistic Reasoning in Intelligent Systems*. San Fransisco: Morgan-Kaufmann Pub..

Reid, D. B. (1979). An algorithm for tracking multiple targets. *IEEE Trans. Automatic Control* **24**, 843–854.

Shachter, R. D. and Kenley, C. R. (1989). Gaussian influence diagrams. *Mgm. Sci.* **35**, 527–550.

Taenzer, E. (1980). Tracking multiple targets simultaneously with a phased array radar. *IEEE Trans. Aerospace Electronic Systems* **41**, 604–614.

Zohar, R. and Geiger, D. (2002a). Estimation in hidden flow models. *Tech. Rep.*, Technion, Israel.

Zohar, R. and Geiger, D. (2002b). Hierarchy matching—preliminary results. *Tech. Rep.*, Technion, Israel.

DISCUSSION

JOSEPH B. KADANE (*Carnegie-Mellon University, USA*)

The nature of this paper changed between the paper in the program and the paper given at the conference. Consequently, almost all my comments concern estimation in hidden flow networks, which is only part of the larger framework of the printed paper.

The motivation given for the hidden flow model is as follows: "Edges represent tracks in a battlefield. Internal nodes correspond to the merger or separation of groups of vehicles. The number of vehicles on each track is constant, and the number of vehicles that enter an internal node equals the number of vehicles that leave it. Various sensors are used to track the scene. Each sensor provides an estimate of the number of vehicles on each track. The goal is to improve the estimation of the number of vehicles on each track by utilizing the flow constraints on internal nodes."

The paper makes three assumptions about the model: Let e_i be the measured flow on edge i, X_i the true flow, and τ_i the precision of the flow. Then $e_i \mid X_i, \tau_i \sim N(X_i, 1/\tau_i)$, with nodes i independent. Second, the prior on the X_i's assumes that the flows in equal the flows out at each internal node. Third, otherwise the prior on the X_i's is uniform.

The motivation leads to some questions about the assumptions. I would worry about measurement bias in the sensors. Is a sensor more likely to miss vehicles than to overcount them? Second, in a real situation, vehicles are likely to be destroyed, get lost, stop working, etc. So while it is quite reasonable to suppose that the number of vehicles out of a node can be no greater than the number that enter, equality seems to me to be a stretch. While this can be repaired formally by having an additional edge for losses, I wonder what kind of sensor will measure this. Finally, I would think that there would usually be informative priors about what flows are plausible. The treatment of time, which standing alone looks problematic, is explained in the larger context of the paper in the volume.

The prior on the vector X of nodes is *very* informative about internal nodes and *very* uninformative about external (source and sink) nodes. Both states of belief can be modelled as limiting normal distributions. Together with the observational normals $e = HX_E + w$, with $w \sim N(0, C)$, where X_E are the true flows at the external nodes, H has a single 1 and all the rest 0's for an external node observation, and -1's for sources and $+1$'s for sinks for an internal node observation. Also C is diagonal with i-th diagonal element $1/\tau_i$.

Then the posterior distribution is the precision-weighted average of the prior distribution and the likelihood, which here simplifies to

$$X_E \mid e \sim N((H^tC^{-1}H)^{-1}H^tC^{-1}e, (H^tC^{-1}H)^{-1}).$$

Contrary to the paper, the assumptions of linearity and unbiasedness are unnecessary. The paper concentrates on finding the posterior means and precisions of the flows.

I now discuss the first example from the paper, which has a single internal node, with the constraint $X_p = X_\ell + X_r$. The paper shows that the posterior flow on edge p has mean and precision

$$\frac{X_p^* = e_p\tau_p + (e_\ell + e_r)\left(\frac{\tau_r\tau_\ell}{\tau_r+\tau_\ell}\right)}{\tau_p^*}, \qquad \tau_p^* = \tau_p + \frac{\tau_r\tau_\ell}{\tau_r + \tau_\ell}.$$

How should this be understood? There are two sources of information about X_p. First, there is the direct observation $e_p \sim N(X_p, 1/\tau_p)$. Second, there is the constraint. From the assumptions, $e_\ell + e_r \sim N(X_p, \sigma^2)$, where σ^2 is the variance of the sum of two independent normal observations, so $\sigma^2 = 1/\tau_r + 1/\tau_\ell = (\tau_\ell + \tau_r)/(\tau_r\tau_\ell)$. The formulae above are then seen to be simple consequences of the fact the variances add for sums of independent normals, although this is a bit obscured by the expression in terms of precisions.

The paper also gives a more complicated second example, which can be handled with similar methods. It is not necessary to introduce artificial simple nodes to resolve nodes with multiple inputs and outputs.

The bulk of the paper is devoted to giving a linear algorithm for finding the mean and precision of the posterior distributions of the flow at each node. It does not give covariances, which means that precisions for sums of flows are not available.

In conclusion, the strength of this paper is that it provides a fast algorithm, linear in the number of nodes, for the posterior means and precisions. The assumptions are somewhat distantly related to the application, however.

REPLY TO THE DISCUSSION

This paper describes a new method for tracking groups of objects. The new aspect of this method is the use of flow conservation constraints as a guiding principle rather than kinematics which is the current leading approach. The paper has three logical parts. The first is the general framework and a description of the system architecture, the second is the estimation of flow (say, number of vehicles) on tracks in the traced arena, and the third is the identification of types of objects according to a given hierarchy of possible object types. A system is now being constructed using the elements proposed in this paper and it will take some time before one can determine how successful this approach is for real sensors with real data.

The comments of Professor Kadane concentrate on Section 3, which is the component of the system that deals with estimating the flow. Some concerns have been raised regarding the assumptions made in formulating the flow estimation problem and some concerns about the priors we have used. Finally, there is a technical suggestion claiming to improve the intuition of our estimation formulae.

The main concern raised by the discussant is that "in a real situation, vehicles are likely to be destroyed, get lost, stop working, etc. So while it is quite reasonable to suppose that the number of vehicles out of a node can be no greater than the number that enter, equality seems to me to be a stretch. While this can be repaired formally by having an additional edge for losses, I wonder what kind of sensor will measure this."

The amount of true vehicles remains unchanged in splits and merges of the flow graph. This is not a stretch but a simple unquestionable truth. However, as the discussant suggests, when some vehicles stop moving, then an additional node needs to be added, a sink node, so that flow conservation is kept in the flow graph. Sensors that measure such stopped vehicles are, for example, based on heat detection, vision, radar, and many others. No real vehicles "get lost", some may not be observed, and so the measurements will not adhere to flow conservation, but these measurements will be corrected according to the flow conservation of the true flow in the entire flow graph. This is precisely the power of the proposed method.

The discussant also worries about measurement bias in the sensors. "Is a sensor more likely to miss vehicles than to overcount them?" Indeed, this is a simplifying assumption we have made. The measurement bias highly depends on the sensor used. For example, in the case of a Moving Target Indicator (MTI) radar, based on Doppler's effect, the detection probability (P_d) of objects depends on the radial velocity of the object relative to the MTI scan. Low radial velocity yields low detection probability.

We sketch below the changes we may introduce to overcome this assumption. First, we estimate the probability of detection P_d per observation. For MTI radar, the radial velocity may be estimated, and given a reasonable model for the radar, the detection probability may be estimated as well. We can now replace the sampled vehicle number e_i with an updated estimate that uses P_d and, similarly, update the precision of measurements, which gets higher as P_d increases, according to a radar model. The new adjusted measurements and precisions can now be fed into our Most Probable Flow Estimation algorithm. Alternatively, the problem may be solved quite efficiently using our Minimum Cost Flow formulation (Section 3.2).

In conclusion, using a non-biased model is more generic in the sense that the bias model is highly dependent on the sensor. Still, defining bias models and estimating the biases as well as the flow values in a global fashion remains an interesting problem to be explored within our framework.

The discussant argues that the flow estimation problem, under the prior assumptions we have adopted, computes the posterior distribution given by

$$X_E \,|\, e \sim N((H^t C^{-1} H)^{-1} H^t C^{-1} e, (H^t C^{-1} H)^{-1}).$$

Indeed this is correct and agrees precisely, for the selected prior, with the formula produced by "linear minimum variance unbiased estimation," albeit the quite different setups. We feel comfortable with the prior we selected and in the tests we have performed so far, it proved effective. If required, the incorporation of more complicated priors will be introduced, and then, the full Bayesian approach, as highlighted by the discussant will surely be of importance.

Finally, the discussant argues that the estimation formulae are "simple consequences of the fact the variances add for sums of independent normals, although this is a bit obscured by the expression in terms of precisions," and that "the paper also gives a more complicated second example, which can be handled with similar methods. It is not necessary to introduce artificial simple nodes to resolve nodes with multiple inputs and outputs."

Clearly, a representation using variances or precisions is equivalent. Some good Bayesian books have chosen to represent many of their formulae using precisions, and so did we. The artificial nodes were introduced for the simplicity-of-presentation of the algorithm and for proving its correctness. In an application, it is not required to actually "add these edges" rather to act as if "they were there". We believe this was clear from our example restated below: "Suppose, for example, we wish to compute the most probable flow of edge l_1 in Figure 3a.

Apply Construction I to obtain Figure 3b. Next apply algorithm *bintree-mpfe*. The first iteration yields: $e_{lr} = e_5 - e_6$ with $\tau_{lr} = \tau_5\tau_6/(\tau_5 + \tau_6)$. After the second iteration $e_{lr} = e_4 + e_5 - e_6$ with

$$\tau_{lr} = \frac{\tau_4\tau_5\tau_6}{\tau_4\tau_5 + \tau_4\tau_6 + \tau_5\tau_6}.$$

After the third iteration $e_{lr} = -e_3 + e_4 + e_5 - e_6$ with

$$\tau_{lr} = \frac{\tau_3\tau_4\tau_5\tau_6}{\tau_3\tau_4\tau_5 + \tau_3\tau_4\tau_6 + \tau_3\tau_5\tau_6 + \tau_4\tau_5\tau_6}$$

and finally, after the forth iteration: $e_{lr} = e_2 - e_3 + e_4 + e_5 - e_6$ with

$$\tau_{lr} = \frac{\tau_2\tau_3\tau_4\tau_5\tau_6}{\tau_2\tau_3\tau_4\tau_5 + \tau_2\tau_3\tau_4\tau_6 + \tau_2\tau_3\tau_5\tau_6 + \tau_2\tau_4\tau_5\tau_6 + \tau_3\tau_4\tau_5\tau_6}.$$

Hence, $X_1^* = (\tau_1 e_1 + \tau_{lr} e_{lr})/(\tau_1 \tau_{lr})$, with $\tau_1^* = \tau_1 + \tau_{lr}$."

Note that our derivation does not use the virtual edges and nodes. It is also interesting to note that the denominator of τ_{lr} is a sum of $O(|E|)$ factors, each consisting of $O(|E|)$ terms. Consequently, direct computation of τ_{lr} could require $O(|E|^2)$ operations in contrast to $O(|E|)$ operations using Construction I and our algorithm *bintree-mpfe*. In other words, although a closed form formula is attainable for the elimination process in polytrees, its efficient computation uses a process that imitates our algorithm.

It is important to observe that trying to solve this simple network using general approaches yields an $O(|E|^3)$ time complexity and an $O(|E|^2)$ space complexity, in contrast to the linear time and space complexity of *polytree-mpfe*.

CONTRIBUTED PAPERS

BAYESIAN STATISTICS 7, pp. 443–451
J. M. Bernardo, M. J. Bayarri, J. O. Berger, A. P. Dawid,
D. Heckerman, A. F. M. Smith and M. West (Eds.)

Bayesian Modeling of Hospital Bed Occupancy Times using a Mixed Generalized Erlang Distribution

M. CONCEPCIÓN AUSÍN   ROSA E. LILLO
Universidad Carlos III de Madrid, Spain
causin@est-econ.uc3m.es   lillo@est-econ.uc3m.es

FABRIZIO RUGGERI
CNR-IMATI, Italy
fabrizio@iami.mi.cnr.it

MICHAEL P. WIPER
Universidad Carlos III de Madrid, Spain
mwiper@est-econ.uc3m.es

SUMMARY

In this paper we model the distribution of length of stay in hospital of geriatric patients. We assume that length of stay has a mixed generalized Erlang distribution and, given data from patients in a geriatric ward of a London hospital, reversible jump methods are used to estimate the predictive distribution of length of stay in hospital. We also address the problem of optimizing the number of beds in hospital with respect to average cost per unit of time where costs are based on lost demand, the number of unoccupied beds and the different types of patients in the hospital.

Keywords: MIXED GENERALIZED ERLANG DISTRIBUTION; M/G/C LOSS SYSTEM; REVERSIBLE JUMP MCMC.

1. INTRODUCTION

The distribution of the time that geriatric patients spend in hospital seems to have a complex behavior. Patients are admitted and subject to an initial acute care which may be followed by a number of different stages of treatment. Most patients are discharged (or die) after a relatively short period of time. However, some patients may remain in hospital over a long time period receiving continuous attention. This suggests that the distribution of lengths of stay in hospital is likely to be very heterogeneous.

Figure 1 shows boxplots of the distribution of the lengths of stay of 1092 geriatric patients at St. George's Hospital in London over the period 1965–1984. The data are a subset of the sample analyzed by *e.g.*, Taylor *et al.* (2000) and may be downloaded from the Royal Statistical Society, `http://www.blackwellpublishers.co.uk/rss`. We can observe from the boxplots that the data are very highly skewed to the left with a number of large outliers.

The diverse patterns of duration of stay in hospital require distinct resources and suitable organization and consequently, for hospital planning, it is important to capture the shape of the distribution of the length of stay. As patients undergo various stages of treatment in hospital, this suggests the use of phase-type distributions (see Neuts, 1981), for modelling the distribution of stay. Such models essentially assume that the total time spent by a patient in hospital can be decomposed into a number of different, exponential, phases (see also Faddy, 1990,1994).

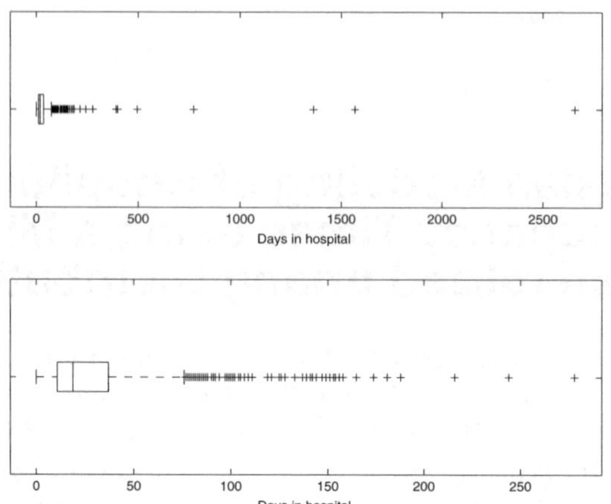

Figure 1. *Boxplots of days spent in hospital for all patients* (top) *and for patients staying under 300 days* (bottom).

A number of authors have used classical techniques to fit a variety of phase-type models to (versions of) the St. George's Hospital data. For example, Harrison and Millard (1991) and Gorunescu *et al.* (1999) considered exponential mixture models and Faddy and McClean (1999) used the more general class of mixed, generalized Erlang (MGE) distributions although they assumed a fixed number of phases. One of the main difficulties with such classical models is that they are not designed to deal with situations where the model dimension, *e.g.*, the number of phases in the MGE model, is unknown. However, in such problems, Bayesian inference can often be carried out using variable dimension MCMC methods. In Section 2, we also assume an MGE model and we show how Bayesian inference can be undertaken using reversible jump techniques under the assumption that the number of phases are unknown.

One of the motivations for studying the length of stay in hospital is to attempt to optimize the number of beds, c. The decision of how many beds to put in a ward will typically be based on both the costs of maintaining beds and the (opportunity) costs of losing prospective patients because the ward is full, assuming that in this case, prospective patients are lost rather than put on a waiting list. Gorunescu *et al.* (1999) analyze this problem using classical techniques. In Section 3, we assess the optimization of the number of beds assuming a slightly more complicated cost function that also takes into account the different costs for short and long stay patients.

In Section 4, we finish with a brief discussion and some possible extensions to other fields.

2. MODELING AND INFERENCE FOR TIME SPENT IN HOSPITAL

In this section, we consider modelling and inference for the time spent in hospital by a patient, T. We assume that T follows a MGE distribution, that is

$$T = \begin{cases} X_1 & \text{with probability } P_1 \\ X_1 + X_2 & \text{with probability } P_2 \\ \vdots & \vdots \\ X_1 + \ldots + X_L & \text{with probability } P_L \end{cases} \tag{1}$$

where $X_r \sim \text{Ex}(\mu_r)$ for $r = 1, \ldots, L$ and $\sum_{r=1}^{L} P_r = 1$.

Thus, from (1), the density of T has a mixture form;

$$f(t \mid L, \boldsymbol{P}, \boldsymbol{\mu}) = \sum_{r=1}^{L} P_r f_r(t \mid \boldsymbol{\mu})$$

where $f_r(t \mid \boldsymbol{\mu})$ is the density function of a sum of r exponentials, or a generalized Erlang distribution

$$f_r\left(t \mid L, \boldsymbol{\mu}\right) = \sum\nolimits_{j=1}^{r} \left(\prod\nolimits_{s \neq j} \left(\frac{\mu_s - \mu_j}{\mu_s \mu_j}\right)^{-1}\right) \mu_j^{2-r} e^{-\mu_j t}. \tag{2}$$

see *e.g.*, Johnson and Kotz (1970). Note that in (2) it is assumed that all μ_r, $r = 1, \ldots, L$ are distinct. Alternative formulae are available in the case where there is some repetition; see Johnson and Kotz (1970).

This distribution is a phase-type distribution of order L and contains the exponential ($L = 1$) and Erlang ($P_L = 1, \mu_1 = \cdots = \mu_L$) distributions as special cases. By increasing the number of phases, it is possible to approximate any (strictly continuous) density function over the positive real line using an MGE distribution. In our case, we will assume that all parameters, including L, are unknown.

Given that we observe the times spent in hospital of n patients, $\boldsymbol{t} = \{t_1, ..., t_n\}$, independently, we now wish to carry out Bayesian inference for this model. The likelihood function takes a very complicated form but can be simplified by introducing the latent variables; Z_i is the phase in which the i-th patient leaves the hospital and X_{ir} is the time spent by the i-th patient in phase r, for $r = 1, \ldots, Z_i$ where $\sum_{r=1}^{Z_i} X_{ir} = T_i$. Then, we have

$$f(t_i, z_i, x_{i1}, \ldots, x_{iz_i} \mid \boldsymbol{\theta}) = P_{z_i} \prod\nolimits_{r=1}^{z_i} \mu_r \exp(-\mu_r x_{ir})$$

where $\boldsymbol{\theta} = (L, \boldsymbol{P}, \boldsymbol{\mu})$.

In order to carry out Bayesian inference, we also need to define prior distributions for the model parameters $\boldsymbol{\theta}$. We assume the following prior dependence structure:

$$f(\boldsymbol{\theta}) = f(L) f(\boldsymbol{P} \mid L) f(\boldsymbol{\mu} \mid L).$$

For the number of phases, L, many prior distributions can be considered. Here, we assume a translated Poisson distribution; $L - 1 \sim \text{Po}(1)$. This choice has the advantage of penalizing overparameterization by giving low probability *a priori* to large numbers of phases, L.

Conditional on L, we use semi-conjugate prior distributions for $(\boldsymbol{P}, \boldsymbol{\mu})$; a Dirichlet distribution for the weights, $\boldsymbol{P} \mid L \sim \text{Di}(1, \ldots, 1)$ and independent exponential distributions, $\mu_r \sim \text{Ex}(0.01)$, for $r = 1, ..., L$. These prior distributions are used mainly for convenience. A possible extension would be to use a hierarchical structure for the prior on $\boldsymbol{\mu}$.

2.1. *Sampling from the Posterior*

Given the data and prior distributions defined earlier, the posterior distribution can be sampled via typical MCMC methods for mixture distributions, using a Gibbs sampler to sample the parameter distributions conditional on L and a reversible jump method to let the chain move through the posterior distribution of L.

Given the latent variables, it is straightforward to show that the conditional posterior of $\boldsymbol{P}$ is still a Dirichlet distribution and the conditional distributions of the μ_r are gammas.

The latent variables can be sampled by sampling the two components of

$$f(z_i, x_{i1}, \ldots, x_{iz_i} \mid t_i, \boldsymbol{\theta}) = f(x_{i1}, \ldots, x_{iz_i} \mid t_i, z_i, \boldsymbol{\theta}) f(z_i \mid t_i, \boldsymbol{\theta}), \quad i = 1, \ldots, n.$$

The conditional posterior $f(z_i \mid t_i, \boldsymbol{\theta})$ is easily derived from (2). It is slightly more complicated to sample from $f(x_{i1}, \ldots, x_{iz_i} \mid t_i, z_i, \boldsymbol{\theta})$. This distribution is a product of z_i exponentials restricted to the subspace $\sum_{r=1}^{z_i} x_{ir} = t_i$. Assuming, without loss of generality, that $\min\{\mu_r\} = \mu_j$ and assuming that $\mu_r \neq \mu_j$ for all $r \neq j$, we have

$$f\left(x_{i1}, \ldots, x_{ij-1}, x_{ij+1}, \ldots, x_{i,z_i} \mid t_i, z_i, \boldsymbol{\theta}\right) \propto \prod_{\substack{r=1 \\ r \neq j}}^{z_i} (\mu_r - \mu_j) \exp\left\{-(\mu_r - \mu_j) x_{ir}\right\}$$

defined over the region $\boldsymbol{R} = \left\{x_{i1} + \cdots + x_{ij-1} + x_{ij+1} + \cdots + x_{i,z_i-1} \leq t_i\right\}$. (A similar formula is available in the case where $\boldsymbol{\mu}$ has multiple minima.) This density can be sampled straightforwardly by, for example, rejection sampling (see *e.g.*, Ripley, 1987).

In order to sample over different numbers of phases L, we can follow the split-combine procedure for mixtures of unknown number of components introduced by Richardson and Green (1997). The proposed moves change the order, L, by one unit in such a manner that in the MGE model, any phase can be split into two or any two consecutive phases can be combined into one. The corresponding changes to the parameters are done such that the marginal distribution of T is preserved. If a combine move is proposed, the latent variables $\left(z_i, x_{i1}, \ldots, x_{iz_i}\right)$ (for $i = 1, \ldots, n$) are changed in the natural way and if a split move is proposed, they are modified analogously to the Gibbs allocation technique described earlier.

2.2. *Results for the St. George's Hospital Data*

Figure 2 shows a histogram of the observed data truncated onto times less than 300 days. The predictive density (solid line) calculated from an MCMC run of 100,000 iterations in equilibrium is overlaid. This is compared with a classical density estimate (dashed line) based on the use of an EM-algorithm by Asmussen *et al.* (1996) to estimate the MGE model with $L = 5$ phases. This method was used in an earlier study of these data by Faddy and McClean (1999). Both density estimates appear very similar. This result is to be expected here as fairly uninformative priors have been used within the Bayesian model.

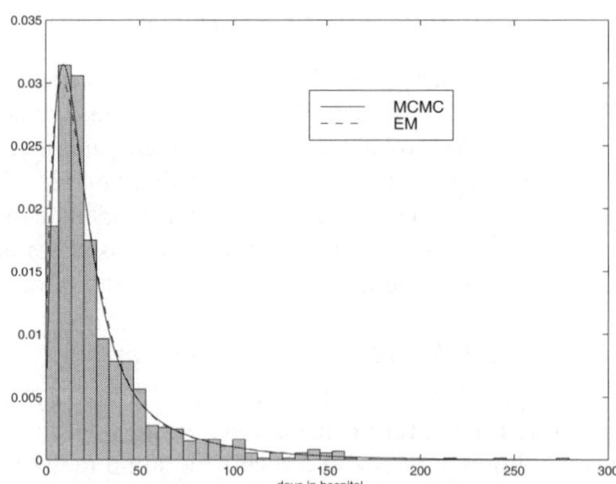

Figure 2. *Densities estimated using the MCMC algorithm and the EM algorithm for a fixed number of phases equal to 5.*

The total computation time for the MCMC algorithm was approximately 1.5 h on a Unix workstation. For the EM algorithm, the computation times are sensitive to the value of L chosen.

For $L = 1$, computation is almost immediate whereas for $L = 10$, this can take 20 min or more. Thus, the overall computation times for the two methods are comparable.

Notice that one advantage of the use of the variable dimension MCMC approach as opposed to the use of the EM method is that the number of phases L does not have to be fixed a priori. Also the uncertainty in the model is reflected directly in the Bayesian approach via the posterior distribution for L, see Figure 3, which means that the predictive density is an average over the various different models. A final advantage of the Bayesian approach is that it gives a natural way of electing an optimum number of beds in the hospital, and of assessing the uncertainty in this selection, via the use of decision theory methods (see Section 3).

Figure 3 shows the posterior probabilities for the total number of phases L. We can observe that the highest probabilities are assigned to $L = 5$ and $L = 6$. This corresponds well with the results of Faddy and McClean (1999). It should be noted that there is little sensitivity to the choice of prior for L in the value of the posterior mode of L. Experiments using different translated Poisson priors have produced very similar results. For example, varying the prior mean of L between 2 and 4, we still find that most of the posterior probability mass is concentrated on L values of 4–8, although the posterior mode varies between 5 and 7. Small changes in the prior distributions of the remaining parameters appear to have little effect. Furthermore, as we would expect, the predictive density estimation of the time spent in hospital by a patient is unaffected by these changes to the priors.

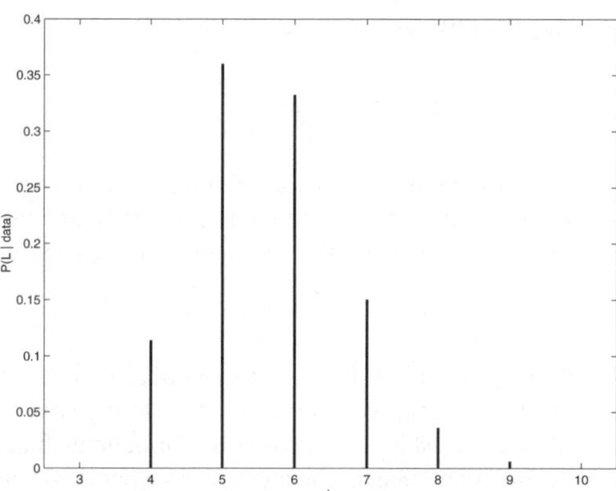

Figure 3. *Probabilities for different numbers of phases L.*

Finally, note that the mean number of days spent by a patient in hospital is estimated from the MCMC output to be around 37.3 (s.d. $= 4$).

3. OPTIMIZING THE NUMBER OF BEDS IN HOSPITAL

In this section, our aim is to formulate a cost function which can be used to optimize the number of beds in the hospital, c. We assume here that patients arrive at the hospital according to a simple Poisson process with rate λ and that when no beds are available, these patients do not join a waiting list but instead are lost to the system, being admitted to other hospitals, for example. The most important costs to be considered will be due to having insufficient beds or having too many empty beds and we therefore need to estimate the probabilities that various numbers of beds are occupied.

Under these conditions, the number of patients in the hospital can be modelled as a M/G/c loss system, that is, a queueing system with Poisson arrivals, general service distribution, c servers and fixed capacity c, that is, no queueing; (see *e.g.*, Tijms, 1990). Here the general service time is identified with the length of the stay and its distribution is approximated by the MGE model, introduced in Section 2. The capacity of the system is equal to the number of beds, c and subsequently there is no queueing.

The average number of patients arriving per unit of time is λ, and the mean time that patients stay in hospital (service time) is

$$\mathrm{E}[T \mid \boldsymbol{\theta}] = \sum_{r=1}^{L} \left(1 - \sum_{j=1}^{r-1} P_j\right) \frac{1}{\mu_r}.$$

The *offered load*, a of the system is defined to be the mean number of arrivals in a service (patient stay) time, that is $a = \lambda \mathrm{E}[T \mid \boldsymbol{\theta}]$.

The stationary distribution of the number of busy servers (beds), N, in a M/G/c loss system is given by, (see *e.g.*, Tijms, 1990),

$$P(N = j \mid a) = \frac{a^j / j!}{\sum_{k=0}^{c} a^k / k!}, \qquad k = 0, ..., c.$$

This is a truncated Poisson distribution defined on $[0, c]$. The probability that an arriving patient finds all c beds occupied is given by $B(c, a)$ where

$$B(c, a) = \frac{a^c / c!}{\sum_{k=0}^{c} a^k / k!} \tag{3}$$

(see *e.g.*, Cohen, 1976). This formula, known as *Erlang's loss formula*, corresponds to the fraction of prospective patients that is lost due to having insufficient beds.

The mean number of busy beds, can also be obtained (see *e.g.*, Tijms, 1990),

$$\mathrm{E}[N \mid a] = a\,[1 - B(c, a)]\,. \tag{4}$$

We can now deal with the problem of deciding the optimal number of beds in hospital. We will consider three possible costs. First, we assume an opportunity cost, $\pi > 0$, for each patient that is lost because no beds are available. The proportion of patients that cannot be admitted is given by the Erlang loss formula and hence, the average cost per day caused by lost patients is

$$\pi \lambda B\,(c, a)\,.$$

Second, we assume a holding cost, $h > 0$, for each empty bed per day. The average number of empty beds is equal to the total number of beds, c minus the average number of busy beds, given by (4). Thus, the average cost per unit of time due to unoccupied beds is given by

$$h\,\{c - a[1 - B(c, a)]\}$$

Finally, we consider a patient cost, r, for each patient in hospital (busy bed) per day. We consider different costs associated with the pattern of length of stay. Assume, for example, different costs, r_S, r_M, r_L, for "short-stay" (S), "medium-stay" (M) and "long-stay" (L) patients, respectively. This will typically be the case in practice when some patients will enter the hospital for urgent and expensive treatment such as operations, etc., and may then leave relatively quickly when cured, whereas others may enter because of general poor health, when they may spend a long period in hospital but without needing intensive treatment. By using the service time distribution, T, we can easily estimate the proportion of patients of every class, p_S, p_M and p_L. From (4), the average number of beds occupied by each type of patient at any given time will by given by $p_k a[1 - B(c, a)]$; for $k = S, M$ or L. Then, the average cost per day due to occupied beds will be,

$$ra[1 - B(c, a)]$$

where $r = r_S p_S + r_M p_M + r_L p_L$.

Combining these cost functions we have that the average cost function per unit of time given that there are c beds is

$$g(c) = \pi\lambda B(c, a) + h\{c - a[1 - B(c, a)]\} + ra[1 - B(c, a)]$$

Note that as c grows, the Erlang loss formula $B(c, a)$ approaches zero and thus, the cost function will approach be approximately linearly increasing for large c. It can also be shown that a sufficient, but not necessary condition for the cost function to have a unique minimum (or two equal, successive minima) is that $\pi\lambda + ha - ra > 0$.

Finally, in the case where we have a sample of patient stay times and the model parameters are unknown, as described in Section 2, it is straightforward to calculate the expected cost function given the MCMC output in the usual way. Results for the St. George's Hospital data are presented in the next section.

3.1. *Results for the St. George's Hospital Data*

There is no direct information on the arrival process of patients in the St. George's Hospital data. Instead, we assume here that the arrival parameter λ is known. Supposing that $\lambda = 1.5$, which approximately corresponds to the values used by Gorunescu *et al.* (1999), then recalling from Section 2.2 that the predictive mean estimate of the length stay in hospital was around 37.3, then the offered load can be estimated to be $\mathrm{E}[a \mid \boldsymbol{t}] = 37.3 \times 1.5 = 55.95$, which corresponds to the average number of patients who arrive during a mean length of stay in hospital.

Table 1. *Optimal numbers of beds for different interarrival rates.*

λ	Optimal c	$\mathrm{E}[g(c)]$	$s.d.[g(c)]$
1.0	38	174.26	15.22
1.5	57	259.31	23.28
2.0	75	344.08	31.37

Figure 4 illustrates the estimated mean cost function (solid line), the median (thick solid line), and an 80% predictive interval (dashed lines), where we assume the following costs: $h = 1, \pi = 200, r_S = 4, r_M = 2$ and $r_L = 5$ units. Note that the values for h and π are based on those of Gorunescu *et al.* (1999).

As can be observed, the optimal number of beds is estimated to be 57 when the minimum cost is 259.31 units. However, the maximum variance is reached near the minimum number of beds; specifically when $c = 58$ and there is some uncertainty in the range 55–60.

It is also important to explore the sensitivity of these results to different interarrival rates of patients. As can be shown, the optimal value of c increases with λ as does the average cost and its variance. Table 1 illustrates the optima for λ between 1 and 2. There is a reasonably large degree of sensitivity to λ.

As we noted earlier in Section 2.2, the predictive distribution of the stay in hospital is insensitive to the prior distribution for the number of phases L. The same result applies to

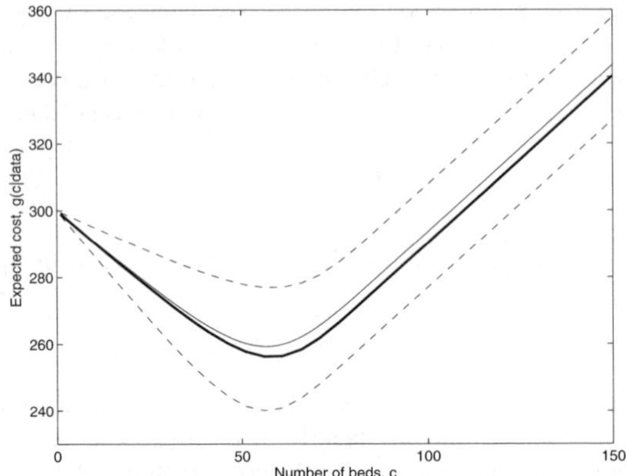

Figure 4. *Mean and median cost functions and 80% predictive interval.*

the predicted cost function and so the election of the optimal number of beds is not strongly influenced by this prior. Note, however, that we could also check the sensitivity to the chosen values of the cost parameters h, π, etc., could also be explored, although this is not done here.

4. DISCUSSION

In this paper, we have developed a Bayesian methodology to make inference about the distribution of the length of stay of patients in hospital and to enable us to optimize the number of beds in the hospital. A number of modifications and extensions are possible.

First, although the MGE distribution seems to fit the data reasonably well, it would also be possible to consider alternatives which might be more flexible, *e.g.*, mixtures of gamma or Erlang distributions (see *e.g.*, Wiper *et al.*, 2000), or a general phase-type model (see *e.g.*, Bladt *et al.*, 2001). Another possibility is to consider the effects of covariates such as the age of patients and year of admission on the number of days spent in hospital (see also Faddy and McClean, 1999).

Second, the assumption that all patients who cannot find a bed are lost may be overly restrictive. It is possible to consider alternative systems to model the throughput of patients such as a simple M/G/c queueing model. In this case, we could assume a waiting list of patients and compute the stationary distribution of the system. For queueing systems with general service distributions, this is complex but in the case where the service time distribution is phase-type, this can be carried out using matrix–geometric methods; see *e.g.*, Ausín *et al.* (2002). In this case, the cost function should be modified to take account of waiting times.

Finally, note that this procedure could be applied to other types of data, for example, stays in hotels and teletraffic data problems.

ACKNOWLEDGEMENTS

Some of the work for this paper was carried out while Conchi Ausín and Mike Wiper were visiting CNR-IMATI in November to December 2001. Conchi Ausín, Rosa Lillo and Mike Wiper also acknowledge support from the Spanish Ministry of Science and Technology through grant BEC2000-0167.

REFERENCES

Ausín, M. C., Wiper, M. P. and Lillo, R. E. (2002). Bayesian estimation for the M/G/1 queue using a phase type approximation. *J. Statist. Plann . Inference* (to appear).

Asmussen, S., Nerman, O. and Olsson, M. (1996). Fitting phase type distributions via EM algorithm. *Scand. J. Statist.* **23**, 419–441.

Bladt, M., Gonzalez, A. and Lauritzen, S. L. (2001). The estimation of phase-type related functionals through Markov chain Monte Carlo methods. *Scand. Actuarial J.* (to appear).

Cohen, J. W. (1976). *On Regenerative Processes in Queueing Theory*. Berlin: Springer.

Faddy, M. J. (1990). Compartmental models with phase-type residence time distributions. *Appl. Stochastic Models Data Anal.* **6**, 121–127.

Faddy, M. J. (1994). Examples of fitting structured phase-type distributions. *Appl. Stochastic Models Data Anal.* **10**, 247–255.

Faddy, M. J. and McClean, S. I. (1999). Analyzing data on lengths of stay of hospital patients using phase-type distributions. *Appl. Stochastic Models Business Industry* **15**, 311–317.

Gorenescu F., McClean S. I. and Millard, P. H. (1999). Using a M/PH/C queue to optimize hospital bed occupancy. *Proceedings of the Applied Stochastic Models and Data Analysis Conference, Lisbon*, 106–111.

Harrison, G. W. and Millard, P. H. (1991). Balancing acute and long term care: The mathematics of throughput in departments of geriatric medicine. *Methods Inform. Med.* **30**, 221–228.

Johnson, N. L. and Kotz, S. (1970). *Distributions in Statistics: Continuous Univariate Distributions*. New York: Wiley.

Neuts, M. F. (1981). *Matrix–Geometric Solutions in Stochastic Models*. Baltimore: John Hopkins University Press.

Richardson, S. and Green, P. (1997). On Bayesian analysis of mixtures with an unknown number of components. *J. R. Statist. Soc. B* **59**, 731–792.

Ripley, B. D. (1987). *Stochastic Simulation*. New York: Wiley.

Taylor, G. J., McClean, S. I. and Millard, P. H. (2000). Stochastic Models of Geriatric Patient Bed Occupancy Behavior. *J. R. Statist. Soc. A* **163**, 39–48.

Tijms, H. C. (1990). *Stochastic Modelling and Analysis: A Computational Approach*. Chichester: Wiley.

Wiper, M. P., Ríos Insua, D. and Ruggeri, F. (2001). Mixtures of gamma distributions with applications. *J. Comput. Graph. Statist.* **10**, 440–454.

BAYESIAN STATISTICS 7, pp. 453–463
J. M. Bernardo, M. J. Bayarri, J. O. Berger, A. P. Dawid,
D. Heckerman, A. F. M. Smith and M. West (Eds.)

The Variational Bayesian EM Algorithm for Incomplete Data: With Application to Scoring Graphical Model Structures

MATTHEW J. BEAL and ZOUBIN GHAHRAMANI
Gatsby Computational Neuroscience Unit, UCL, UK
m.beal@gatsby.ucl.ac.uk   zoubin@gatsby.ucl.ac.uk

SUMMARY

We present an efficient procedure for estimating the marginal likelihood of probabilistic models with latent variables or incomplete data. This method constructs and optimizes a lower bound on the marginal likelihood using variational calculus, resulting in an iterative algorithm which generalizes the EM algorithm by maintaining posterior distributions over both latent variables *and parameters*. We define the family of conjugate-exponential models—which includes finite mixtures of exponential family models, factor analysis, hidden Markov models, linear state-space models, and other models of interest—for which this bound on the marginal likelihood can be computed very simply through a modification of the standard EM algorithm. In particular, we focus on applying these bounds to the problem of scoring discrete directed graphical model structures (Bayesian networks). Extensive simulations comparing the variational bounds to the usual approach based on the Bayesian Information Criterion (BIC) and to a sampling-based gold standard method known as Annealed Importance Sampling (AIS) show that variational bounds substantially outperform BIC in finding the correct model structure at relatively little computational cost, while approaching the performance of the much more costly AIS procedure. Using AIS allows us to provide the first serious case study of the tightness of variational bounds. We also analyze the performance of AIS through a variety of criteria, and outline directions in which this work can be extended.

Keywords: MARGINAL LIKELIHOOD; LATENT VARIABLES; VARIATIONAL METHODS; GRAPHICAL MODELS; ANNEALED IMPORTANCE SAMPLING; STRUCTURE SCORING; BAYES FACTORS.

1. INTRODUCTION

Statistical modelling problems often involve a large number of random variables and it is often convenient to express the conditional independence relations between these variables graphically. Such graphical models are an intuitive tool for visualizing dependencies between the variables. Moreover, by exploiting the conditional independence relationships, they provide a backbone upon which it has been possible to derive efficient message-propagating algorithms for conditioning and marginalizing variables in the model given new evidence (Pearl, 1988; Lauritzen and Spiegelhalter, 1988; Heckerman, 1996; Cowell *et al.*, 1999). Many standard statistical models, especially Bayesian models with hierarchical priors, can be expressed naturally using probabilistic graphical models. This representation can be helpful developing both sampling methods (e.g. Gibbs sampling) and exact inference methods (e.g. junction-tree algorithm) for these models.

An important and difficult problem in Bayesian inference is computing the marginal likelihood of a model. This problem appears under several guises: as computing the Bayes factor (the ratio of two marginal likelihoods; Kass and Raftery, 1995), or computing the normalizing constant of a posterior distribution (known in statistical physics as the "partition function" and in machine learning as the "evidence"). The marginal likelihood is an important quantity because it allows us to select among several model structures. It is a difficult quantity to compute because it involves integrating over all parameters and latent variables, which is usually such a high dimensional and complicated integral that most simple approximations fail catastrophically.

In this paper, we describe the use of variational methods to approximate the marginal likelihood and posterior distributions of complex models. Variational methods, which have been used extensively in Bayesian machine learning for several years, provide a lower bound on the marginal likelihood which can be computed efficiently. In the next subsections we review Bayesian approaches to learning model structure. In Section 2 we turn to describing variational methods applied to Bayesian learning, deriving the variational Bayesian EM algorithm and comparing it to the EM algorithm for maximum *a posteriori* (MAP) estimation. In Section 3, we focus on models in the conjugate-exponential family and derive the basic results. Section 4 introduces the specific problem of learning the conditional independence structure of directed acyclic graphical models with latent variables. We compare variational methods to BIC and annealed importance sampling (AIS). We conclude with a discussion of other variational methods in section 5, and areas for future work in the general area of learning model structure.

1.1. *Bayesian Learning of Model Structure*

We use the term "model structure" to denote a variety of things. (1) In probabilistic graphical models, each graph implies a set of conditional independence statements between the variables in the graph. For example, in directed acyclic graphs (DAGs) if two variables X and Y are d-separated by a third Z (see Pearl, 1988, for a definition), then X and Y are conditionally independent given Z. The model structure learning problem is inferring the conditional independence relationships that hold given a set of (complete or incomplete) observations of the variables. (2) A special case of this problem is input variable selection in regression. Selecting which input (i.e. explanatory) variables are needed to predict the output (i.e. response) variable in the regression can be equivalently cast as deciding whether each input variable is a parent of the output variable in the corresponding directed graph. (3) Many statistical models of interest contain discrete nominal latent variables. A model structure learning problem of interest is choosing the cardinality of the discrete latent variable. Examples of this problem include deciding how many mixture components in a finite mixture model or how many hidden states in a hidden Markov model. (4) Other statistical models contain real-valued vectors of latent variables, making it necessary to do inference on the dimensionality of the latent vector. Examples of this include choosing the intrinsic dimensionality in a probabilistic principal components analysis (PCA) or factor analysis (FA) model or in a linear-Gaussian state-space model.

An obvious problem with maximum likelihood methods is that the likelihood function will generally be higher for more complex model structures, leading to overfitting. Bayesian approaches overcome overfitting by treating the parameters $\boldsymbol{\theta}_i$ from a model m_i (out of a set of models) as unknown random variables and averaging over the likelihood one would obtain from different settings of $\boldsymbol{\theta}_i$:

$$p(\boldsymbol{y} \,|\, m_i) = \int p(\boldsymbol{y} \,|\, \boldsymbol{\theta}_i, m_i) p(\boldsymbol{\theta}_i \,|\, m_i) \, d\boldsymbol{\theta}_i \,. \tag{1}$$

$p(\boldsymbol{y} \,|\, m_i)$ is the *marginal likelihood*[1] for a data set $\boldsymbol{y}$ assuming model m_i, where $p(\boldsymbol{\theta} \,|\, m_i)$ is the prior distribution over parameters. Integrating out parameters penalizes models with

[1] In the machine learning community this is sometimes referred to as the *evidence* for model m_i.

more degrees of freedom since these models can *a priori* model a larger range of data sets. This property of Bayesian integration has been called Ockham's razor, since it favors simpler explanations (models) for the data over complex ones (Jefferys and Berger, 1992; MacKay, 1995)[2] The overfitting problem is avoided simply because no parameter in the pure Bayesian approach is actually *fit* to the data.

Given a prior distribution over model structures $p(m_i)$ and a prior distribution over parameters for each model structure $p(\boldsymbol{\theta} \mid m_i)$, observing the data set $\boldsymbol{y}$ induces a posterior distribution over models given by Bayes rule:

$$p(m_i \mid \boldsymbol{y}) = \frac{p(m_i)p(\boldsymbol{y} \mid m_i)}{p(\boldsymbol{y})}, \qquad p(\boldsymbol{\theta} \mid \boldsymbol{y}, m_i) = \frac{p(\boldsymbol{y} \mid \boldsymbol{\theta}, m_i)p(\boldsymbol{\theta} \mid m_i)}{p(\boldsymbol{y} \mid m_i)}. \tag{2}$$

Our uncertainty over parameters is quantified by the posterior $p(\boldsymbol{\theta} \mid \boldsymbol{y}, m_i)$. The density at a new data point y is obtained by averaging over both the uncertainty in the model structure and in the parameters, $p(y \mid \boldsymbol{y}) = \sum_{m_i} \int p(y \mid \boldsymbol{\theta}, m_i, \boldsymbol{y}) p(\boldsymbol{\theta} \mid m_i, \boldsymbol{y}) p(m_i \mid \boldsymbol{y}) \, d\boldsymbol{\theta}$, which is the *predictive distribution*. Although in theory we should average over all possible structures, in practice, constraints on storage and computation or ease of interpretability may lead us to select a most probable model structure by maximizing $p(m_i \mid \boldsymbol{y})$. We focus on methods for computing probabilities over structures.

1.2. *Practical Bayesian Approaches*

For most models of interest it is computationally and analytically intractable to perform the integrals required for (1) and (2) exactly. Not only do these involve very high dimensional integrals but for models with parameter symmetries (such as mixture models) the integrand can have exponentially many well-separated modes. Focusing on (1), we briefly outline several methods that have been used to approximate this integral before turning to variational methods.

In the Bayesian statistics community *Markov chain Monte Carlo* (MCMC) is the method of choice for approximating difficult high-dimensional expectations and integrals. While there are many MCMC methods which result asymptotically in samples from the posterior distribution over parameters, for models with symmetries it is hard to get mixing between modes, which can be crucial in obtaining valid estimates of the marginal likelihood. Several methods that have been proposed for estimating marginal likelihoods are the candidate method (Chib, 1995), bridge sampling and path sampling (Gelman and Meng, 1998) and the closely related Annealed Importance Sampling (Neal, 2001). For large-scale problems, sampling methods are often not the method of choice because they can be slow and the posterior distribution over parameters is stored as a set of samples, which can be inefficient from a memory standpoint. In this paper, we chose Annealed Importance Sampling (AIS) as the gold standard with respect to which we compare faster non-sampling based approximations.

Another approach to Bayesian integration is the *Laplace approximation* which makes a local Gaussian approximation around a MAP parameter estimate (Kass and Raftery, 1995; MacKay, 1995). This is based on the fact that in the large data limit, given some regularity conditions, the posterior approaches a Gaussian around the MAP estimate. However, for models with symmetries these regularity conditions do not hold and the posterior can approach a mixture of exponentially many Gaussians. The normality assumption can be inaccurate for small data sets (for which, in principle, the advantages of Bayesian integration over MAP are largest).

[2] However, it is important to keep in mind that a realistic model of the data might need to be complex. It is therefore often advisable to use the most "complex" model for which it is possible to do inference, ideally setting up priors that allow the limit of infinitely many parameters to be taken, rather than to artificially limit the number of parameters in the model (Neal, 1996; Rasmussen and Ghahramani, 2001).

Local Gaussian approximations are also poorly suited to bounded, constrained, or positive parameters such as the mixing proportions of a mixture model, so it is often advisable to reparameterize to make Normality reasonable. Finally, the Gaussian approximation requires computing or approximating the determinant of the Hessian matrix at the MAP estimate, which can be computationally costly, that is, $O(d^3)$ for d parameters.

An even more drastic but much less costly approximation to the marginal likelihood is given by the Bayesian Information Criterion (BIC; Schwarz, 1978) which can be derived from the Laplace approximation by dropping all terms that do not scale with the number of data points, n. For a model with d well-determined parameters, using the MAP parameter estimate $\hat{\boldsymbol{\theta}}$ the BIC approximation to the log marginal likelihood is:

$$\log p(\boldsymbol{y} \mid m_i) \approx \log p(\boldsymbol{y} \mid \hat{\boldsymbol{\theta}}, m_i) - \tfrac{d}{2} \log n.$$

2. MARGINAL LIKELIHOOD VIA A VARIATIONAL PRINCIPLE

We review how variational methods can be used to approximate the integrals required for Bayesian learning. While examples of these methods have been used for several years in the machine learning community they are less known to the Bayesian statistics community. Moreover, here we present the general framework for a large family of models, we investigate a novel application of the framework to scoring the structures of discrete graphical models, and provide the first serious assessments of how tight the variational bounds are and how well they compare to sampling methods. We focus on models with latent or hidden variables, considering a general incomplete data setting.

2.1. *Lower Bounding the Marginal Likelihood*

Let $\boldsymbol{y}$ denote the observed variables, $\boldsymbol{x}$ denote the latent variables, and $\boldsymbol{\theta}$ denote the parameters. The log marginal likelihood of a data set $\boldsymbol{y}$ can be lower bounded by introducing any distribution over both latent variables and parameters which has support where $p(\boldsymbol{x}, \boldsymbol{\theta} \mid \boldsymbol{y}, m)$ does, and then appealing to Jensen's inequality (due to the concavity of the logarithm function):

$$\begin{aligned}\log p(\boldsymbol{y} \mid m) = \log \int p(\boldsymbol{y}, \boldsymbol{x}, \boldsymbol{\theta} \mid m)\, d\boldsymbol{x}\, d\boldsymbol{\theta} &= \log \int q(\boldsymbol{x}, \boldsymbol{\theta}) \frac{p(\boldsymbol{y}, \boldsymbol{x}, \boldsymbol{\theta} \mid m)}{q(\boldsymbol{x}, \boldsymbol{\theta})}\, d\boldsymbol{x}\, d\boldsymbol{\theta} \\ &\geq \int q(\boldsymbol{x}, \boldsymbol{\theta}) \log \frac{p(\boldsymbol{y}, \boldsymbol{x}, \boldsymbol{\theta} \mid m)}{q(\boldsymbol{x}, \boldsymbol{\theta})}\, d\boldsymbol{x}\, d\boldsymbol{\theta}.\end{aligned}$$

Maximizing this lower bound with respect to the free distribution $q(\boldsymbol{x}, \boldsymbol{\theta})$ results in $q(\boldsymbol{x}, \boldsymbol{\theta}) = p(\boldsymbol{x}, \boldsymbol{\theta} \mid \boldsymbol{y}, m)$ which when substituted above turns the inequality into an equality. This does not simplify the problem since evaluating the true posterior distribution $p(\boldsymbol{x}, \boldsymbol{\theta} \mid \boldsymbol{y}, m)$ requires knowing its normalizing constant, the marginal likelihood. Instead we use a simpler, factorized approximation to $q(\boldsymbol{x}, \boldsymbol{\theta}) \approx q_{\boldsymbol{x}}(\boldsymbol{x}) q_{\boldsymbol{\theta}}(\boldsymbol{\theta})$:

$$\log p(\boldsymbol{y} \mid m) \geq \int q_{\boldsymbol{x}}(\boldsymbol{x}) q_{\boldsymbol{\theta}}(\boldsymbol{\theta}) \log \frac{p(\boldsymbol{y}, \boldsymbol{x}, \boldsymbol{\theta} \mid m)}{q_{\boldsymbol{x}}(\boldsymbol{x}) q_{\boldsymbol{\theta}}(\boldsymbol{\theta})}\, d\boldsymbol{x}\, d\boldsymbol{\theta} = \mathcal{F}_m(q_{\boldsymbol{x}}(\boldsymbol{x}), q_{\boldsymbol{\theta}}(\boldsymbol{\theta}), \boldsymbol{y}). \tag{3}$$

The quantity $\mathcal{F}$ is a functional of the free distributions $q_{\boldsymbol{x}}(\boldsymbol{x})$ and $q_{\boldsymbol{\theta}}(\boldsymbol{\theta})$.

2.2. *Variational Bayesian EM*

The variational Bayesian algorithm iteratively maximizes $\mathcal{F}$ in Eq. (3) with respect to the free distributions, $q_{\boldsymbol{x}}(\boldsymbol{x})$ and $q_{\boldsymbol{\theta}}(\boldsymbol{\theta})$. We use elementary calculus of variations to take functional derivatives of the lower bound with respect $q_{\boldsymbol{x}}(\boldsymbol{x})$ and $q_{\boldsymbol{\theta}}(\theta)$, each while holding the other fixed. This results in the following update equations where the superscript (t) denotes the iteration number.

$$\begin{aligned} q_{\boldsymbol{x}}^{(t+1)}(\boldsymbol{x}) &\propto \exp\left[\int \log p(\boldsymbol{x},\boldsymbol{y}\,|\,\boldsymbol{\theta},m)\, q_{\boldsymbol{\theta}}^{(t)}(\boldsymbol{\theta})\, d\boldsymbol{\theta}\right], \\ q_{\boldsymbol{\theta}}^{(t+1)}(\boldsymbol{\theta}) &\propto p(\boldsymbol{\theta}\,|\,m)\, \exp\left[\int \log p(\boldsymbol{x},\boldsymbol{y}\,|\,\boldsymbol{\theta},m)\, q_{\boldsymbol{x}}^{(t+1)}(\boldsymbol{x})\, d\boldsymbol{x}\right]. \end{aligned} \tag{4}$$

Clearly, $q_{\boldsymbol{\theta}}(\boldsymbol{\theta})$ and $q_{\boldsymbol{x}_i}(\boldsymbol{x}_i)$ are coupled, so we iterate these equations until convergence. Readers familiar with the EM algorithm (Dempster *et al.,* 1977) may note the similarity between this iterative algorithm and EM. We call this procedure the *Variational Bayesian EM Algorithm* for reasons which will become clearer in the following sections (see also Attias, 2000, and Ghahramani and Beal, 2001).

Re-writing (3), it is easy to see that maximizing $\mathcal{F}$ is equivalent to minimizing the KL divergence between $q_{\boldsymbol{x}}(\boldsymbol{x})\; q_{\boldsymbol{\theta}}(\boldsymbol{\theta})$ and the joint posterior $p(\boldsymbol{x},\boldsymbol{\theta}\,|\,\boldsymbol{y},m)$:

$$\log p(\boldsymbol{y}\,|\,m) - \mathcal{F}_m(q_{\boldsymbol{x}}(\boldsymbol{x}), q_{\boldsymbol{\theta}}(\boldsymbol{\theta}), \boldsymbol{y}) = \int q_{\boldsymbol{x}}(\boldsymbol{x})\; q_{\boldsymbol{\theta}}(\boldsymbol{\theta}) \log \frac{q_{\boldsymbol{x}}(\boldsymbol{x})\; q_{\boldsymbol{\theta}}(\boldsymbol{\theta})}{p(\boldsymbol{\theta},\boldsymbol{x}\,|\,\boldsymbol{y},m)}\, d\boldsymbol{x}\, d\boldsymbol{\theta} = \mathrm{KL}(q\|p)\,.$$

Note that whilst this factorization of the posterior distribution over latent variables and parameters may seem drastic, one can think of it as replacing stochastic dependencies between $\boldsymbol{x}$ and $\boldsymbol{\theta}$ with deterministic dependencies between relevant moments of the two sets of variables.

Variational methods for lower bounding probabilities have been explored by several researchers in the past decade. Hinton and van Camp (1993) proposed an early approach for Bayesian learning of one-hidden layer neural networks using the restriction that $q_{\boldsymbol{\theta}}(\boldsymbol{\theta})$ is Gaussian. Neal and Hinton (1998) presented a generalization of EM, which made use of Jensen's inequality to allow partial E-steps. Jordan *et al.* (1998) review variational methods in a general context. Variational Bayesian methods have been applied to various models with latent variables (Waterhouse *et al.* , 1995; MacKay, 1997; Bishop, 1999; Attias, 2000; Ghahramani and Beal, 2000). The structural EM algorithm for scoring discrete graphical models (Friedman, 1998) is closely related to the variational method described here except that in (4) the distribution over $\boldsymbol{\theta}$ is replaced by the MAP estimate.

3. CONJUGATE-EXPONENTIAL MODELS

We consider a particular class of graphical models with latent variables, which we call *conjugate-exponential* (CE) models. We explicitly apply the variational method to these parametric families, resulting in a simple generalization of EM[3] . Conjugate-exponential models satisfy two conditions:

Condition 1. *The complete data likelihood is that of an exponential family:*

$$p(\boldsymbol{x},\boldsymbol{y}\,|\,\boldsymbol{\theta}) = f(\boldsymbol{x},\boldsymbol{y})\; g(\boldsymbol{\theta}) \exp\left\{\boldsymbol{\phi}(\boldsymbol{\theta})^t \boldsymbol{u}(\boldsymbol{x},\boldsymbol{y})\right\},$$

where $\boldsymbol{\phi}(\boldsymbol{\theta})$ is the vector of natural parameters, and $\boldsymbol{u}$ and f and g are the functions that define the exponential family.

Condition 2. *Given hyperparameters η and $\boldsymbol{\nu}$, the parameter prior is conjugate to the complete data likelihood:*

$$p(\boldsymbol{\theta}\,|\,\eta,\boldsymbol{\nu}) = h(\eta,\boldsymbol{\nu})\; g(\boldsymbol{\theta})^{\eta} \exp\left\{\boldsymbol{\phi}(\boldsymbol{\theta})^t \boldsymbol{\nu}\right\},$$

[3] This section follows the exposition in Ghahramani and Beal (2001), which also includes several general results for directed and undirected graphs.

Theorem. (Conjugate-Exponential Models). *Given an iid data set* $\boldsymbol{y} = \{\boldsymbol{y}_1, \cdots \boldsymbol{y}_n\}$, *if the model satisfies conditions (1) and (2), then at every iteration of the variational Bayesian EM algorithm and at the maxima of* $\mathcal{F}(q_{\boldsymbol{x}}(\boldsymbol{x}), q_{\boldsymbol{\theta}}(\boldsymbol{\theta}), \boldsymbol{y})$:

(a) $q_{\boldsymbol{\theta}}(\boldsymbol{\theta})$ *is conjugate with parameters* $\tilde{\eta} = \eta + n$, $\tilde{\boldsymbol{\nu}} = \boldsymbol{\nu} + \sum_{i=1}^{n} \overline{\boldsymbol{u}}(\boldsymbol{y}_i)$:

$$q_{\boldsymbol{\theta}}(\boldsymbol{\theta}) = h(\tilde{\eta}, \tilde{\boldsymbol{\nu}}) g(\boldsymbol{\theta})^{\tilde{\eta}} \exp\left\{\boldsymbol{\phi}(\boldsymbol{\theta})^t \tilde{\boldsymbol{\nu}}\right\} \tag{5}$$

where $\overline{\boldsymbol{u}}(\boldsymbol{y}_i) = \mathrm{E}_{q_{\boldsymbol{x}_i}}(\boldsymbol{u}(\boldsymbol{x}_i, \boldsymbol{y}_i))$, *using* $\mathrm{E}_{q_{\boldsymbol{x}_i}}$ *to denote expectation under the variational posterior over the latent variable(s) associated with the i-th datum.*

(b) $q_{\boldsymbol{x}}(\boldsymbol{x}) = \prod_{i=1}^{n} q_{\boldsymbol{x}_i}(\boldsymbol{x}_i)$ *with*

$$q_{\boldsymbol{x}_i}(\boldsymbol{x}_i) = p(\boldsymbol{x}_i \mid \boldsymbol{y}_i, \overline{\boldsymbol{\phi}}) \propto f(\boldsymbol{x}_i, \boldsymbol{y}_i) \exp\left\{\overline{\boldsymbol{\phi}}^t \boldsymbol{u}(\boldsymbol{x}_i, \boldsymbol{y}_i)\right\} \tag{6}$$

where $\overline{\boldsymbol{\phi}} = \mathrm{E}_{q_{\boldsymbol{\theta}}}(\boldsymbol{\phi}(\boldsymbol{\theta}))$, *the expectation of the natural parameter.*

Proof. Substitute the parametric forms from the definition of the CE family into the variational extrema given in (4), revealing forms for $q_{\boldsymbol{x}}(\boldsymbol{x})$ and $q_{\boldsymbol{\theta}}(\boldsymbol{\theta})$ according to (5) and (6) respectively. For CE models these forms are then closed under iterations of variational Bayesian EM, ensuring the Theorem continues to hold through to convergence to a local maximum of the lower bound on the marginal likelihood.

3.1. *Comparison of Variational Bayesian EM and EM for MAP estimation*

It is instructive to compare (4) with the EM algorithm for MAP estimation. We use an alternative derivation of EM due to Neal and Hinton (1998):

EM for MAP estimation	Variational Bayesian EM
Goal: maximize $p(\boldsymbol{\theta} \mid \boldsymbol{y}, m)$ w.r.t. $\boldsymbol{\theta}$	**Goal:** lower bound $p(\boldsymbol{y} \mid m)$
E Step: compute	**VB-E Step:** compute
$q_{\boldsymbol{x}}^{(t+1)}(\boldsymbol{x}) = p(\boldsymbol{x} \mid \boldsymbol{y}, \boldsymbol{\theta}^{(t)})$	$q_{\boldsymbol{x}}^{(t+1)}(\boldsymbol{x}) = p(\boldsymbol{x} \mid \boldsymbol{y}, \overline{\boldsymbol{\phi}}^{(t)})$
M Step:	**VB-M Step:**
$\boldsymbol{\theta}^{(t+1)} = \arg\max_{\boldsymbol{\theta}} \int q_{\boldsymbol{x}}^{(t+1)}(\boldsymbol{x}) \log p(\boldsymbol{x}, \boldsymbol{y}, \boldsymbol{\theta})\, d\boldsymbol{x}$	$q_{\boldsymbol{\theta}}^{(t+1)}(\boldsymbol{\theta}) \propto \exp\left[\int q_{\boldsymbol{x}}^{(t+1)}(\boldsymbol{x}) \log p(\boldsymbol{x}, \boldsymbol{y}, \boldsymbol{\theta})\, d\boldsymbol{x}\right]$

The Variational Bayesian EM algorithm reduces to the ordinary EM algorithm if we restrict the parameter density to a point estimate (i.e. Dirac delta function), $q_{\boldsymbol{\theta}}(\boldsymbol{\theta}) = \delta(\boldsymbol{\theta} - \boldsymbol{\theta}^*)$. The VB-E step has about the same time complexity as the E step, and is in all ways identical except that it is re-written in terms of the expected natural parameters, $\overline{\boldsymbol{\phi}}$. In particular, we can make use of all relevant propagation algorithms such as junction tree, Kalman smoothing, or belief propagation. The VB-M step computes a *distribution* over parameters (in the conjugate family) rather than a point estimate. Both algorithms monotonically increase an objective function.

4. VARIATIONAL SCORING OF MODEL STRUCTURES

We apply the variational method to the non-trivial problem of learning the conditional independence structure of directed acyclic graphical models with latent variables. We compare our VB bounds for the marginal likelihood to the standard method of sc oring graphs based on BIC (Schwarz, 1978), and also to a sampling-based gold standard AIS (Neal, 2001).

We consider a small graph consisting of six discrete variables: two binary-valued latent (hidden) variables and four observed variables each of cardinality five. We restrict ourselves to all bipartite structures in which latent variables are parents of the observed variables. We are interested in how successful different scoring methods are at learning from data the true graph structure (i.e. which latent variables are parents of which observed variables). Since all variables are discrete, by placing independent Dirichlet priors on all parameters the model becomes conjugate-exponential.

Data were generated from the graph shown in Figure 1 (top); we call this the "true" structure. We instantiated a setting of this graph's parameters under the prior (once only), and generated incrementally larger data sets from the model.[4] We chose this particular structure because it contains enough links to induce non-trivial correlations amongst the observed variables, whilst it has few enough nodes that we can exhaustively evaluate the marginal likelihood of all possible alternative structures.[5]

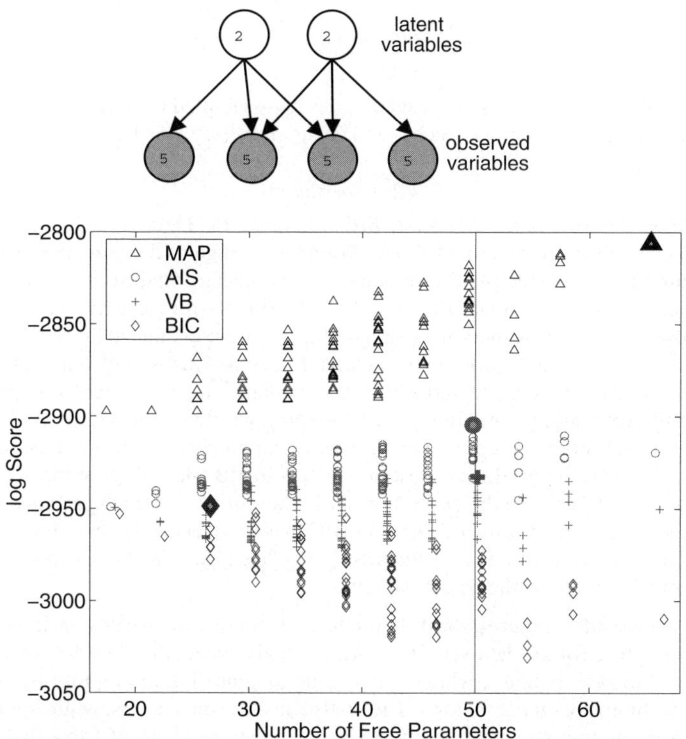

Figure 1. (Top:) *The true structure used to generate the data—one of 136 possible distinct structures accounting for permutations of latent variables.* (Bottom) *Marginal likelihood estimates at* $n = 480$ *for all structures using MAP, AIS, VB and BIC scores, plotted against the number of free parameters in the structure (staggered for each number of parameters for clarity). The true structure has 50 parameters (vertical line); the highest scoring structure for each method is shown with a bold symbol.*

[4] Experiments averaging results over multiple true structures and parameter settings would have been prohibitive as our sampling runs took over 300 CPU hours.

[5] Exhaustive enumeration is of academic interest only—in practice one would embed different structure scoring methods in a greedy model search outer loop (Friedman, 1998) to find probable structures.

4.1. *Annealed Importance Sampling*

We estimate $\mathcal{Z}_1 \equiv \int p(\boldsymbol{\theta} \,|\, m) p(\boldsymbol{y} \,|\, \boldsymbol{\theta}, m)\, d\boldsymbol{\theta}$ by computing

$$\frac{\mathcal{Z}_1}{\mathcal{Z}_0} = \frac{\mathcal{Z}_{\tau(1)}}{\mathcal{Z}_{\tau(0)}} \frac{\mathcal{Z}_{\tau(2)}}{\mathcal{Z}_{\tau(1)}} \cdots \frac{\mathcal{Z}_{\tau(T)}}{\mathcal{Z}_{\tau(T-1)}}$$

where $0 = \tau(0) \leq \tau(1) \cdots \leq \tau(T) = 1$, and $\mathcal{Z}_\tau \equiv \int p(\boldsymbol{\theta} \,|\, m) p(\boldsymbol{y} \,|\, \boldsymbol{\theta}, m)^\tau d\boldsymbol{\theta}$. The sequence $\tau(0) \ldots \tau(T)$ is an annealing schedule of inverse temperatures. Each ratio $\mathcal{Z}_{\tau(t)}/\mathcal{Z}_{\tau(t-1)}$ is estimated by running a Metropolis sampler for

$$\boldsymbol{\theta}^{(i)} \sim p(\boldsymbol{\theta} \,|\, m) p(\boldsymbol{y} \,|\, \boldsymbol{\theta}, m)^{\tau(t-1)} / \mathcal{Z}_{\tau(t-1)},$$

and computing the importance estimate:

$$\frac{1}{N} \sum_i p(\boldsymbol{y} \,|\, \boldsymbol{\theta}^{(i)}, m)^{\tau(t)-\tau(t-1)}.$$

This estimate of $\mathcal{Z}_{\tau(t)}/\mathcal{Z}_{\tau(t-1)}$ is unbiased if each step is sampled from equilibrium; we approximate this by taking one sample at each τ and changing τ very slowly.

4.2. *Experiments*

Scoring all possible structures with MAP, BIC, VB, and AIS. There are 136 distinct structures with the basic architecture described above. For a large range of data set sizes, we ran EM on each structure to compute the MAP estimate of the parameters, and from it computed the BIC score. We also ran the variational Bayesian EM algorithm with the same initial conditions to obtain the lower bound on the marginal likelihood. Finally, we estimated the marginal likelihood using AIS, annealing from the prior to the posterior in 16,384 steps, with a nonlinear annealing schedule in τ tuned to reduce the variance in the estimate, and a Metropolis proposal tuned to give reasonable acceptance rates. In Figure 1 (bottom), we have shown the MAP, AIS, VB and BIC scores for each structure, ordered by number of parameters, for 480 data points. This data set size was chosen as it was the smallest in which both VB and AIS gave the highest score to the true structure. We can see the general upward trend for MAP, which prefers more complex structures and the general downward trend for BIC, which (over)penalizes complexity. AIS lies above the VB lower bound for all structures as we would expect. At 480 data points VB appears to be close to AIS and finds the correct structure.

Rank of the True Model. Figure 2 (top) shows the rank out of 136 given by each scoring method to the true structure for 20 data set sizes. All methods eventually find the correct structure, although the AIS rank is noisy, which may be due to annealing too rapidly (we examine the effect of annealing time on AIS below). BIC finds it at 1120 data points, while VB does so after 480; moreover, the true structure is almost always ranked better by VB than BIC. Especially for small amounts of data, AIS outperforms both VB and BIC; nevertheless, it finds the true structure only after 480 data points.

Performance of AIS with Sampling Time. We ran AIS on the 480 point data set for varying duration annealing schedules, ranging from 64 samples to over 260,000. Figure 2 (bottom) shows annealing runs starting at 10 random initial parameters sampled from the prior, and also at one starting from the true parameters that generated the data (marked with a $*$); also shown are the BIC and VB scores. We see that all runs converge for sufficiently long annealing schedules. It takes roughly 1000 samples to approach the BIC score and about 5000 to pass the VB lower bound.

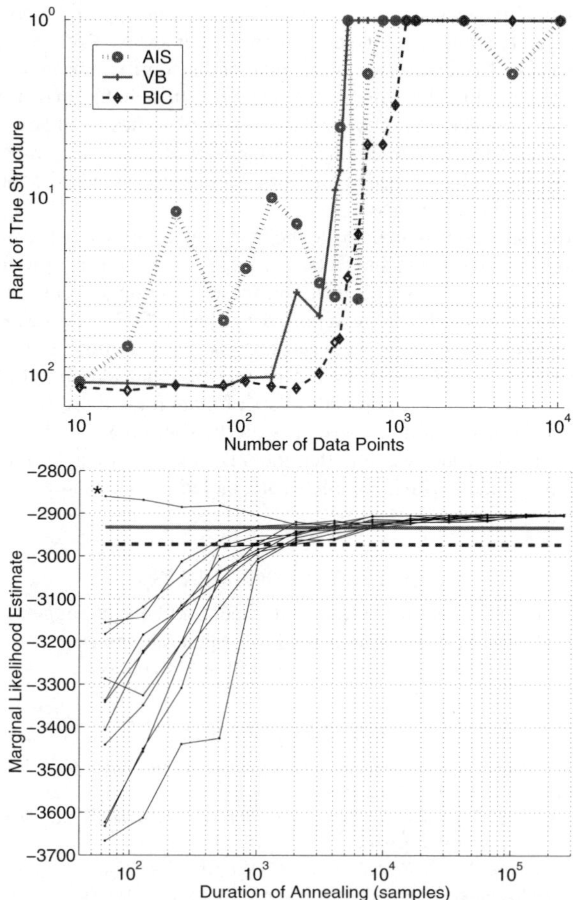

Figure 2. (Top:) *Rank given to the true structure by each scoring method for varying data set sizes (higher in plot is better).* **(Bottom:)** *AIS estimates of the marginal likelihood for $n = 480$, for different initial conditions of the sampler against different duration annealing schedules ($*$ indicates setting initial parameters to true values). Also shown are the BIC score (dashed) and the VB lower bound (solid).*

Computation Time. Scoring all 136 structures at 480 data points on a 1 GHz Pentium III processor took: 200 s with BIC, 575 s with VB, and 55,000 s (15 h) with AIS (using 16,384 samples as in the main experiments). Referring back to Figure 2 (bottom), we can infer that in this example, not only was VB about 100 times faster than AIS, but it also locked on to the true structure more reliably than AIS.

5. CONCLUSIONS AND EVOLVING DIRECTIONS FOR STRUCTURE SCORING

This paper has presented the variational lower bound method for handling intractable marginal likelihoods. A promising future direction is to explore the Bethe and Kikuchi family of variational methods (Yedidia *et al.,* 2001), which may be more accurate but do not provide the assurance of being a bound. These re-express the negative log marginal likelihood as a "free energy" from statistical physics, and then approximate the entropy of the posterior distribution over latent variables and parameters neglecting higher order terms. There are several procedures for minimizing the Bethe free energy as a functional of the approximate posterior distribution to obtain estimates of the marginal likelihood. Interestingly, the belief propagation algorithm, even when run on multiply connected graphs (i.e. "loopy" graphs), has fixed points at the stationary points of the Bethe free energy. While belief propagation on loopy graphs is not guaranteed to converge, it often works well in practice, and has become the standard approach to decoding state-of-the-art error-correcting codes.

We have also explored approximations to the marginal likelihood based on the Cheeseman–Stutz score as outlined in Chickering and Heckerman (1996), and corrections to the BIC criterion as outlined in Geiger *et al.* (1996). Results on these, and suggestions for how they can be combined with variational methods are reported in Beal (2003). A reasonable comparison should take into account the tradeoff between accuracy and computational complexity. We found in this paper that the VB lower bound on the marginal likelihood is a good criterion for model selection which can be computed orders of magnitude more quickly than sampling-based criteria. We have had similar experiences on the usefulness of VB for model selection in a variety of other models, and we note that the VB framework can be readily applied to all conjugate-exponential models with incomplete data, including, for example, graphical Gaussian models.

ACKNOWLEDGEMENTS

We thank Carl Edward Rasmussen for extensive input and helpful suggestions, especially regarding AIS. Part of this work was done while Z. Ghahramani was visiting the Center for Automated Learning and Discovery at Carnegie Mellon University.

REFERENCES

Attias, H. (2000). A variational Bayesian framework for graphical models. *Advances in Neural Information Processing Systems* **12**. Cambridge, MA: MIT Press.

Beal, M. J. (2003). *Variational Algorithms for Approximate Bayesian Inference*. Ph.D. Thesis, Gatsby Computational Neuroscience Unit, University College London, UK.

Bishop, C. M. (1999). Variational PCA. *Proc. Ninth Int. Conf. on Artificial Neural Networks*, ICANN.

Chib, S. (1995). Marginal likelihood from the Gibbs output. *J. Am. Statist. Ass.* **90**, 1313–1321.

Chickering, D. M. and Heckerman, D. (1996). Efficient approximations for the marginal likelihood of Bayesian networks with hidden variables. *Proc. 12th Conf. on Uncertainty in Artificial Intelligence*, UAI 96. San Fransisco: Morgan-Kaufmann Pub., 158–168.

Cowell, R. G., Dawid, A. P., Lauritzen, S. L. and Spiegelhalter, D. J. (1999). *Probabilistic Networks and Expert Systems*. New York: Springer.

Dempster, A. P., Laird, N. M. and Rubin, D. B. (1977). Maximum likelihood from incomplete data via the EM algorithm. *J. R. Statist. Soc. B* **39**, 1–38.

Friedman, N. (1998). The Bayesian structural EM algorithm. *Proc. 14th Conf. on Uncertainty in Artificial Intelligence*, UAI 96. San Fransisco: Morgan-Kaufmann Pub..

Geiger, D., Heckerman D. and Meek C. (1996). Asymptotic model selection for directed networks with hidden variables. *Proc. 12th Conf. on Uncertainty in Artificial Intelligence (UAI '96)*. San Fransisco: Morgan-Kaufmann Pub..

Gelman, A. and Meng, X. (1998). Simulating normalizing constants: From importance sampling to bridge sampling to path sampling. *Statist. Sci.* **13**, 163–185.

Ghahramani, Z. and Beal, M. J. (2000). Variational inference for Bayesian mixtures of factor analyzers. *Advances in Neural Information Processing Systems* **12**. Cambridge, MA: MIT Press.

Ghahramani, Z. and Beal, M. J. (2001). Propagation algorithms for variational Bayesian learning. *Advances in Neural Information Processing Systems* **13**. Cambridge, MA: MIT Press.

Heckerman, D. (1996). A tutorial on learning with Bayesian networks. *Tech. Rep.*, MSR-TR-95-06.Microsoft Research `ftp.research.microsoft.com/pub/tr/TR-95-06.PS`

Hinton, G. E. and van Camp, D. (1993). Keeping neural networks simple by minimizing the description length of the weights. *Sixth ACM Conference on Computational Learning Theory, Santa Cruz.*

Jefferys, W. H. and Berger, J. O. (1992). Ockham's razor and Bayesian analysis. *Ann. Statist.* **80**, 64–72.

Jordan, M. I., Ghahramani, Z., Jaakkola, T. S., and Saul, L. K. (1998). An introduction to variational methods in graphical models. *Learning in Graphical Models* (M. I. Jordan ed). Dordrecht: Kluwer.

Kass, R. E. and Raftery, A. E. (1995). Bayes factors. *J. Am. Statist. Ass.* **90**, 773–795.

Lauritzen, S. L. and Spiegelhalter, D. J. (1988). Local computations with probabilities on graphical structures and their application to expert systems. *J. R. Statist. Soc. B* **50**, 154–227.

MacKay, D. J. C. (1995). Probable networks and plausible predictions: A review of practical Bayesian methods for supervised neural networks. *Network: Comput. Neural Systems* **6**, 469–505.

MacKay, D. J. C. (1997). Ensemble learning for hidden Markov models. *Tech. Rep.*, University of Cambridge, UK.

Neal, R. M. (1996). *Bayesian Learning in Neural Networks*. New York: Springer.

Neal, R. M. (2001). Annealed importance sampling. *Statist. Comput.* **11**, 125–139.

Neal, R. M. and Hinton, G. E. (1998). A view of the EM algorithm that justifies incremental, sparse, and other variants. *Learning in Graphical Models* (M. I. Jordan ed). Dordrecht: Kluwer, 355–369.

Pearl, J. (1988). *Probabilistic Reasoning in Intelligent Systems: Networks of Plausible Inference*. San Fransisco: Morgan-Kaufmann Pub.

Rasmussen, C. E. and Ghahramani, Z. (2001). Occam's razor. *Advances in Neural Information Processing Systems* **13**, Cambridge, MA: MIT Press.

Schwarz, G. (1978). Estimating the dimension of a model. *Ann. Statist.* **6**, 461–464.

Waterhouse, S., MacKay, D. J. C. and Robinson, T. (1995). Bayesian methods for mixtures of experts. *Advances in Neural Information Processing Systems* **7**. Cambridge, MA: MIT Press.

Yedidia, J., Freeman, W. T. and Weiss, Y. (2001). Generalized belief propagation. *Advances in Neural Information Processing Systems* **13**. Cambridge, MA: MIT Press.

BAYESIAN STATISTICS 7, pp. 465–476
J. M. Bernardo, M. J. Bayarri, J. O. Berger, A. P. Dawid,
D. Heckerman, A. F. M. Smith and M. West (Eds.)

Intrinsic Estimation

JOSÉ M. BERNARDO and MIGUEL A. JUÁREZ
Universitat de València, Spain
jose.m.bernardo@uv.es miguel.juarez@uv.es

SUMMARY

In this paper, the problem of parametric point estimation is addressed from an objective Bayesian viewpoint. Arguing that pure statistical estimation may be appropriately described as a precise decision problem, where the loss function is a measure of the divergence between the assumed model and the estimated model, the information-based *intrinsic discrepancy* is proposed as an appropriate loss function. The *intrinsic estimator* is then defined as that minimizing the expected loss with respect to the appropriate *reference* posterior distribution. The resulting estimators are shown to have attractive *invariance* properties. As demonstrated with illustrative examples, the proposed theory either leads to new, arguably better estimators, or provides a new perspective on well-established solutions.

Keywords: INTRINSIC DISCREPANCY; INTRINSIC LOSS; LOGARITHMIC KL-DIVERGENCE; POINT ESTIMATION; REFERENCE ANALYSIS; REFERENCE PRIORS.

1. INTRODUCTION

It is well known that, from a Bayesian viewpoint, the final result of *any* problem of statistical inference is the posterior distribution of the quantity of interest. However, in more than two dimensions, the description (either graphical or analytical) of the posterior distribution is difficult and some "location" measure is often required for descriptive purposes. Moreover, there are many situations where a point estimate of the quantity of interest is specifically needed (and often even legally required) as part of the statistical report; simple examples include quoting the optimal dose of a drug per kilogram of body weight, or estimating the net weight of a canned food.

The typical Bayesian approach to point estimation formulates the problem as a decision problem, where the action space is the set of possible values of the quantity of interest. For each loss function and prior distribution on the model parameters, the *Bayes estimator* is obtained as that which minimizes the corresponding posterior expected loss. It is well known that the solution may dramatically depend both on the choice of the loss function and on the choice of the prior distribution.

In practice, in most situations where point estimation is of interest, an *objective* point estimate of the quantity of interest is actually required: objective in the very precise sense of exclusively depending on the assumed probability model (*i.e.*, on the conditional distribution of the data given the parameters) and the available data. Moreover, in purely inferential settings (where interest focuses on the actual mechanism which governs the data), this estimate is typically required to be *invariant* under one-to-one transformations of either the data or the parameter space. In this paper, an information-theory based loss function is combined with reference analysis to propose an objective Bayesian approach to point estimation which satisfies these desiderata.

In Section 2, the standard Bayesian formulation of point estimation as a decision problem is recalled and its conventional "automatic" answers are briefly discussed. Section 3 presents the proposed methodology. In Section 4, a number of illustrative examples are discussed. Section 5 contains some final remarks and suggests areas for additional research.

2. THE FORMAL DECISION PROBLEM

Let $\{p(\boldsymbol{x} \,|\, \boldsymbol{\theta}), \boldsymbol{x} \in \mathcal{X}, \boldsymbol{\theta} \in \Theta\}$ be a *probability model* assumed to describe the probabilistic behavior of the observable data $\boldsymbol{x}$, and suppose that a point estimator $\boldsymbol{\theta}^e = \boldsymbol{\theta}^e(\boldsymbol{x})$ of the parameter $\boldsymbol{\theta}$ is required. It is well known that this can be formulated as a decision problem under uncertainty, where the action space is the class $\mathcal{A} = \{\boldsymbol{\theta}^e \in \Theta\}$ of possible parameter values. In a purely inferential setting, the optimal estimate $\boldsymbol{\theta}^*$ is supposed to identify the best proxy, $p(\boldsymbol{x} \,|\, \boldsymbol{\theta}^*)$, to the unknown probability model, $p(\boldsymbol{x} \,|\, \boldsymbol{\theta}^a)$, where $\boldsymbol{\theta}^a$ stands for the *actual* (unknown) value of the parameter.

Let $l(\boldsymbol{\theta}^e, \boldsymbol{\theta}^a)$ be a *loss function* measuring the consequences of estimating $\boldsymbol{\theta}^a$ by $\boldsymbol{\theta}^e$. In a purely inferential context, $l(\boldsymbol{\theta}^e, \boldsymbol{\theta}^a)$ should measure the consequences of using the model $p(\boldsymbol{x} \,|\, \boldsymbol{\theta}^e)$ instead of the true, unknown model $p(\boldsymbol{x} \,|\, \boldsymbol{\theta}^a)$. For any loss function $l(\boldsymbol{\theta}^e, \boldsymbol{\theta}^a)$ and (possibly improper) prior $p(\boldsymbol{\theta})$, the *Bayes estimator* $\boldsymbol{\theta}^b = \boldsymbol{\theta}^b(\boldsymbol{x})$ of the parameter $\boldsymbol{\theta}$ is that minimizing the corresponding posterior loss, so that

$$\boldsymbol{\theta}^b(\boldsymbol{x}) = \arg \min_{\boldsymbol{\theta}^e \in \Theta} \int_{\Theta} l\,(\boldsymbol{\theta}^e, \boldsymbol{\theta})\, p\,(\boldsymbol{\theta} \,|\, \boldsymbol{x})\; d\boldsymbol{\theta}, \qquad (1)$$

where $p\,(\boldsymbol{\theta} \,|\, \boldsymbol{x}) \propto p(\boldsymbol{x} \,|\, \boldsymbol{\theta})\, p(\boldsymbol{\theta})$ is the posterior distribution of the parameter vector $\boldsymbol{\theta}$.

A number of conventional loss functions have been proposed in the literature, and their associated Bayes estimators are frequently quoted in Bayesian analysis:

Squared error loss. If the loss function is quadratic, of the form $(\boldsymbol{\theta}^e - \boldsymbol{\theta}^a)^t W (\boldsymbol{\theta}^e - \boldsymbol{\theta}^a)$, where W is a (known) symmetric positive definite matrix, then the Bayes estimator is the *posterior mean* $\mathrm{E}[\boldsymbol{\theta} \,|\, \boldsymbol{x}]$, provided it exists.

Zero-one loss. If the loss function takes the value zero if $\boldsymbol{\theta}^e$ belongs to a ball of radius ϵ centered at $\boldsymbol{\theta}^a$, and the value one otherwise, then the Bayes estimator tends towards the *posterior mode* $\mathrm{Mo}[\boldsymbol{\theta} \,|\, \boldsymbol{x}]$ as $\epsilon \to 0$, provided the mode exists and is unique.

Absolute error loss. If θ is *one-dimensional*, and the loss function is of the form $c\,|\theta^e - \theta^a|$, for some $c > 0$, then the Bayes estimator is the *posterior median* $\mathrm{Me}[\theta \,|\, \boldsymbol{x}]$.

Neither the posterior mean nor the posterior mode are invariant under one-to-one transformations of the parameter of interest. Thus, $\boldsymbol{\theta}^e$ can be declared to be the best estimator of $\boldsymbol{\theta}^a$, while $\phi(\boldsymbol{\theta}^e)$ is declared *not* to be the best estimator for $\phi^a = \phi(\boldsymbol{\theta}^a)$; this is unattractive in a scientific, purely inferential context, where interest is explicitly focused on identifying the actual probability model $p(\boldsymbol{x} \,|\, \boldsymbol{\theta}^a) = p(\boldsymbol{x} \,|\, \phi^a)$. The one-dimensional posterior median *is* invariant, but is not easily generalizable to more than one dimension. Naturally, if besides the likelihood and the prior, the functional form of the loss function is consistently transformed, invariance would be achieved, but this is rarely done in purely inferential problems. Hence, it is suggested that *invariant loss functions* should be used. More precisely it is argued that, in a purely inferential context, the loss function $l(\boldsymbol{\theta}^e, \boldsymbol{\theta})$ should *not* be chosen to measure the discrepancy between $\boldsymbol{\theta}^e$ and $\boldsymbol{\theta}^a$, but to directly measure the discrepancy between the models $p(\boldsymbol{x} \,|\, \boldsymbol{\theta}^e)$ and $p(\boldsymbol{x} \,|\, \boldsymbol{\theta}^a)$ which they label. This type of *intrinsic* loss is typically invariant under reparametrization, and therefore produces invariant estimators.

An appropriate choice of the loss function is, however, only part of the solution. To obtain an objective Bayes estimator, an *objective prior* must be used. It is argued that reference analysis may successfully be used to provide an adequate prior specification.

3. INTRINSIC ESTIMATION

3.1. *The Loss Function*

Conventional loss functions typically depend on the particular metric used to index the model, being defined as a measure of the distance between the parameter and its estimate. We claim that, in a purely inferential context, one should rather be interested in the discrepancy between the models labelled by the true value of the parameter and its estimate. A loss function of the form $l(\boldsymbol{\theta}_1, \boldsymbol{\theta}_2) = l\{p(\boldsymbol{x} \mid \boldsymbol{\theta}_1), p(\boldsymbol{x} \mid \boldsymbol{\theta}_2)\}$ is called an *intrinsic loss* (Robert, 1996).

Bernardo (1979a) and Bernardo and Smith (1994, Ch. 3) argue that scientific inference is well described as a formal decision problem, where the terminal loss function is a proper scoring rule. One of the most extensively studied of these is the *directed logarithmic divergence* (Gibbs, 1902; Shannon, 1948; Jeffreys, 1948; Good, 1950; Kullback and Leibler, 1951; Chernoff, 1952; Savage, 1954; Huzurbazar, 1955; Kullback, 1959; Jaynes, 1983). If $p(\boldsymbol{x} \mid \boldsymbol{\theta}_1)$ and $p(\boldsymbol{x} \mid \boldsymbol{\theta}_2)$ are probability densities with the same support $\mathcal{X}$, the directed logarithmic divergence of $p(\boldsymbol{x} \mid \boldsymbol{\theta}_2)$ from $p(\boldsymbol{x} \mid \boldsymbol{\theta}_1)$ is defined as

$$k_{\mathcal{X}}(\boldsymbol{\theta}_2 \mid \boldsymbol{\theta}_1) = \int_{\mathcal{X}} p(\boldsymbol{x} \mid \boldsymbol{\theta}_1) \, \log \frac{p(\boldsymbol{x} \mid \boldsymbol{\theta}_1)}{p(\boldsymbol{x} \mid \boldsymbol{\theta}_2)} \, d\boldsymbol{x}. \tag{2}$$

The directed logarithmic divergence (often referred to as Kullback–Leibler divergence) is *non-negative*, and it is *invariant* under one-to-one transformations of either $\boldsymbol{x}$ or $\boldsymbol{\theta}$. It is also *additive* in the sense that, if $\boldsymbol{x} \in \mathcal{X}$ and $\boldsymbol{y} \in \mathcal{Y}$ are conditionally independent given $\boldsymbol{\theta}$, then the divergence $k_{\mathcal{X},\mathcal{Y}}(\boldsymbol{\theta}_2 \mid \boldsymbol{\theta}_1)$ of $p(\boldsymbol{x}, \boldsymbol{y} \mid \boldsymbol{\theta}_2)$ from $p(\boldsymbol{x}, \boldsymbol{y} \mid \boldsymbol{\theta}_1)$ is simply $k_{\mathcal{X}}(\boldsymbol{\theta}_2 \mid \boldsymbol{\theta}_1) + k_{\mathcal{Y}}(\boldsymbol{\theta}_2 \mid \boldsymbol{\theta}_1)$; in particular, if data $\boldsymbol{x}$ are assumed to be a random sample $\boldsymbol{x} = \{x_1, \ldots, x_n\}$ from $p(x \mid \boldsymbol{\theta})$, then the divergence of $p(\boldsymbol{x} \mid \boldsymbol{\theta}_2)$ from $p(\boldsymbol{x} \mid \boldsymbol{\theta}_1)$ is simply n times the divergence of $p(x \mid \boldsymbol{\theta}_2)$ from $p(x \mid \boldsymbol{\theta}_1)$. Under appropriate regularity conditions, there are many connections between the logarithmic divergence and Fisher's information (see, *e.g.*, Stone, 1959; Bernardo and Smith, 1994, Ch. 5; Schervish, 1995, p. 118). Furthermore, $k_{\mathcal{X}}(\boldsymbol{\theta}_2 \mid \boldsymbol{\theta}_1)$ has an attractive interpretation in information-theoretical terms: it is the expected amount of information (in natural units, *nits*) necessary to recover $p(\boldsymbol{x} \mid \boldsymbol{\theta}_1)$ from $p(\boldsymbol{x} \mid \boldsymbol{\theta}_2)$.

However, the Kullback–Leibler divergence is not symmetric and it diverges if the support of $p(\boldsymbol{x} \mid \boldsymbol{\theta}_2)$ is a strict subset of the support of $p(\boldsymbol{x} \mid \boldsymbol{\theta}_1)$. To simultaneously address these two unwelcome features we propose to use the symmetric *intrinsic discrepancy* $\delta_{\mathcal{X}}(\boldsymbol{\theta}_1, \boldsymbol{\theta}_2)$, introduced in Bernardo and Rueda (2002), and defined as $\delta_{\mathcal{X}}(\boldsymbol{\theta}_1, \boldsymbol{\theta}_2) = \min\{k_{\mathcal{X}}(\boldsymbol{\theta}_1 \mid \boldsymbol{\theta}_2), k_{\mathcal{X}}(\boldsymbol{\theta}_2 \mid \boldsymbol{\theta}_1)\}$. To simplify the notation, the subindex $\mathcal{X}$ will be dropped from both $\delta_{\mathcal{X}}(\boldsymbol{\theta}_2 \mid \boldsymbol{\theta}_1)$ and $k_{\mathcal{X}}(\boldsymbol{\theta}_2 \mid \boldsymbol{\theta}_1)$ whenever there is no danger of confusion.

Definition 1. (*Intrinsic Discrepancy Loss*). Let $\{p(\boldsymbol{x} \mid \boldsymbol{\theta}), \boldsymbol{x} \in \mathcal{X}(\boldsymbol{\theta}), \boldsymbol{\theta} \in \Theta\}$ be a family of probability models for $\boldsymbol{x} \in \mathcal{X}(\boldsymbol{\theta})$, where the sample space may depend on the parameter value. The intrinsic discrepancy, $\delta_{\mathcal{X}}(\boldsymbol{\theta}_1, \boldsymbol{\theta}_2)$, between $p(\boldsymbol{x} \mid \boldsymbol{\theta}_1)$ and $p(\boldsymbol{x} \mid \boldsymbol{\theta}_2)$ is defined as

$$\min \left\{ \int_{\mathcal{X}(\boldsymbol{\theta}_1)} p(\boldsymbol{x} \mid \boldsymbol{\theta}_1) \log \left[\frac{p(\boldsymbol{x} \mid \boldsymbol{\theta}_1)}{p(\boldsymbol{x} \mid \boldsymbol{\theta}_2)} \right] d\boldsymbol{x}, \int_{\mathcal{X}(\boldsymbol{\theta}_2)} p(\boldsymbol{x} \mid \boldsymbol{\theta}_2) \log \left[\frac{p(\boldsymbol{x} \mid \boldsymbol{\theta}_2)}{p(\boldsymbol{x} \mid \boldsymbol{\theta}_1)} \right] d\boldsymbol{x} \right\}$$

provided one of the two integrals is finite.

The intrinsic discrepancy inherits a number of attractive properties from the directed logarithmic divergence. Indeed, it is non-negative and vanishes if, and only if, $p(\boldsymbol{x} \mid \boldsymbol{\theta}_1) = p(\boldsymbol{x} \mid \boldsymbol{\theta}_2)$

almost everywhere; it is invariant under one-to-one transformations of either $\boldsymbol{x}$ or $\boldsymbol{\theta}$; if the available data $\boldsymbol{x}$ consist of a random sample $\boldsymbol{x} = \{x_1, \ldots, x_n\}$ from $p(x \mid \boldsymbol{\theta})$, then the intrinsic divergence between $p(\boldsymbol{x} \mid \boldsymbol{\theta}_1)$ and $p(\boldsymbol{x} \mid \boldsymbol{\theta}_2)$ is simply n times the intrinsic divergence between $p(x \mid \boldsymbol{\theta}_1)$ and $p(x \mid \boldsymbol{\theta}_2)$. However, in contrast with the directed logarithmic divergence, the intrinsic discrepancy is *symmetric* and, if $p(\boldsymbol{x} \mid \boldsymbol{\theta}_1)$ and $p(\boldsymbol{x} \mid \boldsymbol{\theta}_2)$ have nested supports, so that $p\,(\boldsymbol{x} \mid \boldsymbol{\theta}_1) > 0$ iff $\boldsymbol{x} \in \mathcal{X}(\boldsymbol{\theta}_1)$, $p\,(\boldsymbol{x} \mid \boldsymbol{\theta}_2) > 0$ iff $\boldsymbol{x} \in \mathcal{X}(\boldsymbol{\theta}_2)$, and either $\mathcal{X}(\boldsymbol{\theta}_1) \subset \mathcal{X}(\boldsymbol{\theta}_2)$ or $\mathcal{X}(\boldsymbol{\theta}_2) \subset \mathcal{X}(\boldsymbol{\theta}_1)$, then the intrinsic discrepancy is typically finite, and reduces to a directed logarithmic divergence. More specifically, $\delta\,(\boldsymbol{\theta}_1, \boldsymbol{\theta}_2) = k\,(\boldsymbol{\theta}_1 \mid \boldsymbol{\theta}_2)$ when $\mathcal{X}(\boldsymbol{\theta}_2) \subset \mathcal{X}(\boldsymbol{\theta}_1)$, and $\delta\,(\boldsymbol{\theta}_1, \boldsymbol{\theta}_2) = k\,(\boldsymbol{\theta}_2 \mid \boldsymbol{\theta}_1)$ when $\mathcal{X}(\boldsymbol{\theta}_1) \subset \mathcal{X}(\boldsymbol{\theta}_2)$.

3.2. *The Prior Function*

Under the Bayesian paradigm, the outcome of any inference problem (the posterior distribution of the quantity of interest) combines the information provided by the data with relevant available prior information. In many situations, however, either the available prior information on the quantity of interest is too vague to warrant the effort required to have it formalized in the form of a probability distribution, or it is too subjective to be useful in scientific communication or public decision-making. It is, therefore, important to be able to identify the mathematical form of a "relatively uninformative" prior function, *i.e.*, a function (not necessarily a probability distribution) that, when formally used as a prior distribution in Bayes theorem, would have a minimal effect, relative to the data, on the posterior inference. More formally, suppose that the probability mechanism which has generated the available data $\boldsymbol{x}$ is assumed to be $p(\boldsymbol{x} \mid \boldsymbol{\theta})$, for some $\boldsymbol{\theta} \in \Theta$, and that the quantity of interest is some real-valued function $\phi = \phi(\boldsymbol{\theta})$ of the model parameter $\boldsymbol{\theta}$. Without loss of generality, it may be assumed that the probability model is of the form $p(\boldsymbol{x} \mid \phi, \boldsymbol{\lambda})$, $\phi \in \Phi$, $\boldsymbol{\lambda} \in \Lambda$, where $\boldsymbol{\lambda}$ is some appropriately chosen nuisance parameter vector. What is then required is to identify that joint prior function $\pi_\phi(\phi, \boldsymbol{\lambda})$ which would have a *minimal effect* on the corresponding marginal posterior distribution of the quantity of interest ϕ,

$$\pi(\phi \mid \boldsymbol{x}) \propto \int_\Lambda p(\boldsymbol{x} \mid \phi, \boldsymbol{\lambda})\, \pi_\phi(\phi, \boldsymbol{\lambda})\, d\boldsymbol{\lambda},$$

a prior which, to use a conventional expression, "would let the data speak for themselves" about the likely values of ϕ. Note that, within a given probability model $p(\boldsymbol{x} \mid \boldsymbol{\theta})$, the prior which could be described as "relatively uninformative" about the value of $\phi = \phi(\boldsymbol{\theta})$ will typically depend on the particular quantity of interest, $\phi = \phi(\boldsymbol{\theta})$.

Much work has been done to formulate priors which make the idea described above mathematically precise. Using an information-theoretic approach, Bernardo (1979b) introduced an algorithm to derive *reference* distributions; this is possibly the most general approach available. The reference prior $\pi_\phi(\boldsymbol{\theta})$ identifies a *possible* prior for $\boldsymbol{\theta}$, namely that describing a situation were relevant knowledge about the quantity of interest $\phi = \phi(\boldsymbol{\theta})$ (beyond that universally accepted) may be held to be negligible compared to the information about that quantity which repeated experimentation from a particular data generating mechanism $p(\boldsymbol{x} \mid \boldsymbol{\theta})$ might possibly provide. More recent work containing many refinements to the original formulation include Berger and Bernardo (1989, 1992), Bernardo and Smith (1994, Ch. 5) and Bernardo (1997). Bernardo and Ramón (1998) offers a simple introduction to reference analysis.

Any statistical analysis obviously contains a fair number of subjective elements; these include (among others) the data selected, the model assumptions and the choice of the quantities of interest. Reference analysis may be argued to provide "objective" Bayesian inferences in precisely the same sense that conventional statistical methods claim to be "objective"; namely in the sense that the solutions provided depend exclusively on the model assumptions and on the observed data.

In any decision problem, the quantity of interest is that function of the parameters which enters the loss function. Formally, in a decision problem with uncertainty about $\boldsymbol{\theta}$, actions $\{a \in \mathcal{A}\}$, and loss function $l(a, \phi(\boldsymbol{\theta}))$, the quantity of interest is $\phi = \phi(\boldsymbol{\theta})$. We have argued that, in point estimation, an appropriate loss function is the intrinsic discrepancy $l(\boldsymbol{\theta}^e, \boldsymbol{\theta}) = \delta(\boldsymbol{\theta}^e, \boldsymbol{\theta})$. It follows that, to obtain an objective (reference) intrinsic estimator, one should minimize the expected intrinsic loss with respect to the reference posterior distribution $\pi_\delta(\boldsymbol{\theta} \mid \boldsymbol{x})$, derived from the reference prior $\pi_\delta(\boldsymbol{\theta})$ obtained when the quantity of interest is the intrinsic discrepancy $\delta = \delta(\boldsymbol{\theta}^e, \boldsymbol{\theta})$; thus, one should minimize

$$d(\boldsymbol{\theta}^e \mid \boldsymbol{x}) = \mathrm{E}[\delta \mid \boldsymbol{x}] = \int_\Theta \delta(\boldsymbol{\theta}^e, \boldsymbol{\theta})\, \pi_\delta(\boldsymbol{\theta} \mid \boldsymbol{x})\, d\boldsymbol{\theta}. \tag{3}$$

Definition 2. (*Intrinsic Estimator*). Let $\{p(\boldsymbol{x} \mid \boldsymbol{\theta}), \boldsymbol{x} \in \mathcal{X}(\boldsymbol{\theta}), \boldsymbol{\theta} \in \Theta\}$ be a family of probability models for some observable data $\boldsymbol{x}$, where the sample space may possibly depend on the parameter value. The intrinsic estimator,

$$\boldsymbol{\theta}^*(\boldsymbol{x}) = \arg \min_{\boldsymbol{\theta}^e \in \Theta} d(\boldsymbol{\theta}^e \mid \boldsymbol{x}) = \arg \min_{\boldsymbol{\theta}^e \in \Theta} \int_\Theta \delta\,(\boldsymbol{\theta}^e, \boldsymbol{\theta})\; \pi_\delta\,(\boldsymbol{\theta} \mid \boldsymbol{x})\; d\boldsymbol{\theta}$$

is that minimizing the *reference* posterior expectation of the intrinsic discrepancy.

Reference distributions are known to be invariant under piecewise invertible transformations of the parameter (Datta and Ghosh, 1996) in the sense that, for any such transformation $\boldsymbol{\omega}(\boldsymbol{\theta})$ of $\boldsymbol{\theta}$, the reference posterior of $\boldsymbol{\omega}$, $\pi(\boldsymbol{\omega} \mid \boldsymbol{x})$ is that obtained from $\pi(\boldsymbol{\theta} \mid \boldsymbol{x})$ by standard probability calculus. Since the intrinsic discrepancy is itself invariant, it follows that (for any dimensionality) the intrinsic estimator is *invariant* under piecewise invertible transformations; thus, for any such transformation $\boldsymbol{\omega}(\boldsymbol{\theta})$ of the parameter vector, one has $\boldsymbol{\omega}^*(\boldsymbol{x}) = \boldsymbol{\omega}(\boldsymbol{\theta}^*(\boldsymbol{x}))$.

3.1. *A Simple Example: Bernoulli Data*

Let data $\boldsymbol{x} = \{x_1, \ldots, x_n\}$ consist of n conditionally independent Bernoulli observations with parameter θ, so that $p(x \mid \theta) = \theta^x(1-\theta)^{1-x}$, $x \in \{0, 1\}$. It is easily verified that the directed logarithmic divergence of $p(x \mid \theta_2)$ from $p(x \mid \theta_1)$ is

$$k(\theta_2 \mid \theta_1) = \theta_1 \log[\theta_1/\theta_2] + (1-\theta_1)\log[(1-\theta_1)/(1-\theta_2)]$$

Moreover, it is easily shown that $k(\theta_2 \mid \theta_1) < k(\theta_1 \mid \theta_2)$ iff $\theta_1 < \theta_2 < 1 - \theta_1$; thus, the intrinsic discrepancy between $p(\boldsymbol{x} \mid \theta^e)$ and $p(\boldsymbol{x} \mid \theta)$, represented in the left pane of Figure 1, is

$$\delta(\theta^e, \theta) = n \begin{cases} k(\theta \mid \theta^e) & \theta \in (\theta^e, 1-\theta^e), \\ k(\theta^e \mid \theta) & \text{otherwise.} \end{cases}$$

Since $\delta(\theta^e, \theta)$ is a piecewise invertible function of θ, the δ-reference prior is just the θ-reference prior and, since Bernoulli is a regular model, this is Jeffreys prior, $\pi(\theta) = \mathrm{Be}(\theta \mid \frac{1}{2}, \frac{1}{2})$. The corresponding reference posterior is the Beta distribution $\pi(\theta \mid \boldsymbol{x}) = \mathrm{Be}(\theta \mid r + \frac{1}{2}, n - r + \frac{1}{2})$, with $r = \sum x_i$, and the reference expected posterior intrinsic discrepancy is the concave function

$$d\,(\theta^e, \boldsymbol{x}) = \int_0^1 \delta\,(\theta^e, \theta)\ \mathrm{Be}(\theta \mid r + \tfrac{1}{2}, n - r + \tfrac{1}{2})\, d\theta.$$

The intrinsic estimator is its unique minimum $\theta^*(\boldsymbol{x}) = \arg\min_{\theta^e \in (0,1)} d\,(\theta^e, \boldsymbol{x})$, which is easily computed by one-dimensional numerical integration. A very good approximation is given by

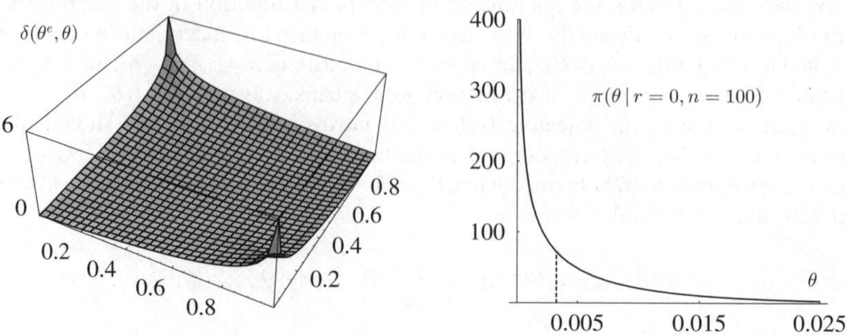

Figure 1. *Intrinsic discrepancy and reference posterior density for a Bernoulli parameter.*

the arithmetic average of the Bayes estimators which corresponds to using $k(\theta \mid \theta^e)$ and $k(\theta^e \mid \theta)$ as loss functions,

$$\theta^*(\boldsymbol{x}) \approx \frac{1}{2}\left(\frac{r+1/2}{n+1} + \frac{\exp[\psi(r+1/2)]}{\exp[\psi(r+1/2)] + \exp[\psi(n-r+1/2)]}\right), \tag{4}$$

where $\psi(.)$ is the digamma function.

As a numerical illustration, suppose that, to investigate the prevalence of a rare disease, a random sample of size $n = 100$ has been drawn and that no affected person has been found, so that $r = 0$. The reference posterior is $\text{Be}(\theta \mid 0.5, 100.5)$ (shown in the right pane of Figure 1), and the exact intrinsic estimator (shown with a dashed line) is $\theta^*(\boldsymbol{x}) = 0.00324$. The approximation (4) yields $\theta^*(\boldsymbol{x}) \approx 0.00318$. The posterior median is 0.00227. We note in passing that the MLE estimator, $\hat{\theta} = r/n = 0$, is obviously misleading in this case.

4. FURTHER EXAMPLES

To illustrate the above methodology and to compare the resulting estimators with those derived by conventional methods, a few more examples will be discussed.

4.1. *Uniform model,* $\text{Un}(x \mid 0, \theta)$

Consider first a simple non-regular example. Let $\boldsymbol{x} = \{x_1, \ldots, x_n\}$, be a random sample from the uniform distribution $\text{Un}(x \mid 0, \theta) = \theta^{-1}$, $0 < x < \theta$. It is immediately verified that $t = \max\{x_1, \ldots, x_n\}$ is a sufficient statistic. The directed logarithmic divergence of $\text{Un}(x \mid 0, \theta_2)$ from $\text{Un}(x \mid 0, \theta_1)$ is

$$k(\theta_1 \mid \theta_2) = n \begin{cases} \log(\theta_1/\theta_2) & \theta_1 \geq \theta_2, \\ \infty & \theta_1 < \theta_2; \end{cases}$$

thus the intrinsic discrepancy between $p(\boldsymbol{x} \mid \theta^e)$ and $p(\boldsymbol{x} \mid \theta)$ is

$$\delta(\theta^e, \theta) = n \begin{cases} \log(\theta^e/\theta) & \theta \leq \theta^e, \\ \log(\theta/\theta^e) & \theta \geq \theta^e, \end{cases}$$

shown in the left pane of Figure 2. Since the intrinsic discrepancy $\delta(\theta^e, \theta)$ is a piecewise invertible function of θ, the δ-reference prior is also the θ-reference prior. Since the sample space

$\mathcal{X}(\theta) = (0, \theta)$ depends on the parameter θ, this is not a regular problem and, hence, Jeffreys prior is not defined. The general expression for the reference prior in one-dimensional continuous problems with an asymptotically sufficient, consistent estimator $\tilde{\theta} = \tilde{\theta}(\boldsymbol{x})$ is (Bernardo and Smith, 1994, p. 312)

$$\pi(\theta) \propto p^*(\theta \,|\, \tilde{\theta})\Big|_{\tilde{\theta}=\theta} \tag{5}$$

where $p^*(\theta \,|\, \tilde{\theta})$ is any asymptotic approximation to the posterior distribution of θ (an expression that reduces to Jeffreys prior in regular problems).

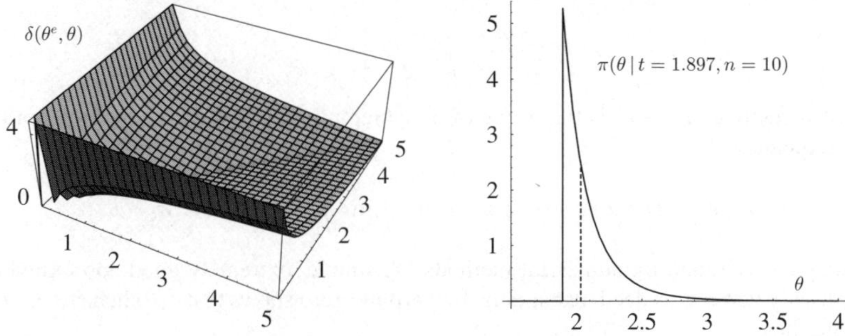

Figure 2. *Intrinsic discrepancy and reference density for the parameter θ of a uniform* $\text{Un}(\cdot \,|\, 0, \theta)$ *model.*

In this problem, the likelihood function is $L(\theta \,|\, \boldsymbol{x}) = \theta^{-n}$, if $\theta > t$, and zero otherwise, where $t = \max\{x_1, \ldots, x_n\}$. Hence, an asymptotic posterior is $p^*(\theta \,|\, t, n) \propto \theta^{-n}, \theta > t$. Computing the missing proportionality constant yields $p^*(\theta \,|\, t, n) = (n-1)t^{n-1}\theta^{-n}$. Since t is a sufficient, consistent estimator of θ, Eq. (5) may be used to obtain the θ-reference prior as $\pi(\theta) \propto t^{n-1}\theta^{-n}|_{t=\theta} = \theta^{-1}$. The corresponding posterior is the Pareto distribution $\pi(\theta \,|\, \boldsymbol{x}) = \text{Pa}(\theta \,|\, n, t) = n\, t^n\, \theta^{-(n+1)}$, $\theta > t$. The reference expected posterior intrinsic discrepancy is then easily found to be

$$d\,(\theta^e, \boldsymbol{x}) = \int_t^\infty \delta(\theta^e, \theta)\; \text{Pa}(\theta \,|\, n, t)\, d\theta = 2\Big(\frac{t}{\theta^e}\Big)^n - n\, \log\Big(\frac{t}{\theta^e}\Big) - 1,$$

which is minimized at $\theta^e = 2^{1/n}\, t$. Hence, the intrinsic estimator is $\theta^*(\boldsymbol{x}) = 2^{1/n}\, t$, which is actually the median of the reference posterior.

As an illustration, a random sample of size $n = 10$ was simulated from a Uniform distribution $\text{Un}(x \,|\, 0, \theta)$ with $\theta = 2$, yielding a maximum $t = 1.897$. The corresponding reference posterior, $\text{Pa}(\theta \,|\, 10, 1.897)$, is shown in the right pane of Figure 2. The intrinsic estimator, $\theta^*(\boldsymbol{x}) = 2.033$ is indicated with a dashed line.

4.2. *Normal Variance with Known Mean*

Combining the invariance properties of the intrinsic discrepancy with a judicious choice of the parametrization often simplifies the required computations. As an illustration, consider estimation of the normal variance.

Let $\boldsymbol{x} = \{x_1, \ldots, x_n\}$ be a random sample from a Normal $\text{N}(x \,|\, \mu, \sigma^2)$ distribution with known mean μ, and let s_x^2 be the corresponding sample variance, so that $ns_x^2 = \sum_j (x_j - \mu)^2$. The required intrinsic discrepancy is $n\, \delta\{N(x \,|\, \mu, \sigma^2), N(x \,|\, \mu, \sigma_e^2)\}$; letting $y = (x - \mu)/\sigma_e$ and using the fact that the intrinsic discrepancy is invariant under one-to-one transformations of the data, this may be written as $n\,\delta\{N(y\,|\,0,\sigma^2/\sigma_e^2), N(y\,|\,0,1)\}$. The last discrepancy has a simple expression in terms of $\theta = \theta(\sigma_e) = \log(\sigma/\sigma_e)$; specifically,

$$\delta\{N(x\,|\,\mu,\sigma^2), N(x\,|\,\mu,\sigma_e^2)\} = \delta(\theta) = 2|\theta| + 2\exp(-2|\theta|) - 1, \tag{6}$$

a symmetric function around zero, which is represented in the left pane of Figure 3. Moreover, since δ is a piecewise invertible function of θ (and hence of σ) and this is a regular problem, the reference prior for δ is the same as the reference prior for σ, the corresponding Jeffreys prior, $\pi(\sigma) = \sigma^{-1}$; in terms of θ, this transforms into the uniform prior $\pi_\delta(\theta) = 1$. The corresponding reference posterior is easily found to be

$$\pi(\theta\,|\,\boldsymbol{x},\sigma_e) = 2e^{-2\theta}\,\mathrm{Ga}\Big(\lambda\;\Big|\;\frac{n}{s},\frac{ns_y^2}{2}\Big)\Big|_{\lambda=e^{-2\theta}}, \qquad ns_y^2 = \frac{ns_x^2}{\sigma_e^2}.$$

The intrinsic estimator $\sigma^*(\boldsymbol{x})$ is that value of σ_e which minimizes the expected reference posterior discrepancy:

$$\sigma^*(\boldsymbol{x}) = \arg\min_{\sigma_e>0}\, d(\sigma_e\,|\,\boldsymbol{x}) = \arg\min_{\sigma_e>0} \int_{\Re} \delta(\theta)\,\pi(\theta\,|\,\boldsymbol{x},\sigma_e)\,d\theta, \tag{7}$$

which may easily found by numerical methods. A simple, extremely good approximation to $\sigma^*(\boldsymbol{x})$ is easily derived. Indeed, expanding $\delta(\theta)$ around zero shows that $\delta(\theta)$ behaves as θ^2 near the origin; thus,

$$d(\sigma_e\,|\,\boldsymbol{x}) = \int_{\Re} \delta(\theta)\,\pi(\theta\,|\,\boldsymbol{x},\sigma_e)\,d\theta \approx \int_0^\infty (\log\sigma - \log\sigma_e)^2\,\pi(\sigma\,|\,\boldsymbol{x})\,d\sigma,$$

and the last integral is minimized by σ_e^* such that

$$\log\sigma_e^* = \mathrm{E}[\log\sigma\,|\,\boldsymbol{x}] = \frac{1}{2}\Big[\log\frac{ns_x^2}{2} - \psi\Big(\frac{n}{2}\Big)\Big],$$

where $\psi(.)$ is the digamma function which, for moderate values of x, is well approximated by $\log x - (2x)^{-1}$. Since intrinsic estimation is an invariant procedure under reparametrization, this provides an approximation to the intrinsic estimator of the standard deviation given by

$$\sigma^*(\boldsymbol{x}) \approx s_x\,\frac{n+1/2}{n}; \quad s_x = \sqrt{\frac{\sum_{j=1}^n (x_j-\mu)^2}{n}}, \tag{8}$$

which is slightly larger than the MLE estimator $\hat\sigma = s_x$.

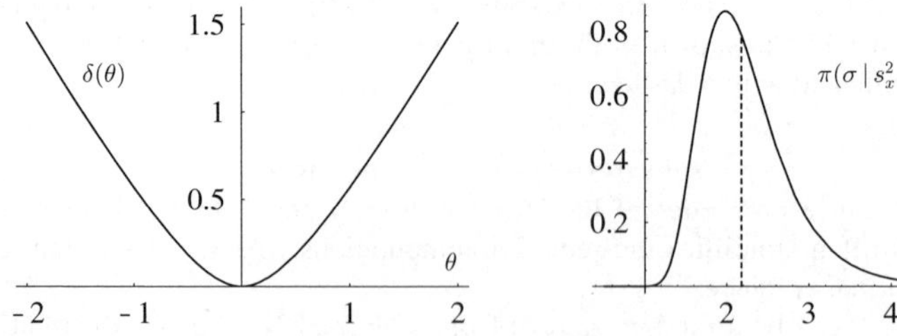

Figure 3. *Intrinsic discrepancy between $N(x\,|\,\mu,\sigma^2)$ and $N(x\,|\,\mu,\sigma_e^2)$ in terms of $\theta = \log(\sigma/\sigma_e)$, and marginal reference posterior of the standard deviation σ, given a simulated random sample of size $n = 10$ from a Normal* $\mathrm{N}(x\,|\,0,2^2)$ *distribution.*

As a numerical illustration, a sample of size $n = 10$ was simulated from a Normal distribution, $\mathrm{N}(x \,|\, 0,\, 2^2)$, yielding $s_x^2 = 4.326$. The corresponding reference posterior, the square root of an inverted gamma with parameters 5 and 21.63, is shown in the right pane of Figure 3. The exact intrinsic estimator, obtained from (7), is $\sigma^*(\boldsymbol{x}) = 2.178$, indicated with a dashed line. The approximation (8) yields $\sigma^*(\boldsymbol{x}) \approx 2.184$. The MLE is $\hat{\sigma} = s_x = 2.080$.

4.3. *Multivariate Mean Vector*

Let data consist of the mean vector $\overline{\boldsymbol{x}}$ from k-variate normal $\mathrm{N}_k(\overline{\boldsymbol{x}} \,|\, \boldsymbol{\mu}, n^{-1}\boldsymbol{I})$. The directed logarithmic divergence of $p(\overline{\boldsymbol{x}} \,|\, \boldsymbol{\mu}^e)$ from $p(\overline{\boldsymbol{x}} \,|\, \boldsymbol{\mu})$ is symmetric in this case, and hence equal to the intrinsic discrepancy

$$\delta(\boldsymbol{\mu}^e, \boldsymbol{\mu}) = \frac{n}{2}(\boldsymbol{\mu}^e - \boldsymbol{\mu})^t(\boldsymbol{\mu}^e - \boldsymbol{\mu}) = \frac{n}{2}\,\phi,$$

where $\phi = (\boldsymbol{\mu}^e - \boldsymbol{\mu})^t(\boldsymbol{\mu}^e - \boldsymbol{\mu})$. Thus, in this problem, the intrinsic discrepancy loss is a quadratic loss in terms of the parameter vector $\boldsymbol{\mu}$.

The intrinsic discrepancy is a linear function of $\phi = \|\boldsymbol{\mu}^e - \boldsymbol{\mu}\|$. Changing to centered generalized polar coordinates, it is found (Bernardo, 1979b; Ferrándiz, 1985; Berger *et al.*, 1998) that the reference posterior density for ϕ is

$$\pi(\phi \,|\, \overline{\boldsymbol{x}}) = \pi(\phi \,|\, t) \propto p(t \,|\, \phi)\, \pi(\phi) \propto \chi^2(nt \,|\, k, n\phi)\, \phi^{-1/2},$$

where $t = (\boldsymbol{\mu}^e - \overline{\boldsymbol{x}})^t(\boldsymbol{\mu}^e - \overline{\boldsymbol{x}})$. Note that this is *very different* from the posterior for ϕ which corresponds to the usual uniform prior for $\boldsymbol{\mu}$, known to lead to Stein's (1959) paradox. The expected reference posterior intrinsic loss may then be expressed in terms of the hypergeometric ${}_1F_1$ function as

$$d(\boldsymbol{\mu}^e, \overline{\boldsymbol{x}}) = \frac{n}{2}\,\mathrm{E}[\phi \,|\, \overline{\boldsymbol{x}}] = \frac{1}{2}\frac{{}_1F_1(3/2,\, k/2,\, nt/2)}{{}_1F_1(1/2,\, k/2,\, nt/2)} = d(nt, k),$$

which only depends on the data through

$$t = t(\boldsymbol{\mu}^e, \overline{\boldsymbol{x}}) = \|\boldsymbol{\mu}^e - \overline{\boldsymbol{x}}\|\,.$$

The expected intrinsic loss $d(nt, k)$ increases with nt for any dimension k and attains its minimum at $t = 0$, that is, when $\boldsymbol{\mu}^e = \overline{\boldsymbol{x}}$. The behavior of $d(nt, k)$ as a function of nt is shown in the left pane of Figure 4 for different values of k. It follows that, if the model is multivariate normal and there is *no further assumption on exchangeability* of the μ_j's, then the intrinsic estimator $\boldsymbol{\mu}^*$ is simply the sample mean $\overline{\boldsymbol{x}}$. The expected intrinsic loss of the Bayes estimator, namely $\boldsymbol{\mu}^* = \overline{\boldsymbol{x}}$, is

$$d(\boldsymbol{\mu}^*, \overline{\boldsymbol{x}}) = d(0, k) = \tfrac{1}{2}\,\cdot$$

Shrinking towards the overall mean $\boldsymbol{x}_0$, leading to ridge-type estimates of the general form

$$\tilde{\boldsymbol{\mu}}(\alpha) = \alpha\, \boldsymbol{x}_0 + (1 - \alpha)\overline{\boldsymbol{x}}, \qquad 0 < \alpha < 1,$$

will only increase the expected loss. Indeed,

$$d(\tilde{\boldsymbol{\mu}}(\alpha), \overline{\boldsymbol{x}}) = \frac{1}{2}\frac{{}_1F_1(3/2,\, k/2,\, nr_\alpha/2)}{{}_1F_1(1/2,\, k/2,\, nr_\alpha/2)}\,, \qquad r_\alpha = \frac{\alpha^2}{k}\sum_{i \neq j}(\mu_i - \mu_j)^2,$$

is an increasing function of nr_α and, hence, an increasing function of α. It follows that, with respect to the reference posterior, all ridge estimators have a *larger* expected loss than the sample mean. Similarly, the James–Stein estimator (James and Stein, 1961),

$$\tilde{\boldsymbol{\mu}}_{js} = (1 - (k-2)||\overline{\boldsymbol{x}}||^{-1})\overline{\boldsymbol{x}}, \quad k > 2,$$

which shrinks towards the origin rather than towards the overall mean, corresponds to $r_\alpha = 1$ and, hence, also has a larger expected loss than the sample mean. The expected intrinsic losses (quadratic in this case) of these estimators, for $k = 3$ and the particular random sample $\overline{\boldsymbol{x}} = \{0.72, -0.71, 1.67\}$, simulated from $\mathrm{N}_3(\overline{\boldsymbol{x}} \,|\, 0, \boldsymbol{I}_3)$, are compared in the right pane of Figure 4.

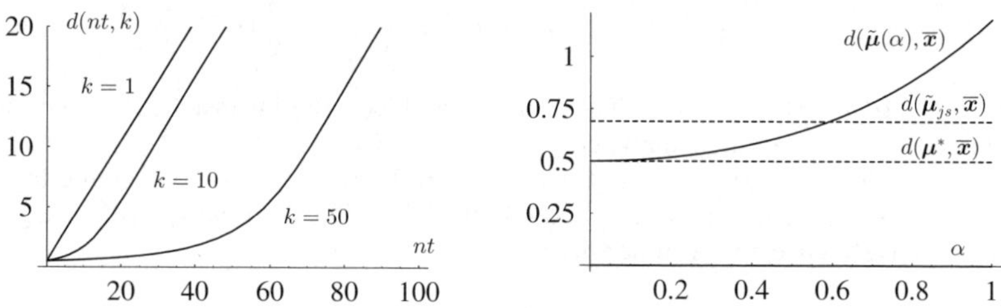

Figure 4. *Reference expected posterior losses in estimating a multivariate normal mean.*

The preceding analysis suggests that the frequent practice of shrinking towards either the origin or the overall mean may be inappropriate, unless there is information which justifies an exchangeability assumption for the μ_i's; in this case, a hierarchical model should be developed, and the intrinsic estimator will indeed be a ridge-type estimator. However, with a plain multivariate normal assumption, shrinking will only increase the (reference) expected loss. Thus, do not shrink without a good reason!

5. FINAL REMARKS

The intrinsic discrepancy, based on the theory of information, introduced in Bernardo and Rueda (2002) for densities which either have the same or nested supports, and further explored in this paper, has been shown to have many attractive properties. It is *symmetric*; it is *invariant*; it is typically finite for *non-regular problems*, and it is *calibrated* in natural information units. Indeed, the intrinsic divergence may be used to define a new type of *convergence* which is natural to consider in Bayesian statistics:

Definition 3. (*Intrinsic Convergence*). The sequence of probability densities $\{p_i\}_{i=1}^{\infty}$ converges *intrinsically* to the probability density p if, and only if, $\lim_{i\to\infty} \delta\{p_i, p\} = 0$.

Exploring the properties of this new definition of convergence will be the subject of future research. Further work is also needed to extend this definition to situations in which the densities are defined over arbitrary supports.

Intrinsic estimators were obtained by minimizing the reference posterior expectation of the intrinsic loss,

$$d(\boldsymbol{\theta}^e \,|\, \boldsymbol{x}) = \int_\Theta \delta(\boldsymbol{\theta}^e, \boldsymbol{\theta})\, \pi_\delta(\boldsymbol{\theta} \,|\, \boldsymbol{x})\, d\boldsymbol{\theta}.$$

Conditional on the assumed model, the positive statistic $d(\boldsymbol{\theta}_0 \,|\, \boldsymbol{x})$ is a natural measure of the *compatibility* of any $\boldsymbol{\theta}_0 \in \Theta$ with the observed data $\boldsymbol{x}$. Consequently, the intrinsic statistic

$d(\boldsymbol{\theta}_0 \mid \boldsymbol{x})$ is a natural test statistic which finds immediate applications in *precise hypothesis testing*, leading to BRC, the Bayesian reference criterion (Bernardo, 1999; Bernardo and Rueda, 2002).

We have focused on the use of the intrinsic discrepancy in reference problems, where no prior information is assumed on the parameter values. However, because of its nice properties, the intrinsic discrepancy is an eminently reasonable loss function to consider in problems where prior information (possibly in the form of a hierarchical model) is, in fact, available.

ACKNOWLEDGEMENTS

The authors gratefully acknowledge the comments received from Professor Dennis Lindley and from an anonymous referee. The research of José M. Bernardo was partially funded with grants GV01-7 of the Generalitat Valenciana, and BMF2001-2889 of the Ministerio de Ciencia y Tecnología, Spain. The research of Miguel Juárez was supported by CONACyT, Mexico.

REFERENCES

Berger, J. O. and Bernardo, J. M. (1989). Estimating a product of means: Bayesian analysis with reference priors. *J. Am. Statist. Ass.* **84**, 200–207.

Berger, J. O. and Bernardo, J. M. (1992). On the development of reference priors. *Bayesian Statistics 4* (J. M. Bernardo, J. O. Berger, A. P. Dawid and A. F. M. Smith, eds). Oxford: Oxford University Press, 35–60 (with discussion).

Berger, J. O., Philippe, A. and Robert, C. P. (1998). Estimation of quadratic functions: Uninformative priors for non-centrality parameters. *Statist. Sinica* **8**, 359–376.

Bernardo, J. M. (1979a). Expected information as expected utility. *Ann. Statist.* **7**, 686–690.

Bernardo, J. M. (1979b). Reference posterior distributions for Bayesian inference. *J. R. Statist. Soc. B* **41**, 113–147 (with discussion). Reprinted in *Bayesian Inference* (N. G. Polson and G. C. Tiao, eds). Brookfield, VT: Edward Elgar, 1995, 229–263.

Bernardo, J. M. (1997). Non-informative priors do not exist. *J. Statist. Plann . Inference* **65**, 159–189 (with discussion).

Bernardo, J. M. (1999). Nested hypothesis testing: The Bayesian reference criterion. *Bayesian Statistics 6* (J. M. Bernardo, J. O. Berger, A. P. Dawid and A. F. M. Smith, eds). Oxford: Oxford University Press, 101–130 (with discussion).

Bernardo, J. M. and Ramón, J. M. (1998). An introduction to Bayesian reference analysis: Inference on the ratio of multinomial parameters. *J. R. Statist. Soc. D* **47**, 101–135.

Bernardo, J. M. and Rueda, R. (2002). Bayesian hypothesis testing: A reference approach. *Internat. Statist. Rev.* **70**, 351–372.

Bernardo, J. M. and Smith, A. F. M. (1994). *Bayesian Theory*. Chichester: Wiley.

Chernoff, H. (1952). A measure of asymptotic efficiency for tests of a hypothesis based on the sum of observations. *Ann. Math. Statist.* **23**, 493–507.

Datta, G. S. and Ghosh, M. (1996). On the invariance of uninformative priors. *Ann. Statist.* **24**, 141–159.

Ferrándiz, J. R. (1985). Bayesian inference on Mahalanobis distance: an alternative approach to Bayesian model testing. *Bayesian Statistics 2* (J. M. Bernardo, M. H. DeGroot, D. V. Lindley and A. F. M. Smith, eds), Amsterdam: North-Holland, 645–654.

Gibbs, J. W. (1902). *Elementary Principles of Statistical Mechanics*. Reprinted, 1981. Woodbridge, CT: Ox Bow Press.

Good, I. J. (1950). *Probability and the Weighing of Evidence*. New York: Hafner Press.

Huzurbazar, V. S. (1955). Exact forms of some invariants for distributions admitting sufficient statistics. *Biometrika* **42**, 533–537.

James, W. and Stein, C. (1961). Estimation with quadratic loss. *Proc. Fourth Berkeley Symp.* **1** (J. Neyman and E. L. Scott, eds). Berkeley: Univ. California Press, 361–380.

Jaynes, E. T. (1983). *Papers on Probability, Statistics and Statistical Physics* (R. D. Rosenkrantz, ed). Dordrecht: Reidel.

Jeffreys, H. (1948). *Theory of Probability* (3rd edn, 1961). Oxford: Oxford University Press.

Kullback, S. (1959). *Information Theory and Statistics*, 2nd edn, 1968. New York: Dover.

Kullback, S. and Leibler, R. A. (1951). On information and sufficiency. *Ann. Math. Statist.* **22**, 79–86.

Robert, C. P. (1996). Intrinsic losses. *Theory and Decision* **40**, 191–214.

Savage, L. J. (1954). *The Foundations of Statistics.* New York: Dover.

Schervish, M. J. (1995). *Theory of Statistics*. Berlin: Springer.

Shannon, C. E. (1948). A mathematical theory of communication. *Bell Systems Tech. J.* **27**, 379–423 and 623–656. Reprinted in *The Mathematical Theory of Communication* (Shannon, C. E. and Weaver, W., 1949). Urbana: University of Illinois Press.

Stein, C. (1959). An example of wide discrepancy between fiducial and confidence intervals. *Ann. Math. Statist.* **30**, 877–880.

Stone, M. (1959). Application of a measure of information to the design and comparison of experiments. *Ann. Math. Statist.* **30**, 55–70.

BAYESIAN STATISTICS 7, pp. 477–484
J. M. Bernardo, M. J. Bayarri, J. O. Berger, A. P. Dawid,
D. Heckerman, A. F. M. Smith and M. West (Eds.)

Robust Analysis of Salamander Data, Generalized Linear Model with Random Effects

S. T. BORIS CHOY, JENNIFER S. K. CHAN and CARRIE H. K. YAM
The University of Hong Kong, Hong Kong
bchoy@hkuspace.hku.hk jchan@hkustask.hku.hk cyam@graduate.hku.hk

SUMMARY

The salamander mating data reported by McCullogh and Nelder (1989) have been analyzed by many researchers using both classical and Bayesian approaches. The study design in the data is crossed rather than nested. This imposes great computation difficulties when using the classical approach because the likelihood function cannot be factorized. Moreover, the likelihood function will become very complicated when non-normal random effects distributions are adopted. Hence, previous researches have always assumed normal distributions for the random effects. In this paper, we adopt a Bayesian approach and choose the Student-t distribution as an alternative to the normal distribution in order to allow for the possible thick tail behavior of the random effects and hence to provide a robust analysis. To identify possible outlying random effects, we represent the Student-t distribution as a scale mixtures of normal distribution within the Bayesian framework and the models are implemented using the Markov chain Monte Carlo method, in particular, the Gibbs sampler. Finally, we highlight the novel and interesting features of this data set.

Keywords: GENERALIZED LINEAR MIXED MODEL; ROBUST; SCALE MIXTURE OF NORMALS (SMN); MARKOV CHAIN MONTE CARLO; GIBBS SAMPLER; METROPOLIS-HASTINGS ALGORITHM.

1. INTRODUCTION

In the past, classical linear models were widely used for normal data in statistical inference. They were then extended to the generalized linear models (GLMs) by McCullagh and Nelder (1989) to handle data from the exponential family of distributions. In the GLM setting, a monotonic differentiable link function is specified to relate the mean of an exponential family to the linear structure of the covariates. This extension has immense impacts on both theoretical and practical aspects of classical statistics. Over the past twenty years, the GLMs have been improved and modified to the generalized linear mixed models (GLMMs) by the inclusion of random effects terms into the linear predictor of the GLMs so as to handle the auto correlation and overdispersion of the data.

One of the well-known data set which is modelled by the GLMMs is the salamander mating data reported by McCullagh and Nelder (1989). The responses are outcomes of mating (successful or unsuccessful) from a total of 20 female and 20 male salamanders. By the experimental design, the male and female random effects are crossed rather than nested. This hinders us from factorizing the likelihood function that involves integrals of 20 dimensions and is beyond the capacity of numerical approximation.

Many estimation methods have been proposed to overcome the difficulties in evaluating the likelihood function. For example, McCullagh and Nelder (1989) used the estimating equation approach. Breslow and Lin (1995) adopted the penalized quasi-likelihood (PQL) approach though the PQL estimates are biased. Lin and Breslow (1996), therefore, proposed the biased corrected PQL estimates to improve the asymptotic performance of the PQL estimates. Moreover, Shun (1997) developed a modified Laplace approximation approach in estimation. Kuk (1999), however, pointed out that the modified Laplace expansion is too technical and is problem-specific, and therefore, suggested a Laplace importance sampling approach. Apart from the classical approaches mentioned above, Bayesian approaches with simulation-based methodologies can provide an efficient means to the estimation problems of the GLMs and GLMMs. See Smith and Robert (1993) and Tierney (1994) for more details about the Bayesian computation methodologies. Karim and Zeger (1992) adopted the Gibbs sampling method within a Bayesian framework to analyze the salamander mating data. On the other hand, previous models mainly used a logit link function for binary data. McCulloch (1994) and Chan and Kuk (1997) adopted a probit link function. McCulloch (1994) proposed the Monte Carlo EM algorithm for parameter estimation and Chan and Kuk (1997) extended the method to correlated random effects.

However, most of the current works on the analysis of the salamander mating data using the GLMMs assume normal random effects. In practice, it may be more appropriate to use a wider class of random effects distributions to considerably widen the scope of applications. For example, Choy and Smith (1997) suggested the use of scale mixtures of normal (SMN) distributions, which include the Student-t, symmetric stable, exponential-power and Laplace distributions, in statistical analysis for robustness consideration because of their heavy-tailed behavior. See also West (1984, 1987) and Fernández and Steel (2000) for more details about the SMN distributions and their applications.

The main objective of this paper is to provide a robust analysis on the salamander data through a full Bayesian approach. A logit link function is considered and the random effects are modelled with the Student-t distribution, which is expressed into a SMN distribution. We shall demonstrate that the SMN representation for the Student-t distribution can simplify the Bayesian computation and enable us to detect potential outlying random effects in the salamander data. To allow for more flexibility, we further assume that the degrees of freedom of the Student-t distribution are random and are assigned suitable prior distributions within a Bayesian framework.

This paper is organized as follows. In Section 2, we briefly describe the salamander mating data. In Section 3, the GLMM with the Student-t distribution for the random effects is presented. We provide a full Bayesian analysis for the salamander mating data where the analysis is carried out by WinBUGS (Window version of Bayesian Using Gibbs Sampling). The results of the analysis are given in Section 4 and a conclusion is presented in Section 5.

2. SALAMANDER MATING DATA

The salamander mating experiments were conducted by S. Arnold and P. Verrell of the Department of Ecology and Evolution at the University of Chicago. The salamanders came from two populations: Rough Butt (RB) and Whiteside (WS). The objective of this experiment was to investigate whether there were barriers to interbreeding in the salamanders from these two geographically isolated populations.

There were totally three experiments. The first experiment was done in the summer of 1986. The second experiment was carried out in the fall of the same year with the same set of animals. The third experiment was done in the fall of the same year but with a different set of salamanders. Each experiment consists of 20 female and 20 male salamanders and each female salamander was paired up with six male salamanders, in which three male salamanders come from RB and the other three from WS. Outcomes are the results of matings (1 if the mating is successful, 0 if the mating is unsuccessful). There are totally 120 observations in either experiment. For example, the first female salamander was paired up with the 1st, fourth and fifth male salamanders from RB and the 11th, 14th and 15th male salamanders from WS. Table 1 gives the numbers of successful matings of female and males salamanders in each experiment.

Table 1. *Observed number of successful matings of female and male salamanders in each experiment.*

i	1	2	3	4	5	6	7	8	9	10	11	12	13	14	15	16	17	18	19	20
Expt. 1																				
Female i	5	6	5	5	6	5	1	3	4	2	4	1	1	4	1	1	5	3	5	5
Male i	5	3	5	3	4	2	1	2	2	1	4	5	6	5	4	3	1	5	4	4
Expt. 2																				
Female i	3	4	5	1	2	3	4	2	5	3	2	2	3	3	0	5	2	0	6	4
Male i	4	0	5	5	2	2	1	3	2	1	2	3	4	4	3	4	1	5	4	4
Expt. 3																				
Female i	5	1	5	3	3	5	1	5	3	5	2	3	3	3	4	1	2	2	2	2
Male i	2	3	1	3	1	2	2	5	1	5	1	1	6	4	4	6	3	4	1	5

Obviously, there exists a crossed structure in the design of experiment. The outcomes of an experiment are not independent because of the repeated measures (6 times) on the same animal. Furthermore, the outcomes are correlated since a female salamander would share the same male salamander with other female salamanders. Therefore, we incorporate random effects into the GLMMs to deal with the problems of heterogeneity and clustering. For robustness consideration, we model the random effects using the Student-t distribution whose probability density function is expressed into the following normal scale mixtures form

$$t_\alpha(x|\theta,\sigma^2) = \int_0^\infty N\left(x\,\middle|\,\theta,\frac{\sigma^2}{\lambda}\right) Ga\left(\lambda\,\middle|\,\frac{\alpha}{2},\frac{\alpha}{2}\right) d\lambda$$

where θ, σ^2 and α are the location parameter, scale parameter and degrees of freedom of the Student-t distribution, $N(\cdot|\cdot,\cdot)$ is the normal density function, and $Ga(\cdot|a,b)$ is the gamma density function with mean a/b. This representation enables us to introduce an auxiliary variable λ to get the following joint density

$$f(x,\lambda|\theta,\sigma^2) = N\left(x\,\middle|\,\theta,\frac{\sigma^2}{\lambda}\right) Ga\left(\lambda\,\middle|\,\frac{\alpha}{2},\frac{\alpha}{2}\right)$$

and this results in the following hierarchical structure:

$$X|\theta,\sigma^2,\lambda \sim N\left(\theta,\frac{\sigma^2}{\lambda}\right), \qquad \lambda \sim Ga\left(\frac{\alpha}{2},\frac{\alpha}{2}\right),$$

which is extremely useful in Gibbs sampling algorithm.

3. MODELLING THE DATA

The GLMM for analyzing the salamander mating data is presented as follows. Let $Y_{ij}, i = 1,\ldots,20$, $j = 1,\ldots,20$, be the binary response of mating (1 = success, 0 = failure) of the ith female salamander with the jth male salamander. The fixed effects denote the populations that the male and female salamanders belong to. WSF_i equals to 1 if the i-th female salamander came from Whiteside. Otherwise, WSF_i equals to 0. Similarly, WSM_j equals to 1 if the j-th male salamander came from Whiteside. Otherwise, WSM_j equals to 0. $WSF_i \times WSM_j$ is the interaction between the male and female population effects. The random effects a_i and b_j represent the effects of the i-th female salamander and j-th male salamanders involved in the mating. A logit link function is used in the GLMM framework and we have

$$\text{logit}(p_{ij}) = \beta_0 + \beta_1 WSF_i + \beta_2 WSM_j + \beta_3 WSF_i \times WSM_j + a_i + b_j$$

where $p_{ij} = Pr(Y_{ij} = 1|a_i, b_j)$ and $\boldsymbol{\beta} = (\beta_0, \beta_1, \beta_2, \beta_3)$ is the vector of fixed effects parameters. The random effects a_i are assumed to have a Student-t distribution with location 0, scale σ_f^2 and λ_f degrees of freedom while b_i are assumed to have a Student-t distribution with location 0, scale σ_m^2, and α_m degrees of freedom. With the SMN representation for the Student-t distribution, the random effects distributions for a_i and b_j are expressed as

$$a_i|\sigma_f^2, \lambda_{f_i} \sim \mathrm{N}\left(0, \frac{\sigma_f^2}{\lambda_{f_i}}\right) \qquad b_i|\sigma_m^2, \lambda_{m_i} \sim \mathrm{N}\left(0, \frac{\sigma_m^2}{\lambda_{m_i}}\right)$$

$$\lambda_{f_i}|\alpha_f \sim \mathrm{Ga}\left(\frac{\alpha_f}{2}, \frac{\alpha_f}{2}\right) \qquad \lambda_{m_i}|\alpha_m \sim \mathrm{Ga}\left(\frac{\alpha_m}{2}, \frac{\alpha_m}{2}\right)$$

To complete a Bayesian framework for the GLMM, we assign a multivariate normal prior to $\boldsymbol{\beta}$, conjugate inverse gamma priors to σ_f^2 and σ_m^2 and gamma priors to the degrees of freedom, α_f and α_m, i.e.

$$\boldsymbol{\beta} \sim \mathrm{N}(\boldsymbol{\mu}, \Sigma) \qquad \sigma_f^2 \sim \mathrm{IG}(c_1, d_1) \qquad \sigma_m^2 \sim \mathrm{IG}(c_2, d_2)$$

$$\alpha_f \sim \mathrm{Ga}(c_3, d_3) \qquad \alpha_m \sim \mathrm{Ga}(c_4, d_4)$$

where $\boldsymbol{\mu}, \Sigma, c_k$ and d_k, $k = 1, 2, 3, 4$ are specified.

Using the SMN representation for the Student-t distribution, the above GLMM can be easily implemented by the Markov chain Monte Carlo methods, such as Gibbs sampler. For those non-standard full conditional distributions, Metropolis–Hastings or rejection sampling algorithms can be used for random variate generation. However, the analysis of the salamander mating data in this paper relies on the software WinBUGS (Spiegelhalter *et. al.*, 1997).

4. RESULTS

The Bayesian GLMM described in the previous section is fitted to the salamander mating data of each of the three experiments. For illustrative purposes, we adopt vague priors for β and non-informative priors for $\sigma_f^2, \sigma_m^2, \alpha_f$ and α_m to reflect our ignorance about the parameters of the GLMM. For the WinBUGS, we shall then set large numbers to the diagonal entries of Σ and $c_k = 0.001$ and $d_k = 0.001$ for $k = 1, 2, 3, 4$. The Gibbs sampler was run for 35000 iterations with the first 5000 iterations discarded as the burn-in period. We then pick up simulated values every 30 iterations to reduce autocorrelation and to mimic a random sample of 1,000 simulated data from the joint posterior distribution. The dependency of the simulated values can be checked by plotting the auto correlation functions.

4.1. *Parameter Estimations*

The posterior estimates of the parameters of interest can be easily obtained from the 1000 simulated values of the Gibbs output. For each of the three salamander mating experiments, the boxplots of the degrees of freedom, α_f and α_m, for the Student-t random effects distributions are given in Figure 1 and Table 2 displays their posterior medians and inter-quartile ranges. In the first experiment, the posterior medians of α_f and α_m are around 13 and 10, respectively. In

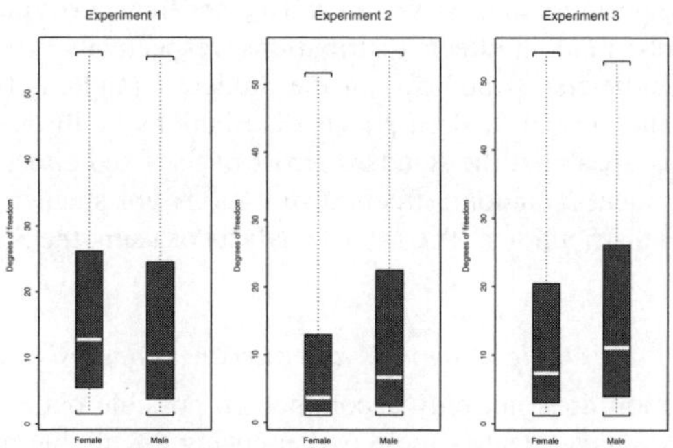

Figure 1. *Box plots of the posterior sample of the degrees of freedom, α_f and α_m, in each experiment.*

Table 2. *Posterior medians (inter-quartile ranges in parentheses) of the degrees of freedom α_f and α_m for the Student-t random effects distributions.*

	Experiment 1	Experiment 2	Experiment 3
α_f	12.8 (20.7)	3.7 (11.2)	7.4 (17.4)
α_m	9.9 (20.7)	6.7 (20.0)	11.1 (22.1)

Table 3. *Posterior means (with standard errors in parentheses) of β, σ_f and σ_m under the normal and Student-t random effects distributions.*

Parameters	β_0	β_1	β_2	β_3	σ_f	σ_m
Student-t						
Experiment 1	1.52	-3.37	-0.51	3.68	1.52	0.38
	(0.84)	(1.19)	(0.77)	(1.17)	(0.57)	(0.38)
Experiment 2	0.60	-2.77	-0.82	4.29	1.43	0.97
	(0.83)	(1.30)	(0.87)	(1.33)	(0.77)	(0.61)
Experiment 3	0.99	-3.59	-0.74	4.20	0.54	1.54
	(0.85)	(1.06)	(1.16)	(1.20)	(0.44)	(0.65)
Normal						
Experiment 1	1.48	-3.33	-0.45	3.62	1.61	0.46
	(0.80)	(1.17)	(0.73)	(1.10)	(0.58)	(0.43)
Experiment 2	0.65	-2.89	-0.85	4.28	1.90	1.24
	(0.97)	(1.31)	(0.94)	(1.36)	(0.77)	(0.65)
Experiment 3	1.14	-3.65	-0.90	4.29	0.60	1.78
	(0.81)	(1.12)	(1.08)	(1.31)	(0.51)	(0.62)

the second experiment, the values are around 4 and 7, respectively. We would expect that the random effects distributions for the salamander experiment in the fall would have heavier tails than those in the summer. However, in the third experiment held in fall using another set of salamanders, the estimates of α_f and α_m are around 7 and 11, respectively, which shows that the other set of female and male salamanders may have behaved differently than the salamanders in the second experiment. Nonetheless, our analysis here has demonstrated that Student-t distribution is appropriate than the normal distribution in modelling the random effects of the salamander mating experiments.

Table 3 exhibits the posterior means of β, σ_f and σ_m and their corresponding standard errors, for normal and Student-t random effects distributions, respectively. Obviously, the posterior means of the scale parameters α_f and α_m, of the Student-t random effects distributions are smaller than those of the normal random effects distributions in all three salamander mating experiments. Generally speaking, the standard errors of the fixed effects parameters, $\beta_k, k = 0, \ldots, 3$, under the Student-$t$ random effects distributions are smaller than those under the normal random effects distributions. The robust analysis of using the Student-t distribution is clearly illustrated.

4.2. *Outlier diagnosis and robustness analysis*

The Student-t distribution does not only accommodate possible outliers to provide a robust analysis but also enables us to identify these outliers using the mixing parameters of its SMN representation. Consider the SMN representation of the Student-t distribution

$$t_\alpha(x|\theta, \sigma^2) = \int_0^\infty N\left(x \,\middle|\, \theta, \frac{\sigma^2}{\lambda}\right) Ga\left(\lambda \,\middle|\, \frac{\alpha}{2}, \frac{\alpha}{2}\right) d\lambda.$$

If an observation is a possible outlier, then a way to accommodate this observation is to inflate the variance of the normal distribution in the above mixture form by diminishing the magnitude of the mixing parameter λ. Therefore, to identify possible outlying random effects in the salamander mating experiments, we can compare the posterior means of $\lambda_{f_i}, i = 1, \ldots, 20$, and $\lambda_{m_j}, j = 1, \ldots, 20$. The smaller the values of λ_{f_i} and λ_{m_j}, the more outlying the random effects are.

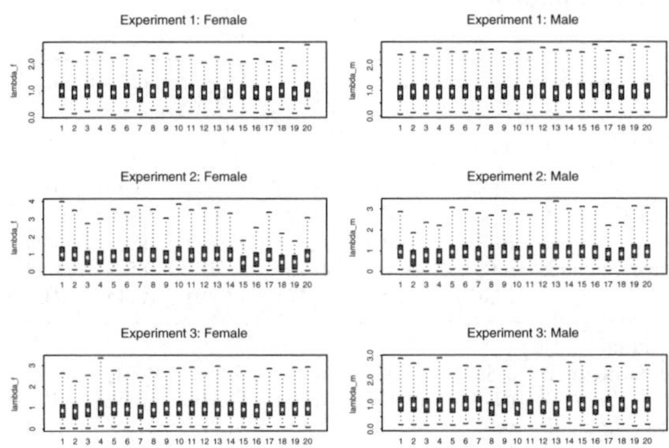

Figure 2. *Box plots of the posterior sample of the degrees of freedom, α_f and α_m, in each experiment.*

Figure 2 plots the boxplots of the mixing parameters, λ_{f_i} and λ_{m_j}, of female and male random effects for each experiment and small values are associated with outlying random effects. In the first experiment, the seventh female salamander may produce an outlying random effects, though it is not obvious.

When we inspect the data carefully, we reveal that the seventh female salamander has only one successful mating out of six occasions. Although the 12th, 13th, 15th and 16th female salamanders, which come from WS, have also one successful mating, the seventh female salamander is identified to be a potential outlier because it comes from RB which has generally a high successful rate of matings. For the male salamanders, there are no obvious outlying random effects.

In the second experiment, there are three possible outlying random effects which are associated with the 15th, 18th and 19th female salamanders. The corresponding λ_f values are comparatively small. In fact, both the 15th and 18th female salamanders failed all matings while the 19th female salamander succeeded in all matings. For the male salamanders, the second salamander is identified to provide an outlying random effect because it did not mate with any of the female salamanders in all six occasions. Two more possible outlying random effects come from the third and fourth male salamanders, both of which have five successful matings out of six occasions. Moreover, it is interesting to note that even though the same set of salamanders were used in the first and the second experiments, the female and male salamanders with outlying effects are not the same in these two experiments. In fact, the random effects in the second experiment are more diversified. Since the two experiments were conducted in different seasons, this indicates the presence of significant seasonal effects. See Chan and Kuk (1997) for the results which show significant seasonal effects in the data. In the third experiment, there are no obvious outlying random effects.

4.3. *Goodness-of-Fit*

The goodness-of-fit of the proposed models can be assessed by the estimated proportions of successful matings for female and male salamanders from different geographically isolated populations, WS and RB. Let π_{ij} be the proportions of successful matings between a female from population i and a male from population j where $i, j = W$ or R corresponding to WS or RB, respectively. For example, π_{RW} is the proportion of successful matings between a female living in RB and a male living in WS.

Table 4. *Observed and expected proportions of successful matings by combinations of geographical areas of the female and male salamanders for the three experiments by two random effects models with normal and Student-t random effects. Percentages of error averaged over combinations of geographical areas of the female and male salamanders are for the the three experiments by two random effects models are also reported.*

	Observed proportion			Expected proportion					
				Student-t			Normal		
	Expt.1	Expt.2	Expt.3	Expt.1	Expt.2	Expt.3	Expt.1	Expt.2	Expt.3
π_{WW}	0.700	0.667	0.633	0.710	0.667	0.635	0.719	0.671	0.644
π_{WR}	0.233	0.233	0.167	0.219	0.234	0.157	0.211	0.207	0.156
π_{RW}	0.667	0.467	0.533	0.667	0.475	0.533	0.675	0.470	0.541
π_{RR}	0.730	0.600	0.667	0.742	0.582	0.680	0.742	0.596	0.682
Average % error				0.023	0.013	0.021	0.037	0.033	0.030

Table 4 presents the observed and expected proportions for each experiment. Both normal random effects and Student-t random effects distributions are considered for comparative study. The averages of percentage error in each experiment show that a better fit is associated with the Student-t random effects distributions.

5. CONCLUSION

For the salamander mating data, we have shown that the GLM with Student-t random effects distributions can accommodate over-dispersion, capture outlying random effects and hence provide a robust analysis that the usual GLM with normal random effects distributions cannot offer. Furthermore, the SMN representation for the Student-t density enables an efficient Gibbs sampling algorithm and also provides a global diagnosis for outlying random effects. This is the first attempt among previous research work on the salamander mating data to identify the possible male and female salamanders with outlying effects in mating in the three experiments. To provide more flexibility for the GLMMs, we assume that the degrees of freedom of the random effects, α_f and α_m, are unknown and let the data speak for themselves by using non-informative priors.

In fact, there is room for further improvement. For example, other distributions, such as the exponential power and symmetric stable distributions, can be used for the random effects. Gibbs sampler algorithm may be simplified by using the SMN representation of the stable distribution and the scale mixtures of uniform representation of the exponential power distribution. These open a new area for further research.

ACKNOWLEDGEMENT

This work was supported by a research grant from the Research Grants Council of HKSAR (Project No. HKU 7257/99H). The authors would like to thank the Associate Editor and a referee for comments.

REFERENCES

Breslow, N. E. and Lin, X. (1995). Bias correction in generalized linear mixed models with a single component of dispersion. *Biometrika* **82**, 81–91.

Chan, J. S. K. and Kuk, A. Y. C. (1997). Maximum likelihood estimation for probit-linear mixed models with correlated random effects. *Biometrics* **53**, 86–97.

Choy, S. T. B. and Smith, A. F. M. (1997). Hierarchical models with scale mixtures of normal distribution. *Test* **6**, 205–211.

Fernández, C. and Steel, M. F. J. (2000). Bayesian regression analysis with scale mixtures of normals. *Econometric Theory* **16**, 80–101.

Karim, M. R. and Zeger, S. L. (1992). Generalized linear models with random effects; salamander mating revisited. *Biometrics* **48**, 631–644.

Kuk, A. Y. C. (1999). Laplace importance sampling for generalized linear mixed models. *J. Statist. Comput. Simulation* **63**, 143–158.

Lin, X. and Breslow, N. E. (1996). Bias correction in generalized linear mixed models with multiple components of dispersion. *J. Am. Statist. Ass.* **91**, 1007–1016.

McCullagh, P. and Nelder, J. A. (1989). *Generalized Linear Models*, 2nd edn. London: Chapman and Hall.

McCulloch, C. E. (1994). Maximum likelihood variance components estimation for binary data. *J. Am. Statist. Ass.* **89**, 330–335.

Shun, Z. (1997). Another look at the salamander mating data: a modified Laplace approximation approach. *J. Am. Statist. Ass.* **92**, 341–349.

Spiegelhalter, A. D., Thomas, A, Best, N. G. and Gilks, W. (1997). *BUGS 0.6: Bayesian Inference using Gibbs Sampling*. Cambridge, UK: MRC Biostatistical Unit.

Smith, A. F. M. and Roberts, G. O. (1993). Bayesian computation via the Gibbs sampler and related Markov Chain Monte Carlo Methods. *J. R. Statist. Soc. B* **55**, 3–23.

Tierney, L. (1994). Markov Chain for exploring posterior distributions. *Ann. Statist.* **22**, 1701–1762 (with discussion).

West, M. (1984). Outlier models and posterior distributions in Bayesian linear regression. *J. R. Statist. Soc. B* **46**, 431–439.

West, M. (1987). On scale mixtures of normal distributions. *Biometrika* **74**, 646–648.

BAYESIAN STATISTICS 7, pp. 485–492
J. M. Bernardo, M. J. Bayarri, J. O. Berger, A. P. Dawid,
D. Heckerman, A. F. M. Smith and M. West (Eds.)

A Relationship Between Randomized Manipulation and Parameter Independence

ALIREZA DANESHKHAH and JIM Q. SMITH
The University of Warwick, UK
a.daneshkhah@warwick.ac.uk j.q.smith@warwick.ac.uk

SUMMARY

We argue that a Bayesian will usually need to specify a joint prior density of the conditional probabilities of a Causal Bayesian network (CBN). We show that in order to do this, under very mild conditions it will be necessary to demand that this joint prior density exhibits the properties of local and global independence. To make the connection between prior independence and causality, it is first necessary to strengthen slightly the assumptions of factorization invariance under manipulation which induces randomized intervention. We introduce the hypercausal BN (HCBN) that asserts a set of factorizations of densities which are invariant to a class of "do" operations larger than those considered by Pearl. We show that if a BN is assumed to be hypercausal, then the prior distributions on the probabilities of the idle system must exhibit local and global independence.

Keywords: CAUSAL BAYESIAN NETWORK; DIRICHLET DISTRIBUTION; ESSENTIAL GRAPH; HYPERCAUSALITY; LOCAL AND GLOBAL INDEPENDENCE; RANDOMIZED MANIPULATION.

1. INTRODUCTION

In the recent past various authors (Spirtes *et al.*, 1993; Pearl, 1995; Lauritzen, 2001), have described causation in a Bayesian Network (BN) as a hypothesis of an invariance of a collection of factorization of a density under the manipulation of its nodes. Thus, in 1995, Pearl defines a Causal Bayesian network (CBN) for $x = (x_i, i \in I = \{1, \ldots, k\})$ as follows.

Let $p(x)$ be a probability distribution on a set X of variables representing the conditional independence relations coded in a BN G in an idle system: *i.e.*, a system where all nodes are simply observed and not manipulated, and let $p(x \mid do(V = v))$ denote the distribution resulting from the manipulation $do(V = v)$ that intervenes on a subset V of variables and sets them to values v. Denote by p_* the set of all distributions $p(x \mid do(V = v))$, $V \subseteq X$ including $p(x)$, which represents no intervention (*i.e.*, $V = \emptyset$).

Definition 1. A BN G is said to be a $Causal\ Bayesian\ Network$ (CBN) compatible p_* if and only if the following three conditions hold for every $p(. \mid do(V = v)) \in p_*$:

(i) $p(x \mid do(V = v))$ is Markov relative to G;
(ii) $p(x_i \mid do(V = v)) = 1$ for all $X_i \in V$ whenever x_i is consistent with $V = v$, and is otherwise zero;
(iii) $p(x_i \mid pa_i, do(V = v)) = p(x_i \mid pa_i)$ for all $X_i \notin V$ whenever pa_i is consistent with $V = v$, and is otherwise zero.

This definition tells us how to read, from a single BN, a whole collection of new probability distributions which assert not only what will happen if we do not intervene in the system but also what will happen if we manipulate each of the nodes. The authors above argue that it is often but not always the case that in practice when we assert a BN is valid, we also believe it is causal in the sense defined above. This BN, therefore, does not simply describes how things are, but also asserts what will happen if the system is controlled or manipulated. One consequence of this extension is that unlike in the general BN, an unambiguous meaning is given to the directionality of the edges in the BN. Each directed CBN in Markov equivalent BNs corresponds to a different class of factorizations. The hypothesis of the CBN enables us to compute the distribution $p(x \mid do(V = v))$ resulting from any intervention $do(V = v)$ as a truncated factorization

$$p(x \mid do(V = v)) = \prod_{\{i \mid X_i \notin V\}} p(x_i \mid pa_i) \ \forall x \text{ consistent with v.}$$

When G is a CBN with respect to p_*, the following two properties must hold.

Property 1. For all i,

$$p(x_i \mid pa_i) = p.(x_i \mid do(PA_i = pa_i))$$

Property 2. For all i and for every subset S of variables disjoint for $\{X_i, PA_i\}$, we have

$$p(x_i \mid do(S = s, PA_i = pa_i)) = p(x_i \mid do(PA_i = pa_i))$$

where PA_i denote to parent set of its child X_i.

Now although Pearl (2000) does not focus on this issue, in practice, the conditional probabilities in a BN usually need to be estimated (*e.g.*, Jordan, 1998). It is therefore natural, from a Bayesian perspective, to ask what constraints, if any, need to imposed on the prior probabilities on the idle system to ensure that its marginal mass function is consistent with the CBN hypothesis. One property that one might want to demand is that, whether we were to learn the value of a probabilities θ from some extraneous source or set this probability to a fixed values: *e.g.*, by randomization, it should be legitimate to "plug-in" this value: that is, to *manipulate* the value of θ to this known value and retain the integrity of the BN.

To make a connection between prior independence and causality, it is first necessary to strengthen slightly the assumptions of factorization invariance under manipulation which induces randomized intervention. In section 2, we introduce the Hypercausal Bayesian Network (HCBN) that asserts a set of factorizations of densities which are invariant to a class of "*do*" operations larger than those considered by Pearl. This requires us to develop the ideas of Koster (2000) and Lauritzen (2001) about randomized intervention. We show that if a BN is assumed to be hypercausal and we wish to learn about the probabilities defining it, then the prior distributions on the probabilities of the idle system (*i.e.*, the BN without intervention) must exhibit local and global parameter independence. On the other hand if we assume that the idle system exhibits local and global independence, it is extendable in a natural way, to a HCBN, and in particular to a CBN. In Section 3, we briefly discuss a concept of *multicausality* which asserts hypercausality for all BNs in the equivalence class of an essential graph (see Anderson *et al.*, 1997, for details of the essential graph).

2. CAUSALITY AND PARAMETER INDEPENDENCE ASSUMPTIONS

In this section, we explore the relationships between Pearl's manipulation concept and local and global independence. We use the following notation. Let the vector $X = (X_1, X_2, \ldots, X_k)$, of nodes of a BN G have its components X_i, $1 \leq i \leq k$, listed in an order compatible with G and their corresponding vectors of probabilities $\underline{\theta}_1, \ldots, \underline{\theta}_k$ compatibly with the partial order of the

BN. So we list the parameters in the order $\underline{\theta}_1, \ldots, \underline{\theta}_k$, where $\underline{\theta}_i = \{\theta_{i|pa(i)}\}$. The components of $\underline{\theta}_i$ is taken in some arbitrary but fixed order within the vector. In general the prior joint density of $\underline{\theta}$ can be written as

$$p(\underline{\theta}) = \prod_{i=1}^{k} p(\underline{\theta}_i \mid \underline{\theta}^{i-1}), \quad \underline{\theta}^{i-1} = \{\underline{\theta}_1, \ldots, \underline{\theta}_{i-1}\}$$

We use the standard notation that let $\underline{\theta}_A$ represent the subset of $\underline{\theta}$ whose indices $i \in A$, A is a subset of $\{1, \ldots, k\}$. Here k is the total number of conditional probabilities needed to define G, or equivalently the number of components of $\underline{\theta}$. Each component of $\underline{\theta}_i$ will be

$$\theta_{i(j)|pa_{i(l)}} = p(X_i = x_{i(j)} \mid PA_i = pa_{i(l)})$$

where $\theta_{i(j)|pa_{i(l)}}$ denotes the parameter associated with the level j of i-th variable and the level l of its parents. Therefore, $\underline{\theta}_i = \{\theta_{i(j)|pa_{i(l)}}, \ 1 \leq j \leq n_i, \ 1 \leq l \leq m_i\}$, where the each component of $\underline{\theta}_i$ is positive and for the fixed l (the fixed level of parent or for the components in the same strata), $\Sigma_{j=1}^{n_i} \theta_{i(j)|pa_{i(l)}} = 1$. Here n_i and m_i denote the numbers of the states of the i-th variable and its parents set, respectively.

The vector $\underline{\theta}$ is said to exhibit *local* and *global* independence, if all its components are mutually independent, that is, the parameters $\underline{\theta} = (\underline{\theta}_j, \ 1 \leq j \leq k)$ associated with the nodes of a BN G are globally independent if the joint prior density factorizes as $p(\underline{\theta}) = \prod_{i=1}^{k} p(\underline{\theta}_i)$, and the parameters associated with the node i, $1 \leq i \leq k$ are locally independent if the corresponding joint prior density factorizes as $p(\underline{\theta}_{i(j)}) = \prod_l p(\theta_{i(j)|pa_{i(l)}})$ for $1 \leq j \leq n_i$ (see Spiegelhalter and Lauritzen, 1990), for details).

2.1. *Randomization and Cause*

Whether in an industrial or medical context, causal effects are usually investigated by performing randomized trials in designed experiments. For example, medical trials will manipulate otherwise idle systems, usually using stratified (or contingent) randomized experiments and may deduce the beneficial effect of one treatment A over another B. This benefit is typically interpreted as causal: if I were to administer (*i.e.*, manipulate a client to have) treatment A then the effects would be better (on average over strata) than if I were to administer treatment B (see *e.g.*, Rubin, 1978). It has therefore been recognized by Koster (2000) and Lauritzen (2001) that it is important to develop an algebra for causation for randomized experiments on BNs.

Example 1. Let a BN have three nodes: patient's sex (S), dose of treatment (D), and patient's recovery (R). The unmanipulated factorization associated with the BN

$$S \longrightarrow D \longrightarrow R$$

is given by

$$p(s, d, r) = p(s)p(d \mid s)p(r \mid d)$$

We want to study the effect of $\hat{d}(s) = do(D = g(s))$ (*i.e.*, set different values of D given different values of s through g, where g is an appropriate function) on the BN above. The effect of this experiment is not to allow nature to choose D as a function of sex, but to instead substitute a new (stratified) randomized selection $\pi(d \mid s)$. When we do this on the BN above, it is natural to assert that the joint mass function represents the factorization

$$p_{\hat{d}(s)}(s, d, r) = p(s)\pi(d \mid s)p(r \mid \hat{d}(s))$$

where $\pi(d \mid s)$ denotes this contingent randomized strategy. The densities on the probabilities in an uncertain BN with local and global independence will satisfy

$$p_{\hat{d}(s)}(\theta_s, \theta_d, \theta_r) = p(\theta_s)\pi(s)p(\theta_{r|\hat{d}(s)})$$

where $\pi(s) = p(\theta_{\hat{d}(s)})$ is degenerated on g.

This motivates the following definitions.

Definition 2. The contingent randomized intervention, $do(\underline{\theta}_A = \underline{\theta}^*_A)$ on a BN G whenever the contingent $pa_{i(l)}$ arises, manipulates X_i to a value $\hat{x}_{i(j)}$ according to the set randomized probabilities $\theta^*_{i(j)|pa_{i(l)}} = p(X_i = \hat{x}_{i(j)} \mid PA_i = pa_{i(l)})$ for each $i \in A$. When several interventions are employed simultaneously the effect of the manipulation is calculated in an order compatible with G.

In fact, in the definition above, we set X_i to $\hat{x}_{i(j)}$ in the l-th strata associated with variable i.

The effect of a contingent manipulation is two-fold. First it changes the distribution of probabilities in the system, substituting a randomized distribution for an idle one. Second, it manipulates a particular X_i to take a different value from the one it might have taken in the idle system.

This natural assumption about the effect of a contingent manipulation conditional on the probability vector $\underline{\theta}$ is analogous to the invariance of factorization formula of Spirtes *et al.* (1993) and Pearl (1995) for non-random manipulation, and is consistent with Robins's (1987) definition in a non-graphical context.

Definition 3. Call a Bayesian network *Contingently Causal* (CCBN), if under the contingent manipulation $do(X_i = \hat{x}_{i(j)} \mid PA_i = pa_{i(l)}) = do(\theta_{i(j)|pa_{i(l)}} = \theta^*_{i(j)|pa_{i(l)}})$, then this implies that, for all configurations of X consistent with $\{X_i = \hat{x}_{i(j)} \mid PA_i = pa_{i(l)}\}$ and $pa_{i(l)}$,

$$\begin{aligned} &p(x \mid \underline{\theta}_{\hat{i}}, do(\theta_{i(j)|pa_{i(l)}} = \theta^*_{i(j)|pa_{i(l)}})) \\ &= \frac{p(x \mid \underline{\theta})}{p(X_i = \hat{x}_{i(j)} \mid \underline{\theta}, PA_i = pa_{i(l)})} = \Big\{ \prod_{v=1,k,v\neq i} \underline{\theta}_v \Big\} \times \theta^*_{i(j)|pa_{i(l)}} \end{aligned}$$

For all configurations, $pa_{i(l)}$ not consistent with $\{X_i = \hat{x}_{i(j)} \mid PA_i = pa_{i(l)}\}$, let $p(x \mid \underline{\theta}_{\hat{i}}, do(\theta_{i(j)|pa_{i(l)}} = \theta^*_{i(j)|pa_{i(l)}})) = 0$.

Here $\underline{\theta}_{\hat{i}}$ denotes the $\underline{\theta}$ vector with the i-th component missing and

$$p\Big(x \mid \underline{\theta}_{\hat{i}}, do(\theta_{i(j)|pa_{i(l)}} = \theta^*_{i(j)|pa_{i(l)}})\Big)$$

denotes the mass function of X given the mentioned manipulation above. Note that this contingent manipulation manipulates X_i to a value $\hat{x}_{i(j)}$ when $PA_i = pa_{i(l)}$, but otherwise keeps the system idle.

A question arising from the definition above is: how should a Bayesian modify a prior density $p(\underline{\theta})$ on $\underline{\theta}$ after an intervention $do(\underline{\theta}_A = \underline{\theta}^*_A)$? Well, if $\underline{\theta}^*_A$ is determined "at random" then it must surely be uninformative about $\underline{\theta}_{\bar{A}}$, where $\bar{A}$ stands for the complement of A in $\{1, \ldots, k\}$. So the original prior margin of $\underline{\theta}_{\bar{A}}$ and the margin after this randomized manipulation should exactly coincide. Denote the density of $\underline{\theta}_{\bar{A}}$ after contingent intervention $do(\underline{\theta}_{\bar{A}} = \underline{\theta}^*_{\bar{A}})$ by $p(\underline{\theta}_{\bar{A}} \mid do(\underline{\theta}_{\bar{A}} = \underline{\theta}^*_{\bar{A}}))$.

Definition 4. Call a contingent randomized intervention $do(\underline{\theta}_A = \underline{\theta}^*_A)$ on an uncertain BN, $Bayes\ faithful$ if

$$p(\underline{\theta}_{\bar{A}} \mid do(\underline{\theta}_A = \underline{\theta}^*_A)) = p(\underline{\theta}_{\bar{A}})$$

where $p(\underline{\theta}_{\bar{A}})$ is the Bayesian' prior marginal density on $\underline{\theta}_{\bar{A}}$.

2.2. *Hypercausality*

We can now define an analogous definition of causality that includes beliefs by asserting than invariance after intervention of non-ancestral probabilities holds in the extended BN which includes probabilities θ in its graph as well as X (see Example 2). Let $A_u = \{1, \ldots, u\}, 1 \leq u \leq k$.

Definition 5. Say an uncertain BN is a $Hypercausal\ BN$ (HCBN) if it is a CCBN and for all Bayes faithful contingent intervention $do(\underline{\theta}_{A_u} = \hat{\underline{\theta}}_{A_u}),\quad 1 \leq u \leq k$ defined above, $p(\underline{\theta}_{\bar{A}_u} \mid do(\underline{\theta}_{A_u} = \hat{\underline{\theta}}_{A_u})) = p(\underline{\theta}_{\bar{A}_u} \mid \hat{\underline{\theta}}_{A_u})$. Where $\bar{A}_u$ denotes the complement of A_u.

Note that Pearl (1995) defines and discusses the intervention $do(X_i = \hat{x}_i)$ for BNs with known probabilities. This can be thought of as a degenerate form of contingent randomized intervention on $\hat{\theta}_i$ on a CCBN as

$$\hat{\theta}_{i(j)|pa_{i(l)}} = \begin{cases} 1, & \text{if } x_{i(j)} = \hat{x}_i \text{ for all } j \text{ and } l; \\ 0, & \text{otherwise.} \end{cases}$$

Koster (2000) gives a generalization of this definition, setting

$$\hat{\theta}_{i(j)\,|\,pa_{i(l)}} = \begin{cases} \theta^*(x_i), & \text{if } x_{i(j)} = \hat{x}_i \text{ for all } j \text{ and } l\,; \\ 0 & \text{otherwise.} \end{cases}$$

where $0 \leq \theta^*(.) \leq 1$.

It is easily checked that his intervention formula under CCBN (and hence CBN) and the Bayes faithfulness assumptions coincides with ours, conditional on $\hat{\theta}_{i(j)|pa_{i(l)}}$. However we have seen in the example above that we may want to use different randomizations for different configurations of parents of X_i and there is absolutely no reason within this framework not to extend his definition to include this case.

Before we can define an uncertain analog of a CBN, we first need to define unambiguously what is meant by "observing the probability θ_i." It is most natural to follow Pearl and Example 2 and to define this conditioning as a limiting proportion of a sequence of observed margins.

So assume a selection mechanism, acting on a random sequence $\{X_{i(j)} = t\}_{l \geq 1}$ of observations respecting the idle BN, which records $\{X_{i(j)} = t\}_{t \geq 1}$ only for the m parent $pa_{i(l)}$ configuration values associated with indices $i \in A$, where m is the number of the components of A.

Call the selected subsequence $\{w_A(s)\}_{s \geq 1}$, where $w_A(s) = \{w_i(s_i) : i \in A\}$, where s_i indexes the s-th observation of the i-th variable. We can now define

$$p(\underline{\theta}_{\bar{A}} \mid \underline{\theta}_A) = \lim_{r \to \infty} p(\underline{\theta}_{\bar{A}} \mid \{w_A(s)\}_{1 \leq s \leq r})$$

where $\inf s_i \to \infty$ as $r \to \infty$. It is easily checked that this definition gives us the familiar definition of conditioning.

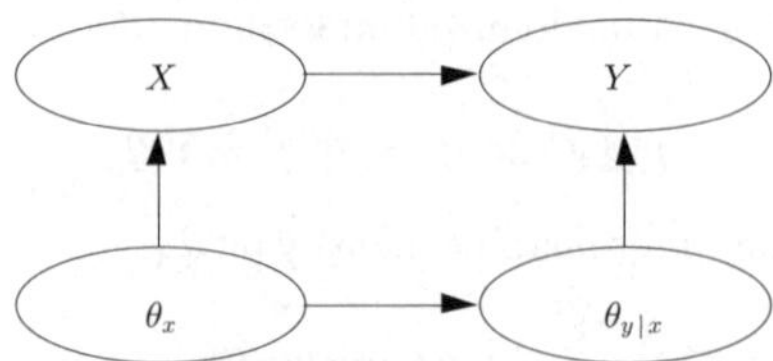

Figure 1. *The BN with the dependent parameters of Example 2.*

Example 2. It is clear that there must be a close link between causality and parameter independence. Consider the graph G on two binary random variables X and Y, where the parameters θ_x and $\theta_{y|x}$ are dependent (Figure 1).

This implies that if we learn the value of θ_x through,—for example, observing an extremely large, independent data set on the X margin—then $\theta_{Y|X=x} = p(Y = 1 \mid X = 1)$ will, in general, change because of dependence above. On the other hand, if we manipulate that system to take a randomized sample on X so that $p(X = 1) = \theta_x$, then by definition this 'randomization' should leave $\theta_{Y|X=x}$ unchanged and have the same margin it had in the idle system. So if there is a prior dependence between θ_x and $\theta_{y|x}$, then we cannot expect to be able to identify a randomly manipulated system with one learnt from an idle system. Note that, the reason we introduced contingent randomization was to be able to consider the separate manipulation of all components $\underline{\theta}_i$ of the probability vector $\underline{\theta}$ to values other than zero and one as considered in standard rules. This heuristic argument motivates the following theorem.

Theorem 1. *G is a HCBN if and only if it exhibits local and global independence*

Proof. By the definition of Bayes faithfulness on this CCBN

$$p(\underline{\theta}_{\bar{A}_u} \mid do(\underline{\theta}_{A_u} = \hat{\underline{\theta}}_{A_u})) = p(\hat{\underline{\theta}}_{A_u})$$

so by the definition of an HCBN, equivalently, the probabilities in the idle system satisfy

$$p(\underline{\theta}_{\bar{A}_u}) = p(\underline{\theta}_{\bar{A}_u} \mid \hat{\underline{\theta}}_{A_u}), \quad 1 \leq u \leq k$$

Hence, by definition of A_u iff $\underline{\theta}_k \perp \underline{\theta}_1, \ldots, \underline{\theta}_{u-1}$, $1 \leq u \leq k-1$; but this is verified iff $\perp_{i=1}^{k} \underline{\theta}_i$, *i.e.*, iff G exhibits local and global independence. □

Corollary *An HCBN exhibits the property that for all Bayes faithful contingent interventions on* $\underline{\theta}_A$, $A \subseteq \{1, \ldots, k\}$,

$$p\Big(\underline{\theta}_{\bar{A}} \mid do(\underline{\theta}_A = \hat{\underline{\theta}}_A)\Big) = p(\underline{\theta}_{\bar{A}} \mid \hat{\underline{\theta}}_A).$$

3. MULTICAUSAL ESSENTIAL GRAPHS

It has been known for a long time that two BNs are Markov equivalent if and only if their mixed pattern (Verma and Pearl,1990, 1992) or their mixed essential graphs (Anderson *et al.*, 1997) agree. It follows that BNs with the same essential graphs will be indistinguishable from each other from an observational study of an idle system. On the other hand the authors mentioned above have argued that various causal deductions can be made within the pattern (or essential graph): the deduced directed edges in observed estimates of these mixed graphs on idle (*i.e.*, unmanipulated) systems allowing us to make these causal assertions.

Now suppose, on the basis of observations of analogous idle system to the one under study, a Bayesian is confident in asserting a particular essential graph H as valid hypothesis for another analogous unmanipulated system. The components of conditional probabilities of this new system are, however, uncertain and the researcher wants to make prior assumptions which are consistent with $every$ HCBN consistent with the given essential graph. In Daneshkhah and Smith (2002), we introduce a concept of multicausality which asserts hypercausality for all BNs in the equivalence class of an essential graph. This Bayesian, therefore, makes the strongest statements that could be made about invariance to manipulation consistent with the causal information embedded in the essential graph.

We show that this Bayesian's prior density on the conditional probabilities in the model must satisfy a generalization of the Geiger and Heckerman (GH) condition (see Theorem 2 in Geiger and Heckerman, 1997). In the special case when the essential graph is undirected, the family of priors must be chosen from degenerate Hyper-Dirichlet family (see Dawid and Lauritzen, 1993).

So what we have established is that if a Bayesian is prepared to make bold enough causal assertions within a single uncertain BN then this not only introduces independence relationships between parameters, but can also characterize prior families of distributions on these parameters. We believe this is a very helpful way of thinking about this class of models. Note that this multicausal essential graph characterization concerns a single hypothesis. It is not an assertion about a common prior to be used for CBN for the model selection as is more typical in, for example, Heckerman *et al.* (1995), Cowell *et al.* (1999) and Cooper and Yoo (1999).

4. DISCUSSION

The assumption of local and global independence is made almost universally in practice when BNs are estimated. For example, Geiger and Heckerman (1997), Heckerman *et al.* (1995) and Cowell *et al.* (1999) implicitly make most models extendable to HCBN. Addressing model selection, Cooper and Yoo (1999) incorporate experimental data, and again assume local and global independence.

However, the implication of using such priors for each model in the selection set is fierce. For example, suppose we believe that an expert might not be calibrated and that there is a possibility of some systematic bias in her probability assessments about $\underline{\theta}$. Then, in Example 2, by learning from marginal information on X that θ_x was over specified, we may legitimately want to adjust the expert's assessments of $\theta_{y|x}$ downwards under the assumption that the expert consistently tends to overestimate. In this rather common scenario it is important that, for any candidate model, the prior on the probabilities vector, $\underline{\theta}$ does not exhibit local and global independence in the idle system.

In general, then, local and global independence and hence hypercausality corresponds to a strong assumption of modularity in the idle system. It is only valid when it is reasonable to assume that we can simulate the behavior of the system by a composite family of simulated modules, one for each $p(X_i \mid PA_i(X) = x)$ which randomize over a prior distribution for $\theta_{i|PA_i}$. But if this level of modularity does hold, then the system is at least compatible with a causal extension of the type described above.

REFERENCES

Andersson, S. A., Madigan, D. and Perlman, M. D. (1997). A characterization of Markov equivalence classes for acyclic. *Ann. Statist.* **24**, 505–541.

Cooper, G. F. and Yoo. C. (1999). Causal Discovery from a Mixture of Experimental and Observational Data. *Proc. 5th Conference on Uncertainty in Artificial Intelligence* (K. B. Laskey and H. Prade, eds). San Fransisco: Morgan-Kaufmann.

Cowell, R. G., Dawid, A. P., Lauritzen, S. L. and Spiegelhalter, D. J. (1999). *Probabilistic Networks and Expert Systems*, Berlin: Springer.

Daneshkhah, A. and Smith, J. Q. (2002). Multicausal Prior Families, Randomization and Essential Graphs. *Proc. 1st European Workshop on Probabilistic Graphical Models* (J. A. Gamez and A. Salmerón eds). Cuenca, Spain, 25–34.

Dawid, A. and Lauritzen, S. L. (1993). Hyper Markov laws in the statistical analysis of decomposable graphical models. *Ann. Statist.* **21**, 1272–1317.

Geiger, D. and Heckerman, D. (1997). A characterization of the Dirichlet distribution through global and local parameter independence. *Ann. Statist.* **25**, 1344–1369.

Heckerman, D., Geiger, D. and Chickering, D. M. (1995). Learning Bayesian networks: The combination of knowledge and statistical data. *Machine Learning* **20**, 197–243.

Jordan, M. I. (1998). *Learning in Graphical Models*. Dordrecht: Kluwer.

Koster, J. T. A. (2000). Graphs, causality and structural equation models (unpublished). www.knaw.nl/09public/rm/koster.pdf

Lauritzen, S. L. (2001). Causal inference from graphical models. *Complex Stochastic Systems* (O. E. Barndorff-Nielsen, D. R. Cox and C. Kluppelberg, eds). London: Chapman and Hall, 63–108.

Pearl, J. (1995). Causal diagrams for experimental research. *Biometrika* **28**, 669-710.

Pearl, J. (2000). *Causality: Models, Reasoning and Inference*. Cambridge: Cambridge University Press.

Robins, J. M. (1986). A new approach to causal inference in mortality studies with a sustained exposure period-applications to control of the healthy worker survivor effect. *Math. Modelling* **7**, 1393–1512.

Rubin, D. B. (1978). Bayesian inference for causal effects: The role of randomized. *Ann. Statist.* **6**, 34–58.

Spiegelhalter, D. J. and Lauritzen, S. L. (1990). Sequential updating of conditional probabilities on directed graphical structures. *Networks* **20**, 579–605.

Spirtes, P., Glymour, C. and Scheines, R. (1993). *Causation, Prediction, and Search*. New York: Springer.

Verma, T. and Pearl, J. (1990). Equivalence and synthesis of causal models. *Uncertainty in Artificial Intelligence* **6** (P. Bonissone, M. Henrion, L. N. Kanal and J. F. Lemmer eds). Amsterdam: Elsevier, 255–268.

Verma, T. and Pearl, J. (1992). An algorithm for deciding if a set of observed independence has a causal explanation. *Uncertainty in Artificial Intelligence* **8**(D. Dubois, M. P. Wellman, B. D'Ambrosio and P. Smets, eds). New York: Wiley, 124–144.

BAYESIAN STATISTICS 7, pp. 493–500
J. M. Bernardo, M. J. Bayarri, J. O. Berger, A. P. Dawid,
D. Heckerman, A. F. M. Smith and M. West (Eds.)

Markov Random Field Extensions using State Space Models

CLAUS DETHLEFSEN
Aalborg University, Denmark
dethlef@math.auc.dk

SUMMARY

We elaborate on the link between state space models and (Gaussian) Markov random fields. We extend the Markov random field models by generalizing the corresponding state space model. It turns out that several non-Gaussian spatial models can be analyzed by combining approximate Kalman filter techniques with importance sampling. We illustrate the ideas by formulating a model for edge detection in digital images, which then forms the basis of a simulation study.

Keywords: EXTENDED KALMAN SMOOTHING; IMAGE RESTORATION; EDGE DETECTION; GAUSSIAN MIXTURES; LATTICE DATA.

1. INTRODUCTION

The class of state space models is very broad and comprises structural time series models, ARIMA models, cubic spline models, and, as demonstrated by Lavine (1999), also Markov random field models. The Kalman filter techniques are powerful tools for inference in such sequential models. Basic references on state space model methodology are Harvey (1989), West and Harrison (1997) and Durbin and Koopman (2001). In the past decade, there has been a certain interest in developing Markov chain Monte Carlo (MCMC) methods for the analysis of complex state state space models (see Carlin *et al.*, 1992; Carter and Kohn, 1994; Frühwirth-Schnatter, 1994; de Jong and Shephard, 1996). Our approach is not based on MCMC but on iterated extended Kalman smoothing, which may be combined with importance sampling for exact simulation (see Durbin and Koopman, 2001). Using this method, we avoid the MCMC problems of ensuring that the Markov chain is mixing well and assessing whether the chain has converged. Writing Markov random field models as state space models, following Lavine (1999), makes it possible to use Kalman filter techniques to extend and analyze more complex Markov random field models. We show how to analyze such extensions and we also deliver an illustration, formulating a model for restoring digital images with focus on finding edges in the image. However, the new class of models also have applications within agricultural experiments (see, *e.g.*, Besag and Higdon, 1999) and disease mapping (see, *e.g.*, Knorr-Held and Rue, 2002).

2. GAUSSIAN STATE SPACE MODELS

The Gaussian state space model involves two processes, namely the latent state process $\boldsymbol{\theta}_t = \boldsymbol{G}_t\boldsymbol{\theta}_{t-1} + \boldsymbol{\omega}_t$ with $\boldsymbol{\omega}_t \sim \mathrm{N}_p(\mathbf{0}, \boldsymbol{W}_t)$, and the observation process $\boldsymbol{y}_t = \boldsymbol{F}_t'\boldsymbol{\theta}_t + \boldsymbol{\nu}_t$ with $\boldsymbol{\nu}_t \sim \mathrm{N}_d(\mathbf{0}, \boldsymbol{V}_t)$. The latent process is initialized by $\boldsymbol{\theta}_0 \sim \mathrm{N}_p(\boldsymbol{m}_0, \boldsymbol{C}_0)$. We assume that the disturbances $\{\boldsymbol{\nu}_t\}$ and $\{\boldsymbol{\omega}_t\}$ are both serially independent and mutually independent. The possible time-dependent system matrices $\boldsymbol{F}_t$, $\boldsymbol{G}_t$, $\boldsymbol{V}_t$ and $\boldsymbol{W}_t$ are all considered known for every t. They may also depend on a parameter vector $\boldsymbol{\psi}$ but this is suppressed in the notation.

The Kalman filter recursively yields $p(\boldsymbol{\theta}_t \mid D_t)$, the conditional distribution of $\boldsymbol{\theta}_t$ given all information available, D_t, at current time t,

$$\begin{aligned}
\boldsymbol{\theta}_t \mid D_{t-1} &\sim \mathrm{N}_p(\overbrace{\boldsymbol{G}_t\boldsymbol{m}_{t-1}}^{\boldsymbol{a}_t}, \overbrace{\boldsymbol{G}_t\boldsymbol{C}_{t-1}\boldsymbol{G}_t' + \boldsymbol{W}_t}^{\boldsymbol{R}_t}) \\
\boldsymbol{y}_t \mid D_{t-1} &\sim \mathrm{N}_d(\overbrace{\boldsymbol{F}_t'\boldsymbol{a}_t}^{\boldsymbol{f}_t}, \overbrace{\boldsymbol{F}_t'\boldsymbol{R}_t\boldsymbol{F}_t + \boldsymbol{V}_t}^{\boldsymbol{Q}_t}) \\
\boldsymbol{\theta}_t \mid D_t &\sim \mathrm{N}_p(\underbrace{\boldsymbol{a}_t + \overbrace{\boldsymbol{R}_t\boldsymbol{F}_t\boldsymbol{Q}_t^{-1}}^{\boldsymbol{A}_t}\overbrace{(\boldsymbol{y}_t - \boldsymbol{f}_t)}^{\boldsymbol{e}_t}}_{\boldsymbol{m}_t}, \underbrace{\boldsymbol{R}_t - \boldsymbol{A}_t\boldsymbol{Q}_t\boldsymbol{A}_t'}_{\boldsymbol{C}_t}).
\end{aligned}$$

Assessment of the state vector, $\boldsymbol{\theta}_t$, using all available information, D_n, is called Kalman smoothing and we write $(\boldsymbol{\theta}_t \mid D_n) \sim \mathrm{N}_p(\widetilde{\boldsymbol{m}}_t, \widetilde{\boldsymbol{C}}_t)$. Starting with $\widetilde{\boldsymbol{m}}_n = \boldsymbol{m}_n$ and $\widetilde{\boldsymbol{C}}_n = \boldsymbol{C}_n$, the Kalman smoother is a backwards recursion in time, $t = n-1, \ldots, 1$, with $\widetilde{\boldsymbol{m}}_t = \boldsymbol{m}_t + \boldsymbol{B}_t(\widetilde{\boldsymbol{m}}_{t+1} - \boldsymbol{a}_{t+1})$ and $\widetilde{\boldsymbol{C}}_t = \boldsymbol{C}_t + \boldsymbol{B}_t(\widetilde{\boldsymbol{C}}_{t+1} - \boldsymbol{R}_{t+1})\boldsymbol{B}_t'$, where $\boldsymbol{B}_t = \boldsymbol{C}_t\boldsymbol{G}_{t+1}'\boldsymbol{R}_{t+1}^{-1}$. When p is large, it is often computationally faster to use the mathematically equivalent disturbance smoother (see Koopman, 1993).

The posterior mode of $p(\boldsymbol{\theta} \mid \boldsymbol{y})$ is $\widetilde{\boldsymbol{m}}' = (\widetilde{\boldsymbol{m}}_1', \ldots, \widetilde{\boldsymbol{m}}_n')$. From the definition of conditional densities $\widetilde{\boldsymbol{m}}$ maximizes $p(\boldsymbol{\theta}, \boldsymbol{y})$ and thus also

$$\log p(\boldsymbol{\theta}, \boldsymbol{y}) = \sum_{t=1}^{n} \log p(\boldsymbol{y}_t \mid \boldsymbol{\theta}_t) + \sum_{t=1}^{n} \log p(\boldsymbol{\theta}_t \mid \boldsymbol{\theta}_{t-1}) + \log p(\boldsymbol{\theta}_0). \tag{1}$$

The derivative with respect to $\boldsymbol{\theta}$ is zero at the maximum, so $\widetilde{\boldsymbol{m}}$ solves the following equations:

$$\frac{\partial \log p(\boldsymbol{y}_t \mid \boldsymbol{\theta}_t)}{\partial \boldsymbol{\theta}_t} + \frac{\partial \log p(\boldsymbol{\theta}_t \mid \boldsymbol{\theta}_{t-1})}{\partial \boldsymbol{\theta}_t} + \frac{\partial \log p(\boldsymbol{\theta}_{t+1} \mid \boldsymbol{\theta}_t)}{\partial \boldsymbol{\theta}_t} \cdot \mathcal{I}_{[t \neq n]} = 0,$$

where $\mathcal{I}_{[t \neq n]}$ is an indicator function, which is 1, when $t \neq n$, and zero otherwise. From the definition of the state space model this gives

$$\boldsymbol{F}_t\boldsymbol{V}_t^{-1}(\boldsymbol{y}_t - \boldsymbol{F}_t'\boldsymbol{\theta}_t) - \boldsymbol{W}_t^{-1}(\boldsymbol{\theta}_t - \boldsymbol{G}_t\boldsymbol{\theta}_{t-1}) + \boldsymbol{G}_{t+1}'\boldsymbol{W}_{t+1}^{-1}(\boldsymbol{\theta}_{t+1} - \boldsymbol{G}_{t+1}\boldsymbol{\theta}_t) \cdot \mathcal{I}_{[t \neq n]} = 0. \tag{2}$$

We may now interpret the Kalman smoother as being an algorithm to solve (2) recursively.

The log likelihood function for a vector of hyperparameters $\boldsymbol{\psi}$ is given by

$$l(\boldsymbol{\psi}) = \sum_{i=1}^{n} \log p(\boldsymbol{y}_t \mid \boldsymbol{y}_1, \ldots, \boldsymbol{y}_{t-1}, \boldsymbol{\psi}) = c - \frac{1}{2}\sum_{t=1}^{n}\left\{\log|\boldsymbol{Q}_t| + \|\boldsymbol{y}_t - \boldsymbol{f}_t\|^2_{\boldsymbol{Q}_t^{-1}}\right\}, \tag{3}$$

where $\|\boldsymbol{x}\|^2_{\Sigma} = \boldsymbol{x}'\Sigma\boldsymbol{x}$ and c is a constant. The log likelihood for a given value of $\boldsymbol{\psi}$ can thus be obtained directly from the Kalman filter. The expression can then be maximized numerically yielding the maximum likelihood estimate. Approximate standard errors can be extracted from numerical second derivatives.

3. NON GAUSSIAN STATE SPACE MODELS

Consider a state space model with the observation density $p(\boldsymbol{y}_t \mid \boldsymbol{F}_t'\boldsymbol{\theta}_t)$ which may be non Gaussian. The latent process is specified as $\boldsymbol{\theta}_t = \boldsymbol{G}_t\boldsymbol{\theta}_{t-1} + \boldsymbol{\omega}_t$, where $\boldsymbol{\omega}_t \sim p(\boldsymbol{\omega}_t)$ may be non Gaussian. Conditional on the latent process, the observations are assumed serially independent. For notational convenience, we define $\boldsymbol{\lambda}_t = \boldsymbol{F}_t'\boldsymbol{\theta}_t$.

West *et al.* (1985), Kitagawa (1987) and Carlin *et al.* (1992) analyzed similar models using approximative conjugate analysis, numerical approximation and MCMC techniques, respectively. Another approach to analyze these types of state space models is by particle filtering (see, *e.g.*, Doucet *et al.*, 2000). The exposition here is due to Durbin and Koopman (2001) and is based on maximizing the posterior $p(\boldsymbol{\theta} \mid \boldsymbol{y})$ with respect to $\boldsymbol{\theta}$. This is equivalent to maximizing $\log p(\boldsymbol{\theta}, \boldsymbol{y})$ given by (1). Differentiation with respect to $\boldsymbol{\theta}_t$ and equating to zero, yields

$$\frac{\partial \log p(\boldsymbol{\theta}, \boldsymbol{y})}{\partial \boldsymbol{\theta}_t} = \boldsymbol{F}_t \frac{\partial \log p(\boldsymbol{y}_t \mid \boldsymbol{\lambda}_t)}{\partial \boldsymbol{\lambda}_t} - \frac{\partial \log p(\boldsymbol{\omega}_t)}{\partial \boldsymbol{\omega}_t} + \boldsymbol{G}_{t+1}' \frac{\partial \log p(\boldsymbol{\omega}_{t+1})}{\partial \boldsymbol{\omega}_{t+1}} \cdot \mathcal{I}_{[t \neq n]} = 0.$$

We assume that the densities are sufficiently well behaved so that a unique maximum exists and that it solves the above equations. For a discussion of this point, see Durbin and Koopman (2001).

The strategy employed to find the maximum is to obtain an approximation to the state space model and thus identifying $\widetilde{\boldsymbol{y}}_t$, $\widetilde{\boldsymbol{V}}_t$, $\widetilde{\boldsymbol{W}}_t$ by comparing with (2). The approximation requires an initial value $\widetilde{\boldsymbol{\theta}}$, which is improved by iteration, the new value being the output from the Kalman smoother, $\widetilde{\boldsymbol{m}}$, in the approximating linear state space model. The procedure is called iterated extended Kalman smoothing.

We will consider two methods for approximation, depending on the form of the densities. Both methods ensure a common mode of the approximating model and the original model. The first method also ensures common curvature at the mode; however, it is not always applicable, for which reason the second method must be used.

Method 1: Letting $\widetilde{\boldsymbol{\lambda}}_t = \boldsymbol{F}_t'\widetilde{\boldsymbol{\theta}}_t$, the observation part is linearized as

$$\frac{\partial \log p(\boldsymbol{y}_t \mid \boldsymbol{\lambda}_t)}{\partial \boldsymbol{\lambda}_t} \approx \left.\frac{\partial \log p(\boldsymbol{y}_t \mid \boldsymbol{\lambda}_t)}{\partial \boldsymbol{\lambda}_t}\right|_{\boldsymbol{\lambda}_t = \widetilde{\boldsymbol{\lambda}}_t} + \left.\frac{\partial^2 \log p(\boldsymbol{y}_t \mid \boldsymbol{\lambda}_t)}{\partial \boldsymbol{\lambda}_t \partial \boldsymbol{\lambda}_t'}\right|_{\boldsymbol{\lambda}_t = \widetilde{\boldsymbol{\lambda}}_t} (\boldsymbol{\lambda}_t - \widetilde{\boldsymbol{\lambda}}_t)$$

Comparing this with the first term in (2), we recognize it as a state space model with (linear) observation part specified by

$$\widetilde{\boldsymbol{V}}_t^{-1} = \left.\frac{\partial^2 \log p(\boldsymbol{y}_t \mid \boldsymbol{\lambda}_t)}{\partial \boldsymbol{\lambda}_t \partial \boldsymbol{\lambda}_t'}\right|_{\boldsymbol{\lambda}_t = \widetilde{\boldsymbol{\lambda}}_t}$$

$$\widetilde{\boldsymbol{y}}_t = \widetilde{\boldsymbol{\lambda}}_t - \widetilde{\boldsymbol{V}}_t \left.\frac{\partial \log p(\boldsymbol{y}_t \mid \boldsymbol{\lambda}_t)}{\partial \boldsymbol{\lambda}_t}\right|_{\boldsymbol{\lambda}_t = \widetilde{\boldsymbol{\lambda}}_t}.$$

For example, exponential family distributions can be linearized using Method 1.

Method 2: In the second approach, it is assumed that the log densities are a function of $(\boldsymbol{y}_t - \boldsymbol{\lambda}_t)^2$ or $\boldsymbol{\omega}_t^2$, respectively. The method applies to either or both of the derivatives $\partial \log p(\boldsymbol{y}_t \mid \boldsymbol{\lambda}_t)/\partial \boldsymbol{\lambda}_t$ and $\partial \log p(\boldsymbol{\omega}_t)/\partial \boldsymbol{\omega}_t$, evaluated at $\widetilde{\boldsymbol{\lambda}}_t = \boldsymbol{F}_t'\widetilde{\boldsymbol{\theta}}_t$ and $\widetilde{\boldsymbol{\omega}}_t = \widetilde{\boldsymbol{\theta}}_t - \boldsymbol{G}_t\widetilde{\boldsymbol{\theta}}_{t-1}$, respectively. The latter term, at time $t+1$ and evaluated at $\widetilde{\boldsymbol{\omega}}_{t+1}$, is also needed for insertion in (2).

Since

$$\frac{\partial \log p(\boldsymbol{y}_t \mid \boldsymbol{\lambda}_t)}{\partial \boldsymbol{\lambda}_t} = -2 \frac{\partial \log p(\boldsymbol{y}_t \mid \boldsymbol{\lambda}_t)}{\partial (\boldsymbol{y}_t - \boldsymbol{\lambda}_t)^2} (\boldsymbol{y}_t - \boldsymbol{\lambda}_t)$$

and

$$\frac{\partial \log p(\boldsymbol{\omega}_t)}{\partial \boldsymbol{\omega}_t} = -2\frac{\partial \log p(\boldsymbol{\omega}_t)}{\partial \boldsymbol{\omega}_t^2}(\boldsymbol{\theta}_t - \boldsymbol{G}_t\boldsymbol{\theta}_{t-1}),$$

we see by comparison with (2) that the approximating model is given by

$$\widetilde{\boldsymbol{V}}_t = -\frac{1}{2}\left[\frac{\partial \log p(\boldsymbol{y}_t \mid \boldsymbol{\lambda}_t)}{\partial (\boldsymbol{y}_t - \boldsymbol{\lambda}_t)^2}\bigg|_{\boldsymbol{\lambda}_t = \widetilde{\boldsymbol{\lambda}}_t}\right]^{-1}$$

$$\widetilde{\boldsymbol{W}}_t = -\frac{1}{2}\left[\frac{\partial \log p(\boldsymbol{\omega}_t)}{\partial \boldsymbol{\omega}_t^2}\bigg|_{\boldsymbol{\omega}_t = \widetilde{\boldsymbol{\omega}}_t}\right]^{-1}.$$

For example, t-distributions or Gaussian mixtures can be approximated by Method 2.

4. MARKOV RANDOM FIELDS

Let S denote an $I \times J$ regular, finite lattice. Elements of S are called sites or pixels, and a typical element is denoted s or ij or (i,j), depending on the context. A site s is associated with two random variables, y_s and θ_s, denoting the observed and the latent value at s, respectively. The vector of random variables at a set of sites $A \subseteq S$ is denoted $\boldsymbol{y}_A$ and, at the remaining sites $\boldsymbol{y}_{-A} = \boldsymbol{y}_{S\setminus A}$. The shorthand notation for $\boldsymbol{y}_S$ is $\boldsymbol{y}$. The vector $\boldsymbol{y}_i = \{y_{ij}\}_{j=1\dots J}$ is the variables in the ith row, and $\boldsymbol{y}_{-i}$ is the vector at sites outside the ith row. Similar notation is used for θ and other derived variables.

Let

$$\boldsymbol{T}_l = \begin{bmatrix} 1 & -1 & 0 & \cdots & 0 \\ -1 & 2 & \ddots & \ddots & \vdots \\ 0 & \ddots & \ddots & \ddots & 0 \\ \vdots & \ddots & \ddots & 2 & -1 \\ 0 & \cdots & 0 & -1 & 1 \end{bmatrix}_{l \times l}.$$

Then the prior density is given by Besag (1974),

$$p(\boldsymbol{\theta}) \propto \exp\left(-\frac{1}{2}\boldsymbol{\theta}'\boldsymbol{P}\boldsymbol{\theta}\right), \tag{4}$$

where $\boldsymbol{P}$ is the $IJ \times IJ$ precision matrix

$$\boldsymbol{P} = \tau_1^{-2}\boldsymbol{T}_I \otimes \boldsymbol{I}_J + \tau_2^{-2}\boldsymbol{I}_I \otimes \boldsymbol{T}_J,$$

and τ_1^2 and τ_2^2 are hyperparameters that we assume are known. The prior is improper since the precision matrix is singular. The parameters τ_1^2 and τ_2^2 measure the degree of dependency in the row and column direction, respectively.

The observations $\boldsymbol{y}$ are assumed to be normally distributed given $\boldsymbol{\theta}$,

$$\boldsymbol{y} \mid \boldsymbol{\theta} \sim \mathrm{N}_{IJ}(\boldsymbol{F}'\boldsymbol{\theta}, \Sigma).$$

We will assume that Σ and $\boldsymbol{F}$ are given so that, in the following, the posterior is proper. The posterior distribution of $\boldsymbol{\theta}$ is given by Bayes' Theorem,

$$\boldsymbol{\theta} \mid \boldsymbol{y} \sim \mathrm{N}_{IJ}(\boldsymbol{m}, \boldsymbol{C}), \tag{5}$$

where $\boldsymbol{C} = (\boldsymbol{P}+\boldsymbol{F}\Sigma^{-1}\boldsymbol{F}')^{-1}$ and $\boldsymbol{m} = \boldsymbol{C}\Sigma^{-1}\boldsymbol{y}$. The aim is to assess the posterior distribution, $p(\boldsymbol{\theta} \mid \boldsymbol{y})$, in a computationally attractive way. Note that the matrices to be inverted in the above expression are of size $IJ \times IJ$.

The Markov random field model is equivalent to the following state space model in the sense that their posterior distributions are identical. The state space model is evolving following the rows instead of time.

$$\begin{pmatrix} \boldsymbol{y}_i \\ \boldsymbol{x}_i \end{pmatrix} \Big| \, \boldsymbol{\theta}_i \sim \mathrm{N}\left[\begin{pmatrix} \boldsymbol{F}_i'\boldsymbol{\theta}_i \\ \boldsymbol{H}\boldsymbol{\theta}_i \end{pmatrix}, \begin{bmatrix} \Sigma_i & \boldsymbol{0} \\ \boldsymbol{0} & \tau_1^2 \boldsymbol{I}_{J-1} \end{bmatrix}\right] \tag{6}$$

$$\boldsymbol{\theta}_i \mid \boldsymbol{\theta}_{i-1} \sim \mathrm{N}(\boldsymbol{\theta}_{i-1}, \tau_2^2 \boldsymbol{I}_J) \tag{7}$$

$$p(\boldsymbol{\theta}_1) \propto 1, \tag{8}$$

where

$$\boldsymbol{H} = \begin{bmatrix} 1 & -1 & 0 & \cdots & 0 \\ 0 & 1 & \ddots & \ddots & \vdots \\ \vdots & \ddots & \ddots & -1 & 0 \\ 0 & \cdots & 0 & 1 & -1 \end{bmatrix}.$$

Thus $\boldsymbol{y}_i$ are the observed rows, $\boldsymbol{\theta}_i$ are the corresponding latent variables, and $\boldsymbol{x}_i$ are so-called pseudo observations. The analysis of the model is carried out assuming that the pseudo observations are observed to be zero as this ensures the equivalence of the state space model with the Markov random field model. In other words, $\boldsymbol{\theta} \mid \boldsymbol{x} = \boldsymbol{0}$ corresponds to the Markov random field prior (4), and $\boldsymbol{\theta} \mid \boldsymbol{x} = \boldsymbol{0}, \boldsymbol{y}$ corresponds to the posterior (5). This equivalence was established by Lavine (1999).

5. EXTENSIONS TO MARKOV RANDOM FIELD MODELS

We now extend the Markov random field model (6)–(8). For notational convenience, we assume $\boldsymbol{F}_i = \boldsymbol{I}$ and $\Sigma_i = \sigma^2\boldsymbol{I}$. The model can then be written in coordinate form as

$$y_{ij} \mid \theta_{ij} \sim \mathrm{N}(\theta_{ij}, \sigma^2)$$

$$x_{ij} \mid (\theta_{ij}, \theta_{i,j+1}) \sim \mathrm{N}(\theta_{ij} - \theta_{i,j+1}, \tau_1^2) \tag{9}$$

$$\theta_{ij} \mid \theta_{i-1,j} \sim \mathrm{N}(\theta_{i-1,j}, \tau_2^2). \tag{10}$$

The idea is to substitute any or all of the Gaussian distributions with more general distributions. These may be mixed in any way and this opens up a large class of models. For example, the observations may be Poisson distributed conditional on a Markov random field similar to the model analyzed by Christensen and Waagepetersen (2002). Another example is the fertility model considered by Besag and Higdon (1999) where t-distributions are used to allow for observational outliers and for jumps in the underlying fertility. In both examples, MCMC methods were used to analyze the models. They were reanalyzed in Dethlefsen (2002), using our methodology which does not make use of MCMC. Our approach is approximative; however, using importance sampling as described in Durbin and Koopman (2001), it is possible to assess various functions of interest using exact simulation.

We illustrate our approach by substituting the distributions in (9) and (10) by a mixture of Gaussian distributions, arriving at an image restoration model. Further illustrations are given in Dethlefsen (2002).

5.1. *Application in Digital Image Analysis*

Let the observed digital image $\boldsymbol{y}$ be represented by the gray scale values y_{ij} in an $I \times J$ lattice made up of pixels ij, for $i = 1, \ldots, I$ and $j = 1, \ldots, J$. We assume that y_{ij} is an indirect observation of the noise-free pixel-value θ_{ij}, so that the noise-free image is $\boldsymbol{\theta}$.

Let $\mathcal{M}(\mu, k, v_1^2, v_2^2) = k\mathrm{N}(\mu, v_1^2)+(1-k)\mathrm{N}(\mu, v_2^2)$, be a mixture of two univariate Gaussian distributions. Let $X \sim \mathcal{M}(0, k, v_1^2, v_2^2)$ and denote the density of X by $p(x)$. The log density is a function of x^2, so we use Method 2 to approximate $p(x)$ in the point, ξ. Thus, $p(x)$ is approximated by a Gaussian distribution with zero mean and variance $\widetilde{v}^2$, where

$$\widetilde{v}^2 = \frac{k \exp[-\xi/(2v_1^2)]/v_1 + (1-k)\exp[-\xi/(2v_2^2)]/v_2}{k \exp[-\xi/(2v_1^2)]/v_1^3 + (1-k)\exp[-\xi/(2v_2^2)]/v_2^3}. \tag{11}$$

We now formulate a model for image restoration by replacing (9) and (10) with

$$x_{ij} \,|\, (\theta_{ij}, \theta_{i,j+1}) \sim \mathcal{M}(\theta_{ij} - \theta_{i,j+1}, k, \tau^2, c\tau^2)$$

$$\theta_{ij} \,|\, \theta_{i-1,j} \sim \mathcal{M}(\theta_{i-1,j}, k, \tau^2, c\tau^2),$$

where k is the probability of having variance τ^2, and $c > 1$ is the scaling factor for the variance of a jump. When c approaches 1, the model approaches the Gaussian Markov random field model where edges tend to be blurred due to the smoothing nature of this model. For larger values of c, the model implies that a larger difference in neighboring values is needed in order to classify the jump as an edge.

In the approximating state space model, each pixel is associated with two variances. These are calculated using (11) with ξ substituted by $\mathrm{E}[\theta_{ij} \,|\, \boldsymbol{x} = \boldsymbol{0}, \boldsymbol{y}]$ and $\mathrm{E}[\theta_{ij} - \theta_{i,j+1} \,|\, \boldsymbol{x} = \boldsymbol{0}, \boldsymbol{y}]$, respectively, and improved by iteration. If these variances are both small, the pixel is in a smooth part of the image. If one or both are large, this indicates an edge in either left-right and/or up-down directions.

Simulation Example. To illustrate the methodology, we simulate an image and denoise it using the image restoration model. Implementation was realized using R (see Ihaka and Gentleman, 1996). The programs are available from the web page `www.math.auc.dk/~dethlef`.

The simulated image is built up by four additive parts; (a) an "egg box" function; (b) a solid disc; (c) a solid rectangle and; (d) independent, Gaussian noise. The sum of these four contributions determines the gray scale value in each pixel.

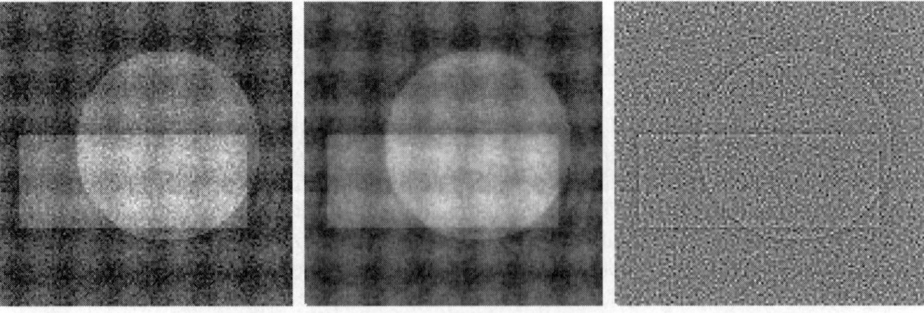

Figure 1. *To the left is shown a simulated image. The middle image shows the posterior mode found by Kalman smoothing using the Gaussian Markov random field model. The right image shows the residual image.*

Figure 1 shows a simulated image (left) of size 128×128 together with the posterior mode (middle) and residuals (right) of the smoothed image obtained by the Kalman smoother using the Gaussian Markov random field model with parameters obtained from maximum likelihood estimation.

The log likelihood is obtained from (3) without the terms resulting from the pseudo-observations. It proved useful to concentrate σ^2 out of the likelihood, leaving only the fraction τ^2/σ^2 to be estimated by the numerical maximization algorithm. For convenience, we worked with the transformed parameter $\log \tau/\sigma$, allowing the maximizer to suggest any real number as input. In all runs, we have chosen $m_{0j} = 128$ and $\boldsymbol{C}_0 = 1000 \cdot \boldsymbol{I}$.

The resulting maximum likelihood estimates were $\widehat{\tau}^2 = 338$ and $\widehat{\sigma}^2 = 117$, and the posterior mode of $\boldsymbol{\theta}$ is displayed in Figure 1 (middle). The result is very smooth and edges are blurred, as expected from the model. The residual image in Figure 1 (right) also indicates that the edges are over-smoothed.

Given a value of $\log \tau/\sigma$, the Kalman smoother took approximately 3 minutes in R on a SUN Enterprise 220R machine. Maximum likelihood estimation of this parameter took approximately 90 min in R running on a SUN Enterprise 220R machine.

Figure 2. *The posterior mode* (left) *after 50 iterations using the iterated extended Kalman smoother. The residual image is shown in the middle and to the right is shown the average of the two variances for each pixel calculated via* (11).

The restored image after 50 iterations of the iterated extended Kalman smoother from the image restoration model is seen in Figure 2 (left) along with the residual image (middle). The parameters chosen were $\sigma^2 = 60$, $\tau^2 = 50$, $k = 0.95$ and $c = 25$. These were chosen by tuning since all attempts to perform maximum likelihood estimation failed due to numeric instabilities. One run with 50 iterations in the model took approximately 5 h in R.

The image to the right in Figure 2 shows the average of the up-down and left-right variances calculated in each iteration by (11). As seen, the edges are now found, although the "egg box" function confuses the image slightly. The edges in the posterior mode of $\boldsymbol{\theta}$ are clearer than the result from the Gaussian Markov random field model. The residual image resembles white noise and indicates a good fit.

6. DISCUSSION

We provide an alternative to MCMC analysis of spatial models. For non-Gaussian state space models, the iterated extended Kalman smoother is capable of finding an approximating Gaussian state space model with the same posterior mode. This allows us to construct Markov random field models with non Gaussian increments. Then, the approximating state space model can be used as importance density as described in Durbin and Koopman (2001), to provide exact sampling of quantities of interest.

Our approach is empirical Bayes in the sense that we have estimated hyperparameters using maximum likelihood estimation. Durbin and Koopman (2001) describe a similar methodology for a full Bayesian estimation of all parameters using iterated extended Kalman smoothing, in combination with importance sampling. Their approach is, however, more computational demanding.

In the image restoration example we experience a weakness of our method. When the lattice is high-dimensional, the iterated extended Kalman smoother is slow. For this reason, we have not employed importance sampling in the example. However, the result from the approximating state space model seems very satisfactory.

We find that the methodology has a great potential and a wide range of applications. This is illustrated in Dethlefsen (2002) by examples from agricultural experiments.

ACKNOWLEDGEMENTS

I am indebted to my Ph.D. supervisor Søren Lundbye-Christensen for inspiring discussions.

REFERENCES

Besag, J. (1974). Spatial interaction and the statistical analysis of lattice systems *J. R. Statist. Soc. B* **36**, 192–236 (with discussion).

Besag, J. and Higdon, D. (1999). Bayesian analysis of agricultural field experiments *J. R. Statist. Soc. B* **61**, 691–746 (with discussion).

Carlin, B. P., Polson, N. G., and Stoffer, D. S. (1992). A Monte Carlo approach to nonnormal and nonlinear state-space modeling. *J. Am. Statist. Ass.* **87**, 493–500.

Carter, C. K. and Kohn, R. (1994). On Gibbs sampling for state space models. *Biometrika* **81**, 541–553.

Christensen, O. F. and Waagepetersen, R. (2002). Bayesian prediction of spatial count data using generalised linear mixed models. *Biometrics* **58** (to appear).

de Jong, P. and Shephard, N. (1995). The simulation smoother for time series models. *Biometrika* **82**, 339–350.

Dethlefsen, C. (2002). *Space Time Problems and Applications*. Ph.D. Thesis, Aalborg University, Denmark.

Doucet, A., Godsill, S. J. and West, M. (2000). Monte Carlo filtering and smoothing with application to time-varying spectral estimation. *Proc. IEEE International Conference on Acoustics, Speech and Signal Processing* **2**, 701–704.

Durbin, J. and Koopman, S. J. (2001). *Time Series Analysis by State Space Methods*. Oxford: Oxford University Press.

Frühwirth-Schnatter, S. (1994), Data Augmentation and Dynamic Linear Models. *J. Time Ser. Anal.* **15**, 183–202.

Harvey, A. C. (1989). *Forecasting, Structural Time Series Models and the Kalman Filter*. Cambridge: Cambridge University Press.

Ihaka, R. and Gentleman, R. (1996). R: A language for data analysis and graphics. *J. Comput. Graph. Statist.* **5**, 299–314.

Kitagawa, G. (1987). Non-gaussian state-space modeling of nonstationary time series *J. Am. Statist. Ass.* **82**, 1032–1063 (with discussion).

Knorr-Held, L. and Rue, H. (2002). On block updating in Markov random field models for disease mapping. *Scand. J. Statist.* (to appear).

Koopman, S. J. (1993). Disturbance smoother for state space models. *Biometrika* **80**, 117–126.

Lavine, M. (1999). Another look at conditionally Gaussian Markov random fields. *Bayesian Statistics 6* (J. M. Bernardo, J. O. Berger, A. P. Dawid and A. F. M. Smith, eds). Oxford: Oxford University Press, 371–387 (with discussion).

West, M. and Harrison, J. (1997). *Bayesian Forecasting and Dynamic Models*. Berlin: Springer.

West, M. and Harrison, J. and Migon, H.S. (1985). Dynamic generalized linear models and Bayesian forecasting *J. Am. Statist. Ass.* **80**, 73–97 (with discussion).

BAYESIAN STATISTICS 7, pp. 501–510
J. M. Bernardo, M. J. Bayarri, J. O. Berger, A. P. Dawid, D. Heckerman, A. F. M. Smith and M. West (Eds.)

Bayesian Estimation of the Grade of Membership Model

ELENA A. EROSHEVA
University of Washington, USA
elena@stat.washington.edu

SUMMARY

The Grade of Membership model first appeared in the context of medical diagnosis problem in the 1970s. As a latent structure model for discrete response data, it deals with individual heterogeneity by introducing a set of extreme profiles and individual membership scores for each extreme profile. Existing estimation methods for the GoM model are maximum-likelihood based. Models closely related to the GoM have recently appeared in the genetics and machine learning literatures.

We consider a Bayesian estimation approach based on the latent class representation of the GoM model. Augmenting data by the latent class indicators allows us to obtain posterior distribution of the model parameters via a Gibbs sampler, when model hyperparameters are known, or via a Metropolis–Hastings algorithm within the Gibbs, when they are unknown. We illustrate the estimation method on a subset of a disability survey data for the case of two extreme profiles.

Keywords: DATA AUGMENTATION; GIBBS SAMPLER; GOM; LATENT CLASS; METROPOLIS–HASTINGS.

1. MODEL FORMULATION

Consider settings where several categorical characteristics are observed for every individual. Latent structure models, which involve parameters for latent or non-observable quantities, can often provide valuable insights for interpreting the underlying phenomenon. This paper reexamines a special latent structure model, the *Grade of Membership* (GoM), which was originally developed in the 1970s by Max Woodbury and is fully described in a monograph by Manton *et al.* (1994).

The GoM model postulates that a population can be characterized by its *extreme profiles* that are defined by their conditional probabilities of response to observed categorical variables, analogously to latent classes. Subject-specific parameters are components of the vector of *GoM scores* that define "proportions" of the individual's membership for each of the extreme profiles. Assume discrete responses are recorded on J dichotomous items for N individuals. Here, we only consider dichotomous items, but the GoM model can deal with polytomous items as well. Let $x_i = (x_{i1}, \ldots, x_{iJ})$ denote a response pattern, where x_{ij} is a binary random variable indicating a response of individual i to item j, $i = 1, \ldots, N$, $j = 1, \ldots, J$. Suppose there are K extreme profiles (or subpopulations) in the population. Assume that each subject can be characterized by a vector of membership (GoM) scores $g_i = (g_{i1}, \ldots, g_{iK})$, one score for each extreme profile. The GoM scores are non-negative and sum to unity over extreme profiles:

$$\sum_k g_{ik} = 1, \quad i = 1, \ldots, N. \tag{1}$$

Extreme profile response probabilities, denoted by λ_{kj}, are probabilities of positive response to question j for a complete member of extreme profile k:

$$\lambda_{kj} = Pr(x_{ij} = 1 \,|\, g_{ik} = 1). \tag{2}$$

To complete the model formulation, Manton *et al.* (1994) assume that: (a) the probability of response of individual i to question j, given the GoM scores, is

$$Pr(x_{ij} = 1 \,|\, g_i) = \sum_{k=1}^{K} g_{ik} \cdot \lambda_{kj}; \tag{3}$$

(b) conditional on the values of GoM scores, the responses, x_{ij}, are independent for different values of j; (c) the responses, x_{ij}, are independent for different values of i; (d) the GoM scores, g_{ik}, are realizations of the components of a random vector with some distribution, D_α, parameterized be vector α.

Assumption (a) postulates that individual response probabilities are convex combinations of response probabilities from K extreme profiles weighted by individual's membership scores. Assumption (b) is known as the *local independence* assumption in psychometrics. Assumption (c) corresponds to individuals being randomly sampled from a population. These first three assumptions are essential for the GoM model. Assumption (d), however, has an ambiguous status in the GoM model literature and it is not used in the GoM estimation procedure described by Manton *et al.* (1994), nor it is implemented in the software package for the GoM model (DSIGoM, 1999).

In this paper, we explicitly employ assumption (d) to develop a Bayesian estimation framework for the GoM model. An equivalent way of writing the model, omitting the subject index, for $j = 1, \ldots, J$, is

$$x_j \,|\, g \sim \text{Bern}\left(\sum_{k=1}^{K} g_k \cdot \lambda_{kj} \right),$$

$$g \sim D_\alpha. \tag{4}$$

Although the standard formulation of the GoM model has hierarchical structure, full conditional distributions are intractable. Existing estimation methods that rely on structure in Eq. (4) are maximum-likelihood based (*e.g.*, Potthoff *et al.* 2000).

In the next section, we present a latent class representation of the GoM model which provides the basis for MCMC algorithms that we describe in Section 3. We consider two cases, when hyperparameters, α, are known and when they are unknown. Our goal is to estimate the response probabilities of the extreme profiles, $\{\lambda_{kj}\}$, and the hyperparameters, α, when unknown. We illustrate our method using a subset of functional disability data from the National Long Term Care Survey in Section 4.

2. LATENT CLASS REPRESENTATION

Latent variable models contain subject-specific quantities that are not observable. For discrete outcomes, these models are called latent class models when latent variables are discrete, and latent trait models when latent variables are continuous. Since membership scores in the GoM model are continuous latent variables, the GoM can be thought of as a latent trait model (Erosheva, 2001). It can be shown, however, that a set of constraints proposed by Haberman (1995) results into a latent class model under which the resulting joint distribution of the manifest variables is exactly the same as under the GoM model (Erosheva, 2002). To simplify notation, we omit the subject index in our description of this representation.

The latent class representation of the GoM model relies on J categorical latent variables $z = (z_1, z_2, \ldots, z_J)$, one for each observable discrete variable. Each latent variable z_j can take on K values from $\{1, 2, \ldots, K\}$. In this fashion, a latent vector $z \in \Omega = \{1, 2, \ldots, K\}^J$ defines a latent class. Assume that $z_1, \ldots, z_J$ are exchangeable and that the probability mass function over the latent classes is constrained to be the expected value of the J-fold product of the GoM scores:

$$\pi_z = Pr\,(z_1, z_2, \ldots, z_J) = \mathrm{E}_{D\alpha}\Big(\prod\nolimits_{j=1}^{J}\prod\nolimits_{k=1}^{K} g_k^{z_{jk}}\Big), \quad z \in \Omega = \{1, 2, \ldots, K\}^J, \tag{5}$$

where $z_{jk} = 1$, if $z_j = k$, and $z_{jk} = 0$ otherwise. If $g_1, \ldots, g_K$ random variables are nonnegative such that $\sum_k g_k = 1$, the functional constraints (Equation 5) define a probability measure on Ω (Erosheva, 2002).

To specify the conditional distribution of the manifest variables, given the latent class indicators, z, we employ two assumptions. The direct influence assumption states that the probability distribution of response x_j depends only on the realization of the jth latent variable, z_j, and is independent of other latent realizations z_l, $l \neq j$:

$$\Pr\,(x_j \,|\, z_1, \ldots, z_J) = \Pr\,(x_j \,|\, z_j)\,, \quad j = 1, \ldots, J. \tag{6}$$

The second assumption is the local independence which states that, conditional on the latent class, the responses x_j are independent.

If we denote the conditional probability $Pr\,(x_j = 1 \,|\, z_j = k)$ by λ_{kj}, $k = 1, \ldots, K$, $j = 1, \ldots, J$, then the probability distribution of the observable variables under described latent class model coincides with the distribution under the GoM model, assuming the GoM scores are nonnegative random variables that add up to one (Erosheva, 2002). Then the GoM model is a constrained latent class model where the distribution on the latent classes is constrained and is a function of the GoM scores distribution.

This latent class representation adds another level to the model hierarchy. For $j = 1, \ldots, J$,

$$\begin{aligned} x_j \,|\, z_j &\sim \mathrm{Bern}\Big(\prod\nolimits_{k=1}^{K} \lambda_{kj}^{z_{jk}}\Big), \\ z_j \,|\, g &\sim \mathrm{Mult}(1, g_1, \ldots, g_K), \\ g &\sim D_\alpha. \end{aligned} \tag{7}$$

The latent classification z_j determines the response probability for observable x_j. In this framework, the GoM scores can be thought of as another set of parameters in the model, and α as a vector of hyperparameters.

3. DATA AUGMENTATION AND MARKOV CHAIN MONTE CARLO ALGORITHMS

The latent class representation of the GoM model leads naturally to a data augmentation approach to computing the posterior distribution (Tanner, 1996): we augment the observed data with latent class indicators z_i, $i = 1, \ldots, N$. We assume the GoM scores follow a Dirichlet distribution, which is conjugate to multinomial, and construct MCMC algorithms to obtain samples from the posterior distribution.

Let $\boldsymbol{x}$ denote the matrix of observed responses x_{ij} for all subjects. Augment the observed data for each subject with realizations of the latent classification variables $z_i = (z_{i1}, \ldots, z_{iJ})$.

Denote by $\boldsymbol{z}$ the matrix of latent classifications z_{ij}. As before, let $z_{ijk} = 1$, if $z_{ij} = k$, and $z_{ijk} = 0$ otherwise. Let $\boldsymbol{\lambda}$ denote the matrix of extreme profile response probabilities.

We assume the prior on the extreme profile response probabilities $\boldsymbol{\lambda}$ is independent of the prior on the hyperparameters, α. The joint distribution for the parameters and augmented data $p(\boldsymbol{x}, \boldsymbol{z}, \boldsymbol{g}, \boldsymbol{\lambda}, \alpha)$ can be written as

$$p(\boldsymbol{\lambda})p(\alpha)\left(\prod_{i=1}^{N} \mathrm{Dir}(g_i \mid \alpha)\right)\prod_{i=1}^{N}\prod_{j=1}^{J}\prod_{k=1}^{K}\left(g_{ik}\lambda_{kj}^{x_{ij}}(1-\lambda_{kj})^{1-x_{ij}}\right)^{z_{ijk}}, \tag{8}$$

where

$$\mathrm{Dir}(g_i \mid \alpha) = \frac{\Gamma(\sum_k \alpha_k)}{\Gamma(\alpha_1)...\Gamma(\alpha_K)} g_{i1}^{\alpha_1 - 1} \dots g_{iK}^{\alpha_K - 1}.$$

We further assume that the prior distribution of extreme profile response probabilities treats items and extreme profiles as independent

$$p(\boldsymbol{\lambda}) = \prod_{k=1}^{K}\prod_{j=1}^{J} p(\lambda_{kj}). \tag{9}$$

We take $p(\lambda_{kj})$ to be Beta(η_1, η_2), and for simplicity in what follows we use $\eta_1 = \eta_2 = 1$.

3.1. *Known Dirichlet Distribution of the GoM Scores*

Suppose the hyperparameters α are known, a Gibbs sampler can then be constructed by using complete conditional distributions

$$p(z_i \mid \dots) \propto \prod_{j=1}^{J}\prod_{k=1}^{K}\left(g_{ik}\lambda_{kj}^{x_{ij}}(1-\lambda_{kj})^{1-x_{ij}}\right)^{z_{ijk}}, \tag{10}$$

$$p(\lambda_{kj} \mid \dots) \propto p(\lambda_{kj}) \cdot \prod_{i=1}^{N}\left(\lambda_{kj}^{x_{ij}}(1-\lambda_{kj})^{1-x_{ij}}\right), \tag{11}$$

$$p(g_i \mid \dots) \propto \mathrm{Dir}(g_i \mid \alpha) \cdot \prod_{j=1}^{J}\prod_{k=1}^{K} g_{ik}^{z_{ijk}}. \tag{12}$$

where ... stands for all other variables from $\boldsymbol{x}, \boldsymbol{z}, \boldsymbol{g}, \boldsymbol{\lambda}$. We give graphical representation of the GoM model in Figure 1.

One simulation draw of the Gibbs sampler consists of an *imputation step* and a *posterior step* (Tanner, 1996). During the $(m+1)$-th imputation step, a realization of the latent classifications $z_i^{(m+1)}$ is obtained for every person from the conditional predictive distribution. During the $(m+1)$-th posterior step, realizations of the parameters $\lambda_{kj}^{(m+1)}$ and $g_{ik}^{(m+1)}$ are obtained from the augmented posterior distribution.

- *Imputation step*: Sample $z_{ij},\ i = 1, \dots, N,\ j = 1, \dots, J$:

$$z_{ij}^{(m+1)} \sim Mult\,(1, p_1, \dots, p_K),\ \ p_k = g_{ik}\lambda_{kj}^{x_{ij}}(1-\lambda_{kj})^{1-x_{ij}}. \tag{13}$$

- *Posterior step*: Sample $\lambda_{kj},\ k = 1, \dots, K,\ j = 1, \dots, J$:

$$\lambda_{kj}^{(m+1)} \sim Beta\left(1 + \sum_{i=1}^{N} x_{ij}z_{ijk}, 1 + \sum_{i=1}^{N}(z_{ijk} - x_{ij}z_{ijk})\right) \tag{14}$$

Sample the GoM scores g_i for each individual $i = 1, \ldots, I$:

$$g_i^{(m+1)} \sim Dir_K\left(\alpha_1 + \sum_{j=1}^{J} z_{ij1}, \ldots, \alpha_K + \sum_{j=1}^{J} z_{ijK}\right). \qquad (15)$$

When the distribution of the GoM scores is known, we can use BUGS to obtain a posterior distribution of the model parameters.

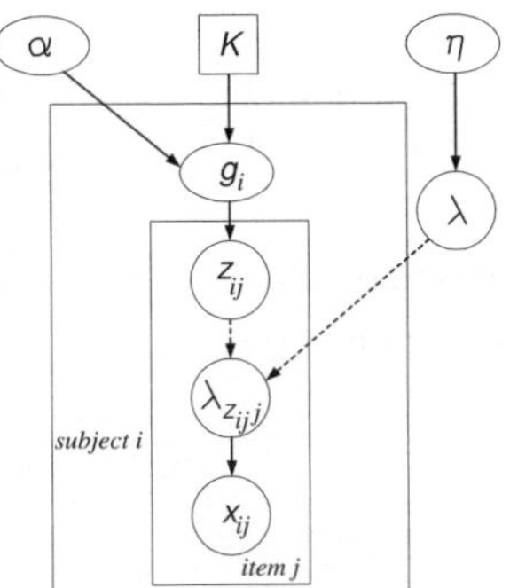

Figure 1. *GoM graphical diagram.*

3.2. *Unknown Dirichlet Distribution of the GoM Scores*

If hyperparameter vector α is unknown, samples from its posterior distribution can be obtained via a Metropolis–Hastings step within the Gibbs sampler (Tierney, 1994).

Consider reparameterizing $\alpha = (\alpha_1, \ldots, \alpha_K)$ with $\xi = (\xi_1, \ldots, \xi_K)$ and α_0, where $\alpha_0 = \sum_k \alpha_k$ and $\xi_k = \alpha_k/\alpha_0$. In this fashion, ξ_k represents the "proportion" of the population that belongs to the kth extreme profile, and α governs the "spread" of the distribution within the convex set determined by the extreme profiles. The closer α_0 is to zero, the more probability is concentrated near the extreme profiles; similarly, for large α_0, more probability is concentrated near the population average. This reparameterization seems reasonable because one can imagine that there exists a prior information about one group of the parameters but not about another.

We also assume that α_0 and ξ are independent since they govern two unrelated qualities of the distribution of the GoM scores, extreme profile proportions and a shape of the distribution. As before, the GoM scores are independent of the structural parameters. Thus, the prior distributions of ξ and α are independent of the prior distribution on $\boldsymbol{\lambda}$. The joint distribution of the parameters and augmented data is

$$p(\boldsymbol{\lambda})p(\alpha_0)p(\xi)\left(\prod_{i=1}^{N} \mathrm{Dir}(g_i \mid \alpha)\right)\prod_{i=1}^{N}\prod_{j=1}^{J}\prod_{k=1}^{K}\left(g_{ik}\lambda_{kj}^{x_{ij}}(1-\lambda_{kj})^{1-x_{ij}}\right)^{z_{ijk}}. \qquad (16)$$

Assuming a proper diffuse Gamma(τ, β) prior for α_0 with shape parameter τ and inverse scale parameter β, we obtain that the full conditional distribution for α_0 is

$$p(\alpha_0 \mid \ldots) \propto \alpha_0^{\tau-1}\exp\left[-\left(\beta - \sum_{k=1}^{K}\xi_k\sum_{i=1}^{N}\log g_{ik}\right)\alpha_0\right]\left[\frac{\Gamma(\alpha_0)}{\Gamma(\xi_1\alpha_0)\ldots\Gamma(\xi_K\alpha_0)}\right]^N, \qquad (17)$$

where ... on the left-hand side stands for all other variables.

We take a proposal distribution for the next draw of α_0, $p(\alpha_0^* \mid \alpha_0^{(m)})$, to be gamma with expected value set at the value of the last draw, $\alpha_0^{(m)}$, and the shape parameter set at a convenient constant $\gamma > 1$. Here, γ is the tuning parameter for the Metropolis–Hastings step. The inverse scale parameter for the proposal distribution is then $\gamma/\alpha_0^{(m)}$. In order to obtain $(m+1)$-th draw of α_0, the Metropolis–Hastings algorithm requires:

(1) draw a candidate point α_0^* from $p(\alpha_0^* \mid \alpha_0^{(m)})$;

(2) calculate the proposal ratio

$$r_{\alpha_0} = \frac{p(\alpha_0^* \mid \ldots) p(\alpha_0^{(m)} \mid \alpha_0^*)}{p(\alpha_0^{(m)} \mid \ldots) p(\alpha_0^* \mid \alpha_0^{(m)})};$$

(3) assign $\alpha_0^{(m+1)} = \alpha_0^*$ with probability $\min\{1, r_{\alpha_0}\}$, otherwise assign $\alpha_0^{(m+1)} = \alpha_0^{(m)}$.

The proposal ratio for the $(m+1)$-th draw of α_0 is

$$r_{\alpha_0} = r_{\alpha_0}(M) \cdot r_{\alpha_0}(H),$$

where

$$r_{\alpha_0}(M) = \left(\frac{\alpha_0^*}{\alpha_0^{(m)}}\right)^{\tau-1} \exp\left[-\left(\beta - \sum_{k=1}^{K} \xi_k \sum_{i=1}^{N} \log g_{ik}\right)(\alpha_0^* - \alpha_0^{(m)})\right] \times \left[\frac{\Gamma(\alpha_0^*)\Gamma(\xi_1 \alpha_0^{(m)}) \ldots \Gamma(\xi_K \alpha_0^{(m)})}{\Gamma(\alpha_0^{(m)})\Gamma(\xi_1 \alpha_0^*) \ldots \Gamma(\xi_K \alpha_0^*)}\right]^N;$$

$$r_{\alpha_0}(H) = \left(\frac{\alpha_0^{(m)}}{\alpha_0^*}\right)^{2\gamma-1} \exp\left[-\gamma(\alpha_0^{(m)}/\alpha_0^* - \alpha_0^*/\alpha_0^{(m)})\right].$$

Here, $r_{\alpha_0}(M)$ is the likelihood component of the proposal ratio and $r_{\alpha_0}(H)$ is the component of the proposal ratio that accounts for nonsymmetric proposal distribution.

In the absence of strong prior opinion about population proportions ξ of the extreme profiles, we take prior distribution $p(\xi)$ being a uniform on the simplex, $\text{Dir}(1, \ldots, 1)$. Then, the full conditional distribution for ξ, up to a constant of proportionality, is:

$$p(\xi \mid \ldots) \propto \exp\left[\alpha_0 \sum_{k=1}^{K} \xi_k \sum_{i=1}^{N} \log g_{ik}\right] \left[\frac{\Gamma(\alpha_0)}{\Gamma(\xi_1 \alpha_0) \ldots \Gamma(\xi_K \alpha_0)}\right]^N \tag{18}$$

where ... on the left-hand side stands for all other variables.

We chose the proposal distribution for ξ be centered at the previous draw and have reasonably small variance for each component. Thus, $\text{Dir}(\xi^* \mid \eta K \xi_1^{(m)}, \ldots, \eta K \xi_K^{(m)})$ sets the variance of the kth component to be $\xi_k^{(m)}(1 - \xi_k^{(m)})/(\eta K + 1)$. Denote the proposal distribution by $p(\xi^* \mid \xi^{(m)})$. The Metropolis–Hastings algorithm has three steps:

(1) draw a candidate point ξ^* from $p(\xi^* \mid \xi^{(m)})$;

(2) calculate the proposal ratio

$$r_\xi = \frac{p(\xi^* \mid \ldots) p(\xi^{(m)} \mid \xi^*)}{p(\xi^{(m)} \mid \ldots) p(\xi^* \mid \xi^{(m)})};$$

(3) assign $\xi^{(m+1)} = \xi^*$ with probability $\min\{1, r_\xi\}$, otherwise assign $\xi^{(m+1)} = \xi^{(m)}$.

The proposal ratio for ξ is

$$r_\xi = \exp\left[\alpha_0 \sum_{k=1}^{K}\sum_{i=1}^{N} \log g_{ik}(\xi_k^* - \xi_k^{(m)})\right]\left[\frac{\Gamma(\xi_1^{(m)}\alpha_0)\ldots\Gamma(\xi_K^{(m)}\alpha_0)}{\Gamma(\xi_1^*\alpha_0)\ldots\Gamma(\xi_K^*\alpha_0)}\right]^N$$
$$\times\ \frac{\Gamma(\eta K\xi_1^{(m)})\ldots\Gamma(\eta K\xi_K^{(m)})}{\Gamma(\eta K\xi_1^*)\ldots\Gamma(\eta K\xi_K^*)}\cdot\frac{(\xi_1^{(m)})^{\xi^*-1}\ldots(\xi_K^{(m)})^{\xi^*-1}}{(\xi_1^*)^{\xi^{(m)}-1}\ldots(\xi_K^*)^{\xi^{(m)}-1}},$$

where η is a tuning parameter which can be set to a suitable constant.

4. ILLUSTRATION

Contingency tables constructed from categorical data on disability in the elderly people often contain only a few very large cell counts but many small counts including counts of one. We illustrate the Bayesian approach to the GoM model on a subset of 16 binary functional disability measures from the analytic file of the National Long Term Care Survey (NLTCS). The subset consists of six measures of *activities of daily living* (ADLs), which include basic activities of hygiene and personal care, and 10 measures of *instrumental activities of daily living* (IADLs), which include basic activities necessary to reside in the community (see Table 1). For every ADL/IADL measure, subjects are classified as being either disabled or healthy on that measure.

We consider the simplest case of the GoM model with two extreme profiles ($K = 2$) for the 16 ADL/IADL measures, pooled across four survey waves, 1982, 1984, 1989, and 1994, with the total sample size of 21,574. Out of possible $2^{16} = 65,536$ response patterns, 3152 distinct responses occurred in the sample. Roughly 55% of the cell counts equal one, 9% are within 5-9, 5% are within 10–19, and only 4% contain counts 20 and above.

To obtain starting values, we fitted the two-class latent class model to our data. We used BUGS (Spiegelhalter *et al.*, 1996) software to obtain posterior distribution of the latent class model parameters, and CODA (Best *et al.*, 1996) to assess convergence. In particular, we examined the Geweke and Heidelberger and Welch diagnostics, trace plots and summary statistics generated by CODA.

We placed a uniform prior on the parameters of the latent class model. We ran the BUGS code for 4500 iterations, and then discarded the first 1000 as a burn-in. Convergence diagnostics indicated that the chain converged. Posterior means (standard deviations) of the latent class probabilities are

$$\Pr(\text{1st class}) = 0.345(1\text{e} - 04), \qquad \Pr(\text{2nd class}) = 0.655(1\text{e} - 04).$$

We assumed the distribution of the GoM scores is unknown. We used the latent class response probabilities as starting values for extreme profile response probabilities, and posterior probabilities of belonging to classes as starting values for the GoM scores. We placed a uniform prior on the hyperparameters (ξ_1, ξ_2) and a Beta$(2, 2)$ prior on α_0. Tuning parameters were set at $\gamma = 50,\ \eta = 10$. A chain of 90,000 samples with thinning parameter $q = 10$ gave perfect univariate convergence diagnostics for all model parameters, after the first 10,000 samples were discarded as a burn-in. The (joint) log-likelihood convergence diagnostics indicated overall convergence of the multivariate posterior distribution. We examined the plots of successive iterations, and found that the label-switching problem is not encountered in this analysis because the two profiles are very well separated on each item. Posterior means (standard deviations) of the hyperparameters are

$$\alpha_0 = 0.521(2e - 02), \quad \xi_1 = 0.433(4e - 02), \quad \xi_2 = 0.567(4e - 02).$$

Table 1. *Posterior mean (standard deviation) of conditional response probabilities for the two-class latent class model and for the two-profile GoM model. The ADL items are:* (1) *eating,* (2) *getting in/out of bed,* (3) *getting around inside,* (4) *dressing,* (5) *bathing,* (6) *using toilet. The IADL items are:* (7) *doing heavy house work,* (8) *doing light house work,* (9) *doing laundry,* (10) *cooking,* (11) *grocery shopping,* (12) *getting about outside,* (13) *traveling,* (14) *managing money,* (15) *taking medicine,* (16) *telephoning.*

j	Class $k = 1$	Class $k = 2$	Profile $k = 1$	Profile $k = 2$
1	0.298 (2e-04)	0.005 (2e-05)	0.319 (7e-03)	0.000 (9e-05)
2	0.656 (2e-04)	0.075 (7e-05)	0.817 (1e-02)	0.000 (5e-04)
3	0.801 (2e-04)	0.193 (1e-04)	0.962 (6e-03)	0.088 (5e-03)
4	0.542 (2e-04)	0.031 (5e-05)	0.641 (1e-02)	0.000 (3e-04)
5	0.847 (1e-04)	0.223 (1e-04)	0.993 (4e-03)	0.128 (5e-03)
6	0.589 (2e-04)	0.068 (7e-05)	0.722 (9e-03)	0.004 (2e-03)
7	0.983 (5e-05)	0.513 (1e-04)	1.000 (2e-04)	0.475 (6e-03)
8	0.594 (2e-04)	0.018 (4e-05)	0.687 (1e-02)	0.000 (1e-04)
9	0.832 (2e-04)	0.103 (9e-05)	0.982 (6e-03)	0.015 (3e-03)
10	0.692 (2e-04)	0.030 (5e-05)	0.812 (1e-02)	0.000 (1e-04)
11	0.925 (1e-04)	0.253 (1e-04)	1.000 (3e-04)	0.161 (6e-03)
12	0.885 (1e-04)	0.380 (1e-04)	0.988 (3e-03)	0.297 (6e-03)
13	0.848 (1e-04)	0.306 (1e-04)	0.949 (4e-03)	0.222 (6e-03)
14	0.522 (2e-04)	0.075 (8e-05)	0.625 (8e-03)	0.018 (3e-03)
15	0.504 (2e-04)	0.056 (6e-05)	0.598 (9e-03)	0.012 (2e-03)
16	0.349 (2e-04)	0.039 (5e-05)	0.404 (7e-03)	0.011 (2e-03)

Conditional response probabilities from both the latent class model and the GoM model (Table 1) suggest that the two classes can be interpreted as "disabled" $(k = 1)$ and "healthy" $(k = 2)$. Most of the conditional response probabilities in the GoM model are closer to the extremes, compared with those in the latent class model. The estimate $\alpha_0 = 0.521$ indicates that the posterior distribution of the GoM scores is bathtub-shaped. Estimated Dirichlet proportions, $\xi_1 = 0.433$ and $\xi_2 = 0.567$, are the cumulative proportions of item responses that elderly make as "disabled" or "healthy". These are closer to 0.5, compared to the "healthy" and "disabled" latent class proportions from the latent class model.

Additional analysis, not presented in this paper, indicated that the two profile GoM model does not provide a sufficiently good fit to the data. Hence, to draw substantive conclusions from these data, we need to examine a number of models with $K > 2$. For further details on the analysis of the NLTCS data, see Erosheva (2002).

5. CONCLUDING REMARKS

The Bayesian approach to the GoM model presented here is based on the assumption that the membership scores are Dirichlet latent variables. Thus, in contrast to traditional GoM estimation methods, which treat the membership scores as fixed unknown individual-specific quantities, our approach allows for population inference. A somewhat related approach by Potthoff *et al.* (2000) also assumes that the membership scores are distributed Dirichlet but it is maximum-likelihood based, and it has been implemented only in low-dimensional special cases with $J \leq 4$, and requires substantial increases in human and computer time for higher dimensions (of J and K). Our Bayesian approach is not limited to specific values of J and K, and, for higher dimensions, requires only computer time increases and adjustments of tuning parameters. Potential problems in high dimensions may include slow convergence, low mixing of MCMC chains, and the problem of label switching, none of which were encountered in our illustration example.

It seems important to point out that the GoM model deals with correlated categorical outcomes. However, instead of modelling correlation explicitly, it does so by introducing latent variables, the membership scores. The assumption is that latent membership scores account for interdependence among observed categorical variables.

Analogous to factor analysis, the GoM model provides a method to analyze interrelationship among observed categorical variables. To analyze dependence of observed categorical variables on a number of covariates, Chib and Greenberg (1998) use multivariate probit models. They use the underlying latent variable approach to obtain simulation of the posterior distribution. The underlying latent variable approach to factor analysis for discrete variables is described in Bartholomew and Knott (1999). In essence, the underlying variable approach assumes that categorical outcomes are incomplete observations of underlying continuous latent variables. The simulation approach presented in this paper does not make such an assumption. For some connections of the GoM model with latent structure models studied in statistical literature, such as factor analysis and the Rasch model, see Erosheva (2002).

Recently, new statistical models in genetics and in machine learning have been published that are remarkably similar to the GoM model. For example, Pritchard *et al.* (2000) developed a clustering model with admixture similar in formulation to the latent class representation of the GoM model, for applications to multilocus genotype data. They presented results obtained from an MCMC algorithm with a fixed Dirichlet distribution on the admixture proportions (these are the membership scores in our terminology).

Developed to study the composition of documents in machine learning, Hofmann's (2001) Probabilistic Latent Semantic Analysis and Blei *et al.* (2001) Latent Dirichlet Allocation models are also similar to different variations of the GoM model. Noting that their model is intractable for exact probabilistic inference, Blei *et al.* (2002) employ a fast variational method to obtain an approximate posterior for the model parameters.

All of these models share the idea of soft classification. They represent individuals as having partial membership in several subpopulations and employ the same conditional probability structure as the GoM model, but they differ in their sampling schemes underlying the data generation process. Understanding connections among these models will allow us to borrow approaches and results across the different literatures.

ACKNOWLEDGEMENTS

This work is based on the author's dissertation research in the Department of Statistics, Carnegie Mellon University and was supported in part by Grant No. 1R03 AG18986-01 from the National Institute on Aging. The author is grateful to Stephen Fienberg, Brian Junker, and Tom Minka for helpful discussions.

REFERENCES

Bartholomew, D. J., and Knott, M. (1999). *Latent Variable Models and Factor Analysis*, 2nd edn. London: Edward Arnold, 214.

Best, N., Cowles, M.K., and Vines, K. (1996). CODA: *Convergence Diagnosis and Output Analysis Software for Gibbs Sampling Output* (version 0.30). Cambridge, UK: MRC.

Blei, D. M., Ng, A. Y., and Jordan, M. I. (2001). Latent Dirichlet Allocation. *Advances in Neural Information Processing Systems* **14**.

Blei, D. M., Jordan, M. I. and Ng, A. Y. (2002). Latent Dirichlet models for applications in information retrieval (this volume).

Chib, S. and Greenberg, E. (1998). Analysis of multivariate probit models. *Biometrika* **85**, 347–361.

DSIGoM (1999). *User Documentation for DSIGoM* (version 1.0). Decision Systems, Inc.

Erosheva, E. A. (2001). Comparing latent structures of the grade of membership, Rasch and latent class models. *Tech. Rep.*, University of Washington, USA.

Erosheva, E. A. (2002). *Grade of Membership and Latent Structure Models with Application to Disability Survey Data.* Ph.D. Thesis, Carnegie Mellon University, USA.

Haberman, S. J. (1995). Review of *Statistical Applications Using Fuzzy Sets*, by K. G. Manton etal *J. Am. Statist. Ass.* **158**, 1131–1133.

Hofmann, T. (2001). Unsupervised learning by probabilistic latent semantic analyis. *Machine Learning* **42**, 177–196.

Manton, K. G., Woodbury, M. A., and Tolley, H. D. (1994). *Statistical Applications Using Fuzzy Sets*. New York: Wiley.

Pritchard, J. K., Stephens, M. and Donnelly, P. (2000). Inference of population structure using multilocus genotype data. *Genetics* **155**, 945–959.

Potthoff, R. G., Manton, K. G., Woodbury, M. A., and Tolley, H. D. (2000). Dirichlet generalizations of latent-class models. *J. Classification* **17**, 315–353.

Tanner, M. A. (1996). *Tools for Statistical Inference. Methods for the Exploration of Posterior Distributions and Likelihood Functions*, 3rd edn. Berlin: Springer.

Tierney, L. (1994). Markov chains for exploring posterior distributions. *Ann. Statist.* **22**, 1701–1728.

Spiegelhalter, D., Thomas, A., Best, N. and Gilks, W. (1996). *BUGS 0.5: Bayesian Inference Using Gibbs Sampling Manual*. Cambridge, UK: MRC.

BAYESIAN STATISTICS 7, pp. 511–518
J. M. Bernardo, M. J. Bayarri, J. O. Berger, A. P. Dawid,
D. Heckerman, A. F. M. Smith and M. West (Eds.)

A Variant Version of the Pólya–Eggenberger Urn Model

LUIS G. ESTEVES SERGIO WECHSLER
Universidade de São Paulo, Brazil
lesteves@ime.usp.br sw@ime.usp.br

PILAR L. IGLESIAS
Pontificia Universidad Católica de Chile, Chile
pliz@mat.puc.cl

ANTÔNIO L. PEREIRA
Universidade de São Paulo, Brazil
alpereir@ime.usp.br

SUMMARY

We present a new version of Pólya urn scheme via the introduction of a prior probability distribution for the initial composition, that is, for the initial numbers of black and white balls in the urn. We also determine conditions for an exchangeable process taking values on $\{0,1\}^\infty$ to be well approximated by a suitable Pólya process with unknown initial composition. The predictivistic features in such processes are also discussed.

Keywords: DE FINETTI'S REPRESENTATION THEOREM; EXCHANGEABLE PROCESSES; MIXTURES; PARAMETERS; PÓLYA URN SCHEME; URN MODELS.

1. PÓLYA–EGGENBERGER URN MODELS

Pólya–Eggenberger urn schemes are central among models developed to represent forms of contagion not only for their pioneering aspect but also for their generality and mathematical properties. We begin by reviewing the Pólya–Eggenberger urn model.

Let us consider an urn containing initially a white balls and b black balls. Pólya and Eggenberger (1923) described the following procedure of successive draws of balls from the urn: a ball is equiprobably drawn from the urn and it is then returned to the urn together with c additional balls of its color (the particular case $c = 1$ was first studied by Markov, 1906). According to this sequential scheme, drawing of a ball of a specific color in a certain stage of the process increases the probability of this color in the next stage (Feller, 1957, named this phenomenon "aftereffect"). This characteristic of the process made Pólya consider it a model for specific forms of "contagion." Formally, let $\{X_n\}_{n\geq 1}$ be a stochastic process taking values on $\{0,1\}^\infty$ with probability measure $\mathbb{P}$ defined by $\forall n \in \mathrm{N}$, $\forall (x_1,\ldots,x_n) \in \{0,1\}^n$,

$$\mathbb{P}(X_1 = x_1,\ldots,X_n = x_n) = \frac{\Gamma\left(\frac{a}{c}+t_n\right)\Gamma\left(\frac{b}{c}+n-t_n\right)\Gamma\left(\frac{a+b}{c}\right)}{\Gamma\left(\frac{a}{c}\right)\Gamma\left(\frac{b}{c}\right)\Gamma\left(\frac{a+b}{c}+n\right)}, \tag{1.1}$$

where $t_n = \sum_{i=1}^{n} x_i$, $\Gamma(\cdot)$ is the gamma function, and $a, b, c \in \mathrm{N}$. The process $\{X_n\}_{n \geq 1}$ describes the evolution of a Pólya–Eggenberger urn model with initial configuration (a, b), that is, with a white and b black balls initially in the urn, and with c balls being added to the urn at each stage. X_n denotes the indicator of a white ball being drawn in the n-th stage. It is easily seen that a Pólya–Eggenberger process is exchangeable and, therefore, satisfies De Finetti's Representation Theorem with De Finetti's measure given by a Beta density function with parameters a/c and b/c.

Generalizations of Pólya–Eggenberger urn model have been developed in various directions. Among these alternatives, we mention: (a) replacement not only of c additional balls of the same color of the drawn ball but also of d balls of the opposite color; (b) replacement of balls of both colors, as in (a), depending on the stage of the process and on the color of the last ball drawn and (c) replacement of random numbers C and D of balls to the urn. In each of the aforementioned variations the resulting process is not exchangeable in general. Variants of Pólya–Eggenberger urn models considering more than two colors are also contemplated in the literature (for a detailed review of all such variants see Johnson and Kotz, 1977). We now present another version of the Pólya–Eggenberger urn model.

Let us again consider an urn containing black and white balls. Let us suppose, however, that the numbers of white and black balls, A and B, respectively, are now unknown. It is then natural, under the subjectivistic standpoint, to consider a probability measure on $(\mathrm{N}^2, \mathcal{P}(\mathrm{N}^2))$ which quantifies personal uncertainty about the initial composition of the urn. As to the process evolution, we suppose that, in each stage, a ball is uniformly chosen from the urn and then returned to it together with another ball of the same color (Pólya–Eggenberger urn model with $c = 1$). We call this alternative process a Pólya–Eggenberger urn model with unknown (or random) initial composition. In this case, with $\{((a_k, b_k), p_k)\}_{k \in \mathrm{N}}$, $(a_k, b_k) \in \mathrm{N}^2$, $p_k \geq 0$, $k \in \mathrm{N}$, and $\sum_{k \in \mathrm{N}} p_k = 1$ defining the prior distribution for the initial composition of the urn, we obtain, $\forall n \in \mathrm{N}$, $\forall (x_1, \ldots, x_n) \in \{0,1\}^n$,

$$\mathbb{P}(X_1 = x_1, \ldots, X_n = x_n) = \sum_{k=1}^{\infty} p_k \cdot \text{Pólya}\left((x_1, \ldots, x_n) \,|\, (a_k, b_k)\right),$$

where

$$\text{Pólya}\left((x_1, \ldots, x_n) \,|\, (a, b)\right) = \frac{\Gamma\left(a + \sum_{i=1}^{n} x_i\right) \Gamma\left(b + n - \sum_{i=1}^{n} x_i\right) \Gamma(a+b)}{\Gamma(a)\Gamma(b)\Gamma(a+b+n)}.$$

More general probability measures on $(\mathbb{R}_+^2, \mathcal{B}(\mathbb{R}_+^2))$ could have been included in the family of distributions above. Nevertheless, as such probability measures should incorporate uncertainty about the pair (A, B), which takes values only on N^2, we did not allow for this possibility. We present some properties of Pólya–Eggenberger urn models with unknown initial composition in the sequel.

Proposition 1.1. *For every probability measure on* $(\mathrm{N}^2, \mathcal{P}(\mathrm{N}^2))$ *for the pair* (A, B), *the resulting Pólya–Eggenberger urn model with unknown initial composition is exchangeable.*

We omit the easy proof. As the usual variations of the Pólya–Eggenberger model do *not* yield exchangeable processes, the random initial composition version does not correspond to any of them. De Finetti's measure of the Pólya–Eggenberger model with random initial composition is now obtained.

Proposition 1.2. *Let $\{X_n\}_{n\geq 1}$ be the sequence of indicators in a Pólya–Eggenberger urn model with unknown initial composition and prior $\{((a_k, b_k); p_k)\}_{k\in \mathrm{N}}$, $(a_k, b_k) \in \mathrm{N}^2$, $p_k \geq 0$, $k \in \mathrm{N}$, with $\sum_{k\in\mathrm{N}} p_k = 1$. The De Finetti measure of this process is then given by the probability density function*

$$g(\theta) = \sum_{k\in\mathrm{N}} p_k \mathrm{Beta}\,(\theta; (a_k, b_k)), \tag{1.2}$$

where

$$\mathrm{Beta}\,(\theta; (a, b)) = \frac{\Gamma(a+b)}{\Gamma(a)\Gamma(b)}\theta^{a-1}(1-\theta)^{b-1}\mathrm{I\!I}_{[0,1]}(\theta).$$

2. MAIN RESULT

We will now establish conditions for a $\{0,1\}^\infty$-valued exchangeable process to be approximated, in total variation distance, by a suitable Pólya–Eggenberger urn model with unknown initial composition. Two lemmas which play a fundamental rôle in the proof of our main result are now reviewed.

Lemma 2.1. *(S. Bernstein). Let $f : [0,1] \to \mathrm{I\!R}$ be a continuous function. Then, the sequence of Bernstein polynomials*

$$P_n(x) = \sum_{k=0}^{n} \binom{n}{k} f\left(\frac{k}{n}\right) x^k (1-x)^{n-k}, \quad n = 1, 2, \ldots$$

converges uniformly to f as $n \to \infty$.

Marsden (1974) presents a proof of Lemma 2.1 based essentially on simple Probability Calculus and argues that consideration of a scenario involving a game with prizes provides an intuitive explanation of it based on the Law of Large Numbers.

Lemma 2.2. *Let $\{\alpha_n\}_{n\geq 1}$, $\alpha_n \in [0,1]$, be a sequence of numbers and $\psi(t)$ a distribution function with $\psi(0) = 0$ and $\psi(1) = 1$ such that*

$$\alpha_n = \int_0^1 t^n \, d\psi(t), \;\; \forall\, n \in \mathrm{N}. \tag{2.1}$$

Then, $\psi(t) = \int_0^t \varphi(u)\, du$, with $\varphi \geq 0$ bounded, if, and only if, there is $M > 0$ such that, $\forall n \in \mathrm{N}$,

$$\max_{v\in\{0,\ldots,n\}} \left\{ \binom{n}{v} \int_0^1 t^v (1-t)^{n-v} \, d\psi(t) \right\} \leq \frac{M}{n+1}.$$

Lemma 2.2 establishes sufficient conditions for Hausdorff's problem of moments to admit an absolutely continuous solution possessing bounded density function. The proof of this lemma is found in Shohat and Tamarkin (1943). We finally state our main result.

Theorem 2.1. *Let $\{X_n\}_{n\geq 1}$ be an exchangeable process taking values on $\{0,1\}^\infty$. Let us suppose that there is $M > 0$ such that, $\forall n \in \mathrm{N}$,*

$$\max_{v\in\{0,\dots,n\}} P\Big(\sum_{i=1}^{n} X_i = v\Big) \leq \frac{M}{n+1} \cdot$$

We will then have, for every $\varepsilon > 0$, a natural number $m_0 = m_0(\varepsilon)$ and a prior $\{p_0,\dots,p_{m_0}\}$ with $p_k \geq 0$, $k = 0,\dots,m_0$, and $\sum_{k=0}^{m_0} p_k = 1$, such that $\forall n \in \mathrm{N}$ and $\forall(x_1,\dots,x_n) \in \{0,1\}^n$,

$$\Big|P(x_1,\dots,x_n) - \sum_{k=0}^{m_0} p_k \cdot \textit{Pólya}\,((x_1,\dots,x_n)\,|\,(k+1, m_0-k+1))\Big| < \varepsilon, \qquad (2.2)$$

where P denotes the probability measure of the process $\{X_n\}_{n\geq 1}$.

Proof. As $\{X_n\}_{n\geq 1}$ is an exchangeable process taking values on $\{0,1\}^\infty$, De Finetti's Representation Theorem yields the existence of a distribution function ψ such that, $\forall n \in \mathrm{N}$, $\forall(x_1,\dots,x_n) \in \{0,1\}^n$,

$$P(x_1,\dots,x_n) = \int_0^1 \theta^{\sum_{i=1}^n x_i}(1-\theta)^{n-\sum_{i=1}^n x_i}\, d\psi(\theta).$$

In particular, for the vector $1_n = (1,\dots,1)_{1\mathrm{x}n}$, we have

$$\underbrace{P(1_n)}_{\alpha_n} = \int_0^1 \theta^n \, d\psi(\theta), \quad \forall n \in \mathrm{N}.$$

Since, by hypothesis, there is $M > 0$ such that

$$\max_{v\in\{0,\dots,n\}} P\left(\sum_{i=1}^{n} X_i = v\right) = \max_{v\in\{0,\dots,n\}} \binom{n}{v} \int_0^1 \theta^v (1-\theta)^{n-v}\, d\psi \leq \frac{M}{n+1}, \quad \forall n \in \mathrm{N},$$

it follows, by Lemma 2.2, that ψ is absolutely continuous and that $\exists\, \varphi : [0,1] \to \mathrm{I\!R}_+$, bounded, such that

$$\psi(t) = \varphi(t)\, dt\,.$$

In this case, there is a sequence of nonnegative simple functions $\{\varphi_n\}_{n\geq 1}$, such that $\varphi_n \xrightarrow{u} \varphi$. Thus, $\forall \varepsilon > 0$, $\exists\, \varphi_{n_0} \geq 0$, simple, such that

$$\int_0^1 |\varphi - \varphi_{n_0}|\, d\lambda_1 < \frac{\varepsilon}{4}\,, \qquad (2.3)$$

where λ_1 denotes the Lebesgue measure on $(\mathrm{I\!R}, \mathcal{B}(\mathrm{I\!R}))$. Also, there are: a step function $S : [0,1] \to \mathrm{I\!R}_+$ such that

$$\int_0^1 |\varphi_{n_0} - S|\, d\lambda_1 < \frac{\varepsilon}{4}\,, \qquad (2.4)$$

a bounded continuous function $C : [0,1] \to \mathrm{I\!R}_+$ such that

$$\int_0^1 |S - C|\, d\lambda_1 < \frac{\varepsilon}{4}\,, \qquad (2.5)$$

and a sequence of Bernstein polynomials $\{B_m\}_{m\geq 1}$ such that $B_m \xrightarrow{u} C$. Let us then evaluate

$$\Big|P(x_1,\ldots,x_n) - \sum_{k=0}^{m} p_k^{(m)} \cdot \text{Pólya}\,((x_1,\ldots,x_n)\,|\,(k+1, m-k+1))\Big|,$$

with

$$p_k^{(m)} = \frac{C(\frac{k}{m})}{\sum_{j=0}^{m} C(\frac{j}{m})}, \qquad k = 0,\ldots,m.$$

We have

$$\begin{aligned}
&\Big|P(x_1,\ldots,x_n) - \sum_{k=0}^{m} p_k^{(m)} \cdot \text{Pólya}\,((x_1,\ldots,x_n)\,|\,(k+1, m-k+1))\Big| \\
&= \Big|\int_0^1 \theta^{\sum_{i=1}^n x_i}(1-\theta)^{n-\sum_{i=1}^n x_i}\varphi(\theta)\,d\theta \\
&\quad - \sum_{k=0}^{m} p_k^{(m)} \cdot \int_0^1 \theta^{\sum_{i=1}^n x_i}(1-\theta)^{n-\sum_{i=1}^n x_i}\,\text{Beta}\,(\theta;(k+1,m-k+1))\,d\theta\,\Big|,
\end{aligned}$$

where the last equality follows from De Finetti's Representation Theorem. By Tonelli's Theorem, we obtain

$$\begin{aligned}
&\Big|P(x_1,\ldots,x_n) - \sum_{k=0}^{m} p_k^{(m)} \cdot \text{Pólya}\,((x_1,\ldots,x_n)\,|\,(k+1, m-k+1))\Big| \\
&= \Big|\int_0^1 \theta^{\sum_{i=1}^n x_i}(1-\theta)^{n-\sum_{i=1}^n x_i}\varphi(\theta)\,d\theta \\
&\quad - \int_0^1 \theta^{\sum_{i=1}^n x_i}(1-\theta)^{n-\sum_{i=1}^n x_i}\Big(\sum_{k=0}^{m} p_k^{(m)} \cdot \text{Beta}\,(\theta;(k+1,m-k+1))\Big)\,d\theta\Big| \\
&= \Big|\int_0^1 \theta^{\sum_{i=1}^n x_i}(1-\theta)^{n-\sum_{i=1}^n x_i}[\varphi(\theta) - \frac{P_m(\theta)}{b_m}]\,d\theta\Big|,
\end{aligned}$$

where

$$\begin{aligned}
&\Big|P(x_1,\ldots,x_n) - \sum_{k=0}^{m} p_k^{(m)} \cdot \text{Pólya}\,((x_1,\ldots,x_n)\,|\,(k+1, m-k+1))\Big| \\
&= \Big|\int_0^1 \theta^{\sum_{i=1}^n x_i}(1-\theta)^{n-\sum_{i=1}^n x_i}\Big[\varphi(\theta) - \frac{P_m(\theta)}{b_m}\Big]\,d\theta\Big| \\
&\leq \int_0^1 \theta^{\sum_{i=1}^n x_i}(1-\theta)^{n-\sum_{i=1}^n x_i}\Big|\varphi(\theta) - \frac{P_m(\theta)}{b_m}\Big|\,d\theta\,.
\end{aligned}$$

Since $b_m \to 1$, as $m \to \infty$, we have $P_m(\theta)/b_m \xrightarrow{u} C(\theta)$, and, therefore, for all $\varepsilon > 0$, there exists $m_0 = m_0(\varepsilon) \in \mathrm{N}$ such that if $m \geq m_0$, then

$$\int_0^1 \Big|\frac{P_m(\theta)}{b_m} - C(\theta)\Big|\,d\lambda_1 < \frac{\varepsilon}{4}. \tag{2.6}$$

In this way,

$$\begin{aligned}
&\left|P(x_1,\ldots,x_n) - \sum_{k=0}^{m_0} p_k^{(m_0)} \cdot \text{Pólya}\left((x_1,\ldots,x_n)\,|\,(k+1, m_0-k+1)\right)\right| \\
&\leq \int_0^1 \theta^{\sum_{i=1}^n x_i}(1-\theta)^{n-\sum_{i=1}^n x_i} \left|\varphi(\theta) - \frac{P_{m_0}(\theta)}{b_{m_0}}\right| d\theta \\
&\leq \int_0^1 \left|\varphi(\theta) - \frac{P_{m_0}(\theta)}{b_{m_0}}\right| d\theta \,<\, \varepsilon\,,
\end{aligned}$$

where the last inequality follows from (2.3)–(2.6) and the triangle inequality. Hence, $\forall \varepsilon > 0$, $\exists m_0 = m_0(\varepsilon) \in \mathrm{N}$ and $\{p_0, \ldots, p_{m_0}\}$ with $p_k \geq 0$, $k = 0, \ldots, m_0$, and $\sum_{k=0}^{m_0} p_k = 1$, such that, $\forall n \in \mathrm{N}$ and $\forall (x_1, \ldots, x_n) \in \{0,1\}^n$,

$$\left|P(x_1,\ldots,x_n) - \sum_{k=0}^{m_0} p_k \cdot \text{Pólya}\left((x_1,\ldots,x_n)\,|\,(k+1, m_0-k+1)\right)\right| < \varepsilon\,,$$

concluding the proof. □

Theorem 2.1 states that any $\{0,1\}^\infty$-valued exchangeable process satisfying the condition $\forall n \in \mathrm{N}$, $\max_{v\in\{0,\ldots,n\}} P\left(\sum_{i=1}^n X_i = v\right) \leq M/(n+1)$, for some $M > 0$, can be well approximated by a suitable Pólya–Eggenberger urn model with unknown initial configuration. More precisely, the probability of any event relative to an n-dimensional marginal segment of this process, $n \in \mathrm{N}$, can be estimated, with fixed maximum error $\varepsilon > 0$, by its probability in a Pólya–Eggenberger urn model with unknown initial composition containing, initially, k (where k is unknown) white balls and $m_0 + 2 - k$ black balls with probability $p_{k-1}^{(m_0)}$ specified in the proof of Theorem 2.1, $k = 1, \ldots, m_0 + 1$.

The condition just mentioned points that such approximation is possible whenever the modes of the distributions of $\sum_{i=1}^n X_i$ do not have probability of order higher than $1/n$, that is, whenever the sequence $\{(n+1)P\left(\sum_{i=1}^n X_i = k\right)\}_{n\geq 1}$ is convergent (finite), $\forall k \in \mathrm{N}$. This condition is equivalent to the De Finetti measure of the process $\{X_n\}_{n\geq 1}$ being absolutely continuous with bounded probability density function. Thus, the result of Theorem 2.1 does not hold, for instance, for the exchangeable stochastic process taking values on $\{0,1\}^\infty$, the De Finetti measure of which possesses a probability density function Beta (1/2,1), which is not bounded (as a matter of fact, we have, in this case, $\lim_{n\to\infty}(n+1)P\left(\sum_{i=1}^n X_i = 0\right) = \infty$).

Considering the conditions of Theorem 2.1, with the additional assumption that the De Finetti measure possesses a probability density function itself also continuous, we obtain a much sharper result than (2.2). Under such conditions, we have, $\forall n \in \mathrm{N}$ and $\forall (x_1, \ldots, x_n) \in \{0,1\}^n$,

$$\left|P(x_1,\ldots,x_n) - \sum_{k=0}^{m_0} p_k^{(m_0)} \cdot \text{Pólya}\left((x_1,\ldots,x_n)\,|\,(k+1, m_0-k+1)\right)\right| \leq \frac{\varepsilon}{n+1} << \varepsilon.$$

Furthermore, the distribution of the number of "successes" ($\{X_i = 1\}$) in the first n stages of the process, $\sum_{i=1}^n X_i$, $n \geq 1$, is well approximated by a mixture of Pólya–Eggenberger distributions, that is, $\forall n \in \mathrm{N}$, $\forall t = 0, 1, \ldots, n$, we have

$$\left|P\left(\sum_{i=1}^n X_i = t\right) - \sum_{k=0}^{m_0} p_k \cdot \text{PE}(t\,|\,(k+1, m_0-k+1))\right| < \varepsilon,$$

where PE$(\cdot \mid k+1, m_0 - k+1)$ denotes the Pólya–Eggenberger distribution for the number of white balls drawn in the first n stages of a Pólya–Eggenberger urn process with initial (known) configuration $(k+1, m_0 - k+1)$:

$$\text{PE}(t \mid (k+1, m_0 - k+1)) = \binom{n}{t} \text{Pólya}\,((x_1, \ldots, x_n) \mid (k+1, m_0 - k+1)),$$

with $(x_1, \ldots, x_n) \in \{0,1\}^n$ such that $\sum_{i=1}^n x_i = t$.

We should emphasize that the result of Theorem 2.1 is in a sense qualitative, as it just points the existence of a suitable Pólya–Eggenberger urn model with unknown initial composition that approximates the original exchangeable stochastic process. We did not examine the problem of the optimality of the approximating model (Diaconis and Ylvisaker (1985) suggest modifications in the coefficients of Bernstein polynomials in order to attain a sharper approximation in Lemma 2.1).

3. DISCUSSION

We present an alternative version of the Pólya–Eggenberger urn model via the introduction of a probability measure for the initial composition of the urn. This process is exchangeable with De Finetti measure possessing a mixture of Beta densities as density function. We also state a predictivistic sufficient condition for a $\{0,1\}^\infty$-valued exchangeable process to be well approximated by a suitable Pólya–Eggenberger urn model with unknown initial composition. The condition amounts to an absolutely continuous with bounded density De Finetti measure of the original process. The strengthening of this condition leads to a sharper approximation, of order $1/n$.

Theorem 2.1, even if stating conditions about the observable quantities $\{X_n\}_{n\geq 1}$, does not correspond to an exact predictivistic representation for the Pólya–Eggenberger urn model with $c = 1$. After all, the result in (2.2) only provides approximations for the marginal distributions of the process $\{X_n\}_{n\geq 1}$, not implying the existence of a probability measure $\mu^* : \mathcal{P}(\mathrm{N}^2) \to [0,1]$ such that, $\forall n \in \mathrm{N}$ and $\forall (x_1, \ldots, x_n) \in \{0,1\}^n$,

$$P(x_1, \ldots, x_n) \;=\; \int_{\mathrm{N}^2} \text{Pólya}\,((x_1, \ldots, x_n) \mid (a,b))\, d\mu^*(a,b)\;.$$

In fact, since the sequence of measures $\{\{((k+1, m-k+1), p_k^m)\}_{k=0}^m\}_{m\geq 1}$ in $(\mathrm{N}^2, \mathcal{P}(\mathrm{N}^2))$ is not tight (their respective supports are $\{(a,b) \in \mathrm{N}^2 : a+b = m+2\}$), we cannot guarantee the existence of such a measure μ^*. Even so, the hypotheses of Theorem 2.1 are totally predictivistic, as they take into account conditions on the observable quantities $X_1, X_2, \ldots$ exclusively.

On the other hand, the result has a strong predictivistic flavor, as it involves the uncertainty one feels with respect to the *initial* composition of an urn containing balls of two different types; urn models are very much easier to conceive (if not more natural) than *limits* of relative frequencies or "probabilities" of heads in coin tossing, usual interpretations for the variable θ that figures in De Finetti's Representation Theorem. Experimentations with urns are feasible, verifiable, and allow for diverse probabilistic modelling (Freudenthal, 1960) asserts that "Tossing coins or a dice or playing at cards are not enough flexible. The most general chance instrument is the urn filled with balls of different colors...".) Our point is, in conclusion, that it is easier to think about the initial numbers (A, B) than about limits of frequencies, the former being much more concrete than the latter. Even so,(A, B) is a "parameter" *stricto sensu.*

Questions not considered in this work may be the goal of future investigation. Among others, we mention the obtainment of necessary conditions for the validity of Theorem 2.1 (we have characterized the sufficient condition only), the (possible) derivation of a true predictivistic representation for the Pólya–Eggenberger urn model, that is, the specification of additional conditions for $\{0,1\}^\infty$-valued exchangeable observables to be characterized as mixtures of usual Pólya laws, the extension to more than two colors, and also the specification of conditions in order to approximate the predictive distributions of an exchangeable process in $\{0,1\}^\infty$ by the corresponding predictive distributions of a suitable Pólya–Eggenberger urn model with unknown initial composition (Dalal and Hall, 1983, prove results in this direction for conjugate families of distributions in Bayesian Inference).

REFERENCES

Dalal, S. R. and Hall, W. J. (1983). Approximating priors by mixtures of natural conjugate priors. *J. R. Statist. Soc. B* **45**, 278–286.

Diaconis, P. and Ylvisaker, D. (1985). Quantifying prior opinion. *Bayesian Statistics 2* (J. M. Bernardo, M. H. DeGroot, D. V. Lindley and A. F. M. Smith, eds), Amsterdam: North-Holland, 133–156 (with discussion).

Feller, W. (1957). *An Introduction to Probability Theory and its Applications* **1**. New York: Wiley

Freudenthal, H. (1960). Models in applied probability. *Synthese* **12**, 202–212.

Johnson, N. L. and Kotz, S. (1977). *Urn Models and their Applications*. New York: Wiley

Markov, A. A. (1906). Extension of the law of large numbers to dependent variables. *Izv. Fiz-Mat. Obschch. Kazan Univ., Ser. 2* **15**, 135–256 (in Russian).

Marsden, J. E. (1974). *Elementary Classical Analysis*. San Francisco: W. H. Freeman and Co.

Pólya, G. and Eggenberger, F. (1923). Über die Statistik verketteter Vorgänge. *Z. Angew. Math. Mech.* **3**, 279–289.

Shohat, J. A. and Tamarkin, J. D. (1943). *The Problem of Moments*. New York: American Mathematical Society.

BAYESIAN STATISTICS 7, pp. 519–527
J. M. Bernardo, M. J. Bayarri, J. O. Berger, A. P. Dawid,
D. Heckerman, A. F. M. Smith and M. West (Eds.)

Multi-scale Modelling of 1-D Permeability Fields

MARCO A. R. FERREIRA
UFRJ, Brazil and Duke University, USA
marco@im.ufrj.br

MIKE WEST
Duke University, USA
mw@isds.duke.edu

HERBERT K. H. LEE
UCSC and Duke University, USA
herbie@cse.ucsc.edu

DAVID HIGDON
Duke University and LANL, USA
higdon@isds.duke.edu

ZHUOXIN BI
Petrotel Inc., USA
zbi@petrotel.com

SUMMARY

Permeability plays an important role in subsurface fluid flow studies, being one of the most important quantities for the prediction of fluid flow patterns. The estimation of permeability fields is therefore critical and necessary for the prediction of the behavior of contaminant plumes in aquifers and the production of petroleum from oil fields. In general, there are two types of information that can be used in the estimation of the permeability field: static data and dynamic data. Static data can be regarded as measurements of the permeability field, and are available at different levels of resolution as a result of geological studies, well tests, and laboratory measurements. Dynamic data are obtained through the production from a collection of wells, being the production history in the case of a mature oil field or the result of tracer experiments in the case of aquifers. To incorporate the dynamic data in formal statistical analysis, corresponding likelihood functions for the high-dimensional random field parameters representing the permeability field can be computed with the help of a fluid flow simulator (FFS). The FFS can run at different scales of resolution of the permeability field, lower levels providing faster but less accurate results. In this paper, we incorporate the static information available at the different scales of resolution by using a multi-scale time series model as a prior for 1-D permeability fields. Estimation of the multi-scale permeability field is then performed using an MCMC algorithm with an embedded FFS running at different scales to incorporate the dynamic data. We use simulated data to study the performance of the proposed approach with respect to the recovery of the original permeability field.

Keywords: MULTI-SCALE STOCHASTIC MODELS; SUBSURFACE FLUID FLOW; PERMEABILITY FIELDS; INFORMATION AGGREGATION; MULTI-SCALE TIME SERIES MODELS; INVERSE PROBLEM.

1. INTRODUCTION

One of the main objectives of subsurface hydrology is the prediction of fluid flow patterns, with important applications in the clean-up of aquifers and production of petroleum from oil fields. The subsurface fluid flow is highly dependent on the permeability field, which is defined as how easily fluid moves through the porous medium. Therefore, the estimation of the permeability field is of ultimate importance, but it has to rely on various sources of information providing data at different scales of resolution. In this paper, we incorporate the information available at the different scales of resolution by using the multi-scale time series model introduced by Ferreira *et al.* (2002) as a prior for 1-D permeability fields. The use of this multi-scale model leads to an efficient and coherent procedure for the estimation of permeability fields.

In general subsurface fluid flow studies, there are two types of available information for the estimation of the permeability field: static data and dynamic data. The static information about the permeability field comes from various sources, for example, geological studies, well tests and laboratory measurements. Each of these sources provides information at different levels of aggregation or resolution: geological studies provide information at a very coarse level of resolution, well tests provide information at an intermediate level, and laboratory measurements of core samples give information at a very fine level. A very important aspect of the static information is the dichotomy between level of resolution and coverage of the field: higher resolution data are available for restricted parts of the field, while the coarse-resolution data are generally available for the whole field. In this work, we assume that the static data are direct measurements of the permeability field at the corresponding scales of resolution. The methods proposed so far do not address the incorporation of the information at different scales of resolution in a probabilistically coherent framework. Our use of multi-scale models leads to a framework that integrates the static data available at the various scales of resolution in a coherent manner.

The dynamic data are obtained from the history of oil production or from results of tracer experiments performed in an aquifer. As opposed to the linear relationship with the static data, the permeability field affects the fluid flow and therefore the dynamic data in a highly non-linear manner. When the physical characteristics of the porous media such as permeability field and porosity are known, physical models such as Darcy's Law and the law of conservation of fluid mass can be used to solve the forward problem, that is, to predict the fluid flow and the dynamic data, as discussed in Gelhar (1993). These physical models are described by a system of differential equations that can be numerically solved by computer codes called Fluid Flow Simulators (FFS). For references on FFS, see King and Datta-Gupta (1998) and Vasco and Datta-Gupta (1999). In this work, we use a relatively fast FFS based on streamlines. This simulator computes an approximation for the expected dynamic data for a given permeability field partitioned into discrete grid-blocks; as the resolution of the partition becomes finer, the FFS becomes more accurate but slower.

The use of the dynamic data for the estimation of the permeability field is known in the subsurface hydrology literature as the inverse problem. Earlier references on the inverse problem include Neuman and Yakowitz (1979), Yeh (1986), Oliver (1994), Craig *et al.* (1996), Hegstad and Omre (1997), Oliver *et al.* (1997), Vasco and Datta-Gupta (1999) and Floris *et al.* (1999). In order to incorporate the dynamic data in formal statistical analysis, we propose a framework in which corresponding likelihood functions for the high-dimensional random field parameters representing the permeability field are computed with the help of the FFS embedded in an MCMC scheme. As the MCMC scheme requires several iterations, the velocity of the FFS becomes a critical issue and we need a framework that leads to the use of the faster solutions based on coarser fields.

The inverse problem is ill-posed, *i.e.*, there is an infinite number of permeability field solutions whose expected dynamic data matches exactly the observed dynamic data. Many of those solutions are too rough to be plausible, and proposed approaches generally impose on the permeability field some stochastic regularity constrains. For example, Cunha *et al.* (1996) and Oliver *et al.* (1997) assume that the permeability field follows a Gaussian process model, while Lee *et al.* (2002) assume that the permeabilities follow a Markov random field process. While the Gaussian process model often implies permeability fields that are too smooth, the Markov random field assumption can lead to an undesirably fast decay of the spatial correlation. In this work, we propose as priors for 1-D permeability fields the use of multi-scale models introduced by Ferreira *et al.* (2002). These multi-scale models have the local behavior of Markov random fields, but globally emulate long memory processes being well suited to capture global features of the permeability field.

The organization of the paper is as follows: in Section 2, we present the multi-scale model; in Section 3, we develop the propagation of static information through the different levels of the field; in Section 4, we discuss the incorporation of the dynamic data; in Section 5, we present an application of our proposed framework to a synthetic permeability field; we present conclusions and discussion in Section 6.

2. MULTI-SCALE MODELLING

Our objective is to estimate discretized versions of a true continuous permeability field; we define coarse, intermediate and fine versions of the field denoted respectively by X_0, X_1 and X_2. We use the multi-scale time series model introduced by Ferreira *et al.* (2002) as a prior for the permeability field. More specifically, the prior of the multi-scale permeability field is of the form:

$$p(X_2 \mid X_1, \mu, \phi_2, \sigma_2^2, \tau_2) p(X_1 \mid X_0, \mu, \phi_1, \sigma_1^2, \tau_1) p(X_0 \mid \mu, \phi_0, \sigma_0^2),$$

where

$$p(X_0 \mid \mu, \phi_0, \sigma_0^2) = \mathrm{N}(X_0 \mid \mu \vec{1}_{n_0}, Q_0)$$

$$p(X_1 \mid X_0, \mu, \phi_1, \sigma_1^2, \tau_1) = \frac{\mathrm{N}(X_1 \mid \mu \vec{1}_{n_1}, V_1) \mathrm{N}(X_0 \mid A_1 X_1, \tau_1 I_{n_0})}{\mathrm{N}(X_0 \mid \mu \vec{1}_{n_0}, A_1 V_1 A_1' + \tau_1 I_{n_0})},$$

and

$$p(X_2 \mid X_1, \mu, \phi_2, \sigma_2^2, \tau_2) = \frac{\mathrm{N}(X_2 \mid \mu \vec{1}_{n_2}, V_2) \mathrm{N}(X_1 \mid A_2 X_2, \tau_2 I_{n_1})}{\mathrm{N}(X_1 \mid \mu \vec{1}_{n_1}, A_2 V_2 A_2' + \tau_2 I_{n_1})},$$

where $(Q_0)_{ij} = \sigma_0^2 \phi_0^{|i-j|}/(1-\phi_0^2)$, $(V_1)_{ij} = \sigma_1^2 \phi_1^{|i-j|}/(1-\phi_1^2)$, μ is the mean level of the field, and $A_1 = m_1^{-1} I_{n_1} \otimes \vec{1}_{m_1}$ is the matrix that coarsens the intermediate level X_1 by non-overlapping moving averages. Analogously, $(V_2)_{ij} = \sigma_2^2 \phi_2^{|i-j|}/(1-\phi_2^2)$ and $A_2 = m_2^{-1} I_{n_2} \otimes \vec{1}_{m_2}$.

We denote the available information by: p_{obs} = observed dynamic data; d_0 = measurement of permeability at the coarse scale; d_1 = measurement of permeability at the intermediate scale; d_2 = measurement of permeability at the fine scale. It is assumed that $p_{obs} \mid X_2 \sim \mathrm{N}(f(X_2), \sigma_\epsilon^2 I)$, $d_0 \mid X_0 \sim \mathrm{N}(X_0, S_0)$, $d_1 \mid X_1 \sim \mathrm{N}(X_1, S_1)$, $d_2 \mid X_2 \sim \mathrm{N}(X_2, S_2)$, where S_0, S_1 and S_2 are the known covariance matrices of the measurement errors at the respective levels of resolution and $f(x)$ is the expected dynamic data for the permeability field x that can be computed by the FFS. In this paper, we assume that the approximation error incurred by the FFS is negligible. Typically, S_0, S_1 and S_2 are diagonal matrices or matrices resulting from some geostatistical model applied to the static data.

In summary, the posterior distribution will be proportional to:

$$\begin{aligned} &p(p_{obs} \mid X_2, \sigma_\epsilon^2) p(d_2 \mid X_2) p(d_1 \mid X_1) p(d_0 \mid X_0) p(X_2 \mid X_1, \mu, \phi_2, \sigma_2^2, \tau_2) \\ &\times p(X_1 \mid X_0, \mu, \phi_1, \sigma_1^2, \tau_1) p(X_0 \mid \mu, \phi_0, \sigma_0^2) p(\mu) p(\phi_2, \sigma_2^2, \tau_2) p(\phi_1, \sigma_1^2, \tau_1) p(\phi_0, \sigma_0^2) p(\sigma_\epsilon^2). \end{aligned}$$

In this paper, we propose an MCMC scheme to explore the posterior distribution. Each iteration of the MCMC is performed in a cascade way, from coarser to finer levels. First, all the information about the static data d_0, d_1 and d_2 is sent from the finer to coarser levels by analytically integrating out the finer levels. Then, the dynamic data are used to generate the coarser levels by using the FFS running at those levels. Finally, the finer levels are generated conditional on the generated coarser levels. Besides the faster running time of the FFS at the coarser levels, the MCMC for the coarser levels converges much faster than for the finer levels because the marginal posterior distribution of coarser levels has fewer local maxima than those of finer levels. In our approach, the results at coarser levels guide the simulation at finer levels, and the main consequence is faster convergence at finer levels as compared to convergence of traditional MRF schemes.

The main idea of our MCMC algorithm is the following: we generate the coarsest level X_0 from $p(X_0 \mid p_{obs}, d_0, d_1, d_2)$, that is, we integrate out X_1 and X_2 before generating X_0. Then, we generate X_1 from $p(X_1 \mid X_0, p_{obs}, d_0, d_1, d_2)$ with X_2 integrated out. Finally, we generate X_2 from its full conditional $p(X_2 \mid X_1, X_0, p_{obs}, d_0, d_1, d_2)$. In the subsequent sections, we obtain the expressions of these distributions.

3. PROPAGATING STATIC INFORMATION

In this section, we describe how to send the measurement information from finer to coarser levels. Here, we consider only two levels because the generalization to more levels is straightforward. In short, our objective in this section is to obtain $p(X_0 \mid d_0, d_1)$. For notational simplicity, we omit the dependence of the density functions on the hyperparameters ϕ_0, σ_0^2, μ, ϕ_1, σ_1^2 and τ_1. Hence:

$$p(X_0 \mid d_0, d_1) \propto \int p(d_0 \mid X_0) p(d_1 \mid X_1) p(X_0) p(X_1 \mid X_0) dX_1$$

$$\propto \frac{\mathrm{N}(d_0 \mid X_0, S_0)\mathrm{N}(X_0 \mid \mu\vec{1}_{n_0}, Q_0)}{\mathrm{N}(X_0 \mid \mu\vec{1}_{n_0}, A_1 V_1 A_1' + \tau_1 I_{n_0})} \mathrm{N}(X_0 \mid b_0, B_0)$$

where

$$B_0 = A_1(S_1^{-1} + V_1^{-1})^{-1} A_1' + \tau_1 I_{n_0}, \quad b_0 = A_1(S_1^{-1} + V_1^{-1})^{-1}(S_1^{-1} d_1 + V_1^{-1}\mu\vec{1}_{n_0}).$$

4. INCORPORATING DYNAMIC DATA

In this section, we describe how to incorporate production data from the coarsest to the finest level, after having incorporated the measurement data.

The first step of our MCMC algorithm is the simulation of X_0, conditional on the pressure and the measurement data. We assume that, conditional on the coarse level, the production p_{obs} is independent of the measurements at the finest and intermediate levels. *The main objective of this assumption is the use of the FFS in the generation of each level of resolution.* With this assumption, we can write the conditional distribution of X_0 given p_{obs}, d_0, d_1 and d_2 as:

$$p(X_0 \mid p_{obs}, d_0, d_1, d_2) \propto p(p_{obs}, d_0, d_1, d_2 \mid X_0) p(X_0)$$

$$\propto p(p_{obs} \mid X_0) p(X_0 \mid d_0, d_1, d_2).$$

Note that $p(X_0 \mid d_0, d_1, d_2)$ can be easily obtained using the result outlined in section 3.

Moreover, a good approximation for $p(p_{obs} \mid X_0)$ can be obtained by running the FFS at the coarsest level, that is, $p(p_{obs} \mid X_0) \approx \mathrm{N}(p_{obs} \mid f(X_0), \sigma_{\epsilon 0}^2 I)$. In this way, we avoid the cumbersome computation of the integral in

$$p(p_{obs} \mid X_0) = \int\int p(p_{obs} \mid X_2) p(X_2 \mid X_1) p(X_1 \mid X_0)\, dX_1 dX_2$$

which is very complicated due to the non-linear relationship between p_{obs} and X_2.

In order to obtain the conditional distribution of X_1 given X_0, p_{obs}, d_0, d_1 and d_2, we assume that p_{obs} is independent of d_2 given X_1, implying:

$$p(X_1 \mid X_0, p_{obs}, d_0, d_1, d_2) \propto p(p_{obs} \mid X_1)p(X_1 \mid X_0, d_0, d_1, d_2)$$
$$\propto p(p_{obs} \mid X_1)p(X_1 \mid X_0)p(d_1 \mid X_1) \int p(d_2 \mid X_2)p(X_2 \mid X_1)dX_2.$$

Note that $\int p(d_2 \mid X_2)p(X_2 \mid X_1)dX_2$ is the operation described in section 3 that brings the measurement information about the finest level to the intermediate level. In addition, $p(p_{obs} \mid X_1)$ can be approximated by running the FFS at the intermediate level and assuming that

$$p(p_{obs} \mid X_1) \approx \mathrm{N}(p_{obs} \mid f(X_1), \sigma^2_{\epsilon 1} I).$$

Finally, the conditional distribution of X_2 given X_1, X_0, p_{obs}, d_0, d_1 and d_2 is

$$p(X_2 \mid X_1, X_0, p_{obs}, d_0, d_1, d_2) \propto p(p_{obs} \mid X_2)p(X_2 \mid X_1)p(d_2 \mid X_2).$$

5. APPLICATION

In this section we present an application of our proposed multi-scale framework to a synthetic 1-D permeability field. The design of the field is as follows: The length of the field is equal to 1280 meters and there is only one production well located in the right end of the field. We assume that there are three discretized versions of the field X_0, X_1 and X_2 dividing the field in 8, 32 and 128 grid-blocks respectively. Moreover, we assume that the finest discretized level X_2 is equal to the truth and we are going to perform inference only for the coarser versions X_0 and X_1. In the following subsections we discuss the specification of the prior distribution, the setup of the application, and the results.

5.1. *Prior Specification*

In this section we discuss the specification of the prior distribution for the parameters of the model. In order to reduce the dimension of the prior specification problem, we assume a priori independence of the parameters, allowing us to focus on the specification of the prior for each parameter individually.

There are several aspects peculiar to our approach to the 1-D fluid flow problem that allow us to set informative priors for several parameters of interest. The most important of these characteristics is that the permeability field is modelled on the logarithm scale. Thus, it is easy to think about the amount of relative variability between neighbor grid-blocks within and across different levels of resolution, leading to straightforward specification of the priors for σ_0^2, σ_1^2 and τ_1. The priors for the autoregressive coefficients ϕ_0 and ϕ_1 can be specified by considering the amount of smoothness at each resolution level of the log-permeability field. The overall log-permeability mean μ is assigned a prior with mass in the region where the log-permeability makes physical sense. The variances $\sigma_{\epsilon 0}$ and $\sigma_{\epsilon 1}$ are assigned vague priors. More specifically, the priors used in our particular application are:

$$\phi_0 \sim \mathrm{N}(0.0, 0.5), \quad \sigma_0^2 \sim \mathrm{IG}(3.0, 1.1), \quad \phi_1 \sim \mathrm{N}(0.9, 0.01),$$
$$\sigma_1^2 \sim \mathrm{IG}(57, 0.14), \quad \mu \sim \mathrm{N}(5.0, 1.21), \quad \sigma_{\epsilon 0} \sim \mathrm{IG}(5 \times 10^{-5}, 5 \times 10^{-5}),$$
$$\sigma_{\epsilon 1} \sim \mathrm{IG}(5 \times 10^{-5}, 5 \times 10^{-5}) \quad \tau_1 \sim \mathrm{IG}(57, 0.14).$$

Figure 1. *Original field and coarser versions:* (a) *original field—512 grid-blocks;* (b) *intermediate field—64 grid-blocks;* (c) *coarse field—8 grid-blocks.*

5.2. *Problem Setup*

In this section, we present the setup of the problem, the synthetic permeability field, and the available information for the estimation of the field.

Figure 1 presents the original synthetic permeability field with 512 grid-blocks and coarser versions with 64 and 8 grid-blocks, darker colors corresponding to lower values and lighter colors corresponding to higher values. The i-th grid-block of the synthetic permeability field is equal to $\exp(-3.1710^{-5}i^2 + 0.01625i + 3.91)$, $i = 0, \ldots, 511$, $i = 1$ and $i = 512$ corresponding, respectively, to the left and right ends of the field. The other characteristics of the reservoir such as number of phases, initial pressure, porosity, production rate, cross sectional area, and compressibility are assumed constant and known. Moreover, there is only one production well located in the right end of the field.

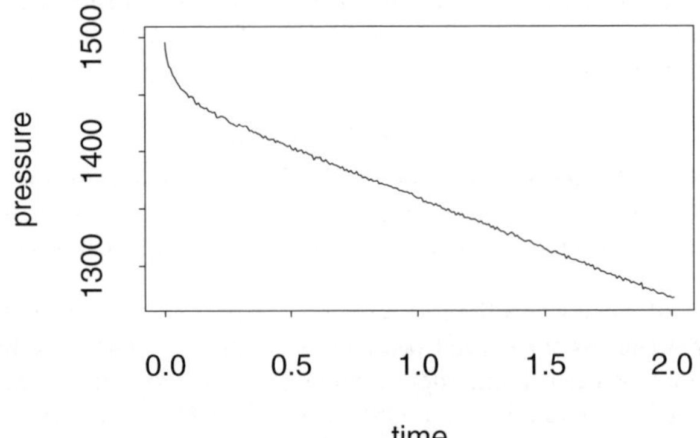

Figure 2. *Dynamic data—pressure curve through time.*

The static data at the coarse level are $d_0 = (4.4, 5.1, 5.6, 5.9, 6.0, 5.7, 5.2, 4.3)$ and $S_0 = 0.1^2 I$. There are static data at the intermediate level only for the grid-blocks 1, 9, 17, 25, 33, 41, 49, 57 and 64 with variance equal to 0.004, with recorded values equal to 3.97, 4.86, 5.50, 5.88, 5.99, 5.85, 5.45, 4.78 and 3.99.

Depending on the production regime, the dynamic data can be either flow rate or pressure at the production well. In this example, it is assumed that the flow rate is kept constant and the dynamic data are given by the pressure curve at the production well. Figure 2 shows this pressure curve computed with the FFS for the first two days of production and perturbed with measurement errors with variance $\sigma_\epsilon^2 = 1.0$.

5.3. *Results*

In this section, we present the results of the application of our multi-scale framework to the estimation of the permeability field outlined in Section 5.2.

Table 1 presents posterior summaries for the hyperparameters of the multi-scale model. The posterior means of ϕ_0, σ_0^2, ϕ_1 and σ_1^2 show the higher degree of smoothness at the intermediate level when compared with the smoothness at the coarse level. The estimate of τ_1 shows that each coarse grid-block is very close to the average of the corresponding intermediate grid-blocks. Finally, the posterior means of $\sigma_{\epsilon 0}^2$ and $\sigma_{\epsilon 1}^2$ show that history matching is more precise at the intermediate level than at the coarse level.

Table 1. *Posterior summaries for the hyperparameters.*

	Mean	Standard deviation
ϕ_0	0.4483	0.3097
σ_0^2	0.4416	0.2207
ϕ_1	0.8630	0.0319
σ_1^2	0.0024	0.0005
τ_1	0.0019	0.0003
$\sigma_{\epsilon 0}^2$	0.9536	0.0874
$\sigma_{\epsilon 1}^2$	0.8915	0.0801
μ	5.0680	0.4810

Figure 3 presents the coarse version of the original field and the posterior mean of the coarse level of resolution. The posterior mean is very close to the coarsest level of the original permeability field indicating that our methodology is working well.

Analogously, Figure 4 presents the intermediate version of the original field and the posterior mean of the intermediate level of resolution. The conclusions are similar to those obtained for the coarse level, the main difference being that the estimation at the intermediate level of resolution recovers a higher degree of detail.

Figure 3. *Coarse level of resolution:* (a) *version of original field;* (b) *posterior mean.*

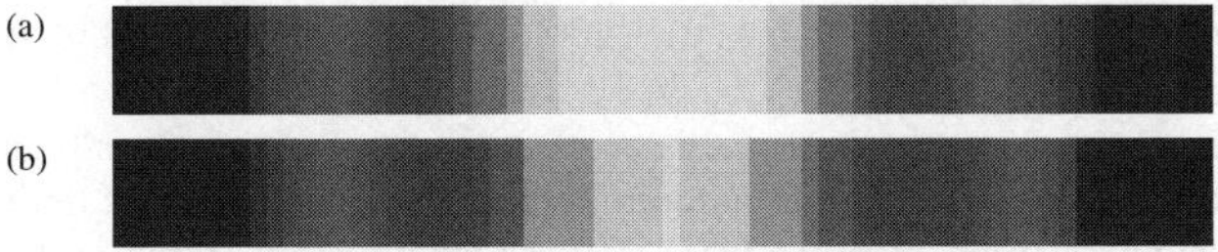

Figure 4. *Intermediate level of resolution:* (a) *version of original field;* (b) *posterior mean.*

As stated in Oliver *et al.* (1997), the main advantage of the Bayesian framework coupled with MCMC technology is the characterization of the uncertainty present in the inference about the permeability field. A useful summary of this uncertainty is the posterior variance of each grid-block. Figure 5 presents the posterior variance across the field at the coarse and intermediate levels of resolution. Figure 5a presents the posterior variance of the log-permeabilities at the coarse level. As the production well is located in the right end of the field, the incorporation of the dynamic data results in smaller posterior variances for grid-blocks closer to that region. In particular, the ratio between the posterior variances of the coarse grid-blocks 1 (left end) and 8 (right end) is about 50, evidencing the reduction in uncertainty due to the proximity of the production well.

Figure 5b presents the posterior variance of the log-permeabilities at the intermediate level. Note that the existence of static data for the grid-blocks 1, 9, 17, 25, 33, 41, 49, 57 and 64 imply smaller posterior variances at these grid-blocks than at their neighbors. Moreover, as the measurement information is equally spaced, the transfer of information along the field causes periodic behavior of the posterior variances. Finally, as in the coarse level of resolution, the posterior variances of the grid-blocks closer to the production well are smaller because of the integration of the dynamic data.

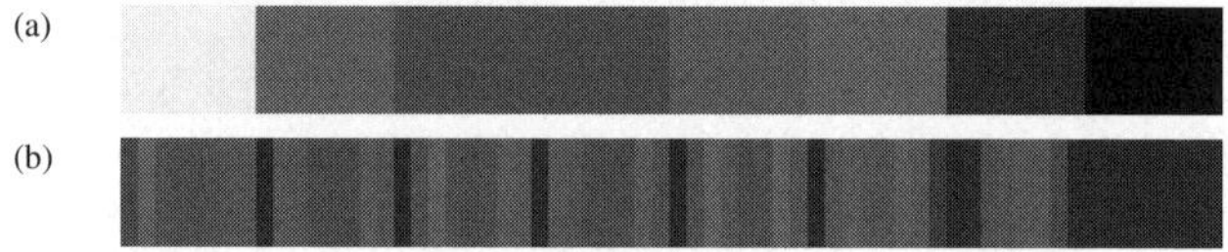

Figure 5. *Posterior variance across the field:* (a) *coarse level.* (b) *intermediate level.*

6. CONCLUSIONS AND DISCUSSION

In this paper, we proposed a 1-D multi-scale model as a prior for permeability fields. We developed an MCMC algorithm with an embedded Fluid Flow Simulator running at each level of resolution to incorporate the dynamic information. The algorithm is performed in a cascade way from coarser to finer levels, coarser levels guiding the realizations at the finer levels, which leads to a more efficient exploration of the posterior distribution of the multi-scale permeability field. The final result is a multi-scale framework that incorporates data at different resolutions and of different types in a coherent and efficient manner.

We applied the proposed methodology to synthetic data. As expected from intuition, analysis of the posterior variances across the field indicates that dynamic data give more information on permeabilities closer to the production well. Finally, the integrated use of data from several different sources allows a very good recovery of the original permeability field.

Even with the restriction to one dimension, there are some potential applications of our proposed methodology to real-world problems. One such example is the estimation of the permeability field of incised valley oil fields, that is, petroleum fields that developed in buried ancient river beds. See Stephen and Dalrymple (2002) for an example of an incised valley oil field.

Ongoing research is the extension of this work to the two-dimensional case. We have already developed theoretical multi-scale random field models, simulation of multi-scale fields from these models, and estimation procedures. Right now we are working on the application of these theoretical developments to the estimation of the permeability field for a contaminated aquifer.

ACKNOWLEDGEMENTS

This work was partially supported by National Science Foundation grants DMS 9873275 and 0233710. Marco A. R. Ferreira's graduate studies were partially supported by a grant from CNPq, Brazil.

REFERENCES

Craig, P. S., Goldstein, M., Seheult, A. H. and Smith, J. A. (1996). Bayes linear strategies for history matching of hydrocarbon reservoirs. *Bayesian Statistics 5* (J. M. Bernardo, J. O. Berger, A. P. Dawid and A. F. M. Smith, eds). Oxford: Oxford University Press, 69–95 (with discussion).

Cunha, L. B., Oliver, D. S., Redner, R. A. and Reynolds, A. C. (1996). A hybrid Markov chain Monte Carlo method for generating permeability fields conditioned to multiwell pressure data and prior information. *SPE Annual Technical Conference and Exhibition*. Denver: SPE.

Ferreira, M. A. R., West, M., Lee, H., and Higdon, D. (2002). A class of multi-scale time series models. *Tech. Rep.*, Duke University, USA.

Floris, F. J. T., Bush, M. D., Cuypers, M., Roggero, F. and Syversveen, A.-R. (1999). Comparison of production forecast uncertainty quantification methods—an integrated study. *1st Symposium on Petroleum Geostatistics*, Toulouse.

Gelhar, L. W. (1993). *Stochastic Subsurface Hydrology*. Englewood Cliffs, NJ: Prentice-Hall.

Hegstad, B. K. and Omre, H. (1997). Uncertainty assessment in history matching and forecasting. *Geostatistics Wollongong '96* **1** (E. Y. Baafi and N. A. Schofield, eds). Dordrecht: Kluwer. 585–596.

King, M. J. and Datta-Gupta, A. (1998). Streamline simulation: a current perspective. *In Situ* **22**, 91–140.

Lee, H., Higdon, D., Bi, Z., Ferreira, M. A. R. and West, M. (2002). Markov random field models for high-dimensional parameters in simulations of fluid flow in porous media. *Technometrics* **44**, 230–241.

Neuman, S. P. and Yakowitz, S. (1979). A statistical approach to the problem of aquifer hydrology: Theory. *Water Resources Res.* **15**, 845–860.

Oliver, D. S. (1994). Incorporation of transient pressure data into reservoir characterization. *In Situ* **18**, 243–275.

Oliver, D. S., Cunha, L. B. and Reynolds, A. C. (1997). Markov chain Monte Carlo methods for conditioning a permeability field to pressure data. *Math. Geol.* **29**, 61–91.

Stephen K. D. and Dalrymple M. (2002). Reservoir simulations developed from an outcrop of incised valley fill strata. *AAPG Bull.* **86**, 797–822.

Vasco, D. W. and Datta-Gupta, A. (1999). Asymptotic solutions for solute transport: a formalism for tracer tomography. *Water Resources Res.* **35**, 1–16.

Yeh, W. W. (1986). Review of parameter identification in groundwater hydrology: the inverse problem. *Water Resources Res.* **22**, 95–108.

BAYESIAN STATISTICS 7, pp. 529–533
J. M. Bernardo, M. J. Bayarri, J. O. Berger, A. P. Dawid,
D. Heckerman, A. F. M. Smith and M. West (Eds.)

Direct Bayes for Interest Parameters

DON A. S. FRASER
University of Toronto, Canada
dfraser@utstat.toronto.edu

NANCY REID
University of Toronto, Canada
reid@utstat.toronto.edu

AUGUSTINE WONG
York University, Canada
august@mathstat.yorku.ca

GRACE Y. YI
University of Waterloo, Canada
yyi@icarus.math.uwaterloo.ca

SUMMARY

For an interest parameter $\psi(\theta)$ the Bayesian method eliminates the nuisance parameter $\lambda(\theta)$ by integrating with respect to a conditional prior for λ given ψ. This conditional prior may be difficult to specify in both the subjective and objective Bayesian contexts. We propose the use of results from likelihood theory that give highly accurate third-order determinations of various marginal distributions in the continuous case. The appropriate conditional prior exists as part of the calculations for the marginal distributions and corresponds to an integration used in the marginalization. To third order, however, the integrated likelihood for the interest parameter can be obtained directly; it needs no model information beyond that commonly available for accurate analyses of likelihood. Simple examples are given where the steps are analytically available to illustrate this direct Bayesian calculation.

Keywords: INTEREST PARAMETER; DIRECT INTEGRATION; NUISANCE ELIMINATION; COMPONENT LIKELIHOOD; BAYESIAN INFERENCE.

1. INTRODUCTION

The elimination of a nuisance parameter λ in order to focus on an interest parameter ψ is of central interest in both Bayesian and frequentist methodologies. In the subjective Bayesian context the needed imputation of a conditional prior for λ given ψ may require excessive attention to issues largely peripheral to the investigation of the interest parameter ψ. And in the objective Bayesian context the direct search for a noninformative conditional prior using Jeffreys or forward or backward reference priors may require model information that is not easily accessible.

We use recent results from likelihood theory that show for continuous models the existence of a third-order prior for λ given ψ that eliminates the nuisance parameter and leads directly to a likelihood for the interest parameter ψ; this interest parameter likelihood $L^*(\psi)$ is recorded below in logarithmic form in (2.3). For discrete models similar results are being developed.

Then with a subjective or objective prior $\pi_1(\psi)$ for just the interest parameter ψ, direct Bayesian inference may be carried out using the posterior density

$$\pi(\psi \mid y^0) \propto \pi_1(\psi) L^*(\psi) \tag{1.1}$$

without any integration; here y^0 denotes the observed data and the dependence of $L^*(\psi)$ on y^0 is suppressed. In effect this indicates the presence of an implicit variable that provides direct measurement information concerning ψ. And it avoids the need for the detailed attention to the nuisance parameter as would be required in a global Bayesian approach.

The direct likelihood $L^*(\psi)$ for ψ does require more information than is provided by the observed likelihood $L(\theta)$, or the related profile likelihood $L_P(\psi)$. But other approaches such as Jeffreys or forward or backward reference priors also require more than observed likelihood. The additional information here is given by the sample space gradient of likelihood at the observed data; this is in contrast to the sample space averages required in other approaches. The model information used is expanded from "at the observed data" to include "to first derivative at the observed data."

2. METHODOLOGY

An important minimum amount of information concerning ψ is provided by the easily calculated profile likelihood

$$L_P(\psi) = \sup_{\psi(\theta)=\psi} L(\theta) = L(\hat{\theta}_\psi),$$

where $\hat{\theta}_\psi$ is the constrained maximum likelihood value. However in the presence of nuisance parameters this profile likelihood can be quite misleading (Cox and Reid, 1987).

For higher precision and accuracy recent likelihood theory uses, either explicitly or implicitly, additional model information in the form of an exponential reparametrization $\varphi(\theta)$ (Fraser etalc 1999). If the model is a full exponential model, then $\varphi(\theta)$ is a canonical parameterization of that model. More generally, $\varphi(\theta)$ is a canonical parameterization of a closely approximating exponential model at the observed data.

Let $\ell(\theta; y)$ be the loglikelihood and suppose the model is continuous. If the dimension of y is the same as the dimension say p of the parameter θ the reparametrization is given by

$$\varphi'(\theta) = \nabla\ell(\theta; y^0) = \frac{\partial}{\partial y}\ell(\theta; y)\,|_{y^0},$$

written here as a row vector; $\nabla\ell(\theta; y^0)$ is the gradient of the loglikelihood evaluated at the data point y^0. In the special case of a full exponential model this extracts a version of the canonical parameter. More generally, when the dimension of y is greater than p the reparametrization is given explicitly as

$$\varphi'(\theta) = \nabla\ell(\theta; y^0)V, \tag{2.1}$$

where $V = (v_1, \ldots, v_p)$ records p linearly independent vectors tangent to an exact or approximate ancillary $a(y)$ at the observed y^0. Some details on the general calculation of such vectors may be found in Appendix A; for some recent views on ancillary conditioning see Fraser (2003).

Let $\hat{\jmath}_{\varphi\varphi}$ be the observed information given by $-\ell_{\varphi\varphi}(\varphi; y) = -(\partial/\partial\varphi)(\partial/\partial\varphi')\ell(\varphi; y)$ evaluated at the observed maximum likelihood value $\hat{\varphi}$. And consider a special linear recalibration $\bar{\varphi}(\theta) = \hat{\jmath}^{1/2}_{\varphi\varphi}\varphi(\theta)$ that gives an identity observed information matrix, $\hat{\jmath}_{\bar{\varphi}\bar{\varphi}} = I$. Such a reparametrization affects the nuisance information only through the Jacobian $X = \partial\varphi(\theta)/\partial\lambda'$, evaluated at the observed $\hat{\theta}_\psi$. Note that

$$|\,j_{(\lambda\lambda)}(\hat{\theta}_\psi)\,| = |\,j_{\lambda\lambda}(\hat{\theta}_\psi)\,|\,|\,X'\hat{\jmath}_{\varphi\varphi}X\,|^{-1}, \tag{2.2}$$

where $j_{\lambda\lambda}(\hat{\theta}_\psi)$ is the observed nuisance information obtained by evaluating the Hessian $-\ell_{\lambda\lambda}(\psi, \lambda; y) = -(\partial/\partial\lambda)(\partial/\partial\lambda')\ell(\psi, \lambda; y)$ at the constrained maximum likelihood value $\hat{\theta}_\psi = (\psi, \hat{\lambda}_\psi)$, and the parentheses enclosing λ in $\jmath_{(\lambda\lambda)}(\hat{\theta}_\psi)$ are to indicate that the nuisance parameter has been reexpressed in terms of the special $\bar{\varphi}(\theta)$ parameterization.

The third order loglikelihood for ψ is then

$$\begin{aligned}\ell^*(\psi) &= \ell_{\mathrm{P}}(\psi) + \frac{1}{2}\log|\jmath_{(\lambda\lambda)}(\hat{\theta}_\psi)|, \\ &= \ell_{\mathrm{P}}(\psi) + \frac{1}{2}\log|j_{\lambda\lambda}(\hat{\theta}_\psi)| - \frac{1}{2}\log|X'\hat{\jmath}_{\varphi\varphi}X|, \end{aligned} \tag{2.3}$$

where $\ell_{\mathrm{P}}(\psi)$ is the logarithm of the profile likelihood $L_{\mathrm{P}}(\psi)$. As presented this can be evaluated without a reparametrization of the model. For background, see Fraser and Reid (1996, 2000, 2002) and Fraser (2002).

The direct Bayes assessment of an interest parameter $\psi(\theta)$ requires the prior for the interest parameter together with the observed full likelihood; it also requires the Jacobian of a reparametrization $\varphi(\theta)$ that implicitly contains substantial information as to how the basic variables provide measurement information on the parameter. This extra information has been instrumental in highly accurate frequentist based inference methods and in the Bayesian context provides an alternative to the sample space averages inherent in the Jeffreys and reference prior analyses (Fraser and Reid, 1996). Thus in addition to the familiar maximum likelihood value $\hat{\theta}$, the constrained maximum likelihood value $\hat{\theta}_\psi$, and the observed nuisance information $j_{\lambda\lambda}(\hat{\theta}_\psi)$, we need the Jacobian of a recalibration of the nuisance information which enters much in the role of a design matrix X; these are typically readily available. The recalibration provides the increase to third-order accuracy from the typical first or second-order accuracy for adjusted profiles.

3. EXAMPLES

In this section, we examine some very simple examples that indicate how the direct Bayes method proceeds in familiar contexts. For larger problems the numerical implementation is computationally routine.

Example 1. *Normal location-scale model.* Consider the simple case of a sample $y_1, \ldots, y_n$ from the $\mathrm{N}(\mu, \sigma^2)$ distribution and suppose we are interested in the component parameter μ. A version of the canonical parameter is $\varphi = (\mu/\sigma^2, 1/\sigma^2)'$ and we note that μ is a ratio of canonical parameters. The profile likelihood for μ is $L_{\mathrm{P}}(\mu) = (1+T^2)^{-n/2}$ where T^2 is the ratio of the usual sums-of-squares, $n(\bar{y}-\mu)^2$ over $s^2 = \Sigma(y_i - \bar{y})^2$. Due to location scale invariance it suffices here to do calculations for just the special data point $(\bar{y}, s^2) = (0, n)$; we then have that the observed information for φ and for $\bar{\varphi} = (\sqrt{n}\mu/\sigma^2, \sqrt{n}/\sqrt{2}\sigma^2)'$ are

$$\hat{\jmath}_{\varphi\varphi} = \begin{pmatrix} n & 0 \\ 0 & n/2 \end{pmatrix},$$

and $\hat{\jmath}_{\bar{\varphi}\bar{\varphi}} = I$. Also we have that the nuisance information is $\jmath_{\sigma^2\sigma^2}(\hat{\theta}_\mu) = n/2\hat{\sigma}_\mu^4$ and the recalibrated nuisance information is $\jmath_{(\sigma^2\sigma^2)}(\hat{\theta}_\mu) = (1+T^2)^2/(1+2\mu^2)$, which is $1 + O(n^{-2})$ in the moderate deviations range about the special data point. This gives $\ell^*(\mu) = \ell_{\mathrm{P}}(\mu)$, which is the profile and corresponds to the very natural Student$(n-1)$ density expression that describes the location μ. Location-scale invariance gives the same expression for $\ell^*(\mu)$ in terms of T at a general data point. The direct Bayes posterior for μ is then obtained by integrating the Student$(n-1)$ density expression for μ combined with the special prior for the parameter μ, a rather natural procedure.

Example 2. *Normal regression model.* As another familiar case consider the normal regression model $y = X\beta + \sigma e$ where e is a sample from the standard normal. To illustrate the theory and without loss of generality in the final expression we take an $n \times r$ canonical design matrix $X = (I, 0)'$ and a standardized data point with $\hat{\beta} = 0$ and $s^2 = \Sigma(y_i - \hat{y}_i)^2 = n$. For an interest parameter $\beta_{(1)} = (\beta_1, \ldots, \beta_d)'$ with $d < r$ we obtain $L^*(\beta_{(1)}) = (1 + T^2)^{-(n-r+d)/2}$ where T^2 is the ratio of the usual sums-of-squares for testing $\beta_{(1)}$; this is quite different from the profile $(1 + T^2)^{-n/2}$ for $\beta_{(1)}$. The posterior for $\beta_{(1)}$ is then obtained by combining the direct prior for that interest parameter with the very natural multivariate Student density expression that has the degrees of freedom appropriate to the corresponding multivariate t-type test.

Example 3. *Models for positive response variables.* Statistical models for positive valued response variables have many important roles in applications. Some examples are the scaled gamma, the inverse gaussian and the log normal. For a general model, consider a positive response $y = \mu z$ where y and z are positive and μ records the scaled reexpression of a standardized error variable z with mean $E(z) = 1$; μ is then the mean of the response y. We also assume that the distribution of z has a shape or concentration parameter β, and for convenience here has exponential family form

$$f(z; \beta) = \exp\{-\beta c(z) + k(\beta)\} f(z) ;$$

in this $c(z)$ provides a distance measure from the center say $z = 1$ of the distribution and is convex with the convenient standardization $c(1) = 1$, $c'(1) = 0$, $c''(1) = 1$.

For analytic illustration here we choose the log normal; the ingredients are:

$$c(z) = 1 + \frac{1}{2}\log^2 z, \qquad k(\beta) = \beta + \frac{1}{2}\log\beta - \frac{1}{8}\beta^{-1}, \qquad f(z) = (2\pi)^{-1/2} z^{-3/2}.$$

With the standardized log normal we have $E(z) = 1$ and z can be written as e^x where x is normal $\left(-\sigma^2/2; \sigma^2\right)$ and $\beta = \sigma^{-2}$.

Consider likelihood for the interest parameter $\psi = \mu$. If we express the lognormal sample in terms of an underlying sample $(x_1, \ldots, x_n)$ from the normal $(\nu; \sigma^2)$ we have $\nu = \log\psi - \sigma^2/2$ and $\sigma^{-2} = \beta$ which give $\psi = \exp\{\nu + \sigma^2/2\} = E\{\exp(x); \nu, \sigma^2\}$. The likelihood from the sample can be expressed in terms of the normal variables $x_i = \log y_i$ and uses $\bar{x}$ and $s^2 = \Sigma(x_i - \bar{x})^2$:

$$\ell(\psi, \sigma^2) = -\frac{n}{2\sigma^2}\left(\bar{x} - \log\psi + \frac{1}{2}\sigma^2\right)^2 - \frac{s^2}{2\sigma^2} - \frac{n}{2}\log\sigma^2 .$$

The full maximum likelihood value is $\hat{\sigma}^2 = s^2/n$; the constrained value is

$$\hat{\sigma}^2_\psi = 2\{1 + \hat{\sigma}^2 + (\bar{x} - \log\psi)^2\}^{1/2} - 2 .$$

With this the log profile likelihood $\ell_{\mathrm{P}}(\psi)$ is readily obtained by substitution in the full log likelihood.

The information adjustment in (2.3) has two terms to be added to the log profile $\ell(\psi)$. The first term simplifies to

$$\frac{1}{2}\log j_{\sigma^2\sigma^2}(\hat{\theta}_\psi) = \frac{1}{2}\log(2 + \hat{\sigma}^2_\psi) - \log\hat{\sigma}^2_\psi .$$

The second term uses $\varphi = (\mu/\sigma^2, 1/\sigma^2)'$ as the canonical parameter with

$$\hat{\jmath}_{\varphi\varphi} = \begin{pmatrix} n\hat{\sigma}^2 & -n\bar{x}\hat{\sigma}^2 \\ -n\bar{x}\hat{\sigma}^2 & n(\bar{x}^2\hat{\sigma}^2 + \hat{\sigma}^4/2) \end{pmatrix} .$$

After some algebra using $\partial\varphi/\partial\sigma^2 = (-\log\psi/\sigma^4, -1/\sigma^4)'$ we obtain

$$|\, X'\jmath_{\varphi\varphi}X \,| = n\hat{\sigma}^2\hat{\sigma}_\psi^{-8}(\log^2\psi - 2\bar{x}\log\psi + \bar{x}^2 + \hat{\sigma}^2/2)$$

which gives the second adjustment term

$$2\log\hat{\sigma}_\psi^2 - \frac{1}{2}\log\{\log^2\psi - 2\bar{x}\log\psi + \bar{x}^2 + \hat{\sigma}^2/2\}.$$

The likelihood for the interest parameter is then

$$L^*(\psi) = \hat{\sigma}_\psi^2 \left(\frac{2 + \hat{\sigma}_\psi^2}{\log^2\psi - 2\bar{x}\log\psi + \bar{x}^2 + \hat{\sigma}^2/2} \right)^{1/2} L_{\mathrm{P}}(\psi).$$

The posterior for ψ is obtained by combining this with a direct prior for the parameter ψ.

For a broad range of familiar examples such as those in Fernández and Steel (1999) and Sun (1997), the posterior densities obtained correspond to those indicated by the available marginal models. For more complicated examples the method can be implemented numerically and produces a posterior density that is easily integrated numerically or by Monte Carlo methods.

REFERENCES

Cox, D. R. and Reid, N. (1987). Parameter orthogonality and approximate conditional inference. *J. R. Statist. Soc. B* **49**, 1–39 (with discussion).

Fernández, C. and Steel, M. F. J. (1999). Reference priors for the location-scale model. *J. Statist. Plann. Inference* **43**, 377 –384.

Fraser, D. A. S. (2002). Likelihood for component parameters. *Biometrika* (to appear).

Fraser, D. A. S. (2003). Ancillaries and conditional inference. *Statist. Sci.* (to appear).

Fraser, D. A. S. and Reid, N. (1996). Bayes posteriors for scalar interest parameters. *Bayesian Statistics 5* (J. M. Bernardo, J. O. Berger, A. P. Dawid and A. F. M. Smith, eds). Oxford: Oxford University Press, 581–585.

Fraser, D. A. S. and Reid, N. (2000). Ancillary information for statistical inference. *Empirical Bayes and Likelihood Inference* (S. E. Ahmed and N. Reid, eds). Berlin: Springer, 185–207.

Fraser, D. A. S. and Reid, N. (2002). Strong matching of frequentist and Bayesian inference. *J. Statist. Plann . Inference* (to appear).

Fraser, D. A. S., Reid, N. and Wu, J. (1999). A simple general formula for tail probabilities for frequentist and Bayesian inference. *Biometrika* **86**, 249–264.

Sun, D. (1997). A note on noninformative priors for Weibull distributions. *J. Statist. Plann . Inference* **61**, 319–338.

APPENDIX A

In the presence of an exact full ancillary $a(y)$ the vectors V are available as a basis for the orthogonal complement of the ancillary gradient $(\partial/\partial y)a'(y)$; this can be obtained as the last p row vectors of A^{-1} where $A = (\partial/\partial y)\{a'(y), \ell_{\theta'}(\theta; y)\}$ evaluated at $(y^0, \hat{\theta}^0)$. And more generally, in the presence of a full n-dimensional pivotal quantity say $z(y)$, the vectors V are available (Fraser and Reid, 2000) as

$$V = -z_{y'}^{-1}(y^0; \hat{\theta}^0) z_{;\theta'}(y^0; \hat{\theta}^0).$$

In the presence of independent scalar coordinates or in the presence of independent vector coordinates each with an associated pivotal quantity, reasonable continuity shows that the conditioning is essentially unique for third-order inference and that only the observed likelihood $\ell^0(\theta)$ and the observed likelihood gradient $\varphi(\theta)$ conforming to the ancillary conditioning are needed for third-order accuracy.

BAYESIAN STATISTICS 7, pp. 535–542
J. M. Bernardo, M. J. Bayarri, J. O. Berger, A. P. Dawid,
D. Heckerman, A. F. M. Smith and M. West (Eds.)

Dynamic Lattice-Markov Spatio-Temporal Models for Environmental Data

LINDA M. GARSIDE and DARREN J. WILKINSON
Newcastle University, UK
l.m.garside@ncl.ac.uk d.j.wilkinson@ncl.ac.uk

SUMMARY

This paper is concerned with the modelling of the latent structure of a Bayesian spatio-temporal model with a view to improving parameter inference, smoothing and prediction. The equilibrium distribution of a time stationary system will be examined, paying particular attention to edge-effects and the effect of grid-coarsening. In order to develop an effective MCMC algorithm, the latent process is integrated out of the model. These techniques will be illustrated using North Atlantic ocean temperature data.

Keywords: COARSE AND FINE MODELS; EDGE-EFFECTS; MARGINAL UPDATING; MCMC.

1. INTRODUCTION

This paper concerns Bayesian inference for dynamic lattice-Markov models with a latent process of Spatio-Temporal Auto-Regressive (STAR) form (Cressie 1991; Pfeifer and Deutsch, 1980), observed irregularly, and with error. Attention focuses on the 2+1-D case. A point at location s and time $t+1$ is influenced by the point s and 8 neighbors at time t. Well-known "edge-effects" destroy the spatial stationarity of finite lattices, and this can lead to biased inferences. In order to reduce bias, the autoregressive weights at the edges are adjusted using multivariate Normal methods since the "usual" solution of embedding the area of interest into a much larger lattice is computationally prohibitive.

The problem of estimating the model parameters from data is explored. Gibbs and block-Gibbs sampling methods suffer from poor mixing due to the high dependence between the parameters and the latent process. Integrating out the latent process leads to an MCMC scheme with good mixing properties. Other approaches to hierarchical modelling in a similar context can be found in Wikle *et al.* (1998). Our technique is applied to Atlantic sea temperatures measured at a depth of 100 m, within the region 20 to 30 degrees latitude and -80 to -20 degrees longitude, for 1978–1988. The data are collected by ships trawling the area and thus occurs irregularly within space and time.

The method described above is computationally intensive. One method for reducing the computation time is to consider modelling on a coarser grid (Higdon *et al.,* 2003); here a grid based on one in four spatial locations is considered. The coarse model has second-order time dependence, but the parameters have the same interpretation as for the fine model.

2. THE MODEL

Consider a Dynamic Linear Model (DLM) which can evolve through space and time (West and Harrison, 1997). The model is described using two equations:

(i) System equation

$$\theta^{(t+1)} \mid \theta^{(t)} \sim \mathrm{N}(G\theta^{(t)}, W), \qquad W = \tau_s^{-1} I;$$

(ii) Observation equation

$$X^{(t)} \mid \theta^{(t)} \sim \mathrm{N}(F_t\theta^{(t)}, V), \qquad V = \tau_o^{-1} I,$$

where $\theta^{(t)} = (\theta_1^t, \theta_2^t, \ldots, \theta_n^t)$ is the vector of latent points in space at time t. Point $\theta_i^{(t)}$ represents location i at time t and G is the matrix which captures the evolution of the the points in space through time. Corresponding to each vector $\theta^{(t)}$ we have a vector of observations $X^{(t)}$, related by the observation matrix F_t. Typically F_t will consist of a subset of rows of the identity matrix corresponding to the points in space for which data is available at time t.

2.1. *2+1-D Model*

Consider the case where we have data on a two-dimensional regular square lattice in space evolving through discrete time. Initialize the latent structure as a grid of independent spatial observations at time $t = 0$. For $t > 0$, the latent points at time $t + 1$ depend on a linear combination of the points at time t. In this model, the vector of latent points $\theta^{(t)}$ is defined to be

$$\theta^{(t)} = \left(\theta_{1,1}^{(t)}, \ldots, \theta_{1,n}^{(t)}, \theta_{2,1}^{(t)}, \ldots, \theta_{2,n}^{(t)}, \ldots, \theta_{m,n}^{(t)}\right).$$

The obvious first-order (five-neighbor) model suffers badly from lattice co-ordinate system artifacts. These can be reduced by considering a second-order (nine-neighbor) model with carefully chosen weights. This corresponds to a dynamic space-time version of the "second-order" neighbor system described by Besag (1974). The model is defined by

$$\begin{aligned} \theta_{i,j}^{(t+1)} \mid \theta^{(t)} \sim N\Big(\alpha\Big\{ & p_{11}^2\theta_{i-1,j-1}^{(t)} + 2p_{11}p_{12}\theta_{i-1,j}^{(t)} + p_{12}^2\theta_{i-1,j+1}^{(t)} \\ & + 2p_{11}p_{12}\theta_{i,j-1}^{(t)} + (2p_{11}p_{22} + 2p_{12}p_{21})\theta_{i,j}^{(t)} + 2p_{12}p_{22}\theta_{i,j+1}^{(t)} \\ & + p_{21}^2\theta_{i+1,j-1}^{(t)} + 2p_{21}p_{22}\theta_{i+1,j}^{(t)} + p_{22}^2\theta_{i+1,j+1}^{(t)}\Big\}, \tau_s^{-1}\Big), \end{aligned} \tag{1}$$

where $\sum_{i,j} p_{ij} = 1$. This model, therefore, has five free parameters. The three free p_{ij} parameters control the degree of anisotropy, α is an autoregressive parameter, and τ_s controls the degree of innovation noise. Stationarity of this system for $\alpha \in [0, 1)$ is established in Pfeifer and Deutsch (1980). The precise form of the weight structure is motivated by considering an additional "hidden" layer at time $t+1/2$, which is also a square lattice, but perfectly offset from the lattices at times t and $t + 1$. If a first-order (four-neighbor) model with edge weights p_{ij} is imposed on the full lattice structure, marginalizing out the hidden layer leads to model (1).

2.2. *Correcting for Edge-effects*

This paper considers only the isotropic case of the model defined by (1) with weights $p_{ij} = 1/4$ giving

$$\begin{aligned} \theta_{ij}^{(t+1)} \mid \theta^{(t)} \sim N\Big(\alpha\Big\{ & \frac{1}{16}\Big[\theta_{i-1,j-1}^{(t)} + \theta_{i-1,j+1}^{(t)} + \theta_{i+1,j-1}^{(t)} + \theta_{i+1,j+1}^{(t)}\Big] \\ & + \frac{1}{8}\Big[\theta_{i-1,j}^{(t)} + \theta_{i,j-1}^{(t)} + \theta_{i,j+1}^{(t)} + \theta_{i+1,j}^{(t)} + \frac{1}{4}\theta_{i,j}^{(t)}\Big\}, \tau_s^{-1}\Big). \end{aligned}$$

The above equation only defines the distribution of the internal nodes. Along the edges and at the corners of the system some neighbors are missing. Standard approaches either simply drop terms corresponding to missing neighbors, or drop terms and then re-scale the remaining weights so that they sum to one. Both of these strategies lead to prominent "edge-effects" which affect the time stationary distribution of the process in space. In the purely spatial context, such effects are often reduced by embedding the region of interest into a much larger lattice. However, in the spatio-temporal context, such an approach is typically computationally infeasible. In the context of a purely spatial Gaussian Markov random fields, more sophisticated approaches to edge weight adjustments have been examined in detail by Besag and Kooperberg (1995). Here we adopt a related approach, better suited to the context of dynamic spatio-temporal models. For a given value of α, the process is linear Gaussian and the time-stationary joint distribution may be computed using numerical linear algebra. From this, the joint distribution of any edge point and its neighbors may be computed, leading to the conditional distribution of the point given available neighbors. This process only needs to be carried out for one typical edge node and one typical corner node, but must be re-computed for each possible α. In practice, the conditional distribution is computed for 100 different values of α in the range $[0, 1)$, and linear interpolation is used for other values of α. Note that the conditional distribution does not need to be re-computed for every possible (α, τ_s) pair as the conditional distributions re-scale in the obvious way for varying τ_s. Figure 1 shows an example of the time-stationary spatial variance surface for the three described modelling strategies with $\alpha = 0.99$ on a 10×10 spatial grid. The first plot represents the model with no edge adjustments (dropping terms corresponding to non-existent neighbors), the second plot has had the edge weights adjusted in a simple way (drop terms and re-scale edge weights to sum to one), whereas the third plot shows the results for adjusting both the edge weights and correcting the variance according to the time stationary distribution. This latter plot is much "flatter" than the previous two, and hence closer to the desired spatial stationarity.

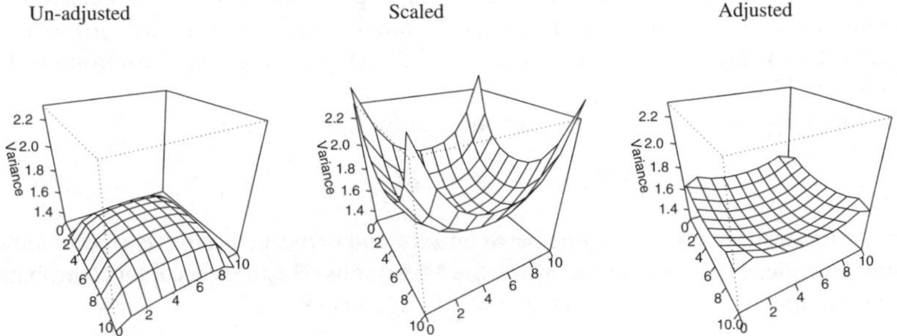

Figure 1. *Spatial variance surfaces for time-stationary distributions based on un-adjusted, scaled and adjusted edge weights.*

2.3. *Coarsening the Model*

For given parameter values, the model described in the previous subsection performs well for spatio-temporal smoothing and interpolation on reasonably fine lattices. However, as the model underlying a good MCMC scheme for parameter inference, it is computationally demanding. One way to reduce the computing time required is to consider a coarser model resulting from considering only one in every four points in space. Unfortunately, if the original fine model is marginalized to the coarser lattice the resulting process is not Markovian, of any order, but can be well approximated by a Markov process of order 2. Each point in space and time $t+2$ is regarded as depending on the nearest neighbor at time $t+1$ and the nine neighbors at time t, as shown in Figure 2.

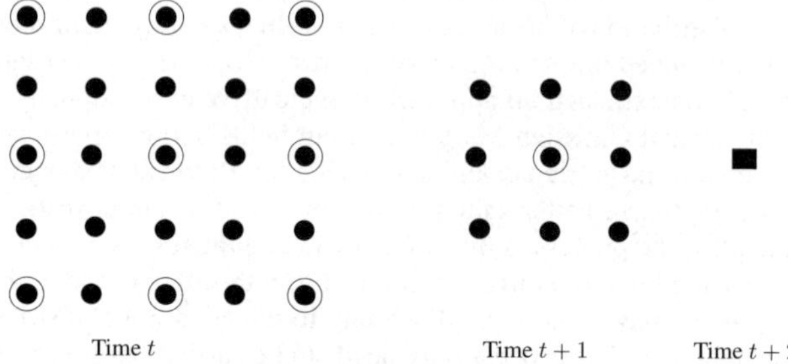

Figure 2. *Circled nodes represent the neighbors in a 2+1-D coarse model.*

As is the case for the edge-effect corrections, analysis of the stationary distribution can uncover the appropriate weights and conditional variances for the coarse model in a manner which makes parameters consistent and share interpretations between the coarse and fine scales.

3. PARAMETER ESTIMATION

In this model we have three unknown parameters, τ_s, τ_o, and α, and latent process $\theta = \{\theta^{(t)} \mid t = 1, 2, \ldots, T\}$. Naive MCMC techniques (such as univariate Gibbs samplers) perform particularly poorly for problems of this type due to the very high dimensional nature of the missing data. A two-block data augmentation scheme which alternately samples $\sigma \mid \theta, X$ and $\theta \mid \sigma, X$, where $\sigma = (\tau_s, \tau_o, \alpha)$, performs much better than the naive techniques (Wilkinson and Yeung, 2002a), but still suffers from poor mixing due to the high posterior dependence between σ and θ (Papaspiliopoulos *et al.,* 2003). Fortunately, a sampler with good mixing properties can be constructed by integrating θ out of problem and directly constructing a Metropolis–Hastings scheme for $\sigma \mid X$.

The following independent priors are adopted for the parameters:

$$\tau_i \sim \text{Ga}(b_i, d_i), \quad i \in \{s, o\}, \qquad \alpha \sim \text{Be}(\gamma, \delta).$$

The beta prior ensures that α remains between zero and one (the intuitively plausible range for temporal dependence). Using these priors, the Metropolis–Hastings acceptance probability for the proposed update $\sigma^* = (\tau_s^*, \tau_o^*, \alpha^*)$ is $\min\{1, A\}$, where

$$A = \frac{[\tau_s^*][\tau_o^*][\alpha^*][X \mid \sigma^*] f(\sigma \mid \sigma^*)}{[\tau_s][\tau_o][\alpha][X \mid \sigma] f(\sigma^* \mid \sigma)}, \tag{2}$$

and $f(\sigma^* \mid \sigma)$ is the proposal density for the parameters. If a symmetric density is used (2) becomes

$$A = \frac{[\tau_s^*][\tau_o^*][\alpha^*][X \mid \sigma^*]}{[\tau_s][\tau_o][\alpha][X \mid \sigma]}.$$

To obtain a rapidly mixing chain, normal random walk updates on the log-precisions work well, and so the density of the log of a gamma is used in the the acceptance probability. The marginal log-likelihood ratio can be calculated in a variety of ways (*e.g.*, using the Kalman filter or causal tree propagation), but is computationally demanding whatever strategy is adopted. The authors recommend using sparse matrix techniques, such as those implemented in the C software library GDAGsim (Wilkinson, 2001). The proposal is constructed by independently sampling

$$\log \tau_o^* \mid \tau_o \sim \mathrm{N}(\log \tau_o, g^{-1}),$$

$$\log \tau_s^* \mid \tau_s \sim \mathrm{N}(\log \tau_s, g^{-1}),$$

$$\alpha^* \mid \alpha \sim \mathrm{N}(\alpha, e),$$

where e and g are Metropolis–Hastings tuning parameters. For a more detailed discussion of the MCMC techniques which can be used for large linear models, see Wilkinson and Yeung (2002b).

4. EXAMPLE: ATLANTIC OCEAN TEMPERATURE DATA

We illustrate these techniques with spatio-temporal data obtained from the National Oceanographic Data Center (`http://www.nodc.noaa.gov/`). The data used in this example were taken from a rectangular region of the Atlantic ocean 20 to 30 degrees latitude and -80 to -20 degrees longitude. Ocean station data (NANSEN) was considered for the time period 1979–1988 inclusive. Observations were indexed according to their time and location for depths of approximately 100 m. The location of the data in space and time is illustrated graphically in Figure 3. Similar data have been analyzed previously by Lavine and Loizier (1999) and examined at one location over time. They also looked at data from a different spatial location between 1905 and 1988, as in Higdon (1998) and Stroud *et al.* (2001).

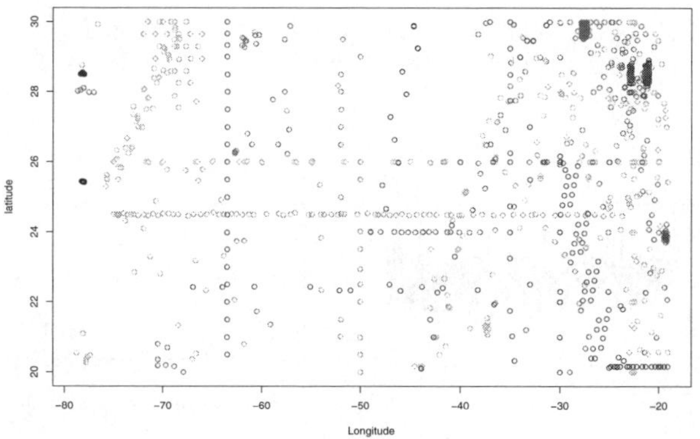

Figure 3. *Sites in space for which data is available (shade indicates time)*

In this example, we consider a fixed α of 0.9 as there is very little information regarding α in this data set. Expert subjective priors for all three parameters would improve on this analysis, and should reduce the confounding between α and τ_s. For illustrative purposes, we adopt vague proper prior distributions for the precisions

$$\tau_s \sim \mathrm{Ga}(1,\, 0.001), \qquad \tau_o \sim \mathrm{Ga}(1,\, 0.001)$$

and consider a $\mathrm{N}(0, 400^{-1})$ random walk update on each component independently. We model the data on a $12 \times 60 \times 10$ grid, but carry out MCMC on the coarsened version of the model, which uses a $6 \times 30 \times 10$ grid. Point estimates (posterior means) of the parameters are computed from the MCMC run, and these are then input into the fine model for smoothing/interpolation.

The results from a typical MCMC run are given in Tables 1 and 2 and Figures 4 and 5. Note, in particular, the rapid decay in autocorrelations and convergence to stationarity signifying a rapidly mixing chain. The posterior mean parameter estimates are $\log(\tau_s) = -0.56$, and $\log(\tau_o) = -0.87$, based on 9,000 iterations. Due to the consistency between the coarse and fine formulations, these parameter estimates can be used as plug-in estimates for the parameters of a fine model for the purposes of smoothing and interpolation. This allows the computation of posterior mean temperature surfaces on the fine scale, as shown in Figures 6, 7 and 8. The height represents the posterior mean temperature conditional on the plug-in parameter estimates from the coarse model MCMC run.

Table 1. *Posterior means and associated standard errors from an MCMC run of 10,000 with 1000 iterations discarded as burn-in and no thinning (conditional on $\alpha = 0.9$).*

	Mean	SD	Naive SE	Time-series SE
$\log(\tau_s)$	-0.5573	0.09848	0.0010381	0.001836
$\log(\tau_o)$	-0.8682	0.04005	0.0004222	0.000759

Table 2. *Posterior quartiles for the parameters from an MCMC run of 10,000 iterations with 1000 iterations discarded as burn-in and no thinning (conditional on $\alpha = 0.9$).*

	2.5%	25%	50%	75%	97.5%
$\log(\tau_s)$	-0.7477	-0.6243	-0.5582	-0.4911	-0.3613
$\log(\tau_o)$	-0.9499	-0.8941	-0.8677	-0.8424	-0.7886

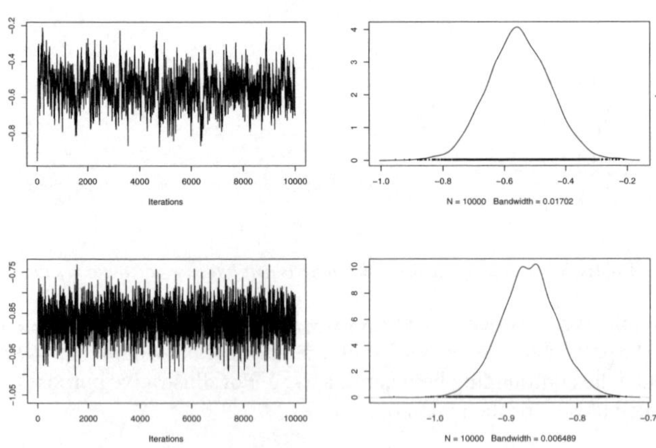

Figure 4. *Trace and density plots for $\log(\tau_s)$ and $\log(\tau_o)$ with α fixed at 0.9, based on 10,000 iterations and no thinning.*

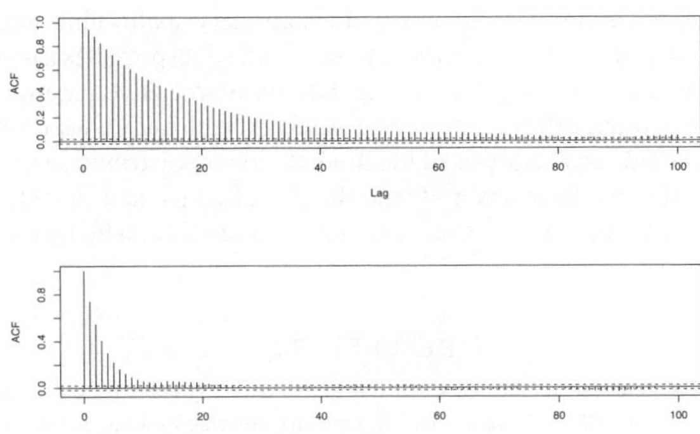

Figure 5. *Autocorrelation plots for* $\log(\tau_s)$ *and* $\log(\tau_o)$ *with* α*=0.9.*

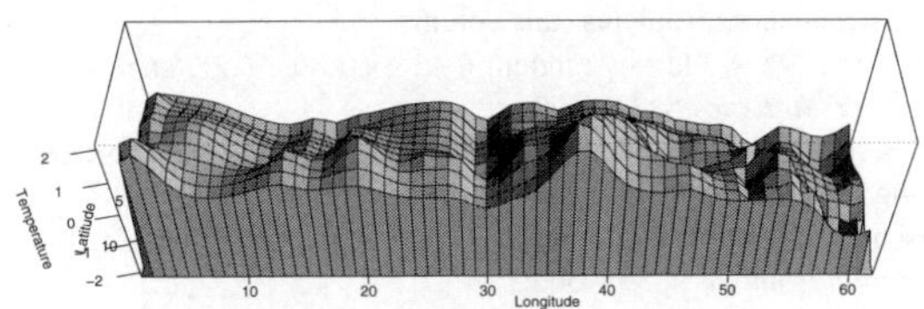

Figure 6. *Surface plot for 1984.*

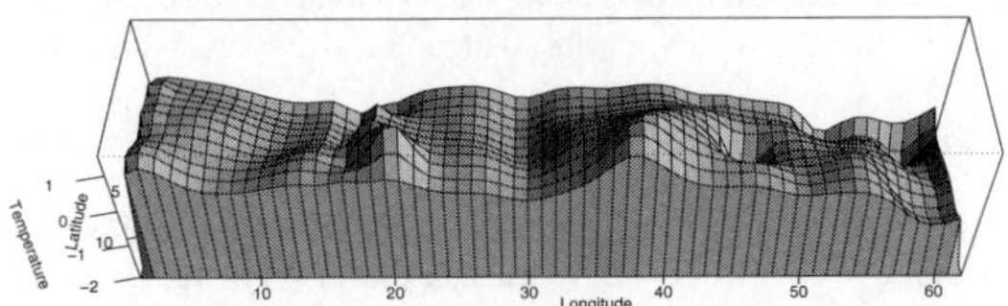

Figure 7. *Surface plot for 1986.*

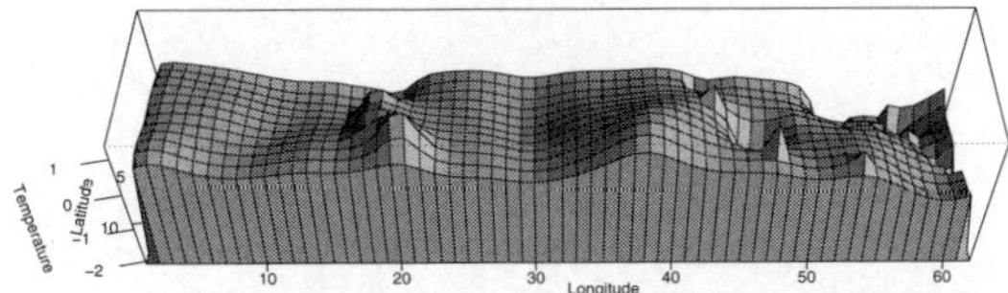

Figure 8. *Surface plot for 1988.*

The surface plots (Figures 6–8) show how the temperature field varies both in space and in time. The original data set had far more data points associated with the earlier years in the study, and this could account for the increased smoothness in the plots for later time periods.

5. CONCLUSIONS

Using a STAR model to capture the dynamics of a stationary spatio-temporal process and then using the DLM formulation to relate the process to available data provides a powerful and flexible framework for inference. Because of the computationally intensive nature of spatio-temporal models it is desirable to reduce the edge-effects associated with lattice-Markov systems, and this can be carried out via numerical analysis of the time-stationary distribution of the STAR process. Similar computational considerations motivate the use of coarsened models constructed so as to remain consistent with the original fine formulation, and this can be tackled in a similar way to the edge-effects.

REFERENCES

Besag, J. (1974). Spatial interaction and the statistical analysis of lattice systems. *J. R. Statist. Soc. B* **36**, 192–236.

Besag, J. and Kooperberg, C. (1995). On conditional and intrinsic autoregressions. *Biometrika* **82**, 733–746.

Cressie, N. A. C. (1991). *Statistics for Spatial Data.* Chichester: Wiley.

Higdon, D. (1998). A process-convolution approach to modelling temperatures in the North Atlantic ocean. *Environ. Ecol. Statis.* **5**, 173–190.

Higdon, D., Lee, H. and Holloman, C. (2003). Markov chain Monte Carlo-based approaches for inference in computationally intensive inverse problems (this volume).

Lavine, M. and Loizier, S. (1999). A Markov random field spatio-temporal analysis of ocean temperature. *Environmental and Ecological Statistics* **6**, 249-273.

Papaspiliopoulos, O., Roberts, G. O. and Sköld, M. (2003). Non-centered parameterizations for hierarchical models and data augmentation (this volume).

Pfeifer, P. E. and Deutsch, S. J. (1980). Stationarity and invertibility regions for low order STARMA models. *Comm. Statist. Simulation Comput.* **9**, 551–562.

Stroud, J. R., Muller, P. and Sanso, B. (2001). Dynamic models for spatio-temporal data. *J. R. Statist. Soc. B* **63**, 673–689.

West, M. and Harrison, J. (1997). *Bayesian Forecasting and Dynamic Models*, 2nd edn. New York: Springer.

Wikle, C. K., Berliner, M. L. and Cressie, N. (1998). Hierarchical Bayesian space-time models. *Environ. Ecol. Statist.* **5**, 117–154.

Wilkinson, D. J. (2001). GDAGsim 0.2 User Guide. *Tech. Rep.*, University of Newcastle. UK. `www.staff.ncl.ac.uk/d.j.wilkinson/software/gdagsim`

Wilkinson, D. J. and Yeung, S. K. H. (2002a). Conditional simulation from highly structured Gaussian systems, with application to blocking-MCMC for the Bayesian analysis of very large linear models. *Statist. Comput.* **12**, 287–300.

Wilkinson, D. J. and Yeung, S. K. H. (2002b). A sparse matrix approach to Bayesian computation in large linear models. *Comput. Statist. Data Anal.* (to appear).

BAYESIAN STATISTICS 7, pp. 543–552
J. M. Bernardo, M. J. Bayarri, J. O. Berger, A. P. Dawid,
D. Heckerman, A. F. M. Smith and M. West (Eds.)

Lymphoscintigraphy of Upper Limbs: A Bayesian Framework

PETR GEBOUSKÝ   MIROSLAV KÁRNÝ
Akademie Věd České Republiky, Czech Republic
gebousky@utia.cas.cz   school@utia.cas.cz

ANTHONY QUINN
University of Dublin, Ireland
aquinn@tcd.ie

SUMMARY

Lymphoscintigraphy is a sensitive diagnostic technique in nuclear medicine. One of its principal applications is the investigation of upper limb lymphedema, a complication of breast cancer. Typically, only two or three snapshots of the distribution of the radioactive tracer in the limb can be obtained, implying a sparse data problem. Hence, traditional inferences of important physiological indicators are completely unreliable. In this paper, the Bayesian paradigm, exploiting available prior information in conjunction with a simplified model of the diffusion dynamics, is used to obtain reliable quantitative evaluations for the first time.

Keywords: DYNAMIC MODEL IDENTIFICATION; BAYESIAN ESTIMATION; SPARSE DATA; UPPER LIMB LYMPHEDEMA; QUANTITATIVE LYMPHOSCINTIGRAPHY.

1. INTRODUCTION

The power of the Bayesian paradigm is evident in inference problems based on a few measurements only. Nuclear medicine is a rich and important application domain where problems of this type are the norm. Diagnostics for lymphedema of the upper limbs, based on *quantitative* scintigraphy, is a case in point. The contributions of the paper to this area are: (a) quantitative lymphoscintigraphy, unavailable until now, is developed and tested; (b) standard prior-regularized Bayesian tools are shown to solve a surprisingly wide range of real tasks; and (c) the modelling role of the Bayesian paradigm is underlined.

Secondary lymphedema—excessive arm swelling caused by a damaged lymphatic system—is a frequently occurring side-effect of breast cancer treatment. It can be simply corrected when diagnosed at a very early stage. Late stages are hard to cure and often lead to full disability. Lymphoscintigraphy—*i.e.*, imaging of the time-dependent dispersion of the injected radionuclide—is used frequently to complement clinical inspections (Fruhling and Bourgeois, 1983; Weissleder and Weissleder, 1988; Verlooy *et al.*, 1997). It visualizes the regional lymph drainage system in a sensitive and non-invasive way (Henkin *et al.*, 1996). Structural and functional information is yielded at the injection depot, along the extremity, and over regional lymph nodes. Qualitative morphological inspection is sufficient for late disease stages. Diagnostics for the critical early stages are poorly supported, with results depending enormously on the skill of the inspecting expert. Quantitative evaluation is currently unavailable. The main reason is that the condition of the patient, as well as economic and time factors, allow only two or three scans of the extremities. A Bayesian attack, one which supplements the sparse data consistently with a rich set of prior knowledge, appears to be the only viable remedy.

2. LYMPHOSCINTIGRAPHIC DATA

In a standardized inspection, about 25 MBq of a $^{99\mathrm{m}}$Tc-labelled solution is injected into an interstitial space of each hand. The injected radio-tracer is carried by lymphatic flow from the interstitial space, through the limb, and finally to the liver.

The scintigraphic images of the radio-tracer visualize the accumulation and flow of the serum through the lymphatic system of the limbs. Its time evolution reflects the state of a patient's lymphatic system. Each image is accumulated over one minute. Images are taken immediately after injection, and then after about 30 and 180 min, respectively. Background is subtracted, correction made for physical decay, and regions of interest (ROI) specified. Figure 1 shows typical ROIs around the axillary nodes, and the upper and lower halves of the arm. Local relative activities for each image are aggregated into the total count over each ROI, and then normalized with respect to the total count in the complete first image, where the latter reflects the administered activity. This calibrates the aggregated data available for quantitative analysis.

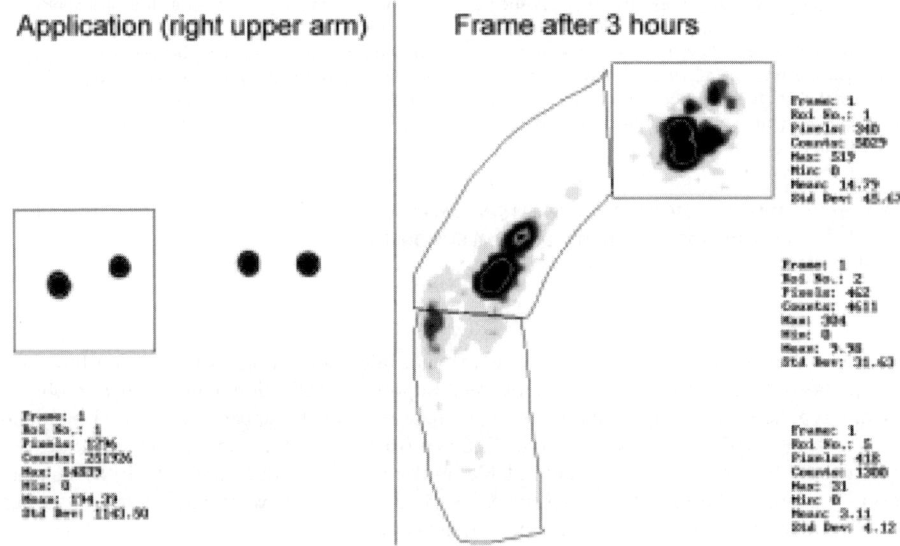

Figure 1. *Typical scintigraphic images of the upper arm, including definition of ROIs.*

3. A PARAMETRIC SYSTEM MODEL

Let t be the number of minutes since the administration time at $t = 0$. Hence, real time is $\tau(t) = t\Delta$, $\Delta = 1$ (min). The actual sampling times are $\mathcal{T} = (t_1, \ldots, t_N)$. Their number, N, is small, typically two or three. The injected tracer is a known input into the lymphatic limb system. The *relative* amount of the injected tracer is a *unit impulse*, *i.e.*, $u_t = \delta_t$ where $\delta_0 = 1$ and $\delta_t = 0$, for $t > 0$. The discrete-time scalar *impulse response*, x_t, of the lymphatic system at the chosen ROI, known as a *time activity curve*, is the relative integral count at time $\tau(t)$ over the pixels of the ROI. Causality implies that $x_t = 0$ for $t < 0$.

The dynamic model relating the sequences u_t and x_t (Oppenheim *et al.*, 1999, see Figure 2) is chosen as a cascade of first-order linear models, with a common parameter a for each of the d sections, and with a single lumped gain parameter, b. It is chosen as a compromise between the complex distributed nature of the lymphatic system and the need for a model with few unknown parameters. This cascading of simple sections describes the gradual penetration of serum through the limb.

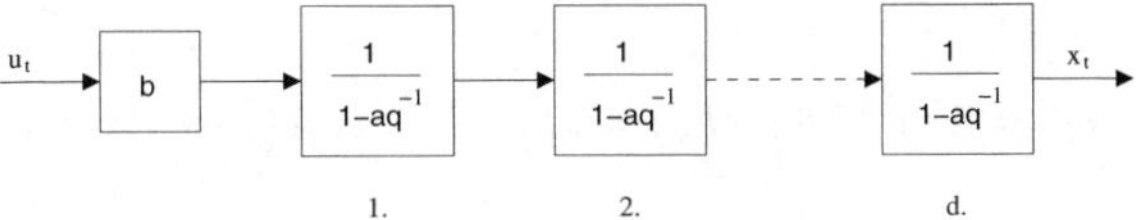

Figure 2. *Parametric cascade model whose impulse response models the lymphoscintigraphy time activity curves (q^{-1} is the backward shift operator, i.e., $q^{-1}x_t = x_{t-1}$).*

Binomial expansion of the system denominator leads to the difference equation:

$$x_t = -\sum_{i=1}^{d} \binom{d}{i} (-a)^i x_{t-i} + bu_t.$$

For $u_t = \delta(t)$, x_t models the *impulse response* of the lymphatic system observed at times t at the particular ROI. Its closed form solution (Oppenheim *et al.*, 1999) is:

$$x_t = b\binom{t+d-1}{t} a^t, \qquad t \geq 0. \tag{1}$$

A rich signal ensemble is generated by the proposed parametric model. It successfully captures —with only three free parameters—the stable, slowly-decaying, non-oscillatory nature of the lymphatic system responses at a particular ROI. In particular, the order parameter, d, allows a rich set of candidate curves to be explored, including those exhibiting a convex nature. Typical curves are illustrated in Figure 3, for $x_0 = b = 1$.

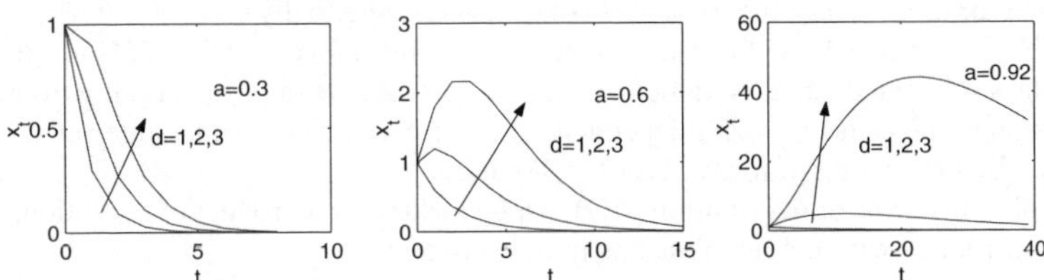

Figure 3. *Typical time activity curves for modelling the time evolution of the lymphatic system at a particular ROI ($b = 1$ in all cases).*

The complete time activity curve is $X = (x_0, x_1, \ldots, x_t, \ldots)'$:

$$X = b\tilde{A}_\vartheta, \quad \tilde{A}_{\vartheta,t+1} = \tbinom{t+d-1}{t} a^t, \quad t = 0, 1, \ldots, \quad \vartheta = (a, d)'. \tag{2}$$

The (noiseless) observations are gathered into the N-vector $X_o = bA_\vartheta$ for the subset of times $\tau(t), t \in \mathcal{T}$ $(A_\vartheta = \tilde{A}_\vartheta(\mathcal{T}))$.

Noisy samples, y_t, of the time activity curves are observed. As emphasized in Section 2, y_t are normalized, aggregated counts for the ROI at time t. Individual counts have a Poisson distribution, but aggregation permits the overall noise process to be approximated well by additive, zero-mean, normal noise e_t, *i.e.*, $y_t = x_t + e_t$. Long inter-sample intervals imply that $e_t, t \in \mathcal{T}$, are approximately independent. The variance r of e_t can be assumed (approximately) constant. Thus, the probability density function (pdf) for the N-vector of observations $Y = (y_1, y_2, \ldots, y_{t_N})'$, is:

$$p(Y \mid \mathcal{T}, \Theta) = \mathcal{N}(Y \mid bA_\vartheta, rI_N) = (2\pi r)^{-N/2} \exp\left(-\frac{1}{2r}||Y - bA_\vartheta||^2\right). \tag{3}$$

$||\cdot||$ denotes the Euclidean norm. The observation model is parameterized by the quadruple $\Theta = (a, b, d, r)'$. The dependence of $p(Y \mid \mathcal{T}, \Theta)$ on the set of actual observation times, $\mathcal{T}$, is expressed explicitly in the notation, to facilitate the later exposition.

4. PRIOR INFORMATION AND BAYESIAN ESTIMATION

Parameters Θ are unknown, and require prior distributions to be elicited. The variance, r, is a property only of the measurement process, not of the patient or the particular ROI. Thus, for its estimation, data from various ROIs and patients can be used. The remaining three parameters are strictly local to the ROI and the patient. They have to be estimated using two or three local patient-specific measurements. This is impossible without prior information, which is rich in this case. This is the key advantage of the Bayesian paradigm in the inference of diagnostically significant quantities from the sparse data. The prior information is expressed through intervals of *a priori* possible Θ, given with comments in the following list:

- $a = \vartheta_1 \quad (0 < a < 1)$: the inspected responses are stable and non-oscillatory; the interval has been shrunk to reflect practically observed slow accumulation dynamics (typically $\vartheta_1 \in (0.9, 0.999)$); specific values depend on model order.
- $d = \vartheta_2 \quad (1 < d \leq \bar{d} \approx 6)$: the parameter describes the penetration rate through the limb and modifies the curve shape; the current limited experience indicates that 6 is the conservative upper bound of the model order.
- $b \quad (0 < b < \bar{b}_\vartheta)$: the response is non-negative and cannot exceed the applied input $u_0 = 1$ (the maximum, $\bar{b}_\vartheta$, as a function of ϑ is evaluated numerically).
- $r \quad (10^{-6} < r < 10^{-4})$: the range corresponds with interval (0.5%, 25%) of the noise in the observed signal; smaller values of r are more probable due to averaging over ROIs; the common variance is *a priori* independent of other parameters. This range can be gradually refined by processing of many sets of ROI data.

The mixed-type prior distribution $p(\Theta)$ can be written as a product of conditional distributions in accordance with the relationships in the above list:

$$p(\Theta) = p(a, b, d, r) = p(b \mid a, d, r)p(a \mid d)p(d)p(r) = p(b \mid \vartheta, r)p(a \mid d)p(d)p(r).$$

The variance r is included in the first factor to simplify evaluations.

The unknown continuous parameter $a = \vartheta_1$ enters the observation model in a nonlinear way (1). To get a feasible solution, it has been discretized. Hence, the pdf $p(a \mid d)$ is replaced by a pf on a discrete grid. The polynomial dependence on a indicates that the influence of a decays exponentially. This motivates selection of a uniform grid on $\exp(a)$. A uniform pf on the *a priori* expected range of the discrete order parameter, *d*, is selected. Since further detailed information is unavailable, the uniform distribution is chosen for both these parameter grids, and justified via the principle of insufficient reason (Jeffreys, 1961).

The remaining pdf's on continuous-valued b and r are expressed in conjugate form (Berger, 1985). This flexible choice simplifies evaluations. For the gain b, the Gaussian pdf is conjugate. Thus:

$$p(b\,|\,\vartheta, r) = \mathcal{N}(b\,|\,\hat{b}_\vartheta, \omega_\vartheta r) = (2\pi\omega_\vartheta r)^{-1/2} \exp\left(-\frac{1}{2\omega_\vartheta r}||b - \hat{b}_\vartheta||^2\right),$$

where $\omega_\vartheta > 0$ and $\hat{b}_\vartheta$ determine this prior. The expected range of b gives $\hat{b}_\vartheta = 0.5\bar{b}_\vartheta$, $\omega_\vartheta = \bar{b}^2_\vartheta/(4\hat{r})$, where $\hat{r}$ is a conservative estimate of the measurement variance r, leading to a relatively flat prior pdf construction. The adopted choice corresponds to one standard deviation on both sides of the mean. The conjugate prior for estimation of $r > 0$ is:

$$p(r) = \left(\frac{(\kappa-2)\varepsilon}{2}\right)^{\frac{\kappa}{2}} \Gamma^{-1}\left(\frac{\kappa}{2}\right) r^{-\frac{\kappa}{2}-1} \exp\left(-\frac{(\kappa-2)\varepsilon}{2r}\right),$$

where $\Gamma(\cdot)$ is the Euler gamma function. This pdf is parameterized by $\varepsilon > 0$ and $\kappa > 0$. It has expectation ε and variance $2\varepsilon^2/(\kappa - 4)$. Use of the Gaussian approximation of this pdf and the physical confidence interval, with half-width equal to one standard deviation, give the choice $\kappa = 7$, $\varepsilon = 3 \times 10^{-5}$, $\hat{r} = 10^{-5}$. A more resolved choice is unnecessary as the interval serves only as an initial conservative guess for specification of the prior pdf on b. Also, it initializes the estimation of common r, which can be improved with each patient.

The Bayesian framework (Berger, 1985; Peterka, 1981) combines the observation model $p(Y\,|\,\mathcal{T}, \Theta)$ (3) and the quantified prior information $p(\Theta)$ into the posterior pdf:

$$p(\Theta\,|\,\mathcal{T}, Y) = \frac{\omega_\vartheta^{-1/2} r^{-(N+\kappa+3)/2} p(\vartheta)}{Cp(Y\,|\,\mathcal{T})} \exp\left[-\frac{1}{2r}\left(||Y - bA_\vartheta||^2 + (\kappa-2)\varepsilon + \frac{||b - \hat{b}_\vartheta||^2}{\omega_\vartheta}\right)\right]. \quad (4)$$

The normalizing constant C is independent of observations Y and measurement times $\mathcal{T}$. The predictive pdf $p(Y\,|\,\mathcal{T})$ in the denominator can be used for determining a good set of sampling time indices $\mathcal{T}$. Introducing the normalized quantities,

$$B_\vartheta = \sqrt{\omega_\vartheta} A_\vartheta, \qquad \hat{B}_\vartheta = \hat{b}_\vartheta / \sqrt{\omega_\vartheta},$$

the predictive pdf is expressed up to another normalizing constant, c, as follows:

$$p(Y\,|\,\mathcal{T}) = c \sum_\vartheta \frac{p(\vartheta)\left(||B_\vartheta||^2 + 1\right)^{-1/2}}{\left(||Y||^2 + \varepsilon + \hat{B}^2_\vartheta - \frac{(Y^{\mathrm{T}} B_\vartheta + \hat{B}_\vartheta)^2}{||B_\vartheta||^2 + 1}\right)^{(N+\kappa)/2}}. \quad (5)$$

Note that the posterior pdf, $p(\Theta\,|\,\mathcal{T}, Y)$, is non-zero only on the grid of the *a priori*-allowed values of $\vartheta = (a, d)$. Correspondingly, the summation in (5) runs over these values only.

5. INFERENCE OF DIAGNOSTICALLY SIGNIFICANT QUANTITIES

The posterior pdf, $p(\Theta\,|\,\mathcal{T}, Y)$, is the starting-point for computing posterior properties of unknown parameters (or functions thereof), several of which have clinical significance.

5.1. *Estimation of the Expected Time Activity Curve, $\hat{X}$*

Construction of this curve has been the principal motivation for the modelling and estimation techniques described in this paper. It follows from (2) that X is a deterministic function of the unknown parameters, Θ. The distribution on the ensemble of possible X is therefore a (highly non-linear) function of the posterior pdf $p(\Theta\,|\,\mathcal{T}, Y)$. From a clinical perspective, the expected value, $\mathrm{E}[X \mid \mathcal{T}, Y]$, is the most important characteristic of the response of the lymphatic system for a particular ROI, and can be calculated directly from (2) and (4):

$$\hat{X} = \mathrm{E}[X \mid \mathcal{T}, Y] = \sum_{\vartheta} \int X(\vartheta, b) p(\vartheta, b, r \mid \mathcal{T}, Y)\, db\, dr = \sum_{\vartheta} \tilde{A}_{\vartheta} \int b p(\vartheta, b \mid \mathcal{T}, Y)\, db. \quad (6)$$

The integral in (6) can be found in analytic form (Peterka, 1981). Hence, the numerical evaluation reduces to a simple summation over the grid of *a priori*-allowed values of ϑ. The covariance matrix of X is computed similarly.

Time activity curves for all ROIs in one limb are shown in Figure 4. The expected value $\hat{X}$ is drawn, together with an envelope formed by intervals with widths equal to the marginal standard deviations. Similar intervals for noisy measurements are also included. The probability with which the time activity curve is expected within this range can be evaluated. Even without this, the time-index of the maximum of the time activity curve provides important qualitative information about the inspected lymphatic system.

5.2. *Residence Time*

The residence time, ξ, is defined as the accumulated activity within a ROI, divided by the administered activity. It is widely accepted in nuclear medicine as a quantitative global characteristic of accumulation kinetics (Stabin, 1996; Heřmanská *et al.,* 1998). With the adopted scaling, the residence time in minutes is found as the area under the time activity curve. The area under the sampled time activity curve is $\xi(a, d, b) = \sum_{t=0}^{\infty} x_t \Delta$, with $\Delta = 1$ (min). This sum converges for the considered stable elementary models with $0 \leq a < 1$. Again the expected residence times, $\mathrm{E}[\xi(a, d, b) \mid \mathcal{T}, Y]$, and standard deviations, σ, are the most instructive characteristics of the pdf $p(\xi \mid \mathcal{T}, Y)$. They are included in the time activity curves in Figures. 4 and 5.

5.3. *Estimates of Model Parameters*

It is expected that the parameter estimates can be used in clinical staging of lymphedema. Point or interval estimates of the elementary time constant, a, and the cascade length, d, are intuitively good indicators of accumulation kinetics. They are described by the analytically obtained marginal posterior distribution $p(\vartheta \mid \mathcal{T}, Y) = p(a, d \mid \mathcal{T}, Y)$. Its discrete nature makes evaluation of moments of interest straightforward. Further research into their clinical significance is underway.

5.4. *Selection of Appropriate Sampling Times $\mathcal{S}$*

This task is critical as the number of measurements is restricted in routine practice. Thus, it is desirable to advance beyond the current empirical choice of standardized sampling times. A collection of patient-specific data, measured more frequently than is the clinical norm (to yield an extended observation set, Y, for each limb of the patient), has provided the aggregated experimental data, $\mathcal{Y}$, needed for Bayesian inference of an optimal sampling grid.

Let us consider a sub-selection, $\mathcal{S} \subset \mathcal{T}$, of sampling times for any one patient limb record within the considered over-sampled experimental set. The predictive pdf for that patient-specific limb data, conditioned on a particular subset of the patient's data, with indices in $\mathcal{S}$, is $p(Y \mid \mathcal{S}) = \int p(Y \mid \Theta, \mathcal{T}) p(\Theta \mid y_t,\ t \in \mathcal{S})\, d\Theta$. This pdf can be evaluated using all data, Y, measured for the particular limb. However, the patient-specific limb data records are mutually independent. Thus, the product of the quantities above—over all sets Y in $\mathcal{Y}$—generates the observation model needed for estimating the unknown optimal sub-selection, $\mathcal{S}$, of sampling times. A prior distribution, $p(\mathcal{S})$ (typically uniform over available choices), and Bayes' rule yield the posterior

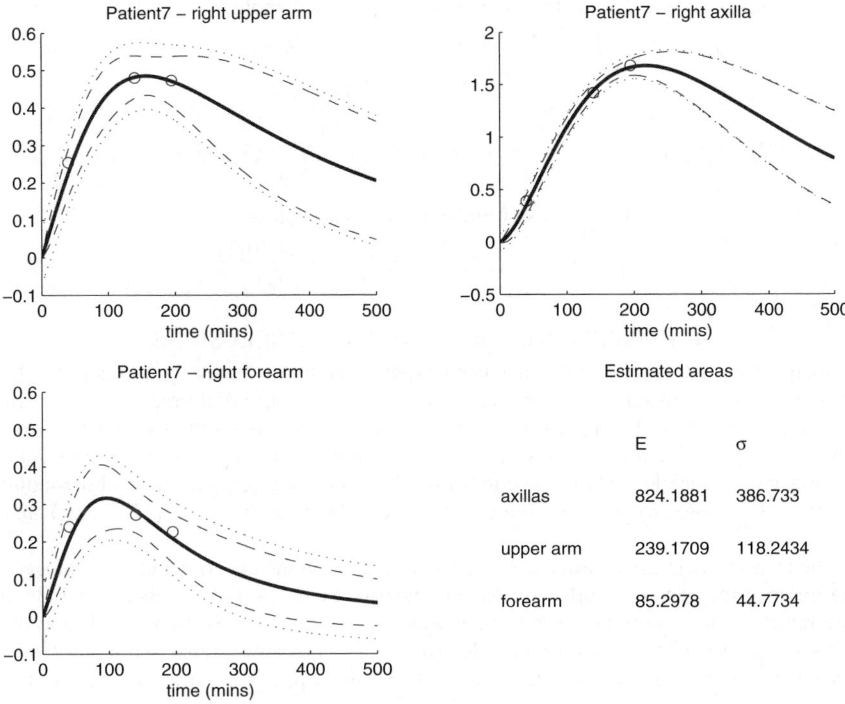

Figure 4. *Example of time activity curves and residence times estimation. Each graph corresponds to a different ROI. The circles denote measured data. Bold solid lines represent $\hat{X}$, the expected time activity curves. Dashed lines represent their uncertainty (mean value $\pm$ marginal standard deviation). Dotted lines represent uncertainty intervals for responses with noise, defined in the same way. Y-axes give % values. Estimates* E *of the expected residence times for each ROI, and their standard deviation σ, are provided in the final quadrant of the figure.*

distribution over the candidate $\mathcal{S}$s, and decision theory yields an optimized choice (Gebouský and Křížová, 2000). This choice reduces to the *Maximum a Posteriori* (MAP) probability estimate of $\mathcal{S}$ if the posterior distribution, $p(\mathcal{S} \mid \mathcal{Y})$, is peaked enough. The results presented below indicate that this simplification is indeed tenable.

5.5. *Comparison of Accumulation on Both Limbs*

Comparison between the responses of a particular patient's upper limbs is a very useful diagnostic aid, since, often, it is known that one limb is healthy, and it can act as a control for evaluation of the other limb. Technically, the need for a quantitative comparison of just two or three scintigraphic images is implied, a task which is hopeless without a Bayesian treatment. The problem is formalized as a test of the hypothesis, H_0, that the accumulation characteristics on both limbs are the same: *i.e.*, that they can be described by a single common model. The alternative, H_1, is defined as the need to model both limbs independently. The Bayesian decision is $\hat{H} = \arg\min_{H \in \{H_0, H_1\}} \sum_j Z(\{Y_1, Y_2\}, H, H_j) p(H_j \mid Y_1, Y_2)$. The loss function, Z, depends generally on the data, Y_1 and Y_2, for the respective limbs. $Z \geq 0$ is a $(2, 2)$ table, typically with zero diagonal entries. Positive entries reflect medical and economic consequences of a bad selection. Bayes' rule and the formulation of the hypotheses imply:

$$p(H \,|\, Y_1, Y_2) \propto p(Y_1, Y_2 \,|\, H)p(H)$$

$$H_0 : p(Y_1, Y_2 \,|\, H_0) = \int_{\Theta} p(Y_1 \,|\, \Theta)p(Y_2 \,|\, \Theta)p(\Theta)d\Theta$$

$$H_1 : p(Y_1, Y_2 \,|\, H_1) = \int_{\Theta_1} p(Y_1 \,|\, \Theta_1)p(\Theta_1)d\Theta_1 \int_{\Theta_2} p(Y_2 \,|\, \Theta_2)p(\Theta_2)d\Theta_2.$$

Then, $\hat{H} = H_0$—*i.e.*, the hypothesis of limb equality is accepted—if:

$$p(H_0 \,|\, Y_1, Y_2) \geq \mathcal{P} = \frac{Z(\{Y_1, Y_2\}, H_0, H_1)}{Z(\{Y_1, Y_2\}, H_0, H_1) + Z(\{Y_1, Y_2\}, H_1, H_0)}. \qquad (7)$$

6. EXPERIMENTAL RESULTS AND CONCLUSIONS

The methodology described above has been experimentally evaluated in all respects. It has been confirmed, for instance, that the pair of measurement times preferred in routine clinical practice (Section 2) has the highest posterior support among the possibilities available from the over-sampled grid (Section 5.4). The choice of the pair of sampling times on the model quality is significant, as is evidenced by the series of results in Figure 5. Optimization of the sampling time pair is therefore essential. In Figure 5, the other inferences discussed in Section 5 are also shown.

The measurement times differ from patient to patient in the experimental set. Hence, they were grouped into sets of similar values, and measurement *intervals*—rather than individual measurement times—were assessed using the method described in Section 5.4. The available intervals, together with optimization results over 22 patients, are summarized in Table 1. The choice was straightforward since the posterior distribution, $p(\mathcal{S}|\mathcal{Y})$, was strongly peaked.

Table 1. *Optimal choice of a pair of measurement times for each ROI (• denote optima).*

Interval	Range(mins)	Forearm	Upper arm	Axillas
$\mathcal{I}_1$	$\langle 30, 50\rangle$	•		
$\mathcal{I}_2$	$\langle 55, 95\rangle$		•	•
$\mathcal{I}_3$	$\langle 115, 150\rangle$	•		
$\mathcal{I}_4$	$\langle 175, 220\rangle$		•	•

Table 1 shows that the delayed data are important for the upper arm and axilla. This corresponds to their increasing distance from the injection site. A compromise in the choice of time intervals has to be made if it is not clear which regions are important for staging the disease. The combination $\{\mathcal{I}_2, \mathcal{I}_4\}$ seems to be the best in this respect.

Table 2 *Comparison of decisions via the proposed Bayesian Quantitative test* (QD) *with Clinical* (CD) *and Visual* (ViD) *decisions on forearm equality.* C^+/C^- *denotes the cases when the limbs are taken as the same/different from the* CD *viewpoint.* V^+, V^- *and* Q^+, Q^- *denote corresponding values for* ViD *and* QD *respectively. The number of cases belonging to the individual groups are quoted.*

Quantitative test (QD)	Clinical difference (CD)		Visual scint. difference (ViD)	
	C^+	C^-	V^+	V^-
Q^+	9	1	7	3
Q^-	0	4	0	4

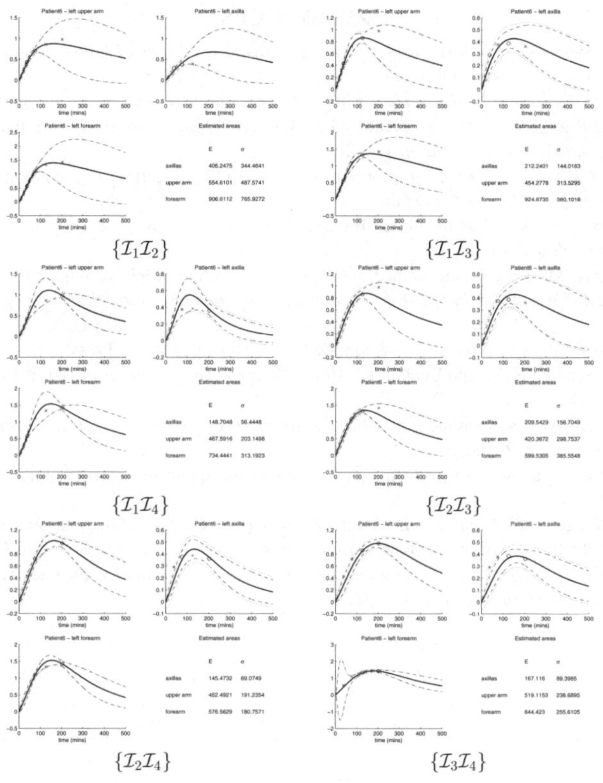

Figure 5. *Examples of time activity curves and residence time estimation from 2 measurements.* $\circ/\times$ *denotes data used/unused for estimation. All possible combinations are demonstrated. Graphical conventions are those of Figure 4.*

The estimates of time activity curves have been welcomed by physicians as a useful semi-quantitative aid for judging hand-staging. Good results have also been achieved in quantitative comparison of the upper limb pairs. Table 2, which gives results for $\mathcal{P} = 0.95$ in (7), illustrates the correspondence in the case of screening tests (Barrett and Swindell, 1981) on the forearms of 14 patients. Limb equality was judged by a clinician who was treating the patients (CD decisions). An independent decision was made by an experienced nuclear medicine expert–who had access to the basic clinical patient data—via visual evaluation of the raw (non-reduced) scintigraphic images (ViD decisions). These medical decisions serve for practical evaluation of the Bayesian quantitative (QD) test (7).

The results must be taken as preliminary, since the amount of available patient data is limited, and definite clinical conclusions are incomplete. Taking into account the discrepancies between experts, however, the high degree of correspondence with the conclusions of both physicians is impressive. It underlines the enormous benefit of the Bayesian approach in lymphoscintigraphy.

ACKNOWLEDGEMENTS

This research was partially supported by GAČR 102/99/1564, AV ČR S1075102.

REFERENCES

Barrett, H. H. and Swindell, W. (1981). *Radiological Imaging. The Theory of Image Formation, Detection, and Processing* **2**. New York: Academic Press.

Berger, B. O. (1985). *Statistical Decision Theory and Bayesian Analysis*. New York: Springer.

Fruhling, J. G. and Bourgeois, P. (1983). Axillary lymphoscintigraphy: current status in the treatment of breast cancer. *CRC Crit. Rev. Oncol. Hematol.* **1**, 1–20.

Gebouský, P. and Křížová, H. (2000). Contribution of Bayesian identification to evaluation of upper limb lymphedema. *Tech. Rep.*, ÚTIA AV ČR, Praha.

Henkin, R. E., Boles, M. A., Dillehay, G. L., Halama, J. R., Karesh, S.M. and Wagner, R.H. (1996). *Nuclear Medicine*. Mosby-Year Book, Inc.

Heřmanská, J., Kárný, M., Jirsa, L. and Němec, J. (1998). Evaluation of biophysical quantities related to the treatment of thyroid diseases. *Quantitative Image Analysis in Functional Scintigraphic Imaging*. Prague: Gamamed. 39–42.

Jeffreys, H. (1991). *Theory of Probability*, 3rd edn. Oxford: Oxford University Press

Oppenheim, A. V., Schafer, R. W. and Buck, J. R. (1999). *Discrete-Time Signal Processing*. Englewood Cliffs, NJ: Prentice-Hall.

Peterka, V. (1981). Bayesian system identification. *Trends and Progress in System Identification* (P. Eykhoff, ed). New York: Pergamon, 239–304.

Stabin, M. G. (1996). Mirdose 3 software for internal dose assessment in nuclear medicine. *J. Nucl. Med.* **37**, 538–546.

Verlooy, H., Biscompte, J. P., Nieuborg, L., Drent, P., Schiepers, C., Mortelmans, L. and De Roo, M. (1997). Noninvasive evaluation of lympho-venous anastomosis in upper limb lymphedema. *Eur. J. Lymphol.* **6**, 27–33.

Weissleder, H. and Weissleder, M. (1988). Lymphedema: Evaluation of qualitative and quantitative lymphoscintigraphy in 238 patients. *Radiology* **167**, 729–735.

BAYESIAN STATISTICS 7, pp. 553–563
J. M. Bernardo, M. J. Bayarri, J. O. Berger, A. P. Dawid,
D. Heckerman, A. F. M. Smith and M. West (Eds.)

Bayesian Analysis of Matched Pairs in the Presence of Covariates

F. JAVIER GIRÓN M. LINA MARTÍNEZ
Universidad de Málaga, Spain
fj_giron@uma.es mlmartinez@uma.es

ELÍAS MORENO FRANCISCO TORRES
Universidad de Granada, Spain
emoreno@ugr.es fdeasis@ugr.es

SUMMARY

In this paper, we address three main questions connected with the statistical analysis of matched pairs in the presence of covariates, from a Bayesian viewpoint. These sort of data appear when measurements are made on the same individuals before and after a treatment, at possibly different dose levels, is applied. The first question refers to the issue of how to construct statistical models for matched pairs data that reflect the dependency of the post-treatment on the pre-treatment observation, the dose level and the values of the covariates. The second one addresses the problem of seeking suitable statistical models for which inferences on the effect of the treatment only depends on some simple summaries of the data, for instance, the differences or ratios of the corresponding paired data. The third issue is concerned with testing the effectiveness or not of the treatment, the relevance of the different dose levels, and the influence of the covariates. Explicit solutions and formulae are derived for the case of an additive linear normal model. These problems are addressed here from a Bayesian default model choice perspective. Comparison of this novel approach to testing with classical and standard (non informative) Bayesian approaches are given, based on a simulation study.

Keywords: ADDITIVE MODELS; COVARIATES; INTRINSIC PRIORS; LOCATION MODELS; NORMAL LINEAR MODEL; MATCHED PAIRS.

1 . INTRODUCTION

The Bayesian analysis of matched pairs when no covariates are considered and the same dose levels of treatment applied to all patients has been fully discussed in Girón *et al.* (2002). Here we extend their model by including auxiliary information provided by a set of covariates $C_1, \ldots, C_k$, which might possibly influence the response to a treatment T applied to the patients at different possible dose levels.

A possible setting for the matched pairs problem in the presence of covariates may be as follows. A study is set up to test whether a new treatment—which can be applied at different dose levels—has some (presumably positive) effect on the patients. There is no assumption about the patients being considered as belonging to a homogeneous group, so that it is assumed that the effect of the treatment may depend on the dose level and, on the other hand, on some characteristics of the patients, closely related to the study, measured by a set of values of a certain set of covariates.

Within this setting, the data for the matched pairs problem consist of a set of the form $\boldsymbol{D} = ((\boldsymbol{c}_1, l_1, x_1, y_1), \ldots, (\boldsymbol{c}_n, l_n, x_n, y_n))$, where x_i is the pre-treatment value of the quantity of interest under study, *i.e.*, *prior* to the treatment, y_i is the post-treatment value of the quantity of interest, *i.e.*, *after* the treatment T is applied at a dose level l_i, and $\boldsymbol{c}_i = (c_{i1}, \ldots, c_{ik})^t$ are the values of the covariates $C_1, \ldots, C_k$ for every individual $i = 1, \ldots, n$.

Analyzing these data in the existing literature is generally done by modelling the difference vector $\boldsymbol{d} = (d_1, \ldots, d_n)^t$, where $d_i = y_i - x_i$, that is the difference between the post-treatment and pre-treatment data, through a statistical model which takes into account the values of the dose levels of the treatment and the values of the covariates; thus taking for granted that $\boldsymbol{d} = \boldsymbol{y} - \boldsymbol{x}$ is a sufficient statistics.

In Section 2 we provide a general formulation based on a hierarchical model and give a brief account of the possible hypothesis testing problems of interest.

The important additive model is introduced in Section 3, where a general theorem guaranteeing that the difference vector $\boldsymbol{d}$ is a sufficient statistics for the model is given.

In Section 4, the additive linear normal model, as an important subclass of the additive model, is further developed in some detail. It turns out that all interesting testing problems involved are particular cases of the general problem of testing whether some subsets of coefficients are null. This problem is addressed from a default Bayesian model selection viewpoint, for which intrinsic priors are derived and an explicit formula for the Bayes factor, computed using these priors, is given.

Some simulation results comparing this approach to testing, based on the posterior probability of the null hypothesis, with the usual one based on the p-values obtained from the ANOVA table are presented in Section 5. Finally, some comments on the main issues of the paper are given in Section 6.

2. A GENERAL MODEL FOR MATCHED DATA IN THE PRESENCE OF COVARIATES

The following model is an extension of the simplified model for matched data—when there are no covariates and the dose level is the same for all individuals—given by Girón *et al.*, 2002). We proceed in a similar fashion as in Section 2 of that paper so as to make this section as self-contained as possible.

Suppose that before applying the treatment, the data $\boldsymbol{x} = (x_1, \ldots, x_n)$ can be regarded as a random sample from a statistical model $f_1(x_i \mid \boldsymbol{c}_i, \boldsymbol{\theta}, \boldsymbol{\phi})$ where $\boldsymbol{c}_i$ is a (presumably low dimension) vector of covariates associated to a generic x_i, $\boldsymbol{\theta}$ is a vector parameter associated to the covariates, and $\boldsymbol{\phi}$ represents a vector of (possible) nuisance parameters.

After applying the treatment at a dose level l_i to a generic x_i, the effect of which we denote by the scalar δ_i, the new observed quantity y_i, conditional on the measurement of x_i and the effect parameter δ_i, is such that the conditional distribution of y_i given x_i and δ_i follows a model whose density is represented by $y_i \mid x_i \sim f_2(y_i \mid m(x_i, \delta_i), \boldsymbol{\psi})$, where $m(x_i, \delta_i)$ is a specified real function, and $\boldsymbol{\psi}$ represents a vector of nuisance parameters. The effect parameter δ_i is, in general, a function of the dose level l_i and the covariates $\boldsymbol{c}_i$ associated to the i-th observation, *i.e.*, $\delta_i = h(\boldsymbol{\beta}, \boldsymbol{c}_i, l_i)$, where $\boldsymbol{\beta}$ is a vector parameter associated to both the covariates and the dose level l_i.

Then, a general sampling model for matched pairs $(\boldsymbol{x}, \boldsymbol{y}) = ((x_1, y_1), \ldots, (x_n, y_n))$ with covariates $\boldsymbol{c}_i$ and dose level l_i, for $i = 1, \ldots, n$, is a hierarchical model of the form

$$x_i \sim f_1(x_i \mid \boldsymbol{c}_i, \boldsymbol{\theta}, \boldsymbol{\phi}), \tag{2.1a}$$

$$y_i \mid x_i \sim f_2(y_i \mid m(x_i, \delta_i), \boldsymbol{\psi}). \tag{2.1b}$$

where $\boldsymbol{\delta} = (\delta_1, \ldots, \delta_n)$, and f_1 and f_2 are probability densities.

A more general model might consider the δ_i's as random variables so that a third stage

$$\delta_i \sim f_3(\delta_i \,|\, \boldsymbol{\beta}, \boldsymbol{c}_i, l_i, \boldsymbol{\varphi}), \tag{2.1c}$$

would be added to preceding model, where $\boldsymbol{\varphi}$ denotes a vector of nuisance parameters, and f_3 is a probability density. Analysis of this general approach will be reported elsewhere.

In the Bayesian setting the model is completed by specifying the joint prior distribution of $\boldsymbol{\theta}, \boldsymbol{\phi}, \boldsymbol{\delta}, \boldsymbol{\psi}$ or $\boldsymbol{\theta}, \boldsymbol{\phi}, \boldsymbol{\beta}, \boldsymbol{\psi}$.

In the sequel, in order to simplify the notation, although all inferences are conditional on the set of covariates $\boldsymbol{C} = (\boldsymbol{c}_1, \ldots, \boldsymbol{c}_n)$ and the dose level vector $\boldsymbol{l} = (l_1, \ldots, l_n)$, we do not explicitly mention it unless necessary for clarity of exposition.

Thus, the observed data $(\boldsymbol{x}, \boldsymbol{y}) = ((x_1, y_1), \ldots, (x_n, y_n))$ are a random sample from the joint sampling model given by

$$f(\boldsymbol{x}, \boldsymbol{y} \,|\, \boldsymbol{\theta}, \boldsymbol{\phi}, \boldsymbol{\delta}, \boldsymbol{\psi}) = \prod\nolimits_{i=1}^{n} f_1(x_i \,|\, \boldsymbol{c}_i, \boldsymbol{\theta}, \boldsymbol{\phi}) f_2(y_i \,|\, m(x_i, \delta_i), \boldsymbol{\psi}),$$

or, equivalently, conditioning on the parameter $\boldsymbol{\beta}$, instead of $\boldsymbol{\delta}$

$$f(\boldsymbol{x}, \boldsymbol{y} \,|\, \boldsymbol{\theta}, \boldsymbol{\phi}, \boldsymbol{\beta}, \boldsymbol{\psi}) = \prod\nolimits_{i=1}^{n} f_1(x_i \,|\, \boldsymbol{c}_i, \boldsymbol{\theta}, \boldsymbol{\phi}) f_2(y_i \,|\, m(x_i, h(\boldsymbol{\beta}, \boldsymbol{c}_i, l_i)), \boldsymbol{\psi}).$$

There are several questions of interest in the matched pairs problem with covariates. The most important one refers to testing hypothesis about the effects vector $\boldsymbol{\delta}$ or some (interpretable or meaningful) function of it, such as, for instance, the average of the δ_i's. Of some importance is also the problem of testing whether the different dose levels of the treatment might have some effect on the overall response. One last issue might consist also in investigating the effect of the covariates in the response to the treatment.

In order to take advantage of the structure of the specific problem at hand and the way the effect of the treatment exerts on the data, some restrictions should be imposed both on the relation between the original data x_i and the effect parameter δ_i, say $m(x_i, \delta_i)$, for $i = 1, \ldots, n$, and on the form of the hierarchical model, that is, the form of the densities f_1 and f_2. Thus, for example, the effect of the treatment may be considered additive if the conditional distribution of the post-treatment datum y_i depends on x_i and δ_i through $x_i + \delta_i$, that is, $m(x_i, \delta_i) = x_i + \delta_i$. In this case, testing whether the *average effect level* of the treatment is null would be equivalent to testing whether $\bar{\delta} = \sum_{i=1}^{n} \delta_i / n = 0$.

Intuitively, the different forms for $m(x_i, \delta_i)$ and the form of the function $h(\boldsymbol{\beta}, \boldsymbol{c}_i, l_i)$ relating the individual effects δ_i to the covariates and the dose levels should be used in conjunction with some statistical models for the pre and post-treatment data such as location models for additive models, and scale models for multiplicative models. Nevertheless, it is only the real problem at hand that dictates the choice of both the statistical and the matched paired models.

From the viewpoint of model selection, the testing problems mentioned above for this very general model are of enormous complexity so that some restrictions on both the hierarchical model (2.1) and the form of the functions $h(\boldsymbol{\beta}, \boldsymbol{c}_i, l_i)$ and $m(x_i, \delta_i)$ are necessary in order to obtain models that are both realistic and computationally feasible.

The next two sections present some results in this direction.

3. THE ADDITIVE MODEL

An additive model is a particular case of the general model (2.1), where $m(x_i, \delta_i) = x_i + \delta_i$ for $i = 1 \ldots, n$. Therefore, it is a hierarchical model of the form

$$x_i \sim f_1(x_i \,|\, \boldsymbol{c}_i, \boldsymbol{\theta}, \boldsymbol{\phi}), \tag{3.1a}$$
$$y_i \,|\, x_i \sim f_2(y_i \,|\, x_i + \delta_i, \boldsymbol{\psi}), \tag{3.1b}$$
$$\delta_i = h(\boldsymbol{\beta}, \boldsymbol{c}_i, l_i). \tag{3.1c}$$

where x_i, y_i, δ_i are real numbers, f_1 and f_2 are densities whose support is the whole real line, and $h(\boldsymbol{\beta}, \mathbf{c}_i, l_i)$ is a real function. The vector parameter $\boldsymbol{\delta} = (\delta_1, \ldots, \delta_n)$ is said to be an *additive effect parameter*.

As remarked in Girón *et al.* (2002), the version of this model without covariates avoids many of the methodological criticisms pointed out by Bickel and Doksum (1977, 211–212); for instance, normality of the data is not necessarily assumed. It also allows for the possibility of the effect of the treatment not to be homogeneous on all the patients.

The following theorem shows that whenever (3.1b) is a location family, irrespective of the form of (3.1a), then full inferences on $\boldsymbol{\beta}$, and consequently on $\boldsymbol{\delta}$, can be obtained by simply conditioning on $\mathbf{d} = \boldsymbol{y} - \boldsymbol{x}$.

Theorem 3.1. *Under model* (3.1), *if for every* $\boldsymbol{\psi}$, $f_2(y_i \,|\, x_i + \delta_i, \boldsymbol{\psi})$ *is a location family with location parameter* $x_i + \delta_i$, *that is* $f_2(y_i \,|\, x_i + \delta_i, \boldsymbol{\psi}) = g_2(y_i - x_i - \delta_i \,|\, \boldsymbol{\psi})$, *where* $g_2(\cdot \,|\, \boldsymbol{\psi})$ *is the generator of the location family, then* $\boldsymbol{x}$ *and* $\mathbf{d}$ *are conditionally independent given* $\boldsymbol{\theta}$, $\boldsymbol{\phi}$, $\boldsymbol{\beta}$, *and* $\boldsymbol{\psi}$. *Further, if a priori* $(\boldsymbol{\theta}, \boldsymbol{\phi})$ *is independent of* $(\boldsymbol{\beta}, \boldsymbol{\psi})$, *then*

$$(\boldsymbol{\beta}, \boldsymbol{\psi}) \,|\, \boldsymbol{x}, \boldsymbol{y} \stackrel{\mathrm{d}}{=} (\boldsymbol{\beta}, \boldsymbol{\psi}) \,|\, \mathbf{d} \qquad \textit{and} \qquad \boldsymbol{\delta} \,|\, \boldsymbol{x}, \boldsymbol{y} \stackrel{\mathrm{d}}{=} \boldsymbol{\delta} \,|\, \mathbf{d}.$$

Proof. As $\mathbf{d} = \boldsymbol{y} - \boldsymbol{x}$, by a simple change of variables, the joint density of the pair $(\boldsymbol{x}, \mathbf{d})$ becomes

$$\begin{aligned} f(\boldsymbol{x}, \mathbf{d} \,|\, \boldsymbol{\theta}, \boldsymbol{\phi}, \boldsymbol{\beta}, \boldsymbol{\psi}) &= \prod_{i=1}^{n} f_1(x_i \,|\, \mathbf{c}_i, \boldsymbol{\theta}, \boldsymbol{\phi}) f_2(y_i \,|\, x_i + h(\boldsymbol{\beta}, \mathbf{c}_i, l_i)), \boldsymbol{\psi}) \\ &= \prod_{i=1}^{n} f_1(x_i \,|\, \mathbf{c}_i, \boldsymbol{\theta}, \boldsymbol{\phi}) g_2(d_i - h(\boldsymbol{\beta}, \mathbf{c}_i, l_i) \,|\, \boldsymbol{\psi}) \\ &= f(\boldsymbol{x} \,|\, \boldsymbol{\theta}, \boldsymbol{\phi}) f(\mathbf{d} \,|\, \boldsymbol{\beta}, \boldsymbol{\psi}), \end{aligned}$$

where $f(\mathbf{d} \,|\, \boldsymbol{\beta}, \boldsymbol{\psi}) = \prod_{i=1}^{n} g_2(d_i - h(\boldsymbol{\beta}, \mathbf{c}_i, l_i) \,|\, \boldsymbol{\psi})$. This proves the first assertion.

From the obvious equation

$$f(\boldsymbol{x}, \boldsymbol{y} \,|\, \boldsymbol{\theta}, \boldsymbol{\phi}, \boldsymbol{\beta}, \boldsymbol{\psi}) = f(\boldsymbol{x}, \mathbf{d} \,|\, \boldsymbol{\theta}, \boldsymbol{\phi}, \boldsymbol{\beta}, \boldsymbol{\psi}).$$

and the prior independence of $(\boldsymbol{\theta}, \boldsymbol{\phi})$ and $(\boldsymbol{\beta}, \boldsymbol{\psi})$, the first part of the theorem implies that

$$f(\boldsymbol{\theta}, \boldsymbol{\phi}, \boldsymbol{\beta}, \boldsymbol{\psi} \,|\, \boldsymbol{x}, \boldsymbol{y}) \propto f(\boldsymbol{\theta}, \boldsymbol{\phi}) f(\boldsymbol{\beta}, \boldsymbol{\psi}) f(\boldsymbol{x} \,|\, \boldsymbol{\theta}, \boldsymbol{\phi}) f(\mathbf{d} \,|\, \boldsymbol{\beta}, \boldsymbol{\psi}) \propto f(\boldsymbol{\theta}, \boldsymbol{\phi} \,|\, \boldsymbol{x}) f(\boldsymbol{\beta}, \boldsymbol{\psi} \,|\, \mathbf{d}).$$

Integrating out the pair $(\boldsymbol{\theta}, \boldsymbol{\phi})$ in this equation results in

$$f(\boldsymbol{\beta}, \boldsymbol{\psi} \,|\, \boldsymbol{x}, \boldsymbol{y}) \propto f(\boldsymbol{\beta}, \boldsymbol{\psi} \,|\, \mathbf{d}),$$

Therefore, both densities are equal a.e. Now, as $\boldsymbol{\delta}$ is a function of $\boldsymbol{\beta}$, the set of covariates $\boldsymbol{C}$, and the dose level vector $\boldsymbol{l} = (l_1, \ldots, l_n)$, then

$$f(\boldsymbol{\delta} \,|\, \boldsymbol{x}, \boldsymbol{y}) = f(\boldsymbol{\delta} \,|\, \mathbf{d}), \quad \text{a.e.}$$

and the second part of the theorem is proved. □

That likelihood based inferences about the vector $\boldsymbol{\beta}$ associated to the covariates and the dose levels, and the effect vector $\boldsymbol{\delta} = (\delta_1, \ldots, \delta_n)$—and the nuisance parameter $\boldsymbol{\psi}$ that might appear in the model f_2, that is in Eq. (3.1b) of the additive model—should be based *only* on the differences between data after and before administering the treatment, that is, on $\mathbf{d}$, is a consequence of the first part of the preceding theorem, which includes the normal case as a very particular case. Moreover, the form of the model for data $\boldsymbol{x}$ is irrelevant to any inferences on these parameters.

From a Bayesian perspective, in order to obtain the posterior distribution of the parameters of interest $\boldsymbol{\beta}$ and/or $\boldsymbol{\delta}$, and possibly of $\boldsymbol{\psi}$, one has to include into the hierarchical model (3.1) an additional independence condition on the parameters of the two models. Otherwise, if there were prior dependence between $(\boldsymbol{\theta}, \boldsymbol{\phi})$ and the pair $(\boldsymbol{\beta}, \boldsymbol{\psi})$, posterior inferences on $(\boldsymbol{\beta}, \boldsymbol{\psi})$ or $(\boldsymbol{\delta}, \boldsymbol{\psi})$ would depend on all available data $(\boldsymbol{x}, \boldsymbol{y})$ or $(\boldsymbol{x}, \boldsymbol{d})$, and not only on $\boldsymbol{d}$.

However, as proved in Girón *et al.* (1997), under rather general conditions, the seemingly very natural independence condition of $\boldsymbol{x} = (x_1, \ldots, x_n)$ and $(\boldsymbol{\beta}, \boldsymbol{\psi})$, for all sample sizes n, implies prior and, also, posterior independence of $(\boldsymbol{\theta}, \boldsymbol{\phi})$ and $(\boldsymbol{\beta}, \boldsymbol{\psi})$ given $(\boldsymbol{x}, \boldsymbol{y})$ or $(\boldsymbol{x}, \boldsymbol{d})$ and then, the second part of Theorem 3.1 holds.

It is interesting to note that Theorem 3.1 holds irrespective of the form of the function $h(\boldsymbol{\beta}, \boldsymbol{c}_i, l_i)$. In the next section, however, we shall restrict the form of $h(\boldsymbol{\beta}, \boldsymbol{c}_i, l_i)$ to be a linear function of the parameter $\boldsymbol{\beta}$.

4. THE ADDITIVE LINEAR NORMAL MODEL

In this section, we consider a particular, but important, case of the additive model (3.1), where f_1 is an arbitrary density, and for every $i = 1, \ldots, n$, $f_2(y_i \mid x_i)$ is assumed to be a normal density, with mean $x_i + \delta_i$, where δ_i is the following linear function

$$\begin{aligned} \delta_i &= \delta_0 + \beta_0 l_i + \beta_1 c_{i1} + \cdots + \beta_k c_{ik} \\ &= \delta_0 + \beta_0 l_i + \boldsymbol{c}_i' \boldsymbol{\beta}, \end{aligned} \tag{4.1}$$

and common variance σ^2. In (4.1), δ_0 is an scalar denoting the intercept, β_0 is the regression coefficient associated to the dose level L, and $\boldsymbol{\beta} = (\beta_1, \ldots, \beta_k)^t$ is the vector of regression coefficients associated to the set of k (with $k < n - 2$) covariates $C_1, \ldots, C_k$.

Now, assuming prior independence of $(\boldsymbol{\theta}, \boldsymbol{\phi})$ and $(\delta_0, \beta_0, \boldsymbol{\beta}, \sigma^2)$, or equivalently, independence of $\boldsymbol{x}$ and $(\delta_0, \beta_0, \boldsymbol{\beta}, \sigma^2)$ for any sample size, from Theorem 3.1 it follows that all Bayesian inferences on $(\delta_0, \beta_0, \boldsymbol{\beta}, \sigma^2)$ should only be based on the difference vector $\boldsymbol{d}$. Furthermore, as it is easy to see, $\boldsymbol{d} = \boldsymbol{y} - \boldsymbol{x}$ conditional on δ_0, β_0, $\boldsymbol{\beta}$ and σ^2 follows the linear normal model

$$\boldsymbol{d} = \boldsymbol{X}\boldsymbol{\alpha} + \boldsymbol{\varepsilon} \tag{4.2}$$

where the design matrix $\boldsymbol{X}$ is

$$\boldsymbol{X} = \begin{pmatrix} 1 & l_1 & c_{11} & \ldots & c_{1k} \\ 1 & l_2 & c_{21} & \ldots & c_{2k} \\ \vdots & \vdots & \vdots & \ddots & \vdots \\ 1 & l_n & c_{n1} & \ldots & c_{nk} \end{pmatrix},$$

$\boldsymbol{\alpha}$ is the column vector made up of the intercept δ_0, and the regression coefficients of the dose level L and the covariates $C_1, \ldots, C_k$, that is, $\boldsymbol{\alpha}^t = (\delta_0, \beta_o, \boldsymbol{\beta}^t)$, and $\boldsymbol{\varepsilon} \sim N_n(0, \sigma^2 \boldsymbol{I})$, where $0 = (0, \ldots, 0)^t$ is a column vector of length n, and $\boldsymbol{I}$ is the identity matrix of dimensions $n \times n$.

If the usual reference prior for $\boldsymbol{\alpha}$ and σ^2, $\pi(\boldsymbol{\alpha}, \sigma^2) \propto 1/\sigma^2$ is used, it is easy to see that the marginal posterior distribution of $\boldsymbol{\alpha}$ given $\boldsymbol{d}$ is a $(k+2)$-variate Student t with location parameter $\tilde{\boldsymbol{\alpha}}$, scale matrix $(ns^2/\nu)(\boldsymbol{X}'\boldsymbol{X})^{-1}$, and degrees of freedom $\nu = n - (k+2)$, where

$$\tilde{\boldsymbol{\alpha}} = (\boldsymbol{X}^t\boldsymbol{X})^{-1}\boldsymbol{X}^t\boldsymbol{d}, \qquad \text{and} \qquad s^2 = \frac{\|\boldsymbol{r}\|^2}{n} = \frac{\boldsymbol{d}^t(\boldsymbol{I} - \boldsymbol{H})\boldsymbol{d}}{n}$$

are the maximum likelihood estimators of $\boldsymbol{\alpha}$ and σ^2 respectively, and $\boldsymbol{r}$ is the vector of residuals, that is, $\boldsymbol{r} = (\boldsymbol{I}-\boldsymbol{H})\boldsymbol{d}$, where $\boldsymbol{H} = \boldsymbol{X}(\boldsymbol{X}^t\boldsymbol{X})^{-1}\boldsymbol{X}^t$ denotes the hat matrix. From this posterior, inferences on δ_0, β_0 or both, such as credible intervals or regions, can be obtained.

As mentioned in Section 2, there are several problems of hypothesis testing in the matched pairs problem. The most important one is: how can we measure the effectiveness or not of the treatment when possibly different dose levels are applied to the individuals in the presence of auxiliary information conveyed by the covariates? There are several possible answers to this question.

In the linear normal model specified by Eq. (4.2) one could test whether the *average effect level* δ is either zero or not by computing the posterior probabilities of $H_0 : \delta = 0$ and $H_1 : \delta \neq 0$.

The impact of the several different dose levels on the response to the treatment is equivalent to testing $H_0 : \beta_0 = 0$ versus $H_1 : \beta_0 \neq 0$.

To test whether the treatment is inefficient in the presence of the covariates is equivalent to testing $H_0 : (\delta_0, \beta_0) = (0, 0)$ versus $H_1 : (\delta_0, \beta_0) \neq (0, 0)$.

4.1. *Testing Whether a Subvector of Regressors is Null as a Model Selection Problem*

Consider the testing problem $H_0 : \boldsymbol{\alpha}_0 = 0$ versus $H_1 : \boldsymbol{\alpha}_0 \neq 0$, where $\boldsymbol{\alpha}_0$ is a subvector of the vector $\boldsymbol{\alpha}$ of length k_0. Partition the $\boldsymbol{\alpha}$ vector as $\boldsymbol{\alpha}^t = (\boldsymbol{\alpha}_0^t, \boldsymbol{\alpha}_1^t)$, and let $\boldsymbol{X} = (\boldsymbol{X}_0|\boldsymbol{X}_1)$ be the corresponding partition of the columns of the design matrix $\boldsymbol{X}$. Thus, $\boldsymbol{X}_1$ is an $n \times k_1$ matrix, where $k_1 = k + 2 - k_0$.

Under the null $H_0 : \boldsymbol{\alpha}_0 = 0$, the sampling model for $\boldsymbol{d}$ is a multivariate normal distribution with mean vector $\boldsymbol{X}_1\boldsymbol{\gamma}_1$ and covariance $\sigma_0^2\boldsymbol{I}$, that is, $\boldsymbol{d} \sim N_n(\boldsymbol{d} \mid \boldsymbol{X}_1\boldsymbol{\gamma}_1, \sigma_0^2\boldsymbol{I})$. Under the alternative $H_1 : \boldsymbol{\alpha}_0 \neq 0$ the sampling model for $\boldsymbol{d}$ is a $N_n(\boldsymbol{d}| \boldsymbol{X}\boldsymbol{\alpha}, \sigma_1^2\boldsymbol{I})$.

Under vague prior information on $(\boldsymbol{\gamma}_1, \sigma_0)$ and $(\boldsymbol{\alpha}, \sigma_1)$, testing H_0 versus H_1 can be formulated as a model choice problem between the two conventional Bayesian models

$$M_0 : N_n(\boldsymbol{d}|\boldsymbol{X}_1\boldsymbol{\gamma}_1, \sigma_0^2\boldsymbol{I}), \qquad \pi_0^N(\boldsymbol{\gamma}_1, \sigma_0) = c_0/\sigma_0^2,$$

and

$$M_1 : N_n(\boldsymbol{d}|\boldsymbol{X}\boldsymbol{\alpha}, \sigma_1^2\boldsymbol{I}), \qquad \pi_1^N(\boldsymbol{\alpha}, \sigma_1) = c_1/\sigma_1^2,$$

where c_0 and c_1 are arbitrary positive constants that cannot be determined because the priors are improper. For the usual 0–1 loss function, model M_0 is chosen if $P(M_0 \mid \boldsymbol{D}) > P(M_1 \mid \boldsymbol{D})$. The posterior probability of each model is computed assuming the conventional default prior $P(M_0) = P(M_1) = 1/2$ on the set of models $\{M_0, M_1\}$. Then, we can write

$$P(M_0 \mid \boldsymbol{D}) = \frac{1}{1 + B_{10}^N(\boldsymbol{D})},$$

where the Bayes factor $B_{10}^N(\boldsymbol{D})$ for comparing model M_1 and H_0 is given by

$$B_{10}^N(\boldsymbol{D}) = \frac{\int N_n(\boldsymbol{d}|\boldsymbol{X}\boldsymbol{\alpha}, \sigma_1^2\boldsymbol{I})\pi_1^N(\boldsymbol{\alpha}, \sigma_1)\, d\boldsymbol{\alpha}\, d\sigma_1}{\int N_n(\boldsymbol{d}|\boldsymbol{X}_1\boldsymbol{\gamma}_1, \sigma_0^2\boldsymbol{I})\pi_0^N(\boldsymbol{\gamma}_1, \sigma_0)\, d\boldsymbol{\gamma}_1\, d\sigma_0}.$$

This expression shows that the usual reference priors cannot be used for model choice because they leave the Bayes factor defined up to the arbitrary multiplicative constant c_1/c_0.

4.2. *Intrinsic Priors*

We note that the sampling model $N_n(\boldsymbol{d} \mid \boldsymbol{X}_1\boldsymbol{\gamma}_1, \sigma_0^2\boldsymbol{I})$ is nested into $N_n(\boldsymbol{d} \mid \boldsymbol{X}\boldsymbol{\alpha}, \sigma_1^2\boldsymbol{I})$. This allows for the application of the intrinsic method to construct intrinsic priors for which the

Bayes factor is well defined (Berger and Pericchi 1996; Moreno *et al.,* 1998). Motivations and justifications for the use of intrinsic priors for model selection have been given by Berger and Pericchi (1996, 1997a, 1997b, 1998), Moreno (1997) and Moreno *et al.* (1998). The method has proven to provide sensible intrinsic priors for a wide variety of testing problems, Berger and Pericchi (1996, 1998), Moreno *et al.* (1998, 1999, 2000), Moreno and Liseo (2002), Moreno *et al.* (2002), Moreno *et al.* (2003), among others.

Lemma 4.1. *The intrinsic prior for parameters* $\boldsymbol{\alpha}$, σ_1 *conditional on an arbitrary but fixed point* (γ_1, σ_0) *is given by*

$$\pi^I(\boldsymbol{\alpha}, \sigma_1 \,|\, \gamma_1, \sigma_0) = N_{k+2}(\boldsymbol{\alpha} \,|\, \tilde{\gamma}_1, (\sigma_0^2 + \sigma_1^2)\boldsymbol{W}^{-1}) \frac{1}{\sigma_0} \left(1 + \frac{\sigma_1^2}{\sigma_0^2}\right)^{-3/2}, \tag{4.3}$$

where $\tilde{\gamma}_1^t = (0^t, \gamma_1^t)$, $\boldsymbol{W} = \boldsymbol{Z}^t \boldsymbol{Z}$ *is a nonsingular matrix and* $\boldsymbol{Z}$ *is a theoretical design matrix of order* $(k+3) \times (k+2)$.

Proof. In order to save space, the proof is omitted. □

By construction $\pi^I(\boldsymbol{\alpha}, \sigma_1 \,|\, \gamma_1, \sigma_0)$ is a probability density for any γ_1, σ_0. The unconditional distribution $\pi^I(\boldsymbol{\alpha}, \sigma_1)$ is obtained by integrating out γ_1 and σ_0 with respect to the conventional prior $\pi_0^N(\gamma_1, \sigma_0) = c_0/\sigma_0^2$. That is,

$$\pi^I(\boldsymbol{\alpha}, \sigma_1) = c_0 \int \pi^I(\boldsymbol{\alpha}, \sigma_1 \,|\, \gamma_1, \sigma_0) \frac{1}{\sigma_0^2} \, d\gamma_1 \, d\sigma_0.$$

The pair of densities $(\pi_0^N(\gamma_1, \sigma_0), \pi^I(\boldsymbol{\alpha}, \sigma_1))$ is called the intrinsic prior, and although they are improper, (i) they are well calibrated since both depend on the same arbitrary constant c_0, and (ii) they are a well established limit of proper priors (Moreno *et al.,* 1998).

4.3. *Design Considerations*

Matrix $\boldsymbol{W}^{-1}$ in (4.3) depends on the regressors of a theoretical training sample of size $k+3$. A way to assess $\boldsymbol{W}^{-1}$ is by using the same underlying idea as that found in the origin of the *Arithmetic Intrinsic Bayes factor* (Berger and Pericchi, 1996) of averaging over all possible minimal training samples of the observed sample. This would give

$$\boldsymbol{W}^{-1} = \frac{1}{L} \sum_{\ell=1}^{L} (\boldsymbol{Z}^t(\ell) \boldsymbol{Z}(\ell))^{-1}, \tag{4.4}$$

where $\{\boldsymbol{Z}(\ell), \ell = 1, ..., L\}$ is the set of submatrices of $\boldsymbol{X}$ of order $(k+3) \times (k+2)$ of full rank.

4.4. *Bayes Factor for Intrinsic Priors*

For the data $\boldsymbol{D}$ and the intrinsic prior derived in Lemma 4.1, the Bayes factor for testing H_0 versus H_1 is given in the next lemma.

Lemma 4.2. *The Bayes factor for comparing models* $M_0 : \{N_n(\boldsymbol{d} \,|\, \boldsymbol{X}_1\gamma_1, \sigma_0^2\boldsymbol{I}), \pi_0^N(\gamma_1, \sigma_0)\}$ *and* $M_1 : \{N_n(\boldsymbol{d} \,|\, \boldsymbol{X}\boldsymbol{\alpha}, \sigma_1^2\boldsymbol{I}), \pi^I(\boldsymbol{\alpha}, \sigma_1)\}$ *turns out to be*

$$B_{10}(\boldsymbol{D}) = |\boldsymbol{X}_1^t \boldsymbol{X}_1|^{1/2} (\boldsymbol{d}^t(\boldsymbol{I} - \boldsymbol{H}_1)\boldsymbol{d})^{(n-k_1+1)/2} I_0,$$

where $\boldsymbol{H}_1 = \boldsymbol{X}_1(\boldsymbol{X}_1^t\boldsymbol{X}_1)^{-1}\boldsymbol{X}_1^t$,

$$I_0 = \int_0^{\pi/2} \frac{d\varphi}{|\boldsymbol{A}(\varphi)|^{1/2}|\boldsymbol{B}(\varphi)|^{1/2}E(\varphi)^{(n-k_1+1)/2}},$$

and

$$\boldsymbol{B}(\varphi) = \sin^2\varphi\boldsymbol{I} + \boldsymbol{X}\boldsymbol{W}^{-1}\boldsymbol{X}^t,$$
$$\boldsymbol{A}(\varphi) = \boldsymbol{X}_1^t\boldsymbol{B}(\varphi)^{-1}\boldsymbol{X}_1,$$
$$E(\varphi) = \boldsymbol{d}^t(\boldsymbol{B}(\varphi)^{-1} - \boldsymbol{B}(\varphi)^{-1}\boldsymbol{X}_1\boldsymbol{A}(\varphi)^{-1}\boldsymbol{X}_1^t\boldsymbol{B}(\varphi)^{-1})\boldsymbol{d}.$$

5. SIMULATION STUDY

To evaluate the performance of the procedure we have carried out an extensive simulation study that will be reported elsewhere. Here we report some results obtained from the simulations concerning the hypothesis $H_0 : \beta_0 = 0$ versus. $H_1 : \beta_0 \neq 0$.

The simulated model includes the treatment and one covariate. The values of these variables were simulated from uniform distributions on the interval (0,10). The sample size was taken $n = 40$, *i.e.*, medium size. Once the treatment and the covariate were simulated we fixed the design matrix $\boldsymbol{X}$. Using this matrix, we simulated $N = 1000$ i.i.d. samples, $\boldsymbol{d}_1, \ldots, \boldsymbol{d}_N$, from model (4.2) under the null $H_0 : \beta_0 = 0$, and another 1000 samples of the same size from some alternatives for $\beta_0 \neq 0$, setting the parameters δ_0 and β_1 as displayed in Table 1. In all cases, the variance was taken as $\sigma^2 = 1$.

Table 1. *Values of the parameters for the simulation study.*

Parameters	H_0	H_1	
β_0	0.0	0.2	0.3
δ_0	1.0	10.0	10.0
β_1	-1.0	-1.0	-1.0

When sampling from the null model, the resulting values of the posterior probabilities of H_0 turned out as expected, *i.e.*, they were close to 1. Figure 1 shows the histograms of the posteriors probabilities of H_0 and the corresponding p-values.

Remark 1. Quite similar histograms are obtained when, under the null, δ_0 and β_1 are set arbitrarily.

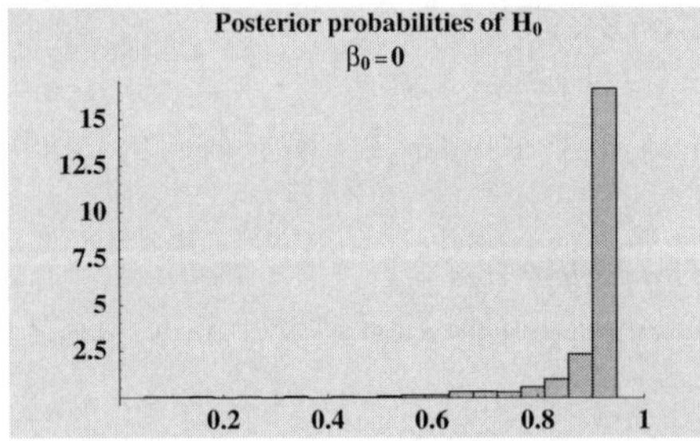

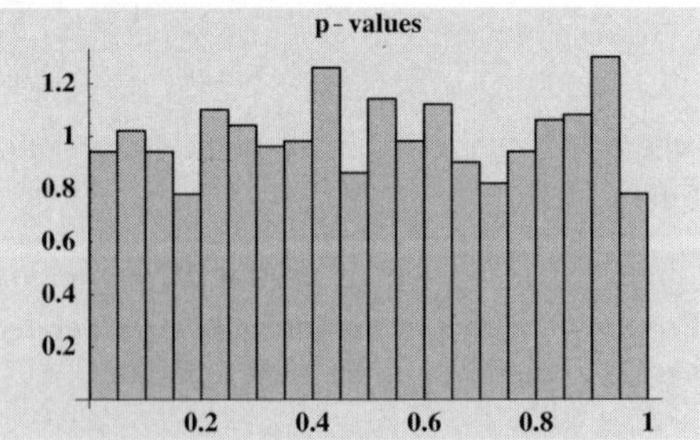

Figure 1. *Histograms of the Posterior probabilities of* H_0 *given* $\boldsymbol{D}$ *and* p*-values when sampling from the null* H_0.

For comparative purposes, the plot of the resulting p-values, and their corresponding posterior probabilities of H_0, is shown in Figure 2. As exemplified by Figure 2, there seems to exist a theoretical relation—a monotone increasing function—between the posterior probabilities of H_0 and the usual p-values associated to this hypothesis. It is apparent from this figure that there are many samples giving posterior probabilities of H_0 higher than 0.5 but, on the other hand, their corresponding p-values are smaller than the usual threshold 0.05.

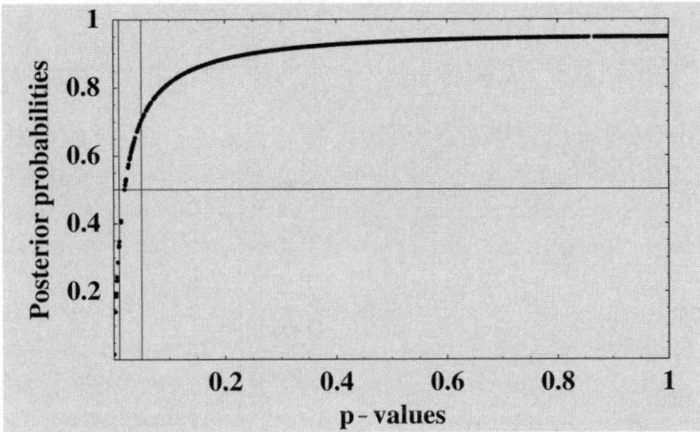

Figure 2. *Plot of the posterior probabilities of H_0 given $\boldsymbol{D}$ against the corresponding p-values.*

It is worth remarking that the "calibration plot or calibration curve," displayed in Figure 2, varies with the sample size n, the design matrix $\boldsymbol{X}$, and the hypothesis to be tested (through the submatrix $\boldsymbol{X}_1$), as is evident from lemma 4.2.

The "calibration plot" can be used to calibrate p-values by means of meaningful values of posterior probabilities of the null hypothesis H_0.

When sampling from the alternative $H_1 : \beta_0 \neq 0$, the resulting values of the posterior probabilities of H_0 decrease to zero as the distance between H_0 and H_1, $D^2 = \beta_0^2/\sigma^2$ increases, regardless of the values of δ_0 and β_1. A similar behavior is also shown by the p-values. In Figure 3, we display the histograms of the posterior probabilities of H_0 and the corresponding p-values for values of the distance $D^2 = 0.04$ and $D^2 = 0.09$, respectively.

6. DISCUSSION

A general hierarchical model is set up for the analysis of matched pairs in the presence of covariates. General conditions are given for the differences between post and pre-treatment observations in the presence of covariates to be a *sufficient statistics* for a general additive model. A fully default Bayesian analysis is carried out and intrinsic priors and Bayes factors for hypothesis testing are derived for the important additive linear normal model.

From the simulation study, it follows that the performance of the intrinsic Bayes factors and their corresponding model posterior probabilities is very satisfactory. This methodology is easily interpretable, takes into account the sample size and the design matrix automatically, and does not present the drawbacks of p-values.

Finally, the model posterior probabilities derived from the intrinsic priors, unlike p-values, address more specifically the issue of *practical significance* (Leonard and Hsu, 1999, pp. 98, 110, 145 and 259).

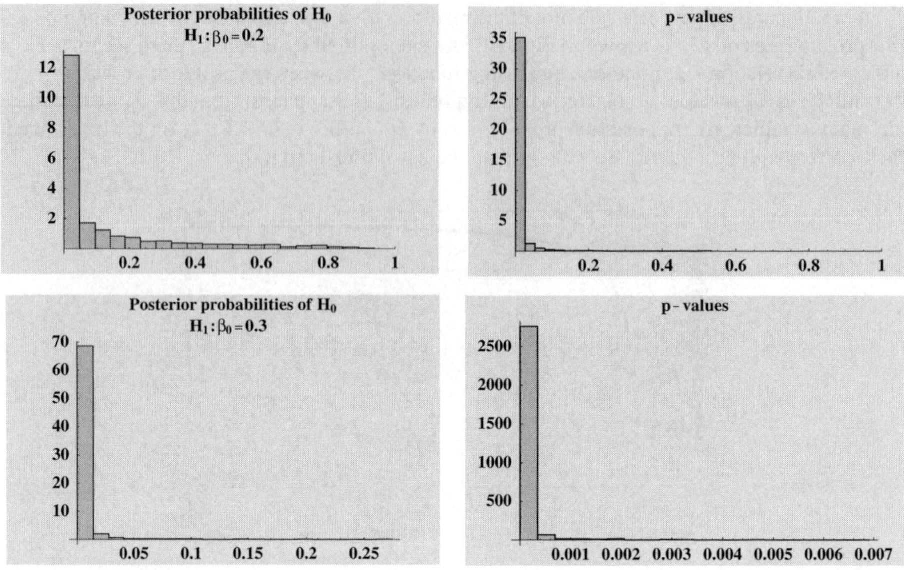

Figure 3. *Histograms of the Posterior probabilities of* H_0 *given* $\boldsymbol{D}$ *and p-values when* $H_1 : \beta_0 = 0.2$ *and* $H_1 : \beta_0 = 0.3$ *are true, respectively.*

ACKNOWLEDGEMENTS

This paper has been partially supported by the *Ministerio de Ciencia y Tecnología* (DGICYT) under projects PB97-1403 C03 and BEC2001-2982, and by *La Consejería de Educación de la Junta de Andalucía.*

REFERENCES

Berger, J. O. and Pericchi, L. R. (1996). The intrinsic Bayes factor for model selection and prediction. *J. Am. Statist. Ass.* **91**, 109–122.

Berger, J. O. and Pericchi, L. R. (1997a). On the justification of default and intrinsic Bayes factor. *Modeling and Prediction* (J. C. Lee *et al.,* eds). New York: Springer, 276–293.

Berger, J. O. and Pericchi, L. R. (1997b). Accurate and stable Bayesian model selection: the median intrinsic Bayes factor. *Sankyà B* **60**, 1–18.

Berger, J. O. and Pericchi, L. R. (1998). On criticism and comparison of default Bayes factors for model selection and hypothesis testing. *Proc. Int. Workshop on Model Selection* (W. Racugno, ed). Bologna: Pitagora, 1–50.

Bickel, P. J. and Doksum, K. A. (1977). *Mathematical Statistics*. San Francisco, CA: Holden-Day

Girón, F. J., Kadane, J. B. and Moreno, E. (1997). Independence issues in imprecise data models: a Bayesian approach. *Compt. Rend. Acad. Sci. Paris Ser. 1* **324**, 1149–1154.

Girón, F. J., Martínez, M. L. and Moreno, E. (2002). Bayesian analysis of matched pairs. *J. Statist. Plann . Inference* (to appear).

Leonard, T. and Hsu, J. S. J. (1999). *Bayesian Methods: An Analysis for Statisticians and Interdisciplinary Researchers*. Cambridge: Cambridge University Press.

Moreno, E. (1997). Bayes factors for intrinsic and fractional priors in nested models: Bayesian robustness. *L_1-Statistical Procedures and Related Topics* (D. Yadolah, ed). Hayward, CA: IMS, 257–270.

Moreno, E., Bertolino, F. and Racugno, W. (1998). An intrinsic limiting procedure for model selection and hypotheses testing. *J. Am. Statist. Ass.* **93**, 1451–1460.

Moreno, E., Bertolino, F. and Racugno, W. (1999). Default Bayesian analysis of the Behrens-Fisher problem. *J. Statist. Plann . Inference* **81**, 323–333.

Moreno, E, Bertolino, F. and Racugno, W. (2000). Bayesian model selection approach to analysis of variance under heteroscedasticity. *J. R. Statist. Soc. D* **46**, 1–15.

Moreno, E. and Liseo, B. (2002). Default priors for testing the number of components of a mixture. *J. Statist. Plann . Inference* (to appear).

Moreno, E., Torres, F. and Casella, G. (2002). Testing equality of regression coefficients in heteroscedastic normal regression models. *Tech. Rep.*, Universidad de Granada, Spain.

Moreno, E., Girón, F. J. and Torres, F. (2003). Intrinsic priors for hypotheses testing in normal regression models. *Rev. Acad. Ciencias Madrid* (to appear).

BAYESIAN STATISTICS 7, pp. 565–575
J. M. Bernardo, M. J. Bayarri, J. O. Berger, A. P. Dawid,
D. Heckerman, A. F. M. Smith and M. West (Eds.)

State Space Models for Density Dependence in Population Ecology

LARA E. JAMIESON   STEPHEN P. BROOKS
University of Cambridge, UK
l.jamieson@statslab.cam.ac.uk   s.p.brooks@statslab.cam.ac.uk

SUMMARY

In this paper, we focus upon the detection and estimation of density dependence in two species of North American ducks on the basis of annual census counts taken over a 47-year period. The data comprise the counts themselves, plus an estimate of the associated standard errors. Classical approaches to investigating density dependence in population ecology from data of this sort tend to suffer from (often extreme) sensitivity to sampling errors. Here, we adopt a Bayesian approach and demonstrate how a state space model can be used to account for both measurement and process error within the analysis. We use (reversible jump) MCMC to both fit and discriminate between the competing models and investigate goodness-of-fit using both cross-validation and Bayesian p-values. We conclude that one of these species of duck appears to exhibit density dependence, whilst the other does not and discuss the implications that this new result has for the future management of these duck populations.

Keywords: REVERSIBLE JUMP MCMC; BAYESIAN P-VALUES; CROSS-VALIDATION; MODEL DISCRIMINATION.

1. INTRODUCTION

As national and international legislation puts increasing pressure on local authorities to identify and protect key wildlife species and their habitats, the ability to design and implement effective management strategies is of paramount importance. In order to design an appropriate management strategy, key factors affecting the population must be understood. In particular, the identification of factors affecting survival and/or population size is an integral part of the management design process and the question as to whether or not population density affects population size is one of the first that must be addressed (see, *e.g.*, Bulmer, 1975; Nicholls *et al.*, 1984; Vickery and Nudds, 1984).

Essentially, if the population exhibits density dependence, this acts as a population stabilizing mechanism which tends to move the population size towards the mean level. When the population size is small, there is a natural pressure on the population to increase its numbers and vice versa. Thus, density dependence increases the population's ability to cope with "shocks" to the system (*i.e.*, rapid increases or, more often, decreases in population size). In terms of management policy, the presence of density dependence within the population indicates an ability to resist perturbations to the system induced by human activity (; Nicholls *et al.*, 1984; Massot *et al.*, 1992). Such activities might vary from the direct effects of hunting to less obvious effects from changes in land use within the population's natural habitat, for example.

Much (almost all) of the work in this area has been classical in nature and focuses upon simple hypothesis tests for the presence (or otherwise) of density dependence. Though the data available for any given population is often rich and varied, much of it is often ignored in order to simplify the classical model fitting and selection process. In this paper we examine data collected from annual aerial surveys (conducted between 1955 and 2001) of two species of North American ducks: the Gadwall (*Anas strepera*) and the Green-winged Teal (*Anas crecca*). Data come in the form of annual population size estimates (based upon the raw counts and then "adjusted for biases" by comparing the aerial counts with random samples on the ground), together with associated standard errors. The raw count data themselves are, unfortunately, not available.

These populations are managed in the sense that, for example, hunting licenses and planning permission for work within key environmental habitats are granted (or not) in order to maintain a stable population size. This "ideal" population level is fixed by the "North American Waterfowl Management Population Plan" which is, in itself, a valuable source of information. The data and population goals are summarized in Figure 1.

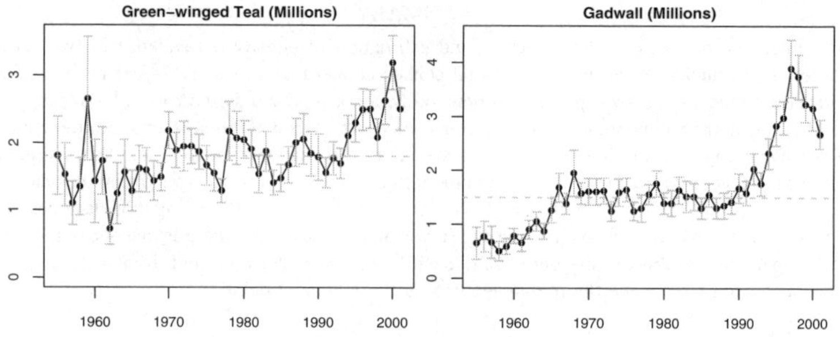

Figure 1. *The population size estimates and associated* 95% *confidence intervals, together with the population goals (dashed) for the two species of duck under study.*

Data in the form of population estimates and standard errors are common in the population ecology literature and data collection authorities are often very protective of the raw data *i.e.*, the original counts upon which the estimates are based. Though Kalman filter-based methods are available (Sullivan, 1992; Newman, 1998), most analyses of the published data simply ignore the standard errors and treat the population estimates as if they were true population sizes. However, whilst also wasting valuable information, this simplifying assumption has a confounding effect upon the estimation of density dependence so that density dependence appears to be present even when it is not (Vickery and Nudds, 1984). The false detection of density dependence can have a substantial effect upon the corresponding management strategy with possibly dire consequences for the species concerned.

In order to make full use of the data available and to avoid this confounding effect, we seek to separate the observation error (associated with the data collection and population size estimation processes) from the system error (associated with the variability within the population itself) so that the density dependence model is fitted to the true underlying (but hidden) population size rather than the estimates provided. This can be most easily achieved through the development of a state space model separating the system and observation processes as follows.

Let N_t denote the true population size at time $t = 1, .., T$. In order to describe the density dependence process, we adopt the model of Dennis and Taper (1994) in which

$$N_{t+1} = N_t \exp(b_0 + b_1 N_t + \epsilon_t) \tag{1}$$

where $\epsilon_t \sim \mathrm{N}(\epsilon_t \mid 0, \sigma^2)$. Note that $b_1 = 0$ in (1) indicates a general trend with $Normal$ errors on a log scale. This basic model can be extended to incorporate density-dependent lags of greater than one year and can be simplified by using the log-transformation $p_t = \log_e N_t$, to give the additive model of order k

$$p_t = p_{t-1} + b_0 + \sum\nolimits_{\tau=1}^{k} b_\tau \exp(p_{t-\tau}) + \epsilon_t, \tag{2}$$

for $t = 1, \ldots, T$ and $k = 0, \ldots, k_{\max}$. Eq. (1) therefore describes the system process and the question of whether or not density dependence exists then boils down to a model determination question: does $k = 0$?

The N_t above denote the true population sizes, which are unknown. However, the data provide us with information as to what these values might be. In particular, the data provide us with estimates $\hat{N}_t$ of the true underlying population size together with associated standard errors s_t. The observation process, which relates the observed data to the true underlying population values, is then described by a simple Gaussian process whereby $\hat{N}_t \mid N_t, s_t^2 \sim \mathrm{N}(\hat{N}_t \mid N_t, s_t^2)$ or equivalently

$$\hat{N}_t \mid p_t, s_t^2 \sim \mathrm{N}(\hat{N}_t \mid e^{p_t}, s_t^2). \tag{3}$$

for $t = 1, \ldots, T$. Note that for $t \leq k$, Eq. (2) involves p-terms with a non-positive index which are not included in (3), since the estimates $\hat{N}_t$ are available only from time $t = 1$ onwards. When we have a large number of observed values we may choose to approximate the likelihood by considering data points from time $t = k$ onwards. However, with models of high order, this involves "ignoring" an increasingly high proportion of the data. Since our data comprise only $T = 47$ observations we shall use the full likelihood and essentially treat the $p_0, \ldots, p_{1-k}$ as additional model parameters to be estimated.

For notational ease we refer to the full set of abundance values $\{p_t, t = 1-k, \ldots, 0, 1, 2, \ldots, T\}$ as $\boldsymbol{P}^k$, the observed series as $\hat{\boldsymbol{N}}$, and associated standard errors as $\boldsymbol{S}$. We shall also let $\boldsymbol{b}^k$ denote the vector of system parameters $b_0, \ldots, b_k$. Hence, the observed data are $\hat{\boldsymbol{N}}$ and $\boldsymbol{S}$; and $\boldsymbol{b}^k$, σ^2 and $\boldsymbol{P}^k$ are the parameters to be estimated under model k.

In the next section, we derive the likelihood for this state space model, discuss priors for the models and their associated parameters, and derive the corresponding posterior distributions. We then discuss the use of (reversible jump) MCMC methods for fitting the models in Section 3, together with model-checking techniques to ensure that the fitted models describe the data adequately. In Section 4, we summarize the results and conclusions of our analysis, before ending with some general discussion.

2. THE BAYESIAN ANALYSIS

The Bayesian analysis involves the combination of the likelihood with the priors to obtain the posterior distribution as the basis for inference. In this section, we derive the form of the likelihood for the state space model described above, before eliciting priors for the model parameters and finally extending the posterior to consider the model determination problem.

2.1. *The Likelihood*

The likelihood comes in two parts: one corresponding to the system and one to the observation equations.

If we decompose $\boldsymbol{P}^k$ into two components $\boldsymbol{P}^k_-$ and $\boldsymbol{P}_-$, where $\boldsymbol{P}^k_-$ denotes the elements $(p_{1-k}, \ldots, 0)$ of $\boldsymbol{P}^k$ and $\boldsymbol{P}_-$ denotes the remainder *i.e.*, $(p_1, \ldots, p_T)$, then we can use (2) to obtain the following likelihood associated with the system process under model k:

$$L_{sys}(\boldsymbol{P}_- \mid \boldsymbol{b}^k, \sigma^2, \boldsymbol{P}^k_-) = (2\pi\sigma^2)^{-T/2} \exp\Big\{ -\frac{1}{2\sigma^2} \sum_{t=1}^{T} \Big(p_t - p_{t-1} - b_0 - \sum_{\tau=1}^{k} b_\tau e^{p_{t-\tau}} \Big)^2 \Big\}. \tag{4}$$

Similarly, (3) can be used to obtain the following likelihood associated with the observed process:

$$L_{obs}(\hat{\boldsymbol{N}} \mid \boldsymbol{P}^k, \boldsymbol{S}) = \frac{1}{(2\pi)^{T/2} \prod_{t=1}^{\prime} s_t} \exp\Big\{ -\frac{1}{2} \sum_{t=1}^{\prime} \Big(\frac{\hat{N}_t - e^{p_t}}{s_t} \Big)^2 \Big\}. \tag{5}$$

Of course, the interpretation of L_{sys} as a likelihood is slightly misleading in that it contains no terms corresponding to the observed data. However, it is intended to represent the likelihood of the $\boldsymbol{P}_-$ vector had it been observed. An alternative interpretation is as a prior for the unobserved $\boldsymbol{P}_-$ conditioning on the remaining parameters. Hence, when we come to specify priors for the model parameters in the next section, we need only specify priors for the $\boldsymbol{P}_-^k$, since those for $\boldsymbol{P}_-$ are essentially already defined through L_{sys}.

2.2. *The Priors*

The North American Waterfowl Management Plan provides suitable guidance for the specification of priors on the elements of $\boldsymbol{P}_-^k$ for each of our two duck species. The plan was signed in 1986 aiming to restore waterfowl numbers to levels of the 1970s. The plan suggests that our *a priori* beliefs about the general level of Gadwall and Green-winged Teal populations should be centered around 1.5 and 1.8 million, respectively, and with corresponding standard deviations of around 2.7×10^5 and 2.2×10^5, reflecting a fair degree of uncertainty on the population level in any given year. Since populations below and above the expected level seem *a priori* equally likely and since the chance of observing populations far from the expected level decreases fairly rapidly with distance, we take independent normal priors with the means and SD's given above for the population levels at times $t = 1 - k, ..., 0$.

With regard to the remaining parameters, very little information is available *a priori* and so we wish to specify reasonably vague priors on these parameters. Here, we take independent priors, with $b_i \sim N(b_i \mid 0, \sigma_b^2)$, $\forall i = 0, ..., k$ and $\sigma^2 \sim \text{Ga}(\sigma^{-2} \mid \alpha, \beta)$, so that $p(\sigma^2) = (\beta^\alpha / \Gamma(\alpha))\, \sigma^{2-(\alpha+1)} e^{-\beta/\sigma^2}$, where $\sigma_b^2 = 100$ and $\alpha = \beta = 10^{-3}$.

Conducting a standard sensitivity study, the posterior means and variances of the model parameters were almost entirely insensitive to changes in the prior parameters within a reasonable range. At the extremes, some sensitivity of the posterior model probabilities was observed, but was not sufficiently great to alter the interpretation of the results, even if the absolute values of the model probabilities did change slightly.

With the likelihoods and priors described above, the posterior for the parameters $\boldsymbol{\theta}^k = (\boldsymbol{b}^k, \sigma^2, \boldsymbol{P}^k)$ under model k is given by

$$\pi(\boldsymbol{\theta}^k \mid \hat{\boldsymbol{N}}, \boldsymbol{S}) = L_{obs}(\hat{\boldsymbol{N}} \mid \boldsymbol{P}^k, \boldsymbol{S}) L_{sys}(\boldsymbol{P}_- \mid \boldsymbol{b}^k, \sigma^2, \boldsymbol{P}_-^k) p(\boldsymbol{P}_-^k) p(\boldsymbol{b}^k) p(\sigma^2).$$

If we wish to extend the posterior distribution to allow for *a priori* uncertainty as to the model, we may specify prior model probabilities placing mass $p(k)$ on each of the models $k = 0, \ldots, k_{\max}$ and we obtain the following joint posterior over both model and parameter space.

$$\pi(\boldsymbol{\theta}^k, k \mid \hat{\boldsymbol{N}}, \boldsymbol{S}) = L_{obs}(\hat{\boldsymbol{N}} \mid \boldsymbol{P}^k, \boldsymbol{S}, k) L_{sys}(\boldsymbol{P}_- \mid k, \boldsymbol{b}^k, \sigma^2, \boldsymbol{P}_-^k) p(\boldsymbol{P}_-^k \mid k) p(\boldsymbol{b}^k \mid k) p(\sigma^2) p(k).$$

For the purposes of our analysis, we will take uniform priors over all models $k = 0, \ldots, k_{\max}$ and set $k_{\max} = 5$ though models of this order, as we shall see, enjoy essentially zero posterior mass.

Bayesian inference is then based upon these posterior distributions and often summarized in the form of posterior means and variances (for example) for parameters of interest. Since these posterior summaries are analytically intractable in this case, we use Markov chain Monte Carlo (MCMC) methods to obtain estimates of the posterior quantities of interest.

3. THE IMPLEMENTATION

Since the number of parameters changes with the model, we distinguish between two forms of updating scheme. The first updates the parameters within the current model, using a mixture of Gibbs sampler (Casella and George 1992) and Metropolis–Hastings (Chib and Greenberg 1995) updates, whilst the second updates the model itself, using reversible jump MCMC (Green 1995; Brooks *et al.* 2002).

3.1. *Updating the Parameters*

With fixed model order, k, the posterior conditional distributions of most of the model parameters are standard distributions *i.e.*, easy to sample from directly, with the exception of $\boldsymbol{P}_-$ and p_0.

From (6) it is clear that

$$\pi(\sigma^2 \mid \boldsymbol{b}^k, \boldsymbol{P}^k, k, \hat{\boldsymbol{N}}, \boldsymbol{S}) \propto L_{sys}(\boldsymbol{P}_- \mid k, \boldsymbol{b}^k, \sigma^2, \boldsymbol{P}_-^k) p(\sigma^2).$$

Hence, it is easy to show that the conditional posterior for σ^2 is an Inverse-Gamma with parameters $\alpha + T/2$ and $\beta + \frac{1}{2}\sum_{t=1}' \left(p_t - p_{t-1} - b_0 - \sum_{\tau=1}^k b_\tau \exp(p_{t-\tau})\right)^2$.

The conditional posterior distributions for the elements of $\boldsymbol{P}_-$ and for p_0 are of non-standard form and so we require Metropolis–Hastings moves to update these parameters. In practice a random walk Metropolis algorithm (Brooks, 1998) with a uniform proposal of width 0.3 seems to work well. However, for the remaining elements of $\boldsymbol{P}_-^k$, the posterior conditional distribution of $N_{-i} = \exp(p_{-i})$ is normal with mean and variance given by

$$\frac{\mu_0\sigma^2 + \sigma_0^2 \sum_{t=1}^{k-i} b_{t+i}(p_t - p_{t-1} - b_0 - \sum_{\tau:\tau\neq t+i} b_\tau e^{(p_{t-\tau})})}{\sigma^2 + \sigma_0^2 \sum_{\tau=(i+1)}^k b_\tau^2} \quad \text{and} \quad \frac{\sigma^2\sigma_0^2}{\sigma^2 + \sigma_0^2 \sum_{\tau=(i+1)}^k b_\tau^2}$$

respectively, where μ_0 and σ_0^2 denote the corresponding prior means and variances.

Finally, we consider the updating of the parameter vector $\boldsymbol{b}^k$. First, we rewrite the likelihood corresponding to the system process in matrix-vector form, so that $\boldsymbol{\epsilon} = \boldsymbol{d} - \boldsymbol{Y}^k\boldsymbol{b}^k$ with

$$\boldsymbol{d} = \begin{pmatrix} p_1 - p_0 \\ \vdots \\ p_t - p_{t-1} \\ \vdots \\ p_T - p_{T-1} \end{pmatrix} \quad \text{and} \quad \boldsymbol{Y}^k = \begin{pmatrix} 1 & e^{p_0} & e^{p_{-1}} & \dots & e^{p_{-K+1}} \\ 1 & e^{p_1} & e^{p_0} & \dots & e^{p_{-K+2}} \\ \vdots & \vdots & \vdots & & \vdots \\ 1 & e^{p_{K-1}} & e^{p_{K-2}} & \dots & e^{p_0} \\ \vdots & \vdots & \vdots & & \vdots \\ 1 & e^{p_{T-1}} & e^{p_{T-2}} & \dots & e^{p_{T-K}} \end{pmatrix}.$$

Since we have independent normal $\mathrm{N}(b_i \mid 0, \sigma_b^2)$ priors on the b_i, the prior for the vector $\boldsymbol{b}^k$ is simply $\mathrm{N}_k(\boldsymbol{b}^k \mid 0, \sigma_b^2\boldsymbol{I}^k)$, where $\boldsymbol{I}^k$ denotes the k-dimensional identity matrix. It is then simple to show that the corresponding posterior conditional for $\boldsymbol{b}^k$ is a multivariate normal with mean $\boldsymbol{\mu}^k$ and covariance matrix $\boldsymbol{C}^k$, where

$$\boldsymbol{\mu}^k = \sigma^{-2}\boldsymbol{C}^k(\boldsymbol{Y}^k)'\boldsymbol{d} \quad \text{and} \quad \boldsymbol{C}^k = (\sigma^{-2}(\boldsymbol{Y}^k)'\boldsymbol{Y}^k + \sigma_b^{-2}\boldsymbol{I}^k)^{-1}.$$

Thus, we are able to update the $\boldsymbol{b}^k$ parameters as a block. This greatly improves MCMC mixing, as the posterior correlations between the elements of $\boldsymbol{b}^k$ can be quite high (see Roberts and Sahu, 1997). The one drawback is the computational expense associated with the inversion required for $\boldsymbol{\mu}^k$. However, for the moderate dimensions that we consider here, the computational burden is minimal.

3.2. *Updating the Model*

Updating the model involves adding or deleting parameters (elements of the $\boldsymbol{b}^k$ and $\boldsymbol{P}^k_-$ vectors) and thus requires a scheme capable of moving between spaces of different dimension. There are various trans-dimensional Markov chain updating schemes, but the most popular is the reversible jump (RJ) MCMC updating scheme first proposed by Green (1995). See also Green (2002) and Brooks *et al.* (2003).

Suppose that we wish to move from a model of order k to a model of order $k+1$. This involves adding new parameters b^{k+1} and p_{-k}. The obvious move is to propose new values for these parameters whilst keeping all of the existing parameters fixed *i.e.*, we propose moving from $\boldsymbol{b}^k$ to $(\boldsymbol{b}^k, b_{k+1})$ etc. However, this produces a very poorly mixing chain, since moving to a model with a higher order dependence invariably changes the effects from previous years. Keeping the values of $b_1, \ldots, b_k$ fixed therefore leads to extremely high rejection rates and the chain rarely moves between models. Clearly, we require a between-model move that proposes new values for the entire $\boldsymbol{b}$-vector. Note that since the interpretation (and hence the value) of the $\boldsymbol{P}^k_-$ vector essentially remains unchanged between models, keeping these fixed for between model moves has no such detrimental effect on the acceptance rates (similarly for σ^2). Hence, we will retain these parameters and only propose new values for the $\boldsymbol{b}$-vector and for p_{-k}.

Suppose that we propose a value for p_{-k} directly from its prior and to replace the current parameter vector $\boldsymbol{b}^k$ by $\boldsymbol{b}^{k+1} \sim \mathrm{N}(\boldsymbol{b}^{k+1} \mid \boldsymbol{\mu}^{k+1}, \boldsymbol{C}^{k+1})$ *i.e.*, the posterior conditional for $\boldsymbol{b}^{k+1}$ in the new model, keeping σ^2 and $\boldsymbol{P}^k_-$ fixed. This proposal corresponds to the so-called "second-order" method of Brooks *et al.* (2003), which aims to maximize the acceptance probability of trans-dimensional jumps by taking a Taylor series expansion of the acceptance ratio about certain canonical jumps. See Brooks *et al.* (2003) for further details and discussion.

With this transition scheme, the usual Jacobian in the reversible jump acceptance ratio is 1 and, using the "Candidate's Identity" (Besag, 1989)

$$\frac{\pi(k, \boldsymbol{b}^k, p_{1-k} \mid \sigma^2, \boldsymbol{P}^k)}{\pi(\boldsymbol{b}^k, p_{1-k} \mid k, \sigma^2, \boldsymbol{P}^k)} = \pi(k \mid \sigma^2, \boldsymbol{P}^k),$$

it is easy to show that the reversible jump MCMC acceptance probability reduces to

$$\alpha\left[(k, \boldsymbol{b}^k), ((k+1), \boldsymbol{b}^{k+1})\right] = \min\left\{1, \frac{\pi(k+1 \mid \sigma^2, \boldsymbol{P}^k)}{\pi(k \mid \sigma^2, \boldsymbol{P}^k)} \frac{q_{k+1,k}}{q_{k,k+1}}\right\},$$

where $q_{k,k'}$ denotes the probability of proposing to go from model k to model k'. Following Godsill (1997), it is fairly straightforward to show that the acceptance probability simplifies to

$$\min\left\{1, \frac{1}{\sigma_b^k} \frac{|\boldsymbol{C}^k|^{-1/2}}{|\boldsymbol{C}^{k+1}|^{-1/2}} \frac{\exp(\frac{1}{2}(\boldsymbol{\mu}^{k+1})^T(\boldsymbol{C}^{k+1})^{-1}\boldsymbol{\mu}^{k+1})}{\exp(\frac{1}{2}(\boldsymbol{\mu}^k)'(\boldsymbol{C}^k)^{-1}\boldsymbol{\mu}^k)} \frac{q_{k+1,k}}{q_{k,k+1}}\right\},$$

by integrating the parameters $\boldsymbol{b}^k$ from the system likelihood given in (4).

Thus, the acceptance probability does not depend upon the current (or proposed new) value of the $\boldsymbol{b}$-vector (though it does depend upon the proposed new value of p_k, through $\boldsymbol{\mu}^{k+1}$

and C^{k+1}). In practice, a new value of p_k is proposed and the move to model $k+1$ is then either accepted or rejected. Only if it is accepted is the new parameter vector b^{k+1} generated. This both improves computational efficiency and provides a rapidly mixing chain which is easily capable of traversing model space.

The Markov chain Monte Carlo scheme described above will provide draws from the posterior distribution. These samples can then be used to estimate posterior means and variances for parameters of interest, by simply calculating the corresponding samples means and variances. Similarly, the posterior model probabilities are estimated by the proportion of time spent in each model. These posterior model probabilities can be used either to discriminate between models or to model average, for example. However, since the posterior model probabilities are known only up to a normalization constant (formed by summing the values over the range of models under consideration) they tell us only about the relative merits of different models and nothing about how well any of the models actually describe the data. In order to be sure that the range of models considered actually describe the data well, we need to implement model checking (or goodness-of-fit-type) methods.

3.3. *Model Checking*

There are two formal model checking procedures commonly used in the literature: the Bayesian p-value, and cross-validation.

Bayesian p-values (Gelman and Meng, 1996) can be used to check the discrepancy between the sample values and the observed. The classical discrepancy statistic is to take $D(x,\vartheta) = \sum_j(\sqrt{x_j}-\sqrt{e_j})^2$, where x is the data or observed values and e_j the expected; j indexes the vectors. In our case ϑ_i is the sample $\boldsymbol{P}^k$ produced by the i-th iteration of the MCMC sampler. Using the observed likelihood of (5), we calculate $D(\boldsymbol{x},\vartheta_i) = L_{obs}(\hat{\boldsymbol{N}} \mid e^{\boldsymbol{P}^k}, \boldsymbol{S})$. We sample $\tilde{\boldsymbol{N}}$ from the observed model using ϑ_i to obtain $D(\boldsymbol{x}_i,\vartheta_i) = L_{obs}(\tilde{\boldsymbol{N}} \mid e^{\boldsymbol{P}^k}, \boldsymbol{S})$. The Bayesian p-value is then the proportion of times that $D(\boldsymbol{x}_i,\vartheta_i) > D(\boldsymbol{x},\vartheta_i)$ which, if the model describes the data well, should be close to $1/2$. See Bayarri and Berger (1998) and Brooks *et al.* (2000) for further details.

Cross-validation (Gelman *et al.,* 1995; Carlin and Louis, 1996) involves treating observed values as if they were missing and investigating how well the model predicts this value. Suppose we treat observation N_{t*} as a missing value and update the remaining parameters as before. For iteration i sample N^i_{t*} using a Metropolis–Hastings step, excluding the $\hat{N}_{t*}$ term. We then sample $\tilde{N}^i_{t*}$ from the model *i.e.*, $\mathrm{N}(\tilde{N}^i_{t*} \mid N^i_{t*}, s^2_{t*})$, and calculate the ratio $\tilde{N}^i_{t*}/\hat{N}_{t*}$. Repeating the same procedure for each $t = 1, ..., T$ we can calculate

$$CV_i = \frac{1}{T}\sum\nolimits'_{t*=1}\frac{\tilde{N}^i_{t*}}{\hat{N}_{t*}}$$

which, if the model describes the data well, should be close to 1.

In the next section, we present the result of our analysis, including parameter estimates and posterior model probabilities. We also employ the model checking procedures described above and interpret the results.

4. RESULTS

Trial runs were used to establish appropriate parameters for the Metropolis–Hastings proposals and to investigate mixing and convergence using standard diagnostic techniques (Cowles and Carlin, 1996; Brooks and Roberts, 1998; Mengersen *et al.,* 1998). The output below was based upon reversible jump MCMC simulations with 6 million iterations after discarding an initial 100,000 iterations as burn-in. Fixed model simulations were also run to gain parameter estimates with low posterior model probability and these were run for 2 million iterations and a burn-in of only 20,000 iterations.

Table 1. *Posterior summary statistics for the parameters of primary interest.* HPDI *denotes the* 95% *highest posterior density interval for the corresponding parameter.*

Parameter		Gadwall Model $k=0$	Gadwall Model $k=1$	Green-winged Teal Model $k=0$	Green-winged Teal Model $k=1$
b_1	Mean	–	-0.030	–	-0.131
	$\boldsymbol{P}(b_1 < 0)$	–	0.84	–	0.97
	HPDI	–	(-0.090,0.030)	–	(-0.278,0.006)
b_0	Mean	0.025	0.073	0.009	0.239
	HPDI	(-0.019,0.069)	(0.032,0.178)	(-0.038,0.056)	(-0.008,0.502)
σ^2	Mean	0.022	0.023	0.026	0.029
	HPDI	(0.009,0.037)	(0.010, 0.039)	(0.010,0.045)	(0.012, 0.049)

Table 1 provides summary statistics for the parameters of primary interest under the models $k = 0$ and $k = 1$. The corresponding density plots for b_1 under the model $k = 1$ are provided in Figure 2.

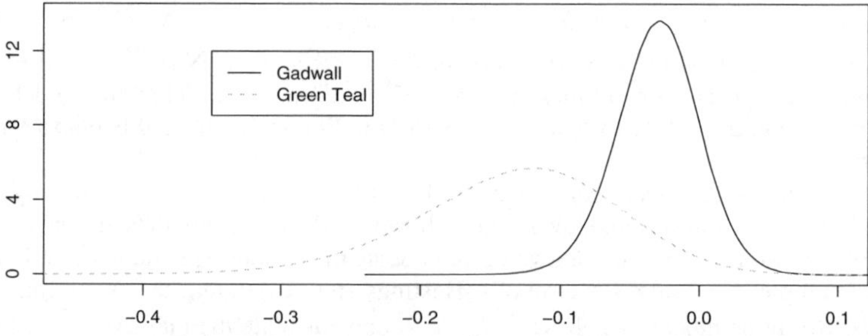

Figure 2. *Density estimates for* b_1*, for the* $k = 1$ *model.*

It is clear from Figure 2 that the posterior distributions for b_1 under the $k = 1$ model differ between the two species, though they have several features in common. As would be expected, the majority of the posterior mass is placed on negative values of b_1, corresponding to the population-stabilizing mechanism discussed earlier. For example, the posterior probability that b_1 is negative for the Green-winged Teal data is 0.97. Essentially, with a negative value for b_1, small populations should grow and large ones shrink, pushing the population towards a stable equilibrium at some species-dependent level. Positive values for b_1 would suggest highly unstable populations that would be unlikely to survive. From Table 1 we see that the estimates for σ^2 are similar for both species (and under both models), with that for the Green-winged Teal slightly larger; reflecting the larger year-to-year variability in the population size observed in Figure 1.

Finally, we observe that the value of b_0 differs quite substantially between the two models—again, as expected. This supports our decision to update the entire $\boldsymbol{b}$-vector when moving

Table 2. *Posterior Model probabilities, cross validation statistics and Bayesian p-values for the two duck species.*

	Gadwall			Green-winged Teal		
k	Model Prob	CV	p-value	Model Prob	CV	p-value
0	0.981	1.002	0.536	0.176	0.985	0.758
1	0.019	1.002	0.520	0.812	0.989	0.686
2	0.000	1.001	0.527	0.011	0.989	0.679
3	0.000	1.001	0.531	0.000	0.989	0.682
4	0.000	1.001	0.534	0.000	0.989	0.681

between different models, since there appears to be little overlap even in the 95% HPDI's between the b_0 values between the two models in Table 1.

Table 2 provides the posterior model probabilities, together with the corresponding Bayesian p-values and cross-validation statistics. We see clear evidence from the posterior model probabilities that the Gadwall is not density dependent (with posterior model probability exceeding 0.98) and some evidence to suggest that the Green-winged Teal population is density dependent with lag 1. The posterior model probability associated with $k = 1$ exceeds 0.80, corresponding to a Bayes factor in excess of 4 when compared with the $k = 0$ model. This conclusion agrees with our observations from the posterior densities of b_1 in which the value of $b_1 = 0$ lay well within the corresponding 95% HPDI for the Gadwall, but not for the Green-winged Teal.

For the Gadwall data set the cross-validation statistics and p-values do not differ greatly between any of the models and suggest that all of the models appear to describe the data well. For the Green-winged Teal data set, the cross-validation statistics are, again, very similar, but suggest that the model tends to very slightly underestimate the missing values. In fact the deviation of the CV statistic from unity is caused almost entirely by a single observation, that corresponding to the year 1959. The p-values are clearly worse for the Green-winged Teal data set than for the Gadwall and this may well be explained by the outlying nature of the observation for 1959 and we shall return this point shortly. The model $k = 0$ clearly performs least well and this is consistent with the majority of the posterior mass being placed on higher order (basically just $k = 1$) models. Thus, overall, the models appear to describe the data well, though investigation of the apparently poorer fit to the Green-winged Teal data set is worthy of further investigation and can, in fact, be explained by examining the posterior estimates of the true underlying population sizes.

Figure 3 provides the model-averaged posterior medians (the means are the same) and associated 95% HPDI's for the true population sizes with the Green-winged Teal data. We can see that the Bayesian estimates (and intervals) match the original data pretty closely, but that the Bayesian estimation procedure has a mild smoothing effect on the estimates, as we might perhaps expect. However, there is one year (1959) where the Bayesian estimate and the original data point differ dramatically. Interestingly, this point corresponds to a year where the population appears to have very nearly doubled from the corresponding figure from the previous year. In addition, the associated standard error for this year is considerably wider than that observed for any other year, suggesting that there may well have been a substantial discrepancy between the aerial counts upon which the original population size estimates were based and the ground-based counts used to "correct" them. This increased uncertainty in the original population estimate provides the state space model with the flexibility to place the estimate of the true underlying population far from the corresponding data point so that it is consistent with the system-level process which describes the remainder of the data so well.

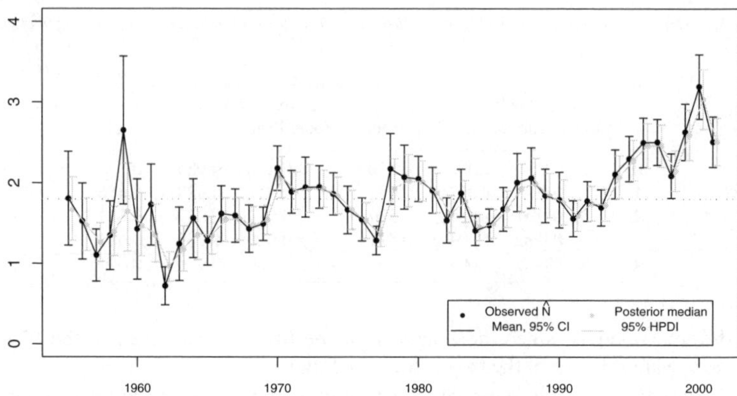

Figure 3. *The Bayesian estimates of the true population size for the Green-winged Teal data, together with the original data. N.B. The Bayesian estimates are displaced slightly to the right for clarity.*

Fitting the system model directly to the raw data and ignoring the standard errors provided can give dramatically different results. The most dramatic example we have observed comes from a study of canvasback ducks where the posterior probability of density dependence rises from 43 to 97% when the errors are ignored. See Jamieson and Brooks (2003) for further details and discussion.

5. DISCUSSION

In this paper we discuss a Bayesian procedure for fitting a suitable state space model to two sets of census count data. This is the first analysis of these data to properly incorporate the standard errors provided with the population estimates and remove the confounding problem associated with analyzing the raw data as if they were the true population levels. Further, this is the first analysis to detect the density-dependent nature of the Green-winged Teal data set which has major implications for the Waterfowl Population Management Plan. In particular, with much discussion as to whether or not the human exploitation of various species of North American waterfowl increases overall mortality, the presence of a density-dependent population process suggests that (within reasonable bounds) the Green-winged Teal population is likely to compensate naturally for any negative effect that human activity might have. However, the Gadwall population does not appear to possess the same self-stabilizing property and so more restrictive management regimes may be required in order to ensure the continued success of this particular species.

The application of reversible jump MCMC methodology greatly simplifies the model determination procedure, which, previously, had consisted of *adhoc* interpretations of classical confidence intervals for the parameter b_1. (These were *adhoc* in the sense that the data were treated as the true population sizes rather than estimates.) In addition, the estimation of the posterior model probabilities allows us to provide model-averaged inference. In this paper, we provide model-averaged estimates of the true underlying population sizes, but of perhaps greater value would be model-averaged predictions for the future development of the populations under study. These would fully account for the uncertainty in the raw data, parameter uncertainty and finally model uncertainty and would give a far more realistic picture of what might happen to these populations in the future.

REFERENCES

Bayarri, M. and Berger, J. (1998). Quantifying surprise in the data and model verification. *Bayesian Statistics 6* (J. M. Bernardo, J. O. Berger, A. P. Dawid and A. F. M. Smith, eds). Oxford: Oxford University Press, 53–82 (with discussion).

Besag, J. (1989). A candidate's formula—A curious result in Bayesian prediction. *Biometrika* **76**, 183.

Brooks, S. P. (1998). Markov chain Monte Carlo method and its application. *The Statistician* **47**, 69–100.

Brooks, S. P., Catchpole, E. A. and Morgan, B. J. T. (2000). Bayesian animal survival estimation. *Statist. Sci.* **15**, 357–376.

Brooks, S. P., Giudici, P. and Roberts, G. O. (2003). Efficient construction of reversible jump proposal distributions. *J. R. Statist. Soc. B* (to appear). (with discussion).

Brooks, S. P. and Roberts, G. O. (1998). Diagnosing convergence of Markov chain Monte Carlo algorithms. *Statist. Comput.* **8**, 319–335.

Bulmer, M. G. (1975). The statistical analysis of density dependence. *Biometrics* **31**, 901–911.

Carlin, B. P. and Louis, T. A. (1996). *Bayes and Empirical Bayes Methods for Data Analysis*. London: Chapman and Hall.

Casella, G. and George, E. I. (1992). *Explaining the Gibbs sampler,J. Am. Statist. Ass.* **46** 167–174.

Chib, S. and Greenberg, E. (1995). Understanding the Metropolis–Hastings algorithm. *Am. Statist.* **49**, 327–335.

Cowles, M. K. and Carlin, B. P. (1996). Markov chain Monte Carlo convergence diagnostics: A comparative review. *J. Am. Statist. Ass.* **91**, 883–904.

Dennis, B. and Taper, M. L. (1994). Density dependence in time series observations of natural populations: Estimation and testing. *Ecological Monographs* **64**, 205–224.

Gelman, A., Carlin, J. B., Stern, H. S. and Rubin, D. B. (1995). *Bayesian Data Analysis*. London: Chapman and Hall.

Gelman, A. and Meng, X. (1996). Model checking and model improvement. *Markov Chain Monte Carlo in Practice* (W. R. Gilks, S. Richardson and D. J. Spiegelhalter, eds). London: Chapman and Hall, 189–201.

Godsill, S. J. (1997). On the relationship between MCMC model uncertainty methods. *Tech. Rep.*, Cambridge University, UK.

Green, P. J. (1995). Reversible jump Markov chain Monte Carlo computation and Bayesian model determination. *Biometrika* **82**, 711–732.

Green, P. J. (2002). Efficient construction of reversible jump MCMC proposal distributions. *Highly Structured Stochastic Systems*. Oxford: Oxford University Press (to appear).

Jamieson, L. E. and Brooks, S. P. (2003). Investigating the population dynamics of North American ducks. *Tech. Rep.*, University of Cambridge, UK.

Massot, M., Clobert, J., Pilorge, T., Lecomte, J. and Barbault, R. (1992). Density dependence in the common lizard: Demographic consequences of a density manipulation. *Ecology* **73**, 1742–1756.

Mengersen, K. L., Robert, C. P. and Guihenneuc-Jouyaux, C. (1998). MCMC convergence diagnostics: A reviewww. *Bayesian Statistics 6* (J. M. Bernardo, J. O. Berger, A. P. Dawid and A. F. M. Smith, eds). Oxford: Oxford University Press 415–440.

Newman, K. B. (1998). State-Space modelling of animal movement and mortality with application to salmon. *Biometrics* **54**, 1290–1314.

Nicholls, J. D., Conroy, M. J., Anderson, D. R. and Burnham, K. P. (1984). Compensatory mortality in waterfowl populations: A review of the evidence and implications for research and management. *Trans. North Ame. Wildlife Nat. Reserve Conf.* **49**, 535–554.

Roberts, G. O. and Sahu, S. K. (1997). Updating schemes, covariance structure, blocking and parameterization for the Gibbs sampler. *J. R. Statist. Soc. B* **59**, 291–318.

Sullivan, P. J. (1992). A Kalman Filter approach to catch-at-length analysis. *Biometrics* **48**, 237–257.

Vickery, W. and Nudds, T. (1984). Detection of density dependent effects in annual duck censuses. *Ecology* **65**, 96–104.

BAYESIAN STATISTICS 7, pp. 577–585
J. M. Bernardo, M. J. Bayarri, J. O. Berger, A. P. Dawid,
D. Heckerman, A. F. M. Smith and M. West (Eds.)

A Marginal Ergodic Theorem

MICHAEL LAVINE
Duke University, USA
michael@stat.duke.edu

SUMMARY

In recent years, there have been several papers giving examples of Markov Chain Monte Carlo (MCMC) algorithms whose invariant measures are improper (have infinite mass) and which therefore are not positive recurrent, yet which have subchains from which valid inference can be derived. These are nonergodic (not having a limiting distribution) Markov chains (MCs) that can be written, possibly after transformation, as $Z = \{Z(n); n \geq 0\} = \{(X(n), Y(n)); n \geq 0\}$ for which the subchain $X(n)$ is ergodic (has a limiting distribution). This paper gives a *marginal ergodic theorem* which (a) gives a formula for bounding the lim inf and lim sup as $n \to \infty$ of the distribution of $X(n)$ and (b) often allows for direct calculation of the limiting distribution, should one exist.

Keywords: MARKOV CHAIN MONTE CARLO, ERGODIC THEOREM, IMPROPER DISTRIBUTION.

1. BACKGROUND

In recent years, several papers including Gelfand and Sahu (1999), Hobert (2001), Liu and Wu (1999), Meng and van Dyk (1999) and van Dyk and Meng (2001) have appeared which use MCMC samplers based on improper posteriors to make valid inference about a proper target posterior. The following example is in Hobert (2001). It or a close relative has appeared in Liu and Wu (1999), Meng and van Dyk (1997, 1999) and van Dyk and Meng (2001).

Example 1. Let $U \sim t(\nu, \mu, 1)$, the t distribution with ν degrees of freedom and unknown location parameter μ. (ν is known.) We can express the distribution of U as

$$U \mid \alpha, q \sim \mathrm{N}\Big(\mu, \frac{\alpha}{q}\Big), \quad q \mid \alpha \sim \mathrm{Ga}\Big(\frac{\nu}{2}, \frac{\nu}{2\alpha}\Big), \quad p(\alpha) \propto \alpha^p$$

for some $p < -0.5$. Then the posterior for (μ, α, q) is improper but an MCMC chain based on alternate draws from $[q \mid \mu, \alpha, u]$ and $[\mu, \alpha \mid q, u]$ has a subchain for μ that not only has a proper limiting distribution but yields i.i.d. draws.

In this and similar examples there is a vector of parameters X in a space $\mathcal{X}$ having a proper "target" distribution, often but not necessarily a posterior distribution from a Bayesian analysis. The vector X is augmented with Y to form $Z = (X, Y)$, where $Y \in \mathcal{Y}$ and $Z \in \mathcal{Z}$. In Example 1 $X = \mu$ and $Y = (\alpha, q)$. A Markov chain (MC) $Z(n) = (X(n), Y(n))$ is constructed such that $\lim_{n\to\infty} X(n)$ has the desired target distribution even though $\lim_{n\to\infty} Z(n)$ is irrelevant or might not even exist. The cited papers give sufficient conditions on $Z(n)$ to ensure that $X(n)$ has the desired limiting distribution in various special cases. Our purpose is to find a theorem more in the spirit of the ergodic theorem, which gives a formula for computing the limiting distribution of an MC from knowledge of its local properties.

We use the notation $K(z \mid z')$ or $K(x, y \mid x', y')$ to denote both transition probabilities and their densities. We assume throughout that K is irreducible and aperiodic. See Borovkov (1998), Robert and Casella (1999) or any book on Markov chains for definitions. A measure μ on $\mathcal{Z}$ is said to be an *invariant* measure for K if, for every measurable set B, $\mu(B) = \int K(B \mid z')\mu(dz')$. Invariant σ-finite measures are unique up to multiplication by constants (Borovkov, 1998, p. 5; Robert and Casella, 1999, p. 155).

We deal generally with the situation in which there is a proper target distribution for X, say $\pi_X(x)$. We wish to construct an MC with transition kernel $K(x, y \mid x', y')$ for which the subchain X converges in distribution to the target. Usually, Y is introduced either to simplify or accelerate convergence; but the purpose of Y is irrelevant for this paper. (Acceleration is the point of the series of papers by Meng and van Dyk in 1997, 1999, 2001.) Hobert (2001) shows that if $X(n)$ is a Markov chain and, in addition, μ has a density π that can be written as $\pi(z) = \pi(x, y) = \pi_X(x)\pi_Y(y)$ and the integral

$$\int\int k(x, y \mid x', y')\pi_Y(y')\, dx\, dy'$$

is free of x', then the subchain X is positive recurrent with invariant density π_X. Liu and Wu (1999) and Meng and van Dyk (1999) show that if K is the limit of transition kernels K_i, where K_i has μ_i as its stationary distribution and μ_i has π_X as its X marginal, then K also has π_X as its limiting X marginal. Liu and Wu (1999) also show that if Y is a locally compact transformation group with a prior corresponding to a Haar measure then a particular MC has the desired limiting distribution. In this paper we impose less severe conditions, similar to those needed for the usual ergodic theorem.

To illustrate the delicacy of the situation, consider the following example.

Example 2. Let $\mathcal{X} = \{0, 1\}$, $\mathcal{Y} = \{\ldots, -1, 0, 1, \ldots\}$ and $\mathcal{Z} = \mathcal{X} \times \mathcal{Y}$. Figure 1 shows the points $z \in \mathcal{Z}$.

Let the measure μ assign mass 1 to every point; that is, for every $z \in \mathcal{Z}$, $\mu(z) = 1$. Consider the following two MCs.

$$\begin{array}{ccccccccc} & \bullet & \bullet & \bullet & \bullet & \bullet & \bullet & \bullet & \\ \cdots & & & & & & & & \cdots \\ & \bullet & \bullet & \bullet & \bullet & \bullet & \bullet & \bullet & \end{array}$$

Figure 1. *The points $z = (y, x)$ in Example 2.*

MC 2a. Let $K_{2a}(x, y \mid x', y')$ be determined by *(a)* $y = y' - 1, y'$, or $y' + 1$ with probabilities 0.25, 0.5 and 0.25 and *(b)* $x \sim \text{Br}(0.5)$ independent of x', y' and y. MC 2a has μ as its invariant measure because for each (x, y), $\mu(x, y) = \sum_{x',y'} K(x, y \mid x', y')\mu(x', y')$. There are only six nonzero terms in the summation corresponding to the six possible (x', y') pairs from which (x, y) might have come.

MC 2b. Change variables from (x, y) to (u, v) according to

$$u(x, y) = \begin{cases} 0 & \text{if } x = 0, \\ 1 & \text{if } x = 1, y \text{ is odd}, \\ 2 & \text{if } x = 1, y \text{ is even}. \end{cases}$$

$$v(x, y) = \begin{cases} y & \text{if } x = 0, \\ y/2 & \text{if } x = 1, y \text{ is even}, \\ (y - 1)/2 & \text{if } x = 1, y \text{ is odd}. \end{cases}$$

	-2	-2	-1	-1	0	0	1	1	2	2	
...											...
	-4	-3	-2	-1	0	1	2	3	4	5	

Figure 2. *The points $z = (y, x)$ in Example 2, labelled by v.*

Figure 2 shows the points $z \in \mathcal{Z}$ labelled by the value of v.

Within any set of three points where v is constant, u assumes the values 0, 1 and 2, with $u = 0$ occurring on the bottom row and $u = 1, 2$ on the top row. Let $K_{2b}(u, v \mid u', v')$ be determined by (a) $v = v' - 1, v'$, or $v' + 1$ with probabilities 0.25, 0.5 and 0.25 and (b) $u = 0, 1$ or 2 with probabilities 1/3, 1/3 and 1/3, independent of u', v' and v. MC 2b has μ as its invariant measure because for each (u, v), $\mu(u, v) = \sum_{u',v'} K(u, v \mid u', v')\mu(u', v')$.

MC 2a and MC 2b both have μ as their invariant measure. However, in MC 2a the long run frequency of $x = 0$ is 0.5 and hence the limiting distribution of X is Br(0.5), but in MC 2b the long run frequency of $u = 0$, and hence $x = x(u, v) = 0$, is 1/3 and the limiting distribution of X is Br(1/3).

A given invariant measure μ, if it is improper, does not have a unique limiting distribution. A given transition kernel has, up to proportionality, a unique invariant measure (Robert and Casella, 1999, p. 155); but two transition kernels with different limiting distributions may share the same invariant measure. The point of the example is that nonuniqueness holds even for the limiting marginal distribution of X, even when $\mu = \mu_X \times \mu_Y$ and μ_X is proper. Specifically, we can write μ as a product: $\mu(x, y) = p(y) \times p(x \mid y) = 2 \times .5$ where $p(x \mid y)$ does not depend on y. One might have hoped, therefore, that $p(x \mid y)$ would have a natural interpretation as the marginal distribution of X and that any MC for which μ is invariant would have Br(0.5) as its limiting distribution for $X(n)$. The example shows that this hope is unrealized. The point deserves stress because there are published papers that assert uniqueness.

To drive the point home, consider a more realistic example with a common statistical model, an overparameterized ANOVA, having what appears to be a proper embedded posterior. Similar examples appear in Gelfand and Sahu (1999) and Hobert (2001).

Example 3. Let $D \sim \mathrm{N}(U + V, 1)$. This is the simplest possible overparameterized ANOVA: there is just a single group, D is the data, U is the overall mean, V is the offset for the group. The prior $p(u, v) \propto 1$ yields the improper posterior $p(u, v \mid d) \propto \exp\{-\frac{1}{2}(d - u - v)^2\}$. The posterior is easily Gibbs sampled: $U \mid d, v \sim \mathrm{N}(d - v, 1)$ and $V \mid d, u \sim \mathrm{N}(d - u, 1)$. Let $S = U + V$. It is easily seen that $U(n+1) + V(n+1) \mid U(n) + V(n) \sim \mathrm{N}(d, 1)$, so the chain yields i.i.d. samples from the "obvious" embedded posterior $S \mid d \sim \mathrm{N}(d, 1)$.

Now let $T = (U + V)/(U - V)$. Transforming densities $p(u, v \mid d) \to p(s, t \mid d)$ formally yields $p(s, t \mid d) \propto \exp\{-\frac{1}{2}(s - d)^2\}|s|t^{-2}$, a product density in which $S \not\sim \mathrm{N}(d, 1)$. The nonuniqueness of the S density here is just another version of the nonuniqueness of the X distribution in Example 2.

From now on we concentrate on discrete distributions to make the math and the concepts clearer.

Example 2 can be analyzed by existing theorems. In that example, $X(n)$ is a Markov chain. And, as we saw earlier, μ does factor so Hobert's theorem may apply. For MC 2a, Hobert's condition is

$$\sum_x \sum_{y'} K_{2a}(x, y \mid x', y') = \sum_{y'} K_{2a}(y \mid x', y') = 2$$

which is free of x' and therefore the limiting distribution of $X(n)$ under MC 2a is Br(0.5). On the other hand, for MC 2b Hobert's condition is

$$\sum_x \sum_{y'} K_{2b}(x, y \mid x', y') = \sum_{y'} K_{2b}(y \mid x', y')$$

which does depend on x' and therefore Hobert does not guarantee that MC 2b has the right limiting distribution. On yet another hand,

$$\sum_u \sum_{v'} K_{2b}(u, v \mid u', v') = \sum_{v'} K_{2b}(v \mid u', v')$$

is free of u' so the limiting distribution of $U(n)$ under MC 2b is Mult $(1/3, 1/3, 1/3)$ and therefore the limiting distribution of $X(n)$ is Br(2/3).

Also, μ is the limit of μ_i where μ_i is defined to be μ restricted to $\{(x, y) : |y| \leq i\}$ and K_{2a} is the limit of $K_{2a,i}$ where $K_{2a,i}$ is K_{2a} restricted to $\{(x, y) : |y| \leq i\}$ and suitably modified at the boundary so that μ_i is its stationary distribution. Therefore, Liu and Wu (1999) and Meng and van Dyk (1999) guarantee that K_{2a} has a limiting X marginal of Br(0.5). On the other hand, μ is also the limit of ν_i where ν_i is simply μ restricted to $\{(x, y) : |y| \leq i \text{ if } x = 0 \text{ and } |y| \leq 2i \text{ if } x = 1\}$. But K_{2a} is not the limit of transition kernels having ν_i as their stationary distributions. However, K_{2b} is the limit of transition kernels $K_{2b,i}$ having ν_i as their stationary distributions; so Liu and Wu (1999) and Meng and van Dyk (1999) guarantee that K_{2b} has a limiting X marginal of Br(2/3).

In Examples 1 and 2, $X(n)$ is an MC. Examples 3 and 4 show that $X(n)$ may converge even when it is not Markov and in situations not covered by existing theorems.

Example 4. Let $\mathcal{Z}$ be as in Example 2. Let transition kernel $K_4(x, y \mid x', y')$ be defined by *(a)* $y = y' - 1, y'$, or $y' + 1$ with probabilities .4, .2 and .4 and *(b)* $x \sim$ Br$(1/3)$ if y is even; $x \sim$ Br$(2/3)$ if y is odd. It is easily verified that $X(n)$ is not Markov under K_4 and that $X(n)$ has limiting distribution Br(0.5).

Example 4 can be handled by existing theorems after a transformation of variables. Let

$$\begin{aligned} u(x, y) &= \lfloor y/2 \rfloor \\ v(x, y) &= x \\ w(x, y) &= \begin{cases} 0 & \text{if } y \text{ is even} \\ 1 & \text{if } y \text{ is odd} \end{cases} \end{aligned}$$

where $\lfloor y \rfloor$ is the *floor* of y, the largest integer less than or equal to y. The kernel K_4 is Markov in (v, w), satisfies Hobert's criterion and is the limit of kernels having proper invariant measures. Therefore any of the existing theorems show that the limiting distribution of (v, w) is the measure $\pi_{4,V,W}$ defined by

$$\pi_{4,V,W}(v, w) = \begin{cases} 1/3 & \text{if } w = 0, v = 1, \\ 1/3 & \text{if } w = 1, v = 0, \\ 2/3 & \text{if } w = 0, v = 0, \\ 2/3 & \text{if } w = 1, v = 1. \end{cases}$$

which has marginal $\pi_{4,V} = \text{Br}(0.5)$.

Example 5. Let $\mathcal{Z}$ be as in Example 4. Let transition kernel $K_5(x, y \mid x', y')$ be defined by (a) $y = y' - 1, y'$, or $y' + 1$ with probabilities 0.4, 0.2 and 0.4 and (b) $x \sim \text{Br}(0.5 + 1/(|y| + 2))$.

Example 5 cannot be handled by the existing theorems. Intuition says that K_5 is arbitrarily close to K_{2a} outside of bounded sets so we expect the limiting distribution of $X(n)$ to be Br(0.5). This intuition is correct, as the Marginal Ergodic Theorem will show.

2. A MARGINAL ERGODIC THEOREM

We adopt and adapt the proof of the Ergodic Theorem in Borovkov (1998, Theorem. 2.2) which is based on a renewal argument. See also the numerous references therein. Let $Z(n)$ be a Markov chain in $(\mathcal{Z}, \mathcal{B}_\mathcal{Z})$. When we want to emphasize the initial state we write $Z(n) = Z(z, n)$, the state at time n, where z is the state at time $n = 0$. $Z(n)$ is partitioned, possibly after transformation, into components $Z(n) = (X(n), Y(n)) = (X(z, n), Y(z, n))$. The state space and σ-field are $(\mathcal{X} \times \mathcal{Y}, \mathcal{B}_\mathcal{X} \times \mathcal{B}_\mathcal{Y})$. We assume throughout that $Z(n)$ is aperiodic and irreducible. For n-step transition probabilities from initial state z to a set $B_\mathcal{Z} \in \mathcal{B}_\mathcal{Z}$ we write

$$P(z, n, B_\mathcal{Z}) = \Pr[Z(z, n) \in B_\mathcal{Z}]$$

The usual ergodic theorem requires the following two conditions (see Borovkov, 1998, 11–12.) Let there exist a set $V \in \mathcal{B}_\mathcal{Z}$, a probability measure φ on $(\mathcal{Z}, \mathcal{B}_\mathcal{Z})$, a number $p \in (0, 1)$, and an integer $n_0 \geq 1$ such that

$$\sup_{z \in V} \mathrm{E}\tau(z) < \infty, \qquad \text{(Condition I)}$$

where $\tau(z) = \min\{k \geq 1 : Z(z, k) \in V\}$ is the time at which the Markov chain $Z(n)$, having started in state z, returns to the set V and

$$P(z, n_0, B_\mathcal{Z}) > p\varphi(B_\mathcal{Z}) \qquad \text{(Condition II}')$$

for every $B_\mathcal{Z} \in \mathcal{B}_\mathcal{Z}$ and every $z \in V$. Condition I is *positive recurrence*; Condition II$'$ is a *mixing* condition. Conditions I and II$'$ (along with irreducibility and aperiodicity) are sufficient to imply ergodicity of $Z(n)$. They are implied by the assumption of *Harris recurrence*, the assumption usually seen in the statistics literature on MCMC.

Because we are concerned with limiting distributions on $\mathcal{X}$ and not on all of $\mathcal{Z}$, the marginal ergodic theorem requires the following modification of Condition II$'$. Let there exist a set $V \in \mathcal{B}_\mathcal{Z}$, a probability measure φ on $(\mathcal{X}, \mathcal{B}_\mathcal{X})$, a number $p \in (0, 1)$, and an integer $n_0 \geq 1$ such that

$$P(z, n_0, B_\mathcal{X} \times \mathcal{Y}) \equiv \Pr[X(z, n_0) \in B_\mathcal{X}] > p\varphi(B_\mathcal{X}) \qquad \text{(Condition II)}$$

for every $B_\mathcal{X} \in \mathcal{B}_\mathcal{X}$ and every $z \in V$.

Before stating the main theorem we first describe the extended Markov chain Z^* on which the renewal argument is based. For each $z \in \mathcal{Z}$, let ψ_z be any probability measure on $(\mathcal{Z}, \mathcal{B}_\mathcal{Z})$ satisfying $P(z, 1, B_\mathcal{Z}) > p\psi_z(B_\mathcal{Z})$ for all $B_\mathcal{Z} \in \mathcal{B}_\mathcal{Z}$ and having $\mathcal{X}$-marginal φ; *i.e.*, $\psi_z(B_\mathcal{X} \times \mathcal{Y}) = \varphi(\mathcal{B}_\mathcal{X})$. (The measure having $\mathcal{X}$-marginal φ and conditional $P(Y(n) \mid X(n-1))$ just as in the Z chain will do.) For each $z \in \mathcal{Z}$, define the probability measure

$$Q_z(\cdot) \equiv \frac{1}{1-p} \left(P(z, 1, \cdot) - p\psi_z(\cdot)\right).$$

Define the extended chain Z^* by $Z^*(n) = (\tilde{Z}(n), W(n))$ where *(a)* $\tilde{Z}(n) \in \mathcal{Z}$ and *(b)* $W(n)$ is an i.i.d. sequence of Br(p) random variables, where p is from Condition II. The distribution of $Z^*(n)$ is given by

$$(a) \quad W(n) \sim \text{iid Br}(p)$$
$$(b1) \quad \tilde{Z}((z, w), 1, B_\mathcal{Z}) = Z(z, 1, B_\mathcal{Z})$$

if $z \notin V$ and

$$(b2a) \quad \tilde{Z}((z, 1), 1, B_\mathcal{Z}) = \psi_z(B_\mathcal{Z})$$
$$(b2b) \quad \tilde{Z}((z, 0), 1, B_\mathcal{Z}) = Q_z(B_\mathcal{Z})$$

if $z \in V$. That is, if $z \notin V$ then $\tilde{Z}(n)$ has the same transition probabilities as $Z(n)$, independent of $W(n)$; but if $z \in V$ then $\tilde{Z}(n)$ has transition probabilities either ψ_z or Q_z according to whether $W(n)$ is 1 or 0, respectively. Because $W(n) \sim \mathrm{Br}(p)$, the marginal transition probabilities of $\tilde{Z}(n)$ are

$$p\psi_z(\cdot) + (1-p)Q_z(\cdot) = P(z, 1, \cdot),$$

the same as $Z(n)$, so the distributions of $Z(n)$ and $\tilde{Z}(n)$ coincide.

Let $V^* = \{z^* : \tilde{z} \in V, w = 1\}$. The process Z^* has the following property: $\Pr[X^*(n+1) \in B_{\mathcal{X}} \mid z^*(n)]$ is independent of z^* provided that $z^* \in V^*$. So the trajectory of Z^* is divided into cycles by the times of arrival in V^*. In the usual ergodic theorem, the cycles are actually i.i.d., and hence renewals, because the first step in each cycle is taken according to distribution φ which does not depend on z^*. In the marginal ergodic theorem the cycles are not i.i.d. because the first step in a cycle is taken according to distribution ψ_z which does depend on z. However, every ψ_z has the same $\mathcal{X}$-marginal, and hence so do the first steps of all the cycles.

For each $z' \in V$, put

$$\tau^*(z') = \min\{k \geq 1 : Z^*((z', 1), k) \in V^*\}$$

$$a(z') = \mathrm{E}\tau^*(z') = \sum_{k=1}^{\infty} P_k(z')$$

where $P_k(z') = \Pr[\tau(z') \geq k]$. Let

$$a^\ell = \inf_{z'} a(z') \qquad a^u = \sup_{z'} a(z').$$

Marginal Ergodic Theorem *Let Conditions I and II hold. In addition, suppose that the $\tau^*(z)$ are uniformly summable. (Condition III) That is, for every $\epsilon > 0$ there exists N such that $\sum_N^\infty P_k(z^*) < \epsilon$ for every z. Then the set functions π^ℓ and π^u on X defined by:*

$$\pi^u(B_{\mathcal{X}}) = \frac{1}{a^\ell} \sum_{k=1}^{\infty} \sup_{z' \in V} P_k(z') P(z', k, B_{\mathcal{X}} \times \mathcal{Y})$$

$$\pi^\ell(B_{\mathcal{X}}) = \frac{1}{a^u} \sum_{k=1}^{\infty} \inf_{z' \in V} P_k(z') P(z', k, B_{\mathcal{X}} \times \mathcal{Y})$$

satisfy

$$\liminf_{n \to \infty} P(z, n, B_{\mathcal{X}} \times \mathcal{Y}) \geq \pi^\ell(B_{\mathcal{X}})$$

$$\limsup_{n \to \infty} P(z, n, B_{\mathcal{X}} \times \mathcal{Y}) \leq \pi^u(B_{\mathcal{X}}).$$

Proof. For reasons of space we do not give the proof here. It follows the proof of the Ergodic Theorem on pages 17–25 of Borovkov (1998). Curious readers may email the author for details.

3. DISCUSSION

It is worth examining in more detail the structure of π^ℓ and π^u. In the ergodic theorem there is but a single π defined as

$$\pi(B_{\mathcal{Z}}) = \frac{1}{a} \sum_{k=1}^{\infty} P_k P(k, B_{\mathcal{Z}})$$

$$= \frac{1}{\mathrm{E}\tau^*} \sum_{k=1}^{\infty} \Pr\left[Z^*(1) \notin V^*, \ldots, Z^*(k-1) \notin V^*, Z^*(k) \in B_{\mathcal{Z}}\right]$$

The chain $Z(\cdot)$ is divided into i.i.d. cycles. The distribution of $Z(n)$, $\mathcal{L}[Z(n)]$, is written as $\sum_k \mathcal{L}[Z(n) \mid \gamma(n) = n-k] \Pr[\gamma(n) = n-k]$. The quantities a, τ^*, P_k and $P(k, B_{\mathcal{Z}})$ are free of z because the first step in each cycle is made according to distribution φ which is free of z. For the marginal ergodic theorem, the first step of a cycle is made according to ψ_z which is not free of z, although its X-marginal is. Because the distribution of each cycle depends on the state z' in which the cycle starts, we end up with bounds that are the supremum and infimum with respect to the different possible values of z'.

In any given example there may be several valid choices of n_0, V, p and φ. For the ergodic theorem these choices make no difference to the value of π. For the marginal ergodic theorem it sometimes happens that π^u and π^ℓ do depend on these choices. Though the theorem holds for any valid choice, there may be particular choices that yield tight bounds while others yield loose bounds. In the examples that follow we use this idea to advantage by taking the infimum of π^u and the supremum of π^ℓ over a class of valid n_0's or V's.

Example 6. (*Example 2 continued*). For K_{2a} $X(n) \sim \text{Br}(0.5)$ independent of $Z(n-1)$ and therefore, as we saw earlier, the limiting distribution of $X(n)$ is $\text{Br}(0.5)$. What does the Marginal Ergodic Theorem say? We can take $V = \mathcal{Z}$ and $n_0 = 1$ because *(a)* Condition I is satisfied trivially with $\tau(z) = 1$ for all z and *(b)* Condition II is satisfied with $n_0 = 1$, $\varphi = \text{Br}(0.5)$ and any $p \in (0, 1)$. Then

$$a^u = a^\ell = 1/p$$
$$P_k(z') \equiv \Pr[\tau^* \geq k] = (1-p)^{k-1} \quad \text{for all } z'$$
$$P(z', k, 1) = P(z', k, 0) = 0.5 \quad \text{for all } z', k$$

and thus $\pi^\ell = \pi^u \equiv \pi$ where $\pi(0) = \pi(1) = 0.5$. Because π^ℓ and π^u coincide the marginal ergodic theorem says that $X(n)$ has a limiting distribution equal to π.

Similar reasoning applies to any MC for which $X(n)$ is independent of $Z(n-1)$. For instance, Example 1 yields i.i.d. draws for μ. Therefore π^ℓ and π^u coincide and the marginal ergodic theorem says that there is a limiting distribution π for μ and tells us what it is. Of course π is the distribution of the i.i.d. draws, a fact previously noted by Hobert (2001).

Example 7. (*Example 5 continued*). In Example 5 we can also take $V = \mathcal{Z}$ and $n_0 = 1$. Then, for $k = 1$ and $B_{\mathcal{X}} = \{1\}$,

$$\sup_{z' \in V} P_k(z') P(z', k, B_{\mathcal{X}}) = P_1((0,0)) P((0,0), 1, 1) = 0.2 + (0.8)(\frac{5}{6}) = \frac{13}{15}$$
$$> \inf_{z' \in V} P_k(z') P(z', k, B_{\mathcal{X}}) = P_1((\infty,0)) P((\infty,0), 1, 1) = 1/2.$$

Similar inequalities hold so that for all $k = 1, 2, \ldots$ and $B_{\mathcal{X}} = \{0\}, \{1\}$,

$$\sup_{z' \in V} P_k(z') P(z', k, B_{\mathcal{X}}) > \inf_{z' \in V} P_k(z') P(z', k, B_{\mathcal{X}}).$$

Therefore, π^ℓ and π^u differ, and it seems that the Marginal Ergodic Theorem, while true, fails to give tight bounds in a case where a limiting distribution actually exists. However, for any m, we may also choose $V = \{(x, y) : |y| > m\}$. As m increases,

$$\sup_{z' \in V} P_1(z') P(z', 1, 1) = P_1((m, 0)) P((m, 0), 1, 1) \to 1/2$$
$$= \inf_{z' \in V} P_1(z') P(z', 1, 1)$$

and similarly for other values of k and $B_{\mathcal{X}}$. In fact, for any $B_{\mathcal{X}}$,

$$\lim_{m\to\infty} \pi^u(B_{\mathcal{X}}) - \pi^\ell(B_{\mathcal{X}}) = 0$$

Because the theorem is true for any V satisfying the conditions we see that Example 5 does have a limiting distribution π, that π is determined solely by the set of z's outside any bounded set, and that $\pi = \mathrm{Br}(0.5)$.

Example 8. (*Example 4 continued*). In Example 4 we can also take $V = \mathcal{Z}$ and $n_0 = 1$. Then, for $k = 1$ and $B_{\mathcal{X}} = \{1\}$, $\sup_{z' \in V} P_1(z')P(z', 1, 1)$ occurs when y' is even; the sup is equal to 0.2(1/3) + 0.8(2/3) = 0.6. The inf occurs when y' is odd and is equal to 0.2(2/3) + 0.8(1/3) = 0.4. So again π^ℓ and π^u do not coincide even though there is a limiting distribution. However, we may, if we choose, use a larger value of n_0. We would declare n to be a quasi-renewal time only if $Z^*(n) \in V^*$ and there had been an interval of at least n_0 steps since the previous quasi-renewal. Therefore, $P_k(z') = 0$ for $k < n_0$ and $P(z', k, B_{\mathcal{X}})$ matters only for values of $k \geq n_0$. Since $\lim_{n\to\infty} P((z'), n, B_{\mathcal{X}}) = 0.5$ regardless of z', π^ℓ and π^u converge as n_0 goes to ∞ and the marginal ergodic theorem correctly identifies the limiting distribution.

In all the examples so far, $X(n)$ has had a limiting distribution. In the next example $X(n)$ does not have a limiting distribution. Instead, the long run frequency of $X(n) = 0$ and $X(n) = 1$ varies between two bounds which are correctly identified by π^ℓ and π^u.

Example 9. Let $\mathcal{Z}$ be as in Example 2. Let transition kernel $K_9(x, y \mid x', y')$ be defined by (a) $y = y' - 1, y'$, or $y' + 1$ with probabilities 0.4, 0.2 and 0.4 and (b) $x \sim \mathrm{Br}(1/3)$ if $y < 0$; $x \sim \mathrm{Br}(2/3)$ if $y \geq 0$. In this example we can take $V = \mathcal{Z}$, $n_0 = 1$, $\varphi = \mathrm{Br}(0.5)$ and $p = 2/3$. Then

$$\begin{aligned}
P_k(z') \equiv \Pr[\tau^* \geq k] = (1-p)^{k-1} \qquad & \text{for all } z' \\
a^\ell = a^u = 1/p = 3/2 \qquad & \\
\inf_{z'} P(k, z', 1 \times \mathcal{Y}) = \inf_{z'} P(k, z', 0 \times \mathcal{Y}) = 1/3 \qquad & \text{for all } k \\
\sup_{z'} P(k, z', 1 \times \mathcal{Y}) = \sup_{z'} P(k, z', 0 \times \mathcal{Y}) = 2/3 \qquad & \text{for all } k,
\end{aligned}$$

so

$$\pi^\ell(0) = \pi^\ell(1) = \frac{2}{3}\sum (1/3)^{k-1} 1/3 = 1/3$$
$$\pi^u(0) = \pi^u(1) = \frac{2}{3}\sum (1/3)^{k-1} 2/3 = 2/3$$

and these quantities do not change if we take a smaller V or larger n_0.

In fact, the MC defined by K_9 alternates between spending an increasingly large amount of time in the positive Y-axis and an increasingly large amount of time in the negative Y-axis. In the first case, the sampling frequency of X approaches Br(2/3); in the second it approaches Br(1/3). The marginal ergodic theorem correctly identifies the fact that the long run frequency of X alternates between the two values.

Finally, it should be noted that when $Y(n)$ is defined independently of $X(n)$, then $X(n)$ can be viewed as a *recursive chain* or a *Markov chain in a random environment*. Following Borovkov (1998, p. 166), a sequence $X(n)$ is called a recursive chain governed by the sequence $Y(n)$ if for all $n \geq 0$ and $B \in \mathcal{B}_{\mathcal{X}}$,

$$\Pr[X(n+1) \in B \mid \mathcal{F}_n] = \Pr[X(n+1) \in B \mid X(n), Y(n)] \quad \text{a.s.}$$

where $\mathcal{F}_n$ is the σ-field generated by $X(0), \ldots, X(n), Y(0), \ldots, Y(n)$. There are ergodic theorems for recursive chains. See, for example, Theorem 14.1 (p. 178) and other theorems in Borovkov's Chapter 14. They all require some version of stationarity on the sequence $Y(n)$ and hence are less general, for our purposes, than the Marginal Ergodic Theorem.

REFERENCES

Borovkov, A. A. (1998). *Ergodicity and Stability of Stochastic Processes*, Chichester: Wiley.

Gelfand, A. and Sahu, S. (1999). Identifiability, improper priors and Gibbs sampling for generalized linear models. *J. Am. Statist. Ass.* **94**, 247–253.

Hobert, J. (2001). Stability relationships among the Gibbs sampler and its subchains. *J. Comput. Graph. Statist.* **10**, 185–205.

Liu, J. and Wu, Y. N. (1999). Parameter expansion scheme for data augmentation. *J. Am. Statist. Ass.* **94**, 1264–1274.

Meng, X. L. and van Dyk, D. (1997). The EM algorithm—an old folk song sung to a fast new tune. *J. R. Statist. Soc. B* **60**, 511–567 (with discussion).

Meng, X. L. and van Dyk, D. (1999). Seeking efficient data augmentation schemes via conditional and marginal augmentation. *Biometrika* **86**, 301–320.

Robert, C. and Casella, G. (1999). *Monte Carlo Statistical Methods*. New York: Springer.

van Dyk, D. and Meng, X. L. (2001). The art of data augmentation, *J. Comput. Graph. Statist.* **10**, 1–50.

BAYESIAN STATISTICS 7, pp. 587–596
J. M. Bernardo, M. J. Bayarri, J. O. Berger, A. P. Dawid,
D. Heckerman, A. F. M. Smith and M. West (Eds.)

Exact Bayesian Inference for a Class of Nonlinear Systems with Application to Robotic Assembly

TYNE LEFEBVRE   KLAAS GADEYNE
Katholieke Universiteit Leuven, Belgium
{tine.lefebvre klaas.gadeyne}@mech.kuleuven.ac.be

HERMAN BRUYNINCKX and JORIS DE SCHUTTER
Katholieke Universiteit Leuven, Belgium
{herman.bruyninckx joris.deschutter}@mech.kuleuven.ac.be

SUMMARY

This paper presents a new finite-dimensional Bayesian filter. The filter calculates the exact analytical expression for the posterior probability density function (pdf) of static systems with any kind of nonlinear measurement equation subject to Gaussian measurement uncertainty. The paper also extends this filter to a limited class of dynamic systems. The filter is applied to the estimation of the inaccurately known position and orientation of two mating parts during autonomous robotic assembly. The sufficient statistics of the posterior pdf are obtained by Kalman Filter formulas, making online estimation possible.

Keywords: EXACT BAYESIAN FILTERING; NONLINEAR SYSTEMS; ROBOTIC MANIPULATION.

1. INTRODUCTION

Exact finite-dimensional filters exist only for a small class of systems. The best known example is the Kalman Filter for linear systems (*i.e.*, systems for which both the process equation and the measurement equation are linear) with Gaussian uncertainties. In this case, the posterior pdf is a Gaussian represented by its mean and its covariance matrix. Other examples are the filters of Beneš (1981), which requires the measurement equation to be linear, and Daum (1988), applicable to a more general class of systems with nonlinear process and measurement equations for which the posterior pdf is any exponential distribution.

This paper proposes a new filter that can deal with static systems (parameter estimation) with any kind of nonlinear measurement equation subject to Gaussian measurement uncertainty (Section 2). Section 3 extends the results to a limited class of dynamic systems. The core idea of the new filter is to linearize the process and measurement equation in a higher-dimensional state space. The new filter uses a Kalman Filter to solve the linear estimation problem and returns the "mean vector" and "covariance matrix" of the Gaussian in the higher-dimensional state space. These are sufficient statistics of the posterior pdf over the original state, however in general, they do *not* represent the mean and covariance of this posterior. The posterior represents a subclass of the exponential distributions, hence, the new filter solves a subclass of the estimation problems solved by the Daum filter. Its main advantage compared to the latter is a low computational complexity. In Section 4, the new filter is applied to the autonomous execution of a robotic assembly task for which the position and orientation of the parts to assemble is inaccurately known.

2. STATIC SYSTEMS

This section describes the filter equations for static systems with any kind of nonlinear measurement equation $\boldsymbol{z}_k = \boldsymbol{h}_k(\boldsymbol{x}) + \boldsymbol{\rho}_{m,k}$ where $\boldsymbol{\rho}_{m,k}$ denotes Gaussian measurement uncertainty $\mathrm{N}(\boldsymbol{\rho}_{m,k} \,|\, 0, \boldsymbol{R}_k)$, which is uncorrelated over time. In order to calculate the posterior pdf over $\boldsymbol{x}$, we define[1] a higher-dimensional state vector $\boldsymbol{x}' = \boldsymbol{g}(\boldsymbol{x})$ in which the measurement equations are linear $\boldsymbol{z}_k = \boldsymbol{H}_k\, \boldsymbol{x}' + \boldsymbol{d}_k + \boldsymbol{\rho}_{m,k}$. Theorem 1 derives the analytical posterior over the original state $\boldsymbol{x}$.

Theorem 1 . (*Analytical posterior pdf for static systems*). *The posterior pdf over the state $\boldsymbol{x}$, given the measurements up to time step k $\boldsymbol{Z}_k = \{\boldsymbol{z}_1, \boldsymbol{z}_2, \ldots, \boldsymbol{z}_k\}$ and the measurement equation*

$$\boldsymbol{z}_k = \boldsymbol{h}_k(\boldsymbol{x}) + \boldsymbol{\rho}_{m,k} = \boldsymbol{H}_k\, \boldsymbol{g}(\boldsymbol{x}) + \boldsymbol{d}_k + \boldsymbol{\rho}_{m,k},$$

with Gaussian measurement uncertainty $\mathrm{N}(\boldsymbol{\rho}_{m,k} \,|\, 0, \boldsymbol{R}_k)$ which is uncorrelated over time, can be expressed in analytical form as

$$p(\boldsymbol{x} \,|\, \boldsymbol{Z}_k) \sim p(\boldsymbol{x}) \exp\Big(-\tfrac{1}{2}\,(\boldsymbol{g}(\boldsymbol{x}) - \boldsymbol{\mu}'_k)^t \boldsymbol{P}_k'^{-1}(\boldsymbol{g}(\boldsymbol{x}) - \boldsymbol{\mu}'_k)\Big), \tag{1}$$

where $\sim$ denotes that the pdf equals the right hand side of the equation up to a normalization factor. $p(\boldsymbol{x})$ is the prior pdf over $\boldsymbol{x}$; $\boldsymbol{\mu}'_k$ and $\boldsymbol{P}'_k$ are defined by Kalman Filter equations for the state estimation of $\boldsymbol{x}' = \boldsymbol{g}(\boldsymbol{x})$ (the iteration starts with $\boldsymbol{P}_0'^{-1} = 0$):

$$\boldsymbol{K}_k = \boldsymbol{P}'_{k-1}\boldsymbol{H}^t_k(\boldsymbol{H}_k\boldsymbol{P}'_{k-1}\boldsymbol{H}^t_k + \boldsymbol{R}_k)^{-1}, \tag{2}$$
$$\boldsymbol{\mu}'_k = \boldsymbol{\mu}'_{k-1} + \boldsymbol{K}_k(\boldsymbol{z}_k - \boldsymbol{H}_k\boldsymbol{\mu}'_{k-1} - \boldsymbol{d}_k), \tag{3}$$
$$\boldsymbol{P}'_k = (\boldsymbol{I} - \boldsymbol{K}_k\boldsymbol{H}_k)\boldsymbol{P}'_{k-1}. \tag{4}$$

Note that Equation (4) is equivalent to $\boldsymbol{P}_k'^{-1} = \boldsymbol{P}_{k-1}'^{-1} + \boldsymbol{H}^t_k\boldsymbol{R}_k^{-1}\boldsymbol{H}_k$.

Proof. Define a higher-dimensional state vector $\boldsymbol{x}' = \boldsymbol{g}(\boldsymbol{x})$ in which the measurement equations are linear. The prior pdf $p(\boldsymbol{x})$ is of the form (1) because $\boldsymbol{P}_0'^{-1} = 0$. We now prove Eqs. (1)–(4) by induction: assume $p(\boldsymbol{x} \,|\, \boldsymbol{Z}_{k-1})$ is of the form (1), then

$$p(\boldsymbol{x} \,|\, \boldsymbol{Z}_k) \sim p(\boldsymbol{x} \,|\, \boldsymbol{Z}_{k-1})p(\boldsymbol{z}_k \,|\, \boldsymbol{x}, \boldsymbol{Z}_{k-1}) = p(\boldsymbol{x} \,|\, \boldsymbol{Z}_{k-1})p(\boldsymbol{z}_k \,|\, \boldsymbol{x}),$$

where

$$p(\boldsymbol{x} \,|\, \boldsymbol{Z}_{k-1}) \sim p(\boldsymbol{x}) \exp\Big(-\tfrac{1}{2}\,(\boldsymbol{g}(\boldsymbol{x}) - \boldsymbol{\mu}'_{k-1})^t \boldsymbol{P}_{k-1}'^{-1}\,(\boldsymbol{g}(\boldsymbol{x}) - \boldsymbol{\mu}'_{k-1})\Big),$$
$$p(\boldsymbol{z}_k \,|\, \boldsymbol{x}) \sim \exp\Big(-\tfrac{1}{2}\,(\boldsymbol{H}_k\, \boldsymbol{g}(\boldsymbol{x}) + \boldsymbol{d}_k - \boldsymbol{z}_k)^t \boldsymbol{R}_k^{-1}(\boldsymbol{H}_k\, \boldsymbol{g}(\boldsymbol{x}) + \boldsymbol{d}_k - \boldsymbol{z}_k)\Big).$$

We simplified $p(\boldsymbol{z}_k \,|\, \boldsymbol{x}, \boldsymbol{Z}_{k-1})$ to $p(\boldsymbol{z}_k \,|\, \boldsymbol{x})$ because the measurement uncertainty is uncorrelated over time. Matrix calculations lead to:

$$p(\boldsymbol{x} \,|\, \boldsymbol{Z}_k) \sim p(\boldsymbol{x}) \exp\Big(-\frac{1}{2}(\boldsymbol{g}(\boldsymbol{x}) - \boldsymbol{\mu}'_k)^t \boldsymbol{P}_k'^{-1}(\boldsymbol{g}(\boldsymbol{x}) - \boldsymbol{\mu}'_k)\Big),$$
$$\boldsymbol{P}_k'^{-1} = \boldsymbol{P}_{k-1}'^{-1} + \boldsymbol{H}^t_k\boldsymbol{R}_k^{-1}\boldsymbol{H}_k,$$
$$\boldsymbol{\mu}'_k = \boldsymbol{P}'_k(\boldsymbol{P}_{k-1}'^{-1}\boldsymbol{\mu}'_{k-1} + \boldsymbol{H}^t_k\boldsymbol{R}_k^{-1}(\boldsymbol{z}_k - \boldsymbol{d}_k)).$$

[1] When dealing with a limited number of different measurement equations (different sensors), we can always find an $\boldsymbol{x}' = \boldsymbol{g}(\boldsymbol{x})$ that linearizes them. For example, when only one measurement equation $\boldsymbol{h}(\boldsymbol{x})$ is used, we can take $\boldsymbol{x}' = \boldsymbol{h}(\boldsymbol{x})$, $\boldsymbol{H}_k = \boldsymbol{I}$ and $\boldsymbol{d}_k = 0$; for N different equations $\boldsymbol{h}_i(\boldsymbol{x})$, we can choose $\boldsymbol{x}' = [\,\boldsymbol{h}^t_1(\boldsymbol{x}) \quad \ldots \quad \boldsymbol{h}^t_N(\boldsymbol{x})\,]^t$.

These equations are obtained using the fact that covariance matrices are symmetric:

$$\boldsymbol{P}'_{k-1} = \boldsymbol{P}'^{t}_{k-1}, \qquad \boldsymbol{R}_k = \boldsymbol{R}^t_k, \quad \boldsymbol{P}'_k = \boldsymbol{P}'^{t}_k.$$

Applying the matrix inversion lemma[2] shows that $\boldsymbol{\mu}'_k$ and $\boldsymbol{P}'_k$ are defined by Eqs. (2)–(4). □

A simple example illustrates the filter. Suppose that we want to estimate an angle α, whose value is $90°$. The measurement equation is $z_k = a_k \cos(\alpha) + b_k \sin(\alpha) + \rho$ with measurement uncertainty $\mathrm{N}(\rho|0, \sigma_\rho^2)$. This equation is linearized by taking a higher-dimensional state $\boldsymbol{x}' = \boldsymbol{g}(\alpha) = [\cos(\alpha), \sin(\alpha)]^t$, $z_k = [a_k, b_k]\, \boldsymbol{x}' + \rho$. Figure 1 shows the likelihood over $\boldsymbol{x}'$ after one measurement. The filtering in the higher-dimensional state space is *not* constrained to the subspace of values for which an $\boldsymbol{x}$ exists such that $\boldsymbol{x}' = \boldsymbol{g}(\boldsymbol{x})$. For this example, it means that the likelihood

$$p(z_1 \,|\, \boldsymbol{x}') \sim \exp\Big(-\frac{1}{2}(\boldsymbol{x}' - \boldsymbol{\mu}'_1)^t \boldsymbol{P}'^{-1}_1 (\boldsymbol{x}' - \boldsymbol{\mu}'_1)\Big),$$

plotted on Figure 1 (left), is valid for all values of $\boldsymbol{x}'$ in $\Re^2$, including values which do *not* correspond to the cosine and sine of an angle. The likelihood of the measurement over the original state $\boldsymbol{x}$, being

$$p(z_1 \,|\, \boldsymbol{x}) \sim \exp\Big(-\frac{1}{2}(\boldsymbol{g}(\boldsymbol{x}) - \boldsymbol{\mu}'_1)^t \boldsymbol{P}'^{-1}_1 (\boldsymbol{g}(\boldsymbol{x}) - \boldsymbol{\mu}'_1)\Big),$$

is obtained by looking at the subspace of values $\boldsymbol{x}'$ for which an $\boldsymbol{x}$ exists such that $\boldsymbol{x}' = \boldsymbol{g}(\boldsymbol{x})$. This operation is not always trivial, Section 4 presents two algorithms (Monte Carlo sampling and an Iterated Extended Kalman Filter). Figure 1 (right) shows the likelihood of the measurement over the original state for the sine/cosine example. Figure 2 shows the prior pdf $p(\alpha)$, the likelihood $p(z_1 \,|\, \alpha)$ from Figure 1 (right), and the posterior pdf $p(\alpha \,|\, z_1) \sim p(\alpha)p(z_1 \,|\, \alpha)$ as a function of α.

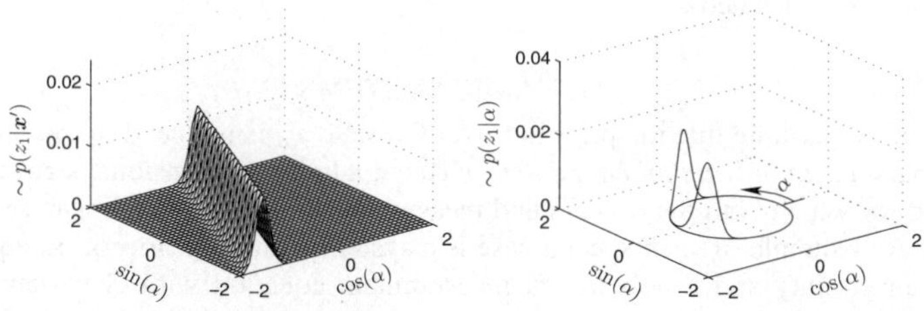

Figure 1. left: *the likelihood $p(z_1 \,|\, \boldsymbol{x}')$ is valid for all values of $\boldsymbol{x}'$ in $\Re^2$, including values which do not correspond to the cosine and sine of an angle.* **right:** *the likelihood $p(z_1 \,|\, \alpha)$ is obtained by looking at the subspace of values $\boldsymbol{x}'$ for which an $\boldsymbol{x}$ exists such that $\boldsymbol{x}' = \boldsymbol{g}(\boldsymbol{x})$.*

[2] The matrix inversion lemma states that $(\boldsymbol{A} + \boldsymbol{BCB}^t)^{-1} = \boldsymbol{A}^{-1} - \boldsymbol{A}^{-1}\boldsymbol{B}(\boldsymbol{B}^t\boldsymbol{A}^{-1}\boldsymbol{B} + \boldsymbol{C}^{-1})^{-1}\boldsymbol{B}^t\boldsymbol{A}^{-1}$.

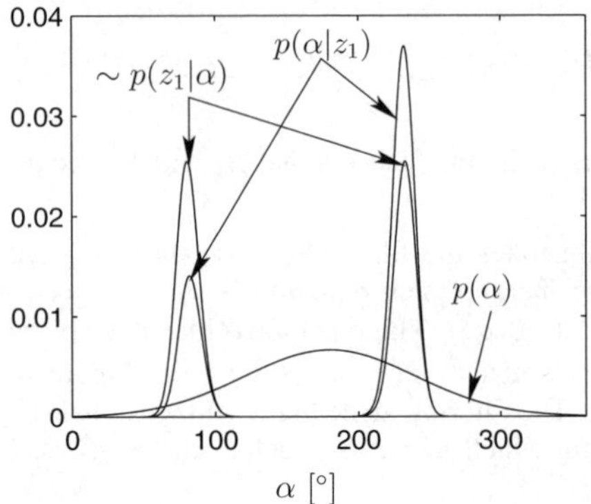

Figure 2. *$p(\alpha \mid z_1) \sim p(\alpha)p(z_1 \mid \alpha)$ where $p(z_1 \mid \alpha)$ is the one obtained from Figure 1 (right).*

Remark 1. If Gaussian, the prior information $\mathrm{N}(\boldsymbol{x} \mid \boldsymbol{\mu}_0, \boldsymbol{P}_0)$ can be introduced in an alternative way by augmenting $\boldsymbol{x}'$: $\boldsymbol{x}' \leftarrow \begin{bmatrix} \boldsymbol{x}' \\ \boldsymbol{x} \end{bmatrix}$ and replacing (1) by

$$p(\boldsymbol{x} \mid \boldsymbol{Z}_k) \sim \exp\Big(-\frac{1}{2}(\boldsymbol{g}(\boldsymbol{x}) - \boldsymbol{\mu}'_k)^t \boldsymbol{P}'^{-1}_k (\boldsymbol{g}(\boldsymbol{x}) - \boldsymbol{\mu}'_k)\Big),$$

$$\boldsymbol{\mu}'_0 = \begin{bmatrix} 0 \\ \boldsymbol{\mu}_0 \end{bmatrix}, \qquad \boldsymbol{P}'^{-1}_0 = \begin{bmatrix} 0 & 0 \\ 0 & \boldsymbol{P}_0^{-1} \end{bmatrix}.$$

Remark 2. In two cases, the dimension of the new state $\boldsymbol{x}'$ is *not higher* than that of the state $\boldsymbol{x}$:

1. If some (combinations) of the original states $\boldsymbol{x}$ are unobserved, the dimension of $\boldsymbol{x}'$ can be lower than that of $\boldsymbol{x}$. In this case, the original estimation problem can equally be described with less parameters. For this new problem description, $\boldsymbol{x}'$ is of equal (see case 2) or higher dimension than $\boldsymbol{x}$.
2. If the original measurement equation is linearizable by applying a one-to-one transformation to the state $\boldsymbol{x}$, then $\boldsymbol{x}'$ has the same dimension of $\boldsymbol{x}$. In this case, by expressing the measurement equation as a function of the transformed state $\boldsymbol{x}'$, the system is linear and no nonlinear filter is needed.

3. DYNAMIC SYSTEMS

In this section, we assume that the prior pdf over the state $\boldsymbol{x}_0$ is included in the prior pdf over $\boldsymbol{x}'_0$ (see Remark 1). *In only a limited number of cases*, a higher-dimensional state $\boldsymbol{x}'_k = \boldsymbol{g}(\boldsymbol{x}_k)$ can be found for which both the process and measurement equations are linear with Gaussian uncertainty. An example of such special case is a system with a linear process equation with *no* process uncertainty and a polynomial measurement equation with Gaussian uncertainty $\mathrm{N}(\boldsymbol{\rho}_{m,k} \mid 0, \boldsymbol{R}_k)$. In this case a state $\boldsymbol{x}'_k$ can always be found which linearizes both equations: for example, for a two-dimensional state $\boldsymbol{x}_k = [\, u_k \quad v_k \,]^t$:

$$\boldsymbol{x}_k = \boldsymbol{a}_0 + \boldsymbol{a}_1 u_{k-1} + \boldsymbol{a}_2 v_{k-1}, \qquad \boldsymbol{z}_k = \boldsymbol{b}_0 + \sum_{i=1}^{i=N} \sum_{j=1}^{j=i} \boldsymbol{b}_{ij} u_k^{i-j} v_k^j + \boldsymbol{\rho}_{m,k},$$

$$\boldsymbol{x}'^t_k = \begin{bmatrix} u_k^N & v_k u_k^{N-1} & \dots & v_k^{N-1} u_k & v_k^N & \dots & u_k^2 & v_k u_k & v_k^2 & u_k & v_k \end{bmatrix}.$$

Theorem 2 derives the analytical posterior $p(\boldsymbol{x}_k \,|\, \boldsymbol{Z}_k)$ over the original state for dynamic systems for which an appropriate $\boldsymbol{x}'$ is found. Note that we are calculating the marginal posterior $p(\boldsymbol{x}_k \,|\, \boldsymbol{Z}_k)$ over the state $\boldsymbol{x}$ at the last time step k and *not* the joint posterior $p(\boldsymbol{x}_1, \ldots, \boldsymbol{x}_k \,|\, \boldsymbol{Z}_k)$ over the state values at all previous time steps.

Theorem 2 . *(Analytical posterior pdf for dynamic systems). Consider an estimation problem with prior uncertainty $N(\boldsymbol{x}_0 \,|\, \boldsymbol{\mu}_{0|0}, \boldsymbol{P}_{0|0})$, measurements $\boldsymbol{Z}_j = \{\boldsymbol{z}_1, \boldsymbol{z}_2, \ldots, \boldsymbol{z}_j\}$, and process and measurement equations*

$$\begin{aligned} \boldsymbol{x}'_k &= \boldsymbol{A}_{k-1}\boldsymbol{x}'_{k-1} + \boldsymbol{c}_{k-1} + \boldsymbol{\rho}_{p,k-1}, \qquad \mathrm{N}(\boldsymbol{\rho}_{p,k-1} \,|\, 0, \boldsymbol{Q}_{k-1}), \\ \boldsymbol{z}_j &= \boldsymbol{H}_j\, \boldsymbol{x}'_j + \boldsymbol{d}_j + \boldsymbol{\rho}_{m,j}, \qquad \mathrm{N}(\boldsymbol{\rho}_{m,j} \,|\, 0, \boldsymbol{R}_j), \end{aligned}$$

where $\boldsymbol{x}'_k = \boldsymbol{g}(\boldsymbol{x}_k)$ and for which the Gaussian process and measurement uncertainties are mutually uncorrelated and uncorrelated over time. For this system, the posterior pdf over the state $\boldsymbol{x}_k$ at time step k, can be expressed in analytical form as

$$p(\boldsymbol{x}_k \,|\, \boldsymbol{Z}_j) \sim \exp\Big(-\frac{1}{2}(\boldsymbol{g}(\boldsymbol{x}_k) - \boldsymbol{\mu}'_{k\,|\,j})^t \boldsymbol{P}'^{-1}_{k\,|\,j}(\boldsymbol{g}(\boldsymbol{x}_k) - \boldsymbol{\mu}'_{k\,|\,j})\Big) \tag{5}$$

where $\boldsymbol{\mu}'_{k\,|\,j}$ and $\boldsymbol{P}'_{k\,|\,j}$ ($j = k-1$, $j = k$) are defined by Kalman Filter equations for the state estimation of $\boldsymbol{x}'_k$:

$$\boldsymbol{\mu}'_{k\,|\,k-1} = \boldsymbol{A}_{k-1}\boldsymbol{\mu}'_{k-1\,|\,k-1} + \boldsymbol{c}_{k-1}, \tag{6}$$

$$\boldsymbol{P}'_{k\,|\,k-1} = \boldsymbol{A}_{k-1}\boldsymbol{P}'_{k-1\,|\,k-1}\boldsymbol{A}^t_{k-1} + \boldsymbol{Q}_{k-1}, \tag{7}$$

$$\boldsymbol{K}_k = \boldsymbol{P}'_{k\,|\,k-1}\boldsymbol{H}^t_k(\boldsymbol{H}_k\boldsymbol{P}'_{k\,|\,k-1}\boldsymbol{H}^t_k + \boldsymbol{R}_k)^{-1}, \tag{8}$$

$$\boldsymbol{\mu}'_{k\,|\,k} = \boldsymbol{\mu}'_{k\,|\,k-1} + \boldsymbol{K}_k(\boldsymbol{z}_k - \boldsymbol{H}_k\boldsymbol{\mu}'_{k\,|\,k-1} - \boldsymbol{d}_k), \tag{9}$$

$$\boldsymbol{P}'_{k\,|\,k} = (\boldsymbol{I} - \boldsymbol{K}_k\boldsymbol{H}_k)\boldsymbol{P}'_{k\,|\,k-1}. \tag{10}$$

If no measurement $\boldsymbol{z}_k$ is available at a certain time step k, then Eqs. (8)–(10) *reduce to $\boldsymbol{\mu}'_{k\,|\,k} = \boldsymbol{\mu}'_{k\,|\,k-1}$, $\boldsymbol{P}'_{k\,|\,k} = \boldsymbol{P}'_{k\,|\,k-1}$. The iteration starts with*

$$\boldsymbol{\mu}'_{0|0} = \begin{bmatrix} 0 \\ \boldsymbol{\mu}_{0|0} \end{bmatrix}, \qquad \boldsymbol{P}'^{-1}_{0|0} = \begin{bmatrix} 0 & 0 \\ 0 & \boldsymbol{P}^{-1}_{0|0} \end{bmatrix}.$$

Proof. Before processing any measurements, $p(\boldsymbol{x}_0)$ is of the form (5). The first part of the proof derives the *process update, i.e.*, Eq. (5) for $j = k-1$ and Eqs. (6)–(7). The second part of the proof focuses on the *measurement update, i.e.*, Eq. (5) for $j = k$ and Eqs. (8)–(10).

Process update. We prove Eqs. (5)–(7) by induction: assume $p(\boldsymbol{x}_{k-1} \,|\, \boldsymbol{Z}_{k-1})$ is of the form (5), then

$$p(\boldsymbol{x}_k \,|\, \boldsymbol{Z}_{k-1}) = \int_{\boldsymbol{x}'_{k-1}} p(\boldsymbol{x}_k \,|\, \boldsymbol{x}'_{k-1}, \boldsymbol{Z}_{k-1}) p(\boldsymbol{x}'_{k-1} \,|\, \boldsymbol{Z}_{k-1})\, d\boldsymbol{x}'_{k-1},$$

with

$$\begin{aligned} p(\boldsymbol{x}_k \,|\, \boldsymbol{x}'_{k-1}, \boldsymbol{Z}_{k-1}) &= p(\boldsymbol{x}_k \,|\, \boldsymbol{x}'_{k-1}) \\ &\sim \exp\Big(-\tfrac{1}{2}\,(\boldsymbol{A}_{k-1}\boldsymbol{x}'_{k-1} + \boldsymbol{c}_{k-1} - \boldsymbol{g}(\boldsymbol{x}_k))^t \boldsymbol{Q}^{-1}_{k-1}(\boldsymbol{A}_{k-1}\boldsymbol{x}'_{k-1} + \boldsymbol{c}_{k-1} - \boldsymbol{g}(\boldsymbol{x}_k))\Big), \\ p(\boldsymbol{x}'_{k-1} \,|\, \boldsymbol{Z}_{k-1}) &\sim \exp\Big(-\tfrac{1}{2}\,(\boldsymbol{x}'_{k-1} - \boldsymbol{\mu}'_{k-1\,|\,k-1})^t \boldsymbol{P}'^{-1}_{k-1\,|\,k-1}(\boldsymbol{x}'_{k-1} - \boldsymbol{\mu}'_{k-1\,|\,k-1})\Big). \end{aligned}$$

We simplified $p(\boldsymbol{x}_k \mid \boldsymbol{x}'_{k-1}, \boldsymbol{Z}_{k-1})$ to $p(\boldsymbol{x}_k \mid \boldsymbol{x}'_{k-1})$ because the process and measurement uncertainty are mutually uncorrelated and uncorrelated over time. After some matrix calculations a factor

$$\exp\Big(-\tfrac{1}{2}\,(\boldsymbol{g}(\boldsymbol{x}_k) - \boldsymbol{\mu}'_{k\,|\,k-1})^t \boldsymbol{P}'^{-1}_{k\,|\,k-1}(\boldsymbol{g}(\boldsymbol{x}_k) - \boldsymbol{\mu}'_{k\,|\,k-1})\Big)$$

is identified which is independent of $\boldsymbol{x}'_{k-1}$. Hence:

$$p(\boldsymbol{x}_k \mid \boldsymbol{Z}_{k-1}) \sim \exp\Big(-\tfrac{1}{2}\,(\boldsymbol{g}(\boldsymbol{x}_k) - \boldsymbol{\mu}'_{k\,|\,k-1})^t \boldsymbol{P}'^{-1}_{k\,|\,k-1}(\boldsymbol{g}(\boldsymbol{x}_k) - \boldsymbol{\mu}'_{k\,|\,k-1})\Big)$$
$$\int_{\boldsymbol{x}'_{k-1}} \exp\Big(-\tfrac{1}{2}\,(\boldsymbol{x}'_{k-1} - \boldsymbol{v})^t \boldsymbol{V}^{-1}(\boldsymbol{x}'_{k-1} - \boldsymbol{v})\Big)\, d\boldsymbol{x}'_{k-1}, \quad (11)$$

where

$$\boldsymbol{V}^{-1} = \boldsymbol{P}'^{-1}_{k-1\,|\,k-1} + \boldsymbol{A}^t_{k-1}\boldsymbol{Q}^{-1}_{k-1}\boldsymbol{A}_{k-1},$$
$$\boldsymbol{v} = \boldsymbol{V}(\boldsymbol{P}'^{-1}_{k-1\,|\,k-1}\boldsymbol{\mu}'_{k-1\,|\,k-1} + \boldsymbol{A}^t_{k-1}\boldsymbol{Q}^{-1}_{k-1}(\boldsymbol{g}(\boldsymbol{x}_k) - \boldsymbol{c}_{k-1})),$$
$$\boldsymbol{P}'^{-1}_{k\,|\,k-1} = \boldsymbol{Q}^{-1}_{k-1} - \boldsymbol{Q}^{-1}_{k-1}\boldsymbol{A}_{k-1}\boldsymbol{V}\boldsymbol{A}^t_{k-1}\boldsymbol{Q}^{-1}_{k-1}, \quad (12)$$
$$\boldsymbol{\mu}'_{k\,|\,k-1} = \boldsymbol{P}'_{k\,|\,k-1}(\boldsymbol{Q}^{-1}_{k-1}\boldsymbol{c}_{k-1} - \boldsymbol{Q}^{-1}_{k-1}\boldsymbol{A}_{k-1}\boldsymbol{V}(\boldsymbol{A}^t_{k-1}\boldsymbol{Q}^{-1}_{k-1}\boldsymbol{c}_{k-1} - \boldsymbol{P}'^{-1}_{k-1\,|\,k-1}\boldsymbol{\mu}'_{k-1\,|\,k-1})). \quad (13)$$

These equations are obtained by using the fact that covariance matrices are symmetric: $\boldsymbol{P}'_{k-1\,|\,k-1} = \boldsymbol{P}'^t_{k-1\,|\,k-1}$, $\boldsymbol{Q}_{k-1} = \boldsymbol{Q}^t_{k-1}$, $\boldsymbol{V} = \boldsymbol{V}^t$ and $\boldsymbol{P}'_{k\,|\,k-1} = \boldsymbol{P}'^t_{k\,|\,k-1}$. If we define n as the dimension of $\boldsymbol{x}'$, then

$$\int_{\boldsymbol{x}'_{k-1}} \exp\Big(-\tfrac{1}{2}\,(\boldsymbol{x}'_{k-1} - \boldsymbol{v})^t \boldsymbol{V}^{-1}(\boldsymbol{x}'_{k-1} - \boldsymbol{v})\Big)\, d\boldsymbol{x}'_{k-1} = \sqrt{(2\pi)^n \det(\boldsymbol{V})}.$$

This result is independent of $\boldsymbol{x}_k$, hence (11) simplifies to

$$p(\boldsymbol{x}_k \mid \boldsymbol{Z}_{k-1}) \sim \exp\Big(-\tfrac{1}{2}\,(\boldsymbol{g}(\boldsymbol{x}_k) - \boldsymbol{\mu}'_{k\,|\,k-1})^t \boldsymbol{P}'^{-1}_{k\,|\,k-1}(\boldsymbol{g}(\boldsymbol{x}_k) - \boldsymbol{\mu}'_{k\,|\,k-1})\Big),$$

i.e., a pdf of the form (5). Applying the matrix inversion lemma[2] to Eqs. (12)-(13) shows that $\boldsymbol{\mu}'_{k|k-1}$ and $\boldsymbol{P}'_{k|k-1}$ are defined by Eqs. (6)-(7).

Measurement update. Given a $p(\boldsymbol{x}_k \mid \boldsymbol{Z}_{k-1})$ of the form (5), we need to derive $p(\boldsymbol{x}_k \mid \boldsymbol{Z}_k)$, knowing that

$$p(\boldsymbol{x}_k \mid \boldsymbol{Z}_k) \sim p(\boldsymbol{x}_k \mid \boldsymbol{Z}_{k-1})p(\boldsymbol{z}_k \mid \boldsymbol{x}_k, \boldsymbol{Z}_{k-1}).$$

The proof is similar to the proof of Theorem 1 and yields Eq. (5) and Eqs. (8)-(10). □

Remark 3. The standard Kalman Filter is a special case of the new filter: if the system has linear process and measurement equations and Gaussian process and measurement uncertainty, then we can choose $\boldsymbol{x}' = \boldsymbol{x}$ in order to "linearize" the system.

4. APPLICATION: ROBOTIC ASSEMBLY IN POORLY STRUCTURED ENVIRONMENTS

4.1. *Introduction*

Figure 3 illustrates the experiment: the robot grasped a cube and assembles it with an object in the environment (a corner with three walls) by executing a sequence of contact formations

(CFs). The position and orientation of the cube and the corner object (the state $\boldsymbol{x}$, also called the "geometrical parameters") are not well known at the beginning of the task. The position and orientation of the cube (frame $\{m\}$) are defined with respect to the robot gripper (frame $\{g\}$), the position and orientation of the corner object (frame $\{e\}$) are defined with respect to a world frame (frame $\{w\}$): the real values of these geometrical parameters do not change during task execution, *i.e.*, the system is *static*. During the task execution, the robot estimates the parameters by processing measurements of (i) the contact force and moment measured at 10 Hz, (ii) the translational and rotational velocity of the robot gripper at 10 Hz and (iii) occasionally, the position and orientation of the gripper. The initial uncertainty on the parameters is Gaussian with covariance $(20°)^2$ for all angles and $(500\ mm)^2$ for all distances. The side length of the cube is $250\ mm$. The system is able to detect contact transitions via a consistency test (Figure 4) and to recognize the CF. The measurement equations are nonlinear in the geometrical parameters.

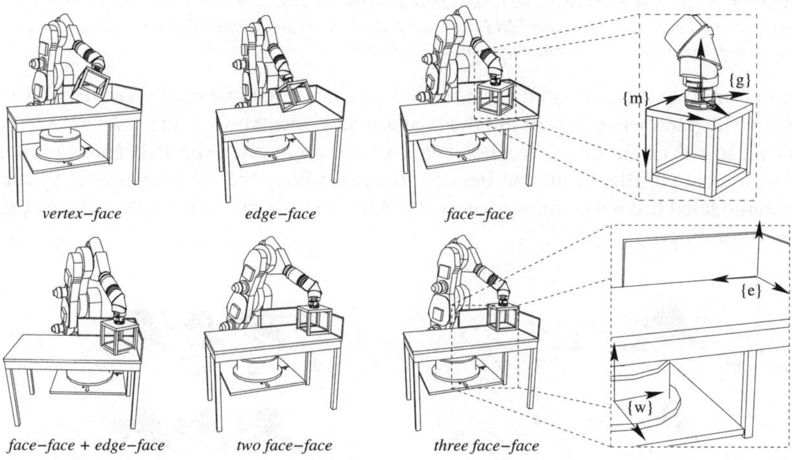

Figure 3. *Assembly of a cube into a corner by a sequence of different contact formations* (CFs), *for example, vertex-face contact, edge-face contact, etc.*

For each CF, the measurement equations are different, this corresponds to different $\boldsymbol{z}_i = \boldsymbol{h}_i(\boldsymbol{x}) + \boldsymbol{\rho}_{m,i}$ of footnote 1. The vector $\boldsymbol{x}'$ is chosen to linearize all these equations. Its dimension is 60, which is higher than 12, the dimension of $\boldsymbol{x}$. For details on the control, the measurement equations, the detection of contact transitions, the recognition of CFs, the inconsistency of the commonly used Extended Kalman Filters, etc., see Lefebvre *et al.* (2002).

4.2. *Results*

Equation (1) describes the (non-Gaussian) analytical posterior pdf over $\boldsymbol{x}$. However, what we want to know is the expected value and covariance of the marginal posterior pdfs over each of the parameters. We compare two methods to obtain these results:

A *Monte Carlo algorithm using a Metropolis–Hasting sampler* returns the posterior sample mean and covariance. Figure 5 shows the sample histograms of the parameters describing the position and orientation of the *corner* object, after having processed 100, 200 and 300 measurements (the

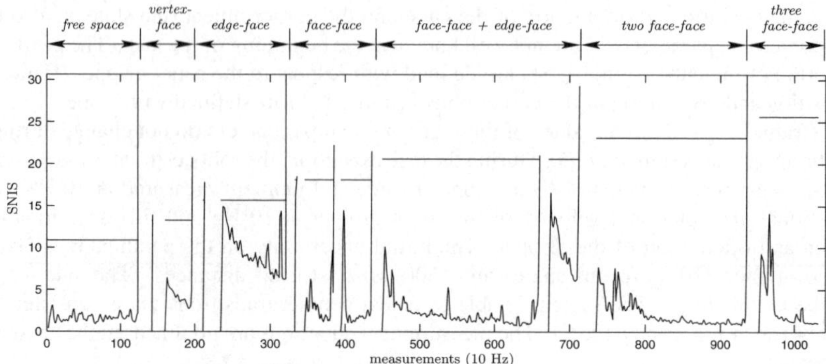

Figure 4. SNIS (*Sum of Normalized Innovations Squared*) consistency test. *The horizontal lines are the confidence boundaries. When the SNIS value exceeds the horizontal line, an inconsistency is reported. The figure shows that each contact transition is detected.*

number of time steps was limited because the method is computationally very expensive). The results for the parameters describing the position and orientation of the *cube* are similar. The sample mean and covariance give a good approximation for the expected value and covariance of the analytical pdf (the deviations between the real values and the estimates of θ_x^e and θ_y^e are due to unmodeled friction). Unfortunately, MCMC methods are computationally too expensive to be used on-line.

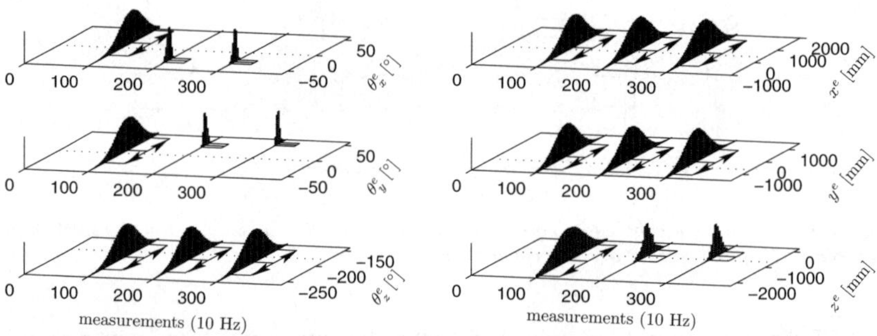

Figure 5. *Sample histograms of the marginal pdf of the three translations x^e, y^e, z^e and three Euler-ZYX angles $\theta_x^e, \theta_y^e, \theta_z^e$ of the corner object, i.e., the translation and rotation between frames $\{e\}$ and $\{w\}$, Figure 3. The pdfs are sampled after having processed 100, 200 and 300 measurements. Intervals corresponding to the sample mean plus/minus twice the standard deviation are indicated with arrows. The real value is plotted with a dotted line.*

An *Iterated Extended Kalman Filter* (IEKF) gives a good approximation of the expected value and covariance *if* the measurements fully observe the system, see Lefebvre *et al.* (2001). At every time step k, an IEKF is constructed with prior pdf $\mathrm{N}(\boldsymbol{x} \mid \boldsymbol{\mu}_{0|0}, \boldsymbol{P}_{0|0})$, measurement $\boldsymbol{\mu}'_k = \boldsymbol{g}(\boldsymbol{x}) + \boldsymbol{\rho}_k$ and measurement uncertainty $\mathrm{N}(\boldsymbol{\rho}_k \mid 0, \boldsymbol{P}'_k)$. The advantage of this method is that it is computationally not very expensive and therefore can be used *on-line*. Figure 6 shows the means and covariances of the marginal posterior pdfs over each of the parameters

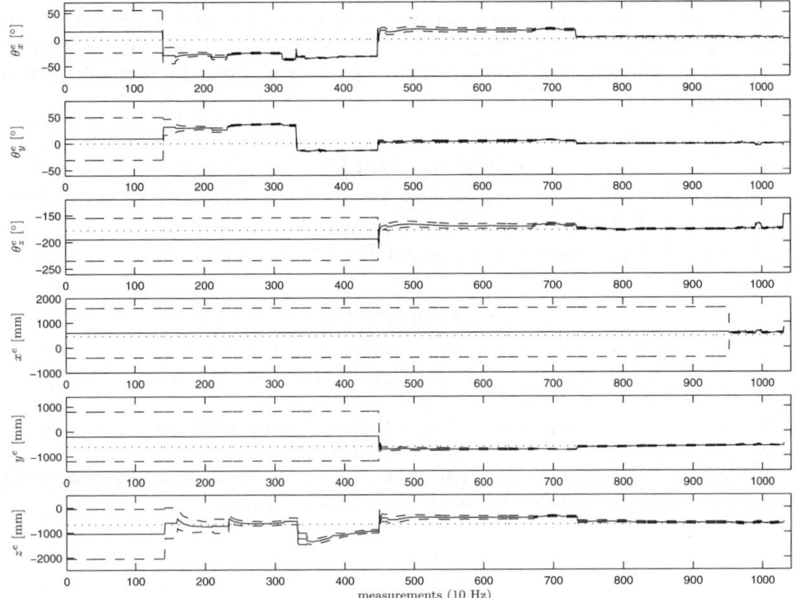

Figure 6. *Results of the Iterated Extended Kalman Filter: means (full lines) of the marginal pdf of the three translations x^e, y^e, z^e and three Euler-ZYX angles θ_x^e, θ_y^e, θ_z^e of the corner object, i.e., the translation and rotation between frames $\{e\}$ and $\{w\}$, Figure 3. The uncertainty on the calculated mean is visualized by twice the standard deviation (dashed lines). The real values are plotted in dotted lines.*

describing the position and orientation of the *corner* object, calculated by an IEKF. The results for the parameters describing the position and orientation of the *cube* are similar.

Figures 5 and 6 show that during the *motion in free space* (first 124 measurements), none of the marginal pdfs change: the motion in free space does not provide any information about the position and orientation of the two objects. After the recognition of the *vertex-face* CF (measurement 141), the uncertainty on several of the geometrical parameter estimates decreases, while for others it stays at the initial uncertainty. For example, the orientation of the corner object around the axis perpendicular to the contacting face (θ_z^e) and its translation inside this face (x^e and y^e) are not estimated, because the *vertex-face* CF does not provide any information about these. θ_z^e and y^e are estimated more accurately from measurement 442 on, *i.e.*, when a contact with one vertical face is made; x^e keeps its initial uncertainty until a *three face-face* contact is recognized at measurement 941. Note that the deviations on the estimation of the parameters θ_x^e and θ_y^e are due to unmodeled contact friction.

Comparing Figures 5 and 6 shows that the IEKF returns estimates (Figure 6, full lines) in the peak of the posterior pdf of Figure 5. However, the IEKF can return underestimated covariances: the intervals between the dashed lines for θ_x^e, θ_y^e and z^e in Figure 6 are smaller than the corresponding uncertainty intervals of Figure 5.

5. CONCLUSION

This paper presented a new finite-dimensional Bayesian filter for static systems with any kind of nonlinear measurement equation subject to Gaussian measurement uncertainty and for a limited class of dynamic systems. The core idea of the filter is to linearize the process and measurement equations in a higher-dimensional state space. The sufficient statistics of the posterior pdf are obtained through Kalman Filtering in this higher-dimensional space. Section 4 applied the new filter to the online estimation of the inaccurately known position and orientation of two parts during assembly.

ACKNOWLEDGEMENTS

T. Lefebvre and H. Bruyninckx are, respectively, Doctoral and Postdoctoral Fellows of the Fund for Scientific Research–Flanders (F.W.O.) in Belgium. Financial support by K. U. Leuven's Concerted Research Action GOA/99/04 is gratefully acknowledged.

REFERENCES

Beneš, V. E. (1981). Exact finite-dimensional filters for certain diffusions with nonlinear drift. *Stochastics* **5**, 65–92.

Daum, F. (1988). New exact nonlinear filters. *Bayesian Analysis of Time Series and Dynamical Models* (J. C. Spall, ed). New York: Marcel Dekker, 199–226.

Lefebvre, T., Bruyninckx, H. and De Schutter, J. (2001). Kalman filters for nonlinear systems: A comparison of performance. *Tech. Rep.*, Katholieke Universiteit Leuven, Belgium.

Lefebvre, T., Bruyninckx, H. and De Schutter, J. (2002). Polyhedral contact formation modeling and identification for autonomous compliant motion. *IEEE Trans. Robotics Automat.* (to appear).

BAYESIAN STATISTICS 7, pp. 597–606
J. M. Bernardo, M. J. Bayarri, J. O. Berger, A. P. Dawid,
D. Heckerman, A. F. M. Smith and M. West (Eds.)

Compatible Priors for Causal Bayesian Networks

VALENTINA LEUCARI and GUIDO CONSONNI
Università di Pavia, Italy
vl@dimat.unipv.it guido.consonni@unipv.it

SUMMARY

We consider discrete causal DAG-models (or Bayesian Networks) wherein the ordering of the variables is fixed across model structures. Given a prior on the parameter space of a model we describe a method for deriving a compatible prior on the parameter space of a submodel. This allows to generate automatically compatible priors for model parameters starting from a single prior relative to the largest entertained model. Our method makes use of a general procedure for constructing compatible priors for causal DAG-models, named reference conditioning, which is invariant within a suitable class of re-parameterizations and is model intrinsic. We show that if the generating prior satisfies global parameter independence, so does the compatible prior; in addition, prior modularity holds. Further results are obtained when the starting prior is product Dirichlet. A simple illustration of the methodology, and comparisons with alternative methods, are presented.

Keywords: GLOBAL INDEPENDENCE; INVARIANCE; LOCAL INDEPENDENCE; PRIOR MODULARITY; PRODUCT DIRICHLET; REFERENCE CONDITIONING.

1. INTRODUCTION

Bayesian Networks (BNs), based on Directed Acyclic Graphs (DAGs), have become increasingly popular to communicate, model and manage efficiently complex systems. They also represent a major tool in probabilistic expert systems. For an excellent account see Cowell *et al.* (1999), to which we refer for further background reading.

There are two ways to consider a BN: either as a belief model or as a "causal" model. In the former case, the focus is on the conditional independence properties embodied by the recursive factorization of the joint distribution, which can be read off from the DAG itself. On the other hand, the causal interpretation is especially appropriate, whenever one can assume the absence of unmeasured confounders, together with an *ordering* of the variables involved (see *e.g.*, Lauritzen and Richardson, 2002). In such models the primitive building blocks are represented by the local conditional distributions.

The structure of the DAG underlying a BN is typically uncertain, so that one may wish to entertain several BNs and eventually perform model determination. If a Bayesian viewpoint is adopted, the issue of specifying a prior distribution on the parameter space of each network (*e.g.*, with the aim of computing a Bayes factor) represents a severe problem, especially when the number of networks is very large. We, therefore, look for methods capable of automatically generating priors on parameter spaces across model structures. Although not logically necessary, it is also natural to require that these priors be related, in some way.

Specifically, assume that there are two models $\mathcal{M}$ and $\mathcal{M}_0$ for the same observable $\boldsymbol{X}$, having parameter space Θ and Θ_0, respectively, with $\Theta_0 \subset \Theta$, so that $\mathcal{M}_0$ is nested within $\mathcal{M}$.

Write $\pi(\cdot)$ and $\pi_0(\cdot)$ for the prior densities on the corresponding parameter spaces. Given $\pi(\cdot)$, the objective is to design a procedure capable of specifying a "compatible" prior $\pi_0(\cdot)$. One motivation for this requirement is that the resulting Bayes factor should be least influenced by dissimilarities between the two priors due to differences in the elicitation processes, and should therefore represent more faithfully the strength of the support that the data lend to each model. The issue of prior compatibility was previously discussed, in particular, by Spiegelhalter *et al.* (1993), Heckerman *et al.* (1995) and Cowell (1996). Cowell *et al.* (1999, Section 11.6) is a very useful review.

Suppose that $\Theta_0 = \{\boldsymbol{\theta} \in \Theta : \boldsymbol{t}(\boldsymbol{\theta}) = \boldsymbol{t}_0\}$, for some function $\boldsymbol{t} := \boldsymbol{t}(\cdot)$ and fixed value $\boldsymbol{t}_0$. Then a "natural" approach is to derive $\pi_0(\cdot)$ by conditioning $\pi(\cdot)$ on $\boldsymbol{t} = \boldsymbol{t}_0$. The drawback of this approach is that different conditioning functions may identify the *same* parameter space Θ_0 while giving rise to different priors $\pi_0(\cdot)$. Motivated by this difficulty, Dawid and Lauritzen (2000) suggested a novel approach, called *Jeffreys conditioning*, to construct $\pi_0(\cdot)$. Subsequently, Roverato and Consonni (2001) (henceforth R&C) argued that Jeffreys conditioning may be inadequate for causal DAG-models and proposed an alternative method, named *reference conditioning*, which incorporates explicitly the causal structure of the DAG.

In this paper we apply reference conditioning to find compatible priors for the parameters of causal BNs.

2. BACKGROUND

2.1. *Causal Bayesian Networks*

A DAG is a pair $\mathcal{D} = (V, E)$ where V is a finite set of vertices and E is a finite set of directed edges, represented by "$\rightarrow$". If $u \rightarrow v$ then u is said to be a parent of v and the set of parents of v is denoted by $\mathrm{pa}(v)$. Random variables or vectors (rvs) X_v and $\boldsymbol{X}_{\mathrm{pa}(v)}$ are associated to v and $\mathrm{pa}(v)$, taking values respectively in $\mathcal{X}_v$ and $\mathcal{X}_{\mathrm{pa}(v)} = \times_{u \in \mathrm{pa}(v)} \mathcal{X}_u$.

Let $\mathcal{M}^{\mathcal{D}}$ represent a DAG-model, *i.e.*, a family of probability distributions for $\boldsymbol{X} = \{X_v, v \in V\}$ parameterized by $\boldsymbol{\theta} = \{\boldsymbol{\theta}_{v|\mathrm{pa}(v)}, v \in V\} \in \Theta$, where $\boldsymbol{\theta}_{v|\mathrm{pa}(v)}$ indexes the conditional distribution of X_v given $\boldsymbol{X}_{\mathrm{pa}(v)}$. The joint density of $\boldsymbol{X}$, relative to a suitable product measure, admits the following factorization according to $\mathcal{D}$

$$p(\boldsymbol{x} \mid \boldsymbol{\theta}) = \prod_{v \in V} p_v(x_v \mid \boldsymbol{x}_{\mathrm{pa}(v)}, \boldsymbol{\theta}_{v|\mathrm{pa}(v)}). \tag{1}$$

We shall only consider DAG-models with a *fixed* vertex ordering wherein each X_v is a discrete random variable, and refer to them as causal Bayesian Networks. If X_v is a d_v-state random variable we let $\mathcal{X}_v = \{0, \ldots, d_v - 1\}$, $\mathcal{X}_v^0 = \{1, \ldots, d_v - 1\}$ and define

$$\theta_{v|\mathrm{pa}(v)}^{y_v|y_{\mathrm{pa}(v)}} = \Pr_{\boldsymbol{\theta}}\{X_v = y_v \mid \boldsymbol{X}_{\mathrm{pa}(v)} = \boldsymbol{y}_{\mathrm{pa}(v)}\}, \quad \boldsymbol{\theta}_{v|\mathrm{pa}(v)}^{y_{\mathrm{pa}(v)}} = (\theta_{v\,|\,\mathrm{pa}(v)}^{1\,|\,y_{\mathrm{pa}(v)}}, \ldots, \theta_{v\,|\,\mathrm{pa}(v)}^{d_v-1\,|\,y_{\mathrm{pa}(v)}}),$$

so that one can write

$$\boldsymbol{\theta}_{v|\mathrm{pa}(v)} = \{\boldsymbol{\theta}_{v|\mathrm{pa}(v)}^{y_{\mathrm{pa}(v)}}, \boldsymbol{y}_{\mathrm{pa}(v)} \in \mathcal{X}_{\mathrm{pa}(v)}\}. \tag{2}$$

We call the collection $\boldsymbol{\theta} = \{\boldsymbol{\theta}_{v|\mathrm{pa}(v)}, v \in V\}$ defined in (2) the *conditional-probability-parameter*, while the components $\boldsymbol{\theta}_{v|\mathrm{pa}(v)}^{y_{\mathrm{pa}(v)}}$ are named *local parameters*. The latter are variation-independent under multinomial sampling.

Two structural properties that are often imposed on the prior $\pi(\cdot)$ for $\boldsymbol{\theta}$ are: i) *global parameter independence* (GPI) and ii) *local parameter independence* (LPI) defined by

$$\text{GPI} : \pi(\boldsymbol{\theta}) = \prod_{v \in V} \pi_v(\boldsymbol{\theta}_{v|\text{pa}(v)}) \qquad \text{LPI} : \pi_v(\boldsymbol{\theta}_{v|\text{pa}(v)}) = \prod_{\boldsymbol{y}_{\text{pa}(v)} \in \mathcal{X}_{\text{pa}(v)}} \pi_v^{y_{\text{pa}(v)}}(\boldsymbol{\theta}_{v|\text{pa}(v)}^{y_{\text{pa}(v)}}).$$

2.2. *Compatible Priors by Means of Conditioning*

Consider the setup described in the Introduction. Let model $\mathcal{M}$ be parameterized by $\boldsymbol{\theta} \in \Theta$, where Θ is an open subset of $\Re^q$, and let $\mathcal{M}_0$ be a submodel of $\mathcal{M}$ identified by $\Theta_0 = \{\boldsymbol{\theta} \in \Theta : \boldsymbol{t}(\boldsymbol{\theta}) = \boldsymbol{t}_0\}$, where $\boldsymbol{t} := \boldsymbol{t}(\cdot)$ is a measurable function on Θ and $\boldsymbol{t}_0$ a suitable (vector-valued) constant. Let $r < q$ be the dimension of Θ_0. Assume that $\boldsymbol{\theta} \sim \pi(\cdot)$, with $\pi(\cdot)$ absolutely continuous with respect to Lebesgue measure. Suppose now that Θ_0 can be equivalently expressed as $\Theta_0 = \{\boldsymbol{\theta} \in \Theta : \boldsymbol{s}(\boldsymbol{\theta}) = \boldsymbol{s}_0\}$ using an alternative function $\boldsymbol{s} := \boldsymbol{s}(\cdot)$. As previously remarked, conditioning $\pi(\cdot)$ on $\boldsymbol{t} = \boldsymbol{t}_0$ may give a result different from conditioning it on $\boldsymbol{s} = \boldsymbol{s}_0$. Since the choice of the conditioning function is typically driven by the parameterization adopted to describe model $\mathcal{M}$ we can conclude, somewhat imprecisely but expressively, that a conditioning procedure is not invariant to model re-parameterizations.

Under Jeffreys conditioning (see Dawid and Lauritzen, 2000), the compatible prior distribution on Θ_0 has density, w.r.t. Lebesgue measure (on $\Re^r$), given by

$$\pi_0(\boldsymbol{\theta}) \propto \pi(\boldsymbol{\theta}) \frac{j_0(\boldsymbol{\theta})}{j(\boldsymbol{\theta})}, \qquad \boldsymbol{\theta} \in \Theta_0,$$

where $j(\cdot), j_0(\cdot)$ represent the densities of the Jeffreys measures under $\mathcal{M}$, respectively $\mathcal{M}_0$. Jeffreys conditioning is invariant to re-parameterization and is model intrinsic, since the procedure only relies on the information matrix of the model.

Assume now that two causal DAG-models $\mathcal{M}^{\mathcal{D}}$ and $\mathcal{M}^{\mathcal{D}_0}$ are under investigation. They share the same vertex ordering and the latter is nested within the former. R&C argue that a causal DAG-model cannot be arbitrarily re-parameterized, and define the class of *$\mathcal{D}$-modular* parameterizations. Loosely speaking, an element in this class can be unambiguously associated to the original local parameterization and in this sense it preserves the modular structure of the DAG-model itself. This leads them to recommend a conditioning procedure invariant within this class. While Jeffreys conditioning is clearly invariant within the class of $\mathcal{D}$-modular parameterizations, it is not the appropriate measure to be used, as exemplified by R&C in the Gaussian case, since it fails to satisfy *prior modularity*. If $\pi(\cdot)$ and $\pi_0(\cdot)$ satisfy GPI, prior modularity (defined in Heckerman *et al.,* 1995) states that $\pi_v(\cdot) = \pi_{0v}(\cdot)$, whenever $\text{pa}_0(v) = \text{pa}(v)$, where $\text{pa}_0(v)$ denotes the set of parents of v under $\mathcal{D}_0$. In other words, parameters corresponding to vertices whose parents are the same under the two models should have the same prior. Arguably Jeffreys conditioning fails to satisfy prior modularity because it is "universally" invariant, while only invariance within the class of $\mathcal{D}$-modular parameterizations is called for in causal DAG-models. One measure which satisfies this requirement, and is model intrinsic, is that based on the reference prior, see Bernardo (1979) and Berger and Bernardo (1992) for the general theory. Invariance of reference priors is dealt with in Yang (1995) and Datta and Ghosh (1996).

If $r(\cdot)$, $r_0(\cdot)$ denote the densities of the reference (prior) measure under $\mathcal{M}^{\mathcal{D}}$, respectively, $\mathcal{M}^{\mathcal{D}_0}$, w.r.t. the local-parameter-grouping, then the compatible prior $\pi_0(\cdot)$ obtained *via reference conditioning* is defined by

$$\pi_0(\boldsymbol{\theta}) \propto \pi(\boldsymbol{\theta}) \frac{r_0(\boldsymbol{\theta})}{r(\boldsymbol{\theta})}, \qquad \boldsymbol{\theta} \in \Theta_0. \tag{3}$$

We now briefly summarize some construction aspects of reference priors, under simplified conditions, in order to let the reader appreciate our results. Consider a statistical model parameterized by $\boldsymbol{\phi} \in \Phi$. Berger and Bernardo (1992) motivate and describe a general algorithm to find reference priors for the parameter $\boldsymbol{\phi}$. Such an algorithm is greatly simplified if the posterior distribution of $\boldsymbol{\phi}$ is asymptotically normal (the so-called *regular case*). Let $\boldsymbol{H}(\boldsymbol{\phi})$ denote the Fisher information matrix in the parameterization $\boldsymbol{\phi}$. We assume that $\boldsymbol{\phi}$ is grouped into s components $(\boldsymbol{\phi}_{(1)}, \ldots, \boldsymbol{\phi}_{(s)})$. Define $\boldsymbol{\phi}_{\sim k}$ to be $\boldsymbol{\phi}$ grouped as above without the component $\boldsymbol{\phi}_{(k)}$. The elements of $\boldsymbol{\phi}$ are usually ordered according to inferential importance; in particular, the parameters of interest should come first. For simplicity we assume that the components $\boldsymbol{\phi}_{(k)}$'s are variation-independent. Under the regular case, if $\boldsymbol{H}(\boldsymbol{\phi}) = \text{diag}\{\boldsymbol{H}_{11}(\boldsymbol{\phi}), \ldots, \boldsymbol{H}_{ss}(\boldsymbol{\phi})\}$ and $\det\{\boldsymbol{H}_{kk}(\boldsymbol{\phi})\} = a_k(\boldsymbol{\phi}_{(k)}) b_k(\boldsymbol{\phi}_{\sim k}), \forall\, k \in \{1, \ldots, s\}$, for some positive functions $a_k(\cdot)$ and $b_k(\cdot)$, then the density—with respect to Lebesgue measure—of the s-group reference prior for $\boldsymbol{\phi}$ is given by

$$r(\boldsymbol{\phi}_{(1)}, \ldots, \boldsymbol{\phi}_{(s)}) \propto \prod_{k=1}^{s} a_k(\boldsymbol{\phi}_{(k)})^{1/2}. \tag{4}$$

We remark that the prior in (4) does not depend on the ordering of the s groups (in general the reference prior depends both on the grouping and the ordering). Strictly speaking the above result holds under a further assumption about the structure of the sequence of compact subsets expanding to Φ, which in our case is most naturally satisfied. We omit details and refer for a proof to Datta and Ghosh (1995); a further useful reference is Gutiérrez-Peña and Rueda (2003).

3. REFERENCE CONDITIONING FOR BAYESIAN NETWORKS

Let $\mathcal{M}^{\mathcal{D}}$ be a causal BN indexed by the conditional-probability-parameter $\boldsymbol{\theta}$ introduced above. Set $\mathcal{D}_0 = (V, E_0 \subset E)$, so that $\mathcal{D}_0$ is obtained from $\mathcal{D}$ by deleting some of the edges while preserving the causal ordering for the vertices, and let $\mathcal{M}^{\mathcal{D}_0}$ denote the corresponding BN. In the following subsections we shall describe the steps required to compute the compatible prior $\pi_0(\cdot)$ given a prior $\pi(\cdot)$ for $\boldsymbol{\theta}$, using reference conditioning.

3.1. *Information Matrix*

We detail computations for $\mathcal{M}^{\mathcal{D}}$ only, since similar calculations hold for $\mathcal{M}^{\mathcal{D}_0}$.

Recall the factorization in (1) and consider a single vertex v. To each configuration $\boldsymbol{y}_{\text{pa}(v)}$ of $\boldsymbol{X}_{\text{pa}(v)}$ there corresponds a local parameter $\boldsymbol{\theta}_{v|\text{pa}(v)}^{\boldsymbol{y}_{\text{pa}(v)}}$, so that the conditional distribution of X_v given $\boldsymbol{X}_{\text{pa}(v)} = \boldsymbol{y}_{\text{pa}(v)}$ can be written as

$$p_v(x_v \,|\, \boldsymbol{y}_{\text{pa}(v)}, \boldsymbol{\theta}_{v|\text{pa}(v)}^{\boldsymbol{y}_{\text{pa}(v)}}) = \Big(1 - \sum_{y_v \in \mathcal{X}_v^0} \theta_{v|\text{pa}(v)}^{y_v|\boldsymbol{y}_{\text{pa}(v)}}\Big)^{1_{\{0\}}(x_v)} \prod_{y_v \in \mathcal{X}_v^0} \Big(\theta_{v|\text{pa}(v)}^{y_v|\boldsymbol{y}_{\text{pa}(v)}}\Big)^{1_{\{y_v\}}(x_v)}$$

where $1_A(\cdot)$ is the indicator function of the set A.

Setting $S_v^0(\boldsymbol{\theta}_{v|\text{pa}(v)}^{\boldsymbol{y}_{\text{pa}(v)}}) = \sum_{y_v \in \mathcal{X}_v^0} \theta_{v|\text{pa}(v)}^{y_v|\boldsymbol{y}_{\text{pa}(v)}}$, the joint density becomes

$$p(\boldsymbol{x} \,|\, \boldsymbol{\theta}) = \prod_{v \in V} \prod_{\boldsymbol{y}_{\text{pa}(v)} \in \mathcal{X}_{\text{pa}(v)}} \Big[p_v(x_v \,|\, \boldsymbol{y}_{\text{pa}(v)}, \boldsymbol{\theta}_{v|\text{pa}(v)}^{\boldsymbol{y}_{\text{pa}(v)}})\Big]^{1_{\{\boldsymbol{y}_{\text{pa}(v)}\}}(\boldsymbol{x}_{\text{pa}(v)})} \tag{5}$$

whcih may also be written as

$$\prod_{v \in V} \prod_{\boldsymbol{y}_{\text{pa}(v)} \in \mathcal{X}_{\text{pa}(v)}} \Big[(1 - S_v^0(\boldsymbol{\theta}_{v|\text{pa}(v)}^{\boldsymbol{y}_{\text{pa}(v)}}))^{1_{\{0\}}(x_v)} \prod_{y_v \in \mathcal{X}_v^0} (\theta_{v|\text{pa}(v)}^{y_v|\boldsymbol{y}_{\text{pa}(v)}})^{1_{\{y_v\}}(x_v)}\Big]^{1_{\{\boldsymbol{y}_{\text{pa}(v)}\}}(\boldsymbol{x}_{\text{pa}(v)})}.$$

From Eq. (5) it appears that the local parameters $\boldsymbol{\theta}_{v|\mathrm{pa}(v)}^{y_{\mathrm{pa}(v)}}$s are orthogonal, yielding the following result.

Lemma 1 *Let $\mathcal{M}^{\mathcal{D}}$ be a BN with conditional-probability-parameter $\boldsymbol{\theta}$. Then*
(i) *The Fisher information matrix relative to $\boldsymbol{\theta}$ is*

$$\boldsymbol{H}(\boldsymbol{\theta}) = \mathrm{diag}\{\boldsymbol{H}_{v|\mathrm{pa}(v)}^{y_{\mathrm{pa}(v)}}, v \in V, \boldsymbol{y}_{\mathrm{pa}(v)} \in \mathcal{X}_{\mathrm{pa}(v)}\}$$

where $\boldsymbol{H}_{v|\mathrm{pa}(v)}^{y_{\mathrm{pa}(v)}} = \left(\Pr_{\boldsymbol{\theta}}\{\boldsymbol{X}_{\mathrm{pa}(v)} = \boldsymbol{y}_{\mathrm{pa}(v)}\}\right) \boldsymbol{G}_{v|\mathrm{pa}(v)}^{y_{\mathrm{pa}(v)}}$ *is a square matrix of order* $\dim(\boldsymbol{\theta}_{v|\mathrm{pa}(v)}^{y_{\mathrm{pa}(v)}}) = d_v - 1$ *and* $\boldsymbol{G}_{v|\mathrm{pa}(v)}^{y_{\mathrm{pa}(v)}}$ *is the Fisher information matrix for a Multinomial family with one trial and cell probabilities* $\boldsymbol{\theta}_{v|\mathrm{pa}(v)}^{y_{\mathrm{pa}(v)}}$.
(ii) *The determinant of each matrix block is*

$$\det(\boldsymbol{H}_{v|\mathrm{pa}(v)}^{y_{\mathrm{pa}(v)}}) = \left(\Pr_{\boldsymbol{\theta}}\{\boldsymbol{X}_{\mathrm{pa}(v)} = \boldsymbol{y}_{\mathrm{pa}(v)}\}\right)^{d_v-1} \left[(1 - S_v^0(\boldsymbol{\theta}_{v|\mathrm{pa}(v)}^{y_{\mathrm{pa}(v)}}))P_v^0(\boldsymbol{\theta}_{v|\mathrm{pa}(v)}^{y_{\mathrm{pa}(v)}})\right]^{-1},$$

where

$$P_v^0(\boldsymbol{\theta}_{v|\mathrm{pa}(v)}^{y_{\mathrm{pa}(v)}}) = \prod_{y_v \in \mathcal{X}_v^0} \theta_{v|\mathrm{pa}(v)}^{y_v|y_{\mathrm{pa}(v)}}.$$

Proof. (i) To each vector $\boldsymbol{\theta}_{v|\mathrm{pa}(v)}^{y_{\mathrm{pa}(v)}}$ there corresponds a single term in the double product in (5) and thus a square block of order $d_v - 1$ in the information matrix

$$\boldsymbol{H}_{v|\mathrm{pa}(v)}^{y_{\mathrm{pa}(v)}} = -\mathrm{E}_{\boldsymbol{\theta}}\Big\{1_{\{\boldsymbol{y}_{\mathrm{pa}(v)}\}}(\boldsymbol{X}_{\mathrm{pa}(v)}) \frac{\partial^2}{\partial \boldsymbol{\theta}_{v|\mathrm{pa}(v)}^{y_{\mathrm{pa}(v)}} \partial (\boldsymbol{\theta}_{v|\mathrm{pa}(v)}^{y_{\mathrm{pa}(v)}})^T} \Big[1_{\{0\}}(X_v) \log(1 - S_v^0(\boldsymbol{\theta}_{v|\mathrm{pa}(v)}^{y_{\mathrm{pa}(v)}}))$$

$$+ \sum_{y_v \in \mathcal{X}_v^0} 1_{\{y_v\}}(X_v) \log \theta_{v|\mathrm{pa}(v)}^{y_v|y_{\mathrm{pa}(v)}}\Big]\Big\}$$

$$= \Pr_{\boldsymbol{\theta}}\{\boldsymbol{X}_{\mathrm{pa}(v)} = \boldsymbol{y}_{\mathrm{pa}(v)}\} \mathrm{E}_{\boldsymbol{\theta}}\Big\{ - \frac{\partial^2}{\partial \boldsymbol{\theta}_{v|\mathrm{pa}(v)}^{y_{\mathrm{pa}(v)}} \partial (\boldsymbol{\theta}_{v|\mathrm{pa}(v)}^{y_{\mathrm{pa}(v)}})^T}$$

$$\times \Big[1_{\{0\}}(X_v) \log(1 - S_v^0(\boldsymbol{\theta}_{v|\mathrm{pa}(v)}^{y_{\mathrm{pa}(v)}})) + \sum_{y_v \in \mathcal{X}_v^0} 1_{\{y_v\}}(X_v) \log \theta_{v|\mathrm{pa}(v)}^{y_v|y_{\mathrm{pa}(v)}}\Big]\Big\}.$$

The expression in square brackets can be regarded as the log-likelihood function relative to the random vector $(1_{\{1\}}(X_v), \ldots, 1_{\{d_v-1\}}(X_v)) \sim \mathrm{Mu}(1, \boldsymbol{\theta}_{v|\mathrm{pa}(v)}^{y_{\mathrm{pa}(v)}})$.
(ii) The determinant of each block is easily recovered recalling that the determinant of the Fisher information matrix for a Multinomial family with n trials and probability vector $(\eta_1, \ldots, \eta_m)$, $\sum_{j=1}^m \eta_i < 1$, is

$$n^m\Big[(1 - \sum_{j=1}^m \eta_j) \prod_{j=1}^m \eta_j\Big]^{-1},$$

(see *e.g.*, Bernardo and Smith, 1994, p. 336). □

We now compute the reference measure for $\boldsymbol{\theta}$ w.r.t the local-parameter-grouping.

Proposition 2 *Let $\mathcal{M}^{\mathcal{D}}$ be a BN with conditional-probability-parameter $\boldsymbol{\theta}$. Then the reference measure for $\boldsymbol{\theta}$, relative to the grouping $\{\boldsymbol{\theta}_{v|\mathrm{pa}(v)}^{y_{\mathrm{pa}(v)}}, v \in V, \boldsymbol{y}_{\mathrm{pa}(v)} \in \mathcal{X}_{\mathrm{pa}(v)}\}$, is order-invariant and has density*

$$r(\boldsymbol{\theta}) \propto \prod_{v \in V} \left[r_v(\boldsymbol{\theta}_{v|\mathrm{pa}(v)})\right]^{1/2} \tag{6}$$

where $r_v(\boldsymbol{\theta}_{v|\mathrm{pa}(v)}) \propto \prod_{\boldsymbol{y}_{\mathrm{pa}(v)} \in \mathcal{X}_{\mathrm{pa}(v)}} \left[(1 - S_v^0(\boldsymbol{\theta}_{v|\mathrm{pa}(v)}^{y_{\mathrm{pa}(v)}}))P_v^0(\boldsymbol{\theta}_{v|\mathrm{pa}(v)}^{y_{\mathrm{pa}(v)}})\right]^{-1}$.

Proof. Vectors representing the local-parameter-groups are variation-independent as well as orthogonal, implying order-invariance of the corresponding reference measure, provided each subset in the sequence expanding to Θ has the structure of a Cartesian product. Moreover $\Pr_{\boldsymbol{\theta}}\{\boldsymbol{X}_{\mathrm{pa}(v)} = \boldsymbol{y}_{\mathrm{pa}(v)}\}$ is independent of $\boldsymbol{\theta}_{v|\mathrm{pa}(v)}^{y_{\mathrm{pa}(v)}}$ for each $v \in V$ and $\boldsymbol{y}_{\mathrm{pa}(v)} \in \mathcal{X}_{\mathrm{pa}(v)}$. The construction in (4) can thus be exploited, w.r.t. such grouping, setting $a_k(\boldsymbol{\phi}_{(k)}) = \left[(1 - S_v^0(\boldsymbol{\theta}_{v|\mathrm{pa}(v)}^{y_{\mathrm{pa}(v)}}))P_v^0(\boldsymbol{\theta}_{v|\mathrm{pa}(v)}^{y_{\mathrm{pa}(v)}})\right]^{-1}$, yielding the following factorization

$$r(\boldsymbol{\theta}) \propto \prod_{v \in V} \prod_{\boldsymbol{y}_{\mathrm{pa}(v)} \in \mathcal{X}_{\mathrm{pa}(v)}} \left[(1 - S_v^0(\boldsymbol{\theta}_{v|\mathrm{pa}(v)}^{y_{\mathrm{pa}(v)}}))P_v^0(\boldsymbol{\theta}_{v|\mathrm{pa}(v)}^{y_{\mathrm{pa}(v)}})\right]^{-1/2}.$$

□

The density $r_0(\cdot)$ of the reference measure under $\mathcal{M}^{\mathcal{D}_0}$ has an analogous expression. We omit details.

3.2. *Compatible priors for Bayesian Networks*

Consider $\mathcal{M}^{\mathcal{D}}$ and $\mathcal{M}^{\mathcal{D}_0}$ as defined at the beginning of Section 3, so that for each $v \in V$, $\mathrm{pa}_0(v) \subseteq \mathrm{pa}(v)$. Set $\mathrm{pa}^*(v) = \mathrm{pa}(v) \backslash \mathrm{pa}_0(v)$ and $V^* = \{v \in V : \mathrm{pa}(v) \neq \mathrm{pa}_0(v)\}$. The parameter space corresponding to $\mathcal{M}^{\mathcal{D}_0}$ is identified through the constraints

$$\theta_{v|\mathrm{pa}^*(v) \cup \mathrm{pa}_0(v)}^{y_v | y_{\mathrm{pa}^*(v)}, y_{\mathrm{pa}_0(v)}} = \theta_{v|\mathrm{pa}_0(v)}^{y_v | y_{\mathrm{pa}_0(v)}},$$

say, for all $\boldsymbol{y}_{\mathrm{pa}^*(v)} \in \mathcal{X}_{\mathrm{pa}^*(v)}$. Let

$$\boldsymbol{\theta}_{v|\mathrm{pa}_0(v)} = \left\{\theta_{v|\mathrm{pa}_0(v)}^{y_v | y_{\mathrm{pa}_0(v)}}, \boldsymbol{y}_v \in \mathcal{X}_v^0, \boldsymbol{y}_{\mathrm{pa}_0(v)} \in \mathcal{X}_{\mathrm{pa}_0(v)}\right\}$$

and set $\boldsymbol{\theta}_0 = \{\boldsymbol{\theta}_{v|\mathrm{pa}(v)}, v \in V \backslash V^*\} \cup \{\boldsymbol{\theta}_{v|\mathrm{pa}_0(v)}, v \in V^*\}$, so that $\boldsymbol{\theta}_0$ represents the conditional probability parameter for model $\mathcal{M}^{\mathcal{D}_0}$. The following theorem states the main result of the paper.

Theorem 3 *Let $\mathcal{M}^{\mathcal{D}}$ be a BN with conditional-probability-parameter $\boldsymbol{\theta}$ having prior density $\pi(\cdot)$. Let $\mathcal{M}^{\mathcal{D}_0}$ be a submodel of $\mathcal{M}^{\mathcal{D}}$ with conditional-probability-parameter $\boldsymbol{\theta}_0$. Then*
(i) *The compatible prior $\pi_0(\cdot)$ obtained via reference conditioning has density*

$$\pi_0(\boldsymbol{\theta}_0) \propto \pi(\boldsymbol{\theta}_0) \prod_{v \in V^*} \frac{1}{r_{0v}(\boldsymbol{\theta}_{v|\mathrm{pa}_0(v)})^{1/2(\prod_{u \in \mathrm{pa}^*(v)} d_u - 1)}} \tag{7}$$

where

$$r_{0v}(\boldsymbol{\theta}_{v|\mathrm{pa}_0(v)}) \propto \prod_{\boldsymbol{y}_{\mathrm{pa}_0(v)} \in \mathcal{X}_{\mathrm{pa}_0(v)}} \left[(1 - S_v^0(\boldsymbol{\theta}_{v|\mathrm{pa}_0(v)}^{y_{\mathrm{pa}_0(v)}}))P_v^0(\boldsymbol{\theta}_{v|\mathrm{pa}_0(v)}^{y_{\mathrm{pa}_0(v)}})\right]^{-1}.$$

(ii) *If $\pi(\cdot)$ satisfies GPI, so does $\pi_0(\cdot)$. Moreover, $\pi_0(\cdot)$ satisfies prior modularity, i.e., the distribution of the parameters $\boldsymbol{\theta}_{v|\mathrm{pa}(v)}^{y_{\mathrm{pa}(v)}}$, $v \in V \backslash V^*$, is the same under $\pi(\cdot)$ and $\pi_0(\cdot)$.*

Proof. (i) Applying Proposition 2 w.r.t. $\mathcal{M}^{\mathcal{D}_0}$ and $\boldsymbol{\theta}_0$ and recalling the factorization in Eq. (6) one obtains

$$r_{0v}(\boldsymbol{\theta}_{v|\mathrm{pa}_0(v)}) = r_v(\boldsymbol{\theta}_{v|\mathrm{pa}(v)}) \quad \text{if } v \in V \backslash V^*$$

$$r_{0v}(\boldsymbol{\theta}_{v|\mathrm{pa}_0(v)}) \propto \prod_{\boldsymbol{y}_{\mathrm{pa}_0(v)} \in \mathcal{X}_{\mathrm{pa}_0(v)}} \left[(1 - S_v^0(\boldsymbol{\theta}_{v|\mathrm{pa}_0(v)}^{y_{\mathrm{pa}_0(v)}}))P_v^0(\boldsymbol{\theta}_{v|\mathrm{pa}_0(v)}^{y_{\mathrm{pa}_0(v)}})\right]^{-1} \quad \text{if } v \in V^*.$$

As a consequence, equation (3) becomes

$$\pi_0(\boldsymbol{\theta}_0) \propto \pi(\boldsymbol{\theta}_0) \left[\prod_{v \in V^*} \frac{r_{0v}(\boldsymbol{\theta}_{v|\mathrm{pa}_0(v)})}{r_v(\boldsymbol{\theta}_{v|\mathrm{pa}_0(v)})}\right]^{1/2} \left[\prod_{v \in V \backslash V^*} \frac{r_{0v}(\boldsymbol{\theta}_{v|\mathrm{pa}(v)})}{r_v(\boldsymbol{\theta}_{v|\mathrm{pa}(v)})}\right]^{1/2}$$

$$\propto \pi(\boldsymbol{\theta}_0) \prod_{v \in V^*} \frac{\prod_{\boldsymbol{y}_{\mathrm{pa}_0(v)} \in \mathcal{X}_{\mathrm{pa}_0(v)}} \left[(1 - S_v^0(\boldsymbol{\theta}_{v|\mathrm{pa}_0(v)}))P_v^0(\boldsymbol{\theta}_{v|\mathrm{pa}_0(v)})\right]^{-1/2}}{\prod_{\boldsymbol{y}_{\mathrm{pa}^*(v)} \in \mathcal{X}_{\mathrm{pa}^*(v)}} \prod_{\boldsymbol{y}_{\mathrm{pa}_0(v)} \in \mathcal{X}_{\mathrm{pa}_0(v)}} \left[(1 - S_v^0(\boldsymbol{\theta}_{v|\mathrm{pa}_0(v)}^{y_{\mathrm{pa}_0(v)}}))P_v^0(\boldsymbol{\theta}_{v|\mathrm{pa}_0(v)}^{y_{\mathrm{pa}_0(v)}})\right]^{-1/2}}$$

$$\propto \pi(\boldsymbol{\theta}_0) \prod_{v \in V^*} \prod_{\boldsymbol{y}_{\mathrm{pa}_0(v)} \in \mathcal{X}_{\mathrm{pa}_0(v)}} \left[(1 - S_v^0(\boldsymbol{\theta}_{v|\mathrm{pa}_0(v)}^{y_{\mathrm{pa}_0(v)}}))P_v^0(\boldsymbol{\theta}_{v|\mathrm{pa}_0(v)}^{y_{\mathrm{pa}_0(v)}})\right]^{1/2(\prod_{u \in \mathrm{pa}^*(v)} d_u - 1)}.$$

(ii) Suppose GPI holds for $\pi(\cdot)$. Then one can write

$$\pi_0(\boldsymbol{\theta}_0) \propto \Big(\prod_{v \in V \backslash V^*} \pi_v(\boldsymbol{\theta}_{v|\mathrm{pa}(v)})\Big)$$

$$\times \prod_{v \in V^*} \{\pi_v(\boldsymbol{\theta}_{v|\mathrm{pa}_0(v)}) \prod_{\boldsymbol{y}_{\mathrm{pa}_0(v)} \in \mathcal{X}_{\mathrm{pa}_0(v)}} [(1 - S_v^0(\boldsymbol{\theta}_{v|\mathrm{pa}_0(v)}^{y_{\mathrm{pa}_0(v)}}))P_v^0(\boldsymbol{\theta}_{v|\mathrm{pa}_0(v)}^{y_{\mathrm{pa}_0(v)}})]^{1/2(\prod_{u \in \mathrm{pa}^*(v)} d_u - 1)}\}$$

from which it appears immediately that GPI is satisfied by $\pi_0(\cdot)$. Furthermore prior modularity holds since priors for vertices $v \in V \backslash V^*$ are unaffected by the conditioning procedure. □

We remark that formula (7) is computationally inexpensive. Furthermore notice that $\pi_0(\cdot)$ may be improper even if $\pi(\cdot)$ is proper.

Suppose now that a prior distribution is assigned to $\boldsymbol{\theta}$ satisfying GPI and LPI with local densities

$$\pi_v^{y_{\mathrm{pa}(v)}}(\boldsymbol{\theta}_{v|\mathrm{pa}(v)}^{y_{\mathrm{pa}(v)}}) \propto (1 - S_v^0(\boldsymbol{\theta}_{v|\mathrm{pa}(v)}^{y_{\mathrm{pa}(v)}}))^{\alpha_{v\,|\,\mathrm{pa}(v)}^{0\,|\,y_{\mathrm{pa}(v)}} - 1} \prod_{y_v \in \mathcal{X}_v^0} (\theta_{v|\mathrm{pa}(v)}^{y_v|y_{\mathrm{pa}(v)}})^{\alpha_{v|\mathrm{pa}(v)}^{y_v|y_{\mathrm{pa}(v)}} - 1}$$

that is, a Dirichlet with strictly positive hyperparameters $\alpha_{v|\mathrm{pa}(v)}^{y_v|y_{\mathrm{pa}(v)}}$. Such a prior is called *product Dirichlet*. The total precision associated to vertex v is denoted by $\alpha_{v|\mathrm{pa}(v)}^{+|+} = \sum_{y_v \in \mathcal{X}_v} \sum_{\boldsymbol{y}_{\mathrm{pa}(v)} \in \mathcal{X}_{\mathrm{pa}(v)}} \alpha_{v|\mathrm{pa}(v)}^{y_v|y_{\mathrm{pa}(v)}}$. The following result provides an expression for the compatible prior $\pi_0(\cdot)$.

Proposition 4. *Assume that $\pi(\cdot)$ is product Dirichlet. If the compatible prior $\pi_0(\cdot)$ is obtained by means of reference conditioning then*
(i) *$\pi_0(\cdot)$ is product Dirichlet (so that GPI and LPI hold) and satisfies prior modularity.*
(ii) *The marginal distribution of $\boldsymbol{\theta}_{v|\mathrm{pa}_0(v)}^{y_{\mathrm{pa}_0(v)}}$, $v \in V^*$, under $\pi_0(\cdot)$ has hyperparameters*

$$\alpha_{v|\mathrm{pa}_0(v)}^{y_v|y_{\mathrm{pa}_0(v)}} = \Big(\sum_{y_{\mathrm{pa}^*(v)} \in \mathcal{X}_{\mathrm{pa}^*(v)}} \alpha_{v|\mathrm{pa}(v)}^{y_v|y_{\mathrm{pa}^*(v)}} \Big) - \frac{1}{2} \Big(\prod_{u \in \mathrm{pa}^*(v)} d_u - 1 \Big)$$

.
(iii) *The total precision associated to vertex $v \in V^*$ under $\pi_0(\cdot)$ is*

$$\alpha_{v|\mathrm{pa}_0(v)}^{+|+} = \alpha_{v|\mathrm{pa}(v)}^{+|+} - \frac{1}{2} \Big(\prod_{u \in \mathrm{pa}^*(v)} d_u - 1 \Big) \Big(\prod_{u \in \mathrm{pa}_0(v)} d_u \Big) d_v.$$

Proof. For (i) and (ii), compute equation (7) for the product Dirichlet prior. For (iii), use (ii) and the definition of total precision. □

We remark that the total precision associated to each vertex $v \in V^*$ is lower under $\pi_0(\cdot)$ than under $\pi(\cdot)$. Moreover we stress the fact that a product Dirichlet structure for $\pi_0(\cdot)$ is *implied* under reference conditioning, unlike in other approaches, see Section 4.

4. EXAMPLES

We illustrate our methodology using two simple BNs and compare our results in Proposition 4 with the method proposed by Spiegelhalter *et al.* (1993) and that suggested by Cowell (1996). Both papers assume that the prior under either model is product Dirichlet. Spiegelhalter *et al.* (1993) use the *expansion and contraction* (EC) algorithm to find a matching prior which preserves total precisions for all vertices. Cowell (1996) considers a fixed ordering, as we do, and finds the closest prior based on a criterion of *minimum prior and expected posterior* Kullback-Leibler divergences (KL). His method admits a few variants depending on the distribution used to compute expectations over all possible future observations, and thus does not lead to a unique answer. In the examples this distribution corresponds to that under the larger model.

Figure 1. *DAGs corresponding to $\mathcal{M}^{\mathcal{D}^i}$, $i = 1, 2$ ($\mathcal{M}^{\mathcal{D}_0^i}$ obtained by dropping all the arrows).*

Table 1. $\mathcal{M}^{\mathcal{D}^1}$, $\mathcal{M}^{\mathcal{D}^1_0}$: *hyperparameters under reference conditioning, EC and KL.*

					π_0	π_0^{EC}	π_0^{KL}			π_0	π_0^{EC}	π_0^{KL}
	α_a^1	α_a^0	$\alpha_{b\vert a}^{1\vert 1}$	$\alpha_{b\vert a}^{1\vert 0}$		α_b^1		$\alpha_{b\vert a}^{0\vert 1}$	$\alpha_{b\vert a}^{0\vert 0}$		α_b^0	
π_1	1.0	8.0	1.0	7.0	7.5	8.9	7.2	9.0	21.0	29.5	29.1	23.8
π_2	1.0	8.0	1.0	7.0	7.5	12.1	7.0	9.0	4.0	12.5	8.9	5.1
π_3	1.0	8.0	6.0	15.0	20.5	20.8	17.7	4.0	12.0	15.5	16.2	13.8
π_4	3.0	4.0	1.0	7.0	7.5	7.0	5.7	9.0	21.0	29.5	31.0	25.1
π_5	3.0	4.0	1.0	7.0	7.5	8.9	5.2	9.0	4.0	12.5	12.5	7.6
π_6	3.0	4.0	6.0	15.0	20.5	21.3	17.2	4.0	12.0	15.5	15.7	12.7

Table 2. $\mathcal{M}^{\mathcal{D}^2_0}$: *hyperparameters and precisions under reference conditioning and EC.*

	α_b^0	α_b^1	α_b^2	α_b^3	α_b^+	α_c^0	α_c^1	α_c^2	α_c^3	α_c^4	α_c^+
π_0	66.0	28.0	14.0	17.0	125.0	91.5	107.5	61.5	153.5	145.5	559.5
π_0^{EC}	75.0	17.2	11.8	25.0	129.0	151.9	56.6	55.1	155.4	168.0	587.0

Example 1. Let $\mathcal{M}^{\mathcal{D}^1}$ be a bivariate BN where $V = \{a, b\}$ and $\text{pa}(b) = \{a\}$, see Figure 1. X_a and X_b are binary "$0 - 1$" rvs, so that $\boldsymbol{\theta} = (\theta_a^1, \theta_{b|a}^{1|1}, \theta_{b|a}^{1|0})$, each component being the probability of value 1. Suppose $\boldsymbol{\theta}$ is assigned a product Beta distribution with $\theta_a^1 \sim \text{Be}(1, 8)$, $\theta_{\text{b|a}}^{1|1} \sim \text{Be}(1, 9)$, $\theta_{\text{b|a}}^{1|0} \sim \text{Be}(7, 21)$. If $\mathcal{M}^{\mathcal{D}^1_0}$ is the independence model wherein $X_a \perp\!\!\!\perp X_b$ parameterized with $\boldsymbol{\theta}_0 = (\theta_a^1, \theta_b^1)$, reference conditioning implies that $\pi_0(\cdot)$ is still product Beta with $\theta_a^1 \sim \text{Be}(1, 8)$, $\theta_{\text{b}}^1 \sim \text{Be}(7.5, 29.5)$. Table 1 reports the hyperparameters for six different prior specifications of $\pi(\cdot)$ under $\mathcal{M}^{\mathcal{D}^1}$ (including the one above, indicated by π_1), together with the corresponding hyperparameters for three types of compatible priors, namely those computed *via* reference conditioning (π_0), expansion-contraction (π_0^{EC}) and Kullback-Leibler minimizations (π_0^{KL}). Clearly only the hyperparameters corresponding to vertex b are shown, since those pertaining to vertex a do not change by prior modularity, which is satisfied by each of the three compatible priors. On the whole it appears that reference conditioning and EC are in closer agreement between themselves than with KL.

Example 2. Consider the complete DAG $\mathcal{D}^2$ of Figure 1 and a model $\mathcal{M}^{\mathcal{D}^2}$ where the corresponding rvs X_a, X_b and X_c take values, respectively, in $\{0, 1, 2\}$, $\{0, 1, 2, 3\}$ and $\{0, 1, 2, 3, 4\}$. A product Dirichlet is specified for the conditional-probability-parameter of such a model with total precisions: $\alpha_a^+ = 18$, $\alpha_{b|a}^{+|+} = 129$, $\alpha_{c|ab}^{+|++} = 587$. Specifically, defining

$$\boldsymbol{\alpha}_{v|\text{pa}(v)}^{y_{\text{pa}(v)}} = (\alpha_{v|\text{pa}(v)}^{0|y_{\text{pa}(v)}}, \ldots, \alpha_{v|\text{pa}(v)}^{d_v - 1|y_{\text{pa}(v)}}),$$

we fix

$$\boldsymbol{\alpha}_a = (9, 2, 7), \quad \boldsymbol{\alpha}_{b|a}^0 = (12, 1, 1, 5), \quad \boldsymbol{\alpha}_{b|a}^1 = (4, 23, 3, 1), \quad \boldsymbol{\alpha}_{b|a}^2 = (51, 5, 11, 12)$$

(to save space we do not report the hyperparameters for the prior on $\boldsymbol{\theta}_{\text{c}|ab}^{y_a, y_b}$). Compatible priors for the independence model $\mathcal{M}^{\mathcal{D}^2_0}$ are computed using both reference conditioning and the expansion-contraction method, yielding the results in Table 2. While the total precisions α_b^+ and α_c^+ are broadly similar under the two approaches, there appear to be some marked differences between specific pairs of hyperparameters.

5. DISCUSSION

Under the product Dirichlet assumption our method allows for varying degrees of total precisions of priors across vertices (for a given prior) and across model structures for each given vertex. This is in contrast to previous works such as Heckerman *et al.* (1995) and Spiegelhalter *et al.* (1993), and similar, in this respect, to Cowell (1996). A further similarity with the latter paper is the assumption of a causal ordering of the variables. However, Cowell's method is rooted in minimizing divergence, while our approach stems from invariance considerations. We also remark that reference conditioning works from the prior on the parameter space of a larger model to that of a nested model. This is not required in the alternative approaches mentioned above. On the other hand, while the latter impose *a priori* the structural properties of the compatible prior, such as GPI or product Dirichlet, such features become *consequences* of the structure of the generating prior within the reference conditioning framework.

ACKNOWLEDGEMENT

Research partially supported by MIUR, Italy, grant 2001138888.

REFERENCES

Berger, J. O. and Bernardo, J. M. (1992). On the development of the reference prior method. *Bayesian Statistics 6* (J. M. Bernardo, J. O. Berger, A. P. Dawid and A. F. M. Smith, eds). Oxford: Oxford University Press, 35–60 (with discussion).

Bernardo, J. M. (1979). Reference posterior distributions for Bayesian inference. *J. R. Statist. Soc. B* **41**, 113–147, (with discussion).

Bernardo, J. M. and Smith, A. F. M. (1994). *Bayesian Theory*. Chichester: Wiley.

Cowell, R. G. (1996). On compatible priors for Bayesian networks. *IEEE Trans. Pattern Anal. Machine Intelligence* **18**, 901–911.

Cowell, R. G., Dawid, A. P., Lauritzen S. L. and Spiegelhalter, D. J. (1999). *Probabilistic Networks and Expert Systems*. New York: Springer.

Datta, G. S. and Ghosh, M. (1995). Some remarks on non-informative priors. *J. Am. Statist. Ass.* **90**, 1357–1363.

Datta, G. S. and Ghosh, M. (1996). On the invariance of noninformative priors. *Ann. Statist.* **24**, 141–159.

Dawid, A. P. and Lauritzen, S. L. (2000). Compatible prior distributions. *Bayesian Methods with Applications to Science, Policy and Official Statistics* (E. George, ed). Brussels: European Community, 109–118.

Gutiérrez-Peña, E. and Rueda, R. (2003). Reference priors for exponential families. *J. Statist. Plann. Inference* **110**, 33–54.

Heckerman, D., Geiger, D. and Chickering, D. (1995). Learning Bayesian networks: The combination of knowledge and statistical data. *Machine Learning* **20**, 197–243.

Lauritzen, S. L. and Richardson, T. S. (2002). Chain graph models and their causal interpretation. *J. R. Statist. Soc. B* **64**, 321–361. (with discussion).

Roverato, A. and Consonni, G. (2001). Compatible prior distributions for DAG models. *Tech. Rep.*, Università di Pavia, Italy.

Spiegelhalter, D. J. , Dawid, A. P. , Lauritzen, S. L. and Cowell, R. G. (1993). Bayesian analysis in expert systems. *Statist. Sci.* **8**, 219–283 (with discussion).

Yang, R. (1995). Invariance of the reference prior under reparametrization. *Test* **4**, 83–94.

BAYESIAN STATISTICS 7, pp. 607–617
J. M. Bernardo, M. J. Bayarri, J. O. Berger, A. P. Dawid,
D. Heckerman, A. F. M. Smith and M. West (Eds.)

On the Application of Logistic Regression Modelling in Microarray Studies

BART J. A. MERTENS
Leiden University, The Netherlands
b.mertens@lumc.nl

SUMMARY

This paper demonstrates how a fully Bayesian hierarchical modelling implementation renders the logistic regression model suited for the analysis of microarray data. We make a point to provide special justification for the use of logistic regression and Bayesian modelling in this context. Results from the methodology on a well-known data set are shown. We provide a brief discussion with reference to alternative approaches.

Keywords: MICROARRAYS; LOGISTIC REGRESSION; HIERARCHICAL MODELLING; ALL/AML DATA; HIGH-DIMENSIONAL DATA; SHRINKAGE REGRESSION; DIMENSION REDUCTION; CONTINUUM LEAST SQUARES REGRESSION.

1. INTRODUCTION

1.1. *Microarrays and High-Dimensional Data Analysis*

Microarrays (sometimes also referred to as "DNA chips") simultaneously record sample responses on a large collection of genes, thus yielding data of extremely high dimensionality. The methodology is to a considerable extent driven by advances in technology. It is envisaged that the chips will eventually be produced in large numbers from standard "matrices" or templates (which are in themselves costly to produce), thus allowing economies of scale. Evaluation of the chip responses is largely automated. At present, microarrays (and sample preparation) are still expensive. Hence, sample sizes from microarray experiments are small, which exacerbates the singularity problems which are encountered with such high-dimensional data.

1.2. *Microarrays and Medical Science*

From a medical point of view there are a number of reasons why microarray technology is of interest. The first and most simple of all is to allow rapid classification or identification based on tissue samples which may be easily collected from patients, such as blood samples (patient-specific inference) for example. Specific examples are in the rapid screening of samples or populations, patient diagnosis, and the like. A more complex purpose is to understand disease processes. Specifically, there is an interest to identify genes involved in specific disease pathways (gene-specific inference). Studies of discriminant design are of special interest when such inference is required, as we can exploit the class distinction in seeking gene sets with good discriminating ability. Finally, the methodology could eventually be used to improve patient treatment assignment or perform treatment optimization, which are medical decision-making problems. This is therefore a more complex problem as compared to the previous two inference tasks, as it would ideally also have to incorporate medical costs of treatment (both from an institutional and patient perspective). We will not consider this latter problem in this paper.

1.3. *An Example Data Set*

We consider the MIT dataset on the discrimination of acute lymphoid leukemia (ALL) from acute myeloid leukemia (AML). The dataset contains microarray measurements on 72 patients in total, of which 47 are of acute lymphoid type and 25 of acute myeloid type. Samples were hybridized to single color Affymetrix arrays, yielding measurements on 7129 genes for each patient from each array. The data and the medical problem generating it were first described in the by now well-known paper by Golub *et al.* (1999). [Note this example does not live up to all the qualifications described above, as the samples are from bone marrow, which is rather more difficult to obtain from patients as compared to blood samples.]

2. LOGISTIC REGRESSION MODELLING

A great variety of methods has by now been proposed and also applied to the analysis of microarray data, ranging from crude multiple testing procedures based on naive applications of two-sample t-testing procedures and p-value evaluation, discriminant dimension reduction methods, and a great collection of machine learning type procedures, to name but a few. Indeed, the abundance of "methods" applied to this problem shows little evidence of much systematic thought as to how methodology should account for the problems posed by this particular type of high dimensional data. In particular, there is little regard for the users of the outcomes of the analyses or for how the methodology will take its place in the greater genome research project. We, therefore, want to provide some justification for proposing logistic regression and in particular also for the need of a fully probabilistic implementation, before discussing any models or results themselves.

2.1. *Logistic Regression Modelling for Microarray Data*

From a medical sciences and statistics perspective, there a special reasons to be interested in logistic regression based approaches to the analysis of microarray data.

1. The data we are interested in is of discriminant type. Logistic regression modelling can (at least from a conceptual point of view) be easily extended to other such settings as regression (of a continuous trait) or even survival prediction for individual patients.
2. The method is relatively well-known within the medical profession. Indeed, logistic regression almost invariably is the first statistical methodology medical science people will encounter which explicitly uses and requires the formulation of probabilities within a statistical model. Most medical personnel and biomedical sciences students need to study the method as a requirement for their degree studies. The method is also popular among epidemiologists and thus provides common ground.
3. From a (purely) methodological statistical point of view, the method is semi-parametric in nature and will emulate classical linear discriminant results under a general class of elliptical distributions. Modelling class membership in terms of the gene responses also avoids the need to make explicit assumptions about the correlation structure, which could be a somewhat painful endeavor with such high dimensions and small sample sizes.

2.2. *Fully Bayesian Analysis for the Logistic Model*

The appeal of the logistic model becomes even stronger if placed within a fully Bayesian context.

1. From a purely "technical" point of view, we obviously have a singularity problem to solve as a large collection of genes will be evaluated within each sample. Exchangeable modelling of the regression coefficients allows borrowing of information across genes to alleviate the problem.
2. At the time of writing, we have little knowledge of which genes will be involved in the relevant disease processes. DNA chips are not tailor made to a specific medical problem and study design. As we discussed before, array production is supposed to achieve an economy of scale and hence genes are present by design of the chip, not of the statistical experiment. It is appealing, therefore, to deal with the large collection of genes as if a sample from a larger population. This again leads naturally to exchangeability based modelling of the a priori structure.
3. Bayesian modelling through exchangeability leads to shrinkage estimation. This is generally a good idea if (patient-specific) prediction (of class membership, but similar comment would extend to other such problems) is at stake.
4. Microarray experiments are presently so small in sample size that common sense suggests we may not be able to discriminate genes with any great certainty. If future follow-up experiments become available to alleviate this problem, then present day results may be easily incorporated through the prior structure of the same models.
5. Indeed, it is one of the aims of genome research to develop more precise ideas of disease pathways. If this were to ensue, prior structure will again provide a natural formulation to incorporate such information, as it develops.

We present an analysis through hierarchical modelling first and an alternative through Bayesian variable selection subsequently.

2.3. *Hierarchical Logistic Regression*

We consider a sequence of four hierarchical model specifications, each of which is a generalization of the previous model. The basic model (1) is written as

$$\log(p/(1-p)) = \alpha + \boldsymbol{x}^t\boldsymbol{\beta}$$

where $\alpha \sim \mathrm{N}(0, 10^2)$ and $\boldsymbol{\beta}^t = (\beta_1, ..., \beta_p)$ with $\beta_i \sim \mathrm{N}(0, 0.1^2)$. We denote the sequence of gene responses generated from any DNA chip as $\boldsymbol{x}^t = (x_1, ..., x_p)$ and p is the probability that an array with observed expression $\boldsymbol{x}$ is an AML case. In practice, this model will have little use as it requires specification of the a priori variation. We present it here partly as a way to evaluate stability of the model estimates (in comparison to the other subsequent models). We may now expand the model to that of "logistic ridge regression" (model 2) by allowing the between-gene variation to be estimated from the data:

$$\beta_i \sim \mathrm{N}(0, \sigma^2)$$

and $\sigma^2 \sim \mathrm{Ig}(1e-3, 1e-3)$. In most applications this model will suffice, particularly with small sample size and will provide an estimate of the between-gene variation of regression coefficients.

Modelling the *a priori* structure may also be of particular interest from an applied point of view as we may speculate that genes differ in mean response and particularly, in variation. Bayesian hierarchical modelling easily allows expanding the model to accommodate such differences between genes. We, therefore, generalize by first allowing variability in the prior mean (model 3):

$$\beta_i \sim \mathrm{N}(\mu, \sigma^2)$$

and $\mu \sim \mathrm{N}(0, 10^2)$, $\sigma^2 \sim \mathrm{Ig}(1e-3, 1e-3)$. This model is of little interest (except again as a way to check stability of model estimates) as we already incorporate random variability of an intercept term in the previous model. Hence, we generalize to allow gene-specific variation (model 4):

$$\beta_i \sim \mathrm{N}(\mu_i, \sigma_i^2)$$

and $\mu_i \sim \mathrm{N}(0, 10^2)$, $\sigma_i^2 \sim \mathrm{Ig}(1e-3, 1e-3)$ (similar ideas on assuming gene-specific variation in the microarray context can be found in work by West *et al.*, 2002, and Ibrahim *et al.*, 2002).

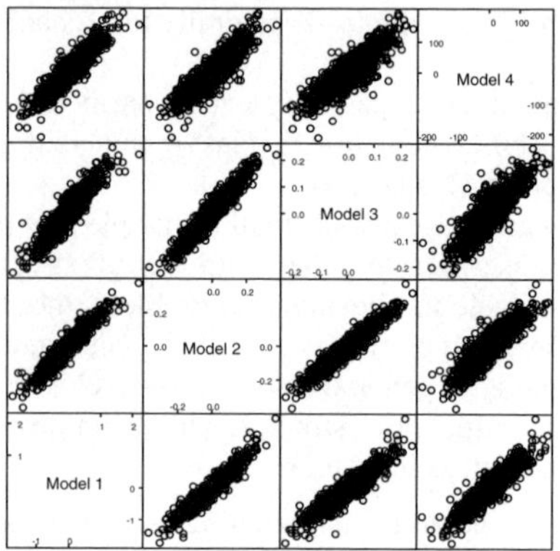

Figure 1. *Scatterplot matrix of posterior mean β coefficients for the four hierarchical logistic regression models. (All estimates are multiplied by a factor 1e4.)*

The above four models were estimated through full (ARS) Gibbs sampling. The predictor data were mean-centered and scaled first. For each model, 2000 burn-in samples were generated first and then discarded, following which 10000 more simulations were drawn. Posterior summaries were calculated by keeping track of means and standard deviations through an updating rule, thus avoiding the need to store vast matrices of simulated data. Following these 12000 samples, we always reinitialized the updating rule (but not the Gibbs chain) and sampling was continued for a further 4000 draws in order to check for stability of estimates, in comparison with results based on estimates from the previous 10,000 simulations. For the fourth model, sampling was continued until 20,000 simulations were available in total, following the initial 2000 burn-in samples and this sampling procedure was repeated from two distinct starting points. The first of these put all starting values for parameters β_i and μ_i equal to zero, with all σ_i equal to one. The second starting from a random draw from independent standard normal distributions for parameters β_i and μ_i and draws from the uniform distribution $\mathrm{Un}(0, 10)$ for parameters σ_i. Comparison of results from these calculations, both within and between models could uncover no evidence to demonstrate lack of convergence.

Figure 1 shows a scatterplot matrix of the posterior mean β regression coefficients across the models. All models roughly calibrate the same regression parameters, keeping the slightly differing amounts of shrinkage in mind. The between-gene variability was generally found to be extremely small. For example, we found that $\overline{\sigma}$=0.024 (median = 0.021, s.d. = 0.016) for the logistic ridge regression model. A density estimate based on simulated posterior standard deviations σ for this model (Figure 4) is unimodal and strictly increasing at zero with 95% HPD interval ranging from 0.0035 to 0.0544. Results from other models are similar (e.g. $\overline{\sigma}$=0.018

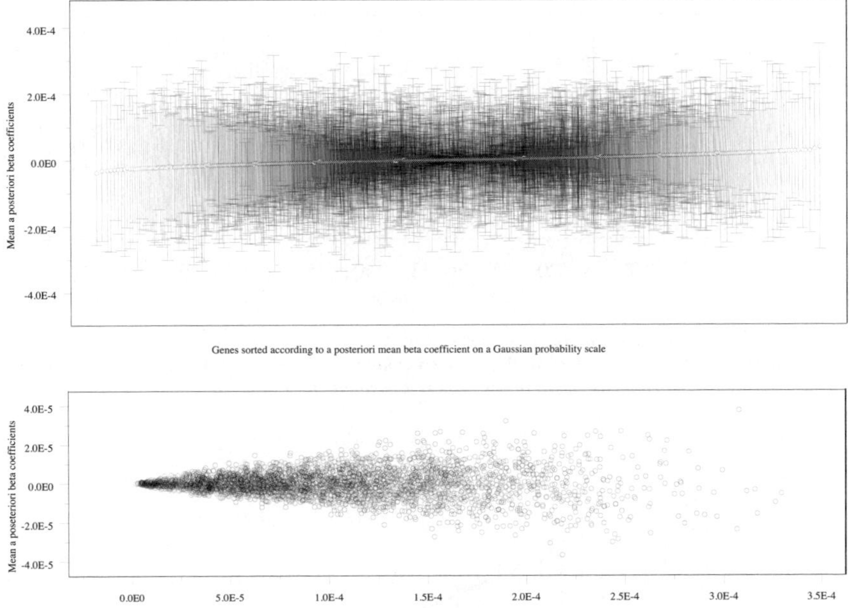

Figure 2. *Posterior mean of β coefficients for the logistic ridge regression model (second model). The top picture shows error bars representing the posterior standard deviation, sorts the genes in increasing order of magnitude of the associated posterior mean regression parameter and plotted using a Gaussian probability scale. The bottom picture plots posterior mean regression coefficients versus associated standard deviations.*

(s.d. = 0.01) for the third model). The small between-gene variability implies that genes will barely be rankable with respect to calculated regression coefficients. This is confirmed by plotting mean gene regression coefficients versus standard deviations or plotting largest/smallest gene regression coefficients with standard deviations (see Figure 2, which plots results for all genes, while using a normal probability axis to improve readability for the most extreme effects in the top plot). Other versions of the above plots (such as plotting scaled regression coefficients defined as mean effect/standard deviation of effect) lead to the same conclusions. Obviously, this may create an inferential headache for the medical profession, but merely adequately reflects the consequences of a combination of high dimensionality and small sample size for these data.

With respect to classification, all models classify most samples correctly (cutoff at 50% and based on the logistic regression prediction equation using the posterior mean of the regression coefficients), with the exception of sample 68. Group separation was particularly good for the fourth model which classified all samples correctly and with a high degree of confidence (not shown). Figure 3 shows boxplots of the posterior densities of classification probabilities across samples for the logistic ridge regression model. Separation is particularly good for the first group and less so for the last, which is again likely due to the differing sample sizes.

We should note that the good classification performance of the model(s) is not in contradiction to the problems in finding specific genes which can separate groups, as apparent in Figure 2. In particular, we should not use these results to naively infer that such genes do not exist as this would clearly confuse gene-specific hypotheses with a general hypothesis as to the

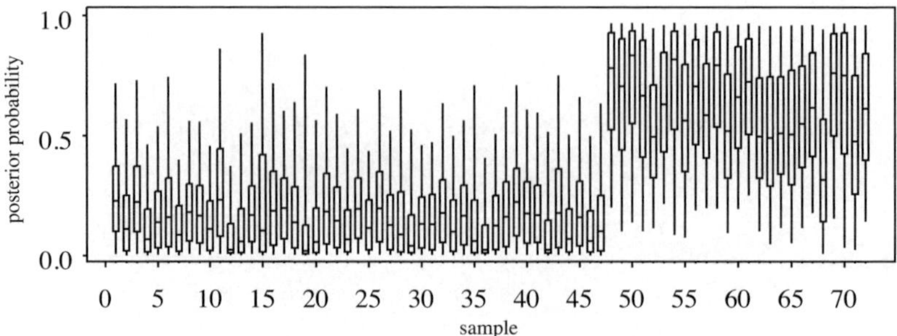

Figure 3. *Boxplots representing the posterior densities of fitted class probabilities (for logistic ridge regression). The first 47 samples are ALL, the remaining 25 samples are AML. Only sample 68 is misclassified, using a predictor based on the posterior mean of the β coefficients.*

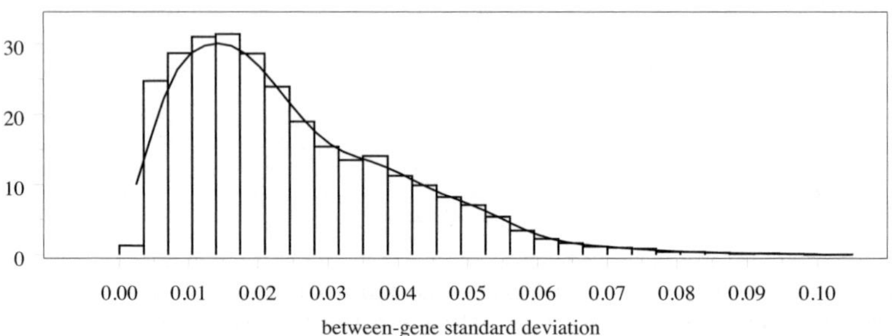

Figure 4. *Histogram of simulated posterior between-gene standard deviations σ for the logistic ridge regression model, with superimposed density estimate.*

presence of any separating gene in the complete set. Instead, a general hypothesis as to the presence of "discriminating genes" could be based on an investigation of the posterior density of σ in the ridge regression model (Figure 4), for example. Because of the unimodal nature of this density, there does appear to be some evidence of the presence of such genes. However, problems in identifying specific genes result from the flatness of the likelihood, caused by the high dimensionality and small sample size. As a consequence, the regression model tends to average across genes, as there are many competing models with distinct parameter estimates which can separate the classes well, which in turn and because of the high dimension causes the small scales apparent in Figures 1 and 2.

2.4. *Stochastic Search Gene Selection*

It is unlikely that problems in the identification of specific genes will be avoided as long as sample sizes of microarray experiments remain small. As an alternative to the above analysis, we may explore a Bayesian gene (variable) selection method within the logistic model (see also George and McCullogh, 1993). The model may be written as

$$\log(p/(1-p)) = \alpha + \boldsymbol{x}^t\boldsymbol{\beta}, \qquad \alpha \sim \mathrm{N}(0, 10^2)$$

and $\boldsymbol{\beta}^t = (\beta_1, ..., \beta_p)$ with

$$P(\beta_i \mid \gamma_i) = (1-\gamma_i)\mathrm{N}(0, \tau^2) + \gamma_i \mathrm{N}(0, c^2\tau^2)$$

and $\tau = 0.05$, $c = 100$ and $P(\gamma_i) = 0.05$.

We could have chosen to estimate the fraction of genes involved in the disease process itself as well, which would have been an interesting exercise. However, for the sake of the argument and to simplify the implementation, we have chosen to keep the fraction fixed and specified *a priori*. Results may then be thought of as "identifying the top 5% of genes with greatest impact."

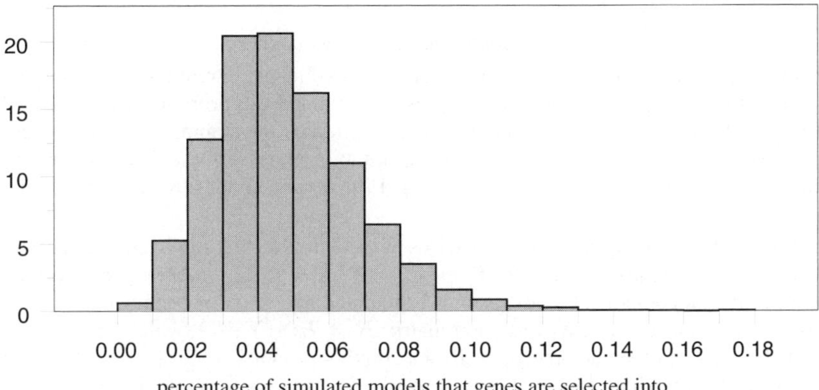

Figure 5. *Histogram which represents the distribution of the number of models that genes are typically selected into, expressed as a percentage of the number of logistic regression models simulated, using stochastic search gene selection.*

The number of genes selected to the model lies between 150 and 200, which is consistent with $P(\gamma_i) = 0.05$. All genes are selected at least once to the model. There are no genes which are present in at least 50% of all simulated models. The maximum number of times a gene is selected out of all models is 18%, for a single gene only. Hence, we get similar results as for the previous analyses: in effect we can not discriminate well between the genes. This in contrast to the classification results which are effectively identical to those above. Figure 5 shows a histogram of the number of models genes are selected to, expressed as percentages of the number of model generated.

3. APPROXIMATE METHODS

It is of interest to compare the above methods with some related approximate methods. We consider penalized likelihood estimation and continuum least squares regression in particular.

3.1. *Penalized Likelihood*

We have already given arguments in favor of a fully Bayesian treatment of the logistic regression approach to the analysis of the microarray data we are considering here, which go beyond considerations relating only to the ill conditioned nature of the data. Nevertheless, smoothing techniques based on penalized likelihood could also be applied in this context, which is popular in certain applied fields and could give results similar to those presented here for the logistic ridge regression model. More importantly, the computational costs of such procedures are a fraction of what is needed to evaluate the fully Bayesian models we presented above, which is a very important argument which can promote use of such "approximate methods" in practice. We will compare results from our second hierarchical model to those based on a penalized estimation for the logistic model which optimizes the likelihood

$$L - \lambda \sum_j \beta_j^2/2$$

and where L is the Bernoulli likelihood (see Eilers *et al.*, 2001). Similar methodology of an empirical Bayes nature is also discussed in papers by Fearn *et al.* (1999) and Khan and Raftery (1996).

3.2. *Continuum Least Squares*

Continuum regression is a regression-based approach to dimension-reduction which derives an oblique rotation of the principal component axes. The method was proposed and developed first by Stone and Brooks (1990, 1992) and has been inspired by developments in the related field of chemometrics. It was quickly shown that continuum regression yields estimators closely related to those from ridge regression (Sundberg, 1993) (same applies to the penalized method of course).

Let $\boldsymbol{X}$ and $\boldsymbol{Y}$ be the mean-centered matrices of predictors (n by p) and outcome measures (n by 1), respectively. We also put $\boldsymbol{c} = \boldsymbol{X}^t\boldsymbol{Y}$ and $\boldsymbol{C} = \boldsymbol{X}^t\boldsymbol{X} = \boldsymbol{Q}\Lambda\boldsymbol{Q}^t$, where $\boldsymbol{Q}$ is the orthogonal matrix of principal component loadings and Λ the diagonal matrix of component eigenvalues. Continuum least squares derives a rotation matrix $\boldsymbol{R} = \boldsymbol{R}(\alpha; \boldsymbol{Y}, \boldsymbol{X})$, which depends on both the predictor data and the outcome, as well as a "mixing proportion" $0 \leq \alpha \leq 1$. The rotation is applied to the principal component axes to give rotated components $\boldsymbol{V} = \boldsymbol{XQR}$. We write $\boldsymbol{A} = \boldsymbol{QR} = (\boldsymbol{a}_1, .., \boldsymbol{a}_r)$ (r denotes the rank). The rotation directions are now derived sequentially to optimize the criterion

$$T = \frac{(\boldsymbol{a}^t\boldsymbol{c})^2}{\|Y\|^2\boldsymbol{a}^t\boldsymbol{C}\boldsymbol{a}} \cdot (\boldsymbol{a}^t\boldsymbol{C}\boldsymbol{a})^{\alpha/(1-\alpha)} \propto (\boldsymbol{a}^t\boldsymbol{c})^2(\boldsymbol{a}^t\boldsymbol{C}\boldsymbol{a})^{(\alpha/(1-\alpha)-1)}$$

Clearly, some interesting special cases exist, such as $\alpha = 0$, which implies we optimize correlation (subject to singular value decomposition and thus Moore–Penrose inverse), which gives an "ordinary" least squares solution. Putting $\alpha = 1$ means we only optimize for the variance of the predictors and thus gives principal components. The solution for $\alpha = 0.5$ is well-known among chemometricians as "partial least squares" and optimizes a 50/50 mix of variance and correlation. Rotated components are now entered "top-down" to the regression equation and both the number of component regressors and the optimal α are determined through cross-validation (for example).

3.3. *Comparison*

We have used cross-validation to apply both the penalized likelihood and continuum regression approaches to the MIT ALL/AML data. For continuum regression, we have used regression of class indicators as is often done in the literature. Both give roughly the same answer but shrink less than the fully Bayesian methods ($\lambda = 0.1$ for penalized likelihood, which is equivalent to $\sigma \approx 3$). The difference in shrinkage is also not uniform between the fully Bayesian and approximate methods (Figure 6). This is quite likely due to the order of magnitude difference in shrinkage between the methods and the fact that only the fully Bayesian calculation can average across the distribution of between-gene variation. It is also a crucial problem with these methods that they do not give estimates of the variability or uncertainty of either the estimated coefficients or the calibrated class probabilities. Furthermore, both methods are essentially algorithmic in nature. This applies for continuum regression in particular for which no distributional properties are known.

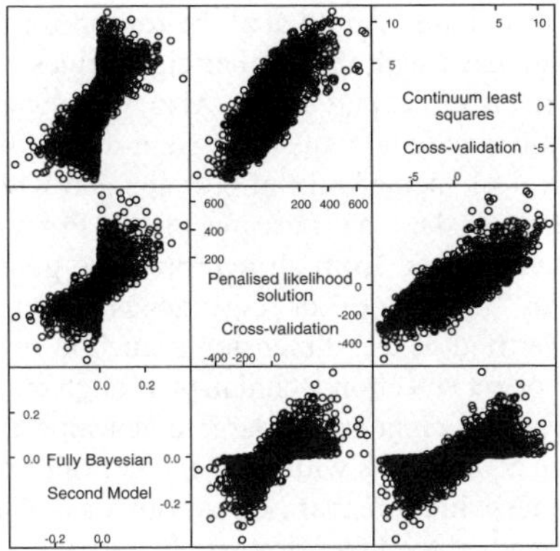

Figure 6. *Scatterplot matrix of posterior mean β coefficients for the "logistic ridge regression" model versus fitted regression parameters in approximate penalized likelihood and continuum least squares solutions (cross-validation based). (All estimates are multiplied by a factor 1e4.)*

4. DISCUSSION

4.1. *Dimension Reduction*

With such extreme dimensionality as encountered in microarray problems, the intuition of many statisticians will be to apply some form of dimension reduction, through principal components or singular value decomposition, for example. This is typically done prior to modelling and the model is then specified with respect to the reduced space. Such a "procedure" is appealing in this context, but entails some unexpected and subtle problems for microarray data. The first point to make is that there should be a substantial amount of correlation in the data if dimension reduction is to be a viable approach to the analysis. Figure 7 shows a histogram of all pooled within-group correlations for the MIT microarray data, based on all available gene pairs (reproduced from Mertens, 2002). Clearly the within-group correlations are typically small, which is what we would expect when scanning across a large collection of genes. In fact, the number of correlations in excess of 50% in absolute value is smaller than 4% of all calculable correlations.

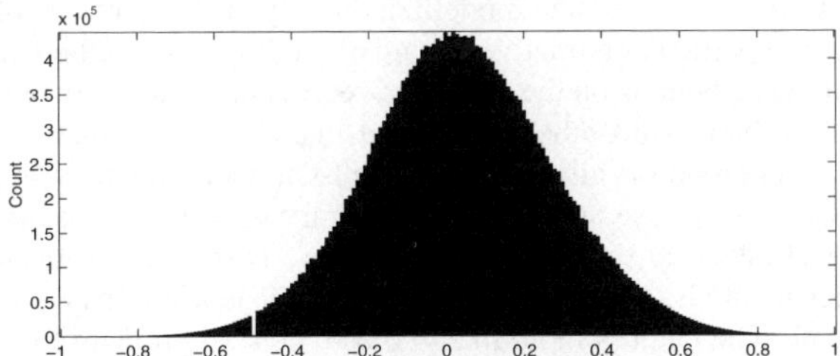

Figure 7. *Histogram of between-gene correlations for the leukemia microarray data, based on the pooled within-group correlations calculated from all 72 observations and considering all distinct gene pairs.*

Clearly, this is a rather crude approach to explore the nature of the singularity problem posed by these data. Nevertheless, it poses the question whether analysis based on initial naive dimension reduction techniques should at all be trusted for these data. When exploring singular value directions associated with the smaller eigenvalues, we should expect to quickly encounter directions which are in effect pure noise. Also, when using principal components for example, we have many choices as to how the dimension reduction is carried out: do we only decompose within-group error or "some" mix of between and within-group error (as singular value decomposition on the whole data matrix seems to do). From a predictive perspective this is unlikely to matter much. However, from an interpretative point of view, any terminology referring to the components as "supergenes" or "eigengenes" should be treated with the greatest suspicion. Frequentists have tried to use dimension reduction in regression-type approaches by applying top-down component selection techniques through cross-validation (for example), which gives an implicit greater weight to the large-eigenvalue eigenvectors. Interpretation problems will still arise when components with smaller eigenvalue enter the prediction equation in such a rigid selection scheme, which are themselves poorly associated with the outcome. West *et al.* (2000) have largely remedied the latter problem in a very interesting development which transfers the component-regression problem to the Bayesian setting. However, the singular value decomposition itself is still fixed *a priori* in this analysis, which entails all the problems of poor definition of the smaller-eigenvalue components and the associated difficulties of interpretation, particularly for such data as we have for microarrays. Furthermore, dimension reduction should ideally also take into account the context within which the decomposition will be used, as is done by continuum least squares, for example. Unfortunately, the latter method is plagued by lack of any model with respect to which the calculations can be interpreted. Presumably, a truly fully Bayesian analysis of the problem, which incorporates the dimension reduction step itself into the Bayesian paradigm could remedy the problem, but this would likely be extremely costly from a computational point of view. Other high-dimensional applications such as found in near infrared spectroscopy suffer (somewhat) less from the above mentioned problems as the correlations across the spectrum are of an entirely different nature, which implies that components are themselves better defined and interpretable.

4.2. *Hierarchical Logistic Regression Modelling*

Clearly, there will be plenty of development on the statistical modelling of microarray data, or indeed other such high-dimensional data types which the genome project is undoubtedly going to produce in the future. In the meantime, this paper has demonstrated how a fully Bayesian implementation of logistic regression makes the method suited to the analysis of the data and leads to inferences which are easily and naturally interpreted by practicians. The methodology allows for inferences and modelling directly on the genes involved, both *a priori* and *a posteriori*. It is a particular attractive feature of the Bayesian method that it allows for the calculation of confidence bounds on the estimated regression parameters or class probabilities, which is a crucial problem with other "penalized likelihood" methods which is difficult to overcome. Logistic regression is also relatively well-known in the medical profession, which should make it more easy to use and explain. Contrary to expectations, we found the Gibbs sampler to work well, even in these large dimensions. Discriminant results are comparable to those from other methods. Gene ranking was complicated however, as the between-gene variability was small. The latter is more likely to be a general problem for microarray studies however, which also relates to sample size. This should, therefore, not be taken as a deficiency of the Bayes approach. Indeed the opposite is true, as it warns us to be careful about ascribing undue interpretation and meaning to individual genes, even when discrimination is easy.

ACKNOWLEDGEMENTS

All computation has been carried out in Matlab (Version 6, release 12). Thanks to Paul Eilers for providing code for the penalized logistic regression model.

REFERENCES

Eilers, P. H. C., Boer, J. M., van Ommen, G. J. and van Houwelingen, H. C. (2001). Classification of microarray data with penalized logistic regression. *Microarrays: Optical Technologies and Informatics* (M. L. Bittner *et al.* eds). San Jose, CA: SPIE, 187–198

Fearn, T., Brown, P. J. and Haque, M. S. (1999). Logistic discrimination with many variables. *Rev. Acad. Ciencias Madrid* **93**, 337–342.

George, E. I. and McCullogh, R. E. (1993). Variable selection via Gibbs sampling. *J. Am. Statist. Ass.* **85**, 398–409.

Golub, T. R., Slonim, D. K., Tamayo, P. *et al.* (1999). Molecular classification of cancer: class discovery and class prediction by gene expression monitoring. *Science* **286**, 531–537.

Ibrahim, J. G., Chen, M.-H. and Gray, R. J. (2002). Bayesian models for gene expression with DNA microarray data. *J. Am. Statist. Ass.* **97**, 88–99.

Kahn, M. J. and Raftery, A. E. (1996). Discharge rates of medicare stroke patients to skilled nursing facilities: Bayesian logistic regression with unobserved heterogeneity. *J. Am. Statist. Ass.* **91**, 29–41.

Mertens, B. J. A. (2002). Microarrays, pattern recognition and exploratory data analysis. *Statist. Med.* (to appear).

Stone, M. and Brooks, R. J. (1990). Continuum regression: cross-validated sequentially constructed prediction embracing ordinary least squares, partial least squares and principal components regression. *J. R. Statist. Soc. B* **52**, 237–269 (with discussion).

Stone, M. and Brooks, R. J. (1992). Corrigendum to Stone and Brooks, 1990. *J. R. Statist. Soc. B* **54**, 906–907.

Sundberg, R. (1993) Continuum regression and ridge regression. *J. R. Statist. Soc. B* **55**, 653–659.

West, M., Nevins, J. R. *et al.* (2000). DNA Microarray data analysis and regression modelling for genetic expression profiling. *Tech. Rep.*, Duke University, USA.

BAYESIAN STATISTICS 7, pp. 619–629
J. M. Bernardo, M. J. Bayarri, J. O. Berger, A. P. Dawid,
D. Heckerman, A. F. M. Smith and M. West (Eds.)

Density Modeling and Clustering Using Dirichlet Diffusion Trees

RADFORD M. NEAL
University of Toronto, Canada
radford@stat.utoronto.ca

SUMMARY

I introduce a family of prior distributions over multivariate distributions, based on the use of a "Dirichlet diffusion tree" to generate exchangeable data sets. These priors can be viewed as generalizations of Dirichlet processes and of Dirichlet process mixtures, but unlike simple mixtures, they can capture the hierarchical structure present in many distributions, by means of the latent diffusion tree underlying the data. This latent tree also provides a hierarchical clustering of the data, which, unlike *ad hoc* clustering methods, comes with probabilistic indications of uncertainty. The relevance of each variable to the clustering can also be determined. Although Dirichlet diffusion trees are defined in terms of a continuous-time process, posterior inference involves only finite-dimensional quantities, allowing computation to be performed by reasonably efficient Markov chain Monte Carlo methods. The methods are demonstrated on problems of modelling a two-dimensional density and of clustering gene expression data.

Keywords: DENSITY MODELLING; HIERARCHICAL CLUSTERING; PRIOR DISTRIBUTIONS.

1. INTRODUCTION

Unknown distributions are encountered when estimating the density of observed data and when modelling the distribution of random effects or other latent variables. Exploratory data analysis can also be viewed in terms of finding features of the data, such as clusters, that are useful in modelling its distribution. A Bayesian model involving an unknown distribution requires a prior distribution over distributions. For such a model to be useful in practice, the prior must be an adequate approximation to our actual prior beliefs about the unknown distribution, and it must be possible to compute the predictive distribution for new data with reasonable efficiency.

The Dirichlet process (Ferguson, 1973) is a simple and computationally tractable prior for an unknown distribution. However, it produces distributions that are discrete with probability one, making it unsuitable for density modelling. This can be avoided by convolving the distribution with some continuous kernel, or more generally, by using a Dirichlet process to define a mixture distribution with infinitely many components, of some simple parametric form (Ferguson, 1983; Antoniak, 1974;). Such Dirichlet process mixture models are not always ideal, however, because they use a prior distribution in which the parameters of one mixture component are independent of the parameters of other components. For many problems, we would expect instead that the components will be hierarchically organized, in ways analogous to the hierarchical grouping of organisms belonging to various species. Even if no obvious hierarchy is present, modelling a complex distribution by a mixture of simple distributions will require that each mode of the distribution be modelled using many of these simple mixture components, which will have similar parameters, and therefore form clusters themselves. Since Dirichlet process mixture models do not capture this hierarchical structure, inference using these models will be inefficient—it will take more data than it should to force the model to create appropriate components, since the model does not "know" that these components will likely resemble other components.

The prior distribution over distributions I discuss here can be seen as a hierarchical generalization of Dirichlet process mixture models. The latent structure underlying the data for this model is what I call a "Dirichlet diffusion tree," whose terminal nodes (leaves) are the data points, and whose non-terminal nodes represent groupings of data points in a hierarchy. This latent tree structure generalizes the one-level grouping of data points into clusters that underlies a Dirichlet process mixture model. For some problems, this structure may be merely a device for obtaining a more suitable prior distribution over distributions. For exploratory applications, however, the latent diffusion tree structure may be of intrinsic interest, as it provides a hierarchical clustering of the data.

Most commonly used hierarchical clustering methods are not based on any probabilistic model of the data. Williams (2000) discusses hierarchical Bayesian mixture models that are similar in objective to the models described here, but which have a finite (though unknown) number of components. Polya trees are another generalization of the Dirichlet process that produces distributions that have hierarchical structure, and which can be continuous. However, Polya tree priors have an unfortunate dependence on an arbitrary set of division points, at which discontinuities in the density functions occur. Walker *et al.* (1999) review these and other related priors over distributions, including the "reinforced random walks" of Coppersmith and Diaconis, which resemble the diffusion trees discussed here, but differ in crucial respects.

Inference involving Dirichlet diffusion trees will not be as computationally easy as inference for simple Dirichlet process or Polya tree models, or as simple hierarchical clustering. Markov chain methods are needed to sample from the posterior distribution over trees. I describe here some techniques for constructing suitable Markov chain samplers, and demonstrate that they work well in a simple two-dimensional density modelling problem, and on a more difficult problem of clustering tumor cells based on gene expression data.

2. THE DIRICHLET DIFFUSION TREE PRIOR

I will define the Dirichlet diffusion tree prior over distributions by giving a procedure for randomly generating a data set of n points, each a vector of p real numbers, in which the data points are drawn independently from a common distribution drawn from the prior. The procedure generates these random data sets one point at time, with each point being drawn from its conditional distribution given the previously generated points, in a manner analogous to the "Polya urn" procedure for defining a Dirichlet process prior (Blackwell and MacQueen, 1973). Such a procedure will be consistent with the data points being independently drawn from some unknown distribution as long as the distribution over data sets produced by the procedure is exchangeable, since such an exchangeable prior on data sets is equivalent to a prior over distributions by de Finetti's Representation Theorem (Bernardo and Smith, 1994, Section 4.3).

2.1. *Generation of Data Points Using a Diffusion Tree*

Each point in a data set drawn from the Dirichlet diffusion tree prior is generated by following a path of a diffusion process. Paths to different data points are linked in a tree structure, according to when each path diverges from previous paths. Generation of these paths will be described below by a sequential procedure, in which paths are generated for each data point in turn, but we will see in Section 2.3 that the ordering of the data points is in fact immaterial.

The first data point is generated by a Gaussian diffusion process (*i.e.*, Brownian motion), beginning at some origin, which I will here fix at zero. From this origin, the diffusion process operates for a predetermined length of time, which without loss of generality can be fixed at one. If at time t, the process has reached the point $X_1(t)$, the point reached an infinitesimal time, dt, later will be $X_1(t+dt) = X_1(t) + N_1(t)$, where $N_1(t)$ is a Gaussian random variable with mean zero and covariance $\sigma^2 I\, dt$, were σ^2 is a parameter of the diffusion process. The $N_1(t)$ at different times are independent. The end point of the path, $X_1(1)$, is the sum of the infinitesimal increments $N_1(t)$, and is easily seen to have a Gaussian distribution with mean zero and covariance $\sigma^2 I$. This end point is the first point in the data set.

The second point in the data set is also generated by following a path from the origin. This path, $X_2(t)$, initially follows the path leading to the first point. However, the two paths will diverge at some time, T_d, after which the path to the second point is independent of the remainder of the first path. In other words, the infinitesimal increments for the second path, $N_2(t)$, are equal to the corresponding increments for the first path, $N_1(t)$, for $t < T_d$, but thereafter, $N_2(t)$ and $N_1(t)$ are independent. The distribution of the divergence time, T_d, can be expressed in terms of a "divergence function", $a(t)$. At each time t before divergence occurs, the probability that the path to the second data point will diverge from the path to the first data point within the next infinitesimal time period of duration dt is given by $a(t)\, dt$.

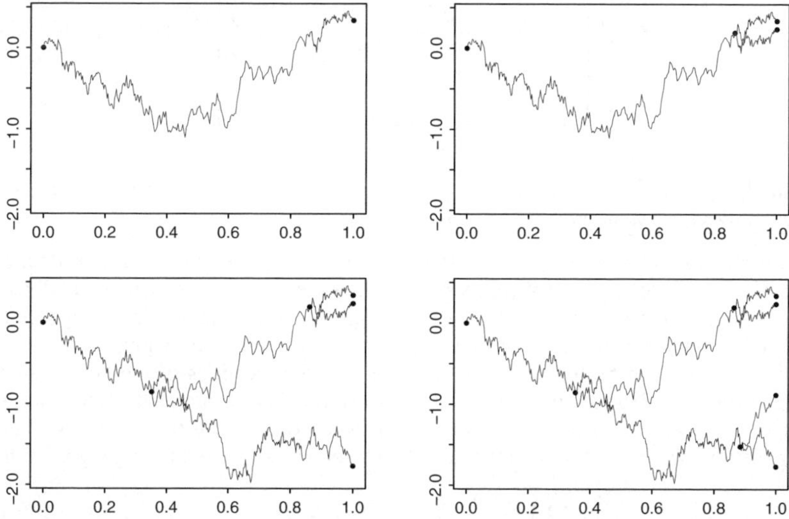

Figure 1. *Generation of a data set of four real numbers from the Dirichlet diffusion tree prior with* $\sigma = 1$ *and* $a(t) = 1/(1-t)$. *Diffusion time is on the horizontal axis, data values on the vertical axis. The upper-left plot shows the path to the first data point* (0.34). *The upper-right plot shows the path to the second data point* (0.24) *diverging from the first path at* $t = 0.86$. *In the lower-left plot, the path to the third data point* (−1.77) *diverges from the first two paths at* $t = 0.35$. *In the lower-right plot, the path to the fourth data point* (−0.87) *follows the path to the third data point until* $t = 0.88$.

In general, the ith point in the data set is obtained by following a path from the origin that initially coincides with the path to the previous $i-1$ data points. If the new path has not diverged at a time when paths to past data points diverged, the new path chooses between these past paths with probabilities proportional to the numbers of past paths that went each way—this reinforcement of previous events is characteristic of "Polya urn" schemes. If at time t, the new path is following a path traversed by m previous paths, the probability that it will diverge from this path within an infinitesimal interval of duration dt is $a(t)dt/m$—note the division by m, which is another aspect of reinforcement of past events. Once divergence occurs, the new path moves independently of previous paths.

Figure 1 illustrates the diffusion tree process for a data set of $n = 4$ data points, each a single real number (*i.e.*, $p = 1$).

2.2. *Probability of Generating a Given Tree of Data Points*

The probability of obtaining a given data set along with its underlying tree can be expressed more easily if the details of the continuous paths taken are suppressed, leaving only the structure of the tree (that is, how data points are hierarchically grouped), the times at which paths diverged, and the locations of these divergence points and of the final data points. Figure 2 shows this less-detailed representation of the example shown in Figure 1.

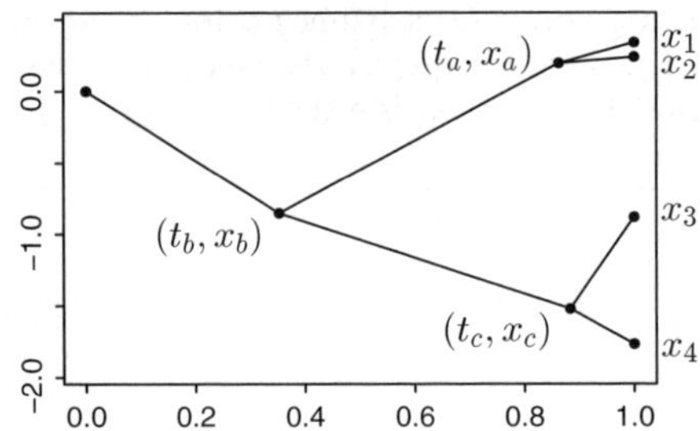

Figure 2. *The data x_1, x_2, x_3, and x_4 generated above and its underlying tree, with details of the paths suppressed except at the divergence points a, b, and c.*

The probability of obtaining a given tree and data set can be written as a product of two factors. The tree factor is the probability of obtaining the given tree structure and divergence times. The data factor is the probability of obtaining the given locations for divergence points and final data points, when the tree structure and divergence times are as given.

The tree factor can be found without reference to the locations of the points, since the divergence function depends only on t. The probability that a new path following a path previously traversed m times will not diverge between time s and time t can be found by dividing the time from s to t into k intervals of duration $(t-s)/k$, and letting k go to infinity:

$$P(\text{no divergence}) = \lim_{k\to\infty} \prod_{i=0}^{k-1} \Big(1 - a(s+i(t-s)/k)[(t-s)/k]/m\Big) = e^{(A(s)-A(t))/m}$$

Here, $A(t) = \int_0^t a(u)\,du$. Using this formula, the probability density for obtaining the tree structure and divergence times for the example in Figure 2 is found to be as follows:

$$e^{-A(t_a)}\, a(t_a) \times e^{-A(t_b)/2}\, (a(t_b)/2) \times e^{-A(t_b)/3}\, (1/3)\, e^{A(t_b)-A(t_c)}\, a(t_c)$$

Given the tree structure and divergence times, the data factor is just a product of Gaussian densities, since the distribution for the location of a point that has diffused for time d starting from location x is Gaussian with mean x and covariance $\sigma^2 Id$. Letting $\mathrm{N}(x \mid \mu, \sigma^2)$ be the Gaussian probability density function, we can write the data factor for the example in Figure 2 as follows:

$$\mathrm{N}(x_b \mid 0,\, \sigma^2 t_b) \times \mathrm{N}(x_a \mid x_b,\, \sigma^2(t_a-t_b)) \times \mathrm{N}(x_1 \mid x_a,\, \sigma^2(1-t_a)) \times \mathrm{N}(x_2 \mid x_a,\, \sigma^2(1-t_a))$$
$$\times\, \mathrm{N}(x_c \mid x_b,\, \sigma^2(t_c-t_b)) \times \mathrm{N}(x_3 \mid x_c,\, \sigma^2(1-t_c)) \times \mathrm{N}(x_4 \mid x_c,\, \sigma^2(1-t_c))$$

2.3. *Proof of Exchangeability*

To show that the procedure for generating a data set using a Dirichlet diffusion tree defines a valid prior over distributions, we must show that the probability density for a data set does not change when the order of the data points is permuted—*i.e.*, that the prior is "exchangeable". I will show this by proving the stronger property that the probability density for producing a data set along with its underlying tree structure and the times and locations of the divergence points is the same for any ordering of data points. Exchangeability follows from this by summing over all possible tree structures and integrating over times and locations of divergences.

The probability density for a data set along with an underlying tree can be written as a product of factors, each pertaining to one segment of the tree. (For example, the tree in Figure 2 has seven segments.) For each segment, $(t_u, x_u) - (t_v, x_v)$, there is a factor in the probability density that corresponds to the density for the diffusion process starting at x_u to move to x_v in time $t_v - t_u$, which is $\mathrm{N}(x_v \,|\, x_u,\, \sigma^2 I(t_v - t_u))$. The product of these factors is the overall data factor in the density. Since the set of segments making up the tree does not depend on the order of the data points, neither will this data factor.

A segment of the tree traversed by more than one path will also be associated with a factor in the overall probability density pertaining to the lack of divergence of these paths before the end of the segment. If such a segment ends before $t = 1$, there will be another a factor for the probability density for one path to diverge at the end of this segment, along with factors for the probabilities of later paths taking the branches they did. The product of such factors for all segments is the overall factor relating to the tree structure. We need to show that this factor does not actually depend on the ordering of the data points, even though it is expressed in an order-dependent way above.

Consider a segment, $(t_u, x_u) - (t_v, x_v)$, that was traversed by $m > 1$ paths. The probability that the $m-1$ paths after the first do not diverge before t_v is $\prod_{i=1}^{m-1} e^{(A(t_u) - A(t_v))/i}$, which does not depend on the order of the data points. If $t_v = 1$, this is the whole factor for this segment. Otherwise, suppose that the first $i-1$ paths do not diverge at t_v, but that path i does diverge at t_v. (Note that i will be at least two, and that some path must diverge at t_v, as otherwise the segment would not end at that time.) The probability density for this divergence is $a(t_v)/(i-1)$. Subsequent paths take one or the other branch at this divergence point, with probabilities proportional to the number that have gone each way previously. Suppose that n_1 paths in total go the way of the first path, and n_2 go the way of path i. (Note that $n_1 \geq i-1$, and $n_1 + n_2 = m$.) The probability that path j (with $j > i$) goes the way of the first path will be $c_1/(j-1)$, where c_1 is the number of paths that went that way before, which will vary from $i-1$ to $n_1 - 1$. The probability that path j goes the other way will be $c_2/(j-1)$, where c_2 is the number of paths that went the other way before, which will vary from 1 to $n_2 - 1$. The product of these branching probabilities for all $j > i$ will be

$$\prod_{c_1=i-1}^{n_1-1} c_1 \cdot \prod_{c_2=1}^{n_2-1} c_2 \bigg/ \prod_{j=i+1}^{m} (j-1) = \frac{(n_1-1)!}{(i-2)!} \cdot (n_2-1)! \cdot \frac{(i-1)!}{(m-1)!}$$

$$= (i-1) \cdot \frac{(n_1-1)!\,(n_2-1)!}{(m-1)!}\,.$$

When this is multiplied by the probability density of $a(t_v)/(i-1)$ for path i diverging at time t_v, the two factors of $i-1$ cancel, leaving a result that does not depend on i, and hence is independent of the order of the data points. This argument can be modified to handle the case where $a(t)$ has an infinite peak, allowing more than one path to diverge at the same time.

3. PROPERTIES OF DIRICHLET DIFFUSION TREE PRIORS

The properties of a Dirichlet diffusion tree prior vary with the choice of divergence function, $a(t)$. I will investigate these properties here by looking at data sets generated from these priors. I also investigate when the distributions produced are continuous and absolutely continuous. More details on the properties of Dirichlet diffusion tree priors are found in Neal (2001).

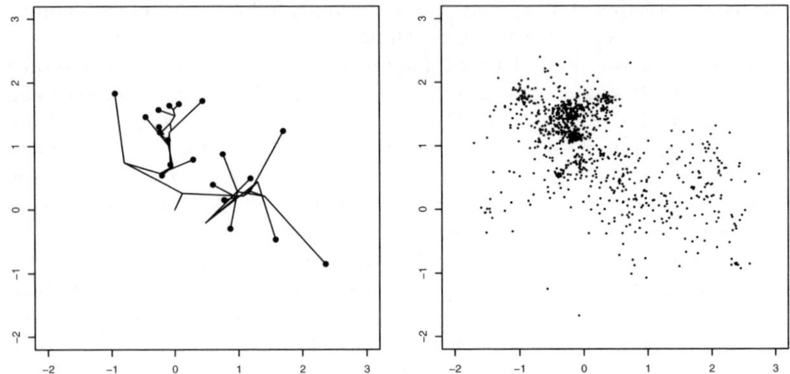

Figure 3. *Generation of a two-dimensional data set from the Dirichlet diffusion tree prior with* $\sigma = 1$ *and* $a(t) = 1/(1-t)$. *The plot on the left shows the first 20 data points generated, along with the underlying tree structure. The right plot shows 1000 data points obtained by continuing the procedure beyond these 20 points.*

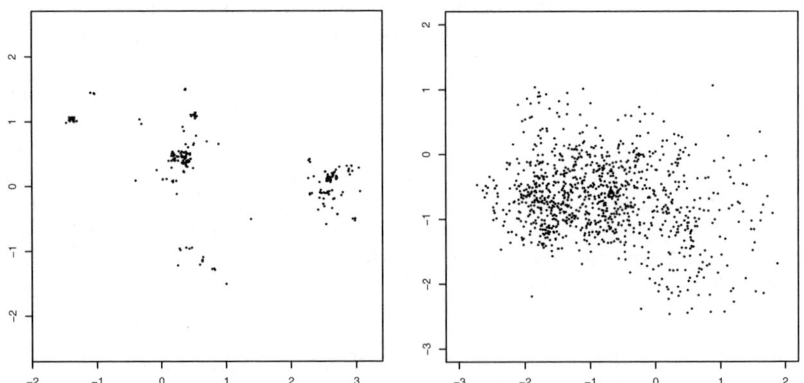

Figure 4. *Two data sets of 1000 points drawn from Dirichlet diffusion tree priors with* $\sigma = 1$. *For the data set on the left, the divergence function used was* $a(t) = (1/4)/(1-t)$. *For the data set on the right,* $a(t) = (3/2)/(1-t)$.

Divergence functions of the form $a(t) = c/(1-t)$ have integrals that diverge only logarithmically as $t \to 1$: $A(t) = \int_0^t a(u)\,du = -c\log(1-t)$. Distributions drawn from such a prior will be continuous—that is, the probability that two data points will be identical is zero.

These distributions will, however, show some degree of tight clustering, varying with c, due to the possibility that divergence will not occur until quite close to $t = 1$. Figure 3 shows the generation of a two-dimensional data set of 1000 points, drawn from this prior with $c = 1$. We see here the hierarchical structure that the Dirichlet diffusion tree prior can produce—there are not just multiple modes in this data, but further structure within each mode. This hierarchy is more obvious in the data set shown on the left of Figure 4, produced from the prior with $c = 1/4$, which has tighter clusters. As seen in the right of Figure 4, when $c = 3/2$, the distributions generated are smoother, and do not exhibit any obvious hierarchical structure.

Priors with different characteristics can be obtained using a divergence function of the form $a(t) = b + d/(1-t)^2$. This divergence function can produce well-separated clusters that exhibit a clear hierarchical structure, but with the points within each cluster being smoothly distributed, due to the rapid increase in $a(t)$ as t approaches one.

The distributions produced by a Dirichlet diffusion tree prior will be continuous (with probability one) when the divergence function, $a(t)$, is such that $A(t) = \int_0^1 a(t)\,dt$ is infinite. However, it does not follow that distributions drawn from such a prior will be absolutely continuous, which is what is required for them to have density functions. I have investigated empirically when Dirichlet diffusion tree priors produce absolutely continuous distributions by looking at distances to nearest neighbors in a large sample from the distribution. For a distribution with a continuous density function, the density in the region near where the two nearest neighbors of a data point are located will be approximately constant. From this one can derive that if r is the ratio of the distance to the nearest other data vector divided by the distance to the second-nearest other data vector, the distribution of r^p will be uniform over $(0, 1)$. The empirical distribution of r^p for a large sample, therefore, provides evidence of whether or not the distribution from which the sample came is absolutely continuous.

The empirical results I have obtained (Neal 2001) lead me to conjecture that Dirichlet diffusion tree priors for p-dimensional distributions with $a(t) = c/(1-t)$ produce distributions that are not absolutely continuous when $c < p/2$, but which are absolutely continuous when $c > p/2$. I conjecture that, in contrast, all Dirichlet diffusion tree priors with divergence functions of the form $a(t) = b + d/(1-t)^2$ with $d > 0$ produce absolutely continuous distributions.

4. MODELS BASED ON DIRICHLET DIFFUSION TREES

In practice, the smoothness and other characteristics of an unknown distribution will seldom be known exactly, and it will, therefore, be appropriate to give higher-level prior distributions to the parameters of the divergence function, allowing these characteristics to be inferred from the data. The variance of the diffusion process for a variable would also usually be a hyperparameter, which can adapt to the scale of that variable; different variables might well be given different variances.

If the data were observed with some amount of noise, or the data values were rounded, it would be appropriate to regard the Dirichlet diffusion tree as defining the prior for the distribution of the unrounded, noise-free values, not for the data actually observed. For some problems, a heavy-tailed noise distribution may be appropriate, to prevent outliers from having an undue effect on the clustering. When the data is categorical, unobserved latent values can be introduced that determine the probabilities of the observed data (*e.g.*, via a logistic model). The distribution of these vectors of latent values could then be given a Dirichlet diffusion tree prior, indirectly defining a prior for the joint distribution of the categorical values.

A model in which the amount of noise for each variable is controlled by a hyperparameter may be useful even when there actually is no appreciable noise in the measurements, since with such a model, some or all of the variation of some variables can be explained as "noise" that is unrelated to variation in other variables. If the posterior distribution for the noise hyperparameter for a variable is concentrated near a large value, the model will have effectively learned that this variable is not relevant to the clustering of data items.

5. MARKOV CHAIN SAMPLING FOR DIRICHLET DIFFUSION TREE MODELS

The finite representation of a Dirichlet diffusion tree shown in Figure 2 is used for Markov chain Monte Carlo computations. The state of the Markov chain used to sample from the posterior distribution will consist of at least the *structure* of the tree (the hierarchical organization of data points) and the *divergence times* for non-terminal nodes. Depending on the model and sampling method used, the Markov chain state may also include the *locations* of non-terminal nodes, *latent vectors* underlying the data (*e.g.*, the noise-free values of observations), and *hyperparameters* such as diffusion variances, noise levels, and parameters of the divergence function.

For models of real data without noise, the locations of terminal nodes are fixed at the data points. For models with latent vectors, the terminal nodes are fixed to these latent vectors, if they are represented explicitly. Neither latent vectors nor terminal node locations need be represented explicitly when the data are modelled as having Gaussian noise, since their distributions conditional on the structure of the tree and the divergence times will be Gaussian, and hence easily integrated over. The locations of non-terminal nodes can also be integrated over, since their joint distribution given the data or latent vectors is Gaussian. This can be done in time proportional to the number of data points, by exploiting the tree structure, as is done in a similar context by Williams (2000). If non-terminal node locations are instead kept as part of the Markov chain state, it takes only linear time to draw new values for them from their joint distribution (conditional on the tree structure, hyperparameters, and latent vectors).

It is easy to sample for the diffusion and noise variance hyperparameters when divergence times and locations for all nodes are known. Node locations can be temporarily sampled for this purpose if they are not being retained in the state. Metropolis updates could be done for the parameters of the divergence function, though I instead use slice sampling (Neal, 2003).

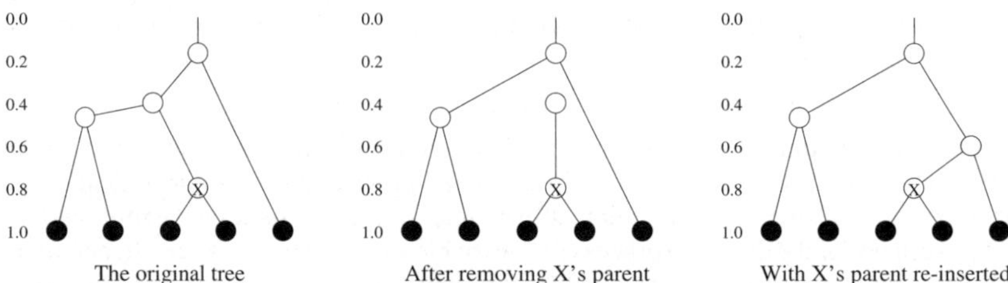

Figure 5. *Modifying a tree with a parent move. Divergence time is shown on the vertical axis, location on the horizontal axis. Black circles are the terminal nodes.*

The most crucial issue is how to sample for different tree structures, a problem that is similar to sampling for phylogenetic trees (see, *e.g.*, Mau *et al.*, 1999). Figure 5 illustrates one approach that I have used, based on Metropolis–Hastings updates that propose to move the parent of a randomly chosen terminal or non-terminal node to a new location in the tree. After the parent of the chosen node is temporarily removed from the tree, a new position for it in the tree is found by simulating the data generation process described in Section 2.1, up to the time where the new path diverges from previous paths. (If this time is later than the child node's divergence time, the process is repeated.) This proposed position in the tree is then accepted or rejected based on the likelihood, integrating over any missing node locations. It is important that at least the location of the parent node being moved be integrated over. This is easily done, and if non-terminal node locations are being retained, a new location for the parent can easily be sampled after the update (whether the proposal was accepted or not). If non-terminal node locations are not being retained, the change in likelihood resulting from the proposed move can be computed in time proportional to the depth of the tree, which typically grows logarithmically with the number of data points. I have also tried another approach to modifying the tree structure, in which a non-terminal node's position in the tree is updated by slice sampling. Best results are obtained when both approaches are combined.

The models and Markov chain sampling methods described above are implemented as part of my software for "flexible Bayesian modelling", which is available from my web page, `http://www.cs.utoronto.ca/~radford`. This software was used for the two examples that follow.

6. MODELING A TWO-DIMENSIONAL DENSITY

I tested the ability of Dirichlet diffusion trees to model a bivariate distribution using an artificial data set, generated by a somewhat complex procedure that produces an absolutely continuous distribution that is not of any simple form. (For details of the distribution and of the models and sampling procedures, see the software documentation, in which this is an example.) The data set of 500 points is shown on the left of Figure 6, along with a simple kernel density estimate, produced by the `kde2d` function in R's MASS library, with default parameter settings.

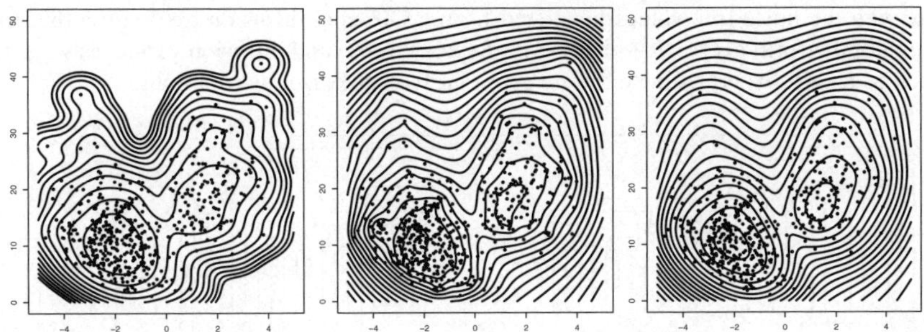

Figure 6. *The two-dimensional density modelling example. The plots show the data points along with contours of the natural log of the estimated density, spaced 0.5 apart. The left plot shows a simple kernel density estimate; the middle and right plots show Dirichlet diffusion tree estimates with different divergence functions.*

These data were modelled directly (with no noise) using two Dirichlet diffusion tree models, with divergence functions of $a(t) = c/(1-t)$ and $a(t) = b + d/(1-t)^2$. For both models, the two variables had separate hyperparameters controlling diffusion variances, which were given fairly vague prior distributions. The parameters of the divergence function (c for the first model, b and d for the second) were also given fairly vague priors.

The middle plot in Figure 6 shows the density estimate found using $a(t) = c/(1-t)$. This estimate was computed by simulating the generation of a 501st data point from 75 trees taken from the posterior distribution. For each of these trees, 500 paths were simulated that would end at each of the 500 data points if no divergence were to occur, with the time and location where divergence did occur defining a Gaussian distribution for the generated point. (This stratified sampling scheme improves the approximation, compared to generating 500 independent paths.) The final density estimate is an equal mixture of these 75×500 Gaussian distributions.

The posterior distribution of c for this model had mean 1.5 and standard deviation 0.2, so according to my conjecture, the distribution produced will be absolutely continuous. The actual distribution from which the data was drawn is indeed absolutely continuous, but with a density that is smoother than the estimate found with this divergence function. Even when $c \approx 1.5$ (as in the right of Figure 4), the prior with $a(t) = c/(1-t)$ favors distributions with some small-scale clustering, producing peaks in the density estimate around each data point.

The right plot in Figure 6 shows a density estimate found in the same way using a Dirchlet diffusion tree with $a(t) = b + d/(1-t)^2$. It displays a smoothness that is more appropriate for this data.

7. CLUSTERING GENE EXPRESSION DATA

I have also applied Dirichlet diffusion tree models to data on gene expression in leukemia cells gathered by Golub *et al.* (1999). These data contains levels of expression for 3571 genes in the tumors of 72 leukemia patients. (Expression levels for more genes were measured, but were discarded as being unreliable. The remaining data was preprocessed to avoid spurious variation, and scaled so the variance of each variable was one.) Each tumor was classified as one of three types—AML, ALL-B, or ALL-T—based on data other than gene expression. The aim of this test is to see whether by clustering the gene expression data we can rediscover these three known types.

I first tried modelling this this data using a subset of only 200 genes, randomly selected from the total of 3571. The expression levels of these genes were modelled as having Gaussian noise, with the noise-free values generated from a Dirichlet diffusion tree with a divergence function of the form $a(t) = b + d/(1-t)^2$. Separate noise and diffusion variances were given to each gene. All hyperparameters were given fairly vague prior distributions.

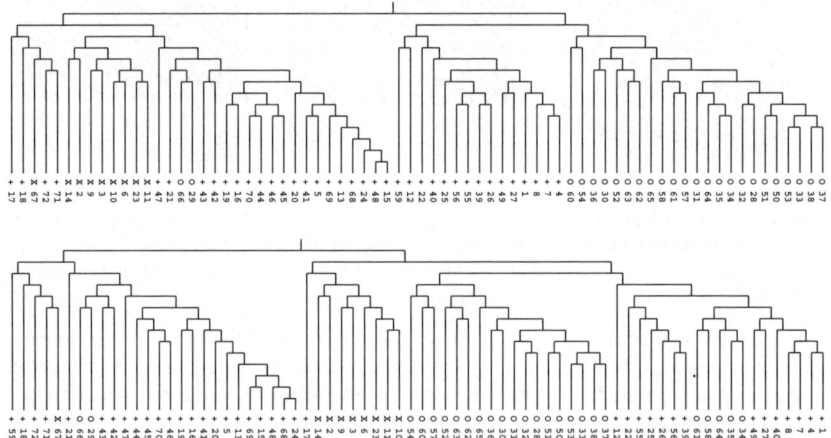

Figure 7. *Two hierarchical clusterings of the 72 tumors, drawn from the posterior distribution. Tumor types are shown by* o = AML, + = ALL-B, *and* x = ALL-T.

I ran two replicated Markov chain simulations, each taking approximately 90 minutes on a 1.7 GHz Pentium 4 processor. The state for these Markov chains did not include the locations of non-terminal nodes; these were instead integrated over. The two chains produced consistent results. Both appear to have converged after about one-third of the run, judging by trace plots of hyperparameters. I also ran chains in which non-terminal node locations were explicitly represented. This sped up tree updates by a factor of eight, but reduced their acceptance rate, with the result that efficiency was comparable to when these locations were integrated over.

Figure 7 shows dendrograms derived from two diffusion trees sampled from the posterior distribution. The clustering found is mostly consistent with the pre-existing classification, showing that the Dirichlet diffusion tree method produces useful results on problems of this scale. The posterior means of the diffusion and noise standard deviations for different genes varied by a factor of five, showing that the model has discovered that some genes are more relevant to the clustering than others.

When all 3571 genes are used, each Markov chain update takes much longer (increasing in proportion to the number of genes), and the chain shows some signs of not completely converging. The results produced are reasonable, but correspond to the pre-existing classification less closely than was the case using only 200 genes. I am presently investigating whether annealing methods can improve convergence on this very high dimensional problem.

ACKNOWLEDGEMENTS

I thank Sandrine Dudoit for providing the program for preprocessing the gene expression data. This work was supported by the Natural Sciences and Engineering Research Council of Canada.

REFERENCES

Antoniak, C. E. (1974). Mixtures of Dirichlet processes with applications to Bayesian nonparametric problems. *Ann. Statist.* **2**, 1152–1174.

Bernardo, J. M. and Smith, A. F. M. (1994). *Bayesian Theory*. Chichester: Wiley.

Blackwell, D. and MacQueen, J. B. (1973). Ferguson distributions via Pólya urn schemes. *Ann. Statist.* **1**, 353–355.

Ferguson, T. S. (1973). A Bayesian analysis of some nonparametric problems. *Ann. Statist.* **1**, 209–230.

Ferguson, T. S. (1983). Bayesian density estimation by mixtures of normal distributions. *Recent Advances in Statistics* (H. Rizvi and J. Rustagi, eds). New York: Academic Press.

Golub, T. R., Slonim, D. K., Tamahyo, P., Huard, C., Gaasenbeek, M., Mesirov, J. P., Coller, H., Loh, M. L., Downing, J. R., Caligiuri, M. A., Bloomfield, C. D. and Lander, E. S. (1999). Molecular classification of cancer: class discovery and class prediction by gene expression monitoring. *Science* **286**, 531–537.

Mau, B., Newton, M. A., and Larget, B. (1999). Bayesian phylogenetic inference via Markov chain Monte Carlo methods. *Biometrics* **55**, 1–12.

Neal, R. M. (2001). Defining priors for distributions using Dirichlet diffusion trees. *Tech. Rep.*, University of Toronto, Canada.

Neal, R. M. (2003). Slice sampling. *Ann. Statist.* (to appear, with discussion).

Walker, S. G., Damien, P., Purushottam, W. L., and Smith, A. F. M. (1999). Bayesian nonparametric inference for random distributions and related functions. *J. R. Statist. Soc. B* **61**, 485–527 (with discussion).

Williams, C. K. I. (2000). A MCMC approach to hierarchical mixture modelling. *Advances in Neural Information Processing Systems 12* (S. A. Solla *et al.*, eds). Cambridge, MA: MIT Press.

BAYESIAN STATISTICS 7, pp. 631–639
J. M. Bernardo, M. J. Bayarri, J. O. Berger, A. P. Dawid,
D. Heckerman, A. F. M. Smith and M. West (Eds.)

Outlier Robust Estimation of a Finite Population Total

LAWRENCE I. PETTIT and ROGER A. SUGDEN
Goldsmiths College, UK
l.pettit@gold.ac.uk r.sugden@gold.ac.uk

SUMMARY

We discuss a Bayesian approach to estimating a finite population total when the mean of the observed sample values is assumed to be proportional to a known covariate. Observations have one of two possible variances, a standard form or an inflated form. Observations with the larger variance are termed outliers. We learn about the proportion of outliers and parameters of the model from the sample. Inference about the mean and variance of the population total is made using the predictive distribution. The frequentist properties of the procedure are analyzed for a particular finite population by examining the bias of the estimator and coverage of the confidence interval. The performance of the estimator is shown to be superior to the Horvitz–Thompson estimator and a simple Bayes estimator which allows only the standard variance.

Keywords: BIAS; COVERAGE; OUTLIER; PREDICTION.

1. INTRODUCTION

A number of authors have examined the problem of robustly estimating a population total. Rubin (1983) discusses Bayesian estimation of a total where there is no covariate information by transforming the sample to normality. In a model-based approach, Chambers (1986) and Chambers and Kokic (1993) apply classical robust techniques, in particular M-estimation. They introduce the idea of a working model for individual responses from which outliers deviate, but no particular outlier model or class of models is specified. In contrast, Box and Tiao (1968) give both a working model for the majority of observations and an outlier model for the remainder specifying variance inflation. Their main concern is Bayesian inference on the parameters of the infinite (super-) population generating the data. See Freeman (1980) for a review of Bayesian approaches to outliers and Pettit and Smith (1985) for a different approach. Rao (1978) also considers a shifted mean model in a model-based, but non-Bayesian, approach to finite population estimation.

A feature of finite population problems is the need to predict responses of the unsampled units. Ericson (1969) shows how this is straightforward in a Bayesian framework if we assume all observations follow the same working model so that, in the regression context, some function of the responses and the covariates is *a priori* exchangeable over the units. Exchangeability does not preclude the existence of outliers, of course.

Bolfarine and Zacks (1992, p. 140) robustify Bayesian inference on the finite population mean by assuming all units follow the *same* working model, and consider a small finite set of possible models for the population. Suppose, for example, that our working model is the known density f_1 and our alternative model is the known density f_2 (possibly including variance inflation). If the responses are $Y_1, \ldots, Y_N$ with realizations $y_1, \ldots, y_n$ then the joint prior density of the responses is

$$(1-\alpha)\prod_{i=1}^{N} f_1(y_i) + \alpha \prod_{i=1}^{N} f_2(y_i)$$

where $(1-\alpha)$ is the prior belief in the first model for the finite population. This is inappropriate for outlier problems since we wish to model the outliers as coming from a distinct distribution.

Our approach is more in the spirit of Box and Tiao (1968), who in this example would consider a joint prior density

$$\prod_{i=1}^{N}[(1-\alpha)f_1(y_i) + \alpha f_2(y_i)]$$

so that units are exchangeable but with a finite mixture for the marginal prior. Mixing is done at unit rather than population level. Our approach extends Box and Tiao (1968) in that α, the probability that a particular unit is an outlier, is unknown and assigned a prior distribution. Moreover they took a fixed variance inflation factor and did not consider prediction.

We have chosen to work with analytic results. An obvious alternative would be to use a Markov chain Monte Carlo method (see, *e.g.*, Verdinelli and Wasserman, 1991). In practice for samples larger than considered here we would probably do so. However, this would add simulation errors which are difficult to quantify and we consequently decided to obtain exact results.

2. MODELS

We suppose we have a finite population of N identifiable units labelled $1, 2, \ldots, N$. Attached to the ith unit is the known value of a concomitant x_i and an unknown response Y_i, assumed to be measured without error for sampled units. We take a sample of size n, with labels in $S \subset \{1, 2, \ldots, N\}$ drawn according to a non-informative sample design $p(S \mid \underline{x})$. We assume a linear regression model through the origin but allow for two types of units through the assumed variances. Thus responses for some units ("non-outliers") are distributed as

$$Y \sim N(\beta x, \sigma^2 x_0^{g-g_0} x^{g_0})$$

and the rest ("outliers") have a distribution which is

$$Y \sim N(\beta x, \sigma^2 x^g).$$

Note that $g > g_0$ are specified by the analyst and that x_0, which is introduced to preserve scales of measurement, is chosen to be no larger than the smallest x_i value. This ensures that outliers have a larger variance than the non-outliers. It may help the reader to think of $g_0 = 1$ and $g = 2$ representing that the standard deviation of Y is proportional to the square root of x for non-outliers or x for outliers. In practice we have found that our choices for x_0, g and g_0 are not independent. See the example later.

We introduce an indicator variable λ, unobserved in the sample and unknown, to show which observations are outliers. Thus, $\lambda_i = 0$ means that unit i is an outlier and $\lambda_i = 1$ means unit i is not an outlier. We suppose the prior on the λ's is the product of N independent Bernoulli distributions and with the further assumption that the λ's are exchangeable we have that

$$\Pr[\lambda_i = 0] = \alpha \qquad i = 1, \ldots, N$$

where α is an unknown parameter. We assign a beta prior to α with parameters a and b and mean $a/(a+b)$.

To complete the prior specification we assume the following conjugate priors for β and σ^2:

$$\beta \,|\, \sigma^2 \sim N(b_0, \sigma^2/h)$$
$$1/\sigma^2 = v \sim \Gamma(\nu_0/2, \nu_0 S_0^2/2)$$

where $\Gamma(c, d)$ represents a gamma distribution with mean c/d.

3. POSTERIOR CALCULATIONS

We define the following sequence of models which refer only to outliers in the sample, not the unsampled population

Model	Description
M_0	No outliers in sample S
$\vdots$	
$M_j(S^{(j)})$	There are j outliers in the sample S with identifying labels $S^{(j)} \subset S$
$\vdots$	
M_n	All sample observations are outliers

The prior implied for j, the number of outliers in the sample, by the priors for λ and α is beta-binomial

$$\Pr[j] = \binom{n}{j} \frac{\beta(a+j, b+n-j)}{\beta(a,b)} \qquad j = 0, 1, \ldots, n$$

where $\beta(a, b)$ is the beta function and hence the prior weight on model $M_j(S^{(j)})$ is

$$\pi[M_j(S^{(j)})] = \frac{\beta(a+j, b+n-j)}{\beta(a,b)}.$$

Note that we do not learn from the sample about α except through the model $M_j(S^{(j)})$ as j, if it was observed, would be sufficient for α.

It follows that the posterior for α conditional on the model $M_j(S^{(j)})$ is a beta distribution with mean $(a+j)/(a+b+n)$.

The posterior distributions for $\beta \,|\, \sigma^2$ and σ^2 conditional on a particular model are derived using standard Bayesian calculations. For notational convenience we suppose that in model $M_j(S^{(j)})$ the the outliers are the first j observations. We find that

$$\beta \,|\, \sigma^2, \text{data}, M_j(S^{(j)}) \sim N\left[b(S^{(j)}), \sigma^2/n(S^{(j)})\right]$$

where

$$n(S^{(j)})b(S^{(j)}) = \left(\sum_{i \le j} \frac{y_i}{x_i^{g-1}} + \frac{1}{x_0^{g-g_0}} \sum_{i>j} \frac{y_i}{x_i^{g_0-1}} + hb_0 \right)$$

and

$$n(S^{(j)}) = \sum_{i \le j} x_i^{2-g} + \sum_{i>j} \frac{x_i^{2-g_0}}{x_0^{g-g_0}} + h$$

is the equivalent sample size for β. Also the posterior of $1/\sigma^2$ is a Gamma distribution with parameters $\frac{1}{2}(\nu_0+n)$ and $\frac{1}{2}(v_0+n)S_j^2$ where

$$(\nu_0+n)S_j^2 = \nu_0 S_0^2 + \left\{h\text{RSS}_0(S^{(j)}) + \Big(\sum_{i\le j} x_i^{2-g} + \sum_{i>j}\frac{x_i^{2-g_0}}{x_0^{g-g_0}}\Big)\,\text{RSS}_j(S^{(j)})\right\}/n(S^{(j)}),$$

$$\text{RSS}_0(S^{(j)}) = \sum_{i\le j}\frac{(y_i-b_0x_i)^2}{x_i^g} + \frac{1}{x_0^{g-g_0}}\sum_{i>j}\frac{(y_i-b_0x_i)^2}{x_i^{g_0}},$$

$$\text{RSS}_j(S^{(j)}) = \sum_{i\le j}\frac{(y_i-\hat\beta_jx_i)^2}{x_i^g} + \frac{1}{x_0^{g-g_0}}\sum_{i>j}\frac{(y_i-\hat\beta_jx_i)^2}{x_i^{g_0}},$$

and

$$(n(S^{(j)})-h)\hat\beta_j = n(S^{(j)})b(S^{(j)}) - hb_0.$$

To find the posterior probability of each model $\pi^*[M_j(S^{(j)})]$ we need to find the marginal distribution of the data $p(\text{data}\mid M_j(S^{(j)}))$ since

$$\pi^*[M_j(S^{(j)})] = \frac{\pi[M_j(S^{(j)})]p(\text{data}\mid M_j(S^{(j)}))}{\sum \pi[M_j(S^{(j)})]p(\text{data}\mid M_j(S^{(j)}))}$$

where the summation is over the 2^n possible models.

Derivation of the marginal distribution of the data for a given model is obtained by integration. The required expression is

$$\frac{h^{\frac{1}{2}}\Gamma(\frac{n+\nu_0}{2})(\nu_0S_0^2)^{\nu_0/2}\{(\nu_0+n)S_j^2\}^{-(n+\nu_0)/2}}{\pi^{\frac{n}{2}}\Gamma(\frac{\nu_0}{2})\{n(S^{(j)})\prod_{i\le j}x_i^g\prod_{i>j}x_i^{g_0}\}^{1/2}x_0^{(g-g_0)(n-j)/2}}.$$

4. INFERENCE FOR THE POPULATION

We suppose that the main aim of the analysis is to make predictions about some feature of the population. This is in contrast to most Bayesian work on outliers where the main focus is inference on the posterior distribution of β and σ^2. If we are interested in the population total $T=\sum_{i=1}^N Y_i$ then we need to find the predictive distributions of the unsampled units Y_u, $u\notin S$ since we know the total for the sampled units.

Predictive inferences for unsampled units conditional on a particular model are fairly straightforward. First note that the mean of an unsampled unit Y_u with corresponding covariate value x_u is the same for outliers and non-outliers. Thus

$$E[Y_u|\text{data}, M_j(S^{(j)})] = x_u b(S^{(j)}).$$

The variance of Y_u is given by the sum of terms representing the expectation of the variance of Y_u conditioning on the value of λ for Y_u and the variance of the expectation. It follows that it is given by

$$\left[\frac{x_u^{g_0}}{a+b+n}\{(a+j)x_u^{g-g_0}+(b+n-j)x_0^{g-g_0}\}+\frac{x_u^2}{n(S^{(j)})}\right]E[\sigma^2\mid\text{data}, M_j(S^{(j)})].$$

Inference about the population total is slightly more complicated. The posterior mean of T is

$$\begin{aligned}\mathrm{E}[T \mid \text{data}] &= \sum_{i\in S} y_i + \sum_{u\notin S} E[Y_u \mid \text{data}] \\ &= \sum_{i\in S} y_i + \sum_{u\notin S} \sum_{M_j(S^{(j)})} E[Y_u \mid \text{data}, M_j(S^{(j)})]\pi^*[M_j(S^{(j)})] \\ &= \sum_{i\in S} y_i + \sum_{u\notin S} \sum_{M_j(S^{(j)})} x_u b(S^{(j)})\pi^*[M_j(S^{(j)})]\end{aligned}$$

and can thus be calculated from quantities already found.

The posterior variance of T is given by

$$\begin{aligned}\mathrm{Var}[T \mid \text{data}] &= \sum_{u\notin S} x_u^g \sum_{M_j(S^{(j)})} \frac{a+j}{a+b+n} E[\sigma^2 \mid \text{data}, M_j(S^{(j)})]\pi^*[M_j(S^{(j)})] \\ &+ \sum_{u\notin S} x_u^{g0} x_0^{g-g0} \sum_{M_j(S^{(j)})} \frac{b+n-j}{a+b+n} E[\sigma^2 \mid \text{data}, M_j(S^{(j)})]\pi^*[M_j(S^{(j)})] \\ &+ \Big(\sum_{u\notin S} x_u\Big)^2 \sum_{M_j(S^{(j)})} [E(\beta \mid \text{data}, M_j(S^{(j)})) - E(\beta \mid \text{data})]^2 \pi^*[(M_j(S^{(j)})] \\ &+ \Big(\sum_{u\notin S} x_u\Big)^2 \sum_{M_j(S^{(j)})} \frac{E[\sigma^2 \mid \text{data}, M_j(S^{(j)})]}{n(S^{(j)})} \pi^*[(M_j(S^{(j)})].\end{aligned}$$

5. A NUMERICAL EXAMPLE

5.1. *Introduction*

We take the "Counties 70" population described in Royall and Cumberland (1981) as a suitable illustration of our procedures. Scatter plots of Y (total population) against x (no. of households) for $N = 304$ counties revealed many counties close to a regression line through the origin with little or no increasing spread, but also a proportion of counties at a considerable distance with increasing spread. For such a population the model of Section 2 may be appropriate.

For $n = 8$, we will examine by simulation the sampling properties of procedures which specify a nominal 95% interval for the unknown population total T

$$\hat{T} \pm t_{n-1}(0.025)\sqrt{\hat{V}} \tag{1}$$

where $\hat{T}$ is a sample estimator for T based on a sample s and $\hat{V}$ is a variance estimator that is a sample estimator of the sampling variance of that estimator. We will also examine the sampling properties of $\hat{T}$ and $\hat{V}$ themselves. For the Bayes procedures, $\hat{T}$ is the posterior mean of T and $\hat{V}$ is the posterior variance.

Studies of the Horvitz–Thompson estimator together with the Sen–Yates–Grundy variance estimator under πPS sampling, usually show under coverage (*i.e.*, considerably less than 95%) for (1) and considerable lack of symmetry in the sampling distribution for $\hat{T}$, although both $\hat{T}$ and $\hat{V}$ are unbiased with respect to the sampling design. A recent study (Sugden *et al.*, 1996) gave only 87.8% coverage for samples of size 8 under Chao's πPS scheme, where the sampling standard deviation of the estimator was approximately 5.92×10^5, 5.3% of the true value (1.1243112×10^7) of the population total. Under systematic πPS sampling the coverage was even worse, only 84.3%, but the estimator had a smaller sampling standard deviation of 5.36×10^5.

5.2. *Choice of Prior Parameters*

We will study the effects on the sampling properties of varying a limited number of the nine prior parameters: $\{b_0, x_0, \nu_0, S_0^2, h, g_0, g, a, b\}$. We should emphasize that part of the aim of the study is to investigate the robustness of the prior specification on the coverage and other sampling properties. We have therefore chosen a wide range of prior values, not necessarily reflecting substantive prior knowledge. The method we employ to choose the parameters may be useful for thinking about such problems.

We fix the prior mean b_0 of the slope parameter β at 4 and the lower limit x_0 (for which outliers and non-outliers have the same variance and are hence indistinguishable) at 480, just less than the minimum of the known x_i values. We also fix the prior "degrees of freedom" ν_0 to be equal to 6. This gives σ^2 a prior mean of $(3S_0^2)/2$ and the precision $v = 1/\sigma^2$ a coefficient of variation (variance/mean^2) of $2/\nu_0 = 1/3$. The ratio S_0^2/h is fixed by requiring that the expected prior variance of β is $1/4$, so $h = 6S_0^2$.

The prior mean of the outlier probability α is also fixed at 0.1, or 10%, but we allow for varying the prior uncertainty about this mean by taking both $a = 1, b = 9$ and $a = 0.1, b = 0.9$, giving prior standard deviations of 0.09 and 0.21, respectively. This represents considerable uncertainty about the proportion of outliers in the population. In the simulation we also consider a "simple Bayes" procedure where outliers are assumed to be absent, corresponding to the choice $a = 0, b > 0$.

It remains to determine our prior beliefs about the variation about the regression line and the relative standard deviations of outliers and non-outliers. Given σ^2, (or v), the ratio of standard deviations at x is $(x/x_0)^{(g-g_0)/2}$. By taking this ratio to be 5, 10 and 20 at $x = 30,000$ we obtain three values for $g - g_0$: 0.778415, 1.11366 and 1.44891, respectively. The shape of the variance function for non-outliers is determined by g_0. A value of zero corresponds to constant variance and we also take 0.25 and 0.5 as alternatives to reflect a slightly increasing variance. Finally, S_0^2 is determined by taking the square root of the prior expected (common) variance at $x = x_0$ to be 5,000, 10,000 and 20,000.

For the full Bayes procedure this gives a 2×3^3 design in the simulation and 3×3 corresponding factor combinations for the simple Bayes. We believe this represents an adequate preliminary study of robustness to prior specifications.

5.3. *Results and Discussion*

Table 1 shows that the Bayesian estimators have a small sampling bias, interestingly of mainly opposite sign to the simple Bayes procedure. This bias mainly results from the choice of prior mean and prior standard deviation for the slope parameter β. There is little variation with other prior specifications.

Table 2 also has little variation, but shows that our Bayes procedure has a clear advantage for almost any prior over both the simple Bayes and the Horvitz–Thompson strategy where the sampling standard deviation is around 5%.

The relative sample bias of both Bayes procedures (bias/s.d.) is quite large and we might expect this would adversely affect the coverage of nominal 95% confidence intervals. However, this is not the case. Table 4 shows acceptable coverage probabilities, unsurprisingly increasing with S_0^2, for most prior specifications and in many cases considerably superior to Horvitz–Thompson. Coverage also increases with g and the variability of outliers. There is little

Table 1. *Sampling bias as % of T.*

s.d. at $x = x_0 = 480$			
g_0	5,000	10,000	20,000
Simple Bayes			
0.00	+2.7	+2.5	+1.8
0.25	+2.4	+1.8	+0.5
0.50	+1.7	+0.6	−1.1
Bayes: $a = 1, b = 9$ (top) $a = 0.1, b = 0.9$ (bottom)			
$g - g_0 = 0.778415$			
0.00	−2.1	−2.2	−1.7
	−1.6	−2.0	−1.4
0.25	−2.3	−1.8	−1.3
	−2.1	−1.6	−0.6
0.50	−2.0	−1.4	−1.3
	−1.8	−0.8	−1.2
$g - g_0 = 1.11366$			
0.00	−2.8	−2.6	−1.7
	−2.7	−2.4	−1.3
0.25	−2.7	−1.9	−1.0
	−2.5	−1.5	−0.2
0.50	−2.1	−1.1	−1.2
	−1.8	−0.4	−1.1
$g - g_0 = 1.44891$			
0.00	−2.9	−2.5	−1.4
	−2.9	−2.3	−2.4
0.25	−2.7	−1.6	−0.5
	−2.5	−1.2	+0.1
0.50	−1.9	−0.7	−1.1
	−1.5	+0.0	−1.1

Table 2. *Sampling s.d. as % of T.*

s.d. at $x = x_0 = 480$			
g_0	5,000	10,000	20,000
Simple Bayes			
0.00	8.7	8.3	7.3
0.25	8.0	7.2	5.3
0.50	7.0	5.4	3.1
Bayes: $a = 1, b = 9$ (top) $a = 0.1, b = 0.9$ (bottom)			
$g - g_0 = 0.778415$			
0.00	3.6	3.6	3.7
	3.8	3.8	3.9
0.25	3.4	3.6	3.3
	3.5	3.8	3.9
0.50	3.5	3.3	2.8
	3.6	3.9	3.0
$g - g_0 = 1.11366$			
0.00	3.1	3.7	3.9
	3.0	3.8	4.2
0.25	3.4	3.8	3.7
	3.5	4.0	4.4
0.50	3.6	3.6	2.9
	3.8	4.3	3.0
$g - g_0 = 1.44891$			
0.00	3.1	4.0	4.2
	3.2	4.1	4.1
0.25	3.7	4.0	4.1
	3.8	4.4	4.8
0.50	3.9	4.1	3.0
	4.1	4.7	3.1

variation though with prior uncertainty on the proportion of outliers. Many of the coverages might be regarded as unacceptably large (in many cases 100%), so that the intervals are "too wide". This is due to the large uncertainty in the prior specification and can be "explained" by Table 3, which shows in these cases that the variance estimator is upwardly biased in the sense that it has a larger expectation than the actual variance of the Bayes estimator. However, it is generally true that even an unbiased variance estimator does not guarantee good coverage.

It is possible, of course, to simulate the exact posterior distribution of T rather than calculate mean and variance for substitution into (1). This distribution could be used for each sample to calculate an asymmetric interval based on tail areas. We did this for an arbitrary sample and found the simulated posterior distribution was almost normal (in comparison with the sampling distribution of the Horvitz–Thompson estimator which is generally skewed). Similarly the sampling distribution of the posterior mean as an estimator of T was also found to be pleasingly symmetrical, indeed normal. We expect, therefore, that these results give a good indication of what may be obtained in practice with similar populations.

Table 3. *Ratio of the (sampling) expectation of the posterior variance to the (sampling) variance of the posterior mean.*

s.d. at $x = x_0 = 480$			
g_0	5000	10000	20000
Simple Bayes			
0.00	0.23	0.28	0.50
0.25	0.27	0.43	1.25
0.50	0.41	1.01	5.06
Bayes: $a = 1, b = 9$ (top) $a = 0.1, b = 0.9$ (bottom)			
$g - g_0 = 0.778415$			
0.00	0.84	1.17	1.89
	0.94	1.16	1.74
0.25	1.11	1.57	3.76
	1.16	1.50	2.64
0.50	1.41	2.98	7.67
	1.38	2.25	5.74
$g - g_0 = 1.11366$			
0.00	1.04	1.24	2.12
	1.25	1.15	1.74
0.25	1.19	1.72	4.15
	1.15	1.50	2.34
0.50	1.50	3.26	10.2
	1.37	2.03	6.16
$g - g_0 = 1.44891$			
0.00	1.25	1.58	3.21
	1.35	1.34	1.29
0.25	1.48	2.60	6.33
	1.28	1.76	2.47
0.50	2.25	4.98	21.8
	1.67	2.14	8.55

Table 4. *Coverage of nominal 95% interval.*

s.d. at $x = x_0 = 480$			
g_0	5,000	10,000	20,000
Simple Bayes			
0.00	62.7	75.0	93.7
0.25	72.7	91.3	100.0
0.50	88.4	99.8	100.0
Bayes: $a = 1, b = 9$ (top) $a = 0.1, b = 0.9$ (bottom)			
$g - g_0 = 0.778415$			
0.00	82.4	94.2	99.7
	84.3	92.2	99.2
0.25	90.2	98.9	100.0
	89.5	98.2	100.0
0.50	97.3	100.0	100.0
	96.1	100.0	100.0
$g - g_0 = 1.11366$			
0.00	82.1	95.8	100.0
	82.3	91.4	99.5
0.25	91.7	99.3	100.0
	88.3	98.3	100.0
0.50	98.2	100.0	100.0
	96.1	100.0	100.0
$g - g_0 = 1.44891$			
0.00	89.3	98.0	100.0
	85.7	93.2	92.0
0.25	96.8	100.0	100.0
	90.5	99.0	100.0
0.50	99.9	100.0	100.0
	97.9	100.0	100.0

REFERENCES

Bolfarine, H. and Zacks, S. (1992). *Prediction Theory for Finite Populations*. Berlin: Springer.

Box, G.E.P. and Tiao, G.C. (1968). A Bayesian approach to some outlier problems. *Biometrika* **55**, 119–129.

Chambers, R. L. (1986). Outlier robust finite population estimation. *J. Am. Statist. Ass.* **81**, 1063–1069.

Chambers, R. L. and Kokic, P. N. (1993). Outlier robust sample survey inference. *Bull. Int. Statist. Inst.* **49-2**, 55–72.

Ericson, W. A. (1969). Subjective Bayesian models in sampling finite populations. *J. R. Statist. Soc. B* **31**, 195–233 (with discussion).

Freeman, P.R. (1980) On the number of outliers in data from a linear model. *Bayesian Statistics* (J. M. Bernardo, M. H. DeGroot, D. V. Lindley and A. F. M. Smith, eds). Valencia: University Press, 349–365 (with discussion).

Pettit, L. I. and Smith, A. F. M. (1985). Outliers and influential observations in linear models. *Bayesian Statistics 2* (J. M. Bernardo, M. H. DeGroot, D. V. Lindley and A. F. M. Smith, eds), Amsterdam: North-Holland, 473–494 (with discussion).

Rao, J. N. K. (1978). Sampling designs involving unequal probabilities of selection and robust estimation of a finite population total. *Contributions to Survey Sampling and Applied Statistics* (H. A. David, ed). New York: Academic Press, 69–87.

Royall, R. M. and Cumberland, W.G. (1981). An empirical study of the ratio estimator and estimators of its variance. *J. Am. Statist. Ass.* **76**, 66–88 (with discussion).

Rubin, D. B. (1983). A case study of the robustness of Bayesian methods of inference: Estimating the total in a finite population using transformations to normality. *Scientific Inference, Data Analysis and Robustness* (G. E. P. Box, T. Leonard and C,-F. Wu, eds). New York: Academic Press, 213–243.

Sugden, R. A., Smith, T. M. F. and Brown, R. P. (1996). Chao's list sequential scheme for unequal probability sampling. *J. Appl. Statist.* **23**, 413–421.

Verdinelli, I and Wasserman, L. (1991). Bayesian analysis of outlier problems using the Gibbs sampler. *Statistics and Computing* **1**, 135–139.

BAYESIAN STATISTICS 7, pp. 641–650
J. M. Bernardo, M. J. Bayarri, J. O. Berger, A. P. Dawid,
D. Heckerman, A. F. M. Smith and M. West (Eds.)

Bayesian Inference for Derivative Prices

NICHOLAS G. POLSON
University of Chicago, USA
ngp@gsb.uchicago.edu

JONATHAN R. STROUD
University of Pennsylvania, USA
stroud@wharton.upenn.edu

SUMMARY

This paper develops a methodology for parameter and state variable inference using both asset and derivative price information. We combine theoretical pricing models and asset dynamics to generate a joint posterior for parameters and state variables and provide an MCMC simulation strategy for inference. There are several advantages of our inferential approach. First, more precise parameter estimates are obtained when both asset and derivative price information are used. Second, we provide a diagnostic tool for model misspecification based on agreement of the state and parameter estimates with and without derivative price information. Furthermore, the time series properties of the state variables can also be used to evaluate model fit. We illustrate our methodology using daily equity index options on the Standard and Poor's (S&P 500) index from 1998–2002.

Keywords: DERIVATIVE PRICING; MCMC; STOCHASTIC VOLATILITY; SMOOTHING; OPTION PRICING; LEVERAGE EFFECT.

1. INTRODUCTION

Bayesian inference using both asset and derivative price information presents a number of challenges. While many authors have considered the problem of parameter inference in asset pricing models, few have considered the problem of also incorporating derivative price information. The advantages of using additional derivative price information is twofold. First, more precise parameter and state variable estimates are obtained. Second, these estimates can be used to provide model misspecification diagnostics.

Financial asset pricing theory determines derivative prices as a conditional expectation of a discounted payoff given the current state variables and parameter values. The expectation is taken with respect to a risk-neutral pricing kernel, which is derived from the underlying asset price dynamics by a transformation of the underlying parameters. For closed form derivative pricing under a wide range of models we refer the reader to Duffie *et al.* (2000). See Johannes and Polson (2002) for further discussion. The joint inference problem, then, is to combine this information with the usual asset pricing model for the dynamics of the underlying asset and state variables.

Our empirical work focuses on the commonly used leverage stochastic volatility model for equity index options. This model uses a number of features for explaining asset and derivative prices: an unobserved mean reverting stochastic volatility state variable and a vector of parameters including a correlation between the underlying asset returns and volatility. Closed form derivative pricing is available and this includes an extra parameter governing the market price of volatility risk.

Figure 1 provides motivation for the additional market price of volatility risk parameter by comparing implied volatilities with historical filtered volatilities (see Polson *et al.*, 2002) using daily S&P 500 equity index return data for 1990–2001. The implied volatilities are consistently higher than the filtered ones due to a number of reasons. The main factor is that option pricing accounts for the market price of volatility risk. Option pricing is determined as a conditional expectation under the risk neutral measure and the market price of volatility risk effectively raises the level of volatility under which pricing occurs. This is discussed further in Section 2.2.

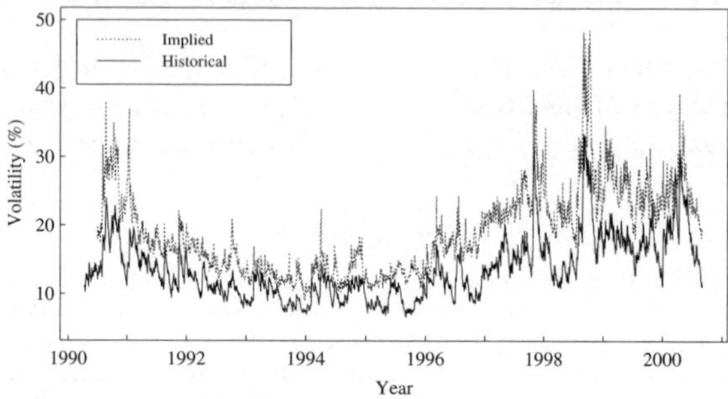

Figure 1. *Implied and historical volatilities for the S&P 500 Index.*

Related literature on combining derivative and asset price information is contained in Chernov and Ghysels (2000) and Johannes and Polson (2002). For option pricing applications, see Heston and Nandi (2000) and Eraker (2002) for S&P 500 options data, Florentina *et al.* (2002) for the IBEX 35 Spanish index, Lin *et al.* (2002) for the FTSE 100. Forbes *et al.* (2002) provides a simulation study and Lewis (2000) a current review of option pricing methods under stochastic volatility.

The main reason for combining information is to provide more precise parameter estimates. In addition, it provides a diagnostic tool for model misspecification. This is due to the fact that, theoretically, the estimated parameters implicit in derivatives prices should be consistent with those using asset price data alone. This insight also yields a model misspecification diagnostic by examining the consistency of the time series properties of the estimated state variables.

The rest of the paper is outlined as follows. Section 2 defines the basic problem and provides the necessary likelihood functions for combining derivative and asset price information. We explicitly study the leverage stochastic volatility model and show how to perform inference using MCMC methods. Section 3 provides an application to S&P 500 equity index options. Finally, Section 4 concludes.

2. BAYESIAN INFERENCE FOR DERIVATIVE PRICING

This section describes the general problem in three stages. First, we describe the inference problem when both asset and derivative prices are observed. Second, we provide an example of inference for equity index options using a leverage stochastic volatility model. Finally, we describe how to implement an MCMC simulation strategy to provide inference for unobserved state variables and parameters given both asset and derivative prices.

2.1. *The Problem*

To describe the basic problem, consider observations on both asset and derivative prices. Let Y^S denote the asset price returns and let Y^D denote the panel of derivative prices. The dynamics of the asset prices Y^S are determined via a likelihood $p(Y^S \mid X, \Theta)$ where X denotes a state variable and Θ a parameter vector. We assume that these evolve according to a model for the asset price dynamics which we denote by the joint distribution $p(X, \Theta) = p(X \mid \Theta)p(\Theta)$.

Inference using only the asset price information Y^S then leads to a posterior distribution for state and parameters given by

$$p(X, \Theta \mid Y^S) \;\propto\; p(Y^S \mid X, \Theta)\, p(X, \Theta)$$

The corresponding marginal distributions $p(\Theta \mid Y^S)$ and $p(X \mid Y^S)$ provide the usual inferential summaries.

The problem of incorporating the additional derivative price information can be described as follows. First, from a theoretical asset pricing model, the derivative price is determined as a function of the state variable X_t, parameter Θ and current asset price Y_t^S. We denote this function by $F(Y_t^S, X_t, \Theta, \Lambda)$, where Λ denotes additional market price of risk parameters. These parameters are used in derivative pricing, but do not affect the asset price or state dynamics. We assume that the observed derivative price Y_t^D is equal to the theoretical value plus a pricing error:

$$Y_t^D \;=\; F(Y_t^S, X_t, \Theta, \Lambda) \;+\; \epsilon_t^D.$$

This leads to a likelihood for the observed derivative prices denoted by $p(Y^D \mid Y^S, \Theta, \Lambda)$.

In order to calculate $F(Y_t^S, X_t, \Theta, \Lambda)$ we need the risk-neutral dynamics of the Y^S process. This is determined by transforming the original state dynamics with the market price of risk parameters. See Duffie *et al.* (2000) and Johannes and Polson (2002) for a number of examples involving equity and interest rate models. More specifically, $F(Y_t^S, X_t, \Theta, \Lambda)$ is given by conditional expectation under $\mathrm{E}^Q(\cdot | X_t, \Theta, \Lambda)$, namely

$$F(Y_t^S, X_t, \Theta, \Lambda) \;=\; E^Q\left(e^{-\int_t^T r_s ds} g(S_T) \mid X_t, \Theta, \Lambda\right)$$

where $g(S_T)$ denotes the payout function for the underlying derivative.

The joint inference problem is then the computation of the joint posterior distribution of state variables and parameters. The parameters include those driving the state evolution Θ and those that are used to determine the risk neutral pricing measure Λ. Our data $Y = (Y^D, Y^S)$ represents a panel of derivative prices in Y^D and the underlying series Y^S and the joint posterior of interest is given by

$$p(X, \Theta, \Lambda \mid Y) \;\propto\; p(Y \mid X, \Theta, \Lambda)\, p(X, \Theta)\, p(\Lambda)$$

where $p(Y \mid X, \Theta, \Lambda)$ denotes the likelihood function, $p(X, \Theta)$ is as described above, and $p(\Lambda)$ denotes a prior on the market price of risk parameters.

The likelihood can be further decomposed as

$$p(Y \mid X, \Theta, \Lambda) \;=\; p(Y^D \mid Y^S, X, \Theta, \Lambda)\, p(Y^S \mid X, \Theta)$$

where $p(Y^S \mid X, \Theta)$ describes the evolution of the underlying asset price. We now describe an application to equity index options using a leverage stochastic volatility model.

2.2. *Stochastic Volatility Option Pricing*

A common approach to equity option pricing is to use a leverage stochastic volatility model due to Heston (1993). Suppose that the underlying equity price S_t evolves according to a stochastic volatility model with square-root dynamics for variance V_t, and correlated errors. The logarithmic asset price $Y_t^S = \log S_t$ then satisfies

$$dY_t^S \;=\; \mu(V_t)\, dt + \sqrt{V_t}\, dW_t^s$$

$$dV_t = \kappa(\theta - V_t)\,dt + \sigma_v\sqrt{V_t}\,dW_t^v$$

for some drift $\mu(V_t)$. Here (W_t^s, W_t^v) are a pair of Brownian motions with correlation ρ. This correlation, or so-called leverage effect (Black, 1976), is important to explain the empirical fact that volatility increases faster as equity prices drop.[1] The parameters (κ, θ) govern the speed of mean reversion and the long-run mean of volatility and σ_v measures the volatility of volatility.

Equity option prices Y_t^D are determined as a function of $F(Y_t^S, V_t, \Theta, \Lambda)$. Consider a derivative written on the underlying equity with a payoff $g(S_T)$ at some future time T. Let r_s denote the instantaneous short rate. Using standard arbitrage arguments (Duffie, 2000; Duffie *et al.*, 2000), the price of a call option is given by the conditional expectation

$$F(Y_t^S, V_t, \Theta, \Lambda) = E^Q\left(e^{-r\tau}(S_T - K)_+ \,|\, V_t, \Theta, \Lambda\right)$$

where the option payout function is $(S_T - K)_+$ at time T and exercise price K; E^Q is taken with respect to the risk-neutral dynamics; and $h(x)_+ = \max\{h(z), 0\}$. The instantaneous short rate r is assumed constant over the time interval $\tau = T - t$. The risk-neutral dynamics are given by the evolution

$$dS_t = r_t S_t dt + \sqrt{V_t} S_t\,dW_t^s$$
$$dV_t = \kappa^*(\theta^* - V_t)dt + \sigma_v\sqrt{V_t}\,dW_t^v$$

where

$$\kappa^* = \kappa + \lambda_v \text{ and } \theta^* = \frac{\kappa}{\kappa + \lambda_v}\theta\,.$$

Here λ_v is the market price of volatility risk. Heston (1993) motivates this choice of a market price of volatility risk which is proportional to V_t, namely $\lambda(V_t, t) = \lambda_v V_t$. Typically, the market price of risk parameter λ_v is negative which leads to option pricing being implemented with θ transformed higher to $(\kappa/\kappa + \lambda_v)\,\theta$.

One of the advantages of this flexible model is that analogous to Black Scholes there is a closed form pricing equation. Specifically, there exists a pair of probabilities $P_j(V_t, \Theta, \Lambda), j = 1, 2$, such that the call price $F(Y_t^S, V_t, \Theta, \Lambda)$ is given by:

$$F(Y_t^S, V_t, \Theta, \Lambda) = S_t P_1(V_t, \Theta, \Lambda) - Ke^{-r\tau}P_2(V_t, \Theta, \Lambda)$$

Specifically, $P_j(V_t, \Theta, \Lambda) = Pr_j\left(\ln(S_T/K) \geq 0 \,|\, V_t, \Theta, \Lambda\right)$, where these probabilities[2] can be determined by inverting a characteristic function

$$P_j(V_t, \Theta, \Lambda) = \frac{1}{2} + \frac{1}{\pi}\int_0^\infty Re\left[\frac{e^{-i\phi\ln(K)}f_j(V_t, \Theta, \Lambda)}{i\phi}\right]d\phi$$

where

$$f_j(V_t, \Theta, \Lambda) = e^{C(\tau;\phi) + D(\tau;\phi)V_t + i\phi\ln(S_t)},$$

and the coefficients $C(\tau; \phi)$ and $D(\tau; \phi)$ are defined in Heston (1993). The correlation and volatility of volatility parameters affect the skewness and kurtosis of the underlying return distribution and Heston (1993) describes how the correlation ρ affects the probabilities $P_j(V_t, \Theta, \Lambda)$. For example, as is typically the case, when the correlation is negative then the underlying risk neutral distribution has a skewed left tail which in turn decreases the price of out-of-the-money call options while increasing the out-of-the-money put options.

---

[1] While this model has been shown to provide very feasible dynamics for asset prices and for patterns in option prices (Heston, 1993) there is still further empirical evidence for jumps in returns (Bates, 1996, 2000; Bakshi *et al.*, 1997; Eraker, 2002) as well as volatility (see Eraker etalc 2002).

[2] The probabilities P_j are determined under the respective measures $dY_t^S = [r \pm \frac{1}{2}V_t]\,dt + \sqrt{V_t}\,dW_T^S$ and $dV_t = (a_j - b_j v_t)\,dt + \sigma_v\sqrt{V_t}\,dW_T^V$ where a_j and b_j are defined in Heston (1993).

2.3. *MCMC Implementation*

This section describes how to sample from the joint distribution of state variables and parameters using MCMC methods. First, we describe how to calculate the necessary likelihood functions with and without the use of derivatives prices. Then we illustrate how to use MCMC methods on the stochastic volatility Heston model described in Section 2.2. First, we use an Euler time discretisation of the continuous time model and write the observed daily log returns, $Y_t^S = \log(S_{t+1}/S_t)$, as evolving according to

$$Y_t^S = \mu + \sqrt{V_t}\epsilon_t^s$$
$$V_{t+1} - V_t = \kappa(\theta - V_t) + \sigma_v \sqrt{V_t}\epsilon_t^v.$$

For simplicity we assume that the drift μ is constant and that the unobserved state variables consists of a vector of volatilities $V = (V_0, \ldots, V_T)$ and the parameters are given by $\Theta = (\kappa, \theta, \sigma_v, \rho)$.

Incorporating the derivative price information leads to an observation equation consisting of a panel of N prices for T time periods of the form $Y_{i,t}^D = F_i(Y_t^S, V_t, \Theta, \Lambda) + \sigma_D \epsilon_{i,t}^D$ for $i = 1, \ldots, N$ and $t = 1, \ldots, T$. This gives an extra likelihood term

$$p(Y^D \mid Y^S, V, \Theta, \Lambda) \propto \sigma_D{}^{-NT} \exp\left(-\frac{1}{2\sigma_D^2} \sum_{i=1}^{N} \sum_{t=1}^{T} \left(Y_{i,t}^D - F_i(Y_t^S, V_t, \Theta, \Lambda) \right)^2 \right).$$

Combining this information with the asset dynamics gives a joint posterior distribution

$$p(X, \Theta, \Lambda \mid Y) \propto p(Y^D \mid Y^S, X, \Theta, \Lambda)\, p(Y^S \mid X, \Theta)\, p(X, \Theta)\, p(\Lambda)$$

Inference is then determined by sampling from the joint posterior distribution of the parameters and volatility states. For example, in the leverage SV model, we sample in a block algorithm defined by the set of conditionals

Volatility states:	$p(V_t \mid V_{-t}, \sigma_D^2, \mu, \lambda_v, \kappa, \theta, \rho, \sigma_v, Y)$
Stock price drift:	$p(\mu \mid V, \sigma_D^2, \lambda_v, \kappa, \theta, \rho, \sigma_v, Y)$
Pricing variance:	$p(\sigma_D^2 \mid V, \mu, \lambda_v, \kappa, \theta, \rho, \sigma_v, Y)$
Market price of risk:	$p(\lambda_v \mid V, \mu, \sigma_D^2, \kappa, \theta, \rho, \sigma_v, Y)$
Volatility parameters:	$p(\kappa, \theta \mid V, \mu, \sigma_D^2, \lambda_v, \rho, \sigma_v, Y)$
Correlation and volatility of volatility:	$p(\rho, \sigma_v \mid V, \mu, \sigma_D^2, \lambda_v, \kappa, \theta, Y)$

In constract to the problem of inference with only asset prices, many of these conditionals are unavailable in closed form when option prices are included due to the nonlinear pricing function $F(Y_t^S, V_t, \Theta, \Lambda)$ in the derivative likelihood. We, therefore, use a Metroplis–Hastings algorithm to sample the parameters that are affected by this, namely $(\lambda_v, \kappa, \theta, \rho, \sigma_v)$, and the volatilities V_t. To update the parameters, we used the block scheme indicated above with a normal proposal distribution centered at the current value with an estimated posterior covariance determined from a short pilot sample and a scale factor chosen to give acceptance probabilities around 30–50%. For the volatilities we used a univariate truncated normal centered at the current value with a variance chosen to yield acceptance rates of around 50%. Finally, the full conditional distributions for μ and σ_D^2 are known in closed form, being normal and inverse-gamma distributions respectively. We found that the algorithm converges fairly rapidly when the pricing error variance is fixed. However, we also found that convergence is much slower when σ_D^2 is unknown. In this case, we used a large Monte Carlo sample size of $M = 100,000$ and we found that this provides reasonable convergence diagnostics.

For our prior specification $p(\Theta, \Lambda)$ we use diffuse priors[3] where appropriate and normal priors for the transformed parameters $(\mu, \kappa, \kappa\theta)$. Inference is performed using the marginal posterior distribution $p(\Theta \mid Y)$. We use posterior means and standard deviations throughout as summaries.

3. APPLICATION: S&P 500 EQUITY OPTIONS

To illustrate our methodology we use daily data from the S&P 500 equity index returns and option prices from 1998 to 2002. The option prices are specified in the form of implied volatilities for different levels of the option Δ, specifically 0.25, 0.50, and 0.75, corresponding to out-of-the-money, at-the-money, and in-the-money options, respectively. For the purpose of illustration, we focus on short-term options with 1 month to expiration, giving a panel of $N = 3$ options at each time period.

For our option pricing model we use the leverage stochastic volatility model described above. The goal of our analysis is to provide more precise parameter estimates and to show how model diagnostics can be provided based on agreement of the posteriors with and without derivative prices.

Table 1 provides posterior estimates and standard deviations (in parentheses) for the mean annualized volatility $E(\sqrt{V_t})$, the speed of mean reversion κ, the volatility of volatility σ_v, the leverage effect ρ, the drift μ, and the market price of risk parameter λ_v. We consider two scenarios for the pricing errors: first, $\sigma_D = 5\%$, representing a typical bid–ask spread; and second, we leave σ_D unknown. The posterior mean for the pricing error was 3%. Our results show that there are little differences between these two cases.

Table 1. *Posterior Estimates of Model Parameters With and Without Derivative Information.*

Parameter	Y^S	Y^S, Y^D	Y^S, Y^D
		$\sigma_D = .05$	σ_D unknown
$E(\sqrt{V_t})$	21.60	21.90	21.82
	(1.46)	(1.15)	(0.86)
κ	0.045	0.046	0.055
	(0.012)	(0.005)	(0.005)
σ_v	0.047	0.048	0.055
	(0.004)	(0.002)	(0.001)
ρ	-0.70	-0.78	-0.74
	(0.08)	(0.02)	(0.01)
μ	.0014	0.0011	0.0010
	(.0004)	(.0004)	(0.0004)
λ_v	–	-0.010	-0.017
	–	(0.005)	(0.005)

We computed inference with and without derivative prices. As theory would suggest, our parameter estimates are very similar in both cases. However, the use of the derivative data

[3] Specifically, in our application we use the following prior distributions: $\mu \sim \mathrm{N}(0, 0.001^2)$ for the drift parameter; $\kappa\theta \sim \mathrm{N}(0.000004, 0.00002^2)$ and $\kappa \sim \mathrm{N}(0.02, 0.02^2)$ for the long run volatility mean and speed of mean reversion; $\rho \sim \mathrm{Un}(-1, 1)$ for the leverage effect; $\sigma_v^{-2} \sim \mathrm{Ga}(2, 0.0001)$ for the volatility of volatility; $\lambda_v \sim \mathrm{N}(0, 0.03^2)$ for the market price of risk parameter; and $V_0 \sim \mathrm{N}(0.0002, 0.00002^2)$ for the initial volatility.

provides more precise parameter estimates in some cases by a factor of eight. For example, the leverage effect is notoriously hard to estimate with asset price data alone (see Jacquier *et al.*, 2002). We find that, with equity price data alone, the posterior mean is $E(\rho \mid Y^S) = -0.70$ and the posterior standard deviation is $SD(\rho \mid Y^S) = 0.08$. With the additional derivative pricing information, we find $E(\rho \mid Y^S, Y^D) = -0.74$ and $SD(\rho \mid Y^S, Y^D) = 0.01$.[4]

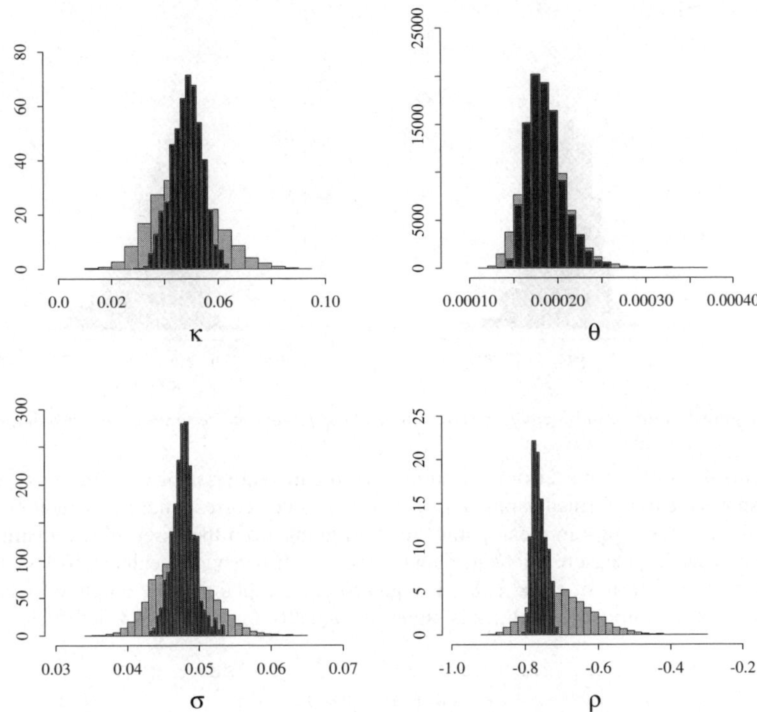

Figure 2. *Posterior histograms of* Θ, *leverage model,* $\sigma_D = 0.05$. *Without option data* (light); *with option data* (dark).

Figure 2 shows the posterior histograms for the four parameters $(\kappa, \theta, \sigma_v, \rho)$ with and without derivative price information. For the derivative pricing case we assumed a 5% pricing error. Notice that, as theory would suggest, the posterior means are in close agreement for the two cases, suggesting no model misspecification. As described above, the most significant reduction for the posterior SDs is for the leverage effect ρ.

Second, we provide estimates of the underlying state variables V_t. Comparing the state variables with and without derivative prices provides a useful model diagnostic. If the theoretical model is a true representation of the derivative prices and asset dynamics then these estimates should agree. Differences in sub-periods can be suggestive of periods of market stress or liquidity and credit risk events.

[4] These estimates are also consistent with Nandi's (1998) estimates for the S&P 500 of -0.79. There are a number of other estimates for different datasets and financial series; for example, Bakshi *et al.* (1997) finds an estimate of -0.64 and Eraker (2002) finds a lower estimate of -0.54 due to the inclusion of jumps.

To illustrate this, Figure 3 compares the estimated state variables $\hat{V}^S$ and $\hat{V}^D$ with and without the additional derivative price information. Notice that although there is close agreement for most of the period, there are a number of discrepancies. For example, the period from August 26 to October 20, 1998, shows clear differences in the estimated state volatilities. Although this could be due to estimation error, this suggests the need for other state variables such as jumps to explain the external market factors occurring in this period, such as the Russian GKO bond and LTCM crisis.

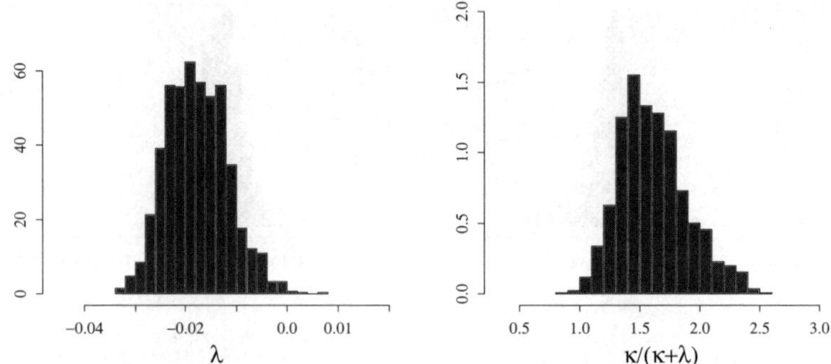

Figure 3. *Posterior histograms of market price of risk parameters, leverage model, unknown σ_D.*

Figure 4 shows the posterior distributions for the market price of volatility risk, $p(\lambda_v \mid Y)$. It also shows the transformed parameter $\kappa/(\kappa + \lambda_v)$, which corresponds to the factor multiplies that multiplies the long-run mean parameter θ in going from the asset price dynamics to the risk-neutral pricing measure. Not surprisingly, there is strong evidence that λ_v is less than zero, and the posterior mean of $\kappa/(\kappa + \lambda_v)$ is equal to 1.65, explaining the empirical fact that the mean level of the implied volatilities is higher than that of the estimated volatilities.

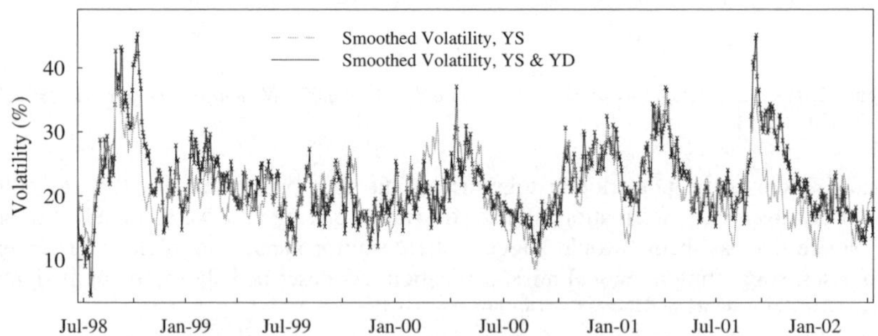

Figure 4. *Annualized estimated volatilities, $\hat{V}_t^S$ and $\hat{V}_t^D$, for the leverage model with $\sigma_D = .05$.*

To provide an illustration of model misspecification we consider a stochastic volatility model without a leverage effect, *i.e.*, $\rho = 0$. Figure 5 illustrates the parameter posterior distributions $p(\Theta \mid \rho = 0, Y^S)$ and $p(\Theta \mid \rho = 0, Y^S, Y^D)$ under the no leverage model with and without the extra derivative price information. If the model is correctly specified the estimates should be identical, the only difference being the tighter standard errors when the derivative prices are

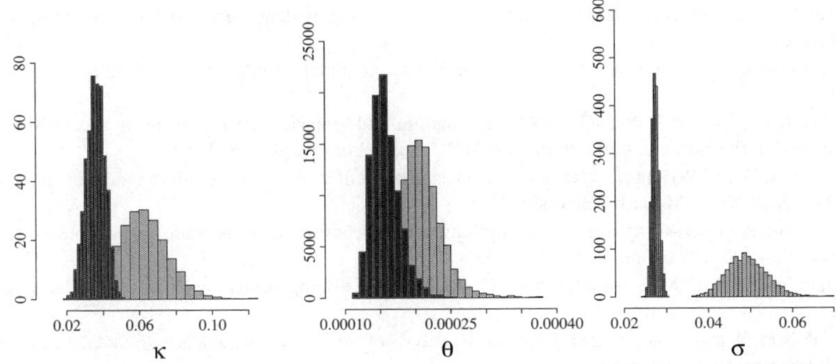

Figure 5. *Posterior histograms of* Θ*, no leverage model,* $\sigma_D = 0.05$*. Without option data* (light)*; with option data* (dark).

added. Notice the dramatic differences in parameter estimates. This clearly indicates that the model without a leverage effect is misspecified.

4. DISCUSSION

This paper develops techniques for Bayesian inference that combines information from derivative and asset price data. Combining such information is increasingly common in practice as it provides more precise parameter estimates. It has the additional feature of drawing inference about market price of risk parameters which are useful in derivative pricing. Furthermore, the estimated parameters and state variables can be used to provide model misspecification diagnostics. These can be identified by finding discrepancies in the estimated parameters with and without the additional derivative price information, and also by comparing the time series properties of the estimated state variables.

There are a number of interesting directions for further study. First, these methods can clearly be extended to multi-factor models and can be applied in a number of different situations such as term-structure and exchange rates modelling. Second, extending this approach to incorporate sequential state and parameter learning is an important direction. One difficulty that arises in practice is that the theoretical asset pricing framework needs to be extended to allow for sequential parameter learning and estimation risk. Recent work in nonlinear filtering methods such as Johannes *et al.* (2002) and Polson *et al.* (2002) are applicable for sequential state and parameter learning.

REFERENCES

Bakshi, G., Cao, C. and Chen, Z. (1997). Emprirical performance of alternative option pricing models. *J. Fin.* **5**, 2003–2049.

Bates, D. (1996). Jumps and stochastic volatility: Exchange rate processes implicit in deutsche mark options. *Rev. Fin. Studies* **9**, 69–107.

Bates, D. (2000). Post-'87 Crash fears in the S&P 500 Futures option market. *J. Econometrics* **94**, 181–238.

Black, F. (1976). Studies of Stock Price Volatility Changes. *1976 Meetings ASA*, 177–181.

Chernov, M. and Ghysels, E. (2000). A study towards a unified approach to the joint estimation of objective and risk neutral measures for the purposes of option evaluation. *J. Financial Economics*, **56**, 407–458.

Duffie, D. (2000). *Dynamic Asset Pricing Theory*. Princeton: Princeton University Press

Duffie, D., Pan, J. and Singleton, K. (2000). Transform anaplysis and asset pricing for affine jump-diffusions. *Econometrica* **68**, 1343–1376.

Eraker, B. (2002). Do stock prices and volatility jump? Reconciling evidence from spot and option prices. *Tech. Rep.*, Duke University, USA.

Eraker, B., Johannes, M. and Polson, N. (2002). The impact of jumps in returns and volatility. *J. Finance* (to appear).

Florentini, G., Leon, A. and Rubin, G. (2002). Estimation and empirical performance of Heston's stochastic volatility model: the case of a thinly traded market. *J. Empirical Fin.*, **9**, 225–255.

Forbes, C., Martin, G. and Wright, J. (2002). Bayesian estimation of a stochastic volatility model using option and spot prices. *Tech. Rep.*, Monash University, USA.

Heston, S. (1993). A closed-form solution for options with stochastic volatility with applications to bond and currency options. *Rev. Fin. Studies*, **6**, 327–343.

Heston, S. and Nandi, S. (2000). A closed-form GARCH option pricing valuation model. *Rev. Fin. Studies*, **13**, 585–625.

Jacquier, E., Polson, N. and Rossi, P. (2002). Bayesian analysis of stochastic vol models with fat tails and correlated errors. *J. Econometrics* (to appear).

Johannes, M. and Polson, N. (2002). MCMC methods for financial econometrics. *Handbook of Financial Econometrics* (Y. Aït-Sahalia and L. Hansen, eds). Princeton: Princeton University Press. (to appear).

Johannes, M., Polson, N. and Stroud, J. (2002). Nonlinear filtering for Stochastic Differential Equations with Jumps. *Tech. Rep.*, University of Chicago, USA.

Lewis, A. (2000). *Option Valuation under Stochastic Volatility*. Financial Press.

Lin, N. Y.-N., Strong, N. and Xu, X. (2002). Pricing FTSE 100 options under stochastic volatility. *J. Futures Markets* **21**, 127–211.

Nandi, S. (1998). How important is the correlation between assets and volatility in a stochastic volatility model? Empirical evidence from pricing and hedging in the S&P index options market. *J. Banking Finance* **22**, 589–610.

Polson, N., Stroud, J. and Müller, P. (2002). Practical filtering with sequential parameter learning. *Tech. Rep.*, University of Chicago, USA.

BAYESIAN STATISTICS 7, pp. 651–660
J. M. Bernardo, M. J. Bayarri, J. O. Berger, A. P. Dawid,
D. Heckerman, A. F. M. Smith and M. West (Eds.)

Gaussian Processes to Speed up Hybrid Monte Carlo for Expensive Bayesian Integrals

CARL EDWARD RASMUSSEN
Gatsby Unit, University College London, UK
edward@gatsby.ucl.ac.uk

SUMMARY

Hybrid Monte Carlo (HMC) is often the method of choice for computing Bayesian integrals that are not analytically tractable. However, the success of this method may require a very large number of evaluations of the (un-normalized) posterior and its partial derivatives. In situations where the posterior is computationally costly to evaluate, this may lead to an unacceptable computational load for HMC. I propose to use a Gaussian Process model of the (log of the) posterior for most of the computations required by HMC. Within this scheme only occasional evaluation of the actual posterior is required to guarantee that the samples generated have exactly the desired distribution, even if the GP model is somewhat inaccurate. The method is demonstrated on a 10 dimensional problem, where 200 evaluations suffice for the generation of 100 roughly independent points from the posterior. Thus, the proposed scheme allows Bayesian treatment of models with posteriors that are computationally demanding, such as models involving computer simulation.

Keywords: GAUSSIAN PROCESS; MARKOV CHAIN MONTE CARLO; HYBRID MONTE CARLO; NONPARAMETRIC MODELS; DERIVATIVE PROCESS; QUADRATURE; DESIGN.

1. INTRODUCTION

Evaluation of integrals is at the heart of Bayesian inference: integrating wrt. the posterior to make predictions, and integrating the likelihood wrt. the prior to obtain the marginal likelihood for model comparison. The later is often more difficult than the former, since the prior is typically fairly broad and the likelihood very peaked. In this paper, we consider the former, and will refer to the posterior wrt. which we wish to integrate as the *target* density.

For many interesting statistical models, direct analytical integration is not possible and Markov chain Monte Carlo (MCMC) methods are often used. Simple forms of MCMC may require a very large number of evaluations of the target density in order to produce reliable results. One of the main causes is that the target density often exhibits strong dependencies between variables, such that the regions of high target density form a narrow extended manifold; situations where this manifold is broken up into several distinct regions (multi-modality) is even more difficult and will not be specifically dealt with here.

Strong dependencies between variables in the target density are probably typical to most Bayesian inference problems, and extended directions within the manifold corresponds to possible families of explanations for the observed data. Integration wrt. these uncertainties is one of the main virtues of the Bayesian treatment, contrasting with paradigms based on point estimates.

The strong dependencies between variables make it difficult to explore the target density while remaining within regions of high probability. In situations where the posterior itself is computationally demanding to evaluate, this makes simple MCMC schemes impractical. Computationally demanding posteriors arise in several contexts: 1) hierarchical models where the lower levels of the hierarchy have been integrated out to reduce the number of parameters typically results in complex posteriors over the remaining parameters; and 2) models that use computer simulation as an integral part of their specification, such as is typical for many geophysical or meteorological applications. In these situations evaluation of the target density at a single point can in itself be computationally demanding. In such circumstances it is obvious that we need to use the information gained from each evaluation as effectively as possible.

The new method proposed in this paper is an elaboration of the Hybrid Monte Carlo (HMC) method of Duane *et al.* (1987), using a Gaussian Process (GP) model to guide the search and avoid too many references to the target density function. Gaussian process models have previously been suggested for evaluating integrals by O'Hagan (1991) under the name of Bayes–Hermite quadrature, see also the excellent study by O'Hagan (1992). In this methodology an "importance" density is introduced, and the Gaussian process is used to model the integrand times the ratio of the posterior and importance density. For analytical tractability the importance density must be Normal, although this requirement has been relaxed by Kennedy (1998). As we will see, the current approach attempts to tackle exactly the same problems, but in a very different way.

The Hybrid Monte Carlo (HMC) method of Duane *et al.* (1987) is a fairly generally applicable approach to sampling more efficiently from highly correlated densities by suppressing (inefficient) random walk behavior. Even though HMC is often much faster than sampling schemes relying on random walks, it may still require a large number of evaluations of the target density. In this paper, I propose an extension to HMC which allows a further reduction in the number of target evaluations. The key idea is to use an approximate model of the target density for most of the computations and only occasionally require the evaluation of the true target; this is done in such a way that the generated samples will have exactly the desired distribution; see Neal (1996, Sec. 3.5) and Liu (2001, Ch. 9) for related ideas.

The remainder of the paper is structured in the following way. In Section 2 the motivation for seeking to suppress random walks is discussed, and the Hybrid Monte Carlo approach is outlined. In Section 3 some results for Gaussian process models are reviewed including conditioning on observed values of function derivatives. The new algorithm is given in Section 4 and its performance is demonstrated on a couple of simple problems. Conclusion and further discussion are given in Section 5.

2. RANDOM WALK METHODS VERSUS HYBRID MONTE CARLO

In highly correlated distributions, sampling methods relying on random walks are inefficient in exploring the target density. Using steps of size s, it will take an expected L^2/s^2 steps to move some distance L using random walk, since the trajectory will often fold back on itself. This is illustrated in a simple bivariate Normal case in Figure 1, for the very commonly used methods of Gibbs sampling and Metropolis sampling.

Since the typical step sizes are confined to be of the order of the narrowest direction in order to get good acceptance probabilities,[1] the quadratic exploration time of the random walk will lead to poor performance as the strength of the correlation increases. The situation is similar for the Gibbs sampling algorithm. The most straight forward solution to this problem would be to make a variable transformation to eliminate the strong dependencies, but for complex distributions

[1] Strictly speaking, this holds only for $D > 2$ dimensions; in $D = 2$ dimensions the acceptance rate falls only linearly with the proposal width, so the optimal width is the longer of the two directions. So Figure 1 is slightly misleading in this respect.

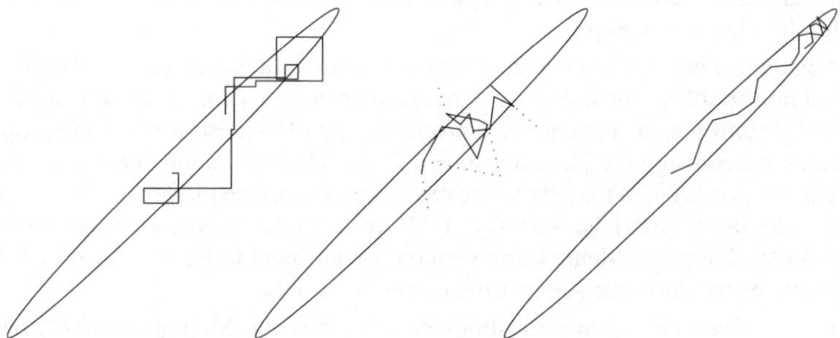

Figure 1. *Examples of sampling from a correlated bivariate Normal density; the one standard deviation contour of the density is drawn. Sample trajectories for three different sampling algorithms are shown.* Left*: 20 (half) iterations of Gibbs sampling;* middle*: 20 Metropolis proposals (rejected proposals are indicated with dotted lines);* right*: 20 sets of leapfrog iterations for Hybrid Monte Carlo. The three trajectories represent comparable computational burdens, although the HMC requires evaluation of derivatives in addition to density values.*

encountered in Bayesian inference this can be very difficult. Other approaches include the ordered overrelaxation scheme of Neal (1998b) and the Hybrid Monte Carlo algorithm, (HCM) which can be seen as a more elaborate way of generating Metropolis proposals.

The HMC algorithm is reviewed in a statistical setting by Neal (1993). In HMC a fictitious dynamical system is introduced in which the variables θ are interpreted as positions, and are augmented by a set of "momentum" variables, ϕ. Metropolis proposals are made by simulating the dynamical system through fictitious time. Because of momentum in the physical system, the random walks are suppressed. The energies of the physical system is related to the probability via the usual duality:

$$E_{\text{pot}} = -\log p(\theta), \quad E_{\text{kin}} = \frac{1}{2}\sum_i \phi_i^2, \quad H = E_{\text{pot}} + E_{\text{kin}}, \quad p(\theta, \phi) = \exp(-H). \tag{1}$$

Thus, the potential energy E_{pot} is related to the target distribution $p(\theta)$, which we wish to sample from. The kinetic energy is related to the square of the momentum. Note that the joint probability of θ and ϕ factors into independent parts; thus we can obtain samples from $p(\theta)$ by sampling from the joint and ignoring the momenta.

The evolution of the system through fictitious time at constant total energy H is governed by Hamilton's equations:

$$\frac{d\theta_i}{dt} = \frac{\partial H}{\partial \phi_i} = \phi_i, \qquad \frac{d\phi_i}{dt} = -\frac{\partial H}{\partial \theta_i} = -\frac{\partial E_{\text{pot}}}{\partial \theta_i}. \tag{2}$$

Unfortunately, for most target densities Eq. (2) are coupled nonlinear differential equations (involving the partial derivative of the log posterior wrt. the parameters) which we cannot solve in closed form. Instead we approximate the solution by iterating first order steps:

$$\phi_i(t+\varepsilon) := \phi_i(t) - \varepsilon\frac{\partial E_{\text{pot}}(\theta(t))}{\partial \theta_i}, \qquad \theta_i(t+\varepsilon) := \theta_i(t) + \varepsilon\phi_i(t+\varepsilon). \tag{3}$$

In practice we use the slightly modified "leapfrog" scheme, and take first a half step for the momentum, then iterate full steps for position and momentum, and finally another half step for the momentum; this is done to make the approximation reversible (which is needed for detailed balance for the Metropolis step).

The full algorithm for HCM becomes iteration of the following steps: (1) Gibbs sample for the momentum variables, whose density is the standard normal (Eq. 1); (2) simulate the physical system for L leapfrog steps to generate a proposal; and (3) accept or reject the proposed state in a Metropolis step using the joint densities $p(\theta, \phi)$. This procedure generates samples from the joint density $p(\theta, \phi)$, and we get the desired samples from the marginal $p(\theta)$ by ignoring the momenta. In Figure 1 (right) the success of HMC in suppressing random walks is shown. Note that the so-called Langevin Monte Carlo method is equivalent to HMC with $L = 1$, but this of course eliminates the desirable suppression of random walks.

If we had solved Hamilton's equations exactly, then the Metropolis step would always accept, since the total energy is conserved by the dynamics. However, since we only followed the dynamics approximately, the energy may change and this may lead to rejections. Notice however, that the samples obtained have exactly the desired distribution, although the dynamics were approximate.

2.1. *Using Hybrid Monte Carlo*

When using the HMC procedure in practice, one needs to address two main issues to ensure that the algorithm is performing well. First, we need to chose a step size ε for the leapfrog iterations. If using a single step size[2] as in Eq. (3) this is fairly easily done by monitoring the resulting acceptance rate. Secondly, once good step sizes have been established, we need to determine the appropriate length L of the simulation. This can usually be done with some experimentation, for example, by monitoring the auto covariance function for parameters and increasing L until roughly independent samples are obtained. Too short trajectories will cause a failure to suppress random walks and too long trajectories will be wasteful of computation.

Once reasonable values for ε and L have been determined, the HMC algorithm can be run to produce the desired samples. In Figure 1 the bivariate Normal has one direction which is 10 times as long as the perpendicular direction. Using a simple Metropolis scheme we would expect the step sizes to be confined by the narrowest direction, and thus expect roughly 10^2 steps to be necessary to generate roughly independent samples. In this example, we see that HMC is able to traverse the distribution using 20 evaluations (of the posterior and its partial derivatives), thus providing improved performance. However, we may still need a large number of evaluations to achieve adequate samples. The key idea in this paper is that we can use a model of the target instead of the target itself for the dynamical simulation, and thereby save greatly on the number of evaluations necessary. It turns out that Gaussian processes are a very convenient tool to model the posterior.

3. GAUSSIAN PROCESS REGRESSION

We can use a noise free[3] Gaussian process to specify a distribution over smooth functions. Consider a set of n pairs of D dimensional covariate x and dependent variables y: $\{x^{(c)}, y^{(c)} \mid c = 1, \ldots, n\}$. We specify that the dependent variables have a joint Normal distribution, with a covariance matrix that depends on the covariates. For example:

[2] It is possible that the appropriate step sizes vary widely between different coordinates of θ; near optimal individual step sizes may not be easy to estimate.

[3] In practice, I add a tiny bit of noise (by adding a small multiple of the identity to the covariance matrix) to improve numerical conditioning.

$$p(\boldsymbol{y}\,|\,\boldsymbol{x}) \sim \mathrm{N}(\boldsymbol{y}\,|\,0, \Sigma), \quad \Sigma_{pq} = C(x^{(p)}, x^{(q)}) = w_0 \exp\Big(-\frac{1}{2}\sum_{d=1}^{D}(x_d^{(p)} - x_d^{(q)})^2/w_d^2\Big), \quad (4)$$

where x_d is the d-th covariate and boldface is used to indicate aggregation over the entire set of n observations. Note, that here $\Sigma_{pq} = \mathrm{Cov}(y^{(p)}, y^{(q)}) = C(x^{(p)}, x^{(q)})$, the covariance between the dependent variables, is a function C of the covariates. This defines a zero mean Gaussian process with a stationary (Gaussian) covariance function, see O'Hagan (1978) and Neal (1998a) for further background on GP models. The covariance is parameterized by hyperparameters: length scales w_d and overall scale w_0. Obviously there are many other possible choices for covariance functions, but here we will stick with Eq. (4).

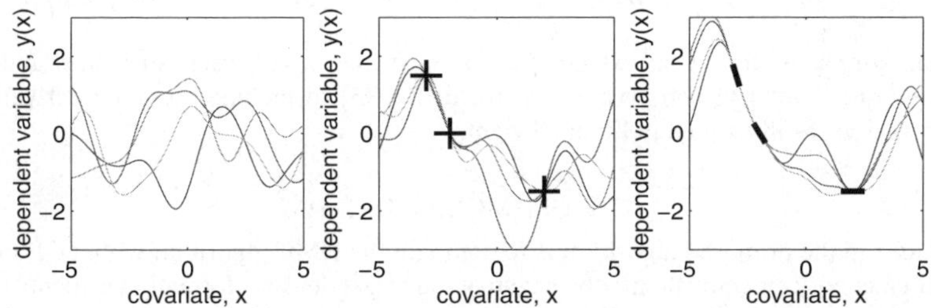

Figure 2. *Examples of random functions drawn from a Gaussian process with Gaussian covariance function (with $w_0 = 1$ and $w_1 = 1$).* Left: *four functions drawn at random;* middle: *four functions drawn at random conditional on three observations indicated with + signs;* right: *same observations as the middle plot, but now the four random functions are also conditional on the derivative at those points, indicated by small tangent segments. It is seen how more information about the function confines the distribution over functions, and how the uncertainty about the functions varies with x.*

Having observed the values of some y variables, we are interested in predicting the y^* corresponding to a new value of the covariate, x^* which is obtained by conditioning the joint on the observed values, yielding a Normal:

$$p(y^*\,|\,x^*, \boldsymbol{y}, \boldsymbol{x}) = \mathrm{N}(y^*\,|\,\mu, \sigma^2), \quad \begin{aligned}\mu &= C(x^*, \boldsymbol{x})C(\boldsymbol{x}, \boldsymbol{x})^{-1}\boldsymbol{y}\\ \sigma^2 &= C(x^*, x^*) - C(x^*, \boldsymbol{x})C(\boldsymbol{x}, \boldsymbol{x})^{-1}C(\boldsymbol{x}, x^*).\end{aligned} \quad (5)$$

A priori we may not have any clear preferences for particular values of the hyperparameters. In principle, we should therefore set a prior over these and integrate out the hyperparameters as in Neal (1998a). For simplicity I will simply optimize the likelihood wrt. the hyperparameters. The log likelihood L and its partial derivatives, Williams and Rasmussen (1996), are given by:

$$\begin{aligned} L = \log p(\boldsymbol{y}|\boldsymbol{x}, \boldsymbol{w}) &= -\frac{1}{2}\log|\Sigma| - \frac{1}{2}\boldsymbol{y}^t\Sigma^{-1}\boldsymbol{y} - \frac{n}{2}\log(2\pi),\\ \frac{\partial L}{\partial w_i} &= -\frac{1}{2}\mathrm{tr}(\Sigma^{-1}\frac{\partial \Sigma}{\partial w_i}) + \frac{1}{2}\boldsymbol{y}^t\Sigma^{-1}\frac{\partial \Sigma}{\partial w_i}\Sigma^{-1}\boldsymbol{y}. \end{aligned} \quad (6)$$

The likelihood can conveniently be optimized with a standard numerical tool such as a conjugate gradient method (subject to local maxima caveats).

3.1. *Gaussian Process Derivatives*

Since differentiation is a linear operator, the derivative of a Gaussian process is again a Gaussian process. For example, the mean of the derivative is equal to the derivative of the mean for predictions:

$$E[\frac{\partial y^*}{\partial x^*_d}] = \frac{\partial E[y^*]}{\partial x^*_d} = \frac{\partial \mu}{\partial x^*_d} = \frac{\partial C(x^*, \boldsymbol{x})}{\partial x^*_d} C(\boldsymbol{x}, \boldsymbol{x})^{-1} \boldsymbol{y}, \tag{7}$$

and we can similarly find the variance of the distribution of the derivative. Equally importantly for the present purposes, we wish to observe values for the partial derivatives and condition on these to make inference about the target function. The covariance between function values and derivatives, and between derivatives are given by:

$$\text{Cov}\Big(y^{(p)}, \frac{\partial y^{(q)}}{\partial x^{(q)}_d}\Big) = \frac{\partial C(x^{(p)}, x^{(q)})}{\partial x^{(q)}_d}, \qquad \text{Cov}\Big(\frac{\partial y^{(p)}}{\partial x^{(p)}_d}, \frac{\partial y^{(q)}}{\partial x^{(q)}_f}\Big) = \frac{\partial^2 C(x^{(p)}, x^{(q)})}{\partial x^{(p)}_d \partial x^{(q)}_f}, \tag{8}$$

so we can simply write out an extended covariance matrix (between all values and partial derivatives) and extend the observation vector in Eq. (5) to include observed derivatives, to make predictions as illustrated in Figure 2 right.

4. THE GPHMC ALGORITHM

The key idea in the proposed algorithm is to combine the HMC algorithm with a GP model of the potential energy, or equivalently the negative log target density; formally we identify x with θ and the dependent variable y with minus the log (un-normalized) target density. The zero mean GP is used to model the observations with the empirical mean subtracted. The GPHMC algorithm is initialized by evaluating the target function at a number, say D of random locations (perhaps sampled from the prior). The algorithm now proceeds in two phases:

Exploratory phase. The goal in this phase is to gather information about the target density. We use HMC with long trajectories, $L = 1000$. We set the potential energy for HMC to $E_{\text{pot}} = \mu - \sigma$ which causes the dynamical simulation to seek out regions of high target density and regions in which the GP model is uncertain about the shape of the target density. The value of the uncertainty is monitored along the trajectory, and if ever the uncertainty reaches a threshold $\sigma = 3$, further simulation seems unlikely to yield meaningful information, so the simulation is stopped, and the target density (and its partial derivatives) are evaluated. In this way the algorithm seeks out good placements of design points, a feature which is crucial if a low number of evaluations are to suffice. The next trajectory is then simulated, either starting from the new point, or from the beginning of the previous trajectory (as to avoid restarting at points of very low probability), based on a Metropolis criterion. Between each trajectory the hyperparameters of the covariance function are refined by 20 iterations of conjugate gradient optimization of the likelihood, Eq. (6) including the newly gathered information.[4]

Sampling phase. The goal of this phase is to generate the desired samples from the target distribution using the GP model generated in the previous phase. We use HMC with $E_{\text{pot}} = \mu$ and $L = 1000$. The end point of the trajectory is subject to a Metropolis acceptance step, requiring a single evaluation of the target density; this ensures that the samples generated will have exactly the desired distribution, although the GP model of the target density may still be

[4] When a reasonable number of data points have been collected, the initial (random) data points are gradually removed, since they will tend to have very low target density values, which renders them unimportant for inference, and they may lead to numerical instabilities for the GP model.

imperfect. The rejection rate indicates how well the GP model has captured the target density; too many rejections calls for an extension of the exploratory phase.

It should be noted that in the new algorithm, the setting of the step sizes ε and trajectory lengths L is much less critical than for the original HMC algorithm. The step size is set low enough that long trajectories get accepted fairly early on in the exploratory phase. The more difficult L parameter, which must be adjusted carefully for optimal performance in HMC, can now simply be set to some quite large value (such as $L = 1000$), since the simulation no longer refers to the actual target density, but uses the GP model, which is assumed much cheaper to evaluate. This property may make the new scheme easier to use in practice.

The computational overhead of the proposed method (excluding the evaluation of the target function and its derivatives) consist of two major components. Firstly, we need to follow the dynamics for L iterations which requires initially a single inversion of $C(\boldsymbol{x}, \boldsymbol{x})$ which is cubic in the number of (retained) constraints $c = i(D + 1)$, where i is the number of iterations. In the sampling phase, the entire dynamics can now be followed in time $\mathcal{O}(Lc)$; in the exploratory phase, where also the uncertainty of the GP is required, the complexity is $\mathcal{O}(Lc^2)$. Secondly, optimization of hyperparameters Eq. (6) requires $\mathcal{O}(c^3)$. As discussed in Section 5, this considerable overhead makes the method most suitable in situations where the target function is very costly to compute.

4.1. *A Demonstration*

In this section I will demonstrate the viability of the proposed approach, by applying it to two problems with known density. The first toy example is a $D = 2$ dimensional target density proportional to $\exp(-8(x_1^2 + x_2^2 - 1)^2)$, see Figure 3. The GPHMC method is run for 100 iterations in each of the two phases. In the sampling phase there were no rejections, confirming that the model of the target density is quite accurate, in fact 100 exploratory iterations seems to be overkill in this example. The generated samples seem to be essentially independent judging by the auto covariance function. In conclusion we have generated 100 independent samples from the desired distribution, evaluating the target function 200 times and its partial derivatives 100 times.

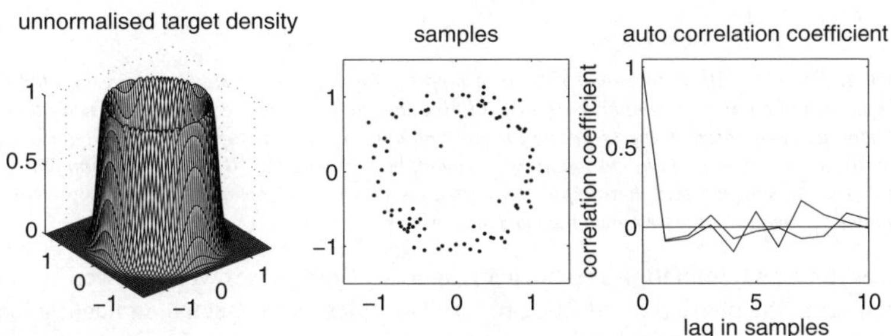

Figure 3. *The $D = 2$ dimensional example. Left: the (un-normalized) target density, middle: the 100 generated samples, right: the auto covariance for each of the two dimensions as a function of the iteration lag number.*

The second example is more challenging in $D = 10$ dimensions, with a Normal target density whose covariance matrix has an eigenvector $(1, 1, \ldots, 1)^t$ with a corresponding eigenvalue of 1.0, and all other eigenvalues are 0.01. Thus, the target density has 9 confined directions and a single direction which is 10 times wider, and all variables are (positively) correlated.

Using a standard density will make it possible to determine whether the generated samples are adequate, and does not make it any easier for the non-parametric procedure. Notice also, that the parameterization of the covariance function does not allow the GP to discover directly that there is one long direction and 9 confined ones (by setting the w hyperparameters) since I have chosen the directions to be diagonal rather than axis aligned. Further, the GP is using a stationary covariance function to model the log of the target density (which is quadratic), which is not tailored to the situation. Of course in any practical situation we should try to use a covariance function which reflects our beliefs about the actual target function at hand; however, we typically wo not have strong knowledge of this, and also the covariance functions that are convenient to handle in practice are limited. All in all, I have attempted to create an example which has a realistic amount of difficulty, while being easy to evaluate.

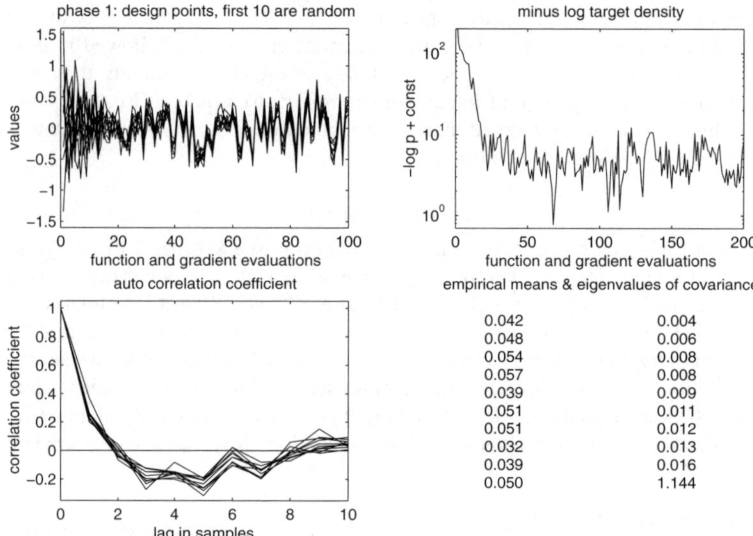

Figure 4. *The* $D = 10$ *dimensional Normal example.* Top left*: the design points generated by the procedure (the first 10 are random* $\mathrm{N}(0, 1)$*); all 10 dimensions are plotted on top of each other; it is seen how the correlation in the target density is gradually recovered;* top right*: negative log density gradually falls as regions of high probability are found;* bottom left*: the 10 auto correlation function for each dimension suggests that the samples generated are close to independent;* bottom right*: empirical mean and eigenvalues of the empirical covariance matrix.*

The results of the simulation are shown in Figure 4. Both phases were allowed 100 iterations. During the sampling phase 5 of the 100 proposed samples were rejected, indicating that the GP approximation to the log target density is not perfect, but still adequate for a high acceptance rate. Also in this example, we generated 100 roughly independent samples from the target density using only 200 evaluations of the density, and 100 evaluations of its partial derivatives. The empirical means are within the tolerance expected based on 100 independent samples, and the eigenvalues of the empirical covariance of the 100 samples also seem typical (when informally compared to draws from the corresponding Wishart distribution). So there are reasons to believe that the procedure produced the correct results in this example.

We can compare this performance to simpler MCMC schemes. The generation of 100 roughly independent samples using simple Metropolis would take on the order of 10,000 evaluations (100 samples times the squared ratio of widest to most confined direction). The HMC procedure with a step size resulting in around 20% rejections, needs roughly $L = 20$ (found through experimentation) resulting in 100 roughly independent samples requiring about 2000 evaluations. In conclusion, this example shows that GPHMC is about an order of magnitude faster than HMC, which itself is about an order of magnitude faster than simple Metropolis. The speed up gained for any task is of course dependent on the magnitude of the dependencies under the target density.

5. CONCLUSIONS, DISCUSSION AND FUTURE DIRECTIONS

A new version of the HMC algorithm has been proposed, which is suitable for doing integrals over distributions that are computationally difficult to evaluate. The way the algorithm has been presented here we also require evaluation of the partial derivatives, which may in some cases be comparatively easy to compute; for example, this is the case for GP models themselves Eq. (6), where evaluation of the likelihood is cubic in the number of constraints, but once this has been done each of the partial derivatives take only quadratic time. If derivatives are not easy to compute, it is of course possible to run the algorithm using only function values.

A simple example in $D = 10$ dimensions has shown the viability of the new method. It is natural to wonder how large problems the method can be used for. This depends on how many constraints are necessary to pin down the target density. Even a Normal distribution requires $\mathcal{O}(D^2)$ constraints; since we are doing a non-parametric fit, and because of the sequential nature of the selection of design points, we probably cannot get close to the globally optimal design, and would expect to need at least a small multiple of D^2 constraints. In the $D = 10$ example, at the end of the exploration phase we have retained 75 (out of the 100, since we gradually remove the initial points which have low density), giving $75 \times 11 = 825$ constraints. Inversion of the corresponding 825×825 covariance matrix takes on the order of one minute on a modern laptop computer. Since the computation times scales cubically in the number of constraints, the overall computation times is expected to be something like D^6 for direct implementations, which unfortunately precludes sizes much larger than $D = 15$ or so.

Interesting future work should compare this new proposed method to the related Bayes–Hermite quadrature of O'Hagan (1991). In particular, it would seem worth investigating whether more appropriate covariance functions for log densities or ratios of densities could be handled. The current use of a stationary covariance with a fixed mean seems inappropriate, but perhaps it is not a bad model in regions of high density, which is the regions that are important to model well. Further understanding and development of these ideas should enable Bayesian treatment of computationally complex models, especially those involving computer simulations.

REFERENCES

Duane, S., Kennedy, A. D., Pendleton, B. J., and Roweth, D. (1987). Hybrid Monte Carlo. *Phys. Lett. B* **55**, 2774–2777.

Kennedy, M. (1998). Bayesian quadrature with non-normal approximating functions. *Statist. Comput.* **8**, 365-375.

Liu, J. S. (2001). *Monte Carlo Strategies in Scientific Computing*. New York: Springer

Neal, R. M. (1993). Probabilistic inference using Markov chain Monte Carlo Methods. *Tech. Rep.*, University of Toronto, Canada.

Neal, R. M. (1996) *Bayesian Learning for Neural Networks*. New York: Springer

Neal, R. M. (1998a). Regression and classification using Gaussian process priors. *Bayesian Statistics 6* (J. M. Bernardo, J. O. Berger, A. P. Dawid and A. F. M. Smith, eds). Oxford: Oxford University Press, 475–501 (with discussion).

Neal, R. M. (1998b) Suppressing random walks in Markov chain Monte Carlo using ordered overrelaxation. *Learning in Graphical Models* (M. I. Jordan, ed). Dordrecht: Kluwer, 205–225.

O'Hagan, A. (1978). Curve fitting and optimal design for prediction. *J. R. Statist. Soc. B* **40**, 1–42 (with discussion).

O'Hagan, A. (1991). Bayes–Hermite quadrature. *J. Statist. Plann. Inference* **29**, 245–260.

O'Hagan, A. (1992). Some Bayesian numerical analysis. *Bayesian Statistics 4* (J. M. Bernardo, J. O. Berger, A. P. Dawid and A. F. M. Smith, eds). Oxford: Oxford University Press, 345–365 (with discussion).

Williams, C. K. I., and Rasmussen, C. E. (1996). Gaussian processes for regression. *Advances in Neural Information Processing Systems* **8** (D. S. Touretzky, M C. Mozer and M. E. Hasselmo, eds). Cambridge, MA: MIT Press, 514–520.

BAYESIAN STATISTICS 7, pp. 661–670
J. M. Bernardo, M. J. Bayarri, J. O. Berger, A. P. Dawid,
D. Heckerman, A. F. M. Smith and M. West (Eds.)

Objective Bayesian Comparison of Laplace Samples from Geophysical Data

ABEL RODRÍGUEZ   GISELLE ÁLVAREZ
Universidad Simón Bolívar, Venezuela
abel@cesma.usb.ve   giselle@cesma.usb.ve

BRUNO SANSÓ
University of California Santa Cruz, USA and *Universidad Simón Bolívar, Venezuela*
bruno@ams.ucsc.edu

SUMMARY

We consider a Bayesian reference analysis for comparing two samples which follow Laplace distributions as well as the analysis of the performance of a Bayesian classification procedure for such a model. The methodology is motivated by an oil exploration problem where the geological structure of a reservoir has to be described. The aim is to characterize the lithology of the reservoir using well registers and seismic traces. We use a continuous wavelet transformation (CWT) of the original data and estimate the power of the transformation that links the energy of the CWT to its scale. For both well logs and seismic traces, we found that such an estimate follows a Laplace distribution with location depending on whether the lithology corresponds to gravel or sand. In this work we explore the comparison problem from an objective Bayesian viewpoint through the use of the Intrinsic Bayes Factor.

Keywords: LAPLACE DISTRIBUTIONS; INTRINSIC BAYES FACTORS; BAYESIAN CLASSIFICATION.

1. INTRODUCTION

In this paper, we consider the problem of comparing two populations of independent samples from Laplace or double exponential distributions using a non-subjective Bayesian approach. In particular, we employ the Intrinsic Bayes Factor (IBF) presented by Berger and Pericchi (1996). We also consider the implications of the results obtained in this comparison on the behavior of a Bayesian classification algorithm.

This study is motivated by an oil exploration problem where the geological structure of a reservoir has to be described using seismic traces and γ-ray logs of wells. This kind of data is shown to have a Laplace distribution under suitable transformations which include the use of continuous wavelet transformations (CWT).

However, the results presented here have much broader implications. Laplace distributions have been proposed as a way to obtain statistical procedures that are more robust to outliers than those based on normal distributions. Some interesting applications are those of Manly (1976) who uses the Laplace distribution to model fitness functions, Easterling (1978) who considers a model for steam generator inspection as exponential responses with Laplace errors and Bagchi *et al.* (1983) who use the Laplace distribution while modelling demand during lead time for slow-moving items.

The paper is organized as follows: first, we describe the motivating problem and the data at hand. Then we pose the identification of differences in the lithology as a comparison of models based on Laplace distributions, obtain the predictive distributions required for the calculation of the IBF and perform those calculations for our problem. Finally, we show a Bayesian approach that can be used for the classification of new observations and provide ROC curves that can be used to asses the misclassification probability of the method if the existing data are used as the training sample.

2. DESCRIPTION OF THE DATA

Seismic traces and γ-ray logs are two types of data commonly available in the area of geophysical exploration. Well logs are signals of high frequency recorded at well locations that can be helpful to differentiate among rock types. In our study we analyze a type of radioactivity log, the γ-ray log. However, well logs are only available at locations where wells have been drilled. Previous to drilling at a specific location, an exhaustive seismic exploration of the reservoir is done. Seismic traces are signals of low frequency, detected at the surface, that register the motions caused by vibratory sources in the earth. These records are generally available in the whole area and not only at the wells.

Our data consist of γ-ray logs for 17 wells, 9 of them known to be located in lithologies that are predominantly gravel and the rest known to be located in lithologies that are predominantly sand. The wells were drilled in a reservoir located in Western Venezuela, covering an area of 100 Km^2 for which seismic data are available. We are interested in detecting differences in the well logs and seismic traces arising from the lithologies and, in case these differences exist, use the seismic data as training samples for a Bayesian classification algorithm.

3. A MODEL FOR THE ENERGY OF SIGNALS

The application of statistical methods on γ-ray logs and seismic traces for characterization of oil reservoir lithology has been an uprising idea in the last years. Some examples are the works of Addy (1998) where neural networks are employed or that of Dumay and Fournier (1988), who apply cluster and factor analysis. Both works base their methods on the raw data.

A different approach, presented in Jiménez *et al.* (1999) treats these signals as "time series" indexed on depth, which allow the authors to perform a discrete wavelet decomposition of the signals. However this method has a reduced discrimination power since the discrete transformation provides only a limited number of points for low frequencies. To solve this we use the continuous wavelet transform, which is defined for any square integrable function $f(t)$ as a function of two variables (in our case, depth and frequency),

$$\mathcal{F}(a,b) = \int f(x)\overline{\psi_{a,b}(x)}\,dx$$

where, $\overline{\psi_{a,b}(x)}$ denotes the conjugate of $\psi_{a,b}(x)$. The function $\psi_{a,b}(x)$ is obtained by translations and dilations of the mother wavelet ψ, that is $\psi_{a,b}(x) = \psi\left((x-b)/a\right)$, for details see Vidakovic (1999). We use the Haar's base in the application we present in this paper, but other options (like the "Mexican hat" and the Gaussian) have provided similar results.

There is a link between the energy of $\mathcal{F}$ and the regularity of f. In fact, if $f \in C^{\alpha}$ (Hölder space with exponent α) then the localized energy is proportional to a power of the scale, that is,

$$(\mathcal{F}(a,b))^2 \propto a^{2\alpha+1}. \tag{1}$$

Thus, after applying the CWT to the original signal, an estimate of α is obtained for each depth b by regressing the points $\log\{[\mathcal{F}(a,b)]^2\}$ on $\log(a)$. This procedure, presented originally in

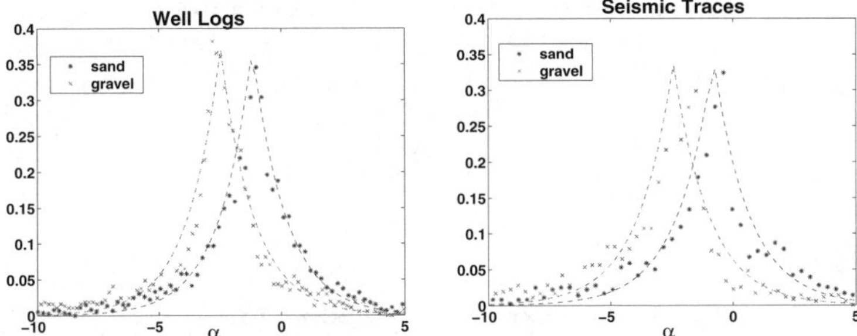

Figure 1. *Empirical distribution of the power parameter α. On the left panel the estimated densities of the coefficients corresponding to sand and gravel wells corresponding to γ-ray logs. An analogous plot is shown on the right panel for seismic traces around the same wells.*

Katul *et al.* (1999), can be considered as a generalization of the one proposed in Jiménez *et al.* (1999).

If we assume that there is no change in the lithology with depth, the values of α just obtained can be though as an iid sample of the energy of the signal. Now, for the data at hand, this distribution happens to have a peaky density that resembles the Laplace distribution (see Figure 1). To assess the validity of this assumption we considered the more general family of power exponential distributions presented in Box and Tiao (1973), defined by

$$p(\alpha_i \mid \mu, \sigma, \beta) = w(\beta)\sigma^{-1} \exp\Big[-c(\beta)\Big|\frac{\alpha_i - \mu}{\sigma}\Big|^{2/(1+\beta)}\Big]$$

where

$$c(\beta) = \left[\frac{\Gamma[\frac{3}{2}(1+\beta)]}{\Gamma[\frac{1}{2}(1+\beta)]}\right]^{1/(1+\beta)} \qquad w(\beta) = \frac{\Gamma[\frac{3}{2}\,(1+\beta)]^{1/2}}{(1+\beta)\{\Gamma[\frac{1}{2}\,(1+\beta)]\}^{3/2}}\,.$$

In this definition $-\infty < \mu < \infty$ and $0 < \sigma < \infty$ are, respectively, location and scale parameters, while $-1 < \beta \leq 1$ is a measure of kurtosis. Changes on β produce a wide range of shapes, from smooth densities like the normal ($\beta = 0$) to peaky ones, like the Laplace ($\beta = 1$). We consider independent reference priors for each parameter to obtain the posterior distribution of μ, σ and β. Figure 2 shows the marginal distribution of β in our case. We observe a high concentration around one, which confirms our selection of the Laplace distribution. Since this same behavior has also been present in resistivity logs and in data obtained from different locations, we are tempted to assume that this is a general distributional result for these types of signals.

Conditional on $\beta = 1$, the posterior mean for μ and σ for each lithology are shown in Table 1.

Table 1. *Estimated parameters for gravel and sand in well logs (left) and seismic traces (right) conditional on $\beta = 1$ (Laplace Model).*

Well Logs

Lithology	$E(\mu \mid \boldsymbol{\alpha})$	$E(\sigma \mid \boldsymbol{\alpha})$	n
Sand	-0.6500	1.5543	4096
Gravel	-2.4550	1.5953	4608

Seismic Traces

Lithology	$E(\mu \mid \boldsymbol{\alpha})$	$E(\sigma \mid \boldsymbol{\alpha})$	n
Sand	-0.8000	1.7013	1024
Gravel	-2.4002	1.7300	1152

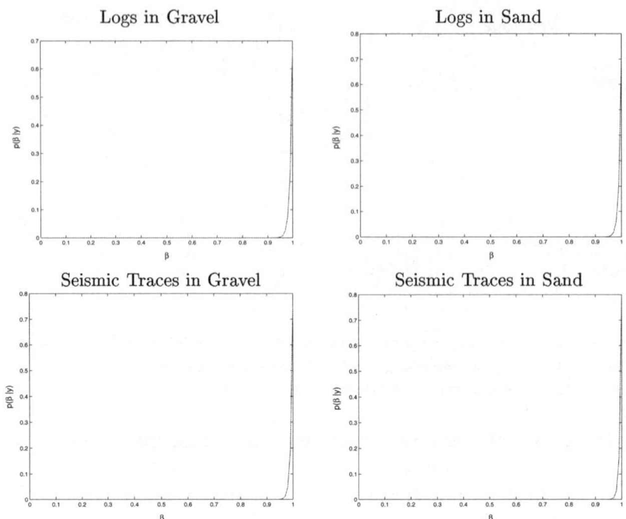

Figure 2. *Plot of $\pi(\beta \mid \boldsymbol{\alpha})$ (up to a multiplicative constant) for γ-ray logs (upper couple) and seismic traces (lower couple) for the two lithologies involved in our problem. In all cases $\beta = 1$ is clearly favored by the data.*

4. LITHOLOGY DISCRIMINATION

It is apparent from Figure 1 and Table 1 that the location of the densities change with the lithology, pointing at the possibility of posing the problem of identifying differences between lithologies as one of comparing Laplace populations. Indeed, it is important to verify if the difference is just on the mean or is also due to the variance. Let $\boldsymbol{x} \in R^{n_x}$ be the vector of power coefficients for either γ-ray logs or seismic traces located in gravel and $\boldsymbol{y} \in R^{n_y}$ the vector of power coefficients for logs in sand. Let $\boldsymbol{z}' = (\boldsymbol{x}', \boldsymbol{y}')$ and $n = n_x + n_y$.

Based on the previous results we assume that $x_i \sim \text{La}(x_i \mid \mu_1, \sigma_1)$ $i = 1, \ldots, n_x$ and $y_i \sim \text{La}(x_i \mid \mu_2, \sigma_2)$ $i = 1, \ldots, n_y$. So we need to consider the following hypotheses

$$M_1: \begin{array}{c} \mu_1 = \mu_2 = \mu \\ \sigma_1 = \sigma_2 = \sigma \end{array} \quad M_2: \begin{array}{c} \mu_1 \neq \mu_2 \\ \sigma_1 = \sigma_2 = \sigma \end{array} \quad M_3 \begin{array}{c} \mu_1 = \mu_2 = \mu \\ \sigma_1 \neq \sigma_2 \end{array} \quad M_4: \begin{array}{c} \mu_1 \neq \mu_2 \\ \sigma_1 \neq \sigma_2 \end{array}$$

where M_2, M_3 or M_4 imply that power coefficients calculated from signals that correspond to different lithologies have different Laplace distributions. This result could be used for for classification purposes.

5. MODEL COMPARISONS

5.1. *The Intrinsic Bayes Factor*

Denote by

(i) $M_1, \ldots, M_k$ the models under consideration.
(ii) $f_i(\boldsymbol{z} \mid \theta_i)$ the density of $\boldsymbol{z}$ under model M_i, $\theta_i \in \Theta_i$.
(iii) $\pi(\theta_i)$ the prior distribution of θ_i and p_i the prior probability of model M_i.
(iv) $m_j(\boldsymbol{z})$ the marginal density of $\boldsymbol{z}$ under M_j.

The key to Bayesian model selection are the Bayes factors:

$$B_{ji} = \frac{\int_{\Theta_j} f_j(\boldsymbol{z} \mid \theta_j)\pi_j(\theta_j)\, d\theta_j}{\int_{\Theta_i} f_i(\boldsymbol{z} \mid \theta_i)\pi_i(\theta_i)\, d\theta_i} = \frac{m_j(\boldsymbol{z})}{m_i(\boldsymbol{z})}$$

which can be combined with the prior probabilities p_i to get posterior probabilities for each model given the data

$$p(M_i \mid \boldsymbol{z}) = \Big[1 + \sum\nolimits_{j \neq i} \frac{p_j}{p_i} B_{ji}(\boldsymbol{z})\Big]^{-1}.$$

In our case we have $k = 4$ different models, each associated with one of the hypotheses presented in the previous section. The corresponding likelihoods, under the simplifying assumption of independence, are

$$f_1(\boldsymbol{z} \mid \mu, \sigma) = \Big(\frac{1}{\sqrt{2}\sigma}\Big)^{nx+ny} \exp\Big\{-\sqrt{2}\Big[\sum_{i=1}^{nx}\Big|\frac{x_i - \mu}{\sigma}\Big| + \sum_{i=1}^{ny}\Big|\frac{y_i - \mu}{\sigma}\Big|\Big]\Big\}$$

$$f_2(\boldsymbol{z} \mid \mu_1, \mu_2, \sigma) = \Big(\frac{1}{\sqrt{2}\sigma}\Big)^{nx+ny} \exp\Big\{-\sqrt{2}\Big[\sum_{i=1}^{nx}\Big|\frac{x_i - \mu_1}{\sigma}\Big| + \sum_{i=1}^{ny}\Big|\frac{y_i - \mu_2}{\sigma}\Big|\Big]\Big\}$$

$$f_3(\boldsymbol{z} \mid \mu, \sigma_1, \sigma_2) = \Big(\frac{1}{\sqrt{2}\sigma_1}\Big)^{nx}\Big(\frac{1}{\sqrt{2}\sigma_2}\Big)^{ny} \exp\Big\{-\sqrt{2}\Big[\sum_{i=1}^{nx}\Big|\frac{x_i - \mu}{\sigma_1}\Big| + \sum_{i=1}^{ny}\Big|\frac{y_i - \mu}{\sigma_2}\Big|\Big]\Big\}$$

$$f_4(\boldsymbol{z} \mid \mu_1, \mu_2, \sigma_1, \sigma_2) = \Big(\frac{1}{\sqrt{2}\sigma_1}\Big)^{nx}\Big(\frac{1}{\sqrt{2}\sigma_2}\Big)^{ny} \exp\Big\{-\sqrt{2}\Big[\sum_{i=1}^{nx}\Big|\frac{x_i - \mu_1}{\sigma_1}\Big| + \sum_{i=1}^{ny}\Big|\frac{y_i - \mu_2}{\sigma_2}\Big|\Big]\Big\}$$

On the other hand, given the nature of our problem, we have very little prior knowledge about the power coefficients. Thus, we assign equal prior probabilities to all models involved (that is, $p_i = 1/4$ for all i) and use Jeffrey's priors for the parameters; these correspond to

$$\pi_1(\mu, \sigma) \propto \frac{1}{\sigma} \quad \pi_2(\mu_1, \mu_2, \sigma) \propto \frac{1}{\sigma^2} \quad \pi_3(\mu, \sigma_1, \sigma_2) \propto \frac{1}{\sigma_1\sigma_2} \quad \pi_4(\mu_1, \mu_2, \sigma_1, \sigma_2) \propto \frac{1}{\sigma_1\sigma_2}.$$

However, Bayes factors are not well defined when improper prior distributions are used. So, we resort to the IBF, which uses a minimal subset of the sample to train the prior distribution (making it proper) and the rest of the sample for model comparison. The dependence on the training sample is eliminated by taking averages over all possible samples. A complete description can be found in Berger and Pericchi (1996).

Let $\boldsymbol{z}(l)$ be the l-th minimal training sample (in our case this is formed by 2 observations from $\boldsymbol{x}$ and 2 observations from $\boldsymbol{y}$) and let L be the total number of training samples. The Arithmetic Intrinsic Bayes Factor (AIBF) is defined as

$$B_{ji}^{AI}(\boldsymbol{z}) = B_{ji}^{N}(\boldsymbol{z})\frac{1}{L}\sum\nolimits_{l=1}^{L} B_{ij}^{N}(\boldsymbol{z}(l))$$

and the Geometric Intrinsic Bayes Factor (GIBF) is defined as

$$B_{ji}^{GI} = B_{ji}^{N}(\boldsymbol{z})\Big(\prod_{l=1}^{L} B_{ij}^{N}(\boldsymbol{z}(l))\Big)^{1/L}.$$

We chose the GIBF for our application since it is coherent for multiple comparisons.

5.2. *Predictive Distributions for the Laplace Likelihood*

In order to calculate the Bayes factor we need to obtain the predictive distributions for each of the models. These are provided by the following propositions. Their proofs are rather technical and are omitted for reasons of space.

Proposition 1. *The predictive distribution for model M_1 is:*

$$m_1(\boldsymbol{z}) = \frac{\Gamma(n-1)}{2^n} \sum_{i=0}^{n} w_i$$

with

$$w_i = \begin{cases} \frac{1}{2i-n}\left[\frac{1}{S_i^{n-1}} - \frac{1}{S_{i+1}^{n-1}}\right] & i \neq \frac{n}{2} \\ (n+1)\left(z_{(n/2+1)} - z_{(n/2)}\right)\frac{1}{S_{n/2}^{n}} & i = \frac{n}{2} \end{cases}$$

where $z_{(i)}$ denotes the i-th order statistic of $\boldsymbol{z}$ and $S_i = n\bar{z}_{(n)} - 2i\bar{z}_{(i)} + 2iz_{(i)} - nz_{(n)}$ with $\bar{z}_{(k)} = k^{-1}\sum_{i=1}^{k} z_{(k)}$ and $\bar{z}_{(0)} = 0$.

Proposition 2. *The predictive distribution for model M_2 is:*

$$m_2(\boldsymbol{z}) = \frac{\Gamma(n-1)}{2^n} \sum_{i=0}^{n_x}\sum_{j=0}^{n_y} w_{ij}$$

where w_{ij} is piecewise defined as

$$\begin{cases} \frac{1}{(2i-n_x)(2j-n_y)}\left(\frac{1}{Q_{i+1,j+1}^{n-1}} - \frac{1}{Q_{i+1,j}^{n-1}} - \frac{1}{Q_{i,j+1}^{n-1}} + \frac{1}{Q_{i,j}^{n-1}}\right) & i \neq \frac{n_x}{2}, j \neq \frac{n_y}{2} \\ \left[y_{\left(\frac{n_y}{2}+1\right)} - y_{\left(\frac{n_y}{2}\right)}\right]\left[\frac{n-1}{2i-n_x}\left(\frac{1}{Q^n_{i,\frac{n_y}{2}}} - \frac{1}{Q^n_{i+1,\frac{n_y}{2}}}\right)\right] & i \neq \frac{n_x}{2}, j = \frac{n_y}{2} \\ \left[x_{\left(\frac{n_x}{2}+1\right)} - x_{\left(\frac{n_x}{2}\right)}\right]\left[\frac{n-1}{2j-n_y}\left(\frac{1}{Q^n_{\frac{n_x}{2},j}} - \frac{1}{Q^n_{\frac{n_x}{2},j+1}}\right)\right] & i = \frac{n_x}{2}, j \neq \frac{n_y}{2} \\ n(n-1)\left(y_{\left(\frac{n_y}{2}+1\right)} - y_{\left(\frac{n_y}{2}\right)}\right)\left(x_{\left(\frac{n_x}{2}+1\right)} - x_{\left(\frac{n_x}{2}\right)}\right)\frac{1}{Q^{n+1}_{n_x/2,n_y/2}} & i = \frac{n_x}{2}, j = \frac{n_y}{2} \end{cases}$$

with

$$Q_{i,j} = S_i^x + S_j^y, \quad S_i^x = n_x\bar{x}_{(n_x)} - 2i\bar{x}_{(i)} + 2ix_{(i)} - n_x x_{(n_x)}$$

and

$$S_j^y = n_y\bar{y}_{(n_y)} - 2j\bar{y}_{(j)} + 2jy_{(j)} - n_y y_{(n_y)}.$$

Proposition 3 *The predictive distribution for model M_3 is:*

$$m_3(\boldsymbol{z}) = \frac{\Gamma(n_x)\Gamma(n_y)}{2^n} \sum_{i=0}^{n} \int_{z_i}^{z_{i+1}} \frac{d\mu}{R_x^{n_x}(\mu)R_y^{n_y}(\mu)}$$

where $R_x(\mu) = \sum_{j=1}^{n_x} |\mu - x_j|$ and $R_y(\mu) = \sum_{j=1}^{n_y} |\mu - y_j|$.

A closed form expression of $m_3(\boldsymbol{z})$ can be obtained using partial fractions, nevertheless, the numerical error involved in solving the linear equations associated with partial fractions for large n_x and n_y is larger than that of using numerical integration, which is thus our preferred method.

Proposition 4 *The predictive distribution for model M_4 can be written in terms of the predictive distribution of M_1 in the following way:*

$$m_4(\boldsymbol{z}) = m_1(\boldsymbol{x})m_1(\boldsymbol{y}).$$

5.3. *Results*

Due to the sample sizes involved, the number of training samples L is in the order of 10^{12} for seismic traces and 10^{14} for well logs. Since this makes the calculations infeasible in terms of computing time, we randomly chose 12,000 training samples.

Table 2. *Model probabilities obtained using the* GIBF.

$P(M_i \mid \boldsymbol{z})$	Well logs	Seismic traces
1	$8.42 \cdot 10^{-11}$	$2.71 \cdot 10^{-7}$
2	0.97	0.96
3	$8.42 \cdot 10^{-11}$	$3.38 \cdot 10^{-7}$
4	0.02	0.03

Table 2 shows the posterior probabilities of each model using the GIBF. The numbers clearly indicate that there is a difference between the location of the distribution of power parameters obtained in gravel and those obtained in sand. We note that the calculation time using Fortran code on a Pentium IV computer was under 20 s for seismic traces and under 6 min for well logs. The calculations were repeated 12 times to study its sensitivity to the training samples, showing no important changes and, thus, no dependence on the reduced set of training samples employed. Increasing the number of training samples did not substantially change the results.

6. BAYESIAN CLASSIFICATION

The predictive densities calculated in the previous section can be used to obtain a Bayesian classification procedure. Let $\boldsymbol{w}$ be a new sample of power coefficients whose lithology is unknown. Given $\boldsymbol{x}$ and $\boldsymbol{y}$, vectors of power coefficients of sand and gravel respectively, we can calculate the probability of each type of lithology as

$$P(\text{sand}) = \frac{m(\boldsymbol{w} \mid \boldsymbol{x})}{m(\boldsymbol{w} \mid \boldsymbol{x}) + m(\boldsymbol{w} \mid \boldsymbol{y})}, \qquad P(\text{gravel}) = \frac{m(\boldsymbol{w} \mid \boldsymbol{y})}{m(\boldsymbol{w} \mid \boldsymbol{x}) + m(\boldsymbol{w} \mid \boldsymbol{y})} = 1 - P(\text{sand}),$$

where

$$m(\boldsymbol{w} \mid \boldsymbol{x}) = \frac{m_1(\boldsymbol{w}, \boldsymbol{x})}{m_1(\boldsymbol{x})} \qquad m(\boldsymbol{w} \mid \boldsymbol{y}) = \frac{m_1(\boldsymbol{w}, \boldsymbol{y})}{m_1(\boldsymbol{y})}.$$

Then, we classify the lithology of $\boldsymbol{w}$ as sand if $P(\text{sand}) > P(\text{gravel})$ and as gravel in the other case. In order to quantify the misclassification error we need to calculate the distribution of a random variable, say B, defined as

$$B = \log\left[m(\boldsymbol{W} \mid \boldsymbol{X})\right] - \log\left[m(\boldsymbol{W} \mid \boldsymbol{Y})\right]$$

which depends on the distribution of $\boldsymbol{X}, \boldsymbol{Y}$ and $\boldsymbol{W}$. If we assume that there is a true model for the data, the following simplification holds:

Proposition 5. *Let* $\boldsymbol{X} = (X_1, \ldots, X_{nx})$, $\boldsymbol{Y} = (Y_1, \ldots, Y_{ny})$, $\boldsymbol{W} = (W_1, \ldots, W_{nw})$ *with* $X_i \sim \text{La}(X_i \mid \mu_1, \sigma_1)$, $Y_i \sim \text{La}(Y_i \mid \mu_2, \sigma_2)$ *and* $W_i \sim \text{La}(W_i \mid \mu_1, \sigma_1)$. *The distribution of* B *depends on* μ_1, μ_2, σ_1 *and* σ_2 *only through the value of*

$$r = \frac{\sigma_1}{\sigma_2} \qquad \text{and} \qquad d = \frac{|\mu_1 - \mu_2|}{\sigma_1\sqrt{1 + r^2}}.$$

The proof of this proposition can be found in the appendix. It can be seen that Proposition 5 holds not only for the classification of Laplace distributions but for any other location-scale family. We shall refer to d as the normalized distance between densities.

Obtaining an analytical expression for the distribution of B as a function of d and r is a difficult task. However, for the purpose of our example we have used simulation to obtain random samples that correspond to $r = 1$, which is the case favored by our IBF calculations, and to the same sample sizes as that of the seismic traces. Plotting the median and 80% quantile of the misclassification probability $P(B < 0)$, we obtained what can be considered as a Bayesian variant of the ROC curve.

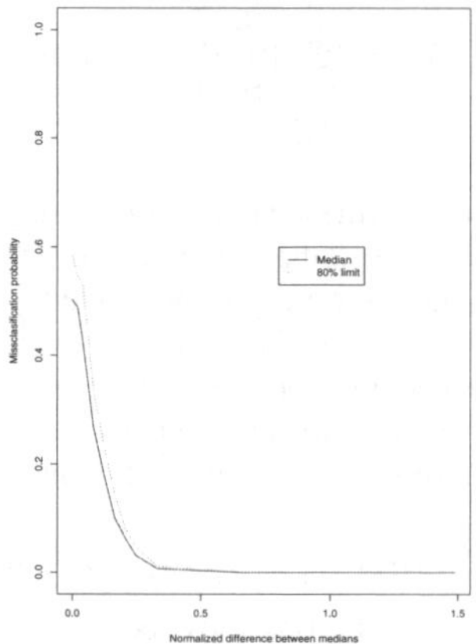

Figure 3. *Simulated response curve for classification of seismic traces with $r = 1$.*

The resulting curve is presented in Figure 3. From Table 1 we see that $d \approx 0.65$, which implies that our method produces a negligible classification error.

7. CONCLUSIONS

We have used the Intrinsic Bayes Factor to compare two Laplace models in the absence of prior knowledge for the parameters. Explicit calculations of the terms involved have been performed, obtaining formulae for the marginals under all the considered models which can be used to obtain fast and accurate results in a broad set of applications. Also, we presented results that justify the construction of Bayesian ROC curves for the classification of samples in location-scale families.

As an application, we have provided theoretical background for the implementation of Bayesian methods in oil reservoir exploration problems, showing that there are substantial differences between the power coefficients of both well logs and seismic traces obtained in sand and in gravel. This empirical fact raises an interesting physical theoretical question. We also prove that, despite the reduced sample sizes, lithology classifications based on seismic traces instead of γ-ray logs are accurate. Álvarez *et al.* (2003) have implemented similar ideas using BIC obtaining results along the lines presented in this paper.

ACKNOWLEDGEMENTS

We specially thank Reinaldo Michelena and Juan Ramón Jiménez from INTEVEP for their support and collaboration. This research was partially funded by grant G97-000592 of CONICIT, Venezuela.

REFERENCES

Addy, S. K. (1998). Neural network classification method helps seismic interval interpretation. *Oil and Gas* **96**, 47–58.

Álvarez, G., Sansó, B., Jiménez, J. R., Michelena, R. (2003). Lithologic characterization of a reservoir using continuous wavelet transforms. *IEEE Trans. Geosci. Remote Sensing* (to appear).

Bagchy, U., Hayya, J. C. and Ord, J. K. (1993). The Hermite distribution as a model of demand during lead time for slow-moving items. *Decision Sciences* **14**, 447–466.

Berger, J. O. and Pericchi, L. R. (1996). The intrinsic Bayes factor for model selection and prediction. *J. Am. Statist. Ass.* **91**, 109–122.

Box, G. E. P. and Tiao, G. (1973). *Bayesian Inference in Statistical Analysis*. New York: Wiley.

Dumay, J. and Fournier, F. (1998). Multivariate statistical analyses applied to seismic facies recognition. *Geophysics* **53**, 1151–1159.

Easterling, R. G. (1978). Exponential responses with double exponential measurement error: A model for steam generator inspection. *Proc. DOE Statistical Symposium, US Department of Energy*, 90–110.

Jiménez, J. R., Peinado, A. and Michelena, R. (1999). Facies recognition using wavelet based fractal analysis on compressed seismic data. *Tech. Rep.*, Intevep, Venezuela.

Katul, G., Vidakovic, B. and Albertson, J. (1999). Estimating global and local scaling exponents in turbulent flows using continuous wavelet transformations. *Tech. Rep.*, Duke University, USA.

Manly, B. (1976). Some examples of double exponential fitness functions. *Heredity* **36**, 229–234.

Vidakovic, B. (1999). *Statistical Modeling by Wavelets*. New York: Wiley.

APPENDIX 1. PROOF OF PROPOSITION 5

Proposition 5 is a straight corollary the following theorem:

Theorem 1. *Let* $f(x \mid \mu, \sigma) = \sigma^{-1} g\left((x-\mu)/\sigma\right)$ *for a given density g, that is, f belongs to a location-scale family. Let*

$$X_i \sim f(X_i \mid \mu, \sigma), \quad Y_i \sim f(Y_i \mid \mu + d, r\sigma), \quad W_i \sim f(W_i \mid \mu, \sigma)$$

for fixed values of r and d, and let $X_i^*, Y_i^*, W_i^* \sim g$. *Then*

$$\begin{aligned} B &= \log\left[m(\boldsymbol{W} \mid \boldsymbol{X})\right] - \log\left[m(\boldsymbol{W} \mid \boldsymbol{Y})\right] \\ B^* &= \log\left[m(\boldsymbol{W}^* \mid \boldsymbol{X}^*)\right] - \log\left[m(\boldsymbol{W}^* \mid r\boldsymbol{Y}^* + d)\right], \end{aligned}$$

where $m(\cdot)$ *denotes the predictive density, have the same distribution.*

Proof. Notice that X_i^*, Y_i^*, W_i^* are standardized versions of X_i, Y_i, W_i. Thus, we have the following relationships for the predictive densities:

$$m(\boldsymbol{W} \mid \boldsymbol{X}) = \frac{\int \frac{1}{\sigma} \prod_{i=1}^{nx} \frac{1}{\sigma} g\left(\frac{X_i - \mu}{\sigma}\right) \prod_{i=1}^{nw} \frac{1}{\sigma} g\left(\frac{W_i - \mu}{\sigma}\right) d\mu\, d\sigma}{\int \frac{1}{\sigma} \prod_{i=1}^{nx} \frac{1}{\sigma} g\left(\frac{X_i - \mu}{\sigma}\right) d\mu\, d\sigma} = \left(\frac{1}{r}\right)^{nw} m(\boldsymbol{W}^* \mid \boldsymbol{X}^*)$$

$$\begin{aligned} m(\boldsymbol{W} \mid \boldsymbol{Y}) &= \frac{\int \frac{1}{\sigma} \prod_{i=1}^{ny} \frac{1}{\sigma} g\left(\frac{Y_i - \mu}{\sigma}\right) \prod_{i=1}^{nw} \frac{1}{\sigma} g\left(\frac{W_i - \mu}{\sigma}\right) d\mu d\sigma}{\int \frac{1}{\sigma} \prod_{i=1}^{ny} \frac{1}{\sigma} g\left(\frac{Y_i - \mu}{\sigma}\right) d\mu d\sigma} \\ &= \left(\frac{1}{r}\right)^{nw} m(\boldsymbol{W}^* \mid r\boldsymbol{Y}^* + d); \end{aligned}$$

thus,

$$\frac{m(\boldsymbol{W} \mid \boldsymbol{X})}{m(\boldsymbol{W} \mid \boldsymbol{Y})} = \frac{m(\boldsymbol{W}^* \mid \boldsymbol{X}^*)}{m(\boldsymbol{W}^* \mid r\boldsymbol{Y}^* + d)}.$$

The former implies that the moment generating functions of B and B^* are equal, in fact

$$\begin{aligned}\phi_B(t) &= \int \left[\frac{m(\boldsymbol{W} \mid \boldsymbol{X})}{m(\boldsymbol{W} \mid \boldsymbol{Y})}\right]^t f(\boldsymbol{X}, \boldsymbol{Y}, \boldsymbol{W})\, d\boldsymbol{X}\, d\boldsymbol{Y}\, d\boldsymbol{W} \\ &= \int \left[\frac{m(\boldsymbol{W}^* \mid \boldsymbol{X}^*)}{m(\boldsymbol{W}^* \mid r\boldsymbol{Y}^* + d)}\right]^t g(\boldsymbol{X}^*, \boldsymbol{Y}^*, \boldsymbol{W}^*)\, d\boldsymbol{X}^*\, d\boldsymbol{Y}^*\, d\boldsymbol{W}^* = \phi_{B^*}(t),\end{aligned}$$

which implies that both variables follow the same distribution. □

BAYESIAN STATISTICS 7, pp. 671–680
J. M. Bernardo, M. J. Bayarri, J. O. Berger, A. P. Dawid,
D. Heckerman, A. F. M. Smith and M. West (Eds.)

The Markov Modulated Poisson Process and Markov Poisson Cascade with Applications to Web Traffic Modeling

STEVEN L. SCOTT
University of Southern California, USA
sls@usc.edu

PADHRAIC SMYTH
University of California, Irvine, USA
smyth@ics.uci.edu

SUMMARY

A Markov modulated Poisson Process (MMPP) is a Poisson process whose rate varies according to a Markov process. The nonhomogeneous MMPP developed in this article is a natural model for point processes whose events combine irregular bursts of activity with predictable (e.g. daily and hourly) patterns. We show how the MMPP may be viewed as a superposition of unobserved Poisson processes that are activated and deactivated by an unobserved Markov process. The MMPP is a continuous time model which may also be viewed as a discretely indexed nonstationary hidden Markov model by viewing intervals between events as a sequence of dependent random variables. The HMM representation allows one to probabilistically reconstruct the latent Markov and Poisson processes using a set of forward–backward recursions. The recursions allow MMPP parameters to be estimated either by an EM algorithm or by a rapidly mixing Markov chain Monte Carlo algorithm which uses the recursions for data augmentation. The Markov–Poisson cascade (MPC) is an MMPP whose underlying Markov process obeys certain restrictions which uniquely order the event rates for the observed process. The ordering avoids a possible label switching issue without slowing down the rapidly mixing algorithms we use to implement the model. We apply the MPC to a data set containing click rate data for individual computer users browsing through the World Wide Web. Because the complete data posterior distribution for the MPC is a product of exponential family distributions we are able to incorporate data from multiple users into a hierarchical model using existing methods from hierarchical Poisson regression.

Keywords: HIDDEN MARKOV MODEL; POINT PROCESS; BURST; FORWARD–BACKWARD RECURSIONS; CONTINUOUS TIME; NONHOMOGENEOUS POISSON PROCESS.

1. INTRODUCTION

The Markov modulated Poisson process (MMPP) is a doubly stochastic Poisson process (Cox, 1955; Gutiérrez-Peña and Nieto-Barajas, 2003) whose rate varies according to a Markov process. This article decomposes the MMPP into a superposition of latent Poisson processes which are activated and deactivated by a latent Markov process. The result is a natural model for point process data where events combine predictable patterns with irregular bursts of activity. The MMPP is most frequently seen in queuing theory (Olivier and Walrand, 1994; Du, 1995) but it has other interesting applications. Davison and Ramesh (1996) applied a discretized MMPP to a binary time series of precipitation data by numerically optimizing the discretized MMPP likelihood. Scott (1998) used the MMPP to model criminal intrusions on a telephone network. Other uses of the MMPP exist in environmental, medical, industrial, and sociological research. Inference for MMPP parameters has received little attention because most applications of the MMPP assume known model parameters. Turin (1996) proposed an EM algorithm for finding maximum likelihood estimates of MMPP parameters. Scott (1999) provided a Bayesian method for inferring the parameters of a stationary two state MMPP. This article extends Scott (1999) to the nonstationary case with an arbitrary number of states. To our knowledge this is the first treatment of inference for the nonhomogeneous MMPP. We show how the MMPP can be viewed as a superposition of latent Poisson processes, which in turn may be expressed as a nonhomogeneous, discretely indexed hidden Markov model (HMM) by partitioning time into intervals between observed events. Expressing the MMPP as an HMM allows one to probabilistically reconstruct the latent Markov and Poisson processes using a set of forward–backward recursions. The recursions allow MMPP parameters to be estimated through familiar latent variable methods such as the EM algorithm or MCMC data augmentation. The Markov–Poisson cascade (MPC) is a special case of MMPP that enforces an ordering of the state space of the underlying Markov process. The MPC maintains the ordering in a natural way, so that a potential label switching issue is avoided without slowing down our rapidly mixing MCMC algorithms or modifying the specification of model parameters as in Robert and Titterington (1998).

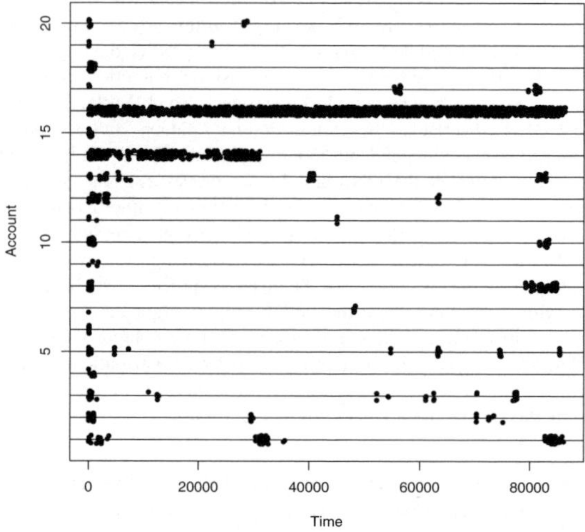

Figure 1. *Mouse click events for the first 20 computers in our data set. Time is measured in seconds from midnight.*

Our motivating example is a data set containing click rate (page request) data produced by 1025 computer users as they navigated through the World Wide Web over a 24 h period. Figure 1 plots page request times for the first 20 users in our data set. The data are a convenience sample from a much larger data set collected by a software tool that automatically logs all page requests within each user's Web browser (with the user's permission). The data are unusual because they were collected on the "client side" at the individuals' computers. It is much easier, and thus more common, for Web traffic studies to be conducted using data obtained from Web servers configured to record the request time, the requested page, and the IP address of the requesting computer ("server side data"). Collecting data on individual users' computers eliminates several technical issues associated with server side data (Cooley *et al.,* 1999). Among the difficulties are that an IP address does not uniquely identify a particular individual, the Web server can only see a user's behavior at one particular site, and some requested pages may not be recorded at the server side because they are provided by the user's memory cache instead of by the Web server.

Computer scientists are interested in page request click rates for academic and commercial reasons. Academically, it is hoped that a better understanding of how individuals use the Web will lead to improved design of Web sites, Web interfaces, and internet traffic management schemes. Commercially, online merchants are interested in click rates because they contain information about an individual's interest in the content of a Web page. Very rapid click rates suggest that an individual is searching for something which he has not found on the current page. Moderate click rates suggest the individual is absorbing content (e.g. reading or listening to music). When a user's click rate falls to zero his session has ended, so resources devoted to him should be directed elsewhere. Click rates can be combined with other information about the content which is being browsed in order to make inference about a user's state of mind. The hidden Markov model representation developed in Section 2 provides a straightforward method of incorporating additional page characteristics into the MMPP, leading to a more elaborate biometric model. For a more detailed discussion of client side Web traffic modeling consult Catledge and Pitkow (1995), Cunha *et al.* (1995), Tauscher and Greenberg (1997), and Cockburn and McKenzie (2001).

The remainder of the article proceeds as follows. Section 2 presents a theory of inference for the MMPP and the MPC. Section 3 applies the model to the mouse click data. Section 4 provides a concluding discussion.

2. MODEL

2.1. *Decomposing the MMPP into Latent Poisson Processes*

Let $N(t)$ be a point process formed by the superposition of unobserved components $N_0(t), \ldots, N_{M-1}(t)$, each of which may sometimes be inactive. Each $N_m(t)$ is a Poisson process whose rate is $\lambda_m(t)$ when $N_m(t)$ is active and 0 when $N_m(t)$ is inactive. We assume that $\lambda_m(t)$ is a parametric or slowly varying nonparametric function supplied by the modeler. An unobserved continuous-time Markov process $A(t)$ determines which component processes are active at time t. Active component processes are independent of one another, and depend on $A(t)$ only through activation/deactivation. Because a variety of models could be used for $A(t)$, depending on physical considerations, we defer discussion of our preferred model until Section 2.4, following material which is independent of the choice of A. For now we assume only that if $A(t) \neq A(t')$ then different subsets of $N_0, \ldots, N_{M-1}$ are active at t and t'.

Heuristically, one imagines that $A(t)$ governs the beginnings and endings of "activity bursts" with intensity $\sum_m \lambda_m(t) I_A^m(t)$, where $I_A^m(t) = 1$ if $A(t)$ indicates process m is active and $I_A^m(t) = 0$ otherwise. Define an activity burst of level m to be an interval (t_0, t_1) such that $N_m(t)$ is active for $t_0 < t < t_1$, $N_m(t)$ is inactive just before t_0 and just after t_1, and events are generated at t_0 and t_1. By defining an activity burst as beginning and ending with its first and last events, we force $A(t)$ to remain constant between events, which eliminates the possibility of pathological "bursts" containing no events. Forcing activity bursts to begin and end with events is mathematically equivalent to assuming that each $A(t)$ transition produces an event. One may think of events produced by $A(t)$ as coming from Poisson processes $B_m(t)$ and $D_m(t)$, which produce the events corresponding to the birth and death of N_m. The current state of $A(t)$ determines whether $B_m(t)$ or $D_m(t)$ are inactive at time t. In particular, N_m cannot be born if it is already alive, and it cannot die if it is already dead. For most applications interest will focus on the case where $B_m(t)$ and $D_m(t)$ have rate functions, denoted $\beta_m(t)$ and $\delta_m(t)$, which are much smaller than $\lambda_m(t)$. The rates $\beta_m(t)$ and $\delta_m(t)$ derive from the "infinitesimal rates" of the generator matrix for $A(t)$. The MMPP is $N = \sum_m (B_m + N_m + D_m)$. Conditional on

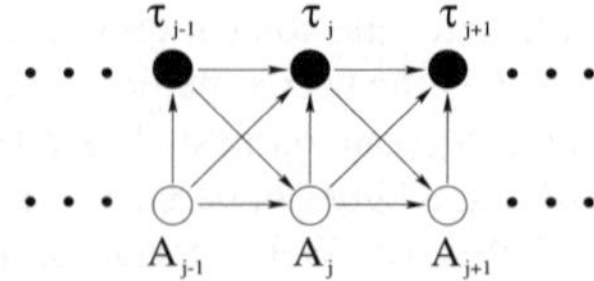

Figure 2. *DAG for the MMPP. Closed and open circles represent observed and unobserved quantities, respectively.*

$A(t) = A$, $N(t)$ is a Poisson process with rate

$$\theta_A(t) = \sum_{m=0}^{M-1} I_A^m(t)\lambda_m(t) + I_B^m(t)\beta_m(t) + I_D^m(t)\delta_m(t) \tag{1}$$

where $I_B^m(t)$ and $I_D^m(t)$ indicate whether $B_m(t)$ and $D_m(t)$ are active at time t.

2.2. *Expressing the MMPP as a Hidden Markov Model*

The MMPP may be expressed as a nonstationary hidden Markov model where the observed data are the event times $\boldsymbol{\tau} = (\tau_1 < \cdots < \tau_n)$ produced by N, and the hidden Markov chain is the sequence $\{A_j = A(\tau_j)\}$. Expressing the MMPP as an HMM allows recursive procedures for HMM's to be used for calculating likelihood, calculating the marginal posterior distribution of each A_j given $\boldsymbol{\tau}$, and estimating model parameters using latent variable methods.

One intuitively understands the MMPP as an HMM because $\{A_j\}$ is a discretization of a continuous-time Markov process, and each τ_j is produced by a Poisson process whose rate is known over the interval (τ_{j-1}, τ_j) given (A_{j-1}, A_j). The formal proof that the MMPP is an HMM follows immediately from a well known fact about Poisson processes presented below as Theorem 1. Note that dependence on model parameters is notationally suppressed for probability calculations in this section.

Theorem 1. *Let $N_0, \ldots, N_M$ be independent Poisson processes with rate functions $\lambda_0(t)$, $\ldots$, $\lambda_M(t)$. Let $\Lambda_m^{(t_0)}(t) = \int_{t_0}^t \lambda_m(u)\,du$. Let T represent the time of the first event generated by any of $N_0, \ldots, N_M$ after time t_0, and let Y denote the index of the first process to produce an event. Then*

$$Pr(T > t, Y = m) = \int_t^\infty \lambda_m(u) \exp\left(-\sum_{r=0}^{M} \Lambda_r^{(t_0)}(u)\right) du.$$

Write τ_j^k for $(\tau_j, \ldots, \tau_k)$ and similarly for other vectors. Theorem 1 implies that $p(\tau_j, A_j \mid \tau_1^{j-1}, A_1^{j-1})$ depends only on (τ_{j-1}, A_{j-1}), a relationship illustrated in Figure 2. Figure 2 is not the traditional DAG associated with a hidden Markov model, but it is sufficiently close that the standard HMM recursions need only be slightly modified. The most important recursive procedure for HMM's is the forward–backward recursion, which calculates the marginal posterior distribution of each (A_{j-1}, A_j) transition conditional on $\boldsymbol{\tau}$. The forward recursion calculates $p_{jrs} = p(A_{j-1} = r, A_j = s \mid \tau_1^j)$. The backward recursion updates these distributions so that they condition on all $\boldsymbol{\tau}$. Theorem 1 implies the forward recursion for the MMPP is

$$p_{jrs} \propto p(\tau_j, A_{j-1} = r, A_j = s \mid \tau_1^{j-1}) = \theta_{rs}^*(\tau_j) \exp\left[-\Theta_r^{(\tau_{j-1})}(\tau_j)\right] \pi_{j-1}(r). \tag{2}$$

The proportionality in (2) is reconciled by $\sum_r \sum_s p_{jrs} = 1$. One recursively computes $\pi_{j-1}(r) = p(A_{j-1} = r \mid \tau_1^{j-1})$ as $\pi_j(s) = \sum_r p_{jrs}$. The symbol $\Theta_r^{(\tau_{j-1})}(\tau_j) = \int_{\tau_{j-1}}^{\tau_j} \theta_r(t)\,dt$ is the expected number of events in (τ_{j-1}, τ_j) given $A(\tau_{j-1}) = r$. If the (r, s) transition implies the birth or death of process m then $\theta_{rs}^*(\tau_j) = \beta_m(\tau_j)$ or $\delta_m(\tau_j)$, respectively. If no births or deaths took place then $\theta_{rs}^*(\tau_j) = \sum_m I_A^m(\tau_{j-1})\lambda_m(\tau_j)$, which is the sum of all λ_m's from processes which were active during (τ_{j-1}, τ_j).

The standard HMM backward recursion applies to the MMPP because the dependence among the A_j's in Figure 2, given $\boldsymbol{\tau}$, is first order Markov. Let $p'_{jrs} = p(A_{j-1} = r, A_j = s \mid \boldsymbol{\tau})$ and $\pi'_j(s) = p(A_j = s \mid \boldsymbol{\tau})$ denote the updated probabilities. The backward recursion begins with $\pi_n(s) = \pi'_n(s)$. Then $p'_{jrs} = p_{jrs}\pi'_j(s)/\pi_j(s)$ where $\pi'_j(s)$ is computed from p'_{j+1rs} in the previous step of the recursion.

2.3. *Parameter Estimation and Posterior Sampling*

The traditional role of the forward–backward recursion is to implement the E-step of an EM algorithm (Baum *et al.,* 1970, Dempster *et al.,* 1977). The complete data likelihood for the MMPP is

$$L_{com} = \prod_{j=1}^{n} \prod_{m=0}^{M-1} \lambda_m(\tau_j)^{y_{jm}} \beta_m(\tau_j)^{b_{jm}} \delta_m(\tau_j)^{d_{jm}} \exp\Big(- \int_{\tau_{j-1}}^{\tau_j} \theta_A(t)\,dt \Big) \qquad (3)$$

where y_{jm}, b_{jm} and d_{jm} are 0/1 indicators revealing which of N_m, B_m, or D_m produced the event at τ_j. Eq. (3) is log-linear in the missing indicators

$$y_{jm}, \quad b_{jm}, \quad d_{jm}, \quad I_A^m(\tau_{j-1}), \quad I_B^m(\tau_{j-1}), \quad I_D^m(\tau_{j-1}),$$

so the E-step of EM simply replaces each indicator with its conditional probability given $\boldsymbol{\tau}$. All required probabilities are available from the forward–backward recursions. The only probability which requires any calculation to extract is

$$p(y_{jm} = 1 \mid \boldsymbol{\tau}) \propto \sum_r p(A_{j-1} = r, A_j = r \mid \tau) I_{A_j=r}^m(\tau_j)\lambda_m(\tau_j).$$

The forward–backward recursions may also be used to sample the missing data directly from its conditional distribution given $\boldsymbol{\tau}$ in a single Gibbs step (Chib, 1996; Scott, 2002). First draw A_n from $\pi_n(s)$. Then draw each A_j given $A_{j+1} = s$ from the distribution proportional to either p_{jrs} or p'_{jrs}. If $A_{j-1} \neq A_j$ then event j is either a birth or a death. If $A_{j-1} = A_j$ then draw $(y_{j0}, \ldots, y_{jM-1})$ from the distribution proportional to $I_{A_j}^m(\tau_j)\lambda_m(\tau_j)$.

Equation (3) says that if $\lambda_m(t)$, $\beta_m(t)$ and $\delta_m(t)$ have distinct parameters, then those parameters are independent in the complete data likelihood. Thus, the parameter estimation or simulation step for an EM or data augmentation algorithm simply involves repeated inference for the parameters of an ordinary Poisson process.

2.4. *Identifiability and The Markov Poisson Cascade*

Depending on physical considerations, there are several state spaces and transition probabilities one could use for $A(t)$. We propose the following model as a default, as it seems to balance physical, computational, and identifiability considerations.

Let $\mathcal{S} = \{0, \ldots, M-1\}$ be the state space, where $A(t) = m$ implies $N_0, \ldots, N_m$ are active and $N_{m+1}, \ldots, N_{M-1}$ are inactive. If $A(t) = m$ then all possible death processes $D_1, \ldots, D_m$

are active, but B_{m+1} is the only active birth process. We call an MMPP with this $A(t)$ a Markov–Poisson cascade (MPC) because the death of a process which is low in the hierarchy immediately deactivates all processes above it. Processes which are terminated by the death of a "lower" process do not generate events. If $A(t) = s$ then Eq. (1) for the MPC becomes

$$\theta_s(t) = \beta_{s+1}(t) + \sum_{m=0}^{s} (\lambda_m(t) + \delta_m(t)) .$$

When paired with a prior distribution enforcing $\lambda_m(t) > \beta_m(t)$ (which simply means that events inside a burst occur more frequently than the bursts themselves) the MPC avoids three identifiability issues associated with the MMPP: label switching, state collapsing, and role reversal. Label switching is an intrinsic feature of all finite mixture models. It occurs because the complete data likelihood is invariant to a permuation of the state labels. The MPC prevents label switching by enforcing $\theta_s(t) < \theta_{s+1}(t)$, provided $\beta_{s+1}(t) + \lambda_s(t) + \delta_s(t) > \beta_s(t)$. This is a very mild restriction which obtains with high posterior probability even under a weak prior. Note that one need not order the individual λ_m's, β_m's or δ_m's. The ordering is automatically enforced on θ_m.

State collapsing and role reversal occur when M is too large to be supported by the data. State collapsing occurs when the model eliminates a redundant state by setting $\lambda_m \approx 0$, which allows N_m to be active without producing any events. Parameters are more interpretable if unnecessary states are left unused. By preventing N_{m+1} from activating unless N_m is active, but allowing lower states to die, the MPC diminishes the prior activation probability of states at the top of the hierarchy. This causes unneccessary states to filter to the top where they remain unused.

Role reversal is a consequence of the symmetry between $\lambda_m(t)$ and $(\beta_m(t), \delta_m(t))$ in (3). Role reversal occurs because a long sequence of events produced by N_m has essentially the same likelihood as an alternating sequence of N_m births and deaths. Scott (1999) observed that a prior forcing $\beta_m < \lambda_m$ prevents role reversal.

3. APPLICATION

We assume user i in the click rate dataset follows an MPC with parameters $\phi_i = \{\lambda_{im}(t), \beta_{im}(t), \delta_{im}(t) : m = 0, \ldots, M-1\}$. For convenience we set $M = 3$, where the three states indicate respectively, the absence of a Web session, a session with a slow click rate, and rapid clicking. In a serious application we would allow M to depend on i using any of several Bayesian methods for model selection or model averaging. We force $\beta_0(t) = \delta_0(t) = 0$ so that N_0 remains active as a baseline to catch isolated events. Each user's click stream almost certainly contains strong daily and hourly patterns, but these patterns are inestimable because only a single day has been observed. Therefore, we assume all rates are constant, e.g. $\lambda_{im}(t) = \lambda_{im}$. Had a much longer time window been observed we could incorporate daily and hourly patterns into the click rates using the timing model proposed by Lambert *et al.* (2001), for example.

The familiar exponential family distributions underlying the MPC make it easy to embed MPC parameters in a hierarchical model. We assume the prior distribution

$$p(\phi_i, \ldots, \phi_n) = \prod_i \prod_m \mathrm{Ga}(\lambda_{im} \mid a_{\lambda m}, b_{\lambda m}) \mathrm{Ga}(\beta_{im} \mid a_{\beta m}, b_{\beta m}) \mathrm{Ga}(\delta_{im} \mid a_{\delta m}, b_{\delta m}), \quad (4)$$

where $\mathrm{Ga}(\cdot \mid a, b)$ is the gamma distribution with mean a/b and variance a/b^2. The hyperparameters in (4) are interpretable as prior event counts and observation times. For example, $a_{\beta m}$ is a prior number of births for N_m, and $b_{\beta m}$ is a prior amount of time spent waiting for N_m

to be born. Following Christiansen and Morris (1997), we assume an improper uniform prior on each a/b and assume each $p(a) = z_0/(z_0 + a)^2$, which is a proper normalized distribution with no moments. Christiansen and Morris show this prior has good frequency properties in the context of Poisson regression. The only "tuning parameter" is z_0, which we set to the relatively uninformative value of 0.10.

Let $\boldsymbol{z}_i$ denote the missing indicators required to compute L_{com} for user i and let $\boldsymbol{\alpha}$ denote the set of $(a_{..}, b_{..})$ pairs in (4). We used an MCMC algorithm which cycles between sampling from $p(\boldsymbol{z}_i \mid \phi_i)$, $p(\phi_i \mid \boldsymbol{z}_i, \boldsymbol{\alpha})$ for each i, and from $p(\boldsymbol{\alpha} | \phi_1, \ldots, \phi_n)$ with $n = 1025$. We ran the algorithm for 5000 iterations and removed the first 1000 as burn-in. Figure 3 shows the remaining 4000 draws of ϕ_i for a sample account. The time series plots and autocorrelation functions in Figure 3 indicate rapid mixing attributable to the forward–backward recursions used in the data augmentation step.

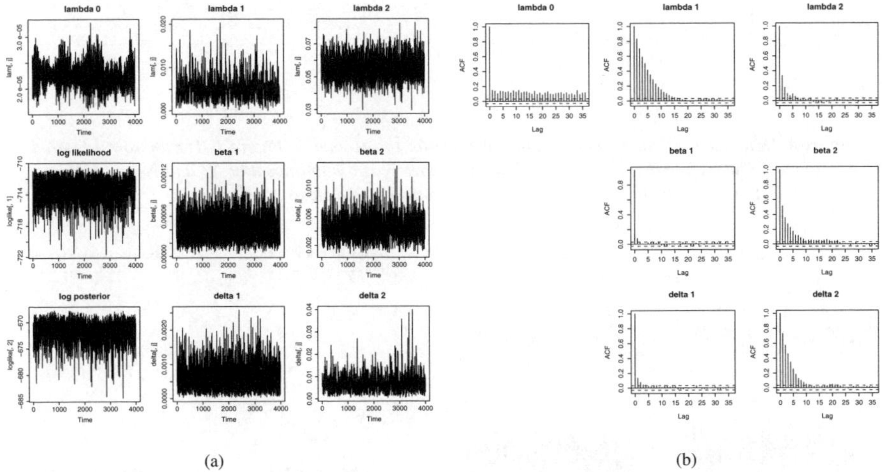

(a) (b)

Figure 3. (a) *Time series plots and* (b) *autocorrelations of 4000 MCMC draws from the posterior distribution of MPC parameters for a sample user.*

Figure 4 plots the data for the sample account, along with the posterior probability that N_1 and N_2 are active each event time. This account was selected because it showed several stray events in addition to the more typical bursts corresponding to Web sessions. We were concerned that the stray events might be interpreted as low intensity Web sessions. Instead, the posterior distribution of $A(t)$ in Figure 4 behaves as hoped. Stray events are attributed to N_0, while N_1 bursts persist across moderate gaps. Level 2 bursts switch more rapidly. The smallest definitive gap between level 2 bursts for this account is the four minute interval containing time 15500. The probabilities in Figure 4 are Bayesian in the sense that they average over the posterior distribution of ϕ_i. A corresponding plot created by running the forward–backward recursions conditional on a point estimate of ϕ_i is not shown because it is nearly identical. This suggests that an online merchant who lacks the time to implement an MCMC algorithm may safely base predictions on quicker empirical Bayes calculations.

The hierarchical model specification allows us to coherently estimate aggregate summaries of Web sessions by examining the posterior distributions of hyperparameters. Figure 5 shows MCMC sample paths for the prior parameters of $\{(a_{\lambda m}, b_{\lambda m}) : m = 0, 1, 2\}$. All three (a, b) pairs have moved from their initial values of (1,1000), suggesting that there is information

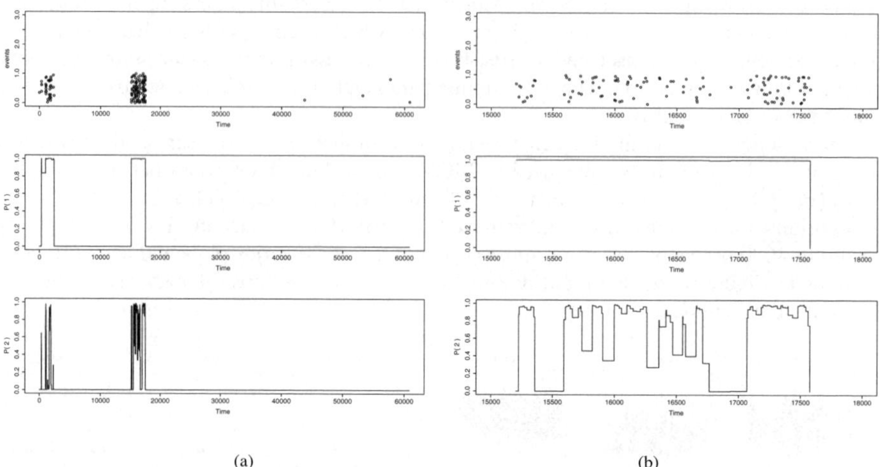

Figure 4. *Data and estimated hidden Markov process for the account in Figure 3. The top row is jittered event times. The middle and bottom rows plot the respective probabilities that N_1 and N_2 are active at time t. Panel* (a) *plots the full observation window. Panel* (b) *is a close-up of panel* (a).

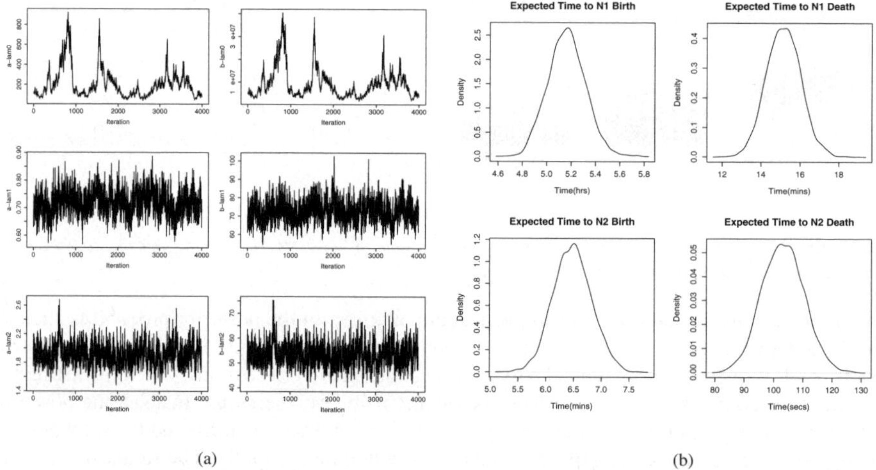

Figure 5. (a) *Sample paths for $a_{\lambda m}$ and $b_{\lambda m}$.* (b) *Posterior distribution of overall expected waiting time until births or deaths of sessions of different levels. Note the different time scales.*

about rate parameters across several accounts. The only parameters that had trouble mixing were $(a_{\lambda 0}, b_{\lambda 0})$, because of the considerable probability mass close to $\lambda_0 = 0$. The posterior medians for $a_{\lambda 0}$ and $b_{\lambda 0}$ are 148.5 and 6,660,925, or about two clicks per day over 78 days. This means the "borrowed information" for λ_0 is 78 times the information in an individual account. Posterior shrinkage for λ_1 and λ_2 is much less, with prior exposure times of about 80 and 50 s, respectively.

Figure 5 also shows the posterior distribution of $b_{\beta m}/a_{\beta m}$ and $b_{\delta m}/a_{\delta m}$ for $m = 1, 2,$

which are the prior expected waiting times until the birth or death of process m. The top row of Figure 5b indicates that the average time between Web sessions is slightly over 5 h, while the average duration of a session is about 15 min. The bottom row suggests that within a Web session, a burst of rapid clicking can be expected about every six minutes, and the burst is expected to last about a minute and a half. The inferences about the average session duration are interesting becasue little seems to be known about distributions of session lengths from past studies (e.g., Catledge and Pitkow, 1995; Cockburn and MacKenzie, 2001). These inferences are apparently the first to come from a model based method of determining Web session boundaries.

4. DISCUSSION

The MPC is a flexible model for point processes subject to irregular bursts of activity. This article has shown how to estimate MPC parameters using either a rapidly mixing MCMC sampler or an EM algorithm. The key to either approach is a set of forward–backward recursions for probabilistically restoring the latent Markov process. The same recursions allow for the filtering of future observations once MPC parameters have been estimated. The computational speed offered by the forward–backward recursions means that the MPC can be a viable model even in applications where MCMC is not feasible due to time constraints. The MPC is a well defined model, free of label switching issues. Finally, because the full conditional distributions underlying the MPC are all familiar models from the exponential family, it is straightforward to incorporate MPC parameters into a hierarchical model for a collection of users.

ACKNOWLEDGEMENTS

The work of P. Smyth was supported in part by the National Science Foundataion under grants IIS-9703120 and IIS-0083489, by a gift from Microsoft Research, and by an IBM Faculty Partnership award.

REFERENCES

Baum, L. E., Petrie, T., Soules, G., and Weiss, N. (1970). A maximization technique occurring in the statistical analysis of probabilistic functions of Markov chains. *Ann. Statist.* **41**, 164–171.

Catledge, L. D. and Pitkow, J. E. (1995). Characterizing browsing strategies in the World-Wide Web. *Comput. Networks ISDN Syst.* **27**, 1065–1073.

Chib, S. (1996). Calculating posterior distributions and modal estimates in Markov mixture models. *J. Econometrics* **75**, 79–97.

Christiansen, C. L. and Morris, C. N. (1997). Hierarchical Poisson regression modeling. *J. Am. Statist. Ass.* **92**, 618–632.

Cockburn, A. and McKenzie, B. J. (2001). What do web users do? An empirical analysis of web use. *Int. J. Human–Computer Studies* **56**, 903–922.

Cooley, R., Mobasher, B., and Srivastava, J. (1999). Data preparation for mining World Wide Web browsing patterns. *Knowledge Information Systems* **1**, 5–32.

Cox, D. R. (1955). Some statistical methods connected with series of events. *J. R. Statist. Soc. B* **17**, 129–164, (with discussion).

Cunha, C., Bestavros, A., and Crovella, M. (1995). Characteristics of WWW client-based traces. *Tech. Rep.*, NEC, USA. `citeseer.nj.nec.com/cunha95characteristics.html`

Davison, A. C. and Ramesh, N. I. (1996). Some models for discretized series of events. *J. Am. Statist. Ass.* **91**, 601–609.

Dempster, A. P., Laird, N. M., and Rubin, D. B. (1977). Maximum likelihood from incomplete data via the EM algorithm. *J. R. Statist. Soc. B* **39**, 1–22.

Du, Q. (1995). A monotonicity result for a single-server queue subject to a Markov-modulated Poisson process. *J. Appl. Probab.* **32**, 1103–1111.

Gutiérrez-Peña, E. and Nieto-Barajas, L. E. (2003). Bayesian nonparametric inference for mixed Poisson processes (this volume).

Lambert, D., Pinheiro, J., and Sun, D. (2001). Estimating millions of dynamic timing patterns in real-time. *J. Am. Statist. Ass.* **96**, 316–330.

Olivier, C. and Walrand, J. (1994). On the existence of finite-dimensional filters for Markov-modulated traffic. *J. Appl. Probab.* **31**, 515–525.

Robert, C. P. and Titterington, D. M. (1998). Reparameterization strategies for hidden Markov models and Bayesian approaches to maximum likelihood estimation. *Statist. Comput.* **8**, 145–158.

Scott, S. L. (1998). *Bayesian Methods and Extensions to the Two State Markov Modulated Poisson Process*. Ph.D. Thesis, Harvard University, USA.

Scott, S. L. (1999). Bayesian analysis of a two state Markov modulated Poisson process. *J. Comput. Graph. Statist.* **8**, 662–670.

Scott, S. L. (2002). Bayesian methods for hidden Markov models: Recursive computing in the 21st century. *J. Am. Statist. Ass.* **97**, 337–351.

Tauscher, L. and Greenberg, S. (1997) Revisitation patterns in World Wide Web navigation. *Human Factors in Computing Systems: Proceedings of the CHI '97 Conference*. New York: ACM, 399–406.

Turin, W. (1996). Fitting probabilistic automata via the EM algorithm. *Comm. Statist. Stochastic Models.* **12**, 405–424.

BAYESIAN STATISTICS 7, pp. 681–690
J. M. Bernardo, M. J. Bayarri, J. O. Berger, A. P. Dawid,
D. Heckerman, A. F. M. Smith and M. West (Eds.)

Modelling Bivariate Extremes in a Region

ELIZABETH L. SMITH and DAVID WALSHAW
Newcastle University, UK
elizabeth.smith2@ncl.ac.uk david.walshaw@ncl.ac.uk

SUMMARY

Practitioners of extreme value methodology have been slow to accept the Bayesian paradigm, and the initial work that *has* been carried out in recent years has reflected the history of the classical approach, in concentrating solely on univariate problems. In this paper we take a first step towards balancing the substantial frequentist literature on *multivariate* extreme value inference by considering problems of bivariate inference from a Bayesian point of view. We relate the bivariate case to inference problems for extremes of environmental variables recorded at a number of locations in a spatial region. We show how inference for bivariate extreme value models can be implemented using an MCMC scheme, and compare two popular model families. We then select one of these families for use in a practical example involving rainfall data. We employ prior information on marginal behavior of extremes constructed from carefully elicited expert beliefs, while prior beliefs about the dependence parameter relate the strength of dependence inversely to the distance between locations, thus exploiting the spatial aspect inherent in the inference problem. We briefly discuss how our ongoing work in this area will lead to a spatial model which enables inference at a particular location of interest to be improved through the model for bivariate extremal dependencies with other locations. We conclude with a pointer to inference for max-stable process models, which are being developed by the authors to address the problems involved in truly multivariate inference.

Keywords: BAYESIAN INFERENCE; EXTREME VALUES; MULTIVARIATE INFERENCE; SPATIAL MODELS; ENVIRONMENTAL VARIABLES.

1. INTRODUCTION

Inference on the extremes of environmental processes is essential for design-specification in civil engineering. The area of extreme value modelling has been rather late in accepting Bayesian ideas, despite the fact that the inference problems that arise are particularly in need of the benefits bestowed by taking the Bayesian approach. In recent years a number of authors have attempted to rectify the situation (*e.g.*, Coles and Powell, 1996; Coles and Tawn, 1996a; Walshaw, 1999, 2000), but thus far only univariate extreme value inference has received any significant consideration. Almost without exception, the large body of literature on multivariate extremes views inference from a frequentist standpoint.

The aim of this paper is to take a first step towards the goal of building a comprehensive Bayesian strategy for inference on multivariate extremes of environmental variables, extending to the infinite-dimensional problems posed by models for extremes over continuous spatial *regions*.

Figure 1 shows the relative locations of a network of 11 sites within a 40 km by 40 km square region in Southwest England, at which detailed daily rainfall records have been kept for many years. Coles and Tawn (1996a) carried out a sophisticated Bayesian analysis of extremes from one of these locations. However, in a (frequentist) multivariate analysis of extremes from

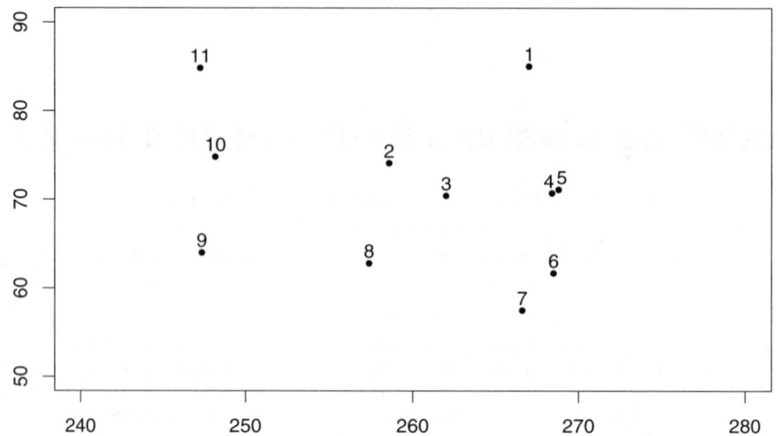

Figure 1. *Map of the network of 11 sites in Southwest England. Spatial plot of the network of sites (latitude versus longitude, in kilometers).*

all 11 sites, Coles and Tawn (1996b) demonstrated the importance of dependence across sites in terms of modelling the extremal behavior of rainfall in this region.

In this paper, we consider models for the bivariate distribution of extremes at pairs of sites. Using models based on limiting joint distributions, we show how posterior inferences for joint extremal behavior can be made using an MCMC scheme. We compare two such models and choose one for analysis of all pairs of sites. Here, informative priors for the marginal behavior of extremes at individual sites are based on modifications of carefully elicited expert beliefs, while a prior specification for the dependence between sites utilizes a model which relates the strength of dependence inversely to the distance separation between sites. This paper is a presentation of ongoing work towards the goal of Bayesian inference for extremes at a location of interest which utilizes the bivariate dependencies with all other sites in the region. For true multivariate inference for extremes at a number of sites, models must be based on higher–dimensional multivariate extreme value theory. We conclude by outlining the authors' ongoing work on Bayesian inference in this situation, based on the theory of max-stable processes.

2. THEORETICAL BACKGROUND

Consider a sequence of identically distributed bivariate random variables $\{(X_i, Y_i); i = 1, 2, \ldots\}$. We are interested in the limiting distribution of

$$\boldsymbol{M} = (\max_{i=1,\ldots,n} \{X_i\}, \max_{i=1,\ldots,n} \{Y_i\}) \tag{1}$$

as $n \to \infty$. Here the (X_i, Y_i) are meant to represent bivariate realizations of some environmental variable, *e.g.*, the total rainfall on day i at two distinct locations from those shown in Figure 1. The vector of componentwise maxima (1), can, by appropriate choice of n, be made to represent a potential quantity of interest for modelling, *e.g.*, the bivariate componentwise annual maximum. Note that this is not, in general, one of the individual observations (X_i, Y_i). Under weak conditions on serial dependence, the central result of univariate extreme value theory determines the form of the marginal distributions: the complete class of limiting distributions for linearly normalized maxima of univariate i.i.d. random variables Z_i under weak conditions is the generalized extreme value (GEV) distribution with distribution function:

$$F(z) = \exp\Big\{ - \Big[1 + \xi\Big(\frac{z - \mu}{\sigma}\Big)\Big]^{-1/\xi} \Big\}, \tag{2}$$

valid on $\{z : [1 + \xi(z - \mu)/\sigma] > 0\}$ (Leadbetter *et al.*, 1983). No analogous parametric family for the complete bivariate limiting class for (1) exists, however it is possible to characterize the family through a characterization of the dependence structure. The central result of bivariate extreme value theory is as follows. If we standardize the margins of (2) by putting $\mu = \sigma = \xi = 1$, then we obtain $\tilde{F}(z) = \exp(-1/z)$ valid for $z > 0$ (the unit Fréchet distribution), and the explicit limiting result, including the normalization, is given by

$$\Pr(\max_{i=1,\ldots,n} \{Z_i\}/n \leq z) = \tilde{F}(z) = \exp(-1/z)$$

valid on $\{z : z > 0\}$. Now consider an independent sequence of bivariate random variables $(\tilde{X}_i, \tilde{Y}_i)$ with marginal distributions $\tilde{F}$, and let

$$\tilde{\boldsymbol{M}}_n = (\tilde{M}_n^X, \tilde{M}_n^Y) = \left(\max_{i=1,\ldots,n} \{\tilde{X}_i\}/n, \max_{i=1,\ldots,n} \{\tilde{Y}_i\}/n \right).$$

Then if

$$\Pr(\tilde{M}_n^X \leq x, \tilde{M}_n^Y \leq y) \xrightarrow{\mathrm{d}} G(x, y)$$

where G is a non-degenerate bivariate distribution function, then

$$G(x, y) = \exp\{-V(x, y)\}, \qquad x > 0, y > 0, \tag{3}$$

where

$$V(x, y) = 2 \int_0^1 \max\left(\frac{\omega}{x}, \frac{1 - \omega}{y}\right) dH(\omega)$$

and H is a distribution function on $[0, 1]$ satisfying the mean constraint

$$\int_0^1 \omega \, dH(\omega) = 1/2. \tag{4}$$

The family of distribution functions G is the *complete class of bivariate extreme value distributions.* Note that this class is in one-to-one correspondence with the set of distribution functions H on $[0, 1]$ satisfying the mean constraint (4). More background to this result can be found in Tawn (1988).

3. MODEL BUILDING

3.1. *Models for Dependence*

Parametric modelling involves identifying parametric sub-families of G, which equates to specifying parametric families for H on $[0, 1]$ with mean equal to 0.5 for every value of the parameter. In practice it is not easy to create models with a useful range of dependence. For example, an early attempt by Gumbel (1960) allowed for only *negative* dependence between variables! However, more recently a number of more practical models have been developed. In this paper we consider the two existing differentiable parametric models identified by Tawn (1988).

The *logistic model* has distribution function given by

$$G(x,y) = \exp\left\{-\left(x^{-1/\alpha} + y^{-1/\alpha}\right)^{\alpha}\right\}, \qquad x > 0, y > 0, \tag{5}$$

for a parameter $\alpha \in (0,1]$. This is obtained non-trivially by appropriate choice of H in (4); see Tawn (1988) for details. This family is popular because of its flexibility. In particular, the limits $\alpha = 1$ and $\alpha \to 0$ correspond to complete independence and perfect dependence respectively, so the full range of positive dependence is available with this model.

The *mixed model* is most easily expressed in terms of its survivor function

$$\overline{G}(x,y) = \exp\left\{-(x+y) + \frac{\beta xy}{x+y}\right\}.$$

for a parameter $\beta \in [0,1]$. Here $\beta = 0$ corresponds to independence, but perfect dependence is not possible. In fact the maximum correlation possible under this model is obtained from an expression given by Tawn (1988) as 0.473. Note that both the logistic model and the mixed model are symmetric, that is to say the variables X and Y are exchangeable.

3.2. *Incorporating the Marginal Distributions*

A full bivariate model for extremal behavior will be based on one of the parametric models, and will be completely specified by a set of GEV parameters for each marginal component, and the appropriate dependence parameter. For example, the logistic model would be completely specified by the vector

$$\boldsymbol{\theta} = (\mu_X, \sigma_X, \xi_X, \mu_Y, \sigma_Y, \xi_Y, \alpha)$$

where the sub-vectors (μ_X, σ_X, ξ_X) and (μ_Y, σ_Y, ξ_Y) determine the marginal GEV distributions (2), and α is the dependence parameter in (5) after transformation to unit Fréchet margins.

4. INFERENCE

We consider the situation where the data available for modelling are componentwise maxima from two locations, $\boldsymbol{M}_r = (M_n^X, M_n^Y)_r, \;\; r = 1, \ldots, t$, taken from t successive time periods of equal length n. This corresponds with the situation where annual maxima are available from the sites of interest. Adopting one of the models in Section 3.1 (thereby assuming exchangeability of maxima between locations), the likelihood can be calculated by substituting the transformed marginal maxima into the appropriate distribution or survivor function. We give the details for the logistic model, although the development for the mixed model is similar. We have from (5),

$$G(\tilde{x}, \tilde{y}) = \exp\left\{-(\tilde{x}^{-1/\alpha} + \tilde{y}^{-1/\alpha})^{\alpha}\right\}, \qquad \tilde{x} > 0, \; \tilde{y} > 0, \tag{6}$$

where

$$\tilde{x} = k_x^{(1/\xi_x)} \quad \text{and} \quad k_x = \left[1 + \xi_x\left(\frac{x - \mu_x}{\sigma_x}\right)\right],$$

with an analogous expression for $\tilde{y}$. Here, from (3) and (6), $V(\tilde{x}, \tilde{y}) = (\tilde{x}^{-1/\alpha} + \tilde{y}^{-1/\alpha})^{\alpha}$. The joint density for a bivariate componentwise maximum (x, y) is then given by

$$g(x,y) = \{V_x(\tilde{x}, \tilde{y}) V_x(\tilde{x}, \tilde{y}) - V_{xy}(\tilde{x}, \tilde{y})\} \exp\{-V(\tilde{x}, \tilde{y})\},$$

where V_x, V_y and V_{xy} represent the derivatives of V (not shown). The prior is most naturally specified as a product of joint priors for each of (μ_X, σ_X, ξ_X) and (μ_Y, σ_Y, ξ_Y), and a univariate prior for α. It is easy to obtain a sample from the approximate posterior distribution for the complete parameter vector $\boldsymbol{\theta}$ using MCMC methods.

In this section, we demonstrate the procedure using vague priors. This is enlightening from the point of view of assessing convergence behavior, and comparing the performances of the logistic and mixed models. We fit the model to annual maximum data from locations 1 and 2 in Figure 1, using the 45 bivariate componentwise maxima available from 1916 to 1973 (some years are missing at each site). Here we use independent univariate normal priors with large variances for each of the six parameters $\mu_X, \phi_X, \xi_X, \mu_Y, \phi_Y, \xi_Y$, where $\phi_X = \log \sigma_X$ and $\phi_Y = \log \sigma_Y$ are more convenient to work with, since the original scale parameters σ_X and σ_Y are constrained to be positive. For the dependence parameter α we use a $U(0, 1)$ prior. It is easy to calculate the univariate conditionals for each parameter in $\boldsymbol{\theta}$ given the other six parameters, and MCMC is implemented as a Gibbs sampler with a random walk Metropolis step, with normal innovations for each parameter.

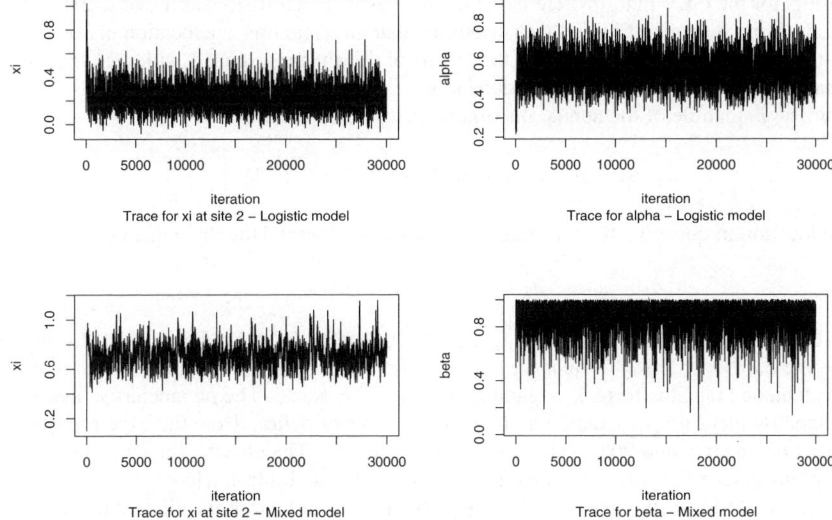

Figure 2. *Trace plots of ξ_Y and the dependence parameter for the logistic and mixed models.*

Figure 2 shows typical sample traces for each of the logistic and mixed models. We show the parameter ξ_Y (for site 2) for each case, and the respective dependence parameters α and β for each model. Convergence appears to occur rapidly for both models. There is a suggestion of poorer mixing of the chain for the mixed model, borne out by the sample autocorrelation functions (not shown). More seriously for the mixed model, the trace for β indicates that the posterior density remains substantial as $\beta \to 1$. Given that this limit corresponds to only moderate dependence between extremes at the two locations, it appears that the mixed model may be too restrictive to represent the levels of extremal dependence which exist between locations in this region. By contrast, the posterior density for α in the logistic model appears to be well contained within the interior of $[0, 1]$, suggesting that the capacity of this model to represent the full range of positive extremal dependence is sufficient to overcome the limitations of the mixed model. This, together with the good convergence and mixing properties exhibited by the time-series plots, leads us to choose the logistic model for the purpose of serious application to the spatial data collected at the 11 locations shown in Figure 1.

5. APPLICATION TO SPATIAL DATA

We wish to fit the logistic model to data from each pair of sites from the 11 locations in Figure 1 using informative priors. Given our knowledge about the locations in Figure 1, it is reasonable to assume that annual maxima from the 11 locations are exchangeable. The number of annual maximum pairs ranges from 28 to 60. Although the assumption of an identical distribution for the daily totals from which the maxima are obtained is unlikely to be realistic because of seasonal variations, many analyses both of data from these locations, and rainfall data generally, demonstrate that the GEV is nevertheless a very good model for the marginal behavior of annual maxima, and that there is no evidence for a long-term trend in extremal behavior of rainfall over the relevant part of the 20th century.

5.1. *Prior Information for the GEV Margins*

We follow Coles and Tawn (1996a) in the use of expert knowledge to construct informative joint priors for the GEV margins. Here, the hydrologist Duncan Reed was asked to express his beliefs about extreme quantiles in the distribution of annual maxima at a location in South West England. It is clear from the paper that this is one of the sites in Figure 1, although there is no information to identify *which* of the stations it is.

The $1-p$ quantile of the annual maximum distribution is given by

$$q_p = \mu + \sigma\left[\{-\log(1-p)\}^{-1/\xi} - 1\right]/\xi. \tag{7}$$

For the location in question, the information requested concerned the three quantities

$$\tilde{q}_1 = q_{p_1}, \quad \tilde{q}_2 = q_{p_2} - q_{p_1}, \quad \tilde{q}_3 = q_{p_3} - q_{p_2}.$$

where the quantiles chosen were $(p_1, p_2, p_3) = (0.1, 0.01, 0.001)$. The use of differences ensures the correct ordering of quantiles. It is assumed that the priors on these quantities are independent, and take the form $\tilde{q}_i \sim \text{gamma}(\lambda_i, \nu_i);\ \ i = 1, 2, 3$. The parameters λ_i and ν_i are determined by measures of location and variability in prior belief. Here the expert was asked to express his median and 90% quantiles for each of the $\tilde{q}_i$. The elicited values, together with the solutions in terms of the gamma parameters are shown in Table 1. Here we have corrected an apparent error for $\tilde{q}_3$ in Coles and Tawn (1996a) by assuming that the elicited quantiles are correct, rather than the gamma parameters.

Table 1. *Elicited prior medians and 90% quantiles for distributions of $\tilde{q}_i$ with associated gamma parameters for the prior distribution.*

Quantile	Median (mm)	90% quantile (mm)	λ_i	ν_i
$\tilde{q}_1$	59	72	38.9	0.67
$\tilde{q}_2$	43	70	7.1	0.16
$\tilde{q}_3$	100	120	41.0	0.41

This prior forms the key component of our prior knowledge about rainfall extremes for the locations in Figure 1, even though we do not know to which site it corresponds. The rainfall climate is relatively homogeneous over the small scale of the region from which the data have been obtained. However the expert did take into account features of the local terrain for the site he was dealing with, together with information about the mean rainfall at the location.

Our prior beliefs about annual maxima for each of the 11 locations are expressed by an appropriate increase in the uncertainty about the quantities $\tilde{q}_i$. We modify the gamma priors by inflating the variance by a factor c, while keeping the mean unchanged. In order to get a feeling for an appropriate choice for c, we experiment with various values and investigate the modified medians and 90% quantiles in the corresponding priors for each $\tilde{q}_i$. We then choose c to reflect the authors' beliefs about the uncertainty associated with the quantities $\tilde{q}_i$. We settle on a value of $c = 3$, giving the medians, the 90% quantiles, and the parameters for our prior gamma distributions for the $\tilde{q}_i$ as shown in Table 2.

Table 2. *Modified prior medians and 90% quantiles for distributions of $\tilde{q}_i$ with associated gamma parameters for the prior distribution.*

Quantile	Median (mm)	90% quantile (mm)	λ_i	ν_i
$\tilde{q}_1$	57	79	13.0	0.22
$\tilde{q}_2$	38	83	2.4	0.05
$\tilde{q}_3$	98	136	13.7	0.14

The joint prior for $(q_{p_1}, q_{p_2}, q_{p_3})$ is then obtained as

$$f(q_{p_1}, q_{p_2}, q_{p_3}) \propto q_{p_1}^{\lambda_1 - 1} \exp(-\nu_1 q_{p_1}) \prod_{i=2}^{3} (q_{p_i} - q_{p_{i-1}})^{\lambda_i - 1} \exp\left\{-\nu_i (q_{p_i} - q_{p_{i-1}})\right\}, \quad (8)$$

on $0 \le q_{p_1} \le q_{p_2} \le q_{p_3}$. Substituting the quantile expression (7) into (8), and multiplying by the Jacobian of the transformation $(q_{p_1}, q_{p_2}, q_{p_3}) \to (\mu, \sigma, \xi)$, leads directly to an expression for the prior in terms of the GEV parameters.

5.2. *Prior Information for the Dependence Parameter α*

Although no expert opinion is available on the extremal dependence between sites, prior knowledge exists in the form of the authors' experience in fitting the logistic model to bivariate annual maxima for rainfall data. This is most easily expressed in terms of the relationship between the dependence parameter α and the distance between sites, d. We use the functional form

$$\alpha = 1 - \exp(-\tau d^{1/2}), \quad (9)$$

which has the desired property of varying continuously between perfect dependence for $d = 0$ and complete independence as $d \to \infty$. The choice of the exponent $d = 1/2$ is somewhat arbitrary, but gives a functional form which reflects the authors' experience in terms of the likely decay of the strength of dependence with increasing distance. Prior information is now expressed as a distribution for the parameter τ. We choose a gamma(0.9, 3) distribution, having mean 0.3 and variance 0.1, which gives distributions for α over the range of distances between sites in Figure 1 which reflect the authors' prior beliefs about the strength of dependence for these distance separations.

5.3. *Posterior Inference*

We fit the logistic model to the sets of bivariate annual maxima available for each of the 55 pairs of sites in Figure 1. Once again the univariate conditionals are easily calculated, and the MCMC scheme with a Metropolis step is used to draw approximate samples from the joint posterior densities. We notice that the performance of the MCMC sampler is considerably improved

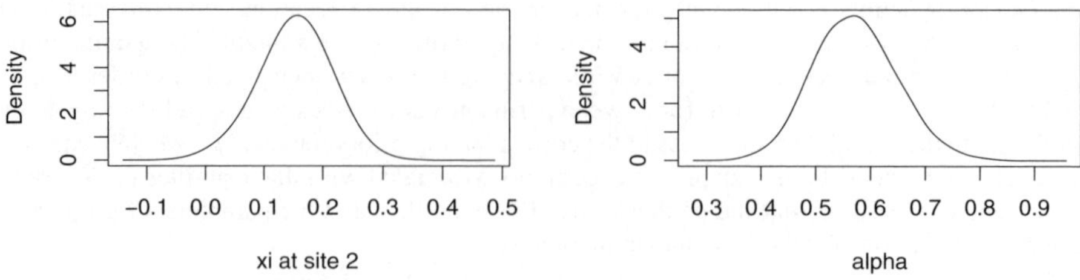

Figure 3. *Posterior densities for* ξ_Y (left) *and* α (right).

now that we are no longer using vague priors, and convergence is apparently achieved within a thousand or so iterations for all location pairs.

Figure 3 shows posterior densities, again obtained from the fit at sites 1 and 2, for the parameter ξ_Y at site 2 and the dependence parameter α. The plots are based on samples of size 20,000, obtained after the first 10,000 iterations have been discarded. This "burn-in" period appears more than sufficient to ensure convergence of the chain. The samples obtained can be used to construct posterior densities for any functions of the parameters; for example return levels would be of particular interest to engineers. Here, however, we focus on posterior inference for the dependence parameter α. Figure 4 is a plot of the posterior mean of α against the distance separation d. Investigation of the sensitivity of inferences to prior specification demonstrates that the overall pattern of the plot is robust; in fact a very similar pattern was obtained when using the non-informative priors described in Section 4, although it was impossible to achieve convergence for several pairs of sites (including the pair (4,5), which are very close together). One feature of interest is the relatively wide scatter of posterior means for α around any given fixed distance d. The relatively small standard errors for the posterior means support the contention that the strength of dependence is not determined by distance separation alone; features of local terrain also appear to play an important rôle.

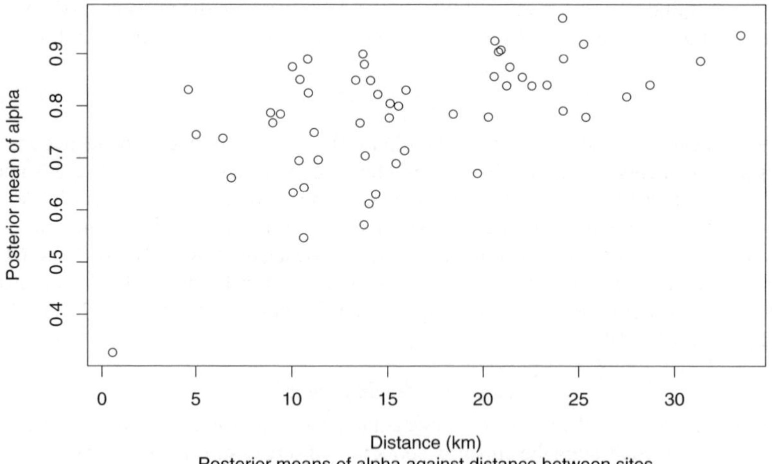

Figure 4. *Plot of posterior means of* α *against the distance between the sites.*

6. DISCUSSION

In this paper, we have shown how to model bivariate extremes obtained from environmental time-series recorded at a network of locations scattered over a spatial region. We have fitted a particular model family to rainfall data collected for such a network, using informative priors which incorporate both the expert beliefs of a hydrologist, and the authors' experience with fitting the chosen model to rainfall data. Posterior inferences have been obtained separately for all pairs of sites. The immediate goal of this ongoing work is to employ the bivariate logistic model as the central component of a spatial model for extremal dependence. This model will be fitted simultaneously to all pairs from a given set of locations, with the full parameter vector consisting of marginal GEV parameters μ_i, σ_i and ξ_i for each location i, and a dependence parameter α_{ij} for each location pair (i, j). We propose to replace the deterministic relationship between α and d, expressed by (9), with a random effects model for τ which allows the pair-specific value τ_{ij} to express terrain adjustments to the strength of dependence. This will allow us to make posterior inferences on the mean value of τ without constraining the value of α, conditional on τ, to be entirely determined by the distance between sites. The principal achievement of this simplistic spatial model will be to allow the inferences on extremes at a particular location of interest to be improved by supplementing the data by information recorded at other sites. This information will be transferred through the bivariate dependencies of extremes at location pairs. For a close-knit network of sites such as the one analyzed in this paper, the dependence is generally strong, and the improvements in precision for posterior inferences are likely to be considerable. This is of great importance for engineers concerned with designing for long-period return levels, since there is typically great uncertainty associated with such quantities.

The longer term aim of the work underway is Bayesian inference for truly multivariate extremes. Engineers are often interested in designing against multivariate, sometimes infinite-dimensional events; for example, the best determiner of possible flooding is the amount of rainfall falling over a continuous *region* in space. Models have been developed for such events, based on the theory of *max-stable processes*, which contain all multivariate extreme value distributions as a sub-family (Coles, 1993; Coles and Tawn, 1996b). However, a frequentist approach was taken to inference. We believe that here too, there is much to be gained by taking the Bayesian line. In addition to the benefits already described, we hope to employ data augmentation to avoid artifacts such as the arbitrary partitioning of sites into those used to estimate dependence, and those used for interpolation of the max-stable process.

ACKNOWLEDGEMENTS

The authors wish to thank Stuart Coles, Jonathan Tawn and the Institute of Hydrology for the rainfall data used in this paper, and Duncan Reed for his expert prior beliefs.

REFERENCES

Coles, S. G. (1993). Regional modelling of extreme storms via max-stable processes. *J. R. Statist. Soc. B* **55**, 797–816.

Coles, S. G. and Powell, E. A. (1996). Bayesian methods in extreme value modelling—a review and new developments. *Internat. Statist. Rev.* **64**, 119–136.

Coles, S. G. and Tawn, J. A. (1996a). A Bayesian analysis of extreme rainfall data. *Appl. Statist.* **45**, 463–478.

Coles, S. G. and Tawn, J. A. (1996b). Modelling extremes of the areal rainfall process. *J. R. Statist. Soc. B* **58**, 329–347.

Gumbel, E. J. (1960). Bivariate exponential distributions. *J. Am. Statist. Ass.* **55**, 698–707.

Leadbetter, M. R., Lindgren, G. and Rootzén, H. (1983). *Extremes and Related properties of Random Sequences and Series*. Berlin: Springer.

Tawn, J. A. (1988). Bivariate extreme value theory: models and estimation. *Biometrika* **75**, 397–415.
Walshaw, D. (1999). Extremes of mixed environmental processes. *Bayesian Statistics 6* (J. M. Bernardo, J. O. Berger, A. P. Dawid and A. F. M. Smith, eds). Oxford: Oxford University Press, 849–858.
Walshaw, D. (2000). Modelling extreme winds in regions prone to hurricanes. *Appl. Statist.* **49**, 51–62.

BAYESIAN STATISTICS 7, pp. 691–700
J. M. Bernardo, M. J. Bayarri, J. O. Berger, A. P. Dawid,
D. Heckerman, A. F. M. Smith and M. West (Eds.)

Bayesian Object Matching with Hierarchical Priors and Markov Chain Monte Carlo

TONI TAMMINEN and JOUKO LAMPINEN
Helsinki University of Technology, Finland
toni.tamminen@hut.fi jouko.lampinen@hut.fi

SUMMARY

We present an object matching system where Bayesian inference is used for estimating the probability distribution of object feature locations in a digital image. The representation of the object contains two parts: the likelihood part which defines the probability of perceiving a pixel image corresponding to the object detail, and a hierarchical prior part which defines the distributions of the feature locations relative to the other features. In this study the objects are human faces. The likelihood part is based on Gabor filters, which are a type of optimal bandpass digital filters. A distortion-tolerant measure of feature similarity is produced by performing statistical analysis on the filter response vectors, and the actual likelihood of observing an image detail given the feature location is computed by analyzing the distribution of the similarities. The prior part defines *a priori* distributions for the relative locations of the features, which correspond to their average locations on the object. The prior is hierarchical in nature, that is, a hyperprior is set for the width of the prior distribution, which allows the system to automatically determine the feature location variance appropriate for the image at hand. The object matching is carried out as Bayesian inference: the goal is to find the posterior probability distribution of the possible matches given the image and the hierarchical prior. Markov Chain Monte Carlo methods, especially Gibbs sampling, are used to draw samples from the posterior distributions of the feature locations, allowing simple estimation of the essential distribution statistics.

Keywords: OBJECT MATCHING; GABOR FILTERS; MARKOV CHAIN MONTE CARLO.

1. INTRODUCTION

Computer vision contains extremely difficult inference tasks. Information is inherently lost in the projection from 3D to 2D, and thus there are many more unknown attributes in the 3D objects than can be estimated from an image. First of all, the image contains no direct information of any dimensions in the view direction. Second, for any image there is an infinite number of combinations of surface textures and shapes that produce exactly the observed image. Furthermore, considering recognition, many objects or classes of objects have considerable internal variations (such as elastic distortions of one person's face, and variations between all objects in the general class "face") making it difficult to decide when two images display the same object. In scene analysis, where the image can contain several objects, additional tasks beside recognition are detection and segmentation of objects, dealing with background clutter, and occlusions, all of which contain their own specific problems.

An important problem in computer vision is object matching, in which the aim is to locate objects present in a scene and fit models to them. In the following we describe an object matching system based on Bayesian inference. The goal of the inference is to find the posterior probability distribution of the possible matches given the image and the object prior. We use Markov chain

Monte Carlo (MCMC) methods, especially Gibbs sampling, to estimate the posterior probability of the matches. The studied problem is the matching of human faces. The prior on the face shape (*i.e.*, , prior on the locations of the corresponding points) is built up in bootstrap fashion: initially the prior is very generic, with the dislocations of nearby points from mean shape are assumed to be uncorrelated. From a set of fitted images we estimate the spatial covariance structure, which can be used as the prior in the next stage. Refining the prior makes the fitting task simpler as the extra prior knowledge reduces the search space to be covered.

2. BACKGROUND

The underdetermined nature of almost any computer vision problem means that there is no unique solution for estimating object attributes from an image. The standard approach in computer vision is to restrict the search to plausible explanations and to use constrained optimization. However, in the recent years the Bayesian approach has gained increasing interest due to its ability to mathematically account for uncertainty and prior assumptions in the model (basically starting with Geman and Geman, 1984).

Methods similar to our feature matching and object shape models include *e.g.*, Wiskott *et al.* (1999) and Cootes *et al.* (2001). Wiskott *et al.* employ Gabor filters and elastic grids for object matching and recognition. Our system differs from this mainly in that our underlying framework is Bayesian instead of the traditional error-minimization approach. Furthermore, we use learned object models as priors to aid the matching process, whereas Wiskott *et al.* measure object similarity one-to-one.

Active Appearance Models (Cootes *et al.*, 2001) are statistical models incorporating both shape and texture variation. The models are matched to novel images by iteratively minimizing an error term composed of texture and shape error. Our model of object shape is similar to the one used by Cootes *et al.*, but we find that the use of Gabor filters achieves more robust matching than the texture-based approach. The face shape model is also reminiscent of linear object class representation of Vetter and Poggio (1995).

3. THE PROBABILITY MODEL

In Bayesian fashion (Gelman *et al.*, 1995) we begin by defining the joint distribution of all the variables in the model. In our case the only observable variable is the perceived image $\boldsymbol{I}$, and the unobservable variables are the locations of the features $\boldsymbol{x}$ and possible hyperparameters $\boldsymbol{\xi}$. Thus, given the model assumptions M, the joint distribution of all the variables is

$$p(\boldsymbol{x}, \boldsymbol{\xi}, \boldsymbol{I} \,|\, M).$$

Conditioning on the observed variables, applying Bayes's formula, and marginalizing over the hyperparameters yields the posterior distribution of the feature locations

$$p(\boldsymbol{x} \,|\, \boldsymbol{I}, M) \propto \int p(\boldsymbol{I} \,|\, \boldsymbol{x}, \boldsymbol{\xi}, M) p(\boldsymbol{x} \,|\, \boldsymbol{\xi}, M) p(\boldsymbol{\xi} \,|\, M) \, d\boldsymbol{\xi}.$$

The model consists of three parts: the likelihood $p(\boldsymbol{I} \,|\, \boldsymbol{x}, \boldsymbol{\xi}, M)$, the prior $p(\boldsymbol{x} \,|\, \boldsymbol{\xi}, M)$ and the hyperprior $p(\boldsymbol{\xi} \,|\, M)$. The likelihood measures the probability of perceiving the image $\boldsymbol{I}$ given the feature locations, hyperparameters, and model assumptions. The likelihood hyperparameters would in theory be related to the nature of the image and the features. For example, likelihood should vary considerably according to whether the perceived image is a photograph or a drawing. However, at the present stage of development we ignore these effects and write likelihood as $p(\boldsymbol{I} \,|\, \boldsymbol{x}, M)$.

The prior term $p(\boldsymbol{x} \mid \boldsymbol{\xi}, M)$ encodes the shape variations of the object. This requires either a good *a priori* model for the objects, or more practically, a learnt model of the object variations. We aim to build this in two steps: at first we use a simple prior which applies to any object class to get a set of samples of the object shape. The shape variations are then estimated from the samples to form the actual shape prior. Here the hyperprior $p(\boldsymbol{\xi} \mid M)$ allows us to leave some of the prior variables open by specifying a prior on them.

In the first stage we assume that the dislocations of the features with respect to the mean locations are independent. The conditional posterior for each feature location, needed for Gibbs sampling, is then

$$p(x_i \mid \boldsymbol{x}_{-i}, \boldsymbol{I}, M) \propto \int p(\boldsymbol{I} \mid x_i, \boldsymbol{x}_{-i}, \boldsymbol{\xi}, M)\, p(x_i \mid \boldsymbol{x}_{-i}, \boldsymbol{\xi}, M) p(\boldsymbol{\xi} \mid \boldsymbol{x}_{-i}, M)\, d\boldsymbol{\xi}, \qquad (1)$$

where $\boldsymbol{x}_{-i}$ denotes all features excluding the ith one. Eq. 1 can be simplified by noticing that the hyperprior and the likelihood do not depend on $\boldsymbol{x}_{-i}$, and that in our model the likelihood is also independent of $\boldsymbol{\xi}$, arriving at

$$p(x_i \mid \boldsymbol{x}_{-i}, \boldsymbol{I}, M) \propto \int p(\boldsymbol{I} \mid x_i, M) p(x_i \mid \boldsymbol{x}_{-i}, \boldsymbol{\xi}, M) p(\boldsymbol{\xi} \mid M) d\boldsymbol{\xi}.$$

Before proceeding with the descriptions of the specific parts of the model, the distinction between "features" and "image details" must be made clear. We use the term "feature" to mean both a certain characteristic of an object (for example, the nose is a feature of the human face), and the mathematical representation of these characteristics. "Image details" are parts of the perceived image, *i.e.*, spatially organized sets of pixel values. It is natural to order the features into a (planar or 3D) labelled graph with the features as the node labels or alternatively simply treated as vectors of coordinates.

4. THE LIKELIHOOD

4.1. *The Gabor Filter*

As outlined in the previous section, in this study the likelihood is proportional to the probability of observing an image detail in a certain place given that the corresponding feature is located there. To measure this, we need three things: a transformation from image details to features, a measure of similarity between features, and a method to assess the likelihood based on the similarities.

We use Gabor filtering as the transformation from image details to features. Gabor filters are optimally localized bandpass filters, in terms of the width of the passband and the extent of the impulse response (Daugman 1988), making them suitable for extracting frequency contents from as small areas as possible. Another justification for their use is that there is evidence that the mammal brain employs Gabor-like structures in its image processing areas (Jones and Palmer 1987).

The basic form of the 2D Gabor filter is

$$h(x, y) = \frac{f^2}{2\pi\sigma_x\sigma_y} \exp\Big(- f^2(\frac{x^2}{2\sigma_x^2} + \frac{y^2}{2\sigma_y^2})\Big) \exp\Big(i(f\cos(\theta)x + f\sin(\theta)y)\Big),$$

where f is the central frequency of the filter, θ the directional angle of the filter and σ_x^2 and σ_y^2 the spatial-domain variances of the filter. The coefficient $f^2/(2\pi\sigma_x\sigma_y)$ compensates for

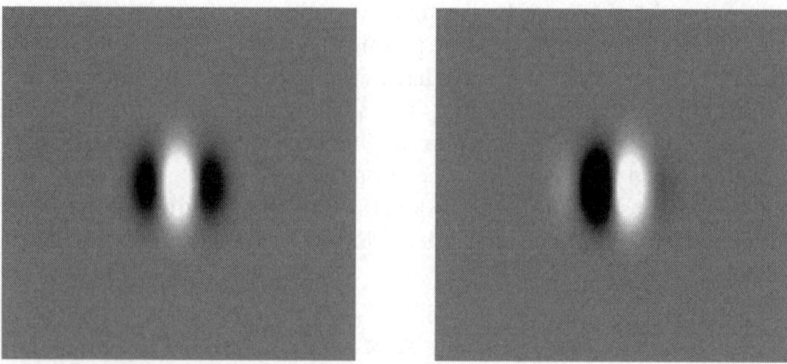

Figure 1. *Real and imaginary parts of a two-dimensional Gabot filter.*

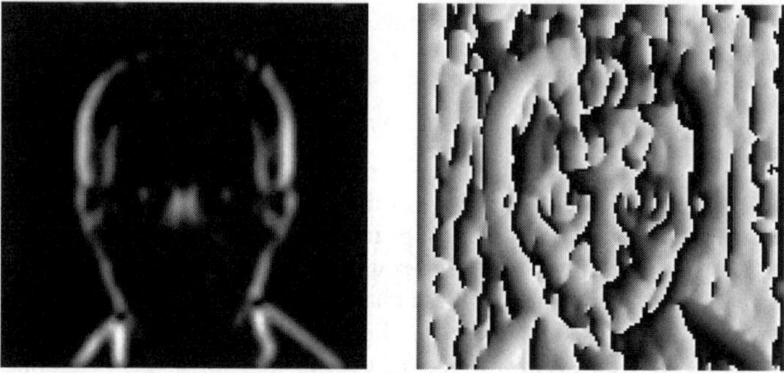

Figure 2. *Sample amplitude and phase responses of a vertical ($\theta = \frac{\pi}{2}$) Gabor filter applied on a facial image.*

the frequency-related decrease of the power spectrum in natural images (Daugman 1988). The Fourier transform or the transfer function of the filter is the Gaussian

$$F(\omega_x, \omega_y) = \exp\left[-\frac{1}{2f^2}\Big(\sigma_x^2(\omega_x - f\cos(\theta))^2 + \sigma_y^2(\omega_y - f\sin(\theta)\Big)^2 \right].$$

The Gabor filter is illustrated in Figures 1 and 2.

4.2. *Comparing the Filter Responses*

Comparison of the outputs of Gabor filter banks in two image locations is usually based only on the amplitude responses of the filters. Typically, the amplitudes of the filter outputs are stacked as a vector (often called a Gabor-jet), and common vector similarity measures are used. For example, Euclidean distance (often after normalization of the length to remove the effect of image contrast), inner product, and angle between the vectors are popular (Lades *et al.*, 1993).

Neglecting the information about phase, however, causes some limitations in the feature space. Consider some fiducial details in the image, which give highly distinctive Gabor filter amplitudes (such as an edge, corner, human eye, etc.) The Gabor jet (amplitude) vector contains only indirect information of the distance from such a detail. For example, the only clue to the distance from an edge is the faster decay of outputs of the smaller receptive field filters. Then robust matching of point chosen from close to some distinct detail is difficult.

In automatic matching of two objects there is need to find point correspondences for arbitrary points in the images, so the reference locations are not necessarily fiducial points of the Gabor filter outputs. In this study we apply a similarity measure sensitive to the filter phases, with amplitude matching based on subspaces instead of single Gabor jets. Another similarity measure sensitive to the phase of the Gabor filters has been presented in Wiskott *et al.* (1999). Our approach differs from this mainly in our use of Gabor jet subspaces, which improves feature localization performance, as will be demonstrated later on.

To model the variation of the Gabor filter responses on the corresponding points in various images, the features associated with the object locations are represented as linear vector subspaces in the Gabor jet space formed with principal component analysis. Thus we have a base $\boldsymbol{P}_i$ for each feature, with the matching error E defined as the angle between a test Gabor jet $\boldsymbol{G}$ and the subspace $\boldsymbol{P}_i$ (Eq. 2). The use of the angle as the error function allows for contrast-invariance (a vital property in feature matching), and by using subspaces to represent the features we attempt to produce features that generalize better to new images than Gabor jets from a single example image. Eighteen example images were used to generate the subspaces.

Also included in the matching error is a phase error term, Eq. (3). The phase information associated to each feature consists of the medians of the phases for all frequencies f, denoted by $\Phi_{f,i}$, of the orientation giving on average the strongest response in the images used to compile the feature subspace. Absolute value of phase is used to ensure that the type of the edge (light/dark or dark/light) does not matter. As mentioned previously, incorporating phase information allows the accurate matching of features not directly associated with an edge and also more accuracy in localization of edge features.

If the similarity measure is completely contrast-invariant, background clutter and faint edge artifacts from image compression can sometimes cause faulty matches. To eliminate these we have an additive *ad hoc* term in the error function which measures the energy of the Gabor jets by summing the squares of the jet vector lengths over the neighborhood of the feature. Squared error is then used to measure the difference between the test and reference energies (Eq. 4). This term is given a low weight in the final error function due to its heuristic nature.

The different error terms are combined to form a similarity measure by scaling and using an exponential function:

$$E_i^{amp} = \arccos\left(\frac{\boldsymbol{G}^t\boldsymbol{P}_i\boldsymbol{P}_i^t\boldsymbol{G}}{||\boldsymbol{G}||\,||\boldsymbol{P}_i\boldsymbol{P}_i^t\boldsymbol{G}||}\right) \tag{2}$$

$$E_i^{phase} = \sum_f ||\Phi_{f,i} - |\phi_{f,i}|||^2 \tag{3}$$

$$E_i^{ener} = (\sum_{area} ||\boldsymbol{G}_i||^2 - \sum_{area} ||\boldsymbol{G}_{ref}||^2)^2 \tag{4}$$

$$E_i^{tot} = \alpha_{amp}E_i^{amp} + \alpha_{phase}E_i^{phase} + \alpha_{ener}E_i^{ener}$$

$$S_i = \exp(-\beta E_i^{tot})$$

where $\phi_{f,i}$ denotes the phases of the filters respective to those stored in the feature $\Phi_{f,i}$, $\boldsymbol{G}_i$ and $\boldsymbol{G}_{ref}$ are the perceived and reference Gabor amplitude jets, respectively, and α_{amp}, α_{phase}, α_{dev} and β are scaling factors. The similarity measure is illustrated and compared to the aforementioned Wiskott's match criterion in Figure 3.

To evaluate the performance of the similarity measure we compared it with a state-of-the-art method presented in Wiskott *et al.* (1999) for matching Gabor filter features. Wiskott's criterion also incorporates phase information to find more accurate matching locations. The measures are compared qualitatively in Figure 3. Quantitatively performance was measured

(a)

(b)

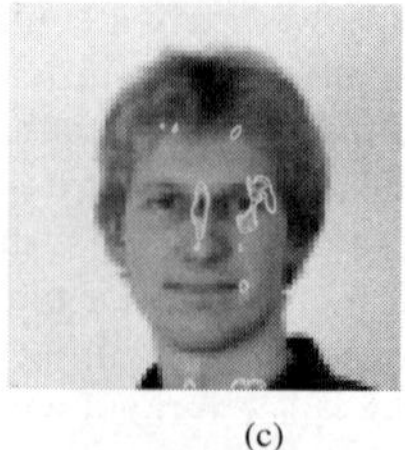
(c)

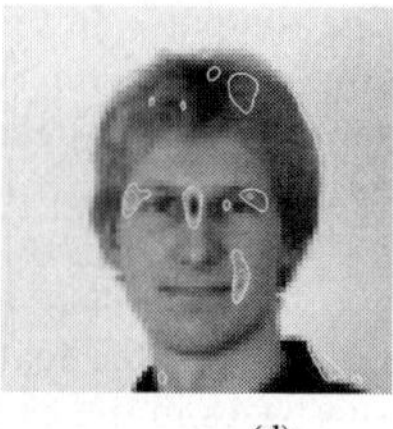
(d)

Figure 3. *The similarity measure. From the left:* (a) *similarity with feature subspace without phase information;* (b) *Wiskott similarity without phase information;* (c) *our combined similarity measure; and* (d) *Wiskott similarity with phase information. The target feature is in all cases the point between the eyes, with dark contours signifying higher similarity. The results with our measure are quite similar to the ones obtained by the Wiskott criterion. Note how the use of phase information allows for much more precise location of the feature.*

by using the IMM-DTU annotated face image database (Stegmann, 2002). Three distances from the selected feature locations were: the distance from the local maximum of the similarity distribution closest to the correct location, the distance from the maximum in a rectangular area around the correct location, and the distance from the global maximum. Our model was formed in a leave-one-out cross-validation way (one image was left out and the model formed with the others), while the Wiskott criterion was computed with respect to two or five randomly chosen images and the largest similarity for each pixel was chosen, resembling the Elastic Bunch Graph Matching framework (Wiskott *et al.*, 1999). With two images the computational complexity was approximately equivalent between the methods, while with five the computational load with Wiskott's measure was considerably heavier. The results were averaged over features and images, and are shown in table 1.

Table 1. *Comparison between the similarity measure presented in this study and the measure presented in Wiskott et al. (1999).* W *stands for Wiskott's criterion and 5 and 2 denote the number of training images used. All distances are in pixels.*

Distance from	Our measure	W 5	W 2
Local max	2.84	3.48	4.34
Neighborhood max	5.34	5.10	5.33
Global max	71.1	52.6	56.3

It can be seen that the performance of the criterion proposed here is roughly similar to Wiskott's criterion for the best match inside a window around the correct location, and clearly more accurate for the closest local maximum. The error of the global maximum has no much practical importance, as in scene analysis clutter and background (*e.g.*, other faces) can always cause spurious matches that the shape model has to filter out.

5. THE OBJECT PRIOR

As described previously, in the first stage of training the system we do not have a realistic prior for the feature locations, and thus are forced to use simpler priors. We assume that the locations of the features are independently Gaussian distributed around the reference locations given by a sample graph or mean of several samples. In practice a reference grid $\boldsymbol{r}$ is computed from the example objects used in the feature construction. This reference grid is then translated and scaled by a function f according to the other features $\boldsymbol{x}_{-i}$ to produce the mean of the prior of the ith feature, *i.e.*, the prior distribution on feature location x_i is

$$x_i \sim \mathrm{N}(f(\boldsymbol{x}_{-i}, \boldsymbol{r}), \sigma^2).$$

A sample prior is shown in Figure 4. The function f can depend on all the other features or, more realistically, only on the ones close to x_i. Note that the features need not necessarily be ordered into a grid—any graph is possible.

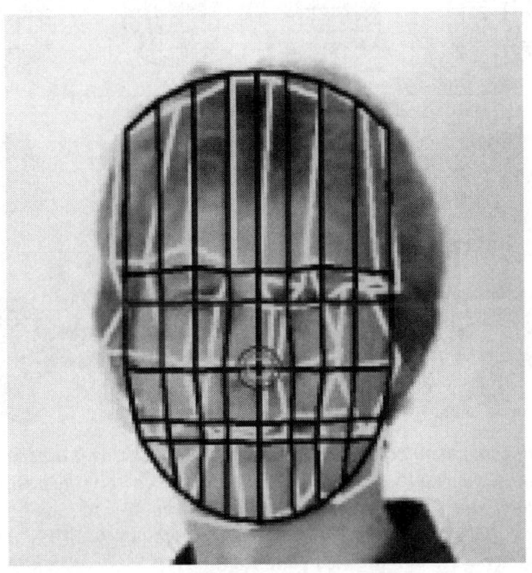

Figure 4. *Illustration of the prior. The light grid is a sample from the posterior distribution, while the dark grid is the scaled reference grid. The contours show the prior for the location of the nose feature.*

The variance hyperparameter σ determines the allowed distortion of the reference grid. In ML approaches, the elasticity or stiffness of the object model is often hand-chosen, and simulated annealing with elaborate cooling schemes is applied (see, *e.g.*, , Wu and Bhanu, 1997) in order to start the matching with more elastic grid to avoid local minima. In the Bayesian approach, we integrate σ out to get the posterior of feature locations. For Gibbs sampling, a convenient hyperprior for the variance σ^2 is the inverse-gamma distribution

$$\sigma^2 \sim \mathrm{Ig}(\alpha, \nu).$$

The full conditional distributions of σ^2 and x_i are straightforward to derive (see, *e.g.*, Neal, 1996):

$$x_i | \boldsymbol{x}_{-i}, \sigma, \boldsymbol{r} \sim \mathrm{N}(f(\boldsymbol{x}_{-i}, \boldsymbol{r}), \sigma^2)$$

$$\sigma^2 | \boldsymbol{x}, \boldsymbol{r}, \alpha, \nu \sim \mathrm{Ig}\left(\frac{\alpha^2\nu + \sum_{i=1}^{n}(x_i - f(\boldsymbol{x}_{-i}, \boldsymbol{r}))^2}{\nu + n}, \nu + n\right).$$

5.1. *The Covariance Prior*

The hierarchical prior described above is not very realistic in the sense that it does not encompass information about the covariance of the feature locations. An estimate of this covariance structure can be obtained by matching a number of objects using the previous prior and applying principal component analysis to find the most significant variations, if necessary (*i.e.*, if the full covariance matrix cannot be estimated accurately enough). To lessen the effect of object pose changes it is sensible to add a mirrored duplicate for each sample so that the shape-related principal components become symmetrical, as can be seen from Figure 5. This kind of prior can be used also as a proposal distribution in Metropolis–Hastings sampling so that the proposed distortions are sensible for the objects.

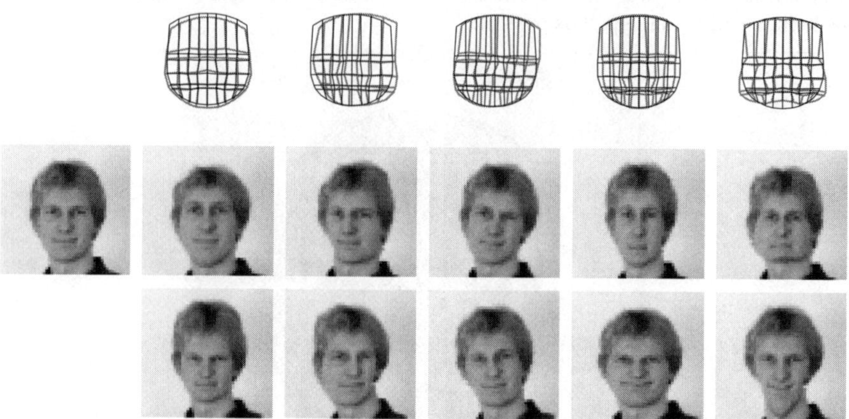

Figure 5. *The principal component grids. In the upper row, the dark grids are the mean or reference grids and the lighter grids show the results of moving the same distance into (the positive) directions specified by the eigenvectors, computed from 38 real images. In the middle and lower rows, the face on the left has been morphed according to the principal components, both into positive and negative directions. It can be seen that components 2 and 3 are related to rotations of the head, while components 1, 4, and 5 are shape-related.*

To make this type of prior more efficient, it would be useful to determine different subspaces of the variations related to pose changes, head shape changes over different persons, and changes due to facial expression for one person.

6. MCMC SAMPLING AND RESULTS

Combining the likelihood and prior parts presented above, we can use Gibbs sampling to draw samples from the posterior. The actual matches are then produced by taking medians of the samples. The starting value for the sampler with the simple prior was such that all features were located at the center point of the image. About 250 samples were used for each image, after a burn-in of the same length. The low number of samples seemed sufficient because the σ parameter converges rather rapidly to a low level (1–2 pixels), after which the chain naturally does not move around very much. Furthermore, repeated simulations with different starting values tended to produce the same results, which reassured us that no significant multimodality was present. Note that the likelihoods for each feature are highly multimodal, but the "correct" solution in the experiments was the only local maximum of the joint posterior for all features. Total time needed for the matching of a single object is about 1 min, most of which is used in the computation of the likelihoods.

The performance of the system was evaluated with the IMM-DTU database (Stegmann, 2002) consisting of 37 annotated facial images. Faces were matched to each image in the database using a model constructed from the other 36 images and the matches were compared with the annotated positions. A set of samples produced with the hierarchical grid prior is shown in Figure 6, and a set of matches is illustrated in Figure 7. Quantitatively the results can be compared to the ones obtained with the AAM framework in Stegmann (2002) by measuring the mean point-to-point error and the mean point-to-curve error (distance from the closest curve point). The results are shown in Table 2.

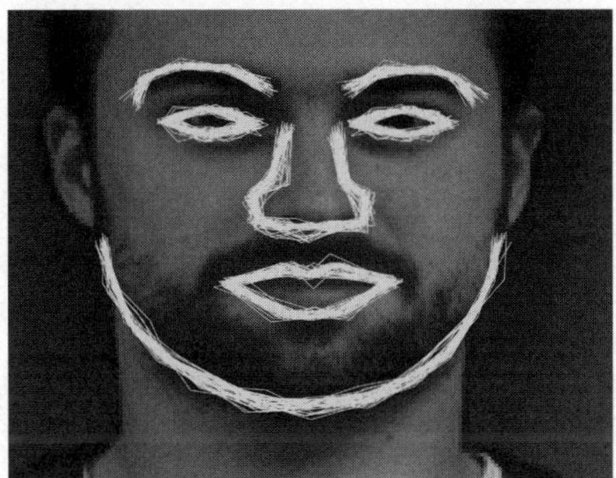

Figure 6. *A set of samples drawn from the posterior with Gibbs sampling.*

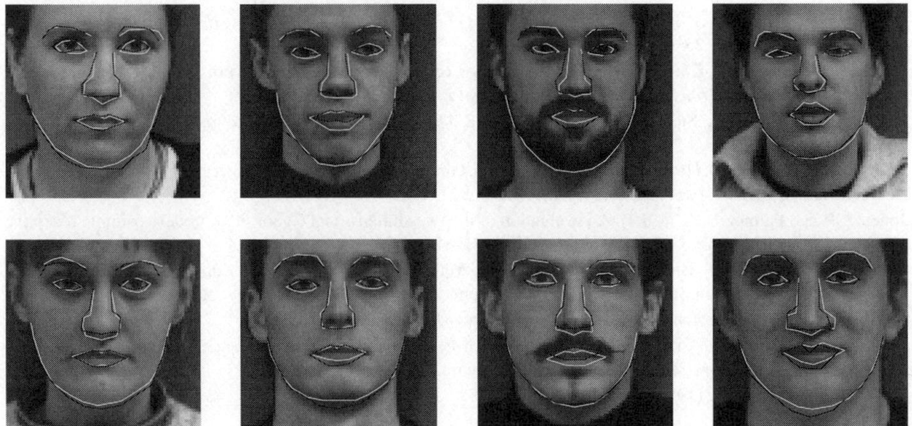

Figure 7. *A set of matches. The dark lines connect the medians of the obtained samples, while the light lines are the annotated positions.*

Table 2. *Comparison of performance between our system and the AAM system of Stegmann (2002).*

Error in pixels	Our system	AAM
Mean point-to-point	5.59	5.54
Mean point-to-curve	2.86	2.93

The results are encouraging—even though our system is still somewhat in development, it performs equally well as the best version of the AAM systems implemented in Stegmann (2002). For face location, our system far surpasses the AAM framework. Cootes *et al.* (2001) report ca. 15% convergence for displacements of 50 pixels (25% of face width). Our system locates the face with 100% accuracy in this (DTU-AAM) database with initial displacements over 100 pixels and initial scale set to zero.

7. CONCLUSION

We have presented an object matching system which employs Gabor filters, hierarchical priors and MCMC sampling in a Bayesian framework. In the face matching task studied in the paper we model the spatial variability of different faces both by a simple grid model with uncorrelated dislocations and by a model of linear correlations. We show how the posterior distribution for any feature can be computed from the locations of the other features and the likelihood function measuring the point correspondence. Gibbs sampling is used for drawing samples from the posterior distributions of the model parameters, allowing the auxiliary (nuisance) parameters to be integrated out from the posterior distributions.

In the future the most critical area of development is the object covariance prior, which needs both practical and theoretical work to make it more applicable. Also, the used grids and object priors should be extended to three dimensions to improve matching performance and to allow us to deal with occlusions and severe changes in pose.

REFERENCES

Cootes, T. F., Edwards, G. J., and Taylor, C. J. (2001). Active appearance models. *IEEE Trans. Pattern Anal. Machine Intelligence* **23**, 681–685.

Daugman, J. G. (1988). Complete discrete 2-D Gabor transforms by neural networks for image analysis and compression. *IEEE Trans. Acoustics, Speech, Signal Process.* **36**, 1169–1179.

Gelman, A., Carlin, J. B., Stern, H. S., and Rubin, D. R. (1995). *Bayesian Data Analysis*. London: Chapman and Hall.

Geman, S. and Geman, D. (1984). Stochastic relaxation, Gibbs distributions and the Bayesian restoration of images. *IEEE Trans. Pattern Anal. Machine Intelligence* **6**, 721–741.

Jones, J. P. and Palmer, L. A. (1987). An evaluation of the two-dimensional Gabor filter model of simple receptive fields in cat striate cortex. *J. Neurophysiol.* **58**, 1233–1258.

Lades, M., Vorbruggen, J., Buhmann, J., Lange, J., v. d. Malsburg, C., Wurtz, R., and Konen, W. (1993). Distortion invariant object recognition in the dynamic link architecture. *IEEE Trans. Comput.* **42**, 300–310.

Neal, R. M. (1996). *Bayesian Learning for Neural Networks*. Berlin: Springer.

Stegmann, M. B. (2002). Analysis and segmentation of face images using point annotations and linear subspace techniques. *Tech. Rep.*, Technical University, Denmark.

Vetter, T. and Poggio, T. (1995). Linear object classes and image synthesis from a single example image. *Tech. Rep.*, MIT, USA.

Wiskott, L., Fellous, J.-M., Krüger, N., and von der Malsburg, C. (1999). Face recognition by elastic bunch graph matching. *Intelligent Biometric Techniques in Fingerprint and Face Recognition* (J. L. Jain, U. Halici, I. Hayashi, and S. B. Lee, eds.). Boca Raton: CRC Press, 355–396.

Wu, X. and Bhanu, B. (1997). Gabor wavelet representation for 3-D object recognition. *IEEE Trans. Image Process.* **6**, 47–64.

BAYESIAN STATISTICS 7, pp. 701–710
J. M. Bernardo, M. J. Bayarri, J. O. Berger, A. P. Dawid,
D. Heckerman, A. F. M. Smith and M. West (Eds.)

Expected Utility Estimation via Cross-Validation

AKI VEHTARI and JOUKO LAMPINEN
Helsinki University of Technology, Finland
aki.vehtari@hut.fi jouko.lampinen@hut.fi

SUMMARY

We discuss practical methods for the assessment, comparison and selection of complex hierarchical Bayesian models. A natural way to assess the goodness of the model is to estimate its future predictive capability by estimating expected utilities. Instead of just making a point estimate, it is important to obtain the distribution of the expected utility estimate in order to describe the associated uncertainty. We synthesize and extend the previous work in several ways. We give a unified presentation from the Bayesian viewpoint emphasizing the assumptions made and propose practical methods to obtain the distributions of the expected utility estimates. We discuss the properties of two practical methods, the importance sampling leave-one-out and the k-fold cross-validation. We propose a quick and generic approach based on the Bayesian bootstrap for obtaining samples from the distributions of the expected utility estimates. These distributions can also be used for model comparison, for example, by computing the probability of one model having a better expected utility than some other model. We discuss how the cross-validation approach differs from other predictive density approaches, and the relationship of cross-validation to information criteria approaches, which can also be used to estimate the expected utilities. We illustrate the discussion with one toy and two real world examples.

Keywords: EXPECTED UTILITY; CROSS-VALIDATION; MODEL ASSESSMENT; MODEL COMPARISON; PREDICTIVE DENSITIES; INFORMATION CRITERIA.

1. INTRODUCTION

1.1. *Expected Utilities*

In prediction and decision problems, it is natural to assess the predictive ability of the model by estimating the expected utilities, that is, the relative values of consequences of using the model (Good, 1952; Bernardo and Smith, 1994).

The posterior predictive distribution of output $y^{(n+1)}$ for the new input $x^{(n+1)}$ given the training data $D = \{(x^{(i)}, y^{(i)}); i = 1, 2, \ldots, n\}$ is obtained by

$$p(y^{(n+1)} \mid x^{(n+1)}, D, M) = \int p(y^{(n+1)} \mid x^{(n+1)}, \theta, D, M) p(\theta \mid x^{(n+1)}, D, M)\, d\theta.$$

where θ denotes all the model parameters and hyperparameters of the prior structures and M is all the prior knowledge in the model specification. We assume that knowing $x^{(n+1)}$ does not give more information about θ, that is, $p(\theta \mid x^{(n+1)}, D, M) = p(\theta \mid D, M)$.

We would like to estimate how good our model is by estimating the quality of the predictions the model makes for future observations from the same process that generated the given set of training data D. Given a utility function u, the expected utility is obtained by taking the expectation

$$\bar{u} = \mathrm{E}_{(x^{(n+1)}, y^{(n+1)})} \left[u \left(y^{(n+1)}, x^{(n+1)}, D, M \right) \right].$$

The expectation could also be replaced by some other summary quantity, such as the α-quantile. Note that considering the expected utility for the next sample is equivalent to taking the expectation over all future samples. Preferably, the utility u would be application-specific, measuring the expected benefit or cost of using the model. For example, Draper and Fouskakis (2000) discuss an example in which monetary utility is used for data collection costs and the accuracy of predicting mortality rate in health policy problem. Examples of generic utilities are the absolute error

$$u = \mathrm{abs}(\mathrm{E}_{y^{(n+1)}}[y^{(n+1)} \mid x^{(n+1)}, D, M] - y^{(n+1)})$$

and the predictive likelihood

$$u = p(y^{(n+1)} \mid x^{(n+1)}, D, M)$$

The predictive likelihood measures how well the model estimates the whole predictive distribution and is thus especially useful in model comparison. It is also useful in non-prediction problems, in which the goal is to get scientific insight in modelled phenomena.

1.2. *Cross-Validation Predictive Densities*

Expected utilities can be estimated using cross-validation (CV) predictive densities. As the distribution of $(x^{(n+1)}, y^{(n+1)})$ is unknown, we approximate it by using the samples we already have, that is, we assume that the distribution can be reasonably well approximated using the (weighted) training data $\{(x^{(i)}, y^{(i)}); i = 1, 2, \ldots, n\}$. To simulate the fact that the future observations are not in the training data, the ith observation $(x^{(i)}, y^{(i)})$ in the training data is left out, and then the predictive distribution for $y^{(i)}$ is computed with a model that is fitted to all of the observations except $(x^{(i)}, y^{(i)})$. By repeating this for every point in the training data, we get a collection of leave-one-out cross-validation (LOO-CV) predictive densities

$$\{p(y^{(i)} \mid x^{(i)}, D^{(\setminus i)}, M); i = 1, 2, \ldots, n\}$$

where $D^{(\setminus i)}$ denotes all the elements of D except $(x^{(i)}, y^{(i)})$. To get the expected utility estimate, these predictive densities are compared to the actual $y^{(i)}$'s using the utility u, and the expectation is taken over i

$$\bar{u}_{\mathrm{LOO}} = \mathrm{E}_i \left[u(y^{(i)}, x^{(i)}, D^{(\setminus i)}, M) \right]$$

If the future distribution is expected to be different from the distribution of the training data, observations could be weighted appropriately. By appropriate modifications of the algorithm, the cross-validation predictive densities can also be computed for data with a nested structure or other finite range dependencies. Vehtari and Lampinen (2002a) discuss these issues and assumptions made on future data distributions in more detail.

For simple models, the LOO-CV-predictive densities may be computed quickly using analytical solutions, but models that are more complex usually require a full model fitting for each of the n predictive densities. When using the Monte Carlo methods we have to sample from $p(\theta \mid D^{(\setminus i)}, M)$ for each i. If sampling is slow (*e.g.*, when using MCMC methods), importance sampling LOO-CV (IS-LOO-CV) or the k-fold-CV can be used to reduce the computational burden.

1.3. *Previous Work*

Cross-validation methods for model assessment and comparison have been proposed by several authors: for early accounts see Stone (1974) and Geisser (1975), and for a more recent review see Gelfand *et al.* (1992). The cross-validation predictive density dates at least to Geisser and Eddy (1979), and reviews of cross-validation and other predictive densities appear in Gelfand and Dey (1994) and Gelfand (1996). Bernardo and Smith (1994, Ch. 6) also discuss briefly how cross-validation approximates the formal Bayes procedure of computing the expected utilities.

2. METHODS

2.1. *Importance Sampling Leave-One-Out Cross-Validation*

In IS-LOO-CV, instead of sampling directly from $p(\theta \mid D^{(\backslash i)}, M)$, samples from the full posterior $p(\theta \mid D, M)$ are reused and the full posterior is used as the importance sampling density for the leave-one-out posterior densities (Gelfand *et al.,* 1992; Gelfand, 1996). Additional computation time compared to sampling from the full posterior distribution is negligible.

The reliability of the importance sampling can be estimated by examining the expected variability of the importance weights. We propose to use a heuristic measure of effective sample sizes based on an approximation of the variance of importance weights computed as $m_{\text{eff}}^{(i)} = 1/\sum_{j=1}^{m}(w_j^{(i)})^2$, where $w_j^{(i)}$ are normalized weights (Kong *et al.,* 1994). See further discussion of estimating the reliability of the IS-LOO-CV in Vehtari and Lampinen (2002a). If there is reason to suspect the reliability of the importance sampling, we suggest using predictive densities from the k-fold-CV, discussed in the next section.

2.2. *k-Fold Cross-Validation*

k-fold CV is a robust way of obtaining CV predictive densities for complex hierarchical Bayesian models. In k-fold-CV, we sample only from k (*e.g.*, $k = 10$) k-fold-CV distributions $p(\theta \mid D^{(\backslash s(i))}, M)$ and get a collection of k-fold-CV predictive densities

$$\{p(y^{(i)} \mid x^{(i)}, D^{(\backslash s(i))}, M); i = 1, 2, \ldots, n\}$$

where $s(i)$ is a set of data points as follows: the data are divided into k groups so that their sizes are as nearly equal as possible and $s(i)$ is the set of data points in the group where the ith data point belongs. In the case of data with nested structure, the grouping needs to respect the hierarchical nature of the data, and in the case of non-structured finite range dependency, the group size should be selected according to the range of the dependency.

Since the k-fold-CV predictive densities are based on smaller training data sets $D^{(\backslash s(i))}$ than the full data set D, the expected utility estimate is slightly biased. This bias can be corrected using a first order bias correction (Burman, 1989):

$$\begin{aligned}
\bar{u}_{\text{tr}} &= \mathrm{E}_i[u(y^{(i)}, x^{(i)}, D, M)] \\
\bar{u}_{\text{cvtr}} &= \mathrm{E}_j[\mathrm{E}_i[u(y^{(i)}, x^{(i)}, D^{(\backslash s_j)}, M)]]; \quad j = 1, \ldots, k \\
\bar{u}_{\text{CCV}} &= \bar{u}_{\text{CV}} + \bar{u}_{\text{tr}} - \bar{u}_{\text{cvtr}},
\end{aligned}$$

where $\bar{u}_{\text{tr}}$ is the expected utility evaluated with the training data given the training data, that is, the training error or the expected utility computed with the marginal posterior predictive densities, and $\bar{u}_{\text{cvtr}}$ is the average of the expected utilities evaluated with the training data given the k-fold-CV training sets.

2.3. *Distribution of the Expected Utility Estimate*

Instead of just making a point estimate, it is important to obtain the distribution of the expected utility estimate in order to describe the associated uncertainty. These distributions can also be used to compare models.

Our goal is to estimate the expected utilities given the training data D, but the cross-validation predictive densities

$$p(y \mid x^{(i)}, D^{(\backslash s_j)}, M)$$

are based on training data sets $D^{(\backslash s_j)}$, which are each slightly different. This makes the u_i's slightly dependent in a way that will increase the estimate of the variability of the $\bar{u}$. In the case of LOO-CV, this increase is negligible (unless n is very small), and in the case of k-fold-CV, it is practically negligible with reasonable values of k. If in doubt, this increase could be estimated as proposed by Vehtari and Lampinen (2002a).

We propose to use the Bayesian bootstrap (BB; Rubin, 1981) to obtain samples from the distribution of the expected utility estimate. In this approach it is assumed that the posterior probabilities for the samples z_i of a random variable Z have a Dirichlet distribution and values of Z that are not observed have zero posterior probability. Sampling from the Dirichlet distribution gives BB samples from *the distribution of the distribution of* Z, and thus samples of any parameter of this distribution can be obtained. We first sample from the distributions of each u_i (variability due to Monte Carlo integration) and then from the distribution of the $\bar{u}$ (variability due to the approximation of the future data distribution). From obtained samples, it is easy to compute, for example, credible intervals (CI), highest probability density intervals (HDPI), and kernel density estimates. The approach can handle arbitrary summary quantities and gives a good approximation also in non-Gaussian cases.

2.4. *Model Comparison with Expected Utilities*

The distributions of the expected utility estimates can be used to compare models, for example, by plotting the distribution of $\bar{u}_{M_1-M_2} = \mathrm{E}_i[u_{M_1,i} - u_{M_2,i}]$ or computing the probability $p(\bar{u}_{M_1-M_2} > 0)$. Following the simplicity postulate (Jeffreys, 1961), it is useful to start from simpler models and then test if more complex models would give significantly better predictions. Note that comparison of point estimates instead of distributions could easily lead to selection of unnecessarily complex models.

An extra advantage of comparing the expected utilities is that even if there is high probability that one model is better, it might be discovered that the difference between the expected utilities is still practically negligible. For example, it is possible that using a statistically better model would save only a negligible amount of money.

3. RELATIONS TO OTHER APPROACHES

3.1. *Other Predictive Approaches*

For notational convenience we omit the x and consider expected predictive log-likelihood $\mathrm{E}_{y^{(n+1)}}[\log p(y^{(n+1)} \mid D, M)]$, which can be estimated using the cross-validation predictive densities as $\mathrm{E}_i[\log p(y^{(i)} \mid D^{(\backslash i)}, M)]$.

Relations to other predictive densities can be illustrated by comparing the equations in Table 1, where D_* is an exact replicate of D, and $y^{(s_i)}$ is a set of data points so that

$$y^{(s_1)} = \emptyset, \quad y^{(s_2)} = y^{(1)}, \quad y^{(s_i)} = y^{(1,\ldots,i-1)}; i = 3, \ldots, n.$$

Table 1. *Relations to other predictive densities*

Cross-validation (Expected utility)	$\frac{1}{n}\sum_{i=1}^{n} \log p(y^{(i)} \mid D^{(\setminus i)}, M) = \mathrm{E}_i \left[\log p(y^{(i)} \mid D^{(\setminus i)}, M)\right]$
Marginal posterior (Training error):	$\frac{1}{n}\sum_{i=1}^{n} \log p(y^{(i)} \mid D_*, M)$
Posterior (Posterior BF):	$\frac{1}{n}\sum_{i=1}^{n} \log p(y^{(i)} \mid y^{(s_i)}, D_*, M) = \frac{1}{n} \log p(D \mid D_*, M)$
Prior (Bayes Factor):	$\frac{1}{n}\sum_{i=1}^{n} \log p(y^{(i)} \mid y^{(s_i)}, M) = \frac{1}{n} \log p(D \mid M)$

Next we discuss the relations and differences in more detail. Other less interesting possibilities not discussed here are marginal prior, partial, intrinsic, fractional and prequential predictive distributions.

Marginal posterior predictive densities. Estimating the expected utilities with the marginal posterior predictive densities measures the goodness of the predictions as if the future data would be exact replicates of the training data. Accordingly this estimate is often called the training error. It is well known that this underestimates the generalization error of flexible models as it does not correctly measure the out-off-sample performance. However, if the effective number of parameters is relatively small (i.e., if $p_{\text{eff}} \ll n$), marginal posterior predictive densities may be useful approximations to cross-validation predictive densities (see section 3.2), and in this case may be used to save computational resources. In case of flexible non-linear models (see, *e.g.*, section 4) p_{eff} is usually relatively large.

The marginal posterior predictive densities are also useful in *Bayesian posterior analysis* advocated, for example, by Gelman *et al.* (1996). In the Bayesian posterior analysis, the goal is to compare posterior predictive replications to the data and examine the aspects of the data that might not accurately be described by the model. Thus, the Bayesian posterior analysis is complementary to the use of the expected utilities in model assessment. To avoid using the data twice, we have also used the cross-validation predictive densities for such analysis. This approach has also been used in some form by Gelfand *et al.* (1992), Gelfand (1996), and Draper (1996).

Posterior predictive densities. Comparison of the joint posterior predictive densities leads to the posterior Bayes factor $p(D \mid D_*, M_1)/p(D \mid D_*, M_2)$ (Aitkin, 1991). Comparing the above equations it is obvious that use of the joint posterior predictive densities would produce even worse estimates for the expected utilities than the marginal posterior predictive densities; thus this method is to be avoided.

Prior predictive densities. Comparison of the joint prior predictive densities leads to the Bayes factor $p(D \mid M_1)/p(D \mid M_2)$ (Kass and Raftery, 1995). In an expected utility sense

$$\frac{1}{n} \log p(D \mid M) = \frac{1}{n} \sum_{i} \log p(y^{(i)} \mid y^{(s_i)}, M)$$

is an average of predictive log-likelihoods with the number of data points used for fitting ranging from 0 to $n - 1$. As the learning curve is usually steeper with smaller number of data points, this is less than the expected predictive log-likelihood with $n/2$ data points. As there are terms which are conditioned on none or very few data points, the prior predictive approach is sensitive to prior changes. With more vague priors and more flexible models the few first terms dominate the expectation unless n is very large.

This comparison shows that the prior predictive likelihood of the model can be used as a lower bound for the expected predictive likelihood (favoring less flexible models). Vehtari and Lampinen (2002b) discuss problems with a very large number of models, where it is not computationally possible to use cross-validation for each model. They propose to use variable dimension MCMC methods to estimate the posterior probabilities of models, which can be used to obtain relative prior predictive likelihood values: in this way it is possible to select a smaller set of models for which expected utilities are estimated via cross-validation.

3.2. INFORMATION CRITERIA

Information criteria such as AIC, NIC, DIC (Akaike, 1973; Murata *et al.,* 1994; Spiegelhalter *et al.,* 2002) also estimate expected utilities (except the BIC by Schwarz, 1978, which is based on prior predictive densities). AIC and DIC are defined using deviance as utility. NIC is defined using arbitrary utilities, and a generalization of DIC using arbitrary utilities was presented by Vehtari (2002). Given a utility function u, it is possible to use Monte Carlo samples to estimate $E_\theta[\bar{u}(\theta)]$ and $\bar{u}(E_\theta[\theta])$, and then compute an expected utility estimate as

$$\bar{u}_{\mathrm{DIC}} = \bar{u}(E_\theta[\theta]) + 2(E_\theta[\bar{u}(\theta)] - \bar{u}(E_\theta[\theta])).$$

Information criteria estimate expected utilities using asymptotic approximations, which will not necessarily be accurate with complex hierarchical models and finite data. The CV approach uses full predictive distributions, obtained by integrating out the unknown parameters, while information criteria use *plug-in* predictive distributions (maximum likelihood, maximum a posteriori or posterior mean), which ignore the uncertainty about parameter values and model. The cross-validation approach is less sensitive to parametrization than information criteria, as it deals directly with predictive distributions. With appropriate grouping k-fold-CV can also be used when there are finite range dependencies in the data, while use of information criteria is limited to cases with more strict restrictions on dependencies (Vehtari, 2003). In the case of information criteria, the distribution of the estimate is not so easy to estimate and usually only point estimates are used. Even if an information criterion is used for models for which assumptions of the criterion hold, this may lead to selection of unnecessarily complex models, as more complex models with possibly better but not significantly better criterion values may be selected.

Spiegelhalter *et al.* (2002) divided the estimation of the expected deviance (DIC) into model fit and complexity parts, where the latter is called the effective number of parameters p_{eff}. In the CV approach, an estimate of p_{eff} is not needed, but it can be estimated in the case of independent data by the difference of the marginal posterior predictive log-likelihood and the expected predictive log-likelihood (Vehtari, 2001, Ch. 3.3)

$$p_{\mathrm{eff,CV}} = \sum_{i=1}^{n} \log p(y^{(i)} \mid D, M) - \sum_{i=1}^{n} \log p(y^{(i)} \mid D^{(\setminus i)}, M).$$

When k-fold-CV is used, the second term is replaced with the bias corrected estimate (Section 2.2).

4. ILLUSTRATIVE EXAMPLES

As illustrative examples, we use multi layer perception (MLP) neural networks and Gaussian processes (GP) with Markov Chain Monte Carlo sampling (Neal, 1996, 1999; Lampinen and Vehtari, 2001) in one toy problem (MacKay's robot arm) and two real world problems: concrete quality estimation and forest scene classification (see Vehtari and Lampinen, 2001, for details of the models, priors and MCMC parameters). The MCMC sampling using the FBM [1] software and Matlab-code partly derived from the FBM and Netlab[2] toolbox. Importance weights for MLP networks and GPs were computed as described in Vehtari (2001, Ch. 3.2.2).

4.1. *Toy Problem: MacKay's Robot Arm*

The task is to learn the mapping from joint angles to position for an imaginary robot arm. Two real input variables, x_1 and x_2, represent the joint angles and two real target values, y_1 and y_2, represent the resulting arm position in rectangular coordinates. The relationship between inputs and targets is $y_1 = 2.0\cos(x_1) + 1.3\cos(x_1 + x_2) + e_1, y_2 = 2.0\sin(x_1) + 1.3\sin(x_1 + x_2) + e_2$, where e_1 and e_2 are independent Gaussian noise variables of standard deviation 0.05. As training data sets, we used the same data sets that were used by MacKay (1992). We used an 8-hidden-unit MLP and a GP with normal (N) residual model.

Figure 1 shows the different components that contribute to the uncertainty in the estimate of the expected utility. The variability due to having slightly different training sets in 10-fold-CV and the variability due to the Monte Carlo approximation are negligible compared to the variability due to not knowing the true noise variance. Figure 1 also demonstrates the comparison of models using paired comparison of the expected utilities. The two models are almost indistinguishable on grounds of predictive utility.

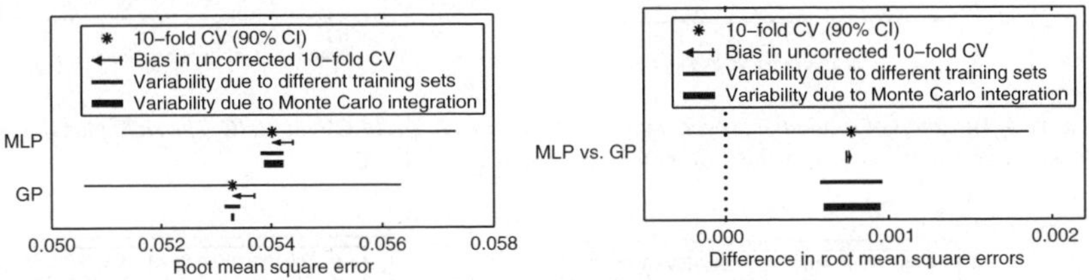

Figure 1. *The left plot shows the different components that contribute to the uncertainty, and bias correction for the expected utility (root mean square error) for MLP and GP. The right plot shows same information for the expected difference of root mean square errors.*

4.2. *Case Study I: Concrete quality estimation*

The goal of this project was to develop a model for predicting the quality properties of concrete, as a part of a large quality control program of the industrial partner of the project (Järvenpää, 2001).

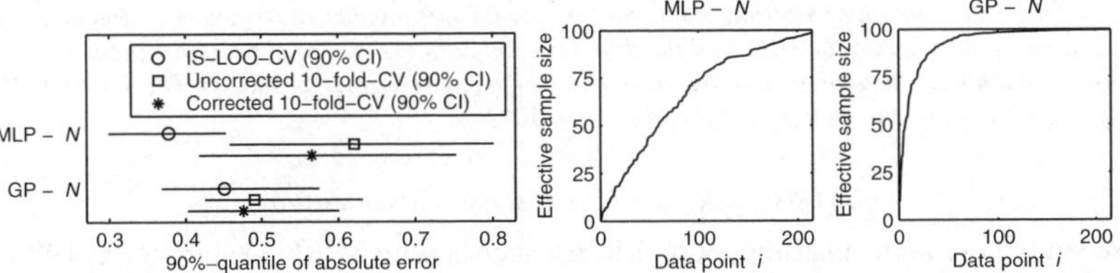

Figure 2. *The left plot shows the comparison of IS-LOO-CV and k-fold-CV with and without bias correction. The right plot shows the effective sample sizes of the importance sampling $m_{\text{eff}}^{(i)}$ for each data point i (sorted in increasing order). Small effective sample sizes imply that IS-LOO-CV does not work.*

[1] www.cs.toronto.edu/~radford/fbm.software.html

[2] www.ncrg.aston.ac.uk/netlab/

In the study, we had 27 explanatory variables and 215 samples. Here we report results for the volume percentage of air in the concrete. We tested 10-hidden-unit MLP networks and GP models with Normal (N), Student's t_ν, input dependent Normal (in.dep.-N) and input dependent t_ν residual models.

Figure 2 shows some results comparing IS-LOO-CV and k-fold-CV. IS-LOO-CV fails as importance sampling does not work well in this problem. k-fold-CV without bias correction gives overly pessimistic estimates. Figure 3 shows results comparing k-fold-CV and DIC in estimating the expected utilities and the effective number of parameters for four different noise models. DIC gives more optimistic estimates, which is probably due to using the plug-in predictive distribution and ignoring the uncertainty about the parameter values. Furthermore, since DIC provides only point estimates, it is harder to know whether the difference between models is significant

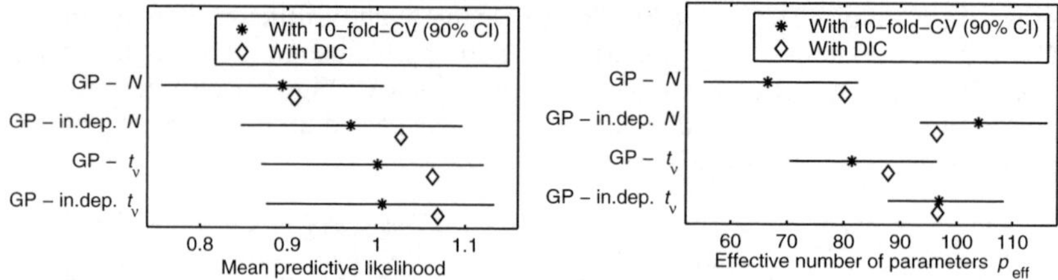

Figure 3. *The left plot shows the expected utility estimates with k-fold-CV and DIC. The right plot shows the effective number of parameters estimates with k-fold-CV and DIC.*

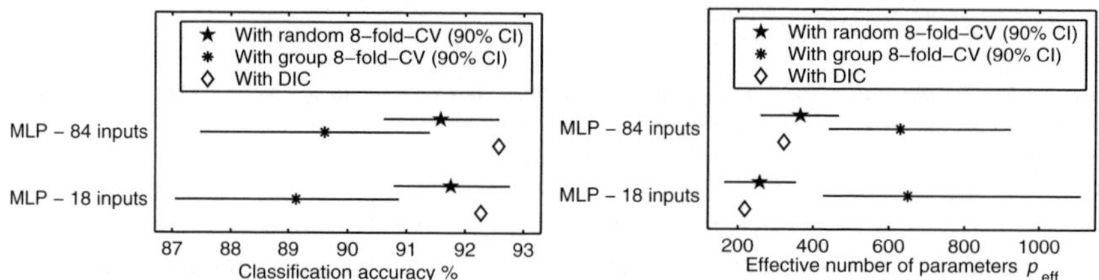

Figure 4. *The left plot shows the expected mean predictive likelihoods computed with the 8-fold-CV and the DIC. The right plot shows the estimates for the effective number of parameters computed with the random 8-fold-CV and the DIC, and the difference between the marginal posterior predictive log-likelihood and the expected predictive log-likelihood computed with group 8-fold-CV. The 84-input MLP had $p_{\mathrm{total}} = 1721$ and the 18-input MLP had $p_{\mathrm{total}} = 401$.*

4.3. *Case Study II: Forest scene classification*

The case study here is the classification of forest scenes with MLP (Vehtari *et al.*, 1998). The final objective of the project was to assess the accuracy of estimating the volumes of growing trees from digital images. To locate the tree trunks and to initialize the fitting of the trunk contour model, a classification of the image pixels to tree and non-tree classes was necessary. Training data was 4800 samples from 48 images (100 pixels from each image) with 84 different Gabor and statistical features as input variables. We tested two 20-hidden-unit MLPs with a logistic likelihood model. The first MLP used all 84 inputs and the second MLP used a reduced set of 18 inputs selected using the reversible jump MCMC method (Vehtari, 2001).

The training data has a nested structure as textures and lighting conditions are more similar in different parts of one image than in different images. If LOO-CV is used or data are divided randomly in the k-fold-CV, the training and test sets may have data points from the same image, which would lead to over-optimistic estimates. This is caused by the fact that instead of having 4800 independent data points, we have 48 sample images which each have 100 highly dependent data points. This increases our uncertainty about the future data. To get a more accurate estimate of the expected utility for new unseen images, the training data set has to be divided by images.

Figure 4 shows the expected classification accuracy and the effective number of parameters estimated via cross-validation and (generalized) DIC. The 8-fold-CV with random data division and the DIC give overly optimistic estimates of the expected classification accuracy. The random 8-fold-CV also underestimates the uncertainty in the estimate. The DIC and the random 8-fold-CV give similar estimates of the effective number of parameters, which supports the argument that DIC assumes independent data points. If there are dependencies in the data, it is not possible to explain the difference between the model fit (the marginal posterior predictive log-likelihood) and the expected predictive log-likelihood with the effective number of parameters because this difference may be larger than the total number of parameters in the model. For example, in the case of the 18-input MLP with total of 401 parameters, this difference is about 650.

ACKNOWLEDGEMENTS

The authors would like to thank D. Draper and D. Spiegelhalter for helpful comments on the manuscript.

REFERENCES

Aitkin, M. (1991). Posterior Bayes factors *J. R. Statist. Soc. B* **53**, 111–142 (with discussion).

Akaike, H. (1973). Information theory and an extension of the maximum likelihood principle. *2nd Int. Symp. on Inform. Theory* (B. N. Petrov and F. Csaki, eds). Budapest: Academiai Kiado. Reprinted in *Breakthroughs in Statistics* **1** (S. Kotz and N. L. Johnson, eds). Berlin: Springer, 610–624.

Bernardo, J. M. and Smith, A. F. M. (1994). *Bayesian Theory*. Chichester: Wiley

Burman, P. (1989). A comparative study of ordinary cross-validation, v-fold cross-validation and the repeated learning-testing methods. *Biometrika* **76**, 503–514.

Draper, D. (1996). Discussion of a paper by A Gelman *et al. Statist. Sinica* **6**, 760–767.

Draper, D. and Fouskakis, D. (2000). A case study of stochastic optimization in health policy: problem formulation and preliminary results. *J. Global Optim.* **18**, 399–416.

Geisser, S. (1975). The predictive sample reuse method with applications. *J. Am. Statist. Ass.* **70**, 320–328.

Geisser, S. and Eddy, W. F. (1979). A predictive approach to model selection. *J. Am. Statist. Ass.* **74**, 153–160.

Gelfand, A. E. (1996). Model determination using sampling-based methods. *Markov Chain Monte Carlo in Practice* (W. R. Gilks, S. Richardson and D. J. Spiegelhalter, eds). London: Chapman and Hall, 145–162.

Gelfand, A. E. and Dey, D. K. (1994). Bayesian model choice: Asymptotics and exact calculations. *J. R. Statist. Soc. B* **56**, 501–514.

Gelfand, A. E., Dey, D. K. and Chang, H. (1992). Model determination using predictive distributions with implementation via sampling-based methods *Bayesian Statistics 4* (J. M. Bernardo, J. O. Berger, A. P. Dawid and A. F. M. Smith, eds). Oxford: Oxford University Press, 147–167 (with discussion).

Gelman, A., Meng, X.-L. and Stern, H. (1996). Posterior predictive assessment of model fitness via realized discrepancies. *Statist. Sinica* **6**, 733–807 (with discussion).

Good, I. J. (1952). Rational decisions. *J. R. Statist. Soc. B* **14**, 107–114.

Jeffreys, H. (1961). *Theory of Probability*, 3rd edn. Oxford: Oxford University Press

Järvenpää, H. (2001). *Quality Characteristics of Fine Aggregates and Controlling their Effects on Concrete*. Helsinki: The Finnish Academies of Technology.

Kass, R. E. and Raftery, A. E. (1995). Bayes factors. *J. Am. Statist. Ass.* **90**, 773–795.

Kong, A., Liu, J. S. and Wong, W. H. (1994). Sequential imputations and Bayesian missing data problems. *J. Am. Statist. Ass.* **89**, 278–288.

Lampinen, J. and Vehtari, A. (2001). Bayesian approach for neural networks—Review and case studies. *Neural Networks* **14**, 7–24.

MacKay, D. J. C. (1992). A practical Bayesian framework for backpropagation networks. *Neural Comput.* **4**, 448–472.

Murata, N., Yoshizawa, S. and Amari, S.-I. (1994). Network information criterion—Determining the number of hidden units for an artificial neural network model. *IEEE Trans. Neural Networks* **5**, 865–872.

Neal, R. M. (1996). *Bayesian Learning for Neural Networks*. New York: Springer

Neal, R. M. (1999). Regression and classification using Gaussian process priors. *Bayesian Statistics 6* (J. M. Bernardo, J. O. Berger, A. P. Dawid and A. F. M. Smith, eds). Oxford: Oxford University Press, 475–501 (with discussion).

Rubin, D. B. (1981). The Bayesian bootstrap. *Ann. Statist.* **9**, 130–134.

Schwarz, G. (1978). Estimating the dimension of a model. *Ann. Statist.* **6**, 461–464.

Spiegelhalter, D., Best, N. G., Carlin, B. P. and van der Linde, A. (2002). Bayesian measures of model complexity and fit. *J. R. Statist. Soc. B* **64**, 583–639 (with discussion).

Stone, M. (1974). Cross-validatory choice and assessment of statistical predictions *J. R. Statist. Soc. B* **36**, 111–147 (with discussion).

Vehtari, A. (2001). *Bayesian Model Assessment and Selection Using Expected Utilities.* Ph.D. Thesis, Helsinki University of Technology, Finland. `lib.hut.fi/Diss/2001/isbn9512257653`.

Vehtari, A. (2002). Discussion of a paper by Spiegelhalter *et al. J. R. Statist. Soc. B* **64**, 620.

Vehtari, A. (2003). Discussion of a paper by Carlin *et al.* (this volume).

Vehtari, A., Heikkonen, J., Lampinen, J. and Juujärvi, J. (1998). Using Bayesian neural networks to classify forest scenes. *Intelligent Robots and Computer Vision XVII: Algorithms, Techniques, and Active Vision* (D. P. Casasent, ed.). Bellingham: SPIE, 66–73.

Vehtari, A. and Lampinen, J. (2001). On Bayesian model assessment and choice using cross-validation predictive densities. *Tech. Rep.*, Helsinki University of Technology, Finland.

Vehtari, A. and Lampinen, J. (2002a). Bayesian model assessment and comparison using cross-validation predictive densities. *Neural Comput.* **14**, 2439–2468.

Vehtari, A. and Lampinen, J. (2002b). Bayesian input variable selection using posterior probabilities and expected utilities. *Tech. Rep.*, Helsinki University of Technology, Finland.

BAYESIAN STATISTICS 7, pp. 711–719
J. M. Bernardo, M. J. Bayarri, J. O. Berger, A. P. Dawid,
D. Heckerman, A. F. M. Smith and M. West (Eds.)

A Method for Sequential Optimization in Bayesian Analysis

MIGUEL VIRTO
Centro de Investigación Militar Operativa, Spain
mvirto@fi.upm.es

JACINTO MARTÍN
Universidad de Extremadura, Spain
jrmartin@unex.es

DAVID RÍOS INSUA
Universidad Rey Juan Carlos, Spain
drios@escet.urjc.es

ARMINDA MORENO DÍAZ
Universidad Politécnica de Madrid, Spain
amoreno@fi.upm.es

SUMMARY

Bayesian analysis frequently leads to sequential optimization problems under uncertainty. Some examples include optimal sequential design, the analysis of influence diagrams and decision trees and many Operations Research related problems in areas like maintenance or reliability. We introduce here a method to deal with such problems, within a stochastic dynamic programming context. The basic approach is a Metropolis augmented probability simulation algorithm. Once introduced, we describe how to exploit the dynamic structure of the problem in a Bellman-like approach and describe our basic algorithm. Implementation details are discussed together with an illustrative example in preventive maintenance. An appendix includes theoretical results.

Keywords: STOCHASTIC OPTIMIZATION; SEQUENTIAL OPTIMIZATION; MCMC METHODS.

1. INTRODUCTION

In sequential decision-making problems, the decision maker chooses consecutive actions, according to the system status and his preferences, to form a decision policy. This scheme is present in a wide range of applications within the industry, management and services. The mathematical models considered include stochastic dynamic programming, Markov and semi-markov decision processes, stochastic control and various heuristic approaches, many of them having their origin in Bellman's (1957) *optimality principle*.

We shall consider systems in which decisions are made sequentially, obtaining an immediate utility after each decision, which also modifies the environment for future decisions. As the actual decision path will depend on the way the random aspects play out, *solving* this sort of problems will involve providing a decision rule which maximizes expected utility, for every possible state. Following Esogbue (1989), we shall focus on discrete, markovian consecutive problems. Unfortunately, real problems render dynamic programming unusable in this context because of the *dimensionality curse*, in spite of various attempts to mitigate such problem, see Morin (1977) for a review.

Our aim here is to explore the applicability of Markov Chain Monte Carlo methods (MCMC) to solve stochastic dynamic programming problems. After formulating the type of problems we shall be dealing with, Section 3 introduces a general strategy based on the augmented probability simulation method (Bielza *et al.*, 1999). To do so, we provide a Metropolis Hastings version of such algorithm, which we extend to sequential problems. Section 4 applies the proposed method to a maintenance problem. We discuss implementation details in Section 5 and provide a theoretical justification in the Appendix.

2. STOCHASTIC DYNAMIC PROGRAMMING PROBLEMS

The systems we consider exhibit the following features:

(i) At each discrete time t, the system is in some observable state θ_t that belongs to a discrete set Θ_t. We assume a finite horizon T.
(ii) For each state $\Theta_t = \theta_t$, there is an action space $\mathcal{D}_t(\theta_t)$, not necessarily finite, where we make our decisions a_t.
(iii) Consequences of actions provide a utility $u_t(a_t, \theta_{t+1})$ when moving randomly from Θ_t to Θ_{t+1}, due to decision a_t. The utility function is assumed to be separable.
(iv) The system satisfies the markovian property with respect to state space Θ_t.
(v) A policy π specifies a sequence of decisions for each stage and state. For finite horizon T, we must find the optimal policy π^*, maximizing expected utility.

We shall consider finite horizon problems in which Bellman's equation is most appropriate. When we move from state θ_t, at time t, to θ_{t+1}, at time $t+1$, choosing alternative a_t, Bellman's equation would be

$$v_t(\theta_t) = \max_{a_t} \left[\int (u_t(a_t, \theta_{t+1}) + v_{t+1}(\theta_{t+1}))\, p_t(\theta_{t+1})\, d\theta_{t+1} \right], \tag{1}$$

with $v_{T+1}(\theta) = 0$, for each θ. If state variables are discrete, or we discretize them, we may apply backwards induction to solve (1)

$$v_t(\theta_t) = \max_{a_t} \left[\sum_{\theta_{t+1}} (u_t(a_t, \theta_{t+1}) + v_{t+1}(\theta_{t+1}))\, p_t(\theta_{t+1}) \right],$$

starting with $v_{T+1}(\theta) = 0$, and then proceeding with $t = T, T-1, T-2, ..., 1$. As we know, in many important problems, computational requirements will overwhelm us due to the *dimensionality curse*.

White (1993) reviews several solution procedures, including linear programming (Manne, 1960), policy improving algorithms (Howard, 1960) and value improving algorithms (Bellman, 1957). Many other methods aim at mitigating the dimensionality curse such as Denardo (1982), Lovejoy (1986), Johnson *et al.* (1993) or Bertsekas and Tsitsiklis (1996). Stochastic programming methods may aid in some circumstances (Birge and Louveaux, 1997). We shall explore here a novel method.

3. A METROPOLIS AUGMENTED PROBABILITY SIMULATION ALGORITHM

3.1. *The Basic Approach*

The basic algorithm we shall use is a Metropolis augmented probability simulation algorithm. Suppose we have to choose an alternative a from a set $\mathcal{A}$. There is uncertainty about a state θ, measured by a posterior distribution $p_a(\theta)$, after alternative a has been chosen. Preferences over consequences are modelled through a utility function $u(a, \theta)$. We aim at finding the maximum expected utility alternative a^* satisfying

$$\Psi(a^*) = \max_{a \in \mathcal{A}} \left\{ \Psi(a) = \int u(a, \theta)\, p_a(\theta) d\theta \right\}.$$

We start from the augmented simulation method in Bielza *et al.* (1999). To do so, define an augmented artificial distribution on $\Theta \times A$, with h such that

$$h(\theta, a) \propto u(\theta, a)\, p_a(\theta)$$

where the function $u(\theta, a)$ must be positive and $h(\theta, a)$ must be integrable. The marginal over a is then proportional to the expected utility

$$h(a) = \int h(\theta, a)\, d\theta \propto \int u(\theta, a)\, p_a(\theta)\, d\theta = \Psi(a).$$

Therefore, the optimal alternative a^* will be the mode of the marginal in the space of alternatives. As a consequence, we may use the following strategy:

1. Generate a sample from distribution $h(\theta, a)$.
2. Transform it to a sample from the marginal $h(a)$.
3. Identify the marginal mode.

To generate the sample from $h(\cdot, \cdot)$, we shall use a Metropolis–Hastings algorithm (Metropolis *et al.*, 1953).

3.2. *An Algorithm for Dynamic Programming Problems*

We shall solve the problem applying the augmented probability simulation strategy, with a variant to estimate modes which resembles the dynamic programming principle. Note that we could use MCMC methods sequentially, as suggested in Müller (1999) for optimal sequential design.

Alternatively, we shall study the possibility of jointly simulating all the stages and study the output of the MCMC method. Let a history be a vector of realized states and actions and let $u(\theta_1, a_1, ..., \theta_T, a_T, \theta_{T+1})$ be the utility of such history, which we assume positive. Then, the artificial distribution $u(\theta, a)\, p_a(\theta)$ will be

$$\begin{aligned} &u(\theta_1, a_1, \theta_2, a_2, \ldots, \theta_T, a_T, \theta_{T+1})\, p_a(\theta_1, \theta_2, ..., \theta_{T+1}) \\ &\qquad = u(\theta_1, a_1, \theta_2, a_2, \ldots, \theta_T, a_T, \theta_{T+1})\, p(\theta_1)\, p_{a_1}(\theta_2 \,|\, \theta_1)\ \ldots\ p_{a_T}(\theta_{T+1} \,|\, \theta_T). \end{aligned}$$

The basic idea takes into account that, given a Monte Carlo approximation to the artificial distribution, integrating out the randomness after the last decision is equivalent to accumulating certain frequencies. Then, once the simulation is undertaken, we must apply a backwards method with the frequencies of the generated histories. We apply Metropolis–Hastings over complete histories $h_{T+1} = (\theta_1, a_1, \theta_2, a_2, ..., \theta_T, a_T, \theta_{T+1})$ with the following algorithm, where fr designates frequency. q is a candidate generating distribution from which we generate histories and $u(h)$ is the global utility of history h.

1. `Start from an initial history` h^0, `do` $i = 0$

   1.1. `Compute` $u(h^0) = u(a_1, \theta_2) + u(a_2, \theta_3) + \ldots + u(a_T, \theta_{T+1})$

2. `Until convergence:`

   2.1. `Generate a history` h^c `from distribution` $q(h \,|\, h^i)$

   2.2. `Evaluate` $u(h^c)$

   2.3. `Compute`

$$\alpha = \min \left\{ 1, \frac{u(h^c)\, p(\theta^c)}{u(h^i)\, p(\theta^i)} \frac{q(a^i, \theta^i \,|\, a^c, \theta^c)}{q(a^c, \theta^c \,|\, a^i, \theta^i)} \right\}$$

2.4. Do $h^{i+1} = h^c$ with probability α or $h^{i+1} = h^i$ with probability $(1-\alpha)$

3. Once convergence has been achieved, with sample $\{h^1, h^2, ..., h^n\}$

3.1. Determine, for all θ_T

$$a_T^*(\theta_T) = \ arg \max_{a_T} \left\{ \sum_h fr(h = (\ldots, \theta_T, a_T)) \right\}$$

3.2. By backwards induction, determine for all θ_t

$$a_t^*(\theta_t) = \ arg \max_{a_t} \left\{ \sum_h fr(h = (\ldots, \theta_t, a_t, \theta_{t+1}, a_{t+1}^*, \ldots, a_T^*)) \right\}$$

for $t = T-1, ..., 1$

Steps 1 and 2 describe a Metropolis–Hastings algorithm that converges, under appropriate conditions, to the target distribution $u(\theta, a)\, p_a(\theta)$. In step 2.1, we need to design the candidate generating distribution $q(h^c \mid h^i)$. The evaluation step 2.2 will be simple as the utility function is separable. As a result of the first two steps, we obtain a sample of different histories $\{h^1, h^2, ..., h^n\}$ with frequencies $fr(h^i)$. Step 3 finds the modes of the marginal distribution on decisions, but we do it sequentially, first integrating out last component θ_{T+1} to determine $a_T^*(\theta_T)$. Note that this optimal decision need not be unique. It may be also convenient to keep trace of several alternatives, if their frequencies are similar. By discarding those histories without optimal a_T, we find out by backwards induction $a_{T-1}^*(\theta_{T-1})$ and continue with $a_{T-2}^*(\theta_{T-2})$, ..., and, finally, $a_1^*(\theta_1)$.

4. AN EXAMPLE IN PREVENTIVE MAINTENANCE

Many standard OR stochastic problems fit within the described framework. Here we shall pay attention to preventive maintenance problems. The basic data in these problems are the expected failures which help in forecasting replacements, and, consequently, required stocks. We consider an example in Kaufmann (1970). To better illustrate the method, we have reduced the number of stages in the original problem from five to three. We assume we are managing an equipment for four years. Starting with a new equipment we must make maintenance decisions at the end of first, second and third years. The state variable θ is the equipment age. At the beginning of each period, we decide whether to keep the equipment as it is, with a small maintenance cost, or replace it with a new equipment, selling the current one. The resale price is always smaller than the acquisition cost of a new one. Each time a failure takes place, there is a cost of 35 units. Given the failure rate as a function of equipment's age, the buying costs of new machines and the residual value of an equipment, one can construct the cost matrix of each decision. Decision $a = 0$ means buying a new equipment and $a = i$ means buying an equipment of age i. An age of 6 or more is equivalent to a broken machine. If the decision coincides with our equipment age, we keep it, with the subsequent maintenance cost.

Assuming we also have a fixed income of 270 m.u. when the system is operating, we aim at maximizing profits. Although this example may be solved by backwards induction, we adapt the proposed algorithm to illustrate the method. In this case both the decision and state spaces are discrete. We shall implement an independent chain MCMC algorithm, see Tierney (1994). As the first state is $\theta_1 = 1$, a history will de described as $h = (a_1, \theta_2, a_2, \theta_3, a_3, \theta_4)$.

We choose the candidate generating distribution

$$q(h) = q(\theta_1, a_1, \theta_2, a_2, \theta_3, a_3, \theta_4) = q_1(\theta_1)\, q_2(a_1)\, q_1(\theta_2)\, q_2(a_2)\, q_1(\theta_3)\, q_2(a_3)\, q_1(\theta_4)$$

with $q_1(\theta) = p(\theta)$, the incumbent probability model describing the evolution of the system, and, for $q_2(a)$, the discrete uniform distribution. In such a way, we simplify α as we have an independent chain. We should mention that the expected utilities vary between 10 and 271 mu. The probabilities p vary between 0.008 and 1, and the q's between 1.11E-4 and 0.125. These ranges favor MCMC convergence.

Table 1. *Accumulated frequencies to obtain optimal policy.*

St. / Dec.	0	1	2	3	4	5
Third stage						
1	229992	**234458**	0	0	0	0
2	195287	186681	**196235**	0	0	0
3	**79037**	77796	78121	78378	0	0
4	**7518**	7147	6917	6968	7463	0
5	9693	9648	8950	9146	**9783**	8257
6	96575	91306	90957	88595	**98211**	86880
Second stage						
1	116182	**116845**	0	0	0	0
2	**101321**	96427	99021	0	0	0
3	0	0	0	0	0	0
4	0	0	0	0	0	0
5	0	0	0	0	0	0
6	**16955**	15332	15464	15568	16525	15602
First stage						
1	116845	**118276**	0	0	0	0

To ensure convergence, we may replicate until the less frequent decision a_3 reaches a predetermined minimum level. In our case, we have fixed that decisions like $a_3 = 2$ when $\theta_3 = 4$ appear more than 5000 iterations. We have also discarded an initial burn-in period of 10,000 iterations. After 15 min running a C-code algorithm in a Sun SparcStation 5 (two million iterations) we obtain the frequency distribution, for the 82 possible histories.

From this frequency distribution we determine optimal decisions aggregating frequencies as indicated in the algorithm. For example, being in state 1 in the third stage, $\theta_3 = 1$, the frequency of decision $a_3 = 1$ is obtained accumulating frequencies of histories ending with the combination $\theta_3 = 1$ and $a_3 = 1$ and whatever value for θ_4, that is,

$$\begin{aligned} fr(\theta_3 = 1; a_3 = 1) &= fr(01011) + fr(12011) + fr(16011) \\ &= 116182 + 101321 + 16955 = 234458. \end{aligned}$$

History 01011 means buying a new machine at the end of first year, $a_1 = 0$. The machine is working properly at the end of the second year, θ_2=1, but you decide to change it again, $a_2 = 0$. At the end of the third year, it is still operating and you decide to keep it as $a_3 = 1$. This process of accumulating frequencies leads to Table 1. Optimal decisions are in boldface. Eventually, given the randomness involved, we might be getting close to an optimal solution.

5. FREQUENCY BALANCE

Even clustering them, it is clear from Table 1 the imbalance between frequencies among histories. For the third period, the decisions we might choose while staying in state 4 have frequencies around 7000, whereas in state 1, frequencies are over 200,000. We may correct this introducing state dependent $k(\theta_t^j)$ or decision dependent $k(a_t^j)$ factors. Using the example to better illustrate the idea, the algorithm will be slightly modified as follows:

1. `Start from an initial history` h^0 `, do` $i = 0$
   1.1. `Evaluate` $u(h^0)$
2. `Until convergence:`
   2.1. `Generate a history` h^c `from distribution` $q(h^c \mid h^i)$
   2.2. `Evaluate` $u(h^c)$
   2.3. `Compute`

$$\alpha = \min \left\{ 1, \ \frac{u(h^c)\, k(\theta^c)\, q(a^i \mid \theta^i)}{u(h^i)\, k(\theta^i)\, q(a^c \mid \theta^c)} \right\}$$

   2.4. `Do` $h^{i+1} = h^c$ `with probability` α or $h^{i+1} = h^i$ `with probability` $(1 - \alpha)$
3. `With the sample` $\{h^1, h^2, ..., h^n\}$
   3.1. `Compute`

$$a_3^*(\theta_3) = \ arg \ \max_{a_3} \left\{ \sum_h \frac{fr(h = (\cdot, \cdot, \cdot, \cdot, \theta_3, a_3))}{k(\theta_3)} \right\} \ for \ all \ \theta_3$$

   3.2. `Then`

$$a_2^*(\theta_2) = \ arg \ \max_{a_2} \left\{ \sum_h \frac{fr(h = (\cdot, \cdot, \theta_2, a_2, \cdot, a_3^*))}{k(\theta_2)} \right\} \ for \ all \ \theta_2$$

   3.3. `Finally`

$$a_1^*(\theta_1 = 1) = \ arg \ \max_{a_1} \left\{ \sum_h \frac{fr(h = (\theta_1, a_1, \cdot, a_2^*, \cdot, a_3^*))}{k(\theta_1)} \right\}$$

The k's included in step 2.3 modify the α probabilities to promote or prevent transitions to certain states. Step 3 removes the effect of the constants. We link the k's with utilities rather than with probabilities.

Closer inspection of frequency distribution of histories shows that those describing machine failure are less usual. Thus, we decide to favor histories having an equipment failure in the first period, introducing a factor $k(\theta_2 = 6) = 4$. We obtain a new table with the observed frequencies for the 82 histories. Clustering by periods yields the solution of the problem in Table 2 for two million iterations.

For instance, to compute $fr(\theta_1 = 1, a_1 = 1)$, we proceed as follows:

$$fr(\theta_1 = 1, a_1 = 1) = fr(12a_2^*) + \frac{fr(16a_2^*)}{4} = 62830 + \frac{41587}{4} = 73226.$$

Table 2 offers an alternative optimal solution. For instance, in stage 3 and state $\theta = 4$, $a = 0$ is optimal as Kaufmann mentions, but $a = 4$ is optimal as well.

Table 2. *New clustering of frequencies.*

St./ Dec.	0	1	2	3	4	5
Third Stage						
1	172970	**176086**	0	0	0	0
2	145876	139278	**147041**	0	0	0
3	**69560**	68453	69262	68521	0	0
4	18368	17741	17459	17065	**18599**	0
5	24084	22989	22541	22129	**24385**	20435
6	126011	118213	119024	114871	**127257**	111782
Second Stage						
1	71669	**72278**	0	0	0	0
2	**62830**	59733	61567	0	0	0
3	0	0	0	0	0	0
4	0	0	0	0	0	0
5	0	0	0	0	0	0
6	**41587**	38046	37950	38030	41149	38089
First Stage						
1	72278	**73227**	0	0	0	0

6. DISCUSSION

We have described a simulation method to cope with sequential decision problems embedded in a stochastic dynamic programming framework. We have also developed a way in which we can control, in a very specific manner, a history's frequency distribution, dissolving its effect once the simulation is performed. This approach is related with importance sampling. We have illustrated the method with an application to maintenance problems identifying new optimal solutions.

We have not faced the challenge of reducing dimensionality of the problem as the proposed algorithm works with the whole sequence of stages, affecting the computational burden. However, there are other potential difficulties in stochastic dynamic programming such as the explicit knowledge of probability transition matrices, required by classical resolution methods. This issue could be even more critical and demanding than the dimensionality itself. The algorithm proposed here is specially suitable when this explicit knowledge is not available due to complex probabilistic environments changing in time.

Moreover, the method can be used in continuous problems. If the set of states is finite, we do not need to modify the algorithm. We only have to estimate modes from continuous distributions. If the set of states is not finite, we may apply our method after using a grid in the state space, as in Müller (1999).

Although we have considered θ observable, the method would work even with partially observed processes. This would be an issue in future comparative studies.

ACKNOWLEDGEMENTS

Research supported by grants from CICYT, CAM, Junta de Extremadura, URJC and MCyT DPI2001-3731.

REFERENCES

Bellman, R. E. (1957). *Dynamic Programming*. Princeton: Princeton University Press.

Bertsekas, D. P. and Tsitsiklis, J. N. (1996). *Neuro-Dynamic Programing*. Athena Scientific.

Bielza, C., Müller, P. and Ríos-Insua, D. (1999). Decision analysis by augmented probability simulation. *Manage. Sci.* **45**, 7, 995–1007.

Birge, J. R. (1985). Decomposition and partitioning methods for multi-stage stochastic linear programs. *Oper. Res.* **33**, 987–1007.

Birge, J. R. and Louveaux, M. (1997). *Introduction to Stochastic Programming*. Berlin: Springer.

Denardo E. V. (1982). *Dynamic Programing*. Englewood Cliffs, NJ: Prentice-Hall.

Esogbue, A. O. (1989). A taxonomic treatment of dynamic programming models of water resources systems. *Dynamic Programming for Optimal Water Resources Analysis* (A. O. Esogbue, ed). Englewood Cliffs, NJ: Prentice-Hall, 27–72.

Johnson, S. A., Stedinger, C. A., Shoemaker, C. A., Li, Y. and Tejada-Guibert, J. A. (1993). Numerical solution of continuous-state dynamic programs using linear and spline interpolation. *Oper. Res.* **41**, 484–500.

Howard, R. A. (1960). *Dynamic Programming and Markov Processes*. Cambridge: Cambridge University Press.

Kaufmann, (1970). *Méthodes et Modéles de la Recherche Opérationnelle*. Paris: Dunod.

Lovejoy, W. S. (1986). Policy Bounds for Markov Decision Processes. *Oper. Res.* **34**, 630–637.

Manne, A. S. (1960). Linear programming and sequential decisions. *Manage. Sci.* **6**, 259–267.

Metropolis, N., Rosenbluth, A. W., Rosenbluth, M. N., Teller, A. H. and Teller, E. (1953). Equations of state calculations by fast computing machines. *J. Chem. Phys.* **21**, 1087–1092.

Morin, T. L. (1977). Computational advances in dynamic programing. *Dynamic Programing and Its Applications* (M. L. Puterman ed). New York: Academic Press.

Müller, P. (1999). Simulation based optimal design. *Bayesian Statistics 6* (J. M. Bernardo, J. O. Berger, A. P. Dawid and A. F. M. Smith, eds). Oxford: Oxford University Press, 459–474 (with discussion).

Tierney, L. (1994). Markov chains for exploring posterior distributions. *Ann. Statist.* **22**, 1701–1762.

White, D. J. (1993). *Markov Decision Processes*. Chichester: Wiley.

APPENDIX: THEORETICAL JUSTIFICATION

Recall that histories are vectors of realized states and actions $h_t{=}(\theta_1, a_1, \theta_2, a_2, ..., a_{t-1}, \theta_t)$. Let A_t be the vector of actions $(a_1, a_2, ..., a_t)$. Assume we know $u(\theta_1, a_1, \theta_2, a_2, ..., \theta_T, a_T, \theta_{T+1})\, p_a(\theta_1, \theta_2, ..., \theta_{T+1})$ for each $(\theta_1,\theta_2, ..., \theta_{T+1})$ and $(a_1,a_2,...,a_T)$. We may write the target distribution as

$$\begin{aligned} u(h_{T+1})\, p_{A_T}(\theta_1, \theta_2, ..., \theta_{T+1}) &= u(h_{T+1})\, p_{A_{T-1}}(\theta_1, \theta_2, ..., \theta_T)\, p_{A_T}(\theta_{T+1} \,|\, \theta_1, \theta_2, ..., \theta_T) \\ &= u(h_{T+1})\, p_{A_{T-1}}(\theta_1, \theta_2, ..., \theta_T)\, p_{a_T}(\theta_{T+1} \,|\, \theta_T) \end{aligned}$$

where that decomposition stems from the conditional probability definition and the markovian property. Considering the integral in θ_{T+1} we shall have

$$\int u(h_{T+1})\, p_{A_{T-1}}(\theta_1, \theta_2, ..., \theta_T)\, p_{a_T}(\theta_{T+1} \,|\, \theta_T)\, d\theta_{T+1}.$$

As utilities are additive, the decomposition $u(h_{T+1}) = u(h_T) + u(a_T, \theta_{T+1})$ leads to

$$\left(u(h_T) + \int u(a_T, \theta_{T+1})\, p_{a_T}(\theta_{T+1} \,|\, \theta_T)\, d\theta_{T+1} \right) p_{A_{T-1}}(\theta_1, \theta_2, ..., \theta_T) \tag{2}$$

Once history h_T is fixed, the utility $u(h_T)$ and the probability $p_{A_{T-1}}(\theta_1, \theta_2, ..., \theta_T)$ would be constant and the integral would represent the expected utility at last stage, making decision a_T. Denoting

$$\begin{aligned} U_T(a_T, h_T) =& u(h_T)\, p_{A_{T-1}}(\theta_1, \theta_2, ..., \theta_T) \\ &+ p_{A_{T-1}}(\theta_1, \theta_2, ..., \theta_T) \int u(a_T, \theta_{T+1})\, p_{a_T}(\theta_{T+1} \,|\, \theta_T)\, d\theta_{T+1} \end{aligned}$$

this utility depends on decision a_T and it comprises two terms: the first one would be constant and depends on the history until previous decision, the other one is proportional to the expected utility of this decision. The optimal U_T will coincide with the maximum expected utility of last decision. Let $a_T^* = a_T^*(h_T)$ be the decision that maximizes this expected utility and define

$$E^*(\theta_T) = \max_{a_T} \int u(a_T, \theta_{T+1})\, p_{a_T}(\theta_{T+1} \,|\, \theta_T)\, d\theta_{T+1}$$
$$= \int u(a_T^*, \theta_{T+1})\, p_{a_T}(\theta_{T+1} \,|\, \theta_T)\, d\theta_{T+1}.$$

Now, in (2), we may consider only the histories satisfying $a_T = a_T^*$, i.e.,

$$\{u(h_T) + E^*(\theta_T)\}\, p_{A_{T-1}}(\theta_1, \theta_2, ..., \theta_T).$$

We may now separate θ_T from the joint distribution and decompose the utility to get

$$\{u(h_{T-1}) + u(a_{T-1}, \theta_T) + E^*(\theta_T)\}\, p_{A_{T-2}}(\theta_1, \theta_2, ..., \theta_{T-1})\, p_{a_{T-1}}(\theta_T \,|\, \theta_{T-1}).$$

Integrating out θ_T, we are led to

$$U(a_{T-1}, h_{T-1}) = \Big(u(h_{T-1}) + \int u(a_{T-1}, \theta_T)\, p_{a_{T-1}}(\theta_T \,|\, \theta_{T-1})\, d\theta_T +$$
$$+ \int E^*(\theta_T)\, p_{a_{T-1}}(\theta_T \,|\, \theta_{T-1})\, d\theta_T\Big)\, p_{A_{T-2}}(\theta_1, ..., \theta_{T-1})$$

where, once the previous history h_{T-1} has been fixed, we shall have a function of the decision a_{T-1} whose maximum coincides with the maximum of the expected utility until the end, allowing us to determine the optimal decision $a_{T-1}^* = a_{T-1}^*(h_{T-1})$.

Defining again

$$E^*(\theta_{T-1}) = \max_{a_{T-1}} \Big(\int u(a_{T-1}, \theta_T)\, p_{a_{T-1}}(\theta_T \,|\, \theta_{T-1})\, d\theta_T$$
$$+ \int E^*(\theta_T)\, p_{a_{T-1}}(\theta_T \,|\, \theta_{T-1})\, d\theta_T\Big)$$

we would continue the process in stages $T-2, T-3$,..., until stage 1 when we would have obtained

$$U(a_1) = \int \{u(a_1, \theta_2) + E^*(\theta_2)\}\, p_{a_1}(\theta_2) d\theta_2$$

which allows us to determine the first optimal decision a_1^*.

As the target distribution of the MCMC algorithm is

$$u(\theta_1, a_1, \theta_2, a_2, ..., \theta_T, a_T, \theta_{T+1})\, p_a(\theta_1, \theta_2, ..., \theta_{T+1})$$

the history frequencies obtained $fr(\theta_1, a_1, \theta_2, a_2, ..., \theta_T, a_T, \theta_{T+1})$ provide an approximation to this distribution. Then, accumulating the frequencies of all histories coinciding in all components except θ_{T+1}, we shall have

$$fr(\theta_1, a_1, \theta_2, a_2, ..., \theta_T, a_T) = \sum_{\forall \theta_{T+1}} fr(\theta_1, a_1, \theta_2, a_2, ..., \theta_T, a_T, \theta_{T+1})$$

which estimates integral (2), that is, $\widehat{U}_T(a_T, h_T)$, and the history with maximum frequency $fr^*(\theta_1, a_1, \theta_2, a_2, ..., \theta_T, a_T)$ will be an approximation to the optimal decision $a_T^*(h_T)$. Being the process markovian, all histories until time t ending up in the same state θ_t will contribute to establish the same optimal alternative $a_t^*(h_t)$, in spite of the fact that their utilities might be different. That is why we can find an approximation to $a_T^*(\theta_T)$ accumulating the frequencies of all histories ending up in the same θ_T. Repeating the same process at later stages, we shall be able to find the optimal decisions a_t^*, as pointed out in the algorithm.

BAYESIAN STATISTICS 7, pp. 721–732
J. M. Bernardo, M. J. Bayarri, J. O. Berger, A. P. Dawid, D. Heckerman, A. F. M. Smith and M. West (Eds.)

Modelling Gene Expression Data over Time: Curve Clustering with Informative Prior Distributions

JON C. WAKEFIELD CHUAN ZHOU
University of Washington, Seattle, USA
jonno@u.washington.edu czhou@u.washington.edu

STEVE G. SELF
Fred Hutchinson Cancer Research Center, Seattle, USA
sgs@hivnet.fhcrc.org

SUMMARY

The aim of many microarray experiments is to discover genes that exhibit similar behavior, that is, co-express. A common approach to analysis is to apply generic clustering algorithms that produce a single cluster allocation for each gene. Such a strategy does not account for the experimental context, and provides no measure of the uncertainty in this classification. The model introduced in this paper is specifically tailored to the application in hand, so allowing the incorporation of prior information, and the probabilistic quantification of cluster membership. In this paper, we are interested in the situation in which the experiments are indexed by a variable that has a natural ordering such as time, temperature or dose level, and propose a four-stage hierarchical model for the analysis of such data. This model assumes that each gene follows one of a number of underlying trajectories, where the number may be assumed unknown, and the specific form of the trajectory depends on the experimental context. An initial filter based on a Bayes factor reduces the dimensionality of the data. The model is illustrated using two experiments carried out on yeast, one during sporulation, and another during cell-cycle. A summary measure that we emphasize is the posterior probability of co-expression which, in the context of our model, corresponds to the probability that two or more genes fall in the same group.

Keywords: BAYES FACTORS; BIRTH–DEATH MARKOV CHAIN MONTE CARLO; CLUSTERING; CO-EXPRESSION; MICROARRAY EXPERIMENTS; MIXTURE MODELS.

1. INTRODUCTION

Recently, there has been a huge interest in the analysis of gene expression data from DNA microarray experiments (e.g. Collins, 1999). This technology allows the simultaneous recording of activity, as measured by messenger RNA (mRNA) levels that reflect the level of transcription from DNA to RNA, in a large number of genes. The resultant data allow the possibility for gaining a greater understanding of the transcription dynamics on a genome-wide scale, under varying experimental conditions.

The data we analyze in Sections 4 and 5 arise from cDNA microarrays in which a fluorescently labelled cDNA sample is obtained for each time point and, to control for the unknown amount of DNA in the sample, is added to another fluorescently labelled sample from a reference experiment with another dye. The combined sample is hybridized to arrays containing templates on each gene in the genome. The amount of signal recorded reflects the extent of mRNA expression for each gene and at each time point, relative to the reference. While acknowledging their importance, we do not consider many of the generic issues that are pertinent to the analysis of data from microarray experiments such as signal extraction, array and dye effects, normalization, and background subtraction.

There are various scientific questions that may be assessed via microarray experiments, in this paper we are interested in the situation in which the experiments may be indexed by an ordered variable such as time, temperature, or the dose level of a toxin. In our scenario, the aim is to gain insight into those genes that behave similarly over the course of the experiment. By comparing genes of unknown function with profiles that are similar to genes of known function, clues to function may be obtained. Hence, *co-expression* of genes is of interest. A variety of approaches to this problem have been proposed. By far the most common is to apply generic supervised or unsupervised clustering algorithms to the data. For example, Eisen *et al.* (1998) use hierarchical clustering with distance measured via correlation, Chu *et al.* (1998) cluster to *known* profiles again using correlation, and Tamayo *et al.* (1999) use self-organizing maps. These approaches, though useful as a starting point are deficient in a number of respects. First, the data are clustered on the basis of the raw measurements and so classifications can be sensitive to outlying observations. To overcome this, often the raw data are "filtered" to remove aberrant observations, though this procedure has an *ad hoc* flavor (specific recipes are given in Section 3). Second, no measure of the certainty of the classification is given. A remedy to this latter problem has been proposed by Kerr and Churchill (2000), who describe a method for accessing the uncertainty by bootstrapping the residuals from an analysis of variance model that includes the gene-time effects of interest; the proportion of bootstrap samples that cluster to the original classification then gives a measure of the reliability. Finally, clustering algorithms are generic and are in no way tuned to the application in hand (except perhaps for the measure of dissimilarity used), and in particular do not allow the incorporation of prior information.

In this paper, we propose a model-based approach to this problem in which, heuristically speaking, we explicitly model the trajectory as a function of the ordering variable (e.g. time) and a gene-specific set of parameters. We then cluster on the basis of the latter, with our probabilistic framework providing quantitative measures of classification. The structure of this paper is as follows. In Section 2 we describe our modelling framework, and in Section 3 an initial screen using Bayes factors that is applied to reduce the dimensionality of the data. In Sections 4 and 5, we apply our model to two data sets, both collected in yeast, one during sporulation, and one during the cell-cycle. These analyses are not intended to represent substantive contributions to yeast genetics but have been selected to highlight posterior probabilities of co-expression (the sporulation data), and the use of prior information of a specific (periodic) form (cell-cycle data). Section 6 contains a concluding discussion.

2. MODEL DESCRIPTION

2.1. *Clustering Model*

Let Y_{it} denote, for gene i, the log-ratio of mRNA expression level measured during experiment t, relative to a reference experiment, $i = 1, ..., N, t = 1, ..., T$. The experiments may be indexed by any ordered variable, but for descriptive purposes we suppose that the variable is time and denote the actual sampling times by X_t, where for simplicity we have assumed that the design is the same for each gene. We then model these data via the following four-stage hierarchical model.

Stage one: For the observed data we have

$$y_{it} = f(\boldsymbol{\theta}_i, X_t) + e_{it},$$

where $e_{it} \sim_{iid} \mathrm{N}(0, \sigma_e^2)$ and $f(\boldsymbol{\theta}_i, X_t)$ denotes the form of the trajectory and depends on a gene-specific set of parameters, $\boldsymbol{\theta}_i$, and the experiment time, X_t.

Stage two: Conditional on C known trajectories, we introduce trajectory membership indicators Z_i that reflect the underlying trajectory that gene i follows so that $\boldsymbol{\theta}_i = \boldsymbol{\theta}^c$ if $Z_i = c, c = 1, ..., C$. We model $\boldsymbol{\theta}^c \mid \boldsymbol{\phi}, C,\ c = 1, ..., C$, as arising from a distribution that depends on unknown parameters $\boldsymbol{\phi}$. This is a prior for the collection of trajectories.

Stage three: We assume that

$$\Pr(Z_1 = z_1, ..., Z_N = z_N \mid \boldsymbol{\pi}, C) = \prod_{i=1}^{N} \Pr(Z_i = z_i \mid \boldsymbol{\pi}, C),$$

where $\Pr(Z_i = z_i \mid \boldsymbol{\pi}, C) = \pi_{z_i}$, $z_i = 1, ..., C$, and $\boldsymbol{\pi} = (\pi_1, ..., \pi_C)$. In the examples of this paper we take the prior for $\boldsymbol{\pi} \mid C$ to be $\mathrm{Di}(1, ..., 1)$.

Stage four: Finally, we place prior distributions on σ_e^2, $\boldsymbol{\pi} \mid C$ and $\boldsymbol{\phi}$, and possibly C (if the latter is assumed unknown).

We now provide an interpretation of the random variables Z_i that represent the gene expression curve that gene i is following. The change in expression levels describing transcription from DNA to RNA that occurs within the cell nucleus is determined by, amongst other things, the regulatory proteins that are responsible for gene i and on the timings at which these proteins are acting (either to activate or suppress activity). The curve membership indicators can therefore be viewed as summarizing the proteins that are relevant for gene i, and if two genes lie in the same cluster it is evidence of shared transcription factors.

Our approach to assume a mixture model is crucially different to the "model-based" clustering approach of Yeung *et al.* (2001) who analyze similar data but simply assume that the data arise from a mixture of T-dimensional normal distributions and hence do not acknowledge the time-ordering of the data (the analysis would be unchanged if the time ordering were permuted). In particular it would be desirable to allow serial dependence, within such an approach, but the MCLUST software (Fraley and Raftery, 1998) that is used by Yeung *et al.* (2001) does not allow for this possibility. Medvedovic and Sivaganesan (2002) also describe a Bayesian hierarchical model for microarray data, but again do not consider specific curve forms.

2.2. *Computation*

For fixed C, samples may be generated from the model described in the previous section using a Metropolis-Hastings Markov chain Monte Carlo (MCMC) algorithm. For unknown C either reversible jump MCMC (Green, 1995), or birth–death MCMC (Stephens, 2000a) may be used. We use the latter since it is relatively straightforward to implement in our context. The algorithm obtains samples from the posterior $p(C, \boldsymbol{\pi}, \boldsymbol{\theta}, \boldsymbol{\phi}, \sigma_e^2 \mid \boldsymbol{y})$ by simulating a continuous time marked point process to sample points $\boldsymbol{\theta}^c$ with associated "marks" π_c, with a Gibbs sampler for $\boldsymbol{\phi}, \sigma_e^2$ (and possibly $\boldsymbol{\theta}$). In our examples convergence was diagnosed informally by examination of trace plots of parameters and comparison of results from multiple chains initiated at different starting points. We also ran fixed C analyses and compared the results to the BDMCMC algorithm.

We follow Stephens (2000a) quite closely and take the birth rate to be mean of the Poisson prior that we specify for C. For a birth we simulate $\pi \sim \mathrm{Be}(1, C)$, and the new trajectory vector from the second stage prior, which in our two examples is a normal distribution. The death rate for component c, d_c, is the likelihood for the collection $\boldsymbol{\pi}, \boldsymbol{\theta}$ with $\pi_c, \boldsymbol{\theta}^c$ removed, divided by the likelihood for $\boldsymbol{\pi}, \boldsymbol{\theta}$. Component c is selected to die with probability $d_c / \sum_{c'} d_{c'}$.

As discussed in detail in Richardson and Green (1997) (and the accompanying discussion), and Stephens (2000b) there is a fundamental non-identifiability associated with mixture problems in that the posterior contains $C!$ modes of equal height that are indistinguishable from the data alone. To identify parameters uniquely some form of "labelling" must be carried out. In Sections 4 and 5 we describe specific labelling methods, but the fundamental question of co-expression of genes i and i' may be answered without resolving this problem since the quantity $\Pr(Z_i = c, Z_{i'} = c \mid \boldsymbol{y})$ is invariant to re-labelling.

3. INITIAL FILTERING

The method we have proposed may be computationally prohibitive for a large number of genes (depending on the functional form of the trajectories), and so we describe an initial screen to reduce the dimensionality of the data. In the microarrays literature, similar procedures are carried out but in a relatively *ad hoc* manner. For example, Eisen *et al.* (1998) only include genes for study if their expression levels deviate from those at time zero by at least a factor of three in at least two time points, while Tamayo *et al.* (1999) exclude yeast genes that do not show a relative change of two units and an absolute change of 35 units; values of 3 and 100 are used for human genes, though no justification for these values was given. We define genes as "un-interesting" if their mRNA levels remain at a constant level over the time course of the experiment.

We propose a very simple approach in which we compare the models $M_0 : \mu_1 = \mu_2 = \ldots = \mu_T = \mu$ and M_1 : not M_0, via the Bayes factor $p(\boldsymbol{y}_i \mid M_1)/p(\boldsymbol{y}_i \mid M_0)$, where

$$I_1 = p(\boldsymbol{y}_i \mid M_1) = \prod_{t=1}^{T} \int_{\mu_t} \int_{\sigma_e^2} p(y_{it} \mid \mu_t, \sigma_e^2) \times \pi(\mu_t)\pi(\sigma_e^2)\, d\mu_t\, d\sigma_e^2, \tag{1}$$

and

$$I_0 = p(\boldsymbol{y}_i \mid M_0) = \prod_{t=1}^{T} \int_{\sigma_e^2} p(y_{it} \mid \mu_t = \mu, \sigma_e^2) \times \pi(\mu)\pi(\sigma_e^2)\, d\mu\, d\sigma_e^2. \tag{2}$$

We assume that $Y_{it} \mid \mu_t, \sigma_e^2 \sim_{iid} N(\mu_t, \sigma_e^2)$, with priors $\pi(\mu)$ and $\pi(\sigma_e^{-2})$ given, respectively, by $\mu_t \sim_{iid} N(m_0, v_0)$, $t = 1, \ldots, T$, and $\sigma_e^{-2} \sim Ga(a_0, b_0)$. Due to the normalization we choose $m_0 = 0$ and take $v_0 = 2^2$ to reflect the range of variation observed in similar experiments (e.g. Eisen *et al.*, 1998). The data of Chu *et al.* (1998) includes data at $t = 0$, relative to another $t = 0$ experiment that therefore act as a "reference" and reflect measurement error, σ_e^2, only. Hence we place a highly informative prior on σ_e^2 for the calculation of Bayes factors. The reference data were analyzed under the model $\bar{Y}_0 \mid \mu_0, \sigma_e^2 \sim N(\mu_0, \sigma_e^2/N)$ with the improper prior $\pi(\mu_0, \sigma_e^2) \propto \sigma_e^{-2}$ which leads to the posterior $\sigma_e^{-2} \mid \boldsymbol{y}_0 \sim Ga\{(N-1)/2, (N-1)s^2/2\}$, where s^2 is the unbiased estimator of σ^2. Since N is large the posterior distribution is highly concentrated.

Importance sampling based on sampling from the prior is relatively efficient given our informative priors, and results in (1) and (2) being estimated, respectively, by

$$\hat{I}_1 = S^{-1} \sum_{s=1}^{S} \prod_{t=1}^{T} p(y_{it} \mid \mu_t^{(s)}, \sigma_e^{2(s)}),$$

and $\hat{I}_0 = S^{-1} \sum_{s=1}^{S} \prod_{t=1}^{T} p(y_{it} \mid \mu^{(s)}, \sigma_e^{2(s)})$, where $\mu_t^{(s)}, \mu^{(s)} \sim_{iid} \pi(\mu)$, $\sigma_e^{2(s)} \sim_{iid} \pi(\sigma_e^2)$. It would be desirable to analyze all of the data together with the clustering model described in

Section 4.2, with a "zero" cluster for the un-interesting genes, but the increase in computational burden is unlikely to be worthwhile in terms of identifying interesting clusters.

4. EXAMPLE 1: SPORULATION DATA

4.1. *Data Description*

We first describe the experiment of Chu *et al.* (1998) which contained data on seven microarrays with a reference at time $t = 0$, and activity during sporulation measured at times $X_t = \{0.5, 2, 5, 7, 9, 11.5\}$ so that $T = 6$. Chu *et al.* (1998) were interested in genes that exhibited similar profiles, and to this end they created seven "characteristic curves" by averaging, via a visual inspection, genes contained in each profile. Figure 1 gives the sets of genes in each profile, along with the mean trajectories. After an initial screen in which 80% of genes were eliminated they clustered the remaining genes to each profile (using correlation as the distance measure). This approach provides a list of genes that appear to conform to each profile, but does not give a measure of the uncertainty of this classification, in common with other distance-based clustering procedures (e.g. Eisen *et al.*, 1998; Tamayo *et al.*, 1999).

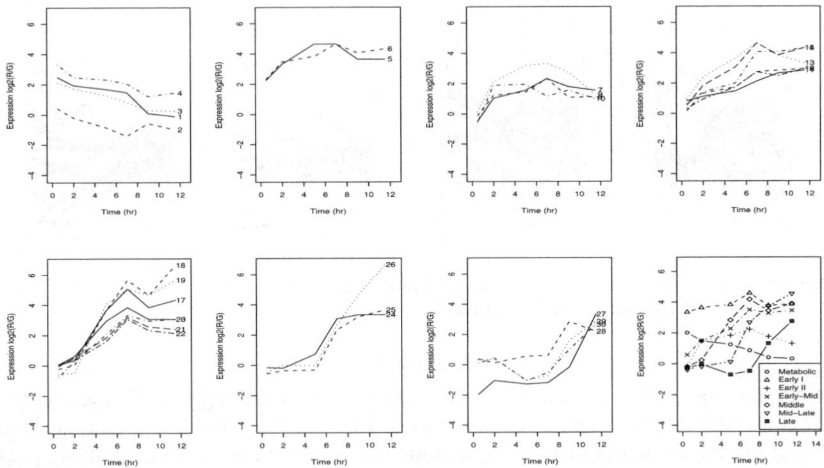

Figure 1. *Hand-picked genes in each of seven groups (from Chu et al., 1998), along with mean trajectories* (panel 8).

4.2. *Model Description*

In this experiment Y_{it} represents the $\log_2$-ratio of expression for gene i at time t, relative to time $t = 0$, $i = 1, ..., 6118$, $t = 1, ..., 6$. We reiterate that the aim is to discover structure over time, and in particular to determine genes that follow common trajectories. In the absence of further information, we would expect the trajectories to be smooth after the onset of sporulation (that is following time $X_1 = 0.5$) which occurs rapidly. To this end we assume a first-order random walk model so that at stage one we have

$$y_{it} \mid \theta_{it}, \sigma_e^2 \sim_{iid} N(\theta_{it}, \sigma_e^2),$$

with at stage two $\theta_{it} = \theta_t^c$ if $Z_i = c$ and

$$\theta_t^c = \theta_{t-1}^c + u_t,$$

for $t = 2, ..., T$, and $u_t \sim_{iid} N(0, \Delta_t \tau^2)$, where $\Delta_t = X_t - X_{t-1}$ so that observations closer in time are more likely to be similar. For the first time point we have $\theta_1^c \sim_{iid} N(0, \tau_1^2)$. With reference to the model described in Section 4.2, $\boldsymbol{\phi} = (\tau_1^2, \tau^2)$ and, conditional on C, the data are modelled as arising from the C trajectories $\boldsymbol{\theta}^1 = (\theta_1^1, ..., \theta_T^1), ..., \boldsymbol{\theta}^C = (\theta_1^C, ..., \theta_T^C)$. Collapsing over stages two and three shows that we are modelling the data as arising from a mixture of C underlying smooth trajectories, *i.e.*, $\boldsymbol{\theta}_i = \sum_{c=1}^{C} \pi_c \boldsymbol{\theta}^c$.

We now discuss the fourth stage prior which here requires specification for $\sigma_e^2, \tau_1^2, \tau^2$ and C. The prior for σ_e^2 is identical to that described in Section 3. For specification of priors for each of the variances τ_1^2 and τ^2 we pick a "most likely" and an "upper value" for the standard deviation. These values are then converted to the inverse variance scale, and we pick the parameters of the gamma distribution to line up the mode with the most likely point, and the 95% point of the distribution with the upper value (which requires a numerical search). We choose the modal value for τ_1^2 to be 2^2 (which is consistent with $\mu_0 \sim N(0, 2^2)$), and for τ^2 (which is a conditional variance) we take the most likely value to be 1.5^2 (so that in one unit time interval we expect the trajectory to be within ± 3.0 with probability 0.95); we take 3^2 and 2^2 as upper values for τ_1^2 and τ^2, respectively. Figure 2 shows four simulations from this prior with fixed $C = 10$ and 20 genes within each cluster. The prior for C was Poisson with mean 15.

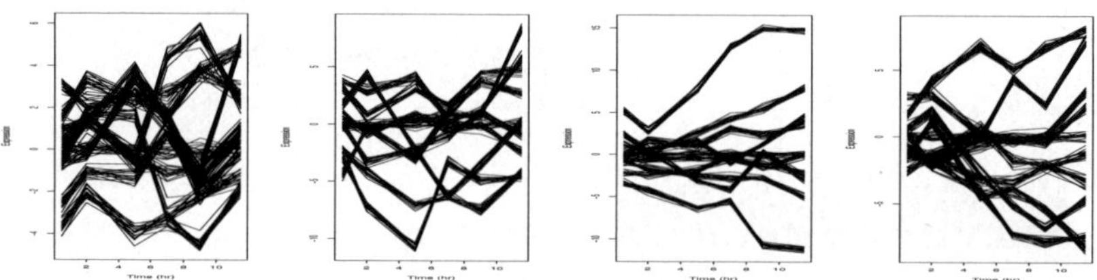

Figure 2. *Four simulations from the random walk prior, with measurement error added. There are $C = 10$ groups, each of which contain 20 genes.*

4.3. *Analysis*

We applied the initial filter described in Section 3 to the $N = 6118$ genes. Figure 3 displays the 100 "top" and 100 "bottom" genes in terms of $\Pr(M_1 \mid \boldsymbol{y})$. We see the same qualitative behavior of the trajectories as that observed in the prior simulations of Figure 2, though in the latter the groups are less distinct. We chose to analyze the 1104 genes for which $\Pr(M_1 \mid \boldsymbol{y}) > 0.999$. We varied the size of the variance on the normal prior for μ_t and found that though the resultant Bayes factors showed sensitivity to this choice, the ranking on the genes with highest $\Pr(M_1 \mid \boldsymbol{y})$ was robust.

We found that the posterior for C was highly sensitive to the prior choices, adding to the need for informative prior distributions. Figure 4 shows the trace plot of C versus iteration number, and the marginal posterior of C, following a burn-in of 40,000 iterations. For 80,000 iterations, the computer time was approximately 4 h on a Unix cluster.

For illustration, Figure 5 shows the curves that result from values of $C = 10$, 12, 14, 16 (these may be compared with Figure 4B of Chu *et al.*, 1998). Re-labelling was carried out on the basis of the mean at the second time-point (since these were relatively distinct) and showed good agreement with the decision theoretic approach of Stephens (2000b). As C increases the extreme trajectories remain relatively constant. The posterior medians of σ_e were 0.50, 0.47, 0.45, 0.44 for $C = 10, 12, 14, 16$, respectively. The prior median of the informative prior on σ_e was 0.25 which suggests there is some model misspecification (or that the measurement error

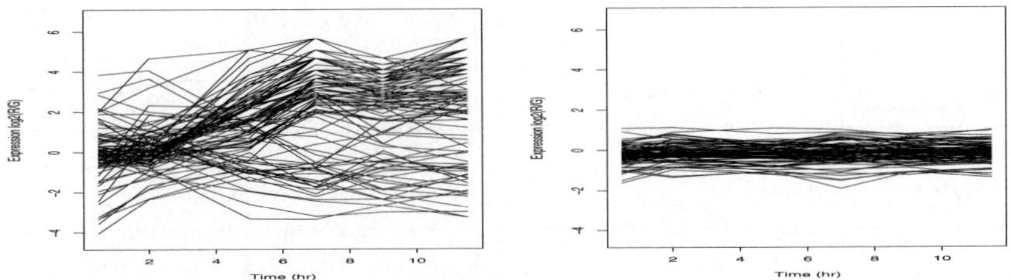

Figure 3. *Expression levels versus time for 100 genes with the highest* (left panel), *and the lowest* (right panel) *values of* $\Pr(M_1 \mid \boldsymbol{y})$ *where* M_1 *is the model of non-constant level.*

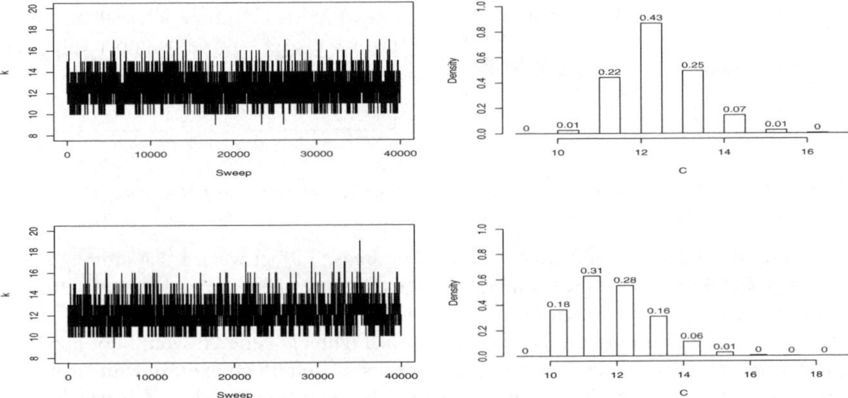

Figure 4. *Trace plot of the number of clusters C and the posterior distribution of C (after a burn-in of 40,000 iterations), from BDMCMC analysis of the sporulation data* (top row), *and the cell-cycle data* (bottom row).

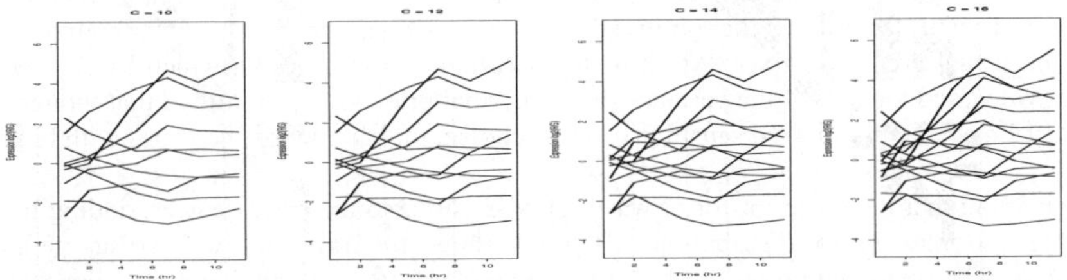

Figure 5. *Mean expression levels as a function of time, for numbers of cluster* $C = 10, 12, 14, 16$.

increases after the zero time point). For the $C = 12$ case we plot, in Figure 6, the mean curves along with the genes clustered to these curves.

In Figure 7 we summarize the pairwise probabilities based on the BDMCMC analysis, averaging across all C. For illustration, the genes we examine are 30 hand-picked genes highlighted by Chu *et al.* (1998), and reproduced in Figure 1. If the cluster membership was consistent with this figure then we would see blocks of shaded areas for genes in the same

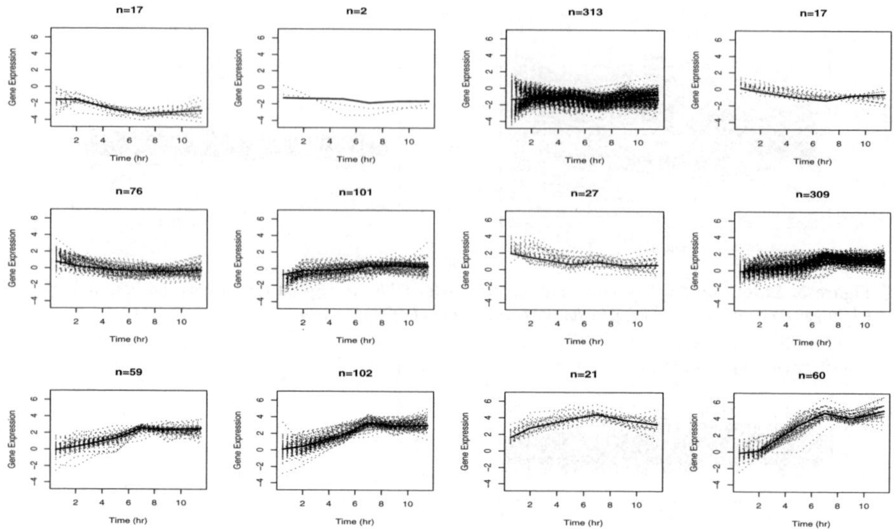

Figure 6. *Posterior profiles and genes classified to each of these profiles (using MAP classification), conditional on $C = 12$ clusters.*

group (close to the diagonal), and white in the other areas. In fact we see that although there is a greater probability of co-expression close to the diagonal, there are both genes within the same hand-picked collection which do not appear to co-express, and genes in other groups that co-express. Concentrating on the sixth group in Figure 1, gene 26 would not appear to co-express with any of the other 29 genes, while genes 24 and 25 co-express with each other and also with genes 11, 12, 16, 20, 21 and 22. Hence, we see that our model offers new insights into co-expression.

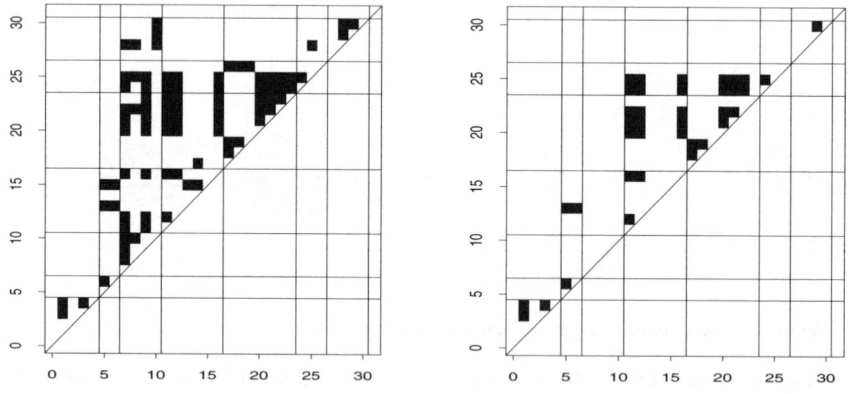

Figure 7. *Heat map showing pairwise probabilities of common cluster membership of the 30 genes in Figure 1. The solid lines separate the different groups. On the left the shaded squares denote those pairwise probabilities greater than 0.5, while on the right the cut-off probability is 0.8.*

5. EXAMPLE 2: CELL-CYCLE DATA

5.1. *Data Description*

Spellman *et al.* (1998) describe a number of microarray experiments that were carried out to create a comprehensive list of yeast genes whose expression levels vary periodically within the cell cycle. For illustration we analyze one of the data sets that measured levels on $N = 4489$ genes that were synchronized via α factor, and randomly select 800 genes. Expression levels were measured every 7 min for 140 min (so that $T = 20$ measurements were recorded in total). The expression levels of the 800 genes are shown in Figure 9, after application of our model. We take as our objective the identification of genes that display similar periodicity.

5.2. *Model Description*

In this example our model formulation is strongly driven by prior information. Specifically we assume that the data arise from a mixture of functions with periodic structure, in particular:

$$Y_{it} = R_i \cos(\omega t + \phi_i) + e_{it} = A_i X_{1t} + B_i X_{2t} + e_{it},$$

where R_i is the amplitude and ϕ_i the phase of gene i, $X_{1t} = \sin(\omega t)$ and $X_{2t} = \cos(\omega t)$, and $\omega = 2\pi/p$ with $p = 66$ minutes as the known period (obtained from Spellman *et al.*, 1998). It may seem more natural to model in terms of amplitude and phase, but because of the irregular constraints on the collection (R_i, ϕ_i) we prefer to formulate our mixture model in terms of $\boldsymbol{\theta}_i = (A_i, B_i)$. Figure 8 gives the least squares estimates for (A_i, B_i) (left panel), and $\log R_i, \log[(\pi/2 - \phi_i)/(\pi/2 + \phi_i)]$ (right panel) and clearly shows the irregularity of the joint distribution of the latter which does not allow us to simply parameterize in terms of functions of the phase and amplitude.

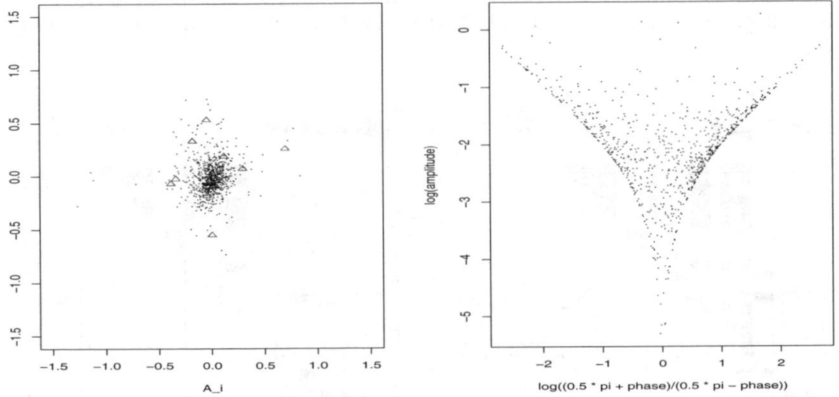

Figure 8. *Least squares estimates* $\hat{A}_i, \hat{B}_i$ *from the model* $\mathrm{E}[Y_{it} \mid A_i, B_i] = A_i \sin(\omega t) + B_i \cos(\omega t)$, *for gene* i (left panel), *and* $\log(\hat{R}_i), \log\{(\pi/2 + \hat{\phi}_i)/(\pi/2 - \hat{\phi}_i)\}$ (right panel), $i = 1, ..., 800$.

At the second stage of the model we assume that $\boldsymbol{\theta}_i$ arise from a mixture of bivariate normal distributions, *i.e.*, $\boldsymbol{\theta}_i = \boldsymbol{\theta}^c$ if $Z_i = c$, and $\boldsymbol{\theta}^c \mid \boldsymbol{m}, \boldsymbol{V} \sim_{iid} N(\boldsymbol{m}, \boldsymbol{V})$, $c = 1, ..., C$, where we assume that $\boldsymbol{m}, \boldsymbol{V}$ are known. We estimate these from the data (along the lines followed by Richardson and Green, 1997, and Stephens, 2000a). In particular, we take $\boldsymbol{m} = (-0.215, -0.005)$, which correspond to the means of the least squares estimates, and $\boldsymbol{V}$ with elements $(4.45\ 0\ 0\ 2.13)$ which are the ranges of the least squares estimates squared. Given that our model here is exploratory we are not troubled by the mild dependence of the prior on the data. The priors for σ_e^2 and C were as for the sporulation data. Figure 9 shows simulations from our prior distribution. We are modelling the A_i, B_i pairs as arising from a mixture of bivariate normal distributions and so, in this example we would not want to filter out the constant genes since this would leave a "hole" close to zero. For these data (although we acknowledge the random error around each curve in our first stage distribution) we have effectively reduced the dimensionality of the data from 20 to 2.

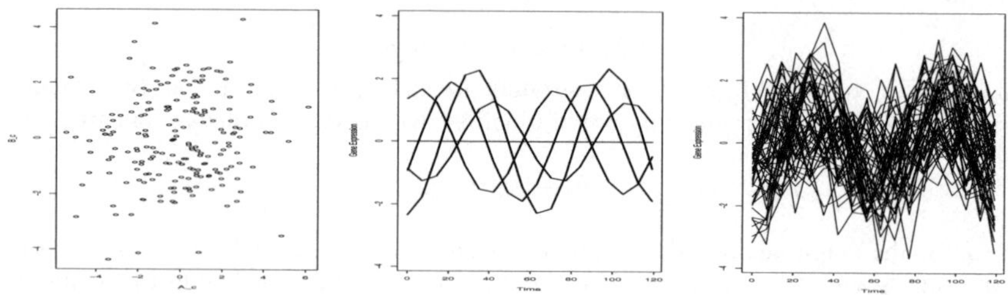

Figure 9. *200 simulations of (A_c, B_c) for the cell-cycle parameters (left); five trajectories without random error (center); 55 simulations from 5 groups, 4 simulated plus zero group with random error (right).*

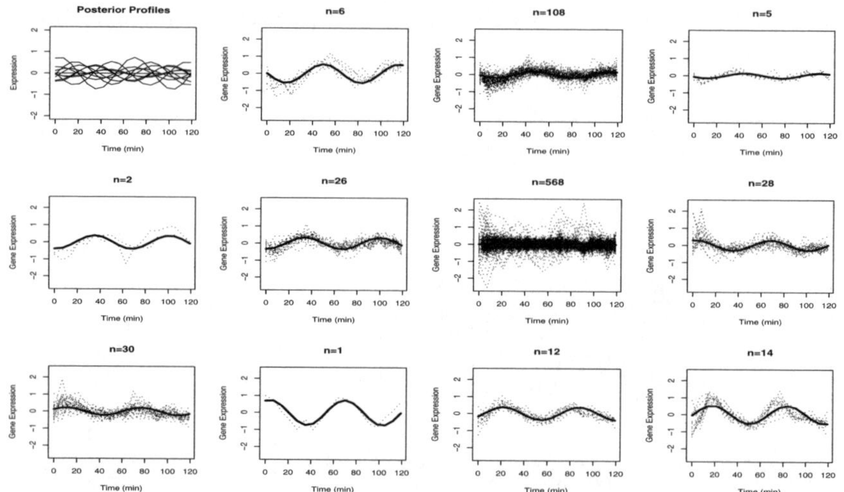

Figure 10. *Posterior profiles and genes classified to each of these profiles (using MAP classification), conditional on $C = 11$ clusters.*

5.3. *Analysis*

The bottom row of Figure 4 shows the behavior of C as a function of iteration number, and the posterior distribution of C. We see that the posterior for the latter is concentrated between 10 and 15. The computational overhead was a little less in this example, due to the use of a linear model that reduces a number of the required calculations. For the $C = 11$ analysis we relabelled on the basis of B^c (which were relatively distinct); again we found good agreement with the decision theoretic approach of Stephens (2000b).

The means $\boldsymbol{\theta}^c = (A^c, B^c)$, $c = 1, ..., 11$, are displayed in the left hand panel of Figure 8. In Figure 10, we show the collection of profiles, conditional on $C = 11$. We classified each gene to a profile based on $\Pr(Z_i \mid \boldsymbol{y})$ and these classifications are displayed in Figure 10. The largest cluster is in the third panel of the second row and corresponds to the "zero" cluster which has a flat profile. We see that the classifications look reasonable, though there is clearly some model misspecification. In particular a number of the trajectories look to have attenuated profiles as time increases.

6. DISCUSSION

In this paper, we have introduced a model that allows a quantitative description of gene co-expression, in contrast to clustering techniques that are currently used. A systematic comparison with these more traditional techniques is unfortunately beyond the scope of this paper. In investigations not reported here, we have found that the number of clusters, and sometime the co-expression probabilities, may be highly sensitive to the prior distributions and we would have far less faith in our quantitative conclusions if the priors we had used were not based on biological and experiment-specific information.

Although we have emphasized that posterior probabilities of co-expression are invariant to component re-labelling, for other summaries such as the reporting of mean trajectories, the problem remains. An alternative to the procedure followed for the examples, and is closer to the context is to label on the basis of known marker genes.

In our model formulation of Section 2 we assumed *a priori* that the trajectory indicator variables Z_i were independent. In practice there will often be substantial information available to place collections of genes in the same cluster with high probability. We are currently working on how to extend our independence model, using ideas from Markov random field modelling.

More substantively, our eventual aim is to combine expression and sequence data. Models for the latter have been extensively developed, see, for example, Liu *et al.* (1999). A Bayesian framework is ideally suited to such an endeavor, since it allows a natural combination of multiple data sources, and the incorporation of prior information, which is likely to be essential in complex problems such as these.

ACKNOWLEDGEMENTS

The authors would like to thank Matthew Stephens for useful discussions, and in particular for advice on birth–death MCMC and label-switching.

REFERENCES

Chu, S., DeRisi, J., Eisen, M., Mulholland, J., Botstein, D., Brown, P., and Herskowitz, I. (1998). The transcriptional program of sporulation in budding yeast. *Science* **282**, 699–705.

Collins, F. (1999). The chipping forecast. *Nat. Genet.* **21**, 1–60 .

Eisen, M., Spellman, P., Brown, P., and Botstein, D. (1998). Cluster analysis and display of genome-wide expression patterns. *Proc. Natl. Acad. Sci. USA* **95**, 14863–14868.

Fraley, C. and Raftery, A. (1998). How many clusters? Which clustering method? Answers via model-based cluster analysis. *Comput. J.* **41**, 578–588.

Green, P. (1995). Reversible jump Markov chain Monte Carlo computation and Bayesian model determination. *Biometrika* **82**, 711–732.

Kerr, K. and Churchill, G. (2000). Bootstrapping cluster analysis: Assessing the reliability of conclusions from microarray experiments. *Proc. Natl. Acad. Sci. USA* **98**, 8961–8965.

Liu, J., Neuwald, A., and Lawrence, C. (1999). Markov structures in biological sequence alignments. *J. Am. Statist. Ass.* **94**, 1–15.

Medvedovic, M. and Sivaganesan, S. (2002). Bayesian infinite mixture model-based clustering of gene expression profiles. *Bioinformatics* **18**, 1194–1206.

Richardson, S. and Green, P. (1997). On Bayesian analysis of mixtures with an unknown number of components, *J. R. Statist. Soc. B* **59**, 731–792 (with discussion).

Spellman, P., Sherlock, G., Zhang, M., Iyer, V., Anders, K., Eisen, M., Brown, P., Botstein, D., and Futcher, B. (1998). Comprehensive identification of cell cycle-regulated genes of yeast saccharomyces cerevisiae by microarray hybridization. *Mol. Biol. Cell* **9**, 3273–3297.

Stephens, M. (2000a). Bayesian analysis of mixture models with an unknown number of components—an alternative to reversible jump methods. *Ann. Statist.* **28**, 40–74.

Stephens, M. (2000b). Dealing with label-switching in mixture models. *J. R. Statist. Soc. B* **62** , 795–809.

Tamayo, P., Slonim, D., Mesirov, J., Zhu, Q., Kitareewan, S., and Dmitrovsky, E . (1999). Interpreting patterns of gene expression with self-organizing maps: methods and application to hematopoietic differentiation. *Proc. Natl. Acad. Sci. USA* **96**, 2907–2912.

Yeung, K., Fraley, C., Murua, A., Raftery, A., and Ruzzo, W. (2001). Model-based clustering and data transformations for gene expression data. *Bioinformatics* **17**, 977–987.

BAYESIAN STATISTICS 7, pp. 733–742
J. M. Bernardo, M. J. Bayarri, J. O. Berger, A. P. Dawid,
D. Heckerman, A. F. M. Smith and M. West (Eds.)

Bayesian Factor Regression Models in the "Large p, Small n" Paradigm

MIKE WEST
Duke University, USA
mw@isds.duke.edu

SUMMARY

I discuss Bayesian factor regression models with many explanatory variables. These models are of particular interest and applicability in problems of prediction, but also for elucidating underlying structure in predictor variables. One key motivating application here is in studies of gene expression in functional genomics. I first discuss empirical factor (principal components) regression, and the use of general classes of shrinkage priors, with an example. These models raise foundational questions for Bayesians, and related practical issues, due to the use of design-dependent priors and the need to recover inferences on the effects of the original, high-dimensional predictors. I then discuss latent factor models for high-dimensional variables, and regression approaches in which low-dimensional latent factors are the predictor variables. These models generalize empirical factor regression, provide for more incisive evaluation of factor structure underlying high-dimensional predictors, and resolve the modelling and practical issues in empirical factor models by casting the latter as limiting special cases. Finally, I turn to questions of prior specification in these models, and introduce *sparse latent factor models* to induce sparsity in factor loadings matrices. Embedding such sparse latent factor models in factor regressions provides a novel approach to variable selection with very many predictors. The paper concludes with an example of sparse factor analysis of gene expression data and comments about further research.

Keywords: DIMENSION REDUCTION, GENE EXPRESSION ANALYSIS, HIGH-DIMENSIONAL COVARIATES, LATENT FACTOR MODELS, SHRINKAGE PRIORS.

1. EMPIRICAL FACTOR REGRESSION MODELS

1.1. *SVD Regression*

Begin with the linear model $\boldsymbol{y} = \boldsymbol{X}\boldsymbol{\beta} + \boldsymbol{\epsilon}$ where $\boldsymbol{y}$ is the n-vector of responses, $\boldsymbol{X}$ is the $n \times p$ matrix of predictors, $\boldsymbol{\beta}$ is the p-vector regression parameter, and $\boldsymbol{\epsilon} \sim \mathrm{N}(\boldsymbol{\epsilon} \,|\, 0, \sigma^2 \boldsymbol{I})$ is the n-vector error term. Of key interest are cases when $p >> n$, when $\boldsymbol{X}$ is "long and skinny." The standard empirical factor (principal component) regression is best represented using the reduced singular-value decomposition (SVD) of $\boldsymbol{X}$, namely $\boldsymbol{X} = \boldsymbol{F}\boldsymbol{A}$ where $\boldsymbol{F}$ is the $n \times k$ factor matrix (columns are factors, rows are samples) and $\boldsymbol{A}$ is the $k \times p$ SVD "loadings" matrix, subject to $\boldsymbol{A}\boldsymbol{A}' = \boldsymbol{I}$ and $\boldsymbol{F}'\boldsymbol{F} = \boldsymbol{D}^2$ where $\boldsymbol{D}$ is the diagonal matrix of k positive singular values, d_i, arranged in decreasing order. This reduced form assumes factors with zero singular values have been ignored without loss; $k \leq n$ with equality only if all singular values are positive. Now the regression transforms via $\boldsymbol{X}\boldsymbol{\beta} = \boldsymbol{F}\boldsymbol{\theta}$ where $\boldsymbol{\theta} = \boldsymbol{A}\boldsymbol{\beta}$ is the k-vector of regression parameters for the factor variables, representing a possibly massive dimension reduction from p to k parameters.

Inherently, observing $\boldsymbol{y}$ provides information only on the underlying factor regression parameters $\boldsymbol{\theta}$, so prior specification directly in factor space has become common. Generalized shrinkage (or ridge regression) priors on $\boldsymbol{\theta}$ have become popular as MCMC methods now permit their routine use. A particularly flexible class of such priors assumes independent T distributions for the elements $\theta_1, \ldots, \theta_k$ of $\boldsymbol{\theta}$, so allowing for varying degrees of shrinkage in each of the orthogonal factor dimensions. A particular example has $\theta_i \sim \mathrm{N}(\theta_i \,|\, 0, c_i/\phi_i)$ where $\phi_i \sim \mathrm{Ga}(\phi_i \,|\, r/2, r/2)$ independently, for some $r > 0$; the tuning parameter r is the degree-of-freedom parameter for the implied T distribution for θ_i that follows on marginalization over the random precision ϕ_i. Here the c_i are weights that may be used, for example, to indicate the prior view that higher-order factors are expected to play lesser roles in the regression—often, though not necessarily, the case. In the first example below this is the case, and the model uses $c_i = \rho i^{-2}$, with scale factor ρ to be estimated, for example. This also allows for removal of factors in prior specification, by setting a c_i to zero. Analyses typically also adopt an inverse gamma prior for the error variance σ^2.

These models are easily implemented using MCMC, with complete conditional posteriors of generally standard forms. The conditional posterior for $\boldsymbol{\theta}$ given the ϕ_i is multivariate normal, and the ϕ_i are conditionally independent gamma variates given $\boldsymbol{\theta}$. The example below utilizes this to generate sequences of posterior samples for $\boldsymbol{\theta}$ and the ϕ_i.

1.2. *Coherence and Inference on Original Regression Parameters*

A basic modelling issue arises from the explicit design, and sample size, dependence of the empirical factor model. The key $\boldsymbol{\theta}$ parameter is directly defined as a function of $\boldsymbol{X}$ and $\boldsymbol{\beta}$, so parameter definition changes as the sample size and design changes; the specification of priors over these design-dependent parameters must be coherent with respect to changing n and $\boldsymbol{X}$, and the question is raised of whether this can be assured. This question is answered in Section 2.

A related practical issue is that of inference on the original regression parameters $\boldsymbol{\beta}$. The framework has priors and posteriors for $\boldsymbol{\theta}$, but leaves open questions of "inverting" the dimension-reducing map to make inferences on $\boldsymbol{\beta}$. The many-one map $\boldsymbol{\theta} = \boldsymbol{A}\boldsymbol{\beta}$ has multiple generalized inverses $\boldsymbol{\beta} = \boldsymbol{A}'\boldsymbol{\theta} + \boldsymbol{b}$ for all p-vectors $\boldsymbol{b}$ such that $\boldsymbol{A}\boldsymbol{b} = 0$. Again, this issue is fully resolved in Section 2. Here, simply note that, for predictive purposes, the choice of $\boldsymbol{b}$ is irrelevant. A canonical choice of generalized inverse is the standard "least-norm" inverse based on $\boldsymbol{b} = 0$, *i.e.*, $\boldsymbol{\beta}^* = \boldsymbol{A}'\boldsymbol{\theta}$. The analysis in the example below uses $\boldsymbol{\beta}^*$; again, Section 2 explains and justifies this properly. Posterior samples for $\boldsymbol{\theta}$ trivially imply samples for $\boldsymbol{\beta}^*$ which may be summarized for inference.

One interesting connection to make is that the prior on $\boldsymbol{\beta}^*$ implied by the prior on $\boldsymbol{\theta}$ of Section 1.1 is a generalization of the g-prior of Zellner (1986). Given conditional $\mathrm{N}(\theta_i \,|\, 0, c_i/\phi_i)$ priors, the implied prior for $\boldsymbol{\beta} = \boldsymbol{\beta}^*$ is singular normal with density proportional to

$$\exp\{-\boldsymbol{\beta}'\boldsymbol{A}'\boldsymbol{G}\boldsymbol{A}\boldsymbol{\beta}/2\}$$

, where $\boldsymbol{G}$ is diagonal with elements ϕ_i/c_i. This makes explicit the design-dependency of the prior and also the individual scaling in each of the singular factor dimensions. The Zellner g-prior corresponds to taking $\boldsymbol{G} = g\boldsymbol{D}^2$ for some positive scalar $g > 0$. I, therefore, refer to this approach as defining a class of *generalized singular g-priors* (or gsg-priors).

1.3. *Prediction*

Prediction is practically straightforward, though raises foundational questions related to the design-dependency issue. Technically, response values to be predicted are simply treated as missing values to be imputed, and the MCMC analysis is trivially extended to sample these values at each iteration. That is, $\boldsymbol{y}$ is partitioned into a vector of training samples, $\boldsymbol{y}_t$, and a vector of validation cases $\boldsymbol{y}_v$ to be predicted; the design matrix is conformably partitioned as $[\boldsymbol{X}_t; \boldsymbol{X}_v]$. The MCMC imputes $\boldsymbol{y}_v$ from the implied (normal) conditional posterior. This requires that the model is specified and analyzed conditional on all predictor values, including $\boldsymbol{X}_v$, and the empirical factor regression model is based on decomposition of the full $\boldsymbol{X}$ matrix. As a result, the factors $\boldsymbol{F}$ are evaluated based on predictors $\boldsymbol{X}_v$ as well as $\boldsymbol{X}_t$; thus $\boldsymbol{X}_v$ forms part of the model and prior structure even though the corresponding responses are missing. One message is that required predictor values must be contemplated prior to analysis, or analysis fully repeated if new predictions are required.

Again, this apparent issue is interpreted and resolved in Section 2 where the empirical model is understood to arise from a more elaborate latent factor regression model. In the example now, this approach is simply adopted for prediction of validation samples.

1.4. *Example: Analysis of Biscuit Dough Data*

This example concerns biscuit dough data analyzed in Brown *et al.* (1999), and originally in Osborne *et al.* (1984). The study aims to predict biscuit dough constituents based on spectral characteristics of dough measured using near infrared (NIR) spectroscopy. Hence, the predictor for each dough sample is a reflectance spectrum on a grid of wavelengths. The analysis here uses the same data as Brown *et al.* (although their framework is multivariate); these authors utilize a decision-theoretic variable selection method, rather than shrinkage priors or factor models. The response chosen here is fat content of dough samples, the predictors are $p = 300$ NIR reflectance measures at equally spaced wavelengths over 1202–2400 nanometers (nm), with 39 training samples and 39 validation cases to be predicted. The fat content response is standardized, and the predictors are centered; the centered spectra are graphed in Figure 1.

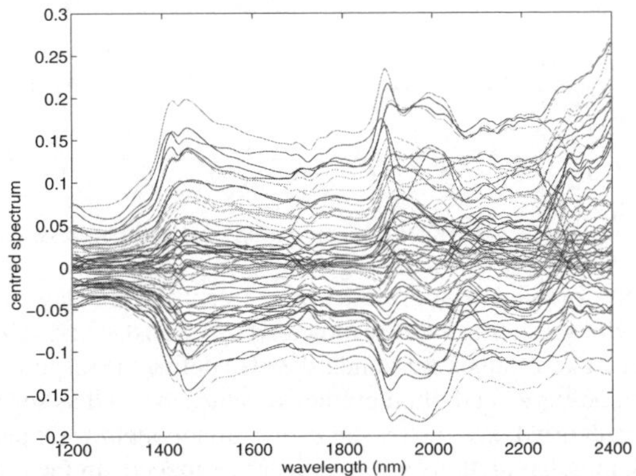

Figure 1. *Centered spectral predictors of 78 biscuit dough samples.*

Several analyses have been studied, varying the tuning parameters r and c_i. A summary of analysis with $r = 5$ and $c_i = \rho i^{-2}$ is given here; analyses with r between 1 and 10 give substantially similar results, and $r = 5$ is marginally optimal as measured simply by the mean square prediction error computed in the validation sample. The prior for σ^{-2} is unit mean exponential, reflecting the known range of the standardized response data but otherwise representing a relatively diffuse prior, and ρ has a diffuse prior. The singular values of the 300×78 matrix $\boldsymbol{X}$ decay rapidly and are truncated to zero, with the corresponding factors dropped, past the point where 99.995% of the total variation, as measured by the cumulative sum of squares of the d_i, is accounted for; this leads to $k = 16$ factors in the model, with $c_i = 0$ for $i > 16$. Adding more factors, at the expense of increased computation, does not in this example improve predictions. Further, a reduced number is consistent with the inherent smoothness of the predictor variable, and will generally be experienced when predictors are curves. The MCMC analysis summarized has 20,000 samples selected from a run of 100,000 by choosing every fifth sample, and following a burn-in of 1,000 samples. Convergence is swift and clean, consistent with experiences with a range of other examples and MCMC experiments.

Figures 2 and 3 graph the approximate posterior means of $\boldsymbol{\theta}$ and $\beta^* = \boldsymbol{A}'\boldsymbol{\theta}$, respectively, the former with equal-tails 90% posterior intervals. Figure 3 uses asterisks to mark the ten β_i^* values with largest absolute posterior means. Of these, there is a small cluster at just over 1700nm, at values 1718, 1722, 1726, 1730 and 1734. These are noteworthy since, as remarked by Brown *et al.* , this is a region where fat is known to have a characteristic absorbance; Brown *et al.* identify the point 1718 nm too. In Figure 1 it is possible to discern small wiggles in the spectra in this wavelength region.

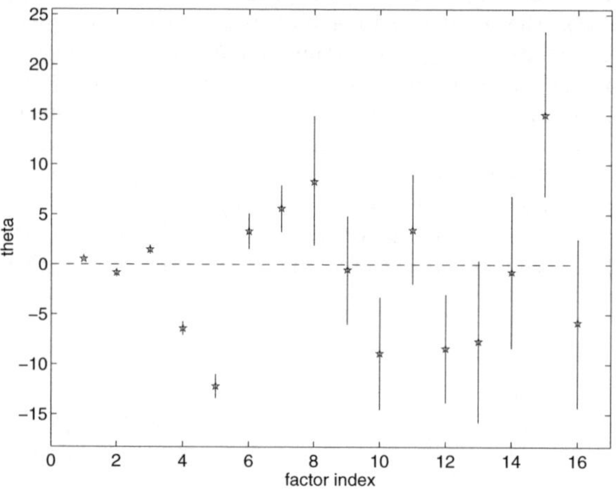

Figure 2. *Biscuit dough analysis: Estimates of empirical factor regression coefficients* $\boldsymbol{\theta}$.

Figure 4 addresses prediction and model assessment via display of data plotted against fitted and predicted values; training data are indicated by asterisks, and validation data by circles (recall the original fat content values are standardized to define the response variable). The close concordance between observed and fitted/predicted values (as well as several other exploratory residual analyses, not reported) give no reason to question model fit. Of particular interest is the fact that the out-of-sample predictions are very accurate indeed. In fact, on the basis of simple mean square prediction errors, this analysis improves on results of Brown *et al.* , albeit only marginally. There is sensitivity to the tuning parameters r and the c_i, and some experimentation is needed to investigate this; the results reported are based on such experimentation and rough optimization over these values with respect to the predictive assessment.

2. LATENT FACTOR REGRESSION MODELS

Formal latent factor models aim to partition variation in the predictor variables into multiple components that reflect common patterns, and separate these from variation that is idiosyncratic to each variable, or "noise." Here I note the theoretical structure of standard linear, latent factor models and define a class of *factor regression models* that naturally relate underlying latent structure in high-dimensional predictors to responses. I show that the empirical model of

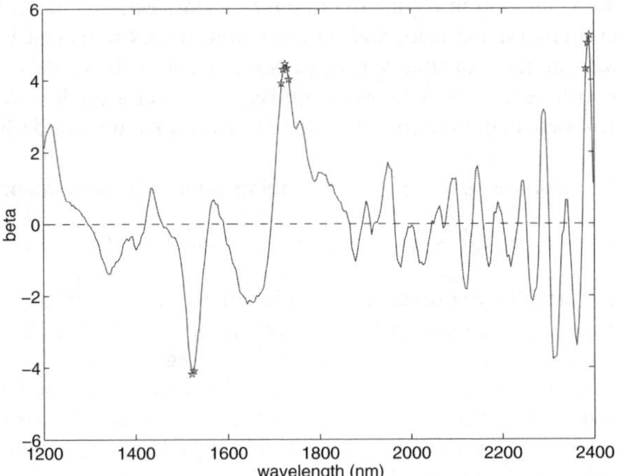

Figure 3. *Biscuit dough analysis: Estimates of wavelength regression coefficients β^*.*

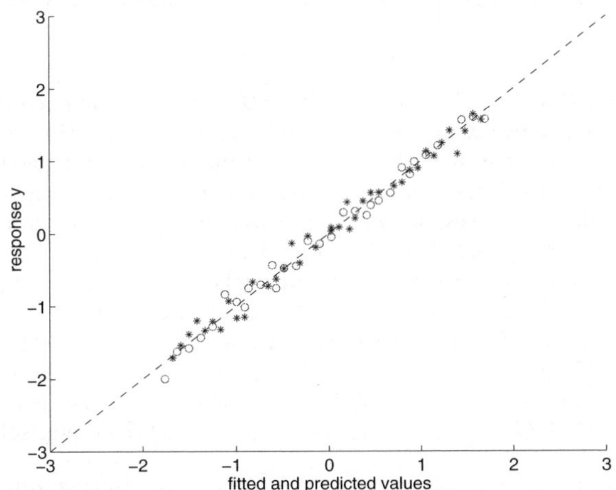

Figure 4. *Biscuit dough analysis: Response versus fitted and predicted values.*

Section 1 arises as a limiting case, and how this clarifies the design-dependency issues discussed in Section 1.

Write $\boldsymbol{x}_i'$ for the i-th row of $\boldsymbol{X}$ and consider the latent factor model (*e.g.*, Aguilar and West, 2000; Lopes and West, 2003)

$$\boldsymbol{x}_i = \boldsymbol{B}\boldsymbol{\lambda}_i + \boldsymbol{\nu}_i \tag{1}$$

where

$$\boldsymbol{\lambda}_i \sim \mathrm{N}(\boldsymbol{\lambda}_i \,|\, 0, \Delta^2) \qquad \text{and} \qquad \boldsymbol{\nu}_i \sim \mathrm{N}(\boldsymbol{\nu}_i \,|\, 0, \Psi^2). \tag{2}$$

Here $\boldsymbol{\lambda}_i$ is a k-vector of uncertain latent factors for case i, $\boldsymbol{B}$ is a $p \times k$ factor loadings matrix parameter, and $\boldsymbol{\nu}_i$ is a vector of idiosyncratic noise terms; both Δ and Ψ are diagonal. The number of factors is fixed and $k << p$. With appropriate identifying constraints on $\boldsymbol{B}$, this is an estimable model that attributes common structure in $\boldsymbol{X}$ to underlying k-dimensional factors, and isolates variation that is purely idiosyncratic in the $\boldsymbol{\nu}_i$ terms. MCMC based Bayesian analysis, and aspects of identification and prior specification, appear in Lopes and West (2003) and, in more elaborate factor models in time series, in Aguilar and West (2000). These two papers provide copious references to a large literature on Bayesian factor models. Mackay and Miskin (2001) is a recent, independent contribution that is also partly motivated by gene expression studies.

Assume that the responses $\boldsymbol{y} = (y_1, \ldots, y_n)'$ relate directly to the k latent factors; for each i,

$$y_i = \boldsymbol{\lambda}_i'\boldsymbol{\theta} + \epsilon_i \qquad \text{where} \qquad \epsilon_i \sim \mathrm{N}(\epsilon_i \,|\, 0, \sigma^2). \tag{3}$$

The original design variables $\boldsymbol{x}_i$ provide information on the latent variables through (1), but do not enter the regression; y_i is conditionally independent of $\boldsymbol{x}_i$ given $\boldsymbol{\lambda}_i$. One implication is that idiosyncratic variation in $\boldsymbol{X}$ now has no influence on the regression.

In Section 3, I discuss and exemplify analysis of the latent factor model alone; space here precludes full development of analysis and examples of the linked factor regression models, but it is of key interest to consider a theoretical limiting special case under the natural prior specification $\boldsymbol{\theta} \sim \mathrm{N}(\boldsymbol{\theta} \,|\, 0, \boldsymbol{G}^{-1})$ with diagonal precision matrix $\boldsymbol{G}$.

Equations (1)–(3) imply a joint normal distribution for $(y_i, \boldsymbol{x}_i, \boldsymbol{\lambda}_i)$, and hence an implied conditional distribution for y_i given only $\boldsymbol{x}_i$ and the model parameters. After some algebra, this is

$$(y_i \,|\, \boldsymbol{x}_i) \sim \mathrm{N}(y_i \,|\, \boldsymbol{x}_i'\boldsymbol{\beta}, \sigma^2 + \boldsymbol{\theta}'\boldsymbol{C}\boldsymbol{\theta}) \tag{4}$$

where $\boldsymbol{\beta} = \Psi^{-2}\boldsymbol{B}\boldsymbol{C}\boldsymbol{\theta}$ with $\boldsymbol{C}^{-1} = \Delta^{-2} + \boldsymbol{B}'\Psi^{-2}\boldsymbol{B}$. Clearly the implied regression of y_i on $\boldsymbol{x}_i$ is linear, with a theoretically implied and unique extension of the (low-dimensional) factor regression parameter $\boldsymbol{\theta}$ to the (high-dimensional) predictor regression parameter $\boldsymbol{\beta}$. Further, under the specific prior for $\boldsymbol{\theta}$, there is a unique (singular normal) prior implied for $\boldsymbol{\beta}$.

Consider now the special case in which $\Psi = s\boldsymbol{I}$, and in which identification is enforced by assuming $\boldsymbol{B}$ to be orthogonal. Then $\boldsymbol{C}$ is diagonal with elements $s^2d_i^2/(s^2 + d_i^2)$ where the d_i are the elements of the diagonal matrix Δ. Now take the limit as $s \to 0$, so that the latent factors explain essentially all the variation in the predictors. Then (1) reduces to $\boldsymbol{x}_i = \boldsymbol{B}\boldsymbol{\lambda}_i$ or, in matrix form, $\boldsymbol{X} = \Lambda\boldsymbol{B}'$ where Λ has rows $\boldsymbol{\lambda}_i'$. Assuming $n \geq k$, this recovers the SVD decomposition of $\boldsymbol{X}$ with $\boldsymbol{B} = \boldsymbol{A}'$, and this limiting special case of the latent factor model defines the empirical factor model. In this limit, it also easily follows that $\boldsymbol{\beta} = \boldsymbol{A}'\boldsymbol{\theta}$. Under the chosen prior $\mathrm{N}(\boldsymbol{\theta} \,|\, 0, \boldsymbol{G}^{-1})$ with $\boldsymbol{G}$ diagonal, it follows that $\boldsymbol{\beta}$ has precisely the gsg-prior of Section 1.2.

Hence, this special limiting case of a formal latent factor model leads to the SVD regression and the gsg-prior. The tie-up is exact, and explains away the issues of design, and sample size, dependence of the parameter and prior, and of recovering inferences on the regression in the original predictor variables. Inference on $\boldsymbol{\beta}$ flows directly from that on $\boldsymbol{\theta}$. All predictor values for validation cases must be included in the analysis of training data as they inform, under (1), on parameters of the latent factor model and therefore, indirectly, on values of the latent factors underlying the training data.

3. SPARSE FACTOR MODELS

3.1. *Motivating Applications in Gene Expression Profiling*

Original motivation for this work comes from gene expression analysis in which predictors are genes and p may range up to 30,000. Some of our initial studies (Spang *et al.*, 2001; West *et al.*, 2000, 2001) involved binary regression, with a probit model constructed by treating the y_i as latent and observing only indicators of $y_i > 0$. Logistic and other variants are also standard extensions (Albert and Johnson, 1999, Ch. 3). This development of generalized shrinkage priors with singular factors enabled the use of high-dimensional predictors in gene expression analysis and, in part, underlies the interest in more formal and flexible factor models.

In gene expression profiling, the predictor variables are recorded expression levels of individual genes, and the responses are clinical or physiological outcomes. Inherently, multiple biological factors underlie patterns of gene expression variation, so latent factor approaches are natural – we imagine that latent factors reflect individual biological functions (gene networks or pathways). This is also a motivating context for *sparse* models. Each biological factor involves a number of genes, perhaps a few to a few hundred, but not all genes; so each column of $\boldsymbol{B}$ will have many zeros. Similarly, a given gene may play roles in one or a small number of biological pathways, but will not be involved in all; so each row of $\boldsymbol{B}$ will have many zeros. It is therefore substantively appropriate to use priors that induce sparsity in $\boldsymbol{B}$.

3.2. *Bayesian Specification of Sparse Factor Models*

A Bayesian approach to defining sparse factor structure uses priors on the elements B_{ij} (gene i, factor j) of $\boldsymbol{B}$ that induce zeros with high probability. Within column (factor) j, take the B_{ij} to be independent with priors

$$\pi_j \delta_0(B_{ij}) + (1 - \pi_j)\mathrm{N}(B_{ij} \,|\, 0, 1)$$

where $\delta_0(\cdot)$ is the unit point mass at zero, and where π_j has a prior heavily concentrated near 1. Note that the unit scale of the normal component is convenient; the arbitrary scale of factor j is already accommodated in the variance parameter d_j^2. Assuming specified priors on all model parameters, MCMC analysis extends existing approaches that utilize normal priors on $\boldsymbol{B}$ (Aguilar and West, 2000; Lopes and West, 2003) to incorporate these new mixture priors, and include sampling of the π_j. The analysis is inherently parallelizable; MCMC sequences through columns of $\boldsymbol{B}$ (*i.e.*, through factors), and within each column the (many) elements B_{ij} are, *a posteriori*, conditionally independent given values of the prior and model parameters, latent factors and data. Hence, for fixed j, the set of p values B_{ij} $(i = 1, \ldots, p)$, may be sampled efficiently, in parallel. Some consideration of identification constraints is needed; we may use the popular lower triangular method (Aguilar and West, 2000; Lopes and West, 1999, and references therein) that simply fixes $k(k+1)/2$ selected elements of $\boldsymbol{B}$ at 0 or 1, and then use the mixture prior for the remaining (many) elements.

3.3. *Example: Factors in Breast Cancer Gene Expression Data*

Some illustration comes from analysis of expression levels of $p = 6128$ genes measured (using Affymetrix DNA microarrays) on $n = 49$ breast cancer tumor samples. The data comes from the study reported in West *et al.* (2000, 2001) and Spang *et al.* (2001), where full details of the data and context may be found. Here I discuss some aspects of a new analysis using the sparse latent factor model, with $k = 25$ factors. Additional prior specifications include priors on the variances d_j^2 of the latent factors and the idiosyncratic variances ψ_i^2 (the elements of the diagonal matrix Ψ); these are each specified via independent $\mathrm{Ga}(\cdot \,|\, 0.01, 0.01)$ priors on reciprocal variances. Finally, the use of a prior on π_j that heavily favors very high values is critical in enforcing very many zeros in the loadings matrix; here, with $p = 6128$ and $k = 25$, analysis utilizes $\pi_j \sim \mathrm{Be}(\pi_j \,|\, 999, 1)$.

Figure 5 displays posterior means of the values of four of the factors, chosen as those four with largest values of the posterior means of the d_j^2 and plotted from the top down. The values of the factors (vertical axis) are plotted against sample number (horizontal axis). The first factor is essentially zero apart from on samples (tumors) $7, 8, 11$ and 46, and the second essentially zero but for cases 7 and 8. These four cases had been much explored in earlier analyses, and, relative to most of the data, have quite apparent differences in large numbers of genes. The second factor shows that cases 7 and 8 share a common pattern of covariation not exhibited by 11 and 46. These four cases, and particularly 7 and 8, are in fact questionable due to concerns about the quality of the DNA microarray hybridization; in analyses in West *et al.* (2001) cases 7 and 8 had been held out due to these data quality concerns. It is of interest here to note that the full data analysis using the sparse factor model is itself capable of identifying these questionable cases, and protecting inferences on other factors from their effects.

The third factor plotted is of key interest in connection with comparison of oestrogen receptor (ER) status of tumors. We had earlier developed binary factor regression models to predictively discriminate ER status based on a selected set of about 50–100 genes (West *et al.,* 2000, 2001; Spang *et al.,* 2001). That analysis format is effective but involves the pre-selection of smaller numbers of discriminatory genes, and one motivating interest in sparse factor models is the potential to automate variable selection at the factor loading level. This potential is realized and evident here. The third factor is color coded: red indicates ER positive tumors, blue ER negative. Factor 3 evidently separates the two groups quite well, with four cases $(16, 31, 40, 43)$ in the mid-ground. Earlier analysis using gene screening had identified these four cases, and follow-on investigations reversed the ER status determination of number 31, so the analysis is in fact discriminating cases very well based on this factor alone.

For comparison, Figure 6 displays the four dominant empirical SVD factors (principal components) from the 6128 genes. Factors 1 and 2 bear resemblance to those from the model-based analysis, though the identification of questionable samples via two "outlier factors" is significantly obscured by the confounding of idiosyncratic noise in gene expression levels—a key inherent and limiting feature of empirical factor analyses with many variables. The third empirical factor certainly relates to the ER discrimination, but again the signal is obscured and much less clearly defined than it is in the model-based analysis.

4. ADDITIONAL COMMENTS

Sparse factor regression models offer a promising framework for dimension reduction in predictor space and for regression variable selection with many predictors. If a small number of latent factors associate with the response (as in the breast cancer ER study, where a single factor is primarily implicated) then a sparse factor model implies that only those genes with non-zero loadings on those factors are relevant; variable selection is then induced, automatically. In the ER study, only 60 genes have posterior probability of non-zero values exceeding 0.5; most of these genes show up in our prior studies and those of other groups exploring ER pathways in breast cancer.

Full analysis of the sparse factor model combined with binary regression has been explored in cross-validation studies of both ER and lymph node (LN) status with this breast cancer data, comparing with results in West *et al.* (2001). This prior work uses gene selection/screening and SVD factor regression. The cross-validation predictions are very similar, perhaps even slightly better with the sparse factor model. I stress that this model uses all 6128 genes whereas our prior published analysis selects 100 based on correlation with ER or LN status, so removing noise via ad-hoc preliminary variable selection. The comparability of predictions is very strong evidence for the efficacy of the sparse modelling approach in dealing formally—and automatically—with that most critical and challenging variable selection problem. Further development and experience with this approach is needed; a key need is the development of efficient software for distributed processing to address the very challenging computational demands of model fitting.

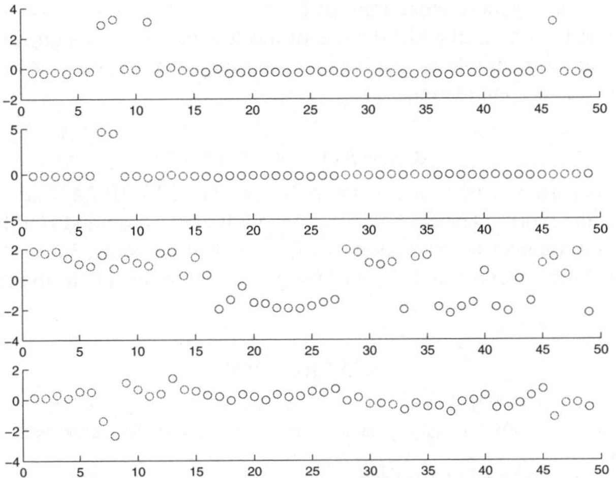

Figure 5. *Four factors in sparse factor analysis of gene expression data.*

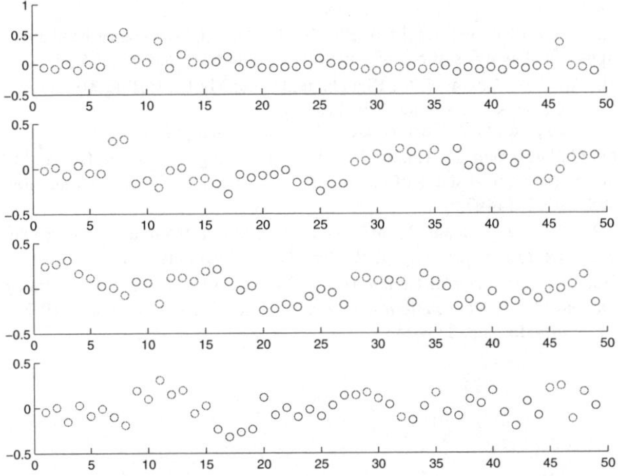

Figure 6. *Four factors in SVD of gene expression data.*

Additional questions concern the incorporation of substantive, informative prior information into these large-scale models. This raises questions of both how flexible the current models are in terms of the scope for customizing them to incorporate specific prior information, and of how they might be generalized. A further key question is the identification, or estimation, of the number of factors. In some studies, simply increasing k and exploring posterior estimates of factors and their variances suffices (Aguilar and West, 2000), though the problem remains an open research area and utilizing formal approaches is a challenge (Lopes and West, 1999). This question is discussed also by Mackay and Miskin (2001). In the new sparse factor model introduced here, using too few factors confounds higher-order structure in the factors being estimated and, in particular, induces likelihood functions that very strongly suggest non-zero factor loadings for many gene-factor combinations than might be expected on scientific grounds; this seems to be simply an artifact of the use of too few factors. Though simple to diagnose, this problem is far from simple to resolve as fitting large numbers of factors is significantly challenging in terms of computation.

ACKNOWLEDGEMENTS

This research was partially supported by the NSF (grants DMS-0102227 and DMS-0112340). Some aspects of the work relate to collaborations with Ming Liao and Hedibert Lopes. I am grateful to Marina Vanucci for provision of the biscuit dough data, for useful conversations with Jim Berger, Merlise Clyde and Rainer Spang, and for comments from the editors and two referees.

REFERENCES

Albert, J. and Johnson, V. E. (1999). *Ordinal Data Models*. New York: Springer-Verlag.

Aguilar, O. and West, M. (2000), Bayesian dynamic factor models and portfolio allocation. *J. Business Econom. Statist.* **18**, 338–357.

Brown, P. J., Fearn, T. and Vannucci, M. (1999) The choice of variables in multivariate regression: a non-conjugate Bayesian decision theory approach. *Biometrika*, **86**, 635–648.

Lopes, H. and West, M. (2003), Model uncertainty in factor analysis. *Statist. Sinica* (to appear).

Mackay, D. J. C. and Miskin, J. (2001), Latent variable models for gene expression data (unpublished). `www. inference.phy.ca.ac.uk/mackay`

Osborne, B. G., Fearn, T., Miller, A. R. and Douglas, S. (1984). Applications of near infrared reflectance spectroscopy to compositional analysis of biscuits and biscuit doughs. *J. Sci. Food Agric.,* **35**, 99–105.

Spang, R., Zuzan, H., West. M., Nevins, J. R., Blanchette, C. and Marks, J. R. (2001). Prediction and uncertainty in the analysis of gene expression profiles. *Silico Biology* **2**. `http://www.bioinfo.de/ isb/gcb01/talks/spang/index.html`.

West, M., Blanchette, C., Dressman, H., Huang, E., Ishida, S., Spang, R., Zuzan, H., Marks, J. R. and Nevins, J. R. (2001). Predicting the clinical status of human breast cancer utilizing gene expression profiles. *Proc. Natl. Acad. Sci. USA* **98**, 11462–11467.

West, M., Nevins, J. R., Marks, J. R, Spang, R. and Zuzan, H. (2000). DNA microarray data analysis and regression modeling for genetic expression profiling. *Tech. Rep.*, Duke University, USA.

Zellner, A. (1986). On assessing prior distributions and Bayesian regression analysis with g-prior distributions. *Bayesian Inference and Decision Techniques: Essays in Honor of Bruno de Finetti* (P. K. Goel and A. Zellner, eds). Amsterdam: North-Holland, 233–243.

BAYESIAN STATISTICS 7, pp. 743–750
J. M. Bernardo, M. J. Bayarri, J. O. Berger, A. P. Dawid,
D. Heckerman, A. F. M. Smith and M. West (Eds.)

A Bayesian Analysis of Smooth Transitions in Trend

PINGPING ZHENG
Lancaster University, UK
pingping.zheng@lancaster.ac.uk

JOHN M. MARRIOTT
Nottingham Trent University, UK
john.marriott@ntu.ac.uk

SUMMARY

Many real time series exhibit changes in trend. However, because the observed changes are frequently not abrupt, models that specify step changes do not adequately represent the observed behavior. Smooth transition models allow for a smooth change from one linear regime to the next.

This paper is concerned with parameter estimation and prediction for a linear trend model with a smooth transition in both the intercept and slope. The smooth transition is modelled using a logistic transition function. We present a Bayesian approach to inference for this model that allows for an autocorrelated error structure and uses the exact likelihood. We develop a slice sampler for the analysis.

We apply the analysis to a published macroeconomic data set which exhibits a slowly changing time trend.

Keywords: GIBBS SAMPLER; NONLINEAR TREND; STRUCTURAL CHANGE; SLICE SAMPLER; SMOOTH TRANSITION.

1. INTRODUCTION

For many years research in econometric time series analysis has been concerned with the question of whether or not an observed series should be modelled as a non-stationary process with an autoregressive unit root or as a stationary process around a deterministic trend. In recent years many researchers have investigated models that allow for a permanent step change in the level (and possibly also the slope) of the stationary process with a linear trend.

Much of the literature on this subject has been concerned with testing for the presence of a unit root when there is a structural break. There are a large number of reports in the literature of Bayesian approaches to testing for unit roots although they mostly neglect the possibility of a structural break. Two exceptions are De Jong (1996) and Marriott and Newbold (2000).

An alternative approach to modelling changes in trend is suggested by Leybourne *et al.* (1998). They argue that the assumption of an instantaneous deterministic structural change is not realistic for real time series. When a structural change occurs the different variables that influence any macroeconomic time series would react at different rates, the resulting time path would therefore be expected to exhibit a gradual change in the deterministic component. For this reason Leybourne *et al.* (1998) use smooth transition models to represent deterministic structural change.

Smooth transition models were first considered by Bacon and Watts (1971) and Maddala (1977). Teräsvirta and Anderson (1992), Teräsvirta (1994) and Greenaway *et al.* (1997) used smooth transition models to describe structural changes and model growth. Leybourne *et al.* (1998) and Sollis *et al.* (1999) tested the null hypothesis of a unit root against the alternative of a smooth transition for economic time series and concluded that "a smooth transition in linear trend constitutes a plausible generator of economic time series and warrants serious consideration."

In this paper, we consider the smooth transition model

$$\begin{aligned} y_t =& \alpha_1 + \beta_1 t + \alpha_2 S_t(\gamma,\tau) + \beta_2 t S_t(\gamma,\tau) + v_t, \qquad t = 0,1,\ldots,T, \\ v_t =& \rho v_{t-1} + \varepsilon_t, \end{aligned} \tag{1}$$

where the ε_t are iid N(0, σ^2), $|\rho| < 1$, and $S_t(\gamma,\tau)$ is the logistic smooth transition (LST) function,

$$S_t(\gamma,\tau) = \frac{1}{1+\exp\{-\gamma(t-\tau T)\}}, \tag{2}$$

with $\tau \in (0,1)$, $\gamma > 0$. This model corresponds to the most general of those discussed by Leybourne *et al.* (1998, Model C) and can be used to describe both increasing and decreasing behavior in observed time series. The parameter τ determines the midpoint of the transition, which is at τT, while the parameter γ determines the speed of the transition. For moderate sample sizes, for example $T+1 = 100$, the sample paths arising from $\gamma \geq 1$ exhibit very steep transitions that resemble step changes.

In the classical literature estimation of (1) has used nonlinear least squares and Granger and Teräsvirta (1993, Ch. 7) point out that the estimate of γ can converge very slowly.

We use a Bayesian approach, based upon the full likelihood, that employs Markov chain Monte Carlo (MCMC) methods to sample from the posterior densities and then use these samples to make inferences for the parameters of the model. In this way, we avoid complex analytic integrals which would otherwise be needed. In particular, we use the MCMC method, often referred to as the slice sampler, to carry out sampling from the posterior densities.

In Section 2, we derive the likelihood and present a Bayesian analysis for (1). The conditional posterior distributions for the parameters γ, τ and ρ are seen to be non-standard and in Section 3 we develop the slice sampler used in the analysis. Section 4 provides an illustration of our approach through its application to the modelling of an economic time series.

2. THE BAYESIAN ANALYSIS

In order to use Bayes theorem to make inferences about the model, we need to obtain the likelihood function.

We first eliminate v_t from the model, to obtain

$$\begin{aligned} y_t - \rho y_{t-1} =& (1-\rho)\alpha_1 + ((1-\rho)t + \rho)\beta_1 \\ &+ (S_t - \rho S_{t-1})\alpha_2 + (tS_t - \rho(t-1)S_{t-1})\beta_2 + \varepsilon_t, \end{aligned}$$

for $t = 1,2,\ldots,T$ where we have written $S_t = S_t(\gamma,\tau)$ for the smooth transition function. We now define a transformation from y to x, by $x_t = y_t - \rho y_{t-1}$ for $t > 0$ and we also write $x_0 = \sqrt{1-\rho^2}\,(\alpha_1 + \alpha_2 S_0) + \tilde{v}_0$ to complete the transformation, which has the Jacobian $J = \sqrt{1-\rho^2}$. We note that $\tilde{v}_0 = \sqrt{1-\rho^2} v_0$ has marginal distribution $\mathrm{N}(0,\sigma^2)$, independent of ε_t, $t > 0$.

We also write

$$X = \begin{pmatrix} \sqrt{1-\rho^2} & 0 & \sqrt{1-\rho^2}S_0 & 0 \\ 1-\rho & 1 & S_1 - \rho S_0 & S_1 \\ 1-\rho & 2-\rho & S_2 - \rho S_1 & 2S_2 - \rho S_1 \\ \vdots & \vdots & \vdots & \vdots \\ 1-\rho & (1-\rho)T+\rho & S_T - \rho S_{T-1} & TS_T - \rho(T-1)S_{T-1} \end{pmatrix}$$

so that the model can be written as

$$x = X\beta + \epsilon,$$

where $\varepsilon = (\tilde{v}_0, \varepsilon_1, \ldots, \varepsilon_T)'$ and $\beta = (\alpha_1, \beta_1, \alpha_2, \beta_2)'$. This formulation of the model allows us to exploit some of the standard results from Zellner (1987, chapter 3). If we now write the parameter vector as $\theta = (\beta', \gamma, \tau, \rho, \sigma)'$, the distribution for x can be written as

$$p(x \mid \theta) = \left(\frac{1}{\sqrt{2\pi}\sigma}\right)^{T+1} \exp\left\{-\frac{1}{2\sigma^2}\left[\nu s^2 + (\beta - \hat{\beta})' X' X (\beta - \hat{\beta})\right]\right\},$$

where $\nu = T - 3$, $\hat{\beta} = (X'X)^{-1}X'x$, and $\nu s^2 = (x - X\hat{\beta})'(x - X\hat{\beta})$.

The likelihood is then

$$p(y \mid \theta) = \sqrt{1-\rho^2}\left(\frac{1}{\sqrt{2\pi}\sigma}\right)^{T+1} \exp\left\{-\frac{1}{2\sigma^2}\left[\nu s^2 + (\beta - \hat{\beta})' X' X (\beta - \hat{\beta})\right]\right\}.$$

If we condition on the parameters γ, τ, and ρ, a joint conjugate prior for β and σ is provided by:

$$p(\beta \mid \sigma) = \frac{|V|^{-1/2}}{(\sqrt{2\pi}\sigma)^4} \exp\left\{-\frac{1}{2\sigma^2}(\beta - \beta^0)' V^{-1} (\beta - \beta^0)\right\},$$

where V is a 4×4 positive definite symmetric constant matrix, β^0 is a 4 dimensional constant vector; and

$$p(\sigma) = \frac{2}{\Gamma(b/2)}\left(\frac{a}{2}\right)^{b/2} \frac{1}{\sigma^{b+1}} \exp\left\{-\frac{a}{2\sigma^2}\right\}, \qquad \sigma > 0,$$

where $a, b > 0$ are constants. In our example below we follow Monahan (1983) and choose his "diffuse" prior for which $V = k_\beta^2 I_4$ with $k_\beta = 16$, $a = b = 1/128$ and I_4 is the 4×4 unit matrix.

In choosing to fit the smooth transition model to data, we assume *a priori* that the transition happens in the time span of the data, that is, $\tau \in (0, 1)$, and in the absence of any other information we adopt a uniform prior for τ. For the parameter γ we use a gamma distribution, $\mathrm{Ga}(\gamma \mid a_1, b_1)$. In what follows we choose $a_1 = 4$ so that the prior mean for γ is $4b_1$ and the standard deviation is $2b_1$. The prior for γ covers the interval $(0, 10b_1)$ with probability of about 0.99. In our experience, values of γ larger than 0.4 should not be considered unless there is evidence of an abrupt transition during the time of measuring the data. We take this into account in our choice of a value for b_1 in the example presented below. For the autoregressive parameter ρ we choose a uniform prior on $(-1, 1)$. We further assume that (β, σ), γ, τ and ρ are *a priori* independently distributed, so that we have the following joint prior distribution for the parameter vector θ

$$p(\theta) \propto \frac{1}{\sigma^4} \exp\left\{-\frac{1}{2\sigma^2}(\beta - \beta^0)' V^{-1} (\beta - \beta^0)\right\} \times \frac{1}{\sigma^{b+1}} \exp\left\{-\frac{a}{2\sigma^2}\right\} \mathrm{Ga}(\gamma \mid a_1, b_1).$$

The posterior for θ is then

$$p(\theta \mid y) \propto \frac{|V_1|^{1/2}}{\sigma^4} \exp\left\{-\frac{1}{2\sigma^2}(\beta - \beta^1)' V_1 (\beta - \beta^1)\right\} \times \frac{(\nu s_1^2)^{(T+b+1)/2}}{\sigma^{T+b+2}} \exp\left\{-\frac{\nu s_1^2}{2\sigma^2}\right\} \mathrm{Ga}(\gamma \mid a_1, b_1) \times |V_1|^{-1/2} (\nu s_1^2)^{-(T+b+1)/2} \sqrt{1-\rho^2}, \tag{3}$$

where $V_1 = X'X + V^{-1}$, $\beta^1 = V_1^{-1}(X'X\hat{\beta} + V^{-1}\beta^0)$, and $\nu s_1^2 = a + \nu s^2 + (\beta^1 - \hat{\beta})'X'X(\beta^0 - \hat{\beta})$. We can now approach the problem of posterior inference for θ. We notice that the full conditional distributions for β and σ are multivariate normal and inverted gamma respectively. However V_1, β^1 and νs_1^2 are complicated functions of ρ, γ and τ and the full conditional distributions for these remaining parameters are not standard. For this reason we develop a slice sampler for the Bayesian analysis of this model.

3. THE SLICE SAMPLER

As was mentioned previously we propose the use of a slice sampler for the Bayesian analysis of the smooth transition model (1). Slice samplers are a form of *auxiliary variable* technique which introduce auxiliary random variables to facilitate the design of an MCMC sampling algorithm. Swendsen and Wang (1987) first used this technique for the Ising model and Besag and Green (1993) introduced the slice sampler into the statistical literature. Roberts and Rosenthal (1999) discussed the convergence of slice sampler Markov chains and Damien *et al.* (1999) demonstrated the use of auxiliary variables for sampling non-standard densities for a variety of examples. More details of slice samplers are discussed in Neal (2003).

Suppose $f(\theta)$ is a density which is given by

$$f(\theta) \propto \pi(\theta) \prod_{i=1}^{m} l_i(\theta), \tag{4}$$

where $\pi(.)$ is a density of known form and the $l_i(.)$ are non-negative functions, then a Gibbs sampler is constructed by introducing *auxiliary variables* $u = (u_1, u_2, \cdots, u_m)$ as follows (given that the current state of the vector θ is θ_n):

(1) sample m independent uniform random variables $u_{n+1,1}, u_{n+1,2}, \ldots, u_{n+1,m}$, with $u_{n+1,i} \sim \text{Un}(0, l_i(\theta_n))$;
(2) sample $\theta_{n+1} \sim \pi(\theta)$ conditional on the set $L = \{\theta : l_i(\theta) > u_{n+1,i}\}$,

where $\text{Un}(a, b)$ denotes the uniform distribution on the interval (a, b).

In order to construct a slice sampler for the smooth transition model (1) we observe that the joint posterior density (3) can be factorized as

$$p(\theta \mid y) \propto \pi(\theta) \prod_{i=1}^{3} l_i(\theta),$$

which has the same form as (4) with

$$\pi(\theta) \propto \frac{|V_1|^{1/2}}{(\sqrt{2\pi}\sigma)^4} \exp\left\{-\frac{1}{2\sigma^2}(\beta - \beta^1)'V_1(\beta - \beta^1)\right\} \times \frac{(\nu s_1^2)^{(T+b+1)/2}}{\sigma^{T+b+2}} \exp\left\{-\frac{\nu s_1^2}{2\sigma^2}\right\} \text{Ga}(\gamma \mid a_1, b_1),$$

$l_1(\theta) = |V_1|^{-1/2}$, $l_2(\theta) = (\nu s_1^2)^{-(T+b+1)/2}$ and $l_3(\theta) = \sqrt{1 - \rho^2}$.

In order to sample directly from $\pi(\theta)$ we employ the following factorization:

$$\pi(\theta) = p(\gamma)\, p(\tau)\, p(\rho)\, p(\sigma \mid \gamma, \tau, \rho) p(\beta \mid \gamma, \tau, \rho, \sigma),$$

where

$$p(\gamma) = \text{Ga}(\gamma \mid a_1, b_1)$$
$$p(\tau) = I_{\{0<\tau<1\}}$$
$$p(\rho) = 0.5 I_{\{-1<\rho<1\}}$$
$$p(\sigma \mid \gamma, \tau, \rho) \propto \frac{(\nu s_1^2)^{(T+b+1)/2}}{\sigma^{T+b+2}} \exp\left\{-\frac{\nu s_1^2}{2\sigma^2}\right\},$$
$$p(\beta \mid \gamma, \tau, \rho, \sigma) = \frac{|V_1|^{1/2}}{(\sqrt{2\pi}\sigma)^4} \exp\left\{-\frac{1}{2\sigma^2}(\beta - \beta^1)' V_1 (\beta - \beta^1)\right\},$$

and $I_{\{\cdot\}}$ is the indicator function.

The sampling of θ from $\pi(\theta)$ can then be carried out in the following steps:

(1) first, sample γ from the gamma distribution $\text{Ga}(\gamma \mid a_1, b_1)$, and then sample τ and ρ as $\text{Un}(0,1)$ and $\text{Un}(-1,1)$ respectively;
(2) sample σ from the inverted gamma distribution $p(\sigma \mid \gamma, \tau, \rho)$;
(3) finally sample β from the multivariate normal distribution $p(\beta \mid \gamma, \tau, \rho, \sigma)$.

The slice sampler is, therefore, implemented by iterating between

(1) sampling θ from $\pi(\theta)$ by rejection sampling, conditional on θ being in $L(u_1, u_2, u_3) = \{\theta : l_i(\theta) \geq u_i, i = 1, 2, 3\}$ and
(2) independently sampling new values of u_i from $\text{Un}(0, l_i(\theta))$, $i = 1, 2, 3$.

We notice that

$$\{\theta : l_3(\theta) \geq u_3\} = \left\{\theta : -\sqrt{1 - u_3^2} \leq \rho \leq \sqrt{1 - u_3^2}\right\}$$

and also that $l_i(\theta), i = 1, 2$ are functions of γ, τ and ρ only. Therefore we do not need to sample σ and β until we have proper samples for ρ, γ and τ which satisfy $l_i(\theta) \geq u_i$, $i = 1, 2, 3$. This leads to the adjusted scheme

(1) sample ρ from $U\left(-\sqrt{1 - u_3^2}, \sqrt{1 - u_3^2}\right)$;
(2) sample γ and τ independently, conditional on being in

$$L(u_1, u_2) = \{(\gamma, \tau, \rho) : l_i(\theta) \geq u_i, i = 1, 2\},$$

by means of rejection sampling;
(3) sample σ from $p(\sigma \mid \gamma, \tau, \rho)$;
(4) sample β from $p(\beta \mid \gamma, \tau, \rho, \sigma)$;
(5) sample new values of u_i from $\text{Un}(0, l_i(\theta))$ independently for $i = 1, 2, 3$.

In the following section we illustrate our approach using an economic time series that has recently been discussed in the literature.

4. AN EXAMPLE

The economic time series we have chosen to illustrate our approach concerns industrial production for the UK from 1780 to 1913. This series was examined by Crafts *et al.* (1989), who discuss the economic context of the data, and Newbold and Agiakloglou (1991). For series such as this, the primary interest of economic historians has been in trend growth rates and the behavior of observed data around such trend growth paths. For this reason the series considered here consists of the natural logarithms of 134 values of an index of aggregate annual production for the UK from 1700 to 1913. Crafts *et al.* (1989) argue that the series exhibits gradual change, rather than several break points, and Leybourne *et al.* (1998) found evidence of a smooth transition in trend when testing this series for the presence of a unit autoregressive root. The data are plotted as the unbroken line in Figure 2.

We now return to our choice of prior for γ, the parameter governing the speed of transition. We decided to use $\mathrm{Ga}(\gamma \mid a_1, b_1)$ for this prior and suggested $a_1 = 4$. Because economists believe that there is a *gradual* smooth transition in this UK industrial production series, we selected a small value for b_1, $b_1 = 0.025$, giving a prior mean for γ of 0.1 with prior standard deviation 0.05.

We used both the Raftery and Lewis (1992a, 1992b) and the Gelman and Rubin (1992) diagnostics to examine the convergence of the Markov chain. For the Raftery and Lewis diagnostic, we chose the initial values $\gamma = 0.1$, $\tau = 0.5$ and $\rho = 0$. For the Gelman and Rubin diagnostic, we generated six parallel chains, with dispersed values for γ, τ and ρ. In both instances we monitored the 2.5th and 97.5th percentiles. Both approaches indicated that convergence was fast, after a burn-in of at most 200 cycles. We chose to use a burn-in of 1000 cycles and then ran a single chain for a further 50,000 cycles for the inference.

Figure 1 shows plots of the posterior densities for the model parameters, Table 1 gives their posterior means and standard deviations and Table 2 presents quantiles of the posterior distributions.

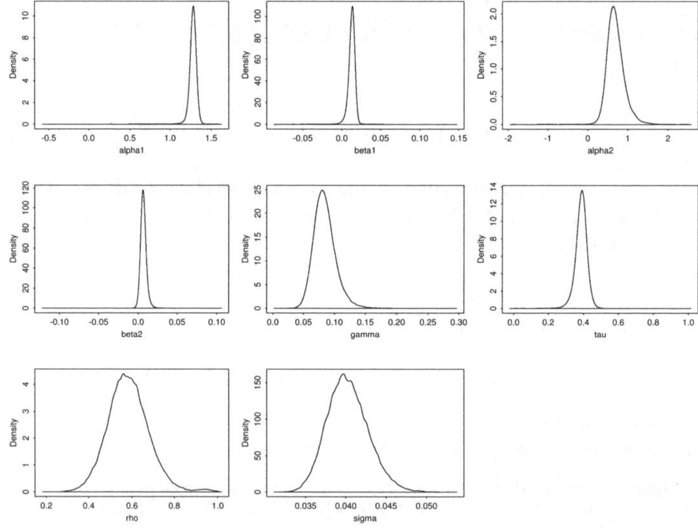

Figure 1. *Posterior densities for the UK industrial production series.*

It is clear from Tables 1 and 2 that there is strong support for both the linear trend and the smooth transition components with the 95% posterior probability intervals for all four of the parameters, α_1, β_1, α_2 and β_2, lying above zero. Similar consideration of the posterior distribution of ρ shows strong evidence of positive serial correlation in the error term. Using these results we can also obtain 95% posterior probability intervals for τ and γ. The interval for τ places the center of the transition between 1821 and 1841 with the posterior mean occurring in 1832. The corresponding interval for γ, 0.054 to 0.125, shows the transition speed to be fairly slow.

Table 1. *Posterior moments for the UK industrial production series.*

	α_1	β_1	α_2	β_2
mean	1.28	0.0128	0.693	0.00692
sd	0.0477	0.00447	0.222	0.00412
	γ	τ	ρ	σ
mean	0.0842	0.387	0.585	0.0403
sd	0.0183	0.0426	0.097	0.00255

Table 2. *Posterior quantiles for the UK industrial production series.*

	$p = P(X \leq x)$						
p	0.025	0.05	0.25	0.5	0.75	0.95	0.975
α_1	1.19	1.21	1.26	1.29	1.31	1.34	1.36
β_1	0.00271	0.00531	0.0106	0.0132	0.0156	0.0189	0.02
α_2	0.328	0.386	0.55	0.67	0.811	1.08	1.2
β_2	4.98e-05	0.00115	0.00433	0.00656	0.00897	0.0135	0.0158
γ	0.0538	0.0582	0.072	0.0824	0.0941	0.116	0.125
τ	0.308	0.329	0.369	0.39	0.409	0.44	0.451
ρ	0.407	0.435	0.521	0.58	0.643	0.747	0.788
σ	0.0356	0.0363	0.0385	0.0401	0.0419	0.0447	0.0456

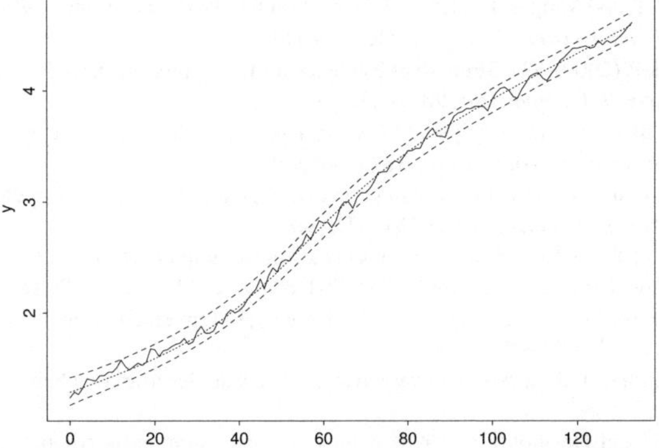

Figure 2. *95% predictive probability intervals for UK industrial production.*

From a Bayesian point of view, any fit of the model (1) to the data should be based on the posterior predictive distribution for each time point, and in particular on the value of the mean $\mathrm{E}(\tilde{y}_t \mid y)$. The output from the sampler was also used to construct samples from the posterior predictive distribution of $\tilde{y}_t$, $t = 1, 2, \ldots, T$. The resulting values of $\mathrm{E}(\tilde{y}_t \mid y)$, together with approximate 95% posterior predictive probability limits, are plotted in Figure 2. The nature of the smooth change in the trend is clearly shown in the picture.

5. CONCLUSION

As was implied by Sollis *et al.* (1999), the smooth transition model (1) should be more widely entertained as a possible data generating mechanism when considering the behavior of economic time series. Hitherto this particular nonlinear model has been largely ignored in the Bayesian literature.

We have presented a Bayesian analysis for the smooth transition model and have shown how to undertake posterior inference for the parameters from which posterior predictive inference of interest to economists can be obtained.

An application of our approach to the UK industrial production time series has shown that this type of model can be seriously considered as a sensible choice for modelling the behavior of the series.

REFERENCES

Bacon, D. W. and Watts, D. G. (1971). Estimating the transition between two intersecting straight lines. *Biometrika* **58**, 525–34.

Besag, J. and Green, P. J. (1993). Spatial Statistics and Bayesian Computation. *J. R. Statist. Soc. B* **55**, 25–37.

Crafts, N. F. R., Leybourne, S. J. and Mills, T. C. (1989). Trends and cycles in British industrial production, 1700–1913. *J. R. Statist. Soc. A* **152**, 43–60.

Damien, P., Wakefield, J. and Walker, S. (1999). Gibbs sampling for Bayesian non-conjugate and hierarchical models by Using Auxiliary Variables. *J. R. Statist. Soc. B* **61**, 331–344.

De Jong, D. N. (1996). A Bayesian search for structural breaks in U.S. GNP. *Adv. Econometrics B* **11**, 109–146.

Gelman, A. and Rubin, D. (1992). Inference from Iterative Simulation Using Multiple Sequences. *Statist. Sci.* **7**, 457–511.

Granger, C. W. J. and Teräsvirta, T. (1993). *Modelling Nonlinear Economic Relationships.* Oxford: Oxford University Press.

Greenaway, D., Leybourne, S. and Sapsford, D. (1997). Modelling growth (and liberalisation) using smooth transitions analysis. *Economic Inquiry* **35**, 798–814.

Leybourne, S., Newbold, P. and Vougas, D. (1998). Unit roots and smooth transitions. *J. Time Ser. Anal.* **19**, 83–97.

Maddala, G. S. (1977). *Econometrics.* New York: McGraw-Hill.

Marriott, J. and Newbold, P. (2000). The Strength of Evidence for Unit Autoregressive Roots and Structural Breaks: A Bayesian Perspective. *J. Econometrics* **98**, 1–25.

Monahan, J. F. (1983). Fully Bayesian Analysis of ARMA Time Series Models. *J. Econometrics* **21**, 307–331.

Neal, R. M. (2003). Slice sampling. *Ann. Statist.* **31** (to appear).

Newbold, P. and Agiakloglou, C. (1991). Looking for evolving growth rates and cycles in British industrial production, 1700–1913. *J. R. Statist. Soc. A* **154**, 341–348.

Raftery, A. and Lewis, S. (1992a). How many iterations in the Gibbs sampler? *Bayesian Statistics 4* (J. M. Bernardo, J. O. Berger, A. P. Dawid and A. F. M. Smith, eds). Oxford: Oxford University Press, 763–773.

Raftery, A. and Lewis, S. (1992b) One long run with diagnostics: Implementation strategies for Markov chain Monte Carlo. *Statist. Sci.* **7**, 493–497.

Roberts, G. O. and Rosenthal, J. S. (1999). Convergence of slice sampler Markov chains. *J. R. Statist. Soc. B* **61**, 643–660.

Sollis, R., Leybourne, S. and Newbold, P. (1999). Unit roots and asymmetric smooth transitions. *J. Time Ser. Anal.* **20**, 671–677.

Swendsen, R. H. and Wang, J. S. (1987). Nonuniversal critical dynamics in Monte Carlo simulations. *Phys. Rev. Lett.* **58**, 86–88.

Teräsvirta, T. (1994). Specification, estimation, and evaluation of smooth transition autoregressive models. *J. Am. Statist. Ass.* **89**, 208–218.

Teräsvirta, T. and Anderson, H. M. (1992). Characterising nonlinearities in business cycles using smooth transition autoregressive models. *J. Appl. Economet.* **7**, 119–S136.

Zellner, A. (1987). *An Introduction to Bayesian Inference in Econometrics*. Melbourne, FL: Krieger.